Number Theory, Analysis and Geometry

Dorian Goldfeld • Jay Jorgenson • Peter Jones
Dinakar Ramakrishnan • Kenneth A. Ribet
John Tate
Editors

Number Theory, Analysis and Geometry

In Memory of Serge Lang

Springer

Editors
Dorian Goldfeld
Department of Mathematics
Columbia University
New York, NY 10027
USA
goldfeld@columbia.edu

Jay Jorgenson
Department of Mathematics
City University of New York
New York, NY 10031
USA
jjorgenson@mindspring.com

Peter Jones
Department of Mathematics
Yale University
New Haven, CT 06520
USA
jones@math.yale.edu

Dinakar Ramakrishnan
Department of Mathematics
California Institute of Technology
Pasadena, CA 91125
USA
dinakar@caltech.edu

Kenneth A. Ribet
Department of Mathematics
University of California at Berkeley
Berkeley, CA 94720
USA
ribet@math.berkeley.edu

John Tate
Department of Mathematics
Harvard University
Cambridge, MA 02138
USA
tate@math.utexas.edu

ISBN 978-1-4614-1259-5 e-ISBN 978-1-4614-1260-1
DOI 10.1007/978-1-4614-1260-1
Springer New York Dordrecht Heidelberg London

Library of Congress Control Number: 2011941121

Printed on acid-free paper

Springer is part of Springer Science+Business Media (www.springer.com)

Serge Lang (Photo provided courtesy of Kenneth A. Ribet.)

Preface

Serge Lang was an iconic figure in mathematics, both for his own important work and, perhaps even more crucially, for the indelible impact he left on the field, and on his students and colleagues. It would be difficult to find a mathematician who came of age in the past forty years, who had not been exposed to Serge's articles, monographs, and textbooks. Serge's writing shaped the mathematical perspectives of all who came in contact with them. Many were challenged by the glimpses of open problems and conjectures that Serge interweaved with his expositions of established subjects. Serge's exposition invariably transcended our discipline's preference for brevity and perfection, which often obscures the intuition underlying the subject. Serge was never one to conform.

One of Serge's uplifting qualities was his openness to new areas of mathematics and his concurrent willingness, even eagerness, to learn novel concepts and techniques. He was constantly reinventing himself, while sharing his accumulated wisdom with students and young mathematicians. Over the course of his career, he traversed a tremendous amount of mathematical ground. As he moved from subject to subject, he found analogies that led to important questions in such areas as number theory, arithmetic geometry, and the theory of negatively curved spaces. Lang's conjectures will keep many mathematicians occupied far into the future.

This memorial volume contains articles in a variety of areas of mathematics, attempting to represent Serge's breadth of interest and impact. We are happy to publish here (for the first time) Serge's final paper, *The heat kernel, theta inversion, and zetas on* $\Gamma\backslash G/K$, written jointly with one of us (J. Jorgenson). Except for that one article, which was left in the form it assumed just before Serge's passing, every other entry here was thoroughly refereed. We thank all the authors for their contributions to the volume and for their willingness to put up and comply with our demands for revision. Thanks also to the anonymous referees for their excellent and timely work.

We, the editors, are pleased to be a part of this production, especially since we were all fortunate enough to know Serge personally. We thank Stacey Croomes, the math administrator at Caltech, for her invaluable help in organizing the receipt of the articles, the refereeing process, and the revisions. We are grateful to Ann Kostant and Elizabeth Loew of Springer for their enthusiasm and helpful advice during the many months of editorial preparation. It took a village to produce this volume.

Columbia University — Dorian Goldfeld
Yale University — Peter Jones
The City College of New York — Jay Jorgenson
Caltech — Dinakar Ramakrishnan
University of California, Berkeley — Kenneth A. Ribet
Harvard University — John Tate

Contents

Publications of Serge Lang: from 2000 and beyond

The five volumes of Serge Lang's *Collected Papers* from 1952 to 1999 were published by Springer-Verlag and noted below. His additional publications from 2000 can be found on the website of the American Mathematical Society's MathSciNet and are listed here.

Titles of books and journal articles are in italics.

[2000a] *Collected Papers.* Vol. I. 1952–1970. Springer-Verlag, New York, 2000. xxiv+525 pp.

[2000b] *Collected Papers.* Vol. II. 1971–1977. Springer-Verlag, New York, 2000. xvi+590 pp.

[2000c] *Collected Papers.* Vol. III. 1978–1990. Springer-Verlag, New York, 2000. xvi+393 pp.

[2000d] *Collected Papers.* Vol. IV. 1990–1996. Springer-Verlag, New York, 2000. xvi+471 pp.

[2001a] *Collected Papers.* Vol. V. 1993–1999. With Jay Jorgenson. Springer-Verlag, New York, 2001. xvi+426 pp.

[2001b] (with Jay Jorgenson). Guinand's theorem and functional equations for the Cramér functions. *J. Number Theory* **86** (2001), no. 2, 351–367.

[2001b] (with Jay Jorgenson). *Spherical inversion on* $\mathrm{SL}_n(\mathbf{R})$. Springer Monographs in Mathematics. Springer-Verlag, New York, 2001. xx+426 pp.

[2001c] (with Jay Jorgenson). The ubiquitous heat kernel. *Mathematics Unlimited—2001 and Beyond*, Springer, Berlin, 2001, pp. 655–683.

[2001d] Comments on non-references in Weil's works. *Gaz. Math.* No. 90 (2001), 46–52. 01A80.

[2002a] Short Calculus. The original edition of *A First Course in Calculus*, [Addison-Wesley, Reading, MA, 1964.] Undergraduate Texts in Mathematics. Springer-Verlag, New York, 2002. xii+260 pp.

[2002b] *Algebra.* Revised third edition. Graduate Texts in Mathematics, Vol. 211. Springer-Verlag, New York, 2002. xvi+914 pp.

[2002c] Comments on non-references in Weil's works. *Mitt. Dtsch. Math.-Ver.* 2002, no. 1, 49–56.

[2002d] *Introduction to Differentiable Manifolds.* Second Edition. Universitext. Springer-Verlag, New York, 2002. xii+250 pp.

[2002e] Comments on non-references in Weil's work. (Italian) Translated from Mitt. Dtsch. Math.-Ver. 2002, no. 1, 49–56 [MR1900622 (2003c: 01058)]. *Lett. Mat. Pristem* No. **44** (2002), 29–37.

[2002f] Comments on Chow's work [see 1410978]. Reprinted from *Notices Amer. Math. Soc.* 43 (1996), no. 10, 1117–1124. *Nankai Tracts Math.*, 5, *Contemporary Trends in Algebraic Geometry and Algebraic Topology* (Tianjin, 2000), World Sci. Publ., River Edge, NJ, 2002, pp 243–250.

[2003a] (with Jay Jorgenson). Analytic continuation and identities involving heat, Poisson, wave and Bessel kernels. *Math. Nachr.* **258** (2003), 44–70.

[2003b] (with Jay Jorgenson). *Spherical inversion on* $\mathrm{SL}_2(\mathbf{C})$. *Heat kernels and analysis on manifolds, graphs, and metric spaces* (Paris, 2002), 241–270, Contemp. Math., 338, Amer. Math. Soc., Providence, RI, 2003.

[2005a] (with Jay Jorgenson). A Gaussian space of test functions. *Math. Nachr.* **278** (2005), no. 7-8, 824–832.

[2005b] On the AMS Notices publication of Krieger's translation of Weil's 1940 letter [MR2125268]. *Notices Amer. Math. Soc.* 52 (2005), no. 6, 612–622.

[2005c] (with Jay Jorgenson). $\mathrm{Pos}_n(\mathrm{R})$ and Eisenstein series. Lecture Notes in Mathematics, 1868. Springer-Verlag, Berlin, 2005. viii+168 pp.

[2008] (with Jay Jorgenson). *The heat kernel and theta inversion on* $\mathrm{SL}_2(\mathbf{C})$. Springer Monographs in Mathematics. Springer, New York, 2008. x+319 pp.

[2009] (with Jay Jorgenson). *Heat Eisenstein series on* $\mathrm{SL}_n(\mathbf{C})$. Mem. Amer. Math. Soc. 201 (2009), no. 946, viii+127 pp.

[2010] (with Jay Jorgenson). *Heat kernel, theta inversions, and zetas on* $\Gamma\backslash G/K$. In this volume.

Additionally, the following articles about Lang appeared after September 2005.

[2006a] Marc Hindry: La géométrie diophantienne, selon Serge Lang. (French) [Diophantine geometry according to Serge Lang] *Gaz. Math.* No. 1**108** (2006), 17–32.

[2006b] David E. Rohrlich: Serge Lang. *Gaz. Math.* No. **108** (2006), 33–34.

[2006c] Michel Waldschmidt: Les contributions de Serge Lang à la thorie des nombres transcendants. (French) [Serge Lang's contributions to the theory of transcendental numbers] *Gaz. Math.* No. **108** (2006), 35–46.

[2006d] Jay Jorgenson and Steven G. Krantz: Serge Lang, 1927–2005. *Notices Amer. Math. Soc.* 53 (2006), no. 5, 536–553.

[2006e] Obituary: Serge Lang (1927–2005). (Spanish) *Lect. Mat.* **27** (2006), no. 2, 166–167.

[2007] Jay Jorgenson and Steven G. Krantz: The mathematical contributions of Serge Lang. *Notices Amer. Math. Soc.* 54 (2007), no. 4, 476–497.

Introduction

John Tate

This introduction is meant as a brief account of Serge Lang's life and his enormously varied contributions to mathematics. Much more about this remarkable man can be found in two articles in the *Notices of the AMS* by Jay Jorgenson and Steven G. Krantz. The first of these, "Serge Lang, 1927–2005" (May 2006) contains a fuller account of Lang's life than we can give here, and includes memories of Serge by twenty-two of his friends, students, colleagues, and even some whom Serge might have seen as adversaries. Read together, these short pieces give a vivid image of Lang. The second article, "Mathematical Contributions of Serge Lang" (April 2007), contains an overview of his research, followed by discussions of its different aspects by seven colleagues in the various fields. These articles, and conversations with Lang's friends Dick Gross, Dinakar Ramakrishnan, and Ken Ribet, have been of great help to me in writing this introduction.

Lang spent his childhood in Saint-Germain-en-Laye, a western suburb of Paris famous for its chateau and long terrace with a view over the valley of the Seine and Paris in the distance. Lang's teen years were spent in quite different surroundings. After emigrating with his family to Los Angeles, he attended Beverly Hills High and Caltech, graduating in 1946 with a BA in physics.

He then did a year and a half of military service with the U.S. Army in Europe. This was of great help to him in his future career, for he served in a clerical position in which he learned to type at incredible speed.

Next, Serge enrolled in the graduate philosophy program at Princeton. Fortunately for mathematics he was disappointed by the quality of the philosophy seminars, and managed to switch to the math graduate program the following year. I don't know why he chose mathematics, but the Princeton math program was an outstanding one and student morale was very high.

J. Tate
Department of Mathematics, Harvard University, Cambridge, MA 02138, USA
e-mail: tate@math.utexas.edu

Though he was not particularly well prepared, Serge plunged right in. For example, knowing little number theory, he attended Emil Artin's seminar on class field theory during his first year and was fascinated. Math, especially algebra and number theory, were the subjects for him! He soon became one of Artin's Ph.D. students along with me and a few others that Artin had brought with him from Indiana. We all felt very fortunate to have Artin as our advisor. In the foreword to his collected works, Lang writes, "I take this opportunity to express once more my appreciation for having been Artin's student. I could not have had a better start in my mathematical life."

Lang got his Ph.D in 1951 with a thesis on quasi-algebraic closure in which he proved that a field complete in a discrete valuation with algebraically closed residue field is quasi-algebraically closed, and that the same is true for several kinds of dense subfields of such a field. Another result in his thesis is a key ingredient in the proof of the Ax–Kochen theorem,[1] which can be viewed as a corrected version of Artin's conjecture that p-adic fields have property C_2. For each degree d, this is true for all but a finite set of primes p, but not necessarily for all, as Artin had guessed.

Lang stayed in Princeton for two more years, with postdoc positions at the university and at the Institute for Advanced Study. Then, after two years of an instructorship at the University of Chicago, where he interacted with Weil and his circle, he accepted a permanent position at Columbia University. He stayed there for the next fifteen years except for a Fulbright Fellowship year in Paris during 1957–58. In 1970, Lang resigned his position at Columbia in protest of the university's handling of student unrest during 1968–69. After visiting professorships at Princeton and Harvard, he accepted a permanent position at Yale. He retired from Yale in 2005, a few months before his death. That is a bare-bones account of Lang's life and the way he got into mathematics.

Serge Lang contributed to mathematics in so many ways that it's hard to know where to begin. Let's start with some remarks on his research, with no attempt to cover it completely. He published his *Collected Papers* in five volumes with Springer in 2000. They contain all of his research papers through 1999, together with reprints of a few of his Springer Lecture Notes and two of his books that were out of print. There are also some interesting accounts of some special topics on which Lang held strong views, especially in Volume IV.

In a brief foreword in Volume 1, Lang gives his own classification of his work into periods as follows:

1.	1951–1954	Thesis on quasi-algebraic closure and related matters.
2.	1954–1962	Algebraic geometry and abelian (or group) varieties; geometric class field theory.[2]

[1] As Deligne pointed out, I misstated the Ax–Kochen theorem in my *Notices* article on Lang's early work, by interchanging "prime p" and "degree d." I hope this "senior moment" misled no one.

[2] This work was the beginning of higher-dimensional class field theory and earned Lang a Cole Prize in 1959.

3. 1963–1975	Transcendental numbers and Diophantine approximation on algebraic groups.
4. 1970	First paper on analytic number theory—jump to Jorgenson–Lang.
5. 1975	$SL_2(\mathbf{R})$—jump to Jorgenson–Lang.
6. 1972–1977	Lang–Trotter Frobenius distributions.
7. 1973–1981	Modular curves, Kubert–Lang modular units.
8. 1974, 1982–1991	Diophantine geometry, complex hyperbolic spaces, and Nevanlinna theory.
9. 1985, 1988	Riemann–Roch and Arakelov theory.
10. 1992–2000+	Jorgenson–Lang (analytic number theory and connections with spectral analysis, heat kernel, differential geometry, Lie groups, and symmetric spaces).

The arrangement of the 2007 *Notices* article fits quite well with Lang's scheme. Here is a list of the authors in the order of their appearance, followed by the periods of Lang's work they discuss. Tate 1,2; Buium 1,2,3; Waldschmidt 3; Rohrlich 6,7; Vojta 8,9; Jorgenson 10; and Kim, who wrote on the theme of the fundamental group in Lang's work, rather than on a specific period.

Lang wrote over 70 research papers and proved many important theorems, but of at least equal significance were his conjectures, his points of view, and his way of looking at things. Waldschmidt expresses this well (loc. cit.), writing about Lang's work on transcendental numbers:

> With his outstanding insight and his remarkable pedagogical gifts, Lang comes into the picture and contributes to the subject in at least two very different ways: on the one hand, he simplifies the arguments (sometimes excessively) and produces the first very clear proofs which can be taught easily; on the other hand, he introduces new tools, like group varieties, which put the topic closer to the interests of many a mathematician.

Waldschmidt concludes his article as follows:

> Among the contributions of Lang to transcendental number theory (also to Diophantine geometry), the least are not his many conjectures which shed a new light on the subject. On the contrary, he had a way of considering what the situation should be, which was impressive. Indeed, he succeeded in getting rid of the limits from the existing results and methods. He made very few errors in his predictions, especially if we compare them with the large number of conjectures he proposed. His description of the subject will be a guideline for a very long time.

As Vojta points out, the title of Lang's magnum opus, *Fundamentals of Diophantine Geometry* suggests that Serge's outlook on number theory was decidedly geometric. Mazur puts this beautifully in concluding his memory of Serge article in the *Notices*:

> Over the decades of mathematics, Lang was led, more specifically, by an over-arching vision, which he pursued through the agency of various fields of mathematics. The vision, baldly put, is that *geometry* is an extraordinarily striking dictator of qualitative *dipohantine* behavior. The still open *Conjecture of Lang* in higher dimensions continues to serve as a

guiding principle to the way in which the grand subjects of geometry and number theory meet, just as Serge himself served as an inspiror of generations of mathematicians, and a spokesman for intellectual honesty.

The conjecture of Lang to which Mazur refers is easy to state. A projective algebraic variety V defined over a number field $F \subset \mathbf{C}$ is *Mordellic* if and only if the corresponding complex space $V(\mathbf{C})$ is *hyperbolic*. Here Mordellic means that for each finite extension field E of F, the set $V(E)$ of points of V with coordinates in E is finite. Hyperbolic meant for Lang, when he made the conjecture in 1974, that the Kobayashi semidistance on $V(\mathbf{C})$ is actually a distance, but we now know, thanks to Brody (1978), that this property is equivalent to there being no nonconstant holomorphic map $\mathbf{C} \to V(\mathbf{C})$. A Riemann surface is hyperbolic if and only if its genus is ≥ 2, so that for curves, Lang's conjecture is equivalent to the famous Mordell conjecture proved by Faltings in 1983. In higher dimensions the conjecture is still open, though it has been proved for closed subvarieties of abelian varieties.

In the 1980s Lang thought deeply about the Mordellic–hyperbolic relationship and introduced plausible variants of the above conjecture which have turned out to have very interesting unexpected implications, such as the existence of a bound $B(g, F)$ depending only on g and F for the number of rational points on a curve of genus $g \geq 2$ defined over a number field F.

Serge led a regular life. During the winter holidays he visited his sister in Los Angeles. He spent the early summer in Europe and July–August in Berkeley, where he had an apartment. In Europe he spent a month in one place, Paris in the early years, Bonn later in his life, but also visited regularly other mathematical centers, Zurich, Berlin, Moscow.... In Berkeley he interacted with the large community of resident and visiting mathematicians.

Lang was an effective communicator, an excellent source of mathematical news. Dick Gross likens his gathering and distributing information to the cross-pollination of a bumblebee. Serge kept in touch with friends in many places, not only in person, by traveling, but by phone. If he had a question or thought for anyone anywhere in the world, he just picked up the phone. When Yale was considering an offer to Serge, I remember warning the department that if he accepted, its phone bill would at least double, but that in fact, the phoning he would be doing was just one more reason for making the offer.

A more important reason for Yale's doing so is that Lang was an excellent and caring teacher. This was recognized by his being awarded the Dylon Hixon Prize for Teaching Excellence in the Natural Sciences at Yale. We were reminded of the esteem in which his former students held him by the testimony of so many of them at the memorial meeting for Lang at Yale in February 2006. One of them, Anthony Petrello, announced the establishment of a Yale fund for an annual prize in Lang's honor which he was launching with a large seed contribution, and the promise of matching funds.

Another memorial to Lang is the Serge Lang Undergraduate Lecture Series at Berkeley. There, when students returned to classes at the end of August, Serge often gave talks to the Math Undergraduate Student Association (MUSA). His talks over the years were incorporated into his 1999 volume "Math talks for Undergraduates." These talks became formalized, and each year from 1999 to 2005 Serge gave a MUSA lecture at 4pm on the first day of classes. The lecture on August 25, 2005, "Weierstrass–Dirac Families," shortly before his death, was one of Serge Lang's last mathematics talks. In response to MUSA's request to somehow continue this tradition, the Department of Mathematics inaugurated the Serge Lang Undergraduate Lecture Series, each fall inviting someone to give a lecture for undergraduates. Ken Ribet made this happen and Anthony Petrello contributed a major part of the initial funding.

Serge was known not only for his support of his students, but also of his younger colleagues at the start of their careers. John Coates thought one of Serge's most remarkable qualities was his unstinting support of young mathematicians. Barry Mazur, after recounting his first encounter with Serge, writes:

> And Serge did this sort of thing through the decades, with many of the young; he would proffer to them gracious, yet demanding, invitations to engage as a genuine colleague—not teacher to student, but mathematician to mathematician; he did all this naturally, and with extraordinary generosity and success.

Lang was awarded the 1999 AMS Leroy P. Steele Prize for Mathematical Exposition "for his many mathematics books." The amount of mathematical knowledge that has been made accessible to students of all ages all over the world by Lang's more than 40 books is amazing to contemplate. Their range both in subject and in level is astonishingly broad. Most were new and modern for their time with Lang's insistence on functoriality and axiomatization. He was almost unique in the way he regularly learned new topics throughout his life, topics often not close to his main interests (algebra and number theory), and wrote textbooks on them (such as "Differentiable Manifolds" and $SL_2(\mathbf{R})$), thereby influencing new generations of students pursuing those fields. He brought excitement to his books that challenged readers to rise above themselves by tackling them.

Lang's Algebra is a classic, still the best reference book in algebra in print today. The first edition in 1965 has been kept up-to-date with new editions and revisions. The latest is the corrected fourth printing of the revised third edition (2004). For Lang, the important thing in a book is its timeliness, and its global aspects, such as arrangement of topics, and degree of abstraction. He did not worry much about an occasional error in a proof, and was widely criticized for this. Given the short time he spent writing a book, there are relatively few of these oversights, and when he became aware of one he was highly attentive to its correction in the next edition or the next printing. These oversights could be useful. I remember a few times first recognizing a student who turned out to be very strong when s/he came for help in understanding one of Lang's faulty arguments.

Lang was driven to publish. In addition to his own writing, he saw to the publication of at least two books which were not his own, namely *Class Field Theory* by Artin and me, and the Collected Papers of Emil Artin. He should really have been included among the authors of the former, for the main part of it is essentially Serge's rewriting of his own notes from the 1951–52 Princeton seminar, and the book's publication would never have happened without Serge's persistent prodding and attending to the details. In the case of Artin's papers, edited by Serge Lang and John T. Tate, it was Lang who did all the work in collecting the material and preparing it for publication. He assigned me the trivial task of writing a paragraph on Artin's conjectures to justify my sharing the editorship. In the first printing, Addison-Wesley made the mistake of including Lang's and my names on the cover, below Artin's. Lang hit the ceiling, insisting that the whole printing be redone. Fortunately a solution was found. Our names are masked by a decorative strip which was added to the cover, although they are discernible if you know they are there.

To touch on some of Serge's personal characteristics: He did things fast. He typed fast, ate fast, drove fast, walked fast. I still remember trying to keep up with him on walks across campus. On a wintry day he used only earmuffs and gloves (no coat) to keep warm. He seemed to have a hummingbird-like metabolism. Besides being fast, he did not waste time. He hated small talk.

Serge cared about quality. His possessions and gifts were chosen thoughtfully. He had impeccable taste and sought out fine things. I think of his leather jacket, his gloves, his silk scarf and clothing generally, but also of his appliances, his collection of rugs, and his furniture. In the summer of 1958, after a year in Paris, we each went through Copenhagen to purchase Danish furniture. Serge chose from the best collections, I from the mid-range.

Except for an occasional sip of a vintage wine, Lang did not drink, but he insisted on supplying a fine cognac for Yale's (and often Harvard's as well) winter holiday party eggnog. The batter for the crepes suzettes he served was beaten by hand with a whisk, never with a mechanical mixer.

It is hard to imagine that Serge's many mathematical activities left time for much else, but he was an accomplished musician, playing piano and lute, and in later years he spent much time and energy exposing cases of what he saw as scientific, editorial or bureaucratic irresponsibility, by compiling and distributing, at considerable personal expense for photocopying and postage, collections[3] of relevant original documents, in his aim to create transparency and promote clear understanding of the facts of a case. He also supported many people and causes he found to be worthy, in ways ranging from enthusiastic encouragement to assistance in funding.

In Serge's battles with the establishment, his positions were almost always fundamentally sound, though the extremes to which he went and the vehemence with which he pursued his points of view may have prevented some from realizing this. Serge could be difficult. To him things were black or white. To compromise was not his way.

Serge Lang devoted his life to advancing mathematics, to teaching, and to fighting for honesty in science and politics.

[3]These collections were known as "files" to the people on Lang's extensive "cc lists."

Raynaud's group-scheme and reduction of coverings

Dan Abramovich with an Appendix by Jonathan Lubin

In grateful memory of Serge Lang

Abstract Let $Y_K \to X_K$ be a Galois covering of smooth curves over a field of characteristic 0, with Galois group G. We assume K is the fraction field of a discrete valuation ring R with residue characteristic p. Assuming $p^2 \nmid G$ and the p-Sylow subgroup of G is normal, we consider the possible reductions of the covering modulo p. In our main theorem we show the existence, after base change, of a twisted curve $\mathcal{X} \to \text{Spec}(R)$, a group scheme $\mathcal{G} \to \mathcal{X}$ and a covering $Y \to \mathcal{X}$ extending $Y_K \to X_K$, with Y a stable curve, such that Y is a $\mathcal{G}$-torsor.

In case $p^2 | G$ counterexamples to the analogous statement are given; in the appendix a strong counterexample is given, where a non-free effective action of α_p^2 on a smooth 1-dimensional formal group is shown to lift to characteristic 0.

Key words Algebraic curves • Galois coverings

Mathematics Subject Classification (2010): 14H25, 14H30

1 Introduction

1.1 Reduction of coverings of degree divisible by p

Let R be a discrete valuation ring of mixed characteristics, with spectrum $S = \text{Spec}\, R$. Denote the generic point η with fraction field K, and the special

D. Abramovich (✉) • J. Lubin
Department of Mathematics, Brown University, Box 1917, Providence, RI, 02912
e-mail: abrmovic@math.brown.edu; lubinj@math.brown.edu

D. Goldfeld et al. (eds.), *Number Theory, Analysis and Geometry: In Memory of Serge Lang*, DOI 10.1007/978-1-4614-1260-1_1,

point s with residue field k of characteristic $p > 0$. Consider a generically smooth, stable pointed curve $Y \to S$ with an action of a finite group G of order divisible by p. Denote $X = Y/G$. We assume that G acts freely on the complement of the marked points in Y_η; it then follows that G respects the branches of all nodes of Y_s.

In situations where the order of G is prime to the residue characteristic, the reduced covering $Y_s \to X_s$ is an admissible G-covering, and a nice complete moduli space of admissible G-coverings exists. An extensive literature exists describing that situation, see e.g., [H-M, Mo, Ek, W, ℵ-C-V]. However, in our case where the residue characteristic divides the order of G, interesting phenomena occur (see e.g., [ℵ-Oo]). The situation was studied by a number of people; we will concern ourselves with results of Raynaud [Ra] and, in a less direct way, Henrio [He] and more recently Maugeais [Ma]. Related work of Saidi [Sa1, Sa2, Sa3], Wewers and Bouw [W1, W2, W3, Bo, B-W1, B-W2], Romagny [Ro] and others provides additional inspiration. In [ℵ-O-V2, Section 5] the curve Y is replaced by something which could be much more singular, and therefore the results are somewhat orthogonal to the situation here.

Thus, in our case where $p \mid |G|$, the covering $Y \to X$ is no longer generically étale on each fiber. It is natural to consider some sort of group-scheme degeneration $\mathcal{G} \to X$ of G, in such a way that Y might be considered something like an admissible $\mathcal{G}$-covering. In our main theorem we show this is the case under appropriate assumptions:

Theorem 3.2.2. Assume $p^2 \nmid |G|$ and the p-Sylow subgroup of G is normal.
There exist

(1) a twisted curve $\mathcal{X} \to X$,
(2) a finite flat group scheme $\mathcal{G} \to \mathcal{X}$,
(3) a homomorphism $G_{\mathcal{X}} \to \mathcal{G}$ which is an isomorphism on $\mathcal{X}_K$,
(4) a lifting $Y \to \mathcal{X}$ of $Y \to X$, and
(5) an action of $\mathcal{G}$ on Y through which the action of G factors,

such that $Y \to \mathcal{X}$ is a principal $\mathcal{G}$-bundle.
The formation of $\mathcal{G}$ commutes with any flat and quasi-finite base change $R \subset R'$.

It is important to note that, unlike the characteristic 0 case, X is not a stable pointed curve in general.

1.2 Background

Raynaud ([Ra], Proposition 1.2.1) considered such a degeneration locally at the generic points of the irreducible components of X_s, in the special case where $|G| = p$; our first goal, see Theorem 3.1.1 below, is to work out its extension to the smooth locus of X, and slightly more general groups, where as above $p^2 \nmid |G|$ and the p-Sylow subgroup of G is normal. The case where $p^2 \mid |G|$ remains a question which I find very interesting. See Example 2.1.7 and the appendix by J. Lubin for a negative result in general, the discussion of question 2.1.5 for positive results in the literature, and Remark 2.1.8 for a positive result for small ramification.

One still needs to understand the structure of $Y \to X$ at the nodes of X_s and Y_s. Henrio, working p-adic analytically, derived algebraic data along X_s, involving numerical combinatorial invariants and differential forms, which in some sense classify $Y_s \to X_s$. Our second goal in this note is to present a different approach to such degenerations at a node, modeled on *twisted curves*, i.e. curves with algebraic stack structures. Borrowing a metaphor from A. Ogus, these twisted curves have served well in the past as a sort of "magic powder" one sprinkles over the "bad locus" of certain structures, which brings about a hidden good property. The point here is that, just as in [ℵ-C-V], the introduction of twisted curves allows one to replace $Y \to X$ by something that is actually a principal bundle. Unlike the case of residue characteristics prime to $|G|$, the twisted curves will in general be Artin stacks rather than Deligne–Mumford stacks. See [Ol], [ℵ-O-V2].

1.3 *Towards a proper moduli space*

The main theorem should be thought of as a first step in constructing a nice proper moduli space of degenerate coverings in mixed characteristics - it gives a special case of the valuative criterion for properness. In joint work with M. Romagny we plan to complete this task. Foundations have only recently been developed in [ℵ-O-V, ℵ-O-V2].

1.4 *Brief introduction to twisted curves*

A twisted pointed curve over a base scheme S is a diagram as follows:

$$\begin{array}{ccc}
\Sigma_i^{\mathcal{C}} & \hookrightarrow & \mathcal{C} \\
\downarrow & & \downarrow \\
\Sigma_i^{C} & \hookrightarrow & C \\
& & \downarrow \\
& & S.
\end{array}$$

Here we follow [ℵ-O-V2, Section 2]:

- $C \to S$ is a usual n-pointed nodal curve, with sections $\Sigma_i^C, i = 1, \ldots, n$;
- $\mathcal{C}$ is an algebraic stack with finite diagonal having C as its coarse moduli space;

- $\Sigma_i^{\mathcal{C}} \subset \mathcal{C}$ are its markings, each of which a gerbe banded by some $\boldsymbol{\mu}_{r_i}$ over $\Sigma_i^C \simeq S$;
- $\mathcal{C} \to C$ is an isomorphism away from nodes and markings of C;
- at a marking of C, where the strict henselization C^{sh} is described by $(\mathrm{Spec}_S\, \mathcal{O}_S[x])^{\mathrm{sh}}$ and Σ_i^C is the vanishing locus of x, the twisted curve $\mathcal{C}$ is described as
$$\left[\, (\mathrm{Spec}_S\, \mathcal{O}_S[u])^{\mathrm{sh}} \,/\, \boldsymbol{\mu}_{r_i} \,\right],$$
where $\boldsymbol{\mu}_{r_i}$ acts on u via the standard character and $u^{r_i} = x$, and $\Sigma_i^{\mathcal{C}}$ is the quotient of the vanishing locus of u;
- at a node of C, where the strict henselization C^{sh} is described by $(\mathrm{Spec}_S\, \mathcal{O}_S[x, y]/\,(xy - f))^{\mathrm{sh}}$ with $f \in (\mathcal{O}_S)^{\mathrm{sh}}$, the twisted curve $\mathcal{C}$ is described as
$$\left[\, (\,\mathrm{Spec}_S\, \mathcal{O}_S\, [u, v] \,/\, (uv - g)\,)^{\mathrm{sh}} \,/\, \boldsymbol{\mu}_r \,\right]$$
for some r, where $\boldsymbol{\mu}_r$ acts via
$$(u, v) \mapsto (\zeta_r u, \zeta_r^{-1} v),$$
and $u^r = x, v^r = y$ and $g^r = f$.

Of course the description on the level of strict henselization descends to some étale neighborhoods. In case p divides r_i or r, the twisted curve $\mathcal{C}$ is not a Deligne–Mumford stack, and it is a little bit of a miracle, following from [ℵ-O-V2, Proposition 2.3], that one can use such a nice description locally in the étale topology (or on strict henselizations) rather than the f.p.p.f. topology. The reader is warned that transition isomorphisms between the étale local charts are in general not given in étale neighborhoods but rather in smooth or f.p.p.f. charts.

Near a marking Σ_i^C, the twisted curve is determined, uniquely up to a unique isomorphism, by the choice of r_i. In fact locally in the Zariski topology of C we can write $\mathcal{C} = \sqrt[r_i]{(C, \Sigma_i^C)}$, see [ℵ-G-V] for notation and proof. Near a node, $\mathcal{C}$ is still uniquely determined by r, but not up to a unique isomorphism - in fact $\mathrm{Aut}_C\, \mathcal{C}$ has a factor $\boldsymbol{\mu}_r$ for each such twisted node of index r, see [ℵ-C-V].

1.5 Acknowledgements

Thanks to Angelo Vistoli for help, and to F. Andreatta, A. Corti, A.J. de Jong, M. Rosen and N. Shepherd-Barron for patient ears and useful comments. I also heartily thank J. Lubin, who pointed me in the direction of Example 2.1.7, and in particular saved me from desperate efforts to prove results when $p^2 \mid |G|$. I am indebted to M. Romagny, whose beautiful computation of a key example in residue characteristic 2 clarified the situation at hand and led to a big improvement in the results obtained.

Thanks to the referee for a careful reading, helpful suggestions, and for pointing out important developments in recent papers.

Research partially supported by NSF grant DMS-0070970, a Forscheimer Fellowship and Landau Center Fellowship

2 Extensions of group-schemes and their actions in dimension 1 and 2

2.1 Raynaud's group-scheme

Raynaud (see [Ra], Proposition 1.2.1, see also Romagny, [Ro]) considers the following construction: let U be integral and let V/U be a finite flat G-invariant morphism of schemes, with G finite. Assume that the action of G on the generic fiber of V/U is faithful. We can view this as an action of the constant group scheme G_U on V, and we consider the schematic image $\mathcal{G}$ of the associated homomorphism of group-schemes

$$G_U \to \mathrm{Aut}_U V.$$

Since, by definition, the image $G_U \to U$ is finite, we have that $\mathcal{G} \to U$ is finite as well. The scheme $\mathcal{G} \to U$ can also be recovered as the closure of the image of the generic fiber of G_U, which is, by the faithfulness assumption, a subscheme of $\mathrm{Aut}_U V$. By definition $\mathcal{G}$ acts faithfully on V.

Definition 2.1.1. We call the scheme $\mathcal{G}$ the effective model of G acting on V/U.

Note that a priori we do not know that $\mathcal{G}$ is a group-scheme. It is however automatically a flat group-scheme if U is the spectrum of a Dedekind domain. This follows because, in that case, the image of $\mathcal{G} \times_U \mathcal{G} \to \mathrm{Aut}_U V$ is also flat, and therefore must coincide with $\mathcal{G}$.

Also note that, if s is a closed point of U whose residue characteristic is prime to the order of G, then the fiber of $\mathcal{G}$ over s is simply G. So this effective model is only of interest when the residue characteristic divides $|G|$.

The following is a result of Raynaud, see [Ra], Proposition 1.2.1. The statement here is slightly extended as Raynaud assumes $|G| = p$:

Proposition 2.1.2. *Let U be the spectrum of a discrete valuation ring, with special point s of residue characteristic p and generic point η. Let $V \to U$ be a finite and flat morphism, and assume that the fiber V_s of V over s is reduced (but not assuming geometrically reduced). Assume given a finite group G, with normal p-Sylow subgroup,* such that *$p^2 \nmid |G|$, and an action of G on V such that $V \to U$ is G-invariant, and such that the generic fiber $V_\eta \to \{\eta\}$ is a principal homogeneous space. Let $\mathcal{G} \to U$ be the effective model of G acting on V/U.*

Then V/U is a principal bundle under the action of $\mathcal{G} \to U$.

Remark 2.1.3. An analogous construction in a wider array of cases is given in Romagny's [Ro1, Theorem A]. Romagny does not aim to construct a principal bundle; on the other hand he shows that an effective model for an action exists even if V/U and G/U are not finite, under very mild hypotheses.

Proof. As in Raynaud's argument, it suffices to show that the stabilizer of the diagonal of $V_s \times_U V_s$ inside the group-scheme $V_s \times_U \mathcal{G}$ is trivial. Since G acts transitively on the closed points t_i of V_s sending the stabilizer on t_i to that over t_j, it is enough to show that one of these stabilizers, say over $t \in V_s$, is trivial. But this stabilizer P is a group-scheme over the residue field $k(t)$ with degree $\deg P \mid p$, and if nontrivial it is of degree exactly p. In such a case it must coincide with the pullback of the unique p-Sylow group-subscheme of $\mathcal{G}$; therefore that p-sylow acts trivially, contradicting the fact that $\mathcal{G}$ acts effectively. □

Remark 2.1.4. In case the inertia group is not normal, Raynaud passes to an auxiliary cover, which encodes much of the behavior of $V \to U$.

Question 2.1.5. *What can one say about the action of $\mathcal{G}$ on V in case the order of G (and the degree of $V \to U$) is divisible by p^2, but the inertia group is still normal? Specifically, what happens if $|G| = p^2$?*

In the latter case, consider a subgroup $P \subset G$ of order p. It can be argued, as in Raynaud's proof, that the effective model $\mathcal{P} \to U$ of P acts freely on V, and thus $V \to V/P$ is a principal $\mathcal{P}$-bundle. Similarly, if $\mathcal{Q}$ is the effective model of G/P acting on V/P, then $V/P \to U$ is a principal $\mathcal{Q}$-bundle. At the same time, we have an action of the effective model $\mathcal{G}$ of G on V/P, but it is not necessarily the case that $\mathcal{G}/\mathcal{P} \to \mathcal{Q}$ is an isomorphism.

While the statement of Question 2.1.5 is somewhat vague, two definite answers can already be given. First, if one concentrates on effective models of the action in the sense of Romagny, a great deal can be said. The recent work of Tossici [To1,To2] concentrates on the case where $G_K = \mathbb{Z}/p^2\mathbb{Z}$ and $\mathcal{O}_U$ contains a primitive root of unity of order p^2. The paper [To1] describes explicitly the possible models $\mathcal{G}$ of G_K; in [To2] an explicit description of the effective model of G_K acting on V is provided. I think it would be of interest to see if results as in Theorems 3.1.1 and 3.2.2 can be obtained for more general effective models such as these.

Second, in general no model of $\mathcal{G}$ will act freely on V. This is the case even for some of the prettiest actions one can consider. This makes giving a complete answer to the previous Question 2.1.5 tricky, and underscores the importance of work such as Tossici's.

As probably the simplest example, consider an action of $\mathcal{G}_0 = (\alpha_p)^2 = \operatorname{Spec} k[a,b]/(a^p, b^p)$ on $k(t)$. Examples of liftings of a non-free action of the type

$$t \mapsto t + a + f(t)b$$

for any residue characteristic have been written down by Romagny (personal communication) and Saidi (see [Sa4]). The case of

$$t \mapsto t + a + t^p b \tag{1}$$

is particularly appealing, as it involves torsion and endomorphisms of a formal group. I therefore ask

Question 2.1.6. *Can one lift the action (1) to characteristic 0?*

A formal positive answer in arbitrary residue characteristics is given by Jonathan Lubin in the appendix. Here I discuss explicitly the case of residue characteristic 2, where this action can be obtained as a reduction of an action of $(\mathbb{Z}/2\mathbb{Z})^2$ on a smooth curve. I concentrate on the local picture (making it global is not difficult):

Example 2.1.7. Let $R = \mathbb{Z}_2[\sqrt{2}]$. Consider the group-scheme Y/R defined by

$$t * t' = t + t' + \sqrt{2}\, t\, t'$$

This is an additive reduction of the multiplicative group. The reduction of the subgroup μ_2 is given as

$$\operatorname{Spec} R[a] \Big/ \left(a\,(a + \sqrt{2})\right),$$

reducing to α_2. It acts on Y by translation via the addition law as above:

$$t \mapsto t + a + \sqrt{2}\, a\, t.$$

The reduction of the action of $\mathbb{Z}/2\mathbb{Z}$ by inversion is the same group-scheme, again reducing to α_2, which we write as

$$\operatorname{Spec} R[b] \Big/ \left(b\,(b + \sqrt{2})\right).$$

This time the action is given by

$$t \mapsto \left(1 + \sqrt{2}\, b\right) t - \frac{b\, t^2}{1 + \sqrt{2}\, t}.$$

Since 2-torsion is fixed by inversion, these actions commute. Explicitly, the action of the product is given by

$$\begin{aligned} t \mapsto a &+ \left(1 + \sqrt{2}\, b\right) t - \frac{b\, t^2}{1 + \sqrt{2}\, t} \\ &+ \sqrt{2}\, a \left(\left(1 + \sqrt{2}\, b\right) t - \frac{b\, t^2}{1 + \sqrt{2}\, t} \right). \end{aligned}$$

The reduction modulo $\sqrt{2}$ is given by

$$t \mapsto t + a + t^2 b,$$

as required.

Remark 2.1.8. Raynaud's arguments do work when $p^2 \mid |G|$ if the p-Sylow group-scheme of $\mathcal{G}$ has only étale and cyclotomic Jordan–Hölder factors. This is because, in that case, there are no nonconstant group subschemes in the reduction. In particular this works whenever the absolute ramification index over $\mathbb{Z}_p$ is $< p$.

2.2 Extension from codimension 1 to codimension 2

Consider now the case where U a Gorenstein noetherian scheme, $\dim U = 2$, and V/U finite flat and G-invariant as above. Consider the S2-saturation $\mathcal{G}' \to \mathcal{G}$ of the effective model $\mathcal{G}$ of the G action on V/U. In Section 6.1.2 of [Va] Vasconcelos considers such saturation (S_2-ification in his terminology). His Proposition 6.21 on page 318 applies in our situation, and gives the existence and a characterization of the S2-saturation. We have

Lemma 2.2.1. *If $\mathcal{G}' \to U$ is flat, then $\mathcal{G}'$ is a group-scheme acting on V.*

Proof. We claim that the rational map $\mathcal{G}' \times_U \mathcal{G}' \dashrightarrow \mathcal{G}'$ induced by multiplication in $\mathrm{Aut}_U V$ is everywhere defined. Indeed the graph of this map is finite over $\mathcal{G}' \times_U \mathcal{G}'$ and isomorphic to it over the locus where $\mathcal{G}' \to \mathcal{G}$ is an isomorphism, whose complement has codimension ≥ 2. Now $\mathcal{G}'$ is S2 and of dimension 2, hence Cohen–Macaulay. Pulling back to $\mathcal{G}'$ the flat Cohen–Macaulay $\mathcal{G}' \to U$ we get that $\mathcal{G}' \times_U \mathcal{G}'$ is Cohen–Macaulay, in particular S2. This implies that the graph of $\mathcal{G}' \times_U \mathcal{G}' \dashrightarrow \mathcal{G}'$ is isomorphic to $\mathcal{G}' \times_U \mathcal{G}'$, and therefore the map is regular. The same works for the map defined by the inverse in $\mathrm{Aut}_U V$. This makes $\mathcal{G}'$ a group-scheme, and the map $\mathcal{G}' \to \mathrm{Aut}_U V$ into a group-homomorphism. □

This applies, in particular, when U is regular:

Lemma 2.2.2. *If U is regular, the S2-saturation $\mathcal{G}'$ of the effective model $\mathcal{G}$ is a finite flat group-scheme acting on V.*

Proof. Again $\mathcal{G}'$, being 2-dimensional and S2, is Cohen–Macaulay, and being finite over the nonsingular scheme U, it is finite *and flat* over U (indeed its structure sheaf, being saturated, is locally free over the nonsingular 2-dimensional scheme U). The result follows from Lemma 2.2.1. □

When the action on the generic fiber is free, we have more:

Proposition 2.2.3. *Let U be a Cohen–Macaulay integral scheme with* $\dim U = 2$. *Let $V \to U$ be a G invariant, finite, flat and Cohen–Macaulay morphism, and*

assume the action of G on the generic fiber is free. Let $\mathcal{G} \to U$ be the effective model of the action. Assume that for every codimension -1 point ξ, the action of the fiber $\mathcal{G}_\xi$ on V_ξ is free.

Then

(1) $\mathcal{G} \to U$ is a flat group-scheme, and
(2) The action of $\mathcal{G}$ on V is free.

Note that, by Raynaud's result 2.1.2, the assumptions hold when $U = V/G$ is local of mixed characteristics $(0, p)$, the fibers V_ξ are reduced, the p-Sylow of G is normal and $p^2 \nmid |G|$.

Proof. Consider the S2-saturation $\mathcal{G}'$ of $\mathcal{G}$. Since $V \to U$ is flat and Cohen–Macaulay, the same is true for $V \times_U V \to V$ and for $\mathcal{G}' \times_U V \to \mathcal{G}'$. Since $\mathcal{G}'$ and V are Cohen–Macaulay, we have that $V \times_U V$ and $\mathcal{G}' \times_U V$ are Cohen–Macaulay, hence S2, as well. The morphism $\mathcal{G}' \times_U V \to V \times_U V$ induced by the action $\mathcal{G}' \to \mathrm{Aut}_U V$ is finite birational and restricts to an isomorphism in codimension 1. By the S2 property it is an isomorphism. In particular we have that $\mathcal{G}' \times_U V \to V$ is flat, and since $V \to U$ is faithfully flat we have that $\mathcal{G}' \to U$ is flat. By Lemma 2.2.1 we have that $\mathcal{G}' \to U$ is a finite flat group-scheme acting on V, and the isomorphism $\mathcal{G}' \times_U V \to V \times_U V$ shows that the action is free, in particular $\mathcal{G}' \to \mathcal{G}$ is an isomorphism. □

3 Curves

3.1 The smooth locus

The main case of interest for us is the following:

Let R be a complete discrete valuation ring of mixed characteristic, with fraction field K of characteristic 0, residue field k of characteristic $p > 0$, and spectrum S. Assume $Y \to S$ is a stable pointed curve with smooth generic fiber, G a finite group acting on Y over S, and denote

$$X = Y/G.$$

We assume that the closure of the locus of fixed points of G in Y_K forms a disjoint union of marked sections of the smooth locus Y_{sm}. Hence for every node $y \in Y$, the stabilizer in G of y keeps the branches of Y at y invariant. We denote the complement of the closure in Y_{sm} of the fixed locus of the generic fiber by Y_{gen}, and the image in X by X_{gen} — the so called *general locus*.

Note that the morphism $Y_{\mathrm{sm}} \to X_{\mathrm{sm}}$ is flat.

The propositions above give:

Theorem 3.1.1. *Assume $p^2 \nmid |G|$ and the p-Sylow subgroup of G is normal.*

There exist

(1). a finite flat group-scheme $\mathcal{G}_{\text{sm}} \to X_{\text{sm}}$*,*
(2). a homomorphism $G_{X_{\text{sm}}} \to \mathcal{G}_{\text{sm}}$ *which is an isomorphism over* X_K*, and*
(3). an action of $\mathcal{G}_{\text{sm}}$ *on* Y_{sm} *through which the action of* G *factors,*

such that $Y_{\text{gen}} \to X_{\text{gen}}$ *is a principal* $\mathcal{G}_{\text{sm}}$*-bundle.*

The formation of $\mathcal{G}_{\text{sm}}$ *commutes with any flat and quasi-finite base change* $R \subset R'$.

Proof. Let $\mathcal{G}_{\text{sm}} \to X_{\text{sm}}$ be the S2-saturation of the effective model of the action of G on Y_{sm}. As X_{sm} is smooth we can apply Lemma 2.2.2; therefore $\mathcal{G}_{\text{sm}} \to X_{\text{sm}}$ is a finite flat group-scheme acting on Y_{sm}, giving (1) and (3). Part(2) applies since over K the group G does not degenerate.

The assumptions on G mean we can apply Proposition 2.1.2, so the action of $\mathcal{G}_{\text{sm}}|X_{\text{gen}}$ on Y_{gen} is free in codimension-1. We can therefore apply Proposition 2.2.3, and obtain that $Y_{\text{gen}} \to X_{\text{gen}}$ is a principal bundle.

The formation of $\mathcal{G}_{\text{sm}}$ clearly commutes with base change when restricted to the locus where it acts freely, and also over X_K. As it is flat and $S2$, its formation also commutes with base change across the remaining codimension-2 locus. □

It would be really interesting to see what happens for other groups G.

3.2 The structure of Y and G over nodes and markings of X

What can be done about the singular points and markings of X and Y? It is easy to see that even in the case of characteristic 0, the cover $Y \to X$ is not a principal bundle in general; it is already not a principal bundle at the fixed points of Y_K, and rarely a principal bundle at the nodes. However, the behavior of $Y \to X$ at the nodes is very interesting. My suggested approach here is to follow the method of [ℵ-V1, ℵ-V2, ℵ-C-V, Ol, ℵ-O-V2] using twisted curves. Let us first consider the cover $Y \to X$ itself and investigate its structure from this point of view.

Consider first a node $P \in X$ where étale locally X^{sh} is described by the equation $xy = \pi^m$, with π a uniformizer in S. Similarly, take a node $Q \in Y$ over P with local equation $st = \pi^n$. Say the local degree of $Y \to X$ at Q is d, so without loss of generality we can write $x = s^d \mu$ and $y = t^d \nu$, where μ and ν are units on Y^{sh}. Comparing the Cartier divisors of x, y, s, t and π on Y^{sh}, we get that $m = dn$, and $\mu\nu = 1$. Note that, since G acts transitively on the points of Y lying over $P \in X$, the degree d is independent of the choice of Q, and we may denote it d_P, to indicate its dependence on P.

Consider now the twisted curve $\mathcal{X}$ having index d_P at each node P. Recall from above that it has local description

$$\left[\left(\operatorname{Spec}_S \mathcal{O}_S\,[u,v] \,/\, (uv - \pi^n) \right)^{\text{sh}} \,/\, \boldsymbol{\mu}_d \right].$$

Write $Z = (\mathrm{Spec}_S\, \mathcal{O}_S\,[u, v] \,/\, (uv - \pi^n))^{\mathrm{sh}}$. We stress again that up to a non-unique isomorphism $\mathcal{X}$ does not depend on the choice of local coordinates.

Lemma 3.2.1. *There is a lifting, unique up to a unique isomorphism, of $Y \to X$ to a finite flat Cohen–Macaulay morphism $Y \to \mathcal{X}$.*

Proof. Recall that the coordinate s on Y^{sh} is related to x via $x = s^d \mu$ with μ a unit. Consider the $\boldsymbol{\mu}_d$-cover $P \to Y^{\mathrm{sh}}$ given by

$$P = \mathrm{Spec}\, \mathcal{O}_{Y^{\mathrm{sh}}}[w]/(w^d - \mu),$$

using the same unit μ, where $\boldsymbol{\mu}_d$ acts via $w \mapsto \zeta_d w$. Define a morphism $P \to Z$ via $u = sw$ and $v = t/w$. This morphism is clearly equivariant, giving a morphism $Y^{\mathrm{sh}} \to [Z/\mu_p] = \mathcal{X}^{\mathrm{sh}}$. Since $(sw)^p = x = u^p$ and $(t/w)^p = y = v^p$ this lifts the given map $Y \to X$. It is a tedious but straightforward exercise to show that the morphism on strict henselization descends to give the required morphism $Y \to \mathcal{X}$. The uniqueness statement follows from the fact that $\mathcal{X}$ is a separated stack. To check that $Y \to \mathcal{X}$ is flat it suffices to show P $\to$ Z flat. This follows from the local criterion for flatness: the fiber over $u = v = 0$ is given by $s = t = 0, w^d = c$ where c is the constant coefficient of μ at $s = t = 0$. This is a scheme of degree precisely d as required. Since Y and $\mathcal{X}$ (or, for that matter, P and Z) are Cohen–Macaulay, the morphism is Cohen–Macaulay. □

We now have our main theorem:

Theorem 3.2.2. *Assume $p^2 \nmid |G|$ and the p-Sylow subgroup of G is normal.*
There exist

(1) a twisted curve $\mathcal{X} \to X$,
(2) a finite flat group scheme $\mathcal{G} \to \mathcal{X}$,
(3) a homomorphism $G_{\mathcal{X}} \to \mathcal{G}$ which is an isomorphism on $\mathcal{X}_K$,
(4) a lifting $Y \to \mathcal{X}$ of $Y \to X$, and
(5) an action of $\mathcal{G}$ on Y through which the action of G factors,

such that $Y \to \mathcal{X}$ is a principal $\mathcal{G}$-bundle.
The formation of $\mathcal{G}$ commutes with any flat and quasi-finite base change $R \subset R'$.

Proof. There are two issues we need to resolve here: the construction of $\mathcal{G}$ at the nodes, and the construction of $\mathcal{X}$ and $\mathcal{G}$ at the markings.

First we need to extend $\mathcal{G}$ over nodes. We have that $Y \to \mathcal{X}$ is flat and Cohen–Macaulay at the nodes; by Theorem 3.1.1 we have that $Y_{\mathrm{gen}} \to X_{\mathrm{gen}}$ is a principal bundle under $\mathcal{G}_{\mathrm{gen}}$. By Proposition 2.2.3 the effective model $\mathcal{G} \to \mathcal{X}$ of the action of $G_{\mathcal{X}}$ on the $\mathcal{X}$-scheme Y is a finite flat group-scheme over $\mathcal{X}$, and away from the markings Y is a principal bundle.

Next, we deal with the markings: the local picture of $Y_K \to X_K$ at a marking is $Y^{\mathrm{sh}} = (\mathrm{Spec}\, R[s])^{\mathrm{sh}}$ and $X^{\mathrm{sh}} = (\mathrm{Spec}\, R[x])^{\mathrm{sh}}$ where $x = s^d$, and the stabilizer in G of $s = 0$ on Y is identified with $\boldsymbol{\mu}_d$, acting via $s \mapsto \zeta s$. We give $\mathcal{X}$ the unique structure of a twisted curve with index d along this marking; locally around the marking $s = 0$ we have $\mathcal{X}^{\text{sh}} = [(\text{Spec}\, \mathcal{O}_S[u])^{\text{sh}}/\boldsymbol{\mu}_d]$. The discussion above shows that $Y_K \to \mathcal{X}_K$ is a principal G-bundle. Applying Proposition 2.2.3 again we obtain that $\mathcal{G}$ is a group-scheme and Y is a principal bundle. □

So, in view of the characteristic 0 discussion in [ℵ-C-V], we might call $\mathcal{Y} \to \tilde{\mathcal{X}}$ a twisted $\mathcal{G}$-bundle.

This suggests an approach to lifting covers from characteristic p to characteristic 0, by breaking it in two stages: (1) lifting group-schemes over $\mathcal{X}$, and (2) lifting the covers. Recent work of Wewers [W3] seems to support such an approach.

Appendix A. Lifting a non-free action on a formal group

by Jonathan Lubin

The Question. In characteristic $p > 0$, consider the substitution $t \mapsto a + t + bt^p$, where $a^p = b^p = 0$. This clearly defines a group-scheme of rank p^2, isomorphic to $\alpha_p \times \alpha_p$, and an action of the groups-cheme on a curve, in this case the affine line. Question 2.1.6 asked whether this group-scheme and this action can be lifted to characteristic zero, over a suitably ramified extension of $\mathbb{Z}_p$.

The Answer. It's a partial yes, in that the example presented here shows an action not on the affine line but on the formal version of this, the formal spectrum of $\mathfrak{O}[[t]]$, where $\mathfrak{O}$ is the ring of integers in a well-chosen ramified extension of $\mathbb{Q}_p$. But if the question is whether there is any example of an action of $\alpha_p \times \alpha_p$ on a genuine algebraic curve in characteristic zero, then I must plead ignorance.

In general, if R is a ring and f and g are power series in one variable over R, then it makes no sense to compose the series, $f \circ g$, unless g has zero constant term. Yet, there are situations where R has a suitable complete topology, when $f \circ g$ can make sense even when $g(0) \neq 0$. Let us detail one fairly general such situation:

If $(\mathfrak{o}, \mathfrak{m})$ is a complete local ring, then on the category of complete local $\mathfrak{o}$-algebras (R, M) we define a group functor denoted $\mathfrak{B}$ or $\mathfrak{B}_{\mathfrak{o}}$, such that $\mathfrak{B}(R)$ is the set of power series

$$f(t) = \sum_{j \geq 0} c_j t^j \quad \in \quad R[[t]]$$

for which $c_0 \in M$ and $c_1 \notin M$. Our desire is that $\mathfrak{B}_{\mathfrak{o}}(R)$ should be a group under composition of power series, and indeed the condition on c_0 guarantees that composition will be well defined, while the condition on c_1 guarantees that the series will have an inverse in $\mathfrak{B}(R)$. One sees now that if κ is the characteristic-p field of definition in Question 2.1.6, and if R is the local κ-algebra $\kappa[a, b]/(a^p, b^p)$, then the series $a + t + bt^p$ is an element of $B_\kappa(R)$. The relation

$$(a + t + bt^p) \circ (a' + t + b't^p) = (a + a') + t + (b + b')t^p$$

shows that the group-scheme that's being described is finite and isomorphic to $\alpha_p \times \alpha_p$.

The Method. We take a formal group F of finite height that has a subgroup of order p as well as a group of automorphisms of order p. Now, finite groups of automorphisms of a formal group of finite height are always étale, but by taking a slight blowup of F, we convert the automorphism subgroup to a local group-scheme, without going so far as to make the above group of torsion points of F étale as well. Then, allowing ourselves a slight abuse of language, our desired lifting consists of all substitutions

$$t \mapsto a \tilde{+} t \tilde{+} [b]_{F'}(t),$$

where a is a torsion point of the blow-up of F, and $1 + b$ is a p-th root of 1. In the displayed formula, F' is the blown-up version of F, the tilde over the plus-sign indicates addition with respect to F', and as usual, $[b]_{F'}(t)$ is the endomorphism whose first-degree term is bt. I suppose that very confident people may be able to look at the preceding explanation and say, Of Course, No Problem, End of Story. But I'm not so confident, and the rest of this paper is devoted to filling in the gaps and making sure, to my own satisfaction at least, that everything is on the up and up. To those confident readers, everything from here on may thus be unnecessary, though the summary 1–7 at the end of this note may be an aid to flagging assurance.

A.1 Some algebra

Let $\zeta = \zeta_p$ be a primitive p-th root of 1 in an algebraic extension of $\mathbb{Q}_p$, and let $\mathfrak{o} = \mathbb{Z}_p[\zeta]$. Let also $\pi = \zeta - 1$, a prime element of $\mathfrak{o}$, and let k be the fraction field of $\mathfrak{o}$. In the ring $\mathfrak{o}[T]/(T^p - 1)$, let us call Γ the image of T, and let us consider $\Delta = \frac{\Gamma - 1}{\pi}$. Then the minimal polynomial for Δ is

$$T^p + \frac{p}{\pi}T^{p-1} + \frac{p(p-1)}{2\pi^2}T^{p-2} + \cdots + \frac{p(p-1)}{2\pi^{p-2}}T^2 + \frac{p}{\pi^{p-1}}T, \qquad (*)$$

in which the coefficient of T is a unit in $\mathfrak{o}$ congruent to -1 modulo π. Let us call B the ring $\mathfrak{o}[\Delta]$; we need to establish a few facts about it. I will use capital Greek letters for elements of B, lower case Greek letters for elements of $\mathfrak{o}$.

Lemma A.1.1. *The ring B is isomorphic to $\mathfrak{o} \oplus \mathfrak{o} \oplus \cdots \oplus \mathfrak{o}$, with p factors. In B, every element Θ satisfies the condition that $\Theta^p - \Theta \in \pi B$.*

Proof. The minimal polynomial for Δ, described above, is $T^p - T$ modulo π. By Hensel's lemma it splits into dinstinct linear factors over the complete local ring $\mathfrak{o}$, so that the first part of the statement is verified. Since each element $\beta \in \mathfrak{o}$ has the property that $\beta^p - \beta \in \pi\mathfrak{o}$, the corresponding property holds for elements of B as well. It may be of interest to note that this is not true of the subring $\mathfrak{o}[\Gamma]$ of B.

A.2 Endomorphisms of the fundamental formal group

We start with the polynomial $f(t) = \pi t + t^p \in \mathfrak{o}[[t]]$, which has associated to it a unique formal group $F(x, y) \in \mathfrak{o}[[x, y]]$ for which $f \in \mathrm{End}_{\mathfrak{o}}(F)$, as proved in [LT]. The following is hardly surprising:

Lemma A.2.1. *For each $\Theta \in B$, there is a unique series $[\Theta]_F(t) \in B[[t]]$ such that $[\Theta]'_F(0) = \Theta$ and $f \circ [\Theta]_F = [\Theta]_F \circ f$; this series is an element of* $\mathrm{End}_B(F)$. *In particular for $\Theta = \pi$ we have $[\pi]_F = f$.*

This may be proved by using either of the halves of Lemma A.1.1; if one wishes to use the fact that $\Theta^p - \Theta$ is always in πB, then the proof of the first Lemma in [LT] goes through word-for-word. The statement $[\pi]_F = f$ follows by uniqueness.

The endomorphism ring $\mathrm{End}_B(F)$ contains in particular the series $[\Delta]_F$ and $[\Gamma]_F$; the p-fold iterate of the latter series is the identity series t. And since $\Gamma = 1 + \pi\Delta$, our periodic series $[\Gamma]_F(t)$ may also be written as

$$F\left(t,\ ([\pi]_F \circ [\Delta]_F)\,(t)\right).$$

If β is an element of $\ker([\pi]_F)$, then the series $\tau_\beta(t) = F(t, \beta)$ commutes with both

$$[\zeta]_F(t) \quad = \quad F\left(t,\ [\pi]_F(t)\right)$$

and

$$[\Gamma]_F(t) \quad = \quad F\left(t,\ ([\Delta]_F \circ [\pi]_F)\,(t)\right).$$

If only $B = \mathfrak{o}[\Delta]$ had not been an étale $\mathfrak{o}$-algebra, we could have taken $\ker([\pi]_F) \times \mathrm{Spec}(B)$ as our desired lifting of $\alpha_p \times \alpha_p$. After all, the points of $\ker([\pi]_F)$ are the β's mentioned above, and the points of $\mathrm{Spec}(B)$ are essentially the p-th roots of unity ξ, and the substitution

$$t \mapsto F\left(\beta, [\xi]_F(t)\right)$$

would be our lifting of the substitution mentioned in the introduction. There is the additional problem that in case $p = 2$, F is of height one and so $\ker[\pi]$ is not a lifting of α_p, but the étaleness of the other factor is a much bigger obstacle.

Because of the form of $f(t) = [\pi]_F(t) = \pi t + t^p$, not only F but also all the B-endomorphisms $[\Theta]_F$ have the property that the only nonzero terms are in degrees congruent to 1 modulo $p - 1$. Any such series can be written, that is, in the form $\sum_{j\ge 0} H_j$ where each H_j is a form or monomial of degree $1 + j(p - 1)$. For want of a better term, I will call any series with this last property $(p - 1)$-lacunary.

Now I want to let $\mathfrak{O}$ be any complete local $\mathfrak{o}$-algebra in which π is no longer indecomposable, $\pi = \lambda\mu$ where both λ and μ are nonunits. Minimally, one may take $\lambda = \mu = \sqrt{\pi}$ and $\mathfrak{O} = \mathfrak{o}[\sqrt{\pi}]$. Or we may let $\mathfrak{O}$ be the ring of integers in any properly ramified algebraic extension K of k, and λ any element of K with valuation $0 < v(\lambda) < v(\pi) = 1$. Or, generically, we can take $\mathfrak{O} = \mathfrak{o}[[\lambda, \pi/\lambda]]$, a ring that can be described alternatively as $\mathfrak{o}[[\lambda,\mu]]/(\lambda\mu-\pi)$ or as the set of all doubly infinite Laurent series $\sum_{j\in\mathbb{Z}}\alpha_j\lambda^j$ in the indeterminate λ and with coefficients $\alpha_j\in\mathfrak{o}$ satisfying the additional condition that $j+v(\alpha_j)\geq 0$, where v is the (additive) valuation on $\mathfrak{o}$ and k normalized so that $v(\pi)=1$.

If G is a $(p-1)$-lacunary series in one or more variables, I will call the λ-blowup of G, denoted by $G^{(\lambda)}$, the series formed from G in the following way: if $G=\sum_{j\geq 0}H_j$, each H_j being homogeneous of degree $1+j(p-1)$, then $G^{(\lambda)}=\sum_{j\geq 0}\lambda^j H_j$.

When we apply the above operation to F and its endomorphisms, here is what happens: $F^{(\lambda)}$ becomes a formal group whose reduction modulo the maximal ideal of $\mathfrak{O}$ is just the additive formal group $x+y$. The maps $\mathrm{End}_{\mathfrak{o}}(F)\to\mathrm{End}_{\mathfrak{O}}(F^{(\lambda)})$ and $End_B(F)\to\mathrm{End}_{B\otimes_{\mathfrak{o}}\mathfrak{O}}(F^{(\lambda)})$ that take $g(t)$ to $g^{(\lambda)}(t)$ are injections. For any $\Theta\in B$, I will write $[\Theta]^{(\lambda)}$ for $[\Theta]_{F^{(\lambda)}}=([\Theta]_F)^{(\lambda)}$; then since $[\pi]^{(\lambda)}(t)=\pi t+\lambda t^p=\lambda(\mu t+t^p)$, the new formal group $F^{(\lambda)}$ has at least one nontrivial finite subgroup, namely the set of roots of $\mu t+t^p$, under the group law furnished by $F^{(\lambda)}$, and they certainly are the geometric points of Spec $(\mathfrak{O}[t]/(\mu t+t^p))$, but this is not the kernel of $[\pi]^{(\lambda)}$, since the standard construction of kernel in that case leads to something that's not flat. Rather, if we call $g(t)=\mu t+t^p$, then the finite group-scheme we're talking about is the kernel of $g: F^{(\lambda)}\to F^{(\lambda^2)}$.

Seeing just how a group scheme lifting α_p acts on $F^{(\lambda)}$ is a little trickier and more unusual. Our aim is to show that the automorphism $[\Gamma]^{(\lambda)}(t)$ of $F^{(\lambda)}$ lies in $\mathfrak{O}[\Delta'][[t]]$, where $\Delta'=\lambda\Delta$ has the $\mathfrak{O}$-minimal polynomial

$$T^p+\frac{p\lambda}{\pi}T^{p-1}+\frac{p(p-1)\lambda^2}{2\pi^2}T^{p-2}+\cdots+\frac{p(p-1)\lambda^{p-2}}{2\pi^{p-2}}T^2+\frac{p\lambda^{p-1}}{\pi^{p-1}}T \quad (**)$$

$$=T^p+\frac{p}{\mu}T^{p-1}+\frac{p(p-1)}{2\mu^2}T^{p-2}+\cdots+\frac{p(p-1)}{2\mu^{p-2}}T^2+\frac{p}{\mu^{p-1}}T;$$

Note that this polynomial is congruent to T^p modulo the maximal ideal $\mathfrak{M}$ of $\mathfrak{O}$.

Now recall that $\Gamma=1+\Delta\pi$, so that the series $[\Gamma](t)$, which is periodic of period p with respect to substitution of series, whether we are talking about automorphisms of the original F or of the blown-up $F^{(\lambda)}$, can be written

$$[\Gamma](t) \;=\; F\left(t,\,([\Delta]\circ[\pi])(t)\right).$$

Since every element of B is an $\mathfrak{o}$-linear combination of $\{1,\Delta,\ldots,\Delta^{p-1}\}$, we may write

$$[\Delta]_F(t) \;=\; \Delta t+\sum_{j\geq 1}C_j t^{j(p-1)+1}\in B[[t]],$$

where, as remarked, each coefficient C_j is an $\mathfrak{o}$-linear combination of the powers of Δ, up to Δ^{p-1}. It follows that $[\Delta]^{(\lambda)}$, the corresponding endomorphism of F, has the form

$$[\Delta]_{F^{(\lambda)}}(t) \;=\; \Delta t+\sum_{j\geq 1}C_j\lambda^j t^{j(p-1)+1}\in\mathfrak{O}[[t]],$$

where the C_j's are the same in both displayed formulas. Now, what of $[\Delta]^{(\lambda)} \circ [\pi]^{(\lambda)} = [\Delta]^{(\lambda)}(\pi t + \lambda t^p)$? Making the indicated substitution gives

$$\begin{aligned}\Delta\,(\pi t + \lambda t^p) &+ \sum_{j\geq 1} C_j \lambda^j (\pi t + \lambda t^p)^{j(p-1)+1} \\ &= \Delta'\,(\mu t + t^p) + \sum C_j \lambda^{jp+1}(\mu t + t^p)^{j(p-1)+1}.\end{aligned}$$

But now because $C_j \in B = \mathfrak{o}[\Delta]$, we also have $C_j \lambda^{jp+1} \in \lambda\mathfrak{O}[\Delta'] = \lambda\mathfrak{O}[\lambda\Delta]$, since the j's all are at least 1. This shows that $[\pi\Delta]^{(\lambda)}(t)$ is a power series with coefficients in $\mathfrak{O}[\Delta']$, and indeed, modulo $\mathfrak{M}$, this series is just $\Delta' t^p$. Finally, when we add this series and the series t by means of the formal group $F^{(\lambda)}(x, y) \equiv x + y \mod \mathfrak{M}$, the result, namely $[\Gamma]^{(\lambda)}(t)$, has coefficients in $B' = \mathfrak{O}[\Delta']$, and is congruent modulo $\mathfrak{M}$ to $t + \Delta' t^p$. One more remark is necessary, the obvious one that if $\mu\alpha + \alpha^p = 0$, then $[\Delta\pi]^{(\lambda)}(\alpha) = 0$ and $[\Gamma]^{(\lambda)}(\alpha) = \alpha$.

In summary, this is what we now have:

(1) The ring $\mathfrak{o}$ is $\mathbb{Z}_p[\zeta]$, where $\zeta = \zeta_p$ is a primitive p-th root of unity, and we use the prime element $\pi = \zeta - 1$.
(2) The ring $\mathfrak{O}$ is any suitably ramified extension of $\mathfrak{o}$, the minimal example being $\mathfrak{O} = \mathfrak{o}[\sqrt{\pi}]$. This $\mathfrak{O}$ is the ring over which our liftings and action are defined, and we identify in it elements $\lambda, \mu \in \mathfrak{O}$ with $\lambda\mu = \pi$.
(3) The formal group F over $\mathfrak{o}$ has $\pi t + t^p$ as an endomorphism and thus has $\mathfrak{o}$ as its ring of "absolute" endomorphisms (over the ring of integers of any algebraic extension field of the fraction field of $\mathfrak{o}$). Allowing for abuse of language, there is a unique $\mathfrak{o}$-subgroup-scheme of F of rank p, namely $\ker[\pi]_F = \mathrm{Spec}(A)$, where $A = \mathfrak{o}[[t]]/\,([\pi]_F(t))$.
(4) The finite $\mathfrak{o}$-algebra B is $\mathfrak{o}[\Delta]$, where the minimal polynomial for Δ over $\mathfrak{o}$ is given in formula $(*)$. Algebraically, B is $\mathfrak{o}^{\oplus p}$, and when we call $\Gamma = 1 + \pi\Delta \in B$, we have $\Gamma^p = 1$. The scheme $\mathrm{Spec}(B)$ is a finite étale group-scheme of order p; the element $\Gamma \in B$ is a generic p-th root of unity, and the operation of the étale group-scheme on the formal-affine line is $t \to [\Gamma]_F(t) = F\,(t, ([\pi] \circ [\Delta])(t))$.
(5) We use $\lambda \in \mathfrak{O}$ to form a sort of blowup of F, which we call $F^{(\lambda)}$ and which is described on the preceding page. This formal group has the subgroup-scheme $\mathrm{Spec}(A')$, where $A' = \mathfrak{O}[[t]]/(\mu t + t^p)$, and this group-scheme acts on the formal-affine line by the substitution $t \to F(\lambda)(a, t)$ when a is any root of $\mu t + t^p$.
(6) We define $\Delta' = \lambda\Delta \in B \otimes_{\mathfrak{o}} \mathfrak{O}$, and note that Δ' has the minimal polynomial over $\mathfrak{O}$ given by $(**)$ on the preceding page. Call $B' = \mathfrak{O}[\Delta']$. The periodic power series $[\Gamma]^{(\lambda)}(t)$, originally defined to be in $B \otimes_{\mathfrak{o}} \mathfrak{O}[[t]]$, actually is in $B'[[t]]$ and as an element of this ring, it becomes $t + \Delta' t^p$ in $B' \otimes_{\mathfrak{O}} \mathfrak{O}/\mathfrak{M}[[t]]$.
(7) Since the series $[\Gamma]^{(\lambda)}(t)$ and the $F^{(\lambda)}(a, t)$ mentioned in (5) commute, we do indeed have a finite group-scheme, namely $\mathrm{Spec}(A' \otimes_{\mathfrak{O}} B')$, acting on the formal-affine line in such a way that over $\mathfrak{O}/\mathfrak{M}$, the action is $t \mapsto a + t + \Delta' t^p$.

References

[ℵ-C-V] D. Abramovich, A. Corti and A. Vistoli, *Twisted bundles and admissible covers,* Comm. Algebra **31** (2003), 3547–3618.

[ℵ-G-V] D. Abramovich, T. Graber and A. Vistoli *Gromov–Witten theory of Deligne–Mumford stacks*, Amer. J. Math. 130 (2008), no. 5, 1337–1398.

[ℵ-O-V] D. Abramovich, M. Olsson and A. Vistoli, *Tame stacks in positive characteristic.* Ann. Inst. Fourier (Grenoble) 58 (2008), no. 4, 1057–1091.

[ℵ-O-V2] D. Abramovich, M. Olsson and A. Vistoli, *Twisted stable maps to tame Artin stacks*, J. Alg. Geom S 1056–3911 (2010) 00569-3; published electronically.

[ℵ-Oo] D. Abramovich and F. Oort, *Stable maps and Hurwitz schemes in mixed characteristic.* Advances in algebraic geometry motivated by physics (Lowell, MA, 2000), E. Previato, ed., 89–100, Contemp. Math., 276, Amer. Math. Soc., Providence, RI, 2001.

[ℵ-V1] D. Abramovich and A. Vistoli, *Compactifying the space of stable maps*, J. Amer. Math. Soc. 15 (2002), no. 1, 27–75.

[ℵ-V2] ——, *Twisted stable maps and quantum cohomology of stacks*, Intersection theory and moduli, 97–138, ICTP Lect. Notes, XIX, Abdus Salam Int. Cent. Theoret. Phys., Trieste, 2004.

[Bo] I. I. Bouw, *Reduction of the Hurwitz space of metacyclic covers*, Duke Math. J. 121 (2004), no. 1, 75–111.

[B-W1] I. I. Bouw and S. Wewers, *Reduction of covers and Hurwitz spaces*, J. Reine Angew. Math. 574 (2004), 1–49.

[B-W2] ——, *Stable reduction of modular curves*, Mathematisches Institut, Georg-August-Universität Göttingen: Seminars 2003/2004, 37–41, Universitätsdrucke Göttingen, Gttingen, 2004.

[Ek] T. Ekedahl, *Boundary behaviour of Hurwitz schemes.* The moduli space of curves (Texel Island, 1994), 173–198, Prog. Math., 129, Birkhäuser Boston, Boston, MA, 1995.

[He] Y. Henrio, *Arbres de Hurwitz et automorphismes d'ordre p des disques et des couronnes p-adiques formels*, preprint math.AG/0011098.

[H-M] J. Harris and D. Mumford, *On the Kodaira dimension of the moduli space of curves,* Invent. Math. **67** (1982), no. 1, 23–88.

[LT] J. Lubin and J. Tate, *Formal complex multiplication in local fields,* Annals of Math. 81 (1965), 380–387.

[Ma] S. Maugeais, *On a compactification of a Hurwitz space in the wild case*, preprint math. AG/0509118.

[Mo] S. Mochizuki, *The geometry of the compactification of the Hurwitz scheme,* Publ. Res. Inst. Math. Sci. **31** (1995), no. 3, 355–441.

[Ol] M. Olsson, *On (log) twisted curves*, Compos. Math. 143 (2007), no. 2, 476–494.

[O-T] F. Oort and J. Tate, *Group schemes of prime order*, Ann. Sci. École Norm. Sup. (4), **3** (1970), 1–21.

[Ra] M. Raynaud, *Spécialisation des revêtements en caractéristique* $p > 0$. Ann. Sci. École Norm. Sup. (4) **32** (1999), no. 1, 87–126.

[Ro] M. Romagny, *Sur quelques aspects des champs de revêtements de courbes algébriques*, Ph.D. Thesis, University of Grenoble, 2002.

[Ro1] ——, Effective models of group schemes, preprint arXiv:0904.3167.

[Sa1] M. Saidi, *Torsors under finite flat group schemes of rank p with Galois action*, Math. Z. 245 (2003), no. 4, 695–710.

[Sa2] ——, *Wild ramification and a vanishing cycle formula*, J. Algebra 273 (2004), no. 1, 108–128.

[Sa3] ——, *Galois covers of degree p and semi-stable reduction of curves in mixed characteristics.* Publ. Res. Inst. Math. Sci. 43 (2007), no. 3, 661–684.

[Sa4] ——, *On the existence of a torsor structure for Galois covers*, preprint math.AG/0403389

[To1] Dajano Tossici, *Models of $\mathbb{Z}/p^2\mathbb{Z}$ over a d.v.r. of unequal characteristic, Models of $\mu_{p^2,k}$ over a discrete valuation sing; with appendix by Xavier Caruso*. J. Aligebra 323 (2010), no. 7, 1908–1957.

[To2] Dajano Tossici, *Effective models and extension of torsors over a discrete valuation ring of unequal characteristic*, Int. Math. Res. Not. IMRN 2008, Art. ID rnn111, 68 pp.

[Va] Wolmer Vasconcelos, *Integral closure. Rees algebras, multiplicities, algorithms.* Springer Monographs in Mathematics. Springer-Verlag, Berlin, 2005.

[W] S. Wewers, *Construction of Hurwitz spaces*, Institut für Experimentelle Mathematik preprint No. 21 (1998).

[W1] S. Wewers, *Reduction and lifting of special metacyclic covers*, Ann. Sci. École Norm. Sup. (4) 36 (2003), no. 1, 113–138.

[W2] ——, *Three point covers with bad reduction*, J. Amer. Math. Soc. 16 (2003), no. 4, 991–1032.

[W3] ——, *Formal deformation of curves with group scheme action*, Ann. Inst. Fourier (Grenoble) 55 (2005), no. 4, 1105–1165.

The modular degree, congruence primes, and multiplicity one

Amod Agashe*, **Kenneth A. Ribet, and William A. Stein**

To Serge Lang,
our friend and teacher,
someone who always knew a fact from a hole in the ground

Abstract The modular degree and congruence number are two fundamental invariants of an elliptic curve over the rational field. Frey and Müller have asked whether these invariants coincide. We find that the question has a negative answer, and show that in the counterexamples, multiplicity one (defined below) does not hold. At the same time, we prove a theorem about the relation between the two invariants: the modular degree divides the congruence number, and the ratio is divisible only by primes whose squares divide the conductor of the elliptic curve. We discuss the ratio even in the case where the square of a prime does divide the conductor, and we study analogues of the two invariants for modular abelian varieties of arbitrary dimension.

---

*Agashe was partially supported by National Science Foundation Grant No. 0603668. Stein was partially supported by National Science Foundation Grant No. 0653968.

A. Agashe (✉)
Department of Mathematics, Florida State University, Tallahassee, Florida 32312, U.S.A.
e-mail: agashe@math.fsu.edu.

K.A. Ribet
Department of Mathematics, UC Berkeley Mail Code 3840, 970 Evans Hall Berkeley, CA 94720-3840
e-mail: ribet@math.berkeley.edu

W.A. Stein
University of Washington; Office: 423 Padelford Seattle, WA 98195-4350
e-mail: wstein@gmail.com

D. Goldfeld et al. (eds.), *Number Theory, Analysis and Geometry: In Memory of Serge Lang*, DOI 10.1007/978-1-4614-1260-1_2,

Key words elliptic curves • abelian varieties • modular degree • congruence primes • multiplicity one

Mathematics Subject Classification (2010): 11G05, 11610, 11G18, 11F33

1 Introduction

Let E be an elliptic curve over $\mathbf{Q}$. By [BCDT01], we may view E as an abelian variety quotient over $\mathbf{Q}$ of the modular Jacobian $J_0(N)$, where N is the conductor of E. We assume that the kernel of the map $J_0(N) \to E$ is connected, i.e., that E is an *optimal quotient* of $J_0(N)$ (this can always be done by replacing E by an isogenous curve if needed). The *modular degree* m_E is the degree of the composite map $X_0(N) \to J_0(N) \to E$, where we map $X_0(N)$ to $J_0(N)$ by sending $P \in X_0(N)$ to $[P] - [\infty] \in J_0(N)$.

Let $f_E = \sum a_n q^n \in S_2(\Gamma_0(N), \mathbf{C})$ be the newform attached to E. The *congruence number* r_E of E is the largest integer such that there is an element $g = \sum b_n q^n \in S_2(\Gamma_0(N))$ with integer Fourier coefficients b_n that is orthogonal to f_E with respect to the Petersson inner product, and congruent to f_E modulo r_E (i.e., $a_n \equiv b_n \pmod{r_E}$ for all n).

Section 2 is about relations between r_E and m_E. For example, $m_E \mid r_E$. In [FM99, Q. 4.4], Frey and Müller asked whether $r_E = m_E$. We give examples in which $r_E \neq m_E$, and show that in these examples, there is a maximal ideal $\mathfrak{m}$ of the Hecke algebra $\mathbf{T}$, such that $J_0(N)[\mathfrak{m}]$ has dimension more than two over $\mathbf{T}/\mathfrak{m}$ (this is the failure of multiplicity one alluded to above). We then conjecture that for any prime p, $\mathrm{ord}_p(r_E/m_E) \leq \frac{1}{2}\mathrm{ord}_p(N)$, and prove this conjecture when $\mathrm{ord}_p(N) \leq 1$.

In Section 3, we consider analogs of the modular degree and the congruence number for certain modular abelian varieties that are not necessarily elliptic curves; these include optimal quotients of $J_0(N)$ and $J_1(N)$ of any dimension associated to newforms. Section 3 may be read independently of Section 2. In Sections 4 and 5 we prove the main theorem of this paper (Theorem 3.6), and also give some examples of failure of what we call multiplicity one for differentials (see Definition 5.13).

Acknowledgments The authors are grateful to M. Baker, F. Calegari, B. Conrad, J. Cremona, G. Frey, H. W. Lenstra, Jr. and B. Noohi for discussions and advice regarding this paper. We would especially like to thank B. Conrad for the material in the appendix and for his suggestions concerning a number of technical facts that are inputs to our arguments. The first author is also grateful to the Max-Planck-Institut für Mathematik for its hospitality during a visit when he partly worked on this paper.

2 Elliptic curves

In Section 2.1, we discuss relationships between the modular degree and the congruence number of an elliptic curve. In Section 2.2 we recall the notion of multiplicity one and give new examples in which it fails.

2.1 Modular degree and congruence number

Let N be a positive integer and let $X_0(N)$ be the modular curve over $\mathbf{Q}$ that classifies isomorphism classes of elliptic curves with a cyclic subgroup of order N. The Hecke algebra $\mathbf{T}$ of level N is the subring of the ring of endomorphisms of $J_0(N) = \mathrm{Jac}(X_0(N))$ generated by the Hecke operators T_n for all $n \geq 1$. Let f be a newform of weight 2 for $\Gamma_0(N)$ with integer Fourier coefficients, and let I_f be kernel of the homomorphism $\mathbf{T} \to \mathbf{Z}[\ldots, a_n(f), \ldots]$ that sends T_n to a_n. Then the quotient $E = J_0(N)/I_f J_0(N)$ is an elliptic curve over $\mathbf{Q}$. We call E the *optimal quotient* associated to f. Composing the embedding $X_0(N) \hookrightarrow J_0(N)$ that sends ∞ to 0 with the quotient map $J_0(N) \to E$, we obtain a surjective morphism of curves $\phi_E : X_0(N) \to E$. Recall that the *modular degree* m_E of E is the degree of ϕ_E.

Let $S_2(\mathbf{Z})$ denote the group of cuspforms of weight 2 on $\Gamma_0(N)$ with integral Fourier coefficients, and if G is a subgroup of $S_2(\mathbf{Z})$, let $G^\perp$ denote the subgroup of $S_2(\mathbf{Z})$ consisting of cuspforms that are orthogonal to f with respect to the Petersson inner product. The *congruence number* of E (really, that of f) is the positive integer r_E defined by either of the following equivalent conditions:

(i) r_E is the largest integer r such that there exists $g \in (\mathbf{Z}f)^\perp$ with $f \equiv g \bmod r$.
(ii) r_E is the order of the quotient group $\frac{S_2(\mathbf{Z})}{\mathbf{Z}f + (\mathbf{Z}f)^\perp}$.

We say that a prime is a *congruence prime for* E if it divides the congruence number r_E. Congruence primes have been studied by Doi, Hida, Ribet, Mazur and others (see, e.g., [Rib83, §1]), and played an important role in Wiles's work [Wil95] on Fermat's last theorem. Frey and Mai-Murty have observed that an appropriate asymptotic bound on the modular degree is equivalent to the abc-conjecture (see [Fre97, p.544] and [Mur99, p.180]). Thus, results that relate congruence primes and the modular degree may be of great interest.

Theorem 2.1. *Let E be an elliptic curve over $\mathbf{Q}$ of conductor N, with modular degree m_E and congruence number r_E. Then $m_E \mid r_E$ and if $\mathrm{ord}_p(N) \leq 1$, then $\mathrm{ord}_p(r_E) = \mathrm{ord}_p(m_E)$.*

Thus any prime that divides the modular degree of an elliptic curve E is a congruence prime for E, and if p is a congruence prime for E such that p^2 does not divide the conductor of E, then p divides the modular degree of E. The divisibility $m_E \mid r_E$ was first discussed in [Zag85, Th. 3], where it is attributed to the second author (Ribet); however in [Zag85] the divisibility was mistakenly written in the opposite direction. For some other expositions of the proof that $m_E \mid r_E$, see [AU96, Lem 3.2] and [CK04]. We generalize this divisibility and prove it in Theorem 3.6(a). The second part of Theorem 2.1, i.e., that if $\mathrm{ord}_p(N) \leq 1$, then $\mathrm{ord}_p(r_E) = \mathrm{ord}_p(m_E)$, follows from the more general Theorem 3.6(b) below.

Table 1 Differing Modular Degree and Congruence Number

Curve	m_E	r_E	Curve	m_E	r_E	Curve	m_E	r_E
54B1	2	6	99A1	4	12	128A1	4	32
64A1	2	4	108A1	6	18	128B1	8	32
72A1	4	8	112A1	8	16	128C1	4	32
80A1	4	8	112B1	4	8	128D1	8	32
88A1	8	16	112C1	8	16	135A1	12	36
92B1	6	12	120A1	8	16	144A1	4	8
96A1	4	8	124A1	6	12	144B1	8	16
96B1	4	8	126A1	8	24			

Note that [AU96, Prop. 3.3–3.4] implies the weaker statement that if $p \nmid N$, then $\mathrm{ord}_p(r_E) = \mathrm{ord}_p(m_E)$, since [AU96, Prop. 3.3] implies

$$\mathrm{ord}_p(r_E) - \mathrm{ord}_p(m_E) = \mathrm{ord}_p(\#\mathcal{C}) - \mathrm{ord}_p(c_E) - \mathrm{ord}_p(\#\mathcal{D}),$$

and by [AU96, Prop. 3.4], $\mathrm{ord}_p(\#\mathcal{C}) = 0$. Here c_E is the Manin constant of E, which is an integer (e.g., see [ARS06]), and $\mathcal{C}$ and $\mathcal{D}$ are certain groups.

Frey and Müller [FM99, Ques. 4.4] asked whether $r_E = m_E$ in general. After implementing an algorithm to compute r_E in Magma [BCP97], we quickly found that the answer is no. The counterexamples at conductor $N \leq 144$ are given in Table 1, where the curve is given using the notation of [Cre97].

For example, the elliptic curve 54B1, given by the equation $y^2 + xy + y = x^3 - x^2 + x - 1$, has $r_E = 6$ and $m_E = 2$. To see explicitly that $3 \mid r_E$, observe that the newform corresponding to E is $f = q+q^2+q^4-3q^5-q^7+\cdots$ and the newform corresponding to $X_0(27)$ is $g = q - 2q^4 - q^7 + \cdots$, so $g(q) + g(q^2)$ appears to be congruent to f modulo 3. To prove this congruence, we checked it for 18 Fourier coefficients, where the sufficiency of precision to degree 18 was determined using [Stu87].

It is unclear whether there is a bound on the possible primes p that occur. For example, for the curve 242B1 of conductor $N = 2 \cdot 11^2$ we have

$$m_E = 2^4 \neq r_E = 2^4 \cdot 11.$$

We propose the following replacement for Question 4.4 of [FM99]:

Conjecture 2.2. *Let E be an optimal elliptic curve of conductor N and p be any prime. Then*

$$\mathrm{ord}_p\left(\frac{r_E}{m_E}\right) \leq \frac{1}{2}\mathrm{ord}_p(N).$$

We verified Conjecture 2.2 using Sage [S$^+$09] for every optimal elliptic curve quotient of $J_0(N)$, with $N \leq 557$.

If $p \geq 5$, then $\mathrm{ord}_p(N) \leq 2$, so a special case of the conjecture is

$$\mathrm{ord}_p\left(\frac{r_E}{m_E}\right) \leq 1 \qquad \text{for any } p \geq 5.$$

2.2 Multiplicity one and its failure

We say that a maximal ideal $\mathfrak{m}$ of $\mathbf{T}$ satisfies *multiplicity one* if $J_0(N)[\mathfrak{m}]$ is of dimension two over $\mathbf{T}/\mathfrak{m}$. The reason one calls this "multiplicitly one" is that if the canonical two-dimensional representation $\rho_\mathfrak{m}$ over $\mathbf{T}/\mathfrak{m}$ attached to $\mathfrak{m}$ (e.g., see [Rib90, Prop. 5.1]) is irreducible, then $J_0(N)[\mathfrak{m}]$ is a direct sum of copies of $\rho_\mathfrak{m}$ [Rib90, Thm. 5.2], and a maximal ideal $\mathfrak{m}$ of $\mathbf{T}$ satisfies *multiplicity one* precisely if the multiplicity of $\rho_\mathfrak{m}$ in this decomposition is one. Even if $\rho_\mathfrak{m}$ is reducible, the definition of multiplicity one given above is relevant (e.g., see [Maz77, Cor. 16.3]). The notion of multiplicity one, originally found in Mazur [Maz77], has played an important role in several places (e.g., in Wiles's proof of Fermat's last theorem: see Thm. 2.1 in [Wil95]).

In [MR91, §13], the authors find examples of failure of multiplicity one in which if p is the residue characteristic of $\mathfrak{m}$, then $p^3 \mid N$, and $\rho_\mathfrak{m}$ is modular of level N/p^2. Kilford [Kil02] found examples of failure of multiplicity one where N is prime and the residue characteristic of $\mathfrak{m}$ is 2. See also [Wie07] and [KW08] for examples of failure of multiplicity one in the $\Gamma_1(N)$ context. We now give examples of failure of multiplicity one where the square of the residue characteristic of $\mathfrak{m}$ divides the level (the residue characteristic is often odd).

Proposition 2.3. *Suppose E is an optimal elliptic curve over $\mathbf{Q}$ of conductor N and p is a prime such that $p \mid r_E$ but $p \nmid m_E$. Then there is a maximal ideal $\mathfrak{m}$ of $\mathbf{T}$ with residue characteristic p such that $\dim_{\mathbf{T}/\mathfrak{m}} J_0(N)[\mathfrak{m}] > 2$, i.e., multiplicity one fails for $\mathfrak{m}$.*

The proposition follows from the more general Proposition 5.9 below. It follows from the proposition above that any example in Table 1 where simultaneously a prime divides r_E but does not divide m_E provides an example of failure of multiplicity one. In such examples, the associated representation $\rho_\mathfrak{m}$ may or may not be irreducible. For example, for the elliptic curve 54B1 and $p = 3$, we have $\mathrm{ord}_3(r_E) = 1$ but $\mathrm{ord}_3(m_E) = 0$, so there is a maximal ideal $\mathfrak{m}$ with residue characteristic 3 such that multiplicity one fails for $\mathfrak{m}$. The curve 54B1 has rational 3-torsion, so $\rho_\mathfrak{m}$ is reducible. On the other hand, for the elliptic curve 99A1, we have $\mathrm{ord}_3(r_E) = 1$ but $\mathrm{ord}_3(m_E) = 0$, so again there is a maximal ideal $\mathfrak{m}$ with residue characteristic 3 such that multiplicity one fails for $\mathfrak{m}$. Moreover, $J_0(99)$ is isogenous to a product of elliptic curves, none of which admit a rational 3-isogeny. Hence $\rho_\mathfrak{m}$ is irreducible.

The notion of multiplicity one at a maximal ideal $\mathfrak{m}$ is closely related to Gorensteinness of the completion of $\mathbf{T}$ at $\mathfrak{m}$ (e.g., see [Til97]). Kilford [Kil02] found examples of failure of Gorensteinness (and multiplicity one) at the prime 2 for certain prime levels. In the examples as above where multiplicity one fails for some maximal ideal, it would be interesting to do computations (e.g., as in [Kil02]) to see if the completion of the Hecke algebra at the maximal ideal is Gorenstein or not.

3 Modular abelian varieties of arbitrary dimension

For $N \geq 4$, let Γ be either $\Gamma_0(N)$ or $\Gamma_1(N)$. Let X be the modular curve over $\mathbf{Q}$ associated to Γ, and let J be the Jacobian of X. Let A and B be abelian subvarieties of J such that $A + B = J$, $A \cap B$ is finite, and every endomorphism of J over $\mathbf{Q}$ preserves A and B. In this section, we generalize the notions of the congruence number and the modular degree to subvarieties A as above, and state a theorem relating the two numbers, which we prove in Sections 4 and 5.

We first give a general example of A and B as above. Up to isogeny, J is the product of factors $J_f^{e(f)}$ where f runs over the set of newforms of level dividing N, taken up to Galois conjugation, and $e(f)$ is the number of divisors of $N/N(f)$, where $N(f)$ is the level of f. Here J_f is the standard abelian subvariety of J attached to f by Shimura [Shi94, Thm. 7.14]. Let A' be the sum of $J_f^{e(f)}$ for some set of f's (taken up to Galois conjugation), and let B' be the sum of all the other $J_f^{e(f)}$'s. Clearly $A' + B' = J$. The J_f's are simple (over $\mathbf{Q}$), hence $A' \cap B'$ is finite. In view of the following lemma, A' and B' provide an example of A and B respectively as above. Note that by $\mathrm{End}(J)$ we mean the ring of endomorphisms of J defined over $\mathbf{Q}$.

Lemma 3.1. $\mathrm{End}(J)$ *preserves* A' *and* B'.

Proof. Suppose $\mathrm{End}(J)$ does not preserve A' (the case of B' is symmetric). Then since the J_f's are simple, that means that some abelian subvariety J_g of A' is isogenous to some abelian subvariety J_h of B', where $g \neq h$. Pick a prime ℓ. If f is a newform, then let ρ_f denote the canonical absolutely irreducible ℓ-adic representation attached to f. Now $\overline{\mathbf{Q}}_\ell \otimes V_\ell(J_f)^{ß}$ is a direct sum of copies of $\rho_{\sigma(f)}$ as σ ranges over all embeddings into $\overline{\mathbf{Q}}$ of the field generated by the Fourier coefficients of f. Thus the above implies that there are distinct newforms g' and h' (of some level dividing N) such that $\rho_{g'} \cong \rho_{h'}$. Now each ρ_f satisfies $\mathrm{tr}(\rho_f(\mathrm{Frob}_p)) = a_p(f)$ for all $p \nmid N\ell$. Thus for all $p \nmid N\ell$, we have $a_p(g') = a_p(h')$. By the multiplicity one theory (e.g., see [Li75, Cor. 3, pg. 300]), this means that $g' = h'$, a contradiction. $\square$

We now give a more specific example, which will include the case of elliptic curves. Recall that $\mathbf{T}$ denotes the Hecke algebra. If $f = \sum a_n(f)q^n \in S_2(\Gamma)$ is a newform and $I_f = \ker(\mathbf{T} \to \mathbf{Z}[\ldots, a_n(f), \ldots])$, then $A_f = J/I_f J$ is the *newform quotient* associated to f. It is an abelian variety over $\mathbf{Q}$ of dimension equal to the degree of the field $\mathbf{Q}(\ldots, a_n(f), \ldots)$. Let ϕ_2 denote the quotient map $J \to A$. If C is an abelian variety, then we denote its dual abelian variety by $C^\vee$. There is a canonical principal polarization $\theta : J \cong J^\vee$. Dualizing ϕ_2, we obtain a closed immersion $\phi_2^\vee : A_f^\vee \to J^\vee$, which when composed with $\theta^{-1} : J^\vee \cong J$ gives us an injection $\phi_1 : A_f^\vee \hookrightarrow J$. One slight complication is that the isomorphism θ does not respect the action of $\mathbf{T}$, because if T is a Hecke operator on J, then on $J^\vee$ it acts as

$W_N T W_N$, where W_N is the Atkin–Lehner involution (see e.g., [DI95, Rem. 10.2.2]). However, on the new quotient J^{new}, the action of the Hecke operators commutes with that of W_N, so since the quotient map $J \to A_f$ factors through J^{new}, the Hecke action on $A_f^\vee$ induced by the embedding $A_f^\vee \to J^\vee$ and the action on $A_f^\vee$ induced by the injection $\phi_1 : A_f^\vee \to J$ are the same. Hence $A_f^\vee$ is isomorphic to $\phi_1(A_f^\vee)$ as a $\mathbf{T}$-module, and $\phi_1(A_f^\vee) = J_f$ (this follows from the characterization of J_f in [Shi94, Thm. 7.14]). For simplicity, we will often denote $\phi_1(A_f^\vee) = J_f$ by just $A_f^\vee$. Let ϕ be the composite map $A_f^\vee \xrightarrow{\phi_1} J \xrightarrow{\phi_2} A_f$; then ϕ is a polarization (induced by dual of the polarization of J). Thus $A_f^\vee + I_f J = J$ and $A_f^\vee \cap I_f J$ is finite. Hence, in view of Lemma 3.1, $A_f^\vee$ and $I_f J$ provide an example of A and B as in the beginning of this section.

The *exponent* of a finite group G is the smallest positive integer n such that every element of G has order dividing n (i.e., such that for all $x \in G, nx = 0$).

Definition 3.2. *The* modular exponent $\tilde{n}_A$ *of A is the exponent of $A \cap B$ and the* modular number n_A *of A is its order.*

Note that the definition is symmetric with respect to A and B. In fact, the definition depends on both A and B, unlike what the notation may suggest—we have suppressed the dependence on B for ease of notation, with the understanding that there is a natural choice of B (e.g., this is the case in the examples we gave above). If f is a newform, then by the modular exponent/number of A_f, we mean that of $A = A_f^\vee$, with $B = I_f J$. In this situation, since ϕ is a polarization, n_{A_f} is a perfect square (e.g., see [AS05, Lemma 3.14]). When A_f is an elliptic curve, ϕ is multiplication by the modular degree m_E. Hence $A \cap B = \ker(\phi)$ is $(\mathbf{Z}/m_E\mathbf{Z})^2$, and so for elliptic curves, the modular exponent is equal to the modular degree and the modular number is the square of the modular degree.

If R is a subring of $\mathbf{C}$, let $S_2(R) = S_2(\Gamma; R)$ denote the subgroup of $S_2(\Gamma; \mathbf{C})$ consisting of cups forms whose Fourier expansions at the cusp ∞ have coefficients in R. There is a $\mathbf{T}$-equivariant bilinear pairing $\mathbf{T} \times S_2(\mathbf{Z}) \to \mathbf{Z}$ given by $(t, g) \mapsto a_1(t(g))$, which is perfect by [AU96, Lemma 2.1] (see also [Rib83, Theorem 2.2]). The action of $\mathbf{T}$ on $\mathrm{H}_1(J, \mathbf{Z})$ is a faithful representation that embeds $\mathbf{T}$ into $\mathrm{Mat}_{2d}(\mathbf{Z}) \cong \mathbf{Z}^{(2d)^2}$. But $\mathbf{Z}$ is Noetherian, so $\mathbf{T}$ is finitely generated over $\mathbf{Z}$, and hence so is $S_2(\mathbf{Z})$. Let $\mathbf{T}_A$ be the image of $\mathbf{T}$ in $\mathrm{End}(A)$, and let $\mathbf{T}_B$ be the image of $\mathbf{T}$ in $\mathrm{End}(B)$ (since $\mathbf{T} \subset \mathrm{End}(J)$, $\mathbf{T}$ preserves A and B). Since $A + B = J$, the natural map $\mathbf{T} \to \mathbf{T}_A \oplus \mathbf{T}_B$ is injective, and moreover, its cokernel is finite (since $A \cap B$ is finite).

Let $S_A = \mathrm{Hom}(\mathbf{T}_A, \mathbf{Z})$ and $S_B = \mathrm{Hom}(\mathbf{T}_B, \mathbf{Z})$ be subgroups of $S_2(\mathbf{Z})$ obtained via the pairing above. Let $\mathrm{Ext}^1 = \mathrm{Ext}^1_{\mathbf{Z}}$ denote the first Ext functor in the category of $\mathbf{Z}$-modules.

Lemma 3.3. *There is a canonical isomorphism of* $\mathbf{T}$*-modules*

$$\mathrm{Ext}^1((\mathbf{T}_A \oplus \mathbf{T}_B)/\mathbf{T}, \mathbf{Z}) \cong S_2(\mathbf{Z})/(S_A + S_B).$$

The groups $S_2(\mathbf{Z})/(S_A + S_B)$ *and* $(\mathbf{T}_A \oplus \mathbf{T}_B)/\mathbf{T}$ *are isomorphic.*

Proof. Apply the $\mathrm{Hom}(-, \mathbf{Z})$ functor to the short exact sequence

$$0 \to \mathbf{T} \to \mathbf{T}_A \oplus \mathbf{T}_B \to (\mathbf{T}_A \oplus \mathbf{T}_B)/\mathbf{T} \to 0$$

to obtain a three-term exact sequence

$$0 \to \mathrm{Hom}(\mathbf{T}_A \oplus \mathbf{T}_B, \mathbf{Z}) \to \mathrm{Hom}(\mathbf{T}, \mathbf{Z}) \to \mathrm{Ext}^1((\mathbf{T}_A \oplus \mathbf{T}_B)/\mathbf{T}, \mathbf{Z}) \to 0. \quad (1)$$

The perfect $\mathbf{T}$-equivariant bilinear pairing $\mathbf{T} \times S_2(\mathbf{Z}) \to \mathbf{Z}$ given by $(t, g) \mapsto a_1(t(g))$ transforms (1) into an exact sequence

$$0 \to S_A \oplus S_B \to S_2(\mathbf{Z}) \to \mathrm{Ext}^1((\mathbf{T}_A \oplus \mathbf{T}_B)/\mathbf{T}, \mathbf{Z}) \to 0$$

of $\mathbf{T}$-modules, which proves the first claim in the lemma. Finally note that if G is any finite abelian group, then $\mathrm{Ext}^1(G, \mathbf{Z}) \approx G$ as groups, which gives the second result of the lemma. □

Definition 3.4. *The exponent of either of the isomorphic groups* $S_2(\mathbf{Z})/(S_A + S_B)$ *and* $(\mathbf{T}_A \oplus \mathbf{T}_B)/\mathbf{T}$ *is the* congruence exponent $\tilde{r}_A$ *of* A *and the order of the groups is the* congruence number r_A.

Note that this definition is also symmetric with respect to A and B, and again, the definition depends on both A and B, unlike what the notation may suggest — we have suppressed the dependence on B with the implicit understanding that B has been chosen (given A). If f is a newform, then by the congruence exponent/number of A_f, we mean that of $A = A_f^\vee$, with $B = I_f J$. In this situation, $\mathbf{T}_A = \mathbf{T}/I_f$ and $S_A = S_2(\mathbf{Z})[I_f]$. Recall that a subgroup H of an abelian group G is said to be *saturated* (in G) if G/H is torsion-free. Now $\mathrm{Hom}(\mathbf{T}_B, \mathbf{Z})$ is the unique saturated Hecke-stable complement of $S_2(\mathbf{Z})[I_f]$ in $S_2(\mathbf{Z})$, hence must equal $S_2(\mathbf{Z})[I_f]^\perp$, where we recall that $S_2(\mathbf{Z})[I_f]^\perp$ denotes the orthogonal complement of $S_2(\mathbf{Z})[I_f]$ in $S_2(\mathbf{Z})$ with respect to the Petersson inner product. Thus the congruence exponent $\tilde{r}_{A_f}$ is the exponent of the group

$$\frac{S_2(\mathbf{Z})}{S_2(\mathbf{Z})[I_f] + S_2(\mathbf{Z})[I_f]^\perp}, \quad (2)$$

and the congruence number r_{A_f} is its order. In particular, our definition of r_{A_f} generalizes the definition in Section 2.1 when A_f is an elliptic curve.

Remark 3.5. If R is a subring of $\mathbf{C}$, then $S_2(\mathbf{Z}) \otimes_{\mathbf{Z}} R = S_2(R)$ (see, e.g., the discussion in [DI95, §12]). Thus the analog of the group displayed in (2) with $\mathbf{Z}$ replaced by an algebraic integer ring (or even $\overline{\mathbf{Z}}$) gives a torsion module whose annihilator ideal meets $\mathbf{Z}$ in the ideal generated by the congruence exponent.

The following generalizes Theorem 2.1:

Theorem 3.6. *Let A and B be as in the first paragraph of Section 3. Then:*

(a) $\tilde{n}_A \mid \tilde{r}_A$.
(b) Let $\Gamma = \Gamma_0(N)$*. If* $p \nmid N$*, then* $\mathrm{ord}_p(\tilde{r}_A) = \mathrm{ord}_p(\tilde{n}_A)$*. If* $f \in S_2(\Gamma_0(N), \mathbf{C})$ *is a newform, then* $\mathrm{ord}_p(\tilde{r}_{A_f}) = \mathrm{ord}_p(\tilde{n}_{A_f})$ *whenever* $p^2 \nmid N$.

We give the proof of part (a) of this theorem in Section 4 and of part (b) in Section 5. The two sections may be read independently of each other.

Remark 3.7. Let $f \in S_2(\Gamma, \mathbf{C})$ be a newform. When A_f is an elliptic curve, Theorem 3.6 implies that the modular degree divides the congruence number (since for an elliptic curve, the modular degree and modular exponent are the same), and that $n_{A_f} \mid r_{A_f}^2$ (since for an elliptic curve, the modular number is the square of the modular exponent). In general, for a higher dimensional newform quotient, the divisibility $n_{A_f} \mid r_{A_f}^2$ need not hold. For example, there is a newform of degree 24 in $S_2(\Gamma_0(431))$ such that

$$n_{A_f} = (2^{11} \cdot 6947)^2 \nmid r_{A_f}^2 = (2^{10} \cdot 6947)^2.$$

Note that 431 is prime and mod 2 multiplicity one fails for $J_0(431)$ (see [Kil02]).

4 Proof of Theorem 3.6(a)

Since $\mathrm{End}(J)$ preserves A and B, we have a map $\mathrm{End}(J) \to \mathrm{End}(A) \oplus \mathrm{End}(B)$; moreover, since $A + B = J$, this map is injective. We have the following commutative diagram with exact rows:

$$\begin{array}{ccccccccc}
0 & \longrightarrow & \mathbf{T} & \longrightarrow & \mathbf{T}_A \oplus \mathbf{T}_B & \longrightarrow & \dfrac{\mathbf{T}_A \oplus \mathbf{T}_B}{\mathbf{T}} & \longrightarrow & 0 \\
 & & \downarrow & & \downarrow & & \vdots & & \\
0 & \longrightarrow & \mathrm{End}(J) & \longrightarrow & \mathrm{End}(A) \oplus \mathrm{End}(B) & \longrightarrow & \dfrac{\mathrm{End}(A) \oplus \mathrm{End}(B)}{\mathrm{End}(J)} & \longrightarrow & 0.
\end{array} \tag{3}$$

The first two vertical maps are clearly injections, and the rightmost vertical map is defined naturally so that the diagram is commutative. Let

$$e = (1, 0) \in \mathbf{T}_A \oplus \mathbf{T}_B,$$

and let e_1 and e_2 denote the images of e in the groups $(\mathbf{T}_A \oplus \mathbf{T}_B)/\mathbf{T}$ and $(\mathrm{End}(A) \oplus \mathrm{End}(B))/\mathrm{End}(J)$, respectively. Since $A \cap B$ is finite (in addition to the fact that $A + B = J$), the two quotient groups on the right side of (3) are finite, so e_1 and e_2 have finite order.

Lemma 4.1. *The element* $e_2 \in (\mathrm{End}(A) \oplus \mathrm{End}(B))/\mathrm{End}(J)$ *defined above has order* $\tilde{n}_A$.

Proof. By the *denominator* of any $\varphi \in \mathrm{End}(J) \otimes \mathbf{Q}$, we mean the smallest positive integer n such that $n\varphi \in \mathrm{End}(J)$. Let $\pi_A, \pi_B \in \mathrm{End}(J) \otimes \mathbf{Q}$ be projection onto A and B, respectively. Let n be the order of e_2, so n is the denominator of π_A, which equals the denominator of π_B (since $\pi_A + \pi_B = 1_J$, so that $\pi_B = 1_J - \pi_A$). We want to show that n is equal to $\tilde{n}_A$, the exponent of $A \cap B$.

Let i_A and i_B be the embeddings of A and B into J, respectively. We view $n\pi_A$ and $n\pi_B$ as morphisms $J \to A$ and $J \to B$, respectively. Let $\varphi = (n\pi_A, n\pi_B) \in \mathrm{Hom}(J, A \times B)$; then $\varphi \circ (i_A + i_B) = [n]_{A \times B}$. We have an exact sequence

$$0 \to A \cap B \xrightarrow{x \mapsto (x,-x)} A \times B \xrightarrow{i_A + i_B} J \to 0.$$

Let Δ be the image of $A \cap B$. Then by exactness,

$$[n]\Delta = (\varphi \circ (i_A + i_B))(\Delta) = \varphi \circ ((i_A + i_B)(\Delta)) = \varphi(\{0\}) = \{0\},$$

so n is a multiple of the exponent $\tilde{n}_A$ of $A \cap B$.

To show the opposite divisibility, consider the commutative diagram

$$\begin{array}{ccccccccc} 0 & \longrightarrow & A \cap B & \xrightarrow{x \mapsto (x,-x)} & A \times B & \longrightarrow & J & \longrightarrow & 0 \\ & & \big\downarrow [\tilde{n}_A] & & \big\downarrow ([\tilde{n}_A],0) & & \vdots\,\psi & & \\ 0 & \longrightarrow & A \cap B & \xrightarrow{x \mapsto (x,-x)} & A \times B & \longrightarrow & J & \longrightarrow & 0, \end{array}$$

where the middle vertical map is $(a, b) \mapsto (\tilde{n}_A a, 0)$ and the map ψ exists because $[\tilde{n}_A](A \cap B) = 0$. But $\psi = \tilde{n}_A \pi_A$ in $\mathrm{End}(J) \otimes \mathbf{Q}$. This shows that $\tilde{n}_A \pi_A \in \mathrm{End}(J)$, i.e., that $\tilde{n}_A$ is a multiple of the denominator n of π_A. □

Lemma 4.2. *The element* $e_1 \in (\mathbf{T}_A \oplus \mathbf{T}_B)/\mathbf{T}$ *has order* $\tilde{r}_A$.

Proof. We want to show that the order r of e_1 equals the exponent of $M = (\mathbf{T}_A \oplus \mathbf{T}_B)/\mathbf{T}$. Since e_1 is an element of M, the exponent of M is divisible by r. To obtain the reverse divisibility, consider any element x of M. Let $(a, b) \in \mathbf{T}_A \oplus \mathbf{T}_B$ be such that its image in M is x. By definition of e_1 and r, we have $(r, 0) \in \mathbf{T}$, and since $1 = (1, 1) \in \mathbf{T}$, we also have $(0, r) \in \mathbf{T}$. Thus $(\mathbf{T}r, 0)$ and $(0, \mathbf{T}r)$ are both subsets of $\mathbf{T}$ (i.e., are in the image of $\mathbf{T}$ under the map $\mathbf{T} \to \mathbf{T}_A \oplus \mathbf{T}_B$), so $r(a, b) = (ra, rb) = (ra, 0) + (0, rb) \in \mathbf{T}$. This implies that the order of x divides r. Since this is true for every $x \in M$, we conclude that the exponent of M divides r. □

Proof of Theorem 3.6(a). Since e_2 is the image of e_1 under the rightmost vertical homomorphism in (3), the order of e_2 divides that of e_1. Now apply Lemmas 4.1 and 4.2. □

5 Proof of Theorem 3.6(b)

Let $\mathbf{T}'$ be the saturation of $\mathbf{T} = \mathbf{Z}[\ldots, T_n, \ldots]$ in $\mathrm{End}(J)$, i.e.,

$$\mathbf{T}' = \mathrm{End}(J) \cap (\mathbf{T} \otimes \mathbf{Q}).$$

The quotient $\mathbf{T}'/\mathbf{T}$ is a finitely generated abelian group because both $\mathbf{T}$ and $\mathrm{End}(J)$ are finitely generated over $\mathbf{Z}$. Since $\mathbf{T}'/\mathbf{T}$ is also a torsion group, it is finite.

In Section 5.1, we introduce two ideals R and S of the Hecke algebra that are generalizations of the notions of the congruence exponent and the modular exponent respectively. We will see that $R \subset S$ and show that there is a natural injection $S/R \hookrightarrow \mathbf{T}'/\mathbf{T}$. In Section 5.2, we will prove that $\mathbf{T}$ and $\mathbf{T}'$ agree locally at a maximal ideal of $\mathbf{T}$ under the condition that we call "multiplicity one for differentials"; we also give examples where this condition does not hold. Theorem 3.6(b) itself is proved at the end of Section 5.1, by applying the results of Section 5.1 and a proposition that is proved in Section 5.2 to show that $R = S$ locally at a prime p such that $p \nmid N$ (when A is the dual of newform quotient, the condition that $p \nmid N$ can be replaced by $p^2 \nmid N$).

5.1 *The congruence and intersection ideals*

In this section, we work in slightly more generality, and take A and B to be as in the first paragraph of Section 3 (so Γ can be $\Gamma_1(N)$, and A need not be the dual of a newform quotient). Let $\pi_A : \mathbf{T} \to \mathbf{T}_A$ and $\pi_B : \mathbf{T} \to \mathbf{T}_B$ be the natural projection maps.

Definition 5.1. *With the setup as above, we define the* congruence ideal *as* $R = \pi_A(\ker(\pi_B)) \subset \mathbf{T}_A$, *and the* intersection ideal *as* $S = \mathrm{Ann}_{\mathbf{T}_A}(A \cap B)$.

Lemma 5.2. *We have* $R \subset S$.

Proof. By definition, R consists of restrictions to A of Hecke operators that vanish on B, while S consists of restrictions to A of Hecke operators that vanish on $A \cap B$. The lemma follows since the image in $\mathbf{T}_A$ of an operator that vanishes on B also vanishes on $A \cap B$. □

Remark 5.3. By Lemma 5.2, we have a surjection $\mathbf{T}_A/R \to \mathbf{T}_A/S$. Note that π_A induces an isomorphism

$$\frac{\mathbf{T}}{\ker(\pi_A) + \ker(\pi_B)} \xrightarrow{\simeq} \frac{\mathbf{T}_A}{R},$$

and we have an isomorphism

$$\frac{\mathbf{T}}{\ker(\pi_A) + \ker(\pi_B)} \xrightarrow{\simeq} \frac{\mathbf{T}_A \oplus \mathbf{T}_B}{\mathbf{T}}$$

obtained by sending $t \in \mathbf{T}$ to $(\pi_A(t), 0) \in \mathbf{T}_A \oplus \mathbf{T}_B$. Hence by Definition 3.4, the exponent of $\mathbf{T}_A/R$ is $\tilde{r}_A$ and its order is r_A. Also, $\tilde{n}_A$ is the exponent of $A \cap B$, and one expects that it is also the exponent of $\mathbf{T}_A/S$ (certainly multiplication by $\tilde{n}_A$ annihilates $\mathbf{T}_A/S$), which would give another proof that $\tilde{n}_A \mid \tilde{r}_A$. Instead of pursuing this question, we record the following result, which will be needed later.

Proposition 5.4. *If p is a prime such that the localizations of R and S at p coincide, then* $\mathrm{ord}_p(\tilde{r}_A) \leq \mathrm{ord}_p(\tilde{n}_A)$.

Proof. Under the hypothesis, the surjection $\mathbf{T}_A/R \to \mathbf{T}_A/S$ is an isomorphism locally at p. The lemma follows from the observations above that $\tilde{r}_A$ is the exponent of $\mathbf{T}_A/R$ and that $\tilde{n}_A$ annihilates $\mathbf{T}_A/S$. □

Lemma 5.5. *There is a natural inclusion $S/R \hookrightarrow \mathbf{T}'/\mathbf{T}$ of $\mathbf{T}$-modules.*

Proof. We have

$$\mathbf{T} \otimes \mathbf{Q} \cong (\mathbf{T}_A \otimes \mathbf{Q}) \oplus (\mathbf{T}_B \otimes \mathbf{Q}) \subset (\mathrm{End}(A) \otimes \mathbf{Q}) \oplus (\mathrm{End}(B) \otimes \mathbf{Q}) \cong \mathrm{End}(J) \otimes \mathbf{Q},$$

which we use to view $\mathbf{T}$ and $\mathbf{T}_A$ as sitting inside $\mathrm{End}(J) \otimes \mathbf{Q}$. Also, the groups $\mathrm{End}(J)$ and $\mathbf{T}'$ sit naturally in $\mathrm{End}(J) \otimes \mathbf{Q}$. By definition, $R = \mathbf{T}_A \cap \mathbf{T}$. Since an endomorphism of $A \times B$ factors through $A \times B \to J$ if and only if it kills $A \cap B$ embedded in $A \times B$ via $x \mapsto (x, -x)$, we have that $S = \mathbf{T}_A \cap \mathrm{End}(J)$ and this equals $\mathbf{T}_A \cap \mathbf{T}'$ (since a suitable multiple of any element of $\mathbf{T}_A$ lands in $\mathbf{T}$, when both are viewed as subgroups of $\mathbf{T} \otimes \mathbf{Q} \subset \mathrm{End}(J) \otimes \mathbf{Q}$). Hence we have $R = S \cap \mathbf{T}$ with intersection taken inside $\mathbf{T}' \subset \mathrm{End}(J) \otimes \mathbf{Q}$. Thus

$$S/R = S/(S \cap \mathbf{T}) \cong (S + \mathbf{T})/\mathbf{T} \hookrightarrow \mathbf{T}'/\mathbf{T}.$$ □

If $\mathfrak{m}$ is a maximal ideal of $\mathbf{T}$, then we say that two Hecke modules, with one contained in the other, *agree locally at* $\mathfrak{m}$ if their localizations at $\mathfrak{m}$ are the same. Let I_A denote the kernel of the map $\mathbf{T} \to \mathbf{T}_A$. As an immediate consequence of Lemma 5.5, we have:

Proposition 5.6. *If $\mathfrak{m}$ is a maximal ideal of $\mathbf{T}$ containing I_A that is not in* $\mathrm{Supp}_{\mathbf{T}}(\mathbf{T}'/\mathbf{T})$*, then the corresponding maximal ideal $\mathfrak{m}/I_A$ of $\mathbf{T}_A$ is not in the support of S/R, i.e.: if $\mathbf{T}$ and $\mathbf{T}'$ agree locally at $\mathfrak{m}$, then R and S also agree locally at $\mathfrak{m}/I_A$.*

Remark 5.7. The ring

$$\mathbf{T}'' = \mathrm{End}(J) \cap (\mathbf{T}_A \times \mathbf{T}_B) = \mathbf{T}' \cap (\mathbf{T}_A \times \mathbf{T}_B)$$

is often of interest, where the intersection is taken in $\mathrm{End}(J) \otimes \mathbf{Q}$. We proved above that there is a natural inclusion $S/R \hookrightarrow \mathbf{T}'/\mathbf{T}$. This inclusion yields an isomorphism $S/R \xrightarrow{\sim} \mathbf{T}''/\mathbf{T}$, as is clear from the "if and only if" statement in the proof of Lemma 5.5. The ideals R and S are equal if the rings $\mathbf{T}$ and $\mathbf{T}''$ coincide. Even when $\mathbf{T}'$ is bigger than $\mathbf{T}$, its subring $\mathbf{T}''$ may be not far from $\mathbf{T}$.

The following lemma and proposition will not be used in the proof of Theorem 3.6(b), but they are of interest from the point of view of multiplicity one.

Lemma 5.8. *Let p be a prime and let $\mathfrak{m}$ be a maximal ideal of $\mathbf{T}$ with residue characteristic p. Suppose $\mathfrak{m}$ satisfies the multiplicity one condition (i.e., $J[\mathfrak{m}]$ is of dimension two over $\mathbf{T}/\mathfrak{m}$). Then the completions of $\mathbf{T}$ and $\mathbf{T}'$ at $\mathfrak{m}$ are isomorphic.*

Proof. As in [Maz77, p.92], consider the Tate module $\mathrm{Ta}_{\mathfrak{m}}(J)$, which is the Pontryagin dual of the $\mathfrak{m}$-divisible group associated to $J(\overline{\mathbf{Q}})$. Since $J[\mathfrak{m}]$ is of dimension two over $\mathbf{T}/\mathfrak{m}$, it follows that $\mathrm{Ta}_{\mathfrak{m}}(J)$ is free of rank 2 over $\mathbf{T}_{\mathfrak{m}}$, where the subscript denotes completion (see, e.g., [Til97, p. 332-333]). If r is an element of $\mathbf{T}'_{\mathfrak{m}}$, then r operates $\mathbf{T}_{\mathfrak{m}}$-linearly on $\mathrm{Ta}_{\mathfrak{m}}(J)$, and thus may be viewed as a 2×2 matrix with entries in $\mathbf{T}_{\mathfrak{m}}$. Further, some non-zero integer multiple of r operates on $\mathrm{Ta}_{\mathfrak{m}}(J)$ as an element of $\mathbf{T}_{\mathfrak{m}}$, i.e., as a scalar. Thus r must be a scalar to start with, i.e., actually lies in $\mathbf{T}_{\mathfrak{m}}$. Hence $\mathbf{T}'_{\mathfrak{m}} = \mathbf{T}_{\mathfrak{m}}$ as claimed. □

Proposition 5.9. *Let p be a prime such that all maximal ideals $\mathfrak{m}$ of $\mathbf{T}$ with residue characteristic p that contain I_A satisfy multiplicity one. Then $\mathrm{ord}_p(\tilde{r}_A) = \mathrm{ord}_p(\tilde{n}_A)$.*

Proof. This follows from Lemma 5.8, Lemma 5.5, Proposition 5.4, and Theorem 3.6(a). □

Proposition 5.10. *Let $\Gamma = \Gamma_0(N)$. Let p be a prime such that $p^2 \nmid N$, and let $\mathfrak{m}$ be a maximal ideal of $\mathbf{T}$ with residue characteristic p. If $p|N$, then assume that $I_f \subseteq \mathfrak{m}$ for some newform f. Then $\mathbf{T}$ and $\mathbf{T}'$ agree locally at $\mathfrak{m}$.*

Since the proof of this proposition is rather technical, we have postponed it to Section 5.2. Admitting this proposition, we may now finish the proof of Theorem 3.6(b).

Proof of Theorem 3.6(b). Recall that A and B are abelian subvarieties of $J = J_0(N)$ such that $A + B = J$, $A \cap B$ is finite, and every endomorphism of J over $\mathbf{Q}$ preserves A and B.

We first want to show that if a prime p does not divide N, then $\mathrm{ord}_p(\tilde{r}_A) = \mathrm{ord}_p(\tilde{n}_A)$. In view of Theorem 3.6(a) and Proposition 5.4, it suffices to check that R and S coincide locally at p. By Proposition 5.6, it suffices to check that $\mathbf{T}$ and $\mathbf{T}'$ are locally equal at all maximal ideals that divide p. If $p \nmid N$, then this follows from Proposition 5.10, which proves part of Theorem 3.6(b).

It remains to show that if $f \in S_2(\Gamma_0(N), \mathbf{C})$ is a newform and $p \parallel N$, then $\mathrm{ord}_p(\tilde{r}_{A_f}) = \mathrm{ord}_p(\tilde{n}_{A_f})$. Note that the Hecke algebra $\mathbf{T}$ acts on S/R through its quotient $\mathbf{T}_{A_f^\vee} = \mathbf{T}/\mathrm{Ann}_{\mathbf{T}} A_f^\vee$ since the action of $\mathbf{T}$ on R and on S factors through this quotient. Thus, in view of Theorem 3.6(a) and Proposition 5.4, it suffices to check that R and S coincide locally at maximal ideals of $\mathbf{T}$ that divide p and contain $\mathrm{Ann}_{\mathbf{T}} A_f^{\vee} = I_f$ (the equality follows since I_f is saturated). But this follows from Proposition 5.6 and Proposition 5.10. □

5.2 Multiplicity one for differentials

This section is devoted to the proof of Proposition 5.10 as well as a discussion of the notion of multiplicity one for differentials (Definition 5.13). In this section, we take $\Gamma = \Gamma_0(N)$.

Let p be a prime such that $p^2 \nmid N$. Let $M_0(N)$ denote the compactified coarse moduli scheme associated to $\Gamma_0(N)$ (as in [DR73, § IV.3]) over $\mathbf{Z}_p$, and let $X_0(N)_{\mathbf{Z}_p}$ denote its minimal regular resolution obtained by suitable blow-up of the points $j = 0, 1728$ in characteristic dividing N, when they are supersingular (cf. [Maz77, p.63]). Let $\Omega_{X_0(N)/\mathbf{Z}_p}$ denote the relative dualizing sheaf of $X_0(N)_{\mathbf{Z}_p}$ over $\mathbf{Z}_p$ (it is the sheaf of regular differentials as in [MR91, §7]). We denote by $X_0(N)_{\mathbf{F}_p}$ the special fiber of $X_0(N)_{\mathbf{Z}_p}$ at the prime p and by $\Omega_{X_0(N)/\mathbf{F}_p}$ the relative dualizing sheaf of $X_0(N)_{\mathbf{F}_p}$ over $\mathbf{F}_p$.

The usual Hecke operators and the Atkin–Lehner involutions (corresponding to primes dividing N) of $J_0(N)$ over $\mathbf{Q}$ extend uniquely to act on the base change to $\mathbf{Z}_p$ of the Néron model of $J_0(N)$, which we denote by $J_{\mathbf{Z}_p}$. The natural morphism $\mathrm{Pic}^0_{X_0(N)/\mathbf{Z}_p} \to J_{\mathbf{Z}_p}$ identifies $\mathrm{Pic}^0_{X_0(N)/\mathbf{Z}_p}$ with the identity component of $J_{\mathbf{Z}_p}$ (see, e.g., [BLR90, §9.4–9.5]). Passing to tangent spaces along the identity section over $\mathbf{Z}_p$, we obtain an isomorphism $H^1(X_0(N)_{\mathbf{Z}_p}, \mathcal{O}_{X_0(N)_{\mathbf{Z}_p}}) \cong \mathrm{Tan}(J_{\mathbf{Z}_p})$. Using Grothendieck duality, one gets an isomorphism $\mathrm{Cot}(J_{\mathbf{Z}_p}) \cong H^0(X_0(N)_{\mathbf{Z}_p}, \Omega_{X_0(N)/\mathbf{Z}_p})$, where $\mathrm{Cot}(J_{\mathbf{Z}_p})$ is the cotangent space at the identity section (cf. [Maz78, p. 140]). Now the Hecke operators and the Atkin–Lehner involutions act on $\mathrm{Cot}(J_{\mathbf{Z}_p})$, and hence via the last isomorphism above, we get an action of the Hecke operators and the Atkin–Lehner involutions on $H^0(X_0(N)_{\mathbf{Z}_p}, \Omega_{X_0(N)/\mathbf{Z}_p})$. Following the proof of Prop. 3.3 on p. 68 of [Maz77], specialization induces an isomorphism

$$\mathrm{H}^0(X_0(N)_{\mathbf{F}_p}, \Omega_{X_0(N)/\mathbf{F}_p}) \cong \mathrm{H}^0(X_0(N)_{\mathbf{Z}_p}, \Omega_{X_0(N)/\mathbf{Z}_p}) \otimes_{\mathbf{Z}_p} \mathbf{F}_p.$$

In this way, we get an action of the Hecke operators and the Atkin–Lehner involutions on $H^0(X_0(N)_{\mathbf{F}_p}, \Omega_{X_0(N)/\mathbf{F}_p})$ as well.

The following lemma is implicit in [Maz77, p. 95].

Lemma 5.11 (Mazur). *Let $\mathfrak{m}$ be a maximal ideal of $\mathbf{T}$ of residue characteristic p (recall that $p^2 \nmid N$). Suppose*

$$\dim_{\mathbf{T}/\mathfrak{m}} \mathrm{H}^0(X_0(N)_{\mathbf{F}_p}, \Omega_{X_0(N)/\mathbf{F}_p})[\mathfrak{m}] \leq 1.$$

Then $\mathbf{T}$ and $\mathbf{T}'$ agree locally at $\mathfrak{m}$.

Proof. Let M denote the group $H^1(X_0(N)_{\mathbf{Z}_p}, \mathcal{O}_{X_0(N)})$, where $\mathcal{O}_{X_0(N)}$ is the structure sheaf of $X_0(N)$. As explained in [Maz77, p. 95], we have an action of $\mathrm{End}_{\mathbf{Q}} J_0(N)$ on M, and the action of $\mathbf{T}$ on M via the inclusion $\mathbf{T} \subset \mathrm{End}_{\mathbf{Q}} J_0(N)$ is faithful, so likewise for the action by $\mathbf{T}'$. Hence we have an injection $\phi : \mathbf{T}' \hookrightarrow \mathrm{End}_{\mathbf{T}} M$. Suppose $\mathfrak{m}$ is a maximal ideal of $\mathbf{T}$ that satisfies the hypotheses of the lemma. To prove that $\mathbf{T}_{\mathfrak{m}} = \mathbf{T}'_{\mathfrak{m}}$ it suffices to prove the following claim: □

Claim: The map $\phi|_{\mathbf{T}}$ is surjective locally at $\mathfrak{m}$.

Proof. It suffices to show that M is generated by a single element over $\mathbf{T}$ locally at $\mathfrak{m}$, and in turn, by Nakayama's lemma, it suffices to check that the dimension of the $\mathbf{T}/\mathfrak{m}$-vector space $M/\mathfrak{m}M$ is at most one. Now $M/\mathfrak{m}M$ is dual to $H^0(X_0(N)_{\mathbf{F}_p}, \Omega_{X_0(N)/\mathbf{F}_p})[\mathfrak{m}]$. Since we are assuming that $\dim_{\mathbf{T}/\mathfrak{m}} H^0 (X_0(N)_{\mathbf{F}_p}, \Omega_{X_0(N)/\mathbf{F}_p})[\mathfrak{m}] \leq 1$, we have $\dim_{\mathbf{T}/\mathfrak{m}}(M/\mathfrak{m}M) \leq 1$, which proves the claim.

Remark 5.12. Note that Lemma 5.8 may provide an alternate route to the conclusion of the previous lemma (sometimes one can prove multiplicity one for a maximal ideal without relying on multiplicity one for differentials, e.g., see [Dia97]). Observe that in the proofs of Lemmas 5.11 and 5.8, all we needed was (locally) a non-zero free $\mathbf{T}$-module (of finite rank, say) that is attached functorially to J. In Lemma 5.11, the module we used was $H^1(X_0(N)_{\mathbf{Z}_p}, \mathcal{O}_{X_0(N)})$; locally, it is free because its reduction modulo $\mathfrak{m}$ is of the same dimension as its generic rank (namely 1). In Lemma 5.8, we used the $\mathfrak{m}$-adic Tate module, whose reduction mod $\mathfrak{m}$ is of the same dimension as its generic rank (namely 2).

Definition 5.13. *If* $\mathfrak{m}$ *is a maximal ideal of the Hecke algebra* $\mathbf{T}$ *of residue characteristic* p*, we say that* $\mathfrak{m}$ *satisfies* multiplicity one for differentials *if*

$$\dim_{\mathbf{T}/\mathfrak{m}}(\mathrm{H}^0(X_0(N)_{\mathbf{F}_p}, \Omega_{X_0(N)/\mathbf{F}_p})[\mathfrak{m}]) = 1.$$

The above condition, which first appeared in [Maz77], plays an important role in several places, including Wiles's proof of Fermat's last theorem (see [Wil95, Lemma 2.2]). It has been used to prove multiplicity one for $\mathfrak{m}$ (as in Section 2.2) and Gorensteinness of the completion of $\mathbf{T}$ at $\mathfrak{m}$ (under certain hypotheses; see, e.g., [Til97]).

5.2.1 Failure of multiplicity one for differentials

In this section, we digress to discuss examples of failure of multiplicity one for differentials. The reader interested in the proof of Proposition 5.10 may jump to Section 5.2.2 below.

By Lemma 5.11, if $p^2 \nmid N$ and if the multiplicity one condition for differentials holds at $\mathfrak{m}$, then $\mathbf{T}$ and $\mathbf{T}'$ agree locally at $\mathfrak{m}$. It is thus of interest to compute the quotient group $\mathbf{T}'/\mathbf{T}$ for various N. We compute this index in Sage [S+09]. and

Table 2 Nonzero Quotients $\mathbf{T}'/\mathbf{T}$ for $N \leq 325$

44	C_2	160	$C_2^3 \oplus C_4 \oplus C_8$	245	C_7^2
46	C_2	162	C_3^4	248	$C_2^7 \oplus C_4 \oplus C_8$
54	C_3	164	C_2^3	250	C_5^8
56	C_2	166	C_2	252	$C_2^2 \oplus C_6^3 \oplus C_{12}$
60	C_2	168	$C_2^5 \oplus C_4$	254	C_2^2
62	C_2	169	C_{13}	256	$C_2^3 \oplus C_4^2 \oplus C_8^2 \oplus C_{16}$
64	C_2	171	C_3^2	260	C_2^6
68	C_2	172	C_2^3	261	C_3^4
72	C_2	174	C_2	262	C_2^2
76	C_2	175	C_5	264	$C_2^7 \oplus C_4^3$
78	C_2	176	$C_2^2 \oplus C_4^2 \oplus C_8$	268	C_2^5
80	C_4	180	$C_2 \oplus C_6^2$	270	$C_3^9 \oplus C_6^2$
84	C_2	184	$C_2^5 \oplus C_4 \oplus C_8$	272	$C_2^3 \oplus C_4^4 \oplus C_8$
88	$C_2 \oplus C_4$	186	C_2^2	275	C_5^4
92	$C_2^2 \oplus C_4$	188	$C_2^4 \oplus C_4^2$	276	$C_2^7 \oplus C_4^2$
94	C_2^2	189	C_3^5	278	C_2
96	C_2^3	190	C_2^3	279	C_3^4
99	C_3^2	192	$C_2^3 \oplus C_4^3 \oplus C_8$	280	$C_2^7 \oplus C_4^3$
104	C_2^2	196	C_{14}	282	C_2^2
108	$C_3^2 \oplus C_6$	198	C_3^4	284	$C_2^6 \oplus C_4^3$
110	C_2	200	$C_2^3 \oplus C_{10}$	286	C_2^4
112	$C_2 \oplus C_4$	204	C_2^5	288	$C_2^7 \oplus C_4^3 \oplus C_{12} \oplus C_{24}$
116	C_2^2	206	C_2^2	289	C_{17}^2
118	C_2	207	C_3^4	290	C_2
120	$C_2^3 \oplus C_4$	208	$C_2^2 \oplus C_4^3$	292	C_2^5
124	$C_2^2 \oplus C_4$	210	C_2	294	C_7^4
125	C_5^2	212	C_2^4	296	$C_2^6 \oplus C_4^2$
126	$C_3 \oplus C_6$	214	C_2	297	$C_3^8 \oplus C_9$
128	$C_2 \oplus C_4 \oplus C_8$	216	$C_3 \oplus C_6^5 \oplus C_{12}$	300	$C_2^2 \oplus C_{10}^3$
132	C_2^3	220	$C_2^5 \oplus C_4$	302	C_2^3
135	C_3^3	224	$C_2^5 \oplus C_4^2 \oplus C_8$	304	$C_2^4 \oplus C_4^4 \oplus C_8$
136	$C_2^2 \oplus C_4$	225	C_5	306	C_3^6
140	C_2^3	228	C_2^5	308	C_2^7
142	C_2^3	230	C_2^2	310	C_2^3
144	$C_2^3 \oplus C_4$	232	$C_2^4 \oplus C_4^2$	312	$C_2^{11} \oplus C_4^2 \oplus C_8$
147	C_7	234	$C_3^2 \oplus C_6^2$	315	C_3^6
148	C_2^2	236	$C_2^5 \oplus C_4$	316	$C_2^6 \oplus C_4^2$
150	C_5	238	C_2^4	318	C_2^4
152	$C_2^3 \oplus C_4$	240	$C_2^7 \oplus C_4^3 \oplus C_8$	320	$C_2^6 \oplus C_4^3 \oplus C_8^3 \oplus C_{16}$
153	C_3	242	C_{11}^2	322	C_2^2
156	$C_2^3 \oplus C_4$	243	$C_3^4 \oplus C_9^2$	324	$C_3^7 \oplus C_6^3 \oplus C_{18}$
158	C_2^2	244	C_2^4	325	C_5^3

obtain Table 2, where the first column contains N for $N \leq 325$ and the second column contains the quotient group $\mathbf{T}'/\mathbf{T}$, where C_n denotes a cyclic group of order n.

In each case in which a prime p divides $[\mathbf{T}' : \mathbf{T}]$ but $p^2 \nmid N$, Lemma 5.11 implies that there is some maximal ideal $\mathfrak{m}$ of $\mathbf{T}$ of residue characteristic p for which multiplicity one for differentials does not hold. For example, when $N = 46$, we find that $[\mathbf{T}' : \mathbf{T}] = 2$, and $2^2 \nmid N$; thus there is a maximal ideal $\mathfrak{m}$ of $\mathbf{T}$ of residue characteristic 2 for which multiplicity one for differentials does not hold.

In Table 2, we observe that whenever p divides $[\mathbf{T}' : \mathbf{T}]$, then $p = 2$ or $p^2 \mid N$. This raises the question: is it true that if p is odd and $p^2 \nmid N$, then multiplicity one for differentials holds for maximal ideals $\mathfrak{m}$ of $\mathbf{T}$ of residue characteristic p? Lemma 5.20 below gives an affirmative answer in one direction (the other direction is usually easy), but under the hypothesis that if $p \mid N$ then U_p acts as a non-zero scalar on $H^0(X_0(N)_{\mathbf{F}_p}, \Omega_{X_0(N)/\mathbf{F}_p})[\mathfrak{m}]$.

5.2.2 Proof of Proposition 5.10

The main point is to prove that the hypothesis

$$\dim_{\mathbf{T}/\mathfrak{m}} \mathrm{H}^0(X_0(N)_{\mathbf{F}_p}, \Omega_{X_0(N)/\mathbf{F}_p})[\mathfrak{m}] \leq 1$$

of Lemma 5.11 holds for suitable maximal ideals $\mathfrak{m}$. This is achieved in Lemma 5.20 below, whose proof requires an Eichler–Shimura type relation for U_p (Lemma 5.15 below). We obtain this relation by modifying the argument in [Wil80, §5], which is in the $\Gamma_1(N)$ context, to the $\Gamma_0(N)$ situation. Let L denote the maximal unramified extension of $\mathbf{Q}_p$ and let $\mathcal{O}_L$ denote the ring of integers of L. For the sake of completeness, we state below a lemma that is well known (e.g., it is used implicitly in [Wil80, p. 18]); the proof was indicated to us by F. Calegari.

Lemma 5.14. *Let E be an elliptic curve over $\mathcal{O}_L$ with good ordinary reduction. Then the subgroup schemes of E of order p are p copies of $\mathbf{Z}/p\mathbf{Z}$ and one copy of μ_p.*

Proof. Let $G = E[p]$, and consider its connected-étale sequence

$$0 \to G^0 \to G \to G^{\mathrm{et}} \to 0.$$

Now G^0 is in the kernel of the reduction map, and we know that the reduction of $E[p]$ has non-trivial order. Hence G^{et} is non-trivial. By Cartier duality, G^0 is also non-trivial. Hence G^{et} is a $\mathbf{Z}/p\mathbf{Z}$ and by duality, G^0 is a μ_p. Thus one of the subgroup schemes of E of order p is a copy of μ_p. Let H be any other subgroup scheme of E of order p. Then H^0 has to be trivial, since otherwise $H = H^0$ is a non-trivial subgroup scheme of $G^0 = \mu_p$, hence is equal to $G^0 = \mu_p$, which has already been accounted for. Thus H is étale, and hence is a copy of $\mathbf{Z}/p\mathbf{Z}$. The lemma follows, since there are $p + 1$ subgroup schemes of order p in $E[p]$, hence in E. □

We assume that $p \| N$ until just after the proof of Lemma 5.18. Let $M = N/p$. We will use the superscript h to denote the subscheme of $M_0(N)$ obtained by removing the supersingular points in characteristic p. Following [DR73, VI.6.9] and [DR73, § V.2], the $\overline{\mathbf{F}}_p$-valued points of $M_0(N)^h$ are in one-to-one correspondence with isomorphism classes of triples consisting of

(a) a generalized elliptic curve E over $\overline{\mathbf{F}}_p$, whose smooth locus we denote E^{sm},
(b) a subgroup of $E^{\mathrm{sm}}[p]$ isomorphic to $\boldsymbol{\mu}_p$ or to $\mathbf{Z}/p\mathbf{Z}$, and
(c) a subgroup $\mathbf{Z}/M\mathbf{Z}$ of $E^{\mathrm{sm}}[M]$,

such that the subgroup generated by the subgroups in (b) and (c) above meets every irreducible component of every geometric fiber of E over $\overline{\mathbf{F}}_p$. Also, $M_0(N)_{\overline{\mathbf{F}}_p}$ has two irreducible components, which may be described according as whether the subgroup in (b) is isomorphic to $\boldsymbol{\mu}_p$ or to $\mathbf{Z}/p\mathbf{Z}$. As mentioned earlier, $X_0(N)_{\overline{\mathbf{F}}_p}$ is obtained from $M_0(N)_{\overline{\mathbf{F}}_p}$ by suitable blowups and consists of two copies of $X_0(M)_{\overline{\mathbf{F}}_p}$ identified at supersingular points, along with some copies of $\mathbf{P}^1$ (see the description of $X_0(N)_{\overline{\mathbf{F}}_p}$ on p. 175–177 of [Maz77] for details). One of the copies of $X_0(M)_{\overline{\mathbf{F}}_p}$ corresponds to the irreducible component of $M_0(N)_{\overline{\mathbf{F}}_p}$ where the subgroup in (b) is isomorphic to $\mathbf{Z}/p\mathbf{Z}$; we denote this copy by C_0. The other copy of $X_0(M)_{\overline{\mathbf{F}}_p}$ corresponds to the irreducible component of $M_0(N)_{\overline{\mathbf{F}}_p}$ where the subgroup in (b) is isomorphic to $\boldsymbol{\mu}_p$, and contains the cusp ∞; we denote this copy by C_1. We denote the copies (if any) of $\mathbf{P}^1$ by $C_2, \ldots, C_r$, where r is one less than the total number of irreducible components of $X_0(N)_{\overline{\mathbf{F}}_p}$.

The usual endomorphisms U_p and W_p of $J_0(N)$ over $\mathbf{Q}$ can be extended by base change to L, and extend uniquely to act on the Néron model of $J_0(N)$ over $\mathcal{O}_L$. Since the formation of Néron models is compatible with completions and unramified base change, this action is compatible with the already-defined action on the Néron model of $J_0(N)$ over $\mathbf{Z}_p$. The identity component of the special fiber of the Néron model of $J_0(N)$ over $\mathcal{O}_L$ is $\mathrm{Pic}^0_{X_0(N)/\overline{\mathbf{F}}_p}$, whose maximal abelian variety quotient is $\prod_{i=0}^r \mathrm{Pic}^0_{C_i/\overline{\mathbf{F}}_p}$ (cf. [DR73, I.3.7] and [BLR90, §9.2, Example 8]). Thus we get an action of U_p and W_p on $\mathrm{Pic}^0_{X_0(N)/\overline{\mathbf{F}}_p}$ and on $\prod_{i=0}^r \mathrm{Pic}^0_{C_i/\overline{\mathbf{F}}_p}$. Let Frob_p denote the Frobenius morphism on $C_{0/\overline{\mathbf{F}}_p}$.

Lemma 5.15. *The endomorphisms U_p and W_p of $\prod_{i=0}^r \mathrm{Pic}^0_{C_i/\overline{\mathbf{F}}_p}$ satisfy $U_p = \mathrm{Frob}_p + (p-1)W_p$ on $\mathrm{Pic}^0_{C_0/\overline{\mathbf{F}}_p}$.*

Proof. The proof is a modification of the proof of Theorem 5.3 in [Wil80], along with some details borrowed from the proof of Theorem 5.16 in B. Conrad's appendix to [RS01].

It suffices to check the desired identity on a Zariski dense subset of $\mathrm{Pic}^0{}_{C_0/\overline{\mathbf{F}}_p}(\overline{\mathbf{F}}_p) = J(C_0)(\overline{\mathbf{F}}_p)$, where $J(C_0)$ is the Jacobian of C_0. If g is the genus of C_0, then fixing a base point, we get a surjection $C_0^g \to J(C_0)$. Hence if U is any dense open subset of $C_0(\overline{\mathbf{F}}_p)$, then U^g hits a Zariski dense subset of $J(C_0)(\overline{\mathbf{F}}_p)$. Taking U to be the ordinary locus of $C_0(\overline{\mathbf{F}}_p)$, it thus suffices to prove the desired identity on divisors of the form $(Q)-(Q')$, where the elliptic curves corresponding to $Q, Q' \in C_0(\overline{\mathbf{F}}_p)$ are ordinary.

Let $\mathcal{M}_0(N)$ denote the algebraic stack over $\mathcal{O}_L$ associated to $\Gamma_0(N)$ by [DR73, IV.3.3, IV.4.2], whose associated coarse moduli scheme is $M_0(N)$ (over $\mathcal{O}_L$). Let $\pi : \mathcal{M}_0(N) \to M_0(N)$ denote the associated natural map. If $k = \overline{\mathbf{F}}_p$ or an algebraic closure of L, then π is an isomorphism on k-valued points, and so we will often identify points on $M_0(N)(k)$ with points on $\mathcal{M}_0(N)(k)$. Let Q be an ordinary point on $C_0(\overline{\mathbf{F}}_p)$. Then Q is given by a triple $(\overline{E}, \overline{C}, \overline{D})$, where $\overline{E}$ is an ordinary elliptic curve over $\overline{\mathbf{F}}_p$, $\overline{C}$ is a subgroup isomorphic to $\mathbf{Z}/p\mathbf{Z}$, and $\overline{D}$ is a subgroup isomorphic to $\mathbf{Z}/M\mathbf{Z}$. We can choose a Weierstrass model $E \hookrightarrow \mathbf{P}^2_{\mathcal{O}_L}$ lifting $\overline{E}$; then E is canonically an elliptic curve by [KM85, Chap. 2]. By Lemma 5.14 and its proof, there is a subgroup C of E isomorphic to $\mathbf{Z}/p\mathbf{Z}$ that lifts $\overline{C}$. Also, as argued in [RS01, p. 219], there is a subgroup D of E isomorphic to $\mathbf{Z}/M\mathbf{Z}$ that lifts $\overline{D}$. Then (E, C, D) gives a point on $\mathcal{M}_0(N)(\mathcal{O}_L)$ (cf. [DR73, V.1.6]), whose image in $M_0(N)(\mathcal{O}_L)$ corresponds to a point P in $X_0(N)(\mathcal{O}_L)$ (since E has ordinary reduction). We will use a bar to denote specialization. Thus we have $Q = \overline{P}$. Similarly, given another point $Q' \in C_0(\overline{\mathbf{F}}_p)$, we will denote the corresponding associated quantities by a prime superscript (thus P' in $X_0(N)(\mathcal{O}_L)$ denotes a lift of Q', etc.). As mentioned in the previous paragraph, it suffices to prove the relation claimed in the lemma for elements of the form $(Q) - (Q')$ in $\mathrm{Pic}^0_{C_0/\overline{\mathbf{F}}_p}(\overline{\mathbf{F}}_p)$. Viewing P and P' as relative effective Cartier divisors of degree one, we see that $U_p((Q) - (Q'))$ is the image of $U_p((P) - (P'))$ under specialization, i.e., $U_p((Q) - (Q')) = \overline{U_p((P) - (P'))}$.

We next compute $U_p((P) - (P'))$. Now $\mathrm{Pic}^0{}_{X_0(N)/\mathcal{O}_L}$ is the identity component of $J_0(N)_{\mathcal{O}_L}$, and we have $J_0(N)_{\mathcal{O}_L}(\mathcal{O}_L) = J_0(N)(L) \subseteq J_0(N)(\overline{L})$, where $\overline{L}$ is an algebraic closure of L. Denoting base change to $\overline{L}$ by a subscript $\overline{L}$, we have

$$\begin{aligned} U_p((E_{\overline{L}}, C_{\overline{L}}, D_{\overline{L}}) &- (E'_{\overline{L}}, C'_{\overline{L}}, D'_{\overline{L}})) \\ &= \sum_{A_{\overline{L}}} (E_{\overline{L}}/A_{\overline{L}}, (C_{\overline{L}} + A_{\overline{L}})/A_{\overline{L}}, (D_{\overline{L}} + A_{\overline{L}})/A_{\overline{L}}) \\ &\quad - \sum_{A'_{\overline{L}}} (E'_{\overline{L}}/A'_{\overline{L}}, (C'_{\overline{L}} + A'_{\overline{L}})/A'_{\overline{L}}, (D'_{\overline{L}} + A'_{\overline{L}})/A'_{\overline{L}}), \qquad (4) \end{aligned}$$

where $A_{\overline{L}}$ runs through the subgroups of $E_{\overline{L}}$ of order p except $C_{\overline{L}}$ (and similarly for $A'_{\overline{L}}$). Enlarging L by a finite extension if needed (which does not change the residue field $\overline{\mathbf{F}}_p$) we may assume that there are $p + 1$ subgroups of order p in E_L. Their scheme-theoretic closures in E over $\mathcal{O}_L$ are the subgroups schemes mentioned in Lemma 5.14. If A is a subgroup scheme of E of order p, then we denote the quotient map $E \to E/A$ by α_A. Consider the Cartier divisors corresponding to $U_p((P) - (P'))$ and to

$$\begin{aligned} &\Big(\pi(E/\mu_p, \alpha_{\mu_p}(C), \alpha_{\mu_p}(D)) + \sum_{B} \pi(E/B, \mathrm{cl}(\alpha_B(C)), \alpha_B(D))\Big) \\ &\quad -\Big(\pi(E'/\mu'_p, \alpha_{\mu'_p}(C'), \alpha_{\mu'_p}(D')) + \sum_{B'} \pi(E'/B', \mathrm{cl}(\alpha_{B'}(C')), \alpha_{B'}(D'))\Big), \end{aligned}$$

where B runs through the subgroups of E isomorphic to $\mathbf{Z}/p\mathbf{Z}$ except for C, and $\mathrm{cl}(\alpha_B(C))$ denotes the Zariski closure of $\alpha_B(C)$ in E/B (and similarly with prime superscripts). These two divisors coincide since they induce the same $\overline{L}$-point by (4).

Passing to special fibers, and noting that the special fiber of the Néron model of E/A is given by $\overline{E}/\overline{A}$, we find that

$$U_p((Q)-(Q')) = \overline{U_p((P)-(P'))}$$

$$= \Big((\overline{E}/\overline{\mu_p}, \overline{\alpha}_{\mu_p}(\overline{C}), \overline{\alpha}_{\mu_p}(\overline{D})) + \sum_B (\overline{E}/\overline{B}, \overline{\mathrm{cl}(\alpha_B(C))}, \overline{\alpha}_B(\overline{D}))\Big) \quad (5)$$

$$-\Big((\overline{E'}/\overline{\mu'_p}, \overline{\alpha}_{\mu'_p}(\overline{C'}), \overline{\alpha}_{\mu'_p}(\overline{D'})) + \sum_{B'} (\overline{E'}/\overline{B'}, \overline{\mathrm{cl}(\alpha_{B'}(C'))}, \overline{\alpha}_{B'}(\overline{D'}))\Big), \quad (6)$$

where B again runs through the subgroups of E isomorphic to $\mathbf{Z}/p\mathbf{Z}$ except for C (and a similar statment holds with prime superscripts).

Let F_p denote the relative Frobenius map $\overline{E} \to \overline{E}^{(p)}$ over $\overline{\mathbf{F}}_p$. Now μ_p is in the kernel of F_p, and since the quotient map $\overline{\alpha}_{\mu_p}$ has the same degree as F_p, there is an isomorphism $\phi : \overline{E}/\overline{\mu_p} \overset{\cong}{\to} \overline{E}^{(p)}$ such that $F_p = \phi \circ \overline{\alpha}_{\mu_p}$. Also ϕ induces an isomorphism $\overline{\alpha}_{\mu_p}(\overline{C}) \overset{\cong}{\to} \overline{C}^{(p)}$ and $\overline{\alpha}_{\mu_p}(\overline{D}) \overset{\cong}{\to} \overline{D}^{(p)}$. Thus the first term in (5) is identified with $(\overline{E}^{(p)}, \overline{C}^{(p)}, \overline{D}^{(p)})$, which is the image under Frob_p of $\overline{P} = (\overline{E}, \overline{C}, \overline{D})$. Similarly, the first term in (6) is $\mathrm{Frob}_p(\overline{P'})$.

As for the sum over B in (5), note that in each term, we are quotienting by a group B which is isomorphic to $\mathbf{Z}/p\mathbf{Z}$, and hence $\mathrm{cl}(\alpha_B(C))$ is of μ_p-type. In a manner similar to the computation of the action of U_p, we find that

$$W_p((\overline{E}, \overline{C}, \overline{D}) - (\overline{E'}, \overline{C'}, \overline{D'}))$$

$$= (\overline{E}/\overline{C}, \overline{E}[p]/\overline{C}, (\overline{D}+\overline{C})/\overline{C}) \quad (7)$$

$$- (\overline{E'}/\overline{C'}, \overline{E'}[p]/\overline{C'}, (\overline{D'}+\overline{C'})/\overline{C'}). \quad (8)$$

Considering that $\overline{P} = (\overline{E}, \overline{C}, \overline{D})$, with $\overline{C}$ isomorphic to $\mathbf{Z}/p\mathbf{Z}$, we see that $\overline{E}[p]/\overline{C}$ is isomorphic to μ_p. Also, if B is as in the sum in (5), then $\overline{B}$ is a $\mathbf{Z}/p\mathbf{Z}$, but there is only one copy of $\mathbf{Z}/p\mathbf{Z}$ in $\overline{E}$, since E has good ordinary reduction; hence $\overline{B} = \overline{C}$. Thus each of the terms in the sum over B in (5) is the term in (7). A similar statment holds with prime superscripts (viz., each of the terms in the sum over B' in (6) is the term in (8)).

The lemma now follows from the previous two paragraphs. □

Since we are assuming that $p||N$, the curve $X_0(N)_{\overline{\mathbf{F}}_p}$ has ordinary double point singularities, and so the differentials in $H^0(X_0(N)_{\overline{\mathbf{F}}_p}, \Omega_{X_0(N)/\overline{\mathbf{F}}_p})$ may be identified with meromorphic differentials $(\omega_i)_{i=0,\dots,r}$ on $\coprod_{i=0}^r C_i$ whose only possible poles are at points on $\coprod_{i=0}^r C_i$ lying over an intersection point of two components in $X_0(N)_{\overline{\mathbf{F}}_p}$ and where the sum of the residues at the points lying over an intersection point is zero; such differentials are called *regular differentials* (see [Con00, §5.2] for the justification that the relative dualizing sheaf under Grothendieck duality is indeed the sheaf of regular differentials). By a *holomorphic differential* in $H^0(X_0(N)_{\overline{\mathbf{F}}_p}, \Omega_{X_0(N)/\overline{\mathbf{F}}_p})$, we mean a regular differential all of whose corresponding ω_i have no poles at all (i.e., for all i, $\omega_i \in H^0(C_i, \Omega_{C_i/\overline{\mathbf{F}}_p})$). The subspace of holomorphic differentials may be identified with $\prod_{i=0}^r H^0(C_i, \Omega_{C_i/\overline{\mathbf{F}}_p})$ (which we will often do implicitly), and we let i_1 denote the corresponding injection $\prod_{i=0}^r H^0(C_i, \Omega_{C_i/\overline{\mathbf{F}}_p}) \hookrightarrow H^0(X_0(N)_{\overline{\mathbf{F}}_p}, \Omega_{X_0(N)/\overline{\mathbf{F}}_p})$.

In a manner similar to the description in the third paragraph of Section 5.2, Grothendieck duality gives an isomorphism

$$\Theta : H^0(X_0(N)_{\mathcal{O}_L}, \Omega_{X_0(N)/\mathcal{O}_L}) \xrightarrow{\cong} \mathrm{Cot}(\mathrm{Pic}^0_{X_0(N)/\mathcal{O}_L}), \tag{9}$$

where Cot denotes the cotangent space at the identity section. Since we have an action of U_p and W_p on $\mathrm{Pic}^0_{X_0(N)/\mathcal{O}_L}$ (by viewing it as the identity component of the Néron model of $J_0(N)$ over $\mathcal{O}_L$), we may use Θ to get an action of these operators on $H^0(X_0(N)_{\mathcal{O}_L}, \Omega_{X_0(N)/\mathcal{O}_L})$. As before, Prop. 3.3 on p. 68 of [Maz77] implies that base change to $\overline{\mathbf{F}}_p$ gives an isomorphism

$$\mathrm{H}^0(X_0(N)_{\overline{\mathbf{F}}_p}, \Omega_{X_0(N)/\overline{\mathbf{F}}_p}) \cong \mathrm{H}^0(X_0(N)_{\mathcal{O}_L}, \Omega_{X_0(N)/\mathcal{O}_L}) \otimes_{\mathcal{O}_L} \overline{\mathbf{F}}_p. \tag{10}$$

From this, we get an action of U_p and W_p on $H^0(X_0(N)_{\overline{\mathbf{F}}_p}, \Omega_{X_0(N)/\overline{\mathbf{F}}_p})$.

Corollary 5.16. *The endomorphisms U_p and W_p of $\mathrm{H}^0(X_0(N)_{\overline{\mathbf{F}}_p}, \Omega_{X_0(N)/\overline{\mathbf{F}}_p})$ preserve the subspace $\prod_{i=0}^r H^0(C_i, \Omega_{C_i/\overline{\mathbf{F}}_p})$, and satisfy $U_p = \pm\mathrm{Frob}_p^* + (p-1)W_p$ on $H^0(C_0, \Omega_{C_0/\overline{\mathbf{F}}_p})$, where Frob_p^* denotes pullback by Frob_p and where we have a possible sign ambiguity $\pm$ (which will not affect us later).*

Proof. The proof is based on the following diagram; we describe below some of the maps in it that have not been defined yet.

$$\begin{array}{ccc}
H^0(X_0(N)_{\mathcal{O}_L}, \Omega_{X_0(N)/\mathcal{O}_L}) & \xrightarrow{\ \Theta\ } & \mathrm{Cot}(\mathrm{Pic}^0_{X_0(N)/\mathcal{O}_L}) \\
\downarrow{\scriptstyle \pi_1} & & \downarrow{\scriptstyle \pi_2} \\
H^0(X_0(N)_{\overline{\mathbf{F}}_p}, \Omega_{X_0(N)/\overline{\mathbf{F}}_p}) & \xrightarrow{\ \theta\ } & \mathrm{Cot}(\mathrm{Pic}^0_{X_0(N)/\overline{\mathbf{F}}_p}) \\
\uparrow{\scriptstyle i_1} & & \uparrow{\scriptstyle i_2} \\
\prod_{i=0}^r H^0(C_i, \Omega_{C_i/\overline{\mathbf{F}}_p}) & \xrightarrow{\ \theta'\ } & \prod_{i=0}^r \mathrm{Cot}(\mathrm{Pic}^0_{C_i/\overline{\mathbf{F}}_p}).
\end{array}$$

Firstly, Cot always denotes the cotangent space at the identity section. The map π_1 is obtained by base change to $\overline{\mathbf{F}}_p$. By (10), π_1 is surjective. The map π_2 is obtained by observing that $\mathrm{Pic}^0_{X_0(N)/\overline{\mathbf{F}}_p}$ is the identity component of the special fiber of the Néron model of $J_0(N)$ over $\mathcal{O}_L$, and hence maps to the identity component of the Néron model of $J_0(N)$ over $\mathcal{O}_L$, which is $\mathrm{Pic}^0{}_{X_0(N)/\mathcal{O}_L}$. The map θ is obtained using Grothendieck duality. The compatibility of Grothendieck duality under base change (see [Con00]) implies that the top square in the diagram above commutes.

Now we have already defined actions of U_p and W_p on $\mathrm{Pic}^0_{X_0(N)/\mathcal{O}_L}$ and on $\mathrm{Pic}^0_{X_0(N)/\overline{\mathbf{F}}_p}$ (just before Lemma 5.15). Thus we get actions of U_p and W_p on $\mathrm{Cot}(\mathrm{Pic}^0_{X_0(N)/\mathcal{O}_L})$ and on $\mathrm{Cot}(\mathrm{Pic}^0_{X_0(N)/\overline{\mathbf{F}}_p})$. From the definitions of these actions we see that π_2 is compatible with the actions on its domain and codomain. Recall that we used the isomorphism Θ to induce actions of U_p and W_p on $H^0(X_0(N)_{\mathcal{O}_L}, \Omega_{X_0(N)/\mathcal{O}_L})$ and then used formula (10) to get actions on $H^0(X_0(N)_{\overline{\mathbf{F}}_p}, \Omega_{X_0(N)/\overline{\mathbf{F}}_p})$. Thus Θ and π_1 are also compatible with the actions of U_p and W_p on their domain and codomain. Let $\omega \in H^0(X_0(N)_{\overline{\mathbf{F}}_p}, \Omega_{X_0(N)/\overline{\mathbf{F}}_p})$, and let $\Omega \in H^0(X_0(N)_{\mathcal{O}_L}, \Omega_{X_0(N)/\mathcal{O}_L})$ be such that $\pi_1(\Omega) = \omega$. Then $\theta(U_p(\omega)) = \theta(\pi_1(U_p(\Omega))) = \pi_2(\Theta(U_p(\Omega))) = \pi_2(U_p(\Theta(\Omega))) = U_p(\pi_2(\Theta(\Omega))) = U_p(\theta(\pi_2(\Omega))) = U_p(\theta(\omega))$. Thus we see that the isomorphism θ is compatible with the action of U_p (and similarly for W_p) on its domain and codomain.

Now we turn to the bottom square in the diagram above. As mentioned earlier, the injection i_2 arises because $\prod_{i=0}^r \mathrm{Pic}^0_{C_i/\overline{\mathbf{F}}_p}$ is the maximal abelian variety quotient of the identity component $\mathrm{Pic}^0_{X_0(N)/\overline{\mathbf{F}}_p}$ of the special fiber of the Néron model of $J_0(N)$ over $\mathcal{O}_L$. The map θ' is the isomorphism coming from Serre duality.

Next, by [Con00, §5.2], the Grothendieck duality isomorphism θ is the same as the isomorphism coming from the duality theory of Rosenlicht (as in [Ser88, Chap. IV]), perhaps up to multiplication by -1. Assume for the moment that there is no sign ambiguity, so that θ is indeed the isomorphism coming from the duality theory of Rosenlicht. One can check that the Serre duality isomorphism θ' is induced by the Rosenlicht duality isomorphism θ via the inclusions i_1 and i_2 by looking at the proof of the two dualities in [Ser88, Chaps. II and IV]. Note that in [Ser88], the curve X over the field k (notation as in loc. cit.) is assumed to be irreducible. This hypothesis is needed in loc. cit. (for our purposes) only to show that $H^1(X, k(X)) = 0$ (p. 12, loc. cit.); the latter condition holds so long as X is reduced (see top of p. 165 in [AK70], as well as the bottom of p. 138 and top of p. 132 therein), which is true in our case (taking $X = X_0(N)_{\overline{\mathbf{F}}_p}$ and $k = \overline{\mathbf{F}}_p$) . We remark that our contention that the Serre duality isomorphism θ' is induced by the Rosenlicht duality isomorphism θ via the inclusions i_1 and i_2 also follows from Section 6 (an appendix provided to us by Brian Conrad), by taking $C = X_0(N)_{\overline{\mathbf{F}}_p}$ and C' to be any of the C_i in Section 6. In any case, we conclude that the bottom square in the diagram above commutes as well, perhaps up to multiplication by -1.

Now the action of U_p and W_p on $\prod_{i=0}^r \mathrm{Pic}^0_{C_i/\overline{\mathbf{F}}_p}$ was defined by identifying it as the maximal abelian variety quotient of $\mathrm{Pic}^0_{X_0(N)/\overline{\mathbf{F}}_p}$. Thus we see that i_2 is compatible with the action of U_p and W_p on its domain and codomain. Considering that moreover the isomorphism θ is compatible with the action of U_p (and W_p) and the bottom square in the diagram above commutes, perhaps up to multiplication by -1, we see that U_p and W_p preserve $\prod_{i=0}^{r} H^0(C_i, \Omega_{C_i/\overline{\mathbf{F}}_p})$. Now since θ is compatible with the action of U_p and W_p on its domain and codomain, so is θ'. Thus we may use the isomorphism θ' to translate the identity in Lemma 5.15 from the right to the left of θ' to get the desired identity in the corollary, where the $\pm$ ambiguity in front of $\mathrm{Frob}_p{}^*$ is really due to the sign ambiguity about the compatibility of the action of U_p and W_p on the two sides of the isomorphism θ'. □

Remark 5.17. We defined the action of the Hecke operators and the Atkin–Lehner involution in characteristic p from their definition in characteristic 0 in a somewhat indirect manner via the Néron mapping property, Grothendieck duality, etc (cf. beginning of Section 5.2). This has made our proofs rather complicated, since we have to show several compatibilities (as in the previous Corollary 5.16 and the upcoming Lemma 5.18). After this article was written, B. Conrad pointed out to us that one can define the action of the Hecke operators on suitable Artin stacks over $\mathbf{Z}$ for $\Gamma_0(N)$-structures (see [Con07]) in such a way that the definition agrees with the usual definition of the Hecke operators over $\mathbf{Q}$. This naturallly defines the action of the Hecke operators on objects related to $X_0(N)$ such as differentials, Picard groups, etc., in characteristic p and these definitions are automatically "compatible" with the corresponding definitions in characteristic zero. This alternative method would have been a less complicated way to proceed.

By [Maz77, Prop. II.3.3] we have an isomorphism

$$\mathrm{H}^0(X_0(N)_{\overline{\mathbf{F}}_p}, \Omega_{X_0(N)/\overline{\mathbf{F}}_p}) \cong \mathrm{H}^0(X_0(N)_{\mathbf{F}_p}, \Omega_{X_0(N)/\mathbf{F}_p}) \otimes_{\mathbf{F}_p} \overline{\mathbf{F}}_p,$$

using which we may identify $\mathrm{H}^0(X_0(N)_{\mathbf{F}_p}, \Omega_{X_0(N)/\mathbf{F}_p})$ as a subspace of $\mathrm{H}^0(X_0(N)_{\overline{\mathbf{F}}_p}, \Omega_{X_0(N)/\overline{\mathbf{F}}_p})$. Just before Corollary 5.16, we defined an action of U_p (and W_p) on $\mathrm{H}^0(X_0(N)_{\overline{\mathbf{F}}_p}, \Omega_{X_0(N)/\overline{\mathbf{F}}_p})$.

Lemma 5.18. *The action of U_p (respectively W_p) on $\mathrm{H}^0(X_0(N)_{\overline{\mathbf{F}}_p}, \Omega_{X_0(N)/\overline{\mathbf{F}}_p})$ preserves the subspace $\mathrm{H}^0(X_0(N)_{\mathbf{F}_p}, \Omega_{X_0(N)/\mathbf{F}_p})$, and agrees with the action of U_p (respectively W_p) on this subspace that we defined earlier in the third paragraph of Section 5.2.*

Proof. We have the following diagram, obtained by the obvious base changes:

$$\begin{array}{ccccc}
H^0(X_0(N)_{\overline{\mathbf{F}}_p}, \Omega_{X_0(N)/\overline{\mathbf{F}}_p}) & \leftarrow & H^0(X_0(N)_{\mathcal{O}_L}, \Omega_{X_0(N)/\mathcal{O}_L}) & \stackrel{\Theta}{\rightarrow} & \mathrm{Cot}(\mathrm{Pic}^0_{X_0(N)/\mathcal{O}_L}), \\
\uparrow & & \uparrow & & \uparrow \\
H^0(X_0(N)_{\mathbf{F}_p}, \Omega_{X_0(N)/\mathbf{F}_p}) & \leftarrow & H^0(X_0(N)_{\mathbf{Z}_p}, \Omega_{X_0(N)/\mathbf{Z}_p}) & \stackrel{\Theta'}{\rightarrow} & \mathrm{Cot}(\mathrm{Pic}^0_{X_0(N)/\mathbf{Z}_p}),
\end{array}$$

where the map Θ' is the isomorphism coming from Grothendieck duality as discussed in the third paragraph of Section 5.2. Now the action of U_p and W_p on $\mathrm{Cot}(\mathrm{Pic}^0_{X_0(N)/\mathcal{O}_L}) = \mathrm{Cot}(J_0(N)_{\mathcal{O}_L})$ (where $J_0(N)_{\mathcal{O}_L}$ is the Néron model of $J_0(N)$ over $\mathcal{O}_L$) was obtained by base changing from $\mathbf{Z}_p$. Considering that the formation of Néron models is compatible with completions and unramified base change, we see that the rightmost vertical map above is compatible under the action of U_p and W_p. Also, the action of U_p and W_p on $H^0(X_0(N)_{\mathcal{O}_L}, \Omega_{X_0(N)/\mathcal{O}_L})$ (respectively on $H^0(X_0(N)_{\overline{\mathbf{F}}_p}, \Omega_{X_0(N)/\overline{\mathbf{F}}_p})$) was obtained via the isomorphism Θ (respectively Θ'). Thus the rightmost two horizonal maps above are also compatible under the action of U_p and W_p on their domain and codomain. Finally, the compatibility of Grothendieck duality under base change (see [Con00]) implies that the right square in the diagram above commutes. Arguing as in the third paragraph of the proof of Corollary 5.16, one sees then that the middle vertical map above is compatible under the action of U_p and W_p.

Now the already-defined action of U_p and W_p on $\mathrm{H}^0(X_0(N)_{\mathbf{F}_p}, \Omega_{X_0(N)/\mathbf{F}_p})$ in the third paragraph of Section 5.2 is obtained via the lower leftward pointing arrow in the diagram above, and the action of U_p and W_p on $\mathrm{H}^0(X_0(N)_{\overline{\mathbf{F}}_p}, \Omega_{X_0(N)/\overline{\mathbf{F}}_p})$ is obtained via the upper leftward pointing arrow in the diagram above. Thus the leftmost two horizontal arrows are compatible under the action of U_p and W_p on their domain and codomain. Repeated applications of [Maz77, Prop. II.3.3] show that the left square also commutes. Using all this, we see that the action of U_p (respectively W_p) on $\mathrm{H}^0(X_0(N)_{\mathbf{F}_p}, \Omega_{X_0(N)/\mathbf{F}_p})$ viewed as a subspace of $\mathrm{H}^0(X_0(N)_{\overline{\mathbf{F}}_p}, \Omega_{X_0(N)/\overline{\mathbf{F}}_p})$ agrees with the action of U_p (respectively W_p) on $\mathrm{H}^0(X_0(N)_{\mathbf{F}_p}, \Omega_{X_0(N)/\mathbf{F}_p})$ defined in the third paragraph of Section 5.2, and in particular that U_p and W_p preserve this subspace. □

We now revert to the assumption that p is a prime such that $p^2 \nmid N$ (in particular p may not necessarily divide N). The Tate curve over $\mathbf{F}_p[[q]]$ gives rise to a morphism from Spec $\mathbf{F}_p[[q]]$ to the smooth locus of $X_0(N)_{\mathbf{F}_p} \to$ Spec $\mathbf{F}_p$. Since the module of completed Kähler differentials for $\mathbf{F}_p[[q]]$ over $\mathbf{F}_p$ is free of rank 1 on the basis dq, we obtain a map

$$q\text{-exp} : H^0(X_0(N)_{\mathbf{F}_p}, \Omega_{X_0(N)/\mathbf{F}_p}) \to \mathbf{F}_p[[q]].$$

If $p \nmid N$, then by a *holomorphic differential* in $H^0(X_0(N)_{\mathbf{F}_p}, \Omega_{X_0(N)/\mathbf{F}_p})$, we mean any differential in $H^0(X_0(N)_{\mathbf{F}_p}, \Omega_{X_0(N)/\mathbf{F}_p})$.

Lemma 5.19. *Recall that p is a prime such that $p^2 \nmid N$, and $\mathfrak{m}$ is a maximal ideal of $\mathbf{T}$ with residue characteristic p. If $p|N$, then assume that U_p acts as a non-zero scalar on $H^0(X_0(N)_{\mathbf{F}_p}, \Omega_{X_0(N)/\mathbf{F}_p})[\mathfrak{m}]$. Then the map q-exp restricted to homomorphic differentials in $\mathrm{H}^0(X_0(N)_{\mathbf{F}_p}, \Omega_{X_0(N)/\mathbf{F}_p})[\mathfrak{m}]$ is injective.*

Proof. The essential argument is quite standard, going back to Mazur, so we only sketch the ideas. For some of the details, we refer the reader to the proof of Lemma 4.2 in [ARS06]. If $p \nmid N$, the injectivity follows from the q-expansion principle. So suppose that $p \parallel N$, and let $M = N/p$. Recall that $X_0(N)_{\overline{\mathbf{F}}_p}$ is obtained from $M_0(N)_{\overline{\mathbf{F}}_p}$ by suitable blowups at supersingular points and consists of two copies of $X_0(M)_{\overline{\mathbf{F}}_p}$ identified at supersingular points, along with some copies of $\mathbf{P}^1$. Suppose $\omega \in H^0(X_0(N)_{\mathbf{F}_p}, \Omega_{X_0(N)/\mathbf{F}_p})[\mathfrak{m}]$ is a holomorphic differential that is in the kernel of q-exp. Then the q-expansion principle implies that ω vanishes on the copy of $X_0(M)_{\overline{\mathbf{F}}_p}$ containing the cusp ∞, i.e., on C_1. By Corollary 5.16 and Lemma 5.18, we have $U_p(\omega|_{C_0}) = \pm \mathrm{Frob}_p{}^*(\omega|_{C_0}) + (p-1)W_p(\omega|_{C_0})$. But pullback by Frob_p is the trivial map and W_p swaps C_0 and C_1, so $U_p(\omega|_{C_0}) = (p-1)(\omega|_{C_1}) = 0$. Now by hypothesis, U_p acts as multiplication by a non-zero scalar, hence ω is trivial on C_0. Thus ω is trivial on both copies of $X_0(M)_{\overline{\mathbf{F}}_p}$. One can show that then ω is trivial on the copies of $\mathbf{P}^1$ as well (see the proof of Lemma 4.2 in [ARS06]). Thus ω is trivial on $X_0(N)_{\overline{\mathbf{F}}_p}$, hence on $X_0(N)_{\mathbf{F}_p}$. □

Lemma 5.20. *We continue our hypotheses that p is a prime such that $p^2 \nmid N$, $\mathfrak{m}$ is a maximal ideal of $\mathbf{T}$ with residue characteristic p, and if $p|N$, then U_p acts as a non-zero scalar on $H^0(X_0(N)_{\mathbf{F}_p}, \Omega_{X_0(N)/\mathbf{F}_p})[\mathfrak{m}]$. Then*

$$\dim_{\mathbf{T}/\mathfrak{m}} \mathrm{H}^0(X_0(N)_{\mathbf{F}_p}, \Omega_{X_0(N)/\mathbf{F}_p})[\mathfrak{m}] \leq 1.$$

Proof. The idea behind the proof is the same as in the proof of Lemma 2.2 in [Wil80, p. 485-487], which in turn builds on ideas from p. 94–95 of [Maz77]. However, parts of our arguments are somewhat different, and may be considered alternatives to some of the methods in the works cited in the previous sentence.

If $\omega \in \mathrm{H}^0(X_0(N)_{\mathbf{F}_p}, \Omega_{X_0(N)/\mathbf{F}_p})$ and $n \geq 1$, then let $a_n(\omega)$ denote the coefficient of q^n in q-exp(ω). We have a pairing $\mathrm{H}^0(X_0(N)_{\mathbf{F}_p}, \Omega_{X_0(N)/\mathbf{F}_p}) \times \mathbf{T} \to \mathbf{F}_p$ that takes (ω, T) to $a_1(T\omega)$. This induces a map

$$\psi : \mathrm{H}^0(X_0(N)_{\mathbf{F}_p}, \Omega_{X_0(N)/\mathbf{F}_p})[\mathfrak{m}] \to \mathrm{Hom}_{\mathbf{F}_p}(\mathbf{T}/\mathfrak{m}, \mathbf{F}_p),$$

which is a homomorphism of $\mathbf{T}/\mathfrak{m}$-vector spaces.

Claim 1: If $\omega \in \ker(\psi)$, then q-exp(ω) is trivial.

Proof. Following the proof of Prop. 3.3 on p. 68 of [Maz77], we have

$$\mathrm{H}^0(X_0(N)_{\mathbf{F}_p}, \Omega_{X_0(N)/\mathbf{F}_p}) \cong \mathrm{H}^0(X_0(N)_{\mathbf{Z}_p}, \Omega_{X_0(N)/\mathbf{Z}_p}) \otimes_{\mathbf{Z}_p} \mathbf{F}_p, \quad (11)$$

and

$$\mathrm{H}^0(X_0(N)(\mathbf{C}), \Omega_{X_0(N)(\mathbf{C})/\mathbf{C}}) \cong \mathrm{H}^0(X_0(N)_{\mathbf{Z}_p}, \Omega_{X_0(N)/\mathbf{Z}_p}) \otimes_{\mathbf{Z}_p} \mathbf{C}. \quad (12)$$

The definition of the action of the Hecke operators on $\mathrm{H}^0(X_0(N)_{\mathbf{Z}_p}, \Omega_{X_0(N)/\mathbf{Z}_p})$ defined in the third paragraph of Section 5.2 shows that this action is compatible with the action of the Hecke operators on $\mathrm{H}^0(X_0(N)(\mathbf{C}), \Omega_{X_0(N)(\mathbf{C})/\mathbf{C}})$ under (12). Also, the action of the Hecke operators on $\mathrm{H}^0(X_0(N)_{\mathbf{F}_p}, \Omega_{X_0(N)/\mathbf{F}_p})$ was defined in the third paragraph of Section 5.2 via their action on $\mathrm{H}^0(X_0(N)_{\mathbf{Z}_p}, \Omega_{X_0(N)/\mathbf{Z}_p})$ using (11), so these actions are clearly compatible under (11). Now

$$\mathrm{H}^0(X_0(N)(\mathbf{C}), \Omega_{X_0(N)(\mathbf{C})/\mathbf{C}}) \cong \mathrm{H}^0(J_0(N)(\mathbf{C}), \Omega_{J_0(N)(\mathbf{C})/\mathbf{C}}) \cong S_2(\Gamma_0(N), \mathbf{C}),$$

and thus $a_1(T_n\omega) = a_n(\omega)$ for $\omega \in \mathrm{H}^0(X_0(N)(\mathbf{C}), \Omega_{X_0(N)(\mathbf{C})/\mathbf{C}})$. Hence, by (11), (12), and the discussion above, we also have the formula $a_1(T_n\omega) = a_n(\omega)$ for $\omega \in \mathrm{H}^0(X_0(N)_{\mathbf{F}_p}, \Omega_{X_0(N)/\mathbf{F}_p})$.

Thus if $\omega \in \ker(\psi)$, then $a_n(\omega) = a_1(T_n\omega) = 0$ for all $n \geq 1$, i.e., q-exp(ω) is trivial, as was to be shown. □

Claim 2: The $\mathbf{T}/\mathfrak{m}$-dimension of the subspace of holomorphic differentials in $\mathrm{H}^0(X_0(N)_{\mathbf{F}_p}, \Omega_{X_0(N)/\mathbf{F}_p})[\mathfrak{m}]$ is at most 1.

Proof. If ω is a holomorphic differential in $\mathrm{H}^0(X_0(N)_{\mathbf{F}_p}, \Omega_{X_0(N)/\mathbf{F}_p})[\mathfrak{m}]$ and $\psi(\omega) = 0$, then by Claim 1, q-exp(ω) is trivial, and hence by Lemma 5.19, ω is trivial. This proves that ψ is injective when restricted to the subspace of holomorphic differentials. Now the group $\mathrm{Hom}_{\mathbf{F}_p}(\mathbf{T}/\mathfrak{m}, \mathbf{F}_p)$ has the same size as $\mathbf{T}/\mathfrak{m}$, which completes the argument because ψ embeds the subspace of holomorphic differentials in $\mathrm{H}^0(X_0(N)_{\mathbf{F}_p}, \Omega_{X_0(N)/\mathbf{F}_p})[\mathfrak{m}]$ into $\mathrm{Hom}_{\mathbf{F}_p}(\mathbf{T}/\mathfrak{m}, \mathbf{F}_p)$, which has dimension 1 as a $\mathbf{T}/\mathfrak{m}$-vector space. □

Claim 2 proves the lemma in the case when $p \nmid N$. We now prove that $\dim_{\mathbf{T}/\mathfrak{m}} \mathrm{H}^0(X_0(N)_{\mathbf{F}_p}, \Omega_{X_0(N)/\mathbf{F}_p})[\mathfrak{m}] \leq 1$ when $p || N$, which will finish the proof of the lemma. Following the proof of Lemma 2.2 in [Wil95], we break the argument into two cases:

Case I: There is no non-zero holomorphic differential in

$$\mathrm{H}^0(X_0(N)_{\mathbf{F}_p}, \Omega_{X_0(N)/\mathbf{F}_p})[\mathfrak{m}].$$

Suppose ω_1 and ω_2 are two differentials in $\mathrm{H}^0(X_0(N)_{\mathbf{F}_p}, \Omega_{X_0(N)/\mathbf{F}_p})[\mathfrak{m}]$. Then we can find a pair $(\mu, \lambda) \in (\mathbf{T}/\mathfrak{m})^2$ with $(\mu, \lambda) \neq (0, 0)$ such that $\mu\psi(\omega_1) - \lambda\psi(\omega_2) = 0$, i.e., $\psi(\mu\omega_1 - \lambda\omega_2) = 0$. Hence by Claim 1, q-exp$(\mu\omega_1 - \lambda\omega_2) = 0$. Viewing $\mu\omega_1 - \lambda\omega_2$ as an element of $\mathrm{H}^0(X_0(N)_{\overline{\mathbf{F}}_p}, \Omega_{X_0(N)/\overline{\mathbf{F}}_p})$, we see that $\mu\omega_1 - \lambda\omega_2$ vanishes on C_1 (recall that C_1 is the copy of $X_0(N/p)_{\overline{\mathbf{F}}_p}$ that contains the cusp ∞) by the "q-expansion principle" (see the proof of Lemma 4.2 in [ARS06] for details). Now $C_2, \ldots, C_r$ (the copies of $\mathbf{P}^1$) arise as chains that link C_1 and C_0 (recall that C_0 is the copy of $X_0(N/p)_{\overline{\mathbf{F}}_p}$ that does not contain the cusp ∞) and each of $C_2, \ldots, C_r$ has at most two points of intersection, with all intersection points being ordinary double points (see the description of $X_0(N)_{\overline{\mathbf{F}}_p}$ on p. 175–177 of [Maz77] for details). Taking into consideration the definition of regular differentials and the residue theorem we see that $\mu\omega_1 - \lambda\omega_2$ is holomorphic on the curves among $C_2, \ldots, C_r$ that intersect C_1 (for details, see the proof of Lemma 4.2 in [ARS06] in a similar situation). Now a curve among $C_2, \ldots, C_r$ that does *not* intersect C_1 intersects exactly one curve among $C_2, \ldots, C_r$ that *does* intersect C_1. Hence by repeating the argument above, $\mu\omega_1 - \lambda\omega_2$ is holomorphic on each curve in $C_2, \ldots, C_r$ that does *not* intersect C_1 as well. Thus $\mu\omega_1 - \lambda\omega_2$ is holomorphic on all of $X_0(N)_{\overline{\mathbf{F}}_p}$ except perhaps on C_0. But the only possible poles of $\mu\omega_1 - \lambda\omega_2$ on C_0 are over points of intersection with other components, and again, considering the definition of regular differentials, we see that there are no such poles, i.e., $\mu\omega_1 - \lambda\omega_2$ is holomorphic on C_0 as well. Thus $\mu\omega_1 - \lambda\omega_2$ is holomorphic everywhere and is an element of $\mathrm{H}^0(X_0(N)_{\mathbf{F}_p}, \Omega_{X_0(N)/\mathbf{F}_p})[\mathfrak{m}]$. Hence it is trivial by the hypothesis of this case. Thus ω_1 and ω_2 are linearly dependent. Since ω_1 and ω_2 were arbitrary, this shows that $\dim_{\mathbf{T}/\mathfrak{m}} \mathrm{H}^0(X_0(N)_{\mathbf{F}_p}, \Omega_{X_0(N)/\mathbf{F}_p})[\mathfrak{m}] \leq 1$ in this case.

Case II: There is a non-zero holomorphic differential

$$\omega \in \mathrm{H}^0(X_0(N)_{\mathbf{F}_p}, \Omega_{X_0(N)/\mathbf{F}_p})[\mathfrak{m}].$$

By Lemma 5.19, q-exp(ω) is non-trivial, and so by Claim 1, $\psi(\omega) \neq 0$. Let $\omega' \in \mathrm{H}^0(X_0(N)_{\mathbf{F}_p}, \Omega_{X_0(N)/\mathbf{F}_p})[\mathfrak{m}]$. Then there is a $\lambda \in \mathbf{T}/\mathfrak{m}$ such that $\psi(\omega') - \lambda\psi(\omega) = 0$, i.e., $\psi(\omega' - \lambda\omega) = 0$. As in the proof of Case I, we conclude that $\omega' - \lambda\omega$ is holomorphic; in particular ω' is holomorphic. Thus every differential in $\mathrm{H}^0(X_0(N)_{\mathbf{F}_p}, \Omega_{X_0(N)/\mathbf{F}_p})[\mathfrak{m}]$ is holomorphic. Then by Claim 2, $\dim_{\mathbf{T}/\mathfrak{m}} \mathrm{H}^0(X_0(N)_{\mathbf{F}_p}, \Omega_{X_0(N)/\mathbf{F}_p})[\mathfrak{m}] \leq 1$ in this case as well. □

(Proof of Proposition 5.10). Recall that the hypotheses of Proposition 5.10 are that p is a prime such that $p^2 \nmid N$, $\mathfrak{m}$ is a maximal ideal of $\mathbf{T}$ with residue characteristic p such that if $p|N$, then $I_f \subseteq \mathfrak{m}$ for some newform f. We wish to show that then $\mathbf{T}$ and $\mathbf{T}'$ agree locally at $\mathfrak{m}$.

If $p \nmid N$, then the result follows from Lemmas 5.11 and 5.20. If f is a newform and $p|N$, then U_p acts as ± 1 on f, and hence $U_p \pm 1 \in I_f$. Thus if $p|N$ and $I_f \subseteq \mathfrak{m}$ for some newform f, then U_p acts as a non-zero scalar (± 1) on $H^0(X_0(N)_{\mathbf{F}_p}, \Omega_{X_0(N)/\mathbf{F}_p})[\mathfrak{m}]$ (note that the action of U_p on regular differentials was defined compatibly with the usual action of U_p on complex differentials, i.e., on cuspforms; cf. the proof of Claim 1 in the proof of Lemma 5.20). The proposition follows again from Lemmas 5.11 and 5.20. □

6 Duality theory: an appendix by Brian Conrad

Let k be a field and let C be a proper reduced k-scheme with pure dimension 1. Assume that C is generically smooth, and let $C' \subseteq C$ be a non-empty reduced closed subscheme with pure dimension 1 (so C' is also generically smooth). The case of most interest to us is when C is a geometrically connected and semistable curve and C' is a smooth geometrically irreducible component. The inclusion $C' \to C$ induces a natural map of k-groups $\mathrm{Pic}_{C/k} \to \mathrm{Pic}_{C'/k}$, and on tangent spaces at the identity this is the canonical pullback map

$$\theta : \mathrm{H}^1(C, \mathcal{O}_C) \to \mathrm{H}^1(C', \mathcal{O}_{C'})$$

(as we see by computing with dual numbers over k). Each of C and C' satisfies Serre's condition (S_1) by reducedness, so each is Cohen–Macaulay. Thus, by Serre duality we can identify the map of cotangent spaces with the map $\mathrm{H}^0(C', \omega_{C'/k}) \to \mathrm{H}^0(C, \omega_{C/k})$ dual to θ. We wish to give a concrete description of this latter map. To do this, we first review some basic definitions and identifications in duality theory.

In what follows we use Grothendieck's approach to duality theory, which has the merit of permitting more localization operations than in Serre's approach. Since C and C' are Cohen–Macaulay with pure dimension 1, their relative dualizing complexes over k are naturally identified with $\omega_{C/k}[1]$ and $\omega_{C'/k}[1]$ respectively [Con00, 3.5.1]. Since (by construction) the formation of the relative dualizing complex is compatible with Zariski-localization on the source, we have canonical isomorphisms $\omega_{C'/k}|_{C'^{\mathrm{sm}}} \simeq \Omega^1_{C'^{\mathrm{sm}}/k}$ and $\omega_{C/k}|_{C^{\mathrm{sm}}} \simeq \Omega^1_{C^{\mathrm{sm}}/k}$ that coincide on the open locus $U = C^{\mathrm{sm}} \cap C'$ that is dense in C' (and supported in C'^{sm}). If we let $j : C^{\mathrm{sm}} \to C$ and $j' : C'^{\mathrm{sm}} \to C'$ denote the canonical dense open immersions then, by [Con00, 5.2.1] the natural maps

$$\omega_{C'/k} \to j'_*(\Omega^1_{C'^{\mathrm{sm}}/k}), \;\; \omega_{C/k} \to j_*(\Omega^1_{C^{\mathrm{sm}}/k})$$

are injective. By construction this is compatible with the natural isomorphism $\omega_{C/k}|_U \simeq \omega_{C'/k}|_U$. Letting $\eta : \mathrm{Spec}(K) \to C$ and $\eta' : \mathrm{Spec}(K') \to C'$ denote the canonical maps from the schemes of generic points, $\omega_{C'/k}$ maps isomorphically onto a coherent subsheaf of $\eta'_*(\Omega^1_{K'/k})$ and likewise for $\omega_{C/k}$ in $\eta_*(\Omega^1_{K/k})$; these image subsheaves are the so-called sheaves of *regular differentials*, and a classical result of Rosenlicht describes these images explicitly using residues when k is algebraically closed [Con00, 5.2.3]. We will not require Rosenlicht's result for the statement or proof of the theorem below.

Using Grothendieck's theory of relative trace maps, the canonical closed immersion $\iota : C' \to C$ over k induces a trace morphism $\mathrm{Tr}_\iota : \iota_*(\omega_{C'/k}) \to \omega_{C/k}$ whose formation commutes with Zariski-localization on C, so over the dense open $U = \iota^{-1}(C^{\mathrm{sm}}) \subseteq C'$ it induces the natural isomorphism $\omega_{C'/k}|_U \simeq \omega_{C/k}|_U$, or equivalently it is the identity map on $\Omega^1_{U/k}$. Hence, Tr_ι is compatible with the canonical inclusions $\omega_{C'/k} \hookrightarrow \eta'_*(\Omega^1_{K'/k})$ and $\omega_{C/k} \hookrightarrow \eta_*(\Omega^1_{K/k})$. In particular, the map Tr_ι is compatible with the natural identification of meromorphic 1-forms on C' with meromorphic 1-forms on C (i.e., compatible with the injection $\Omega^1_{K'/k} \hookrightarrow \Omega^1_{K/k}$).

Having summarized some inputs from duality theory, we can now state the result we want to prove.

Theorem 6.1. *The pullback* $\mathrm{H}^1(C, \mathcal{O}_C) \to \mathrm{H}^1(C', \mathcal{O}_{C'})$ *is dual to the natural map*

$$\mathrm{H}^0(C', \omega_{C'/k}) = \mathrm{H}^0(C, \iota_*(\omega_{C'/k})) \to \mathrm{H}^0(C, \omega_{C/k}).$$

Proof. Let $\mathrm{Tr}_C : \mathrm{H}^1(C, \omega_{C/k}) \to k$ and $\mathrm{Tr}_{C'} : \mathrm{H}^1(C', \omega_{C'/k}) \to k$ be the canonical trace maps, so our problem is to prove that for $s \in \mathrm{H}^1(C, \mathcal{O}_C)$ and $\xi' \in \mathrm{H}^0(C', \omega_{C'/k}) \subseteq \Omega^1_{K'/k}$,

$$\mathrm{Tr}_{C'}(\xi' \cup s|_{C'}) = \mathrm{Tr}_C(\mathrm{Tr}_\iota(\xi') \cup s)$$

in k. By the functoriality of Grothendieck's trace map, $\mathrm{Tr}_{C'} = \mathrm{Tr}_C \circ \mathrm{H}^1(\mathrm{Tr}_\iota)$ as maps $\mathrm{H}^1(C', \omega_{C'/k}) \to k$. Thus, it suffices to show that the map $\mathrm{H}^1(C', \omega_{C'/k}) \to \mathrm{H}^1(C, \omega_{C/k})$ induced by Tr_ι carries $\xi' \cup s|_{C'}$ to $\mathrm{Tr}_\iota(\xi') \cup s$. We may view dualizing sheaves as subsheaves $\omega_{C/k} \subseteq \eta_*(\Omega^1_{K/k})$ and $\omega_{C'/k} \subseteq \eta'_*(\Omega^1_{K'/k})$ in terms of which we have seen that the abstract trace map Tr_ι is induced by the natural inclusion $\Omega^1_{K'/k} \subseteq \Omega^1_{K/k}$.

To do the computation we work with Čech theory. Let $\{U_n\}$ be an ordered finite open affine cover of C and let $U'_n = U_n \cap C'$, so $\{U'_n\}$ is an open affine cover of C'. The cohomology class s corresponds to a Čech 1-cocycle $\{s_{n,m}\}_{n<m}$ with $s_{n,m} \in \mathcal{O}_C(U_n \cap U_m)$, so s' corresponds to $\{s'_{n,m}\}$ with $s'_{n,m} = s_{n,m}|_{U'_n \cap U'_m}$. Identifying ξ' with an element of $\Omega^1_{K'/k}$, $\xi' \cup s|_{C'} \in \mathrm{H}^1(C', \omega_{C'/k})$ corresponds to $\{s'_{n,m}\xi'\}_{n<m}$ and $\mathrm{Tr}_\iota(\xi') \cup s \in \mathrm{H}^1(C, \omega_{C/k})$ corresponds to $\{s_{n,m}\xi'\}_{n<m}$, where ξ' is viewed in $\Omega^1_{K/k}$ in the natural way. The product $s_{n,m}\xi'$ at the generic points of $U_n \cap U_m$ vanishes at generic points not in C', so the required equality is clear even at the level of Čech 1-cocycles. □

References

[AK70] A. Altman and Steven Kleiman, *Introduction to Grothendieck duality theory*, Lecture Notes in Mathematics, Vol. 146, Springer-Verlag, Berlin, 1970.

[ARS06] A. Agashe, K. Ribet and W. Stein, *The Manin Constant*, Pure and Applied Mathematics Quarterly, Special issue: In honor of John H. Coates (2006), to appear.

[AS05] A. Agashe and W. A. Stein, *Visible evidence for the Birch and Swinnerton-Dyer conjecture for modular abelian varieties of analytic rank zero*, Math. Comp. **74** (2005), no. 249, 455–484, with an appendix by J. Cremona and B. Mazur.

[AU96] A. Abbes and E. Ullmo, *À propos de la conjecture de Manin pour les courbes elliptiques modulaires*, Compositio Math. **103** (1996), no. 3, 269–286.

[BCP97] W. Bosma, J. Cannon, and C. Playoust, *The Magma algebra system. I. The user language*, J. Symbolic Comput. **24** (1997), no. 3–4, 235–265, Computational algebra and number theory (London, 1993).

[BCDT01] C. Breuil, B. Conrad, F. Diamond, and R. Taylor, *On the modularity of elliptic curves over* **Q**: *wild 3-adic exercises*, J. Amer. Math. Soc. **14** (2001), no. 4, 843–939 (electronic).

[BLR90] S. Bosch, W. Lütkebohmert, and M. Raynaud, *Néron models*, Springer-Verlag, Berlin, 1990.

[Con00] B. Conrad, *Grothendieck duality and base change*, Lecture Notes in Mathematics, Vol. 1750, Springer-Verlag, Berlin, 2000.

[Con07] B. Conrad, *Arithmetic moduli of generalized elliptic curves*, J. Inst. Math. Jussieu **6** (2007), no. 2, 209–278.

[CK04] A. C. Cojocaru and E. Kani, *The modular degree and the congruence number of a weight 2 cusp form*, Acta Arith. **114** (2004), no. 2, 159–167.

[Cre97] J. E. Cremona, *Algorithms for modular elliptic curves*, second ed., Cambridge University Press, Cambridge, 1997, available at `http://www.maths.nott.ac.uk/personal/jec/book/`.

[DI95] F. Diamond and J. Im, *Modular forms and modular curves*, Seminar on Fermat's Last Theorem, Providence, RI, 1995, pp. 39–133.

[Dia97] F. Diamond, *The Taylor-Wiles construction and multiplicity one*, Invent. Math. **128** (1997), no. 2, 379–391.

[DR73] P. Deligne and M. Rapoport, *Les schémas de modules de courbes elliptiques*, Modular functions of one variable, II (Proc. Internat. Summer School, Univ. Antwerp, Antwerp, 1972) (Berlin), Springer, 1973, Lecture Notes in Math., Vol. 349 pp. 143–316.

[Fre97] G. Frey, *On ternary equations of Fermat type and relations with elliptic curves*, Modular forms and Fermat's last theorem (Boston, MA, 1995), Springer, New York, 1997, pp. 527–548.

[FM99] G. Frey and M. Müller, *Arithmetic of modular curves and applications*, Algorithmic algebra and number theory (Heidelberg, 1997), Springer, Berlin, 1999, pp. 11–48.

[Kil02] L. J. P. Kilford, *Some non-Gorenstein Hecke algebras attached to spaces of modular forms*, J. Number Theory **97** (2002), no. 1, 157–164.

[KM85] N. M. Katz and B. Mazur, *Arithmetic Moduli of Elliptic Curves*, Princeton University Press, Princeton, N.J., 1985.

[KW08] L. J. P. Kilford and Gabor Wiese, *On the failure of the Gorenstein property for Hecke algebras of prime weight*, Experiment. Math. **17** (2008), no. 1, 37–52. MR MR2410114 (2009c:11075).

[Li75] W-C. Li, *Newforms and functional equations*, Math. Ann. **212** (1975), 285–315.

[Maz77] B. Mazur, *Modular curves and the Eisenstein ideal*, Inst. Hautes Études Sci. Publ. Math. (1977), no. 47, 33–186 (1978).

[Maz78] B. Mazur, *Rational isogenies of prime degree (with an appendix by D. Goldfeld)*, Invent. Math. **44** (1978), no. 2, 129–162.

[MR91] B. Mazur and K. A. Ribet, *Two-dimensional representations in the arithmetic of modular curves*, Astérisque (1991), no. 196–197, 6, 215–255 (1992), Courbes modulaires et courbes de Shimura (Orsay, 1987/1988).

[Mur99] M. R. Murty, *Bounds for congruence primes*, Automorphic forms, automorphic representations, and arithmetic (Fort Worth, TX, 1996), Amer. Math. Soc., Providence, RI, 1999, pp. 177–192.

[Rib75] K. A. Ribet, *Endomorphisms of semi-stable abelian varieties over number fields*, Ann. Math. (2) **101** (1975), 555–562.

[Rib81] K. A. Ribet, *Endomorphism algebras of abelian varieties attached to newforms of weight* 2, Seminar on Number Theory, Paris 1979–80, Progr. Math., Vol. 12, Birkhäuser Boston, Mass., 1981, pp. 263–276.

[Rib83] K. A. Ribet, *Mod p Hecke operators and congruences between modular forms*, Invent. Math. **71** (1983), no. 1, 193–205.

[Rib90] K. A. Ribet, *On modular representations of* $\mathrm{Gal}(\overline{\mathbf{Q}}/\mathbf{Q})$ *arising from modular forms*, Invent. Math. **100** (1990), no. 2, 431–476.

[RS01] K. A. Ribet and W. A. Stein, *Lectures on Serre's conjectures*, Arithmetic algebraic geometry (Park City, UT, 1999), IAS/Park City Math. Ser., Vol. 9, Amer. Math. Soc., Providence, RI, 2001, pp. 143–232.

[Ser88] J-P. Serre, *Algebraic groups and class fields*, Graduate Texts in Mathematics, Vol. 117, Springer-Verlag, New York, 1988, Translated from the French.

[Shi94] G. Shimura, *Introduction to the Arithmetic Theory of Automorphic Functions*, Princeton University Press, Princeton, NJ, 1994, Reprint of the 1971 original, Kan Memorial Lectures, 1.

[S$^+$09] W. A. Stein et al., *Sage Mathematics Software (Version 4.2)*, The Sage Development Team, 2009, `http://www.sagemath.org`.

[Stu87] J. Sturm, *On the congruence of modular forms*, Number theory (New York, 1984–1985), Springer, Berlin, 1987, pp. 275–280.

[Til97] J. Tilouine, *Hecke algebras and the Gorenstein property*, Modular forms and Fermat's last theorem (Boston, MA, 1995), Springer, New York, 1997, pp. 327–342.

[Wie07] G. Wiese, *Multiplicities of Galois representations of weight one*, Algebra Number Theory **1** (2007), no. 1, 67–85, With an appendix by Niko Naumann.

[Wil80] A. Wiles, *Modular curves and the class group of* $\mathbf{Q}(\zeta_p)$, Invent. Math. **58** (1980), no. 1, 1–35.

[Wil95] A. J. Wiles, *Modular elliptic curves and Fermat's last theorem*, Ann. of Math. (2) **141** (1995), no. 3, 443–551.

[Zag85] D. Zagier, *Modular parametrizations of elliptic curves*, Canad. Math. Bull. **28** (1985), no. 3, 372–384.

Le théorème de Siegel–Shidlovsky revisité

Daniel Bertrand

À la mémoire de Serge Lang

Abstract We give a new proof of the Siegel–Shidlovsky theorem, which is based on a new version of Shidlovsky's lemma and on M. Laurent's interpolation determinants. We also establish a dual version of the lemma, and yet another proof of the theorem when the monodromy around 0 is trivial (as in the Lindemann–Weierstrass case).

Key words Algebraic independence

Mathematics Subject Classification (2010): 11J81

1 Introduction

Dans cet article, nous dirons à la suite de Serge Lang [L2] qu'une fonction entière holonome f est une E-fonction si les coefficients de son développement de Taylor à l'origine $\Sigma_{n\geq 0} a_n \frac{z^n}{n!}$ engendrent un corps de nombres K_f, et si la hauteur du point $(a_0, \ldots, a_n)$ de $\mathbf{P}_n(K_f)$ croît au plus géométriquement avec n. Nous n'aborderons pas la question, toujours ouverte, de comparer cette notion à la définition initiale de Siegel (voir à ce propos [R] et [A2.I]). Étant donné un corps de nombres K, nous dirons avec Shidlovsky [Sh] que f est une KE-fonction si, de plus, $K_f \subset K$. Ces définitions s'étendent sans difficulté aux solutions entières de systèmes différentiels

D. Bertrand (✉)
Institut de Mathématiques de Jussieu, Paris
e-mail: bertrand@math.jussieu.fr

D. Goldfeld et al. (eds.), *Number Theory, Analysis and Geometry: In Memory of Serge Lang*, DOI 10.1007/978-1-4614-1260-1_3,

linéaires, et permettent, à l'aide des résultats de [A2.I], de parler de KE-vecteurs horizontaux d'un fibré à connexion sur $\mathbf{P}_1/K$.

Soient $n > 0$ un entier, K un corps de nombres, de degré κ sur $\mathbf{Q}$, $|.|_\infty$ la valeur absolue attachée à un plongement complexe de K (sous-entendu dans les extensions des scalaires de K à $\mathbf{C}$ qui suivent), $\mathcal{M}$ un fibré de rang n sur $\mathbf{P}_1/K$, muni d'une connexion à singularités ∇, et S la réunion des points $0, \infty$ et de l'ensemble des singularités de ∇. On suppose que le point $\gamma = 1$ de $\mathbf{P}_1(K)$ n'appartient pas à S, et on note M_1 la fibre de $\mathcal{M}$ en 1, et $\hat{\mathcal{M}}_1$ le complété formel en 1 du faisceau associé à $\mathcal{M}$. Soit par ailleurs $\mathcal{E} \in \hat{\mathcal{M}}_1 \otimes_K \mathbf{C}$ une section horizontale formelle de $\nabla_\mathbf{C}$ au voisinage de 1. On note $n(\mathcal{E})$ le rang du plus petit sous-fibré de $\mathcal{M}_\mathbf{C}$ dont $\mathcal{E}$ soit une section locale, et $r_1(\mathcal{E})$ la dimension du plus petit K-sous-espace vectoriel W_1 de M_1 dont $\mathcal{E}(1)$ soit un point complexe. En d'autres termes, les coordonnées $\mathcal{E}_1, \ldots, \mathcal{E}_n$ de $\mathcal{E}$ relativement à un repère de $\mathcal{M}$ au-dessus d'un voisinage affine du point 1 forment une solution d'un système différentiel

$$\frac{d}{dz}\begin{pmatrix}\mathcal{E}_1\\ \vdots \\ \mathcal{E}_n\end{pmatrix} = A(z)\begin{pmatrix}\mathcal{E}_1\\ \vdots \\ \mathcal{E}_n\end{pmatrix}, A(z) \in gl_n\big(K(z) \cap K[[z-1]]\big); \qquad (*)$$

elles engendrent dans $\mathbf{C}((z-1))$ un $\mathbf{C}(z)$-espace vectoriel de dimension $n(\mathcal{E})$, et leurs valeurs en 1 engendrent dans $\mathbf{C}$ un K-espace vectoriel de dimension $r := r_1(\mathcal{E})$. Dans ces conditions, on a:

Théorème 1 (Théorème de Siegel–Shidlovsky). *on suppose que $\mathcal{E}$ s'étend par prolongement analytique en 0 en un KE-vecteur. Alors, $r_1(\mathcal{E}) \geq \frac{n(\mathcal{E})}{\kappa}$.*

Par passage aux puissances symétriques, on en déduit de la façon habituelle l'égalité des degrés de transcendance de $\mathbf{C}(z)(\mathcal{E}_1, \ldots, \mathcal{E}_n)$ sur $\mathbf{C}(z)$, et de $\mathbf{Q}(\mathcal{E}_1(1), \ldots, \mathcal{E}_n(1))$ sur $\mathbf{Q}$.

Le théorème de Siegel–Shidlovsky généralise le classique théorème de Lindemann–Weiestrass sur les valeurs de la fonction exponentielle en des points algébriques. Il a connu un regain d'intérêt ces dernières années, à la suite de la preuve adélique que Bézivin et Robba [BR] ont donnée de son avatar originel, puis de son extension au cas général par Y. André [A2.II] et, tout récemment, du raffinement suivant, qu'avait conjecturé Serge Lang ([L2], p.100), et que F. Beukers a déduit de la preuve de [A2] : on peut, dans la conclusion du théorème 1, supprimer le facteur $1/\kappa$, de sorte que $r_1(\mathcal{E})$ est en fait *égal* à sa valeur maximale $n(\mathcal{E})$, même si $K \neq \mathbf{Q}$.

Dans cet article, nous donnons une nouvelle preuve du théorème 1 lui-même, au moyen des *déterminants d'interpolation* de M. Laurent. À partir de paramètres L, T_0, T_1 dont la signification est précisée au §3 ci-dessous, il s'agit donc.

(i) de construire un "morphisme d'évaluation" ϕ de l'espace $\Gamma(L)$ des sections d'un fibré sur $\mathbf{P}_1/K$, à valeur dans un K-espace vectoriel $Ev(T_0, T_1)$;
(ii) de montrer (lemme de zéros; voir §2) que ϕ est injectif dès que les dimensions de ces espaces le permettent (à une constante près), c'est-à-dire ici dès que

$T_0 + rT_1 - nL >> 0$, d'où en prenant des bases, un déterminant mineur $\Delta \in K$, d'ordre $\sim nL$, non nul;

(iii) et de majorer les différentes valeurs absolues de Δ, avec un soin particulier (méthode de M. Laurent) pour $|\Delta|_\infty$. La formule du produit, et un choix convenable de L, T_0 et T_1, fournissent alors l'inégalité recherchée.

Une construction de ce type a déjà été proposée par A. Sert [S] pour le théorème de Lindemann–Weierstrass. Mais sa construction,"duale" de la nôtre (les rôles des paramètres nL et $T_0 + rT_1$ y sont inversés), s'appuie sur un critère de surjectivité pour ϕ. Nous y revenons au §6 de l'article.

Le point (iii) de notre nouvelle preuve fait l'objet du §4. À ce propos, rappelons, selon J-B. Bost [Bo], que tout l'art d'une preuve de transcendance consiste à choisir judicieusement des métriques sur les espaces source et but de l'application ϕ, ou de façon moins canonique, des bases de ces espaces. Nous proposons au §5 de l'article une base de Γ(L) différente de celle du §4, qui, jointe au *théorème de pureté* d'André [A 2.I], devrait fournir encore une autre preuve du théorème 1, mais que je ne sais faire aboutir pour l'instant que dans le cas du théorème de Lindemann–Weierstrass[1]. Ainsi, la preuve du §5 suit mot à mot le schéma précédent; seule varie la technique d'évaluation de la hauteur de Δ au point (iii). Et si le coeur lui en dit, le lecteur pourra déduire du §6 des variations de ces démonstrations, en remplaçant simplement la condition d'injectivité de ϕ au point (ii) par la condition de surjectivité $nL - (T_0 + rT_1) \gg 0$.

On peut finalement voir ce travail comme un pont jeté entre, d'une part, la preuve usuelle [L2], [Sh] du théorème 1, qui démarre par une construction auxiliaire au point 0, et dont, à la fin du §4, nous reprenons le choix de paramètres $T_0 >> T_1$ pour conclure, et d'autre part, la preuve d'Y. André [A2], qui démarre par une construction auxiliaire au point 1, et dont, à la fin du §5, nous reprendrons le choix de paramètres $T_1 >> T_0$. Dans chacune des preuves présentées ici, les constructions sont au contraire globales, et les points 0 et 1 jouent des rôles parallèles, en reflet des outils sur lesquels elles reposent, à savoir :

- une version du *lemme de zéros* de Shidlovsky relative à *plusieurs* points: voir le §2, ainsi que l'énoncé dual (*lemme d'annulation*) du §6;
- le *lemme d'interpolation* de M. Laurent, énoncé au cours du §3, qui tient simultanément compte des points 0 et 1, et fournit aux deux preuves la même majoration de $|\Delta|_\infty$.

2 Le lemme de zéros

Le lemme de zéros dont nous aurons besoin se déduit de la relation de Fuchs généralisée (pour une démonstration plus proche de celle du lemme de Shidlovsky, voir [B2], §4.i). Nous allons en donner une formulation générale, dans un style

[1]Ce texte est une version allégée de [B2], où sont discutées les difficultés liées à son extension au cas général.

familier en théorie de Baker. Avec les notations de l'introduction, soient L un entier > 0, et $\mathcal{M}^*(L)$ le tordu par $\mathcal{O}(L)$ du fibré à connexion dual de $\mathcal{M}$. Il est encore muni d'une connexion à singularités, qu'on note ∇^*; pour $\mathcal{M}$ de type $(a_1, \ldots, a_n)$, où les a_i sont des entiers $\leq L$, ses sections globales s'interprètent comme des n-uplets $s = (s_1, \ldots, s_n)$ de polynômes de degré $\leq L - a_1, \ldots, L - a_n$, à coefficients dans K, et $\nabla^*_{d/dz}(s)$ est représenté par $d/dz(s_1, \ldots, s_n) + (s_1, \ldots, s_n)A$.

Soit $\mathcal{R}$ un ensemble de points de $\mathbf{P}_1(K)$ à distance finie, *éventuellement singuliers* pour ∇. Pour tout $\alpha \in \mathcal{R}$, soit $\hat{\mathcal{W}}_\alpha$ un K-sous-espace vectoriel de $\hat{\mathcal{M}}_\alpha$ formé de sections horizontales pour ∇, c'est-à-dire, dans un repère local, de solutions dans $K^n[[z - \alpha]]$ du système différentiel $(*)$; si $\alpha \notin S$, cela revient à se donner un sous-espace vectoriel W_α de l'espace des "conditions initiales" M_α. On dit qu'une section s de $\mathcal{M}^*(L)$ *s'annule à un ordre* $\geq T$ *le long de* $\hat{\mathcal{W}}_\alpha$ si pour tout $Z \in \hat{\mathcal{W}}_\alpha$, la série formelle $s.Z$ est divisible par $(z - \alpha)^T$, c'est-à-dire si son image dans $\hat{\mathcal{O}}_\alpha[T] := \hat{\mathcal{O}}_\alpha/(z - \alpha)^T\hat{\mathcal{O}}_\alpha$ est nulle.

Théorème 2 (Lemme de zéros). *Il existe un nombre réel* $c(\nabla)$, *effectivement calculable en fonction de* $\mathcal{M}, \nabla$ *et* card $(\mathcal{R})$, *vérifiant la propriété suivante. Soient* $\{T_\alpha, \alpha \in \mathcal{R}; L\}$ *une collection d'entiers* ≥ 0, *et* s *une section non nulle de* $\mathcal{M}^*(L)$ *s'annulant, pour tout* $\alpha \in \mathcal{R}$, *à un ordre* $\geq T_\alpha$ *le long de* $\hat{\mathcal{W}}_\alpha$. *Il existe un sous-fibré* $\mathcal{M}'$ *de* $\mathcal{M}$ *dont la restriction hors de* S *est stable sous* ∇, *sur lequel* s *s'annule, et tel que*

$$\sum_{\alpha \in \mathcal{R}} \dim(\hat{\mathcal{W}}_\alpha/\hat{\mathcal{W}}_\alpha \cap \hat{\mathcal{M}}'_\alpha).T_\alpha \leq rk(\mathcal{M}/\mathcal{M}').L + c(\nabla).$$

[Lorsque $\alpha \notin S$, on pourra remplacer le quotient du terme de gauche par $W_\alpha/W_\alpha \cap M'_\alpha$.]

Démonstration. Soient M le $K(z)$-vectoriel à connexion défini par $\mathcal{M}$, D l'opérateur $\nabla_{d/dz}$ sur M, M' le plus grand sous-espace vectoriel de M stable sous D et contenu dans l'hyperplan $\mathrm{Ker}(s)$ de M, $\overline{M}$ le quotient M/M', muni de la connexion quotient $\overline{D}$, et ν sa dimension. Montrons que le sous-fibré $\mathcal{M}'$ de $\mathcal{M}$ de fibre générique M', sur lequel s s'annule, répond à la question.

Par passage au quotient, s définit une forme linéaire $\overline{s}$ sur $\overline{M}$, qui est un vecteur cyclique pour la connexion duale $\overline{D}^*$ sur $\overline{M}^*$: en effet, l'orthogonal dans $\overline{M}$ du $K(z)[\overline{D}^*]$-module engendré par $\overline{s}$ est stable sous $\overline{D}$, et contenu dans $\mathrm{Ker}(\overline{s})$, donc nul par maximalité de M'. L'annulateur dans $K(z)[d/dz]$ de $\overline{s}$ est donc un opérateur différentiel $\mathcal{L} = P(d/dz)$ de rang $\nu = rk(\mathcal{M}/\mathcal{M}')$.

Pour tout vecteur horizontal $Z \in \hat{\mathcal{W}}_\alpha$ de M, $s.Z$ est solution de l'équation différentielle $\mathcal{L}(s.Z) = (P(D^*)(s)).Z = 0$. De plus, $s.Z = 0$ si et seulement si s annule l'orbite de Z sous l'action du groupe de Galois différentiel d'une extension de Picard–Vessiot $F_\alpha/\mathbf{C}(z)$ de ∇ en α. Comme cette orbite engendre l'espace des vecteurs horizontaux d'un $K(z)$-sous-espace vectoriel M_Z de M stable sous D, on déduit de la maximalité de M' que l'image de M_Z dans $\overline{M}$ est nulle, et la dimension μ_α de l'image de $\hat{\mathcal{W}}_\alpha$ sous s est égale à $\dim(\hat{\mathcal{W}}_\alpha/\hat{\mathcal{W}}_\alpha \cap M'(F_\alpha))$, soit $\mu_\alpha = \dim(\hat{\mathcal{W}}_\alpha/\hat{\mathcal{W}}_\alpha \cap \hat{\mathcal{M}}'_\alpha)$.

Pour tout $\alpha \in \mathcal{R}$, l'équation différentielle $\mathcal{L}y = 0$ admet ainsi dans $K[[z-\alpha]]$ au moins μ_α solutions linéairement indépendantes d'ordre $\geq T_\alpha$. La collection de ses ν exposants en α est donc formée de μ_α entiers $\geq T_\alpha$, et de $\nu - \mu_\alpha$ nombres complexes dont les parties réelles sont minorables en fonction de A (et ≥ 0 si $\alpha \notin S$). De même, les parties réelles de ses exposants à l'infini sont, à l'addition près d'une constante ne dépendant que du type de $\mathcal{M}$ et de A, toutes minorées par $-L$. Enfin, les parties réelles de ses exposants aux autres singularités σ de $\mathcal{L}$ sont ≥ 0 si $\sigma \notin S$, et minorées en fonction de A sinon. La relation de Fuchs généralisée (voir [B1], Thm. 2) entraîne donc

$$-\nu L + \Sigma_{\alpha \in \mathcal{R}}\, \mu_\alpha T_\alpha \leq c(\nabla),$$

où $c(\nabla)$ désigne la somme des opposés des constantes énumérées plus haut, de l'irrégularité globale de $\mathrm{End}(\nabla)$, et de $|\mathcal{R}|n(n-1)$, et le théorème 2 est démontré.

Nous nous restreignons désormais au cas où $\mathcal{R} = \{0, 1\}$, et où $\dim(\hat{\mathcal{W}}_0) = 1$, et nous posons $r = \dim(\hat{\mathcal{W}}_1)$. Pour la preuve du §4, on pourrait suivre l'usage, courant en transcendance, qui consiste à éviter les passages aux quotients auxquels conduisent naturellement les lemmes de zéros grâce à des contraintes numériques sur les paramètres (ici, $L \geq T_1$). Mais cette contrainte n'est pas satisfaite dans la preuve du §5, où la notion suivante s'avère utile: on dit que $\hat{\mathcal{W}}_1$ est *non dégénéré* si pour tout sous-fibré $\mathcal{M}' \neq \mathcal{M}$ génériquement stable sous ∇, les quantités $n' = rk(\mathcal{M}/\mathcal{M}')$ et $r' = \dim(\hat{\mathcal{W}}_1/\hat{\mathcal{W}}_1 \cap \hat{\mathcal{M}}'_1)$ vérifient:

$$\frac{r'}{n'} := \frac{\dim(\hat{\mathcal{W}}_1/\hat{\mathcal{W}}_1 \cap \hat{\mathcal{M}}'_1)}{rk(\mathcal{M}/\mathcal{M}')} \geq \frac{\dim(\hat{\mathcal{W}}_1)}{rk(\mathcal{M})} := \frac{r}{n}. \qquad (**)$$

Pour $\alpha = 0$, où $\dim(\hat{\mathcal{W}}_0) = 1$, la condition de non-dégénérescence correspondante revient à demander que $\hat{\mathcal{W}}_0$ ne soit pas inclus dans un sous-fibré à connexion propre de $\mathcal{M}$. En reprenant l'expression $c(\nabla)$ du théorème 2, on en déduit :

Proposition 1 (Corollaire du lemme de zéros). *Soient $\{T_0, T_1, L\}$ trois entiers ≥ 0, et s une section de $\mathcal{M}^*(L)$ s'annulant, pour $\alpha = 0, 1$, à un ordre $\geq T_\alpha$ le long de $\hat{\mathcal{W}}_\alpha$. On suppose que la droite $\hat{\mathcal{W}}_0$ n'est pas incluse dans un sous-fibré à connexion propre de $\mathcal{M}$, et que l'une* ou *l'autre des hypothèses suivantes est satisfaite :*

(i) $T_0 + rT_1 > nL + c(\nabla)$ *et* $L \geq T_1$;
(ii) $T_0 + rT_1 > nL + nc(\nabla)$ *et* $\hat{\mathcal{W}}_1$ *est non dégénéré.*

Alors, $s = 0$.

Démonstration. Dans le cas contraire, il existerait un sous-fibré $\mathcal{M}'$ vérifiant la conclusion du théorème 2. Mais c'est impossible, puisque l'hypothèse faite sur $\hat{\mathcal{W}}_0$ entraîne $\hat{\mathcal{W}}_0 \cap \hat{\mathcal{M}}'_0 = 0$ (comme $\mathcal{M}' \subset \mathrm{Ker}(s)$, il aurait d'ailleurs suffi d'imposer $s(\hat{\mathcal{W}}_0) \neq 0$), et qu'avec les notations n', r' introduites ci-dessus, on peut minorer, dans sa conclusion, l'expression $\dim(\hat{\mathcal{W}}_0/\hat{\mathcal{W}}_0 \cap \hat{\mathcal{M}}'_0).T_0 + \dim(\hat{\mathcal{W}}_1/\hat{\mathcal{W}}_1 \cap \hat{\mathcal{M}}'_1).T_1$

- dans le cas (i), par

$$T_0+rT_1-\dim(\hat{\mathcal{W}}_1\cap\hat{\mathcal{M}}'_1).T_1 \geq T_0+rT_1-rk(\hat{\mathcal{M}}').T_1 > nL+c(\nabla)-rk(\hat{\mathcal{M}}').L,$$

- dans le cas (ii), par

$$n'\left(\frac{1}{n'}T_0+\frac{r'}{n'}T_1\right)\geq n'\left(\frac{1}{n}T_0+\frac{r}{n}T_1\right)>n'L+n'c(\nabla),$$

et donc toujours strictement par $rk(\mathcal{M}/\mathcal{M}')L + c(\nabla)$.

3 Les deux premières étapes, et le lemme d'interpolation de M. Laurent

Nous pouvons maintenant préciser les deux premières étapes de la preuve du théorème 1. Nous reprenons les notations $\mathcal{M}, \nabla, S, \mathcal{E}, n = rk(\mathcal{M}), n(\mathcal{E}), r_1(\mathcal{E}), \kappa = [K : \mathbf{Q}]$ de l'introduction. Le point 0 est en général une singularité de ∇, mais par hypothèse, la section horizontale $\mathcal{E}$ y admet un prolongement analytique en un KE-vecteur, qui appartient donc à $\hat{\mathcal{M}}_0$. Nous choisissons pour $\hat{\mathcal{W}}_0$ le K-sous-espace vectoriel de dimension 1 que $\mathcal{E}$ y engendre. Au point $1 \notin S$, nous notons W_1 le K-sous-espace vectoriel de dimension $r := r_1(\mathcal{E})$ de M_1 engendré par $\mathcal{E}(1)$, et nous choisissons pour $\hat{\mathcal{W}}_1$ le sous-espace horizontal de $\hat{\mathcal{M}}_1$ correspondant à W_1.

Deux réductions nous seront utiles. Tout d'abord, on peut sans perte de généralité supposer que les composantes $\mathcal{E}_1, \ldots, \mathcal{E}_n$ de la solution étudiée sont *linéairement indépendantes sur* $K(z)$, c'est-à-dire que $n(\mathcal{E}) = n$; en effet (cf. [Sh], 4, lemme 2), le saturé $\tilde{\mathcal{M}}$ du fibré à connexion que $\mathcal{E}$ engendre dans $\mathcal{M}$ n'a pas de singularité au point $\gamma = 1 \notin S$. Appliqué à ce saturé, dont le rang $\tilde{n}$ est égal à $n(\mathcal{E})$, la preuve aboutit à l'inégalité $r_1(\mathcal{E}) \geq \frac{\tilde{n}}{\kappa}$, qui est bien la conclusion recherchée. (En d'autres termes, on peut supposer que $\hat{\mathcal{W}}_0$ est non dégénéré.)

La seconde réduction ne s'avère nécessaire que pour la preuve du §5, où la contrainte numérique $L \geq T_1$ du §4 n'est pas vérifiée, mais par souci de symétrie, nous la ferons dans les deux cas. Sans perte de généralité, on peut également supposer que $\hat{\mathcal{W}}_1$ *est non dégénéré*. En effet, si ce n'est pas le cas, il existe un sous-fibré $\mathcal{M}' \neq \mathcal{M}$ génériquement stable sous ∇ tel qu'avec les notations de $(**)$, $\frac{r'}{n'} < \frac{r}{n}$, et on peut choisir $\mathcal{M}'$ de telle sorte qu'il réalise le minimum de ces expressions. Considérons alors le fibré quotient $\overline{\mathcal{M}} = \mathcal{M}/\mathcal{M}'$, qui est muni de la connexion à singularités quotient $\overline{\nabla}$, et relativement auquel $\overline{\hat{\mathcal{W}}_1} := \hat{\mathcal{W}}_1/\hat{\mathcal{W}}_1 \cap \hat{\mathcal{M}}'_1$ est non dégénéré: en effet, ses sous-fibrés sont de la forme $\overline{\mathcal{M}''}$, avec $\mathcal{M}' \subset \mathrm{Ker}(s)$, et les quantités $\overline{n''} := rk(\overline{\mathcal{M}}/\overline{\mathcal{M}''}) = rk(\mathcal{M}/\mathcal{M}'') := n''$ et $\overline{r''} = \dim(\overline{\hat{\mathcal{W}}}_1/\overline{\hat{\mathcal{W}}}_1 \cap \overline{\hat{\mathcal{M}}''_1}) = \dim(\hat{\mathcal{W}}_1/\hat{\mathcal{W}}_1 \cap \hat{\mathcal{M}}''_1) := r''$ vérifient : $\frac{\overline{r''}}{\overline{n''}} = \frac{r''}{n''} \geq \frac{r'}{n'}$ par minimalité de ce dernier quotient. On vérifie de plus, en complétant en une base de $\mathcal{M}^*(\mathbf{P}_1-\infty)$ un système primitif d'équations de $\mathcal{M}'$, que $\overline{\nabla}$ n'a pas de singularité en 1, et que l'image $\overline{\mathcal{E}}$ de $\mathcal{E}$ dans $\overline{\mathcal{M}}$ est encore un KE-vecteur en 0. Le théorème 1, appliqué à $\overline{\mathcal{M}}$ et $\overline{\mathcal{E}}$ et au sous-espace non dégénéré $\overline{\hat{\mathcal{W}}_1}$, fournit la conclusion $\kappa \geq \frac{n'}{r'}$. Comme $\frac{n'}{r'} > \frac{n}{r}$, on a bien $r \geq \frac{n}{\kappa}$.

Ces réductions étant acquises, soient c le plus petit entier supérieur à $nc(\nabla)$, et T_0, T_1, L trois entiers tels que

$$T_0 + rT_1 = nL + c \qquad (***)$$

Dans ces conditions,

(i) la première étape de la preuve consiste à choisir:

- pour $\Gamma(L)$ l'espace des sections de $\mathcal{M}^*(L)$, qui, pour L assez grand, est de dimension $h^0(L) = n(L+1) - \deg(\mathcal{M}) \sim nL$;
- pour $Ev(T_0, T_1)$ l'espace vectoriel $\mathrm{Hom}_K(\hat{\mathcal{W}}_0, \hat{\mathcal{O}}_0[T_0]) \oplus \mathrm{Hom}_K(\hat{\mathcal{W}}_1, \hat{\mathcal{O}}_1[T_1])$, qui est de dimension $\dim(\hat{\mathcal{W}}_0).T_0 + \dim(\hat{\mathcal{W}}_1).T_1 = T_0 + rT_1$;
- pour ϕ l'application K-linéaire de $\Gamma(L)$ dans $Ev(T_0, T_1)$ somme directe sur $\alpha = 0, 1$, des composées des applications canoniques $\Gamma(L) \to \hat{\mathcal{M}}^*_\alpha$, $\hat{\mathcal{M}}^*{}_\alpha \to \mathrm{Hom}(\hat{\mathcal{W}}_\alpha, \hat{\mathcal{O}}_\alpha)$, $\hat{\mathcal{O}}_\alpha \to \hat{\mathcal{O}}_\alpha[T_\alpha]$. Pour s dans $\Gamma(L)$, $\phi(s)$ représente donc la collection des T_0 premiers coefficients de Taylor en 0 de $s(\mathcal{E})$ et des T_1 premiers coefficients de Taylor en 1 de $s(\mathcal{Z})$, où $\mathcal{Z}$ parcourt $\hat{\mathcal{W}}_1$. On trouvera plus bas une expression matricielle de ϕ: une matrice Φ à $T_1 + rT_0$ lignes et $h^0(L) \sim nL$ colonnes.

(ii) la seconde étape consiste à remarquer que puisque $T_0 + rT_1 = nL + c > nL + nc(\nabla)$, le deuxième cas de la proposition 1, joint aux hypothèses $n = n(\mathcal{E})$ et $\hat{\mathcal{W}}_1$ non dégénéré auxquelles nous nous sommes ramenés, entraîne que pour $s \in \Gamma(L)$, $\phi(s)$ ne peut s'annuler que si $s = 0$, autrement dit que ϕ est injective. L'une des coordonnées de $\Lambda^{h_0(L)}\phi$ dans une base de $\Lambda^{h_0(L)}\mathrm{Hom}(\Gamma(L), Ev(T_1; T_0))$ (c'est-à-dire l'un des mineurs d'ordre $h^0(L)$ de Φ) est ainsi un élément Δ de K non nul. Reste

(iii) à évaluer la hauteur de Δ, c'est-à-dire ses valeurs absolues pour les différentes places $v \in \mathcal{V}_K$ de K.

Pour le facteur local en la place ∞, cette dernière étape repose sur l'énoncé suivant, dont on trouvera diverses variantes dans [La], [S], Lemme 4.4.1, [LP], Lemme 7, [H].

Théorème 3 (Lemme d'interpolation de M. Laurent). *Soient* **D** *un disque de* **C** *centré à l'origine,* γ_0, γ_1 *deux points de* **D**, $\mathcal{T}_0, \mathcal{T}_1$ *deux ensembles d'entiers* ≥ 0 *et majorés respectivement par* $T_0 - 1, T_1 - 1$, h *la somme des cardinaux de* $\mathcal{T}_0$ *et de* $\mathcal{T}_1$, *et* $f_1, \ldots, f_h$ *des fonctions analytiques sur* **D**. *Posons* $\tilde{c} = T_0 + T_1 - h$. *Alors, la fonction* $\delta(z) = \det \begin{pmatrix} \left((\partial^t f_i)(\gamma_0 z)\right)_{t\in\mathcal{T}_0, 1\leq i\leq h} \\ \left((\partial^t f_i)(\gamma_1 z)\right)_{t\in\mathcal{T}_1, 1\leq i\leq h} \end{pmatrix}$ *s'annule en* 0 *avec une multiplicité* $\geq T_0T_1 - \tilde{c}(T_0 + T_1)$.

Démonstration. En effectuant des combinaisons linéaires des colonnes, on peut supposer que $f_i(z) \in z^{i-1}\hat{\mathcal{O}}_0$. L'ordre de δ en 0 est alors minoré par $\Sigma_{i=1,\ldots,h}(i-1) - \Sigma_{t\in\mathcal{T}_0} t - \Sigma_{t\in\mathcal{T}_1} t \geq \frac{h(h-1)}{2} - \frac{T_0(T_0-1)}{2} - \frac{T_1(T_1-1)}{2} = T_0T_1 - \tilde{c}(T_0+T_1) + \frac{\tilde{c}(\tilde{c}+1)}{2}$.

Appliqué aux points $\gamma_0 = 0$ et $\gamma_1 = 1$ (dont on notera qu'ils interviennent ici sur un pied d'égalité), cet énoncé fournit, tant au §4 qu'au §5, une majoration très précise des valeurs absolues en ∞ de mineurs d'ordre $\sim T_0 + T_1$ de Δ. Avant de débuter cette estimation et celles des $|\Delta|_v, v \in \mathcal{V}_K$, je signale pour la commodité du lecteur que les paramètres T_0, T_1 seront en fin de preuve choisis de façon proportionnelle à L, et que les termes dominants des majorations sont de la forme $e^{\gamma L^2 \mathrm{Log} L}$. C'est γ qu'il convient de contrôler précisément. D'autre part, je désignerai par $c_0, c_1, \ldots$ des entiers ≥ 2 qui ne dépendent que de $\mathcal{M}, \nabla, \mathcal{E}$, et que l'on pourrait d'ailleurs majorer de façon effective en termes de la matrice $A(z)$ et de $\mathcal{E}$.

4 Conclusion de la première preuve (cas général)

Matriciellement, ϕ s'exprime comme suit: fixons une base $(s_1, \ldots, s_{h^0(L)})$ de $\Gamma(L)$, et considérons la base de l'espace but de ϕ construite à partir d'une base $\mathcal{Z}_1, \ldots, \mathcal{Z}_r$ de $\hat{\mathcal{W}}_1$, de la base $\mathcal{E}$ de $\hat{\mathcal{W}}_0$, et de la base de l'espace $\hat{\mathcal{O}}_\alpha[T_\alpha]$, $\alpha = 0, 1$ formée des monômes $\{\frac{1}{t!}(z-\alpha)^t\ ; t = 0, \ldots, T_\alpha - 1\}$. Alors, ϕ est représentée par la matrice à $T_0 + rT_1$ lignes et $h^0(L)$ colonnes à coefficients dans K:

$$\Phi = \begin{pmatrix} \Phi_0 \\ \Phi_1 \\ \ldots \\ \Phi_r \end{pmatrix}, \begin{pmatrix} \Phi_0 = \left(\partial^t(s_i.\mathcal{E})(0)\right)_{0\leq t\leq T_0-1; 1\leq i\leq h^0(L)} \\ \ldots \\ \Phi_\rho = \left(\partial^t(s_i.\mathcal{Z}_\rho)(1)\right)_{0\leq t\leq T_1-1; 1\leq i\leq h^0(L)} \\ \ldots \qquad (\rho=1,\ldots,r) \end{pmatrix},$$

où ∂ désigne le champ de vecteurs d/dz.

Du côté de l'espace source, identifions le fibré $\mathcal{M}$ à $\mathcal{O}(a_1) \oplus \cdots . \oplus \mathcal{O}(a_n)$, choisissons un repère adapté $(e_1, \ldots, e_n)$ de sections de $\mathcal{M}$ au-dessus de $\mathbf{P}_1 \setminus \infty$, d'où l'écriture $\mathcal{E} = \Sigma_{\iota=1}^n \mathcal{E}_\iota e_\iota = {}^t(\mathcal{E}_1, \ldots, \mathcal{E}_n)$, $\mathcal{Z}_\rho = \Sigma_{\iota=1}^n \mathcal{Z}_{\rho,\iota} e_\iota$ pour $\rho = 1, \ldots, r$, et une décomposition $\Gamma(L) = \oplus_{\iota=1,\ldots,n} \Gamma(\mathcal{O}(L-a_i))e_\iota^*$. Pour tout L', choisissons pour base de $\Gamma(\mathcal{O}(L'))$ les monômes $\{\frac{1}{\ell!}z^\ell\ ; \ell = 0, \ldots, L'\}$ (une autre base fera l'objet du §5). Dans la base $\{\frac{1}{\ell!}z^\ell e_\iota^*; \iota = 1, \ldots, n, \ell = 0, \ldots, L - a_\iota\}$ de $\Gamma(L) = H^0(\mathcal{M} \otimes \mathcal{O}(L))$ indexant ses $h^0(L) = n(L+1) - \Sigma_{\iota=1}^n a_\iota$ colonnes, notre matrice se réécrit:

$$\Phi = \begin{pmatrix} \Phi_0 \\ \Phi_1 \\ \ldots \\ \Phi_r \end{pmatrix}, \begin{pmatrix} \Phi_0 = \left(\partial^t(\frac{1}{\ell!}z^\ell \mathcal{E}_\iota)(0)\right)_{0\leq t\leq T_0-1; 1\leq \iota\leq n, 0\leq \ell\leq L-a_\iota} \\ \ldots . \\ \Phi_\rho = \left(\partial^t(\frac{1}{\ell!}z^\ell \mathcal{Z}_{\rho,\iota})(1)\right)_{0\leq t\leq T_1-1; 1\leq \iota\leq n, 0\leq \ell\leq L-a_\iota} \\ \ldots \qquad (\rho=1,\ldots,r) \end{pmatrix}.$$

Quitte à remplacer L par $L + c_0$ dans les estimations qui suivent, on peut, comme je vais le faire, supposer les a_ι tous nuls (fibré trivial). Soit alors Δ l'un des déterminants mineurs d'ordre maximal $h_0(L) = n(L+1)$ de Φ non nul.

(A) Majoration de $|\Delta|_v$ *(*$v \in \mathcal{V}_K$ *quelconque)*

Elle repose sur une majoration terme à terme des valeurs absolues v-adiques des coefficients de Δ.

A1. Coefficients des lignes de Δ situées dans Φ_0: les prolongements analytiques à l'origine $\mathcal{E}_\iota(z) = \Sigma_{k\geq 0} a_{\iota,k}\frac{z^k}{k!}$ sont par hypothèse des KE-séries. Comme le coefficient

$$\partial^t(\frac{1}{\ell!}z^\ell \mathcal{E}_\iota)(0) = \binom{t}{\ell} a_{\iota,t-\ell} \qquad \text{si } t \geq \ell,$$

et s'annule si $t < \ell$, sa valeur absolue v-adique est majorée par $|d_{T_0-1}|_v^{-1}$ si v est finie, par $c_2^{T_0}$ si v est archimédienne. Ici, d_{T_0-1} désigne un dénominateur commun des $a_{\iota,k}$, $k = 0, \ldots, T_0 - 1$, qu'on majore suivant la définition de Lang par $c_1^{T_0}$. (Le produit c_1c_2 est relié à la taille des transformées de Laplace des séries $\mathcal{E}_\iota(z)$.)

A2. Coefficients des lignes de Δ situées dans les $\Phi_\rho, \rho = 1, \ldots, r$: pour tout $s \in \Gamma(L)$ et tout $\mathcal{Z}$ horizontal, $\partial^t(s.\mathcal{Z}) = \big((\nabla_\partial^*)^t s\big).\mathcal{Z}$. Comme 1 est un point ordinaire, on en déduit que si v est une place finie, le coefficient $\partial^t(\frac{1}{\ell!}z^\ell \mathcal{Z}_{\rho,\iota})(1)$ a une valeur absolue v-adique $\leq |L!|_v^{-1}|c_3|_v^{-T_1}|c_5|_v$; si v est archimédienne, elle est $\leq T_1!c_4^{T_1}c_6$. [2](Le produit c_3c_4 est relié à la taille du "transformé de Fourier" de ∇ ; le produit c_5c_6 à la hauteur de W_1.)

Δ est somme de $h_0(L)!$ monômes où apparaissent les coefficients du premier type au plus T_0 fois, du deuxième type au plus rT_1 fois. Ainsi

$$\prod_{v\in\mathcal{V}_K, v\neq\infty} |\Delta|_v \leq \big((n(L+1))!\,(c_1c_2)^{T_0^2}\,(c_3c_4)^{rT_1^2}(c_5c_6)^{rT_1}(L!T_1!)^{rT_1}\big)^\kappa.$$

Le terme prépondérant de cette majoration est donc $(L!T_1!)^{rT_1\kappa}$, dont le logarithme croît comme $\kappa(L+T_1)T_1\mathrm{Log}L$.

(B) Majoration de $|\Delta|_\infty$ *(méthode de M. Laurent)*

Comme on l'a dit, elle repose sur le théorème 3. Quelques préparatifs avant de l'appliquer à certains mineurs (d'ordre $h \sim T_0 + T_1$) de Δ. Tout d'abord, $\mathcal{E} \in \hat{\mathcal{M}}_1 \otimes$ **C** étant une combinaison linéaire à coefficients complexes $\lambda_1\mathcal{Z}_1 + \cdots + \lambda_r\mathcal{Z}_r$, avec, disons, $\lambda_1 \neq 0$, on ne change guère Δ quand on remplace dans Φ le bloc Φ_1 par la combinaison linéaire correspondante $\Phi_1' = \lambda_1\Phi_1 + \cdots + \lambda_r\Phi_r$. Plus précisément, Δ se déduit de Φ en rayant $c' := T_0 + rT_1 - h^0(L) = c - n$ lignes (cf. $(***)$). Par conséquent, les indices t d'au moins $T_1 - c'$ lignes de Φ_1 intervenant dans Δ

[2]Il est possible que le théorème de pureté d'André permette de se débarrasser des termes en $T_1!$ dans ce calcul archimédien. Je reviens sur ce point au §5. Ici, le gain serait faible puisqu'on prendra $L \geq T_1$.

sont également indices de lignes de Φ_2, de Φ_3,...et de Φ_r intervenant dans Δ. Notons $\mathcal{T}_1$ l'ensemble, de cardinal $\tau_1 \geq T_1 - c'$, de ces indices communs, et $\mathcal{T}_0$ l'ensemble, de cardinal $\tau_0 \geq T_0 - c'$, des indices de ligne de Φ_0 intervenant dans Δ. À division par $\pm\lambda_1^{\tau_1}$ près, Δ est donc égal au déterminant $\tilde{\Delta}$ de la matrice carrée d'ordre $h^0(L) = n(L+1)$:

$$\tilde{\Phi} = \begin{pmatrix} \tilde{\Phi}_0 \subset \Phi_0 \\ \tilde{\Phi}'_1 \subset \Phi'_1 \\ \tilde{\Psi} \subset \Phi \end{pmatrix}, \quad \begin{pmatrix} \tilde{\Phi}_0 = \left(\partial^t (s_i.\mathcal{E})(0)\right)_{t \in \mathcal{T}_0; 1 \leq i \leq h^0(L)} \\ \tilde{\Phi}'_1 = \left(\partial^t (s_i.\mathcal{E})(1)\right)_{t \in \mathcal{T}_1; 1 \leq i \leq h^0(L)} \\ \tilde{\Psi} \end{pmatrix}.$$

où les lignes de $\tilde{\Psi}$ sont, pour au plus $T_0 - \tau_0$ d'entre elles, des lignes de Φ_0, et pour au plus $rT_1 - \tau_1$ d'entre elles, des lignes de $\Phi_1, \Phi_2, \ldots,$ ou Φ_r.

Suivant [S], nous majorons maintenant la valeur absolue du nombre complexe $\tilde{\Delta}$ en calculant son développement de Laplace suivant les mineurs d'ordre maximal de $\tilde{\Psi}$:

B1. Valeur absolue en ∞ des mineurs d'ordre maximal de $\tilde{\Psi}$: la considération de leur format et les calculs précédents sur $|\Delta|_v$, avec $v = \infty$, montrent que chacun d'eux est $\leq (n(L+1) - \tau_0 - \tau_1)!\, c_2^{(T_0-\tau_0)T_0}\, (c_6 c_4^{T_1} T_1!)^{rT_1-\tau_1}$. Comme $T_0 - \tau_0 \leq c'$, ils sont donc majorés par $(n(L+1) - (\tau_0+\tau_1))! c_2^{c'T_0}(c_6 c_4^{T_1} T_1!)^{rT_1}$.

B2. Valeur absolue en ∞ des mineurs d'ordre maximal $h := \tau_0 + \tau_1$ de la matrice $\begin{pmatrix} \tilde{\Phi}_0 \\ \tilde{\Phi}'_1 \end{pmatrix}$: ils sont tous de la forme $\delta(1)$, pour des fonctions $\delta(z)$ du type étudié au lemme d'interpolation : prendre $\gamma_0 = 0, \gamma_1 = 1$, $f_i = s_i.\mathcal{E}$, où i parcourt une partie à h éléments de l'ensemble $[1, h^0(L)]$, et reprendre les notations ci-dessus pour $\mathcal{T}_0, \mathcal{T}_1$; en particulier, $h = T_0 + T_1 - \tilde{c}$, avec $\tilde{c} \leq 2c$. En vertu du lemme de Schwarz, appliqué sur des disques de rayon 1 et $R > 1$, chacun d'eux admet une majoration de la forme

$$|\delta(1)| \leq R^{-T_0T_1+2c(T_0+T_1)} h!\, (\sup_{t<T_0,i} |\partial^t f_i(0)|)^{\tau_0}\, (\sup_{t<T_1,i} |\partial^t f_i|_R)^{\tau_1}.$$

La première parenthèse se majore comme on l'a fait des coefficients de $|\Delta|_v$ en **A1**. Pour la seconde, noter que les E-fonctions $\mathcal{E}_\iota$ sont des fonctions entières de type exponentiel. Si $c_7 \geq 1$ désigne un majorant de leurs types, on déduit des inégalités de Cauchy $|\partial^t f_i|_R \leq \frac{t!}{R^t} |f_i|_{2R}$ que pour $f_i = \frac{1}{\ell!} z^\ell \mathcal{E}_\iota$ et $R = T_1$, ce dernier terme est majoré par $(\sup_{\ell \leq L} \frac{(2T_1)^\ell}{\ell!} e^{2c_7 T_1})^{T_1} \leq e^{4c_7 T_1^2}$. Ainsi $|\delta(1)| \leq T_1^{-T_0T_1} h! L^{2c(T_0+T_1)} c_2^{T_0^2} e^{4c_7 T_1^2}$. En définitive, la formule de Laplace donne, avec $R = T_1$:

$$|\Delta|_\infty \leq T_1^{-T_0T_1} \times (n(L+1))! c_6^{rT_1} L^{2c(T_0+T_1)} c_4^{rT_1^2} c_2^{2T_0^2} e^{4c_7 T_1^2}.(T_1!)^{rT_1}.$$

Le terme prépondérant de cette majoration est donc $T_1^{-T_0T_1}(T_1!)^{rT_1}$, dont le logarithme croît comme $(-T_0 + rT_1)T_1 Log T_1$.

(C) Formule du produit et choix des paramètres

Nous faisons maintenant croître L indéfiniment, et supposons que T_0 et T_1, qui étaient jusqu'à présent soumis à la seule contrainte $T_0 + rT_1 = nL + c$ de $(***)$, croissent linéairement avec L. Comme $\Delta \neq 0$, la formule du produit entraîne :

$$T_0 T_1 \leq r\kappa L T_1 + r(\kappa+1)T_1^2 + O(L^2/\mathrm{Log}L),$$

donc $\frac{T_0}{L} \leq r\kappa + r(\kappa+1)\frac{T_1}{L} + O(\frac{L}{T_1 \mathrm{Log}L})$. Choisissant, à l'instar de Siegel et Shidlovsky, $T_0 = (n-\epsilon)L, T_1 = (\epsilon L + c)/r$ avec ϵ petit, on obtient: $n - \epsilon \leq r\kappa(1 + O(\epsilon))$, et finalement la conclusion $n \leq r\kappa$ souhaitée.

5 Conclusion de la seconde preuve (cas de Lindemann–Weierstrass)

Dans la preuve précédente, l'hypothèse que $\mathcal{E}$ est un vecteur de E-fonctions est intervenue à deux reprises : au point **A1** pour majorer la hauteur des coefficients des séries entières $s_i.\mathcal{E} \in K[[z]]$; au point **B2** pour majorer la croissance à l'infini des fonctions entières $s_i.\mathcal{E}$. Nous n'avons utilisé des sections horizontales $\mathcal{Z}_\rho \in \hat{\mathcal{W}}_1$ de ∇ que la propriété de convergence au voisinage du point 1. Or le théorème de pureté d'Y. André sur les séries Gevrey de type négatif ([A2.I], Théorèmes 4.2 et 4.3.iii) affirme que dès que $n(\mathcal{E}) = n$, toutes les solutions de ∇ relèvent de la théorie des E-fonctions. Plus précisément (voir [A2.I], corollaire 4.4), ∇ admet alors un système fondamental de solutions en 0 de la forme $U(z)z^C$, où $e^{2i\pi C} \in GL_n(\mathbf{C})$ représente la monodromie de ∇ au point 0, et $U(z)$ est une matrice dont tous les coefficients sont des E-fonctions. Nous allons maintenant faire usage de cette propriété, en supposant, pour éviter la présence au voisinage de 1 de produits de E-fonctions par des G-fonctions, que *la monodromie de* ∇ *en* 0 *est triviale*.

Jointe à l'existence d'un E-vecteur solution de ∇, cette hypothèse limite en pratique les applications aux cas couverts par le théorème de Lindemann–Weiesrtrass. Elle permet en tous cas, par torsion par une puissance entière de z, de supposer que ∇ admet une base de solutions constituée de E-vecteurs, et qu'en particulier, *toutes* ses sections horizontales $\mathcal{Z}_\rho \in \hat{\mathcal{W}}_1$ admettent des prolongements analytiques en des fonctions entières *de type exponentiel*.

Mais pour exploiter cette idée, encore faut-il contrôler les hauteurs des coefficients des séries entières $s_i.\mathcal{Z}_\rho \in K[[z-1]]$. Ceci conduit (voir **A$'$2** ci-dessous) à remplacer les monômes $\frac{z^\ell}{\ell!}$ de la base de $\Gamma(\mathcal{O}(L')) = \mathrm{Sym}^{L'}(\Gamma(\mathcal{O}(1))$ choisie au début du §4 par $\{\frac{1}{\ell!}(z-1)^\ell; \ell = 0, \ldots, L'\}$, d'où une nouvelle matrice

$$\Phi = \begin{pmatrix} \Phi_0 \\ \Phi_1 \\ \ldots \\ \Phi_r \end{pmatrix}, \begin{pmatrix} \Phi_0 = \left(\partial^t(\frac{1}{\ell!}(z-1)^\ell \mathcal{E}_\iota)(0)\right)_{0 \leq t \leq T_0 - 1; 1 \leq \iota \leq n, 0 \leq \ell \leq L - a_\iota} \\ \ldots\ldots \\ \Phi_\rho = \left(\partial^t(\frac{1}{\ell!}(z-1)^\ell \mathcal{Z}_{\rho,\iota})(1)\right)_{0 \leq t \leq T_1 - 1; 1 \leq \iota \leq n, 0 \leq \ell \leq L - a_\iota} \\ \ldots \qquad {\scriptstyle (\rho = 1, \ldots, r)} \end{pmatrix}$$

représentant ϕ.

Cette nouvelle matrice se déduit de celle du §4 par multiplication à droite par un élément de $SL_{h^0(L)}$. En rayant les $T_0 + rT_1 - h^0(L)$ mêmes lignes que précédemment, on obtient donc le *même* déterminant mineur $\Delta \in K^*$. Mais de nouvelles estimations apparaissent pour $|\Delta|_v$:

(A′) Majoration de $|\Delta|_v$, $v \in \mathcal{V}_K$ quelconque

A′1. Coefficients des lignes de Δ situées dans Φ_0: si v est une place finie, les factorielles ne se simplifient plus, et on doit multiplier le majorant trouvé en **A1**, par $|L!|_v^{-1}$; celui des places archimédiennes devient $2^L c_2^{T_0}$.

A′2. Coefficients des lignes de Δ situées dans les Φ_ρ, $\rho = 1, \dots, r$: on remarque maintenant que $\partial^t\big((\frac{1}{\ell!}(z-1)^\ell \mathcal{Z}_{\rho,\iota}\big)(1) = \binom{t}{\ell}\, \partial^{t-\ell}\mathcal{Z}_{\rho,\iota}(1)$, et que 1 est un point ordinaire de ∇. On peut donc supprimer le facteur $|L!|_v^{-1}$ de la majoration de **A2** aux places finies. De plus, on peut, aux places v infinies, en supprimer le facteur $T_1!$: en effet, les fonctions induites par $\mathcal{Z}_{\rho,\iota}$ sur $K_v \subset \mathbf{C}$ sont entières de type exponentiel; si $c_8 \geq 1$ désigne un majorant de leurs types, les inégalités de Cauchy, appliquées à un disque de rayon T_1, entrainent que $|\partial^t\big((\frac{1}{\ell!}(z-1)^\ell \mathcal{Z}_{\rho,\iota}\big)(1)|_v \leq e^{T_1+1}e^{c_8 T_1}$.

Le terme prépondérant du produit $\Pi_{v \neq \infty}|\Delta|_v$ devient donc $(L!)^{\kappa T_0} \sim e^{\kappa L T_0 \mathrm{Log} L}$.

(B′) Majoration de $|\Delta|_\infty$ (méthode de M. Laurent)

On reproduit la démonstration du §4.B jusqu'à l'étude du développement de Laplace, qu'on modifie comme suit.

B′1. Valeur absolue en ∞ des mineurs d'ordre maximal de $\tilde{\Psi}$: pour les mêmes raisons qu'en **A′2**, on peut supprimer le facteur $(T_1!)^{rT_1}$ de la majoration de **B1**;

B′2. Valeur absolue en ∞ des mineurs d'ordre maximal $h := \tau_0 + \tau_1$ de la matrice $\begin{pmatrix}\tilde{\Phi}_0\\ \tilde{\Phi}_1'\end{pmatrix}$: pas de changement par rapport à **B2**, si ce n'est que les fonctions f_i s'écrivent maintenant $f_i = \frac{1}{\ell!}(z-1)^\ell \mathcal{E}_\iota$. On peut alors majorer la première parenthèse comme dans **A′1**, et la seconde en appliquant l'inégalité de Cauchy sur un disque de rayon $R = T_1$.

Le terme prépondérant du majorant de $|\Delta|_\infty$ devient donc $T_1^{-T_0 T_1}$.

(C′) Formule du produit et choix des paramètres

La formule du produit donne cette fois

$$T_0 T_1 \leq \kappa T_0 L + O(L^2/\mathrm{Log}L),$$

donc $\frac{T_1}{L} \leq \kappa + O(\frac{L}{T_0 \mathrm{Log} L})$. Choisissant, à l'instar d'André [A2.II], $T_1 = (\frac{n}{r} - \epsilon)L$, $T_0 = \epsilon r L + c$, on aboutit à la conclusion $\frac{n}{r} \leq \kappa$ souhaitée.

Remarque 1. On voit en combinant les deux approches avant de fixer les paramètres que $\frac{T_0 . rT_1}{(T_0 + rT_1)\inf(T_0, rT_1)} \leq \frac{r}{n}\kappa(1 + o(1))$. C'est bien en prenant rT_1 (comme au §4) ou T_0 (comme ici) aussi petit que possible qu'on obtient le meilleur résultat.

Remarque 2. Le choix de paramètres $rT_1 \sim nL$ fait dans ce §5 évoque la méthode de Gel'fond–Dèbes utilisée dans [A1] et [A2.II]. Yves André me signale d'ailleurs que la démarche présentée ici permet de retrouver ses résultats d'indépendance linéaire [A1] sur les valeurs de G-fonctions.

6 L'approche duale

Restreinte au cas de Lindemann–Weierstrass $\mathcal{E}_\iota(z) = e^{\alpha_\iota z}, \iota = 1, \ldots, n$, la preuve du §4 est "duale" de celle de [S], qui l'a fortement inspirée[3]. Plus précisément, A. Sert étudie la fonction e^z aux points $\alpha_1, \ldots, \alpha_n$, et considère une matrice $\hat{\Phi}$ à $n(L+1)$ lignes et $T_0 + rT_1$ colonnes, construite à partir d'une base $\{b_1, \ldots, b_r\}$ de W_1 dans M_1, et des matrices

$$\hat{\Phi}_0 = \left(\frac{1}{\ell!}\partial^\ell z^t(\alpha_\iota)\right)_{1\le\iota\le n, 0\le\ell\le L; 0\le t\le T_0-1},$$

$$\hat{\Phi}_1 = \left(e^{-\alpha_\iota}\frac{1}{\ell!}\partial^\ell(z^t e^z)(\alpha_\iota)\right)_{1\le\iota\le n, 0\le\ell\le L; 0\le t\le T_1-1}.$$

(Pour passer de mes notations à celles de [S], utiliser le dictionnnaire: $\hat{\Phi} \to M, L \to J, T_0 \to S, T_1 \to T, n \to m, r \to n$.) Comme

$$\frac{1}{\ell!}\partial^\ell z^t(\alpha_\iota) = \partial^t\left(\frac{1}{\ell!}z^l e^{\alpha_\iota z}\right)(0)\,,\; e^{-\alpha_\iota}\frac{1}{\ell!}\partial^\ell(z^t e^z)(\alpha_\iota) = \partial^t\left(\frac{1}{\ell!}z^l e^{\alpha_\iota(z-1)}\right)(1),$$

$\hat{\Phi}$ est essentiellement la transposée de la matrice Φ du §4. Mais malgré sa ressemblance avec celui du §2, le "lemme de zéros" utilisé par Sert ([S], Lemme 4.2) n'est pas de même nature : il énonce l'existence d'une constante $\hat{c} = n^2/4$ telle que (pour $L \ge T_1$), $\hat{\Phi}$ est de rang maximal $T_0 + rT_1$, dès que $nL \ge T_0 + rT_1 + \hat{c}$. C'est un *lemme d'interpolation* au sens de D. Masser : une condition suffisante pour que le morphisme d'évaluation ϕ lui-même (et non son transposé) soit surjectif.

Comme me l'ont fait remarquer M. Laurent et S. Fischler, les formules supra (dualité de Fourier–Borel) montrent qu'un lemme d'interpolation s'interprète comme un lemme de zéros lorsqu'on se restreint à l'ensemble des exponentielles-polynômes, c'est-à-dire à des systèmes différentiels à coefficients constants. Je me propose maintenant d'étendre l'énoncé de surjectivité de [S] au cas d'un système différentiel général. Il est possible que la dualité que fournit la transformation de Fourier (voir [A2.I], §6) y conduise, mais j'établirai cette extension par un argument élémentaire, proche de ceux de Masser et de [F]. Pour des raisons expliquées plus bas, je préfère d'ailleurs appeler "lemme d'annulation" cette généralisation.

Pour alléger, je ne traiterai que de la situation rencontrée à la Proposition 1 du §2, et en supposant le fibré $\mathcal{M}$ trivial. On reprend les notations $n{=}rk(\mathcal{M}), \nabla, \mathcal{R}, \hat{\mathcal{W}}_\alpha$ ($\alpha \in \mathcal{R}$), $c(\nabla)$ du §2, avec donc $\mathcal{R} = \{0, 1\}$ et $\dim(\hat{\mathcal{W}}_0) = 1$. On pose $r = \dim(\hat{\mathcal{W}}_1)$, et on note $\mathcal{E}$ (resp.$\{\mathcal{Z}_1, \ldots, \mathcal{Z}_r\}$) une base $\hat{\mathcal{W}}_0$ (resp. $\hat{\mathcal{W}}_1$) sur K. Outre les hypothèses de non-dégénérescence faites à la Proposition 1, on devra

[3]C'est d'ailleurs une mesure d'indépendance algébrique entièrement explicite que [S] obtient dans ce cas. Dans cet ordre d'idées, je ne sais si la méthode du §4 permettra d'établir, à l'instar de [L1], une version quantitative du théorème 1.

ici supposer que $\mathcal{E}(0) \neq 0$, et que 1 est un point ordinaire de ∇, de sorte que $\{\mathcal{Z}_1(1), \dots, \mathcal{Z}_r(1)\}$ forme une base de W_1.

Proposition 2 (Lemme d'annulation). *Il existe une constante $\hat{c}(\nabla)$ effectivement calculable en fonction de $\mathcal{M}$ et de ∇ et vérifiant la propriété suivante. Soient T_0, T_1, L un triplet d'entiers ≥ 0 et $\{a_{0,t}, 0 \leq t \leq T_0 - 1, a_{i,t}, 1 \leq i \leq r, 0 \leq t \leq T_1 - 1\}$ un $(T_0 + rT_1)$-uple d'éléments de K. Supposons que la droite $\hat{\mathcal{W}}_0$ ne soit pas incluse dans un sous-fibré à connexion propre de $\mathcal{M}$, et que $\mathcal{E}(0) \neq 0$. Supposons par ailleurs que 1 soit un point ordinaire de ∇, et que l'une* ou *l'autre des hypothèses suivantes soit satisfaite:*

(i) $L_1 \geq T_1$;
(ii) $\hat{\mathcal{W}}_1$ *est non dégénéré.*

Supposons enfin que $nL \geq T_0 + rT_1 + \hat{c}(\nabla)$. Alors, il existe une section s de $\mathcal{M}^*(L)$ telle que $\partial^t(s.\mathcal{E})(0) = a_{0,t}$ pour tout $t \leq T_0 - 1$ et $\partial^t(s.\mathcal{Z}_i)(1) = a_{i,t}$ pour tout $i = 1, \dots, r, t \leq T_1 - 1$.

Démonstration. Il s'agit de montrer que sous ces hypothèses, les $T_0 + rT_1$ formes linéaires sur $\Gamma(L) := H^0(\mathcal{M}^*(L))$ définies par

$$ev_{0,t}(s) = \partial^t(s.\mathcal{E})(0)(t = 0, \dots, T_0 - 1);$$
$$ev_{i,t}(s) = \partial^t(s.\mathcal{Z}_i)(1)(i = 1, \dots, r; t = 0, \dots, T_1 - 1)$$

sont linéairement indépendantes sur K (c'est-à-dire que les $T_0 + rT_1$ lignes de la matrice Φ du §4 le sont; le lemme de zéros du §2 exprimait, pour $T_0 + rT_1 \geq nL + c(\nabla)$, l'indépendance linéaire de ses $n(L + 1)$ colonnes). Fixons une matrice représentative $A(z)$ de la connection ∇, dont on suppose que le dénominateur $Q(z)$ ne s'annule pas en 1, et est d'ordre θ minimal en 0, et notons $\deg(\nabla)$ le maximum des degrés de Q et des coefficients de QA.

Supposons tout d'abord que $T_1 = 0$, et que 0 n'est pas une singularité de ∇ (autrement dit, que Q ne s'annule pas en 0; la condition $ev_{0,0} \neq 0$ est alors automatique), et construisons un nombre réel $c_0'(\nabla)$ tel que pour tout L et tout $T_0 \leq nL - c_0'(\nabla)$, les éléments $ev_{0,t} := ev_t; 0 \leq t \leq T_0 - 1$ de $\Gamma(L)^*$ sont linéairement indépendants sur K.

Soit $\tau_0 \geq 1$ le plus grand entier tel que les $ev_t; 0 \leq t \leq \tau_0 - 1$ soient linéairement indépendants. Si un élément s de $\Gamma(L)$ s'annule à l'ordre τ_0 (sous entendu : le long de $\hat{\mathcal{W}}_0$), il s'annulera alors à l'ordre $\tau_0 + 1$. Soit L' tel que $\tau_0 + n > n(L' + 1) \geq \tau_0$. Par algèbre linéaire, il existe $\sigma \in \Gamma(L')$ non nul annulant les $ev_t, t \leq \tau_0 - 1$. Pour tout k tel que $L' + k\deg(\nabla) \leq L$, $(\nabla^*_{Q\partial})^k(\sigma)$ est dans $\Gamma(L)$. Notons k_1 la partie entière de $(L - L')/\deg(\nabla)$.

Comme $Q(0) \neq 0$, on déduit de la relation $\forall s, (\nabla^*_{Q\partial}s).\mathcal{E} = Q\partial(s.\mathcal{E})$ et de l'hypothèse sur ev_{τ_0} que σ s'annule à l'ordre $\tau_0 + k_1$. Le lemme de zéros donne $\tau_0 + k_1 \leq nL' + c(\nabla)$. Ainsi, $k_1 \leq c(\nabla)$, et $\tau_0 \geq nL' \geq n\big(L - (c(\nabla) + 1)\deg(\nabla)\big)$. Pour $T_0 < nL - c_0'(\nabla)$ avec $c_0'(\nabla) = n(c(\nabla) + 1)\deg(\nabla)$, on peut donc affirmer que les $ev_t; 0 \leq t \leq T_0 - 1$ sont linéairement indépendants.

Si 0 est une singularité de ∇, on modifie l'argument de la façon suivante, qui revient à tordre le fibré $\mathcal{M}$ par $\mathcal{O}(-c\theta)$. Soit $Q = z^\theta Q_1$, avec $Q_1(0) \neq 0$, et soit c un entier $\geq 2c(\nabla)$. Pour $\tau_0 - c\theta \sim nL'$, il existe $\sigma \in \Gamma(L')$ s'annulant à l'ordre $\tau_0 - c\theta$ le long de $\hat{\mathcal{W}}_0$. Alors, $\tilde{\sigma} := z^{c\theta}\sigma \in \Gamma(L' + c\theta)$ s'annule à l'ordre τ_0. Pour tout $k < c$ tel que $L' + c\theta + k\deg(\nabla) \leq L$, $(\nabla^*_{Q_1\partial})^k(\tilde{\sigma})$ est dans $\Gamma(L)$. Il en découle comme supra que $\tilde{\sigma}$ s'annule à l'ordre $\tau_0 + k$, σ à l'ordre $\tau_0 - c\theta + k$, donc $\tau_0 - c\theta + k < nL' + c(\nabla)$. On en déduit en notant k_1 la partie entière de $(L - L' - c\theta)/deg(\nabla)$ que $\inf(c, k_1) \leq c(\nabla)$, donc $k_1 < c(\nabla)$. Ainsi, $\tau_0 > nL' + c\theta > nL - (n-1)c\theta - c_0'(\nabla)$, et $c_0''(\nabla) = 2nc_0'(\nabla)\theta$ convient.

Montrons maintenant que si $rT_1 < nL - c_1'(\nabla)$ avec $c_1'(\nabla) = 3nc(\nabla)deg(\nabla)$, les $ev_{i,t}, i = 1, \ldots, r, 0 \leq t \leq T_1 - 1$ sont linéairement indépendants. Soit $\tau_1 \geq 1$ le plus grand entier tel que tous les $ev_{i,t}, i = 1, \ldots, r, t < \tau_1$ sont indépendants, tandis que (disons) ev_{1,τ_1} en dépende. Soit encore $c > 2c(\nabla)$ et soit L' tel que $r\tau_1 + (r-1)c \sim nL'$. Par algèbre linéaire, il existe $\sigma \in \Gamma(L')$ non nul s'annulant à l'ordre $\tau_1 + c$ le long des $\mathcal{Z}_i, i = 2, \ldots, r$, et à l'ordre τ_1 le long de $\mathcal{Z}_1$. Par l'argument supra, σ s'annule à l'ordre $\tau_1 + k$ le long de *tous* les $\mathcal{Z}_i$ tant que $k < c$ et que $L' + k\deg(\nabla) \leq L$, et cela impose sous ces contraintes que $r(\tau_1 + k) \leq nL' + c(\nabla) \leq r\tau_1 + (r-1)c + c(\nabla)$, d'où $k \leq (1 - \frac{1}{2r})c < c$. Ainsi, $[\frac{L-L'}{deg(\nabla)}] \leq (2 - \frac{1}{r})c(\nabla)$ et $r\tau_1 \geq nL - c_1'(\nabla)$.

Le cas général enfin: on va vérifier que la constante $\hat{c}(\nabla) := \sup(c_1'(\nabla), c_0'(\nabla))$ répond à la question. Si $T_0 + rT_1 \leq nL - \hat{c}(\nabla)$, on a en particulier $rT_1 \leq nL - c_1'(\nabla)$, donc les rT_1 évaluations de $\Gamma(L)$ le long de $\hat{\mathcal{W}}_1$ sont linéairement indépendantes. Soit τ_0 le plus grand entier tel que les $ev_{0,t}, t \leq \tau_0 - 1$ les complètent en un système libre. Supposons pour simplifier que 0 n'est pas une singularité (suivre le 2e pas pour le cas général). Pour $c = 2c(\nabla)$ et L' tel que $\tau_0 + r(T_1 + c) \sim nL'$, on construit $\sigma \in \Gamma(L')$ s'annulant à l'ordre $T_1 + c$ le long de $\hat{\mathcal{W}}_1$, à l'ordre τ_0 le long de $\hat{\mathcal{W}}_0$. D'où pour tout $k < c$ tel que $L' + k\deg(\nabla) \leq L$, $\tau_0 + k + r(T_1 + c) \leq nL' + c(\nabla)$, imposant $k \leq c(\nabla)$, et $\tau_0 + rT_1 \geq nL - c_0'(\nabla)$.

Remarque 3. Je préfère donner au lemme la dénomination “annulation” (*vanishing lemma*) plutôt qu'“interpolation”, d'abord pour éviter une confusion avec le lemme d'interpolation du §3, mais surtout parce qu'il exprime l'annulation d'un H^1. Plus précisément, soit $\mathcal{J}_{T_0,T_1}$ le sous-faisceau de $\mathcal{M}^*$ des germes de sections s'annulant à un ordre $\geq T_0$ le long de $\hat{\mathcal{W}}_0$, à un ordre $\geq T_1$ le long de $\hat{\mathcal{W}}_1$. Supposons $\mathcal{M}$ trivial, de sorte que $H^1(\mathcal{M}^*(L)) = 0$ dès que $L > 0$. La suite exacte $0 \to \mathcal{J}_{T_0,T_1} \to \mathcal{M}^* \to \mathcal{M}^*/\mathcal{J}_{T_0,T_1} \to 0$ fournit après tensorisation par $\mathcal{O}(L)$ une suite exacte

$$H^0(\mathcal{J}_{T_0,T_1}(L)) \hookrightarrow H^0(\mathcal{M}^*(L)) \to H^0((\mathcal{M}^*/\mathcal{J}_{T_0,T_1})(L)) \to H^1(\mathcal{J}_{T_0,T_1}(L)) \to 0,$$

où l'on reconnaît en deuxième terme $\Gamma(L)$, en troisième l'espace $Ev(T_0, T_1) \simeq \mathbf{C}^{T_0} \oplus \mathbf{C}^{rT_1}$ (ce toujours sous l'hypothèse $\mathcal{E}(0) \neq 0$, qui signifie que $\hat{\mathcal{M}}_0/\hat{\mathcal{O}}_0 \otimes \hat{\mathcal{W}}_0$ est sans torsion – condition automatiquement satisfaite au point ordinaire 1), et en deuxième flèche notre application ϕ.

Un fibré de degré négatif ne peut être engendré génériquement par ses sections globales. En itérant l'action de la connexion sur les sections de $\mathcal{J}_{T_0',T_1'}(L')$, on en déduit des preuves plus naturelles du lemme de zéros et du lemme d'annulation. Par exemple, dans le cas élémentaire où ∇ est irréductible et avec $c(\nabla) = (2\deg(\nabla) + 1)n^2 = \hat{c}(\nabla)$:

- le fibré $\mathcal{J}_{T_0-\theta n,T_1-n}(L + n\deg(\nabla))$ est de degré $\leq nL - T_0 - rT_1 + c(\nabla)$; donc si cette expression est négative, $H^0(\mathcal{J}_{T_0,T_1}(L))$ doit s'annuler: de façon équivalente, ϕ est injective;
- dès que $nL - T_0 - rT_1 \geq \hat{c}(\nabla)$, le fibré $\mathcal{J}_{T_0+\theta n,T_1+n}(L - n\deg(\nabla))$ a une section non nulle, donc $\mathcal{J}_{T_0,T_1}(L)$ est engendré génériquement par son H^0, et son H^1 s'annule: de façon équivalente, ϕ est surjective.

Même dans le cas général, la situation est plus simple que pour les groupes algébriques [F]: si nos fibres sont de dimension n plutôt que 1, notre base, de dimension 1, ne nécessite pas de recourir à la théorie de l'intersection; elle devrait de plus permettre de passer par dualité de Serre de l'un à l'autre des énoncés.

Remarque 4 (en guise de conclusion). Le fait que le théorème de Siegel–Shidlovsky porte sur des connexions à singularités irrégulières a limité son impact en géométrie algébrique. Était-ce dans cette direction qu'allait la suggestion proposée par P. Cartier à S. Lang, et dont personne ne se souvient (voir [W], §1) ? On peut en tous cas espérer que l'ébauche esquissée par Deligne d'une théorie de Hodge irrégulière lui fournira de nouveaux champs d'applications. Ainsi (voir [D], I.§4 ou II.§8), les valeurs $J_0(\lambda) = \mathrm{Res}_0(e^{t+\frac{\lambda}{t}}dt)$ de la fonction de Bessel (resp. de sa dérivée $J_1(\lambda)$) s'interprètent dans cette théorie comme des quotients par $2i\pi$ de périodes de formes de première (resp. deuxième) espèce généralisées le long d'un cycle invariant sous la monodromie, et le théorème de Siegel montre que—comme pour une courbe elliptique définie sur $\mathbf{Q}(\lambda)$—*ces périodes sont linéairement indépendantes sur* $\overline{\mathbf{Q}}$ *quand* λ *est un nombre algébrique*. Leur transcendance, et celle de leurs autres périodes, ne sont pas encore connues.

References

[A1] Y. André: *G-fonctions et transcendance*; J. Crelle, 476, 1996, 95–125.

[A2] Y. André: *Séries Gevrey de type arithmétique, I et II*; Annals of Maths, 151, 2000, 705–740 et 741–756.

[B1] D. Bertrand: *On André's proof of the Siegel–Shidlovsky theorem*; Publ. Keio U., 27, 1999, 51–63.

[B2] D. Bertrand: *Le théorème de Siegel–Shidlovsky revisité* (version préliminaire); Prepubl. IMJ, 370, 2004, http://people.math.jussieu.fr/~preprints/index-2004.html

[Be] F. Beukers: *A refined version of the Siegel-Shidlovskii theorem*; ArXiv.math.NT/0405549, 2004. Annals of Maths 163, 2006, 369–379.

[BR] J-P. Bézivin, P. Robba: *A new p-adic method for proving irrationality and transcendence results*; Annals of Maths 129, 1989, 151–160.

[Bo] J-B. Bost: *Germs of analytic varieties in algebraic varieties: canonical metrics and arithmetic algebraization theorem.*; in *Geometric aspects of Dwork's theory*, A. Adolphson ed., W. de Gruyter, 2004, vol. 2, 371–418.

[D] P. Deligne: *Théorie de Hodge irrégulière*; I (1984); II (2006); in *Correspondance Deligne-Malgrange-Ramis*, Documents mathématiques 5, SMF, 2007.

[F] S. Fischler: *Interpolation on algebraic groups*, Compo. math. 141, 2005, 907–925.

[H] L. Habsieger: *A few remarks related to the four exponentials conjecture*; Publicationes Mathematicae Debrecen, 70, 2007, fasc. 1–2 (11).

[L1] S. Lang: *A transcendence measure for E-functions*; Mathematika 9, 1962 157–161.

[L2] S. Lang: *Introduction to transcendental numbers*; Addison-Wesley, 1966.

[La] M. Laurent: *Linear forms in two logarithms and interpolation determinants*; Acta Arithm., 66, 1994, 181–199.

[LP] M. Laurent, D. Poulakis: *On the global distance between two algebraic points on a curve*; J. Number Th., 104, 2004, 210–254.

[R] J-P. Ramis: *Théorèmes d'indices Gevrey pour les équations différentielles ordinaires*; Memoirs of the AMS, 296, (1984), 95p.

[S] A. Sert: *Une version effective du théorème de Lindemann-Weierstrass par les déterminants d'interpolation*; J. Number Th., 76, 1999, 94–119.

[Sh] A. Shidlovsky: *Transcendental Numbers*, W. de Gruyter, 1989.

[W] M. Waldschmidt: *Les contributions de Serge Lang à la théorie des nombres transcendants*, Gazette des Mathématiciens, Soc. Math. France, 108, avril 2006, 35–46.

Some aspects of harmonic analysis on locally symmetric spaces related to real-form embeddings

Eliot Brenner and Andrew Sinton

Abstract Let $G = \mathrm{SO}_3(\mathbb{C})$, $\Gamma = \mathrm{SO}_3(\mathbb{Z}[\mathbf{i}])$, $K = \mathrm{SO}(3)$, and let X be the locally symmetric space $\Gamma\backslash G/K$. In this paper, we present a relationship between the heat kernel on $\mathrm{SL}_3(\mathbb{C})$ and $\mathrm{SO}_3(\mathbb{C})$. We write down explicit equations defining a fundamental domain for the action of Γ on G/K. The fundamental domain is well adapted for studying the theory of Γ-invariant functions on G/K. We write down equations defining a fundamental domain for the subgroup $\Gamma_{\mathbb{Z}} = \mathrm{SO}(2,1)_{\mathbb{Z}}$ of Γ acting on the symmetric space $G_{\mathbb{R}}/K_{\mathbb{R}}$, where $G_{\mathbb{R}}$ is the split real form $\mathrm{SO}(2,1)$ of G and $K_{\mathbb{R}}$ is its maximal compact subgroup $\mathrm{SO}(2)$. We formulate a simple geometric relation between the fundamental domains of Γ and $\Gamma_{\mathbb{Z}}$ so described. Both the formula for the heat kernel and the fundamental domains are designed to aid in a detailed study of the spectral theory of X and the embedded subspace $X_{\mathbb{R}} = \Gamma_{\mathbb{Z}}\backslash G_{\mathbb{R}}/K_{\mathbb{R}}$.

Key words Kleinian groups • fundamental domains • heat kernel • locally symmetric spaces

Mathematics Subject Classification (2010): 11F55 (Primary), 11F72, 11H55, 11Y35 (Secondary)

E. Brenner (✉)
Courant Institute of Mathematical Sciences, 251 Mercer Street New York, N.Y. 10012
e-mail: epb254@cs.nyu.edu

A. Sinton
NDS Technologies Israel
e-mail: sinton@gmail.com

D. Goldfeld et al. (eds.), *Number Theory, Analysis and Geometry: In Memory of Serge Lang*, DOI 10.1007/978-1-4614-1260-1_4,

General introduction

While Serge Lang and Jay Jorgenson were working towards a derivation of a theta relation and the Selberg zeta function associated to $\mathrm{SL}_n(\mathbb{Z}[\mathbf{i}])\backslash\mathrm{SL}_n(\mathbb{C})/\mathrm{SU}(n)$ from the Selberg trace formula applied to the heat kernel (see the books [20], [18], [17], and [19]) they envisioned a similar "ladder" of trace formulas and zeta functions associated to the groups $\mathrm{SO}_n(\mathbb{Z}[\mathbf{i}])\backslash\mathrm{SO}_n(\mathbb{C})/\mathrm{SO}_n(\mathbb{R})$ (see the introduction to [19]). They suggested to us that we attempt to complete that project, starting with the base case of $\mathrm{SO}_3(\mathbb{Z}[\mathbf{i}])\backslash\mathrm{SO}_3(\mathbb{C})/\mathrm{SO}_3(\mathbb{R})$. This paper consists of two distinct but interrelated parts of this project that have been completed to date. Section 1 constructs the heat gaussian (from which the heat kernel may be derived in an entirely standard way, as on p. 48 of [19]) on $\mathrm{SO}_3(\mathbb{C})$. In analogy with the $\Gamma\backslash\mathrm{SL}_2(\mathbb{C})/\mathrm{SU}(2)$ project, this section has the same purpose (though it follows a different method) as §2.5 of [19]. Section 2 constructs a fundamental domain for the action of the discrete group. In particular, it investigates the group structure and a fundamental domain for $\mathrm{SO}_3(\mathbb{Z}[\mathbf{i}])$ acting on the symmetric space. Also, considered in this section is the "real form" $\mathrm{SO}(2,1)_{\mathbb{Z}}\backslash\mathrm{SO}(2,1)/\mathrm{SO}_3(\mathbb{R})$ (shown to be a hyperbolic surface with one cusp), which can be considered the base case of another ladder $\mathrm{SO}(p,q)_{\mathbb{Z}}\backslash\mathrm{SO}(p,q)/\mathrm{SO}(p+q)$ of locally symmetric spaces. This section corresponds in function to Chapter VI and parts of Chapter IX of [19], although again the method is different, relying directly on an explicit isomorphism of $\Gamma = \mathrm{SO}(3,\mathbb{Z}[\mathbf{i}])$ with a particular Kleinian group (described in §2.1) rather than a direct construction. For much more extensive comments concerning the motivation, context, methods, and possible continuation of the work towards the theta relation, see the introductions preceding each of the two parts.

Acknowledgements We would like to thank the editors for providing this opportunity to honor Serge's memory and legacy, and to thank Jay Jorgenson in particular with whom we spent much time thinking about the ideas related to this paper. The first author gratefully acknowledges the financial support of the Center for Advanced Studies in Mathematics at Ben-Gurion University, which made this collaboration possible.

1 Obtaining the heat kernel by integration over a normal bundle

We take as our starting point knowledge of the heat gaussian via explicit formulas, on $\mathrm{SL}_n(\mathbb{C})/\mathrm{SU}(n)$, for which see [20] in general and [19], §2.5, for the special case of $n = 2$. Our viewpoint will be that one can obtain the heat kernel of the noncompact symmetric space $\mathrm{SO}_n(\mathbb{C})/\mathrm{SO}(n)$ by embedding $\mathrm{SO}_n(\mathbb{C})$ in the larger group $\mathrm{SL}_n(\mathbb{C})$ and integrating the heat gaussian on $\mathrm{SL}_n(\mathbb{C})/\mathrm{SU}(n)$ along certain normal fibers.

Consider the following general setup. Let $G_1 \subset G$ be noncompact real reductive Lie groups. It is possible to choose a maximal compact subgroup K of G so that

$K_1 := G_1 \cap K$ is a maximal compact subgroup of G_1. In our setup we also assume there exists another, not necessarily reductive, group $H \subset G$ such that G has a smooth decomposition $G = HG_1K$. This then yields a product decomposition on the level of symmetric spaces $G/K \simeq Y \times G_1/K_1$, where Y is some quotient of H. Projecting onto the second component, one obtains a projection $\pi : G/K \longrightarrow G_1/K_1$. Viewing the heat kernel as a probability measure μ_t on G/K, the question is whether the push-forward of μ_t under π is equal to the heat kernel on G_1/K_1. If one prefers to think of the heat kernel (or more precisely, heat gaussian) as a family of functions $\mathbf{h}_t$, then the push-forward becomes "integrating over the normals." More precisely, define the function $\mathbf{h}_t^H$ on G_1/K_1 by

$$\mathbf{h}_t^H(g_1) = \int_H \mathbf{h}_t(hg_1)dh. \tag{1}$$

The question is under what conditions on G, G_1 is $\mathbf{h}_t^H$ the heat gaussian on G_1.

There are two main classes of examples where one does in fact obtain the heat kernel on G_1 in this fashion. The specific example $SO_3(\mathbb{C}) \subset SL_3(\mathbb{C})$ in this paper does not belong to either class and seems to be a low-dimensional coincidence. See comments at the end of the paper for further discussion.

The simplest class is that in which G_1 is a Levi component of a parabolic subgroup of G and H is the nilradical of the parabolic. The push-forward is realized as an integral over the orbits of the nilpotent group H that are perpendicular to G_1/K_1. Note that the case of $(2n-1)$-dimensional hyperbolic space sitting inside $2n$-dimensional hyperbolic space has a similar, though distinct, setup.

The second main class is what Flensted-Jensen worked on, where G_1 is a real form of G and $H = K_1^{\mathbb{C}}$, the complexification of a maximal compact subgroup of G_1. Here the push-forward becomes integrals over spaces traced by the H-orbits of points in G_1/K_1. Except for the H-orbit through eK_1, which is isomorphic to the symmetric space H/K_1, these H-orbits are not composed of geodesics. They are rather built of paths that are perpendicular to G_1/K_1 and that maintain a constant distance from symmetric space which is the H-orbit through the identity. In this case there is a subtlety as to how the above integral describes the heat Gaussian: the value of the heat Gaussian at $g_1g_1^* \in G_1/K_1$ is actually determined by the H-orbit through the coset determined by g_1.

The case of $SO_3(\mathbb{C}) \subset SL_3(\mathbb{C})$ examined here is structurally very similar to Flensted-Jensen's case. The decomposition he uses, $G = HG_1K$, depends only on G_1 being the fixed-point set of an involution of G, and not necessarily a real form. This naturally led to the hope of expanding Flensted-Jensen's results to encompass any $G_1 \subset G$ which is the fixed point set of an involution. Unfortunately, it does not seem to work in general, with the present case $SO_3(\mathbb{C}) \subset SL_3(\mathbb{C})$ being a low-dimensional exception. It is noteworthy that in this case, it is the value of the heat Gaussian at $(g_1g_1^*)^2 \in G_1/K_1$ that is determined by the H-orbit through the coset determined by g_1. For further discussion of this phenomenon, see comments at the end of the paper.

Finally, we mention that in both of the above cases, there is also a functorial action in the other direction: the pull-back under π takes eigenfunctions of the invariant differential operators on G_1/K_1 to eigenfunctions on G/K. In the case that G_1 comes from a parabolic subgroup this corresponds to parabolic induction of representations. When G_1 is a real form of G there seems to be some kind of base-change principle at work. It is not yet clear whether the new example here has some similar interpretation, perhaps being related to the symmetric square lifting.

A closed-form formula for the heat kernel on $SO_3(\mathbb{C})$ can already be deduced from the formulas of Gangolli for the heat kernel on a complex Lie group: cf. [12]. The point of this paper is to show how a relationship between the heat kernel on classical groups may arise naturally through the push-forward of the heat kernel under projection along orbits of the linear group and to place this fact in the context of similar phenomena for different projections. Though Serge's original hope of obtaining all heat kernels of symmetric spaces as the "homomorphic image" of the heat kernel of $SL_n(\mathbb{C})$ seems to be too much to ask for, it would be nice to have a more fundamental understanding of precisely what specific conditions are needed on a double coset decomposition $G = HG_1K$ to ensure that the resulting projection does map the heat kernel of G to the heat kernel of G_1.

1.1 Setup and notation

Let $G = SL_3(\mathbb{C})$ and let $\theta(g) = (g^*)^{-1}$ define the standard Cartan involution. Then $K = G^\theta = SU(3)$ is a maximal compact subgroup of G. We let A and N be the subgroups of positive real diagonal matrices and upper triangular unipotent matrices respectively. As is well known, G has the Iwasawa decomposition $G = NAK$, which is unique, and the Cartan decomposition $G = KAK$, in which the A component is unique up to permutation of its entries.

Let J be the symmetric matrix given by antidiag$(1, -1, 1)$. Then J defines a conjugation of G by $g^\sigma = J\bar{g}J^{-1}$, where bar denotes complex conjugation of each entry of g. Note that $J = J^{-1}$. We then obtain a third involution, τ, by composing σ and θ: $g^\tau = g^{\sigma\theta} = J\,{}^t g^{-1} J$. The group fixed by τ, $G_1 = G^\tau$, is then a J conjugate of the standard embedding of $SO_3(\mathbb{C})$. The advantage of working with this conjugate is that its Iwasawa decomposition is compatible with the standard Iwasawa coordinates of G. On the Lie algebra level we have $\tau(X) = -J\,{}^t X J$ and $\mathfrak{g}_1$ is the set of all matrices of the form

$$\begin{pmatrix} a & b & 0 \\ d & 0 & b \\ 0 & d & -a \end{pmatrix}.$$

G_1 has Iwasawa and Cartan decompositions just like G, where N_1, A_1, and K_1 are obtained by intersecting N, A, and K with G_1.

Let $H = G^{\sigma}$, which is isomorphic to $SL_3(\mathbb{R})$. There is a generalized Cartan decomposition $G = HA_1K$ in which the A_1 component is unique up to the action of W_1, the Weyl group of $SO_3(\mathbb{C})$. This result is due to Flensted-Jensen [9], though the main content can already be found in Mostow [22]. Later, we will use this decomposition to compare the Casimir operators of G and G_1. First we need to decompose $\mathfrak{g}$ and $\mathfrak{g}_1$ under the bracket action of $\mathfrak{a}_1$.

It is immediate that $\mathfrak{a}_1$ acts on $\mathfrak{n}_1$, the Lie algebra of N_1, by scalar multiplication. If $Y_1 \in \mathfrak{a}_1$ is the matrix $\operatorname{diag}(t, 0, -t)$, then for any element $Z \in \mathfrak{n}_1$, $[Y, Z] = \alpha_1(Y)Z = tZ$. Furthermore, the corresponding eigenspace is of real dimension 2.

Decomposing all of $\mathfrak{n}$, the Lie algebra of N, under $\mathfrak{a}_1$ is only slightly more complicated. The eigencharacter α_1 corresponds to an eigenspace of real dimension 4, and the eigencharacter $\alpha_2 = 2\alpha_1$ has eigenspace of real dimension 2. Observe that of the α_1 eigenspaces, two are fixed by τ, and two have eigenvalue -1 under τ. For α_2, both eigenspaces have eigenvalue -1 under τ.

1.2 *Integral formulas*

For the main result, we will need to fix a normalization of measures and obtain an integral formula for the generalized Cartan decomposition above. We fix the invariant measure to be 1 on all nondiscrete compact groups and spaces. Let B (resp. B_1) be a positive multiple of the Killing form of $\mathfrak{g}$ (resp. $\mathfrak{g}_1$). Then $B|_{\mathfrak{g}_1} = c'B_1$. The form B (resp. B_1) is positive definite on $\mathfrak{a}$ (resp. $\mathfrak{a}_1$) and we use the exponential map to push forward the Euclidean measure on $\mathfrak{a}$ (resp. $\mathfrak{a}_1$) to a Haar measure da on A (resp. da_1 on A_1). We now define the Haar measure dg on G such that

$$\int_G f(g)dg = \int_K \int_A \int_K f(kak')J(a)dk\, da\, dk'$$

for any continuous f with compact support. Here, $J(a)$ is the standard Jacobian of the Cartan decomposition $G = KAK$, cf. [15] for example. We normalize the measure on G_1 in the same way and note that in this case, the Jacobian J_1 is simply

$$J_1(a_1) = \sinh^2 \alpha_1(\log a_1).$$

The exponent 2 comes from the fact that the α_1-eigenspace has dimension 2.

Fix an arbitrary Haar measure dh on H. Flensted-Jensen computed the Jacobian for the generalized Cartan decomposition above; cf. [10] Theorem 2.6. In our specific case, the relevant integral formula becomes the following.

Proposition 1.1. *Given the above normalization of measures, there exist a function J' and a constant c such that for all continuous functions on G with compact support we have:*

1. $\int_G f(g)dg = \int_H \int_{A_1} \int_K f(ha_1k)J'(a_1)dk da_1 dh.$
2. $J'(a_1) = cJ_1(a_1^4).$

Proof. The first statement is standard abstract nonsense following from the smooth decomposition $G = HA_1K$. That the Jacobian J' does not depend on the H and K coordinates is a simple consequence of the G-bi-invariance of the Haar measure dg. The specific form of J' is immediate from [10], Theorem 2.6, the fact that $\alpha_2 = 2\alpha_1$, and basic hyperbolic trigonometric identities. Note that the constant c will depend on the specific measure dh that was selected.

1.3 Differential operators and the heat kernels

With measures fixed, all that is needed to define the heat kernel is the second-order Casimir operator. Given an element $Z \in \mathfrak{g}$, define the left-invariant differential operator

$$\tilde{Z}(f)(g) = \partial_t f\left(ge^{tZ}\right)|_{t=0},$$

where f is any smooth function on G. The form B introduced above is nondegenerate on $\mathfrak{g}$ and positive definite on $\mathfrak{a}$, so we can use it to identify $\mathfrak{a}$ and its dual. In particular, we define Y_α to be the unique element in $\mathfrak{a}$ such that $B(Y_\alpha, X) = \alpha(X)$ for every $X \in \mathfrak{a}$. We remind the reader that the second-order Casimir operator of $\mathrm{SL}_3(\mathbb{C})$, ω, is defined by

$$\omega = \sum_{i=1}^{\dim \mathfrak{g}} \tilde{Z}_i' \tilde{Z}_i, \tag{2}$$

where $\{Z_i\}$ is any basis of $\mathfrak{g}$ and $\{Z_i'\}$ is the dual basis under B. The second-order Casimir of $\mathrm{SO}_3(\mathbb{C})$, ω_1, is obtained by inserting subscripts 1 everywhere in (2), taking care to use the form B_1 in calculating the dual basis. Note that there is some positive constant c' such that $B|_{\mathfrak{g}_1} = c' B_1$.

For our main result, we will compare the direct images of ω and ω_1 to A_1. Suppose we are given a function f_1 on A_1 that is invariant under the action of W_1. Using the generalized Cartan decomposition $G = HA_1K$, we extend f_1 to a function on G by defining $\tilde{f}_1(ha_1k) = f_1(a_1)$. Given a differential operator D which is right K- and left H-invariant, the function $D\tilde{f}_1$ is again a W_1 invariant function on A_1. Given an H-left-invariant, K-right-invariant differential operator on G, we can now project it to a differential operator on A_1. Given such a differential operator D, we define its A_1 projection $\tilde{D}$ by $\tilde{D} f_1(a_1) = D\tilde{f}_1(a_1)$, which is then a differential operator on A_1; cf. [25] for a more formal treatment.

Applying this to ω, it is a result of van den Ban and Schlichtkrull [25] that the direct image of ω to A_1 is

$$\tilde{\omega} = \frac{1}{c'}\left(\omega_{A_1} + 2(\coth\alpha_1 + \tanh\alpha_1)\tilde{Y}_{\alpha_1} + 2(\tanh\alpha_2)\tilde{Y}_{\alpha_2}\right), \tag{3}$$

where ω_{A_1} is the Casimir (i.e., Laplacian) of A_1 determined by the form B_1 and the Y_{α_i} are the vectors dual to α_i under B_1 for $i = 1, 2$. Since $\coth x + \tanh x = 2\coth 2x$ and $2\tilde{Y}_{\alpha_1} = \tilde{Y}_{\alpha_2}$, we can simplify to obtain

$$\tilde{\omega} = \frac{1}{c'}(\omega_{A_1} + 8\coth 4\alpha_1 \tilde{Y}_{\alpha_1}). \tag{4}$$

We remind the reader of the standard result, cf. [15], that the direct image of ω_1 to A_1 using the $G_1 = K_1 A_1 K_1$ decomposition (i.e., the radial component of ω_1) is given by

$$\tilde{\omega_1} = \omega_{A_1} + 2\coth \alpha_1 \tilde{Y}_{\alpha_1}. \tag{5}$$

We can now state the main theorem.

Theorem 1.2. *Let f be a K_1-bi-invariant function on G_1 and F be an H-left-invariant, K-right-invariant function on G such that for $a_1 \in A_1$,*

$$F(a_1) = f(a_1^4).$$

Then

$$(\omega F)(a_1) = \frac{16}{c'}(\omega_1 f)(a_1^4).$$

Proof. This is essentially freshman calculus. It is easier to work on $\mathfrak{a}_1$, with $a_1 = e^{Y_1}$ and viewing F and f as functions on $\mathfrak{a}_1$. Under this transformation, ω_{A_1} becomes the standard Laplacian on $\mathfrak{a}_1$, which we denote by $\Delta_{\mathfrak{a}_1}$, and $\tilde{Y}_{\alpha_1}$ becomes the standard directional derivative in the Y_{α_1} direction.

As a mental aid for the computation, we write $F = f \circ 4$, where "4" means multiply the variable by 4. Evaluating $(\omega F)(Y_1)$ term by term, we get

$$(\Delta_{\mathfrak{a}_1} F)(Y_1) = (\Delta_{\mathfrak{a}_1} f \circ 4)(Y_1) = 16(\Delta_{\mathfrak{a}_1} f)(4Y_1), \tag{6}$$

$$8\coth 4\alpha_1(Y_1)(\partial_{Y_{\alpha_1}} F)(Y_1) = 32\coth 4\alpha_1(Y_1)(\partial_{Y_{\alpha_1}} f)(4Y_1). \tag{7}$$

Combining (4), (6) and (7) and using the fact that $4\alpha_1(Y_1) = \alpha_1(4Y_1)$, the theorem is immediate.

This relationship between Casimir operators on G and G_1 leads immediately to a relationship between the solutions of their respective heat equations. We remind the reader that the heat Gaussian $\mathbf{h}_t$, $t > 0$ on G is the unique family of positive K-bi-invariant functions such that $\omega\mathbf{h}_t = \partial_t \mathbf{h}_t$, with total integral 1 for all t, and such that $\mathbf{h}_t$ approaches the Dirac distribution as t approaches zero.

Theorem 1.3. *Let $\mathbf{h}_t$ be the heat Gaussian of G. Then the restriction to A_1 of the heat Gaussian of G_1 is given by*

$$\mathbf{h}_{1,t}(a_1^4) = \frac{4}{c}\int_H \mathbf{h}_{\frac{tc'}{16}}(ha_1)dh.$$

Proof. We first verify that $\mathbf{h}_{1,t}$ satisfies the heat equation on G_1. Let

$$g_t(g) = \int_H \mathbf{h}_{\frac{tc'}{16}}(hg)dh$$

denote the left H-average of $\mathbf{h}_{\frac{tc'}{16}}$ considered as a function on all of G. Then

$$(\partial_t \mathbf{h}_{1,t})(a_1^4) = (\partial_t\ g_t)(a_1) = \int_H \partial_t \mathbf{h}_{\frac{tc'}{16}}(ha_1)dh = \frac{c'}{16}\int_H \omega \mathbf{h}_{\frac{tc'}{16}}(ha_1)dh$$
$$= \frac{c'}{16}\omega \int_H \mathbf{h}_{\frac{tc'}{16}}(ha_1)dh = (\omega \mathbf{h}_{1,t})(a_1^4).$$

Let F and f be as in Theorem 1.2. To see that the total integral of $\mathbf{h}_{1,t}$ is equal to 1 for all t, we use the integration formulas associated to the Cartan and generalized Cartan decompositions. Using the fact that $\alpha_2 = 2\alpha_1$, we see that

$$\int_G F(g)dg = c\int_{\mathfrak{a}_1} F(e^{Y_1})\sinh^2\alpha_1(4Y_1)dY_1 = \frac{c}{4}\int_{\mathfrak{a}_1} f(e^{Y_1})\sinh^2\alpha_1(Y_1)dY_1$$
$$= \frac{c}{4}\int_{G_1} f(g_1)dg_1.$$

This proves the theorem. □

We have deliberately left the constants c and c' undetermined so that one can see directly how their choice of normalizations will affect the relationship between heat gaussians. However, the exponent of 4 on the left-hand side of Theorem 1.3 is structural and cannot be modified. Note that the number 4 also makes an appearance in Proposition 1.1 that is critical to the above proof.

The number 4 appears in another context that we have not yet mentioned. Let ρ (resp. ρ_1) be the sum of all the characters occurring in the decomposition of $\mathfrak{a}$ (resp. $\mathfrak{a}_1$) acting on $\mathfrak{n}$ (resp. $\mathfrak{n}_1$). Then

$$\rho|_{\mathfrak{a}_1} = 4\rho_1.$$

Though it is not clear exactly how these two coefficients are related, it is more than coincidence. Indeed, in the case of Flensted-Jensen's work, where G_1 is a real form of G, the exponent of a_1 occurring in the theorem analogous to Theorem 1.3 above (cf. [9] Theorem 6.1) is 2. There is a similar relationship between ρ's given by

$$\rho|_{\mathfrak{a}_1} = 2\rho_1.$$

Finally, in the case that G_1 arises as the semisimple part of a parabolic subgroup of G, no exponent is needed on a_1, and we have

$$\rho|_{\mathfrak{a}_1} = \rho_1.$$

It was stated in the introduction that it is not possible to extend Flensted-Jensen's results to the case that G_1 is the fixed-point set of an arbitrary involution. The next-simplest case to check is that of $G_1 = \mathrm{SO}_5(\mathbb{C}) \subset \mathrm{SL}_5(\mathbb{C}) = G$. Though there is a nice decomposition $G = HG_1K$, the associated projection of differential operators does not yield a relationship between Casimir's, similar to that of Theorem 1.2 so one cannot even get started in relating the heat kernels. We also note that in this case, the restriction of ρ to $\mathfrak{a}_1$ is not a multiple of ρ_1. One encounters similar obstructions for other cases.

2 Fundamental Domains for $\mathrm{SO}_3(\mathbb{Z}[\mathbf{i}])\backslash\mathrm{SO}_3(\mathbb{C})/\mathrm{SO}(3)$, and for $\mathrm{SO}(2,1)_{\mathbb{Z}}\backslash\mathrm{SO}(2,1)/\mathrm{SO}(2)$

The first author has undertaken, in Chapter 1 of [6], a generalization of the classical theory of Ford fundamental domains (see §2.2 of [16]) for Fuchsian groups to a wide class of group actions including, in particular, $\Gamma_n = \mathrm{SL}_n(\mathbb{Z}[\mathbf{i}])$ acting on $G_n = \mathrm{SL}_n(\mathbb{C})/\mathrm{SU}(n)$ and $\mathrm{GL}(n,\mathbb{Z})$ acting on $\mathrm{GL}(n,\mathbb{R})/\mathrm{SO}(n)$. The present section carries out this theory in a very explicit manner in one concrete, low-dimensional case. This endeavor lies sufficiently outside the mainstream of the modern theory of automorphic forms that we find it necessary to preface the exposition with some remarks concerning the envisioned uses of such an explicit fundamental domain.

In the case of $\mathrm{GL}(n,\mathbb{Z})$ acting on $\mathrm{GL}(n,\mathbb{R})/\mathrm{SO}(n)$, the fundamental domains obtained in [6] coincide with the F_n studied by D. Grenier in [13] and [14] (allowing for the isomorphism of the symmetric space G/K with the quadratic model P). For this reason, we adopt the terminology *Grenier domains* for the generalized Ford domains. A major theme of Grenier's work in these articles is that the F_n for different n are best considered as part of an inductive scheme, since F_m for $m < n$ appear both in the definition of F_n and in his construction of the Satake compactifications of the locally symmetric space $\mathrm{GL}(n,\mathbb{Z})\backslash\mathrm{GL}(n,\mathbb{R})/\mathrm{SO}(n)$. The base case of Grenier's inductive scheme is (ignoring the center of $\mathrm{GL}(n,\mathbb{R})$) provided by Dirichlet's classical fundamental domain for $\mathrm{SL}_2(\mathbb{Z})$ acting on the upper half-plane. The results of this paper may be viewed as providing the base case for an inductive scheme of the same type corresponding to the sequence of locally symmetric spaces in (93), below. Note that the base case for this "orthogonal" sequence is considerably more complicated than the base case for Grenier's "general linear" sequence.

We take advantage of the well-known isomorphism

$$\mathrm{SL}_2(\mathbb{C})/\{\pm I\} \xrightarrow{\cong} \mathrm{SO}_3(\mathbb{C}),$$

specified at the beginning of §2.1, to identify the lattice $\mathrm{SO}_3(\mathbb{Z}[\mathbf{i}])$ with a group of fractional linear transformations acting on $\mathbb{H}^3$. The concrete outcome of the present section is to state explicitly what this arithmetic subgroup is in explicit matrix terms (Proposition 2.8) and give an appropriate fundamental domain, $\mathscr{F}(\mathscr{G})$, for the natural action on hyperbolic 3-space (Proposition 2.19). The reader who is solely interested in this result may wish simply to read 2.3, which is self-contained, and see the interactive graphics representation of the fundamental domain $\mathscr{F}(\mathscr{G})$, available on the internet at [3].

Following J.S. Friedman, A.B. Venkov, and A. Selberg, one defines the **three-dimensional vector Selberg zeta function associated to a Kleinian group Γ and a unitary representation χ of Γ** by

$$Z_{\Gamma,\chi}(s) = \prod_{\{\gamma\}} \prod_{k=0}^{\infty} \det(1 - \chi(\gamma) N_0(\gamma)^{-s-k}), \text{ for } \mathrm{Re}s \gg 0. \tag{1}$$

In (1), called an "Euler project" expression, $\{\gamma\}$ ranges over Γ-conjugacy classes of primitive hyperbolic elements in Γ and $N_0(\gamma)$ denotes the length of the closed geodesic on $\Gamma\backslash G/K$ corresponding to γ. The meromorphic continuation of $Z_{\Gamma,\chi}$ (or, more precisely, of its logarithmic derivative Z'/Z) to the entire complex domain is closely related to an explicit form of the Selberg trace formula, worked out, for example, in [11] in parallel to [8]. It is of obvious interest to obtain relations between the $Z_{\Gamma,\chi}$ of the members of a pair of lattices (Γ, Γ'), where Γ and Γ' are related in various ways. For example, in the case of (Γ, Γ') a pair of Fuchsian groups, with $\Gamma' \subseteq \Gamma$ and $[\Gamma : \Gamma'] < \infty$ (with $Z_{\Gamma,\chi}$ defined similarly for Fuchsian groups), [26] gave a formula which is loosely called a "factorization formula," because in the case Γ' normal in Γ, it specializes to a bona fide factorization of $Z_{\Gamma',\chi}$ as the product of the Z_{Γ,χ_i}, where χ_i ranges over the irreducible direct summands of $\mathrm{Ind}_{\Gamma'}^{\Gamma}\chi$. In [5], we extended the factorization formula, now called "Artin formalism", to arbitrary pairs (Γ, Γ') of commensurable Kleinian groups, dropping the assumption of normality. It is clear from the definition (1) that in order to apply the Artin formalism to particular pairs, such as $\left(\mathbf{c}^{-1}(\mathrm{SO}_3(\mathbb{Z}[\mathbf{i}])), \mathrm{PSL}_2(\mathbb{Z}[\mathbf{i}])\right)$, one needs to develop concrete understanding of the relations between the hyperbolic conjugacy classes of the groups in question. Proposition 2.8, below, lays the foundations for that study. In §7 of [4] we derive formulas for the Selberg zeta function of $\mathrm{SO}_3(\mathbb{Z}[\mathbf{i}])$ (resp. $\mathrm{SO}(2,1)_{\mathbb{Z}}$) in terms of $\mathrm{PSL}_2(\mathbb{Z}[\mathbf{i}])$ (resp. $\mathrm{PSL}_2(\mathbb{Z})$).

In §2.6, we discuss the application of fundamental domains to the study of a more general class of spectral zeta functions.

Based on the $\mathrm{SL}_n/\mathrm{GL}_n$ examples in the literature, one can speculate on future applications of exact fundamental domains to traditional problems in number theory. Some diverse examples of applications of Grenier's domain for $\mathrm{GL}_n(\mathbb{Z})$, acting on the space of positive definite real matrices P_n, include the proof in [7] of a bound on the first nontrivial eigenvalue of the Laplacian for the case $n = 3$ and the investigations of [23] into the minima of Epstein's zeta function. D. Hejhal's pioneering work in the 1970s on explicit computer calculations of eigenvalues of the Laplacian acting on Maass forms relied on an explicit reduction algorithm for identifying an element of the quotient $\mathrm{SL}_2(\mathbb{Z})\backslash\mathbf{H}^2$ with a point in the Dirichlet domain. As part of ongoing renaissance of this field of explicit computer computations related to automorphic forms (for a broad survey of which see the speakers and talks of the 2009 Séminaire de Mathématiques Supérieures, "Automorphic Forms and L-Functions: Computational aspects", [24]), several generalizations of this explicit reduction algorithm have been developed and more will be needed. At the moment the subject of explicit computation of automorphic forms for groups of higher-rank remains in its infancy. It seems likely that as the detailed study of automorphic functions on quotients of $\mathrm{SO}_n(\mathbb{C})$ and its real forms becomes more developed, the exact fundamental domains, which the present paper specifies in the "base case" $n = 2$, will play a large role in investigating certain zeta functions associated to these arithmetic quotients.

We mention the relation of Propositions 2.8, 2.19, and 2.27, below, to some results already in the literature. First, M. Babillot, in Lemma 3.2 of [1], constructs a fundamental domain for $\mathrm{SO}(2,1)_{\mathbb{Z}}$ acting naturally on the hyperboloid of one sheet. The method there bypasses results like Propositions 2.8 and 2.19 by embedding $\mathrm{SO}(2,1)_{\mathbb{Z}}$ as a subgroup of a triangle group of index two. The fundamental domain so obtained is used to give a constructive proof that $\mathrm{SO}(2,1)_{\mathbb{Z}}$ acts with finite covolume, so that a general theorem can be applied to solve a lattice-point counting problem. Also, there is a well-developed theory of *splines*, which are models for the arithmetic quotients of $\mathbb{Q}$-rank-one groups, in a way different from, but related to, (Grenier) fundamental domains. For a recent treatment with a general existence theorem and references, see [27]. It would be interesting (and possibly useful for cohomology calculations of the sort undertaken in [28]) to determine precisely the relation of "duality" that seems to exist between the splines and Grenier fundamental domains. However, this is more relevant to higher rank, and therefore, belongs more to the continuation of the study undertaken in [6] than to the study at hand.

The verifications of all the principal propositions of the present paper are elementary, though lengthy, and they are not needed for the envisioned applications of the results. Accordingly, many details of proofs are omitted and the interested reader is referred to the electronically archived preprint [2] for them.

2.1 Representation of $\mathrm{SO}_3(\mathbb{Z}[\mathbf{i}])$ *as a lattice in* $\mathrm{SL}_2(\mathbb{C})$

We begin by establishing some basic notational conventions. Let n be a positive integer and $\mathfrak{o}$ a ring. We will use $\mathrm{Mat}_n(\mathfrak{o})$ to denote the set of all $n \times n$ square matrices with coefficients in $\mathfrak{o}$. We reserve use the Greek letters α, and so on, for the elements of $\mathrm{Mat}_n(\mathfrak{o})$, and the roman letters a, b, c, d, and so on, for the entries of the matrices. We will denote scalar mutliplication on $\mathrm{Mat}_n(\mathfrak{o})$ by simple juxtaposition. Thus, if $\mathfrak{o} = \mathbb{Z}[\mathbf{i}]$, $\ell \in \mathbb{Z}[\mathbf{i}]$, and $\alpha \in \mathrm{Mat}_2(\mathbb{Z}[\mathbf{i}])$, then

$$\alpha = \begin{pmatrix} a & b \\ c & d \end{pmatrix} \quad \text{implies} \quad \ell\alpha = \begin{pmatrix} \ell a & \ell b \\ \ell c & \ell d \end{pmatrix}.$$

The letters p, q, r, s will be reserved to denote a quadruple of elements of $\mathfrak{o}$ such that $ps - rq = 1$. In what follows, we normally have $\mathfrak{o} = \mathbb{Z}[\mathbf{i}]$ whenever α is written with entries p through s. Therefore,

$$\alpha = \begin{pmatrix} p & q \\ r & s \end{pmatrix} \in \mathrm{SL}_2(\mathbb{Z}[\mathbf{i}]),$$

unless stated otherwise.

We will denote a conjugation action of a group on a space V by $\mathbf{c}_V$ when the context makes clear what this action is. For example, if H is a linear Lie group and $\mathfrak{h}$ the Lie algebra of H, then we have

$$\mathbf{c}_{\mathfrak{h}}(h)X = hXh^{-1}, \quad \text{for all } h \in H,\ X \in \mathfrak{h}.$$

Note that the morphism $\mathbf{c}_{\mathfrak{h}}(h)$ is the image under the Lie functor of the usual conjugation $\mathbf{c}_H(h)$ on the group level. Using $\mathrm{SL}(V)$ to denote the group of unimodular transformations of a vector space V, it is easy to see that

$$\mathbf{c}_{\mathfrak{h}} : H \to \mathrm{SL}(\mathfrak{h}) \text{ is a Lie group morphism.} \tag{2}$$

Henceforth, whenever H is a group acting on a Lie algebra $\mathfrak{h}$ by conjugation, we will omit the subscript $\mathfrak{h}$. Thus, we define

$$\mathbf{c} := \mathbf{c}_{\mathfrak{h}}$$

when we are in the situation of (2).

Except in §2.2, we will use the notation $G = \mathrm{SO}_3(\mathbb{C})$, $\Gamma = \mathrm{SO}_3(\mathbb{Z}[\mathbf{i}])$. We use B to denote the half-trace form on $\mathfrak{sl}_2(\mathbb{C})$, the Lie algebra of traceless 2×2 matrices. That is,

$$B(X, Y) = \frac{1}{2}\mathrm{Tr}(XY).$$

We use the notation $\beta' = \{X_1', X_2', Y'\}$ for the "standard" basis of $\mathfrak{sl}_2(\mathbb{C})$, where

$$X_1' = \begin{pmatrix} 0 & 1 \\ 0 & 0 \end{pmatrix}, \qquad X_2' = \begin{pmatrix} 0 & 0 \\ 1 & 0 \end{pmatrix}, \qquad \text{and} \quad Y' = \begin{pmatrix} 1 & 0 \\ 0 & -1 \end{pmatrix}. \tag{3}$$

The following properties of B are verified either immediately from the definition or by straightforward calculations.

B1 B is nondegenerate.

B2 Setting

$$X_1 = X_1' + X_2', \qquad X_2 = \mathbf{i}(X_1' - X_2'), \qquad \text{and} \quad Y = Y', \tag{4}$$

we obtain an orthonormal basis $\beta = \{X_1, X_2, Y\}$ with respect to the bilinear form B.

B3 B is invariant under the conjugation action of $\mathrm{SL}_2(\mathbb{C})$, meaning that

$$B(X,Y) = B(\mathbf{c}(g)Z, \mathbf{c}(g)W), \quad \text{for all } Z, W \in \mathfrak{sl}_2(\mathbb{C}),\ g \in \mathrm{SL}_2(\mathbb{C}).$$

By **B3**, $\mathbf{c}$ is a morphism of $\mathrm{SL}_2(\mathbb{C})$ into G. The content of part (a) of Proposition 2.1 below is that the morphism $\mathbf{c}$ just described is an epimorphism.

As a consequence of **B1** and **B2**, we have that

$$B(x_1^1 X_1 + x_2^1 X_2 + y^1 Y, x_1^2 X_1 + x_2^2 X_2 + y^2 Y) = x_1^1 x_1^2 + x_2^1 x_2^2 + y^1 y^2, \quad x_j^i, y \in \mathbb{C}. \tag{5}$$

For any bilinear form B on a vector space V, we use $\mathrm{O}(B)$ to denote the group of linear transformations of V preserving B, and we use $\mathrm{SO}(B)$ to denote the unimodular subgroup of $\mathrm{O}(B)$. If B is as in (5), then the isomorphism

$$\mathrm{SO}(B) \cong G \tag{6}$$

induced by the identification of the vector space $\mathfrak{sl}_2(\mathbb{C})$ with $\mathbb{C}\langle X_1, X_2, Y\rangle$ puts a system of coordinates on G. Part (b) of Proposition 2.1, below, will describe the epimorphism $\mathbf{c} : \mathrm{SL}_2(\mathbb{C}) \to G$ in terms of these coordinates.

Proposition 2.1. *With G, $\mathbf{c}$ as above, we have*

(a) The map $\mathbf{c}$ induces an isomorphism

$$\mathrm{SL}_2(\mathbb{C})/\{\pm I\} \xrightarrow{\cong} G$$

of Lie groups.

(b) Relative to the standard coordinates on $\mathrm{SL}_2(\mathbb{C})$ and the coordinates on G induced from the orthonormal basis β of $\mathfrak{sl}_2(\mathbb{C})$, as defined in (4), the epimorphism $\mathbf{c} : \mathrm{SL}_2(\mathbb{C}) \to G$ has the following coordinate expression:

$$\mathbf{c}\left(\begin{pmatrix} a & b \\ c & d \end{pmatrix}\right) = \begin{pmatrix} \frac{a^2-c^2+d^2-b^2}{2} & \frac{\mathbf{i}(a^2-c^2+b^2-d^2)}{2} & cd - ab \\ \frac{\mathbf{i}(b^2+d^2-a^2-c^2)}{2} & \frac{a^2+c^2+b^2+d^2}{2} & \mathbf{i}(ab+cd) \\ -ac+bd & \mathbf{i}(ac+bd) & ad+bc \end{pmatrix}. \tag{7}$$

We establish some further notational conventions regarding conjugation mappings. Whenever a matrix group H has a conjugation action $\mathbf{c}_V$ on a *finite-dimensional vector space* V over a field F, each basis β of V naturally induces a morphism

$$\mathbf{c}_{V,\beta} : H \to \mathrm{GL}_N(F), \text{ where } N = \dim V. \tag{8}$$

Let β, β' be two bases of V. Write $\alpha^{\beta\mapsto\beta'}$ for the change-of-basis matrix from β to β'. That is, if β, β' are written as N-entry row vectors, then

$$\beta\alpha^{\beta\mapsto\beta'} = \beta'. \tag{9}$$

Then elementary linear algebra tells us that

$$\mathbf{c}_{V,\beta} = \mathbf{c}_{\mathrm{GL}_N(F)}\left(\left(\alpha^{\beta\mapsto\beta'}\right)^{-1}\right)\mathbf{c}_{V,\beta'} = \mathbf{c}_{\mathrm{GL}_N(F)}\left(\alpha^{\beta'\mapsto\beta}\right)\mathbf{c}_{V,\beta'}. \tag{10}$$

Assuming that c_V is injective, and writing c_V^{-1} for the left inverse of c_V, we calculate from (10) that

$$\mathbf{c}_{V,\beta}\mathbf{c}_{V,\beta'}^{-1} \in \mathrm{Aut}(\mathrm{GL}_N(F)) \text{ is given by } \mathbf{c}_{\mathrm{GL}_N(F)}\left(\alpha^{\beta\mapsto\beta'}\right). \tag{11}$$

In keeping with the practice established after (2), we will omit the subscript $\mathfrak{h}$ when H is a Lie group acting on its Lie algebra by conjugation. Thus, for any basis β of $\mathfrak{h}$,

$$\mathbf{c}_\beta := \mathbf{c}_{\mathfrak{h},\beta}.$$

Generally speaking, whenever we fix a single basis β for $\mathfrak{h}$ we will blur the distinction between $\mathbf{c}$ and $\mathbf{c}_\beta$. For example, in this paper, whenever $H = \mathrm{SL}_2(\mathbb{C})$ and $V = \mathrm{Lie}(H)$, we will write $\mathbf{c}$ to denote both the "abstract" morphism $\mathbf{c}$ of H into $\mathrm{Aut}(V)$ and the linear morphism $\mathbf{c}_\beta$ of H into $\mathrm{GL}_3(\mathbb{C})$, where β is the orthonormal basis for $\mathrm{Lie}(H)$ defined in (4). Whenever the linear morphism into $\mathrm{GL}_3(\mathbb{C})$ is induced by a basis $\beta' \neq \beta$, the notation $\mathbf{c}_{\beta'}$ will be used.

We now turn our attention to the description of the inverse image $\mathbf{c}^{-1}(\Gamma)$ as a subset of $\mathrm{SL}_2(\mathbb{C})/\{\pm I\}$ with respect to the standard coordinates of $\mathrm{SL}_2(\mathbb{C})$. According to Proposition 2.1, this amounts to describing the quadruples

$$(a,b,c,d) \in \mathbb{C}^4, \text{ with } ad - bc = 1, \text{ and the entries of the right side of (7) integers.} \tag{12}$$

Describing the quadruples meeting conditions (12) will be the subject of the remainder of this section, culminating in Proposition 2.8.

Conventions regarding multiplicative structure of $\mathbb{Z}[\mathbf{i}]$. Before stating the proposition, we establish certain conventions we will use when dealing with the multiplicative properties of the Euclidean ring $\mathbb{Z}[\mathbf{i}]$. First, it is well known that $\mathbb{Z}[\mathbf{i}]$ is a Euclidean, hence principal, ring. That $\mathbb{Z}[\mathbf{i}]$ is principal means that all ideals $\mathscr{I}$ of $\mathbb{Z}[\mathbf{i}]$ are generated by a single element $m \in \mathbb{Z}[\mathbf{i}]$, so that every $\mathscr{I}$ is of the form (m). However, there is an unavoidable ambiguity in the choice of generators caused by the presence in $\mathbb{Z}[\mathbf{i}]$ of four units, $\mathbf{i}^j$, for $j \in \{0, \ldots, 3\}$, in $\mathbb{Z}[\mathbf{i}]$. We will adopt the following convention to sidestep the ambiguity caused by the group of units.

Definition 2.2. We refer to the following subset of $\mathbb{C}^\times$ as the **standard subset**:

$$\{z \in \mathbb{C}^\times \mid \mathrm{Re}(z) > 0,\ \mathrm{Im}(z) \geq 0\}. \tag{13}$$

That is, the standard subset of $\mathbb{C}^\times$ is the union of the interior of the first quadrant and the positive real axis. An element of $\mathbb{Z}[\mathbf{i}]$ in the standard subset will be referred to as a **standard Gaussian integer**, or more simply as a **standard integer** when the context is clear.

Because of the units in $\mathbb{Z}[\mathbf{i}]$, each nonzero ideal $\mathscr{I}$ of $\mathbb{Z}[\mathbf{i}]$ has precisely one generator which is a standard integer. Henceforth, we refer to the generator of $\mathscr{I}$ which is a standard integer as the **standard generator** of $\mathscr{I}$. Unless otherwise stated, whenever we write $\mathscr{I} = (m)$, to indicate the ideal $\mathscr{I}$ generated by an $m \in \mathbb{Z}[\mathbf{i}]$, it will be understood that m is standard. Conversely, whenever we write an ideal $\mathscr{I}$ in the form (m), it will be understood that m is the standard generator of $\mathscr{I}$. Thus, for example, since $(1 - \mathbf{i}) = \mathbf{i}^3(1 + \mathbf{i})$ with $1 + \mathbf{i}$ standard, we write $\mathscr{I} =: (1 - \mathbf{i})\mathbb{Z}[\mathbf{i}]$, defined as the ideal of Gaussian integers divisible by $1 - \mathbf{i}$, in the form $\mathscr{I} = (1 + \mathbf{i})$.

Similar comments apply to Gaussian primes, factorization, and greatest common divisor in $\mathbb{Z}[\mathbf{i}]$. By a "prime in $\mathbb{Z}[\mathbf{i}]$" we will always mean a *standard prime*. By "prime factorization" in $\mathbb{Z}[\mathbf{i}]$ we will always mean *factorization into a product of standard primes*, multiplied by the appropriate unit factor. Note that the convention regarding standard primes uniquely determines the unit factor in a prime factorization. For example, since

$$2 = \mathbf{i}^3(1 + \mathbf{i})^2$$

and $(1 + \mathbf{i})^3$ is standard, the above expression is the standard factorization of the Gaussian integer 2, and $\mathbf{i}^3$ is uniquely determined as the *standard unit factor* in the prime factorization of $2 \in \mathbb{Z}[\mathbf{i}]$.

By convention, unless stated otherwise, the "trivial ideal" $\mathbb{Z}[\mathbf{i}]$ will be understood to belong to the set of ideals of $\mathbb{Z}[\mathbf{i}]$. The standard generator of the trivial ideal $\mathbb{Z}[\mathbf{i}]$ is, of course, 1.

To facilitate the statement of Proposition 2.8, we estblish the following conventions. First, we use ω_8 to denote the unique primitive eighth root of unity in the standard set of $\mathbb{C}^\times$. Observe that

$$\omega_8 = \frac{\sqrt{2}}{2}(1 + \mathbf{i}), \quad \text{and} \quad \omega_8^2 = \mathbf{i}. \tag{14}$$

The $\mathrm{SL}_2(\mathbb{Z}[\mathbf{i}])$-space $\mathrm{M}_2^{\mathrm{N}}$.

Definition 2.3. For $N \in \mathbb{Z}[\mathbf{i}]$, $\mathbf{M}_2^{\mathbf{N}}$ will denote the subset of $\mathrm{Mat}_2(\mathbb{Z}[\mathbf{i}])$ consisting of the elements with determinant N. Since the group $\mathrm{SL}_2(\mathbb{Z}[\mathbf{i}])$ acts on $\mathrm{M}_2^{\mathrm{N}}$ by multiplication on the left, $\mathrm{M}_2^{\mathrm{N}}$ is an $\mathrm{SL}_2(\mathbb{Z}[\mathbf{i}])$-space.

It is not difficult to see that the action of $\mathrm{SL}_2(\mathbb{Z}[\mathbf{i}])$ on $\mathrm{M}_2^{\mathrm{N}}$ fails to be transitive, so $\mathrm{M}_2^{\mathrm{N}}$ is not an $\mathrm{SL}_2(\mathbb{Z}[\mathbf{i}])$-homogeneous space. The purpose of the subsequent definitions and results is to give a description of the orbit structure of the $\mathrm{SL}_2(\mathbb{Z}[\mathbf{i}])$-space $\mathrm{M}_2^{\mathrm{N}}$.

Let

$$\Omega_y := \text{a fixed set of representatives of } \mathbb{Z}[\mathbf{i}]/(y), \text{ for all } y \in \mathbb{Z}[\mathbf{i}]. \tag{15}$$

It is clear that for each $y \in \mathbb{Z}[\mathbf{i}]$, there exists a number of possible choices for Ω_y. For the general result, Proposition 2.6, below, the choice of Ω_y does not matter, and we leave it unspecified. However, in the specific applications of Proposition 2.6, where y is always of the form $y = (1 + \mathbf{i})^n$ for n a positive integer, it will be essential to give an Ω_y explicitly, which we now do.

So let $n \in \mathbb{N}$, $n \geq 1$. In the definition of $\Omega_{(1+\mathbf{i})^n}$, we use the "ceiling" notation, defined by

$$\lceil q \rceil = \text{smallest integer} \geq q, \text{ for } q \in \mathbb{Q}.$$

Now set

$$\Omega_{(1+\mathbf{i})^n} = \left\{ r + s\mathbf{i} \text{ with } r, s \in \mathbb{Z}, 0 \leq r < 2^{\lceil \frac{n}{2} \rceil}, 0 \leq s < 2^{n - \lceil \frac{n}{2} \rceil} \right\}. \tag{16}$$

The definition is justified by Lemma 2.4, below.

Lemma 2.4. *For $n \geq 1$ an integer, let $\Omega_{(1+\mathbf{i})^n}$ be defined as* (16). *Then*

$$\Omega_{(1+\mathbf{i})^n} \textit{ is a complete set of representatives of } \mathbb{Z}[\mathbf{i}]/((1+\mathbf{i})^n)) \textit{ for all } n.$$

Definition 2.5. Let $N \in \mathbb{Z}[\mathbf{i}]$ be fixed, and for each $y \in \mathbb{Z}[\mathbf{i}]$ let Ω_y be as in (15). Define the matrix $\alpha^{\mathrm{N}}(m, x) \in \mathrm{M}_2^{\mathrm{N}}$ as follows:

$$\alpha^{\mathrm{N}}(m, x) = \begin{pmatrix} m & x \\ 0 & \frac{N}{m} \end{pmatrix}, \text{ for } m \in \mathbb{Z}[\mathbf{i}],\ m|N,\ x \in \Omega_{\frac{N}{m}}. \tag{17}$$

It is trivial to verify that $\alpha^{\mathrm{N}}(m, x)$ as given by (17) indeed has determinant N, i.e., $\alpha^{\mathrm{N}}(m, x) \in \mathrm{M}_2^{\mathrm{N}}$. The point of Definition 2.5 is given by the following proposition.

Proposition 2.6. *For $N \in \mathbb{Z}[\mathbf{i}] - \{0\}$, let $\mathrm{M}_2^{\mathrm{N}}$ be the $\mathrm{SL}_2(\mathbb{Z}[\mathbf{i}])$-space of matrices with entries in $\mathbb{Z}[\mathbf{i}]$ and determinant N. Define the matrices $\alpha^{\mathrm{N}}(m, x)$ as in* (17). *Then*

$$M_2^{\mathrm{N}} = \dot{\bigcup_{\substack{m \in \mathbb{Z}[\mathbf{i}] |\, m|N, \\ \frac{N}{m} \text{ standard}}}} \ \dot{\bigcup_{x \in \Omega_{\frac{N}{m}}}} \mathrm{SL}_2(\mathbb{Z}[\mathbf{i}]) \alpha^{\mathrm{N}}(m, x), \tag{18}$$

and (18) *gives the decomposition of the $\mathrm{SL}_2(\mathbb{Z}[\mathbf{i}])$-space $\mathrm{M}_2^{\mathrm{N}}$ into distinct $\mathrm{SL}_2(\mathbb{Z}[\mathbf{i}])$-orbits.*

We now make some comments concerning the significance of Proposition 2.6. First, a statement equivalent to Proposition 2.6 is that an arbitrary $\alpha \in M_2^{\mathrm{N}}$ has a uniquely determined product decomposition of the form

$$\alpha = \begin{pmatrix} a & b \\ c & d \end{pmatrix} = \begin{pmatrix} p & q \\ r & s \end{pmatrix} \begin{pmatrix} m & x \\ 0 & \frac{N}{m} \end{pmatrix}, \text{ with } m \in \mathfrak{o},\ m|N,\ \frac{N}{m} \text{ standard},$$
$$x \in \Omega_{\frac{N}{m}},\ pr - qs = 1. \tag{19}$$

The uniqueness is derived from Proposition 2.6 as follows. The second matrix in the product of (19) is uniquely determined by the matrix α because of the disjointness of the union in (18). The first matrix in the product appearing in (19) is therefore also uniquely determined.

The second remark is that Proposition 2.6 may be thought of as the Gaussian-integer version of the decomposition of elements of $\mathrm{Mat}_2(\mathbb{Z})$ of fixed determinant N, sometimes known as the Hecke decomposition. Occasionally we refer to (19) as the *Gaussian* Hecke decomposition, to distinguish it from this *classical* Hecke decomposition in the context of the rational integers.

The proof is the same as that of the classical decomposition except for some care that has to be taken because of the presence of additional units in $\mathbb{Z}[\mathbf{i}]$. For the classical Hecke decomposition, see page 110, §VII.4, of [21], which is the source of our notation for the Gaussian version.

Statement of the Main Result of §2.1. Let Ξ be an arbitrary subset of $\mathrm{SL}_2(\mathbb{Z}[\mathbf{i}])$. Suppose, at first, that Ξ is actually a *subgroup* of $\mathrm{SL}_2(\mathbb{Z}[\mathbf{i}])$. Since $\mathrm{SL}_2(\mathbb{Z}[\mathbf{i}])\alpha^{\mathrm{N}}(m,x)$ is an $\mathrm{SL}_2(\mathbb{Z}[\mathbf{i}])$-space, it is also a Ξ-space. For general subgroups Ξ, however, the action of Ξ on $\mathrm{SL}_2(\mathbb{Z}[\mathbf{i}])\alpha^{\mathrm{N}}(m,x)$ fails to be transitive, i.e., $\mathrm{SL}_2(\mathbb{Z}[\mathbf{i}])\alpha^{\mathrm{N}}(m,x)$ is not a Ξ-homogeneous space. We will now describe the orbit structure of $\mathrm{SL}_2(\mathbb{Z}[\mathbf{i}])\alpha^{\mathrm{N}}(m,x)$ for a specific subgroup Ξ. In order to make the description of the subgroup and some related subsets of $\mathrm{SL}_2(\mathbb{Z}[\mathbf{i}])$ easier, we introduce the epimorphism

$$\mathrm{red}_{1+\mathbf{i}} : \mathrm{SL}_2(\mathbb{Z}[\mathbf{i}]) \to \mathrm{SL}_2(\mathbb{Z}[\mathbf{i}]/(1+\mathbf{i}))$$

by inducing from the reduction map

$$\mathrm{red}_{1+\mathbf{i}} : \mathbb{Z}[\mathbf{i}] \to \mathbb{Z}[\mathbf{i}]/(1+\mathbf{i}).$$

That is, we "extend" $\mathrm{red}_{1+\mathbf{i}}$ from elements to matrices by setting

$$\mathrm{red}_{1+\mathbf{i}}\left(\begin{pmatrix} p & q \\ r & s \end{pmatrix}\right) = \begin{pmatrix} \mathrm{red}_{1+\mathbf{i}}p & \mathrm{red}_{1+\mathbf{i}}q \\ \mathrm{red}_{1+\mathbf{i}}r & \mathrm{red}_{1+\mathbf{i}}s \end{pmatrix}. \tag{20}$$

Since $\Omega_{1+\mathbf{i}} = \{0,1\}$, we may identify $\mathbb{Z}[\mathbf{i}]/(1+\mathbf{i})$ with $\{0,1\}$. Similarly to the convention with $p,q,r,s \in \mathbb{Z}[\mathbf{i}]$, we use $(\overline{p},\overline{q},\overline{r},\overline{s})$ to denote a quadruple of elements of $\mathbb{Z}[\mathbf{i}]/(1+\mathbf{i})$ such that

$$\overline{p}\,\overline{s} - \overline{r}\,\overline{q} = 1.$$

There are two elements of $\mathrm{SL}_2(\mathbb{Z}[\mathbf{i}]/(1+\mathbf{i}))$ of particular interest:

$$\overline{I} := \begin{pmatrix} 1 & 0 \\ 0 & 1 \end{pmatrix}, \quad \overline{S} := \begin{pmatrix} 0 & 1 \\ 1 & 0 \end{pmatrix} \in \mathrm{SL}_2(\mathbb{Z}[\mathbf{i}]/(1+\mathbf{i})). \tag{21}$$

The notation in (21) is chosen to remind the reader that $\overline{I} = \mathrm{red}_{1+\mathbf{i}}(I)$ and $\overline{S} = \mathrm{red}_{1+\mathbf{i}}(S)$, where I, S are the standard generators of $\mathrm{SL}_2(\mathbb{Z})$, as in §VI.1 of [19]. Since $\overline{S}^2 = \overline{I}$, it is easy to see that $\{\overline{I}, \overline{S}\}$ is a subgroup of $\mathrm{SL}_2(\mathbb{Z}[\mathbf{i}]/(1+\mathbf{i}))$. Now define

$$\Xi_{12} = \mathrm{red}_{1+\mathbf{i}}^{-1}(\{\overline{I}, \overline{S}\}). \tag{22}$$

Since $\mathrm{red}_{1+\mathbf{i}}$ is a morphism, Ξ_{12} is a subgroup of $\mathrm{SL}_2(\mathbb{Z}[\mathbf{i}])$.

Also, using the epimorphism $\mathrm{red}_{1+\mathbf{i}}$, we define the following subsets of $\mathrm{SL}_2(\mathbb{Z}[\mathbf{i}])$:

$$\begin{aligned}
\Xi_1 &= \mathrm{red}_{1+\mathbf{i}}^{-1}\left(\left\{\begin{pmatrix}0&1\\1&1\end{pmatrix}, \begin{pmatrix}1&1\\0&1\end{pmatrix}\right\}\right),\\
\Xi_2 &= \mathrm{red}_{1+\mathbf{i}}^{-1}\left(\left\{\begin{pmatrix}1&1\\1&0\end{pmatrix}, \begin{pmatrix}1&0\\1&1\end{pmatrix}\right\}\right).
\end{aligned} \tag{23}$$

(The subscripts on the Ξ of (22) and (23) are chosen in order to remind the reader of the column in which zeros appear in the matrices of $\mathrm{red}_{1+\mathbf{i}}(\Xi)$.) Since $\mathrm{SL}_2(\mathbb{Z}[\mathbf{i}]/(1+\mathbf{i}))$ consists of the elements $\overline{I}, \overline{S}$ and the four elements appearing on the right-hand side of (23), and $\mathrm{red}_{1+\mathbf{i}}$ is an epimorphism, we have

$$\mathrm{SL}_2(\mathbb{Z}[\mathbf{i}]) = \Xi_1 \mathbin{\dot{\bigcup}} \Xi_2 \mathbin{\dot{\bigcup}} \Xi_{12}. \tag{24}$$

Unlike Ξ_{12}, the subsets Ξ_1 and Ξ_2 of $\mathrm{SL}_2(\mathbb{Z}[\mathbf{i}])$ are not subgroups.

All three subsets Ξ in (22) and (23), though, have a description of the following sort, which gives some insight into the reason for Sublemma 2.7, below. For fixed

$$(\overline{p}\ \overline{q}), (\overline{r}\ \overline{s}) \in \left\{\begin{matrix}(1\ 1),\\(1\ 0),\\(0\ 1)\end{matrix}\right\} \subset (\mathrm{SL}_2(\mathbb{Z}[\mathbf{i}]/(1+\mathbf{i})))^2,$$

we have

$$\Xi = \mathrm{red}_{1+\mathbf{i}}^{-1}\left(\left\{\begin{pmatrix}\overline{p}&\overline{q}\\\overline{r}&\overline{s}\end{pmatrix}, \begin{pmatrix}\overline{r}&\overline{s}\\\overline{p}&\overline{q}\end{pmatrix}\right\}\right). \tag{25}$$

For example, we obtain Ξ_{12} by taking

$$(\overline{p}\ \overline{q}) = (1\ 0) \quad \text{and} \quad (\overline{r}\ \overline{s}) = (0\ 1)$$

in (25).

The reason for introducing the subsets Ξ of (23) is that they allow us, in Sublemma 2.7 below, to describe precisely the orbit structure of the Ξ_{12}-space $\mathrm{SL}_2(\mathbb{Z}[\mathbf{i}])\alpha^{\mathrm{N}}(m, x)$.

Sublemma 2.7 *Using the notation of* (17) *and* (23), *we have*

$$\mathrm{SL}_2(\mathbb{Z}[\mathbf{i}])\alpha^{\mathrm{N}}(m,x) = \dot{\bigcup_{\Xi=\Xi_1,\Xi_2,\Xi_{12}}} \Xi\alpha^{\mathrm{N}}(m,x). \tag{26}$$

Each of the three sets in the union (26) *is closed under the action, by left-multiplication, of Ξ_{12} on $\mathrm{SL}_2(\mathbb{Z}[\mathbf{i}])\alpha^{\mathrm{N}}(m,x)$ and equals precisely one Ξ_{12}-orbit in the space $\mathrm{SL}_2(\mathbb{Z}[\mathbf{i}])\alpha^{\mathrm{N}}(m,x)$.*

Proposition 2.8. *Let* $\mathbf{c}$ *be the morphism from* $\mathrm{SL}_2(\mathbf{C})$ *onto* G *as in* (7). *Let* $\Gamma = \mathrm{SO}_3(\mathbb{Z}[\mathbf{i}])$ *be the group of integral points of* G *in the coordinatization of* G *induced by the isomorphism* (6). *Let the subsets* Ξ_1, Ξ_2, Ξ_{12} *of* $\mathrm{SL}_2(\mathbb{Z}[\mathbf{i}])$ *be as defined in* (22) *and* (23). *Let the matrices* $\alpha^{\mathrm{N}}(m,x)$ *be as in* (17). *Let* $\omega_8 \in \mathbf{C}$ *be as in* (14). *Then we have*

$$\mathbf{c}^{-1}(\Gamma) = \dot{\bigcup_{\delta,=0,1}} \left(\frac{1}{\omega_8^{\delta}} \Xi_{12}\alpha^{\mathbf{i}^{\delta}}(\mathbf{i}^{\delta},0) \dot{\bigcup} \left(\dot{\bigcup_{\epsilon=0,1}} \frac{1}{\omega_8^{\delta}(1+\mathbf{i})} \Xi_2\alpha^{2\mathbf{i}^{1+\delta}}(\mathbf{i}^{1+\delta},\mathbf{i}^{\epsilon}) \right) \right). \tag{27}$$

Remarks

(a) We use $\mathbb{Z}[\omega_8]$ to denote the ring generated over $\mathbb{Z}$ by ω_8. By (14) we have $\mathbb{Z}[\mathbf{i}] \subset \mathbb{Z}[\omega_8]$ and $\mathbb{Z}[\omega_8] = \mathbb{Z}[\omega_8,\mathbf{i}]$. It follows from Proposition 2.8 that $\mathbf{c}^{-1}(\Gamma) \subseteq \mathrm{SL}_2(\mathbb{C})$ is in fact a subset of $\mathrm{SL}_2(\mathbb{Q}(\omega))$. More precisely, of the two parts of the right-hand side of (27), we have

$$\frac{1}{\omega_8^{\delta}} \Xi_{12}\alpha^{\mathbf{i}^{\delta}}(\mathbf{i}^{\delta},0) \subseteq \mathrm{SL}_2(\mathbb{Z}[\mathbf{i},\omega_8]) \quad \text{for } \delta \in \{0,\ 1\}, \tag{28}$$

while

$$\left(\dot{\bigcup_{\epsilon=0,1}} \frac{1}{\omega_8^{\delta}(1+\mathbf{i})} \Xi_2\alpha^{2\mathbf{i}^{1+\delta}}(\mathbf{i}^{1+\delta},\mathbf{i}^{\epsilon}) \right) \subseteq \mathrm{SL}_2\left(\mathbb{Z}\left[\mathbf{i},\omega_8,\frac{1}{1+\mathbf{i}}\right]\right) \quad \text{for } \delta \in \{0,\ 1\}. \tag{29}$$

(b) One can easily verify that the set on the left-hand side of (28) is closed under multiplication, while the set on the left-hand side of (29) is not. More precisely, through a rather lengthy calculation, not included here, one verifies that

$$\text{for } (x,y) \text{ a pair of elements of the form of (29), } xy \text{ is } \begin{cases} \text{of form (29)} \\ \text{or} \\ \text{of form (28).} \end{cases} \tag{30}$$

with each possibility in (30) being realized for an appropriate pair (x,y). These calculations amount to a brute-force verification of the fact that the right-hand side of (27) is closed under multiplication. But because Γ is a group and $\mathbf{c}$ a morphism, this fact also follows from Proposition 2.8.

The explicit representation of $\mathbf{c}^{-1}(\Gamma)$ in Proposition 2.8 allows us to read off certain group-theoretic facts relating $\mathbf{c}^{-1}(\Gamma)$ to $\mathrm{SL}_2(\mathbb{Z}[\mathbf{i}])$. In Lemma 2.9 below we use the notation

$$[G : H] \text{ is the index of } H \text{ in } G, \text{ for any group } G \text{ with subgroup } H.$$

Lemma 2.9. *Let* $\mathbf{c}^{-1}(\Gamma)$ *be the subgroup of* $\mathrm{SL}_2(\mathbb{C})$ *described above, given explicitly in matrix form in (27). All the other notation is also as in Proposition 2.8.*

(a) We have

$$\mathbf{c}^{-1}(\Gamma) \cap \mathrm{SL}_2(\mathbb{Z}[\mathbf{i}]) = \Xi_{12}.$$

(b) We have

$$[\mathbf{c}^{-1}(\Gamma) : \Xi_{12}] = 6, \quad [\mathrm{SL}_2(\mathbb{Z}[\mathbf{i}]) : \Xi_{12}] = 3. \tag{31}$$

Explicitly, the six right cosets of Ξ_{12} *in* $\mathbf{c}^{-1}(\Gamma)$ *are the two cosets obtained by letting* δ *range over* $\{0, 1\}$ *in*

$$\frac{1}{\omega_8^\delta}\Xi_{12}\alpha^{\mathbf{i}^\delta}(\mathbf{i}^\delta, 0) = \frac{1}{\omega_8^\delta}\Xi_{12}\begin{pmatrix}\mathbf{i}^\delta & 0\\ 0 & 1\end{pmatrix}$$

and the four cosets obtained by letting δ, ϵ *range over* $\{0, 1\}$ *independently in*

$$\frac{1}{\omega_8^\delta(1+\mathbf{i})}\Xi_{12}\begin{pmatrix}1 & 0\\ 1 & 1\end{pmatrix}\alpha^{2\mathbf{i}^{1+\delta}}(\mathbf{i}^{1+\delta}, \mathbf{i}^\epsilon) = \frac{1}{\omega_8^\delta(1+\mathbf{i})}\Xi_{12}\begin{pmatrix}\mathbf{i}^{1+\delta} & \mathbf{i}^\epsilon\\ \mathbf{i}^{1+\delta} & 2+\mathbf{i}^\epsilon\end{pmatrix}.$$

2.2 *Good Grenier fundamental domains for arithmetic groups* $\Gamma \in \mathrm{Aut}^+(\mathbf{H}^3)$

We begin with the following definition, which is fundamental to everything that follows.

Definition. Let X be a topological space. Suppose that Γ is a group acting topologically on X, *i.e.*, $\Gamma \subseteq \mathrm{Iso}(X)$. A subset $\mathcal{F}$ of X is called an **exact fundamental domain for the action of Γ on X** if the following conditions are satisfied:

FD 1. The Γ-translates of $\mathcal{F}$ cover X, i.e.,

$$X = \Gamma\mathcal{F}.$$

FD 2. Distinct Γ-translates of $\mathscr{F}$ intersect only on their boundaries, i.e.,

$$\gamma_1, \gamma_2 \in \Gamma,\ \gamma_1 \neq \gamma_2 \text{ implies } \gamma_1\mathscr{F} \cap \gamma_2\mathscr{F} \subseteq \gamma_1\partial\mathscr{F},\ \gamma_2\partial\mathscr{F}.$$

Henceforth, we will drop the word **exact** and refer to such an $\mathscr{F}$ simply as a **fundamental domain**.

For the current section, §2.2, only, G, instead of denoting $SO_3(\mathbb{C})$, will denote $SL_2(\mathbb{C})$. Likewise, instead of denoting $SO_3(\mathbb{Z}[\mathbf{i}])$ or $\mathbf{c}^{-1}(SO_3(\mathbb{Z}[\mathbf{i}]))$, Γ will denote an arbitrary subgroup of $SL_2(\mathbb{C})$ satisfying certain conditions to be given below. The main examples to keep in mind are first, $\Gamma = SL_2(\mathbb{Z})$, the integer subgroup of $SL_2(\mathbb{C})$, and second, $\Gamma = \mathbf{c}^{-1}(SO_3(\mathbb{Z}[\mathbf{i}]))$, the inverse image of the integer subgroup of $SO_3(\mathbb{C})$, described explicitly as a group of fractional linear transformations in Proposition 2.8.

Iwasawa decomposition of $SL_2(\mathbb{C})$. For the reader's convenience, we recall only those results in the context of $SL_2(\mathbb{C})$ which we need to proceed. For proofs and the statements for $SL_n(\mathbb{C})$, see the "Notation and Terminology" section of [20]. Let

$$U = \left\{ \begin{pmatrix} 1 & x \\ 0 & 1 \end{pmatrix} \;\middle|\; x \in \mathbb{C} \right\}, \text{ upper triangular unipotent matrices in } SL_2(\mathbb{C}),$$

$$A = \left\{ \begin{pmatrix} y & 0 \\ 0 & y^{-1} \end{pmatrix} \;\middle|\; y \in \mathbb{R}_+ \right\}, \begin{array}{l}\text{diagonal elements of } SL_2(\mathbb{C}) \text{ with positive} \\ \text{diagonal entries,}\end{array}$$

$$K = SU(2) = \{k \in SL_2(\mathbb{C}) \mid kk^* = 1\}.$$

Here x^* denotes the conjugate-transpose $\overline{x}^t$ of x.

We have the ***Iwasawa decomposition***

$$SL_2(\mathbb{C}) = UAK,$$

and the product map $U \times A \times K \to UAK$ is a differential isomorphism.

The Iwasawa decomposition induces a system of coordinates ϕ on the symmetric space $SL_2(\mathbb{C})/K$. The mapping ϕ is a diffeomorphism between $SL_2(\mathbb{C})/K$ and $\mathbb{R}^3$. The details are as follows. The Iwasawa decomposition gives a uniquely determined product decomposition of $gK \in SL_2(\mathbb{C})/K$ as

$$gK = u(g)a(g)K, \text{ where } u(g) \in U,\ a(g) \in A \text{ are uniquely determined by } gK.$$

Define the **Iwasawa coordinates** $x_1(g), x_2(g) \in \mathbb{R}$, $y(g) \in \mathbb{R}^+$ by the relations

$$u(g) = \begin{pmatrix} 1 & x_1(g) + \mathbf{i}x_2(g) \\ 0 & 1 \end{pmatrix}, \quad a(g) = \begin{pmatrix} y(g)^{\frac{1}{2}} & 0 \\ 0 & y(g)^{-\frac{1}{2}} \end{pmatrix}.$$

By the Iwasawa decomposition, the Iwasawa coordinates of g are uniquely determined. We emphasize that while $x_1(g)$ and $x_2(g)$ range over all the real numbers, $y(g)$ ranges over the positive numbers. As functions on G, x_1 x_2, and y are invariant under right-multiplication by K. Thus x_1, x_2, and y induce coordinates on G/K. Now define the coordinate mappings $\phi_i : \mathrm{SL}_2(\mathbb{C})/K \to \mathbb{R}$, for $i = 1, 2, 3$, by

$$\phi_1 = -\log y, \quad \phi_2 = x_1, \quad \phi_3 = x_2, \tag{32}$$

and set

$$\phi = (\phi_1, \phi_2, \phi_3) : G/K \to \mathbb{R}^3.$$

The mapping ϕ is a diffeomorphism of G/K onto $\mathbb{R}^3$, because the Iwasawa coordinate system is a diffeomorphism, as is log. Thus, there exists the inverse diffeomorphism

$$\phi^{-1} : \mathbb{R}^3 \to G/K.$$

By (32), we can write, explicitly,

$$\phi^{-1}(t_1, t_2, t_3) = t_2 + t_3\mathbf{i} + e^{-t_1}\mathbf{j}, \quad \text{for all } t = (t_1, t_2, t_3) \in \mathbb{R}^3. \tag{33}$$

The quaternion model and the coordinate system on $\mathbf{SL_2(\mathbb{C})}/K$. We will use the model G/K as the upper half-space $\mathbf{H}^3$, defined as the following subset of the quaternions:

$$\mathbf{H}^3 = \{x_1 + x_2\mathbf{i} + y\mathbf{j}, \text{ where } x_1,\ x_2 \in \mathbb{R},\ y \in \mathbb{R}^+\}. \tag{34}$$

Recall that $\mathrm{SL}_2(\mathbb{C})$ acts transitively on $\mathbf{H}^3$ by fractional linear transformations. See §VI.0 of [19] for the details of the action. We note the relation

$$g\mathbf{j} = x_1(g) + x_2(g)\mathbf{i} + y(g)\mathbf{j}. \tag{35}$$

As a result of (35) and the Iwasawa decomposition, we may identify $\mathrm{SL}_2(\mathbb{C})/K$ with $\mathbf{H}^3$. So $\phi : G/K \to \mathbb{R}^3$ induces a diffeomorphism

$$\phi : \mathbf{H}^3 \xrightarrow{\cong} \mathbb{R}^3.$$

Because of (35), if g is any element of G such that $g\mathbf{j} = z$, then $\phi(g) = \phi(z)$. Further, beause of the way we set up the coordinates on $\mathbf{H}^3$, $\phi : \mathbf{H}^3 \to \mathbb{R}^3$ is given explicitly by the same formulas as (32).

As explained in, for example, §VI.0 of [19], the kernel of the action of $\mathrm{SL}_2(\mathbb{C})$ on $\mathbf{H}^3$ is precisely the set $\{\pm I\}$, consisting of the identity matrix and its negative.

For any oriented manifold X equipped with a metric, use the notation

$$\mathrm{Aut}^+(X) = \text{ group of orientation-preserving isometric automorphisms of } X.$$

It is a fact that every element of $\mathrm{Aut}^+(\mathbf{H}^3)$ is realized by a fractional linear transformation in $\mathrm{SL}_2(\mathbb{C})$, unique up to multiplication by ± 1. Therefore, the action of $\mathrm{SL}_2(\mathbb{C})$ on $\mathbf{H}^3$ by fractional linear transformations induces an isomorphism

$$\mathrm{SL}_2(\mathbb{C})/\{\pm I\} \cong \mathrm{Aut}^+(\mathbf{H}^3). \tag{36}$$

The stabilizer in Γ of the first j ϕ-coordinates. In all that follows, if $i, j \in \mathbb{N}$, the notation $[i, j]$ is used to denote the interval of *integers* from i to j, inclusive. The interval $[i, j]$ is defined to be the empty set if $i > j$.

Definition 2.10. For $i, j \in \{1, 2, 3\}$, with $i \leq j$, let $\phi_{[i,j]}$ be the **projection of $\mathbf{H}^3$ onto the $[i, j]$ factors of** $\mathbb{R}^3$. In other words, we let

$$\phi_{[i,j]} = (\phi_i, \phi_{i+1}, \ldots, \phi_j).$$

Since ϕ is a diffeomorphism of $\mathbf{H}^3$, $\phi_{[i,j]}$ is a smooth epimorphism of $\mathbf{H}^3$ onto $\mathbb{R}^{i-j+1}$.

If $\mathcal{K}$ is any subset of $\{1, 2, 3\}$, of size $|\mathcal{K}|$, then we can generalize in the obvious way to define the smooth epimorphism

$$\phi_{\mathcal{K}} : \mathbf{H}^3 \to \mathbb{R}^{|\mathcal{K}|}.$$

Let Γ be a group acting by diffeomorphisms of $\mathbf{H}^3$. For $\gamma \in \Gamma$ we also use γ to denote the diffeomorphism of $\mathbf{H}^3$ defined by the left action of γ on $\mathbf{H}^3$. Therefore, for $l \in \{1, \ldots, 3\}$ the composition $\phi_l \circ \gamma$ is the $\mathbb{R}$-valued function on $\mathbf{H}^3$ defined by

$$\phi_l \circ \gamma(z) = \phi_l(\gamma z) \quad \text{for all } z \in \mathbf{H}^3.$$

We use $\Gamma^{\phi_{[1,j]}}$ to denote the subgroup of Γ whose action stabilizes the first i coordinates. In other words, we set

$$\Gamma^{\phi_{[1,j]}} = \{\gamma \in \Gamma \mid \phi_{[1,j]} = \phi_{[1,j]} \circ \gamma\}.$$

We extend the definition of $\Gamma^{\phi_{[1,j]}}$ to $j = 0, 4$ by adopting the conventions

$$\Gamma^{\phi_{[1,0]}} = \Gamma \quad \text{and} \quad \Gamma^{\phi_{[1,4]}} = 1.$$

Note that by definition, we have the descending sequence of groups

$$\Gamma = \Gamma^{\phi_{[1,0]}} \geq \Gamma^{\phi_1} \geq \Gamma^{\phi_{[1,2]}} \geq \Gamma^{\phi_{[1,3]}} \geq \Gamma^{\phi_{[1,4]}} = 1.$$

Note that the penultimate group in this sequence, namely $\Gamma^{\phi_{[1,3]}}$, equals, by definition, the kernel of the action of Γ on $\mathbf{H}^3$. Assuming that $\Gamma \subset \mathrm{SL}_2(\mathbb{C})$, i.e., that Γ consists of fractional linear transformations, we always have

$$\Gamma^{\phi_{[1,3]}} = \Gamma \cap \{\pm 1\}. \tag{37}$$

Because the $\Gamma^{\phi_{[1,j]}}$ form a descending sequence, for $k, j \in \{1,2,3\}$ with $k < j$, we can consider the left cosets of $\Gamma^{\phi_{[1,k]}}$ in $\Gamma^{\phi_{[1,j]}}$. The left cosets are the sets of the form $\Gamma^{\phi_{[1,j]}}\gamma_k$ for $\gamma_k \in \Gamma^{\phi_{[1,k]}}$. Now let $i, j, k \in \{1,2,3\}, l \leq j, k < j$. By the definition of $\Gamma^{\phi_{[1,j]}}$, the function $\phi_l \circ \gamma_k$ depends only on the left $\Gamma^{\phi_{[1,j]}}$-coset to which γ_k belongs. Therefore, for fixed z we may consider $\phi_l \circ \gamma_k(z)$ to be a well-defined function on the set of left cosets $\Gamma^{\phi_{[1,j]}}\gamma_k$ of $\Gamma^{\phi_{[1,k]}}$ in $\Gamma^{\phi_{[1,j]}}$. We may therefore speak of the $\mathbb{R}$-valued function $\phi_l \circ \Gamma^{\phi_{[1,j]}}\gamma_k$.

In what follows we will most often apply the immediately preceding paragraph when $l = j$ and $k = j - 1$. For $\gamma \in \Gamma^{\phi_{[1,j-1]}}$ and Δ an arbitrary subset of $\Gamma^{\phi_{[1,j]}}$, we have

$$\phi_j(\Delta\gamma z) = \{\phi_j(\gamma z)\}. \tag{38}$$

Therefore, by setting

$$\phi_j \circ \Gamma^{[1,j]}\gamma(z) = \phi_j(\gamma z),$$

we obtain a well-defined function

$$\phi_j \circ \Gamma^{\phi_{[1,j]}}\gamma : \mathbf{H}^3 \to \mathbb{R}.$$

The function $\phi_j \circ \Gamma^{\phi_{[1,j]}}\gamma$ depends only on the $\Gamma^{\phi_{[1,j]}}$-coset to which γ belongs.

For $\gamma \in \Gamma^{\phi_{[1,j-1]}}$, the $\mathbb{R}$-valued function $\phi_j \circ \Gamma^{\phi_{[1,j]}}\gamma$ gives the effect of the action of $\gamma \in \Gamma^{\phi_{[1,j-1]}}$ on the jth coordinate of a point. It is clear from the definition that

$$\phi_j = \phi_j \circ \gamma \text{ if and only if } \Gamma^{\phi_{[1,j]}}\gamma \text{ is the identity left coset of } \Gamma^{\phi_{[1,j]}} \text{ in } \Gamma^{\phi_{[1,j-1]}}. \tag{39}$$

Sections of projections and induced actions of Γ. As before, suppose that Γ is a group acting by diffeomorphisms on $\mathbf{H}^3$, and let $\Gamma^{\phi_{[1,j]}}$ for $j \in \{1,2,3\}$ be defined as above.

For any subset $\mathcal{K}$ of the interval of integers $[1,3]$, we let $\mathcal{K}^c = [1,3] - \mathcal{K}$ be the *complement of* $\mathcal{K}$ *in* $[1,3]$.

Definition 2.11. Let f be a real-valued function

$$f : \mathbf{H}^3 \to \mathbb{R}.$$

Let $\mathcal{K}$ a subset of $[1,3]$. We say that f is **independent of the $\mathcal{K}$ coordinates** if for every $x, y \in \mathbf{H}^3$,

$$\phi_{\mathcal{K}^c}(x) = \phi_{\mathcal{K}^c}(y) \text{ implies } f(x) = f(y).$$

In other words, f is independent of the coordinates in $\mathcal{K}$ if and only if f is constant on the fibers of the projection $\phi_{\mathcal{K}^c}$ onto the $\mathbb{R}$-factors indexed by $\mathcal{K}^c$.

For the next observation, we need to introduce the notion of a section of a projection $\phi_{\mathcal{K}}$. It will not really matter which section we use, so for simplicity, we choose the zero section. For a subinterval $[i, j]$ of $\{1,2,3\}$ of size $j - i + 1$, define

$$\sigma^0_{[i,j]} : \mathbb{R}^{j-1+1} \to \mathbf{H}^3$$

by

$$\sigma^0_{[i,j]}(x_1, \dots, x_{j-i+1}) = (\underbrace{0, \dots, 0}_{i-1}, x_1, \dots, x_{j-i+1}, \underbrace{0, \dots, 0}_{3-j}).$$

The map $\sigma^0_{[i,j]}$ is called the **zero section of the projection $\phi_{[i,j]}$**. The terminology comes from the relation

$$\phi_{[i,j]}\sigma^0_{[i,j]} = \mathrm{Id}_{\mathbb{R}^{i-j+1}}, \tag{40}$$

which is immediately verified. The concept of the zero section of the projection can be generalized from the case of a projection associated with an interval $[i, j]$ to that of an arbitrary subset $\mathscr{K}$ of $\{1, 2, 3\}$, in the obvious way, although we will not have any use for this generalization in the present context.

By use of the zero section, we are able to make a useful reformulation of the condition that $f : \mathbf{H}^3 \to \mathbb{R}$ is independent of the first $j - 1$ coordinates. Let $j \in \{2, 3\}$ and f a real-valued function on $\mathbf{H}^3$. Then

$$f \text{ is independent of the first } j-1 \text{coordinates if and only if } f\,\sigma^0_{[j,3]}\phi_{[j,3]} = \sigma^0_{[j,3]}\phi_{[j,3]}\, f. \tag{41}$$

The reformulation (41) allows us to prove the following result.

Lemma 2.12. *Let Δ be a group acting on $\mathbf{H}^3$, and for $j \in \{1, 2, 3\}$, let $\phi_{[j,3]}$ be the projection of $\mathbf{H}^3$ onto the last $3 - j + 1$ coordinates and let $\sigma^0_{[j,3]}$ be the zero section of $\phi_{[j,3]}$. Suppose that for all $l \in [j, 3]$ and $\delta \in \Delta$ the functions $\phi_l \circ \delta$ are independent of the first $j-1$ coordinates. Then Δ has an induced action on $\mathbb{R}^{3-j+1}$ defined by*

$$\delta_{[j,3]}(\mathbf{t}) = \phi_{[j,3]}(\delta\sigma^0_{[j,3]}(\mathbf{t})), \quad \textit{for all } \mathbf{t} = (t_1, \dots, t_{3-j+1}) \in \mathbb{R}^{3-j+1}. \tag{42}$$

It is an immediate consequence of the definitions that for any group $\tilde{\Gamma}$ acting on $\mathbf{H}^3$ by diffeomorphisms, and any subgroup Γ of $\tilde{\Gamma}$, we have, for $1 \le i \le j \le 3$,

$$\Gamma^{\phi_{[i,j]}} = (\tilde{\Gamma})^{\phi_{[i,j]}} \cap \Gamma. \tag{43}$$

Applying (43) to the case of $\tilde{\Gamma} = \mathrm{SL}_2(\mathbb{C})$ and $i = 1$, we deduce that

$$\Gamma^{\phi_{[1,j]}} = \Gamma \cap \mathrm{SL}_2(\mathbb{C})^{\phi_{[1,j]}}, \tag{44}$$

for any subgroup $\Gamma \subseteq \mathrm{SL}_2(\mathbb{C})$. Because of (44) it is very useful to have an explicit expression for $\mathrm{SL}_2(\mathbb{C})^{\phi_1}$. We carry out the calculation using the relations of (32).

Let $z \in \mathbf{H}^3$ with

$$z = x_1 + x_2 + y\mathbf{j},$$

as in (34). Let

$$g \in \mathrm{SL}_2(\mathbb{C}) \text{ with } g = \begin{pmatrix} a & b \\ c & d \end{pmatrix}.$$

Define

$$y(c,d;z) = \frac{y(z)}{\|cz+d\|^2}, \tag{45}$$

where in (45) and from now on, for a quaternion z, $\|z\|^2$ denotes the squared norm of z, so that $\|z\|^2 = z\bar{z}$. Then we have

$$y(gz) = y(c,d;z). \tag{46}$$

For the details of such calculations, see §VI.0 of [19]. Since

$$\phi_1 : \mathbf{H}^3 \to \mathbb{R} \text{ is defined as } -\log y(\cdot),$$

and log is injective, (46) implies that

$$g \in \mathrm{SL}_2(\mathbb{C})^{\phi_1} \text{ if and only if } y(c,d;z) = y(z) \text{ for all } z \in \mathbf{H}^3. \tag{47}$$

By (47) and (45), we have

$$g \in \mathrm{SL}_2(\mathbb{C})^{\phi_1} \text{ if and only if } \|cz+d\|^2 = 1 \quad \text{for all } z \in \mathbf{H}^3. \tag{48}$$

Clearly, the condition $\|cz+d\|^2 = 1$ is satisfied for all $z \in \mathbf{H}^3$ if and only if $c = 0$ and $\|d\| = 1$. We therefore deduce from (48) that

$$\mathrm{SL}_2(\mathbb{C})^{\phi_1} = \left\{ \begin{pmatrix} \omega^{-1} & x \\ 0 & \omega \end{pmatrix} \;\middle|\; x, \omega \in \mathbb{C}, \|\omega\| = 1 \right\}. \tag{49}$$

As a result of (49), we can easily verify that for $\gamma \in \Gamma^{\phi_1}$, $l \in [2,3]$, the functions $\phi_l \circ \delta$ are independent of the first coordinate. So we can apply Lemma 2.12, in this case, with $j = 2$ and deduce the following:

Lemma 2.13. *Let $\Gamma \subseteq \mathrm{Aut}^+(\mathbf{H}^3)$, let $\phi_{[2,3]}$ be the projection of $\mathbf{H}^3$ onto the last two coordinates, and let $\sigma^0_{[2,3]}$ be the zero section of $\phi_{[2,3]}$. Then Γ has an induced action on $\mathbb{R}^2$ defined by*

$$\gamma_{[2,3]}(\mathbf{t}) = \phi_{[2,3]}(\gamma\sigma^0_{[2,3]}(\mathbf{t})), \quad \textit{for all } \mathbf{t} = (t_1,t_2) \in \mathbb{R}^2. \tag{50}$$

The following theorem is a special case of the main result of the first chapter of [6].

Theorem 2.14. *Let Γ be a subgroup of $\mathrm{SL}_2(\mathbb{C})$, acting on $\mathbf{H}^3$ on the left by fractional linear transformations. Suppose that Γ is commensurable with $\mathrm{SL}_2(\mathbb{Z}[\mathbf{i}])$.*

Let $\mathcal{G}$ be a fundamental domain for the induced action of $\Gamma^{\phi_{[2,3]}}/\{\pm 1\}$ on $\mathbb{R}^2$. Assume further that $\mathcal{G} = \overline{\mathrm{Int}(\mathcal{G})}$. Define

$$\mathcal{F}_1 = \{z \in \mathbf{H}^3 \mid \phi_1(z) \le \phi_1(\gamma z), \textit{ for all } \gamma \in \Gamma\}. \tag{51}$$

Set

$$\mathcal{F}(\mathcal{G}) = \phi_{[2,3]}^{-1}(\mathcal{G}) \cap \mathcal{F}_1. \tag{52}$$

(a) We have $\mathcal{F}(\mathcal{G})$ a fundamental domain for the action of $\Gamma/\{\pm 1\}$ on $\mathbf{H}^3$.
(b) We have

$$\mathcal{F}(\mathcal{G}) = \overline{\mathrm{Int}\big(\mathcal{F}(\mathcal{G})\big)}. \tag{53}$$

(c) Further, $\mathrm{Int}\big(\mathcal{F}_1\big)$ and $\mathrm{Int}\big(\mathcal{F}(\mathcal{G})\big)$ have explicit descriptions as follows:

$$\mathrm{Int}\big(\mathcal{F}_1\big) = \{z \in \mathbf{H}^3 \mid \phi_1(z) < \phi_1(\gamma z), \textit{ for all } \gamma \in \Gamma - \Gamma^{\phi_1}\} \tag{54}$$

and

$$\mathrm{Int}\big(\mathcal{F}(\mathcal{G})\big) = \phi_{[2,3]}^{-1}\big(\mathrm{Int}(\mathcal{G})\big) \cap \mathrm{Int}(\mathcal{F}_1). \tag{55}$$

Considering the coordinate system ϕ on $\mathbf{H}^3$ as fixed, we may think of the fundamental domain $\mathcal{F}$ for $\Gamma^{\phi_{[1,3]}}\backslash\Gamma$ to be a function of the fundamental domain $\mathcal{G}$ for the induced action of Γ^{ϕ_1} on $\mathbb{R}^2$. When we wish to stress this dependence of $\mathcal{F}$ on $\mathcal{G}$, we will write $\mathcal{F}(\mathcal{G})$ instead of $\mathcal{F}$.

Definition 2.15. Suppose that $\Gamma \subseteq \mathrm{Aut}^+(\mathbf{H}^3)$ is commensurable with $\mathrm{SL}_2(\mathbb{Z}[\mathbf{i}])$. Let $\mathcal{G}$ be a fundamental domain for the induced action of $\Gamma^{\phi_{[1,3]}}\backslash\Gamma^{\phi_1}$ on $\mathbb{R}^2$ satisfying $\mathcal{G} = \overline{\mathrm{Int}(\mathcal{G})}$. Then the fundamental domain $\mathcal{F}(\mathcal{G})$ for the action of $\Gamma^{\phi_{[1,3]}}\backslash\Gamma$ defined in (52) is called the **good Grenier fundamental domain for the action of Γ on $\mathbf{H}^3$ associated to the fundamental domain $\mathcal{G}$**.

The reference to the fundamental domain $\mathcal{G}$ is often omitted in practice.

Henceforth, we drop the explicit reference to $\Gamma^{\phi_{[1,3]}}$ and speak of a *fundamental domain of $\Gamma^{\phi_{[1,3]}}\backslash\Gamma$* as a *fundamental domain of Γ*. By (37), Γ is at worst a twofold cover of $\Gamma^{\phi_{[1,3]}}\backslash\Gamma$, so this involves only a minor abuse of terminology.

We will give an expression for a good Grenier fundamental domain $\mathcal{F}(\mathcal{G})$ for $\mathbf{c}^{-1}(\mathrm{SO}_3(\mathbb{Z}[\mathbf{i}]))$ in terms of explicit inequalities, in (73), and again as a convex polytope in $\mathbf{H}^3$, in Proposition 2.19, below.

Example: The Picard domain $\mathcal{F}$ for $\mathrm{SL}_2(\mathbb{Z}[\mathbf{i}])$. Define the following rectangle in $\mathbb{R}^2$:

$$\mathcal{G}_{\mathrm{SL}_2(\mathbb{Z}[\mathbf{i}])^{\phi_1}} = \left\{ (t_1, t_2) \in \mathbb{R}^2 \;\middle|\; t_1 \in \left[-\frac{1}{2}, \frac{1}{2}\right],\ t_2 \in \left[0, \frac{1}{2}\right] \right\}. \tag{56}$$

It is easy to verify, from an explicit description of $\mathrm{SL}_2(\mathbb{Z}[\mathbf{i}])^{\phi_1}$ deduced from (49), that $\mathcal{G}_{\mathrm{SL}_2(\mathbb{Z}[\mathbf{i}])^{\phi_1}}$ is a fundamental domain for the action of $\mathrm{SL}_2(\mathbb{Z}[\mathbf{i}])^{\phi_1}/\{\pm 1\}$.

Further, it is obvious that

$$\mathcal{G}_{\mathrm{SL}_2(\mathbb{Z}[\mathbf{i}])^{\phi_1}} = \overline{\mathrm{Int}(\mathcal{G}_{\mathrm{SL}_2(\mathbb{Z}[\mathbf{i}])^{\phi_1}})}.$$

Therefore, Theorem 2.14 applies. We deduce that with $\mathcal{F}_1$, $\mathcal{F}(\mathcal{G}_{\mathrm{SL}_2(\mathbb{Z}[\mathbf{i}])^{\phi_1}})$ defined as in Theorem 2.14, we have

$$\mathcal{F} := \mathcal{F}(\mathcal{G}_{\mathrm{SL}_2(\mathbb{Z}[\mathbf{i}])^{\phi_1}}) \text{ is a good Grenier fundamental domain for } \mathrm{SL}_2(\mathbb{Z}[\mathbf{i}]).$$

The fundamental domain $\mathcal{F}$ is defined in §VI.1 of [19], where, in keeping with classical terminology, $\mathcal{F}$ is called the **Picard domain**.

In order to complete the example, we now give an explicit description of the set $\mathcal{F}_1$, which will allow the reader to see that "our" $\mathcal{F}$ is exactly the same as the Picard domain. It can be shown that $\mathcal{F}_1$ is the subset of $\mathbb{R}^3$ whose image under the diffeomorphism ϕ^{-1} is given as follows:

$$\phi^{-1}(\mathcal{F}_1) = \{z \in \mathbf{H}^3 \mid \|z - m\| \geq 1, \text{ for all } m \in \mathbb{Z}[\mathbf{i}]\}. \tag{57}$$

Of the infinite set of inequalities defining $\mathcal{F}_1$, all except the one with $d = 0$, i.e., $\|z\|^2 \geq 1$, are trivially satisfied on $\phi_{[2,3]}^{-1}(\mathcal{G}_{\mathrm{SL}_2(\mathbb{Z}[\mathbf{i}])^{\phi_1}})$. Thus, from (57), (56), and (52), we recover the description of the Picard domain by finitely many inequalities given in §VI.1 of [19]:

$$\mathcal{F}(\mathcal{G}_{\mathrm{SL}_2(\mathbb{Z}[\mathbf{i}])^{\phi_1}}) = \left\{ z \in \mathbf{H}^3 \;\middle|\; x_1 \in \left[-\frac{1}{2}, \frac{1}{2}\right], \, x_2 \in \left[0, \frac{1}{2}\right] y, \, \|z\|^2 \geq 1 \right\}. \tag{58}$$

2.3 *Explicit description of the fundamental domain for the action of* $\mathrm{SO}_3(\mathbb{Z}[\mathbf{i}])$ *on* $\mathbf{H}^3$

We now proceed to consider the special case of $\mathbf{c}^{-1}(\mathrm{SO}_3(\mathbb{Z}[\mathbf{i}]))$ in Theorem 2.14 above. In keeping with the general practice of the present paper, we will go back to using G to denote $\mathrm{SO}_3(\mathbb{C})$ exclusively, and Γ to denote the group $\mathrm{SO}_3(\mathbb{Z}[\mathbf{i}])$. Since we are always in this section in the setting of subgroups of $\mathrm{SL}_2(\mathbb{C})$, we will abuse notation slightly and use Γ to denote the isomorphic inverse image $\mathbf{c}^{-1}(\Gamma)$ of $\Gamma = \mathrm{SO}_3(\mathbb{Z}[\mathbf{i}])$ in $\mathrm{SL}_2(\mathbb{C})$.

Also, we treat $\mathbb{R}^2$, the image of the projection $\phi_{[2,3]}$, as $\mathbb{C}$, by identifying the point $(t_1, t_2) \in \mathbb{R}^2$ with $t_1 + \mathbf{i}t_2$. Thus, our "new" $\phi_{[2,3]}$ is defined in terms of the "old" ϕ-coordinates by

$$\phi_{[2,3]}(z) = \phi_2(z) + \mathbf{i}\phi_3(z). \tag{59}$$

Proposition 2.16. ***First form of*** $\mathcal{F}_1$. *Let* $\mathcal{F}_1$ *be as defined in* (51). *All other notation has the same meaning as in Theorem 2.14. Then we have*

$$\mathscr{F}_1 = \{z = x(z) + y(z)\mathbf{j} \in \mathbf{H}^3 \mid \|x(z) - d\|^2 + y(z)^2 \geq 2, \textit{ for } d \in 1 + (1+\mathbf{i})\mathbb{Z}[\mathbf{i}]\}, \tag{60}$$

and $\mathrm{Int}(\mathscr{F}_1)$ *is the same as in* (60), *but with strict inequality instead of nonstrict inequality.*

Fundamental domain $\mathscr{G}$ for Γ^{ϕ_1}. In order to complete the explicit determination of a good Grenier fundamental domain $\mathscr{F}$ for Γ, it remains to describe a suitable fundamental domain $\mathscr{G}$ for Γ^{ϕ_1}. Using (44), (49), and the description of Γ in (27), we deduce that

$$\Gamma^{\phi_1} = \left\{ \begin{pmatrix} \omega_8^\delta & \omega_8^\delta b \\ 0 & \omega_8^{-\delta} \end{pmatrix} \,\middle|\, b \in (1+\mathbf{i})\mathbb{Z}[\mathbf{i}],\ \delta \in \{0,1\} \right\}. \tag{61}$$

It follows from (61) that the subgroup of unipotent elements of Γ^{ϕ_1} is

$$(\Gamma^{\phi_1})_U = \begin{pmatrix} 1 & (1+\mathbf{i})\mathbb{Z}[\mathbf{i}] \\ 0 & 1 \end{pmatrix}. \tag{62}$$

We make note of certain group-theoretic properties of Γ^{ϕ_1} and $(\Gamma^{\phi_1})_U$ that are used in determining the fundamental domains. First, we define the following generating elements:

$$R_{\frac{\pi}{2}} = \begin{pmatrix} \omega_8 & 0 \\ 0 & \omega_8^{-1} \end{pmatrix}, \quad T_{1+\mathbf{i}} = \begin{pmatrix} 1 & 1+\mathbf{i} \\ 0 & 1 \end{pmatrix}, \quad \text{and} \quad T_{1-\mathbf{i}} = \begin{pmatrix} 1 & 1-\mathbf{i} \\ 0 & 1 \end{pmatrix}. \tag{63}$$

It is easily verified, using (61) and (62), that

$$(\Gamma^{\phi_1})_U = \langle T_{1+\mathbf{i}}, T_{1-\mathbf{i}} \rangle, \quad \Gamma^{\phi_1} = \langle R_{\frac{\pi}{2}}, T_{1+\mathbf{i}}, T_{1-\mathbf{i}} \rangle. \tag{64}$$

We calculate, from the definition of $R_{\frac{\pi}{2}}$ and (64), that

$$\mathbf{c}(R_{\frac{\pi}{2}})(\Gamma^{\phi_1})_U = (\Gamma^{\phi_1})_U.$$

Since Γ^{ϕ_1} is generated by $(\Gamma^{\phi_1})_U$ and $R_{\frac{\pi}{2}}$, and $R_{\frac{\pi}{2}}$ has order 4, we deduce that

$$(\Gamma^{\phi_1})_U \text{ is normal in } \Gamma^{\phi_1} \text{ with } [\Gamma^{\phi_1} : (\Gamma^{\phi_1})_U] = 4. \tag{65}$$

Let T be any element of $(\Gamma^{\phi_1})_U$. Then we have a more precise version of (65):

$$\begin{aligned} &\text{The group } \langle TR_{\frac{\pi}{2}} \rangle \text{ of order 4 is a set of representatives} \\ &\text{for the coset group } \Gamma^{\phi_1}/(\Gamma^{\phi_1})_U. \end{aligned} \tag{66}$$

Applying (66) to the case $T = T_{1-\mathrm{i}}$, we have the following: The group $\langle T_{1-\mathrm{i}} R_{\frac{\pi}{2}} \rangle$ of order 4 is a set of representatives for the coset group $\Gamma^{\phi_1}/(\Gamma^{\phi_1})_U$. (67)

It is easily verified that the action of $R_{\frac{\pi}{2}}$ on $\mathbb{C}$ is rotation by an angle $\pi/2$ about the fixed point 0. Furthermore, we calculate from (63) that

$$T_{1-\mathrm{i}} R_{\frac{\pi}{2}} = \mathbf{c}(T_1) R_{\frac{\pi}{2}}.$$

Therefore,

$$\text{The action of } T_{1-\mathrm{i}} R_{\frac{\pi}{2}} \text{ on } \mathbb{C} \text{ is rotation by } \pi/2 \text{ about } 1. \tag{68}$$

The following statement is a special case of Lemma 2.2.7 of [2].

Lemma 2.17. *Let $\mathcal{G}_U$ be a fundamental domain for the action of $(\Gamma^{\phi_1})_U$ on $\mathbf{H}^3$ satisfying*

$$T_{1+\mathrm{i}} R_{\frac{\pi}{2}}(\mathcal{G}_U) = \mathcal{G}_U.$$

Let $\mathcal{G}$ be a fundamental domain for the action of $\langle T_{1+\mathrm{i}} R_{\frac{\pi}{2}} \rangle$ on $\mathcal{G}$. Then $\mathcal{G}$ a fundamental domain for the action of Γ^{ϕ_1} on $\mathbf{H}^3$.

In order to define and work with the sets $\mathcal{G}_U$ and $\mathcal{G}$ which will be fundamental domains for the action of $\Gamma_U^{\phi_1}$ and Γ^{ϕ_1}, it is useful to introduce the notion of a convex hull in a totally geodesic metric space.

A metric space (X, d) will be called **totally geodesic** if for every pair of points $p_1, p_2 \in X$, $p_1 \neq p_2$ there is a unique geodesic segment connecting p_1, p_2. In this situation, the (closed) geodesic segment connecting p_1, p_2 will be denoted by $[p_1, p_2]_d$. A point $x \in X$ is said to lie **between p_1 and p_2** when x lies on $[p_1, p_2]_d$. We then say that $\mathcal{S} \subset X$ is **convex** when $p_1, p_2 \in \mathcal{C}$ and p_3 between p_1 and p_2 implies that $p_3 \in \mathcal{S}$. Let $p_1, \dots, p_r$ be r points in X. The points determine a set

$$\mathcal{C}_d(p_1, \dots, p_r),$$

called the **convex closure** of $p_1, \dots, p_r$, described as the smallest convex subset of X containing the set $\{p_1, \dots, p_r\}$.

Obviously, we can apply the notion of convex hull to any set $\mathcal{S}$, rather than a finite set of points. The definition remains the same, namely that $\mathcal{C}_d(\mathcal{S})$ is the smallest convex subset of X containing $\mathcal{S}$. In general we will use the notation

$$\mathcal{C}_d(\mathcal{S}_1, \dots, \mathcal{S}_r) = \mathcal{C}_d\left(\bigcup_{i=1,\dots,r} \mathcal{S}_i\right).$$

In particular, if we apply these notions to $X = \mathbb{R}^2$ with the ordinary Euclidean metric Euc, then the geodesic segment $[p_1, p_2]_{\mathrm{Euc}}$ is just the line segment joining p_1, p_2. Further, provided that not all the p_i are collinear, $\mathcal{C}(p_1, \dots p_r)$ is a closed convex polygon whose vertices are located at a subset of $\{p_1, \dots, p_r\}$.

We first use the notion of convex closure to record an elementary facts concerning the fundamental domains of groups of translations acting on $\mathbb{R}^2$, identified with $\mathbb{C}$ in the usual way. Let $\omega_1, \omega_2 \in \mathbb{C}$ be linearly independent over $\mathbb{R}$. Then $\mathbb{Z}\omega_1 + \mathbb{Z}\omega_2$ is a lattice in $\mathbb{C}$, and it is well known that all lattices in $\mathbb{C}$ are of this form for suitable ω_1, ω_2. Let T denote the group of translations by elements of $\mathbb{Z}\omega_1 + \mathbb{Z}\omega_2$ acting on $\mathbb{C}$. Then we have

$$\mathscr{C}(0, \omega_1, \omega_2, \omega_1 + \omega_2) \text{ is a fundamental domain for the action of } \mathbb{Z}\omega_1 + \mathbb{Z}\omega_2 \text{ on } \mathbb{C}. \tag{69}$$

Now we define the following polygons in $\mathbb{C} \cong \mathbb{R}^2$. Let

$$\mathscr{G}_U = \mathscr{C}_{\mathrm{Euc}}(0, 2, 1 + \mathbf{i}, 1 - \mathbf{i}),$$

and let

$$\mathscr{G} = \mathscr{C}_{\mathrm{Euc}}(1, 2, 1 + \mathbf{i}). \tag{70}$$

The relation between the polygons is that $\mathscr{G}_U$ is a square centered at 1, while $\mathscr{G}$ is an isosceles right triangle inside $\mathscr{G}_U$, with vertices at the center of $\mathscr{G}_U$ and two of the corners of $\mathscr{G}_U$. Therefore, it follows from (68) that we have

$$\mathscr{G}_U = \bigcup_{i=0,1,2,3} (T_{1+\mathbf{i}} R_{\frac{\pi}{2}})^i \mathscr{G}, \text{ with } (T_{1+\mathbf{i}} R_{\frac{\pi}{2}})^i \mathscr{G} \cap \mathscr{G} \subseteq \partial\mathscr{G}, \text{ for } i \not\equiv 0 \mod 4. \tag{71}$$

The relations (69) and (71) lead to the following lemma.

Lemma 2.18. *Let Γ^{ϕ_1} be as given in* (61) *and $(\Gamma^{\phi_1})_U$ as given in* (62).

(a) *The set $\mathscr{G}_U$ is a fundamental domain for the induced action of $(\Gamma^{\phi_1})_U$ on $\mathbb{C} \cong \mathbb{R}^2$.*
(b) *$\mathscr{G}$ is a fundamental domain for the induced action of $\langle T_{1+\mathbf{i}} R_{\frac{\pi}{2}} \rangle$ on $\mathscr{G}_U$.*
(c) *The set $\mathscr{G}$ is a fundamental domain for the induced action of Γ^{ϕ_1} on $\mathbb{C} \cong \mathbb{R}^2$.*

Form of $\mathscr{F}$ in terms of explicit inequalities. Combining Part (c) of Lemma 2.18, Proposition 2.16, and (52), we deduce that

$$\mathscr{F}(\mathscr{G}) = \{z \in \mathbf{H}^3 \mid \phi_{[2,3]}(z) \in \mathscr{C}_{\mathrm{Euc}}(1, 2, 1 + \mathbf{i}),\ \|x(z) - m\|^2 + y(z)^2 \geq 2, \\ \text{for } m \in 1 + (1 + \mathbf{i})\mathbb{Z}[\mathbf{i}]\}.$$

By (59), the first condition in the description of $\mathscr{F}(\mathscr{G})$ above may be replaced by

$$x(z) \in \mathscr{C}_{\mathrm{Euc}}(1, 2, 1 + \mathbf{i}). \tag{72}$$

Let $z \in \mathbb{C}$ satisfy (72). The element $m = 1$ is the element of $1 + (1 + \mathbf{i})\mathbb{Z}[\mathbf{i}]$ closest to $x(z)$. Therefore, for z satisfying (72), the condition

$$\|x(z) - m\|^2 + y(z)^2 \geq 2, \text{for all } m \in 1 + (1 + \mathbf{i})\mathbb{Z}[\mathbf{i}]$$

reduces to $\|x(z) - 1\|^2 + y(z)^2 \geq 2$. So we may rewrite the description of $\mathcal{F}(\mathcal{G})$ in the form

$$\mathcal{F}(\mathcal{G}) = \{z \in \mathbf{H}^3 \; x(z) \in \mathcal{C}_{\mathrm{Euc}}(1, 2, 1 + \mathbf{i}), \; \|x(z) - 1\|^2 + y(z)^2 \geq 2\}. \qquad (73)$$

Additional facts regarding convex hulls and totally geodesic hypersurfaces in $\overline{\mathbf{H}^3}$. We now extend our "geodesic hull" treatment of $\mathcal{F}$ from the boundary into the interior of $\mathbf{H}^3$. We first recall certain additional facts regarding convex hulls and totally geodesic hypersurfaces in $\mathbf{H}^3$.

The description of the geodesics in $\mathbf{H}^2$ is well known, but the corresponding description of the totally geodesic surfaces in $\mathbf{H}^3$ perhaps not as well known, so we recall it here. Henceforth we abbreviate "totally geodesic" by t.g. Although all t.g. surfaces are related by isometries, in our model they have two basic types. The first type is a vertical upper half-plane passing through the origin with angle θ measured counterclockwise from the real axis, which we denote by $\mathbf{H}^2(\theta)$. The second type is an upper hemisphere centered at the origin with radius r, which we will denote by $\mathbb{S}_r^+(0)$. The t.g. surfaces of $\mathbf{H}^3$ are the $\mathbf{H}^2(\theta)$, the $\mathbb{S}_r^+(0)$, and their translates by elements of $\mathbb{C}$. For each of the basic t.g. surfaces, we produce an isometry $g \in \mathrm{Aut}(\mathbf{H}^3)$, necessarily orientation-reversing, such that $\mathrm{Fix}(g)$ is precisely the surface in question. The existence of such a g shows that the surface is a t.g. surface.

We define

$$\overline{\mathbf{H}^3} = \mathbf{H}^3 \cup \mathbb{C} \cup \infty$$

to be the usual closure of $\mathbf{H}^3$ and extend the action of fractional linear transformations and the notion of the convex hull in the usual way. For any subset $\mathcal{S}$ of $\mathbf{H}^3$, $\overline{\mathcal{S}}$ will denote the closure in $\overline{\mathbf{H}^3}$. For $g \in \mathrm{Aut}(\mathbf{H}^3)$, we will likewise use g to denote the extension of g to the closure $\overline{\mathbf{H}^3}$. Henceforth, we will work exclusively in the setting of the closure $\overline{\mathbf{H}^3}$ of $\mathbf{H}^3$. Thus, we will actually identify the closures of the t.g. surfaces.

The basic orientation-reversing isometry of $\overline{\mathbf{H}^3}$ may be denoted by R^*. With $x_1 + x_2\mathbf{i} + y\mathbf{j} \in \overline{\mathbf{H}^3}$, we have

$$R^*(x_1 + x_2\mathbf{i} + y\mathbf{j}) = x_1 - x_2\mathbf{i} + y\mathbf{j}.$$

Clearly, we have $\mathrm{Fix}(R^*) = \overline{\mathbf{H}^2(0)}$. To obtain isometries corresponding to the other vertical planes, let

$$R_\theta = \begin{pmatrix} e^{\mathbf{i}\theta/2} & 0 \\ 0 & e^{-\mathbf{i}\theta/2} \end{pmatrix}.$$

Because $R_\theta \overline{\mathbf{H}^2(0)} = \overline{\mathbf{H}^2(\theta)}$, we have

$$\mathrm{Fix}(\mathbf{c}(R_\theta)R^*) = \overline{\mathbf{H}^2(\theta)}.$$

To define the isometry I such that $\mathrm{Fix}(I)$ is the basic hemisphere $\overline{\mathbb{S}_0^+(1)}$, let $\bar{z}$ denote the conjugate of the quaternion z, i.e., if $z = x_1 + x_2\mathbf{i} + y\mathbf{j}$, then $\bar{z} = x_1 - x_2\mathbf{i} - y\mathbf{j}$. For $z \in \overline{\mathbf{H}^3}$, set

$$I(z) = 1/\bar{z}.$$

We have the equality $z/I(z) = \|z\|^2$. Observe that $\overline{\mathbb{S}_1^+(0)}$ is precisely the set of quaternions in $\overline{\mathbf{H}^3}$ of norm one. Thus, $\mathrm{Fix}(I) = \overline{\mathbb{S}_1^+(0)}$. For the more general hemispheres $\overline{\mathbb{S}_r^+(0)}$, set

$$A(r) = \begin{pmatrix} \sqrt{r} & 0 \\ 0 & \frac{1}{\sqrt{r}} \end{pmatrix}.$$

Then, since $A(r)\overline{\mathbb{S}_1^+(0)} = \overline{\mathbb{S}_r^+(0)}$, we have $\mathrm{Fix}(\mathbf{c}(A(r))I) = \overline{\mathbb{S}_r^+(0)}$.

In order to denote the convex hull in $\overline{\mathbf{H}^3}$, we use the notation $\mathscr{C}_{\mathbf{H}}$. Therefore, if $\mathrm{d}s^2$ is the hyperbolic metric on $\overline{\mathbf{H}^3}$, we have

$$\mathscr{C}_{\mathbf{H}}(p_1, \ldots, p_r) = \mathscr{C}_{\mathrm{d}s^2}(p_1, \ldots, p_r),$$

in terms of our original notational conventions.

Let $p_1, \ldots, p_r \in \overline{\mathbf{H}^3}$, for $r > 3$, not lying on the same totally geodesic surface, such that for each i, $1 \le i \le r$,

$$p_i \notin \mathscr{C}_{\mathbf{H}}(p_1, \ldots, p_{i-1}, p_{i+1}, \ldots, p_r).$$

Then the set $\mathscr{C}_{\mathbf{H}}(p_1, \ldots, p_r)$ will be called the **solid convex polytope with vertices at $p_1, \ldots, p_r$**. It is clear that for any $p_1, \ldots p_r \in \overline{\mathbf{H}^3}$ not lying in the same totally geodesic surface, $\mathscr{C}_{\mathbf{H}}(p_1, \ldots, p_r)$ is a solid convex polytope with vertices consisting of some subset of the r points.

Description of $\mathscr{F}(\mathscr{G})$ as a solid convex polytope.

Proposition 2.19. *The solid convex polytope with four vertices in $\overline{\mathbf{H}^3}$ given by*

$$\mathscr{F}(\mathscr{G}) = \mathscr{C}_{\mathbf{H}}(1 + \sqrt{2}\mathbf{j}, 2 + \mathbf{j}, 1 + \mathbf{i} + \mathbf{j}, \infty) \tag{74}$$

is a good Grenier fundamental domain for the action of $\Gamma = \mathbf{c}^{-1}(\mathrm{SO}_3(\mathbb{Z}[\mathbf{i}]))$ on $\overline{\mathbf{H}^3}$.

In the interactive graphical representation, [3], $\mathscr{F}(\mathscr{G})$ is the solid region inside the triangular prism and above the red sphere.

2.4 $\mathbf{SO(2,1)_{\mathbb{Z}}}$ as a group of fractional linear transformations

We will now use the results of §2.1 and §2.2 to deduce a realization of $\Gamma_{\mathbb{Z}} = \mathrm{SO}(2,1)_{\mathbb{Z}}$ as a group of fractional linear transformations, as well as a description of a fundamental domain for $\Gamma_{\mathbb{Z}}$ acting on $\mathbf{H}^2$ that is in some sense (to be explained precisely below) compatible with the fundamental domain of Γ acting on $\mathbf{H}^3$.

We maintain the notational conventions established in §2.1. In particular, $G = \mathrm{SO}_3(\mathbb{C})$ and $\Gamma = \mathrm{SO}_3(\mathbb{Z}[\mathbf{i}])$. It is crucial, for the moment, that we observe the distinction between G, Γ and their isomorphic images under $\mathbf{c}^{-1}$.

Definition 2.20. Set

$$\Gamma_{\mathbb{Z}} = \mathbf{c}(\mathrm{SL}_2(\mathbb{R}) \cap \mathbf{c}^{-1}(\Gamma)). \tag{75}$$

Remark 2.21. Note that the elements of $\Gamma_{\mathbb{Z}}$ do not have real entries! The naïve approach to the definition of $\Gamma_{\mathbb{Z}}$ would be to take the elements of Γ with real entries, as in the case of $\mathrm{SL}_2(\mathbb{Z}[\mathbf{i}])$ and $\mathrm{SL}_2(\mathbb{Z})$. However, this clearly cannot be the right definition because the resulting discrete group would be contained in SO(3), hence compact, and hence finite. The justification for Definition 2.20 is contained in Proposition 2.22, below.

Recall the orthonormal basis β for $\mathrm{Lie}(\mathrm{SL}_2(\mathbb{C}))$ defined in (4). Define a new basis η by specifying the change-of-basis matrix

$$\alpha^{\beta \mapsto \eta} = \mathrm{diag}(1, -\mathbf{i}, 1). \tag{76}$$

Let $V_{\mathbb{R}}$ be a *real* vector space of dimension 3. Let $\mathrm{SO}(2,1)$ denote the group of unimodular linear automorphisms of $V_{\mathbb{R}}$ preserving a form $B_{\mathbb{R}}$ on $V_{\mathbb{R}}$ of bilinear signature $(2,1)$. For definiteness, we will take

$$V_{\mathbb{R}} = \mathbb{R}\text{-span}(\eta) \subseteq \mathrm{Lie}(\mathrm{SL}_2(\mathbb{C})), \quad B_{\mathbb{R}} = B|_{V_{\mathbb{R}}},$$

where β' is the basis of $\mathrm{Lie}(\mathrm{SL}_2(\mathbb{C}))$ defined at (3), and B is as usual the Killing form on $\mathrm{Lie}(\mathrm{SL}_2(\mathbb{C}))$. From the fact that β is an orthonormal set under B and from (76), it is immediately verified that $B|_{\mathbb{R}}$ has signature $(2,1)$. Note also that

$$V := V_{\mathbb{R}} \otimes \mathbb{C} = \mathrm{Lie}(\mathrm{SL}_2(\mathbb{C})).$$

By considering $\mathrm{SO}_{2,1}$ as a subset of $\mathrm{GL}_3(\mathbb{R})$ we obtain the **standard representation of $\mathbf{SO(2,1)}$**. We define $\mathrm{SO}(2,1)_{\mathbb{Z}}$ to be the matrices with integer coefficients in the standard representation of $\mathrm{SO}(2,1)$.

Recall from (8) the definition of the morphism

$$\mathbf{c}_\eta := \mathbf{c}_{V,\eta} : \mathrm{SL}_2(\mathbb{C}) \to \mathrm{SL}_3(\mathbb{R}).$$

Proposition 2.22. *Let $\Gamma_{\mathbb{Z}}$ be as defined in* (2.20). *Then the restriction of* $\mathbf{c}_\eta$ *to* $V_{\mathbb{R}}$ *provides an isomorphism*

$$\mathbf{c}_\eta : \mathrm{SL}_2(\mathbb{R})/\{\pm I\} \to \mathrm{SO}(2,1)^0 \tag{77}$$

of Lie groups. The isomorphism of (77) *further restricts to an isomorphism of discrete subgroups*

$$\mathbf{c}_\eta : \mathbf{c}^{-1}(\Gamma_{\mathbb{Z}}) \to \mathrm{SO}(2,1)_{\mathbb{Z}}. \tag{78}$$

As a result, $\mathbf{c}_\eta \mathbf{c}^{-1}$ *exhibits an isomorphism*

$$\Gamma_{\mathbb{Z}} \cong \mathrm{SO}(2,1)_{\mathbb{Z}}. \tag{79}$$

The next proposition, Proposition 2.24, is the analogue of Proposition 2.8 for the real form of the complex group. Proposition 2.24 below is, in contrast, almost a triviality to prove at this point, since it can be deduced rather readily from Proposition 2.8.

For Proposition 2.24, it is necessary to recall the Ξ-subsets of $\mathrm{SL}_2(\mathbb{Z}[\mathbf{i}])$ defined in (22) and (23). For each of the three Ξ-subsets, we define

$$(\Xi)_{\mathbb{Z}} = \Xi \cap \mathrm{SL}_2(\mathbb{R}). \tag{80}$$

The following result both justifies this notation and clarifies the meaning of Proposition 2.24, below.

Lemma 2.23. *Each* $(\Xi)_{\mathbb{Z}}$*-group can be given the following description:*

$$\text{For suitable fixed } \begin{pmatrix}\overline{p} & \overline{q}\end{pmatrix}, \begin{pmatrix}\overline{r} & \overline{s}\end{pmatrix} \in \left\{ \begin{matrix} \begin{pmatrix}1 & 1\end{pmatrix}, \\ \begin{pmatrix}1 & 0\end{pmatrix}, \\ \begin{pmatrix}0 & 1\end{pmatrix} \end{matrix} \right\} \subset (\mathrm{SL}_2(\mathbb{Z}[\mathbf{i}])/(2))^2,$$
$$\Xi = \mathrm{red}_2^{-1}\left(\left\{ \begin{pmatrix}\overline{p} & \overline{q} \\ \overline{r} & \overline{s}\end{pmatrix}, \begin{pmatrix}\overline{r} & \overline{s} \\ \overline{p} & \overline{q}\end{pmatrix} \right\}\right). \tag{81}$$

In order to obtain Ξ_{12} *in this manner, we may take, in* (81),

$$\begin{pmatrix}\overline{p} & \overline{q}\end{pmatrix} = \begin{pmatrix}1 & 0\end{pmatrix} \quad \text{and} \quad \begin{pmatrix}\overline{r} & \overline{s}\end{pmatrix} = \begin{pmatrix}0 & 1\end{pmatrix}.$$

Further, we may take

$$\begin{pmatrix}\overline{p} & \overline{q}\end{pmatrix} = \begin{pmatrix}1 & 1\end{pmatrix}, \text{ in order to obtain } \Xi_1 \text{ and } \Xi_2,$$

and

$$\begin{pmatrix}\overline{r} & \overline{s}\end{pmatrix} = \begin{pmatrix}0 & 1\end{pmatrix}, \text{ in order to obtain } \Xi_1,$$

$$\begin{pmatrix}\overline{r} & \overline{s}\end{pmatrix} = \begin{pmatrix}1 & 0\end{pmatrix}, \text{ in order to obtain } \Xi_2.$$

Proposition 2.24. *With $\Gamma_{\mathbb{Z}}$ defined as in Definition 2.20, we have*

$$\mathbf{c}^{-1}(\Gamma_{\mathbb{Z}}) = (\Xi_{12})_{\mathbb{Z}} \bigcup \frac{1}{\sqrt{2}}(\Xi_2)_{\mathbb{Z}} \begin{pmatrix} 1 & -1 \\ 0 & 2 \end{pmatrix}. \tag{82}$$

From (82), we deduce the following analogue of Lemma 2.9

Lemma 2.25. *Let $\mathbf{c}^{-1}(\Gamma_{\mathbb{Z}})$ be the discrete subgroup of $\mathrm{SL}_2(\mathbb{R})$ defined in (75), and given explicitly in matrix form in (82). All the other notation is also as in Proposition 2.24.*

(a) We have

$$\mathbf{c}^{-1}(\Gamma_{\mathbb{Z}}) \cap \mathrm{SL}_2(\mathbb{Z}) = (\Xi_{12})_{\mathbb{Z}}.$$

(b) We have

$$[\mathbf{c}^{-1}(\Gamma_{\mathbb{Z}}) : (\Xi_{12})_{\mathbb{Z}}] = 2, \quad [\mathrm{SL}_2(\mathbb{Z}) : \Xi_{12}] = 3. \tag{83}$$

Explicitly, a representative of the unique nonidentity right coset of $(\Xi_{12})_{\mathbb{Z}}$ in $\mathbf{c}^{-1}(\Gamma)$ is

$$\frac{1}{\sqrt{2}}\begin{pmatrix} 1 & 0 \\ 1 & 1 \end{pmatrix}\begin{pmatrix} 1 & -1 \\ 0 & 2 \end{pmatrix} = \frac{1}{\sqrt{2}}\begin{pmatrix} 1 & -1 \\ 1 & 1 \end{pmatrix}.$$

2.5 *Fundamental domain for* $\mathrm{SO}(2,1)_{\mathbb{Z}}$ *acting on* $\mathbf{H}^2$ *and its relation to that of* $\mathrm{SO}_3(\mathbb{Z}[\mathbf{i}])$

The main point of this section is that, provided the fundamental domain $\mathscr{G}_{\mathbb{R}}$ of the standard unipotent subgroup of $\mathbf{c}^{-1}(\Gamma_{\mathbb{Z}})$ is chosen in a way that is compatible with the choice of $\mathscr{G}$ in (70), then the good Grenier fundamental domain $\mathscr{F}_{\mathbb{R}}(\mathscr{G}_{\mathbb{R}})$ for $\mathbf{c}^{-1}(\Gamma_{\mathbb{Z}})$ corresponding to $\mathscr{G}_{\mathbb{R}}$ will have a close geometric relationship to $\mathscr{F}(\mathscr{G})$. Based on the classical example of Dirichlet's fundamental domain for $\mathrm{SL}_2(\mathbb{Z})$ acting on $\mathbf{H}^2$ and the Picard domain, one might guess that we would have the equality

$$\mathscr{F}_{\mathbb{R}}(\mathscr{G}_{\mathbb{R}}) = \mathscr{F}(\mathscr{G}) \cap \mathbf{H}^2. \tag{84}$$

In fact, this intersection property cannot hold, because of the presence of additional torsion elements (the powers of $\omega_8 I_2$) in $\mathbf{c}^{-1}(\Gamma)$. However, in a sense which will be made precise in Proposition 2.27, below, the next best thing holds. Namely, the intersection of the set consisting of *two* Γ-translates of $\mathscr{F}(\mathscr{G})$ with $\mathbf{H}^2$ equals $\mathscr{F}_{\mathbb{R}}(\mathscr{G}_{\mathbb{R}})$, for the choice of $\mathscr{G}_{\mathbb{R}}$ in (85), below.

In the case of $\Gamma_{\mathbb{Z}} \subset \mathrm{Aut}^+(\mathbf{H}^2)$, commensurable with $\mathrm{SL}_2(\mathbb{Z}[\mathbf{i}])$, we have the obvious analogue of Theorem 2.14, defining a good Grenier fundamental domain for the action of $\Gamma_{\mathbb{Z}}$. In order to distinguish the real case $\Gamma_{\mathbb{Z}} \subset \mathrm{Aut}^+(\mathbf{H}^2)$ from the complex case, we add the subscript $\mathbb{R}$ to the sets $\mathscr{G}$ $\mathscr{F}_1$, $\mathscr{F}(\mathscr{G})$, and so write $\mathscr{G}_{\mathbb{R}}$ $\mathscr{F}_{1,\mathbb{R}}$, $\mathscr{F}_{\mathbb{R}}(\mathscr{G}_{\mathbb{R}})$. In this case, the good Grenier fundamental domain coincides with the classical notion of the *Ford fundamental domain* for a discrete subgroup of $\mathrm{Aut}^{+}(\mathbf{H}^2)$ of finite covolume. See, for example, [16], p. 44. However, we use the terminology **Grenier domain** even in this context, in order to stress the eventual connections with the higher-rank case.

Explicit Descriptions of $\mathcal{G}_{\mathbb{R}}$ and $\mathcal{F}_{\mathbb{R}}(\mathcal{G}_{\mathbb{R}})$.

Lemma 2.26. *(a) We have*

$$(\Gamma_{\mathbb{Z}})^{\phi_1} = \begin{pmatrix} 1 & 2\mathbb{Z} \\ 0 & 1 \end{pmatrix}.$$

(b) The interval

$$\mathcal{G}_{\mathbb{R}} := [0,2] \tag{85}$$

is a fundamental domain for the action of $\Gamma_{\mathbb{Z}}^{\phi_1}$ on $\mathbb{R}$ satisfying

$$\mathcal{G}_{\mathbb{R}} = \overline{\mathrm{Int}\mathcal{G}_{\mathbb{R}}}.$$

(c) With $\mathcal{G}_{\mathbb{R}}$ as defined in (85)*, part (b) implies that*

$$\mathcal{F}_{\mathbb{R}}\,(\mathcal{G}_{\mathbb{R}}) = \{z \in \mathbf{H}^2 \mid 0 \le x(z) \le 2,\ y(z)^2 + (x-1)^2 \ge 2\} = \mathcal{C}_{\mathbf{H}}(\mathbf{i}, 2+\mathbf{i}, \infty). \tag{86}$$

Geometric relation of $\mathcal{F}_{\mathbb{R}}(\mathcal{G}_{\mathbb{R}})$ to $\mathcal{F}(\mathcal{G})$. In order to relate the fundamental domain of a subgroup of $\mathrm{SL}_2(\mathbb{R})$ acting on $\mathbf{H}^2$ to the fundamental domain of a subgroup of $\mathrm{SL}_2(\mathbb{C})$ acting on $\mathbf{H}^3$, we consider $\mathbf{H}^2$ embedded in $\mathbf{H}^3$ as the totally geodesic surface $\mathbf{H}^2(0)$. Note that

$$\mathbf{H}^2(0) = \{x\mathbf{i} + y\mathbf{j} \mid y > 0\},$$

and the actions of $\mathrm{SL}_2(\mathbb{R})$ on $\mathbf{H}^2$ and $\mathbf{H}^2(0)$ are equivariant with the obvious isomorphism

$$\mathbf{H}^2 \xrightarrow{\cong} \mathbf{H}^2(0), \text{ mapping } x + y\mathbf{j} \mapsto x\mathbf{i} + y\mathbf{j}.$$

Under this isomorphism of $\mathrm{SL}_2(\mathbb{R})$-homogeneous spaces, $\mathcal{F}_{\mathbb{R}}(\mathcal{G}_{\mathbb{R}})$ corresponds to

$$\mathcal{C}_{\mathbf{H}}(\mathbf{j}, 2\mathbf{i} + \mathbf{j}, \infty) \text{ in } \mathbf{H}^2(0). \tag{87}$$

Because of the isomorphism, we can safely ignore the distinction between the forms of $\mathcal{F}_{\mathbb{R}}(\mathcal{G}_{\mathbb{R}})$ in (86) and (87).

Because, as can be verified readily,

$$\mathcal{G}_{\mathbb{R}} = \left(\mathcal{G} \cup \mathbf{c}(T_1)\left(R^2_{\frac{\pi}{2}}\right)\mathcal{G}\right) \cap \mathbf{H}^2_{\mathbf{j}}, \tag{88}$$

we cannot hope that we will have the straightforward relation

$$\mathscr{F}_{\mathbb{R}}\left(\mathscr{G}_{\mathbb{R}}\right) = \mathscr{F}\left(\mathscr{G}\right) \cap \mathbf{H}^2_{\mathbf{j}}$$

that we find in the classical case of $\mathrm{SL}_2(\mathbb{Z}[\mathbf{i}])$ and $\mathrm{SL}_2(\mathbb{Z})$. However, we do have the next best possible relation between the fundamental domains.

Proposition 2.27. *We have the relation*

$$\mathscr{F}\left(\mathscr{G}_{\mathbb{R}}\right) = \left(\mathscr{F}(\mathscr{G}) \cup \mathbf{c}(T_1)\left(R^2_{\frac{\pi}{2}}\right)\mathscr{F}(\mathscr{G})\right) \cap \mathbf{H}^2_{\mathbf{j}}.$$

Remark 2.28. We note for possible future reference that $\mathscr{F}_{\mathbb{R}}(\mathscr{G}_{\mathbb{R}})$ is the *normal geodesic projection* of the union of $\mathscr{F}(\mathscr{G})$ and one translate $\mathbf{c}(T_1)\left(R^2_{\frac{\pi}{2}}\right)\mathscr{F}(\mathscr{G})$ of $\mathscr{F}(\mathscr{G})$. This relation between the fundamental domains is connected to the one given in Proposition 2.27, though neither relation implies the other, in general. The relation may be of some use in relating spectral expansions in the complex case to spectral expansions in the real case. We explored one aspect of this topic in §1 of this paper, above.

In the interactive graphical representation [3], the front face of the "cut-off prism" solid amounts to one half of $\mathscr{F}_{\mathbb{R}}(\mathscr{G}_{\mathbb{R}})$.

2.6 *Spectral zeta functions*

This section discusses a potential application of the results of this paper and indicates a future line of investigation building on this work. Jorgenson and Lang, in works such as [17], [19] (see the introduction to the latter work especially), and [20], have laid out and begun to pursue an ambitious program of using heat kernel analysis to associate additive spectral zeta functions to quotients of symmetric spaces. When completed, this theory would subsume the basic theory of the Riemann zeta function and Selberg zeta function (among others), and clarify the relationship between the zeta functions arising at different geometric levels. The main component of the program is obtaining a theta inversion formula.

In [19], which carries out the derivation of the theta inversion formula for the special case of

$$X = \Gamma\backslash G/K = \mathrm{SL}_2(\mathbb{Z}[\mathbf{i}])\backslash\mathrm{SL}_2(\mathbb{C})/\mathrm{SU}(2,\mathbb{C}),$$

the authors compute the regularized trace of an integral operator on functions on X. The kernel of the integral operator is $\mathbf{K}_{t,X}(z,w)$, the heat kernel on X. The trace of such an integral operator is defined to be the integral on the diagonal

$$\int_X \mathbf{K}_{t,X}(z,z)\,\mathrm{d}z.$$

Although this integral is infinite, because of the cusp of X, the integrals over sets X_Y approximating by covering X only up to some fixed finite "distance" in the cusp are finite and diverge logarithmically in Y. That is,

$$\lim_{Y\to\infty}\int_{X_Y}\mathbf{K}_{t,X}(z,z)\,\mathrm{d}z - c_1(t)\log Y \text{ exists as a } \mathbb{C}\text{-valued function of } t, \tag{89}$$

where $c_1(t)$ is a factor, constant in Y, and determined in [19]. For the purposes of such an integration, we can replace X with a suitable fundamental domain $\mathscr{F}$. Similarly, we replace the truncation X_Y of X with a matching truncation $\mathscr{F}_Y$ of $\mathscr{F}$.

To obtain the theta inversion formula, the limit of (89) is computed in two ways. One computation is from the expression of the heat kernel as the periodized heat kernel on the universal covering space $\mathbf{H}^3$. This method of computing the limit in (89) yields

$$e^{-2t}(4t)^{-\frac{1}{2}}\Theta^{\mathrm{NC}}(1/t) + \Theta^{\mathrm{Cus}}(1/t). \tag{90}$$

In (90), $\Theta^{\mathrm{NC}}(1/t)$, the inverted theta series, is defined in terms of invariants of certain Γ-conjugacy classes in Γ, while $\Theta^{\mathrm{Cus}}(1/t)$, the inverted theta integral, is a sum of products composed of special values of $\zeta_{\mathbb{Q}(\mathbf{i})}$, constants similar to Euler's γ, and single integrals whose Gauss transforms are exact. (We refer to §XIV.7, of [19], for exact definitions of $\Theta^{\mathrm{NC}}(1/t)$, $\Theta^{\mathrm{Cus}}(1/t)$ and the other terms in the theta relation.) The other method of computing the limit in (89) is from the expansion of $\mathbf{K}_{t,X}(z,z)\,\mathrm{d}z$ in terms of the spectrum of the Laplacian Δ_X. This second method of computing the limit of (89) yields

$$\theta_{\mathrm{Cus}}(t) + 1 + \theta_{\mathrm{Eis}}(t), \tag{91}$$

where $\theta_{\mathrm{Cus}}(t)$ is the theta series $\sum_{j=1}^{\infty} e^{-\lambda_j t}$ and λ_j are the eigenvalues of Δ_X, and $\theta_{\mathrm{Eis}}(t)$ is what remains as the limit of the integral of the convolution of $\mathbf{K}_{t,X}(z,w)$ with certain Eisenstein series once the term $c_1(t)\log(Y)$ has been subtracted. Setting equal the two expressions (90) and (91), for the same limit (89), we obtain the theta inversion formula for X:

$$e^{-2t}(4t)^{-\frac{1}{2}}\Theta^{\mathrm{NC}}(1/t) + \Theta^{\mathrm{Cus}}(1/t) = \theta_{\mathrm{Cus}}(t) + 1 + \theta_{\mathrm{Eis}}(t). \tag{92}$$

Next, note that there is an infinite sequence of arithmetic quotients

$$X_n = \mathrm{SL}_n(\mathbb{Z}[\mathbf{i}])\backslash \mathrm{SL}_n(\mathbb{C})/\mathrm{SU}(n), \quad n > 1,$$

having $X = X_2$ as its first nontrivial member. Generalizations of (92) to X_n for $n > 2$ are discussed in [20]. In order to obtain exact formulas analogous to

(92), we would have to integrate over a fundamental domain, rather than over an appoximating Siegel set, which is a more common analytic model in the literature.

In the present work, we initiate an extension of the Jorgenson–Lang project to the sequence of arithmetic quotients

$$\mathrm{SO}_n(\mathbb{Z}[\mathbf{i}])\backslash\mathrm{SO}_n(\mathbb{C})/\mathrm{SO}(n) \tag{93}$$

and related arithmetic quotients of real forms of the symmetric spaces. The main results of the present paper are restricted to the group theory (Propositions 2.8 and 2.24) and fundamental domains (Propositions 2.19 and 2.27) in the first case of $n = 2$. Nevertheless, some of the intermediate results are couched in a more general terminology and notation, with a view towards building upwards from the case $n = 2$, to the case of a general n. Thus, our project includes a natural extension and generalization of Grenier's work in [13] and [14] to a different sequence of symmetric spaces.

The identification

$$\mathrm{SL}_2(\mathbb{C})/\{\pm I\} \xrightarrow{\cong} \mathrm{SO}_3(\mathbb{C})$$

allows us to view the theta inversion relation (conjecturally) associated with the case $n = 2$ in (93) as a theta inversion relation associated with a quotient of $\mathrm{SL}_2(\mathbb{C})/K$ by an arithmetic subgroup different from, but still commensurable with, the "standard" arithmetic subgroup $\mathrm{SL}_2(\mathbb{Z}[\mathbf{i}])$. The results of this paper will, it is hoped, enable future investigations to apply the machinery developed in [19] to this "nonstandard" arithmetic subgroup $\mathbf{c}^{-1}(\mathrm{SO}_3(\mathbb{Z}[\mathbf{i}]))$ of $\mathrm{SL}_2(\mathbb{C})$ to obtain the corresponding theta function.

References

1. Babillot, M.: Points entiers et groupes discrets: de l'analyse aux systèmes dynamiques. In: Rigidité, groupe fondamental et dynamique, *Panor. Synthèses*, vol. 13, pp. 1–119. Soc. Math. France, Paris (2002). With an appendix by Emmanuel Breuillard.
2. Brenner, E.: A fundamental domain of Ford type for some integer subgroups of orthogonal groups. Preprint, arXiv:math.NT/0605012.
3. Brenner, E.: Interactive graphics available for download (2011). URL http://megamachine.org/?p=158.
4. Brenner, E., Spinu, F.: Artin formalism, for Kleinian groups, via heat kernel methods (2007). URL http://megamachine.org/?p=158.
5. Brenner, E., Spinu, F.: Artin formalism for Selberg zeta functions of co-finite Kleinian groups. *J. Théor. Nombres*, Bordeaux **21**(1), 59–75 (2009).
6. Brenner, E.P.: Grenier domains for arithmetic groups and associated tilings. ProQuest LLC, Ann Arbor, MI (2005). Thesis (Ph.D.)–Yale University.
7. Catto, S., Huntley, J., Jorgenson, J., Tepper, D.: On an analogue of Selberg's eigenvalue conjecture for $\mathrm{SL}_3(\mathbf{Z})$. *Proc. Amer. Math. Soc.* **126**(12), 3455–3459 (1998)
8. Elstrodt, J., Grunewald, F., Mennicke, J.: *Groups acting on hyperbolic space*. Springer Monographs in Mathematics. Springer-Verlag, Berlin (1998)

9. Flensted-Jensen, M.: Spherical functions on semisimple Lie groups: A method of reduction to the complex case. *J. Funct. Anal.* **30**, 106–146 (1978).
10. Flensted-Jensen, M.: Discrete series for semisimple symmetric space. *Annals of Math.* **111**, 253–311 (1980).
11. Friedman, J.S.: The Selberg trace formula and Selberg zeta-function for cofinite Kleinian groups with finite-dimensional unitary representations. *Math. Z.* **250**(4), 939–965 (2005)
12. Gangolli, R.: Asymptotic behavior of spectra of compact quotients of certain symmetric spaces. *Acta Math.* **121**, 151–192 (1968)
13. Grenier, D.: Fundamental domains for the general linear group. *Pacific J. Math.* **132**(2), 293–317 (1988)
14. Grenier, D.: On the shape of fundamental domains in $\mathrm{GL}(n, \mathbf{R})/\mathrm{O}(n)$. *Pacific J. Math.* **160**(1), 53–66 (1993)
15. Helgason, S.: *Differential Geometry, Lie Groups, and Symmetric Spaces*. Academic Press (1978)
16. Iwaniec, H.: Introduction to the spectral theory of automorphic forms. Biblioteca de la Revista Matemática Iberoamericana. [Library of the Revista Matemática Iberoamericana]. *Revista Matemática Iberoamericana*, Madrid (1995)
17. Jorgenson, J., Lang, S.: Spherical inversion on $\mathrm{SL}(n, \mathbf{R})$. Springer Monographs in Mathematics. Springer-Verlag, New York (2001)
18. Jorgenson, J., Lang, S.: *$Pos_n(R)$ and Eisenstein series,* Lecture Notes in Mathematics, vol. 1868. Springer-Verlag, Berlin (2005)
19. Jorgenson, J., Lang, S.: *The heat kernel and theta inversion on $SL_2(\mathbf{C})$*. Springer Monographs in Mathematics. Springer, New York (2008)
20. Jorgenson, J., Lang, S.: *Heat Eisenstein series on $L_n(\mathbf{C})$*. Mem. Amer. Math. Soc. **201**(946), viii+127 (2009)
21. Lang, S.: *Introduction to modular forms*. Springer-Verlag, Berlin (1976). Grundlehren der mathematischen Wissenschaften, No. 222
22. Mostow, G.D.: Some new decomposition theorems for semi-simple groups. *Mem. Amer. Math. Soc.* pp. 31–54 (1955)
23. Sarnak, P., Strömbergsson, A.: Minima of Epstein's zeta function and heights of flat tori. *Invent. Math.* **165**(1), 115–151 (2006). DOI 10.1007/s00222-005-0488-2. URL http://dx.doi.org/10.1007/s00222-005-0488-2
24. Séminaire de Mathématiques Supérieures: "Automorphic Forms and L-Functions: Computational aspects" (2009). URL http://www.dms.umontreal.ca/~sms/2009/titles_e.p
25. Van Den Ban, E.P., Schlichtkrull, H.: Expansions for Eisenstein integrals on semisimple symmetric spaces. *Ark. Mat* **35**, 59–86 (1997)
26. Venkov, A.B., Zograf, P.G.: Analogues of Artin's factorization formulas in the spectral theory of automorphic functions associated with induced representations of Fuchsian groups. Izv. *Akad. Nauk SSSR Ser. Mat.* **46**(6), 1150–1158, 1343 (1982)
27. Yasaki, D.: On the existence of spines for $\mathbb{Q}$-rank 1 groups. *Selecta Math.* (N.S.) **12**(3-4), 541–564 (2006). DOI 10.1007/s00029-006-0029-x. URL http://dx.doi.org/10.1007/s00029-006-0029-x
28. Yasaki, D.: An explicit spine for the Picard modular group over the Gaussian integers. *J. Number Theory* **128**(1), 207–234 (2008). DOI 10.1016/j.jnt.2007.03.008. URL http://dx.doi.org/10.1016/j.jnt.2007.03.008

Differential characters on curves

Alexandru Buium

Dedicated to the memory of Serge Lang

Abstract The δ-characters of an abelian variety [B95] are arithmetic analogues of the Manin maps [M63]. Given a smooth projective curve X of genus at least 2 embedded into its Jacobian A, one can consider the restrictions to X of the δ-characters of A; the maps so obtained are referred to as δ-characters of X. It is easy to see that the δ-characters of X have a remarkable symmetry property at the origin. The aim of this paper is to prove that this symmetry property completely characterizes the δ-characters of X.

Key words curves • jacobians • local fields • fermat quotients

Mathematics Subject Classification (2010): 11620, 14H25

1 Introduction

In [M58, M63] Manin introduced some remarkable additive characters of the group of points of an Abelian variety defined over a complex function field; these characters are referred to as *Manin maps* and were used by Manin in his proof of the complex function field analogue of the Mordell conjecture [M63]. Manin's result had been conjectured by S. Lang in [Lan56]. In [B95] arithmetic analogues of the Manin maps were introduced; they were called *δ-characters* and in their definition Manin's horizontal differentiation was replaced by a Fermat quotient operator, δ,

A. Buium (✉)
University of New Mexico, Albuquerque NM 87131, USA
e-mail: buium@math.unm.edu

D. Goldfeld et al. (eds.), *Number Theory, Analysis and Geometry: In Memory of Serge Lang*, DOI 10.1007/978-1-4614-1260-1_5,

which on integers acts as $n \mapsto \delta n := \frac{n-n^p}{p}$, where p is a fixed prime. More generally, if $R := \hat{\mathbf{Z}}_p^{ur}$ is the completion of the maximum unramified extension of the ring $\mathbf{Z}_p$ of p-adic integers, then one considers the operator $\delta : R \to R$ defined by

$$\delta x := \frac{\phi(x) - x^p}{p},$$

where $\phi : R \to R$ is the unique lift of the p-power Frobenius map on $k := R/pR$. (Morally one views δ as an arithmetic analogue of a derivation.) Then, if X is a smooth scheme over R and $f : X(R) \to R$ is a function, one says that f is a *δ-function of order r* if for any point in $X(R)$ there is a Zariski open neighborhood $U \subset X$, a closed embedding $u : U \subset \mathbf{A}^N$, and a restricted power series F with R-coefficients in $(r+1)N$ variables such that

$$f(P) = F(u(P), \delta(u(P)), \ldots, \delta^r(u(P))), \quad P \in U(R).$$

Here we view $u(P) \in \mathbf{A}^N(R) = R^N$ and we say F is *restricted* if its coefficients tend to 0. If G is a group scheme over R, then by a *δ-character* one understands a δ-function $\psi : G(R) \to R$ which is also a group homomorphism into the additive group $\mathbf{G}_a(R) = R$. For the construction of δ-characters on Abelian schemes we refer to [B95]; there always exist δ-characters of order 2, there sometimes exist δ-characters of order 1 and, of course, there are no non-zero δ-characters of order 0. Note that δ-characters were used in [B95] to prove an arithmetic analogue of the Manin theorem of the kernel [M63] and they also have a number of applications to dynamical systems and modular forms; cf. [B05] and the bibliography therein.

Now if X is a smooth projective curve over R of genus $g \geq 2$, then one can compose the δ-characters of the Jacobian of X with Abel–Jacobi maps of X into its Jacobian. What one gets are δ-functions $X(R) \to R$ which we shall call *δ-characters* of X. If f is a δ-character (corresponding to an embedding of X into its Jacobian that sends the point P^0 into the origin), then the "formal germ" of the map

$$\begin{aligned} X(R) \times X(R) &\to R \\ (P_1, P_2) &\mapsto f(P_1) + f(P_2), \end{aligned} \tag{1}$$

at (P^0, P^0) satisfies a certain symmetry property which we shall refer to as *δ-symmetry*. The aim of this paper is to prove that, somewhat surprisingly, the converse of the above is also true: if a δ-function $f : X(R) \to R$ vanishing at a point P^0 is such that the "formal germ" of the map (1) at (P^0, P^0) is δ-symmetric, then f is a δ-character. What we have here is that formal behavior of a function at one point guarantees the correct global behavior of the function.

We refer to the body of the paper for the precise definition of δ-symmetry and further discussion of this concept. We content ourselves here with a few remarks:

Remark 1. (1) For δ-functions f of order 0 (i.e., for functions f arising from usual algebraic geometry) on an affine curve X the δ-symmetry of the formal germ of the map in Equation (1) will be automatic; it will be a consequence of the usual fundamental theorem of symmetric functions. However the analogue of the fundamental theorem of symmetric functions fails for δ-functions of order ≥ 1 and consequently δ-symmetry of formal germs of δ-functions will be far from automatic.

(2) As we shall see, there exist δ-functions $X(R) \to R$ which are not p-adic limits of polynomials in δ-characters.

(3) In [B08] it is shown that δ-symmetry, in the special case of modular curves, is closely related to the action of the Hecke operator $T(p)$ on δ-modular forms (in the sense of [B05]).

In Section 2 below we will introduce the main concepts and notation and we will state our main theorem. Section 3 is devoted to the proof of the theorem.

Acknowledgement While writing this paper the author was partially supported by NSF grant DMS 0552314.

2 Main concepts and statement of the theorem

Let $R = \hat{\mathbf{Z}}_p^{ur}$, $k = R/pR$, $\phi, \delta : R \to R$, be as in the introduction. We denote by $K = R[1/p]$ the fraction field of R. We will assume throughout the paper that $p \neq 2$. For X a smooth scheme over R we denote by $\mathcal{O}^r(X)$ the ring of δ-functions $X(R) \to R$ of order r. For G a smooth group scheme over R we denote by $\mathbf{X}^r(G)$ the R-module of δ-characters $G(R) \to R$ of order r. There are natural inclusions $\mathcal{O}^{r-1}(X) \subset \mathcal{O}^r(X)$, $\mathbf{X}^{r-1}(G) \subset \mathbf{X}^r(G)$. Also there are natural maps $\delta : \mathcal{O}^{r-1}(X) \to \mathcal{O}^r(X)$, $f \mapsto \delta \circ f$, and $\phi : \mathbf{X}^{r-1}(G) \to \mathbf{X}^r(G)$, $\psi \mapsto \phi \circ \psi$. Recall from [B95] (or [B05], Definition 3.5) that there exists a projective system of formal schemes

$$\ldots \to J^r(X) \to J^{r-1}(X) \to \ldots \to J^0(X) = \hat{X},$$

called the *p-jet spaces* of X, such that the ring $\mathcal{O}^r(X)$ naturally identifies with the ring $\mathcal{O}(J^r(X))$ of global functions on $J^r(X)$. Here $\hat{}$ stands for p-adic completion; on the other hand, completion with respect to a maximal ideal will be denoted, as a rule, by the superscript *for*. By a *δ-morphism* $f : X \to Y$ of order r between two smooth schemes we understand a morphism of formal schemes $J^r(X) \to Y$. The latter induces unique morphisms of formal schemes $J^{r+s}(X) \to J^s(Y)$ that commute (in the obvious sense) with δ; this allows one to compose δ-morphisms of orders r_1 and r_2 to get a δ-morphism of order $r_1 + r_2$. Note that the set of δ-morphisms $X \to \mathbf{A}^1$ of order r identifies with the set of δ-functions $X(R) \to R$ of order r. We shall repeatedly use the fact ([B05], Corollary 3.16) that if $U \subset X$ is open, $X_0 := X \otimes k$ is irreducible, and $U_0 \neq \emptyset$, then $\mathcal{O}^r(X) \to \mathcal{O}^r(U)$ is injective with torsion-free cokernel.

Definition 1. Let X be a smooth projective curve over R of genus $g \geq 1$ and let $P^0 \in X(R)$ be an R-point. A δ-morphism $f : X \to \mathbf{A}^1$ is called a *δ-character centered at P^0* (of order r) if

$$f = \psi \circ \beta,$$

where $\beta : X \to A := Jac(X)$ is the Abel–Jacobi map $P \mapsto \mathcal{O}(P - P^0)$ and $\psi : A \to \mathbf{G}_a = \mathbf{A}^1$ is a δ–character of A (of order r).

Remark 2. It follows from [B95], pp. 325–326, that the R-module of δ-characters of order r on X centered at P^0 has rank between $(r-1)g$ and rg. Note that any δ-character on X centered at P^0 vanishes at P^0 and actually vanishes on the set of torsion points $X(R)_{tors} := \beta^{-1}(A(R)_{\text{tors}})$. More generally any δ-character on X takes the same value at $P, Q \in X(R)$ whenever $P - Q$ is torsion in the divisor class group. In particular the p-adic closure $\mathcal{C}^r(X)$ of the R-subalgebra of $\mathcal{O}^r(X)$ generated by all δ-characters of X (centered at various points) does not separate points whose difference is torsion in the divisor class group. On the other hand if the genus is $g \geq 2$ and $r \geq 1$, then by [B96], p. 365, $\mathcal{O}^r(X)$ separates the points of $X(R)$. So, for all $r \geq 1$, we have $\mathcal{C}^r(X) \neq \mathcal{O}^r(X)$ provided there exist distinct points $P, Q \in X(R)$ such that $P - Q$ is torsion in the divisor class group.

Let $\mathbf{T} = \{T^1, \dots, T^g\}$ be a g-tuple of variables. (Later in the paper g will be the genus of our curve.) By a (smooth) *local formal scheme* (of relative dimension g) we understand a formal scheme of the form $X^{\text{for}} \simeq Spf\ R[[\mathbf{T}]]$. If $\mathbf{T}', \dots, \mathbf{T}^{(r)}$ are additional g-tuples of variables, then the local formal scheme

$$J^r(X^{\text{for}}) := Spf\ R[[\mathbf{T}, \dots, \mathbf{T}^{(r)}]]$$

will be referred to as the *p-jet space* of X^{for} of order r. One defines the maps

$$\delta : R[[\mathbf{T}, \dots, \mathbf{T}^{(i)}]] \to R[[\mathbf{T}, \dots, \mathbf{T}^{(i+1)}]]$$

by the formula

$$\delta(F(\mathbf{T}, \dots, \mathbf{T}^{(i)})) := \frac{F^{(\phi)}(\mathbf{T}^p + p\mathbf{T}', \dots, (\mathbf{T}^{(r)})^p + p\mathbf{T}^{(i+1)}) - F(\mathbf{T}, \dots, \mathbf{T}^{(i)})^p}{p};$$

the upper (ϕ) means "twisting coefficients by ϕ". A *δ-morphism* $X^{\text{for}} \to Y^{\text{for}}$ of order r is a morphism of formal schemes $J^r(X^{\text{for}}) \to Y^{\text{for}}$. Any such morphism induces unique morphisms $J^{r+i}(X^{\text{for}}) \to J^i(Y^{\text{for}})$ for all $i \geq 0$ such that the corresponding ring homomorphisms commute, in the obvious sense, with δ. This allows one to define, in an obvious way, the composition of two δ-morphisms of local formal schemes.

Let $\mathcal{F} \in R[[\mathbf{T}_1, \mathbf{T}_2]]^g$ be a formal group law in g variables (always assumed commutative); the local formal scheme $G^{\text{for}} := Spf\ R[[\mathbf{T}]]$ equipped with the formal group law $G^{\text{for}} \times G^{\text{for}} \to G^{\text{for}}$ defined by $\mathcal{F}$ will be referred to as a *local formal group*. In particular we denote by $\mathbf{G}_a^{\text{for}} = Spf\ R[[z]]$ the local formal group defined by the formal group law $z_1 + z_2$. By [B05], Proposition 4.39, for any local formal group G^{for} and any $r \geq 0$ we may consider the formal group law in $g(r+1)$ variables

$$(\mathcal{F}, \delta\mathcal{F}, \ldots, \delta^r\mathcal{F}) \in R[[\mathbf{T}, \mathbf{T}', \ldots, \mathbf{T}^{(r)}]]^{g(r+1)}. \qquad (2)$$

(That this tuple is a formal group law follows from an obvious universality property argument.) The formal group law (2) defines a local formal group $J^r(G^{\text{for}})$ which can be referred to as *p-jet space* of G^{for} of order r. By a *δ-character* $G^{\text{for}} \to \mathbf{G}_a^{\text{for}}$ of a local formal group G^{for} we understand a δ-morphism $G^{\text{for}} \to \mathbf{G}_a^{\text{for}}$ which is compatible in the obvious sense with the group laws.

Remark 3. Assume $g = 1$. If $\mathcal{F}$ has coefficients in an unramified extension $\mathcal{O}$ of $\mathbf{Z}_p$ with residue field $\bar{\mathcal{O}} = \mathcal{O}/p\mathcal{O}$ of size p^ν and if $\bar{\mathcal{F}} \in \bar{\mathcal{O}}[[\mathbf{T}_1, \mathbf{T}_2]]$ has either infinite height or height h, then there always exists at least one non-zero δ-character $G^{\text{for}} \to \mathbf{G}_a^{\text{for}}$ of order $\leq \nu h$; this is a consequence of the proof of Proposition 4.26 in [B05].

Definition 2. A δ-morphism

$$f^{\text{for}} : X^{\text{for}} \to \mathbf{A}^{1,\text{for}} := Spf\ R[[z]]$$

from a local formal scheme will be called a *δ-character* if there exists a morphism $\beta^{\text{for}} : X^{\text{for}} \to G^{\text{for}}$ into a local formal group G^{for} and a δ-character $\psi^{\text{for}} : G^{\text{for}} \to \mathbf{G}_a^{\text{for}} = \mathbf{A}^{1,\text{for}}$ such that

$$f^{\text{for}} = \psi^{\text{for}} \circ \beta^{\text{for}}.$$

Let $t_1, \ldots, t_m$ be a tuple of variables and $s_1, \ldots, s_m$ another tuple of variables, $m \geq 2$. Let $S_j := S_j(t_1, \ldots, t_m)$ be the fundamental symmetric polynomials in $t_1, \ldots, t_m$,

$$S_1 = t_1 + \cdots + t_m, \cdots, S_m = t_1 \ldots t_m.$$

Remark 4. The R-algebra homomorphism

$$R[[s_1, \ldots, s_m, \ldots, s_1^{(r)}, \ldots, s_m^{(r)}]] \to R[[t_1, \ldots, t_m, \ldots, t_1^{(r)}, \ldots, t_m^{(r)}]] \qquad (3)$$

$$s_j^{(i)} \mapsto \delta^i S_j,$$

is injective. Indeed it is enough to check the corresponding statement with R replaced by K, the quotient field of R. But

$$K[t_1, \ldots, t_m, \ldots, t_1^{(r)}, \ldots, t_m^{(r)}] = K[t_1, \ldots, t_m, \ldots, \phi^r(t_1), \ldots, \phi^r(t_m)] \qquad (4)$$

and similarly for the s_i's. Also we have an equality of ideals

$$(t_1, \ldots, t_m, \ldots, t_1^{(r)}, \ldots, t_m^{(r)}) = (t_1, \ldots, t_m, \ldots, \phi^r(t_1), \ldots, \phi^r(t_m))$$

so Equation (4) holds with polynomials replaced by power series. So we are led to check that

$$K[[s_1,\dots,s_m,\dots,\phi^r(s_1),\dots,\phi^r(s_m)]] \to K[[t_1,\dots,t_m,\dots,\phi^r(t_1),\dots,\phi^r(t_m)]]$$

is injective. We claim the latter map is faithfully flat and this will prove its injectivity. Now the claim follows from the fact that the inclusion

$$K[s_1,\dots,s_m,\dots,\phi^r(s_1),\dots,\phi^r(s_m)] \to K[t_1,\dots,t_m,\dots,\phi^r(t_1),\dots,\phi^r(t_m)]$$

is finite and flat and the ideal $(t_1,\dots,t_m,\dots,\phi^r(t_1),\dots,\phi^r(t_m))$ is the only maximal ideal above $(s_1,\dots,s_m,\dots,\phi^r(s_1),\dots,\phi^r(s_m))$.

In what follows we shall view the map (3) as an inclusion.

Definition 3. A series in $R[[t_1,\dots,t_m,\dots,t_1^{(r)},\dots,t_m^{(r)}]]$ is *δ-symmetric* if it is in the image of $R[[s_1,\dots,s_m,\dots,s_1^{(r)},\dots,s_m^{(r)}]]$.

Let t be a variable. For any series $F \in R[[t,t',\dots,t^{(r)}]]$ and any $m \geq 2$ one can consider the series

$$\Sigma_m F := \sum_{i=1}^{m} F\left(t_i,t_i',\dots,t_i^{(r)}\right) \in R\left[\left[t_1,\dots,t_m,\dots,t_1^{(r)},\dots,t_m^{(r)}\right]\right].$$

These series are not δ-symmetric in general. For example, the series $\Sigma_2(t')^2$ is not δ-symmetric. On the other hand, for instance, $\Sigma_m t'$ is δ-symmetric for any $m \geq 2$. An obvious class of examples of series F such that $\Sigma_m F$ are δ-symmetric is provided by the series in

$$\sum_{i=0}^{r} \phi^i(tR[[t]]).$$

A less obvious class of examples of series F such that $\Sigma_m F$ are δ-symmetric is provided by Lemma 1 below. Let us also note that if $\Sigma_m F$ is δ-symmetric then $\Sigma_i F$ is δ-symmetric for all $i \leq m$. Also it is worth noting that if $F \in R[t,t',\dots,t^{(r)}]\hat{}$, then one always has

$$\Sigma_m F \in R[s_1,\dots,s_m,\dots,s_1^{(r)},\dots,s_m^{(r)},\Delta^{-1}]\hat{}$$

where

$$\Delta := \prod_{i<j}(t_i - t_j)^2 \in R[s_1,\dots,s_m]$$

is the discriminant polynomial; this is an immediate consequence of [B05], Proposition 3.27. So, morally, what prevents $\Sigma_m F$ from being δ-symmetric is the possible "occurrence of Δ in the denominators".

Definition 4. Let $f^{\text{for}} : X^{\text{for}} \to \mathbf{A}^{1,for} = Spf\ R[[z]]$ be a δ-morphism with X^{for} of relative dimension 1 and let

$$\Sigma_m f^{\text{for}} : (X^{\text{for}})^m \stackrel{(f^{\text{for}})^m}{\longrightarrow} (\mathbf{A}^{1,for})^m \stackrel{+}{\to} \mathbf{A}^{1,\text{for}}$$

be the induced δ-morphism, $m \geq 2$. Fix an isomorphism $X^{\text{for}} \simeq Spf\ R[[t]]$ and consider the series

$$f^{\text{for},*}z \in R[[t, t', \dots, t^{(r)}]].$$

We say that $\Sigma_m f^{\text{for}}$ is *δ-symmetric* if the series

$$\Sigma_m f^{\text{for},*}z \in R[[t_1, \dots, t_m, \dots, t_1^{(r)}, \dots, t_m^{(r)}]]$$

is δ-symmetric. (The definition does not depend on the choice of the isomorphism $X^{\text{for}} \simeq Spf\ R[[t]]$.)

Now if X is a smooth curve over R and $P^0 \in X(R)$ is a point with reduction mod p denoted by $P_0^0 \in X(k)$, then the completion X^{for} of X at P_0^0 is isomorphic to $Spf\ R[[t]]$ so it is a local formal scheme. Fix such an isomorphism and let

$$f : X \to \mathbf{A}^1$$

be a δ-morphism with $f(P^0) = 0$. Then f naturally defines a δ-morphism of local formal schemes

$$f^{\text{for}} : X^{\text{for}} \to \mathbf{A}^{1,\text{for}};$$

cf. [B05], p. 126.

The main result of this paper is the following:

Theorem 1. Let X be a smooth projective curve of genus at least 2 over R and let $f : X \to \mathbf{A}^1$ be a δ-morphism vanishing at some point $P^0 \in X(R)$. The following are equivalent:

(1) f is a δ-character centered at P^0;
(2) f^{for} is a δ-character;
(3) $\Sigma_m f^{\text{for}}$ is δ-symmetric for all $m \geq 2$;
(4) $\Sigma_2 f^{\text{for}}$ is δ-symmetric.

The implications $1 \Rightarrow 2$ and $3 \Rightarrow 4$ are clear. The rest of this paper is devoted to the proof of $2 \Rightarrow 3$ and $4 \Rightarrow 1$. Before proceeding, let us remark that conditions 2, 3 and 4 in the Theorem look a priori much weaker than condition 1; indeed, the R-module of δ-characters of order r on X^{for} has infinite rank whereas the R-module of δ-characters on X centered at P^0 has finite rank. What morally makes the implication $4 \Rightarrow 1$ hold is the additional assumption that f is a global object, defined on the whole of X.

3 Proof of the theorem

The next lemma settles the implication 2 $\Rightarrow$ 3 in Theorem 5:

Lemma 1. *Let* $f^{\text{for}} : X^{\text{for}} \to \mathbf{A}^{1,for}$ *be a* δ*-character where* X^{for} *has relative dimension* 1. *Then* $\Sigma_m f^{\text{for}}$ *is* δ*-symmetric for all* $m \geq 2$.

Proof. Assume we are in the situation of Definition 2 with

$$X^{\text{for}} = Spf\ R[[t]],$$

$$G^{\text{for}} = Spf\ R[[\mathbf{T}]],$$

$$\mathbf{A}^{1,\text{for}} = Spf\ R[[z]].$$

Here $\mathbf{T}$ is a g-tuple of variables. Let the morphism $\beta^{\text{for}} : X^{\text{for}} \to G^{\text{for}}$ be given by

$$\beta^{\text{for},*}\mathbf{T} = \Phi(t) \in R[[t]]^g$$

and let $\psi^{\text{for}} : G^{\text{for}} \to \mathbf{G}_a^{\text{for}}$ be given by

$$\psi^{\text{for},*}z = \Psi = \Psi(\mathbf{T}, \dots, \mathbf{T}^{(r)}) \in R[[\mathbf{T}, \dots, \mathbf{T}^{(r)}]].$$

Let $\mathcal{F}_m \in R[[\mathbf{T}_1, \dots, \mathbf{T}_m]]^g$ be the tuple of series defining the m-fold addition

$$(G^{\text{for}})^m \to G^{\text{for}}.$$

Then (by a universality property argument) the tuple of series

$$(\mathcal{F}_m, \delta\mathcal{F}_m, \dots, \delta^r\mathcal{F}_m)$$

defines the m-fold addition

$$(J^r(G^{\text{for}}))^m \to J^r(G^{\text{for}}).$$

Recall the classical fundamental theorem of symmetric polynomials saying that the symmetric polynomials in $R[t_1, \dots, t_m]$ belong to the ring $R[s_1, \dots, s_m]$. By this theorem applied to the homogeneous components of

$$\mathcal{G} := \mathcal{F}_m(\Phi(t_1), \dots, \Phi(t_m)),$$

we have $\mathcal{G} \in R[[s_1, \dots, s_m]]^g$. Then, using the fact that Ψ defines homomorphism, we get

$$\begin{aligned}\sum_{i=1}^m \Psi(\Phi(t_i), \dots, \delta^r\Phi(t_i)) &= \Psi(\mathcal{F}_m(\Phi(t_1), \dots, \Phi(t_m)), \dots, \delta^r\mathcal{F}_m(\Phi(t_1), \dots, \Phi(t_m))) \\ &= \Psi(\mathcal{G}, \dots, \delta^r\mathcal{G}) \\ &\in R[[s_1, \dots, s_m, \dots, s_1^{(r)}, \dots, s_m^{(r)}]].\end{aligned}$$

Lemma 2. *Let Y be a smooth scheme over R, with irreducible fibers, and let $U \subset Y$ be an open set. Let $Y_0 := Y \otimes k$, $U_0 := U \otimes k$, and assume $Y_0 \backslash U_0$ has codimension ≥ 2 in Y_0. Then $\mathcal{O}^r(Y) = \mathcal{O}^r(U)$ for all $r \geq 0$.*

Proof. For $r = 0$, $\mathcal{O}^0(Y) = \mathcal{O}(\hat{Y})$, where $\hat{}$ means p-adic completion, and the result is well known. (For convenience we recall the argument. We may assume Y is affine. Take $f \in \mathcal{O}(\hat{U})$. Then the image $\bar{f} \in \mathcal{O}(U_0)$ extends uniquely to some $\bar{g}_1 \in \mathcal{O}(Y_0)$. Since Y is affine $\bar{g}_1$ is the image of some $g_1 \in \mathcal{O}(\hat{Y})$. So $f = g_1 + pf_1$, $f_1 \in \mathcal{O}(\hat{U})$. Similarly one finds $g_2 \in \mathcal{O}(\hat{Y})$ and $f_2 \in \mathcal{O}(\hat{U})$ such that $f_1 = g_2 + pf_2$. Continuing we get, in a similar way, $f = g_1 + pg_2 + \ldots + p^{n-1}g_n + p^n f_n$, $g_i \in \mathcal{O}(\hat{Y})$, $f_n \in \mathcal{O}(\hat{U})$. Clearly the restriction of $\sum p^{i-1} g_i \in \mathcal{O}(\hat{Y})$ to $\hat{U}$ is f.) Now assume r arbitrary and recall that $\mathcal{O}^r(Y) = \mathcal{O}(J^r(Y))$, where $J^r(Y)$ is the p-adic completion of some smooth scheme. So it is enough to check that $J^r(U) \otimes k$ has codimension ≥ 2 in $J^r(Y) \otimes k$. This follows immediately from the fact that $J^r(Y) \otimes k$ is a locally trivial bundle over $Y \otimes k$ with fibers affine spaces and $J^r(U) \otimes k$ is the inverse image of $U \otimes k$ in $J^r(Y) \otimes k$; cf. [B05], Proposition 3.13 and Corollary 3.16.

Lemma 3. *Let X be a smooth projective curve over R, let X^n be the n-fold product of X ($n \geq 2$), let $S_n \times X^n \to X^n$ be the natural action of the symmetric group, and let $X^{(n)} = X^n / S_n$ be the n-fold symmetric product of X. Let $\pi = \pi_n : X^n \to X^{(n)}$ be the natural projection. Let $f \in \mathcal{O}^r(X)$ an let $P^0 \in X(R)$ be such that $f(P^0) = 0$. Let $f^n \in \mathcal{O}^r(X^n)$ be defined by*

$$f^n(P_1, \ldots, P_n) = f(P_1) + \cdots + f(P_n), \quad P_i \in X(R).$$

Assume $\Sigma_2 f^{\text{for}}$ is δ-symmetric. Then f^n descends to a function $f^{(n)} \in \mathcal{O}^r(X^{(n)})$ (i.e., $f^n = f^{(n)} \circ \pi$).

Proof. View the k-points of $X^{(n)}$ as effective divisors on $X_0 := X \otimes k$. Let $H \subset X^{(n)}$ be the image of the union $\tilde{H} = \cup X_{ij}$ of all "principal diagonals" $X_{ij} \subset X^n$; so, for $i < j$, X_{ij} is the image of

$$X^{n-1} \to X^n,$$

$$(P_1, \ldots, P_{n-1}) \mapsto (P_1, \ldots, P_{i-1}, P_i, P_{i+1}, \ldots, P_{j-1}, P_i, P_{j+1}, \ldots, P_n).$$

Then $H_0 := H \otimes k$ is the image of $X_0^{(n-1)}$, hence is an irreducible divisor on $X_0^{(n)}$. The projection

$$X^n \backslash \tilde{H} \to X^{(n)} \backslash H$$

is a finite Galois étale morphism so by [B05], Proposition 3.27, f^n descends to a function $f_1^{(n)} \in \mathcal{O}^r(X^{(n)} \backslash H)$. Choose a reduced divisor D_0 on X_0 of degree $n-2$ whose support does not contain P_0^0 and consider the k-point $2P_0^0 + D_0 \in H_0(k)$.

Claim 1. There exists a neighborhood U of $2P_0^0 + D_0 \in X^{(n)}(k)$ in $X^{(n)}$ such that f^n descends to a function $f_2^{(n)} \in \mathcal{O}^r(U)$.

Note that Claim 1 can be used to end the proof of our lemma. Indeed, $f_1^{(n)} \in \mathcal{O}^r(X^{(n)} \backslash H)$ and $f_2^{(n)} \in \mathcal{O}^r(U)$ obviously coincide on the R-points of

$U \cap (X^{(n)} \backslash H)$ so they yield a function $f_3^{(n)} \in \mathcal{O}^r(U \cup (X^{(n)} \backslash H))$. Since H_0 is an irreducible divisor on $X_0^{(n)}$, the complement of $U_0 \cup (X_0^{(n)} \backslash H_0)$ in $X_0^{(n)}$ has codimension ≥ 2. By Lemma 2, $f_3^{(n)}$ extends to a function $f^{(n)} \in \mathcal{O}^r(X^{(n)})$. Since $f^{(n)} \circ \pi = f^n$ on an open set with non-empty fibers over R it follows that the latter equality holds on the whole of X^n.

We will now prove *Claim 1*.

Choose an affine neighborhood U of $2P_0^0 + D_0$ in $X^{(n)}$ such that the divisor $H \cap U$ is given in U by one equation $h \in \mathcal{O}(U)$. By shrinking U we may assume that U has étale coordinates

$$s_1, s_2, t_3, \ldots, t_n,$$

where $s_1 = t_1 + t_2$, $s_2 = t_1 t_2$ such that t_1, t_2 come from an étale coordinate t on X around P_0^0 and $t_3, \ldots, t_n$ come from étale coordinates on X around the points in the support of D_0. Let $A = \mathcal{O}(U)$ and denote by σ the variables s_1, s_2 and by τ the variables $t_3, \ldots, t_n$. (By the way, we may assume the image of h in $R[[\sigma, \tau]]$ equals $s_1^2 - 4s_2$; we will not need this explicit expression for the image of h.) By [B05], Proposition 3.13, we have identifications

$$\mathcal{O}^r(U) = \hat{A}[\sigma', \tau', \ldots, \sigma^{(r)}, \tau^{(r)}]\hat{}, \tag{5}$$

$$\mathcal{O}^r(U \backslash H) = (A_h)\hat{}[\sigma', \tau', \ldots, \sigma^{(r)}, \tau^{(r)}]\hat{}. \tag{6}$$

We may view $f_1^{(n)}$ as an element in the ring (6). On the other hand, by our assumption that $\Sigma_2 f^{\text{for}}$ is δ-symmetric, we may view $f_1^{(n)}$ as an element in the ring

$$R[[\sigma, \tau, \sigma', \tau', \ldots, \sigma^{(r)}, \tau^{(r)}]]. \tag{7}$$

The rings (5), (6), (7) are subrings of the ring

$$(R[[\sigma, \tau]]_h)\hat{}[[\sigma', \tau', \ldots, \sigma^{(r)}, \tau^{(r)}]]. \tag{8}$$

So *Claim 1* will be proved if we prove

Claim 2. The intersection of the rings (6) and (7) inside the ring (8) equals the ring (5).

In order to prove Claim 2 we prove a series of other claims. Let

$$\bar{h} := h \otimes 1 \in \bar{A} := A \otimes k.$$

Claim 3. $k[[\sigma, \tau]] \cap \bar{A}_{\bar{h}} = \bar{A}$.

To check the non-obvious inclusion "$\subset$" consider the maximal ideal $\bar{M} := (\sigma, \tau) \subset \bar{A}$. Then the $\bar{M}$-adic completion of $\bar{A}_{\bar{M}}$ is $k[[\sigma, \tau]]$ so the extension $\bar{A}_{\bar{M}} \subset k[[\sigma, \tau]]$ is faithfully flat. So if $Q(\bar{A})$ is the fraction field of $\bar{A}$, then

$$k[[\sigma, \tau]] \cap Q(\bar{A}) = \bar{A}_{\bar{M}}.$$

In particular

$$k[[\sigma,\tau]] \cap \bar{A}_{\bar{h}} \subset \bar{A}_{\bar{M}} \cap \bar{A}_{\bar{h}} = \bar{A},$$

which ends the proof of Claim 3. (The last equality holds because any element in $\bar{A}_{\bar{M}} \cap \bar{A}_{\bar{h}}$ defines a rational function u on Spec $\bar{A} = U_0 = U \otimes k$ which is regular at the point $\bar{M}$ and also outside the hypersurface $H_0 \cap U_0$; since $H_0 \cap U_0$ is irreducible, u is regular outside a closed subset of codimension ≥ 2, hence is regular on the whole of U_0.)

Claim 4. $R[[\sigma,\tau]] \cap (A_h)\hat{} = \hat{A}$.

The non-obvious inclusion is, again, "$\subset$". Now if $u \in R[[\sigma,\tau]] \cap (A_h)\hat{}$, then its reduction mod p satisfies $\bar{u} \in k[[\sigma,\tau]] \cap \bar{A}_{\bar{h}}$. By Claim 3 there exists $a_1 \in A$ such that $u - a_1 \in p(A_h)\hat{} \cap pR[[\sigma,\tau]]$ so $u - a_1 = u_1$ with $u_1 \in R[[\sigma,\tau]] \cap (A_h)\hat{}$. Repeating the procedure we find, for any n, that

$$u = a_1 + pa_2 + \cdots + p^{n-1}a_n + p^n u_n,$$

where $a_i \in A$, $u_n \in R[[\sigma,\tau]] \cap (A_h)\hat{}$ which clearly implies $u \in A$.

Let us prove Claim 2. Any element F in the intersection of the rings (6) and (7) is an element in the ring

$$(R[[\sigma,\tau]] \cap (A_h)\hat{})[[\sigma',\tau',\ldots,\sigma^{(r)},\tau^{(r)}]]. \tag{9}$$

By Claim 4 the ring (9) equals the ring

$$\hat{A}[[\sigma',\tau',\ldots,\sigma^{(r)},\tau^{(r)}]]. \tag{10}$$

So F is in the intersection of the ring (10) with (6), hence is a series in $\sigma',\tau',\ldots,\sigma^{(r)},\tau^{(r)}$ with coefficients in $\hat{A}$ which tend to 0 in the ring $(A_h)\hat{}$. Since $\bar{A} \to \bar{A}_{\bar{h}}$ is injective, $p(A_h)\hat{} \cap \hat{A} = p\hat{A}$, so $p^n(A_h)\hat{} \cap \hat{A} = p^n\hat{A}$ for any n. So the coefficients of F tend to 0 in $\hat{A}$ and hence F is in the ring (5) which ends our proof.

Lemma 4. *Let Y be a smooth scheme over R, $\mathcal{E}$ a locally free sheaf on Y, and $\mathbf{P}_Y(\mathcal{E})$ the associated projective bundle over Y. Then, for all $r \geq 0$, the natural map $\mathcal{O}^r(Y) \to \mathcal{O}^r(\mathbf{P}_Y(\mathcal{E}))$ is an isomorphism.*

Proof. We may assume $\mathcal{E}$ is trivial so $\mathbf{P}_Y(\mathcal{E}) = Y \times \mathbf{P}^N$. If $Y =$ Spec R then our assertion was proved in [B96]. For arbitrary Y pick a point $Q^0 \in \mathbf{P}^N(R)$ and let $f \in \mathcal{O}^r(Y \times \mathbf{P}^N)$. Denote by $s : Y \to Y \times \mathbf{P}^N$ the map $s(P) = (P, Q^0)$. If $\pi : Y \times \mathbf{P}^N \to Y$ is the first projection, then

$$(f \circ s \circ \pi)(P,Q) = f(P,Q^0) = f(P,Q)$$

because the map

$$\mathbf{P}^N(R) \to R, \quad Q \mapsto f(P,Q)$$

is constant (by the case $Y =$ Spec R of the lemma). We just proved surjectivity of the map $\mathcal{O}^r(Y) \to \mathcal{O}^r(\mathbf{P}_Y(\mathcal{E}))$. Injectivity is obvious.

We are ready to prove the implication $4 \Rightarrow 1$ in Theorem 5.

Proof. Assume condition 4 holds. For $n \geq 1$ let $\beta_n : X^{(n)} \to A = \mathrm{Jac}(X)$ be the Abel–Jacobi map which, at the level of geometric points, sends a divisor D of degree n into (the isomorphism class of) $\mathcal{O}(D - nP^0)$. By Lemma 3 the induced δ-morphism $f^n : X^n \to \mathbf{A}^1$ descends to a δ-morphism $f^{(n)} : X^{(n)} \to \mathbf{A}^1$. Recall that for any $n \geq 2g - 1$ (where g is the genus of X) we have $X^{(n)} = \mathbf{P}_A(\mathcal{E}_n)$ for some locally free sheaf $\mathcal{E}_n$ on A. By Lemma 4, $f^{(n)}$ descends to a δ-morphism $\psi_n : A \to \mathbf{A}^1$. If $\beta = \beta_1 : X \to A$, then for any $P \in X(R)$,

$$
\begin{aligned}
(\psi_n \circ \beta)(P) &= \psi_n(\mathcal{O}(P - P^0)) \\
&= \psi_n(\mathcal{O}((n-1)P^0 + P - nP^0)) \\
&= f^{(n)}((n-1)P^0 + P) \\
&= (n-1)f(P^0) + f(P) \\
&= f(P).
\end{aligned}
$$

So $\psi_n \circ \beta = f$.

We claim that ψ_n is a δ-character and this will end our proof. Fix the integer $n \geq 2g - 1$ in what follows. Recall that $\beta_g : X^{(g)} \to A$ induces an isomorphism $\beta_g^{-1}(U) \to U$ where $U \subset A$ is some open set with non-empty fibers over R. So there is an open set $V \subset U$ with non-empty fibers over R such that if $\pi_g : X^g \to X^{(g)}$ is the canonical projection then $\tilde{V} := \pi_g^{-1}\beta_g^{-1}(V) \to V$ is étale. So any R-point ξ of V lifts to an R-point of X^g, i.e., there exist $P_1, \ldots, P_g \in X(R)$ such that $\beta_g(P_1 + \cdots + P_g) = \xi$. Hence, if $D_\xi := P_1 + \cdots + P_g + (n-g)P^0$, then $\beta_n(D_\xi) = \xi$. Now let $\xi_1, \xi_2 \in V(R)$ be such that $\xi_1 + \xi_2 \in V(R)$. Then

$$
(D_{\xi_1} - nP^0) + (D_{\xi_2} - nP^0) \sim D_{\xi_1+\xi_2} - nP^0,
$$

where $\sim$ means linear equivalence on the geometric generic fiber of X. So $D_{\xi_1} + D_{\xi_2}$ and $D_{\xi_1+\xi_2} + nP^0$ correspond to R-points of $X^{(2n)}$ that map to the same R-point of A via β_{2n}. Hence we have:

$$
\begin{aligned}
\psi_n(\xi_1) + \psi_n(\xi_2) &= f^{(n)}(D_{\xi_1}) + f^{(n)}(D_{\xi_2}) \\
&= f^{(2n)}(D_{\xi_1} + D_{\xi_2}) \\
&= f^{(2n)}(D_{\xi_1+\xi_2} + nP^0) \\
&= f^{(n)}(D_{\xi_1+\xi_2}) + nf(P^0) \\
&= f^{(n)}(D_{\xi_1+\xi_2}) \\
&= \psi_n(\xi_1 + \xi_2).
\end{aligned}
$$

The above identity holds for all R-points of the open set $(V \times V) \cap \mu^{-1}(V)$ where $\mu : A \times A \to A$ is the addition hence it holds on $A \times A$ hence ψ_n is a δ-character and our claim is proved.

References

[B95] Buium, A.: Differential characters of Abelian varieties over p−adic fields, *Invent. Math.* 122, 309–340 (1995).

[B96] Buium, A.: Geometry of p−jets, *Duke Math.* J. 82, 2, 349–367 (1996).

[B05] Buium, A.: *Arithmetic Differential Equations*, Math. Surveys and Monographs 118, AMS, 2005.

[B08] Buium, A.: Differential eigenforms, *J. Number Theory*, 128 (2008), 979–1010.

[Lan56] Lang, S.: Integral points on curves, *Publ. Math. IHES* 6, 27–43 (1960).

[M58] Manin, Yu. I.: Algebraic curves over fields with differentiation, *Izv. Akad. Nauk SSSR, Ser. Mat.* 22, 737–756 (1958).

[M63] Manin, Yu. I.: Rational points of algebraic curves over function fields, Izv. *Akad. Nauk SSSR, Ser. Mat.* 27, 1395–1440 (1963).

Weyl group multiple Dirichlet series of type A_2

Gautam Chinta and Paul E. Gunnells

In memory of Serge Lang

Abstract A *Weyl group multiple Dirichlet series* is a Dirichlet series in several complex variables attached to a root system Φ. The number of variables equals the rank r of the root system, and the series satisfies a group of functional equations isomorphic to the Weyl group W of Φ. In this paper we construct a Weyl group multiple Dirichlet series over the rational function field using n^{th} order Gauss sums attached to the root system of type A_2. The basic technique is that of [10, 11]; namely, we construct a rational function in r variables invariant under a certain action of W, and use this to build a "local factor" of the global series.

Key words Multiple Dirichlet series • Eisenstein series

Mathematics Subject Classification (2010): Primary 11F68; Secondary 11F30

1 Introduction

Weyl group multiple Dirichlet series are Dirichlet series in r complex variables $s_1, s_2, \ldots, s_r$ that have analytic continuation to $\mathbb{C}^r$, satisfy a group of functional equations isomorphic to the Weyl group of a finite root system of rank r, and

---

G. Chinta (✉)
Department of Mathematics, The City College of CUNY, New York, NY 10031, USA
e-mail: chinta@sci.ccny.cuny.edu

P.E. Gunnells
Department of Mathematics and Statistics, University of Massachusetts, Amherst, MA 01003
e-mail: gunnells@math.umass.edu

D. Goldfeld et al. (eds.), *Number Theory, Analysis and Geometry: In Memory of Serge Lang*, DOI 10.1007/978-1-4614-1260-1_6,

whose coefficients are products of nth order Gauss sums. The study of these series was introduced in [2], which also suggested a method for proving their analytic continuation and functional equations.

Recently a complete proof of these expected properties has been given in [12]. In this paper we describe in detail the construction for the root system A_2. There exist alternate constructions of the series defined here. For A_2 and $n \geq 2$ one falls in the stable range, and therefore our result follows from the work of [3]. (In fact, this case was treated earlier in [2].) Nevertheless there are several reasons why a new treatment of A_2 is desirable. First, the methods used here are completely different from those of [3] and give an alternative technique to construct Weyl group multiple Dirichlet series. Second, the technique presented here works for a root system Φ of arbitrary rank and for arbitrary n, with no stability restriction. This is the subject of [12]; one of the main goals of the present paper is an exposition of our method in the simplest nontrivial case, namely $\Phi = A_2$.

With this latter goal in mind we also adopt certain assumptions to make the exposition simpler. For instance, we work over a rational function field to avoid the annoyance of having to deal with Hilbert symbols. We also focus on the *untwisted case* (see §2 for an explanation of this terminology) to avoid some notational complexities. A comparison with the methods of [2, 10, 11] indicates how to extend our methods to an arbitrary global field containing the $2n^{th}$ roots of unity and to arbitrary twists.

We now describe our main result in greater detail. Let $\mathbb{F}$ be a finite field whose cardinality q is congruent to 1 mod $4n$. Let K be the rational function field $\mathbb{F}(t)$, and let $\mathcal{O} = \mathbb{F}[t]$. Let $\mathcal{O}_{\text{mon}} \subset \mathcal{O}$ be the subset of monic polynomials. We let $K_\infty = \mathbb{F}[[t^{-1}]]$ denote the field of Laurent series in t^{-1}.

For $x, y \in \mathcal{O}$ relatively prime, we denote by $\left(\frac{x}{y}\right)$ the nth order power residue symbol. We have the reciprocity law

$$\left(\frac{x}{y}\right) = \left(\frac{y}{x}\right) \tag{1.1}$$

for x, y monic. The reciprocity law takes this particularly simple form because of our assumption that the cardinality of $\mathbb{F}$ is congruent to 1 mod 4.

Let $y \mapsto e(y)$ be an additive character on K_∞ with the following property: if $I \subset K$ is the set of all $y \in K$ such that the restriction of e to $y\mathcal{O}$ is trivial, then $I = \mathcal{O}$. Fix an embedding ϵ from the the nth roots of unity in $\mathbb{F}$ to $\mathbb{C}^\times$. For $r, c \in \mathcal{O}$ we define the Gauss sum $g(r, \epsilon, c)$ by

$$g(r, \epsilon, c) = \sum_{y \bmod c} \epsilon\left(\left(\frac{y}{c}\right)\right) e\left(\frac{ry}{c}\right).$$

We will also use the notation $g_i(r, c) = g(r, \epsilon^i, c)$ and $g(r, c) = g(r, \epsilon, c)$. Note that ϵ^i is not necessarily an embedding.

We are now ready to define our double Dirichlet series. Put

$$Z(s_1,s_2) = (1-q^{n-ns_1})^{-1}(1-q^{n-ns_2})^{-1}(1-q^{2n-ns_1-ns_2})^{-1} \times \sum_{c_1\in\mathcal{O}_{\text{mon}}}\sum_{c_2\in\mathcal{O}_{\text{mon}}} \frac{H(c_1,c_2)}{|c_1|^{s_1}|c_2|^{s_2}}, \tag{1.2}$$

where the coefficient $H(c_1,c_2)$ is defined as follows:

(1) (Twisted multiplicativity) If $gcd(c_1c_2, d_1d_2) = 1$, then

$$\frac{H(c_1d_1,c_2d_2)}{H(c_1,c_2)H(d_1,d_2)} = \left(\frac{c_1}{d_1}\right)\left(\frac{d_1}{c_1}\right)\left(\frac{c_2}{d_2}\right)\left(\frac{d_2}{c_2}\right)\left(\frac{c_1}{d_2}\right)^{-1}\left(\frac{d_1}{c_2}\right)^{-1}. \tag{1.3}$$

(2) ($\mathfrak{p}$-part) If $\mathfrak{p}$ is prime, then

$$\sum_{k,l\geq 0} H(\mathfrak{p}^k,\mathfrak{p}^l)x^ky^l = 1 + g(1,\mathfrak{p})x + g(1,\mathfrak{p})y + g(1,\mathfrak{p})g(\mathfrak{p},\mathfrak{p})xy + g(1,\mathfrak{p})g(\mathfrak{p},\mathfrak{p}^2)xy^2 + g(1,\mathfrak{p})g(\mathfrak{p},\mathfrak{p}^2)x^2y + g(1,\mathfrak{p})^2g(\mathfrak{p},\mathfrak{p}^2)x^2y^2. \tag{1.4}$$

Our main result is

Theorem 1.1. *The double Dirichlet series $Z(s_1,s_2)$ converges absolutely for* $\mathrm{Re}(s_i)$ *sufficiently large and has an analytic continuation to all $(s_1,s_2)\in\mathbb{C}^2$. Moreover, $Z(s_1,s_2)$ satisfies two functional equations of the form*

$$\sigma_1\colon (s_1,s_2)\mapsto(2-s_1,s_1+s_2-1) \text{ and } \sigma_2\colon (s_1,s_2)\mapsto(s_1+s_2-1,2-s_2). \tag{1.5}$$

These two functional equations generate a subgroup of the affine transformations of $\mathbb{C}^2$ isomorphic to the symmetric group S_3.

The precise statement of the functional equations involves a set of double Dirichlet series $Z(s_1,s_2;i,j)$, where $0\leq i,j\leq n-1$, and where $Z(s_1,s_2) = \sum_{i,j} Z(s_1,s_2;i,j)$; we refer to Theorem 4.1 for details. Moreover, one can explicitly write down $Z(s_1,s_2)$ as a rational function in q^{-s_1},q^{-s_2}. For $n=2$, this was first done by Hoffstein and Rosen [16] (in the case of the rational function field with q congruent to 1 mod 4), and later by Fisher and Friedberg [13] (over a general base field). For $n>2$, again working over the rational function field, the A_2 series have been computed by Chinta [8].

As stated above this theorem follows from the work of [2, 3]. In [6], the authors study the harder problem of constructing twisted Weyl group multiple Dirichlet series associated to the root system A_r. They construct such series for A_2 and present a conjectural description of the series associated to A_r for arbitrary r and n. Recently, Brubaker, Bump and Friedberg have given two different proofs of their conjectures [4, 5], thereby giving a complete definition of Weyl group multiple Dirichlet series associated to A_r. In fact, in [4] the authors prove that the series they construct are Fourier–Whittaker coefficients of Eisenstein series on a metaplectic n-fold cover of GL_{r+1}, thereby establishing Conjecture 1.4 of [3], in the case $G = GL_{r+1}$. Consequently, these Weyl group multiple Dirichlet series inherit from the Eisenstein series functional equations and analytic continuation. Contrastingly, the techniques used in [5] make no use of the connection to Eisenstein series other than in the rank 1 case.

This is also the case in our paper. Our method has the advantage that functional equations are essentially built-in to our definition. As in the case of [2, 3, 6, 10, 11] the Weyl group multiple Dirichlet series are completely determined by their $\mathfrak{p}$-parts and the twisted multiplicativity satisfied by the coefficients. Our approach is to show that if the $\mathfrak{p}$-parts (which can be expressed as rational functions in the $|\mathfrak{p}|^{-s_i}$) satisfy certain functional equations, then the global multiple Dirichlet series satisfies the requisite global functional equations. This leads us to define a certain action of W, the Weyl group of the root system Φ, on a certain subring of the field of rational functions in r indeterminates. This approach, first introduced in [7], has been carried out in the quadratic case for an arbitrary simply-laced root system, see [10, 11]. We extend this approach to arbitrary Φ and n in [12]. However, although the basic ideas are clear, the non-obvious group action required on rational functions can appear unmotivated and complicated in the general setting. Therefore, we feel it is worthwhile in this paper to work out in detail the simplest nontrivial case, the rank two root system A_2.

Here is a short plan of the paper. Section 2 describes the Weyl group action on rational functions that leads to a $\mathfrak{p}$-part (1.4) with the desired functional equations. Although the focus of this paper is untwisted A_2, we work more generally at first and state the full action for a general (simply laced) root system. We then specialize to untwisted A_2. Section 3 reviews the Dirichlet series of Kubota; in the current framework, these series are Weyl group multiple Dirichlet series attached to A_1. The main result of this section is Theorem 3.4, which shows that a certain Dirichlet series $E(s, m)$ built from the function $H(c, d)$ from (1.3)–(1.4) satisfies the same functional equations as Kubota's. Finally, in Section 4 we use Theorem 3.4 to complete the proof of Theorem 1.1. The basic idea is that the (one variable) functional equations of the $E(s, m)$ induce a bivariate functional equation in the double Dirichlet series.

Acknowledgements GC wishes to thank the NSF for support of this research through the FRG grant DMS-0652605. GC also gratefully acknowledges the support of the Alexander von Humboldt Foundation.

PG wishes to thank the NSF for support through Grant DMS-0801214.

2 A Weyl group action

Let Φ be an irreducible simply laced root system of rank r with Weyl group W. Choose an ordering of the roots and let $\Phi = \Phi^+ \cup \Phi^-$ be the decomposition into positive and negative roots. Let

$$\Delta = \{\alpha_1, \alpha_2, \ldots, \alpha_r\}$$

be the set of simple roots and let σ_i be the Weyl group element corresponding to the reflection through the hyperplane perpendicular to α_i. We say that i and j are *adjacent* if $i \neq j$ and $(\sigma_i \sigma_j)^3 = 1$. The Weyl group W is generated by the simple reflections $\sigma_1, \sigma_2, \ldots, \sigma_r$, which satisfy the relations

$$(\sigma_i \sigma_j)^{r(i,j)} = 1 \text{ with } r(i,j) = \begin{cases} 3 & \text{if } i \text{ and } j \text{ are adjacent,} \\ 1 & \text{if } i = j, \text{ and} \\ 2 & \text{otherwise,} \end{cases} \tag{2.1}$$

for $1 \leq i, j \leq r$. The action of the generators σ_i on the roots is

$$\sigma_i \alpha_j = \begin{cases} \alpha_i + \alpha_j & \text{if } i \text{ and } j \text{ are adjacent,} \\ -\alpha_j & \text{if } i = j, \text{ and} \\ \alpha_j & \text{otherwise.} \end{cases} \tag{2.2}$$

Define

$$\operatorname{sgn}(w) = (-1)^{\operatorname{length}(w)}$$

where the length function on W is with respect to the generators $\sigma_1, \sigma_2, \ldots, \sigma_r$. Let Λ be the lattice generated by the roots. Any $\alpha \in \Lambda$ has a unique representation as an integral linear combination of the simple roots:

$$\alpha = k_1\alpha_1 + k_2\alpha_2 + \cdots + k_r\alpha_r. \tag{2.3}$$

We denote by

$$d(\alpha) = k_1 + k_2 + \cdots + k_r$$

the usual height function on Λ and put

$$d_j(\alpha) = \sum_{i \sim j} k_i,$$

where $i \sim j$ means that the nodes labeled by i and j are adjacent in the Dynkin diagram of Φ. Introduce a partial ordering on Λ by defining $\alpha \succeq 0$ if each $k_i \geq 0$ in (2.3). Given $\alpha, \beta \in \Lambda$, define $\alpha \succeq \beta$ if $\alpha - \beta \succeq 0$.

Let $A = \mathbb{C}[\Lambda]$ be the ring of Laurent polynomials on the lattice Λ. Hence A consists of all expressions of the form $f = \sum_{\beta \in \Lambda} a_\beta \mathbf{x}^\beta$, where $a_\beta \in \mathbb{C}$ and almost all are zero, and the multiplication of monomials is defined by addition in Λ: $\mathbf{x}^\beta \mathbf{x}^\lambda = \mathbf{x}^{\beta+\lambda}$. We identify A with $\mathbb{C}[x_1, x_1^{-1}, \ldots, x_r, x_r^{-1}]$ via $\mathbf{x}^{\alpha_i} \mapsto x_i$.

Let $\mathfrak{p}$ be a prime in $\mathcal{O}$ of norm p. Let $\widetilde{A}$ be the localization of A at the multiplicative subset of all expressions of the form

$$\{1 - p^{nd(\alpha)}\mathbf{x}^{nd(\alpha)}, 1 - p^{nd(\alpha)-1}\mathbf{x}^{nd(\alpha)} \mid \alpha \in \Phi^+\}.$$

The group W will act on $\widetilde{A}$, and the action will involve the Gauss sums $g_i(1, \mathfrak{p})$.[1] There is one further parameter necessary for the definition. Let $\ell = (l_1, \dots, l_r)$ be an r-tuple of nonnegative integers. The tuple ℓ is called a *twisting parameter*; it should be thought of as corresponding to the weight $\sum(l_j + 1)\varpi_j$, where the ϖ_j are the fundamental weights of Φ. The case $\ell = (0, \dots, 0)$ is called the *untwisted case*. For each choice of ℓ we will define an action of the Weyl group W on $\widetilde{A}$.

We are now ready to define the W-action. First, we define a "change of variables" action on $\widetilde{A}$ as follows. for $\mathbf{x} = (x_1, x_2, \dots, x_r)$ define $\sigma_i \mathbf{x} = \mathbf{x}'$, where

$$x'_j = \begin{cases} p x_i x_j & \text{if } i \text{ and } j \text{ are adjacent,} \\ 1/(p^2 x_j) & \text{if } i = j, \text{ and} \\ x_j & \text{otherwise.} \end{cases} \tag{2.4}$$

One can easily check that if $f_\beta(\mathbf{x}) = \mathbf{x}^\beta$ is a monomial, then

$$f_\beta(w\mathbf{x}) = q^{d(w^{-1}\beta - \beta)}\mathbf{x}^{w^{-1}\beta}. \tag{2.5}$$

Next, write $f \in A$ as

$$f(\mathbf{x}) = \sum_\beta a_\beta \mathbf{x}^\beta.$$

Given integers k, i, j, define

$$f_k(\mathbf{x}; i, j) = \sum_{\substack{\beta_k \equiv i \bmod n \\ d_k(\beta) \equiv j \bmod n}} a_\beta \mathbf{x}^\beta.$$

We define the action of a generator $\sigma_k \in W$ on f as follows:

$$(f|_\ell \sigma_k)(\mathbf{x}) = (p x_k)^{l_k} \sum_{i=0}^{n-1}\sum_{j=0}^{n-1} \big(\mathcal{P}_{ij}(x_k) f_k(\sigma_k \mathbf{x}; i, j - l_k)$$
$$+ \mathcal{Q}_{ij}(x_k) f_k(\sigma_k \mathbf{x}; j + 1 - i, j - l_k)\big), \tag{2.6}$$

[1] We remark that our normalization for Gauss sums follows [3,6] and not [10,11]. See [11, Remark 3.12] for a discussion of this.

where

$$\mathcal{P}_{ij}(x) = (px)^{1-(-2i+j+1)_n} \frac{1-1/p}{1-p^{n-1}x^n},$$

$$\mathcal{Q}_{ij}(x) = -g^*_{2i-j-1}(1,\mathfrak{p})(px)^{1-n} \frac{1-p^n x^n}{1-p^{n-1}x^n},$$

$$g_i^*(1,\mathfrak{p}) = \begin{cases} g_i(1,\mathfrak{p})/p & \text{if } n \nmid i, \\ -1 & \text{otherwise.} \end{cases}$$

Here $(i)_n \in \{0,\dots,n-1\}$ is the remainder upon division of i by n. We extend this action to all of $\widetilde{A}$ first extending (2.6) to all of A by linearity, and then given $f/g \in \widetilde{A}$ by defining

$$\left(\frac{f}{g}\bigg|_\ell \sigma_k\right)(\mathbf{x}) = \frac{(f|_\ell\sigma_k)(\mathbf{x})}{g(\sigma_k\mathbf{x})}.$$

One can show that this action of the generators extends to an action of W on $\widetilde{A}$; in particular the defining relations (2.1) are satisfied.

Now we specialize to the focus of this paper: we set $\Phi = A_2$ and $\ell = (0,0)$. To simplify notation we write x, y for the variables of $\widetilde{A}$. With these simplifications the action of σ_1 on $f \in A$ takes the form

$$(f|\sigma_1)(x,y) = \sum_{i=0}^{n-1}\sum_{j=0}^{n-1}\Big(\mathcal{P}_{ij}(x) f_1\left(\tfrac{1}{p^2x}, pxy; i, j\right) + \mathcal{Q}_{ij}(x) f_1\left(\tfrac{1}{p^2x}, pxy; j+1-i, j\right)\Big); \tag{2.7}$$

the action of σ_2 is similar. An invariant rational function for this action is

$$h(x,y) = \frac{N(x,y)}{(1-p^{n-1}x^n)(1-p^{n-1}y^n)(1-p^{2n-1}x^n y^n)}, \tag{2.8}$$

where the numerator $N(x,y)$ is

$$\begin{aligned} N(x,y) = N^{(\mathfrak{p})}(x,y) &= 1 + g_1(1,\mathfrak{p})x + g_1(1,\mathfrak{p})y + g_1(1,\mathfrak{p})g_1(\mathfrak{p},\mathfrak{p})xy \\ &\quad + pg_1(1,\mathfrak{p})g_2(1,\mathfrak{p})xy^2 + pg_1(1,\mathfrak{p})g_2(1,\mathfrak{p})x^2y \\ &\quad + pg_1(1,\mathfrak{p})^2 g_2(1,\mathfrak{p})x^2y^2. \end{aligned} \tag{2.9}$$

To compare this with (1.4), note that $pg_2(1, \mathfrak{p}) = g_1(\mathfrak{p}, \mathfrak{p}^2)$. Also note that only the numerator of (2.8) appears in (1.4) because the denominator is incorporated in the factors appearing at the front of (1.2).

Let us write $h(x, y)$ as

$$h(x,y) = \sum_{k,l\geq 0} a(\mathfrak{p}^k, \mathfrak{p}^l) x^k y^l \tag{2.10}$$

$$= \sum_{l\geq 0} y^l \left(\sum_{i=0}^{n-1} \sum_{k=i \bmod n} a(\mathfrak{p}^k, \mathfrak{p}^l) x^k \right)$$

$$= \sum_{l\geq 0} \sum_{i=0}^{n-1} y^l h^{(\mathfrak{p},l)}(x;i).$$

The following two lemmas are proved by a direct computation.

Lemma 2.1. *We have* $N^{(\mathfrak{p})}(x,0) = 1 + g_1(1,\mathfrak{p})x$, $N^{(\mathfrak{p})}(0,y) = 1 + g_1(1,\mathfrak{p})y$ *and for* $j = l \bmod n$, *and* $0 \leq i \leq n-1$,

$$h^{(\mathfrak{p},l)}(x;i) = (px)^l P_{ij}(x) h\left(\tfrac{1}{p^2 x}; i\right) + (px)^l Q_{ij}(x) h\left(\tfrac{1}{p^2 x}; l+1-i\right).$$

Lemma 2.2. *Let*

$$f^{(\mathfrak{p},l)}(x;i) = h^{(\mathfrak{p},l)}(x;i) - \delta g_{2i-l-1}(1,\mathfrak{p}) p^{(2i-l-2)_n} x^{(2i-l-1)_n} h^{(\mathfrak{p},l)}(x, l+1-i)$$

where $\delta = 0$ *if* $l - 2i = -1 \bmod n$ *and is 1 otherwise. Then*

$$f^{(\mathfrak{p},l)}(x;i) = (px)^{l-(l-2i)_n} f^{(\mathfrak{p},l)}\left(\tfrac{1}{p^2 x}; i\right).$$

3 Kubota's Dirichlet series

The basic building blocks of the multiple Dirichlet series are the Kubota Dirichlet series constructed from Gauss sums [17, 18]. Let m be a nonzero polynomial in $\mathcal{O}$ and let s be a complex variable. These series are defined by

$$D(s,m) = (1 - q^{n-ns})^{-1} \sum_{d \in \mathcal{O}_{\text{mon}}} \frac{g(m,d)}{|d|^s} \tag{3.1}$$

and

$$D(s,m;i) = (1 - q^{n-ns})^{-1} \sum_{\substack{\deg d = i \bmod n \\ d \in \mathcal{O}_{\text{mon}}}} \frac{g(m,d)}{|d|^s}. \tag{3.2}$$

Kubota proved that these series have meromorphic continuation to $s \in \mathbb{C}$ with possible poles only at $s = 1 \pm 1/n$ and satisfy a functional equation. Actually, Kubota worked over a number field, but the constructions over a function field are identical.

If the degree of m is $nk + j$, where $0 \le j \le n-1$, this functional equation takes the form

$$D(s,m) = |m|^{1-s} \sum_{0 \le i \le n-1} T_{ij}(s) D(2-s, m; i), \tag{3.3}$$

where the $T_{ij}(s)$ are certain quotients of Dirichlet polynomials. For fixed s the T_{ij} depend only on $2i - j$. We will not need to know anything more about the functional equation, but a more explicit description can be found in Hoffstein [15] or Patterson [20].

Given a set of primes S, we define

$$D_S(s,m) = (1 - q^{n-ns})^{-1} \sum_{\substack{(d,S)=1 \\ d \in \mathcal{O}_{\mathrm{mon}}}} \frac{g(m,d)}{|d|^s}. \tag{3.4}$$

If $m_0 = \prod_{\mathfrak{p} \in S} \mathfrak{p}$ we sometimes write $D_{m_0}(s,m)$ for $D_S(s,m)$.

We record some properties of Gauss sums that we will use repeatedly.

Proposition 3.1. *Let $a, m, c, c' \in \mathcal{O}$.*

(i) If $(a,c) = 1$, then $g_i(am,c) = \left(\frac{a}{c}\right)^{-1} g_i(m,c)$.
(ii) If $(c,c') = 1$, then

$$g_i(m,cc') = g_i(m,c) g_i(m,c') \left(\frac{c}{c'}\right)^{2i}.$$

Using this proposition we can relate the functions D_S to the functions $D_{S'}$ for different sets S and S'. This is the content of the following two lemmas.

Lemma 3.2. *Let $\mathfrak{p} \in \mathcal{O}_{\mathrm{mon}}$ be prime of norm p. For an integer i with $0 \le i \le n-1$ and $m_1, m_2, \mathfrak{p}$ all pairwise relatively prime, we have*

$$D_{m_1}(s, m_2\mathfrak{p}^i) = D_{\mathfrak{p}m_1}(s, m_2\mathfrak{p}^i) + \frac{g(m_2\mathfrak{p}^i, \mathfrak{p}^{i+1})}{p^{(i+1)s}} D_{\mathfrak{p}m_1}\left(s, m_2\mathfrak{p}^{(n-i-2)_n}\right).$$

More generally,

$$D(s,m) = \sum_{S_0 \subset S} \left(\prod_{\mathfrak{p} \in S_0} \frac{g(m,\mathfrak{p}^{i+1})}{|\mathfrak{p}|^{(i+1)s}} \right) D_S\left(s, \prod_{\substack{\mathfrak{p} \in S_0^c \\ \mathfrak{p}^i \| m}} \mathfrak{p}^i \cdot \prod_{\substack{\mathfrak{p} \in S_0 \\ \mathfrak{p}^i \| m}} \mathfrak{p}^{(n-i-2)_n}\right).$$

Proof. We prove only the first part of the lemma. For $\mathfrak{p}, m_1, m_2$ as in the statement,

$$\begin{aligned}(1-q^{n-ns})D_{m_1}(s, m_2\mathfrak{p}^i) &= \sum_{\substack{(d,m_1)=1\\ d\in\mathcal{O}_{\mathrm{mon}}}} \frac{g(m_2\mathfrak{p}^i, d)}{|d|^s}\\ &= \sum_{k\geq 0}\sum_{\substack{(d,m_1\mathfrak{p})=1\\ d\in\mathcal{O}_{\mathrm{mon}}}} \frac{g(m_2\mathfrak{p}^i, d\mathfrak{p}^k)}{|d|^s p^{ks}}\\ &= \sum_{k\geq 0}\sum_{\substack{(d,m_1\mathfrak{p})=1\\ d\in\mathcal{O}_{\mathrm{mon}}}} \frac{g(m_2\mathfrak{p}^i, d)g(m_2\mathfrak{p}^i, \mathfrak{p}^k)}{|d|^s p^{ks}}\left(\frac{d}{\mathfrak{p}^{2k}}\right)\\ &= \sum_{\substack{(d,m_1\mathfrak{p})=1\\ d\in\mathcal{O}_{\mathrm{mon}}}} \frac{g(m_2\mathfrak{p}^i, d)}{|d|^s}\left(\sum_{k\geq 0}\frac{g(m_2\mathfrak{p}^i, \mathfrak{p}^k)}{p^{ks}}\left(\frac{d}{\mathfrak{p}^{2k}}\right)\right).\end{aligned}$$

The Gauss sum in the inner sum vanishes unless $k = 0$ or $i + 1$. This proves the lemma. □

Inverting the previous lemma, we obtain

Lemma 3.3. *If* $0 \leq gi \leq n-2$ *and* $m_1, m_2, \mathfrak{p}$ *as above,*

$$D_{\mathfrak{p}m_1}(s, m_2\mathfrak{p}^i) = \frac{D_{m_1}(s, m_2\mathfrak{p}^i)}{1-|\mathfrak{p}|^{n-1-ns}} - \frac{g(m_2\mathfrak{p}^i, \mathfrak{p}^{i+1})}{|\mathfrak{p}|^{(i+1)s}}\frac{D_{m_1}(s, m_2\mathfrak{p}^{n-i-2})}{1-|\mathfrak{p}|^{n-1-ns}},$$

and if $i = n-1$,

$$D_{\mathfrak{p}m_1}(s, m_2\mathfrak{p}^i) = \frac{D_{m_1}(s, m_2\mathfrak{p}^i)}{1-|\mathfrak{p}|^{n-1-ns}}.$$

Now suppose that $N(x, y) = N^{(\mathfrak{p})}(x, y)$ is the polynomial from (2.9). We define a function H on pairs of powers of $\mathfrak{p}$ by setting $H(\mathfrak{p}^k, \mathfrak{p}^l)$ to be the coefficient of $x^k y^l$ in $N(x, y)$:

$$N(x, y) = \sum H(\mathfrak{p}^k, \mathfrak{p}^l)x^k y^l.$$

We extend H to all pairs of monic polynomials by the twisted multiplicativity relation: if $\gcd(cd, c'd') = 1$, then we put

$$H(cc', dd') = H(c, d)H(c', d')\left(\frac{c}{c'}\right)^2\left(\frac{d}{d'}\right)^2\left(\frac{c}{d'}\right)^{-1}\left(\frac{c'}{d}\right)^{-1}. \tag{3.5}$$

In particular, note that

$$H(d, 1) = g(1, d). \tag{3.6}$$

Now consider the Dirichlet series

$$E(s,m) = (1-q^{n-ns})^{-1} \sum_{d\in\mathcal{O}_{\mathrm{mon}}} \frac{H(d,m)}{d^s}. \tag{3.7}$$

That $E(s,m)$ satisfies the same functional equation as $D(s,m)$ is the main result of this section.

Theorem 3.4. *Let $m \in \mathcal{O}_{\mathrm{mon}}$ be a monic polynomial of degree $nk+j$, where $0 \le j \le n-1$. Then*

$$E(s,m) = |m|^{1-s} \sum_{0\le i\le n-1} T_{ij}(s)E(2-s,m;i).$$

Proof. Before tackling the general case, we first consider $m = \mathfrak{p}^l$ for a prime $\mathfrak{p}$ and $l > 0$. Then

$$\begin{aligned}
E(s,\mathfrak{p}^l) &= (1-q^{n-ns})^{-1} \sum_{\substack{d\in\mathcal{O}_{\mathrm{mon}}\\(d,\mathfrak{p})=1}} \sum_{k\ge 0} \frac{H(d\mathfrak{p}^k,\mathfrak{p}^l)}{d^s|\mathfrak{p}|^{ks}}\\
&= (1-q^{n-ns})^{-1} \sum_{\substack{d\in\mathcal{O}_{\mathrm{mon}}\\(d,\mathfrak{p})=1}} \sum_{k\ge 0} \frac{H(\mathfrak{p}^k,\mathfrak{p}^l)g(1,d)}{|\mathfrak{p}|^{ks}d^s}\left(\frac{d}{\mathfrak{p}^{2k-l}}\right), \text{ by (3.5) and (3.6)}\\
&= \sum_{k\ge 0} \frac{H(\mathfrak{p}^k,\mathfrak{p}^l)}{|\mathfrak{p}|^{ks}} D_{\mathfrak{p}}(s,\mathfrak{p}^{(l-2k)_n})\\
&= \sum_{j=0}^{n-1} D_{\mathfrak{p}}(s,\mathfrak{p}^{(l-2j)_n}) \left(\frac{1}{|\mathfrak{p}|^{js}} \sum_{k\ge 0} \frac{H(\mathfrak{p}^{j+nk},\mathfrak{p}^l)}{|\mathfrak{p}|^{nks}}\right)\\
&= \sum_{j=0}^{n-1} D_{\mathfrak{p}}(s,\mathfrak{p}^{(l-2j)_n}) h^{(\mathfrak{p},l)}(|\mathfrak{p}|^{-s};j),
\end{aligned}$$

where $h^{(\mathfrak{p},l)}$ was introduced in (2.10). Using Lemma 3.3 the previous expression becomes

$$\sum_{j=0}^{n-1} D(s,\mathfrak{p}^{(l-2j)_n}) h^{(\mathfrak{p},l)}(|\mathfrak{p}|^{-s};j)$$

$$-\sum_{j=0}^{n-1} \delta_j \frac{g(\mathfrak{p}^{(l-2j)_n},\mathfrak{p}^{(l-2j)_n+1})}{|\mathfrak{p}|^{((l-2j)_n+1)s}} D(s,\mathfrak{p}^{(2j-l-2)_n}) h^{(\mathfrak{p},l)}(|\mathfrak{p}|^{-s};j), \tag{3.8}$$

where $\delta_j = 0$ if $l - 2j \equiv n-1(n)$ and is 1 otherwise. Replace j by $l+1-j$ in the second summation and regroup to conclude

$$E(s,\mathfrak{p}^l) = \sum_{j=0}^{n-1} D(s,\mathfrak{p}^{(l-2j)_n}) f^{(\mathfrak{p},l)}(|\mathfrak{p}|^{-s};j). \tag{3.9}$$

(Note the use of the identity $n-2-(l-2j)_n = (2j-l-2)_n$.) Using the functional equations (3.3) of D and $f^{(\mathfrak{p},l)}$ (Lemma 2.2), we write

$$\begin{aligned}
&E(s,\mathfrak{p}^l)|\mathfrak{p}|^{-(1-s)l}\\
&= \sum_{j=0}^{n-1}\sum_{i=0}^{n-1} T_{i,(l-2j)_n \deg\mathfrak{p}}(s) D(2-s,\mathfrak{p}^{(l-2j)_n};i) f^{(\mathfrak{p},l)}(2-s;j)\\
&= \sum_{i,j=0}^{n-1} T_{i-j\deg\mathfrak{p},(l-2j)_n\deg\mathfrak{p}}(s) D(2-s,\mathfrak{p}^{(l-2j)_n};i-j\deg\mathfrak{p}) f^{(\mathfrak{p},l)}(2-s;j)\\
&= \sum_{i=0}^{n-1} T_{i,l\deg\mathfrak{p}}(s)\left[\sum_{j=0}^{n-1} D(2-s,\mathfrak{p}^{(l-2j)_n};i-j\deg\mathfrak{p}) f^{(\mathfrak{p},l)}(2-s;j)\right]
\end{aligned}$$

$$= \sum_{i=0}^{n-1} T_{i,l\deg\mathfrak{p}}(s) E(2-s,\mathfrak{p}^l;i), \tag{3.10}$$

where the third equality comes from our remark that the T_{ij} depend only on $2i-j$. This is the functional equation we wished to prove, in the special case $m=\mathfrak{p}^l$.

The argument for general m is similar. Let $m = \mathfrak{p}_1^{l_1}\mathfrak{p}_2^{l_2}\cdots\mathfrak{p}_r^{l_r}$ where the $\mathfrak{p}_i$ are distinct primes and the l_i are positive. Then

$$\begin{aligned}
E(s;m) &= (1-q^{n-ns})^{-1}\sum_{d\in\mathcal{O}_{\text{mon}}}\frac{H(d,m)}{|d|^s}\\
&= (1-q^{n-ns})^{-1}\sum_{\substack{d\in\mathcal{O}_{\text{mon}}\\(d,m)=1}}\sum_{k_1,\dots,k_r\ge 0}\frac{H(d\mathfrak{p}_1^{k_1}\cdots\mathfrak{p}_r^{k_r},\mathfrak{p}_1^{l_1}\cdots\mathfrak{p}_r^{l_r})}{|d|^s|\mathfrak{p}_1|^{k_1 s}\cdots|\mathfrak{p}_r|^{k_r s}}\\
&= (1-q^{n-ns})^{-1}\sum_{\substack{d\in\mathcal{O}_{\text{mon}}\\(d,m)=1}}\sum_{k_1,\dots,k_r\ge 0}\frac{H(d,1)H(\mathfrak{p}_1^{k_1},\mathfrak{p}_1^{l_1})\cdots H(\mathfrak{p}_r^{k_r},\mathfrak{p}_r^{l_r})}{|d|^s|\mathfrak{p}_1|^{k_1 s}\cdots|\mathfrak{p}_r|^{k_r s}}\\
&\quad\times\left(\frac{d}{m}\right)^{-1}\left(\frac{d}{\mathfrak{p}_1^{k_1}\cdots\mathfrak{p}_r^{k_r}}\right)^2\prod_{a\ne b}\left(\frac{\mathfrak{p}_a^{k_a}}{\mathfrak{p}_b^{k_b}}\right)\left(\frac{\mathfrak{p}_a^{l_a}}{\mathfrak{p}_b^{l_b}}\right)\left(\frac{\mathfrak{p}_a^{k_a}}{\mathfrak{p}_b^{l_b}}\right)^{-1}
\end{aligned}$$

$$= \prod_{a \neq b} \left(\frac{\mathfrak{p}_a^{l_a}}{\mathfrak{p}_b^{l_b}} \right) \sum_{j_1=0}^{n-1} \cdots \sum_{j_r=0}^{n-1} D_m(s, \mathfrak{p}_1^{(l_1-2j_1)_n} \cdots \mathfrak{p}_r^{(l_r-2j_r)_n})$$

$$\times h^{(\mathfrak{p}_1,l_1)}(s; j_1) \cdots h^{(\mathfrak{p}_r,l_r)}(s; j_r) \prod_{a \neq b} \left(\frac{\mathfrak{p}_a^{j_a}}{\mathfrak{p}_b^{j_b}} \right) \left(\frac{\mathfrak{p}_a^{j_a}}{\mathfrak{p}_b^{l_b}} \right)^{-1}. \tag{3.11}$$

Denote for the moment by $C(j_1) = C(j_1, \ldots, j_r)$ the product of residue symbols

$$C(j_1) = \prod_{a \neq b} \left(\frac{\mathfrak{p}_a^{j_a}}{\mathfrak{p}_b^{j_b}} \right) \left(\frac{\mathfrak{p}_a^{j_a}}{\mathfrak{p}_b^{l_b}} \right)^{-1}. \tag{3.12}$$

Letting $J_i = (l_i - 2j_i)_n$ for $i = 1, \ldots, r$, we have

(3.13)

$$(1 - |\mathfrak{p}_1|^{n-1-ns}) D_m(s, \mathfrak{p}_1^{J_1} \cdots \mathfrak{p}_r^{J_r}) C(j_1) = D_{\mathfrak{p}_2 \cdots \mathfrak{p}_r}(s, \mathfrak{p}_1^{J_1} \cdots \mathfrak{p}_r^{J_r}) C(j_1)$$

$$- \delta_{j_1} \frac{g(\mathfrak{p}_1^{J_1} \cdots \mathfrak{p}_r^{J_r}, \mathfrak{p}_1^{J_1+1})}{|\mathfrak{p}_1|^{(J_1+1)s}} D_{\mathfrak{p}_2 \cdots \mathfrak{p}_r}(s, \mathfrak{p}_1^{(2j_1-l_1-2)_n} \mathfrak{p}_2^{J_2} \cdots \mathfrak{p}_r^{J_r}) C(j_1)$$

by Lemma 3.3. In the second term on the right-hand side, replace j_1 by $l_1 + 1 - j_1$. For $\delta_{j_1} \neq 0$ this gives

(3.14)

$$\frac{g(\mathfrak{p}_1^{(2j_1-l_1-2)_n} \mathfrak{p}_2^{J_2} \cdots \mathfrak{p}_r^{J_r}, \mathfrak{p}_1^{(2j_1-l_1-1)_n})}{|\mathfrak{p}_1|^{((2j_1-l_1-1)_n)s}} D_{\mathfrak{p}_2 \cdots \mathfrak{p}_r}(s, \mathfrak{p}_1^{J_1} \mathfrak{p}_2^{J_2} \cdots \mathfrak{p}_r^{J_r}) C(l_1 - j_1 + 1).$$

The Gauss sum can be written as

$$\left(\frac{\mathfrak{p}_2^{J_2} \cdots \mathfrak{p}_r^{J_r}}{\mathfrak{p}_1^{2j_1-l_1-1}} \right)^{-1} g(\mathfrak{p}_1^{(2j_1-l_1-2)_n}, \mathfrak{p}_1^{(2j_1-l_1-1)_n}), \tag{3.15}$$

and $C(l_1 - j_1 + 1)$ is

$$\left(\frac{\mathfrak{p}_2^{J_2} \cdots \mathfrak{p}_r^{J_r}}{\mathfrak{p}_1^{l_1-j_1+1}} \right)^{-1} \left(\frac{\mathfrak{p}_2^{j_2} \cdots \mathfrak{p}_r^{j_r}}{\mathfrak{p}_1^{l_1}} \right)^{-1} \left(\prod_{\substack{a \neq b \\ a,b \neq 1}} \left(\frac{\mathfrak{p}_a^{j_a}}{\mathfrak{p}_b^{j_b}} \right) \left(\frac{\mathfrak{p}_a^{j_a}}{\mathfrak{p}_b^{l_b}} \right)^{-1} \right). \tag{3.16}$$

Taking the product of (3.15) and (3.16) yields

(3.17)

$$\left(\frac{\mathfrak{p}_2^{J_2}\cdots\mathfrak{p}_r^{J_r}}{\mathfrak{p}_1^{j_1}}\right)^{-1}\left(\frac{\mathfrak{p}_2^{j_2}\cdots\mathfrak{p}_r^{j_r}}{\mathfrak{p}_1^{l_1}}\right)^{-1} g(\mathfrak{p}_1^{(2j_1-l_1-2)_n},\mathfrak{p}_1^{(2j_1-l_1-1)_n})\left(\prod_{\substack{a\neq b\\ a,b\neq 1}}\left(\frac{\mathfrak{p}_a^{j_a}}{\mathfrak{p}_b^{j_b}}\right)\left(\frac{\mathfrak{p}_a^{j_a}}{\mathfrak{p}_b^{l_b}}\right)^{-1}\right)$$

$$= g(\mathfrak{p}_1^{(2j_1-l_1-2)_n},\mathfrak{p}_1^{(2j_1-l_1-1)_n})C(j_1).$$

Therefore, continuing from the last line of (3.11),

(3.18)

$$E(s,m)=\prod_{a\neq b}\left(\frac{\mathfrak{p}_a^{l_a}}{\mathfrak{p}_b^{l_b}}\right)\sum_{j_1=0}^{n-1}\cdots\sum_{j_r=0}^{n-1} D_{m'}(s,\mathfrak{p}_1^{(l_1-2j_1)_n}\cdots\mathfrak{p}_r^{(l_r-2j_r)_n})$$
$$\times\prod_{a\neq b}\left(\frac{\mathfrak{p}_a^{j_a}}{\mathfrak{p}_b^{j_b}}\right)\left(\frac{\mathfrak{p}_a^{j_a}}{\mathfrak{p}_b^{l_b}}\right)^{-1} f^{(\mathfrak{p}_1,l_1)}(s;j_1)h^{(\mathfrak{p}_2,l_2)}(s;j_2)\cdots h^{(\mathfrak{p}_r,l_r)}(s;j_r),$$

where $m'=\mathfrak{p}_2^{l_2}\cdots\mathfrak{p}_r^{l_r}$. Repeating this procedure to remove the primes from m one at a time, we find that up to a constant of modulus one, $E(s,m)$ is equal to

$$\sum_{j_1=0}^{n-1}\cdots\sum_{j_r=0}^{n-1} D(s,\mathfrak{p}_1^{J_1}\cdots\mathfrak{p}_r^{J_r})\left(\prod_{a=1}^{r} f^{(\mathfrak{p}_a,l_a)}(s;j_a)\right)\prod_{a\neq b}\left(\frac{\mathfrak{p}_a^{j_a}}{\mathfrak{p}_b^{j_b}}\right)\left(\frac{\mathfrak{p}_a^{j_a}}{\mathfrak{p}_b^{l_b}}\right)^{-1}. \tag{3.19}$$

We may now apply the functional equations of D and the $f^{(\mathfrak{p}_a,l_a)}$ as in (3.10) to conclude that $E(s,m)$ satisfies the functional equation

$$E(s,m)=|m|^{1-s}\sum_{i=0}^{n-1}T_{i,\deg m}(s)E(2-s,m;i). \tag{3.20}$$

This completes the proof of the theorem. □

For later use, we record the following bound:

Proposition 3.5 *For all $\epsilon>0$, $m\in\mathcal{O}$ and $0\le i<n$,*

$$(s-1-\tfrac{1}{n})(s-1+\tfrac{1}{n})E(s,m;i)\ll_\epsilon\begin{cases}1 & \text{for } Re(s)>\tfrac{3}{2}+\epsilon\\ |m|^{\frac{1}{2}+\epsilon} & \text{for } \tfrac{1}{2}-\epsilon<Re(s)<\tfrac{3}{2}+\epsilon\\ |m|^{1-s+\epsilon} & \text{for } Re(s)<\tfrac{1}{2}-\epsilon\end{cases}$$

Proof. Use the meromorphy and functional equation of $E(s,m)$ together with the convexity principle, cf. [14, Eq. (2.3)] and [19, Propostion 8.4]. □

4 The double dirichlet series

Recall the definition of the double Dirichlet series from (1.2)–(1.4). In this section we show that $Z(s_1,s_2)$ has a meromorphic continuation to $s_1,s_2 \in \mathbb{C}$ and satisfies a group of functional equation isomorphic to W. In [2], the authors show in detail how the analytic continuation of a Weyl group multiple Dirichlet series follows from the functional equations. Therefore we concentrate on establishing the functional equations of $Z(s_1,s_2)$.

Actually we need to consider slightly different series. For integers $0 \le i,j \le n-1$ we define

$$Z(s_1,s_2;i,j) = (1-q^{n-ns_1})^{-1}(1-q^{n-ns_2})^{-1}(1-q^{2n-ns_1-ns_2})^{-1} \times \sum_{\substack{m\in\mathcal{O}_{\mathrm{mon}}\\ \deg m=i \bmod n}} \sum_{\substack{d\in\mathcal{O}_{\mathrm{mon}}\\ \deg d=j \bmod n}} \frac{H(d,m)}{|m|^{s_1}|d|^{s_2}}. \tag{4.1}$$

We further introduce the notation

$$Z(s_1,s_2;i,*) = \sum_j Z(s_1,s_2;i,j)$$

and

$$Z(s_1,s_2;*,j) = \sum_i Z(s_1,s_2;i,j).$$

These series are absolutely convergent for $\mathrm{Re}(s_1),\mathrm{Re}(s_2) > 3/2$. In fact, we can do a little better. Summing over d first yields

(4.2)

$$\begin{aligned}Z(s_1,s_2;i,*) &= (1-q^{n-ns_1})^{-1}(1-q^{n-ns_2})^{-1}(1-q^{2n-ns_1-ns_2})^{-1}\\ &\quad\times \sum_{\substack{m\in\mathcal{O}_{\mathrm{mon}}\\ \deg m=i \bmod n}} \left(\frac{1}{|m|^{s_1}} \sum_{d\in\mathcal{O}_{\mathrm{mon}}} \frac{H(d,m)}{|d|^{s_2}}\right)\\ &= (1-q^{n-ns_1})^{-1}(1-q^{2n-ns_1-ns_2})^{-1} \sum_{\substack{m\in\mathcal{O}_{\mathrm{mon}}\\ \deg m=i \bmod n}} \frac{E(s_2,m)}{|m|^{s_1}}.\end{aligned}$$

By the convexity bound of Proposition 3.5, this representation of $Z(s_1, s_2; i, *)$ is seen to be meromorphic for $\mathrm{Re}(s_1) > 0, \mathrm{Re}(s_2) > 2$. Alternatively, summing over m first we deduce that $Z(s_1, s_2; i, *)$ is meromorphic for $\mathrm{Re}(s_2) > 0, \mathrm{Re}(s_1) > 2$. Let $\mathcal{R}$ be the tube domain that is the union of these three regions of initial meromorphy:

$$\mathcal{R} = \{\mathrm{Re}(s_1), \mathrm{Re}(s_2) > 3/2\} \cup \{\mathrm{Re}(s_1) > 0, \mathrm{Re}(s_2) > 2\} \\ \cup \{\mathrm{Re}(s_2) > 0, \mathrm{Re}(s_1) > 2\}.$$

Let the Weyl group W act on $\mathbb{C}^2$ by

$$\sigma_1 : (s_1, s_2) \mapsto (2 - s_1, s_1 + s_2 - 1), \quad \sigma_2 : (s_1, s_2) \mapsto (s_1 + s_2 - 1, 2 - s_2). \tag{4.3}$$

Let $\mathcal{F}$ be the real points of a closed fundamental domain for the action of W on $\mathbb{C}^2$:

$$\mathcal{F} = \{\mathrm{Re}(s_1), \mathrm{Re}(s_2) \geq 1\}.$$

One can easily see that $\mathcal{R} \smallsetminus \mathcal{F} \cap \mathcal{R}$ is compact. Therefore, by the principle of analytic continuation and Bochner's tube theorem [1], to prove that $Z(s_1, s_2)$ has a meromorphic continuation to $\mathbb{C}^2$ it suffices to show that the functions $Z(s_1, s_2; i, j)$ satisfy functional equations as (s_1, s_2) goes to $(2 - s_1, s_1 + s_2 - 1)$ and $(s_1 + s_2 - 1, 2 - s_2)$. For details, we refer to [2, Section 3].

To prove the σ_2 functional equation, we begin with (4.2) and write

$$\begin{aligned} Z(s_1, s_2; i, *) &= (1 - q^{n-ns_1})^{-1}(1 - q^{2n-ns_1-ns_2})^{-1} \sum_{\substack{m \in \mathcal{O}_{\mathrm{mon}} \\ \deg m = i \bmod n}} \frac{E(s_2, m)}{|m|^{s_1}} \\ &= (1 - q^{n-ns_1})^{-1}(1 - q^{2n-ns_1-ns_2})^{-1} \\ &\quad \times \sum_{\substack{m \in \mathcal{O}_{\mathrm{mon}} \\ \deg m = i \bmod n}} \frac{|m|^{1-s_2}}{|m|^{s_1}} \sum_{j=0}^{n-1} T_{ji}(s_2) E(2 - s_2, m; j), \quad \text{by Thm. 3.4} \\ &= \sum_{j=0}^{n-1} T_{ji}(s_2) Z(s_1 + s_2 - 1, 2 - s_2; i, j) \end{aligned}$$

The σ_1 functional equation is proved similarly.

We conclude that

Theorem 4.1. *The double Dirichlet series has a meromorphic continuation to $s_1, s_2 \in \mathbb{C}$ and is holomorphic away from the hyperplanes*

$$s_1 = 1 \pm \frac{1}{n}, s_2 = 1 \pm \frac{1}{n} \text{ and } s_1 + s_2 = 2 \pm \frac{1}{n}.$$

Furthermore, $Z(s_1, s_2)$ satisfies the functional equations

$$Z(s_1, s_2) = \sum_{i,j} T_{ji}(s_2) Z(s_1 + s_2 - 1, 2 - s_2; i, j)$$

$$= \sum_{i,j} T_{ij}(s_1) Z(2 - s_1, s_1 + s_2 - 1; i, j).$$

References

1. S. Bochner, *A theorem on analytic continuation of functions in several variables*, Ann. of Math. (2) 39 (1938), no. 1, 14–19.
2. B. Brubaker, D. Bump, G. Chinta, S. Friedberg, J. Hoffstein, *Weyl group multiple Dirichlet series I,* in Multiple Dirichlet Series, Automorphic Forms, and Analytic Number Theory, Proc. Sympos. Pure Math., 75, Amer. Math. Soc., Providence, RI, 2006.
3. B. Brubaker, D. Bump, and S. Friedberg, *Weyl group multiple Dirichlet series II: The stable case*, Invent. Math. **165** (2006), 325–355.
4. B. Brubaker, D. Bump, and S. Friedberg. *Weyl group multiple Dirichlet Series, Eisenstein series and crystal bases*, to appear in Annals of Mathematics.
5. B. Brubaker, D. Bump, and S. Friedberg. *Weyl Group Multiple Dirichlet Series: Type A Combinatorial Theory*, to appear in Annals of Mathematics Studies, Princeton University Press.
6. B. Brubaker, D. Bump, S. Friedberg, and J. Hoffstein, *Weyl group multiple Dirichlet series. III. Eisenstein series and twisted unstable A_r*, Ann. of Math. (2) **166** (2007), no. 1, 293–316.
7. G. Chinta, *Mean values of biquadratic zeta functions*, Invent. Math. **160** (2005), 145–163.
8. G. Chinta. *Multiple Dirichlet series over rational function fields*, Acta Arith., 132(4): 377–391, 2008.
9. G. Chinta, S. Friedberg and J. Hoffstein, *Multiple Dirichlet series and automorphic forms,* in Multiple Dirichlet Series, Automorphic Forms, and Analytic Number Theory, Proc. Sympos. Pure Math., 75, Amer. Math. Soc., Providence, RI, 2006.
10. G. Chinta, S. Friedberg, and P. E. Gunnells. *On the p-parts of quadratic Weyl group multiple Dirichlet series*, J. Reine Angew. Math., 623:1–23, 2008.
11. G. Chinta and P. E. Gunnells, *Weyl group multiple Dirichlet series constructed from quadratic characters,* Invent. Math. 167 (2007), no.2, 327–353.
12. G. Chinta and P. E. Gunnells, *Constructing Weyl group multiple Dirichlet series,* J. Amer. Math. Soc. 23 (2010), no. 1, 189–215.
13. B. Fisher and S. Friedberg, *Double Dirichlet series over function fields*, Compos. Math. **140** (2004), no. 3, 613–630.
14. S. Friedberg, J. Hoffstein, and D. Lieman, *Double Dirichlet series and the n-th order twists of Hecke L-series,* Math. Ann. 327 (2003), no. 2, 315–338.
15. J. Hoffstein, *Theta functions on the n-fold metaplectic cover of* SL(2)*—the function field case*, Invent. Math. 107 (1992), no. 1, 61–86.
16. J. Hoffstein and M. Rosen, *Average values of L-series in function fields*, J. Reine Angew. Math. 426 (1992), 117–150.
17. T. Kubota, *Some results concerning reciprocity law and real analytic automorphic functions,* 1969 Number Theory Institute (Proc. Sympos. Pure Math., Vol. XX, State Univ. New York, Stony Brook, N.Y., 1969), pp. 382–395. Amer. Math. Soc., Providence, R.I., 1971.

18. T. Kubota, *Some number-theoretical results on real analytic automorphic forms,* Several Complex Variables, II (Proc. Internat. Conf., Univ. Maryland, College Park, Md., 1970), pp. 87–96. Lecture Notes in Math., Vol. 185, Springer, Berlin, 1971.
19. R. Livné and S. J. Patterson, *The first moment of cubic exponential sums,* Invent. Math. 148, 79–116 (2002).
20. S. J. Patterson. *Note on a paper of J. Hoffstein* Glasg. Math. J., 49(2):243–255, 2007.

On the geometry of the diffeomorphism group of the circle

Adrian Constantin and Boris Kolev

Dedicated to the memory of Professor Serge Lang

Abstract We discuss some of the possibilities of endowing the diffeomorphism group of the circle with Riemannian structures arising from right-invariant metrics.

Key words Geodesic flow • diffeomorphism group of the circle

Mathematics Subject Classification (2010): 35Q35, 58B25

Preamble

Among the numerous contributions of Professor Serge Lang to mathematics the present contribution relates to his wonderful treatise on differential geometry [18]. Advocating a coordinate-free formalism, classical results from (finite-dimensional) Riemannian geometry were masterfully extended to infinite-dimensional manifolds modeled on a Hilbert space. With few exceptions (e.g., the Hopf–Rinow theorem is not any more valid), most classical results are shown to hold in the infinite-dimensional setting. It is only natural to wonder whether one can, in the same spirit, address rigorously these issues in the case of infinite-dimensional Fréchet manifolds.

A. Constantin (✉)
University of Vienna, Fakultät für Mathematik, Nordbergstraße 15, 1090 Wien, Austria
e-mail: adrian.constantin@univie.ac.at

B. Kolev
CMI, 39, rue F. Joliot-Curie, 13453 Marseille cedex 13, France
e-mail: kolev@cmi.univ-mrs.fr

D. Goldfeld et al. (eds.), *Number Theory, Analysis and Geometry: In Memory of Serge Lang*, DOI 10.1007/978-1-4614-1260-1_7,

The motivation for this question is the fact that certain equations of mathematical physics (like the inviscid Burgers equation from gas dynamics, the Euler equation of hydrodynamics, the Camassa–Holm equation from shallow water theory) were shown to be re-expressions of geodesic flow for right-invariant metrics on groups of smooth diffeomorphisms, and for such complex systems the need for more elegant formulations becomes an issue of paramount importance. The infinite-dimensional Lie groups of smooth diffeomorphisms are not Hilbert manifolds but Fréchet manifolds (their topology is not induced by an inner product but by a countable family of seminorms), situation which raises a number of highly non-trivial technical issues since basic analytic results (like the existence and uniqueness of local solutions for ordinary differential equations with a smooth right-hand side, or like the inverse function theorem) are known not to be valid in general within this setting [12]. Nevertheless, rigorous analytical results can be obtained.

We illustrate the approach by considering the example of the diffeomorphism group of the circle. The presented results are to a large extent the fruit of our joint research effort over the last five years, through which we benefited from the generous encouragement and support of Professor Serge Lang.

1 Introduction

In 1966 Arnold [2] showed that the Euler equations of hydrodynamics can be obtained as the geodesic equations for the infinite-dimensional Lie group of smooth volume- and orientation-preserving diffeomorphisms of the fluid domain D with respect to the right-invariant L^2 inner-product, given on the corresponding Lie algebra of divergence-free vector fields tangent to the boundary of D by

$$\langle u, v \rangle = \int_D (u \cdot v)\, d\mu.$$

Subsequently, other equations from mathematical physics were found to have an interpretation as geodesic flows on diffeomorphism groups (see for example [13,14, 20,21]). Our aim is to explain the setting presented by the diffeomorphism group of the circle. Why study Euler equations on the diffeomorphism group of the circle? For two reasons: On the one hand, it is the simplest of the diffeomorphisms groups and it is expected that understanding some mechanisms within this setting can give insight to deal with more ambitious situations. On the other hand, because it is already the configuration space of two famous equations arising in fluid mechanics: the inviscid *Burgers* equation and the *Camassa–Holm* equation.

Before proceeding with this concrete example in Section 3, we first discuss some fundamental aspects in the general case of an abstract Lie group in Section 2. Section 4 is devoted to the study of bi-Hamiltonian structures for Euler's equations on the diffeomorphism group of the circle. A final section deals with the case of the

Virasoro group and the KdV equation: a case which has been considered prior to the case of the diffeomorphism group of the circle but which, from a didactic point of view, has to be studied afterwards because of additional technical difficulties.

2 Invariant metrics on an abstract Lie group

2.1 *Left-invariant metrics*

A left-invariant metric on a Lie group G is determined by its value at the unit element e of the group, that is, by an inner product on the Lie algebra $\mathfrak{g}$, expressed in terms of a symmetric[1] linear operator

$$A : \mathfrak{g} \to \mathfrak{g}^*,$$

called the *inertia operator*. If $t \mapsto g(t)$ is a geodesic[2] with $g(0) = e$, denote by $\dot{g}(t)$ the derivative with respect to t, and let

$$m(t) =< \dot{g}(t), \cdot >_g \in T_g^* G.$$

2.2 *Angular velocities and momenta*

Introducing the *left and right angular velocities* by

$$u_L = L_{g^{-1}} \dot{g}, \qquad u_R = R_{g^{-1}} \dot{g},$$

and the *left and right angular momenta* by

$$m_L = L_g^* m, \qquad m_R = R_g^* m\,,$$

the following relations hold between these four geometrical objects[3]:

$$m_L = A u_L, \qquad u_R = Ad_g\, u_L, \qquad m_R = Ad_g^*\, m_L, \tag{2.1}$$

where L, R stand for left, respectively right translation.

---

[1] A is symmetric if $(Au, v) = (Av, u)$ for all $u, v \in \mathfrak{g}^*$, where the round brackets stand for the pairing of elements of the dual spaces $\mathfrak{g}$ and $\mathfrak{g}^*$.

[2] Notice that geodesics issuing from some $g_0 \in G$ are obtained via left translation by g_0 from geodesics issuing from e.

[3] The coadjoint action of G on $\mathfrak{g}^*$ is defined by $(\mathrm{Ad}_g^*\, m, u) = (m, \mathrm{Ad}_{g^{-1}}\, u)$.

2.3 Noether's theorem

In its initial formulation by Emmy Noether in 1918, the theorem states that to each infinitesimal transformation which leaves the Lagrangian of a variational problem invariant there corresponds a first integral of the motion. In the sixties, Souriau [24] extended this result to the more general setting of *Symplectic Mechanics*. Before giving a (weak[4]) statement of this theorem let us recall the following fact. If a group G acts smoothly on a manifold M, to each vector $v \in \mathfrak{g}$ corresponds a vector field on M defined by

$$X_v(x) = \left.\frac{d}{ds}\right|_{s=0} \exp(sv)(x).$$

Theorem 2.1 (Noether theorem). *Let* (M, ω) *a symplectic manifold such that* $\omega = d\theta$ $(\theta \in \Omega^1(M))$. *Let* G *be a Lie group acting smoothly on* M *and such that* $g^*\theta = \theta$ *for all* $g \in G$. *Then, for each* G*-invariant function* $H \in C^\infty(M)$, *and each* $v \in \mathfrak{g}$, *the function* $\theta(X_v)$ *is a first integral of the Hamiltonian vector field* X_H.

2.4 Euler's equations

A Riemannian metric on a Lie group G induces an isomorphism between TG and T^*G which permits to pullback on TG the *canonical* 1-form of T^*G. We will call it θ. The exterior derivative $\omega = d\theta$ is a *symplectic form* on TG, and the Hamiltonian flow of the function

$$H(X_g) = \frac{1}{2} < X_g, X_g >, \quad X_g \in T_gG$$

corresponds to the geodesic flow of the metric. If the Riemannian metric is invariant by the natural left action of G on itself, so are θ and H. If $\widetilde{X}_v$ is the vector field on $TG \simeq G \times \mathfrak{g}$ induced by $v \in \mathfrak{g}$, we get

$$\theta_{(g,u)}(X_v) = \left(Ad_g^*(Au), v\right),$$

where A is the inertia operator defined previously. In particular

$$\theta_{(g(t),u_L(t))}(X_v) = (m_R(t), v),$$

[4]The strong version does not require ω to be an *exact form*. It only assumes that ω is G-invariant and that the symplectic group action of G on M has a *moment map*.

and an application of *Noether's theorem* leads us to *Euler's first theorem*

$$\frac{dm_R}{dt} = 0. \tag{2.2}$$

This may be considered as a generalization of the *angular momentum conservation law* of a free rigid body.

Taking the time derivative of the third relation of (2.1), we then obtain *Euler's second theorem*

$$\frac{dm_L}{dt} = -\mathrm{ad}^*_{u_L} m_L, \tag{2.3}$$

where $\mathrm{ad}^*_u m$ represents the coadjoint action[5] of $\mathfrak{g}$ over $\mathfrak{g}^*$.

2.5 *The contravariant formulation*

It is useful to have a contravariant formulation of (2.3). With this purpose in mind, let us introduce the bilinear operator $B : \mathfrak{g} \times \mathfrak{g} \to \mathfrak{g}$ defined by

$$< \mathrm{ad}_u\, v, w >=< B(w,u), v >, \qquad u, v, w \in \mathfrak{g}.$$

Equation (2.3) can then be rewritten as a quadratic differential equation on $\mathfrak{g}$,

$$\dot{u}_L = B(u_L, u_L). \tag{2.4}$$

This may be considered as a generalization of the *evolution equation for the angular velocity* of a free rigid body with a fixed point.

Therefore, integration of the geodesic equations may be reduced to two successive quadratures

$$\begin{cases} \dot{g} & = L_g\, u_L\,, \\ \dot{u}_L & = B(u_L, u_L)\,. \end{cases} \tag{2.5}$$

Remark 2.2. The Levi-Civita connection of a left invariant metric is also left-invariant. It is thus completely defined by the knowledge of

$$\left(\nabla_{L_g u} L_g v\right)(e) = \frac{1}{2}[u, v] - \frac{1}{2}\left(B(u,v) + B(v,u)\right).$$

When the metric is bi-invariant, B is skew-symmetric and $\dot{u}_L = 0$. In that case, the geodesics through the unit element are just the one-parameter subgroups of G.

[5] It is defined by $(\mathrm{ad}^*_\omega\, m, u) = -(m, \mathrm{ad}_\omega\, u)$.

2.6 *Right-invariant metrics*

In the case of a right-invariant metric, relations (2.1) become

$$m_R = Au_R, \qquad u_R = \mathrm{Ad}_g\, u_L, \qquad m_R = \mathrm{Ad}_g^*\, m_L, \tag{2.6}$$

and Euler's equations become

$$\frac{dm_L}{dt} = 0, \qquad \frac{dm_R}{dt} = \mathrm{ad}_{u_R}^* m_R. \tag{2.7}$$

The contravariant formulation of second Euler's equation becomes

$$\dot{u}_R = -B(u_R, u_R), \tag{2.8}$$

3 Right-invariant metrics on the diffeomorphism group of the circle

In his landmark treatise [18], Serge Lang succeeded in fursning an elegant in-depth study of differential geometry on manifolds modeled on an abstract Banach space, proving fundamental results corresponding to the classical theorems known in the finite-dimensional case.

In this section, we will deal with the group $\mathrm{Diff}(\mathbb{S}^1)$ of smooth diffeomorphisms of the circle that are orientation-preserving. This group is naturally equipped with a *Fréchet manifold* structure. More precisely, we can cover $\mathrm{Diff}(\mathbb{S}^1)$ with charts taking values in the *Fréchet vector space*[6] $\mathrm{C}^\infty(\mathbb{S}^1)$ and in such a way that the change of charts are smooth maps (see [5] for more details).

Since the composition and the inverse are smooth maps for this structure we say that $\mathrm{Diff}(\mathbb{S}^1)$ is a *Fréchet-Lie group*. Its "*Lie algebra*" $\mathrm{Vect}(\mathbb{S}^1)$ is isomorphic to $\mathrm{C}^\infty(\mathbb{S}^1)$ with the Lie bracket given by

$$[u, v] = u_x v - u v_x.$$

[6] A topological vector space E has a canonical *uniform structure*. When this structure is *complete* and when the topology of E may be given by a countable family of *semi-norms*, we say that E is a Fréchet vector space. In a Fréchet space, such classical results like the *Cauchy–Lipschitz theorem* or the *local inverse theorem* are no longer valid in general as they are in on Banach manifold. the typical example of a Fréchet space is the space of smooth functions on a compact manifold where semi-norms are just the C^k-norms ($k = 0, 1, \ldots$).

3.1 Right-invariant metrics on Diff($\mathbb{S}^1$)

A *right-invariant* metric on Diff($\mathbb{S}^1$) is defined by an inner product $\mathbf{a}$ on the Lie algebra of the group Vect($\mathbb{S}^1$) $=$ $\mathrm{C}^\infty(\mathbb{S}^1)$. If this inner product is *local*,[7] then according to a result of Peetre [22], it is given by

$$\mathbf{a}(u,v) = \int_{\mathbb{S}^1} u\, A(v)\, dx \qquad u, v \in \mathrm{C}^\infty(\mathbb{S}^1),$$

where A is a symmetric linear differential operator. This operator A, that will be assumed invertible, is called the *inertia operator*. The corresponding bilinear operator B defined in Section 2 is then given by

$$B(w,u) = A^{-1}\left[2A(w)u_x + uA(w)_x\right] \qquad u, w \in \mathrm{C}^\infty(\mathbb{S}^1).$$

An important special case is when the inner product is given by the H^k*-Sobolev* norm on $\mathrm{C}^\infty(\mathbb{S}^1)$ $(k \geq 0)$,

$$\mathbf{a}_k(u,v) = \int_{\mathbb{S}^1} \left(uv + u_x v_x + \cdots + u_x^{(k)} v_x^{(k)}\right) dx \qquad u, v \in \mathrm{C}^\infty(\mathbb{S}^1).$$

In that case, the inertia operator is

$$A_k = 1 - \frac{d^2}{dx^2} + \cdots + (-1)^k \frac{d^{2k}}{dx^{2k}}.$$

3.2 Examples

For $k = 0$ (that is for the L^2-metric), the corresponding Euler's equation (2.8) is the *inviscid Burgers equation*

$$u_t + 3uu_x = 0. \tag{3.1}$$

For $k = 1$ (that is for the H^1-metric), the corresponding Euler's equation (2.8) is the *Camassa–Holm equation*

$$u_t - u_{txx} + 3uu_x - 2u_x u_{xx} - uu_{xxx} = 0. \tag{3.2}$$

---

[7] That is, for all $u, v \in \mathrm{C}^\infty(\mathbb{S}^1)$, $\mathrm{Supp}(u) \cap \mathrm{Supp}(v) = \emptyset \Rightarrow \mathbf{a}(u,v) = 0$.

3.3 Regular dual

Since the topological dual of the Fréchet space $\mathrm{Vect}(\mathbb{S}^1)$ is isomorphic to the space of distributions on the circle, we use a subspace of this topological dual, called the *regular dual*, denoted $\mathrm{Vect}^*(\mathbb{S}^1)$ and consisting of the linear functionals of the form

$$u \mapsto \int_{\mathbb{S}^1} mu\,dx$$

for some function $m \in C^\infty(\mathbb{S}^1)$. The *regular* $\mathrm{Vect}^*(\mathbb{S}^1)$ is naturally isomorphic to the space of *quadratic differentials* $m(x)dx^2$ ($m \in C^\infty(\mathbb{S}^1)$) on the circle, the pairing being given by

$$(m, u) = \int_{\mathbb{S}^1} mu\,dx$$

for $u \in \mathfrak{g} = C^\infty(\mathbb{S}^1)$ and $m \in \mathrm{Vect}^*(\mathbb{S}^1)$.

The coadjoint action of $\mathrm{Diff}(\mathbb{S}^1)$ on $\mathrm{Vect}^*(\mathbb{S}^1)$ is given by

$$\mathrm{Ad}^*_\varphi\, m = \frac{m \circ \varphi^{-1}}{(\varphi_x \circ \varphi^{-1})^2}, \qquad m \in \mathfrak{g}^*, \varphi \in \mathrm{Diff}(\mathbb{S}^1),$$

and the coadjoint action of $\mathrm{Vect}(\mathbb{S}^1)$ on $\mathrm{Vect}^*(\mathbb{S}^1)$ by

$$\mathrm{ad}^*_u\, m = -\,(2mu_x + m_x u)\,, \qquad m \in \mathfrak{g}^*, u \in \mathrm{Vect}(\mathbb{S}^1).$$

3.4 The momentum

The conservation of the left momentum

$$m_L = \mathrm{Ad}^*_{\varphi^{-1}}\, m_R$$

leads [5] to the following conservation law

$$(A(u) \circ \varphi) \cdot \varphi_x^2 = A(u_0),$$

along each geodesic curve φ issued from the unit element Id in the direction u_0.

3.5 The Cauchy problem

Since there are no general local existence theorem for evolution equations on a Fréchet space, the first step in the study of a geodesic flow on the diffeomorphism group is to prove the short-time existence and uniqueness. This was achieved by the authors [5] for the right-invariant metric on $\mathrm{Diff}(\mathbb{S}^1)$ generated by the H^k Sobolev norm for $k \geq 1$.

Theorem 3.1 (Constantin and Kolev 2003). *Let $k \geq 1$. For all $T > 0$, there exists a neighborhood of the origin V in $\mathrm{Vect}(\mathbb{S}^1)$ such that for all $u_0 \in V$, there exists a unique geodesic*

$$\varphi \in C^\infty\left([0,T); \mathrm{Diff}(\mathbb{S}^1)\right)$$

for the metric H^k, starting at $\varphi(0) = Id \in \mathrm{Diff}(\mathbb{S}^1)$ in the direction $u_0 \in T_{Id}\mathrm{Diff}(\mathbb{S}^1)$. Moreover, the solution depends smoothly on the initial data $u_0 \in \mathrm{C}^\infty(\mathbb{S}^1)$.

Remark 3.2. For $k = 0$, which corresponds to the *inviscid Burgers equation*, one can show the short-time existence by the *method of characteristics* but the proof below does not apply.

Remark 3.3. The spirit of the proof given in [5] is to study the evolution equation on each Hilbert space H^n obtaining well-posedness on a maximal interval $[0, T_n)$. Thereafter one checks that the decreasing sequence T_n does not go to 0 as $n \to +\infty$, ensuring thus the short-time existence on $C^\infty = \bigcap_{n=0}^{+\infty} H^n$.

Notice that it is advisable to avoid considering directly the Euler equation

$$u_t = -B_k(u,u) = -A_k^{-1}\left[2A_k(u)u_x + uA_k(u_x)\right]. \tag{3.3}$$

The reason is that A_k is a differential operator of degree $2k$, and therefore the right-hand side of (3.3) is a pseudo-differential operator of degree 1 because of the "bad term" $uA_k(u_x)$. Hence the Cauchy problem for this equation has no meaning in H^n.

The following approach was used in [5] to overcome this difficulty. The operator

$$C_k(u) = A_k(uu_x) - uA_k(u_x)$$

is a quadratic differential operator of degree $2k$. Therefore, if $k \geq 1$, the right-hand side of

$$u_t + uu_x = -A_k^{-1}\left[2A_k(u)u_x - C_k(u)\right] \tag{3.4}$$

is a pseudo-differential operator of degree 0. Moreover, $u_t + uu_x$ is just $v_t \circ \varphi^{-1}$, where $v = \varphi_t = u \circ \varphi$ is the *Lagrangian velocity*.

Sketch of proof. The proof is divided into three steps.

Step 1. For each $n \geq 2k+1$, we have a *well-posed Cauchy problem* on the Hilbert manifold $\mathcal{D}_n \times H^n(\mathbb{S})$:

$$\begin{cases} \varphi_t = v, \\ v_t = R_\varphi \circ P_k \circ R_{\varphi^{-1}}(v), \end{cases}$$

where $P_k = -A_k^{-1} \circ Q_k$ and $Q_k(u) = 2A_k(u)u_x - C_k(u)$. In fact, the map

$$\tilde{P}(\varphi, v) \mapsto \left(\varphi, R_\varphi \circ P_k \circ R_{\varphi^{-1}}(v)\right)$$

is a well-defined smooth (C^∞) vector field on the manifold $\mathcal{D}_n \times H^n(\mathbb{S})$ as shown in Lemma 3.3.

Step 2. Let $\varepsilon > 0$. By the Cauchy theorem applied on $\mathcal{D}_n \times H^n(\mathbb{S})$, which is an open set of a Hilbert space, we know that there exists a positive number T_n, such that for each u_0 in the ball $B_n(0, \varepsilon)$ in $H^n(\mathbb{S})$ there exists a unique solution of the Cauchy problem with initial data $\varphi(0) = \mathrm{Id}$ and $v(0) = u_0$, defined on som e time interval $[0, T_n]$ with $T_n > 0$. Each solution of the Cauchy problem $\mathcal{P}_{n+1}$ is itself a solution of the Cauchy problem $\mathcal{P}_n$. What could happen is that the upper bound $T_{n+1}^*(u_0)$ of the maximal time interval of the solution of $\mathcal{P}_{n+1}$ is smaller than $T_n^*(u_0)$. The second step consists in showing that in fact we have

$$T_n^*(u_0) = T_{2k+1}^*(u_0)$$

for each $n \geq 2k + 1$.

Step 3. The previous steps permit us to define for each $n \geq 2k + 1$ a map

$$F_n : H^n(\mathbb{S}) \cap B_{2k+1}(0, \varepsilon) \to C^\infty([0, T_{2k+1}]; \mathcal{D}_n)$$

which associates to each initial data u_0 the solution φ of the Cauchy problem. This map depends smoothly on u_0 as a consequence of the Cauchy theorem since P_k is smooth. Moreover, we have

$$F_{n|H^{n+1}} = F_{n+1}$$

from which we deduce that the *inductive limit*

$$F : C^\infty(\mathbb{S}) \cap B_{2k+1}(0, \varepsilon) \to C^\infty([0, T_{2k+1}]; \mathcal{D})$$

is also a smooth map. □

Lemma 3.3. *For each $k \geq 1$ and each $n \geq 2k + 1$, the operator*

$$\tilde{P}_k(\varphi, v) = \left(\varphi, R_\varphi \circ P_k \circ R_{\varphi^{-1}}(v)\right)$$

is a smooth map from $\mathcal{D}_n \times H^n(\mathbb{S})$ to $H^n(\mathbb{S}) \times H^n(\mathbb{S})$.

Remark 3.4. We cannot conclude directly from the smoothness of P_k that $\tilde{P}_k$ is smooth because neither the composition nor the inversion are smooth maps on $H^n(\mathbb{S}^1)$.

Proof. To prove the smoothness of $\tilde{P}_k$, we write it as the composition $\tilde{P}_k = \tilde{A}_k^{-1} \circ \tilde{Q}_k$, where

$$\tilde{A}_k(\varphi, v) = \Big(\varphi,\ R_\varphi \circ A_k \circ R_{\varphi^{-1}}(v)\Big)$$

and

$$\tilde{Q}_k(\varphi, v) = \Big(\varphi,\ R_\varphi \circ Q_k \circ R_{\varphi^{-1}}(v)\Big).$$

Note first that

$$R_\varphi \circ A_k \circ R_{\varphi^{-1}}(v) = \sum_{p=0}^{k} (-1)^p \,(v \circ \varphi^{-1})^{(2p)} \circ \varphi$$

is a polynomial expression in the variables

$$\frac{1}{\varphi_x}, \varphi_{xx}, \dots, \varphi^{(2k)}, v, v_x, \dots, v^{(2k)}.$$

For example for $k = 1$, we get

$$R_\varphi \circ A_1 \circ R_{\varphi^{-1}}(v) = v + v_x \frac{\varphi_{xx}}{\varphi_x^3} - v_{xx} \frac{1}{\varphi_x^2},$$

and to prove the general case, we let $a_p = (v \circ \varphi^{-1})^{(p)} \circ \varphi$, and use the recurrence relation

$$a_{p+1} = \frac{1}{\varphi_x} a_p'.$$

A similar reasoning for $R_\varphi \circ Q_k \circ R_{\varphi^{-1}}(v)$, where

$$Q_k(u) = 2A_k(u)u_x - \sum_{p=0}^{k} (-1)^p \sum_{i=1}^{2p} C_{2p}^i u^{(i)} u^{(2p-i+1)},$$

shows that it is also a polynomial expression in the variables

$$\frac{1}{\varphi_x}, \varphi_{xx}, \dots, \varphi^{(2k)}, v, v_x, \dots, v^{(2k)}.$$

To conclude that $\tilde{A}_k$ and $\tilde{Q}_k$ are smooth maps from $\mathcal{D}_n \times H^n(\mathbb{S})$ to $\mathcal{D}_n \times H^{n-2k}(\mathbb{S})$, we use the following known facts (see for example [1]):

(1) For $n \geq 1$, $H^n(\mathbb{S})$ is a Banach algebra and hence polynomial maps on $H^n(\mathbb{S})$ are smooth.

(2) For $n \geq 1$, the map $H^n(\mathbb{S}) \to H^{n-1}(\mathbb{S})$, $v \mapsto v_x$ is smooth.
(3) For $n \geq 1$, the map $H^n(\mathbb{S}) \cap \{v > 0\} \to H^n(\mathbb{S})$, $v \mapsto 1/v$ is smooth.

To show that $\tilde{A}_k^{-1} : \mathcal{D}_n \times H^{n-2k}(\mathbb{S}) \to \mathcal{D}_n \times H^n(\mathbb{S})$ is smooth, we compute the derivative of $\tilde{A}_k$ at an arbitrary point (φ, v), obtaining

$$D\tilde{A}_k(\varphi, v) = \begin{pmatrix} Id & 0 \\ * & R_\varphi \circ A_k \circ R_{\varphi^{-1}} \end{pmatrix}.$$

It is clearly a bounded linear operator in view of the preceding analysis. Moreover, it is an invertible operator since A_k is itself invertible. The application of the *local inversion theorem* in Banach spaces achieves the proof. □

3.6 The exponential map

In classical Riemannian geometry, the *exponential chart* and *normal coordinates* play a very special role. This is a key tool in the study of geodesics.

On Diff($\mathbb{S}^1$) the existence of this privileged chart is not ensured automatically. One may find it useful to recall on this occasion that the *group exponential* of Diff($\mathbb{S}^1$) is not a local diffeomorphism.[8] We have a similar negative result for the Riemannian exponential map of the L^2 metric but a positive result for the H^k metric if $k \geq 1$.

Theorem 3.6 (Constantin and Kolev 2002). *The Riemannian exponential map* $\mathfrak{exp}$ *for the* L^2*-metric on* Diff($\mathbb{S}^1$) *is not a local* C^1*-diffeomorphism near the origin.*

Sketch of proof. Assuming $\mathfrak{exp}$ to be a C^1 local diffeomorphism from a neighborhood of zero to a neighborhood of the identity map of Diff($\mathbb{S}^1$), one can show (see [4]) that the derivative $D\mathfrak{exp}(0)$ of $\mathfrak{exp}$ at zero is the identity map, while

$$(D\mathfrak{exp}_v\, w)(x) = \frac{1}{2c} \int_{x-2c}^{x} w(y)\, dy$$

for $v(x) \equiv c$ (constant). But this yields the contradiction

$$D\mathfrak{exp}_{v_n}\, w_n \equiv 0$$

for

$$w_n = \sin(\pi n x) \in \mathrm{C}^\infty(\mathbb{S}^1) \quad \textit{and} \quad v_n = \frac{1}{n}, \qquad n \geq 1,$$

[8] Indeed, this map is not locally surjective. Otherwise, every diffeomorphism sufficiently near to the identity (for the C^∞ topology) would have a square root. However one can build (see [19]) diffeomorphisms arbitrary near to the identity which have exactly 1 periodic orbit of period $2n$. But the number of periodic orbits of even periods of the square of a diffeomorphism is always even. Therefore, such a diffeomorphism cannot have a square root.

since $v_n \to 0$ in $C^\infty(\mathbb{S}^1)$ as $n \to \infty$, while $D\mathfrak{exp}(v)$ is supposed to be invertible in a neighborhood of $0 \in C^\infty(\mathbb{S}^1)$. □

Theorem 3.7 (Constantin and Kolev 2003). *For $k \geq 1$, the Riemannian exponential map $\mathfrak{exp}$ for the H^k-metric on* Diff($\mathbb{S}^1$)*, is a smooth local diffeomorphism near the origin.*

Sketch of proof. The approach relies on two important consequences of the conservation law obtained in Section 3.4. [5], namely that whenever $n \geq 2k + 1$ we have:

(i) Firstly, in the scale provided by the Sobolev spaces H^n, the geodesic $\varphi_{u_0}(t)$ issuing from the identity in the direction of u_0 inherits at each instant $t > 0$ exactly the regularity of u_0 (that is, if $u_0 \notin H^{n+1}$, then $\varphi_{u_0}(t) \notin H^{n+1}$ for $t > 0$);
(ii) Secondly, for $u_0 \in C^\infty(\mathbb{S}^1)$ there is no function $w \in H^n \setminus H^{n+1}$ such that $D\mathfrak{exp}_{u_0}(w) \in H^{n+1}$.

Taking these two facts for granted, we proceed as follows. Since $D\mathfrak{exp}$ is the identity map, the regularity properties of $\mathfrak{exp}$ established in Section 3.5 in conjunction with the inverse function theorem in Hilbert spaces ensure the existence of two open neighborhoods V_{2k+1} and O_{2k+1} of $0 \in H^{2k+1}$, respectively of the identity in H^{2k+1}, such that $\mathfrak{exp} : V_{2k+1} \to O_{2k+1}$ is a smooth diffeomorphism with $D\mathfrak{exp}_{u_0} : H^{2k+1} \to H^{2k+1}$ bijective for every $u_0 \in V_{2k+1}$. We now claim that $\mathfrak{exp}$ is a smooth diffeomorphism from $V = V_{2k+1} \cap C^\infty(\mathbb{S}^1)$ to $O = O_{2k+1} \cap C^\infty(\mathbb{S}^1)$. Indeed, let $u_0 \in V$. The regularity properties of $\mathfrak{exp}$ ensure that $D\mathfrak{exp}_{u_0}$ is a bounded linear operator from H^n to H^n for every $n \geq 2k + 1$. We now prove inductively that it is a bijection. For $n = 2k + 1$ this is so by our choice of V_{2k+1} and O_{2k+1}. If it is true for $2k + 1 \leq j \leq n$, then $D\mathfrak{exp}_{u_0}$ is injective as a bounded linear map from H^{n+1} to H^{n+1} since its extension to H^n is injective. The second fact emphasized above ensures its surjectivity. Using now the inverse function theorem on Hilbert spaces, both $\mathfrak{exp}$ and its inverse are smooth maps from $V \cap H^n$ to $V \cap H^n$. Letting $n \to \infty$ we obtain that $\mathfrak{exp}$ is a smooth diffeomorphism near $0 \in C^\infty(\mathbb{S}^1)$. □

4 Integrability of the Euler equations

4.1 Lie-Poisson structure

On the dual $\mathfrak{g}^*$ of a Lie algebra, there is a canonical *Poisson structure*[9], defined by

$$\{f, g\}_{LP}(m) = m([d_m f, d_m g]), \qquad f, g \in C^\infty(\mathfrak{g}^*).$$

[9]If $\mathfrak{g}$ is the Lie algebra of a Lie group G, this structure corresponds to the reduction of the canonical symplectic structure on T^*G by the left action of G on T^*G.

and called the *Lie-Poisson structure* (see [25]). Each Euler equation on $\mathfrak{g}^*$ is Hamiltonian relatively to this structure.

In the case of the Lie algebra $\mathrm{Vect}(\mathbb{S}^1)$ of vector fields on the circle, this bracket is represented by the family of operators

$$J(m) = -(mD + Dm),$$

where $D = d/dx$. For an *inertia operator* $A : \mathrm{Vect}(\mathbb{S}^1) \to \mathrm{Vect}^*(\mathbb{S}^1)$, the Hamiltonian is given by

$$H_A(m) = \frac{1}{2}\int_{\mathbb{S}^1} m\, A^{-1}(m), \qquad m \in \mathfrak{g}^*,$$

the corresponding Hamiltonian vector field being

$$X_A(m) = -(mD + Dm)(A^{-1}m).$$

4.2 Bi-Hamiltonian structure

One remarkable property shared by *Burgers* and *Camassa–Holm* equation is the existence of an infinite number of conservation laws for these equations. For example, the following functionals are commuting first integrals for the Burgers equation

$$H_k(m) = \int_{\mathbb{S}^1} m^k\, dx, \qquad (k = 1, 2, \ldots).$$

The mechanism which is at the origin of the existence of these conservation laws is known as the *Lenard scheme* or *bi-Hamiltonian formalism* (see [23] for an excellent historical survey). It is characteristic of evolution equations in infinite dimension known as *formally integrable*, in analogy with classical integrable systems (in the sense of *Liouville*) in finite dimension.

Two Poisson brackets $\{,\}_P$ and $\{,\}_Q$ on the same manifold M, defined by *Poisson bi-vectors* P and Q are said *compatible*, if every linear combination of these brackets

$$\{f, g\}_{\lambda,\mu} = \lambda\{f, g\}_P + \mu\{f, g\}_Q, \qquad \lambda, \mu \in \mathbb{R},$$

is itself a Poisson bracket; that is, it is anti-symmetric, bilinear and it satisfies the Jacobi identity.

When a vector field is Hamiltonian relatively to two *compatible* Poisson brackets, we say that this vector field is bi-Hamiltonian. In the good cases, this leads to the existence of a hierarchy $(H_n)_{n\in\mathbb{N}}$ of commuting first integrals (relatively to both structures). These functions are defined recursively by the so-called *Lenard scheme*:

$$P\, dH_n = Q\, dH_{n+1}, \qquad n = 1, 2, \ldots$$

4.3 Affine Poisson structures on $\mathfrak{g}^*$.

On the dual of a Lie algebra $\mathfrak{g}^*$, one obtains a Poisson bracket by choosing a skew-symmetric bilinear functional γ on $\mathfrak{g}$ and letting

$$\{f, g\}_\gamma(m) = \gamma(d_m f, d_m g), \qquad f, g \in C^\infty(\mathfrak{g}^*).$$

This bracket is compatible with the Lie–Poisson bracket if and only if γ is a 2-*cocycle*,[10] that is:

$$\gamma([u, v], w) + \gamma([v, w], u) + \gamma([w, u], v) = 0,$$

for all $u, v, w \in \mathfrak{g}$.

For instance, it happens that the *Burgers equation* is Hamiltonian relatively to the following cocycle of Vect($\mathbb{S}^1$)

$$\gamma(u, v) = \int_{\mathbb{S}^1} uDv\, dx$$

and the *Camassa–Holm equation* relatively to

$$\gamma(u, v) = \int_{\mathbb{S}^1} u(D - D^3)v\, dx.$$

In fact, these examples are essentially unique (see [6], [16]).

Theorem 4.1 (Constantin and Kolev, 2005). *The only continuous, linear, invertible differential operators A : Vect($\mathbb{S}^1$) $\to$ Vect**($\mathbb{S}^1$) *with* constant coefficients, *whose corresponding Euler vector field X_A is bi-Hamiltonian relatively to some modified Lie–Poisson structure are*

$$A = aI + bD^2,$$

where $a, b \in \mathbb{R}$ satisfy $a - bn^2 \neq 0$ for all $n \in \mathbb{Z}$. The second Hamiltonian structure is induced by the operator

$$Q = -DA = -aD - bD^3,$$

where $D = d/dx$ and the Hamiltonian function is

$$H_3(m) = \frac{1}{2}\int_{\mathbb{S}^1} \left(au^3 - bu(u_x)^2\right) dx,$$

where $m = Au$.

---

[10]A special case occurs when this cocycle γ is a coboundary i.e. $\gamma(u, v) = m_0([u, v])$ for some $m_0 \in \mathfrak{g}^*$ *(freezing structure)*.

5 The Virasoro group and Korteweg-de Vries equation

Historically, the bi-Hamiltonian formalism has been introduced[11] at the end of the 1970s for the *Korteweg-de Vries equation*

$$u_t + 3uu_x - cu_{xxx} = 0, \qquad c \in \mathbb{R}.$$

Notice that the expression $3uu_x - cu_{xxx}$ is not quadratic in u and therefore cannot be written as an Euler equation on $\mathrm{Vect}(\mathbb{S}^1)$. However, it has been shown in [14] that it can be written as an Euler equation for the L^2 metric on the *Virasoro group*,[12] which is a central extension of $\mathrm{Diff}(\mathbb{S}^1)$ by $\mathbb{R}$. This equation was already known in the seventies to be bi-Hamiltonian relatively to the two brackets on $\mathrm{C}^\infty(\mathbb{S}^1)$ defined by the operators D and $-(Dm + mD) + cD^3$.

The Lie–Poisson bracket on the *regular dual* $\mathrm{Vir}^* = \mathrm{C}^\infty(\mathbb{S}^1) \oplus \mathbb{R}$ of the Virasoro algebra is represented by matrix

$$J(m,\alpha) = \begin{pmatrix} -Dm - mD + \alpha D^3 & 0 \\ 0 & 0 \end{pmatrix}.$$

The functions $F(m,\alpha)$ on Vir^* which depend only on α are therefore *Casimir functions* for the canonical structure on Vir^*. In particular, the Hamiltonian flow for the L^2 right-invariant metric leaves invariant each hyperplane $\alpha = c$ (constant). The canonical structure induces on the hyperplane $\alpha = c$, which is isomorphic to $\mathrm{C}^\infty(\mathbb{S}^1)$, a Poisson structure represented by the operator $-(Dm + mD) + cD^3$, which gives a geometric explanation for this operator. Notice that for $c = 0$, we recover the canonical Poisson structure on $\mathrm{Vect}(\mathbb{S}^1)$ and the Burgers equation.

A similar approach can be pursued for the *general Camassa–Holm equation*

$$u_t - u_{txx} + 3uu_x - 2u_xu_{xx} - uu_{xxx} + cu_{xxx} = 0, \qquad c \in \mathbb{R}.$$

This equation can be obtained as the Euler equation for the H^1 right-invariant metric on the Virasoro group.

---

[11]By Gel'fand, Dorfman, Magri. See the review [23].

[12]The composition in the Virasoro group $\mathrm{Vir} = \mathrm{Diff}(\mathbb{S}^1) \times \mathbb{R}$ is given by

$$(\phi,\alpha) \circ (\psi,\beta) = \big(\phi \circ \psi, \alpha + \beta + B(\phi,\psi)\big)$$

where

$$B(\phi,\psi) = -\frac{1}{2} \int_0^1 \log(\phi(\psi(x)))_x \, d\log \psi_x(x)$$

is the *Bott cocycle*.

The theorems which were stated for the diffeomorphism group $\mathrm{Diff}(\mathbb{S}^1)$ in Section 3 on the short-time existence of the geodesic flow and on the Riemannian exponential map for H^k metrics are still true on the Virasoro group but the chart property of the Riemannian exponential map is true only for $k \geq 2$ in this case (see [8]).

References

1. Adams RA, *Sobolev spaces*, Academic Press, 1975.
2. Arnold VI, Sur la géométrie différentielle des groupes de Lie de dimension infinie et ses applications à l'hydrodynamique des fluides parfaits, *Ann. Inst. Fourier (Grenoble)* **16** (1966), 319–361.
3. Arnold VI and Khesin BA, *Topological methods in hydrodynamics*, Springer-Verlag, New York, 1998.
4. Constantin A and Kolev B, On the geometric approach to the motion of inertial mechanical systems, *J. Phys. A* **35**(2002), R51–R79.
5. Constantin A and Kolev B, Geodesic flow on the diffeomorphism group of the circle, *Comment. Math. Helv.* **78** (2003), 787–804.
6. Constantin A and Kolev B, Integrability of invariant metrics on the diffeomorphism group of the circle, *J. Nonlinear Sci.* **16** (2006), 109–122.
7. Constantin A, Kolev B and Lenells J, Integrability of invariant metrics on the Virasoro group, *Phys. Lett. A* **350** (2006), 75–80.
8. Constantin A, Kappeler T, Kolev B, and Topalov P, On geodesic exponential maps of the Virasoro group, *Ann. Glob. Anal. Geom.*, **31** (2007), 155–180.
9. Constantin A and McKean HP, A shallow water equation on the circle, *Comm. Pure Appl. Math.* **52** (1999), 949–982.
10. Ebin DG and Marsden J, Groups of diffeomorphisms and the notion of an incompressible fluid, *Ann. of Math.* **92** (1970), 102–163.
11. Euler L, Theoria motus corporum solidorum seu rigidorum ex primiis nostrae cognitionis principiis stabilita et ad onmes motus qui inhuiusmodi corpora cadere possunt accomodata, *Mémoires de l'Académie des Sciences Berlin* (1765).
12. Hamilton R, The inverse function theorem of Nash and Moser, *Bull. Amer. Math. Soc.* **7** (1982), 66–222.
13. Khesin B and Misiolek G, Euler equations on homogeneous spaces and Virasoro orbits. *Adv. Math.* **176** (2003), no. 1, 116–144.
14. Khesin B and Ovsienko V, The super Korteweg-de Vries equation as an Euler equation. *Functional Anal. Appl.* **21** (1988), no. 4, 329–331.
15. Kolev B, Lie groups and mechanics: an introduction, *J. Nonlinear Math. Phys.* **11** (2004), 480–498.
16. Kolev B, Bi-Hamiltonian systems on the dual of the Lie algebra of vector fields of the circle and periodic shallow water equations, *Phil. Trans. Roy. Soc. London*, **365** (2007), 2333–2357.
17. Kouranbaeva S, The Camassa-Holm equation as a geodesic flow on the diffeomorphism group, *J. Math. Phys.* **40** (1999), 857–868.
18. Lang S, *Fundamentals of Differential Geometry*, Springer-Verlag, New York, 1999.
19. Milnor J, Remarks on infinite-dimensional Lie groups, in *Relativity, Groups and Topology*, pp. 1009–1057, (1984), North-Holland, Amsterdam.
20. Misiołek G, A shallow water equation as a geodesic flow on the Bott-Virasoro group, *J. Geom. Phys.* **24** (1998), 203–208.
21. Ovsienko V and Roger C, Looped cotangent Virasoro algebra and non-linear integrable systems in dimension 2+1, *Comm. Math. Phys.* 273 (2007), 357–378.

22. Peetre J, Une caractérisation abstraite des opérateurs différentiels, *Math. Scand.* **7** (1959), 211–218.
23. Praught J and Smirnov RG, Andrew Lenard: a mystery unraveled, *SIGMA* **1** (2005), 7pp.
24. Souriau JM, *Structure of Dynamical Systems* Birkhäuser, Boston, 1997.
25. Vaisman I, *Lectures on the geometry of Poisson manifolds* Birkhäuser, Basel, 1994.

Harmonic representatives for cuspidal cohomology classes

Józef Dodziuk, Jeffrey McGowan, and Peter Perry

Dedicated to the memory of Serge Lang

Abstract We give a construction of harmonic differentials that uniquely represent cohomology classes of a non-compact Riemann surface of finite topology. We construct these differentials by cutting off all cusps along horocycles and solving a suitable boundary value problem on the truncated surface. We then pass to the limit as the horocycle in each cusp recedes to infinity.

Key words harmonic differentials • non-compact Riemann surfaces

Mathematics Subject Classification (2010): 30F30, 58J60

1 Introduction

This paper contains a construction of unique harmonic representatives of de Rham cohomology classes of a noncompact Riemann surface X of finite geometry with precise estimates on the behavior of these forms at infinity. Our interest in this

J. Dodziuk (✉)
Grad School and University Center (CUNY), PHD Program in Mathematics,
365 5th Avenue, New York, NY 10016-4309
e-mail: jdodziuk@gc.cuny.edu

J. McGowan
Central Connecticut State University, Department of Mathematical Sciences,
1615 Stanley Street, New Britain, CT 06050
e-mail: mcgowan@ccsu.edu

P. Perry
Department of Mathematics, University of Kentucky,
715 Patterson Office Tower, Lexington, KY 40506-0027
e-mail: perry@ms.uky.edu

D. Goldfeld et al. (eds.), *Number Theory, Analysis and Geometry: In Memory of Serge Lang*, DOI 10.1007/978-1-4614-1260-1_8,

question was motivated by an approach, used by Katsuda–Sunada [3] and Phillips–Sarnak [5]), to the problem of counting of closed geodesics in a fixed cohomology class. The approach is based on the Selberg trace formula applied to sections of the flat line bundle E_χ over X induced by the character χ of the fundamental group. The characters are parametrized by the elements of the Jacobian torus $H^1(X,\mathbb{R})/H^1(X,\mathbb{Z})$ and the cohomology classes in turn are parametrized, at least for compact surfaces, by the harmonic forms. In this approach, cf.[3] the harmonic forms are also used to "untwist" the Laplacian Δ_χ acting on sections of the bundles E_χ, replacing it by the unitarily equivalent operator L_χ acting on functions on X or equivalently on automorphic functions on the upper half-plane. If $\chi = \chi_\omega$ corresponds to the harmonic form ω, then the operator

$$L_\chi f = \Delta f - 4\pi i \langle df, \omega\rangle + 4\pi^2 \|\omega\|^2 f,$$

where the inner product and the norm are induced by the Poincaré metric of the upper half-plane. Thus the twisted Laplacians appear as a family of operators acting on a fixed Hilbert space and one can apply perturbation theory to this family. This works very well for compact hyperbolic manifolds since Hodge theory gives good control of harmonic forms. The program was successfully carried out by McGowan–Perry [4] for infinite volume hyperbolic manifolds without cusps and of arbitrary dimension. Their proof relied on an unpublished result of Mazzeo about representing cohomology classes in $H^1(X,\mathbb{R})$ for such manifolds (or more generally for conformally compact manifolds) by bounded harmonic forms.

We hope to apply the results of this paper to the case of Riemann surfaces that have both funnels and cusps. We remark that the case of hyperbolic manifolds of finite volume was handled by Epstein [2] by a different method.

2 Statement of the result and an outline of the method

Let $X = \Gamma\backslash\mathbb{H}^2$ be a geometrically finite Riemann surface with n cusps and k funnels. We will describe a construction of harmonic 1-forms on X representing cohomology classes. Every cohomology class is determined by its periods, i.e., its values on cycles. We are going to describe and construct harmonic forms with nice properties representing cohomology classes. We only consider the case where X is noncompact, i.e., $n + k > 0$ since the compact case is fully covered by Hodge theory. Let $\gamma^1, \dots, \gamma^k$ be cycles at bases of cusps $C^1, \dots, C^k$. Fix a cohomology class $c \in H^1(X,\mathbb{R})$ and let $p_i = \langle c, \gamma^i\rangle$.

We have

Theorem 2.1. *For every class c as above there exists a unique form ω with the following properties.*

1. ω is harmonic, i.e., $d\omega = 0$ and $d * \omega = 0$.

2. *Periods of* ω *are determined by* c, *i.e.,* $\int_z \omega = \langle c, z\rangle$ *for every cycle* z *in* X.
3. *Let* X' *be* X *with arbitrarily small neighborhoods of those punctures for which the period* p_i *is not equal to zero removed. The form* ω *is in* L^2 *on* X'.
4. *On every cusp* C^i *choose coordinates* (r,t) *so that the metric on the cusp takes the form* $ds^2 = dr^2 + e^{-2r}dt^2$ *and* γ_i *is determined by* $r = 0$. *The form* $\omega - p_i dt$ *is in* L^2 *on* C^i.
5. ω *has a smooth extension to the surface* $\overline{X}$ *obtained by adding circles at infinity to all funnels of* X *and has the normal component equal to zero along these circles.*

Moreover, for every cusp C^i, *there exists a constant* $A_i > 0$ *such that*

$$|\omega - p_i dt|_{(r,t)} \leq A_i e^r e^{-2\pi(e^r - 1)}.$$

Our construction of ω will proceed as follows. We parametrize all cusps so that they are isometric with

$$C = [-\ln 2, \infty) \times \mathbb{T}^1 \text{ with the metric } dr^2 + e^{-2r}dt^2, \tag{1}$$

where $\mathbb{T}^1 = \mathbb{R}/\mathbb{Z}$. According to [1, Theorem 4.4.6], the cusps are disjoint. For every positive a we denote by X_a the original surface with all cusps cut off at $r = a$. Consider X_R for large R. X_R is a hyperbolic surface with boundary all of whose components have length e^{-R}. The surface X_R has a natural *conformal* compactification $\overline{X_R}$ whose boundary consists of horocycles cutting off the cusps and the circles at infinity added to funnels. The cohomology of $\overline{X_R}$ is naturally isomorphic to the cohomology of X and we use the de Rham-Hodge theory for manifolds with boundary [7, Proposition 4.2, Corollary 5.7] to construct a harmonic form ω_R on $\overline{X_R}$ that represents c and satisfies absolute boundary conditions $(\omega_R)_n = 0$ on the boundary. Since $*$ operator is a conformal invariant, $\omega_R|X_R$ is an L^2 harmonic form on X_R with periods prescribed by c. We now track ω_R as $R \to \infty$ and X_R develops into X. We show that for a subsequence of R, the limiting form ω exists and has the required properties. The convergence will be uniform on $\overline{X_a}$ for every fixed $a > 0$ and the L^2 norms of ω_R will blow up when $R \to \infty$ only in those truncated cusps for which the corresponding period is not equal to 0.

3 Preliminaries

In this section we set up notation, review conformal invariance of certain objects and justify several statements made in the introduction. Recall that for a positive function λ on a Riemannian manifold (X, ds^2), the metric $\lambda^2 ds^2$ is said to be conformally equivalent to ds^2. If X is a surface, then the Hodge $*$ operator acting on differential forms of degree one depends only on the conformal class of the metric.

The $*$ operator is used to define the L^2 norm and harmonicity of forms. Namely, a differential (a form of degree one) ω is harmonic if and only if

$$d\omega = d * \omega = 0 \tag{2}$$

and the inner product of two differentials is given by

$$(\omega, \eta) = \int_X \omega \wedge *\eta. \tag{3}$$

It follows that the harmonicity and the L^2 inner product together with the associated norm are conformally invariant. Moreover, the normal direction at the boundary is clearly a conformal invariant. Thererefore, vanishing of the normal component of a differential at a boundary point is a conformally invariant condition as well. According to the definition (2) above a harmonic function is locally constant. This is too restrictive and we call a function h harmonic if dh is a harmonic differential, i.e., if $d * dh = 0$ which is equivalent to the usual definition of a harmonic function.

Every Riemannian surface (X, ds^2) admits isothermal coordinates. More precisely, every point of X has a coordinate neighborhood U and two function x and y defined on U such that (x, y) are local coordinates on U and the metric ds^2 is conformal to the Euclidean metric $dx^2 + dy^2$. Now a restriction of a harmonic differential ω to U can be expressed as $a(x, y)dx + b(x, y)dy$ and $*\omega|_U = *(adx + bdy) = ady - bdx$. The equations (2) translate to

$$b_x = a_y \quad \text{and} \quad a_x = -b_y$$

which are Cauchy-Riemann equations for the function $f = a - ib$. Thus, locally, we get a one-to-one correspondence between harmonic differentials $adx + bdy$ and holomorphic functions $f = a - ib$ of $z = x + iy$. Under this correspondence

$$\|\omega\|_U^2 = \int_U \omega \wedge *\omega = \int_U (a^2 + b^2)\, dxdy = \int_U |f|^2\, dxdy. \tag{4}$$

As a consequence, if U is a punctured disk, we see that L^2 harmonic differentials on a punctured disk have a removable singularity. We also note that ω can be recovered from f as follows:

$$\omega = \mathrm{Re}(f(z)dz). \tag{5}$$

For a cusp C and an interval I, we will write $C_I = I \times \mathbb{T}^1$, and to simplify the notation we set $C_R = C_{[R,\infty)}$. We now describe a convenient conformal parametrization of cusps. In particular, we will be interested in $C_{[0,\infty)}$ with the metric as in (1). Of course $C_{[0,R]}$ is conformally equivalent to an annulus in the Euclidean plane and $C_{[0,\infty)}$ is equivalent to a punctured disk. We will make this equivalence explicit. If (ρ, θ) are polar coordinates in the plane, we set $\theta = 2\pi t$

and $\rho = \rho(r)$ with $\rho(0) = 1$. We would like to map the cusp so that the base $r = 0$ is mapped onto the unit circle and the horocycle $r = R$ goes onto a small circle around the origin. Thus, we need ρ to be a decreasing function of r. An easy calculation shows that the metric (1) written in terms of ρ and θ is given by

$$ds^2 = \left(\frac{dr}{d\rho}\right)^2 \left(d\rho^2 + \left(\frac{d\rho}{dr}\right)^2 \frac{e^{-2r}}{4\pi^2} d\theta^2\right).$$

The Euclidean metric is expressed in terms of polar coordinates as $d\rho^2 + \rho^2 d\theta^2$. It follows that the two metrics are conformal if and only if

$$\rho^2 = \left(\frac{d\rho}{dr}\right)^2 \frac{e^{-2r}}{4\pi^2}.$$

Taking into account that $\rho(0) = 1$ and $\frac{d\rho}{dr} < 0$ we solve this equation to obtain

$$\rho = e^{-2\pi(e^r - 1)}. \tag{6}$$

Finally, we observe that the expanding ends of X, so called funnels, are compact from the conformal point of view. Namely, let $F = (-\infty, \infty) \times \mathbb{T}^1$ with the hyperbolic Riemannian metric

$$ds^2 = dr^2 + \ell^2 \cosh^2 r \, dt^2. \tag{7}$$

Here ℓ is the length of the geodesic at the "waist" of the funnel. A calculation similar to the one for the cusp shows that F is conformally equivalent to an annulus in the plane, and thus can be compactified by adding boundary circles. The conformal equivalence is given by the equations

$$\theta = 2\pi t, \qquad \rho = e^{(2\pi/\ell) \int_0^r \frac{du}{\cosh u}}. \tag{8}$$

Since the limits of $\rho(r)$ at $\pm\infty$ are finite these equations define the mapping of F onto an annulus that can be compactified by adding boundary circles. We can use this procedure to compactify all funnels of X by adding a circle at infinity to each to obtain the surface with boundary $\overline{X}$.

We make a comment about the behavior of L^2 harmonic differentials near the boundary points of $\overline{X}$. Since every circle in the $z = x + iy$ plane can be mapped by a Möbius transformation onto the real axis, every boundary point of $\overline{X}$ has a neighborhood that is conformally equivalent to the intersection of a disk D centered at the origin intersected with the *closed* lower half-plane. Writing a differential ω as $\omega = adx + bdy$ we see that the requirement that the normal component of ω vanishes on the boundary is equivalent to $b(x, 0) \equiv 0$. It follows that the associated holomorphic function $f = a - ib$ is real on the real axis so that Schwartz's reflection principle applies. Therefore both $f(z)$ and ω extend to D and the L^2 norms of either one on the upper and lower half-disks are equal. In addition, $\omega = dh$ in D for a harmonic function h determined uniquely up to an additive constant and satisfying $h(x, y) = h(x, -y)$. In our conformal identification of a funnel with an annulus we mapped the added circle at infinity into the outer boundary of the annulus in the complex plane. The remark above implies that both ω and f extend analytically with bounded L^2 norms to a larger annulus containing the image of the funnel in its interior. Moreover, if ω is exact on the whole annulus, then its primitive h extends as well and has equal values at points symmetric with respect to the boundary circle.

Finally we recall the following elementary inequality,

$$|f(0)| \leq \frac{1}{R\sqrt{\pi}} \|f\|_{L^2(D_R)} \tag{9}$$

that holds for every holomorphic function f on a closed disk D_R centered at the origin in the complex plane.

4 Proof of the theorem

We first prove the existence of ω. $\|\eta\|_A$ will denote the L^2 norm of a differential η on the set $A \subset X$. Now, for every $a > 0$ consider the harmonic differential ω_a on $\overline{X_a}$ as in the introduction.

Lemma 4.1. *For every $a \geq 0$, there exists a constant $m(a) > 0$ such for all $R \geq a + 2$,*

$$\|\omega_R - \omega_{a+1}\|_{\overline{X_a}} \leq m_1(a).$$

Let C be one of the cusps C_i, $1 \leq i \leq k$, $p = p_i$. For every $R > 2$

$$\|\omega_R - pdt\|_{C_{[0,R]}} \leq m_2,$$

where $m_2 > 0$ is a constant independent of R.

Proof. We prove the second estimate first since the proof in this case is somewhat simpler but contains all the essential aspects. The harmonic form $\omega_R - pdt$ is exact on $C_{[-\ln 2,R]}$ since it is closed and its only period is equal to zero. Moreover, its normal component vanishes on the outer boundary γ_R of $C_{[0,R]}$. We use the notation $\gamma_r = \{r\} \times \mathbb{T}^1$. If $\omega_R - pdt = dg$, then $d * g = 0$ on C_R. Therefore

$$\|\omega_R - pdt\|^2_{C_{[0,R]}} = \int_{\Gamma_0} g * dg$$

by Stokes' formula. Observe that the function g is determined only up to an additive constant. Thus after adding a suitable constant we can assume that $\inf_{C_{[-\ln 2,1]}} g = 1$.

It follows from the Harnack inequality [6, Theorem 21] that $|g|$ is bounded on $C_{[-(1/2)\ln 2,1/2]}$ by a constant independent of R. Now for a harmonic function, bounds on the function imply bounds on derivatives [6, Section 13]. Therefore $|dg|$ is bounded on γ_0 as well. This proves the second inequality in the lemma.

Consider the form $\omega_R - \omega_{a+1}$ on $\overline{X_{a+1}}$. This form is harmonic and has normal component equal to zero along all circles added to compactify the funnels. By the discussion at the end of Section 3 this form has an extension to an L^2 harmonic form ϕ defined on the surface Y obtained by adding annuli to the funnels. The homology of Y is carried by $\overline{X_{a+1}}$ and therefore all periods of ϕ are equal to zero. It follows that ϕ is exact, i.e., $\phi = dh$ for a harmonic function h on Y. Since $\phi = \omega_R - \omega_{a+1}$ on $\overline{X_{a+1}}$, the normal component of dh vanishes on circles that bound funnels. Using Stokes' formula and the fact that $dh \wedge *dh = d(h * dh)$ we see as above that

$$\|\omega_R - \omega_{a+1}\|^2_{\overline{X_a}} = \int_{\overline{X_a}} dh \wedge *dh = \sum_i \int_{\Gamma_i^a} h * dh, \tag{10}$$

where γ_a^i is the cycle $r = a$ in C^i. We now need to estimate h and dh on γ_i^a. The function h is determined only up to an additive constant. Therefore we can assume that $\min h|_{\overline{X_{a+1}}} = 1$. As we saw above, h extends to a harmonic function h_1 on Y that satisfies the same lower bound as h, i.e., $h_1 \geq 1$. By the Harnack inequality there exists a constant $A > 0$ that depends only on the geometry so that $h(p) \leq A$ for every $p \in \overline{X_{a+1/2}}$. Now, for a harmonic function, the bounds on the function on a disk around a point can be translated into bounds on partial derivatives at the center. It follows that we can bound dh uniformly on Γ_a^i by a constant that depends only on the original surface X and on a. This finishes the proof. □

It follows from the first inequality in the lemma that for a fixed a the norms $\|\omega_R\|_{X_a}$ are bounded independently of R. Locally L^2 harmonic forms can be identified with L^2 holomorphic functions. A uniform bound on L^2 norms implies via (9) that the family ω_R is a "normal family." All the estimates apply via the reflection principle up to the boundary of $\overline{X_a}$. A standard diagonal argument shows there exists a sequence $R_j \to \infty$ such that ω_{R_j} converges uniformly on $\overline{X_a}$ for every fixed $a > 1$ to a harmonic form ω. Obviously, ω is harmonic, has prescribed periods, and its normal component on circles at infinity vanishes.

It remains to investigate the behavior of ω in cusps. On a given cusp $C = C^i$, consider the form η_{R_j} which is equal to $\omega_{R_j} - pdt$ extended by zero to $C_{[R_j,\infty)}$. By Fatou's lemma, using the second inequality in Lemma 4.1, its pointwise limit $\omega - pdt$ is in L^2 on $C_{[0,\infty)}$. This finishes the existence part of the proof.

The uniqueness is easy. If ω_1 and ω_2 are two forms satisfying our conditions, then $\eta = \omega_1 - \omega_2$ is an L^2 harmonic differential on $\overline{X}$ with vanishing periods. Singularities of η in cusps are removable. Consider the Riemann surface Z obtained from X by closing the punctures and adding circles at infinity to funnels. The form η extends to Z, is harmonic, has zero periods and satisfies the absolute boundary conditions $\eta_n = 0$ on the boundary of Z. Therefore, $\eta = dh$ on Z, h is a harmonic function and

$$\|\eta\|^2_{\overline{X}_1} = \int_{\overline{X}_1} dh \wedge *dh = \int_{\overline{X}_1} h\Delta h = 0$$

and $\omega_1 - \omega_2 = \eta = 0$.

Finally, we prove the decay estimate of $\omega - p_i dt$ in cusps. We identify C^i with a punctured disk with coordinate $z = \rho e^{i\theta}$ and use the correspondence between L^2 harmonic differentials and L^2 harmonic forms $\omega - pdt = \mathrm{Re}(f(z)dz)$. The function $f(z)$ must have a removable singularity at zero and is therefore bounded. Thus it is enough to estimate the pointwise norm of $dz = d(\rho e^{i\theta}) = e^{i\theta}d\rho + \rho i e^{i\theta}d\theta$ in terms of the hyperbolic metric $ds^2 = dr^2 + e^{2r}dt^2$. A calculation using (1) and (6) yields

$$|d\rho| = 2\pi e^r e^{-2\pi(e^r-1)} \quad \text{and} \quad |d\theta| = 2\pi|dt| = 2\pi e^r.$$

Thus $|dz| \leq |d\rho| + 2\pi e^r \rho$ which in view of (6) implies our estimate.

References

1. Peter Buser. *Geometry and Spectra of Compact Riemann Surfaces*, Progress in Mathematics **106**. Boston, MA: Birkhäuser Boston, Inc., 1992.
2. Charles L. Epstein. Asymptotics for closed geodesics in a homology class, the finite volume case. *Duke Math. J.* **55** #4 (1987), 717–757.
3. Atsushi Katsuda and Toshikazu Sunada. Homology and closed geodesics in a compact Riemann surface. *Amer. J. Math.* **110** #1 (1988), 145–155.
4. Jeffrey McGowan and Peter Perry. Closed geodesics in homology classes for convex co-compact hyperbolic manifolds. In *Proceedings of the Euroconference on Partial Differential Equations and their Applications to Geometry and Physics (Castelvecchio Pascoli, 2000)*, **91**, 197–209, 2002.
5. Ralph Phillips and Peter Sarnak. Geodesics in homology classes. *Duke Math. J.* **55** #2 (1987), 287–297.
6. Murray H. Protter and Hans F. Weinberger. *Maximum principles in differential equations.* Englewood Cliffs, N.J.: Prentice-Hall Inc., 1967.
7. D. B. Ray and I. M. Singer. R-Torsion and the Laplacian on Riemannian Manifolds. *Advances in Math.* **7** (1971), 145–210.

About the ABC Conjecture and an alternative

Machiel van Frankenhuijsen

In memory of Serge Lang

Abstract After a detailed discussion of the ABC Conjecture, we discuss three alternative conjectures proposed by Baker in 2004. The third alternative is particularly interesting, because there may be a way to prove it using the methods of linear forms in logarithms.

Key words ABC Conjecture • error term in the ABC Conjecture • linear forms in logarithms

Mathematics Subject Classification (2010): Primary 11D75; Secondary 11J86

1 Introduction

I met Serge first around 1990 in Utrecht, when he gave a talk there and I was a student in Nijmegen. I asked him a question during the break, and at some point during the second half, he suddenly pointed a finger at me, asking "let's see what they teach in Nijmegen: when you have a meromorphic function with simple poles and integer residues, what do you do?" After the talk I was surprised and happy when he insisted that I join the group to a restaurant. On the way back in the train, we talked about mathematics, and about a week or two later I was even more surprised when a small box full of books arrived. Since then, we have been in contact regularly until Serge passed away.

M. van Frankenhuijsen (✉)
Department of Mathematics, Utah Valley University, Orem, Utah 84058–5999,
e-mail: vanframa@uvu.edu

D. Goldfeld et al. (eds.), *Number Theory, Analysis and Geometry: In Memory of Serge Lang*, DOI 10.1007/978-1-4614-1260-1_9,

During my thesis research, I asked Serge by email if I could visit him at Yale. He immediately called back by phone, saying “No, it's impossible, I cannot find a place for you to stay.” Half an hour later he called again: he had found a place for me.

I fondly remember the lunches while at Yale. One day, I explained to him the theorem of Stewart and Tijdeman that the error term in the ABC Conjecture is of order $h^{1/2}$, and my idea that this should be related to the fact that the zeros of the Riemann zeta function have real part $1/2$. He looked at me for a moment and then said: “This is insight! Now you have to work until you prove it.”

These and other experiences with Serge have been truly inspiring for me. I consider Serge to be one of the greatest teachers that I have had. He is the one who taught me how to do research (“Formulate the theorem and prove it!”), and how to write it up (when I asked him one time how he wrote so many books, he looked at me with a puzzled expression as if to say “just start and never stop!”).

Serge had a great insight into the interconnection of questions of geometry and Diophantine analysis, thus contributing to the field of Diophantine geometry. One of his latest insights was to emphasize the difference between the error terms in the Vojta Height Inequality and the radicalized version of this inequality. This is well explained in the manuscript *Questions about the error term of Diophantine inequalities* [La05], which, I think, has been left unfinished. We were in contact about this shortly before Serge died. The paper [vF06] was initiated by his question whether the implication of the title is true (Serge's intuition shines through the fact that when he contacted me by phone about this question, I was not sure immediately if this could be proved, but soon I realized how the argument of [vF04] could be modified). Other papers were also inspired by Serge's generous sharing of ideas and enormous drive.

In this paper, we explore another of Serge's questions, asked around the same time, about an alternative to the ABC Conjecture proposed by Baker. We first explain the ABC Conjecture, and illustrate it with several diagrams. Then we explain and discuss Baker's alternative.

I sent an early draft of this paper to Serge shortly before he passed away. The draft contained some graphs and a very incomplete text, which quite upset Serge to my regret. The present paper explains much clearer the available data. The reader will see that no firm conclusion can be arrived at. More data, obtained by a larger exhaustive search of abc sums, is needed for a more definite conclusion. Such data is being assembled in the project *Reken mee met ABC* [LPS09].

Acknowledgement We thank the referee for suggesting a number of important improvements to this paper. David Masser and Joseph Oesterlé kindly shared their recollection of the origins of the ABC Conjecture, and Hendrik Lenstra improved the paper by asking some insightful questions.

2 The ABC Conjecture

In 1985, Masser attended a talk by Oesterlé that involved elliptic curves (and possibly Szpiro's conjecture, which is closely related, see (5) below). It reminded him of another recent development, a theorem for polynomials by Mason [Ma84] (see also [Si84, Sto81]), and he formulated the ABC Conjecture in July 1985 at a London conference in honour of Roth's sixtieth birthday. Only the "Open Problems" were published in some form,[1] but in 1986, as a possible approach to Fermat's Last Theorem, Oesterlé formulated the ABC Conjecture in the Bourbaki seminar [O88].

The *height* of an *abc sum* $P\colon a + b = c$ of coprime integers is

$$h(P) = \max\{\log|a|, \log|b|, \log|c|\}, \tag{1}$$

and the *(logarithmic) radical* of P is defined by

$$r(P) = \sum_{p|abc} \log p. \tag{2}$$

With these definitions, the ABC Conjecture can be formulated as follows.

Conjecture 1. There exists a function ψ such that $\lim_{h\to\infty} \psi(h)/h = 0$ and

$$h(P) - r(P) \leq \psi(h(P)) \tag{3}$$

for all abc sums P.

Smirnov explains in [Sm93] that the ABC Conjecture can be interpreted as a Riemann–Hurwitz inequality for rational numbers, interpreted as functions on spec $\mathbf{Z}$ (see also the appendix of [Mas02]).

In [StTi86], using an extension of Baker's theory of bounds for logarithmic forms, Stewart and Tijdeman obtain that the height of every abc sum is bounded by a power of its exponential radical:

$$h(P) \ll e^{15r(P)}. \tag{4}$$

Apart from subsequent improvements of the exponent 15 to eventually $1/3 + \varepsilon$ [StY01], this is still the best known result. With this result, one sees that if the radical is fixed, i.e., if a, b and c are to be formed using only prime factors from a fixed set of prime numbers, then a, b and c are bounded, and hence there exist only finitely

[1] See [GdS07], which is also of interest to non-Dutch readers for a photographic reproduction of the relevant page.

many such abc sums.[2] Equivalently, in every infinite sequence of abc sums, $r(P)$ is unbounded. However, (4) is too weak to imply Fermat's Last Theorem.

It is easy to deduce Fermat's Last Theorem from the ABC Conjecture, provided the function ψ is explicitly known.[3] We will give the argument here, to make the point that, roughly speaking, the ABC Conjecture says that the exponents of the prime factors of a, b and c are at most $3+\varepsilon$ on average. Suppose that $x^n + y^n = z^n$ and $n \geq 4$. Then we have an abc sum P of height $n \log z$ and radical

$$r(P) = \sum_{p|xyz} \log p \leq \log xyz < 3 \log z.$$

By the ABC Conjecture, $(n-3)\log z \leq \psi(n \log z)$. Since $\psi(h) = o(h)$, there exists an h_0 such that $\psi(h) < h/4$ for $h > h_0$. Then, for $z > e^{h_0/4}$ we have $n \log z > h_0$, and we conclude that $n-3 < n/4$, and hence $n < 4$. There remain only finitely many values for z to check, and for each of these values, we need only check exponents $n \leq h_0/\log z$.

We see that the ABC Conjecture does not allow us to prove Fermat's Last Theorem for exponent 3. However, we can get arbitrarily close to exponent 3. Indeed, letting $z \to \infty$ in the above argument, we see that $n - 3 \leq o(1)$. Szpiro conjectured that

$$\limsup \frac{\log(abc)}{r(P)} = 3, \tag{5}$$

where the limsup is over all abc sums P.

Defining the *quality* of an abc sum P as $q(P) = h(P)/r(P)$, one could state the ABC Conjecture alternatively as

$$\limsup q(P) = 1.$$

Also in the paper [StTi86], Stewart and Tijdeman prove that there exist infinitely many abc sums such that the height is larger than the radical. Their result was subsequently improved by the author [vF00] to

$$\text{for infinitely many abc sums } P\text{:} \quad h(P) - r(P) \geq 6.07\frac{\sqrt{h(P)}}{\log h(P)}. \tag{6}$$

---

[2] In other words, (4) implies the classical result of Siegel and Mahler on the S-unit equation. The innovation of Stewart and Tijdeman was to use Baker's theorem on linear forms in logarithms, generalized to p-adic logarithms, to make this result effective.

[3] If ψ is not explicitly known, one would deduce that there could only be finitely many counterexamples to Fermat's Last Theorem, but one would not know when to stop looking for one.

It follows that there exist infinitely many abc sums with $q(P) > 1$. We computed the quality from two tables. The first table, by Benne de Weger, lists all abc sums with $q(P) > 1.2$ up to $c = 2^{32}$ (i.e., up to height 22.18; our diagrams reflect this by a high density of data points up to this height), and the second table lists abc sums with $q(P) > 1.28$ up to a height of 50, compiled by the author from Abderrahmane Nitaj's tables [N09]. The first table is obtained by an exhaustive search, but the second table is not exhaustive.[4] These data indicate that the quality might be bounded by 1.6299, the quality of the abc sum $2+3^{10}\times 109 = 23^5$ of height 15.6775. However, at present, even a proof of a statement such as $q(P) < 1{,}000$ for every P would be a great theorem.

The disadvantage of the quality is that for any bound $q > 1$, one expects only finitely many abc sums of quality larger than q. Thus up to height 22.18, and as Figure 1 suggests, probably well beyond this height, one finds many sums for which $q(P)$ is about 1.2. One would like to discard those and only record the more interesting sums with $q(P) > 1.4$. But for larger heights, probably when $h(P) > 100$, there may not be any point with $q(P) > 1.4$, and already examples with $q(P) > 1.2$ are interesting. Thus one would like to have a criterion that adapts with the height and would tell us which abc sums are interesting and which ones should be regarded as "common" or "too abundant", a criterion that becomes less restrictive as the height increases.

By (6), if a function ψ as in Conjecture 1 exists, then $\psi(h) \geq 6.07\sqrt{h}/\log h$. This provides us with the kind of adaptive criterion as was mentioned at the end of the previous paragraph[5] (see the curved line in Figure 1). Thus, we regard an abc sum as *interesting* if $h - r \geq 6.07\sqrt{h}/\log h$. In terms of the quality, this means that interesting abc sums have a quality of at least

$$q(P) \geq \left(1 - \frac{6.07}{\sqrt{h(P)}\log h(P)}\right)^{-1}.$$

Thus up to height 22.18, interesting abc sums have a quality of at least 1.72 (hence in this sense, there are no interesting abc sums of height less than 22.18). The first interesting abc sum is

$$5^3 + 2^9 \times 3^{17} \times 13^2 = 11^5 \times 17 \times 31^3 \times 137,$$

with a height of 30.0446, and for heights between 30 and 50, a quality of 1.28 already makes an abc sum interesting.

[4] We have omitted from our table all abc sums with $h > 50$, since beyond a height of 50 our table is definitely not exhaustive and therefore useless. By November 2009, the project [LPS09] had resulted in an exhaustive search up to height 29.9337 (i.e., up to $c = 10^{13}$, apparently improved to 10^{20} [N09]). Schulmeiss has found some very large abc sums that satisfy (6), the largest of which has a height of 5,114. Since these sums were not obtained by an exhaustive search, they are less useful to check different versions of the ABC Conjecture.

[5] This criterion is closely related to the "merit", see [GdS07, dS09]. See also (7) below, which contains the same information as an inequality for the merit.

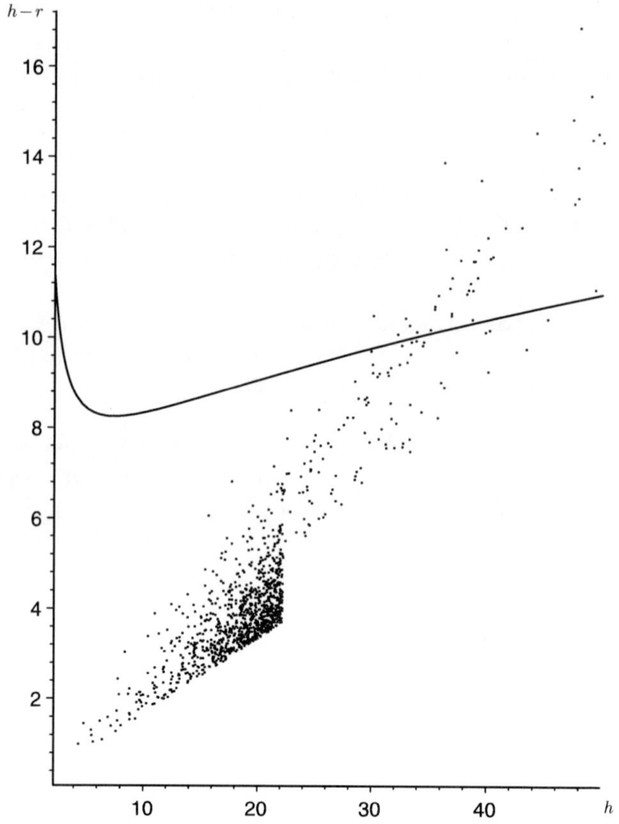

Fig. 1 The points $(h(P), h(P) - r(P))$ and the graph of $6.07\sqrt{h}/\log h$.

In Figure 1, the points $(h(P), h(P) - r(P))$ are compared with the graph of $6.07\sqrt{h}/\log h$. This diagram seems to indicate that $h - r$ grows linearly in h for infinitely many abc sums, which would contradict the ABC Conjecture. However, only a small portion of the sums in the tables satisfy the inequality (6). We expect that as more abc sums become available that satisfy (6), it will become clear that $h(P) - r(P)$ is never much larger than $\sqrt{h(P)}$. Indeed, in [vF95], the author has given a heuristic argument showing that

$$h(P) - r(P) \ll \sqrt{h(P)/\log h(P)} \tag{7}$$

for all abc sums P (see also [StTe]).

Clearly, by (6),

$$\limsup \frac{\log(h(P) - r(P))}{\log h(P)} \geq \frac{1}{2}.$$

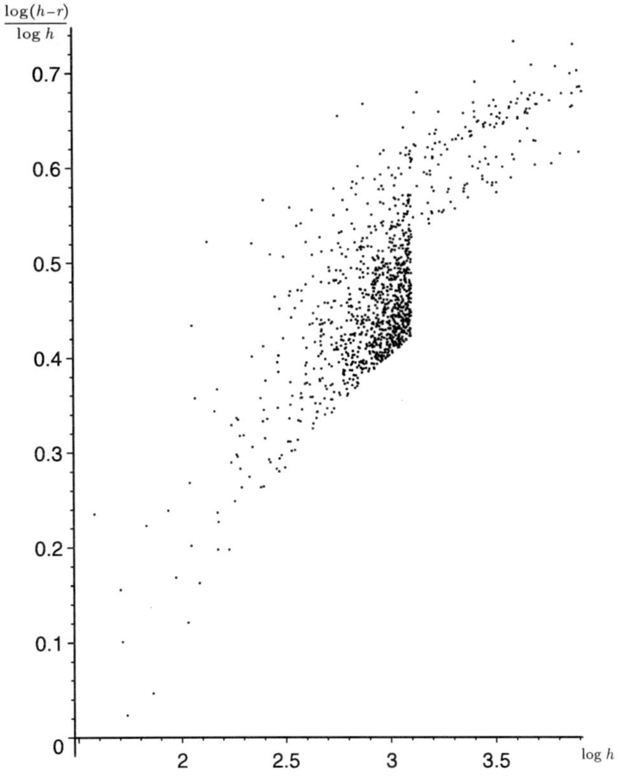

Fig. 2 $\log(h-r)$ compared with $\log h$.

In view of the heuristic inequality (7), we expect that actually, equality holds here.[6] The double logarithmic plot in Figure 2 of the points

$$\left(\log h(P), \frac{\log(h(P)-r(P))}{\log h(P)}\right)$$

suggests that the maximum of $\log(h-r)/\log h$ is 0.733258, for the abc sum

$$19 \times 1307 + 7 \times 29^2 \times 31^8 = 2^8 \times 3^{22} \times 5^4,$$

[6] As alluded to in the introduction, the value $1/2$ may be related to the Riemann Hypothesis. Michel Waldschmidt pointed out to me that the most accessible approach to such a connection may be to construct a sequence of abc sums such that $h(P)-r(P) \geq h(P)^{\theta-\varepsilon}$, given a hypothetical zero of the Riemann zeta function with real part $\theta > 1/2$.

of height 36.1524, $\log h = 3.3877$. It is clear however that the available data is insufficient to reach a definite conclusion, and one expects more definite information to emerge from a complete table of abc examples up to a height of at least about 80.

3 Baker's alternative conjectures

Let $\omega(n)$ be the number of prime factors of the natural number n. By the prime number theorem, $\omega(n)$ is at most of size $\log n / \log\log n$, and by [HW60, Theorem 430], $\omega(n) \sim \log\log n$ on average. Clearly, for an abc sum $a + b = c$, the number of prime factors of abc equals $\omega(abc) = \omega(a) + \omega(b) + \omega(c)$. In [B04], Baker proposes the following conjecture:

Conjecture 2. There exists a constant K such that

$$h(P) - r(P) \leq \omega \log r(P) - (\omega + 1/2) \log \omega + \omega + K$$

for all abc sums $P: a + b = c$ composed of $\omega = \omega(abc)$ different prime numbers.

In Figure 3, we compare Baker's conjecture with the available data from our tables. Clearly, the diagram indicates that the constant K in Conjecture 2 could probably be taken to be -0.8.

Remark 1. By Chen's theorem [C73, C78], there are infinitely many n and prime numbers p, q and r such that $2^n = p + qr$. Thus, $\omega(abc)$ is bounded (by 4) for an infinite sequence of abc sums. In these examples, the radical $r(P) = \log(2pqr)$ is about $2h(P)$. On the other hand, for the examples in our table with $h(P) > r(P)$, especially those that satisfy (6), $\omega(abc)$ is always a relatively large value. Indeed, if Conjecture 2 holds, those abc sums are composed of at least $\omega(abc) = O\left(\sqrt{h}/(\log h)^{3/2}\right)$ primes.

Our formulation of Conjecture 2 is not Baker's original formulation. Baker gives two formulations, one based on $\Theta(R)$, the number of integers $\leq R$ composed of prime factors of R. This is equivalent to Conjecture 2 (see [B96, §7]). Baker states this conjecture in an exponential form involving $\omega!$, where we have used Stirling's formula to replace $\log \omega!$ by $(\omega + 1/2) \log \omega - \omega$. Our formulation seems to exhibit more clearly the relative importance of the different terms. Ignoring all terms of lower order, one could rephrase the conjecture in a weaker form as follows:

Conjecture 3. There exists a constant κ such that for all abc sums $P: a + b = c$ composed of $\omega = \omega(abc)$ different prime numbers,

$$h(P) - r(P) \leq \kappa \omega \log(r(P)/\omega).$$

At present, it seems hopeless to prove the ABC Conjecture itself. It derives its interest mainly from the fact that it allows us to test other conjectures. However, Baker continues by stating his most interesting conjecture. It has the same strength

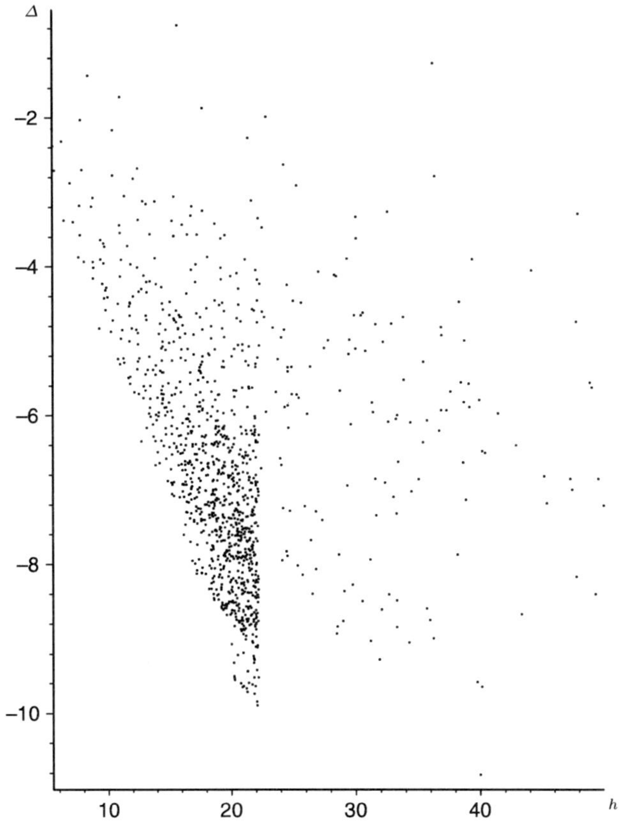

Fig. 3 The size of $\Delta = (h - r) - (\omega \log(r) - (\omega + 1/2)\log(\omega) + \omega)$.

as the ABC Conjecture in the sense that it implies Fermat's Last Theorem, and it may be possible to prove it using the methods of linear forms in logarithms, as Baker explains in [B04, §2] and [B96].

Conjecture 4. There exists a constant κ such that for all abc sums $P{:}\, a + b = c$,

$$h(P) - r(P) \leq \kappa\omega(ab)\log(r(P)/\omega(ab)).$$

Remark 2. The estimate for linear forms in logarithms that is discussed in [B96, §4] involves a product of logarithmic heights. If one could obtain a similar estimate with the product replaced by a sum of the logarithmic heights, one could deduce Conjecture 4. The reader can find an example where a product is successfully replaced by a sum in the Remark on page 37 of [LC90].

For the connection with linear forms in logarithms, it is essential that $\omega(ab)$ in Conjecture 4 only depends on two of the variables. Baker does not specify on which two variables, so permuting a, b and c (and adjusting the signs), the weakest form of Conjecture 4 is obtained when $\omega(ab)$ has the maximal value,

$$\omega(ab) = \omega_{\max} = \max\{\omega(ab), \omega(bc), \omega(ac)\},$$

and the strongest form is obtained for

$$\omega(ab) = \omega_{\min} = \min\{\omega(ab), \omega(bc), \omega(ac)\}.$$

Clearly, Conjecture 4 implies Conjecture 3 with the same value for κ, and Conjecture 4 with $\kappa = 1$ implies Conjecture 2. On the other hand, for the weakest form of Conjecture 4, with $\omega_{\max}$ for $\omega(ab)$, we have[7]

$$\omega(abc) \leq \frac{3}{2}\omega_{\max},$$

so Conjecture 2 implies Conjecture 3 with $\kappa = 3/2$ and $\omega(ab) = \omega_{\max}$.

In Figure 4, we graph $(h - r)/\big(\omega_{\min} \log(r/\omega_{\min})\big)$. Thus one could probably take $\kappa = 2.04$, even in the strongest form of Conjecture 4. The extreme example is the abc sum

$$1 + 2^6 \times 3 \times 5 \times 7 \times 13^4 \times 17 = 239^4,$$

of height 21.9058.

Reasoning as before, each one of the conjectures 2, 3 and 4 imply Fermat's Last Theorem. Indeed, let $x^n + y^n = z^n$. Then

$$\omega(abc) = \omega(xyz) \leq 3 \log z / \log\log z.$$

Also $\omega(ab) \leq 3 \log z / \log\log z$. Using Conjecture 3 or 4, one obtains

$$(n - 3) \log z \leq 3\kappa \log z \log\log\log z / \log\log z.$$

It follows that $n - 3 \leq o(1)$, which implies, as before, that there remain only finitely many cases to check. And if κ is explicitly known, this can be done in a finite search.

[7]If $\omega(c)$ is the least value among $\omega(a)$, $\omega(b)$ and $\omega(c)$, then $\omega_{\max} = \omega(a) + \omega(b)$ and $\omega(abc) = \omega_{\max} + \omega(c) \leq \omega_{\max} + \frac{1}{2}(\omega(a) + \omega(b))$.

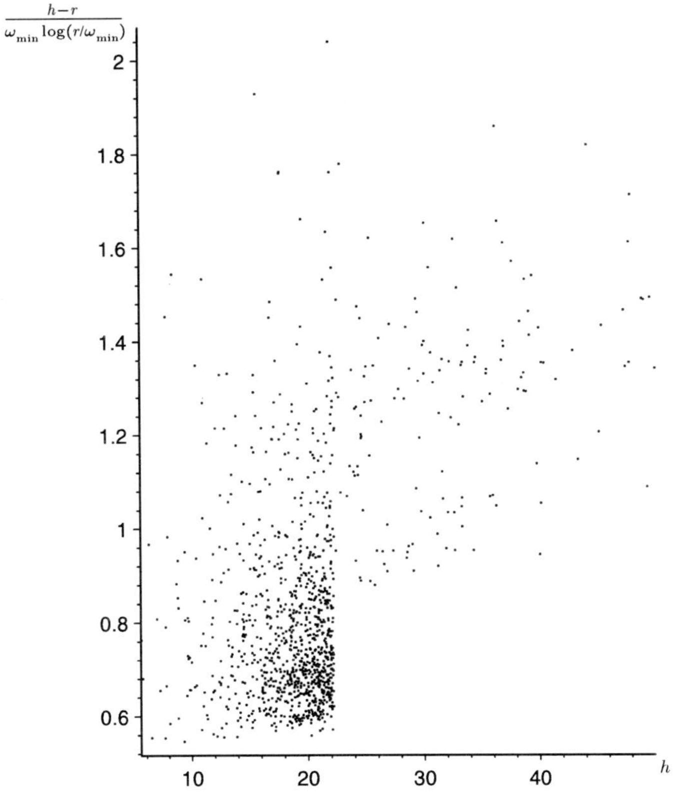

Fig. 4 $h - r$ compared with $\omega_{\min} \log(r/\omega_{\min})$, plotted against the height.

References

[B96] A. Baker, Logarithmic forms and the abc-conjecture, in: *Number theory (Diophantine, computational and algebraic aspects),* Proceedings of the international conference, Eger, Hungary, July 29–August 2, 1996 (Győry, Kalman *et al.*, ed.), de Gruyter, Berlin, 1998, 37–44.

[B04] A. Baker, Experiments on the abc-conjecture, *Publ. Math. Debrecen* **65/3-4** (2004), 253–260.

[C73] J. R. Chen, On the Representation of a Large Even Integer as the Sum of a Prime and the Product of at Most Two Primes, *Sci. Sinica* **16** (1973), 157–176.

[C78] J. R. Chen, On the Representation of a Large Even Integer as the Sum of a Prime and the Product of at Most Two Primes, II, *Sci. Sinica* **21** (1978), 421–430.

[dS09] B. de Smit, http://www.math.leidenuniv.nl/~desmit/abc/, 2009.

[GdS07] G. Geuze, B. de Smit, Reken mee met ABC, *Nieuw Archief voor de Wiskunde* 5/8, no. 1, March 2007.

[HW60] G. H. Hardy and E. M. Wright, *An Introduction to the Theory of Numbers*, 4th ed., Oxford Univ. Press, Oxford, 1960.

[La05] S. Lang, Questions about the error term of Diophantine inequalities, preprint, 2005.

[LC90] S. Lang and W. Cherry, *Topics in Nevanlinna Theory*, Lect. Notes in Math. **1433**, Springer-Verlag, New York, 1990.

[LPS09] H. W. Lenstra, W. J. Palenstijn and B. de Smit, *Reken Mee met ABC,* http://www.rekenmeemetabc.nl/, 2009, and http://abcathome.com/.

[Ma84] R. C. Mason, *Diophantine Equations over Functions Fields*, London Math. Soc. LNS **96**, Cambridge, 1984.

[Mas02] D. Masser, On abc and discriminants, *Proc. Amer. Math. Soc.* **130** (2002), 3141–3150.

[N09] A. Nitaj, http://www.math.unicaen.fr/~nitaj/abc.html, 2009.

[O88] J. Oesterlé, Nouvelles approches du "Théorème" de Fermat, *Sém. Bourbaki* 1987–1988 no. **694**, Astérisque 161–162 (1988), 165–186.

[Si84] J. H. Silverman, The S-unit equation over function fields, *Math. Proc. Cambridge Philos. Soc.* **95** (1984), no. 1, 3–4.

[Sm93] A. L. Smirnov, Hurwitz inequalities for number fields, *St. Petersburg Math. J.* **4** (1993), 357–375.

[StTe] C. L. Stewart, G. Tenenbaum, A refinement of the ABC Conjecture, preprint.

[StTi86] C. L. Stewart and R. Tijdeman, On the Oesterlé-Masser conjecture, *Mh. Math.* **102** (1986), 251–257.

[StY01] C. L. Stewart and K. Yu, On the abc conjecture, II, *Duke Math. J.* **108** (2001), 169–181.

[Sto81] W. W. Stothers, Polynomial identities and hauptmoduln, *Quart. J. Math.* Oxford (2) **32** (1981), 349–370.

[vF95] M. van Frankenhuijsen, *Hyperbolic Spaces and the ABC Conjecture*, thesis, Katholieke Universiteit Nijmegen, 1995.

[vF00] M. van Frankenhuijsen, A lower bound in the ABC Conjecture, *J. of Number Theory* **82** (2000), 91–95.

[vF04] M. van Frankenhuijsen, The ABC conjecture implies Vojta's Height Inequality for Curves, *J. Number Theory* **95** (2002), 289–302.

[vF06] M. van Frankenhuijsen, ABC implies the radicalized Vojta height inequality for curves, *J. Number Theory* **127** (2007), 292–300.

Unifying themes suggested by Belyi's Theorem

Wushi Goldring

Dedicated to the memory of Serge Lang: Teacher and Friend

Abstract Belyi's Theorem states that every curve defined over the field of algebraic numbers admits a map to the projective line with at most three branch points. This paper describes a unifying framework, reaching across several different areas of mathematics, inside which Belyi's Theorem can be understood. The paper explains connections between Belyi's Theorem and (1) The arithmetic and modularity of elliptic curves, (2) abc-type problems and (3) moduli spaces of pointed curves.

Key words Belyi's Theorem • abc • moduli of curves

Mathematics Subject Classification (2010): Primary 11G99; Secondary 11G32, 11Dxx, 11G

1 Introduction

Belyi's Theorem, the main object of our study, states:

Theorem 1.1 (Belyi [Bel80], [Bel02]). *Let X be a connected, smooth, projective curve defined over the field of algebraic numbers $\overline{\mathbf{Q}}$. Then there exists a morphism*

$$\varphi : X \longrightarrow \mathbf{P}^1 \qquad (1.0.1)$$

with

$$\text{Branch Locus } (\varphi) \subset \{0, 1, \infty\}. \qquad (1.0.2)$$

W. Goldring (✉)
Department of Mathematics, Harvard University, Cambridge, MA 02138, USA
e-mail: wushijig@gmail.com

D. Goldfeld et al. (eds.), *Number Theory, Analysis and Geometry: In Memory of Serge Lang*, DOI 10.1007/978-1-4614-1260-1_10,

Belyi's Theorem is an elementary result about curves defined over the algebraic numbers $\overline{\mathbf{Q}}$. One surprising aspect of Belyi's Theorem is that not only is the statement of the theorem elementary, but so are the only two known proofs (both due to Belyi himself). Nothing but the rudiments of the theory of ramification for maps between curves is called upon. These notions were known to Riemann, yet Belyi's Theorem was only discovered in 1979, more than a hundred years later.

This paper describes a unifying framework, reaching across several different areas of mathematics, to understand Belyi's Theorem. The paper explains connections between Belyi's Theorem and (1) the arithmetic and modularity of elliptic curves, (2) abc-type problems and (3) moduli spaces of pointed curves.

First, §2 presents Belyi's two different proofs of his theorem. In each proof we attempt to motivate Belyi's key constructions. Next, §3 is devoted to the connection between Belyi's Theorem and the abc Polynomial Theorem. §4 discusses theorems and conjectures that offer generalizations of Belyi's theorem to characteristic p (§4.1), function fields (§4.2) and higher dimensions (§4.3). In Section 5, we give an exposition of a question of C. Khare, which is an attempt to relate the modularity of elliptic curves to Belyi's Theorem.

The first half of this paper (§§2,3) is completely elementary. The second half (§§4,5) uses a bit of algebraic geometry. Some knowledge of automorphic forms is helpful for Section 5.

2 Belyi's Proofs

Section 2 is devoted to the presentation of Belyi's two proofs of his theorem. For a map $\varphi : C_1 \to C_2$ of smooth, projective curves, we write $B(\varphi)$ for the branch locus of φ and for a point $P \in C_1$, we let $e_\varphi(P)$ denote the ramification index of φ at P. Given X as in Theorem 1.1, we call a map φ satisfying (1.0.1) and (1.0.2) a *Belyi map* for X.

2.1 Outline of section 2 and the proofs

In the interest of making the exposition self-contained, we begin by recording the well-known fact (Lemma 2.1 below) that ramification indices are multiplicative under a composition of morphisms. The basic structure of both of Belyi's proofs is the same. The general strategy is to start with an arbitrary map $h_0 : X \longrightarrow \mathbf{P}^1$ and "refine" it by composing it with a sequence of maps from $\mathbf{P}^1$ to itself, which gradually reduce the complexity of the branch locus, until it becomes $\{0, 1, \infty\}$, the simplest it could possibly be. The complexity of the branch locus is reduced in two steps, which we call the "$\overline{\mathbf{Q}}$ to $\mathbf{Q}$ step" and the "$\mathbf{Q}$ to $\{0, 1, \infty\}$ step", stated and proved as Theorems 2.2 and 2.4 respectively. Belyi's two proofs are the same as far as the "$\overline{\mathbf{Q}}$ to $\mathbf{Q}$ step" is concerned. However the two proofs offer genuinely different approaches to the "$\mathbf{Q}$ to $\{0, 1, \infty\}$ step".

We present Belyi's proofs in the opposite order from that in which they were historically discovered by Belyi. We do this for two reasons. First, we believe that, although it is more sophisticated and harder to understand at first, Belyi's second proof (Proof 1 of Theorem 2.4 below) offers more insight and is more efficient (see also Question 2.7), whilst it is not less elementary than Belyi's first proof (Proof 2 of Theorem 2.4 below). For example, Belyi's second proof of the "$\mathbf{Q}$ to $\{0, 1, \infty\}$ step" requires a single self-map of $\mathbf{P}^1$, whereas the first proof requires (possibly) many such maps. Our second reason for the order of presentation is that Belyi's second proof has gone almost unnoticed[1], while the first proof from [Bel80] is well known and many expositions of it are available. To our knowledge, there is no reference containing Belyi's second proof other than Belyi's original paper [Bel02].

Since Belyi's second proof uses Vandermonde determinants, we have included a subsection which briefly recalls the basic properties of Vandermonde determinants needed in the proof.

Finally, section 2 ends with some further remarks about Belyi's theorem and its proofs.

2.2 *The Two Proofs*

We shall make repeated use of the following well-known property of ramification indices, which we state as a lemma.

Lemma 2.1 (Multiplicativity of Ramification Indices). *Suppose*

$$C_1 \xrightarrow{\varphi} C_2 \xrightarrow{\psi} C_3 \tag{2.2.1}$$

is a composition of maps of smooth projective curves. Then for all $P \in C_1$, one has

$$e_{\psi\circ\varphi}(P) = e_{\psi}(\varphi(P))e_{\varphi}(P). \tag{2.2.2}$$

Theorem 2.2 (Branching Reduction from $\mathbf{P}^1(\overline{\mathbf{Q}})$ to $\mathbf{P}^1(\mathbf{Q})$). *Let $h_0 : X \to \mathbf{P}^1$ be a morphism satisfying $B(h_0) \subset \mathbf{P}^1(\overline{\mathbf{Q}})$. Then there exists a morphism $h_B : \mathbf{P}^1 \to \mathbf{P}^1$ such that the composition*

$$h_B \circ h_0 : X \to \mathbf{P}^1 \text{ satisfies } B(h_B \circ h_0) \subset \mathbf{P}^1(\mathbf{Q}). \tag{2.2.3}$$

Proof. We begin by recursively defining a sequence of polynomials and branch sets. For a finite set $S \subset \overline{\mathbf{Q}}$, we use the notation $m_{S,\mathbf{Q}}(x)$ to denote the minimal polynomial of S over $\mathbf{Q}$ i.e., $m_{S,\mathbf{Q}}(x)$ is the monic polynomial of smallest degree

[1] Apparently one reason for this is that [Bel02] was an MPI preprint for several years that was difficult to access before it was published.

with rational coefficients having all the elements of S as roots. Note that since we do not require the elements of S to be Galois conjugate, the minimal polynomial $m_{S,\mathbf{Q}}(x)$ may be reducible. Put

$$B_0 = B(h_0) - \{\infty\}\,,\, h_1(x) = m_{B_0,\mathbf{Q}}(x), \tag{2.2.4}$$

and define recursively

$$B_i = B(h_i) - \{\infty\}\,,\, h_i(x) = m_{B_{i-1},\mathbf{Q}}(x). \tag{2.2.5}$$

For $i \geq 1$, one can think of the h_i as maps $h_i : \mathbf{P}^1 \to \mathbf{P}^1$, totally ramified at ∞.

Claim 1: For all $i \geq 1$, if $\deg h_i \geq 1$, then $\deg h_{i+1} \leq \deg h_i - 1$. The point here is that the branch sets B_i are Galois stable for $i \geq 1$ i.e., for all $\sigma \in \mathrm{Gal}(\overline{\mathbf{Q}}/\mathbf{Q})$ and $\alpha \in B_i$ one has $\sigma\alpha \in B_i$. Indeed, for all $i \geq 1$ the finite branch sets B_i can be described by means of derivatives as

$$B_i = \{h_i(a) | h_i'(a) = 0\}. \tag{2.2.6}$$

If $\alpha \in B_i$, say $\alpha = h_i(a)$, then

$$\sigma\alpha = \sigma(h_i(a)) = h_i(\sigma a)$$

since $h_i \in \mathbf{Q}[x]$ and

$$h_i'(\sigma a) = \sigma(h_i'(a)) = \sigma(0) = 0$$

since $h_i' \in \mathbf{Q}[x]$ and $h_i(a) \in B_i$.

Since B_i is Galois stable, we have

$$h_{i+1}(x) = \prod_{\alpha \in B_i} (x - \alpha) \tag{2.2.7}$$

for all i. In particular $\deg h_{i+1} = |B_i|$. From (2.2.6) we see that if $\deg h_i \geq 1$, then $|B_i| \leq \deg h_i - 1$, so the claim is true.

By the claim, there is a positive integer ℓ such that h_ℓ is linear, i.e., $\deg h_\ell = 1$. Consider

$$h_B := h_{\ell-1} \circ h_{\ell-2} \circ \cdots \circ h_1 \tag{2.2.8}$$

and put $h = h_B \circ h_0$.

Claim 2: $B(h) \subset \mathbf{P}^1(\mathbf{Q})$. To see this, we use the multiplicativity of ramification indices (Lemma 2.1). Suppose h is ramified at $P \in X$ above $Q \in \mathbf{P}^1$ and $Q \neq \infty$. By Lemma 2.1, there exists i, $0 \leq i \leq \ell - 1$, such that

$$e_{h_i}((h_{i-1} \circ h_{i-2} \circ \cdots \circ h_1 \circ h_0)(P)) > 1. \tag{2.2.9}$$

Case I: $i < \ell - 1$. By definition of B_i, (2.2.9) implies that

$$(h_i \circ h_{i-1} \circ \cdots \circ h_1 \circ h_0)(P) \in B_i, \tag{2.2.10}$$

so

$$(h_{i+1} \circ h_i \circ \cdots \circ h_1 \circ h_0)(P) = 0. \tag{2.2.11}$$

Since $h_i \in \mathbf{Q}[x]$ for all $i \geq 1$, it follows that

$$(h_{\ell-1} \circ \cdots \circ h_{i+1} \circ h_i \circ \cdots \circ h_1 \circ h_0)(P) = h(P) = Q \in \mathbf{Q}, \tag{2.2.12}$$

as desired.

Case II: $i = \ell - 1$. Since h_ℓ is linear and $h_\ell(x) = m_{\mathbf{Q},B_{\ell-1}}(x)$, we must have $|B_{\ell-1}| = 1$ and $B_{\ell-1} \subset \mathbf{Q}$. But $Q \in B_{\ell-1}$ so $Q \in \mathbf{Q}$. □

As an interlude between the two steps of Belyi's proofs, we recall some basic facts about Vandermonde determinants.

Vandermonde determinants. Let K be a field. Given $\alpha_1, \ldots, \alpha_n \in K$, the Vandermonde determinant of $\alpha_1, \ldots, \alpha_n$ is

$$V(\alpha_1, \ldots, \alpha_n) = \begin{vmatrix} 1 & \alpha_1 & \alpha_1^2 & \ldots & \alpha_1^{n-1} \\ 1 & \alpha_2 & \alpha_2^2 & \ldots & \alpha_2^{n-1} \\ \vdots & \vdots & \vdots & \ddots & \vdots \\ 1 & \alpha_n & \alpha_n^2 & \ldots & \alpha_n^{n-1} \end{vmatrix}. \tag{2.2.13}$$

The Vandermonde determinant satisfies *Vandermonde's Identity*:

Lemma 2.3. *One has*

$$V(\alpha_1, \ldots \alpha_n) = \prod_{j<i} (\alpha_i - \alpha_j). \tag{2.2.14}$$

Proof. The proof is by induction on n. Observe that (2.2.14) holds for $n = 2$, as both sides of the equation are equal to $\alpha_2 - \alpha_1$. We may assume that $\alpha_i \neq \alpha_j$ for $i \neq j$ since otherwise both sides of (2.2.14) are 0.

Define a polynomial $v(t)$ of one variable t by $v(t) = V(\alpha_1, \ldots, \alpha_{n-1}, t)$. The degree of $v(t)$ is $n - 1$. Let c be the leading coefficient of $v(t)$. By expanding the determinant defining $v(t)$ along the bottom row, we see that $c = V(\alpha_1, \ldots, \alpha_{n-1})$. Now $v(\alpha_i) = 0$ for $1 \leq i \leq n-1$, since the determinant giving $V(\alpha_1, \ldots, \alpha_{n-1}, \alpha_i)$ has the ith and nth rows equal. Hence $t-\alpha_i$ divides $v(t)$ and since the α_j are distinct, we obtain

$$v(t) = c \prod_{i=1}^{n-1} (t - \alpha_i). \tag{2.2.15}$$

Using induction to express c as a product, (2.2.15) becomes

$$v(t) = \prod_{1 \le k < j \le n-1} (\alpha_j - \alpha_k) \prod_{i=1}^{n-1} (t - \alpha_i). \tag{2.2.16}$$

Substituting α_n into (2.2.16) gives (2.2.14). □

The Vandermonde determinant also satisfies a scalar translation invariance property: For all $\beta \in K$, translating all the entries $\alpha_1, \ldots, \alpha_n$ by β does not affect the Vandermonde determinant. More precisely,

$$V(\alpha_1 + \beta, \ldots, \alpha_n + \beta) = V(\alpha_1, \ldots, \alpha_n). \tag{2.2.17}$$

This translation invariance is an immediate consequence of the Vandermonde Identity (2.2.14), since each term in the factorization of the left-hand side of (2.2.17) is of the form $(\alpha_i + \beta) - (\alpha_j + \beta) = \alpha_i - \alpha_j$. The translation invariance (2.2.17) can also be seen directly from the definition of the Vandermonde determinant, without using the Vandermonde identity, by means of the multilinearity of the determinant. We will make use of the translation invariance property of the Vandermonde determinant at the end of the first proof of Theorem 2.4.

We now come to the second part of Belyi's proofs, the "$\mathbf{Q}$ to $\{0, 1, \infty\}$ step".

Theorem 2.4 (Branching Reduction from $\mathbf{P}^1(\mathbf{Q})$ to $\{0, 1, \infty\}$). *Assume $h : X \to \mathbf{P}^1$ is a morphism with $B(h) \subset \mathbf{P}^1(\mathbf{Q})$. Then there exists a map $g : \mathbf{P}^1 \to \mathbf{P}^1$ such that the composite $g \circ h$ is a Belyi map for X.*

Before embarking on the proofs of this theorem, let us note the following equivalent formulation. Suppose $g : \mathbf{P}^1 \to \mathbf{P}^1$ and put $S = B(h)$. By the multiplicativity of ramification indices, we see that $g \circ h$ is a Belyi map for X if and only if g satisfies the two conditions:

(i) $g(S) \subset \{0, 1, \infty\}$
(ii) $B(g) \subset \{0, 1, \infty\}$.

First proof of Theorem 2.4: This is Belyi's second construction, which uses Vandermonde determinants. We proceed to construct a function $g : \mathbf{P}^1 \to \mathbf{P}^1$ satisfying conditions (i) and (ii). Let $S = B(h) \subset \mathbf{P}^1(\mathbf{Q})$. By composing with a fractional linear transformation (which is everywhere unramified) we may assume

$$S = \{\lambda_1 < \lambda_2 < \cdots < \lambda_n\} \cup \{\infty\} \tag{2.2.18}$$

with $\lambda_i \in \mathbf{Z}$ and $\lambda_1 = 0$.

Consider

$$g(x) = \prod_{i=1}^{m} (x - \lambda_i)^{m_i} \tag{2.2.19}$$

with

$$m_i = (-1)^{i-1} V(\lambda_1, \ldots, \widehat{\lambda_i}, \ldots, \lambda_n) \tag{2.2.20}$$

where the term with a hat $(\widehat{\lambda_i})$ is to be omitted. By choosing g to have the above form, we immediately ensure that g satisfies condition (i), regardless of our choice of exponents m_i[2]. In fact, g maps λ_i to 0 or ∞ according to whether m_i is positive or negative and the fact that g is monic implies that $g(\infty)$ is either 0, 1, or ∞ depending on whether the degree of g is negative, zero, or positive, respectively. Although it is not necessary for the proof, we will determine which of the above cases occurs for each element of S.

The purpose of our particular choice of exponents m_i made in (2.2.20) is to ensure that g satisfy (ii) as well. We shall try to motivate the choice of exponents later in the proof.

Before proving (ii), we determine explicitly the values of g on elements of S. By Vandermonde's Identity (2.2.14) and the fact that the λ_i are increasing, we have that

$$V(\lambda_1, \ldots, \widehat{\lambda_i}, \ldots, \lambda_n) > 0 \tag{2.2.21}$$

for all i since every term in the factorization of $V(\lambda_1, \ldots, \widehat{\lambda_i}, \ldots, \lambda_n)$ is positive. Hence:

$$g(\lambda_i) = \begin{cases} 0 & \text{if} \quad i \equiv 1 \pmod 2 \\ \infty & \text{if} \quad i \equiv 0 \pmod 2. \end{cases} \tag{2.2.22}$$

Next we show $\sum_{i=1}^{n} m_i = 0$, which gives $g(\infty) = 1$. Consider the auxiliary matrix

$$A = \begin{pmatrix} 1 & 1 & \lambda_1 & \lambda_1^2 & \ldots & \lambda_1^{n-2} \\ 1 & 1 & \lambda_2 & \lambda_2^2 & \ldots & \lambda_2^{n-2} \\ \vdots & \vdots & \vdots & \vdots & \ddots & \vdots \\ 1 & 1 & \lambda_n & \lambda_n^2 & \ldots & \lambda_n^{n-2} \end{pmatrix}. \tag{2.2.23}$$

Since the first two columns of A are identical, $\det A = 0$. Expanding $\det A$ along the first column gives $\sum_{i=1}^{n} m_i = 0$.

Now we return to verifying (ii) for g. Consider the logarithmic derivative of g:

$$\frac{g'}{g} = \sum_{i=1}^{n} \frac{m_i}{x - \lambda_i}. \tag{2.2.24}$$

Since the ramification points of g that are not above infinity are the roots of g', the ramification points of g are contained in the set of zeros and poles of g'/g.

Claim. We have the identity

$$\sum_{i=1}^{n} \frac{m_i}{x - \lambda_i} = (-1)^{n-1} \frac{V(\lambda_1, \ldots, \lambda_n)}{\prod_{i=1}^{n}(x - \lambda_i)}. \tag{2.2.25}$$

[2]This is true so long as $m_i \neq 0$ for all i which holds by (2.2.14) and (2.2.18).

The important point is that the numerator on the right-hand side of (2.2.25) is a constant; its value does not matter for proving (ii). The choice of exponents that we made in (2.2.20) is what makes the numerator of the right-hand side of (2.2.25) constant; for an arbitrary choice of exponents, the numerator of the right hand side of (2.2.25) would be a complicated polynomial of degree $n-1$.

Before proving the claim, let us see why it implies that g satisfies (ii). Assuming the claim, we see that g'/g has simple poles at λ_i, a zero of order n at ∞ and no other zeros or poles. Hence any ramification point of g either lies above ∞ or lies in S. By (i) for g, the branch points of g are contained in $\{0,1,\infty\}$, as was to be shown.

We are left with proving the claim. By putting the left-hand side of (2.2.25) over a common denominator, the claim is equivalent to

$$\sum_{i=1}^{n} m_i \prod_{j\neq i}(x-\lambda_j) = (-1)^{n-1} V(\lambda_1,\dots,\lambda_n). \tag{2.2.26}$$

We shall prove (2.2.26) in two steps: First we will show that the constant term of the left hand side is equal to the right-hand side and then we will prove the full identity by using a change of variables trick suggested by B. Mazur. Equality of constant terms in (2.2.26) states that

$$\sum_{i=1}^{n} m_i (-1)^{n-1} \prod_{j\neq i}\lambda_j = (-1)^{n-1} V(\lambda_1,\dots,\lambda_n). \tag{2.2.27}$$

We now use that $\lambda_1 = 0$ to simplify (2.2.27). If $i \neq 1$, then $\prod_{j\neq i}\lambda_j = 0$ since $\lambda_1 = 0$, so (2.2.27) becomes

$$m_1 \prod_{j\neq 1}\lambda_j = V(\lambda_1,\dots,\lambda_n). \tag{2.2.28}$$

Keeping in mind that $\lambda_1 = 0$, expand $V(\lambda_1,\dots,\lambda_n)$ along the first row: All but the first term are zero, so we get

$$V(\lambda_1,\dots,\lambda_n) = \begin{vmatrix} \lambda_2 & \lambda_2^2 & \dots & \lambda_2^{n-1} \\ \vdots & \vdots & \ddots & \vdots \\ \lambda_n & \lambda_n^2 & \dots & \lambda_n^{n-1} \end{vmatrix}. \tag{2.2.29}$$

$$= \prod_{j\neq 1}\lambda_j \begin{vmatrix} 1 & \lambda_2 & \dots & \lambda_2^{n-2} \\ \vdots & \vdots & \ddots & \vdots \\ 1 & \lambda_n & \dots & \lambda_n^{n-2} \end{vmatrix} \tag{2.2.30}$$

by multilinearity of the determinant. This proves (2.2.27).

We regard (2.2.27) as an identity of independent variables $\{\lambda_i\}$[3] and make the change of variables

$$\lambda_i \mapsto -x + \lambda_i \tag{2.2.31}$$

to get

$$\sum_{i=1}^{n}(-1)^{i-1}V(-x+\lambda_1,\ldots,\widehat{-x+\lambda_i},\ldots,-x+\lambda_n)$$
$$\prod_{j\neq i}(-x+\lambda_j) = V(-x+\lambda_1,\ldots,-x+\lambda_n). \tag{2.2.32}$$

We are now in a position to apply the translation invariance property of the Vandermonde determinant explained above. Applying (2.2.17) to (2.2.32) gives

$$\sum_{i=1}^{n}(-1)^{i-1}V(\lambda_1,\ldots,\widehat{\lambda_i},\ldots,\lambda_n)\prod_{j\neq i}(x-\lambda_j) = (-1)^{n-1}V(\lambda_1,\ldots,\lambda_n). \tag{2.2.33}$$

This proves the claim which completes the proof of Belyi's theorem. □

Second proof of Theorem 2.4: This is Belyi's original construction. The idea here is to reduce the size of the branch locus by one at a time, until the size of the branch locus is at most 3. Again set $S = B(h) \subset \mathbf{P}^1(\mathbf{Q})$. We first show by induction on $|S|$ that it suffices to construct g satisfying (i) and (ii) when $|S| = 4$. The crux of the argument is then to treat the case $|S| = 4$ by constructing g explicitly.

Suppose there exists g satisfying (i) and (ii) for all S with $|S| = 4$. Let $s \in S$ and set $U = S - s$. By induction there exists a function $g_U : \mathbf{P}^1 \to \mathbf{P}^1$ satisfying (i) and (ii) for U i.e., $g_U(U) \subset \{0,1,\infty\}$ and $B(g_U) \subset \{0,1,\infty\}$. Let $T = \{0,1,\infty,g_U(s)\}$. By our assumption on $|S| = 4$, there exists $g_T : \mathbf{P}^1 \to \mathbf{P}^1$ satisfying (i) and (ii) for T i.e., $g_T(T) \subset \{0,1,\infty\}$ and $B(g_T) \subset \{0,1,\infty\}$. Now consider $g_S = g_T \circ g_U$. By (i) for g_T and g_U, we have $g_S(S) \subset \{0,1,\infty\}$ and we see from the multiplicativity of ramification indices that $B(g_S) \subset \{0,1,\infty\}$. This completes the induction.

We now show that there exists g satisfying (i) and (ii) when $|S| = 4$. By using a fractional linear transformation, we may assume $S = \{0,1,\infty,s\}$ with $0 < s < 1$. Note that $s \in \mathbf{Q}$. There exist unique positive integers m,n such that $s = \frac{m}{m+n}$ and $(m,n) = 1$. We claim that

$$g(x) = \frac{(m+n)^{m+n}}{m^m n^n}x^m(1-x)^n \tag{2.2.34}$$

[3]In other words, regard (2.2.27) as an identity in $\mathbf{Q}[\lambda_1,\ldots,\lambda_n]$.

satisfies (i) and (ii). We will try to motivate this construction and in the process prove that it works.

To begin with, it makes sense to consider functions given by polynomials of the form

$$f_{c,k,\ell}(x) = cx^k(1-x)^\ell, \tag{2.2.35}$$

with $c \in \overline{\mathbf{Q}}^*$ a non-zero constant, as possible candidates for g. For all $c \neq 0$ and $k, \ell \geq 1$, we have $f_{c,k,\ell}(0) = f_{c,k,\ell}(1) = 0$ and $f_{c,k,\ell}(\infty) = \infty$. For any choice of k, ℓ, there will be a unique choice of c to make $f_{c,k,\ell}(s) = 1$. This takes care of (i). As for (ii), the finite branch values of $f_{c,k,\ell}$ are given by

$$B(f_{c,k,\ell}) - \{\infty\} = \{f_{c,k,\ell}(a) | f'_{c,k,\ell}(a) = 0\}. \tag{2.2.36}$$

But

$$f'_{c,k,\ell}(x) = cx^{k-1}(1-x)^{\ell-1}[k(1-x) - \ell x], \tag{2.2.37}$$

so $f'_{c,k,\ell}$ has just one zero other than zero and one. The value of this exceptional zero depends on the exponents k, ℓ and is given by $x = \frac{k}{k+\ell}$. Hence we choose $k = m$ and $\ell = n$ so that the root of $f'_{c,k,\ell}$ different from 0 and 1 is $\frac{m}{m+n} = s$. The corresponding branch value is then $f_{c,m,n}(s)$. As was said above, there is a choice of c, namely $c = \frac{(m+n)^{m+n}}{m^m n^n}$ so that $f_{c,m,n}(s) = 1$. This is one way to see how to arrive at

$$g(x) = f_{c,k,\ell}(x) \text{ with } c = \frac{(m+n)^{m+n}}{m^m n^n},\ k = m \text{ and } \ell = n. \tag{2.2.38}$$

□

2.3 Remarks

The most interesting point in both of Belyi's proofs of Theorem 2.4 is that the exponents in the functions g satisfying (i) and (ii) are constructed from the elements of S, which are the branch values of h. These constructions use the fact that the branch values of h are integers (or rational numbers) together with the very special property of the ring $\mathbf{Z}$ that its elements are exponents of rational functions. By the latter we mean that, for example, $f(x) = x^\alpha$ is a rational function if and only if $\alpha \in \mathbf{Z}$. This presents a major difficulty in generalizing Belyi's theorem to other situations, such as the function field case, which we discuss in the next section.

The converse of Belyi's theorem is also true:

Theorem 2.5 (Weil [Wei56], Converse of Belyi's theorem). *Suppose X is a connected, smooth, projective curve that admits a map $\varphi : X \to \mathbf{P}^1$ with $B(\varphi) \subset \{0, 1, \infty\}$. Then X is defined over $\overline{\mathbf{Q}}$.*

The converse of Belyi's theorem follows from Weil's descent theory. It is curious that, as far as we know, Weil did not ask whether the converse to Theorem 2.5 is true, thus conjecturing Belyi's theorem, especially that Weil did so much pioneering work in the arithmetic and geometry of covers of curves.

Theorem 2.5 is more clearly understood in terms of descent for the étale fundamental group, (part of) Grothendieck's vast generalization of Weil's theory, which can be found in [Gro71].

Theorem 2.6 (Descent for the étale fundamental group). *Suppose L is an algebraically closed field of characteristic 0 and K is an algebraically closed subfield of L. Let S be a variety over K. Then*

$$\pi_1^{\text{ét}}(S) \cong \pi_1^{\text{ét}}(S \otimes_K L). \tag{2.3.1}$$

Define the Belyi degree of a curve X defined over $\overline{\mathbf{Q}}$ to be the smallest degree of a Belyi map for X. Both proofs of Belyi's Theorem give a bound for the Belyi degree of X in terms of the height of the branch values of the original function h_0. The first proof presented here (Belyi's second proof) is more efficient in the sense that it gives a better bound on the Belyi degree.

Question 2.7. What can be said about the behavior of the function

$$\text{Belyi Degree} : \left\{ \begin{array}{c} \text{connected, smooth, projective} \\ \text{curves defined over } \overline{\mathbf{Q}} \end{array} \right\} \longrightarrow \mathbf{Z}^+? \tag{2.3.2}$$

For example, how close is the bound on the Belyi degree given by Belyi's second proof to the best possible?

We know of only one lower bound for the Belyi degree, which was kindly communicated to us by Zapponi. To state his result, we need the notions of *field of moduli* and *semistable reduction*. Suppose X is a curve defined over $\overline{\mathbf{Q}}$. The Galois group $\text{Gal}(\overline{\mathbf{Q}}/\mathbf{Q})$ acts on the set of isomorphism classes of curves defined over $\overline{\mathbf{Q}}$, and the stabilizer of the isomorphism class of X in $\text{Gal}(\overline{\mathbf{Q}}/\mathbf{Q})$ is of the form $\text{Gal}(\overline{\mathbf{Q}}/K)$ for some number field K. The field of moduli of X is defined to be K. The field of moduli may also be described as being the intersection of all fields of definition for X. The field of moduli of other geometric objects defined over $\overline{\mathbf{Q}}$, such as a cover $f : X \longrightarrow Y$ of varieties, is defined analogously. The field of moduli is not always a field of definition, but in many cases it is, and understanding precisely when it is is a rather subtle problem, see [DD97].

Suppose X is a curve with field of moduli a number field K. Let $\mathfrak{p}$ be a prime of K and let $\mathcal{O}_{K_\mathfrak{p}}$ denote the ring of integers of the completion of K at $\mathfrak{p}$. We say that X has good (resp. semistable) reduction at $\mathfrak{p}$, if X admits a model over Spec $\mathcal{O}_{K_\mathfrak{p}}$ whose special fiber is smooth (resp. a model whose special fiber has at worst nodal singularities and finitely many automorphisms). If X does not have good reduction at $\mathfrak{p}$, then we say X has bad reduction at $\mathfrak{p}$. We say X has good (resp. semistable) reduction over a prime $p \in \mathbf{Z}$ if X has good (resp. semistable) reduction at all primes of K lying above p. Finally, there is the question of how the reduction type of X changes as the base field varies. We shall use only that good and semistable reduction do not change if one passes to a larger field.[4]

Again these definitions extend to covers. However it is important to note that, for example, X may have good reduction at a prime $\mathfrak{p}$ of a field F, while a cover consisting of X and a map from X to $\mathbf{P}^1$, may have bad reduction at $\mathfrak{p}$.

Theorem 2.8 (Zapponi, [Zap08a]). *Let X be a curve defined over $\overline{\mathbf{Q}}$, with field of moduli K. Then the Belyi degree of X is greater or equal to the largest prime $p \in \mathbf{Z}$ over which X has semistable bad reduction.*

Note added in Proof: This result can now be found in a preprint of Zapponi, "On the Belyi degree(s) of a curve defined over a number field", available at http://www.math.jussieu.fr/~zapponi.

The lower bound of Theorem 2.8 cannot be sharp since, for example, there exists infinitely many number fields F and elliptic curves E/F with E having everywhere good reduction over F.

The proof of Theorem 2.8 is a simple application of a theorem of Beckmann [Bec89] relating the primes of bad reduction of X to the Galois groups of Belyi maps for X. In turn, Beckmann's theorem may be obtained as a corollary of Grothendieck's theory of the tame fundamental group. For Grothendieck's theory of the tame fundamental group, see [Ser92] and [Gro71].

Theorem 2.9 (Beckmann, [Bec89]). *Assume (X, φ) is a pair consisting of a curve X and a Belyi map φ for X. Let M be the field of moduli of the pair (X, φ). Let G be the Galois group of the Galois closure of the cover $\varphi : X \longrightarrow \mathbf{P}^1$. For every prime $p \in \mathbf{Z}$, if p does not divide $|G|$, then X has good reduction at primes of M above p. Moreover p is unramified in M.*

We do not give a proof of Beckmann's Theorem, as that would take us too far afield.

Proof of Theorem 2.8. Let φ be a Belyi map for X of degree n. Let K (resp. M) be the field of moduli of X (resp. (X, φ)). Then M is a (possibly non-trivial) finite extension of K. Let G be as in Theorem 2.9 and let $\mathfrak{p}$ be a prime of K, above $p \in \mathbf{Z}$, where X has bad semistable reduction. Then X also has bad semistable reduction at some prime $\wp$ of M, lying above $\mathfrak{p}$. Hence the pair (X, φ) has bad reduction at $\wp$. By Theorem 2.9, p divides $|G|$. But $G \hookrightarrow S_n$, the symmetric group on n letters. Now $|S_n| = n!$, so $|G| | n!$. Hence $p | n!$, so $p \leq n$. □

[4]The deep result here, which we do not use, is the Semistable Reduction Theorem of Deligne-Mumford, which says that every curve acquires semistable reduction over some finite extension of the base field (see [DM69]). However this result is the reason why in Theorem 2.8 we say "semistable bad reduction" rather than simply "bad reduction", which would have been a stronger statement.

3 The *abc* Polynomial Theorem

Let K be an algebraically closed field of characteristic 0. Let $f \in K[x]$. Then the *radical* $R(f)$ of f is the product of the distinct irreducible factors of f. In other words

$$R(f) = \prod_{p|f} p. \tag{3.0.3}$$

We denote the degree of the radical of f by $r(f)$, i.e., $\deg R(f) = r(f)$. The *height* $h(f_1, \ldots, f_n)$ of a finite set $\{f_1, \ldots, f_n\} \subset K[x]$ of polynomials is the maximum of the degrees:

$$h(f_1, \ldots, f_n) = \max\{\deg f_1, \ldots, \deg f_n\}. \tag{3.0.4}$$

The abc Polynomial Theorem states:[5]

Theorem 3.1 (Stothers [Sto81]-Mason [Mas84]). *Suppose $e, f, g \in K[x]$ are not all constant, pairwise relatively prime and satisfy*

$$e + f = g. \tag{3.0.5}$$

Then

$$h(e, f, g) \le r(efg) - 1. \tag{3.0.6}$$

Theorem 3.2. *Equality holds in* (3.0.6) *if and only if f/g is a Belyi map for $\mathbf{P}^1$ and $(f/g)(\infty) \in \{0, 1, \infty\}$.*

Proof. Set $h = h(e, f, g)$ and $\varphi = f/g$. Then $\deg \varphi = h$. Further, we may assume without loss of generality that $\deg e = \deg f = h$. There are then two cases, $\deg g < h$ or $\deg g = h$, which correspond to whether $\varphi(\infty) = \infty$ or not. Our assumptions on the degrees of e, f, g imply that in both cases $\varphi(\infty) \neq 0, 1$. By the Riemann–Hurwitz formula applied to φ,

$$-2 = -2h + \sum_{x \in \mathbf{P}^1} (e_\varphi(x) - 1). \tag{3.0.7}$$

The crucial observation now is that the branching behavior φ over $\{0, 1, \infty\}$ is tied to the radical of efg.

We break up the ramification divisor

$$R = \sum_{x \in \mathbf{P}^1} (e_\varphi(x) - 1) \tag{3.0.8}$$

[5]So called because it is the function field analogue which motivated the famous 'abc Conjecture' of Masser and Oesterlé (see [Oes89]).

as $R = R_{\{0,1,\infty\}} + R_{\mathbf{P}^1-\{0,1,\infty\}}$ with

$$R_{\{0,1,\infty\}} = \sum_{x \in \varphi^{-1}(\{0,1,\infty\})} (e_\varphi(x) - 1) \tag{3.0.9}$$

and

$$R_{\mathbf{P}^1-\{0,1,\infty\}} = \sum_{x \in \varphi^{-1}(\mathbf{P}^1-\{0,1,\infty\})} (e_\varphi(x) - 1). \tag{3.0.10}$$

Now we focus on $R_{\{0,1,\infty\}}$. We have

$$\sum_{x \in \varphi^{-1}(0)} (e_\varphi(x) - 1) = \sum_{x \in \varphi^{-1}(0)} e_\varphi(x) - \sum_{x \in \varphi^{-1}(0)} 1 \tag{3.0.11}$$

$$= h - |\varphi^{-1}(0)| = h - r(f).$$

Similarly, since $\varphi(x) = 1$ if and only if $(g - f)(x) = e(x) = 0$, we have

$$\sum_{x \in \varphi^{-1}(1)} (e_\varphi(x) - 1) = h - r(e). \tag{3.0.12}$$

If $\varphi(x) = \infty$, then either $x = \infty$ or $g(x) = 0$. If $\deg g = h$, then $\varphi(\infty) \neq \infty$, so as above

$$\sum_{x \in \varphi^{-1}(\infty)} (e_\varphi(x) - 1) = h - r(g). \tag{3.0.13}$$

On the other hand, if $\deg g < h$, then $\varphi(\infty) = \infty$, so we have

$$\sum_{x \in \varphi^{-1}(\infty)} (e_\varphi(x) - 1) = h - (r(g) + 1). \tag{3.0.14}$$

The extra contribution of 1 this time comes from $x = \infty$, namely in this case $|\varphi^{-1}(\infty)| = r(g) + 1$.

Putting together these calculations, we see that

$$R_{\{0,1,\infty\}} = \begin{cases} 3h - r(e) - r(f) - r(g) & \text{if} \quad \deg g = h \\ 3h - r(e) - r(f) - r(g) - 1 & \text{if} \quad \deg g < h \end{cases} \tag{3.0.15}$$

By (3.0.7), $2h - 2 = R$ and $R_{\mathbf{P}^1-\{0,1,\infty\}} \geq 0$, so $2h - 2 \geq R_{\{0,1,\infty\}}$. Hence (3.0.15) implies that $2h - 2 \geq 3h - r(e) - r(f) - r(g) - 1$, so

$$h \leq r(e) + r(f) + r(g) - 1.$$

Since e, f, g are pairwise relatively prime, $r(efg) = r(e) + r(f) + r(g)$. This proves (3.0.6). We also see that equality holds in (3.0.6) if and only if $\deg g < h$ and $R_{\mathbf{P}^1-\{0,1,\infty\}} = 0$. Now $\deg g < h$ means that $\varphi(\infty) = \infty$ and $R_{\mathbf{P}^1-\{0,1,\infty\}} = 0$ means that φ is a Belyi map. This proves Theorem 3.2. □

4 Generalizations

In this section, we present some theorems and conjectures which may be viewed as generalizations of Belyi's Theorem in several different directions. In §4.1, two analogues of Belyi's Theorem in characteristic p are given, which we call "The *wild* p-Belyi Theorem" (Theorem 4.1) and "The *tame* p-Belyi Theorem" (Theorem 4.6). It is interesting to note that, regarding the tame p-Belyi question, it is only a theorem for $p > 2$, while it remains an open problem for $p = 2$. Then §4.2 states Question 4.12, which is concerned with generalizing Belyi's Theorem to function fields. Finally, §4.3 considers Belyi's Theorem in the framework of moduli spaces of pointed curves. This is used to explain Braungardt's Question (Question 4.17), which offers a generalization of Belyi's Theorem to higher dimensions.

4.1 Characteristic p

Let k be a (not necessarily finitely generated) perfect field of characteristic $p > 0$ and let $\varphi : C_1 \longrightarrow C_2$ be a map of smooth projective curves defined over k. Recall that there is the dichotomy of *tame* and *wild* ramification in characteristic p. Given $P \in C_1$, the map φ is *tamely ramified* (resp.*wildly ramified*) at P if $p \nmid e_\varphi(P)$ (resp. $p | e_\varphi(P)$). For a divisor D on a curve C, the reduced divisor associated to D (resp. the support of D) will be denoted by D_{red} (resp. $\text{supp}(D)$). We let $D = D_0 - D_\infty$ be the unique decomposition of D as a difference of disjoint effective divisors.

We now state the first of two characteristic p analogues of Belyi's Theorem, which we call "The Wild p-Belyi Theorem". We give two proofs of this result, following [Kat88], [Zap08b]. See also [Kat88] for applications of the "Wild p-Belyi Theorem" to the Langlands correspondence for function fields.

Theorem 4.1 (Wild p-Belyi, [Kat88], [Zap08b]). *Let C/k be a connected, smooth, projective curve. Then there exists a map*

$$\varphi : C \longrightarrow \mathbf{P}^1 \text{ with } B(\varphi) = \{\infty\}. \tag{4.1.1}$$

In other words, C is birational to an étale cover of the affine line $\mathbf{A}^1$.

In case a map φ satisfies (4.1.1), we call φ a *wild p-Belyi map* of C. The tame fundamental group of the affine line $\mathbf{A}^1$ is trivial. Hence, if C is not isomorphic to $\mathbf{P}^1$, then any wild p-Belyi map of C is wildly ramified over ∞, whence the name "*wild* p-Belyi map". On the other hand, this is one of the major differences between characteristic 0 and characteristic p –the full étale fundamental group of the affine line is a huge, non-finitely generated group. This may seem counter-intuitive at least at a superficial level, given our geometric intuition concerning $\mathbf{A}^1(\mathbf{C})$. The Wild p-Belyi Theorem is one manifestation of the fact that, indeed, there are "many" étale covers of the affine line in characteristic p. The study of the (full) fundamental group of the affine line forms an interesting subject in its own right. One of the deepest conjectures in this area was Abhyankar's Conjecture [Abh57], which was proved by Raynaud [Ray94] using many sophisticated techniques from algebraic geometry. For an introduction to the basic problems regarding $\pi_1^{\text{ét}}(\mathbf{A}^1)$ and Abhyankar's Conjecture in particular, see the Bourbaki exposé of Serre [Ser92].[6]

Proof 1 (after [Kat88]). The idea here is to use that, given a finite subgroup G of the $\bar{k}$-points of the additive group scheme $\mathbf{G}_a$, the natural action of G on $\mathbf{A}^1$ satisfies:

1. $\mathbf{A}^1/G \cong \mathbf{A}^1$.
2. The projection map $\pi : \mathbf{A}^1 \longrightarrow \mathbf{A}^1/G$ is a finite étale cover.
3. Over $\bar{k}$, when G is viewed as a subset of $\mathbf{A}^1$, the restriction of π to the complement $\mathbf{A}^1 - G$,

$$\pi|_{(\mathbf{A}^1 - G)} : (\mathbf{A}^1 - G) \longrightarrow \mathbf{G}_m,$$

is a finite étale cover of $\mathbf{G}_m$.

By the Riemann-Roch Theorem, there exists a non-constant function $f \in K(C)$ which has only simple poles.[7] Now f defines a map $f : C \longrightarrow \mathbf{P}^1$ which is unramified above ∞ by construction. Hence f is étale over $\mathbf{A}^1 - B(f)$. Let G be the subgroup of $\mathbf{G}_a(\bar{k})$ generated by $B(f)$. Now we use that char$(k) = p > 0$, to attain that G is finite, since $G = \text{span}_{\mathbf{F}_p}\{B(f)\}$. It follows from (1-3) above that the composite $(\pi \circ f) : C \longrightarrow \mathbf{P}^1$ is étale over $\mathbf{G}_m$. The following lemma allows us to conclude the proof.

Lemma 4.2. *The map*

$$\begin{aligned} \mu &: \mathbf{G}_m \longrightarrow \mathbf{A}^1 \\ \mu &: x \longmapsto (x^p + \tfrac{1}{x}) \end{aligned} \tag{4.1.2}$$

is a finite étale cover of the affine line by the multiplicative group scheme $\mathbf{G}_m$.

Proof. The lemma follows from a simple computation that is left to the reader. □

Indeed, $(\mu \circ \pi \circ f) : C \longrightarrow \mathbf{P}^1$ is étale over $\mathbf{A}^1$, which proves Theorem 4.1. □

---

[6] The only downside to [Ser92] is that it was written before Raynaud [Ray94] proved Abhyankar's conjecture, so it is a bit dated and does not include the most recent developments in the field.

[7] If $n \geq g + 1$ and $P_1, \ldots, P_n$ are distinct points on C, then the divisor $D = P_1 + \cdots P_n$ satisfies $\ell(D) = \deg D - g + 1 + \ell(K - D) \geq \deg D - g + 1 \geq 2$ by Riemann–Roch, so one can find the desired function in $\mathcal{L}(D) = H^0(C, \mathcal{O}_C(D))$.

Proof 2 (after [Zap08b]). Zapponi obtains Theorem 4.1 as a consequence of a more general result relating the existence of maps to $\mathbf{P}^1$ with restricted ramification to the existence of exact rational differentials with restricted divisor type. Zapponi's result is the following:

Theorem 4.3. *Let S be a non-empty finite subset of C. Then there exists a finite étale map $(C - S) \longrightarrow \mathbf{A}^1$ if and only if there exists an exact rational differential ω whose divisor's support is contained in S.*

One source of motivation for such a connection to exist is the following simple fact.

Lemma 4.4. *Suppose C is a connected, smooth, projective curve and $f \in K(C) - K(C)^p$ is a rational function that is not a p-th power. Then the ramification divisor $R(f)$ of the map $f : \varphi : C \longrightarrow \mathbf{P}^1$ defined by f is related to the divisor of the differential df by*

$$(df) = R(f) - R(f)_{\text{red}}. \tag{4.1.3}$$

Theorem 4.1 follows immediately from Theorem 4.3. Indeed, let $f \in K(C)$ be a rational function that is not a pth power, i.e., $f \notin K(C)^p$. Put $\omega = df$ and let S be the support of ω. By construction ω is a non-zero exact differential whose support is contained in S. By Theorem 4.3, there exists a finite étale map $\widetilde{\varphi} : (X - S) \longrightarrow \mathbf{A}^1$. Now $\widetilde{\varphi}$ extends to a map φ satisfying (4.1.1).

To complete Zapponi's proof of Theorem 4.1, it remains to prove Theorem 4.3. Following Zapponi, we first prove a general lemma about exact differentials on curves and then use the lemma to finish off the proof of Theorem 4.3.

Lemma 4.5. *Let $\omega \in \Omega^1_C$ be a non-zero exact rational differential. Then for all $P \in C$ there exists a function $g \in K(C)$, satisfying $dg = \omega$, whose set of poles is contained in the union of $\{P\}$ and the set of poles of ω.*

Proof. Let T be the union of $\{P\}$ and the set of poles of ω. Since ω is exact and non-zero, there exists $g_0 \in K(C) - K(C)^p$ with $dg_0 = \omega$. The argument is now to show, by induction on the degree of the part of $(g_0)_\infty$ disjoint from T, that one can modify g_0 by $h \in K(C)^p$ so that $g_0 + h$ satisfies the theorem. Note that $\mathrm{char}(k) = \mathrm{char}(K(C)) = p$ implies $dh = 0$ for all $h \in K(C)^p$.

Let Z be the set of poles of g_0. If $Z \subset T$, there is nothing to prove. So assume there exists $Q \in Z - T$. Put $n = -\mathrm{ord}_Q(g_0)$, so $n \in \mathbf{Z}^+$. Now n is divisible by p, since otherwise ω would have a pole at Q, contrary to $Q \notin T$. Write $n = pm$, with $m \in \mathbf{Z}^+$ and consider the divisors $D = (2g-1)P + mQ$ and $D - Q$. Both D and $D - Q$ are non-special, since $\deg(D - Q) = [((2g-1) + m) - 1] > 2g - 2$. That is $\ell(\kappa_C - D) = \ell(\kappa_C - (D - Q)) = 0$, with κ_C the canonical class of C. Applying the Riemann–Roch Theorem to D and $D - Q$ and comparing the results yields $\ell(D) - \ell(D - Q) = 1$. Hence there exists $u \in \mathcal{L}(D) - \mathcal{L}(D - Q)$. Therefore there is a non-zero constant $c \in \bar{k}^*$ such that $-\mathrm{ord}_Q(g_0 - cu^p) < n$. Put $h_0 = cu^p$ and $g_1 = g_0 - h_0$. Then $h_0 \in K(C)^p$ and $dg_1 = dg_0 = \omega$. Furthermore,the degree of the part of $(g_1)_\infty$ disjoint from T is strictly smaller than the degree of the part of $(g_0)_\infty$ disjoint from T. By induction, there exist h and g as desired. □

Proof of Theorem 4.3. ($\Longleftarrow$). Suppose $f_0 \in K(C) - K(C)^p$ and $\text{supp}(df_0) \subset S$. Put $\omega = df_0$. Choose $P \in S$ and let T be as in the proof of Lemma 4.5. By Lemma 4.5, there exists $f_P \in K(C) - K(C)^p$ such that $df_P = \omega$ and f_P is regular outside T. We now modify f_P to get a primitive of ω which has a pole at every element of S (f_P may be regular at some point(s) of S). The map to $\mathbf{P}^1$ determined by this primitive will be the map we are looking for.

Let m be a positive integer and consider the divisor $D_1 = mP + \sum_{Q \in (S-P)} Q$. As long as $m \geq 2g + 1$, the Riemann–Roch Theorem guarantees that there exists a function g such that $(g) = D_2 - D_1$, with D_2 an effective divisor. For $c \in \bar{k}^*$, consider the functions $h_c = f_P + cg^p$. For all $Q \in S$, g^p has a pole at Q, so there is at most one value of c for which h_c is regular at Q. Since S is finite and $\bar{k}^*$ is infinite, there exists $c \in \bar{k}^*$ such that h_c has a pole at every element of S. Also, $dh_c = \omega$. Now think of h_c as a map $h_c : C \longrightarrow \mathbf{P}^1$. By construction, $h_c(S) = \{\infty\}$, so h_c is étale over $\mathbf{A}^1$ by Lemma 4.4.

($\Longrightarrow$). Conversely, assume $f : (X - S) \longrightarrow \mathbf{A}^1$ is finite étale. Let t be the usual uniformizing parameter on $\mathbf{A}^1$, so that $(t) = (0) - (\infty)$ on $\mathbf{P}^1$. Let ω be the differential of the pull back of t from $\mathbf{P}^1$ to X by means of f, i.e., $\omega = d(f^*(t))$. By construction ω is exact. Since differentiation commutes with pull backs, $\omega = f^*(dt)$. Hence $\text{supp}(\omega) \subset f^*(\text{supp}(dt)) = f^*(\{\infty\})$. So ω satisfies the properties required by Theorem 4.3. □

Having proved Theorem 4.3, the proof of Theorem 4.1 is now complete. □

We now come to the tame p-Belyi Theorem.

Theorem 4.6 (Tame p-Belyi, [Sai97]). *Let p be an odd prime. Let X be connected, smooth, projective curve defined over $\overline{\mathbf{F}_p}$. Then there exists a tamely ramified morphism*

$$\varphi : X \longrightarrow \mathbf{P}^1 \tag{4.1.4}$$

with

$$B(\varphi) \subset \{0, 1, \infty\}. \tag{4.1.5}$$

If φ is a tamely ramified morphism that satisfies (4.1.4) and (4.1.5), we call φ a *tame p-Belyi map* of C.

Saidi's proof amounts to the observation that Theorem 4.6 is a corollary of the following lemma, which was known to F. Klein and A. Hurwitz, but was first rigorously proved in all characteristics different from two by Fulton [Ful69].

Lemma 4.7 (Fulton [Ful69], p.569, Prop. 8.1). *Let p be an odd prime and let k be an algebraically closed field of characteristic p. Then every connected, smooth, projective curve X over k admits a simply branched morphism to $\mathbf{P}^1$, i.e., there exists*

$$\psi : X \longrightarrow \mathbf{P}^1 \text{ with } e_\psi(P) = 1 \text{ or } 2 \text{ for all } P \in X. \tag{4.1.6}$$

Proof of Theorem 4.6. Take $k = \overline{\mathbf{F}_p}$. Let $\psi : X \longrightarrow \mathbf{P}^1$ be a simply branched cover, as provided by Lemma 4.7. Since p is odd, ψ is tamely ramified. Also, $B(\psi)$

is a finite subset of $\mathbf{P}^1(\overline{\mathbf{F}_p})$, so there exists an integer m such that $B(\psi) \subset \mathbf{F}_{p^m}$. Consider the function f given by

$$f : \mathbf{P}^1 \longrightarrow \mathbf{P}^1$$
$$f : x \longmapsto x^{p^m-1},$$

and form the composite $\varphi = f \circ \psi$. Now f is totally ramified at 0 above 0, at ∞ above ∞ and unramified everywhere else. Since $e_f(0) = e_f(\infty) = \deg f = p^m - 1$, the map f is tame. Also, $f(b) = 1$ for all $b \in B$. By Lemma 2.1, φ satisfies (4.1.5). □

The above proof is not applicable in the case $p = 2$, but the result should still hold for $p = 2$.

Conjecture 4.8 (Tame 2-Belyi). Theorem 4.6 is also true for $p = 2$.

There is only a partial result known in the direction of Conjecture 4.8 (see Theorem 4.10 below). It has been asked whether the proof of Theorem 4.6 could be modified so as to work in characteristic two by using triple ramification instead of double (simple) ramification. More precisely:

Question 4.9. Suppose k is an algebraically closed field of characteristic different from three. Does every curve X over k admit a map

$$\tau : X \longrightarrow \mathbf{P}^1 \text{ with } e_\tau(P) = 1 \text{ or } 3 \text{ for all } P \in X \, ? \qquad (4.1.7)$$

Note that, in case $\mathrm{char}(k) = 2$, if a map τ satisfying (4.1.7) exists for a curve X, then τ is tamely ramified, and so, by the proof of Theorem 4.6, Conjecture 4.8 holds for X. Thus a positive answer to Question 4.9 would imply Conjecture 4.8.

In this direction, Schroer [Sch03] has given a lower bound for the dimension of the locus, in the moduli space $\mathcal{M}_g$ of curves of genus g, of those curves admitting a map τ as in (4.1.7).

Theorem 4.10 (Schroer [Sch03]). *Let k be an algebraically closed field of characteristic $\neq 3$. Define* $\mathrm{Tri}(\mathcal{M}_g)$ *to be the closure of the locus of curves X that admit a map τ as in* (4.1.7). *Then,*

1. $\dim \mathrm{Tri}(\mathcal{M}_{1,1}) = 1$.
2. $\dim \mathrm{Tri}(\mathcal{M}_g) \geq 2g - 3$ *for* $g \geq 2$.

As a final piece of evidence to support Question 4.9, we mention the following result regarding the case $k = \mathbf{C}$.

Theorem 4.11 (Artebani-Pirola, [AP04], p.339, Th. 2). *The Locus* $\mathrm{Tri}(\mathcal{M}_g)(\mathbf{C})$ *is a Zariski open subset of* $\mathcal{M}_g(\mathbf{C})$

The proof of Artebani–Pirola uses complex analysis, so it is not clear whether their method can be modified to prove the same statement for any field k, $\mathrm{char}\, k \neq 3$. Still, we believe that Theorem 4.11 shows at least that Question 4.9 is reasonable.

4.2 The Function Field Case

In this section, we search for a Belyi-type theorem for curves defined over $\overline{\mathbf{Q}(t)}$. Let X be a curve defined over $\overline{\mathbf{Q}(t)}$. By the converse to Belyi's Theorem, if X is generic, then there is no map $\varphi : X \to \mathbf{P}^1$ with $B(\varphi) \subset \{0, 1, \infty\}$. By Belyi's Theorem, there therefore cannot even be such a map with $B(\varphi) \subset \overline{\mathbf{Q}}$. However one can ask:

Question 4.12. For every curve X defined over $\overline{\mathbf{Q}(t)}$, does there exist a map

$$\varphi : X \to \mathbf{P}^1 \text{ with } |B(\varphi)| = 4? \tag{4.2.1}$$

For a more general question along the lines of Question 4.12, see [Par02][8].

We illustrate the nature of this question with two examples, one for which the question can be seen to have a positive answer, the other for which the answer is unknown.

Example 4.13. Elliptic curves over $\overline{\mathbf{Q}(t)}$.

Let $E/\overline{\mathbf{Q}(t)}$ be an elliptic curve. Then the hyperelliptic degree 2 map $E \to \mathbf{P}^1$ (given, for instance, by quotient by the hyperelliptic involution) is branched over $2g(E) + 2 = 4$ points so the answer to Question 4.12 in this case is yes.

Example 4.14. Genus two curves over $\overline{\mathbf{Q}(t)}$.

Let $C/\overline{\mathbf{Q}(t)}$ be a genus 2 curve, which is necessarily hyperelliptic. In this case I do not know the answer to Question 4.12. We can consider again the hyperelliptic degree 2 map $h : C \to \mathbf{P}^1$. This time $|B(h)| = 2g(C) + 2 = 6$. Taking the proof of Belyi's Theorem as our guide, we may ask:

Question 4.15. Does there exist $g : \mathbf{P}^1 \to \mathbf{P}^1$ such that $|B(g \circ h)| = 4$? That is, is there a way of reducing the number of branch points from 6 to 4 in this case?

4.3 Higher Dimensions

The purpose of this section is to explain a question of V. Braungardt [Bra04] (see Question 4.17), which suggests a generalization of Belyi's theorem to higher dimensional varieties.[9] The origin of Braungardt's question is the observation that the space $\mathbf{P}^1 - \{0, 1, \infty\}$ appearing in Belyi's theorem is a moduli space of curves:

---

[8]We thank the referee for bringing this reference to our attention.

[9]A generalization of Belyi's theorem for surfaces in a different direction than the one considered here can be found in [Par02]. Since we believe the generalization in [Par02] is less natural and of a more technical nature, we have chosen not to state it here.

It is the moduli space of genus 0 curves with 4 ordered marked points, since any 3 points in $\mathbf{P}^1$ can be mapped to $\{0, 1, \infty\}$ by a fractional linear transformation, unique up to permutation of $\{0, 1, \infty\}$.

Thus we open this section with a brief discussion of moduli spaces of curves with marked points. Then we state Braungardt's Question. This is followed by an explanation of partial results in low dimension, some remarks on high dimension and a note on fundamental groups of hypersurface complements.

Moduli spaces of curves. Let $\mathcal{M}_{g,n}$ (resp. $\mathcal{M}_{g,[n]}$) denote the moduli space of genus g curves with n ordered marked points (resp.with a subset of cardinality n). The moduli problem describing $\mathcal{M}_{g,n}$ is not representable in the category of schemes when n is sufficiently small compared to g. Moreover, the moduli problem describing $\mathcal{M}_{g,[n]}$ is never represented by a scheme. When the moduli problem is not representable, there does not exist a *fine* moduli space in the category of schemes. The principal obstruction to the representability of a moduli problem is (lack of) rigidity. This means that if the objects being classified by the moduli problem have non-trivial automorphisms, then the moduli problem is not representable in the category of schemes. For all values of g and n other than the three pairs $(g = 1, n = 0)$, $(g = 0, n = 1)$, $(g = 0, n = 2)$, there does exist a scheme which is a *coarse* moduli space for genus g curves with n marked points.

However, our interest in the moduli spaces of curves lies with their covers and fundamental groups. For this aspect of moduli spaces of curves it is essential to work with *fine* moduli, since passing to a *coarse* moduli space often annihilates much of the fundamental group. For this reason we find ourselves forced to view $\mathcal{M}_{g,n}$ (resp. $\mathcal{M}_{g,[n]}$) as a *fine moduli algebraic stack* and to consider $\pi_1(\mathcal{M}_{g,n})$ as the *stack* fundamental group.

The theory of stacks, moduli spaces and their representability in various categories is very deep and sophisticated. Since we want to keep the exposition as self-contained as possible, while at the same time not veering off course to discuss unnecessary technical issues, we proceed as follows. We recall what it means for a map of algebraic stacks to be étale in terms of schemes, which suffices, as far as stacks are concerned, for understanding this paper. Then we illustrate some of the points made above with an example: The moduli space of elliptic curves $\mathcal{M}_{1,1}$, which will also reappear in §5. For more on moduli problems we refer to the three classic works [KM1], [DR73] and [DM69].

A map $\varphi : \mathcal{X} \longrightarrow \mathcal{Y}$ of algebraic stacks is étale if, for every pair (S, α) consisting of a scheme S and a map $\alpha : S \longrightarrow \mathcal{Y}$, the pull-back $S \times_{\mathcal{Y}} \mathcal{X}$ of $\mathcal{X}$ to S is a scheme:

$$\begin{array}{ccc} S \times_{\mathcal{Y}} \mathcal{X} & \longrightarrow & \mathcal{X} \\ \downarrow & & \downarrow \varphi. \\ S & \xrightarrow{\alpha} & \mathcal{Y} \end{array} \tag{4.3.1}$$

For example, one can check that the map $\mathcal{M}_{g,n} \longrightarrow \mathcal{M}_{g,[n]}$, given by forgetting the ordering, is étale as a map of algebraic stacks. For this reason, we will for the most part restrict attention to $\mathcal{M}_{g,[n]}$, rather than $\mathcal{M}_{g,n}$ from this point on.

Example 4.16. **The Moduli space of Elliptic Curves.** In terms of the above, the moduli space of elliptic curves is the moduli space $\mathcal{M}_{1,1}$ of genus 1 curves with 1 marked point (the 1 marked point being the origin, i.e., the identity for the group law). It is the quintessential example of a moduli space, in many ways the *raison d'etre* of moduli theory. Every elliptic curve has a non-trivial automorphism, namely the map called -1 which sends an element x to $-x$. The existence of non-trivial automorphisms of elliptic curves shows that the moduli problem for elliptic curves is not rigid and hence not representable: There does not exist a fine moduli scheme for elliptic curves.

There are two ways to recover a fine moduli space. One way is to consider the moduli space of elliptic curves as a stack, as was mentioned in the general paragraph above. The other way is to impose level structure. This means that one modifies the moduli problem to that of classifying pairs (E, L) where E is an elliptic curve and L is some additional structure arising from E, such as a basis of the N-torsion points of E for some positive integer N. Adding level structure to the moduli problem is a way of "rigidifying" the moduli problem (eliminating non-trivial automorphisms), which is then often sufficient for proving that the modified moduli problem is representable and thus admits a fine moduli scheme.

For example, if instead of classifying elliptic curves, we aim to classify pairs (E, L), with L a basis of the N-torsion of E, then for $N \geq 3$ there exists a fine moduli scheme for this moduli problem: It is the modular curve denoted $Y(N)$.

At the same time, there does exist a coarse moduli scheme for elliptic curves, called the j-line. The j-line is isomorphic to the affine line $\mathbf{A}^1$ by means of the map j which sends an isomorphism class of elliptic curves to its j-invariant. Over the complex numbers, the coarse moduli scheme of elliptic curves is also isomorphic to $SL(2,\mathbf{Z}) \setminus \mathbf{H}$ and the j line can be seen as giving a complex-analytic isomorphism of (open) Riemann surfaces

$$j : SL(2,\mathbf{Z}) \setminus \mathbf{H} \xrightarrow{\sim} \mathbf{C}. \tag{4.3.2}$$

Since the coarse moduli space of elliptic curves is isomorphic to $\mathbf{A}^1$, it is simply connected. By means of (4.3.2), we see that $SL(2,\mathbf{Z}) \setminus \mathbf{H}$ is also simply connected. The (topological) fundamental group of $SL(2,\mathbf{Z}) \setminus \mathbf{H}$ is not $SL(2,\mathbf{Z})$ because the action of $SL(2,\mathbf{Z})$ on the upper half-plane $\mathbf{H}$ is not free. However if we consider $SL(2,\mathbf{Z}) \setminus \mathbf{H}$ as a stack over $\mathbf{C}$, or as an orbifold, then the stack or orbifold fundamental group of $SL(2,\mathbf{Z}) \setminus \mathbf{H}$ is $SL(2,\mathbf{Z})$ (cf. [BN06]).

Braungardt's Question. The main question of this section is:

Question 4.17 (Braungardt, [Bra04]). Is every connected, quasi-projective variety X that is defined over $\overline{\mathbf{Q}}$ birational to a finite étale cover of some moduli space of curves $\mathcal{M}_{g,[n]}$?

There are only a few limited partial results regarding Braungardt's Question, which we now describe.

Dimension 1. Belyi's theorem gives a complete affirmative answer in the case that X is a curve. Indeed, every connected, quasi-projective curve is birational to a connected, smooth projective curve and by Belyi's Theorem every connected smooth projective curve defined over $\overline{\mathbf{Q}}$ is birational to a finite unramified cover of $\mathbf{P}^1 - \{0, 1, \infty\} \cong \mathcal{M}_{0,4}$.

Dimension 2. The moduli spaces $\mathcal{M}_{g,[n]}$ of dimension two are $\mathcal{M}_{0,[5]}$ and $\mathcal{M}_{1,[2]}$. There is an étale cover of algebraic stacks

$$\mathcal{M}_{1,[2]} \xrightarrow{\beta} \mathcal{M}_{0,[5]} \tag{4.3.3}$$

which we now describe. Let $(E; \{q_1, q_2\})$ represent a point $\eta \in \mathcal{M}_{1,[2]}$. Then, up to fractional linear transformation, there is a unique degree 2 map $\varphi_E : E \longrightarrow \mathbf{P}^1$ satisfying $\varphi_E(q_1) = \varphi_E(q_2)$. We define $\beta(\eta) \in \mathcal{M}_{0,[5]}$ to be the point represented by $(\mathbf{P}^1; \{r_1, \dots, r_5\})$, where $B(\varphi) = \{r_1, \dots, r_4\}$ and $r_5 = \varphi(q_1) = \varphi(q_2)$. Therefore in dimension 2 Braungardt's question is equivalent to: *Is every connected, smooth projective surface birational to a finite étale cover of* $\mathcal{M}_{0,[5]}$*?*

Here is a concrete description of $\mathcal{M}_{0,5}$ as a subvariety of $\mathbf{P}^1 \times \mathbf{P}^1$. Let

$$S = [(\mathbf{P}^1 - \{0, 1, \infty\}) \times (\mathbf{P}^1 - \{0, 1, \infty\})] - \Delta, \tag{4.3.4}$$

where Δ denotes the diagonal of the product. Then $\mathcal{M}_{0,5} \cong S$. A point $(a, b) \in S$ corresponds to the 5-pointed projective line $(\mathbf{P}^1; 0, 1, \infty, a, b)$.

An embedding $\mathcal{M}_{0,5} \hookrightarrow \mathbf{P}^2$ is given explicitly in the last paragraph of this section.

Summary of results of Braungardt on his question for surfaces. Braungardt has shown that his question has a positive answer for some special classes of surfaces.

Theorem 4.18 (Braungardt, [Bra04]). *1. All abelian surfaces*[10] *defined over* $\overline{\mathbf{Q}}$ *are birational to finite etale covers of* $\mathcal{M}_{0,[5]}$.

2. *All ruled surfaces, all fiber bundles of fiber genus* ≥ 2 *and all elliptic fibrations defined over* $\overline{\mathbf{Q}}$ *are birational to finite étale covers of* $\mathcal{M}_{0,5}$.
3. *If there is a positive answer to Question 4.12, then every smooth irreducible surface fibered over a curve with connected fibers is birational to a finite étale cover of* $\mathcal{M}_{0,5}$.

From the classification of surfaces, we see that the surfaces not treated in Theorem 4.18 are the non-elliptic K3 surfaces and surfaces of general type. To the best of our knowledge, Braungardt's question is open in these cases.

We only give a proof of the assertion about abelian surfaces. However, note that the proofs of the other parts of Theorem 4.18 are similar in that they use the way the

[10] In this paper abelian surface will always mean algebraic abelian surface.

given surface is related to some curve to reduce to Belyi's theorem for that curve. Indeed, observe that the surfaces covered by Braungardt are exactly those whose geometry is controlled by some underlying curve.

Proof of Theorem 4.18 (1). The essential ingredient in the proof is a well-known dichotomy for abelian surfaces: Every principally polarized abelian surface is either a product of two elliptic curves or is the Jacobian of a genus 2 curve. In fact, every principally polarized abelian surface is the Jacobian of its theta divisor. Furthermore, every abelian surface over an algebraically closed field of characteristic 0 is isogenous to an abelian surface that admits a principal polarization. One can also see that the relevant dimensions happen to coincide. The dimension of $\mathcal{A}_g$, the moduli space of abelian varieties of dimension g is $g(g+1)/2$. On the other hand, for $g \geq 2$, we have $\dim \mathcal{M}_g = 3g - 3$. So $\dim \mathcal{A}_g = \dim \mathcal{M}_g$ if and only if $g = 2$ or $g = 3$. The comparison of dimensions also shows why this proof for abelian surfaces cannot generalize to abelian varieties of dimension greater than 3.

Let A be an abelian surface defined over $\overline{\mathbf{Q}}$. By composing with an isogeny (which is finite and everywhere unramified) we may assume, without loss of generality, that A admits a principal polarization.

Case I: A is a product $E_1 \times E_2$ of two elliptic curves. Since A is defined over $\overline{\mathbf{Q}}$ also E_1 and E_2 are defined over $\overline{\mathbf{Q}}$. Applying Belyi's Theorem to E_1 and E_2 gives Belyi maps $\alpha : E_1 \to \mathbf{P}^1$ and $\beta : E_2 \to \mathbf{P}^1$. Putting together the maps α, β gives a map $\psi : A \to \mathbf{P}^1 \times \mathbf{P}^1$ given by

$$\psi : A \xrightarrow{\alpha \times \beta} \mathbf{P}^1 \times \mathbf{P}^1. \tag{4.3.5}$$

Now, since α and β are Belyi maps, their branch loci are contained in $\{0, 1, \infty\}$. Hence ψ restricts to a finite unramified cover $\psi : \psi^{-1}(S) \to S$, where S is the image of the embedding of $\mathcal{M}_{0,5}$ into the quadric surface $\mathbf{P}^1 \times \mathbf{P}^1$ defined in (4.3.4). Since the complement of S in $\mathbf{P}^1 \times \mathbf{P}^1$ is a proper closed subvariety (it is a union of lines) so is the complement of $\psi^{-1}(S)$ in A. Hence $\psi^{-1}(S)$ is an open subset of A, so $\psi^{-1}(S)$ is birational to A.

Case II: A is the Jacobian of some genus 2 curve C. The Jacobian of C is birational to the symmetric square $\mathrm{Sym}^2(C)$ of C. Since A is defined over $\overline{\mathbf{Q}}$, C is defined over $\overline{\mathbf{Q}}$ too. Let $\varphi : C \to \mathbf{P}^1$ be a Belyi map for C. By functoriality of symmetric powers, there is an induced map $\mathrm{Sym}^2(\varphi) : \mathrm{Sym}^2(C) \to \mathrm{Sym}^2(\mathbf{P}^1)$, whose branch locus is contained in $\mathrm{Sym}^2(\{0, 1, \infty\})$. Hence the map $\mathrm{Sym}^2(\varphi)$ is unramified over the quotient U of $\mathcal{M}_{0,5}$ by the transposition that interchanges the fourth and fifth marked points. Now U is a cover of $\mathcal{M}_{0,[5]}$, so A is birational to a finite unramified covering of $\mathcal{M}_{0,[5]}$. □

Some remarks on dimension at least six. Given the content of Braungardt's Question, it will be convenient to have the following notation. Given a variety V, let $\mathrm{buc}(V)$ denote the set of all varieties W such that W is birational to a finite, unramified cover of V. If V and W are two varieties satisfying both $V \in \mathrm{buc}(W)$

and $W \in \text{buc}(V)$, then we will say that V and W are BUC, written also $V \sim_{\text{BUC}} W$. On the other hand, if V and W satisfy both $V \notin \text{buc}(W)$ and $W \notin \text{buc}(V)$, then we will say that V and W are Ibuc,[11] or that V is Ibuc with W.

In terms of this notation, Braungardt's Question may be rephrased as

$$\bigcup_{\{(g,n) \mid \dim \mathcal{M}_{g,n}=k\}} \text{buc}(\mathcal{M}_{g,[n]}) = \underline{\text{VAR}^k_{\overline{\mathbf{Q}}}}, \tag{4.3.6}$$

where $\underline{\text{VAR}^k_{\overline{\mathbf{Q}}}}$ is the set of varieties of dimension k defined over $\overline{\mathbf{Q}}$.

Conjecture 4.19. 1. The moduli space $\mathcal{M}_3$ of curves of genus 3, is Ibuc with the other moduli spaces $\mathcal{M}_{g,[n]}$ of curves of dimension 6, namely $\mathcal{M}_{0,[9]}$, $\mathcal{M}_{1,[6]}$ and $\mathcal{M}_{2,[3]}$.

2. More generally, assume (g_1, n_1) and (g_2, n_2) are two *distinct* pairs of nonnegative integers satisfying $3g_1 - 3 + n_1 = 3g_2 - 3 + n_2$. Then $\mathcal{M}_{g_1,[n_1]}$ and $\mathcal{M}_{g_2,[n_2]}$ are Ibuc if and only if $\{g_1, g_2\} \not\subset \{0, 1, 2\}$.

The "Only if" part of Conjecture 4.19 is true. This can be seen by using the degree 2 hyperelliptic maps to $\mathbf{P}^1$ that genus 1 and genus 2 curves admit to construct maps between the moduli spaces, of the same dimension, of curves of genus 0,1 and 2 with marked points. This was done above explicitly in the dimension 2 case.

The moduli space of curves $\mathcal{M}_g$ is of general type for sufficiently large g [HM82]. On the other hand, the moduli spaces $\mathcal{M}_{0,n}$ are rational, since $\mathcal{M}_{0,n}$ embeds into $\mathbf{P}^1 \times \cdots \times \mathbf{P}^1$ ($n-3$ copies) as a Zariski open set, via the natural generalization of (4.3.4), namely

$$\begin{aligned} \mathcal{M}_{0,n} &\hookrightarrow \mathbf{P}^1 \times \cdots \times \mathbf{P}^1 \\ (\mathbf{P}^1; p_1, \ldots, p_n) &\mapsto (\gamma(p_4), \ldots, \gamma(p_n)), \end{aligned} \tag{4.3.7}$$

where $\gamma \in \text{Aut}(\mathbf{P}^1)$ is the unique element satisfying $\gamma(p_1) = 0$, $\gamma(p_2) = 1$ and $\gamma(p_3) = \infty$. This shows that, if g is sufficiently large, $\mathcal{M}_{0,n} \notin \text{buc}(\mathcal{M}_g)$. We are not aware of any other results regarding Conjecture 4.19.

As J. Harris pointed out to us, there are several known instances in algebraic geometry where proving the *non-existence* of birational maps has proved to be a daunting task. For example, the general cubic hypersurface in $\mathbf{P}^5$ is conjectured to be *not* rational, yet it has not been proved that a single such hypersurface is irrational.

Suppose for a moment that the answer to Braungardt's Question is "yes" and that Conjecture 4.19 holds true. This assumption has a surprising consequence. Namely, it would entail that to every variety V defined over $\overline{\mathbf{Q}}$ is attached a *discrete invariant*, namely the non-empty finite set

$$\text{mc}(V) = \{(g, n) \in \mathbf{Z}_{\geq 0} \times \mathbf{Z}_{\geq 0} \mid V \in \text{buc}(\mathcal{M}_{g,[n]})\}. \tag{4.3.8}$$

If this is indeed the case, then it would be interesting to characterize the invariant $\text{mc}(V)$ independently of moduli spaces of curves.

[11] "I" for incomparable.

Fundamental groups of hypersurface complements. Let H be a hypersurface in a projective space $\mathbf{P}^m$. It is a basic problem to understand the fundamental group $\pi_1(\mathbf{P}^m - H)$ in terms of the geometry of H. When $m = 1$, H is a finite set of points, and it is classical that, as an abstract group $\pi_1(\mathbf{P}^1(\mathbf{C}) - H)$ is the free group on $|H| - 1$ generators. Although the general problem goes back to Zariski, when m is greater than one, it remains mysterious. As far as we know, the only general result about the fundamental group of the complement of a hypersurface in projective space is the following theorem of Fulton–Deligne in the case $m = 2$.

Theorem 4.20 (Fulton–Deligne, [Ful80], [Del80]). *Let $C \in \mathbf{P}^2(\mathbf{C})$ be a nodal plane curve, i.e., either C is smooth, or the only singularities of C are nodes (also known as ordinary double points). Then the fundamental group $\pi_1(\mathbf{P}^2(\mathbf{C})-C)$ of the complement of C in $\mathbf{P}^2(\mathbf{C})$ is abelian. More precisely, if the irreducible components of C have degrees $d_1, \dots, d_k$, then*

$$\pi_1(\mathbf{P}^2(\mathbf{C}) - C) \cong \mathbf{Z}/d_1\mathbf{Z} \times \cdots \times \mathbf{Z}/d_k\mathbf{Z}. \tag{4.3.9}$$

Fulton [Ful80] proved that the algebraic fundamental group $\pi_1^{\mathrm{alg}}(\mathbf{P}^2(\mathbf{C})-C)$ (which is the profinite completion of the topological $\pi_1(\mathbf{P}^2(\mathbf{C}) - C)$) is abelian. Then Deligne used Fulton's results to deduce (4.3.9) and also gave an exposition of Fulton's work in [Del80].[12]

The connection between fundamental groups of hypersurface complements and Braungardt's question is that in certain cases $\mathcal{M}_{g,n}$ or $\mathcal{M}_{g,[n]}$ can be realized as $\mathbf{P}^m - H$, for some hypersurface H. In fact, for $\mathcal{M}_{0,n}$ this can be done with $m = n - 3$ and H a finite union of hyperplanes. For $\mathcal{M}_{0,4}$, this is the above mentioned isomorphism $\mathcal{M}_{0,4} \cong \mathbf{P}^1 - \{0, 1, \infty\}$. More generally, one has.

Theorem 4.21. *Suppose $n \geq 1$. Let $S = \{p_1, \dots, p_{n+2}\}$ be a collection of $n + 2$ points in general position in $\mathbf{P}^n$. Let H be the union of the $\binom{n+2}{2}$ hyperplanes spanned by subsets of S of cardinality n. Then*

$$\mathcal{M}_{0,n+3} \cong \mathbf{P}^n - H. \tag{4.3.10}$$

In particular, setting $n = 2$ gives:

Corollary 4.22. *$\mathcal{M}_{0,5}$ is isomorphic to the complement, in $\mathbf{P}^2$, of any 6 lines through 4 points, no 3 of which are collinear.*

In view of Corollary 4.22, we see that an affirmative answer to Braungardt's question in dimension 2 would not contradict the Fulton–Deligne Theorem: The union of 6 lines through 4 points in general position is not a nodal curve; its singular locus consists of 4 ordinary triple points and 3 ordinary double points.

---

[12] As is explained in [Ful80] and [Del80], Zariski gave a proof of Theorem 4.20, but it was incomplete because it relied on a flawed argument of Severi.

Proof of Theorem 4.21. Let K be the hyperplane in $\mathbf{P}^n$ spanned by $p_1, \ldots, p_n$ and let L be the line spanned by p_{n+1} and p_{n+2}. Since S is in general position K intersects L in a point; call this point z. Now let $P \in \mathbf{P}^n - H$. We will associate to P a sequence $\varphi(P) = (\varphi_1(P), \ldots, \varphi_n(P))$ of n distinct points lying on $L - \{p_{n+1}, p_{n+2}, z\}$ in such a way that the map

$$\Phi : \begin{array}{ccc} \mathbf{P}^n - H & \longrightarrow & \mathcal{M}_{0,n+3} \\ P & \longmapsto & (L; p_{n+1}, p_{n+2}, z, \varphi_1(P), \ldots, \varphi_n(P)) \end{array} \tag{4.3.11}$$

is an isomorphism.

To define $\varphi_i(P)$, consider the set $S_i(P) = (S - \{p_i\}) \cup \{P\}$. Since S is in general position and $P \notin H$, the set $S_i(P)$ spans a hyperplane $K_i(P)$ in $\mathbf{P}^n$. Using again that $P \notin H$, it follows that $K_i(P)$ intersects L in a point and that that point is distinct from p_{n+1}, p_{n+2} and z. Set $\varphi_i(P)$ to be the unique point of intersection of $K_i(P)$ and L.

Now it is straightforward to check that $\varphi_i(P) \neq \varphi_j(P)$ for all $i \neq j$, so that Φ is well defined, and also that Φ is a bijection.

If $\varphi_i(P) = \varphi_j(P)$ and $i \neq j$, then $\varphi_i(P) \in K_i(P) \cap K_j(P)$, so $(S_i(P) - p_j) \cup \varphi_i(P) = (S_j(P) - p_i) \cup \varphi_j(P)$. This yields an equality of hyperplanes $K_i(P) = \text{span}(S_i(P) - p_j) \cup \varphi_i(P) = \text{span}(S_j(P) - p_i) \cup \varphi_j(P) = K_j(P)$. But $K_i(P) = K_j(P)$ implies $K = K_i(P) = K_j(P)$, so $P \in K$, contradiction.

To see that Φ is injective, note that for all $P \in \mathbf{P}^n - H$, the n hyperplanes $K_i(P)$, $1 \leq i \leq n$, are in general position, so they intersect in a single point, but $P \in K_i(P)$ for all i, so $\cap_{i=1}^n K_i(P) = \{P\}$. Thus $\varphi(P) = \varphi(Q)$ implies $\{P\} = \cap_{i=1}^n K_i(P) = \cap_{i=1}^n K_i(Q) = \{Q\}$, so $P = Q$. Finally, if $(y_1, \ldots, y_n)$ are n distinct points in $L - \{p_{n+1}, p_{n+2}, z\}$, then the n hyperplanes $span\{S_i \cup y_i\}$, $1 \leq i \leq n$, intersect in a point, call it P. By construction $\Phi(P) = (L; p_{n+1}, p_{n+2}, z, y_1, \ldots, y_n)$. Hence Φ is surjective.

It is apparent from the construction of Φ that both Φ and Φ^{-1} are morphisms, so that Φ is in fact an isomorphism.[13] □

5 Modularity of elliptic curves and a question of Khare

In this section we explain a question of Khare ([Kha04]). There are two results that, when viewed in conjunction, lead naturally to Khare's question. The first result is an equivalent formulation of Belyi's Theorem, while the second is the modularity of elliptic curves over $\mathbf{Q}$ (see Lemma 5.1 and Theorem 5.2 below).

[13] An alternative proof of Theorem 4.21 can be gotten by using (4.3.7), followed by a Segre embedding, to embed $\mathcal{M}_{0,n}$ into a large projective space. Then one projects onto a suitably chosen hyperplane. The composite is the desired embedding. We leave the details to the reader.

5.1 Subgroups of $SL(2, \mathbf{Z})$

To state these results we recall some terminology about subgroups of $SL(2, \mathbf{Z})$. Let $N \geq 1$ be an integer. Consider

$$\Gamma(N) = \ker(SL(2, \mathbf{Z}) \to SL(2, \mathbf{Z}/N\mathbf{Z})). \quad (5.1.1)$$

A subgroup Γ of $SL(2, \mathbf{Z})$ is called a *congruence subgroup* if Γ contains $\Gamma(N)$ for some N. An important class of congruence subgroups is given by the subgroups $\Gamma_0(N)$, which are defined by

$$\Gamma_0(N) = \{\begin{pmatrix} a & b \\ c & d \end{pmatrix} \in SL(2, \mathbf{Z}) | N \text{ divides } c\}. \quad (5.1.2)$$

A congruence subgroup is necessarily a finite index subgroup. The converse is not true.[14] Given a finite index subgroup Γ of $SL(2, \mathbf{Z})$, the congruence hull Γ^c of Γ is the intersection of all congruence subgroups containing Γ. The quotient $\Gamma \setminus \mathbf{H}$ has a compactification obtained by adding finitely many points to $\Gamma \setminus \mathbf{H}$. We will denote the compactification of $\Gamma \setminus \mathbf{H}$ by $\overline{\Gamma \setminus \mathbf{H}}$. When $\Gamma = \Gamma_0(N)$, we follow standard notation by writing $\Gamma_0(N) \setminus \mathbf{H} = Y_0(N)$ and $\overline{\Gamma_0(N) \setminus \mathbf{H}} = \overline{Y_0(N)} = X_0(N)$.

5.2 Reformulation of Belyi's theorem as uniformization

The following lemma is an equivalent formulation of Belyi's theorem in terms of quotients of the upper half-plane by finite index subgroups of $SL(2, \mathbf{Z})$. It is easy to prove and has been observed, in various slighly different forms, by many people. Nevertheless, this reformulation of Belyi's theorem plays an important conceptual role, and therefore should not be underestimated, as explained by Mazur in [Maz91].

Let

$$\Gamma_{\text{free}}(4) = \left\{ \begin{pmatrix} a & b \\ c & d \end{pmatrix} \in \Gamma(2) \middle| a \equiv d \equiv 1 \pmod 4 \right\}. \quad (5.2.1)$$

Then $\Gamma_{\text{free}}(4)$ is a level 4 congruence subgroup, i.e., $\Gamma_{\text{free}}(4) \supset \Gamma(4)$.

Lemma 5.1. *Let X be a connected, smooth, projective curve defined over $\overline{\mathbf{Q}}$. Then there exists a finite index subgroup Γ of $\Gamma_{\text{free}}(4)$ such that*

$$X \cong \overline{\Gamma \setminus \mathbf{H}}. \quad (5.2.2)$$

---

[14] Indeed, in a sense which can be made precise, "most" finite index subgroups of $SL(2, \mathbf{Z})$ are not congruence subgroups. (See [LS03] for a discussion of this topic.)

Proof. The equivalence of Lemma 5.1 and Belyi's theorem (Theorem 1.1) is a simple application of the Galois theory of covering spaces, coupled with the observation that $\Gamma_{\text{free}}(4)$ acts freely on $\mathbf{H}$ and $\Gamma_{\text{free}}(4) \backslash \mathbf{H} \cong \mathbf{P}^1 - \{0, 1, \infty\}$.

Let us indicate why the latter holds. One has $-1 \in \Gamma(2)$, $-1 \notin \Gamma_{\text{free}}$ and $[\Gamma(2) : \Gamma_{\text{free}}(4)] = 2$. Hence

$$\Gamma(2) = \{\pm 1\} \times \Gamma_{\text{free}}(4). \tag{5.2.3}$$

The stabilizer of every point in $\mathbf{H}$ for the action of $\Gamma(2)$ is $\{\pm 1\}$, so (5.2.3) implies that Γ_{free} acts freely on $\mathbf{H}$. Now it is elementary and classical that $\Gamma(2)$ has three cusps, which can be mapped to $\{0, 1, \infty\}$, so $\Gamma(2) \backslash \mathbf{H} \cong \mathbf{P}^1 - \{0, 1, \infty\}$. However, by (5.2.3), the action of $\Gamma(2)$ factors through that of $\Gamma_{\text{free}}(4)$. Therefore $\Gamma_{\text{free}}(4) \backslash \mathbf{H} \cong \mathbf{P}^1 - \{0, 1, \infty\}$. In particular, this observation yields that $\pi_1(\mathbf{P}^1 - \{0, 1, \infty\}) \cong \Gamma_{\text{free}}$.[15]

Next, we prove that Belyi's theorem implies Lemma 5.1. Suppose X is a curve defined over $\overline{\mathbf{Q}}$. By Belyi's theorem there exists a Belyi map $\varphi : X \to \mathbf{P}^1$. This means that the restriction $\varphi_{|U} : U \to \mathbf{P}^1 - \{0, 1, \infty\}$ is a finite unramified covering, where $U = \varphi^{-1}(\mathbf{P}^1 - \{0, 1, \infty\})$. Since $\mathbf{P}^1 - \{0, 1, \infty\} \cong \Gamma_{\text{free}}(4) \backslash \mathbf{H}$ and $\pi_1(\mathbf{P}^1 - \{0, 1, \infty\}) \cong \Gamma_{\text{free}}$, the Galois theory of covering spaces implies that $\pi_1(U)$ is a finite index subgroup of $\Gamma_{\text{free}}(4)$ and $\pi_1(U) \backslash \mathbf{H} \cong U$. By compactifying we see that

$$X \cong \overline{\pi_1(U) \backslash \mathbf{H}}. \tag{5.2.4}$$

Conversely, suppose that for some $\Gamma \subset \Gamma_{\text{free}}(4)$ we have (5.2.2). The inclusion $\Gamma \hookrightarrow \Gamma_{\text{free}}(4)$ induces a covering $\Gamma \backslash \mathbf{H} \to \Gamma_{\text{free}}(4) \backslash \mathbf{H}$, which extends uniquely to a Belyi map $X \to \mathbf{P}^1$. □

It is interesting to compare Lemma 5.1 with the classical *uniformization theorem* that every curve of genus at least two is uniformized by the upper half-plane, in the sense that such a curve is a quotient of the upper half-plane by *some* discrete subgroup of $SL(2, \mathbf{R})$.

5.3 Modularity of Elliptic Curves over Q

The following theorem is the Shimura–Taniyama–Weil Conjecture, proved by Wiles [Wil95], Taylor–Wiles [TW95] in the semistable[16] case and completed in full generality by Breuil–Conrad–Diamond–Taylor [BCDT01].

Theorem 5.2 (Modularity of elliptic curves over Q). *Suppose E is an elliptic curve defined over $\mathbf{Q}$. Then there exists an integer N (called the conductor of E) and a non-constant map*

$$X_0(N) \longrightarrow E. \tag{5.3.1}$$

[15] In particular, this argument shows, in a roundabout way, that as an abstract group $\Gamma_{\text{free}}(4)$ is isomorphic to the free group on two generators.

[16] The definition of semistable was given in §2, in the discussion preceding Theorem 2.8.

5.4 Congruence Defects

Following Khare, we define the *congruence defect* $cd(\Gamma)$ of a finite index subgroup Γ of $\Gamma_{\text{free}}(4)$ to be the index of Γ in its congruence hull i.e., $cd(\Gamma) = [\Gamma^c : \Gamma]$. The congruence defect measures how far Γ is from being a congruence subgroup. The congruence defect of Γ is 1 if and only if Γ is a congruence subgroup. Next, we define the congruence defect $cd(E)$ of an elliptic curve E to be

$$cd(E) = \min \left\{ cd(\Gamma) \,\middle|\, \begin{array}{c} \Gamma \subset \Gamma_{\text{free}}(4) \text{ and } \overline{\Gamma \setminus \mathbf{H}} \text{ admits} \\ \text{a non-constant map to } E \end{array} \right\}. \tag{5.4.1}$$

The existence of the congruence defect of an elliptic curve defined over $\overline{\mathbf{Q}}$ follows from the reformulation of Belyi's Theorem (Lemma 5.1).

Question 5.3 (Khare). Let $K/\mathbf{Q}$ be a number field. Does there exist a constant $cd(K)$, the congruence defect of K, such that for every elliptic curve E/K, we have $cd(E) \leq cd(K)$? In other words, is the congruence defect bounded independently of the elliptic curve and solely in terms of the field of definition?

By the modularity of elliptic curves over $\mathbf{Q}$, the congruence defect of $\mathbf{Q}$ exists and is equal to 1. The existence of the congruence defect for any number field different from $\mathbf{Q}$ is not known.

5.5 Modularity over general number fields

Khare's question 5.3 can be seen as a possible generalization of the modularity of elliptic curves over $\mathbf{Q}$ to arbitrary number fields. There is a more standard conjecture, coming from the Langlands correspondence, which offers a generalization of the modularity of elliptic curves over $\mathbf{Q}$ to modularity, or automorphy, of elliptic curves over general number fields.

Conjecture 5.4. Let $K/\mathbf{Q}$ be a number field and $\mathbf{A}_K$ the adeles of K. Let E be an elliptic curve over K. Then there exists a automorphic representation π_E of $GL(2, \mathbf{A}_K)$ which "corresponds" to E in the sense that:

1. The L-functions of E and π_E are equal: $L(E, s) = L(\pi_E, s)$.
2. As ℓ ranges over primes of $\mathbf{Q}$, the (duals of the) Tate modules $T_\ell(E)$ form a compatible system of ℓ-adic Galois representations $\{\rho_\ell\}$ attached to π_E.
3. For all but finitely many primes $\mathfrak{p}$ of K, the characteristic polynomial of $\rho_\ell(\text{Frob}_{\mathfrak{p}})$ is $(x - \alpha_{\mathfrak{p}})(x - \beta_{\mathfrak{p}})$, where $\alpha_{\mathfrak{p}}, \beta_{\mathfrak{p}}$ are the Langlands parameters of the local component $(\pi_E)_{\mathfrak{p}}$ of π_E at $\mathfrak{p}$.

For an introduction to the Langlands correspondence between Galois representations and automorphic representations, see Taylor's ICM talk [Tay02] and its expanded version [Tay04]. In particular, all of the terms used in the statement of the conjecture above are explained in *loc. cit.* For the third condition in the conjecture, see also [Tay94]. Conjecture 5.4 has been proved completely for $K = \mathbf{Q}$ and for totally real fields[17] satisfying certain technical conditions, using the techniques of *modularity lifting theorems*, as described in *loc. cit.* When the current machinery of modularity lifting theorems is perfected—there is reason to believe that this will happen in the next few years—it should allow one to prove, as a corollary, Conjecture 5.4 for all CM fields.[18] Beyond that, if K is not a CM field, then Conjecture 5.4 is wide open and appears to require genuinely new ideas.

In the simplest case of $K = \mathbf{Q}$, both the answer, $cd(\mathbf{Q}) = 1$, to Khare's Question and Conjecture 5.4 are equivalent to the modularity of elliptic curves over $\mathbf{Q}$ (Theorem 5.2). However, it is hard to say whether such a relationship between Khare's question and Conjecture 5.4 persists over any other number field. For example, for number fields K where Conjecture 5.4 is known, it is not known whether this has any implication towards Khare's question for those K. Let K be a number field different from $\mathbf{Q}$. It would be extremely interesting if an affirmative answer to Khare's question (Question 5.3) could be used to say something about Conjecture 5.4.

One significant advantage of Khare's question is that, like (5.3.1), it is geometric, in that it asserts the existence of maps between algebraic varieties. Conjecture 5.4, at least as stated above, lacks the same kind of geometric flavor. When K is totally real, there sometimes exists a geometric equivalent of Conjecture 5.4, in analogy with the $K = \mathbf{Q}$ case. When this analogy works, the modular curves of (5.3.1) are replaced by *Shimura curves*, which can be thought of as arithmetic quotients of the upper half-plane by groups of units of quaternion algebras. It would be very interesting to find a connection, for some totally real fields, between parameterizations of elliptic curves by Shimura curves on the one hand, and the maps involving modular curves afforded by a positive answer to Khare's question.

However, even in the totally real case, one does not always have a parametrization by a Shimura curve. Moreover, when K is not totally real, there is no known idea of how to parameterize elliptic curves over K by Shimura varieties. This makes the possibility of Khare's question providing such a parametrization over arbitrary number fields all the more exciting.

6 Some personal remarks

When I met Lang, I was 16 years old and he was 77. He was giving two talks in one day, on two totally different subjects, with only an hour break in-between, to give the audience a chance to catch its breath. Never again have I seen anyone captivate and

[17]Recall that a number field K is totally real if the images of all its complex embeddings are contained in the real numbers.

[18]Recall that a CM field is a totally imaginary quadratic extension of a totally real field, the prototypical example of CM fields being imaginary quadratic fields.

inspire an audience, generating such electricity in the room, as when Lang lectured on the ABC Conjecture and the Heat Kernel. I walked out of the talks feeling, "I want to work on this stuff!".

The next day Lang made sure that I became his student by showing up at my home in Los Angeles. He wasted no time and started doing mathematics with me on the spot. And so it went on day after day, with Lang continuously surprising me, showing up at my house, calling me early in the morning and late at night to *do* mathematics. That was Lang's approach to mathematics: A unified process of learning, talking and doing research at the same time.

Since Lang passed away, mathematicians often tell me things like "This problem is too hard," and so on. But then I think of Lang. I immediately hear, "Stop fooling around and get back to work!" As I start working, I also hear Lang saying "It's possible, let's do it, let's do it right now!", and then I feel I've got to try, as Lang would have done if he were around.

7 Acknowledgements

This paper is an outgrowth of a talk given at the University of Paris VI and VII in the spring of 2008. I would like to thank Julien Grivaux and Paris VI and VII for inviting me to speak. I thank the anonymous referee for their suggestions and for bringing to my attention a number of interesting references I was unaware of. I am very grateful to Joseph Oesterlé, Barry Mazur, Joe Harris and Leonardo Zapponi for sharing with me their ideas and insights on Belyi's Theorem and many related topics. I thank Jesse Kass and Fred Van der Wyck for helping me with questions in algebraic geometry that arose while this paper was written. I would like to express special thanks to Pierre Deligne, Julien Grivaux and Barry Mazur for reading a draft of this paper and offering me invaluable comments, suggestions and corrections. Finally I would like to thank Serge Lang for initiating me to Belyi's Theorem and the abc Polynomial Theorem.

References

[Abh57] S. Abhyankar. Coverings of algebraic curves. *Amer. J. Math.*, 79:825–856, 1957.

[AP04] M. Artebani and G.-P. Pirola. Algebraic functions with even monodromy. *Proc. Amer. Math. Soc.*, 133:(2)331–341, 2004.

[BCDT01] C. Breuil, B. Conrad, F. Diamond, and R. Taylor. On the modularity of elliptic curves over **Q**: Wild 3-adic exercises. *JAMS*, 14:843–949, 2001.

[Bec89] Beckmann. Ramified primes in the field of moduli of branched coverings of curves. *J. Algebra*, 125:236–255, 1989.

[Bel80] G. V. Belyi. On Galois extensions of a maximal cyclotomic field. *Math. USSR Izvestija*, 14:247–256, 1980.

[Bel02] G. V. Belyi. A new proof of the three point theorem. *Sb. Math.*, 193(3-4):329–332, 2002.

[BN06] K. Behrend and B. Noohi. Uniformization of Deligne-Mumford curves. *J. Reine Angew. Math.*, 599:111–153, 2006.

[Bra04] V. Braungardt. Covers of moduli surfaces. *Compositio Math.*, 140(4):1033–1036, 2004.

[DD97] P. Dèbes and J.-C. Douai. Algebraic covers: Field of moduli versus field of definition. *Ann. Sci. ENS*, 30(3):303–338, 1997.

[Del80] P. Deligne. *Le groupe fondamental du complément d'une courbe plane n'ayant que des points doubles ordinaires est abélien*. In Lecture Notes in Math., Vol. 842, , pages 1–10. Séminaire Bourbaki, 1979-1980.

[DM69] P. Deligne and D. Mumford. The irreducibility of the space of curves of given genus. *Publ. Math. IHES*, 36:75–109, 1969.

[DR73] P. Deligne and M. Rapoport. Les schémas de modules des courbes elliptiques. In *Modular Functions of one variable II*, Vol. 349, Springer Lecture Notes in Math., pages 143–316. Springer-Verlag, 1973.

[Ful69] W. Fulton. Hurwitz schemes and the irreducibility of moduli of algebraic curves. *Ann. Math.*, 90:542–575, 1969.

[Ful80] W. Fulton. On the fundamental group of the complement of a node curve. *Ann. Math.* 111:407–409, 1980.

[Gro71] A. Grothendieck. *Revêtements étales et groupe fondamental* (SGA1), Vol. 224, Lect. Notes in Math., Springer-Verlag, 1971.

[HM82] J. Harris and D. Mumford, On the Kodaira dimension of the moduli space of curves. *Invent. Math.*, 67:23–86, 1982.

[Kat88] N. Katz. Travaux de Laumon, Séminaire Bourbaki, (691):105–132, 1987–1988.

[Kha04] C. Khare. Belyi parametrizations of elliptic curves and congruence defects. *CRM Proc. Lecture Notes*, 36:189–195, 2004.

[KM1] N. Katz and B. Mazur. *Arithmetic moduli of elliptic curves*, Vol. 108, *Annals of Math. Studies*, Princeton Univ. Press, 1985.

[LS03] A. Lubotsky and D. Segal. Subgroup growth, Vol. 212, Progress in Math., Birkhauser-Verlag, 2003.

[Mas84] R. C. Mason. *Diophantine equations over function fields*, Vol. 96, London Math. Soc. Lecture Notes Series. Cambridge University Press, 1984.

[Maz91] B. Mazur. Number theory as gadfly. *Amer. Math. Monthly*, 98(7):593–610, 1991.

[Oes89] J. Oesterlé. Nouvelles approches du "théorème" de Fermat. In *Asterisque*, Vol. 161–162, pages 165–186. Séminaire Bourbaki, 1988/9.

[Par02] K. Paranjape. A geometric characterization of arithmetic varieties. *Proc. Indian. Acad. Sci. (Math)*, 112:1–9, 2002.

[Ray94] M. Raynaud. Revêtements de la drote affine en caractéristique $p > 0$ et conjecture de Abhyankar. *Invent. Math.*, 116:425–462, 1994.

[Sai97] M. Saidi. Revêtements modérés et groupe fondamental de graphes de groupes. *Compositio Math.*, 107:319–338, 1997.

[Sch03] S. Schroer. Curves with only triple ramification. *Ann. Inst. Fourier*, 53(7):2225–2241, 2003.

[Ser92] J.-P. Serre. Revêtements de courbes algébriques. *Astérisque*, Vol. 206, pages 167–182 Séminaire Bourbaki, 1992.

[Sto81] W. Stothers. Polynomial identities and Hauptmoduln. *Quart. J. Math. Oxford (2)*, 32:349–370, 1981.

[Tay94] R. Taylor. l-adic representations associated to modular forms over imaginary quadratic fields. II. *Invent. Math.*, 116:619–643, 1994.

[Tay02] R. Taylor. Galois representations. In *Proc. ICM 2002*, Vol. I, pages 449–474. Higher Ed. Press, Beijing, 2002.

[Tay04] R. Taylor. Galois representations. *Ann. Math. Sci. Toulouse*, 13(1):75–119, 2004.

[TW95] R. Taylor and A. Wiles. Ring-theoretic properties of certain Hecke algebras. *Ann. Math. (2)*, 141(3):553–572, 1995.

[Wei56] A. Weil. The field of definition of a variety. *Amer. J. Math.*, 78:509–524, 1956.
[Wil95] A. Wiles. Modular elliptic curves and Fermat's Last Theorem. *Ann. Math. (2)*, 141(3):443–551, 1995.
[Zap08a] L. Zapponi. A lower bound for the Belyi degree, Private oral communication, December 2008.
[Zap08b] L. Zapponi. On the 1-pointed curves arising as étale covers of the affine line in positive characteristic. *Math Zeit.*, 258(4):711–727, 2008.

On the local divisibility of Heegner points

Benedict H. Gross and James A. Parson

"To Serge Lang"

Abstract We relate the local ℓ-divisibility of a Heegner point on an elliptic curve of conductor N, at a prime p which is inert in the imaginary quadratic field, to the first ℓ-descent on a related abelian variety of level Np.

Key words Elliptic curves over local and global fields • abelian varieties

Mathematics Subject Classification (2010): 11G05, 11607, 11G10

1 Introduction

Heegner points on the modular curve $X_0(N)$, and their images on elliptic curve factors E of the Jacobian, enjoy many remarkable properties. These points are the moduli of level structures with endomorphisms by the ring of integers of an imaginary quadratic field K. Their traces to $E(K)$ have height given by the first derivative at $s = 1$ of the L-function of E over K (cf. [GZ86]), and their ℓ-divisibility in the Mordell–Weil group controls the first ℓ-descent on E over K (cf. [Gro]).

In this paper, we show how their ℓ-divisibility in the local group $E(K_p)$, where p is a prime that is inert in K, often determines a first descent over K on a related

B.H. Gross (✉)
Department of Mathematics, Harvard University, Cambridge, MA 02138
e-mail: gross@math.harvard.edu

J.A. Parson
Hood College, Department of Mathematics, 401 Rosemont Avenue, Frederick, Maryland 21701
e-mail: parson@hood.edu

D. Goldfeld et al. (eds.), *Number Theory, Analysis and Geometry: In Memory of Serge Lang*, DOI 10.1007/978-1-4614-1260-1_11,

abelian variety A over $\mathbb{Q}$. The abelian variety A is associated to a modular form of weight 2 and level Np that is obtained by Ribet's level-raising theorem from the modular form of level N associated to E. This descent result is Theorem 2 below. To prove the descent theorem, we compare the local conditions defining a certain Selmer group for A with those defining the ℓ-Selmer group for E. The conditions agree at places of K prime to p, and at p the condition changes from the unramified local condition to a transverse condition. The parity lemma proved in §5.3 then compares the ranks of the corresponding Selmer groups in terms of the ℓ-divisibility of P in $E(K_p)$ and allows us to understand a first descent on A/K based on Kolyvagin's determination of the first ℓ-descent on E/K.

Some related work on the Selmer group can be found in [BD99, Prop 1.5] and [BD05]; a comparison with the value of the L-function at $s = 1$ is given in [BD99, Thm 1.3].

2 The main theorem

2.1 Heegner points and Kolyvagin's descent

Let E be an elliptic curve over $\mathbb{Q}$ of conductor N. It is now known that E is modular (cf. [BCDT01]): the L-function $L(s, E) = \sum_{n\geq 1} a_n n^{-s}$ of E over $\mathbb{Q}$ is the Mellin transform of a modular form $f(\tau) = \sum_{n\geq 1} a_n e^{2\pi i n\tau}$ of weight 2 on $\Gamma_0(N)$. Together with the isogeny theorem of [Fal83], this implies that there is a dominant morphism

$$\pi : X_0(N) \to E$$

over $\mathbb{Q}$, where $X_0(N)$ is the modular curve classifying elliptic curves with a cyclic N-isogeny (cf. [BSD75]). We will assume π has minimal degree, and that it maps the cusp ∞ to the origin of E. Then π is determined up to sign.

Let K be an imaginary quadratic field where all primes dividing N are split, and choose a factorization $(N) = \mathfrak{n} \cdot \overline{\mathfrak{n}}$ with $\gcd(\mathfrak{n}, \overline{\mathfrak{n}}) = 1$ in the ideals of the ring of integers $\mathcal{O}_K$ of K. The isogeny of complex elliptic curves $\mathbb{C}/\mathcal{O}_K \to \mathbb{C}/\mathfrak{n}^{-1}$ defines a point $x \in X_0(N)(\mathbb{C})$. By the theory of complex multiplication, x is defined over the Hilbert class field H of K. We define

$$P = \mathrm{Tr}_{H/K}\, \pi(x) \qquad \text{in } E(K).$$

Then P is defined up to sign by E and K. For these facts and a general introduction to Heegner points, see [Gro84].

We will assume in the rest of this paper that P has infinite order in $E(K)$. By the main result in [GZ86], this condition holds precisely when $L'(1, E/K) \neq 0$. Since $E(K)$ is finitely generated, P can be divisible only by a finite number of primes ℓ

in $E(K)$. Kolyvagin showed that the group $E(K)$ has rank 1, and he completed the first ℓ-descent at the primes that do not divide P. More precisely, assume that

(a) ℓ is an odd prime that does not divide P in $E(K)$,
(b) the Galois representation

$$\rho : \mathrm{Gal}(\overline{\mathbb{Q}}/\mathbb{Q}) \to \mathrm{Aut}(E[\ell](\overline{\mathbb{Q}}))$$

is surjective.

Recall that the ℓ-Selmer group $\mathrm{Sel}(E/K,\ell)$ is the subgroup of classes in $\mathrm{H}^1(K, E[\ell])$ consisting of classes c whose local restrictions $c_v \in \mathrm{H}^1(K_v, E[\ell])$ lie in the images of the local Kummer maps $E(K_v) \to \mathrm{H}^1(K_v, E[\ell])$. Kolyvagin showed the following:

Theorem 1. *Assume that P has infinite order in $E(K)$ and that conditions (a) and (b) hold. Then the ℓ-Selmer group* $\mathrm{Sel}(E/K,\ell)$ *has dimension* 1 *over $\mathbb{Z}/\ell\mathbb{Z}$, and it is generated by the image of $P \in E(K)$ under the global Kummer map $E(K) \to$* $\mathrm{H}^1(K, E[\ell])$.

A proof is given in [Gro].

2.2 Level-raising and local divisibility of P

Now let $p \neq \ell$ be a prime that is inert in K. Then p does not divide N. Under the conditions of Theorem 1, we will consider the local divisibility of P by ℓ in $E(K_p)$. The following observation motivates the analysis below:

Lemma 1. *If the Heegner point P is not divisible by ℓ in $E(K_p)$, then*

(c) $\quad a_p \equiv \pm(p+1) \pmod{\ell}$.

More precisely, the sign in c) can be taken to be $-\epsilon$, where ϵ is the sign of the functional equation of $L(s, f)$.

Proof. Since P is not divisible by ℓ in $E(K_p)$, it has non-zero image in $E(K_p)/\ell E(K_p)$. As p does not divide $N\ell$, the latter group is isomorphic to $\mathcal{E}(\mathbb{F}_{p^2})/\ell\mathcal{E}(\mathbb{F}_{p^2})$, where $\mathcal{O}_p$ is the ring of integers of K_p and $\mathcal{E}/\mathcal{O}_p$ is the Néron model. Since $\mathcal{E}(\mathbb{F}_{p^2})$ is a finite group, the elliptic curve $\mathcal{E}/\mathbb{F}_{p^2}$ has a rational ℓ-torsion point. Therefore, Frob_p^2 acts on $E[\ell](\overline{\mathbb{Q}})$ with eigenvalues $(1, p^2)$. Since the determinant of Frob_p on $E[\ell](\overline{\mathbb{Q}})$ is p, its eigenvalues on $E[\ell](\overline{\mathbb{Q}})$ are $(\pm 1, \pm p)$. As the trace of Frob_p on $E[\ell](\overline{\mathbb{Q}})$ is equal to a_p modulo ℓ, this completes the proof of (c).

The more precise statement about the sign follows from the formula

$$\overline{P} = -\epsilon P + t,$$

where $t \in E(\mathbb{Q})$ is a torsion point (cf.[Gro84]): by assumption (b), the order of t is prime to ℓ, and so the image of P in $E(K_p)/\ell E(K_p)$ satisfies $\mathrm{Frod}_p(P) = -\epsilon P$. Consequently, $\mathcal{E}/\mathbb{F}_{p^2}$ has a rational ℓ-torsion point in the $-\epsilon$ eigenspace for Frob_p, and the eigenvalues of Frob_p are $-\epsilon$ and $-\epsilon p$. □

From now on, we assume that condition (c) holds, which is automatic when P is not divisible by ℓ in $E(K_p)$ by the lemma. Conditions (b) and (c) are the hypotheses of the level-raising Theorem 1 of [R]. This theorem produces a normalized newform g of level dividing Np that is p-new and that has trivial Nebentypus character. The theorem also constructs a place λ of $\overline{\mathbb{Q}}$ over ℓ, and g and λ have the property that for all rational primes $q \neq p$, one has

$$a_q(f) \equiv a_q(g) \pmod{\lambda}. \tag{2.1}$$

2.3 The Eichler–Shimura construction

Let $h \in S_2(\Gamma_0(M), \mathbb{C})$ be a normalized newform, and let $F = \mathbb{Q}(h)$ be the subfield of $\mathbb{C}$ generated by its Hecke eigenvalues. The Eichler–Shimura construction associates to h a pair (A, i), where A is an abelian variety up to isogeny over $\mathbb{Q}$ of dimension $[F : \mathbb{Q}]$, and where $i : F \to \mathrm{End}^0(A)$ is an isogeny action of F on A that is defined over $\mathbb{Q}$.

Recall the construction: corresponding to the newform h, one has an algebra homomorphism $\mathbb{Q} \otimes_{\mathbb{Z}} \mathbb{T} \to F$, where $\mathbb{T}$ is the Hecke algebra at level M. The object (A, i) is $\mathrm{Hom}_{\mathbb{Q}\otimes\mathbb{T}}(F, J_0(M))$, suitably interpreted, where $J_0(M)$ is considered as an abelian variety over $\mathbb{Q}$ up to isogeny with action of $\mathbb{Q} \otimes \mathbb{T}$ as endomorphisms. Since $J_0(M)$ has good reduction at finite places of $\mathbb{Q}$ not dividing M, so does A. Let ω be a finite place of F over the place w of $\mathbb{Q}$. The ω-adic Tate module $V_\omega(A, i) = F_\omega \otimes_{\mathbb{Q}_w} V_w(A)$ is a 2-dimensional vector space over F_ω, which admits an action of $\mathrm{Gal}(\overline{\mathbb{Q}}/\mathbb{Q})$ that is unramified away from M and w. The Eichler–Shimura relation implies that for any finite place v of $\mathbb{Q}$ prime to M and w, the characteristic polynomial of (arithmetic) Frobenius at v acting on $V_\omega(A, i)$ is $T^2 - a_v(h)T + q_v$, where the Hecke eigenvalue $a_v(h)$ is considered as an element of $\mathbb{Q}(h) = F \subset F_\omega$.

2.4 The theorem

Let (A, i) be the object associated by the Eichler–Shimura construction to the newform g provided by Ribet's level-raising theorem. There is an abelian variety over $\mathbb{Q}$ in the isogeny class A such that the maximal order $R \subset F = \mathbb{Q}(g)$ acts on A compatibly with i (cf.[Shi98], §7.1, Proposition 7). We fix one such abelian variety in the isogeny class, and we write (A, i) for it as well. Many constructions below depend on the action $i : R \to \mathrm{End}(A)$, but since it is fixed throughout, we will generally omit it from the notation.

Let I be the maximal ideal of R induced by the finite place λ of $\overline{\mathbb{Q}}$ provided by Ribet's theorem. The main result below compares the first I-descent on A/K with the first ℓ-descent on E/K. The I-descent structures on A are very similar to the ℓ-descent structures on an elliptic curve. Since I need not be principal, however, one must often tensor with powers of the R-modules I and I^{-1} to define maps whose ℓ-descent analogues would involve multiplying by powers of the generator ℓ of $(\ell) \subset \mathbb{Z}$. For example, for an R-module M, the R-linear analogue of the multiplication-by-ℓ endomorphism is the homomorphism $I \otimes_R M \to M$: if I is principal, then the choice of a generator x, which can also be viewed as an isomorphism $x : R \to I$, converts $I \otimes_R M \to M$ into the endomorphism of multiplication by x on M.

To complete the first I-descent, we define an I-Selmer group of A/K as follows: let $I^{-1} \otimes_R A$ be the K-scheme that represents the functor $T \mapsto I^{-1} \otimes_R A(T) = \mathrm{Hom}_R(I, A(T))$. We show that this functor is representable, and we give more details on the abstract formalism of I-descent in the appendix below. Then $I^{-1} \otimes_R A$ is an abelian variety of the same dimension as A/K, on which R acts as endomorphisms. For example, if $A(\mathbb{C}) = \mathbb{C}^{[F:\mathbb{Q}]}/\Lambda$, then the lattice Λ has a natural R-module structure, and $I^{-1} \otimes_R A(\mathbb{C}) = \mathbb{C}^{[F:\mathbb{Q}]}/I^{-1} \otimes_R \Lambda$. (The embedding of $I^{-1} \otimes_R \Lambda$ as a lattice in $\mathbb{C}^{[F:\mathbb{Q}]}$ is the unique extension of the embedding of its finite-index subgroup Λ.)

The inclusion $R \to I^{-1}$ induces an isogeny $A \to I^{-1} \otimes_R A$ defined over K with kernel $A[I]$, the group scheme of I-torsion sections of A. We thus have the exact sequence

$$0 \longrightarrow A[I] \longrightarrow A \longrightarrow I^{-1} \otimes_R A \longrightarrow 0$$

of group schemes over $\mathrm{Spec}(K)$. Passing to Galois cohomology, we find a global Kummer (boundary) map

$$I^{-1} \otimes_R A(K) \to \mathrm{H}^1(K, A[I]).$$

In the familiar situation of ℓ-descent, the boundary map $E(K) \to \mathrm{H}^1(K, E[\ell])$ maps a point $x \in E(K)$ to the $E[\ell]$-torsor composed of the points $\{\ell^{-1}x\} \subset E(\overline{K})$. In the I-descent formalism, we would replace $x \in E(K)$ with $(\ell^{-1}) \otimes x \in (\ell)^{-1} \otimes E(K)$. The image of $(\ell^{-1}) \otimes x$ under the Kummer map is the fiber over this point of the isogeny $E \to (\ell)^{-1} \otimes E$.

For each place v of K, there are analogous local Kummer maps

$$I^{-1} \otimes_R A(K_v) \to \mathrm{H}^1(K_v, A[I]).$$

The *I-Selmer group of* (A, i), denoted $\mathrm{Sel}(A/K, I)$, is the group of classes $c \in \mathrm{H}^1(K, A[I])$ such that the restriction $c_v \in \mathrm{H}^1(K_v, A[I])$ is in the image of $I^{-1} \otimes_R A(K_v)$ for all places v of K. Evidently the global Kummer map factors through $\mathrm{Sel}(A/K, I) \subset \mathrm{H}^1(K, A[I])$.

Just as for standard ℓ-descent, one can show that $\mathrm{Sel}(A/K, I)$ is a finite group and hence a finite-dimensional R/I-vector space. The kernel of the global Kummer map $I^{-1} \otimes_R A(K) \to \mathrm{Sel}(A/K, I)$ is the image of the natural map

$$R \otimes_R A(K) \to I^{-1} \otimes_R A(K),$$

and so the image of the Kummer map is $I^{-1}/R \otimes_R A(K)$ or, equivalently,

$$I^{-1} \otimes_R (R/I \otimes_R A(K)).$$

The R/I-dimension of $\mathrm{Sel}(A/K, I)$ is thus an upper bound on the rank of $A(K)$ as R-module, just as the $\mathbb{Z}/\ell\mathbb{Z}$-dimension of $\mathrm{Sel}(E/K, \ell)$ is an upper bound on the rank of $E(K)$ as $\mathbb{Z}$-module.

Let us impose the following additional assumptions on p, N, and E:

(d) ℓ is prime to N,

(e) For each prime q dividing N, the dimension of the q-inertia invariants of the modulo-ℓ Galois representation

$$\rho : \mathrm{Gal}(\overline{\mathbb{Q}}/\mathbb{Q}) \to \mathrm{Aut}(E[\ell](\overline{\mathbb{Q}}))$$

is 1, if E has multiplicative reduction at q, and is 0, if E has additive reduction at q,

(f) $p^2 \not\equiv 1 \pmod{\ell}$.

Since ℓ is prime to N, condition (e) is equivalent to the statement that the conductor of ρ in the sense of [Ser87] is equal to the conductor N of $E/\mathbb{Q}$. By level-lowering theorems (cf. [Dia95]), condition (e) means that f is a form of minimal level among those giving rise to the modulo-ℓ representation ρ. It also implies that the form g constructed by level raising has level Np.

We will deduce the following result about the I-descent on A/K from Kolyvagin's result stated in Theorem 1:

Theorem 2. *Assume that $P \in E(K)$ has infinite order and that conditions (a)–(f) hold. Then*

$$\dim_{R/I} \mathrm{Sel}(A/K, I) = 0, \text{ if } P \text{ is not divisible by } \ell \text{ in } E(K_p), \text{ and}$$
$$\dim_{R/I} \mathrm{Sel}(A/K, I) = 2, \text{ if } P \text{ is divisible by } \ell \text{ in } E(K_p).$$

The I-adic Tate module

$$T_I(A, i) = \varprojlim (I^{\otimes n} \otimes_R A[I^n](\overline{\mathbb{Q}}))$$

is a lattice in $V_\lambda(A, i)$ and is thus free of rank 2 over the I-adic completion of R. Therefore, $R/I \otimes_R T_I(A, i) = I \otimes_R A[I](\overline{\mathbb{Q}})$ and hence $A[I](\overline{\mathbb{Q}})$ is

2-dimensional over R/I. By the Eichler–Shimura relation and (2.1), the characteristic polynomials of prime-to-$Np\ell$ Frobenius elements on $R/I \otimes E[\ell](\overline{\mathbb{Q}})$ and $A[I](\overline{\mathbb{Q}})$ agree. Thus, by the Brauer–Nesbitt principle, these Galois modules have isomorphic semi-simplifications. By assumption (b), the Galois module $R/I \otimes E[\ell](\overline{\mathbb{Q}})$ is irreducible, and so the two $(R/I)[\mathrm{Gal}(\overline{\mathbb{Q}}/\mathbb{Q})]$-modules are isomorphic. We choose and fix such an isomorphism for the rest of the paper.

The plan for proving Theorem 2 is to compare the local conditions defining $\mathrm{Sel}(A/K, I) \subset \mathrm{H}^1(K, A[I])$ and

$$R/I \otimes \mathrm{Sel}(E/K, \ell) \subset R/I \otimes \mathrm{H}^1(K, E[\ell]) = \mathrm{H}^1(K, R/I \otimes E[\ell]),$$

by means of the fixed isomorphism $R/I \otimes E[\ell]/\mathbb{Q} = A[I]/\mathbb{Q}$. Using assumptions (a)–(f), we describe these local conditions entirely in terms of the Galois module

$$A[I](\overline{\mathbb{Q}}) = R/I \otimes E[\ell](\overline{\mathbb{Q}}).$$

The local conditions agree at all places v of K except at $v = p$, where the two conditions are transverse. By combining this local analysis with a global parity lemma, proved in §5.3, we deduce Theorem 2 from Theorem 1. In §4 we study the local conditions defining the Selmer groups at places $v \neq p$ of K, and in §5 we study the conditions at $v = p$ and use the parity lemma to prove the theorem. In §6 below, we make explicit the compatibility with the parity predictions of the conjecture of Birch and Swinnerton-Dyer.

3 Néron models of abelian varieties with real multiplication

This section contains preliminary remarks that will be applied in Sections 4 and 5 to Néron models of E and A in order to describe the images of the local Kummer maps entirely in terms of the Galois modules underlying $E[\ell]/\mathbb{Q}$ and $A[I]/\mathbb{Q}$.

3.1 Semi-stable case

Let L be the quotient field of a Henselian discrete valuation ring $\mathcal{O}_L$ with residue field k_L, and let B/L be an abelian variety. Let F_0 be a number field with maximal order R_0. Assume that the dimension of B/L is $[F_0 : \mathbb{Q}]$. Let $j : R_0 \to \mathrm{End}(B/L)$ be a faithful action of R_0. In what follows, we will use $F_0 = \mathbb{Q}$ and $F_0 = F$.

Let $\mathcal{B}/\mathcal{O}_L$ be the Néron model of B/L. We consider first the case when B/L has semi-stable reduction. This condition means that the identity component $\mathcal{B}^0/k_L$

of the special fiber of the Néron model is an extension of an abelian variety by a torus T/k_L. Let $\overline{k_L}$ be a separable closure of k_L. The torus T/k_L is determined by its geometric character lattice $X^*(T/\overline{k_L})$, which is a free $\mathbb{Z}$-module of rank $\dim(T)$, with the induced action of $\mathrm{Gal}(\overline{k_L}/k_L)$ on this lattice. The semi-stable abelian variety B/L has *good reduction*, if the torus component of the special fiber is trivial. In this case, the special fiber is connected and is an abelian variety, and so $\mathcal{B}/\mathcal{O}_L$ is an abelian scheme. If the abelian-variety component of $\mathcal{B}^0/k_L$ is trivial, then B/L has *purely toric reduction*.

Lemma 2. *If B/L has semi-stable reduction, then B/L has either good reduction or purely toric reduction. In the purely toric-reduction case, the functorial action of R_0 on $X^*(T/\overline{k_L})$ makes this lattice an invertible R_0-module.*

Proof. By functoriality of the Néron model, the action j induces a (unital) ring homomorphism $R_0 \to \mathrm{End}(T/\overline{k_L}) = \mathrm{End}(X^*(T/\overline{k_L}))$. Tensoring with $\mathbb{Q}$, we find an F_0-vector-space structure on $\mathbb{Q} \otimes X^*(T/\overline{k_L})$. Therefore, the dimension of T/k_L is a multiple of $[F_0 : \mathbb{Q}] = \dim(B)$. Since $\dim(T) \leq \dim(B/L) = \dim(\mathcal{B}^0/k_L)$, we see that either $T = 0$ or $T = \mathcal{B}^0/k_L$. In the first case, B/L has good reduction. In the second case, B/L has purely toric reduction, and $X^*(T/\overline{k_L})$ is an R_0-lattice in the 1-dimensional F_0-vector space $\mathbb{Q} \otimes X^*(T/\overline{k_L})$. Thus the character lattice is an invertible R_0-module, since R_0 is the maximal order of F_0. □

Consider the purely-toric reduction case of the lemma. Since T is split over $\overline{k_L}$, we have the natural isomorphism

$$T/\overline{k_L} = \underline{\mathrm{Hom}}(X^*(T/\overline{k_L}), \mathbb{G}_m),$$

which is a functorial expression of the fact that $T/\overline{k_L}$ is isomorphic to a product of copies of the multiplicative group $\mathbb{G}_m$ indexed by any basis of $X^*(T/\overline{k_L})$. Thus for any ideal I_0 of R_0, we have

$$T[I_0]/\overline{k_L} = \underline{\mathrm{Hom}}(X^*(T/\overline{k_L}), \mathbb{G}_m)[I_0] = \underline{\mathrm{Hom}}(R_0/I_0 \otimes_{R_0} X^*(T/\overline{k_L}), \mathbb{G}_m).$$

Therefore, we have

$$T[I_0](\overline{k_L}) = \mathrm{Hom}_{\mathbb{Z}}(R_0/I_0 \otimes_{R_0} X^*(T/\overline{k_L}), \overline{k_L}^{\times}).$$

As an abstract group, $\overline{k_L}^{\times}$ is isomorphic to $\prod_{q \neq q_0} \mathbb{Q}_q/\mathbb{Z}_q$, where q runs over primes not equal to the characteristic q_0 of k_L. Consequently, if I_0 is a maximal ideal such that the order of R_0/I_0 is invertible in k_L, then $T[I_0](\overline{k_L})$ is 1-dimensional over R_0/I_0. We state this fact as a lemma for later reference.

Lemma 3. *Assume that B/L has purely toric reduction. Let I_0 be a maximal ideal of R_0 such that the order of R_0/I_0 is invertible in k_L. Then $\mathcal{B}^0[I_0](\overline{k_L})$ is 1-dimensional over R_0/I_0.*

3.2 Component groups in the general case

We now consider abelian varieties B/L with endomorphisms $j : R_0 \to \mathrm{End}(B)$ as above, but we allow arbitrary (not necessarily semi-stable) reduction. Let $\mathcal{B}/\mathcal{O}_L$ be the Néron model of B/L, and let $\mathcal{B}^0/k_L$ be the connected component of the identity of $\mathcal{B}/k_L$. In the general case $\mathcal{B}^0/k_L$ is a successive extension of an abelian variety, a torus, and a connected unipotent group. In this section, we are concerned with the component group scheme $\Phi = \mathcal{B}/\mathcal{B}^0$ of the special fiber. It is a finite, étale R_0-module scheme over k_L. Let $\mathcal{B}^0/\mathcal{O}_L$ be the smooth group scheme whose generic fiber is B/L and whose special fiber is $\mathcal{B}^0/k_L$, i.e., $\mathcal{B}^0/\mathcal{O}_L$ is the complement in $\mathcal{B}/\mathcal{O}_L$ of the non-identity components of $\mathcal{B}/k_L$.

Let $I_0 \subset R_0$ be a maximal ideal such that the order of R_0/I_0 is invertible in $\mathcal{O}_L$ and thus in L and k_L. Let us see how the group $\Phi[I_0](\overline{k_L})$ appears in the Galois module $B[I_0]/L$. Consider the R_0-module extension of $\Phi[I_0](\overline{k_L})$ by $\mathcal{B}^0(\overline{k_L})$ obtained by restricting

$$0 \to \mathcal{B}^0(\overline{k_L}) \to \mathcal{B}(\overline{k_L}) \to \Phi(\overline{k_L}) \to 0$$

to $\Phi[I_0](\overline{k_L})$. We claim that this extension of R_0-modules splits. Since the group $\Phi[I_0](\overline{k_L})$ is a direct sum of copies of R_0/I_0 as R_0-module, it suffices to show that

$$\mathrm{Ext}^1_{R_0}(R_0/I_0, \mathcal{B}^0(\overline{k_L})) = 0. \tag{3.1}$$

To see this vanishing, we consider the homomorphism $\mathcal{B} \to I_0^{-1} \otimes_{R_0} \mathcal{B}$ of smooth group schemes over $\mathcal{O}_L$, as in the formalism of I_0-descent discussed in the appendix. On the relative Lie algebras, this morphism is the natural homomorphism of free $\mathcal{O}_L$-modules

$$\mathrm{Lie}(\mathcal{B}/\mathcal{O}_L) \to \mathrm{Lie}(I_0^{-1} \otimes_{R_0} \mathcal{B}/\mathcal{O}_L) = I_0^{-1} \otimes_{R_0} \mathrm{Lie}(\mathcal{B}/\mathcal{O}_L).$$

The kernel and cokernel are annihilated by I_0; since I_0 contains the order of R_0/I_0, which is a unit in $\mathcal{O}_L$, the map is an isomorphism. Consequently, the homomorphism of smooth group schemes $\mathcal{B} \to I_0^{-1} \otimes_{R_0} \mathcal{B}$ over $\mathcal{O}_L$ is étale. Since the geometric fibers of $\mathcal{B}^0$ and $I_0^{-1} \otimes_{R_0} \mathcal{B}^0$ over $\mathcal{O}_L$ are connected, the homomorphism $\mathcal{B}^0 \to I_0^{-1} \otimes_{R_0} \mathcal{B}^0$ is thus étale and surjective. In particular, $\mathcal{B}^0(\overline{k_L}) \to I_0^{-1} \otimes_{R_0} \mathcal{B}^0(\overline{k_L})$ is surjective.

Consider the projective resolution

$$0 \longrightarrow I_0 \longrightarrow R_0 \longrightarrow R_0/I_0 \longrightarrow 0$$

of the R_0-module R_0/I_0. From the long exact Ext_{R_0}-sequence with coefficients $\mathcal{B}^0(\overline{k_L})$, we find that $\mathrm{Ext}^1_{R_0}(R_0/I_0, \mathcal{B}^0(\overline{k_L}))$ is the cokernel of

$$\mathrm{Hom}_{R_0}(R_0, \mathcal{B}^0(\overline{k_L})) \longrightarrow \mathrm{Hom}_{R_0}(I_0, \mathcal{B}^0(\overline{k_L})).$$

Since the map

$$\mathcal{B}^0(\overline{k_L}) = \mathrm{Hom}_{R_0}(R_0, \mathcal{B}^0(\overline{k_L})) \to \mathrm{Hom}_{R_0}(I_0, \mathcal{B}^0(\overline{k_L})) = I_0^{-1} \otimes_{R_0} \mathcal{B}^0(\overline{k_L})$$

is surjective, we obtain the desired vanishing (3.1).

If we choose a splitting, then we find an R_0/I_0-module isomorphism

$$\mathcal{B}[I_0](\overline{k_L}) = \mathcal{B}^0[I_0](\overline{k_L}) \oplus \Phi[I_0](\overline{k_L}). \tag{3.2}$$

Let $\widetilde{L}$ be a maximal unramified extension of L with residue field $\overline{k}_L$, and let $\mathcal{O}_{\widetilde{L}}$ be the valuation ring of $\widetilde{L}$. By the Néron property, we have $B[I_0](\widetilde{L}) = \mathcal{B}[I_0](\mathcal{O}_{\widetilde{L}})$. Since $\mathcal{B}[I_0]/\mathcal{O}_L$ is étale (as the kernel of an étale homomorphism $\mathcal{B} \to I_0^{-1} \otimes_{R_0} \mathcal{B}$ over $\mathcal{O}_L$) and $\mathcal{O}_L$ is Henselian, the reduction map $\mathcal{B}[I_0](\mathcal{O}_{\widetilde{L}}) \to \mathcal{B}[I_0](\overline{k_L})$ is an isomorphism. Therefore $B[I_0](\widetilde{L}) = \mathcal{B}[I_0](\overline{k_L})$. The R_0/I_0-module $\mathcal{B}[I_0](\widetilde{L})$ is the space of inertia invariants in $\mathcal{B}[I_0](\overline{L})$, where $\overline{L}$ is a separable closure of L containing $\widetilde{L}$. In summary, we have

Lemma 4. *If the order of R_0/I_0 is invertible in $\mathcal{O}_L$, then as R_0/I_0-module, the space of inertia invariants in $B[I_0](\overline{L})$ is $\mathcal{B}^0[I_0](\overline{k_L}) \oplus \Phi[I_0](\overline{k_L})$.*

3.3 Reduction of E and A at finite places of K

We now describe some aspects of the reduction of E and A at primes q of $\mathbb{Q}$ and at finite places v of K using the results of this section and hypotheses (a)–(f). Since N is the conductor of $E/\mathbb{Q}$, if q and v are prime to N, then $E/\mathbb{Q}$ has good reduction at q, and E/K has good reduction at v. Let v be a place of K dividing N and let q be the prime of $\mathbb{Q}$ lying under v. The elliptic curve $E/\mathbb{Q}$ then has either multiplicative or additive reduction at q. Since, by assumption, q is unramified and splits in K, we have $\mathbb{Q}_q = K_v$, and the elliptic curve E/K has the same reduction at v as $E/\mathbb{Q}$ at q. If $E/\mathbb{Q}$ has additive reduction, then by condition (e), there are no non-zero inertia invariants in $E[\ell](\overline{\mathbb{Q}_q})$. Thus by Lemma 4, since ℓ is prime to N, the component group of the reduction at q has no ℓ-torsion. If $E/\mathbb{Q}$ has multiplicative reduction at q, then the inertia invariants in $E[\ell](\overline{\mathbb{Q}_q})$ are 1-dimensional over $\mathbb{Z}/\ell\mathbb{Z}$ by assumption (e). By Lemma 3 and 4, all of these inertia invariants come from the torus part of the reduction, and the component group of the reduction at q again has no non-zero ℓ-torsion.

If q and v are prime to Np, then the abelian variety $A/\mathbb{Q}$ has good reduction at q, and A/K has good reduction at v, since $A/\mathbb{Q}$ is an isogeny factor of $J_0(Np)/\mathbb{Q}$, which has good reduction away from Np. It follows from the construction of the semi-stable model of $J_0(Np)/\mathbb{Q}_p$ over $\mathbb{Z}_p$ (cf. [DR73]) that the new quotient of $J_0(Np)/\mathbb{Q}$ has purely toric reduction at p. Since the eigenform g used to construct $A/\mathbb{Q}$ is new at p, the abelian variety $A/\mathbb{Q}$ thus has purely toric reduction at p. Therefore, A/K has purely toric reduction at places v of K dividing p.

By condition (e), the Galois module $E[\ell](\overline{\mathbb{Q}})$ is ramified at primes q dividing N. Since $R/I \otimes E[\ell](\overline{\mathbb{Q}}) = A[I](\overline{\mathbb{Q}})$ as Galois modules, the Galois module $A[I](\overline{\mathbb{Q}})$ is also ramified at primes q dividing N. Therefore, $A/\mathbb{Q}_q$ has bad reduction at all primes q dividing N. As above, if v is a place of K dividing N and lying over the rational prime q, then q is unramified and split in K. Therefore $K_v = \mathbb{Q}_q$, and so the reduction of A/K at v is the same as the reduction of $A/\mathbb{Q}$ at q. If q divides N but q^2 does not divide N, then $A/\mathbb{Q}$ has semi-stable reduction at q (and hence v), since it is an isogeny factor of $J_0(Np)/\mathbb{Q}$, which has semi-stable reduction at q. Since $A/\mathbb{Q}$ has bad reduction at q, it has purely toric reduction by Lemma 2. The inertia invariants in $A[I](\overline{\mathbb{Q}_q})$ are 1-dimensional over R/I by assumption (e) and the isomorphism $R/I \otimes E[\ell](\overline{\mathbb{Q}}) = A[I](\overline{\mathbb{Q}})$. Therefore, by Lemmas 3 and 4, the component group of the reduction of $A/\mathbb{Q}$ at q has no I-torsion. On the other hand, if q^2 divides N, then $E/\mathbb{Q}$ has additive reduction at q, the inertia invariants in $A[I](\overline{\mathbb{Q}_q})$ are 0-dimensional over R/I by assumption (e) and the isomorphism $R/I \otimes E[\ell](\overline{\mathbb{Q}}) = A[I](\overline{\mathbb{Q}})$. Therefore, by Lemma 4, the component group of the reduction of $A/\mathbb{Q}$ at q has no I-torsion.

4 Proof of Theorem 2: local conditions at $v \neq p$

To prove Theorem 2, we will compare

$$R/I \otimes \mathrm{Sel}(E/K, \ell) \subset R/I \otimes \mathrm{H}^1(K, E[\ell]) = \mathrm{H}^1(K, R/I \otimes E[\ell])$$

with $\mathrm{Sel}(A/K, I) \subset \mathrm{H}^1(K, A[I])$. Here we identify $\mathrm{H}^1(K, R/I \otimes E[\ell])$ with $\mathrm{H}^1(K, A[I])$ using the fixed isomorphism $R/I \otimes E[\ell] = A[I]$ over $\mathbb{Q}$ from §2.4. We compare these two Selmer groups by comparing the local conditions in $\mathrm{H}^1(K_v, A[I])$ that define them. Let $L_v \subset \mathrm{H}^1(K_v, A[I])$ be the R/I-span of the image of

$$E(K_v) \to \mathrm{H}^1(K_v, E[\ell]) \subset \mathrm{H}^1(K_v, A[I]).$$

Then $R/I \otimes \mathrm{Sel}(E/K, \ell)$ consists of the classes $c \in \mathrm{H}^1(K, A[I])$ such that each restriction $c_v \in \mathrm{H}^1(K_v, A[I])$ belongs to L_v. Let $L'_v \subset \mathrm{H}^1(K_v, A[I])$ be the image of the local Kummer map $I^{-1} \otimes_R A(K_v) \to \mathrm{H}^1(K_v, A[I])$, so that $\mathrm{Sel}(A/K, I)$ is the set of classes $c \in \mathrm{H}^1(K, A[I])$ such that $c_v \in L'_v$ for all places v of K.

The remainder of the present section is devoted to proving

Lemma 5. *For $v \neq p$, we have $L_v = L'_v$.*

We prove the lemma by deducing from assumptions (a)–(f) descriptions of L_v and L'_v entirely in terms of the Galois modules $E[\ell]/\mathbb{Q}$ and $A[I]/\mathbb{Q}$. The fixed isomorphism $R/I \otimes E[\ell] = A[I]$ over $\mathbb{Q}$ then allows us to identify the local conditions as in the lemma. In §5 below, we complete the local comparison by analyzing L_p and L'_p, which is the only place at which the defining local conditions for the two Selmer groups differ. With the local comparison in hand, we use a parity lemma based on local and global duality theory in Galois cohomology to deduce Theorem 2 from Theorem 1.

4.1 Types of local conditions

Let $\mathcal{O}_L$ be a discrete valuation ring with quotient field L and residue field k_L. Let V be a locally constant constructible sheaf of abelian groups on the small étale site of $\mathrm{Spec}(L)$. The data of V is carried equivalently by the finite-order $\mathbb{Z}[\mathrm{Gal}(\overline{L}/L)]$-module $V(\overline{L})$, where $\overline{L}/L$ is a separable closure.

The subspace of $\mathrm{H}^1(L, V)$ composed of classes that split over an unramified extension of L is denoted $\mathrm{H}^1_{\mathrm{unr}}(L, V)$. Alternatively, if $j : \mathrm{Spec}(L) \to \mathrm{Spec}(\mathcal{O}_L)$ denotes the inclusion, the unramified classes are

$$\mathrm{H}^1_{\text{ét}}(\mathrm{Spec}(\mathcal{O}_L), j_* V) \subset \mathrm{H}^1_{\text{ét}}(\mathrm{Spec}(L), V) = \mathrm{H}^1(L, V).$$

From either description, it is clear that formation of unramified classes is functorial in V.

Suppose that there is a finite, free group scheme $G/\mathrm{Spec}(\mathcal{O}_L)$ whose restriction to $\mathrm{Spec}(L)$ represents V. One then has the subspace of flat classes valued in $G/\mathcal{O}_L$

$$\mathrm{H}^1_{\mathrm{fl}}(\mathrm{Spec}(\mathcal{O}_L), G) \subset \mathrm{H}^1_{\mathrm{fl}}(\mathrm{Spec}(L), G) = \mathrm{H}^1(L, V).$$

The formation of such a subspace is functorial in the flat model G over $\mathrm{Spec}(\mathcal{O}_L)$.

4.2 Identifying local conditions for I_0-descent on abelian varieties

We return briefly to the general notation of §3: $\mathcal{O}_L$ is a Henselian discrete valuation ring with quotient field L and residue field k_L; one has an abelian variety B/L, equipped with an action of the ring of integers R_0 of a number field F_0 such that $[F_0 : \mathbb{Q}] = \dim(B/L)$. Let $\mathcal{B}/\mathcal{O}_L$ be the Néron model, and let Φ/k_L be the component group scheme of $\mathcal{B}/k_L$. Let $I_0 \subset R_0$ be a maximal ideal such that the order of R_0/I_0 is invertible in L. Then the homomorphism $B \to I_0^{-1} \otimes_{R_0} B$ is an étale isogeny with kernel $B[I_0]$, and we have the Kummer map $I_0^{-1} \otimes_{R_0} B(L) \to \mathrm{H}^1(L, B[I_0])$. The following two lemmas identify the image of the Kummer map under certain restrictions on the reduction of B/L. The cases with $R_0 = \mathbb{Z}$ are entirely standard facts (cf. [Maz72], for instance).

Lemma 6. *If the order of R_0/I_0 is invertible in $\mathcal{O}_L$ and $\Phi[I_0] = 0$, then the image of the Kummer map in $\mathrm{H}^1(L, B[I_0])$ is $\mathrm{H}^1_{\mathrm{unr}}(L, B[I_0])$.*

Proof. Since the order of R_0/I_0 is invertible in $\mathcal{O}_L$, as in §3.2, we find that the homomorphism $\mathcal{B} \to I_0^{-1} \otimes_{R_0} \mathcal{B}$ is étale and that $\mathcal{B}^0 \to I_0^{-1} \otimes_{R_0} \mathcal{B}^0$ is surjective and étale. Since $\Phi[I_0] = 0$ and Φ/k_L is finite étale, the natural map $\Phi \to I_0^{-1} \otimes_{R_0} \Phi$ is an isomorphism, and so $\mathcal{B} \to I_0^{-1} \otimes \mathcal{B}$ is surjective and étale. One thus has the Kummer maps for B/L and $\mathcal{B}/\mathcal{O}_L$, as discussed in the appendix, which are connected by restriction maps in étale cohomology:

$$\begin{array}{ccc} I_0^{-1} \otimes_{R_0} \mathcal{B}(\mathcal{O}_L) & \longrightarrow & \mathrm{H}^1_{\text{ét}}(\mathrm{Spec}(\mathcal{O}_L), \mathcal{B}[I_0]) \\ \downarrow & & \downarrow \\ I_0^{-1} \otimes_{R_0} B(L) & \longrightarrow & \mathrm{H}^1_{\text{ét}}(\mathrm{Spec}(L), B[I_0]). \end{array} \tag{4.1}$$

By the Néron mapping property we have $\mathcal{B}(\mathcal{O}_L) = \mathcal{B}(L)$, and so the left vertical arrow in an isomorphism. Furthermore, the top arrow is surjective, which one sees as follows:

The cokernel of the top arrow is the kernel of

$$\mathrm{H}^1_{\text{ét}}(\mathrm{Spec}(\mathcal{O}_L), \mathcal{B}) \longrightarrow \mathrm{H}^1_{\text{ét}}(\mathrm{Spec}(\mathcal{O}_L), I_0^{-1} \otimes_R \mathcal{B}),$$

and so to see surjectivity, it suffices to check that $\mathrm{H}^1_{\text{ét}}(\mathrm{Spec}(\mathcal{O}_L), \mathcal{B}^0) = 0$ and that $\Phi \to I_0^{-1} \otimes_{R_0} \Phi$ is an isomorphism. The second fact follows from $\Phi[I_0] = 0$, as noted above. To see the first fact, recall that a class in $\mathrm{H}^1(\mathrm{Spec}(\mathcal{O}_L), \mathcal{B}^0)$ can be represented by a right torsor under the étale sheaf represented by $\mathcal{B}^0$. Since the base $\mathcal{O}_L$ is a discrete valuation ring and $\mathcal{B}^0$ is smooth over $\mathcal{O}_L$, this torsor is representable by a scheme over $\mathrm{Spec}(\mathcal{O}_L)$ by a theorem of Raynaud (cf. [Mil80], Chapter III, Theorem 4.3). It suffices therefore to show that any scheme $\mathcal{P}$ over $\mathrm{Spec}(\mathcal{O}_L)$ that is a right $\mathcal{B}^0$-torsor has a section. Since $\mathcal{B}^0/k_L$ is connected, Lang's theorem (cf. [Lan56], Theorem 2) implies that $\mathcal{P}(k_L)$ is non-empty. As $\mathcal{B}^0/\mathrm{Spec}(\mathcal{O}_L)$ and hence $\mathcal{P}/\mathrm{Spec}(\mathcal{O}_L)$ are smooth and $\mathcal{O}_L$ is Henselian, the map $\mathcal{P}(\mathcal{O}_L) \to \mathcal{P}(k_L)$ is surjective. Consequently, the element of $\mathcal{P}(k_L)$ provided by Lang's theorem lifts to a section over $\mathcal{O}_L$.

Since the left vertical arrow in (4.1) is an isomorphism and the top arrow is a surjection, the image of the Kummer map in $\mathrm{H}^1(L, B[I_0])$ is $\mathrm{H}^1_{\text{ét}}(\mathrm{Spec}(\mathcal{O}_L), \mathcal{B}[I_0])$. The Néron mapping property identifies $j_* B = \mathcal{B}$ as sheaves on the small étale site of $\mathrm{Spec}(\mathcal{O}_L)$. Thus

$$\mathcal{B}[I_0] = (j_* B)[I_0] = j_*(B[I_0]),$$

and $\mathrm{H}^1_{\text{ét}}(\mathrm{Spec}(\mathcal{O}_L), \mathcal{B}[I_0]) = \mathrm{H}^1_{\text{unr}}(L, B[I_0])$. □

Alternatively, assume that the order of R_0/I_0 is invertible only in L, but consider only B/L with good reduction. The Néron model $\mathcal{B}/\mathcal{O}_L$ is then an abelian scheme.

Lemma 7. *If B/L has good reduction, then $\mathcal{B}[I_0]/\mathcal{O}_L$ is finite and locally free. Furthermore, the image of the Kummer map $I_0^{-1} \otimes_{R_0} B(L) \to \mathrm{H}^1(L, B[I_0])$ is $\mathrm{H}^1_{\text{fl}}(\mathrm{Spec}(\mathcal{O}_L), \mathcal{B}[I_0])$.*

Proof. As explained in the appendix, since $\mathcal{B}/\mathcal{O}_L$ is an abelian scheme, the homomorphism $\mathcal{B} \to I_0^{-1} \otimes_{R_0} \mathcal{B}$ is flat and surjective. Furthermore, the kernel

$\mathcal{B}[I_0]/\mathcal{O}_L$ is finite and locally free. To complete the proof othe lemma, one can apply the same argument as in Lemma 6, replacing étale cohomology with flat cohomology. □

4.3 *Proof of Lemma 5*

For v the infinite place of K, we have $K_v = \mathbb{C}$, and so $\mathrm{H}^1(K_v, A[I]) = 0$ and $L_v = L'_v = 0$.

Next consider a finite place v of K prime to $p\ell$. Let $\mathcal{E}/\mathcal{O}_v$ and $\mathcal{A}/\mathcal{O}_v$ be the Néron models of E/K_v and A/K_v, respectively. Let Φ_E and Φ_A be the component group schemes of the special fibers of $\mathcal{E}/\mathcal{O}_v$ and $\mathcal{A}/\mathcal{O}_v$, respectively. By the discussion in §3.3, we have $\Phi_E[\ell] = 0$ and $\Phi_A[I] = 0$. Therefore, Lemma 6 implies that

$$L_v = R/I \otimes \mathrm{H}^1_{\mathrm{unr}}(K_v, E[\ell]) \quad \text{and} \quad L'_v = \mathrm{H}^1_{\mathrm{unr}}(K_v, A[I]).$$

Consequently, $L_v = L'_v$.

Finally, let v be a place of K dividing ℓ. Let $\mathcal{E}/\mathbb{Z}_\ell$ and $\mathcal{A}/\mathbb{Z}_\ell$ be the Néron models of $E/\mathbb{Q}_\ell$ and $A/\mathbb{Q}_\ell$, respectively. Since ℓ is prime to Np, these models are abelian schemes. Furthermore, the base changes $\mathcal{E}/\mathcal{O}_v$ and $\mathcal{A}/\mathcal{O}_v$ are the Néron models of E/K_v and A/K_v, respectively. The R/I-vector schemes $R/I \otimes_{\mathbb{Z}} \mathcal{E}[\ell]/\mathbb{Z}_\ell$ and $\mathcal{A}[I]/\mathbb{Z}_\ell$ are finite, free models of $A[I]/\mathbb{Q}_\ell$. Since $\ell > 2$, by Theorem 3.3.3 of [Ray74], the identification of their generic-fiber Galois modules extends uniquely to an isomorphism $R/I \otimes \mathcal{E}[\ell] = \mathcal{A}[I]$ over $\mathbb{Z}_\ell$. By Lemma 7, we have

$$L_v = R/I \otimes \mathrm{H}^1_{\mathrm{fl}}(\mathrm{Spec}(\mathcal{O}_v), \mathcal{E}[\ell]) \quad \text{and} \quad L'_v = \mathrm{H}^1_{\mathrm{fl}}(\mathrm{Spec}(\mathcal{O}_v), \mathcal{A}[I]).$$

Therefore, $L_v = L'_v$.

5 Proof of Theorem 2: local condition at $v = p$ and the parity lemma

5.1 *The conditions at $v = p$*

The $\mathrm{Gal}(\overline{K_p}/K_p)$-module $E[\ell](\overline{K_p})$ is unramified at p, since p is prime to the conductor N of $E/\mathbb{Q}$. By assumption (c), the eigenvalues of Frob_{p^2} on $E[\ell](\overline{K_p})$ are 1 and p^2. By assumption (f), we have $p^2 \not\equiv 1 \pmod{\ell}$, and so $E[\ell](\overline{K_p})$ splits as a sum of the Frob_{p^2}-eigenspaces for the eigenvalues 1 and p^2. Therefore, we have an isomorphism of R/I-vector schemes (or $(R/I)[\mathrm{Gal}(\overline{K_p}/K_p)]$-modules):

$$A[I]/K_p = R/I \oplus R/I(1),$$

where $R/I(1)$ is the R/I-vector scheme $R/I \otimes \mu_\ell$.

Lemma 8. *The spaces* $\mathrm{H}^1(K_p, R/I)$ *and* $\mathrm{H}^1(K_p, R/I(1))$ *are each* 1*-dimensional over* R/I. *Furthermore,*

$$L_p = \mathrm{H}^1(K_p, R/I) \quad \textit{and} \quad L'_p = \mathrm{H}^1(K_p, R/I(1)).$$

Proof. We have $\mathrm{H}^1(K_p, R/I) = \mathrm{Hom}(\mathrm{Gal}(\overline{K_p}/K_p), R/I)$. Since R/I is abstractly a sum of copies of $\mathbb{Z}/\ell\mathbb{Z}$ and $\ell \neq p$, any such homomorphism is tamely ramified. Furthermore, by assumption (f), any tamely ramified homomorphism is unramified. Therefore all classes in $\mathrm{H}^1(K_p, R/I)$ are unramified, and $\mathrm{H}^1(K_p, R/I) \to R/I$, sending a cohomology class to the image of Frob_{p^2} in R/I, is an isomorphism.

By Kummer theory, we have $\mathrm{H}^1(K_p, \mu_\ell) = K_p^\times/(K_p^\times)^\ell$, and the unramified classes are $\mathcal{O}_p^\times/(\mathcal{O}_p^\times)^\ell$. Thus by assumption (f), we have $\mathrm{H}^1_{\mathrm{unr}}(K_p, \mu_\ell) = 0$, and $\mathrm{H}^1(K_p, \mu_\ell)$ is 1-dimensional over $\mathbb{Z}/\ell\mathbb{Z}$, generated by the class of a uniformizer in $K_p^\times/(K_p^\times)^\ell$. Since E/K_p has good reduction we have, by Lemma 6,

$$L_p = R/I \otimes \mathrm{H}^1_{\mathrm{unr}}(K_p, E[\ell]),$$

and so we find that L_p is the summand $\mathrm{H}^1(K_p, R/I)$ of $\mathrm{H}^1(K_p, A[I])$.

To see that $L'_p = \mathrm{H}^1(K_p, R/I(1))$, we will use the rigid-analytic uniformization of A/K_p. Since it is an isogeny factor of the new quotient of $J_0(Np)/\mathbb{Q}$, the abelian variety $A/\mathbb{Q}_p$ has purely toric reduction. Let $\mathcal{A}/\mathbb{Z}_p$ be its Néron model. By Lemma 2, the F-vector space $\mathbb{Q} \otimes_{\mathbb{Z}} X^*(\mathcal{A}^0/\overline{\mathbb{F}}_p)$ is 1-dimensional, and so the group $\mathrm{Gal}(\overline{\mathbb{F}}_p/\mathbb{F}_p)$ acts on it via a continuous $F^\times$-valued character. Since F is totally real, this character must be quadratic. The group scheme $\mathcal{A}^0/\mathbb{F}_{p^2}$ is thus a split torus.

Since $A/\mathbb{Q}_p$ has semi-stable reduction, the (split) torus $\mathcal{A}^0/\mathbb{F}_{p^2}$ is identity component of the special fiber of the Néron model of A/K_p. Let T/K_p be the split torus whose character group is $X^*(\mathcal{A}^0/\mathbb{F}_{p^2})$. Let R act as endomorphisms of T/K_p dual to the action of R on $X^*(\mathcal{A}^0/\mathbb{F}_{p^2})$. One then has the rigid-analytic uniformization $T^{\mathrm{an}} \to A^{\mathrm{an}}$ over K_p (cf. [BL91]), which yields a surjection of $R[\mathrm{Gal}(\overline{K_p}/K_p)]$-modules $T(\overline{K_p}) \to A(\overline{K_p})$. The kernel Λ of $T(\overline{K_p}) \to A(\overline{K_p})$ is a free $\mathbb{Z}$-module of rank $\dim(A)$, on which $\mathrm{Gal}(\overline{K_p}/K_p)$ acts trivially.

Consider the following diagram comparing the Kummer maps for T/K_p and A/K_p, where the vertical maps come from the analytic uniformization $T(\overline{K_p}) \to A(\overline{K_p})$:

$$\begin{array}{ccc} I^{-1} \otimes_R T(K_p) & \longrightarrow & \mathrm{H}^1(K_p, T[I]) \\ \downarrow & & \downarrow \\ I^{-1} \otimes_R A(K_p) & \longrightarrow & \mathrm{H}^1(K_p, A[I]). \end{array} \tag{5.1}$$

Since $\Lambda[I] = 0$, the uniformization induces an injection $T[I] \to A[I]$ over K_p. As $X^*(\mathcal{A}^0/\mathbb{F}_{p^2})$ is locally free of rank 1 over R, the R/I-module scheme $T[I]$

is abstractly $R/I(1)$. Thus the image of $T[I]$ in $A[I]$ is the $R/I(1)$ summand of $A[I] = R/I \oplus R/I(1)$. Consequently, the image of the right vertical arrow in (5.1) is the summand $\mathrm{H}^1(K_p, R/I(1))$ of $\mathrm{H}^1(K_p, A[I])$. The cokernel of the top arrow is contained in $\mathrm{H}^1(K_p, T)$, which vanishes by Hilbert Theorem 90, since T/K_p is a split torus. Finally, the left vertical arrow is surjective, since $\mathrm{H}^1(K_p, \Lambda) = 0$. These three observations imply that the image L'_p of the Kummer map in $\mathrm{H}^1(K_p, A[I])$ is the summand $\mathrm{H}^1(K_p, R/I(1))$. □

5.2 *Comparing the Selmer groups*

To prove Theorem 2, we compare the subspaces $R/I \otimes \mathrm{Sel}(E/K, \ell)$ and $\mathrm{Sel}(A/K, I)$ of $\mathrm{H}^1(K, A[I])$. Let $\mathrm{Sel}_s(A/K, I) \subset \mathrm{Sel}(A/K, I)$ (with s for "strict") be the intersection of these two Selmer groups. By Lemmas 5 and 8, the Selmer group $\mathrm{Sel}_s(A/K, I)$ is the space of classes $c \in \mathrm{H}^1(K, A[I])$ such that for $v \neq p$ the local restriction c_v belongs to $L_v = L'_v$ and such that the local restriction x_p is 0. Since the spaces L_p and L'_p are both 1-dimensional over R/I, the codimension of $\mathrm{Sel}_s(A/K, I)$ in each of $R/I \otimes \mathrm{Sel}(E/K, \ell)$ and $\mathrm{Sel}(A/K, I)$ is at most 1. More precisely, by Theorem 1, the Selmer group $\mathrm{Sel}_s(A/K, I)$ is 1-dimensional, if the Heegner point P is divisible by ℓ in $E(K_p)$: in this case the restriction to p of the image of P in $\mathrm{Sel}(E/K, \ell)$ vanishes, and so $R/I \otimes \mathrm{Sel}(E/K, \ell) = \mathrm{Sel}_s(A/K, I)$. Therefore, if P is divisible by ℓ in $E(K_p)$, the R/I-dimension of $\mathrm{Sel}(A/K, I)$ is either 1 or 2. Similarly, if P is not divisible by ℓ in $E(K_p)$, then $\mathrm{Sel}_s(A/K, I)$ is 0-dimensional; in this case, the R/I-dimension of $\mathrm{Sel}(A/K, I)$ is either 0 or 1.

As explained in §6 below, the conjecture of Birch and Swinnerton-Dyer suggests that the R/I-dimension of $\mathrm{Sel}(A/K, I)$ should be even. In order to finish the proof of Theorem 2, we must exclude the possibility that $\mathrm{Sel}(A/K, I)$ is 1-dimensional over R/I, in harmony with the conjecture. In order rule out the 1-dimensional case, we present in §5.3 a variant of an argument which was shown to us by Benjamin Howard.

5.3 *The parity lemma*

Let k be a finite field of characteristic ℓ and let M be a totally imaginary number field. In what follows, we will take $k = R/I$ and $M = K$. Let V be locally constant constructible étale sheaf of k-vector spaces over $\mathrm{Spec}(M)$, equipped with a perfect, alternating, k-bilinear pairing $V \times V \to k(1)$. In the application to Theorem 2, we will take $V = A[I]$ and use the pairing coming from the Weil pairing on $E[\ell]$. For each finite place v of M one then has the perfect, symmetric, k-bilinear Tate-local-duality pairing $\mathrm{H}^1(M_v, V) \times \mathrm{H}^1(M_v, V) \to k$ mapping $x \times y \mapsto \langle x, y \rangle_v$. The duality pairing is the composition of the product $\mathrm{H}^1(M_v, V) \times \mathrm{H}^1(M_v, V) \to \mathrm{H}^2(M_v, k(1))$

with the reciprocity isomorphism $H^2(M_v, k(1)) \to k$. For each finite place v of M, let $\Lambda_v \subset H^1(M_v, V)$ be a k-subspace. Assume that for almost all v, one has $\Lambda_v = H^1_{\mathrm{unr}}(M_v, V)$. Assume furthermore that each Λ_v is its own annihilator under the Tate pairing. This assumption implies that each $H^1(M_v, V)$ has even dimension over k and that each Λ_v is a totally isotropic subspace of half the dimension of $H^1(M_v, V)$. Thus the Tate pairing on $H^1(M_v, V)$ is a split symmetric bilinear form, and Λ_v is a maximal totally isotropic subspace.

We distinguish one fixed finite place w of M, which will be the place p of K in the proof of Theorem 2. Assume that $H^1(M_w, V)$ is 2-dimensional over k, so that $H^1(M_w, V)$ equipped with the Tate pairing is a hyperbolic plane. There are then exactly two maximal totally isotropic subspaces (lines), namely, the given Λ_w and another subspace Λ'_w. (Recall that k has characteristic $\ell \neq 2$.)

We consider four Selmer groups contained in $H^1(M, V)$, defined by the local conditions Λ_v for $v \neq w$ and differing only in their defining local conditions at w. Let $\mathrm{Sel}_u(V)$ ("unramified") be defined by the local conditions Λ_v at all places v, i.e., $\mathrm{Sel}_u(V) \subset H^1(M, V)$ is the space of classes x such that the restriction x_v belongs to Λ_v for all finite places v of M. Let $\mathrm{Sel}_t(V)$ ("transverse") be defined by the local conditions Λ_v at $v \neq w$ and by Λ'_w at w. Let $\mathrm{Sel}_r(V)$ ("relaxed") be defined by the local conditions Λ_v at $v \neq w$ and no condition at w. Let $\mathrm{Sel}_s(V)$ ("strict") be defined by the local conditions L_v at $v \neq w$ and local vanishing at w.

Lemma 9. *The k-dimensions of $\mathrm{Sel}_u(V)$ and $\mathrm{Sel}_t(V)$ differ by exactly 1; moreover, either*

$$\mathrm{Sel}_u(V) = \mathrm{Sel}_r(V) \quad \textit{and} \quad \mathrm{Sel}_t(V) = \mathrm{Sel}_s(V), \textit{ or}$$

$$\mathrm{Sel}_u(V) = \mathrm{Sel}_s(V) \quad \textit{and} \quad \mathrm{Sel}_t(V) = \mathrm{Sel}_r(V).$$

Proof. Let $x, y \in \mathrm{Sel}_r(V)$. By global class field theory, one has

$$\sum_v \langle x_v, y_v \rangle_v = 0,$$

where $x_v, y_v \in H^1(M_v, V)$ are the restrictions of x and y. Since x, y restrict to elements of the totally isotropic spaces Λ_v for $v \neq w$, one has

$$\langle x_w, y_w \rangle_w = \sum_v \langle x_v, y_v \rangle_v = 0.$$

Therefore, the image of $\mathrm{Sel}_r(V)$ in $H^1(M_w, V)$ under the restriction map is a totally isotropic subspace. Consequently this image is contained in Λ_w or Λ'_w, and

$$\mathrm{Sel}_r(V) = \mathrm{Sel}_u(V) \quad \text{or} \quad \mathrm{Sel}_r(V) = \mathrm{Sel}_t(V).$$

On the other hand, it follows from the global Euler characteristic formula and global duality (cf. [DDT94], Theorem 2.19) that the k-codimension of $\mathrm{Sel}_s(V)$

in $\mathrm{Sel}_r(V)$ is 1: the local conditions for $\mathrm{Sel}_s(V)$ are obtained by relaxing the local condition at w from the self-dual, 1-dimensional Λ_w to the 2-dimensional $\mathrm{H}^1(M_w, V)$, and the dual of the relaxed condition at w is the strict condition at w defining $\mathrm{Sel}_s(V)$. As $\mathrm{Sel}_s(V) = \mathrm{Sel}_u(V) \cap \mathrm{Sel}_t(V)$ in $\mathrm{Sel}_r(V)$, one cannot have $\mathrm{Sel}_r(V) = \mathrm{Sel}_t(V) = \mathrm{Sel}_r(V)$. Therefore, one has either $\mathrm{Sel}_r(V) = \mathrm{Sel}_u(V)$ and $\mathrm{Sel}_s(V) = \mathrm{Sel}_t(V)$ or $\mathrm{Sel}_r(V) = \mathrm{Sel}_t(V)$ and $\mathrm{Sel}_s(V) = \mathrm{Sel}_u(V)$. These relations and the fact that the k-codimension of $\mathrm{Sel}_s(V)$ in $\mathrm{Sel}_r(V)$ is 1 prove the lemma. □

5.4 Completion of the proof of Theorem 2

To finish the proof the theorem, we apply the parity lemma to $A[I]/K$ to compare the subspaces $R/I \otimes \mathrm{Sel}(E/K, \ell)$ and $\mathrm{Sel}(A/K, I)$ of $\mathrm{H}^1(K, A[I])$. The transfer of the Weil pairing on $R/I \otimes E[\ell]$ via the isomorphism $R/I \otimes E[\ell] = A[I]$ to $A[I]$ provides a perfect, alternating pairing $A[I] \times A[I] \to R/I(1)$. The fact that the spaces $L_v \subset \mathrm{H}^1(K_v, A[I])$ are their own annihilators under the Tate pairing follows from Tate local duality for the elliptic curve E/K. Alternatively, one can deduce it directly from the description $L_v = \mathrm{H}^1_{\mathrm{unr}}(K_v, A[I])$ for v prime to ℓ and $L_v = \mathrm{H}^1_{\mathrm{fl}}(\mathrm{Spec}(\mathcal{O}_v, \mathcal{A}[I])$ for v dividing ℓ (cf. [Mil86], Theorem I.2.6, Corollary II.1.10(b), and Theorem III.1.8(b)). As we observed above in Lemma 8, conditions (c) and (f) imply that $A[I]/K_p = R/I \oplus R/I(1)$ and that $\mathrm{H}^1(K_p, A[I])$ is 2-dimensional over R/I. Each 1-dimensional summand R/I and $R/I(1)$ is isotropic for the (alternating) Weil pairing, and so the 1-dimensional summands $\mathrm{H}^1(K_p, R/I)$ and $\mathrm{H}^1(K_p, R/I(1))$ of $\mathrm{H}^1(K_p, A[I])$ are isotropic for the Tate pairing.

We now apply the parity lemma with $M = K$, the étale sheaf $V = A[I]$, equipped with the Weil pairing, and the local conditions $\Lambda_v = L_v$. These structures satisfy the hypotheses of §5.3. We take the distinguished place w to be the place p of K. By the discussion in §5.1, we have $\Lambda'_p = L'_p$. Since for $v \neq p$, we have $L_v = L'_v$, the parity lemma compares the Selmer groups $\mathrm{Sel}_u(A[I]) = R/I \otimes \mathrm{Sel}(E/K, \ell)$ and $\mathrm{Sel}_t(A[I]) = \mathrm{Sel}(A/K, I)$ in $\mathrm{H}^1(K, A[I])$. By Kolyvagin's result stated in Theorem 1, we know that $\mathrm{Sel}(E/K, \ell)$ is 1-dimensional over $\mathbb{Z}/\ell\mathbb{Z}$, generated by the image of the Heegner point P under the Kummer map. Therefore, the parity lemma implies Theorem 2. □

6 Compatibility with the functional equation

In this section we study the signs of the functional equations for L-functions related to f and g. Let $\epsilon = \pm 1$ be the sign of the complete L-function of f, i.e., if $\Lambda(s, f) = (2\pi)^{-s}\Gamma(s)L(s, f)$ is the complete L-function of f, then $\Lambda(s, f) = \epsilon N^{1-s}\Lambda(2 - s, f)$. If W is the Fricke involution of level N, then

$Wf = -\epsilon f$ by Theorem 3.66 of [Shi94]. Let χ be the quadratic Dirichlet character corresponding to $K/\mathbb{Q}$. By [Shi94], Theorem 3.66 and Lemma 3.63 (2), the sign of the complete twisted L-function $\Lambda(s, f, \chi) = (2\pi)^{-s}\Gamma(s)L(s, f, \chi)$ is $\chi(-N) \times \epsilon$. That is,

$$\Lambda(s, f, \chi) = (\chi(-N) \times \epsilon)(r^2 N)^{1-s}\Lambda(2 - s, f, \chi),$$

where r is the conductor of χ. Since the primes dividing N split in K and K is imaginary quadratic, one has $\chi(-N) = \chi(-1) = -1$, and so the sign of $\Lambda(f, s, \chi)$ is $-\epsilon$. Therefore, the sign of $\Lambda(s, f) \times \Lambda(s, f, \chi)$, which is the complete L-function of the base change of f to K, is -1. By the compatibility of the Eichler–Shimura construction with the local Langlands correspondence (cf. [Car86]), the complete L-function $\Lambda(s, E/K) = \Lambda(s, E/\mathbb{Q}) \times \Lambda(s, E/\mathbb{Q}, \chi)$ is $\Lambda(s, f) \times \Lambda(s, f, \chi)$. Thus the functional equation for $\Lambda(s, E/K)$ has sign -1, which is what one would expect from the Birch and Swinnerton-Dyer conjecture and Theorem 1, since $E[\ell](K) = 0$ by condition (b).

Recall that the new level of the newform g is Np by condition (e). Let ϵ' be the sign of the functional equation of $\Lambda(s, g)$. Then, as in the case of f, the sign of the functional equation of $\Lambda(s, g, \chi)$ is $\chi(-Np) \times \epsilon' = \epsilon'$, since the primes dividing N split in K, the field K is imaginary quadratic, and p is inert in $K/\mathbb{Q}$. Consequently, the sign of $\Lambda(s, g) \times \Lambda(s, g, \chi)$, which is the complete L-function of the base change of g to K, is $+1$.

The Birch and Swinnerton-Dyer conjecture for A/K, when combined with the conjecture of Deligne and Gross [D79, Conj 2.7], predicts that the dimension of of the F-vector space $F \otimes_R A(K)$ is equal to the order of vanishing of $L(s, g) \times L(s, g, \chi)$ at $s = 1$. From the sign of the functional equation, one thus expects $F \otimes_R A(K)$ to have even dimension. Theorem 2 implies that the dimension is 0, if ℓ does not divide P in $E(K_p)$, since $I^{-1}/R \otimes_R A(K)$ injects into $\mathrm{Sel}((A, i)/K, I) = 0$. If ℓ does divide P in $E(K_p)$, then by Theorem 2, the group $\mathrm{Sel}((A, i)/K, I)$ is 2-dimensional over R/I. If $Ш(A/K)$ is finite, then one can check using the Cassels–Tate pairing that the dimensions of $\mathrm{Sel}(A/K, I)$ and of $I^{-1}/R \otimes_R A(K)$ have the same parity; thus $F \otimes_R A(K)$ has dimension 2 or 0, according to whether $Ш(A/K)[I] = 0$ or not.

Note that if ℓ does not divide P in $E(K_p)$, then, since the dimension of $F \otimes_R A(K)$ is 0, one finds that the rank of $A/\mathbb{Q}$ is 0. The following lemma shows that this conclusion is compatible with the parity prediction of the Birch and Swinnerton-Dyer conjecture over $\mathbb{Q}$.

Lemma 10. *Suppose that ℓ does not divide P in $E(K_p)$. Then the sign ϵ' of the functional equation of $\Lambda(s, g)$ is $+1$.*

Proof. Owing to the congruence between f and g, we have

$$a_p(f) \equiv a_p(g)(p + 1) \pmod{\lambda}, \tag{6.1}$$

as one sees by comparing the local Galois representations at p associated to f and g. According to Lemma 1, the congruence in (6.1) holds with $a_p(g)$ replaced by $-\epsilon$. Since $p+1$ is invertible modulo ℓ by assumption f), we have, therefore, $a_p(g) \equiv -\epsilon \pmod{\lambda}$. As $a_p(g) = \pm 1$ and $\ell \neq 2$, we conclude that $a_p(g) = -\epsilon$.

Let α be the root in $\overline{\mathbb{Q}_\ell}$ of the Hecke polynomial $T^2 - a_p(f)T + p$ that is congruent to $-\epsilon$ modulo λ, and let β be the other root. Consider the oldform with coefficients in $\overline{\mathbb{Z}_\ell}$

$$h(z) = f(z) - \beta f(pz)$$

in the old space for f at level Np. The form $h(z)$ is an eigenfunction of the Hecke operators T_q for q prime to Np and of U_q for q dividing Np; the eigenvalues at q prime to p agree with those of f, and at p, we have $U_p h = \alpha h$. Since $\alpha \equiv -\epsilon \pmod{\lambda}$, the reductions $\overline{g}, \overline{h} \in S_2(\Gamma_0(Np), \overline{\mathbb{F}}_\ell)$ are eigenfunctions for the Hecke operators T_q for q prime to Np and U_q for q dividing Np with the same eigenvalues. Since both of these cuspidal eigenforms have $a_1 = 1$, we have $\overline{g} = \overline{h}$ by the q-expansion principle.

For a prime q dividing N, let W_q be the Atkin–Lehner involution at q for level N. Since f is a newform with new level N, it is an eigenfunction of all of the W_q. Let $W_q f = \epsilon_q f$. Then the sign ϵ is $-\prod_{q|N} \epsilon_q$.

For each q dividing Np, let W'_q be the Atkin–Lehner involution at q at level Np. Since g is a newform at level Np, it is an eigenfunction for all of the W'_q. One has $W'_p = -U_p$, and so $W'_p g = \epsilon g$. For q dividing N, we have $W'_q h = \epsilon_q h$, and so $\overline{W'_q g} = \overline{W'_q h} = \epsilon_q \overline{g}$. Therefore, $W'_q g = \epsilon_q g$ for q dividing N. The sign ϵ' of the functional equation for $\Lambda(s, g)$ is thus

$$-\epsilon \times \prod_{q|N} \epsilon_q = \epsilon^2 = +1. \qquad \square$$

7 Examples

We now give two examples. The data consist of a 4-tuple (E, K, p, ℓ) satisfying hypothesis (a)–(f). The examples are chosen so that the lifted form g has rational Fourier coefficients. In this case, we have $\mathbb{Q}(g) = \mathbb{Q}$ and $R = \mathbb{Z}$; the abelian variety (A, i) is an elliptic curve, and the I-descent on (A, i) is simply the ℓ-descent A. We wish to thank Noam Elkies and William Stein for help with the computations.

The first example is:

$$E = X_0(57)/\langle W_3, W_{19}\rangle \qquad N = 57$$

$$K = \mathbb{Q}(\sqrt{-59}) \qquad h = 3$$

$$p = 2 \qquad a_2 = -2$$

$$\ell = 5$$

The minimal equation of E is (cf. [BK75]):

$$y^2 + y = x^3 - x^2 - 2x + 2 \qquad \Delta = -3^2.19,$$

and the modular form f associated to E has q-expansion

$$f = q - 2q^2 - q^3 + 2q^4 - 3q^5 + 2q^6 - 5q^7 + \dots.$$

The sign in the functional equation of $L(s, E) = L(s, f)$ is $\epsilon = -1$, and $E(\mathbb{Q})$ is free of rank 1, with generator $e = (2, 1)$. The Heegner point P associated to K is equal to $\pm 2e$. This is *not* divisible by $\ell = 5$ in $E(K_p) = E(K_2)$.

There is a unique newform g of weight 2 and level $114 = Np$ where $W_3' = W_{19}' = +1$ and $W_2' = \epsilon = -1$. It has q-expansion

$$g = q + q^2 - q^3 + q^4 + 2q^5 - q^6 + 0q^7 + \dots$$

congruent (mod 5) to the old form $f(\tau) - 2f(2\tau)$. The elliptic curve A has minimal equation

$$y^2 + xy + y = x^3 + x^2 - 352x - 2431 \qquad \Delta = 2^{20}.3^3.19.$$

The Selmer group $\mathrm{Sel}(A/K, 5) = 0$, and $A(K)$ has rank 0.

The second example is

$$E = X_0(26)/\langle W_2 \rangle \qquad N = 26$$

$$K = \mathbb{Q}(\sqrt{-79}) \qquad h = 5$$

$$p = 3 \qquad a_3 = 1$$

$$\ell = 5$$

The minimal equation of E is (cf. [BK75]):

$$y^2 + xy + y = x^3 - 5x - 8 \qquad \Delta = -2^3.13^3$$

and the modular form f associated to E has q-expansion beginning

$$f = q - q^2 + q^3 + q^4 - 3q^5 - q^6 - q^7 + \dots.$$

The sign in the functional equation of $L(s, E) = L(s, f)$ is $\epsilon = +1$, and $E(\mathbb{Q}) \simeq \mathbb{Z}/3\mathbb{Z}$.

The Heegner point P associated to K has x-coordinate $x(P) = 1700/711$, and generates the free group $E(K)$/torsion. On the other hand, since the denominator of $x(P)$ is divisible by $p = 3$, the point P reduces to the identity (mod 3) and *is* divisible by $\ell = 5$ in $E(K_3)$. We note that E has three points over the field with $p = 3$ elements, and fifteen points over the field with $p^2 = 9$ elements.

In this case, g is the unique newform of weight 2 and level $78 = Np$. It has Fourier expansion beginning

$$g = q - q^2 - q^3 + q^4 + 2q^5 + q^6 + 4q^7 + \dots$$

and is congruent to the old form $f(\tau) + 3f(3\tau)$ (mod 5). The elliptic curve A has minimal equation

$$y^2 + xy = x^3 + x^2 - 19x + 685 \qquad \Delta = -2^{16}.3^5.13$$

and rank 0 over $\mathbb{Q}$. Since P is locally divisible by 5 in $E(K_3)$, the Selmer group $\mathrm{Sel}(A/K, 5)$ has dimension 2 over $\mathbb{Z}/5\mathbb{Z}$. In fact, $A(K)$ has rank 2. In the twisted model over $\mathbb{Q}$:

$$y^2 + xy + y = x^3 - 121830x - 341716424$$
$$\Delta = -2^{16}.3^5.13.79^6$$

the points
$(x, y) = (2732, 139056)$
$(x, y) = (410357, -263076219)$
generate the Mordell–Weil group modulo torsion.

Appendix A. Descent with endomorphisms

In this appendix, we give an expanded discussion of the I-descent used on the abelian variety A/K and its Néron model above. In order to have results general enough to apply easily to group schemes such as the Néron model, we begin very generally with some abstract constructions on sheaves. We then specialize to the sheaves represented by smooth group schemes, explaining how to apply the abstract results to such cases.

A.1 First descent with endomorphisms

Let R be a commutative, unital ring, and let $I \subset R$ be an ideal. Let A be a sheaf of R-modules on some site S, and assume that A is I-injective in the sense that the map of sheaves

$$A = \underline{\mathrm{Hom}}_R(R, A) \to \underline{\mathrm{Hom}}_R(I, A)$$

is surjective. (Note that for any R-module M, the presheaf $T \mapsto \mathrm{Hom}_R(M, A(T))$ is a sheaf.) The kernel of this map is $\underline{\mathrm{Hom}}_R(R/I, A)$. Associating to a sheaf homomorphism $R/I \to A$ the image of the Section 1 identifies $\underline{\mathrm{Hom}}_R(R/I, A)$ with $A[I]$, the sheaf of sections of A killed by elements of I. We thus have an exact sequence of sheaves

$$0 \longrightarrow A[I] \longrightarrow A \longrightarrow \underline{\mathrm{Hom}}_R(I, A) \longrightarrow 0.$$

The boundary map

$$\mathrm{Hom}_R(I, A(S)) \to \mathrm{H}^1(S, A[I])$$

in the associated long exact sequence in cohomology is the *Kummer map*. From the long exact sequence, we see that the kernel of the Kummer map is the image of $\mathrm{Hom}_R(R, A(S)) = A(S)$ in $\mathrm{Hom}_R(I, A(S))$. Similarly, the cokernel of the Kummer map is the kernel of

$$\mathrm{H}^1(S, A) \to \mathrm{H}^1(S, \underline{\mathrm{Hom}}_R(I, A)).$$

If I is invertible as an R-module, then we have

$$\underline{\mathrm{Hom}}_R(I, A) = I^{-1} \otimes_R A,$$

where $I^{-1} = \mathrm{Hom}_R(I, R)$. Here $I^{-1} \otimes_R A$ denotes the presheaf

$$T \mapsto I^{-1} \otimes_R A(T),$$

which is a sheaf since I^{-1} is a flat R-module. Thus we have the exact sequence of sheaves

$$0 \longrightarrow A[I] \longrightarrow A \longrightarrow I^{-1} \otimes_R A \longrightarrow 0.$$

The kernel of the Kummer map

$$I^{-1} \otimes_R A(S) \to \mathrm{H}^1(S, A[I])$$

is the image of $R \otimes_R A(S) = A(S)$, and so the Kummer image is $(I^{-1}/R) \otimes_R A(S)$, or, equivalently, it is $I^{-1} \otimes_R (R/I \otimes_R A(S))$. In this case, we also have

$$\mathrm{H}^1(S, I^{-1} \otimes_R A) = I^{-1} \otimes_R \mathrm{H}^1(S, A).$$

The cokernel of the Kummer map is then the kernel of

$$\mathrm{H}^1(S, A) \to I^{-1} \otimes_R \mathrm{H}^1(S, A),$$

which is $\mathrm{H}^1(S, A)[I]$, the I-torsion of the R-module $\mathrm{H}^1(S, A)$.

If S is the small étale site of $\mathrm{Spec}(K)$ for a number field K, we define $\mathrm{Sel}(A/K, I)$, the I-Selmer group of A, to be the subspace of classes $x \in \mathrm{H}^1(K, A[I])$ whose local restrictions $x_v \in \mathrm{H}^1(K_v, A[I])$ lie in the image of $\mathrm{Hom}_R(I, A(K_v))$ under the local Kummer map for all places v of K. As in the standard case with $R = \mathbb{Z}$ and $I = (\ell)$, the global Kummer map

$$\mathrm{Hom}_R(I, A(K)) \to \mathrm{H}^1(K, A[I])$$

factors through $\mathrm{Sel}(A/K, I)$.

In topology the most familiar analogue of the above formalism is the following: let M be an R-module that has no I-torsion, so that $M \to \mathrm{Hom}_R(I, M)$ is injective. Let N be the cokernel, so that we have the exact sequence of constant sheaves of R-modules on any topological space X

$$0 \longrightarrow M \longrightarrow \underline{\mathrm{Hom}}_R(I, M) \longrightarrow N \longrightarrow 0.$$

Passing to cohomology, we find

$$\begin{array}{ccccc} \dots & \longrightarrow & \mathrm{H}^i(X, M) & \longrightarrow & \mathrm{H}^i(X, \underline{\mathrm{Hom}}_R(I, M)) \longrightarrow \\ \mathrm{H}^i(X, N) & \longrightarrow & \mathrm{H}^{i+1}(X, M) & \longrightarrow & \dots, \end{array}$$

which is a Bockstein-type sequence in the case $R = \mathbb{Z}$ and $I = (\ell)$.

To recast the above discussion in this style, we replace the sheaves A and $\underline{\mathrm{Hom}}_R(I, A)$ with the complexes

$$M = A[-1] \quad \text{and} \quad M' = \underline{\mathrm{Hom}}_R(I, A)[-1],$$

concentrated in degree 1. The I-injectivity of A translates into something like I-torsion freeness of M: the exact sequence of sheaves

$$0 \longrightarrow A[I] \longrightarrow A \longrightarrow \underline{\mathrm{Hom}}_R(I, A) \longrightarrow 0$$

gives rise to a distinguished triangle

$$M \to M' \to A[I] \to M[1].$$

The sheaf $A[I]$ thus plays the same role as the cokernel N in the picture over a topological space X, and the Kummer map is

$$\mathrm{Hom}(I, A(S)) = \mathrm{H}^1(S, M') \to \mathrm{H}^1(S, A[I]).$$

For a concrete example, let S be the fppf site of a field K, and let A is the sheaf represented by an abelian variety over K on which R acts as endomorphisms. Assume that I is invertible as an R-module. Then, as explained in the next section,

$\underline{\mathrm{Hom}}_R(I, A) = I^{-1} \otimes_R A$ is also represented by an abelian variety, on which R acts as endomorphisms. Furthermore, the natural map $A \to I^{-1} \otimes_R A$ of sheaves is surjective. The complexes M and M' are the 1-motives associated to these abelian varieties, and the surjection of sheaves $A \to \underline{\mathrm{Hom}}_R(I, A)$ becomes an injection of motives $M \to M'$. The sheaf $A[I]$ (in degree 0) stands in for the (hypothetical) torsion motive coker$(M \to M')$.

A.2 Smooth R-module schemes

Let R and I be as in §A.1. Assume moreover that I is finitely presented as an R-module. Let X be a scheme and let G/X be an R-module scheme. The choice of a presentation of I produces a scheme representing the fppf sheaf $\underline{\mathrm{Hom}}_R(I, G)$: let $R^{\oplus r} \to R^{\oplus s} \to I \to 0$ be an R-module presentation. Then the kernel of $G^{\oplus r} \to G^{\oplus s}$ represents $\underline{\mathrm{Hom}}_R(I, G)$. We fix a presentation of I and write $\underline{\mathrm{Hom}}_R(I, G)$ for the representing scheme as well as the sheaf. Many scheme-theoretic properties of G/X carry over to $\underline{\mathrm{Hom}}_R(I, G)$. For instance, if G/X is separated (resp. quasi-compact, quasi-separated, locally of finite type, locally of finite presentation, proper, ...), then so is $\underline{\mathrm{Hom}}_R(I, G)/X$. Note finally that if I is invertible as an R-module and G/X has connected geometric fibers, then so does $\underline{\mathrm{Hom}}_R(I, G)$: the surjection $R^{\oplus s} \to I$ splits, and so $\underline{\mathrm{Hom}}_R(I, G)/X$ is a factor of $G^{\oplus r}/X$, which has connected geometric fibers.

Assume from now on that G/X is smooth and that I is invertible as an R-module. It follows from the functorial criterion for smoothness that $\underline{\mathrm{Hom}}_R(I, G)$ is smooth over R. Furthermore, there is a natural isomorphism $\underline{\mathrm{Hom}}_R(I, \mathrm{Lie}(G/X)) = \mathrm{Lie}(\underline{\mathrm{Hom}}_R(I, G)/X)$, and so the relative dimensions of G/X and $\underline{\mathrm{Hom}}_R(I, G)/X$ agree.

The kernel and cokernel of the map on Lie algebras

$$\mathrm{Lie}(G/X) \to \mathrm{Lie}(I^{-1} \otimes_R G/X) = I^{-1} \otimes_R \mathrm{Lie}(G/X)$$

are coherent sheaves on X on which R acts and that are annihilated by I. Thus if I contains the image of $\ell \in \mathbb{Z}$ such that ℓ is invertible on X, this map is an isomorphism, and $G \to I^{-1} \otimes_R G$ is étale. By [SGA70], Exposé VIB, Proposition 3.11, if G/X has connected geometric fibers, the map $G \to \underline{\mathrm{Hom}}_R(I, G)$ is étale and surjective, and so the sheaf on the small étale (or fppf or fpqf) site of X represented by G/X is I-injective.

Alternatively, assume that G/X has connected and semi-abelian geometric fibers, and that I contains the image of any non-zero $\ell \in \mathbb{Z}$. The multiplication-by-ℓ endomorphism of G is then surjective. Since $G \to I^{-1} \otimes_R G$ factors through multiplication by ℓ, the geometric fibers of $I^{-1} \otimes_R G/X$ are connected, and the relative dimensions of G/X and $I^{-1} \otimes_R G/X$ agree, the homomorphism $G \to I^{-1} \otimes_R G$ is also surjective. By [SGA70], Exposé VIB, Proposition 3.11, the homomorphism $G \to I^{-1} \otimes_R G$ is then flat and surjective. Consequently, the sheaf on the fppf (or fpqf) site of X represented by G/X is I-injective.

In either of these cases, since $G \to I^{-1} \otimes_R G$ is flat, the kernel $G[I]/X$ is flat and quasi-finite for dimensional reasons. If G/X is proper, then $G[I]/X$ is moreover finite and locally free.

References

[BCDT01] Christophe Breuil, Brian Conrad, Fred Diamond, and Richard Taylor, *On the modularity of elliptic curves over $\mathbb{Q}$: wild 3-adic exercises*, J. Amer. Math. Soc. **14** (2001), no. 4, 843–939 (electronic).

[BD99] M. Bertolini and H. Darmon, *Euler systems and Jochnowitz congruences*, Amer. J. Math. **121** (1999), no. 2, 259–281.

[BD05] M. Bertolini and H. Darmon, *Iwasawa's main conjecture for elliptic curves over anticyclotomic Z_p-extensions*, Annals of Mathematics **162** (2005), 1–64.

[BK75] B. J. Birch and W. Kuyk (eds.), *Modular functions of one variable. IV*, Springer-Verlag, Berlin, 1975, Lecture Notes in Mathematics, Vol. 476.

[BL91] Siegfried Bosch and Werner Lütkebohmert, *Degenerating abelian varieties*, Topology **30** (1991), no. 4, 653–698.

[BSD75] B. J. Birch and H. P. F. Swinnerton-Dyer, *Elliptic curves and modular functions*, Modular functions of one variable, IV (Proc. Internat. Summer School, Univ. Antwerp, Antwerp, 1972), Springer, Berlin, 1975, pp. 2–32. Lecture Notes in Math., Vol. 476.

[Car86] Henri Carayol, *Sur les représentations l-adiques associées aux formes modulaires de Hilbert*, Ann. Sci. École Norm. Sup. (4) **19** (1986), no. 3, 409–468.

[DDT94] Henri Darmon, Fred Diamond, and Richard Taylor, *Fermat's last theorem*, Current developments in mathematics, 1995 (Cambridge, MA), Internat. Press, Cambridge, MA, 1994, pp. 1–154.

[Dia95] Fred Diamond, *The refined conjecture of Serre*, Elliptic curves, modular forms, Fermat's last theorem (Hong Kong, 1993), Ser. Number Theory, I, Internat. Press, Cambridge, MA, 1995, pp. 22–37.

[D79] P. Deligne, *Valeurs de fonctions L et périodes d'intégrales*, Automorphic forms, and representations, and L-functions, Proceedings of Symposia in Pure Mathematics, volume 33, part 2 (1979) 313–346.

[DR73] P. Deligne and M. Rapoport, *Les schémas de modules de courbes elliptiques*, Modular functions of one variable, II (Proc. Internat. Summer School, Univ. Antwerp, Antwerp, 1972), Springer, Berlin, 1973, pp. 143–316. Lecture Notes in Math., Vol. 349.

[Fal83] G. Faltings, *Endlichkeitssätze für abelsche Varietäten über Zahlkörpern*, Invent. Math. **73** (1983), no. 3, 349–366.

[Gro] Benedict H. Gross, *Kolyvagin's work on modular elliptic curves*, L-functions and arithmetic (Durham, 1989), London Math. Soc. Lecture Note Ser., vol. 153, Cambridge Univ. Press, Cambridge, pp. 235–256.

[Gro84] ———, *Heegner points on $X_0(N)$*, Modular forms (Durham, 1983), Ellis Horwood Ser. Math. Appl.: Statist. Oper. Res., Horwood, Chichester, 1984, pp. 87–105.

[GZ86] Benedict H. Gross and Don B. Zagier, *Heegner points and derivatives of L-series*, Invent. Math. **84** (1986), no. 2, 225–320.

[Lan56] Serge Lang, *Algebraic groups over finite fields*, Amer. J. Math. **78** (1956), 555–563.

[Maz72] Barry Mazur, *Rational points of abelian varieties with values in towers of number fields*, Invent. Math. **18** (1972), 183–266.

[Mil80] James S. Milne, *Étale cohomology*, Princeton Mathematical Series, vol. 33, Princeton University Press, Princeton, N.J., 1980.

[Mil86] J. S. Milne, *Arithmetic duality theorems*, Perspectives in Mathematics, vol. 1, Academic Press Inc., Boston, MA, 1986.

[Ray74] Michel Raynaud, *Schémas en groupes de type* $(p, \dots, p)$, Bull. Soc. Math. France **102** (1974), 241–280.

[R] Kenneth A. Ribet, *Raising the levels of modular representations*, Séminaire de Théorie des Nombres, Paris 1987–88, Progr. Math., vol. 81, Birkhäuser Boston, Boston, MA, 1990, pp. 259–271.

[Ser87] Jean-Pierre Serre, *Sur les représentations modulaires de degré* 2 *de* $\mathrm{Gal}(\overline{\mathbf{Q}}/\mathbf{Q})$, Duke Math. J. **54** (1987), no. 1, 179–230.

[SGA70] *Schémas en groupes. I: Propriétés générales des schémas en groupes*, Séminaire de Géométrie Algébrique du Bois Marie 1962/64 (SGA 3). Dirigé par M. Demazure et A. Grothendieck. Lecture Notes in Mathematics, Vol. 151, Springer-Verlag, Berlin, 1970.

[Shi94] Goro Shimura, *Introduction to the arithmetic theory of automorphic functions*, Publications of the Mathematical Society of Japan, vol. 11, Princeton University Press, Princeton, NJ, 1994, Reprint of the 1971 original, Kanô Memorial Lectures, 1.

[Shi98] ———, *Abelian varieties with complex multiplication and modular functions*, Princeton Mathematical Series, vol. 46, Princeton University Press, Princeton, NJ, 1998.

Uniform estimates for primitive divisors in elliptic divisibility sequences

Patrick Ingram and Joseph H. Silverman

Dedicated to the memory of Serge Lang

Abstract Let P be a nontorsion rational point on an elliptic curve E, given by a minimal Weierstrass equation, and write the first coordinate of nP as A_n/D_n^2, a fraction in lowest terms. The sequence of values D_n is the elliptic divisibility sequence (EDS) associated to P. A prime p is a primitive divisor of D_n if p divides D_n, and p does not divide any earlier term in the sequence. The Zsigmondy set for P is the set of n such that D_n has no primitive divisors. It is known that Z is finite. In the first part of the paper we prove various uniform bounds for the size of the Zsigmondy set, including (1) if the j-invariant of E is integral, then the size of the Zsigmondy set is bounded independently of E and P, and (2) if the abc Conjecture is true, then the size of the Zsigmondy set is bounded independently of E and P for all curves and points. In the second part of the paper, we derive upper bounds for the maximum element in the Zsigmondy set for points on twists of a fixed elliptic curve.

Key words elliptic divisibility sequence • elliptic curve • primitive divisor

Mathematics Subject Classification (2010): 11G05; Secondary 11B37, 14G25, 14H52

P. Ingram
Department of Pure Mathematics, Colorado State University, Fort Collins, CO, USA 80523
e-mail: pingram@math.colostate.edu

J.H. Silverman (✉)
Mathematics Department, Box 1917 Brown University, Providence, RI 02912 USA
e-mail: jhs@math.brown.edu

D. Goldfeld et al. (eds.), *Number Theory, Analysis and Geometry: In Memory of Serge Lang*, DOI 10.1007/978-1-4614-1260-1_12,

1 Introduction

Let $E/\mathbb{Q}$ be an elliptic curve given by a Weierstrass equation

$$y^2 + a_1xy + a_3y = x^3 + a_2x^2 + a_4x + a_6, \qquad a_1, \dots, a_6 \in \mathbb{Z} \tag{1}$$

and let $P \in E(\mathbb{Q})$ be a point of infinite order. For each $n \geq 1$ the nth iterate of P has the form

$$nP = \left(\frac{A_n}{D_n^2}, \frac{B_n}{D_n^3}\right),$$

where the fractions are written in lowest terms and we assume that $D_n > 0$. The sequence $\mathcal{D}_{E,P} = (D_n)_{n\geq 1}$ is called the *elliptic divisibility sequence* associated to E and P. It is a divisibility sequence in the sense that if $m|n$, then $D_m|D_n$, and in fact it satisfies the stronger divisibility relation

$$D_{\gcd(m,n)} = \gcd(D_m, D_n) \qquad \text{for all } m, n \geq 1.$$

(See Section 1.1 for a general statement over Dedekind domains.)

The study of the arithmetic properties of elliptic divisibility sequences was initiated by Morgan Ward in the 1940's [34, 35] and has seen a surge of interest in recent years, see for example [1, 2, 5, 6, 8–13, 18, 23, 31–33].

If $\mathcal{C} = (C_n)_{n\geq 1}$ is any divisibility sequence, one says that a prime p is a *primitive divisor of* C_n if $p|C_n$ but $p \nmid C_1C_2\cdots C_{n-1}$. Primitive divisors of certain divisibility sequences were studied by Zsigmondy [37] in the 19th century. We define the *Zsigmondy set* of a divisibility sequence $\mathcal{C}$ to be

$$Z(\mathcal{C}) = \{n \geq 1 : C_n \text{ does not have a primitive divisor}\}.$$

Zsigmondy was especially interested in divisibility sequences defined by binary linear recurrences satisfying appropriate growth conditions. Bilu, Hanrot, and Voutier [4] recently completed the proof that all of these Lucas and Lehmer sequences satisfy $\max Z(\mathcal{C}) \leq 30$, and there are examples to show that this bound is sharp. They also completely describe all such sequences with $\max Z(\mathcal{C}) \geq 12$. We note that Lucas divisibility sequences are associated to singular elliptic curves, so the material in this paper is, in some sense, a direct generalization of these earlier results.

It is a nontrivial fact that the Zsigmondy set $Z(\mathcal{D}_{E,P})$ of an elliptic divisibility sequence is finite, see [29]. It is natural to ask if there is a uniform bound for $Z(\mathcal{D}_{E,P})$ as there is for the case of binary linear recurrences. The answer is no unless some care is taken, since a simple change of variables $(x, y) \mapsto (u^2x, u^3y)$ allows one to multiply every term of the sequence by a power of u. This is the same trick that allows the creation of elliptic curves with arbitrarily many integer points, and the solution to both problems is the same, namely restrict attention to minimal Weierstrass equations. With this restriction, we prove a reasonably strong uniform bound for the number of elements in the set $Z(\mathcal{D}_{E,P})$, and assuming the abc-conjecture, we show that $\#Z(\mathcal{D}_{E,P})$ is bounded independently of E and P.

Theorem 1. *Let $E/\mathbb{Q}$ be an elliptic curve given by a minimal Weierstrass equation* (1)*, let $P \in E(\mathbb{Q})$ be a point of infinite order, and let $\mathcal{D}_{E,P}$ be the associated elliptic divisibility sequence.*

(a) $\#Z(\mathcal{D}_{E,P})$ is bounded by a constant depending only on the number of primes at which E has split multiplicative reduction.
(b) If the abc-conjecture over $\mathbb{Q}$ is true, then there is an absolute bound for $\#Z(\mathcal{D}_{E,P})$ that is completely independent of E and P.

Remark 1. Primes of split multiplicative reduction necessarily divide the denominator of the j-invariant, so a slightly weaker version of (a) is that $\#Z(\mathcal{D}_{E,P})$ is bounded by a constant depending only on the number of primes dividing the denominator of $j(E)$. So, for example, it is unconditionally true that there is an absolute bound for $\#Z(\mathcal{D}_{E,P})$ as E varies over all elliptic curves with integral j-invariant.

For (b), we prove an unconditional theorem that implies the stated result. We show that $\#Z(\mathcal{D}_{E,P})$ is bounded by a constant depending only on the *Szpiro ratio* of $E/\mathbb{Q}$ defined by

$$\text{Szpiro Ratio}(E/\mathbb{Q}) = \frac{\log|\,\text{Discriminant}\, E/\mathbb{Q}|}{\log|\,\text{Conductor}\, E/\mathbb{Q}|}.$$

It is well known that the (weak) abc-conjecture implies that the Szpiro ratio is bounded independently of E. More precisely, Szpiro has conjectured that for any $\epsilon > 0$ there are only finitely many elliptic curves $E/\mathbb{Q}$ whose Szpiro ratio exceeds $6 + \epsilon$. See [16] for a discussion.

Remark 2. We actually prove a general version of Theorem 1 over number fields. See Theorem 7 in Section 1.3 for the exact statement.

Theorem 1 gives uniform bounds for the size of the Zsigmondy set $Z(\mathcal{D}_{E,P})$, but it does not provide an effective bound for the largest element. Such upper bounds are not known in general for elliptic divisibility sequences, but various partial results are known. For illustrative purposes, we quote a result due to the first author.

Theorem 2. (Ingram [18]) *Let N be a squarefree integer, let E be the elliptic curve $y^2 = x^3 - N^2x$, and let $P \in E(\mathbb{Q})$ be a point of infinite order. Then*

$$Z(\mathcal{D}_{E,P}) \cap 2\mathbb{Z} \subset \{2\} \qquad \textit{and} \qquad Z(\mathcal{D}_{E,P}) \cap 5\mathbb{Z} = \emptyset.$$

Further, if $P \in 2E(\mathbb{Q})$ or if $x(P) < 0$, then $Z(\mathcal{D}_{E,P}) \subset \{1, 2\}$.

The main contribution of [18] is to provide, for fixed $n \geq 3$, an effective method for finding all elliptic divisibility sequences $\mathcal{D}_{E,P}$ arising from curves of the above form, such that $n \in Z(\mathcal{D}_{E,P})$. The problem of finding all such sequences is reduced to that of solving a certain Thue–Mahler equation involving the binary form

$$\prod_{Q}(X - x_Q Y),$$

where Q ranges over points on $E(\bar{\mathbb{Q}})$ of exact order n. Note that this is entirely analogous to the method used in [4], originally outlined by Schinzel [21], wherein the problem of finding all Lucas sequences $\mathcal{C}$ such that $n \in Z(\mathcal{C})$ is reduced to solving a Thue-Mahler equation involving the nth cyclotomic polynomial. Unfortunately, the analogy to [4] does not provide a bound on $\max Z(\mathcal{D}_{E,P})$, but bounds produced by ad hoc methods in [10] can be reduced, using the above observation, to those presented in Theorem 2.

This idea rests on the observation that the points of order n on E vary predictably as E runs over a family of quadratic twists of a fixed curve. Thus, one may show that, for fixed $n \geq 3$, the collection of elliptic divisibility sequences $\mathcal{D}_{E,P}$ such that $n \in Z(\mathcal{D}_{E,P})$, where E runs over the quadratic twists of *any* fixed elliptic curve, is finite and effectively computable. In Section 1.4, we present a result to this effect over number fields. Note that the proof can easily be modified to treat families of curves defined by quartic or sextic twisting, as in [18].

The methods used in [10] to obtain bounds on the largest element of $Z(\mathcal{D}_{E,P}) \cap 2\mathbb{Z}$ make critical use of the existence of rational points of order two on the curves in question, affording one a strong, explicit lower bound on the denominators of points that are divisible by two in the Mordell–Weil group. If one is willing to forsake the effective computability of these bounds, one may generalize these techniques. In the proof of theorem 6 we use Roth's theorem on diophantine approximation to show that

$$\max(Z(\mathcal{D}_{E,P}) \cap p\mathbb{Z})$$

may be bounded for any sufficiently large prime p if E ranges, again, over the quadratic twists of a fixed elliptic curve.

Finally, in Section 1.5 we turn our attention back to elliptic divisibility sequences over $\mathbb{Q}$ with the aim of seeing what certain common conjectures tell us about $\max Z(\mathcal{D}_{E,P})$. We show that if Hall's Conjecture is true, then $Z(\mathcal{D}_{E,P}) \subseteq \{1, 2, 3, 4\}$ for all but finitely many elliptic divisibility sequences arising from (minimal) curves of the form $E : y^2 = x^3 + M$. Analogous results can be obtained for other families with fixed j-invariant if one accepts a generalization of Hall's Conjecture due to Lang. Even with these generous assumptions, a uniform bound on $Z(\mathcal{D}_{E,P})$ seems out of reach.

Acknowledgements The first author's research supported in part by a grant from NSERC of Canada. The second author's research supported by NSA grant H98230-04-1-0064.

1.1 Elliptic divisibility sequences over Dedekind domains

In this section we prove some basic theorems concerning elliptic divisibility sequences over characteristic 0 Dedekind domains R. Let K be the fraction field of R and let E/K be an elliptic curve given by a Weierstrass equation (1) with coefficients $a_i \in R$. For any nonzero point $P = (x_P, y_P) \in E(K)$ we define the *denominator ideal of P* to be the ideal $D_P \subset R$ specified by

$$\operatorname{ord}_{\mathfrak{p}}(D_P) = \frac{1}{2}\max\{0, -\operatorname{ord}_{\mathfrak{p}}(x_P)\} \qquad \text{for all } 0 \neq \mathfrak{p} \in \operatorname{Spec}(R).$$

Here $\operatorname{ord}_{\mathfrak{p}} : K^* \twoheadrightarrow \mathbb{Z}$ is the normalized valuation associated to $\mathfrak{p}$. The *elliptic divisibility sequence* (EDS) associated to a nontorsion point $P \in E(K)$ is the sequence of ideals

$$\mathcal{D}_{E,P} = (D_{nP})_{n\geq 1}.$$

Proposition 1. *Let* $\mathcal{D}_{E,P}$ *be an* EDS *as above and let* $\mathfrak{p} \in \operatorname{Spec}(R)$ *be a nonzero prime ideal. Let* p *be the characteristic of* $R/\mathfrak{p}$ *and let* $e_{\mathfrak{p}} = \operatorname{ord}_{\mathfrak{p}}(p)$ *be the ramification index of* p.

(a) $\operatorname{ord}_{\mathfrak{p}}(D_{\gcd(m,n)P}) = \min\{\operatorname{ord}_{\mathfrak{p}}(D_{mP}), \operatorname{ord}_{\mathfrak{p}}(D_{nP})\}$ *for all* $m, n \geq 1$.

(b) If $\operatorname{ord}_{\mathfrak{p}}(D_{nP}) > e_{\mathfrak{p}}/(p-1)$, *then*

$$\operatorname{ord}_{\mathfrak{p}}(D_{knP}) = \operatorname{ord}_{\mathfrak{p}}(D_{nP}) + \operatorname{ord}_{\mathfrak{p}}(k) \qquad \textit{for all } k \geq 1.$$

Proof. It suffices to prove the proposition after localizing and completing R and K at $\mathfrak{p}$. For each integer $i \geq 1$ let

$$E_i(K) = \{Q \in E(K) : -\operatorname{ord}_{\mathfrak{p}}(x_Q) \geq 2i\}.$$

We also let $E_0(K) = E(K)$, which is not quite standard notation, but suffices for our purposes. Then all of the $E_i(K)$ are subgroups of E, see for example [27, Chap. IV]. We also observe that

$$\operatorname{ord}_{\mathfrak{p}}(D_Q) = \max\{i \geq 0 : Q \in E_i(K)\}.$$

Let m and n be positive integers and let $d = \gcd(m, n)$. Write $d = am + bn$ for some $a, b \in \mathbb{Z}$. Further let

$$i = \operatorname{ord}_{\mathfrak{p}}(D_{mP}), \qquad j = \operatorname{ord}_{\mathfrak{p}}(D_{nP}), \qquad \text{and} \qquad k = \operatorname{ord}_{\mathfrak{p}}(D_{dP}).$$

Thus $mP \in E_i(K)$ and $nP \in E_j(K)$. The fact that the $E_i(K)$ are subgroups of $E(K)$ allows us to conclude that

$$dP = a(mP) + b(nP) \in E_i(K) + E_j(K) = E_{\min\{i,j\}}(K).$$

Hence

$$k = \operatorname{ord}_{\mathfrak{p}}(D_{dP}) \geq \min\{i, j\}.$$

For the opposite inequality, we use the fact that $d \mid m$ to conclude that

$$mP = \frac{m}{d} \cdot dP \in \frac{m}{d} E_k(K) \subset E_k(K),$$

so $i = \mathrm{ord}_{\mathfrak{p}}(D_{mP}) \geq k$. Similarly $j \geq k$, which completes the proof of (a).

In order to prove (b), we use the fact that the subgroup $E_1(K)$ has the structure of a formal group and the $E_i(K)$ form a filtration of subgroups of $E_1(K)$. Further, for $i > e_{\mathfrak{p}}/(p-1)$, there are filtration compatible isomorphisms

$$E_i(K) \longrightarrow \mathfrak{p}^i, \tag{2}$$

where the group structure on $\mathfrak{p}^i$ is simply addition. (See [27, Chap. IV] for proofs of these basic facts.)

Now suppose that $\mathrm{ord}_{\mathfrak{p}}(D_{nP}) > e_{\mathfrak{p}}/(p-1)$. Then we can identify nP with some $z \in R$ satisfying $\mathrm{ord}_{\mathfrak{p}}(z) = \mathrm{ord}_{\mathfrak{p}}(D_{nP})$, and the formal group isomorphism (2) tells us that

$$\mathrm{ord}_{\mathfrak{p}}(D_{knP}) = \mathrm{ord}_{\mathfrak{p}}(kz) = \mathrm{ord}_{\mathfrak{p}}(k) + \mathrm{ord}_{\mathfrak{p}}(z) = \mathrm{ord}_{\mathfrak{p}}(k) + \mathrm{ord}_{\mathfrak{p}}(D_{nP}).$$

This completes the proof of (b).

Definition 1. Let $\mathcal{D}_{E,P}$ be an elliptic divisbility sequence as above. The *rank of apparition of* $\mathcal{D}_{E,P}$ *at the prime* $\mathfrak{p}$ is the smallest integer $r_{\mathfrak{p}} = r_{\mathfrak{p}}(E,P)$ with the property that $\mathfrak{p}$ divides $D_{r_{\mathfrak{p}}P}$.

Definition 2. We say that a prime $\mathfrak{p}$ is *exceptional* for the elliptic divisibility sequence $\mathcal{D}_{E,P}$ if

$$\mathrm{ord}_{\mathfrak{p}}(D_{r_{\mathfrak{p}}P}) \leq \frac{e_{\mathfrak{p}}}{p-1}.$$

We *define a modified rank of apparition* $s_{\mathfrak{p}}$ by

$$s_{\mathfrak{p}} = \min\left\{s \geq 1 : \mathrm{ord}_{\mathfrak{p}}(D_{sP}) > \frac{e_{\mathfrak{p}}}{p-1}\right\}. \tag{3}$$

Thus $\mathfrak{p}$ is exceptional if and only if $s_{\mathfrak{p}} > r_{\mathfrak{p}}$.

Remark 3. If $\mathfrak{p}$ is exceptional, then necessarily

$$1 \leq \mathrm{ord}_{\mathfrak{p}}(D_{r_{\mathfrak{p}}P}) \leq \frac{e_{\mathfrak{p}}}{p-1} \leq \frac{[K:\mathbb{Q}]}{p-1},$$

so $p \leq [K:\mathbb{Q}]+1$. In particular, if K is a number field, then there are only finitely many exceptional primes.

Remark 4. If the given Weierstrass equation for E has good reduction at $\mathfrak{p}$, then $r_{\mathfrak{p}}$ is the order of P in the group $E(\mathbb{F}_{\mathfrak{p}})$.

Proposition 2. *Let $K/\mathbb{Q}$ be a number field of degree d and let $\mathcal{D}_{E,P}$ be an elliptic divisibility sequence. Suppose that $m \in Z(\mathcal{D}_{E,P})$ is in the Zsigmondy set of $\mathcal{D}_{E,P}$.*

Then either $m = s_{\mathfrak{p}}$ *for some exceptional prime* $\mathfrak{p}$ *or else*

$$\log \mathsf{N}_{K/\mathbb{Q}} D_{mP} \le \sum_{\substack{k \mid m \text{ with} \\ m/k \text{ prime}}} \log \mathsf{N}_{K/\mathbb{Q}} D_{kP} + d \log(2d) + d \log(m).$$

Proof. Let $\mathfrak{p}$ be a prime dividing D_{mP}, let $r = r_{\mathfrak{p}}$ be the rank of apparition of $\mathfrak{p}$, and and let $s = s_{\mathfrak{p}}$ be the modified rank of apparition of $\mathfrak{p}$ defined by (3). We consider several cases:

Case 1. $s \nmid m$.
We know that $r|m$, so in particular we see that $s \ne r$. In other words, the prime $\mathfrak{p}$ is exceptional. Using the strong divisibility property Proposition 1(a), we find that

$$\operatorname{ord}_{\mathfrak{p}}\left(D_{\gcd(s,m)P}\right) = \min\left\{\operatorname{ord}_{\mathfrak{p}}(D_{sP}), \operatorname{ord}_{\mathfrak{p}}(D_{mP})\right\}.$$

The assumption that $s \nmid m$ implies that $\gcd(s,m)$ is strictly smaller than s, so the definition of the modified rank of apparition tells us that

$$\operatorname{ord}_{\mathfrak{p}}\left(D_{\gcd(s,m)P}\right) \le \frac{e_{\mathfrak{p}}}{p-1} < \operatorname{ord}_{\mathfrak{p}}(D_s P).$$

We conclude that

$$\operatorname{ord}_{\mathfrak{p}}(D_{mP}) = \operatorname{ord}_{\mathfrak{p}}\left(D_{\gcd(s,m)P}\right) \le \frac{e_{\mathfrak{p}}}{p-1}.$$

Hence the product over all primes in Case 1 satisfies

$$\prod_{\substack{\mathfrak{p} \\ s_{\mathfrak{p}} \nmid m}} \mathfrak{p}^{\operatorname{ord}_{\mathfrak{p}}(D_{mP})} \,\Big|\, \prod_{p} \prod_{\mathfrak{p} | p} \mathfrak{p}^{\lfloor e_{\mathfrak{p}}/(p-1) \rfloor} \,\Big|\, \prod_{p \le [K:\mathbb{Q}]+1} p^{1/(p-1)} \le 2[K:\mathbb{Q}].$$

(For our purposes, it would suffice to know that the penultimate product is bounded by a constant depending only on $[K:\mathbb{Q}]$. We do not actually need the sharp bound of $2[K:\mathbb{Q}]$.)

Case 2. $s|m$ and $s < m$.
In this case we can apply Proposition 1(b) to obtain the estimate

$$\operatorname{ord}_{\mathfrak{p}}(D_{mP}) = \operatorname{ord}_{\mathfrak{p}}(D_{s(m/s)P}) = \operatorname{ord}_{\mathfrak{p}}(D_{sP}) + \operatorname{ord}_{\mathfrak{p}}(m/s).$$

We also note that if $s_{\mathfrak{p}}|m$ with $s_{\mathfrak{p}} \ne m$, then $s_{\mathfrak{p}}$ necessarily divides some divisor k of m having the property that m/k is prime. Hence the product over all primes satisfying Case 2 is bounded by

$$\prod_{\substack{\mathfrak{p} \\ s_{\mathfrak{p}} | m \\ s_{\mathfrak{p}} \ne m}} \mathfrak{p}^{\operatorname{ord}_{\mathfrak{p}}(D_{mP})} \,\Big|\, \prod_{\substack{k|m \text{ with} \\ m/k \text{ prime}}} \prod_{\mathfrak{p}} \mathfrak{p}^{\operatorname{ord}_{\mathfrak{p}}(D_{kP}) + \operatorname{ord}_{\mathfrak{p}}(m/k)} \,\Big|\, m \prod_{\substack{k|m \text{ with} \\ m/k \text{ prime}}} D_{kP}.$$

(In this last expression, the factor of m may be replaced by the squarefree part of m.)

Case 3. $s = m$.
If $\mathfrak{p}$ is exceptional, we have ruled this out by assumption. And if $\mathfrak{p}$ is not exceptional, then $s = r$, and the assumption that m is in the Zsigmondy set implies that $m \neq r$. Hence Case (3) cannot occur.

We now combine the three cases to estimate the norm of D_{mP}. To ease notation, we let $d = [K : \mathbb{Q}]$. Then

$$\begin{aligned}\log \mathsf{N}_{K/\mathbb{Q}} D_{mP} &= \sum_{\mathfrak{p}\in\text{Case 1}} \operatorname{ord}_{\mathfrak{p}}(D_{mP}) \log \mathsf{N}_{K/\mathbb{Q}}\mathfrak{p} + \sum_{\mathfrak{p}\in\text{Case 2}} \operatorname{ord}_{\mathfrak{p}}(D_{mP}) \log \mathsf{N}_{K/\mathbb{Q}}\mathfrak{p} \\ &\leq d \log(2d) + d\log(m) + \sum_{\substack{k\mid m \text{ with} \\ m/k \text{ prime}}} \log \mathsf{N}_{K/\mathbb{Q}} D_{kP}.\end{aligned}$$

This completes the proof of Proposition 2.

We next prove an inequality relating the terms in an elliptic divisibility sequence $\mathcal{D}_{E,P}$ to the heights of the multiples of P. Roughly speaking, the quantity $\log \mathsf{N}_{K/\mathbb{Q}} D_Q$ is the nonarchimedean contribution to the canonical height $\hat{h}_E(Q)$, so we would expect $\log \mathsf{N}_{K/\mathbb{Q}} D_Q$ to be bounded by $\hat{h}_E(Q)$. This is indeed true, and we can make the dependence on E explicit by using standard results relating naive heights to canonical heights. Thus the following result is comparatively elementary compared to Theorem 3 that we prove in Section 1.2.

Proposition 3. *Let E/K be an elliptic curve given by a Weierstrass equation* (1), *let $P \in E(K)$ be a nontorsion point, and define the height of E to be*

$$h(E) = 1 + h\big([1, a_1^{12}, a_2^6, a_3^4, a_4^3, a_6^2]\big).$$

Then

$$\frac{1}{[K : \mathbb{Q}]} \log \mathsf{N}_{K/\mathbb{Q}} D_P \leq \hat{h}_E(P) + O\big(h(E)\big),$$

where the big-O constant is absolute.

Proof. With appropriate normalizations on the absolute values in K, the absolute logarithmic Weil height of $x(P)$ is

$$h\big(x(P)\big) = \sum_{v\in M_K} \max\big\{0, -v(x(P))\big\}.$$

If we sum over only the nonarchimedean places we obtain

$$h\big(x(P)\big) \geq \sum_{v\in M_K^0} \max\big\{0, -v(x(P))\big\} = \frac{2}{[K : \mathbb{Q}]} \log \mathsf{N}_{K/\mathbb{Q}} D_P.$$

Finally, we use a uniform estimate for the difference between the Weil height and the canonical height

$$\hat{h}(P) = \frac{1}{2}h(x(P)) + O(1 + h(E)), \tag{4}$$

see [7, 30, 36] for example, where the big-O constant is absolute.

1.2 A uniform quantitative version of Siegel's integrality theorem for elliptic curves

A famous theorem of Siegel says that an elliptic curve has only finitely many integral points. Siegel actually proved something much stronger. For any point $P \in E(\mathbb{Q})$, we write $x(P) = A_P/D_P^2$ in lowest terms with $D_P \geq 1$ and we set

$$h(P) = \frac{1}{2}\log\max\{|A_P|, D_P^2\}.$$

Then Siegel proved that

$$\lim_{\substack{P\in E(\mathbb{Q})\\ h(P)\to\infty}} \frac{\log D_P^2}{h(P)} = 1. \tag{5}$$

(See [27, IX.3.3].) Using the fact that $h(nP) \sim n^2\hat{h}(P)$, where $\hat{h}$ is the canonical height on E, this shows that elliptic divisibility sequences over $\mathbb{Q}$ grow very rapidly,

$$\lim_{n\to\infty} \frac{\log D_{nP}}{n^2} = \hat{h}(P) > 0. \tag{6}$$

And indeed it is an easy exercise using Siegel's deep result (6) and the elementary estimates given in Proposition 1 to prove that the Zsigmondy set $Z(D_{nP})$ of an elliptic divisibility sequence is finite, see [29] or [27, Exercise 9.4].

Siegel's proof of the finiteness of $E(\mathbb{Z})$ can be used to give an upper bound for $\#E(\mathbb{Z})$, but the bound depends rather badly on the equation defining E. Dem'janenko in a special case and Lang in general [19] made the following conjecture.

Conjecture 1. (Lang–Dem'janenko) Let $E/\mathbb{Q}$ be an elliptic curve given by a minimal Weierstrass equation. Then $\#E(\mathbb{Z})$ is bounded by a constant that depends only on the rank of $E(\mathbb{Q})$.

As in Siegel's work, rather than simply bounding the size of $E(\mathbb{Z})$, one can ask for a uniform bound for the number of points in $E(\mathbb{Q})$ that do not satisfy some inequality related to the limit (5). A bound of this sort was proven by the second author in [28], and in this section we apply the results from [28] to deduce uniform information about Zsigmondy sets of elliptic divisibility sequences.

However, we need to take some care, because an elliptic curve $E/\mathbb{Q}$ and a nontorsion point $P \in E(\mathbb{Q})$ do not completely determine an elliptic divisibility sequence. The reason is that the definition of the associated EDS uses a specific Weierstrass equation for $E/\mathbb{Q}$. One solution is to take a global minimal Weierstrass equation, which works fine over $\mathbb{Q}$, but unfortunately if K has class number larger than 1, then there exist elliptic curves E/K that do not have a global minimal Weierstrass equation [3, 26]. In this case one could work with a Néron model for E/K, but instead we will simply put the dependence on the choice of Weierstrass equation into our notation.

Let K be a number field and let R_K be its ring of integers. For a given 5-tuple of values $\mathbf{a} = (a_1, a_2, a_3, a_4, a_6) \in R_K^5$, let $E_\mathbf{a}$ denote the Weierstrass equation (1) with the given coefficients and define the height of $E_\mathbf{a}$ to be

$$h(E_\mathbf{a}) = 1 + h\big([1, a_1^{12}, a_2^6, a_3^4, a_4^3, a_6^2]\big).$$

The canonical height $\hat{h}$ is a positive definite quadratic form on the Mordell–Weil group $E_\mathbf{a}(K)$ modulo torsion, and we write

$$\lambda(E_\mathbf{a}/K) = \min\{\hat{h}(Q) : Q \in E_\mathbf{a}(K),\ Q \text{ nontorsion}\}.$$

In the language of the geometry of numbers, $\lambda(E_\mathbf{a}/K)$ is the first minimum of the quadratic form $\hat{h}$ on the lattice $E_\mathbf{a}(K)/E_\mathbf{a}(K)_{\text{tors}}$.

We can now state the special case of [28] which we need to give a uniform bound for the size of the Zsigmondy set of an EDS.

Theorem 3. *With notation as above, for all $\epsilon > 0$ and all $d \geq 1$ there is a constant $C = C(\epsilon, d)$ with the following property: Let $K/\mathbb{Q}$ be a number field of degree at most d, let $\mathbf{a} \in R_K^5$ be a 5-tuple so that $E_\mathbf{a}$ is an elliptic curve, let $P \in E_\mathbf{a}(K)$ be a nontorsion point, and let $M \geq 1$. Form the elliptic divisibility sequence $(D_{nP})_{n\geq 1}$ as described in Section* 1.1. *Then the set*

$$\left\{n \geq 1 : \frac{1}{[K:\mathbb{Q}]} \log \mathsf{N}_{K/\mathbb{Q}}(D_{nP}) \leq (1-\epsilon)n^2\hat{h}(P) + Mh(E_\mathbf{a})\right\}$$

has at most $C\sqrt{Mh(E_\mathbf{a})/\lambda(E_\mathbf{a}/K)}$ elements.

Proof. This is a version of Theorem 4.1 in [28], see also [14] for similar results with explicit (albeit huge) constants. We briefly indicate how the results in [28] imply our statement. A direct application of [28, Theorem 4.1] to the family of Weierstrass equations $E \to \mathbb{P}^5$ yields

$$\#\left\{P \in E_\mathbf{a}(K) : \sum_{v\in S} \lambda_{E_\mathbf{a},(O)}(P, v) \geq \epsilon \hat{h}_{E_\mathbf{a}}(P) - Mh(E_\mathbf{a})\right\}$$

$$\leq \#E_\mathbf{a}(K)_{\text{tors}} \cdot C^{1+\#S+\operatorname{rank} E_\mathbf{a}(K)} \left(\frac{Mh(E_\mathbf{a})}{\lambda(E_\mathbf{a}/K)}\right)^{\frac{1}{2}\operatorname{rank} E_\mathbf{a}(K)}. \qquad (7)$$

(We refer the reader to [28] for a complete description of the notation. In particular, the constant C depends only on $[K : \mathbb{Q}]$ and ϵ.) This is not quite what we want, since we are dealing with a rank-one torsion-free subgroup of $E_{\mathbf{a}}(K)$, namely the subgroup generated by a particular point $P \in E_{\mathbf{a}}(K)$. However, the proof in [28] is easily adapted to subgroups of $E(K)$, so in (7) we can replace $E_{\mathbf{a}}(K)$ by the set $\{nP : n \in \mathbb{Z}\}$, which also means that we put $\#E_{\mathbf{a}}(K)_{\text{tors}} = 1$ and rank $E_{\mathbf{a}}(K) = 1$. Further, we take $S = M_K^\infty$ to be the set of archimedean places of K, so we can absorb the dependence on $\#S$ into the constant C. Then for $P \in E_{\mathbf{a}}(K)$ we have the estimate (for a new constant $C = C([K : \mathbb{Q}], \epsilon)$)

$$\#\left\{ n \geq 1 : \sum_{v \in M_K^\infty} \lambda_{E_{\mathbf{a}},(O)}(nP, v) \geq \epsilon \hat{h}_{E_{\mathbf{a}}}(nP) - Mh(E_{\mathbf{a}}) \right\} \leq C \sqrt{\frac{Mh(E_{\mathbf{a}})}{\lambda(E_{\mathbf{a}}/K)}}. \tag{8}$$

The local height function $\lambda_{E_{\mathbf{a}},(O)}$ is given by

$$\lambda_{E_{\mathbf{a}},(O)}(Q, v) = \frac{1}{2} \max\{0, -v(x(Q))\},$$

where the valuations are normalized so that the absolute logarithmic height of $\alpha \in K^*$ is given by the formula $h(\alpha) = \sum_v \max\{0, -v(\alpha)\}$. Hence

$$\begin{aligned} \sum_{M_K^\infty} \lambda_{E_{\mathbf{a}},(O)}(Q, v) &= \frac{1}{2} \sum_{M_K^\infty} \max\{0, -v(x(Q))\} \\ &= \frac{1}{2} h(x(Q)) - \frac{1}{2} \sum_{M_K^0} \max\{0, -v(x)\} \\ &= \frac{1}{2} h(x(Q)) - \frac{1}{[K : \mathbb{Q}]} \log \mathsf{N}_{K/\mathbb{Q}}(D_Q) \\ &= \hat{h}_{E_{\mathbf{a}}}(Q) + O(h(E_{\mathbf{a}})) - \frac{1}{[K : \mathbb{Q}]} \log \mathsf{N}_{K/\mathbb{Q}}(D_Q). \end{aligned}$$

(For the last line we have again used (4). Note that the big-O constant is absolute.) Putting $Q = nP$ yields

$$\sum_{M_K^\infty} \lambda_{E_{\mathbf{a}},(O)}(nP, v) = \hat{h}_{E_{\mathbf{a}}}(nP) - \frac{1}{[K : \mathbb{Q}]} \log \mathsf{N}_{K/\mathbb{Q}}(D_{nP}) + O(h(E_{\mathbf{a}})).$$

Substituting this into (8) gives

$$\begin{aligned} \#\left\{ n \geq 1 : (1 - \epsilon)\hat{h}_{E_{\mathbf{a}}}(nP) + (M - C')h(E_{\mathbf{a}}) \geq \frac{1}{[K : \mathbb{Q}]} \log \mathsf{N}_{K/\mathbb{Q}}(D_{nP}) \right\} \\ \leq C \sqrt{\frac{Mh(E_{\mathbf{a}})}{\lambda(E_{\mathbf{a}}/K)}}, \end{aligned} \tag{9}$$

where $C' \geq 0$ is an absolute constant. Finally, we replace M by $M + C'$, use the inequality $M + C' \leq (1 + C')M$, and increase the value of C accordingly. Then (9) and the canonical height property $\hat{h}(nP) = n^2\hat{h}(P)$ give the desired result, which completes the proof of Theorem 3.

1.3 A uniform Zsigmondy estimate

We have now assembled all the tools needed to prove a uniform bound for the size of the Zsigmondy set of an elliptic divisibility sequence.

Theorem 4. *Continuing with the notation from Sections* 1.2 *and* 1.1, *let* $\mathbf{a} \in R_K^5$ *so that* $E_{\mathbf{a}}$ *is an elliptic curve and let* $P \in E_{\mathbf{a}}(K)$ *be a nontorsion point. Then there is a constant* $C = C\big([K:\mathbb{Q}]\big)$ *such that*

$$\#Z(\mathcal{D}_{E_{\mathbf{a}},P}) \leq C h(E_{\mathbf{a}})/\lambda(E_{\mathbf{a}}/K).$$

Proof. As noted in Remark 3, the number of exceptional primes is bounded by a constant depending only on $[K:\mathbb{Q}]$, so without loss of generality we may discard from $Z(\mathcal{D}_{E_{\mathbf{a}},P})$ all m with the property that $m = s_{\mathfrak{p}}$ for some exceptional prime $\mathfrak{p}$.

Let $m \in Z(\mathcal{D}_{E_{\mathbf{a}},P})$. Then Theorem 3 (with $\epsilon = \frac{1}{4}$ and $M = 1$) says that with at most $O\big(\sqrt{h(E_{\mathbf{a}})/\lambda(E_{\mathbf{a}}/K)}\big)$ exceptions, we have

$$\frac{1}{[K:\mathbb{Q}]}\log \mathsf{N}_{K/\mathbb{Q}}(D_{mP}) \geq \frac{3}{4}m^2\hat{h}(P). \tag{10}$$

In the other direction, Propositions 2 and 3 allow us to estimate

$$\begin{aligned}
&\frac{1}{[K:\mathbb{Q}]}\log \mathsf{N}_{K/\mathbb{Q}} D_{mP} \\
&\quad \leq \sum_{k|m,\,k\neq m} \frac{1}{[K:\mathbb{Q}]}\log \mathsf{N}_{K/\mathbb{Q}} D_{kP} + \log\big(2[K:\mathbb{Q}]\big) + \sqrt{m}\log(m) \\
&\quad \leq \sum_{k|m,\,k\neq m} \big(\hat{h}(kP) + O(h(E_{\mathbf{a}}))\big) + \log\big(2[K:\mathbb{Q}]\big) + \sqrt{m}\log(m) \\
&\quad \leq \sum_{k|m,\,k\neq m} k^2\hat{h}(P) + O(mh(E_{\mathbf{a}})).
\end{aligned} \tag{11}$$

We also have the trivial estimate

$$\sum_{k|m,\,k\neq m} k^2 \leq m^2 \sum_{k|m,\,k\geq 2} \frac{1}{k^2} \leq m^2\big(\zeta(2) - 1\big). \tag{12}$$

Combining (10), (11) and (12) yields

$$\left(\tfrac{7}{4} - \zeta(2)\right) m^2 \hat{h}(P) \le O\big(mh(E_{\mathbf{a}})\big).$$

Since $\frac{7}{4} - \zeta(2) \approx 0.105 > 0$, we find that

$$m \le O\big(h(E_{\mathbf{a}})/\hat{h}(P)\big) \le O\big(h(E_{\mathbf{a}})/\lambda(E_{\mathbf{a}}/K)\big).$$

All of the big-O constants depend only on $[K : \mathbb{Q}]$, so this completes the proof of Theorem 4.

In order to apply Theorem 4, we need some sort of upper bound for the ratio $h(E_{\mathbf{a}})/\lambda(E_{\mathbf{a}}/K)$. The denominator depends only on the K-isomorphism class of $E_{\mathbf{a}}$, while the numerator depends on the particular Weierstrass model. For example, if we let

$$u \star \mathbf{a} = [1, ua_1, u^2a_2, u^3a_3, u^4a_4, u^6a_6],$$

then $E_{u\star\mathbf{a}}$ is K-isomorphic to $E_{\mathbf{a}}$, but it is not hard to see that $h(E_{u\star\mathbf{a}}) = h(E_{\mathbf{a}}) + 12h(u) + O(1)$. Thus in order to obtain a completely uniform upper bound in Theorem 4, we must put some restriction on the choice of $\mathbf{a}$.

Definition 3. Put a partial order on R_K^5 by setting $\mathbf{a} \preceq \mathbf{b}$ if

$$E_{\mathbf{a}} \cong_{/K} E_{\mathbf{b}} \qquad \text{and} \qquad \mathsf{N}_{K/\mathbb{Q}}\operatorname{Disc}(E_{\mathbf{a}}) \le \mathsf{N}_{K/\mathbb{Q}}\operatorname{Disc}(E_{\mathbf{b}}).$$

A Weierstrass equation $E_{\mathbf{a}}$ is called *K-quasiminimal* if $\mathbf{a}$ is a minimal element for this partial order.

It is clear that every elliptic curve has a quasiminimal Weierstrass equation. The following is a natural generalization of a conjecture of Lang [19], which he made based on some preliminary work of Dem'janenko.

Conjecture 2. (Lang) Let $K/\mathbb{Q}$ be a number field. There is a positive constant $C = C(K)$ such that for all K-quasiminimal Weierstrass equations $E_{\mathbf{a}}$ over K we have

$$\lambda(E_{\mathbf{a}}/K) \ge Ch(E_{\mathbf{a}}).$$

Clearly Lang's Conjecture 2 combined with Theorem 4 imply that the size of the Zsigmondy set of an elliptic divisibility sequence on a K-quasiminimal elliptic curve is bounded by a constant depending only on $K/\mathbb{Q}$. We mention two other conjectures that turn out to be related to Lang's conjecture.

Definition 4. The *Szpiro ratio* of an elliptic curve E/K is the quantity

$$\sigma(E/K) = \frac{\log \mathsf{N}_{K/\mathbb{Q}}\operatorname{Disc}(E/K)}{\log \mathsf{N}_{K/\mathbb{Q}}\operatorname{Cond}(E/K)},$$

where Cond(E/K) is the conductor of E/K. (For our purposes, it would suffice to replace Cond(E/K) with the product of the primes at which E/K has bad reduction.)

Conjecture 3. (Szpiro) For any $\epsilon > 0$ there are only finitely many elliptic curves E/K satisfying $\sigma(E/K) \geq 6 + \epsilon$.

It is well known that the abc-conjecture of Masser and Osterlé implies Szpiro's conjecture. Less obvious is the fact that Lang's conjecture is a consequence of Szpiro's conjecture.

Theorem 5. (Hindry–Silverman [16]) *Szpiro's Conjecture* 3 *implies Lang's Conjecture* 2. *More precisely, there is a positive constant* $C = C(K)$ *such that for all* K*-quasiminimal Weierstrass equations* $E_{\mathbf{a}}$ *over* K *we have*

$$\lambda(E_{\mathbf{a}}/K) \geq C^{1+\sigma(E_{\mathbf{a}})} h(E_{\mathbf{a}}).$$

We also quote another partial result on Lang's conjecture in which the constant depends in a mild way on the elliptic curve.

Theorem 6. (Silverman [25]) *With notation as above, let* $\nu(E_{\mathbf{a}})$ *be the number of primes of* K *at which* $E_{\mathbf{a}}$ *has split multiplicative reduction. Then*

$$\lambda(E_{\mathbf{a}}/K) \geq C^{1+\nu(E_{\mathbf{a}})} h(E_{\mathbf{a}}).$$

Combining all of this material yields a number field version of the result (Theorem 1) stated in the introduction.

Theorem 7. *Let* $K/\mathbb{Q}$ *be a number field, let* E/K *be an elliptic curve given by a* K*-quasiminimal Weierstrass equation, let* $\sigma(E/K)$ *be the Szpiro ratio of* E/K, *and let* $\nu(E/K)$ *be the number of primes of* K *at which* E/K *has split multiplicative reduction. Let* $P \in E(K)$ *be a point of infinite order, let* $\mathcal{D}_{E,P}$ *be the associated elliptic divisibility sequence, and consider its Zsigmondy set* $Z(\mathcal{D}_{E,P})$. *There is a constant* $C = C(K)$ *depending only on* K *such that:*

(a) $\#Z(\mathcal{D}_{E,P}) \leq C^{1+\mathrm{n}(E/K)}$.

(b) $\#Z(\mathcal{D}_{E,P}) \leq C^{1+\sigma(E/K)}$.

If Szpiro's conjecture or the abc*-conjecture is true, then* $\#Z(\mathcal{D}_{E,P})$ *is bounded by a constant that depends only on* K.

Proof. The estimate in (a) follows by combining Theorems 4 and 6, and the estimate in (b) follows similarly by combining Theorems 4 and 5. The final statement is clear, since Szpiro's conjecture says that $\sigma(E/K)$ is bounded independently of E, and it is well known that the abc-conjecture implies Szpiro's conjecture. (Note that we actually only need a weak version of either conjecture, i.e., it suffices to have any exponent, we do not need $6+\epsilon$ in Szpiro's conjecture or $1+\epsilon$ in the abc-conjecture.)

1.4 Results for quadratic twists

Theorem 7 bounds $\#Z(\mathcal{D}_{E,P})$ for all curves E/K given by a K-quasiminimal Weierstrass equation with at most a fixed number of primes dividing the denominator of $j(E)$. It is certainly true, then, that $\#Z(\mathcal{D}_{E,P})$ is bounded as E varies over the K-quasiminimal quadratic twists of a fixed curve. It is natural to ask whether, in this specific context, one can bound max $Z(\mathcal{D}_{E,P})$ uniformly. Although a result of this sort seems out of reach at the moment, we can prove a strong bound in the case in which P is in the image of an isogeny. In particular, if we consider, for a sufficiently large prime p, nontorsion points $P \in pE(K)$ as E runs over (minimal) quadratic twists of some fixed curve, we can show that max $Z(\mathcal{D}_{E,P})$ is bounded by a constant depending only on $j(E)$ and p.

Explicit results along these lines, over $\mathbb{Q}$ and with $p = 2$, are given in [10], and we use similar techniques to obtain the more general results described above. In [18], the first author gave sharpened estimates for max $Z(\mathcal{D}_{E,P})$ when $P \in E(\mathbb{Q})$ and $j(E) \in \{0, 1728\}$. Specifically, if we consider minimal $E/\mathbb{Q}$ of the form

$$y^2 = x^3 + Ax \quad \text{or} \quad y^2 = x^3 + A,$$

then for a fixed integer k greater than 3 in the first case and 4 in the second, it was proven that the set of pairs $\{(A, P) : k \in Z(\mathcal{D}_{E,P})\}$ is finite and effectively computable. Thus any bound on max $Z(\mathcal{D}_{E,P})$ for a family of elliptic divisibility sequences arising from these curves may be, with a finite number of exceptions, reduced to a bound of 3 or 4, respectively. Furthermore, this finite set of exceptions is effectively (although often not practically) computable. We will extend these results to families of twists over number fields. For the remainder of the section we will, for simplicity, work with elliptic curves in short Weierstrass form (i.e., with $a_1 = a_2 = a_3 = 0$, in the notation above). We will say that such a model is K-quasiminimal if it has minimal discriminant among K-isomorphic curves in the same form. Such curves might not be K-quasiminimal in the sense above, but will be away from primes dividing 2 or 3.

To prove the next two theorems, we require tools that trace back, in spirit, to the original paper of Ward [35]. Ward considers sequences $(h_n)_{n\in\mathbb{Z}}$ satisfying the relation

$$h_{m+n}h_{m-n} = h_{m+1}h_{m-1}h_n^2 - h_{n+1}h_{n-1}h_m^2, \tag{13}$$

with $h_0 = 0$, $h_1 = 1$, and $h_2 | h_4$ (a definition that makes sense over any integral domain). It is worth noting that the sequences described in [35] do not correspond directly to the sequences discussed here, and we reserve the term *elliptic divisibility sequence* for the latter. For example, if we consider the elliptic divisibility sequence $\mathcal{D}_{E,P}$ defined by the point $(12, 36)$ on the elliptic curve $y^2 = x^3 - 36x$, we have

$$D_P = (1),\ D_{2P} = (2),\ D_{3P} = (23),\ D_{4P} = (140),\ D_{5P} = (52487)\ldots$$

It is clear that we cannot choose generators $h_n \in \mathbb{Z}$ of the ideals D_{nP} which satisfy the recursion (13), in particular because this would require (setting $m = 3$ and $n = 2$)

$$52487 = |h_5| \leq |h_4 h_2^3| + |h_3^3 h_1| = 13287.$$

Although our sequence $\mathcal{D}_{E,P}$ is not necessarily an elliptic divisibility sequence in the sense of Ward, one may associate a Ward-type sequence to it. Recall (from, e.g., [27, p. 105]) the division polynomials of an elliptic curve

$$E : y^2 = x^3 + Ax + B.$$

They are elements of the function field $K(E)$ defined by setting

$$\psi_0 = 0, \quad \psi_1 = 1, \quad \psi_2 = 2y, \quad \psi_3 = 3x^4 + 6Ax^2 + 12Bx - A^2,$$
$$\psi_4 = 4y(x^6 + 5Ax^4 + 20Bx^3 - 5A^2x^2 - 4ABx - 8B^2 - A^3),$$

and then using the recursion (13) to define the other ψ_n. It is an easy exercise to show that ψ_n^2 may be written as a polynomial in x for all n. More precisely, ψ_n (respectively ψ_n/y) may be written as a polynomial in x if n is odd (respectively even). We will abuse notation by writing either $\psi_n(P)$ or $\psi_n(x_P)$ when $\psi_n \in K(x)$.

If $P \in E(K)$, then $(\psi_n(P))_{n\in\mathbb{Z}}$ is a sequence (in K) satisfying (13), by construction. While we would like to consider sequences in R, rather than K, which relate to $\mathcal{D}_{E,P}$, it is difficult to do this if R is not a principal ideal domain. Note that if S_0 is a finite set of primes of R, it is always possible to extend S_0 to a finite set S such that the localization R_S of R at S is a principal ideal domain. If the R_S-ideal $D_P R_S$ is principal, and is generated by t, then a simple induction shows that the sequence

$$h_n = t^{n^2-1}\psi_n(P) \tag{14}$$

is a sequence in R_S satisfying (13). This sequence depends on the choice of t, but only up to multiplication by S-units. The following proposition, due essentially to Ward [35] and Ayad [1, 2], indicates how this sequence relates to $\mathcal{D}_{E,P}$.

Proposition 4. *Let E be an elliptic curve as above, let $P \in E(K)$ be a point of infinite order, let S be a finite set of primes such that R_S is a PID, and let $\mathfrak{p} \in$ Spec(R_S) be a nonzero prime ideal.*

(a) For all n,

$$\mathrm{ord}_{\mathfrak{p}}(h_n) \geq \mathrm{ord}_{\mathfrak{p}}(D_{nP}/D_P).$$

(b) If $\mathrm{ord}_{\mathfrak{p}}(h_n) > \mathrm{ord}_{\mathfrak{p}}(D_{nP}/D_P)$ for any n, then $\mathfrak{p}$ is a prime of bad reduction for E, and P has singular reduction modulo $\mathfrak{p}$.

Proof. To prove (a), we note that if $t \in R_S$ is a generator for $D_P R_S$, and we set

$$k_n = t^{2n^2}\left(x_P\psi_n(P)^2 - \psi_{n+1}(P)\psi_{n-1}(P)\right) \in R_S,$$

for each n, then $x_{nP} = k_n/(th_n)^2$ (see [27, p. 105]). As k_n and h_n are S-integral, and

$$\operatorname{ord}_{\mathfrak{p}}(D_{nP}) = \frac{1}{2}\max\{0, -\operatorname{ord}_{\mathfrak{p}}(x_{nP})\}$$

it follows at once that $\operatorname{ord}_{\mathfrak{p}}(D_{nP}) \leq \operatorname{ord}_{\mathfrak{p}}(th_n)$, and part (a) follows. Part (b) follows from Théorème A of [1].

We now turn our attention to families of quadratic twists of a fixed elliptic curve over K. Let $A, B \in R$ be non-zero, and let

$$E : y^2 = x^3 + Ax + B$$

be a K-quasiminimal elliptic curve. Another elliptic curve

$$E' : y^2 = x^3 + A'x + B'$$

with $A', B' \in R$, is a *quadratic twist* of E if there is a $\overline{K}$-isomorphism from E to E'. Writing down the possible isomorphisms between curves of this form [27, p. 49], we see that we must have $A' = \tau^4 A$ and $B' = \tau^6 B$, for some $\tau \in \overline{K}$. Furthermore,

$$\tau^2 = \frac{AB'}{A'B} \in K.$$

Note that if $S \subseteq \operatorname{Spec}(R)$ is a set of nonzero primes containing all primes dividing $AB(4A^3 + 27B^2)$, and such that R_S is a principal ideal domain, then E' can be K-quasiminimal only if $\tau^2 \in R_S$, and $\operatorname{ord}_{\mathfrak{p}}(\tau^2) \leq 1$ for all $\mathfrak{p} \notin S$. If S is such a set of primes, if $P \in E(K)$, and if h_n is defined as in (14), then

$$\operatorname{ord}_{\mathfrak{p}}(h_n) = \operatorname{ord}_{\mathfrak{p}}(D_{nP}/D_P)$$

for all $\mathfrak{p} \notin S$ and all n, by Proposition 4. This is, however, not the case for sequences $(h_n)_{n\geq\mathbb{Z}}$ and $\mathcal{D}_{E',P}$ corresponding to points on twists of E, since E' may have bad reduction at primes outside of S. The following proposition allows us to restrict the amount by which the orders of these quantities vary at primes of bad reduction for E'.

Proposition 5. *Let E be as above, and let S be a finite set of primes containing all divisors of $AB(4A^3 + 27B^2)$ and all ramified primes, such that R_S is a principal ideal domain. Let E' be a K-quasiminimal twist of E, and let $P \in E'(K)$ be a point of infinite order. If P has bad reduction at $\mathfrak{p} \notin S$, then*

$$\operatorname{ord}_{\mathfrak{p}}(h_n) = \begin{cases} \frac{n^2-1}{2} & \text{if } 2 \nmid n \\ \frac{n^2-4}{2} + \operatorname{ord}_{\mathfrak{p}}(h_2) + \operatorname{ord}_{\mathfrak{p}}(n) & \text{if } 2 \mid n. \end{cases}$$

Proof. Let $\tau \in \overline{K}$ be as above, so that $\tau^2 \in R_S$

$$E' : y^2 = x^3 + \tau^4 Ax + \tau^6 B.$$

The isomorphism $\phi : E \to E'$ is given by

$$\phi(x, y) = \left(x\tau^2, y\tau^3\right).$$

As E' has bad reduction at $\mathfrak{p} \notin S$, and as $\operatorname{Disc}(E') = \tau^{12} \operatorname{Disc}(E)$, we have $\operatorname{ord}_{\mathfrak{p}}(\tau^2) = 1$. Note that P can have singular reduction at $\mathfrak{p}$ only if $\operatorname{ord}_{\mathfrak{p}}(x_P) \geq 1$. If $\operatorname{ord}_{\mathfrak{p}}(x_P) > 1$, then we have $\operatorname{ord}_{\mathfrak{p}}(x_P^3 + \tau^4 Ax_P) > 3$, and so

$$2 \operatorname{ord}_{\mathfrak{p}}(y_P) = \operatorname{ord}_{\mathfrak{p}}(y_P^2) = \operatorname{ord}_{\mathfrak{p}}(x_P^3 + \tau^4 Ax_P + \tau^6 B) = \operatorname{ord}_{\mathfrak{p}}(\tau^6 B) = 3$$

(recalling that $\operatorname{ord}_{\mathfrak{p}}(B) = 0$ for $\mathfrak{p} \notin S$). This is impossible, as $\operatorname{ord}_{\mathfrak{p}}(y_P) \in \mathbb{Z}$, and hence we must have $\operatorname{ord}_{\mathfrak{p}}(x_P) = 1$.

Let $L = K(\tau)$, let $\mathfrak{p} = \mathfrak{q}^2$ in the ring of integers of L, and let $Q = \phi^{-1}(P) \in E(L)$. While Q is not K-rational, it should be noted that $x_Q = x_P/\tau^2 \in K$. As $\operatorname{ord}_{\mathfrak{p}}(x_P) = 1$, we have $\operatorname{ord}_{\mathfrak{p}}(x_Q) = 0$, and in particular $\mathfrak{q} \nmid D_Q$. From this, and the fact that E has good reduction at $\mathfrak{q}$, we see from Proposition 4 that

$$\operatorname{ord}_{\mathfrak{q}}(D_{nQ}) = \operatorname{ord}_{\mathfrak{q}}(\psi_{E,n}(Q))$$

for all n (where $\psi_{E,n}$ is the nth division polynomial for E). In particular,

$$\begin{aligned} \operatorname{ord}_{\mathfrak{q}}(D_{2Q}) &= \operatorname{ord}_{\mathfrak{q}}(y_Q) \\ &= \operatorname{ord}_{\mathfrak{q}}(y_P) - 3 \operatorname{ord}_{\mathfrak{q}}(\tau). \end{aligned}$$

We have seen that

$$\operatorname{ord}_{\mathfrak{q}}(y_P) = 2 \operatorname{ord}_{\mathfrak{p}}(y_P) = \operatorname{ord}_{\mathfrak{p}}(x_P^3 + \tau^4 Ax_P + \tau^6 B) \geq 3,$$

and so (as $\operatorname{ord}_{\mathfrak{q}}(y_P)$ is even)

$$\operatorname{ord}_{\mathfrak{q}}(D_{2Q}) > 0.$$

Because $\mathfrak{p} \notin S$, we must have $e_{\mathfrak{p}} = 1$, and thus $e_{\mathfrak{q}} = 2$ (in the extension $L/\mathbb{Q}$). It follows from Proposition 1(b) that

$$\operatorname{ord}_{\mathfrak{q}}(\psi_{E,n}(Q)) = \operatorname{ord}_{\mathfrak{q}}(D_{nQ}) = \begin{cases} 0 & \text{if } 2 \nmid n \\ \operatorname{ord}_{\mathfrak{q}}(D_{2Q}) + \operatorname{ord}_{\mathfrak{q}}(n) & \text{if } 2 \mid n. \end{cases}$$

An examination of the division polynomials shows that

$$\psi_{E',n}(P) = \tau^{n^2-1} \psi_{E,n}(Q),$$

and so (recalling that $\mathrm{ord}_{\mathfrak{p}}(D_P) = 0$),

$$\begin{aligned}\mathrm{ord}_{\mathfrak{q}}(h_n) &= \mathrm{ord}_{\mathfrak{q}}(\psi_{E',n}(P)) \\ &= \left(n^2 - 1\right) \mathrm{ord}_{\mathfrak{q}}(\tau) + \mathrm{ord}_{\mathfrak{q}}(\psi_{E,n}(Q)).\end{aligned}$$

For n odd, the proposition follows immediately, since $\mathrm{ord}_{\mathfrak{q}}(\tau) = 1$, and $\mathfrak{p} = \mathfrak{q}^2$. For n even, note that

$$\mathrm{ord}_{\mathfrak{q}}(D_{2Q}) = \mathrm{ord}_{\mathfrak{q}}(y_P) - 3\,\mathrm{ord}_{\mathfrak{q}}(\tau) = \mathrm{ord}_{\mathfrak{q}}(h_2) - 3,$$

and so

$$\mathrm{ord}_{\mathfrak{q}}(h_n) = \left(n^2 - 1\right) + \mathrm{ord}_{\mathfrak{q}}(h_2) - 3.$$

Again, we are done as $\mathfrak{p} = \mathfrak{q}^2$.

Theorem 8. *Let E be as above, and fix an integer $k \geq 3$. Then there are only finitely many pairs (P, E') such that E' is a K-quasiminimal twist of E, $P \in E'(K)$ is a point of infinite order, and $k \in Z(\mathcal{D}_{E',P})$.*

Proof. Let S be a set of primes of R which contains all prime divisors of $kAB(4A^3 + 27B^2)$, all ramified primes, and such that R_S is a PID.

By Möbius inversion we write the division polynomials of E as

$$\psi_{E,n} = \prod_{d|n} \psi^*_{E,n},$$

for some functions $\psi^*_{E,n} \in K(E)$, and the same for E'. Note that for $n \geq 3$, $\psi^*_{E',n} \in K(x)$. For functions $f(x) \in K(x)$, we will write

$$\tilde{f}(x, y) = y^{\deg(f)} f(x/y).$$

Now suppose that $P \in E'(K)$ is as in the statement of the theorem, so that D_{kP} has no primitive divisors. As R_S is a principal ideal domain, we will select $s, t \in R_S$ with $(t) = D_P$ and $s = x_P t^2$. Our first observation, from the definition of the division polynomials, is that

$$\tilde{\psi}_{E',n}(s, t^2) = \tilde{\psi}_{E,n}(s, t^2\tau^2)$$

for all n (recall that $\tau^2 \in R_S$ is square-free in R_S). Again, to simplify notation, we will set $h_n = \tilde{\psi}_{E',n}(s, t^2)$.

Let $\mathfrak{p} \notin S$ be a prime of R, and consider $\mathrm{ord}_{\mathfrak{p}}(\tilde{\psi}^*_{E,k}(s, t^2\tau^2))$. There are several cases.

Case 1. P has good reduction at $\mathfrak{p}$ and $r_{\mathfrak{p}} = 1$: In this case, by Proposition 4, we see that

$$\mathrm{ord}_{\mathfrak{p}}(D_{nP}/D_P) = \mathrm{ord}_{\mathfrak{p}}(h_n)$$

for all n. Note that $r_{\mathfrak{p}} = 1$ implies $\mathfrak{p} \mid t$, and hence $\mathfrak{p} \nmid s$. It is an easy exercise [27, p. 105] to show that

$$\psi_{E',m}^2(x) = m^2 x^{m^2-1} + \text{lower order terms}$$

for each m. Recalling that S contains all divisors of k, we have $\operatorname{ord}_{\mathfrak{p}}(d^2 s^{d^2-1}) = 0$ for any $d \mid k$, and so $\operatorname{ord}_{\mathfrak{p}}(h_d) = 0$ for any $d \mid k$. It follows at once that

$$\operatorname{ord}_{\mathfrak{p}}(\tilde{\psi}_k^*(s,t^2)) = 0.$$

Case 2. P has good reduction at $\mathfrak{p}$ and $r_{\mathfrak{p}} > 1$: In this case $\operatorname{ord}_{\mathfrak{p}}(D_P) = 0$, and so Proposition 4 gives us

$$\operatorname{ord}_{\mathfrak{p}}(D_{nP}) = \operatorname{ord}_{\mathfrak{p}}(h_n)$$

for all n. Furthermore, $\mathfrak{p}$ is not a primitive divisor of D_{kP}, and so either $\mathfrak{p} \nmid D_{kP}$, or else $r_{\mathfrak{p}}$ is a proper divisor of k. In the former case it is clear that $\operatorname{ord}_{\mathfrak{p}}(\psi_{E',n}^*(s,t^2)) = 0$, and so we will suppose that we are in the latter case. We have

$$\operatorname{ord}_{\mathfrak{p}}(D_{r_{\mathfrak{p}}mP}) = \operatorname{ord}_{\mathfrak{p}}(D_{r_{\mathfrak{p}}P}) + \operatorname{ord}_{\mathfrak{p}}(m)$$

for all m, and so if μ is the Möbius function, then

$$\begin{aligned}
\operatorname{ord}_{\mathfrak{p}}(\tilde{\psi}_{E',k}^*(s,t^2)) &= \sum_{d \mid k} \mu(d) \operatorname{ord}_{\mathfrak{p}}(h_{\frac{k}{d}}) \\
&= \sum_{d \mid k} \mu(d) \operatorname{ord}_{\mathfrak{p}}(D_{\frac{k}{d}P}) \\
&= \sum_{d \mid (k/r_{\mathfrak{p}})} \mu(d) \operatorname{ord}_{\mathfrak{p}}(D_{\frac{k}{d}P}),
\end{aligned}$$

as $\operatorname{ord}_{\mathfrak{p}}(D_{mP}) = 0$ if $r_{\mathfrak{p}} \nmid m$. Writing $k = k' r_{\mathfrak{p}}$, we obtain

$$\begin{aligned}
\operatorname{ord}_{\mathfrak{p}}(\tilde{\psi}_{E',k}^*(s,t^2)) &= \sum_{d \mid k'} \mu(d) \operatorname{ord}_{\mathfrak{p}}(D_{\frac{r_{\mathfrak{p}}k'}{d}P}) \\
&= \sum_{d \mid k'} \mu(d) \left(\operatorname{ord}_{\mathfrak{p}}(D_{r_{\mathfrak{p}}P}) + \operatorname{ord}_{\mathfrak{p}}(k'/d)\right) \\
&= 0,
\end{aligned}$$

as $\operatorname{ord}_{\mathfrak{p}}(k) = 0$, and as $\sum_{d \mid m} \mu(d) = 0$ for any $m \geq 2$.

Case 3. P has bad reduction at $\mathfrak{p}$: In this case, Proposition 5 ensures that $\operatorname{ord}_{\mathfrak{p}}(t) = 0$, $\operatorname{ord}_{\mathfrak{p}}(s) = 1 = \operatorname{ord}_{\mathfrak{p}}(\tau^2)$. Also, we have

$$\mathrm{ord}_{\mathfrak{p}}(h_n) = \begin{cases} \frac{n^2-1}{2} & \text{if } 2 \nmid n \\ \frac{n^2-4}{2} + \mathrm{ord}_{\mathfrak{p}}(h_2) + \mathrm{ord}_{\mathfrak{p}}(n) & \text{if } 2 \mid n. \end{cases}$$

If k is odd, we have

$$\begin{aligned} \mathrm{ord}_{\mathfrak{p}}(\tilde{\psi}^*_{E',n}(s,t^2)) &= \sum_{d\mid k} \mu(d)\,\mathrm{ord}_{\mathfrak{p}}(h_{\frac{k}{d}}) \\ &= \sum_{d\mid k} \mu(d)\deg(\psi_{E',k/d}) = \deg(\psi^*_{E',n}). \end{aligned}$$

If k is even, we have (recall that $\mathrm{ord}_{\mathfrak{p}}(k) = 0$)

$$\begin{aligned} \mathrm{ord}_{\mathfrak{p}}(\tilde{\psi}^*_{E',n}(s,t^2)) &= \sum_{d\mid k} \mu(d)\,\mathrm{ord}_{\mathfrak{p}}(h_{\frac{k}{d}}) \\ &= \sum_{d\mid k} \mu(d)\deg(\psi_{E',k/d}) + \sum_{\substack{d\mid k \\ k/d \text{ even}}} \mu(d)\left(\mathrm{ord}_{\mathfrak{p}}(h_2) - \frac{3}{2}\right) \\ &= \deg(\psi^*_{E',n}). \end{aligned}$$

The second term in the penultimate line vanishes as $\sum_{d\mid(k/2)} \mu(d) = 0$.

We now have a value for $\mathrm{ord}_{\mathfrak{p}}(\tilde{\psi}^*_{E',n}(s,t^2))$ in each case. To summarize, if we choose a generator $(g) = sR_S + \tau^2 R_S$, we have

$$\mathrm{ord}_{\mathfrak{p}}(\tilde{\psi}^*_{E',n}(s,t^2)) = \deg(\psi^*_{E',n})\,\mathrm{ord}_{\mathfrak{p}}(g)$$

for each $\mathfrak{p} \notin S$. In particular,

$$\tilde{\psi}^*_{E,n}(s/g, t^2\tau^2/g) = g^{-\deg(\psi^*_{E',n})}\tilde{\psi}^*_{E',n}(s,t^2) \tag{15}$$

is an S-unit. But s/g and $t^2\tau^2/g$ are S-integers, and

$$\psi^*_{E,n}(x) = \tilde{\psi}^*_{E,n}(x,1)$$

has at least three distinct roots for $n \geq 3$. Thus (15) defines a Thue–Mahler equation, which has only finitely many solutions. Each solutions traces back to a unique pair $(E', \pm P)$, proving the result.

We now turn our attention to the claim that, for a sufficiently large prime p, $\max Z(\mathcal{D}_{E',P})$ may be bounded for points $P \in pE'(K)$ as E' varies through a family of quadratic twists. Translating the problem back to E, this will require us to obtain a lower bound on sizes of the 'denominators' D_Q of points $Q \in E(\overline{K})$ such that $[K(Q):K] \leq 2$. As mentioned above, Siegel's theorem allows us to conclude that

$$(1-\epsilon)\hat{h}(Q) \leq \frac{1}{[K(Q):\mathbb{Q}]} \log \mathsf{N}_{K(Q)/\mathbb{Q}} D_Q + O(1),$$

but the implied constant depends on the particular field $K(Q)$. This is insufficient for our needs. It turns out, though, that if p is large enough, we can obtain more uniform estimate for points $Q = pQ'$ with $Q' \in E(\overline{K})$ such that $x_{Q'} \in K$.

Proposition 6. *Let E be as above, fix a rational prime $p \geq 3$, and let $\delta > 0$. Let $Q \in E(\overline{K})$ be such that $x_Q \in K$. Then*

$$\left(1 - \frac{[K:\mathbb{Q}](2+\delta)}{p^2}\right)\hat{h}(pQ) \leq \frac{1}{[K(Q):\mathbb{Q}]} \log \mathsf{N}_{K(Q)/\mathbb{Q}} D_{pQ} + O(1),$$

where the implied constant depends only on E, K, δ, and p.

Proof. We begin by noting that there is a map $f : \mathbb{P}^1 \to \mathbb{P}^1$ of degree p^2 such that

$$x_{pQ} = f(x_Q)$$

for all $Q \in E(\overline{K})$. Explicitly, in terms of the division polynomials, we may write

$$f = x - \frac{\psi_{n+1}\psi_{n-1}}{\psi_n^2}.$$

This is, a priori, an element of $K(E)$, but is easily shown to lie in $K(x)$.

Writing $\|x\|_v = |x|_v^{[K_v:\mathbb{Q}_v]}$ for all $v \in M_K$, and letting S denote the set of archimedean places of K, we have

$$\begin{aligned}\frac{1}{[K(Q):\mathbb{Q}]} \log \mathsf{N}_{K(Q)/\mathbb{Q}} D_{pQ} &= \frac{1}{[K:\mathbb{Q}]} \sum_{v \notin S} \frac{1}{2} \max\left\{0, \log \|x_{pQ}\|_v\right\} \\ &= \frac{1}{2} h(x_{pQ}) - \frac{1}{2[K:\mathbb{Q}]} \sum_{v \in S} \max\left\{0, \log \|x_{pQ}\|_v\right\}.\end{aligned}$$

We have (by the basic properties of heights; see Theorems 5.6 and 9.3(e) of [27])

$$h(x_{pQ}) = h(f(x_Q)) = p^2 h(x_Q) = 2p^2\hat{h}(Q) + O(1),$$

where the implied constant depends on E and p.

We will now apply a version of Roth's Theorem to the poles of $f(x)$. Specifically, by Theorem D.9.3 of [17] applied to $\frac{1}{f} \in K(x)$, we find that there is a constant $c > 0$ such that

$$\sum_{v \in S} \min\left\{0, \log \left\|\frac{1}{f(x)}\right\|_v\right\} \geq -[K:\mathbb{Q}]\#S(2+\epsilon)h(x) - \log c$$

for all $x \in K$. In particular, we have

$$\frac{-1}{2[K:\mathbb{Q}]}\sum_{v\in S}\max\left\{0,\log\|f(x_Q)\|_v\right\} \geq -\#S(2+\epsilon)\hat{h}(Q)-c',$$

for some constant c' depending only on E, p, and K. Combining these estimates, and noting that $\#S \leq [K:\mathbb{Q}]$, we see that

$$\frac{1}{2}h(x_{pQ}) - \frac{1}{2[K:\mathbb{Q}]}\sum_{v\in S}\max\left\{0,\log\|x_{pQ}\|_v\right\}$$

$$\geq \left(p^2-[K:\mathbb{Q}](2+\epsilon)\right)\hat{h}(Q)-c'',$$

for some c''. This proves the proposition, as $\hat{h}(pQ) = p^2\hat{h}(Q)$.

Ultimately, under the assumption that the nth term of $\mathcal{D}_{E',P}$ has no primitive divisor, we wish to apply Proposition 2 to derive an upper bound on D_{mP}. The proposition provides no such bound, though, if $m = s_{\mathfrak{p}}$ for some exceptional prime $\mathfrak{p}$. In the proof of Theorem 4 we overcame this by employing the observation that the number of exceptional primes is bounded in terms of $[K:\mathbb{Q}]$. The next proposition shows that, once attention is restricted to a family of quadratic twists, $\max\{s_{\mathfrak{p}} : \mathfrak{p} \text{ is exceptional}\}$ may be similarly constrained.

Proposition 7. *Let E' be an elliptic curve over K, let $P \in E'(K)$, and let $\mathfrak{p}$ be an exceptional prime for $\mathcal{D}_{E',P}$. Then $s_{\mathfrak{p}} \leq M$, for some quantity M depending only on K and $j(E')$.*

Proof. Let $\mathfrak{p}$ be an exceptional prime for $\mathcal{D}_{E',P}$, and let p be the characteristic of $R/\mathfrak{p}$. Let

$$E'_i(K) = \{Q \in E'(K) : -\operatorname{ord}_{\mathfrak{p}}(x_Q) \geq 2i\},$$

as in Proposition 1, and note that $s_{\mathfrak{p}}$ is simply the order of P in $E'(K)/E'_N(K)$, for N the least integer greater than $e_{\mathfrak{p}}/(p-1)$. Note, as per [27, Ch. IV] and the proof of Proposition 1, that $E'_1(K_{\mathfrak{p}})$ is isomorphic to a formal group, and that if z is the coordinate corresponding to $Q \in E'_1(K)$, then $\operatorname{ord}_{\mathfrak{p}}(z) = \operatorname{ord}_{\mathfrak{p}}(D_Q)$. Furthermore, the multiplication-by-p map in the formal group is given by a power series of the form

$$[p]z = pz + O(z^2).$$

Now suppose that $Q \in E'_1(K)$ is not trivial modulo $E'_N(K)$, i.e., that $\operatorname{ord}_{\mathfrak{p}}(D_Q) \leq e_{\mathfrak{p}}/(p-1)$. Then

$$\operatorname{ord}_{\mathfrak{p}}(D_{pQ}) = \operatorname{ord}_{\mathfrak{p}}([p]z) = \operatorname{ord}_{\mathfrak{p}}(z) + \operatorname{ord}_{\mathfrak{p}}(p+O(z)) \geq 2\operatorname{ord}_{\mathfrak{p}}(z),$$

since

$$\operatorname{ord}_{\mathfrak{p}}(z) \leq \frac{e_{\mathfrak{p}}}{p-1} \leq e_{\mathfrak{p}} = \operatorname{ord}_{\mathfrak{p}}(p).$$

By induction, the order of Q in $E'_1(K)/E'_N(K)$ is at most p^n, where n is the least integer greater than $\log_2 N$. This bounds the order of any element of $E'_1(K)/E'_N(K)$ in terms of $e_\mathfrak{p}$ and p.

It now suffices to bound $r_\mathfrak{p}$, which is the order of P in $E'(K)/E'_1(K)$. If E' has good reduction at $\mathfrak{p}$, then

$$E'(K)/E'_1(K) \cong \tilde{E}(R/\mathfrak{p}),$$

where $\tilde{E}$ is the reduction of E' modulo $\mathfrak{p}$. Thus, by Hasse's theorem the order of P in $E'(K)/E'_1(K)$ is at most $(\sqrt{\#(R/\mathfrak{p})}+1)^2$. If E' has bad reduction at $\mathfrak{p}$, let $E'_0(K) \subseteq E'(K)$ be the subgroup of points with nonsingular reduction modulo $\mathfrak{p}$. Hasse's theorem again bounds the size of $E'_0(K)/E'_1(K)$, while the size of $E'(K)/E'_0(K)$ is at most $\max\{4, -\operatorname{ord}_\mathfrak{p}(j(E'))\}$, by a theorem of Kodaira and Néron [27, Theorem 6.1].

Thus, for each $\mathfrak{p}$, we have $s_\mathfrak{p} \le M_\mathfrak{p}$ for some quantity $M_\mathfrak{p}$ depending only on $\mathfrak{p}$ and $j(E')$ (in fact, we have not yet needed the fact that $\mathfrak{p}$ is exceptional). As noted in Remark 3, $\mathfrak{p}$ can be exceptional only if $p \le [K:\mathbb{Q}]+1$. Considering the maximum of the $M_\mathfrak{p}$ over the finitely many primes $\mathfrak{p}$ with $p \le [K:\mathbb{Q}]+1$, we have our bound M.

Theorem 9. *Let E be a K-quasiminimal elliptic curve, let $\epsilon > 0$, and let $p \ge \sqrt{[K:\mathbb{Q}](4+\epsilon)}$ be a rational prime. Then for each K-quasiminimal twist E' of E and each nontorsion point $P \in pE'(K)$,* $\max Z(\mathcal{D}_{E',P}) < C$ *for some constant C which depends on E, K, and p, but not on E' or P.*

Proof. Suppose that D_{mP} has no primitive divisor. By Proposition 7 there is a quantity M, depending only on K and $j(E') = j(E)$, such that $s_\mathfrak{p} \le M$ for all exceptional primes $\mathfrak{p}$. As our goal is to provide a bound on m that depends only on E and K, we will assume that $m > M$. By Proposition 2, then we see that

$$\log |\mathsf{N}_{K/\mathbb{Q}} D_{mP}| \le \sum_{\substack{k|m \text{ with}\\ m/k \text{ prime}}} \log \mathsf{N}_{K/\mathbb{Q}} D_{kP} + d\log(2d) + d\log(m),$$

where $d = [K:\mathbb{Q}]$. Note that

$$\begin{aligned}\sum_{\substack{k|m \text{ with}\\ m/k \text{ prime}}} \log \mathsf{N}_{K/\mathbb{Q}} D_{kP} &\le \sum_{\substack{k|m \text{ with}\\ m/k \text{ prime}}} \frac{d}{2} h(x_{kP}) \\ &\le \sum_{\substack{k|m \text{ with}\\ m/k \text{ prime}}} (d\hat{h}(kP) + O(h(E'))) \\ &\le d\hat{h}(mP) \sum_{q|m} \frac{1}{q^2} + O(\log(m)h(E')).\end{aligned}$$

As before, we have an isomorphism $E \to E'$ by $(x, y) \mapsto (x\tau^2, y\tau^3)$ for some $\tau \in \overline{K}$ with $\tau^2 \in K$. Taking d to be fixed, and noting that $h(E') \ll 1 + h(\tau)$, we have (if $m \in Z(\mathcal{D}_{E',P})$),

$$\log |\mathsf{N}_{K/\mathbb{Q}} D_{mP}| \le d\rho(m)\hat{h}(m^2 P) + O\big(\log(m)(1 + h(\tau))\big),$$

where $\rho(m) = \sum_{q|m} q^{-2}$.

On the other hand, we have $x_{mQ} = x_{mP}/\tau^2$, and so

$$\frac{1}{[K(Q) : K]} \log \mathsf{N}_{K(Q)/\mathbb{Q}} D_{mQ} \le \log \mathsf{N}_{K/\mathbb{Q}} D_{mP} + dh(\tau).$$

Finally, if $P \in pE(K)$, we may write $P = pP'$, for some $P' \in E(K)$, and set $Q' = \phi^{-1}(P')$. It follows from Proposition 6 (applied with $Q = Q'$ and $\delta = \epsilon/2$) that

$$\begin{aligned}\left(1 - \frac{d(2 + \epsilon/2)}{p^2}\right) d\hat{h}(mP) &= \left(1 - \frac{d(2 + \epsilon/2)}{p^2}\right) d\hat{h}(mQ) \\ &\le \frac{1}{[K(Q) : K]} \log \mathsf{N}_{K(Q)/\mathbb{Q}} D_{mQ} + O(1) \\ &\le \log \mathsf{N}_{K/\mathbb{Q}} D_{mP} + O(h(\tau)) \\ &\le d\rho(m)\hat{h}(m^2 P) + O\big(\log(m)h(\tau)\big). \qquad (16)\end{aligned}$$

By Theorem 6, there is a constant $\delta > 0$ depending only on $j(E)$ and K such that $\hat{h}(P) \ge \delta(1 + h(\tau))$ (on the assumption that P is not a point of finite order). Dividing both sides of (16) by $m^2\hat{h}(P)$ then yields

$$1 - \frac{d(2 + \epsilon/2)}{p^2} \le \rho(m) + O\left(\frac{\log(m)}{m^2}\right), \qquad (17)$$

where the implied constant depends only on E, K, and p. But

$$\rho(m) \le \sum_{q \text{ prime}} q^{-2} \le \frac{1}{2^2} + \frac{1}{3^2} + \cdots + \frac{1}{13^2} + \sum_{n=17}^{\infty} \frac{1}{n^2} < \frac{1}{2},$$

and so (17) bounds m, since our condition on p ensures that

$$\frac{d(2 + \epsilon/2)}{p^2} \le \frac{1}{2}.$$

This proves the result.

Finally, we should note that the restriction in this section to curves in short Weierstrass form is not, strictly speaking, necessary. Any elliptic curve over K may be written in short Weierstrass form with a change of variables. If $\mathcal{D}_{E,P}$ is a given elliptic divisibility sequence, we may choose this transformation such that the values of D_{nP} are unchanged at primes not dividing 6. By enlarging the set S in the statement of Theorem 8, then we can treat all K-quasiminimal models of elliptic curves in a given family of quadratic twists (as long as the finiteness in the conclusion of the theorem is now interpreted as finiteness up to this sort of change of variables). Similarly, Theorem 9 may be made independent of finitely many primes (for example, those above 2 and 3) by increasing the set S appearing in the proof of Proposition 6 to include these primes. In this case, however, we must require that $p > \sqrt{\#S(4+\epsilon)}$ (where S contains at least all infinite primes).

1.5 Speculative results over $\mathbb{Q}$

As we have seen in Theorem 7, there is a uniform bound on the size of the set $Z(\mathcal{D}_{E,P})$ if one is prepared to accept the conjecture of Szpiro. It seems, not surprisingly, rather more difficult to establish a uniform bound on the largest element in the set $Z(\mathcal{D}_{E,P})$. However, if one restricts attention to certain families of elliptic curves, then a bound may be obtained under similar assumptions. For simplicity, we work over $\mathbb{Q}$.

Conjecture 4. (Hall [15]) For every $\epsilon > 0$ there exists a constant C_ϵ such that whenever x, y, and $M \neq 0$ are integers satisfying

$$y^2 = x^3 + M,$$

then there is a bound of the form $|x| < C_\epsilon M^{2+\epsilon}$.

Theorem 10. *Suppose Hall's Conjecture 4 holds. Then there is a finite set $\mathcal{E}$ such that if M is a sixth-power free integer, $E : y^2 = x^3 + M$, and $P \in E(\mathbb{Q})$ is nontorsion, then $(M, P) \in \mathcal{E}$ or*

$$Z(\mathcal{D}_{E,P}) \subseteq \{1, 2, 3, 4\}.$$

Proof. In light of Theorem 3 of [18] (the analogue of Theorem 8 in this paper), it suffices to produce a uniform bound on $\max Z(\mathcal{D}_{E,P})$ for all E of the above form. Suppose $Q \in E(\mathbb{Q})$ is a point of infinite order, and let

$$Q = \left(\frac{A_Q}{D_Q^2}, \frac{B_Q}{D_Q^3}\right)$$

as usual. We have $B_Q^2 = A_Q^3 + MD_Q^6$, and thus, for a fixed $\epsilon > 0$, the conjecture of Hall tells us that

$$|A_Q| < C_\epsilon (MD_Q^6)^{2+\epsilon}.$$

It follows that

$$h_x(Q) \leq 6(2+\epsilon)\log|D_Q| + O(\log|M|),$$

where the implied constant depends on ϵ. So we have, for any $Q \in E(\mathbb{Q})$ of infinite order,

$$\log|D_Q| \geq \frac{1}{3(2+\epsilon)}\hat{h}(Q) + O(\log|M|). \tag{18}$$

Now let $P \in E(\mathbb{Q})$. We have by Propositions 2 and 3 that $n \in Z(\mathcal{D}_{E,P})$ only if

$$\log|D_{nP}| \leq \rho(n)n^2\hat{h}(P) + O(\log(n)\log|M|), \tag{19}$$

with explicit constants. Consider (18) with $Q = nP$. Noting that

$$\log|M| \leq O(\hat{h}(P)),$$

we have

$$\frac{1}{3(2+\epsilon)}n^2 < \rho(n)n^2 + O(\log(n)).$$

If we take $\epsilon \leq 1$, we see that n is bounded by some N as long as $\rho(n) < 0.1$, say. The latter condition is ensured if $(n, 6) = 1$. Appealing to Theorem 6, our bound is

$$\max\{N, \max(Z(\mathcal{D}_{E,P}) \cap 2\mathbb{Z}), \max(Z(\mathcal{D}_{E,P}) \cap 3\mathbb{Z})\}.$$

Remark 5. Note that, much as in Theorem 7, we only really need the weaker assumption that Hall's Conjecture holds for sufficiently large ϵ. If (18) holds for any value of ϵ, then (19) yields an upper bound on the values $n \in Z(\mathcal{D}_{E,P})$ such that $\rho(n) < \frac{1}{6(2+\epsilon)}$, for example. But it is easy to check that if $\rho(n) \geq \frac{1}{6(2+\epsilon)}$, then n has a prime divisor $p \leq 6(2+\epsilon)$. Computing a bound as in Theorem 6 for each such p, we obtain a uniform bound on $\max Z(\mathcal{D}_{E,P})$ for $j(E) = 0$.

Extending this idea, one can prove a general result if one accepts two stronger conjectures, both due to Lang. The first was already stated over number fields (Conjecture 2), but we restate it here over $\mathbb{Q}$ for the convenience of the reader, while the other is a genarlization of Conjecture 4.

Conjecture 5. (Lang [19]) There is an absolute constant $C > 0$ such that for every minimal $E/\mathbb{Q}$ and every nontorsion $P \in E(\mathbb{Q})$,

$$\hat{h}(P) > Ch(E).$$

Conjecture 6. (Hall–Lang [20]) There exist absolute constants C_1 and C_2 such that if x, y, A, B are integers with $4A^3 + 27B^2 \neq 0$ and

$$y^2 = x^3 + Ax + B,$$

then $|x| < C_1 \max\{|A|, |B|\}^{C_2}$.

Theorem 11. *Suppose that Conjectures 5 and 6 hold. Then there exist absolute constants M_1 and M_2 such that for all $E/\mathbb{Q}$ and $P \in E(\mathbb{Q})$, if $n \in Z(\mathcal{D}_{E,P})$ then either $n < M_1$ or there is a prime $p < M_2$ such that $p|n$.*

The proof of this theorem is nearly identical to the proof of Theorem 10.

Remark 6. If one restricts attention to a given family of quadratic twists, then Conjecture 5 is known to hold. Thus, if one assumes that Conjecture 6 holds, at least for the given family of twists, one may apply Theorems 8 and 6 to obtain a statement analogous to Theorem 10. That is, one deduces that except for a finite number of exceptions, elliptic divisibility sequences $\mathcal{D}_{E,P}$ arising from this family of twists satisfy $Z(\mathcal{D}_{E,P}) \subseteq \{1, 2\}$. Note also that, under the assumption of Conjecture 6 for all elliptic curves, a uniform version of Theorem 6 would provide a uniform bound on $Z(\mathcal{D}_{E,P})$ for curves $E/\mathbb{Q}$. Such a uniform statement, however, requires a refinement of Roth's theorem that is far beyond current Diophantine analysis.

References

1. Mohamed Ayad. Points S-entiers des courbes elliptiques. *Manuscripta Math.*, 76(3-4): 305–324, 1992.
2. Mohamed Ayad. Périodicité (mod q) des suites elliptiques et points S-entiers sur les courbes elliptiques. *Ann. Inst. Fourier (Grenoble)*, 43(3):585–618, 1993.
3. Ebru Bekyel. The density of elliptic curves having a global minimal Weierstrass equation. *J. Number Theory*, 109(1):41–58, 2004.
4. Yu. Bilu, G. Hanrot, and P. M. Voutier. Existence of primitive divisors of Lucas and Lehmer numbers. *J. Reine Angew. Math.*, 539:75–122, 2001. With an appendix by M. Mignotte.
5. J. Cheon and S. Hahn. Explicit valuations of division polynomials of an elliptic curve. *Manuscripta Math.*, 97(3):319–328, 1998.
6. D. V. Chudnovsky and G. V. Chudnovsky. Sequences of numbers generated by addition in formal groups and new primality and factorization tests. *Adv. in Appl. Math.*, 7(4):385–434, 1986.
7. V. A. Dem'janenko. An estimate of the remainder term in Tate's formula. *Mat. Zametki*, 3:271–278, 1968.
8. Manfred Einsiedler, Graham Everest, and Thomas Ward. Primes in elliptic divisibility sequences. *LMS J. Comput. Math.*, 4:1–13 (electronic), 2001.
9. Graham Everest and Helen King. Prime powers in elliptic divisibility sequences. *Math. Comp.*, 74(252):2061–2071 (electronic), 2005.
10. Graham Everest, Gerald Mclaren, and Thomas Ward. Primitive divisors of elliptic divisibility sequences. MR 2220263, 2005; *J. Number theory*, 118:1, 71–89, 2006.
11. Graham Everest, Victor Miller, and Nelson Stephens. Primes generated by elliptic curves. *Proc. Amer. Math. Soc.*, 132(4):955–963 (electronic), 2004.

12. Graham Everest and Igor E. Shparlinski. Prime divisors of sequences associated to elliptic curves. *Glasg. Math. J.*, 47(1):115–122, 2005.
13. Graham Everest, Alf van der Poorten, Igor Shparlinski, and Thomas Ward. *Recurrence sequences*, Volume 104, *Mathematical Surveys and Monographs*. American Mathematical Society, Providence, RI, 2003.
14. Robert Gross and Joseph Silverman. S-integer points on elliptic curves. *Pacific J. Math.*, 167(2):263–288, 1995.
15. Marshall Hall, Jr. The Diophantine equation $x^3 - y^2 = k$. In *Computers in number theory (Proc. Sci. Res. Council Atlas Sympos. No. 2, Oxford, 1969)*, pages 173–198. Academic Press, London, 1971.
16. M. Hindry and J. H. Silverman. The canonical height and integral points on elliptic curves. *Invent. Math.*, 93(2):419–450, 1988.
17. M. Hindry and J. H. Silverman. *Diophantine Geometry: An introduction*, Vol. 201, *Graduate Texts in Mathematics*. Springer-Verlag, New York, 2000.
18. Patrick Ingram. Elliptic divisibility sequences over certain curves. *J. Number theorey,* 123:2, 473–486, 2007.
19. Serge Lang. *Elliptic curves: Diophantine analysis*, volume 231 of *Grundlehren der Mathematischen Wissenschaften [Fundamental Principles of Mathematical Sciences]*. Springer-Verlag, Berlin, 1978.
20. Serge Lang. Conjectured Diophantine estimates on elliptic curves. In *Arithmetic and Geometry, Vol. I*, Vol. 35, *Progr. Math.*, Birkhäuser Boston, Boston, MA, 1983, 155–171.
21. A. Schinzel. Primitive divisors of the expression $A^n - B^n$ in algebraic number fields. *J. Reine Angew. Math.*, 268/269:27–33, 1974. Collection of articles dedicated to Helmut Hasse on his seventy-fifth birthday, II.
22. Wolfgang M. Schmidt. *Diophantine Approximation*, Vol. 785, Lecture Notes in Mathematics. Springer, Berlin, 1980.
23. Rachel Shipsey. *Elliptic divisibility sequences.* PhD thesis, Goldsmith's College (University of London), 2000.
24. T. N. Shorey and R. Tijdeman. *Exponential Diophantine equations*, Vol. 87 Cambridge Tracts in Mathematics, Cambridge University Press, Cambridge, 1986.
25. Joseph H. Silverman. Lower bound for the canonical height on elliptic curves. *Duke Math. J.*, 48(3):633–648, 1981.
26. Joseph H. Silverman. Weierstrass equations and the minimal discriminant of an elliptic curve. *Mathematika*, 31(2):245–251 (1985), 1984.
27. Joseph H. Silverman. *The Arithmetic of Elliptic Curves*, Vol. 106, *Graduate Texts in Mathematics*. Springer-Verlag, New York, 1986.
28. Joseph H. Silverman. A quantitative version of Siegel's theorem: integral points on elliptic curves and Catalan curves. *J. Reine Angew. Math.*, 378:60–100, 1987.
29. Joseph H. Silverman. Wieferich's criterion and the abc-conjecture. *J. Number Theory*, 30(2):226–237, 1988.
30. Joseph H. Silverman. The difference between the Weil height and the canonical height on elliptic curves. *Math. Comp.*, 55(192):723–743, 1990.
31. Joseph H. Silverman. Common divisors of elliptic divisibility sequences over function fields. *Manuscripta Math.*, 114(4):431–446, 2004.
32. Joseph H. Silverman. p-adic properties of division polynomials and elliptic divisibility sequences. *Math. Ann.*, 332(2):443–471, 2005. Addendum 473–474.
33. Christine Swart. *Elliptic divisibility sequences.* PhD thesis, Royal Holloway (University of London), 2003.
34. Morgan Ward. The law of repetition of primes in an elliptic divisibility sequence. *Duke Math. J.*, 15:941–946, 1948.
35. Morgan Ward. Memoir on elliptic divisibility sequences. *Amer. J. Math.*, 70:31–74, 1948.
36. Horst Günter Zimmer. On the difference of the Weil height and the Néron-Tate height. *Math. Z.*, 147(1):35–51, 1976.
37. K. Zsigmondy. Zur Theorie der Potenzreste. *Monatsh. Math.*, 3:265–284, 1892.

The heat kernel, theta inversion and zetas on $\Gamma \backslash G/K$

Jay Jorgenson and Serge Lang

Abstract Direct and precise connections between zeta functions with functional equations and theta functions with inversion formulas can be made using various integral transforms, namely Laplace, Gauss, and Mellin transforms as well as their inversions. In this article, we will describe how one can initiate the process of constructing geometrically defined zeta functions by beginning inversion formulas which come from heat kernels. We state conjectured spectral expansions for the heat kernel, based on the so-called heat Eisenstein series defined in [JoL 04]. We speculate further, in vague terms, the goal of constructing a type of ladder of zeta functions and describe similar features from elsewhere in mathematics.

Key words Zeta function • heat kernel • spectral expansion

Mathematics Subject Classification (2010): 35K08, 11M36, 11F72

The survey article "Riemann's zeta function and beyond" [GeM 03] and the book *Introduction to the Langlands Program* [BeG 04] describe one direction for the theory of automorphic forms. Its successes and ramifications for three decades have deservedly attracted much attention. We would like to describe here a way of looking at some aspects of analysis on certain spaces of type $\Gamma \backslash G/K$ (which will be defined below) pointing to a different direction with different emphasis. We do not question that both directions interact with each other, but perhaps most effectively after this second direction achieves certain successes of its own. However, this other direction that we are presenting requires a reworking of the foundations of the subject, and explaining from scratch how it evolves inductively from the simplest case

$$\mathbf{SL}_2(\mathbf{Z}[\mathbf{i}]) \backslash \mathbf{SL}_2(\mathbf{C})/K_2,$$

J. Jorgenson (✉) • S. Lang
Department of Mathematics, The City College of New York, 138th and Convent Avenue, New York, NY 10031 USA

D. Goldfeld et al. (eds.), *Number Theory, Analysis and Geometry: In Memory of Serge Lang*, DOI 10.1007/978-1-4614-1260-1_13,

where K_2 is the unitary group SU(2). Actually, even this case is already one level above the most classical case of $2\pi\mathbf{Z}\backslash\mathbf{R}$, i.e., the case of the circle with which we start our exposition.

The heat kernel is present as the ubiquitous and fundamental object [JoL 01b]. As will be seen, one may summarize its manifestation in this survey by saying that the t in the heat kernel $\mathbf{K}_t$ is the same t as in the Poisson theta inversion formula

$$t^{-1/2}\theta_1(1/t) = \theta_2(t).$$

We shall describe a way to zeta objects through theta objects, starting with the most classical case on the circle as above. The procedure takes five steps:

- Start with the heat kernel on G/K.
- Periodize with respect to Γ.
- Determine the eigenfunction expansion on $\Gamma\backslash G/K$.
- Regularize the expansion and integrate over $\Gamma\backslash G$.
- Apply the Gauss transform.

This yields what amounts to the logarithmic derivative, with fudge terms. It may be viewed as preparing the ground for seeing how the functions so obtained fit into ladders (cf. Section 6), whenever one has a sequence of suitable embeddings

$$\cdots \to G_n/K_n \to G_{n+1}/K_{n+1} \to \cdots.$$

The most important case for us will be with $G_n = \mathrm{SL}_n$, or certain subgroups isomorphic to products of certain SL_m $(m < n)$ called parabolics. We shall proceed stepwise. We concentrate on what seems a special case with the groups SL_n; however, we will describe a general theory whereby $\mathrm{SL}_n(\mathbf{C})$ becomes a controlling group for semisimple Lie groups; see Section 6. Other classical groups also play such a controlling role, such as $\mathrm{SO}(p, q)$ and $\mathrm{Sp}(n)$.

Acknowledgements Jorgenson acknowledges support from several PSC-CUNY and NSF grants. We thank Tony Petrello for his support of our joint work, and Mel DelVecchio for her setting the manuscript in TeX. Lang also thanks Petrello for his support of the Yale Mathematics Department.

1 The circle and Riemann's functional equation[1]

The Riemann zeta function $\zeta_{\mathbf{Q}}(s)$ can be defined in two ways:

$$\zeta_{\mathbf{Q}}(s) = \sum n^{-s} \text{ or via the Euler product } \prod(1 - 1/p^s)^{-1}.$$

[1]The present article was completed by Jorgenson and Lang during the summer of 2005 shortly before Lang passed away on September 12, 2005. As such, this article is the last mathematics paper written by Lang. At the time the article was written, it was the intention to describe the future direction that Jorgenson and Lang envisioned for their research.

The first way reflects analytical properties, and the second exhibits a number-theoretic connection with the primes. The analytic connection, which will become relevant in Section 2, is that n can be viewed as the square root of an eigenvalue n^2 for the positive Laplacian on the functions e^{inx}. We return to this later. Here, we want to indicate the way to basic analytic properties of the zeta function.

In general, we define a **theta series** to be a series of the form

$$\theta(t) = \sum a_n e^{-\lambda_n t},$$

where $\{a_n\}$ is a sequence of complex numbers, and $\{\lambda_n\}$ is a sequence of real numbers tending to infinity. In practice, these λ_n are > 0. They increase sufficiently fast so that the series converges absolutely. The notation suggests that often the λ_n will be related to eigenvalues. We consider the universal covering

$$\mathbf{R} \to \mathbf{R}/2\pi\mathbf{Z}.$$

Here $\Gamma = 2\pi\mathbf{Z}$ is the discrete group of translations by integral multiples of 2π. Given a function f on $\mathbf{R}$, its Γ-periodization is the function defined by

$$f^{2\pi\mathbf{Z}}(x) = \sum_{n\in\mathbf{Z}} f(x + 2\pi n).$$

What do we do with a periodic function? Advanced calculus says we expand it in a Fourier series. Are there best possible functions to do this with? The answer depends on what we want to do with them. For some purposes, one wants to deal with non-smooth functions, e.g., the sawtooth function whose Fourier series is $\Sigma e^{inx}/n$. For our purposes, we want a priori better behavior than that. There is essentially a unique class of functions that work out most easily while having extensive applications, e.g., to zeta functions, namely the gaussians $\varphi_c(x) = e^{-x^2/c}$, with some positive constant c, and linear combinations of these, or simply constant factors times these functions. The constant factor is normalized to get the probabilistic condition that the total integral is 1, i.e., for an arbitrary positive function f we let

$$I(f) = \int_{\mathbf{R}} f(x)dx,$$

so $f/I(f)$ is probabilistic. This still leaves a choice for the positive constant c. As far as we are concerned, there is a best way to express this constant, namely $c = 4t$, in which case $I(f) = \sqrt{4\pi t}$, and thus we let

$$\mathbf{g}_t(x) = \frac{1}{\sqrt{4\pi t}} e^{-x^2/4t}. \tag{1}$$

In any case, whether we put $c = 4t$ or not, let $\psi_c = \varphi_c/I(\varphi_c)$. Then $\{\psi_c\}$ is a Weierstrass–Dirac family, or $\{\mathbf{g}_t\}$ is a Weierstrass–Dirac family, which we define as follows. Let $\{g_t\}_{t>0}$ be a family of continuous functions on $\mathbf{R}$. We call $\{g_t\}$ a **Weierstrass–Dirac family** if it satisfies the following conditions:

WD 1. $g_t \geqq 0$.
WD 2. $\int_{\mathbf{R}} g_t(x)dx = 1$.
WD 3. Given $\delta > 0$, we have $\lim_{t\to 0} \int_{d(x,0)\geqq\delta} g_t(x)dx = 0$.

Here, $d(x, 0)$ is the distance between x and 0. But with this notation, generalizations can be immediately formulated, e.g., to euclidean space and onward.

A Dirac family is also called an ***approximation of the identity*** because of the following approximation theorem, whose first manifestation is in [Wei 1885].

Let f be a measurable function on $\mathbf{R}$. Let $\{g_t\}$ be a Weierstrass–Dirac family. Define the convolution

$$(g_t * f)(x) = \int_{\mathbf{R}} g_t(y)f(x-y)dy.$$

Then $g_t * f \to f$ uniformly on every set where f is uniformly continuous.

In particular, we get pointwise convergence where f is continuous.

Now a word about using the parameter t instead of c. The floating constant is uniquely determined by the Dirac family condition and a differential equation, called the ***heat equation***, namely if $\boldsymbol{\omega}_x = (\partial/\partial x)^2$, then

$$(-\boldsymbol{\omega}_x + \partial_t)g_t(x) = 0.$$

This is the structural explanation for formula (1) defining the heat gaussian. We then take the following steps:

- Start with the heat gaussian $\mathbf{g}_t$.
- Periodize it, to get $\mathbf{g}_t^{\Gamma}$.
- Expand the periodization in a Fourier series.

A calculus computation shows that the Fourier series is what occurs on the right of the following relation

$$\frac{1}{\sqrt{4\pi t}} \sum_{n\in\mathbf{Z}} e^{-(x+2\pi n)^2/4t} = \frac{1}{2\pi} \sum_{n\in\mathbf{Z}} e^{-n^2t} e^{inx}.$$

Observe that the left side is an inverted theta series times the power $t^{-\frac{1}{2}}$. The right side is a theta series. Both have variable coefficients depending on x. We then perform a fourth step:

- Put $x = 0$, yielding a ***theta inversion relation***

$$\sum_{n\in\mathbf{Z}} \frac{1}{\sqrt{4\pi t}} e^{-(2\pi n)^2/4t} = \frac{1}{2\pi} \sum_{n\in\mathbf{Z}} e^{-n^2t},$$

also called the ***Poisson inversion formula***. [Note: Because of the special abelian nature of **R**, one gets away with a misleading technical trick to eliminate the variable x by setting $x = 0$.]

Having such an inversion formula, what does one do with it? Riemann applied the ***Mellin transform***, defined by

$$(\mathbf{M}\theta)(s) = \int_0^\infty \theta(t) t^s \frac{dt}{t}.$$

Then easy calculus shows that one gets the zeta function with some fudge factors. The inversion relation (with $t, 1/t$ on the two sides of the relation) becomes the functional equation interchanging s and $1-s$:

$$\mathbf{\Lambda}\zeta(s) = \mathbf{\Lambda}\zeta(1-s) \text{ defining } \mathbf{\Lambda}\zeta(s) = \pi^{-s/2}\mathbf{\Gamma}(s/2)\zeta_{\mathbf{Q}}(s),$$

with fudge factors $\pi^{-s/2}$, $\mathbf{\Gamma}(s/2)$ ($\mathbf{\Gamma}$ boldfaced to distinguish from a discrete subgroup).

We work with another transform, which we call the ***Gauss transform***, defined by

$$\text{Gauss}(\theta)(s) = 2s \int_0^\infty e^{-s^2 t}\theta(t) dt.$$

Never mind the fact that the integrals don't converge for the moment. In the monograph [JoL 94] (reproduced in [LanJo 01] Vol. V) we explain how to regularize this integral. Formally the integral is easily evaluated on functions of the form $e^{-n^2 t}$. The Gauss transform of this term for $n \neq 0$ is $1/(s^2 + n^2)$, and Gauss(θ) is essentially the logarithmic derivative of sine. Thus the Gauss transform has an additive structure (Mittag-Leffler type) in contrast with the multiplicative structure for the ordinary zeta function. The functional equation of Gauss(θ)(s) is given by the obvious symmetry coming from the invariance of the integral under $s \mapsto -s$.

2 The next case: $\mathbf{SL}_2(\mathbf{C})$, eigenfunction expansion

We have a choice: $\mathbf{SL}_2(\mathbf{R})$ or $\mathbf{SL}_2(\mathbf{C})$. From the point of view of freshman calculus, $\mathbf{SL}_2(\mathbf{R})$ is "easier". From any more ambitious point of view, $\mathbf{SL}_2(\mathbf{C})$ is easier, for the following main reason. We want gaussians analogous to those on **R**. For $\mathbf{SL}_2(\mathbf{C})$, the gaussians are "split" in a way that they look like those on **R**. For $\mathbf{SL}_2(\mathbf{R})$, one needs an extra integral which at first looks disagreeable, but which can be explained structurally. We now go into such structures.

We start with the basic question: what is the analogue of Fourier expansion? Here is where the eigenfunction property of $e^{\mathrm{i}nx}$ (or $e^{\mathrm{i}xy}$ for the theory of Fourier transform on **R** itself) becomes central. One does not know in general how to write down a priori elementary functions by a formula that will play the role of these exponentials. But thanks to insights of Bargman, Harish-Chandra, Selberg, Maass, Roelcke and Langlands, one does know that such functions exist as eigenfunctions of a laplacian. We shall now describe how one gets to the eigenfunction expansion.

From complex analysis, everybody knows that the group $G = \mathbf{SL}_2(\mathbf{R})$ acts by fractional linear transformations on the upper half-plane $\mathbf{h}_2$. The subgroup leaving $\mathbf{i}$ fixed is the unitary subgroup K (which is maximal compact). Thus we have an isomorphism of homogeneous spaces

$$G/K \xrightarrow{\approx} \mathbf{h}_2 \text{ sending } z \longmapsto g(z) = (az+b)(cz+d)^{-1}$$

if $g = \begin{pmatrix} a & b \\ c & d \end{pmatrix} \in \mathbf{SL}_2(\mathbf{R})$. The action on the coset space G/K is just translation. The non-commutativity is analyzed via a product decomposition of G. Let

U = subgroup of upper triangular matrices $u(x) = \begin{pmatrix} 1 & x \\ 0 & 1 \end{pmatrix}$ with $x \in \mathbf{R}$.
A = subgroup of diagonal matrices $a = \mathrm{diag}(a_1, a_2)$ with diagonal elements $a_1, a_2 > 0$.
K = real unitary subgroup $K(\mathbf{R})$.

Then the product map

$$U \times A \times K \longrightarrow UAK = G$$

is a bijection (real analytic isomorphism). Each element $g \in G$ has a unique decomposition $g = uak$, called the ***Iwasawa decomposition***, with $u \in U, a \in A, k \in K$.

The group A is isomorphic to the positive reals $\mathbf{R}^+$. For ulterior purposes, we write the isomorphism with notation fitting the additive version on the $\mathbf{R}$-space of real diagonal matrices $H = \mathrm{diag}(h_1, h_2)$ with $h_1, h_2 \in \mathbf{R}$. We let α be the character defined by $\alpha(H) = h_1 - h_2$. Then α is an isomorphism of the diagonal space with $\mathbf{R}$. Its exponential is the isomorphism we want of A with $\mathbf{R}^+$. The map $\chi_\alpha : A \longrightarrow \mathbf{R}^+$ sending

$$a \longmapsto y_a = a^\alpha = e^{\alpha(\log a)} = a_1/a_2 = a_1^2$$

is a group isomorphism of A with $\mathbf{R}^+$. We call y the ***Iwasawa coordinate*** on A.

The two subgroups U and A are used to provide eigenfunction expansions on G/K, thus reducing such expansions to abelian groups.

We shall state the eigenfunction expansion formula, but first we comment on gaussians, for which we want to carry out the explicit steps of Section 1. Being on **R** causes all the possible formulas to be twisted in a somewhat disagreeable manner. Going to $\mathrm{SL}_2(\mathbf{C})$ makes life much easier, and we shall explain below the method for recovering $SL_2(\mathbf{R})$ afterwards. Thus instead of the upper half-plane, we use the upper half 3-space $\mathbf{h}_3$, defined as the set of quaternions

$$z = x + \mathbf{j}y = x_1 + \mathbf{i}x_2 + \mathbf{j}y$$

with $x \in \mathbf{C}$, $y > 0$ and $\mathbf{k}$-coordinate equal to 0. We let $G = SL_2(\mathbf{C})$. Then G acts on $\mathbf{h}_3$ by

$$z \longmapsto (az + b)(cz + d)^{-1}.$$

Brute force shows that the image of the fractional transformation does indeed lie in $\mathbf{h}_3$ (i.e., has $\mathbf{k}$-coordinate 0). The isotropy group of $\mathbf{j}$ is $K(\mathbf{C})$, the usual unitary group over $\mathbf{C}$. When dealing both with $\mathbf{R}$ and $\mathbf{C}$ simultaneously, we have to keep them in the notation, but working with one case for a continued period, we omit this extra notation. We now suppose that we are over $\mathbf{C}$, unless otherwise specified.

We have the three groups:

$U = U(\mathbf{C})$ consisting of elements $u(x)$ with x complex, so $U \approx \mathbf{C}$.
$A =$ same group as for $SL_2(\mathbf{R})$!
$K =$ unitary group.

The product map $U \times A \times K \longrightarrow UAK$is a bijection with $G = SL_2(\mathbf{C})$, again called the Iwasawa decomposition. We have the same character α as before on A, with coordinate y. We call (x, y) the ***Iwasawa coordinates***.

We can now define gaussians. The group G also admits a polar decomposition, $G = KAK$, $g = k_1 b k_2$ with $k_1, k_2 \in K$ and $b \in A$, uniquely determined up to a permutation of the diagonal elements. We define the ***polar height*** σ by

$$\sigma(g)^2 = (\log\ y_b)^2.$$

Then a ***gaussian on G/K*** is a positive multiple of a function having the form

$$\varphi_c(g) = e^{-\sigma^2(g)/2c} \frac{\log y}{(y - y^{-1})/2} = e^{-v^2/2c} \frac{v}{\sinh v}$$
$$= e^{-\sigma^2/2c} \frac{\sigma}{\sinh \sigma},$$

with $c > 0$ with additive variable $v = \log y$, and $y = y_b$ is the polar y. The factor on the right is a Jacobian factor but we do not need to go into this.

We use the ***Iwasawa Haar measure*** given by $d\mu(z) = dx_1 dx_2 dy / y^3$. Then we can define the total integral $I(f)$ for a function f on $G/K = \mathbf{h}_3$, and also the notion of Dirac family, using the distance function on G/K,

$$d(z, w) = \sigma(z^{-1}w)$$

which is G-invariant. We can then define $\psi_c = \varphi_c/I(\varphi_c)$ as we did on $\mathbf{R}$. Then $\{\psi_c\}$ is a Dirac family, using the same definition as on $\mathbf{R}$, except that for the third condition **WD 3**, one takes the distance $d(x, e)$ between x and the unit element of G, or the unit coset eK, because the distance is K-bi-invariant. The approximation theorem is valid. Furthermore, the vector space generated by gaussians is dense in anything one wants, see [JoL 03b]. This space can be used to develop the general spherical inversion theory by explicit formulas [JoL 03a]. Most importantly, it contains the heat kernel discussed below.

As to the laplacian, it is easier here to use the group theory to define a Casimir operator (which would turn out to be a scaling of the laplacian if we used the language and context of Riemannian metrics). Casimir is defined entirely in terms of a G-conjugation invariant scalar product on the Lie algebra $\mathfrak{g}$ of G as follows. The Lie algebra is just the vector space of matrices having trace 0, with the bracket product $[X, Y] = XY - YX$. We use the real trace form for the scalar product of matrices $Z_1, Z_2 \in \mathfrak{g}$, that is, we put

$$B(Z_1, Z_2) = \operatorname{Re} \operatorname{tr}(Z_1 Z_2).$$

Let $Z \in \mathfrak{g}$. One can define a left-invariant differential operator denoted $\tilde{Z}$ or $\mathcal{D}(Z)$ (the directional derivative in the direction of Z) by the formula

$$(\tilde{Z} f)(g) = \frac{d}{dt}\Big|_{t=0} f(g \exp(tZ)).$$

If $\{Z_i\}$ is a basis for $\mathfrak{g}$ over $\mathbf{R}$, and $\{Z_i'\}$ the dual basis, then Casimir is defined by

$$\omega = \sum \tilde{Z}_i \tilde{Z}_i',$$

which is independent of the choice of basis. Given an invariant Riemannian metric on $G/K (= \mathbf{h}_3)$, Casimir is equal to a constant times the laplacian with respect to this metric, but one can also carry out everything in terms of the group structure as we have done above.

The heat equation for Casimir is then $-\omega h_t + \partial_t h_t = 0$. There is a unique heat gaussian (gaussian satisfying the heat equation and the Dirac properties) given by

$$\mathbf{g}_t(g) = (8\pi t)^{-3/2} e^{-2t} e^{-\sigma^2(g)/8t} \frac{\sigma}{\sinh \sigma}, \tag{2}$$

which is in line with the euclidean heat gaussian. The general analysis of which this is a special case was done by Gangolli [Gan 68].

What we have defined above suffices to formulate the first two steps, the periodization being with respect to the group $SL_2(\mathbf{Z}[\mathbf{i}])$ in the complex case, and the group $SL_2(\mathbf{Z})$ in the real case. For the remainder of this section, we set $\Gamma = SL_2(\mathbf{Z}[\mathbf{i}])$

There remains to explain what the eigenfunction expansion is analogous to the Fourier expansion on **R** (or euclidean space). This comes originally from Roelcke [Roe 66] in the context of functional analysis on $SL_2(\mathbf{R})$. Because of the more complicated structure of the group, the expansion has a series like a Fourier series, but it also has a continuous version analogous to the Fourier integral. We give the definitions needed to state the expansion.

A function on $\Gamma\backslash G/K$ is Γ_U-periodic ($\Gamma_U = \Gamma \cap U \approx \mathbf{Z}[\mathbf{i}]$). Hence it has an ordinary Fourier expansion on $\mathbf{C}/\mathbf{Z}[\mathbf{i}]$. We say that a function f is ***cuspidal*** if the constant term of this Fourier series is 0, that is,

$$\int_{\Gamma_U\backslash U} f(ug)d\mu(g) = 0 \text{ for all } g \in G.$$

The cuspidal subspace $L^2_{\text{cus}}(\Gamma\backslash G/K)$ then has an orthonormal basis $\{\psi_j\}$ consisting of eigenfunctions of Casimir, with negative eigenvalues $-\lambda_j$.

On the other hand, to get the continuous part, we use the ***Eisenstein series***, depending on a complex parameter s, and defined by

$$E_s(z) = E(s,z) = \sum_{\Gamma_U\backslash\Gamma} (\gamma z)_A^{s\alpha} = \sum_{\Gamma_U\backslash\Gamma} y(\gamma z)^s,$$

where $(\gamma z)_A$ is the projection on A from the Iwasawa decomposition, so we can apply the character $\chi_{s\alpha}$ to this projection, writing this application exponentially. We can write each term as we did in terms of the Iwasawa y-coordinate. The Eisenstein series is thus the $\Gamma_U\backslash\Gamma$-periodization of the character $\chi_{s\alpha}$. The series is absolutely convergent for $\text{Re}(s) > 2$, and has a meromorphic continuation. The character $\chi_{s\alpha}$ is an eigenfunction of Casimir with eigenvalue $2s(s-2)$, and so is the Eisenstein series.

Convolution on G itself is defined by

$$(f * h)(z) = \int_G f(zw^{-1})h(w)d\mu(w).$$

Given a function of two variables $F(z,w)$, the convolution is defined by

$$(F * h)(z) = \int_G F(z,w)h(w)d\mu(w).$$

We superimpose Γ-periodization on this. For a function f on G, we let

$$f^{\Gamma}(w) = \sum_{\gamma\in\Gamma} f(\gamma w).$$

Then we have the group Fubini theorem

$$\int_G f(w)d\mu(w) = \int_{\Gamma\backslash G} f^\Gamma(w)d\mu_{\Gamma\backslash G}(w) = \int_{\Gamma\backslash G} \sum_{\gamma\in\Gamma} f(\gamma w)d\mu_{\Gamma\backslash G}(w).$$

Lt F be G-invariant (meaning $F(gz, gw) = F(z, w)$ for all $g, z, w \in G$). Define the ***periodization***

$$F^\Gamma(z, w) = \sum_{\gamma\in\Gamma} F(\gamma z, w) = \sum_{\gamma\in\Gamma} F(z, \gamma w).$$

If h is a function on $\Gamma\backslash G$, then

$$\int_G F(z, w)h(w)d\mu(w) = \int_{\Gamma\backslash G} F^\Gamma(z, w)h(w)d\mu_{\Gamma\backslash G}(w),$$

where $\mu_{\Gamma\backslash G}$ is the natural homogeneous space measure induced by $d\mu$ on G, coming from the fact that G and $\Gamma\backslash G$ are locally measure isomorphic.

For convolution purposes, it is useful to deal with the integral kernel in two variables derived from a function in one variable via the group law. Thus we now define the ***heat kernel***

$$\mathbf{K}_t(z, w) = \mathbf{g}_t(z^{-1}w).$$

The heat kernel is right K-invariant in each variable, so defined on $G/K \times G/K$. Its Γ-periodization is given by

$$\mathbf{K}_t^\Gamma(z, w) = \sum_{\gamma\in\Gamma} \mathbf{K}_t(z, \gamma w) = \sum_{\gamma\in\Gamma} \mathbf{K}_t(\gamma z, w) = \sum_{\gamma\in\Gamma} \mathbf{g}_t(z^{-1}\gamma w).$$

Eigenfunctions of Casimir are also eigenfunctions of convolution by the heat kernel. If $\boldsymbol{\omega} f = -\lambda f$, then

$$\mathbf{K}_t * f = e^{-\lambda t} f.$$

In particular, $\boldsymbol{\omega}\chi_{s\alpha} = 2s(s-2)\chi_{s\alpha}$, so

$$\mathbf{K}_t * E_s = e^{2s(s-2)t} E_s.$$

Convolution can of course be defined using any function of two variables, for instance $E(s, z)$, taking the integral with respect to either variable. We have

$$(E * f)(\bar{s}) = \langle f, E_s\rangle_{\Gamma\backslash G} = \int_{\Gamma\backslash G} f(w)\overline{E_s(w)}d\mu_{\Gamma\backslash G}(w),$$

so it is the hermitian scalar product (Eisenstein coefficient up to a constant). Of course, all these formulas hold under assumptions of absolute convergence which have to be specified or proved.

Theorem 2.1. Eigenfunction expansion. *For $f \in C_c^\infty(\Gamma\backslash G/K)$, one has the eigenfunction expansion*

$$f(z) = \sum_{j=1}^{\infty} \langle f, \psi_j \rangle \psi_j + c_0(f) + c_E \int_{\mathrm{Re}(s)=1} (E * f)(\bar{s}) E(s,z) d\,\mathrm{Im}(s).$$

With the Iwasawa measure, the constants are $c_0(f) = \mathrm{res}_2(E * f)$ and $c_E = 1/16\pi$. (Notation: res_2 *means residue at* 2.*)*

We have written the convolution $E * f$, viewing E as an integral kernel function, but of course it has here the interpretation as the scalar product as above. So the Eisenstein series forms a continuous family playing the role of the exponential $e^{\mathbf{i}xy}$ on the line. The integral is actually on the imaginary line $s = 1 + \mathbf{i}r, r \in \mathbf{R}$. The eigenfunction expansion is originally due to Roelcke [Roe 66], in the context of functional analysis and spectral decomposition, so it is usually called spectral expansion. The factor $1/16 = 1/4^2$ cancels a quadruplication due to the units of $\mathbf{Z}[\mathbf{i}]$ in our definition of Eisenstein series.

To avoid disagreeable convergence problems, the eigenfunction expansion was stated for C^∞ functions with compact support, using the space of test functions which go as far as possible in eliminating convergence problems. However, in real life, interesting functions do not have compact support, for instance the gaussians, especially the heat gaussian. However, gaussians have quadratic exponential decay, but Eisenstein series have only linear exponential growth, so no convergence problem arises for their convolutions.

We may now state the eigenfunction expansion for the heat kernel.

Theorem 2.2. *With the Iwasawa measure and $c_0 = 1/\mathrm{vol}(\Gamma\backslash G)$,*

$$\mathbf{K}_t^\Gamma(z,w) = \sum_{j=1}^{\infty} e^{-\lambda_j t} \psi_j(z) \overline{\psi_j(w)} + c_0$$
$$+ c_E \int_{-\infty}^{\infty} e^{-2(r^2+1)t} E(1+\mathbf{i}r, z) E(1-\mathbf{i}r, w) dr.$$

Note that the exponent $-2(r^2+1)$ is the eigenvalue $2s(s-2)$ on the line $1+\mathbf{i}r$ ($r \in \mathbf{R}$), which is the imaginary line at the middle of the critical strip $0 < \mathrm{Re}(s) < 2$. Thus the formula is an expansion of the heat kernel in terms of its eigenfunction components, the first part being discrete, and the second part being continuously dependent on the parameter r, in a manner similar to Fourier series and Fourier integrals. Putting $z = w$ gives the heat kernel on the diagonal. The periodization on the left is then

$$\mathbf{K}_t^\Gamma(z,z) = \sum_{\gamma\in\Gamma} \mathbf{K}_t(\gamma z, z) = \sum_{\gamma\in\Gamma} \mathbf{g}_t(z^{-1}\gamma z).$$

This periodization is therefore expressed in terms of conjugation, and we are led to investigate conjugacy classes.

For any group G and $g \in G$, we let $\mathbf{c}(g)$ be conjugation by g. Let Γ be a subgroup of G and Γ' a subset of Γ which is $\mathbf{c}(\Gamma)$-invariant. Then Γ' is the union of Γ-conjugacy classes. We denote the set of such classes by

$$\mathrm{CC}_\Gamma(\Gamma'),$$

and its elements by $\mathfrak{c}$. Then Γ' is the disjoint union of the classes

$$\mathfrak{c} \in \mathrm{CC}_\Gamma(\Gamma').$$

We apply these general considerations to our concrete case of $\mathrm{SL}_2(\mathbf{C})$ and the heat gaussian, together with the heat kernel. We decompose Γ into two $\mathbf{c}(\Gamma)$-invariant subsets. We define the ***standard cuspidal*** subgroup of Γ to be the subgroup of upper triangular matrices in Γ, denoted by Γ_∞. We define the ***cuspidal*** and ***non-cuspidal*** subsets by

$$\mathrm{Cus}\,\Gamma \text{ or } \Gamma_{\mathrm{Cus}} = \mathbf{c}(\Gamma)\Gamma_\infty \quad \text{and its complement} \quad \mathrm{NC}\Gamma = \Gamma - \Gamma_{\mathrm{Cus}}.$$

The trace $\mathbf{K}_t^\Gamma$ can then be decomposed into two series, non-cuspidal and cuspidal respectively,

$$\mathbf{K}_t^{\mathrm{NC}}(z,z) = \sum_{\gamma\in \mathrm{NCus}\Gamma} \mathbf{K}_t(\gamma z, z) \quad \text{and} \quad \mathbf{K}_t^{\mathrm{Cus}}(z,z) = \sum_{\gamma\in \mathrm{Cus}\,\Gamma} \mathbf{K}_t(\gamma z, z),$$

so we can rewrite the eigenfunction expansion of $\mathbf{K}_t^\Gamma$ in the form:

Theorem 2.2′.

$$\mathbf{K}_t^{\mathrm{NC}}(z,z) + \mathbf{K}_t^{\mathrm{Cus}}(z,z)$$

$$= \sum_{j=1}^{\infty} e^{-\mathrm{i}\lambda_j t}|\psi_j(z)|^2 + c_0 + c_E \int_{-\infty}^{\infty} e^{-2(1+r^2)t}|E(1+\mathbf{i}r, z)|^2 dr.$$

This is the form of the eigenfunction expansion which we shall use in the next section. On the left, the two sums $\mathbf{K}_t^{\mathrm{NC}}$ and $\mathbf{K}_t^{\mathrm{Cus}}$ are inverted theta series with variable coefficients. On the right, we have a theta series with variable coefficients, and a theta integral, also with the variable z as a parameter. The series converge absolutely.

3 The theta inversion formula

We want to integrate the eigenfunction expansion term-by-term over $\Gamma\backslash G$, which has finite volume. Formally, this will give us a theta inversion identity. The main problem is that out of five terms, the integral of two of them is not convergent. Before explaining what to do with them, we first describe what the integration gives in the two cases for which the integral is convergent, namely the non-cuspidal series on the left and the theta series with variable coefficients on the right.

Having picked $\{\psi_j\}$ to be an orthonormal basis of $L^2_{\text{cus}}(\Gamma\backslash G/K)$, using the fact that $\Gamma\backslash G/K$ has finite measure, there is no problem integrating the series, and we get a theta series with constant coefficients,

$$(1) \qquad \theta_{\text{Cus}}(t) = \sum_{j=1}^{\infty} e^{-\lambda_j t},$$

where we remind the reader that $-\lambda_j$ is an eigenvalue of Casimir on $L^2_{\text{cus}}(\Gamma\backslash G/K)$.

The integral of the non-cuspidal trace of the heat kernel is absolutely convergent, and can be computed explicitly, giving an inverted theta series:

$$(2) \qquad \int_{\Gamma\backslash G} \mathbf{K}_t^{\text{NC}}(z,z)d\mu(z) = e^{-2t}(4t)^{-\frac{1}{2}}\Theta^{\text{NC}}(1/t)$$

where $\boldsymbol{\Theta}^{\text{NC}}$ is the non-cuspidal inverted theta series using $\Gamma' = \text{NCus}\Gamma$:

$$(2a) \qquad \boldsymbol{\Theta}^{\text{NC}}(1/t) = \sum_{\mathfrak{c}\in CC_\Gamma(\Gamma')} a_{\mathfrak{c}} e^{-|\log b_{\mathfrak{c}}|^2/4t},$$

with constants $a_{\mathfrak{c}}$, $|\log b_{\mathfrak{c}}|^2$ that are determined explicitly. Cf. [JoL 03], [JoL 04]. Specifically, let $\gamma_{\mathfrak{c}}$ be a representative of $\mathfrak{c}$, and Γ_γ the centralizer of an element $\gamma \in \Gamma$. Then $b_{\mathfrak{c}} = A$-polar component of $\gamma_{\mathfrak{c}}$, and

$$a_{\mathfrak{c}} = \frac{1}{|\text{Tor}(\mathfrak{c})|}\log|\eta_{\mathfrak{c},0}|^2 \frac{(2\pi)^{-\frac{3}{2}}}{|\eta_{\mathfrak{c}} - \eta_{\mathfrak{c}}^{-1}|^2},$$

where

$|\text{Tor}(\mathfrak{c})|$ = order of the torsion group of $\Gamma_{\gamma_{\mathfrak{c}}}$, which is finite;
$\eta_{\mathfrak{c}}$ = eigenvalue of elements in $\mathfrak{c}$;
$\eta_{\mathfrak{c},0}$ = eigenvalue of absolute value >1 for a primitive element of $\Gamma_{\gamma_{\mathfrak{c}}}$ (which is infinite cyclic modulo the torsion group).

This explicit determination can actually take several forms. The form we use in the above reference is in terms of invariants associated with each conjugacy class in the group Γ. There are different ways of expressing these completely in terms of number-theoretic invariants over the rational numbers or over $\mathbf{Z}[\mathbf{i}]$. These two alternatives are reminiscent of theories that establish a bijection between conjugacy classes in a Galois group and objects associated directly with the base (class field theory, covering space theory, etc). This direction deserves a separate treatment elsewhere. It connects with the work of Zagier [Zag 79], [Zag 82], Szmidt [Szm 83], [Szm 87], and Venkov [Ven 73].

This leaves the cuspidal trace and the Eisenstein term to be dealt with. Their integrals over $\Gamma\backslash G$ are divergent. However, both of these divergences are "the same" in the following precise sense. There is a natural fundamental domain $\mathcal{F}$ for $\Gamma\backslash G/K = \Gamma\backslash\mathbf{h}_3$. This domain has a tube structure in terms of the coordinates $(x, y), x \in \mathbf{C}, y > 0$; precisely,

$$-\frac{1}{2} \leqq x_1 \leqq \frac{1}{2}, \quad 0 \leqq x_2 \leqq \frac{1}{2}, \quad x_1^2 + x_2^2 + y^2 \geqq 1.$$

Let $\mathcal{F}(\mathcal{Y})$ be the subset of $\mathcal{F}$ consisting of those points such that $y \leqq Y$ for Y large. Let

$$\mathrm{Eis}_t(Y) = \frac{1}{16\pi} \int\limits_{\mathcal{F}(\mathcal{Y})} \int\limits_{-\infty}^{\infty} e^{-2(1+r^2)t} |E(1 + \mathbf{i}r, z|^2 dr d\mu(z),$$

and

$$\mathrm{Cus}_t(Y) = \int\limits_{\mathcal{F}(\mathcal{Y})} \mathbf{K}_t^{\mathrm{Cus}}(z, z) d\mu(z).$$

Theorem 3.1. $\mathrm{Eis}_t(Y)$ *and* $\mathrm{Cus}_t(Y)$ *have asymptotic expansions of the form*

$$c_1(t) \log Y + c_2(t) + o(1) \textit{ for } Y \to \infty,$$

the factor $c_1(t)$ *being the same for both functions, namely*

$$c_1(t) = \frac{e^{-2t}}{(2\pi t)^{1/2}}.$$

Therefore the term $c_1(t) \log Y$ causing divergence cancels, and we can integrate the rest over all of $\mathcal{F}$ (representing $\Gamma\backslash G/K$) letting $Y \to \infty$. What results are a theta integral and an inverted theta series respectively,

$$\theta_{\mathrm{Eis}}(t) = \lim_{Y\to\infty} [\mathrm{Eis}_t(Y) - c_1(t) \log Y]$$

$$\Theta^{\mathrm{Cus}}(1/t) = \lim_{Y\to\infty} [\mathrm{Cus}_t(Y) - c_1(t) \log Y].$$

These are called the ***regularized*** Eisenstein and cuspidal terms respectively. For the regularized Eisenstein term we get

$$\theta_{\mathrm{Eis}}(t) = -e^{-2t} + \frac{1}{\pi}\int_{-\infty}^{\infty} e^{-2(1+r^2)t}(\phi_{11}'/\phi_{11})(1+\mathbf{i}r)dr, \tag{3}$$

where, putting $\mathbf{F} = \mathbf{Q}(\mathbf{i})$, the function ϕ_{11} is defined by

$$\phi_{11}(s) = \frac{\mathbf{\Lambda}\zeta_{\mathbf{F}}(2-s)}{\mathbf{\Lambda}\zeta_{\mathbf{F}}(s)} \quad \text{and} \quad \mathbf{\Lambda}\zeta_{\mathbf{F}}(s) = \pi^{-s}\mathbf{\Gamma}(s)\zeta_{\mathbf{F}}(s).$$

The use of $\mathbf{\Lambda}$ is classical to denote the "completed" zeta function, with its fudge factors needed for the functional equation. Thus the Riemann–Dedekind zeta function appears!

Finally, the regularized cuspidal inverted theta series is computed. For this, we let

$$u(r) = \begin{pmatrix} 1 & r \\ 0 & 1 \end{pmatrix} \quad \text{and} \quad f_t(r) = \mathbf{g}_t(u(r)).$$

With this, we compute explicit constants a_1, a_2, a_3 and a_r such that

$$\Theta^{\mathrm{Cus}}(1/t) = a_1\int_0^{\infty} f_t(r)rdr + a_2\int_0^{\infty} f_t(r)\log(r)rdr \tag{4}$$

$$+a_3\int_0^{\infty} f_t(r)\log\left(\frac{r^2+4}{r}\right)rdr + a_4.$$

The constants a_1, a_2, a_3, and a_r are explicitly computed using classical known numbers, meaning rational numbers, π, $\log 2$, $\log \pi$, and γ which is the Euler constant, as well as an Euler constant $\gamma_{\mathbf{Q}(\mathbf{i})}$ associated to the number field $\mathbf{Q}(\mathbf{i})$ (see also [Szm 83] and [Ven 73]). As we show, the appearance of the Euler constants comes from the asymptotic expansion of the summation of the appropriate harmonic series, either for the integers or the Gauss integers. An elementary exercise in analytic number theory connects the asymptotic expansion of harmonic series with special values of the corresponding zeta function. On a deeper level, the integrals in (4) can be explicitly evaluated using the general Parseval formula from [JoL 93b], which expresses the series $\Theta^{\mathrm{Cus}}(1/t)$ as integrals of classical functions times (euclidean) gaussinas.

In summary, we thus obtain

Theorem 3.2. *The theta relation holds:*

$$e^{-2t}(4t)^{-\frac{1}{2}}\Theta^{\mathrm{NC}}(1/t) + \Theta^{\mathrm{Cus}}(1/t) = \theta_{\mathrm{Cus}}(t) + 1 + \theta_{\mathrm{Eis}}(t).$$

Readers may compare our explicit theta relation with Arthur's version of Selberg's trace formula [Art 74] in the context of representation theory, and adelization, with test functions having compact support (Assumption 3.5). See also Kubota [Kub 73], Appendix; [Zag 79] and [Szm 83].

We are ready to take the fifth step, which is to take the Gauss transform to yield a function $L(s)$ satisfying a functional equation. Actually, this function is a logarithmic derivative

$$L(s) = Z'/Z(s).$$

If one carries out the above procedure with $\mathrm{SL}_2(\mathbf{R})/K_2(\mathbf{R})$ and a co-compact discrete subgroup Γ, what one gets for Z is essentially the Selberg zeta function for a compact Riemann surface. This follows from a theorem of McKean [McK 72].

It is relevant to note here a comment of Iwaniec [Iwa 95]. He writes down the functional equation of the Selberg zeta function in multiplicative form (10.40)

$$Z(s) = \Psi(s)Z(1-s),$$

and then says on pp. 169:

> *If you wish, the Selberg zeta-function satisfies an analogue of the Riemann hypothesis. However, the analogy with the Riemann zeta function is superficial. First of all, it fails badly when it comes to development into Dirichlet's series. Furthermore, the functional equation (10.40) resists any decent interpretation as a kind of Poisson's summation principle.*

Readers can evaluate for themselves whether "the analogy with the Riemann zeta function is superficial." Whether decent or indecent, our procedure shows the way to get systematically a construction of zeta functions in the most general case of semisimple or reductive Lie groups.

One can then apply the Gauss transform [JoL 94], already mentioned at the end of Section 1, to get a zeta object. We return to this below in higher rank.

4 $\mathbf{SL}_n$ and Eisenstein series

We shall describe conjecturally the direction we are taking. We consider "ladders" of spaces of type $\Gamma\backslash G/K$, with appropriate groups G. To avoid fancier terminology, we describe some conjectures about the way certain spaces $\Gamma_n\backslash G_n/K_n$ fit into each other in the concrete case when

$$G_n = \mathbf{SL}_n(\mathbf{C}), \quad K_n = \text{unitary subgroup}, \quad \Gamma_n = \mathbf{SL}_n(\mathbf{Z}[\mathbf{i}]).$$

We expect these to be both typical examples and also controlling the more general context with semisimple or reductive groups, in a manner which will be discussed in the last section.

Preliminaries

We start with the fact that $\mathbf{SL}_n$ has the same type of Iwasawa decomposition that we saw on $\mathbf{SL}_2$. We let $G_n = \mathbf{SL}_n(\mathbf{C})$ but usually omit the index n. Let

U = subgroup of upper triangular unipotent matrices (1's on the diagonal).
A = subgroup of diagonal matrices with components $a_i > 0$ for $i = 1, \ldots, n$.
K = subgroup of unitary matrices.

Then the product map $U \times A \times K \to UAK = G$ is a bijection, called the ***Iwasawa decomposition***.

As on $\mathbf{SL}_2$, we let B be the real trace form on the Lie algebra $\mathfrak{g}$ of G, consisting of the matrices of trace 0.

In addition, we have the Lie algebras $\mathfrak{n}, \mathfrak{a}, \mathfrak{k}$, where $\mathfrak{n}$ is the vector space of strictly upper triangular matrices, $\mathfrak{a}$ is the space of diagonal matrices, $\mathfrak{k}$ is the space of skew-hermitian matrices, all of them with trace 0. We have the direct sum decomposition

$$\mathfrak{g} = \mathfrak{n} + \mathfrak{a} + \mathfrak{k}.$$

The conjugation action of G on itself induces a conjugation action (functorially) on $\mathfrak{g}$. Its restriction to A induces an action $\mathbf{c}_\mathfrak{n}(A)$ on $\mathfrak{n}$, which is completely reducible into eigenspaces of $\mathbf{R}$-dimension 2, with eigencharacters. In fact, let $E_{ij}\,(i < j)$ be the matrix with (ij)-component 1 and all other components 0. Then $E_{ij}, \mathbf{i}E_{ij}$ (complex case!) are an $\mathbf{R}$-basis of an eigenspace, and the corresponding eigencharacter χ_{ij} is determined by

$$aE_{ij}a^{-1} = (a_i/a_j)E_{ij}, \quad \text{that is} \quad \chi_{ij}(a) = a_i/a_j.$$

The conjugation action of A on $\mathfrak{n}$ corresponds to the regular Lie representation of $\mathfrak{a}$ on $\mathfrak{n}$, that is the homomorphism $\mathrm{reg} : \mathfrak{a} \to \mathrm{End}(\mathfrak{n})$ such that $\mathrm{reg}(H)Z = [H, Z]$. Thus corresponding to the Iwasawa decomposition $G = UAK$, $\mathfrak{g} = \mathfrak{n}+\mathfrak{a}+\mathfrak{k}$, we let

$\mathcal{R}(\mathfrak{n})$ = set of $(\mathfrak{a}, \mathfrak{n})$-characters, also called the ***regular characters***,

these being the characters occurring in the semisimple decomposition of $\mathfrak{n}$ over the Lie-regular action of $\mathfrak{a}$. These characters are the $\alpha = \alpha_{ij}$ such that for a diagonal matrix $H = \mathrm{diag}(h_1, \ldots, h_r)$ we have

$$\alpha_{ij}(H) = h_i - h_j.$$

The multiplicative characters can be indexed by the additive ones, namely,

$$\chi = \chi_\alpha \text{ and } \chi_\alpha(a) = e^{\alpha(\log a)} \text{ so that } \chi_{ij}(a) = a^{\alpha_{ij}}.$$

We let

$\mathcal{S}(\mathfrak{n})$ = subset of what are called the ***simple characters***, $\alpha_{i,i+1}$, $(i = 1, \ldots, r)$.

Every regular character is a linear combination with positive integer coefficients of simple characters. The simple characters form a basis of the dual space $\mathfrak{a}^\vee$.

The theory of $\mathbf{SL}_n$ is built up from subgroups $\mathbf{SL}_m$ with $m < n$ as follows. Given an integer $n \geqq 2$, let $\mathcal{P}$ denote a partition of n, that is

$$n = n_1 + \ldots + n_{r+1}, \text{ letting } r = r_\mathcal{P}$$

with positive integers $n_1, \ldots, n_{r+1}$. If $r = n-1$, we deal with the maximal partition. We consider blocks of $n_i \times n_i$ matrices along the diagonal, with indices i ranging from 1 to $r + 1$. We let

$U_\mathcal{P}$ = subgroup of the unipotent triangular group with nonzero elements strictly above the square blocks, except for the diagonal element equal to 1.

$A_\mathcal{P}$ = subgroup of the diagonal group A with positive diagonal elements that are constant in each block, the whole matrix having determinant 1.

$$G_\mathcal{P} = \prod_{i=1}^{r+1} \mathbf{SL}_{n_i} = \prod_{i=1}^{r+1} G_{n_i} = \text{direct product of the block groups.}$$

$$K_\mathcal{P} = \prod_{i=1}^{r+1} K_{n_i} = \text{unitary subgroup of } G_P.$$

Thus both $G_\mathcal{P}$ and $A_\mathcal{P}$ have product structure, which we may write each as

$$G_\mathcal{P} = \text{diag}(G_{n_1}, \ldots, G_{n_{r+1}}) \quad \text{and} \quad A_\mathcal{P} = \text{diag}(a_1 I_{n_1}, \ldots, a_{r+1} I_{n_{r+1}}).$$

The components a_j $(j = 1, \ldots, r + 1)$ are subject to the determinant condition

$$\prod_{j=1}^{r+1} a_j^{n_j} = 1.$$

We define a ***standard reduced parabolic*** subgroup of H to be a subgroup of the form

$$P = U_\mathcal{P} A_\mathcal{P} G_\mathcal{P} \quad \text{also written} \quad P = U_P A_P G_P.$$

This is a subgroup. Indeed, G_P and A_P centralize each other. Furthermore, A_P and G_P normalize U_P, so P is a subgroup of G. The analysis we are considering on G/K is built up inductively from these subgroups.

Let $\Gamma = \Gamma_n = \mathbf{SL}_n(\mathbf{Z}[\mathbf{i}])$. Then Γ is a discrete subgroup of G, and the homogeneous space

$$\Gamma \backslash G = \mathbf{SL}_n(\mathbf{Z}[\mathbf{i}]) \backslash \mathbf{SL}_n(\mathbf{C})$$

has finite volume for its Haar measure, i.e., the G-invariant measure induced from the action of G.

Let H be a subgroup of G. We use the notation

$$\Gamma_H = \Gamma \cap H.$$

Thus we obtain the subgroups $\Gamma_P, \Gamma_{U_P}, \Gamma_{G_P}$. Note that Γ_{A_P} is the trivial group. For a reduced standard parabolic subgroup P as above, we have the semidirect product decomposition

$$\Gamma_P = \Gamma_{U_P}\Gamma_{G_P}.$$

For $n = 2$, the group U is abelian. For $n > 2$, it is not abelian, but it is decomposable into a Jordan–Hölder sequence of abelian groups, actually vector groups. In any case, $\Gamma_U\backslash U$ and so $\Gamma_{U_P}\backslash U_P$ is compact, and behaves like a "non-abelian torus", even though it is not even a group.

The groups $A_P, G_P, K_{G_P}, U_{G_P}$ are block groups of the same type as A, G, K, U respectiely. We have their Lie algebras

$$\mathfrak{n}_P = \text{Lie}(U_P), \quad \mathfrak{a}_P = \text{Lie}(A_P), \quad \mathfrak{g}_{G_P} = \text{Lie}(G_P), \quad \mathfrak{k}_{G_P} = \text{Lie}(K_{G_P}).$$

One obtains the corresponding relevant characters, for instance,

$\mathcal{R}(\mathfrak{n}_{\mathcal{G}_\mathcal{P}})$ = subset of characters $\alpha \in \mathcal{R}(\mathfrak{n})$ such that $\alpha(\mathfrak{a}_P) = 0$, or equivalently $a^\alpha = 1$ for all $a \in A_P$.

$\mathcal{R}(\mathfrak{n}_\mathcal{P})$ = subset of characters $\alpha \in \mathcal{R}(\mathfrak{n})$ occurring in the $\mathfrak{a}$-semisimple decomposition

$$\mathfrak{n}_P = \bigoplus \mathfrak{n}_{P,\alpha},$$

i.e., such that the α-eigenspaces $\mathfrak{n}_{P,\alpha}$ are $\neq 0$. We have the disjoint union

$$\mathcal{R}(\mathfrak{n}) = \mathcal{R}(\mathfrak{n}_{G_\mathcal{P}}) \cup \mathcal{R}(\mathfrak{n}_\mathcal{P}).$$

We define the traces

$$\tau = \tau_G = \sum_{\alpha\in\mathcal{R}(\mathfrak{n})} m(\alpha)\alpha \quad \text{and} \quad \delta(a) = e^{\tau(\log a)}$$

and

$$\tau_P = \sum_{\alpha\in\mathcal{R}(\mathfrak{n}_\mathcal{P})} m(\alpha)\alpha \quad \text{and} \quad \delta_P(a) = e^{\tau_P(\log a)}.$$

The half traces

$$\rho = \frac{1}{2}\tau \quad \text{and similarly} \quad \rho_P \text{ and } \rho_{G_P}$$

play a significant role, as lying the the middle of a "critical strip" familiar in classical contexts of analytic number theory.

Eisenstein series

We are ready for the main course, which concerns the Eisenstein series. Let P be a reduced standard parabolic of $G = \mathbf{SL}_n(\mathbf{C})$, and $\Gamma = \mathbf{SL}_n(\mathbf{Z}[\mathbf{i}])$. Let χ be a character on A_P. For $\zeta \in \mathfrak{a}^\vee_{P,\mathbf{C}}$ (complexification of the real dual space of $\mathfrak{a}_P$), we let χ_ζ be the character of A_P such that for $a \in A_P$ we have $\chi_\zeta(a) = a^\zeta$. The product map

$$U_P \times A_P \times G_P/K_{G_P} \to G/K \quad \text{given by} \quad (u, a, gK_{G_P}) \mapsto uagK,$$

is a differential isomorphism. Therefore we have its three projections on the three factors, denoted by putting any one of them as an index, for instance z_{A_P}. We define the ***character Eisenstein series*** by

$$\begin{aligned} E_P(\chi_\zeta)(z) = E_P(\zeta, z) &= \sum_{\gamma \in \Gamma_P \backslash \Gamma} (\gamma z)^\zeta_{A_P} \\ &= \sum_{\gamma \in \Gamma_P \backslash \Gamma} \chi_\zeta((\gamma z)_{A_P}). \end{aligned}$$

We must now describe a half space of convergence. We have remarked that the simple characters $\{\alpha_1, \ldots, \alpha_r\}$ $(r = n-1)$ form a basis of $\mathfrak{a}^\vee$. Similarly the simple characters $\alpha_{P,1}, \ldots, \alpha_{P,r_P}$ form a basis of $\mathfrak{a}^\vee_P$. We let $\alpha'_{P,1}, \ldots, \alpha'_{P,r_P}$ be the dual basis of $\mathfrak{a}^\vee_P$ relative to the trace form originally given in $\mathfrak{a}$, restricted to $\mathfrak{a}_P$, and inducing a form on $\mathfrak{a}^\vee_P$ in the natural way. A real character $\xi \in \mathfrak{a}^\vee_P$ is defined to be ***positive***, written $\xi > 0$, if

$$\xi = \sum \sigma_i \alpha'_{P,i} \quad \text{with } \sigma_i > 0 \text{ for all } i.$$

This notion of positivity defines a partial ordering on $\mathfrak{a}^\vee_P$. A complex character $\zeta \in \mathfrak{a}^\vee_{P,\mathbf{C}}$ can be written

$$\zeta = \xi + \mathbf{i}\lambda, \quad \text{where } \xi, \lambda \in \mathfrak{a}^\vee_P \text{ are real characters.}$$

We let $\xi = \mathrm{Re}(\zeta)$ and $\lambda = \mathrm{Im}(\zeta)$. We can write ζ as a linear combination

$$\zeta = \sum s_i \alpha'_{P,\mathbf{i}} \quad \text{with } s_i \in \mathbf{C}.$$

Then $\mathrm{Re}(\zeta) > 0$ is equivalent with $\mathrm{Re}(s_i) > 0$ for all i.

Theorem 4.1. *For $\mathrm{Re}(\zeta) > 2\rho_P$, the Eisenstein series $E_P(\zeta)$ is absolutely convergent, uniformly for z in a compact subset of G/K.*

This theorem for a much wider class of groups (semsimple) is proved in Langlands [Lgld 76], see also [Har 68]. Langlands' monumental work proves the meromorphic continuation and functional equation in this general case, leaving many details to the reader, so access is difficult. The attempt to fill a needed complement [MoW 94] does not really make Langlands' theory easily accessible. One of the participants of the seminar giving rise to this attempt is referred to in the introduction as holding the opinion that "the purpose of the seminar was to render obscure what is not so clear." So foundational material to these theories is not yet in a form that provides relatively easy access, which is one reason why we are concentrating on the basic special case of $\mathbf{SL}_n$. It is not the only reason; see the comments at the end.

Let F_0 be a function on $\Gamma_{G_P}\backslash G_P/K_{G_P}$. Let $\chi = \chi_\zeta$ with $\mathrm{Re}(\zeta) > 2\rho_P$. We define the ***$F_0$-twisted Eisenstein series*** $E_P(F_0, \zeta)$ by the series

$$E_P(F_0,\zeta)(z) = \sum_{\gamma\in\Gamma_P\backslash\Gamma} F_0((\gamma z)_{\mathbf{X}_{G_P}})(\gamma z)_{A_P}^{\zeta},$$

where we abbreviated $G_P/K_{G_P} = \mathbf{X}_{G_P}$. The main twisting function we shall use is the heat gaussian or heat kernel, on which we now comment.

First, on $\mathbf{SL}_2$, the group G_P is the trivial group for the standard parabolic UA, so the only way to define the heat kernel on G_P in this case is the constant 1. Twisting is not visible on $\mathbf{SL}_2$. It is a construction which takes hold in higher dimension.

The heat kernel

We fix the real trace form on the Lie algebra of $\mathbf{SL}_n(\mathbf{C})$. This determines a Riemannian metric on G/K. We can define the Casimir operator exactly as we did on $\mathbf{SL}_2$, via a basis and the dual basis of the Lie algebra, and the corresponding invariant differential operator, again denoted by $\boldsymbol{\omega}$. The ***heat operator*** is then

$$\mathbf{H}_{z,t} = -\boldsymbol{\omega}_z + \partial_t .$$

We may also fix a Haar measure, usually taken to be the one associated with the metric volume form, but it is useful to allow the scalings depending on different choices of Haar measures, as we did on $\mathbf{SL}_2$. The ***heat gaussian*** is then defined as a Weierstrass–Dirac family satisfying the heat equation. A general theorem of Dodziuk [Dod 83] guarantees its existence and uniqueness on a complete Riemannian manifolds with Ricci curvature bounded from below. However, the heat kernel can be reached on spaces G/K via spherical inversion theory. It was so determined and analyzed by Gangolli in the general case of semisimple Lie groups [Gan 68]. We carry this out in a self-contained way for $\mathrm{SL}_2(\mathbf{C})$ in [JoL 03a]. In the complex case, it is obtained from a specific formula for the ***heat gaussian***, which is the K-bi-invariant function such that for all $b \in A$ under Gangolli's normalizations, and a bilinear form $B^\vee$,

$$\mathbf{g}_t(b) = \frac{1}{(4\pi t)^{\dim(G/K)/2}} e^{-|\log b|^2/4t} e^{-\rho^2 t} \prod_{\alpha\in\mathcal{R}(\mathrm{n})} \frac{2\pi}{B^\vee(\alpha,\tau)} \frac{\alpha(\log b)}{\sinh(\alpha(\log b))},$$

where $\rho^2 = B^\vee(\rho, \rho)$, and in the present case, $\dim G/K = n^2 - 1$. The formula is written in such a way that it is the Gangolli formula in the most general case. Different normalizations may introduce a constant in front, as when we pick the Iwasawa Haar measure for SL_2. In any case, gaussians are defined as K-bi-invariant functions given on $b \in A$ by the formula

$$e^{-|\log b|^2/c} \prod_{\alpha \in \mathcal{R}(n)} \frac{\alpha(\log b)}{\sinh(\alpha(\log b))}$$

with $c > 0$, and positive multiples of such functions. We may then first satisfy the Weierstrass-Dirac property by dividing such functions by their total integral, and finally adjust the remaining constant so the heat equation is satisfied, in a manner entirely like the one we described for SL_2, thus obtaining the heat gaussian $\mathbf{g}_t$. One then defines the ***heat kernel*** by the universal recipe

$$\mathbf{K}_t(z, w) = \mathbf{g}_t(z^{-1}w).$$

The heat kernel on a product space being the (tensor) product of the heat kernel on the factors, we have a heat kernel on G_P for each P. We have already commented on the convention in the completely degenerate case of trivial G_P. Since the heat kernel is a function of two variables, we are led to define the two-character ***heat Eisenstein series***

$$E_{P,\Gamma,\mathbf{K}}(t, \zeta_1, \zeta_2, z, w)$$
$$= \sum_{\gamma_1, \gamma_2 \in \Gamma_P \backslash \Gamma} \mathbf{K}^{\Gamma_{G_P}}_{\mathbf{X}_{G_P}, t}((\gamma_1 z)_{\mathbf{X}_{G_P}}, (\gamma_2 w)_{\mathbf{X}_{G_P}})(\gamma_1 z)^{\zeta_1}_{A_P}(\gamma_2 w)^{\zeta_2}_{A_P}.$$

In the simplest case of SL_2, the heat Eisenstein series collapses to the product of two Eisenstein series as defined in Section 2, namely for $P = UA$,

$$E_{P,\Gamma,\mathbf{K},t}(\zeta_1, \zeta_2, z, w) = E(\zeta_1, z)E(\zeta_2, w).$$

There is no $\mathbf{K}$ *or* t.

For simplicity, one may omit the subscripts $\mathbf{X}_{G_P}$ and A_P, viewing the heat kernel $\mathbf{K}_{G_P}$ and the characters as defined on G/K via the projections on $\mathbf{X}_{G_P}$ and A_P respectively.

Let F_P be the function defined by

$$F_P(t, \zeta, \bar{\zeta}, z, w) = \mathbf{K}_{\mathbf{X}_{G_P}, t}(z_{\mathbf{X}_{G_P}}, w_{\mathbf{X}_{G_P}}) z^{\zeta}_{A_P} w^{\bar{\zeta}}_{A_P}.$$

For any eigenfunction F of the heat operator, with eigenvalue λ, we define its ***heated function*** (satisfying the heat equation) by

$$F^{\#} = e^{-\lambda t} F.$$

Theorem 4.2. *The function F_P in the variables (t, z) or (t, w) is an eigenfunction of the Heat operator, with eigenvalue $-ev(\omega, \chi_\zeta)$ (minus the eigenvalue of Casimir on the character χ_ζ). This eigenvalue is explicitly*

$$\lambda_{P,\zeta} = ev(\omega, \chi_\zeta) = B^\vee(\zeta, \zeta) - B^\vee(\zeta, \tau_P).$$

The heated function $F_P^\#$ satisfies the heat equation in the variables (t, z) and (t, w). The Eisenstein series is an eigenfunction of the heat operator, with the same eigenvalue as above, and its heated function

$$E^\#_{P,\Gamma,\mathbf{K}}(t, \zeta, \bar{\zeta}, z, w) = e^{-\lambda_{P,\zeta} t} E_{P,\Gamma,\mathbf{K}}(t, \zeta, \bar{\zeta}, z, w)$$

satisfies the heat equation.

5 Conjectures on eigenfunction expansion, theta inversion and zetas

In this section, we reproduce conjectures from [JoL 03].

By BC we abbreviate the property of being bounded continuous. Let $f \in \mathrm{BC}(\Gamma\backslash G/K)$. Let P be a standard parabolic subgroup of G. We define f to be ***P*-cuspidal** if

$$\int_{\Gamma_{U_P}\backslash U_P} f(uz)du = 0 \text{ for all } z \in G.$$

Let $\xi > 2\pi_P = \tau_P$. We say that f is (P, ξ)-admissible if $f E_{P,\Gamma,K}(\zeta, \bar{\zeta})$ is in $L^1(\Gamma\backslash G/K)$ for $\mathrm{Re}(\zeta) = \xi$. We make the same definition for all ξ after assuming the meromorphic continuation of the Eisenstein series.

Suppose f is (P, ξ)-admissible. The property that f is P-cuspidal is equivalent with the property that for $\mathrm{Re}(\zeta) = \xi$,

$$E_{P,\Gamma,K}(\zeta, \bar{\zeta}) * f = 0.$$

The above property is still phrased within the half space of absolute convergence of the Eisenstein series. We now need to assume the existence of a meromorphic continuation, so that we can work on the critical space $\mathrm{Re}(\zeta) = \rho_P$ itself.

We define $L^2_{\mathrm{dis}}(\Gamma\backslash G/K)$ to be the closure of the subspace of L^2 generated by eigenfunctions of Casimir, and call this subspace the ***discrete part of*** L^2. We define a function to be ***cuspidal*** if it is P-cuspidal for all P, and call the space of such functions the ***cuspidal space***. It is contained in the discrete part of L^2. Its orthonormal complement will be called the ***residual space*** $L^2_{\mathrm{res}}(\Gamma\backslash G/K)$.

For $n = 2$, the discrete part of L^2 is the cuspidal subspace plus the constants. For $n > 2$, it involves more, namely a bigger residual part, for which the only available basic reference is the "jungle" of Langlands [Lgld 76], §7.

Gelfand–Piatetski–Shapiro proved that convolution on the cuspidal space with an L^1 function is a compact operator, and Borel–Garland [BoG 83] extended this to the full discrete part.

Conjecture 5.1 *Suppose that for all ζ with $\mathrm{Re}(\zeta) = \rho_P, t > 0, z \in G$ we have*

$$(E_{P,\Gamma,\mathbf{K}} * f)(t, \zeta, \bar{\zeta}, z) = 0.$$

Then f is in the P-cuspidal + discrete subspace.

The convolution is on $\Gamma\backslash G$. The integral implicit in this convolution is taken over the second $\Gamma\backslash G$ variable. As for the test function, one needs it in a space which will include the heat kernel, e.g. the space of gaussians.

Pushing the need for analytic continuation, we define the ***anti-discrete kernel***

$$\begin{aligned} J_{P,\Gamma,\rho_P,t}(z, w) &= \int_{\mathrm{Re}(\zeta)=\rho_P} E^{\#}_{P,\Gamma,\mathbf{K}}(t, \zeta, \bar{\zeta}, z, w)\, d\,\mathrm{Im}(\zeta) \\ &= \int_{\mathrm{Re}(\zeta)=\rho_P} e^{-\lambda_{P,\zeta} t} E_{P,\Gamma,\mathbf{K}}(t, \zeta, \bar{\zeta}, z, w)\, d\,\mathrm{Im}(\zeta). \end{aligned}$$

Conjecture 5.2 *The map $t \mapsto J_{P,\Gamma,\rho_P,t}(z, w)$ satisfies the semigroup property under convolution on $\Gamma\backslash G$. For f in the appropriate space (containing the gaussians, possibly (P, ξ)-admissible for $\rho_P \leqq \xi \leqq (2 + \varepsilon)\rho_P$) the function*

$$f - \lim_{t\to 0} c_P J_{P,\Gamma,\rho_P,t} * f$$

is in the P-cuspidal subspace + residual space, for some constant c_P.

Considering the anti-Eisenstein product

$$\prod (I - J_P)$$

taken over all P, we are led to:

Conjecture 5.3 *There exist real numbers c'_P such that the function*

$$f - \sum_P c'_P \lim_{t\to 0} J_{P,\Gamma,\rho_P,t} * f$$

is in the cuspidal + residual = discrete subspace.

We are now in a position to state the conjectural form of the eigenfunction expansion for the heat kernel. Let $\{\psi_j\}$ be an orthonormal basis for the discrete part of $L^2(\Gamma\backslash G/K)$ consisting of Casimir eigenfunctions, so $\omega\psi_j = \lambda_j\psi_j$.

Conjecture 5.4 *For $X = G/K$, there exist constants c''_P such that*

$$\mathbf{K}_{\mathbf{X}}^{\Gamma}(t, z, w) = \sum_{j \neq 0} e^{-\lambda_j t} \psi_j(z)\overline{\psi_j(w)}$$

$$+ \; c_0 + \sum_P c''_P \int_{\mathrm{Re}(\zeta)=\rho_P} e^{-\lambda_{P,\zeta} t} E_{P,\Gamma,\mathbf{K}}(t, \zeta, \bar{\zeta}, z, w) d\,\mathrm{Im}(\zeta).$$

In the above formulation, we took ψ_0 for $j = 0$ to be a constant function. The value of the constants c_0 and c''_P depend on various normalizations. However, no matter what, c_0 has to be $1/\mathrm{vol}(\Gamma\backslash G)$.

Conjecture 5.4 is the conjectured formula for the eigenfunction expansion of the heat kernel.

We have reached the stage of Theorem 2.2 on $\mathrm{SL}_n(\mathbf{C})$. Then one has to carry out the regularization, the cancellation procedure between the cuspidal part $\mathbf{K}_t^{\mathrm{Cus}}$ of the heat kernel and the Eisenstein integral, in order to reach the theta inversion formula, thus completing the first four steps listed in the case of SL_2.

The fifth step will consist in taking the Gauss transform to yield zeta objects with an additive structure. This will exhibit the ladder structure, in which the fudge terms of the functional equation of the zeta object L_n at level n will be mostly the zeta objects L_m at lower levels $m < n$, including the vey bottom object which will be ϕ'_{11}/ϕ_{11} (cf. Section 3 (3)), thus bringing in explicitly the number theoretic zeta. The other fudge terms will be gamma type functions, powers of π, exponentials. The general theory of the Gauss transform from [JoL 94] Chapter V will apply.

In this way, the theory of explicit formulas for regularized series merges with the theory of eigenfunction expansions on reductive Lie groups, starting with what first appears as a special case, the SL_n ladder. This broader context was the principal motivation for the axiomatization that we started in [JoL 93], [JoL 94] to make the entire set up (regularized products or series and explicit formulas) applicable simultaneously to the cases of classical analytic number theory and the geometric cases that arise from groups like SL_n (reductive groups).

Note that at levels higher than $n = 2$, the main contributions to the fudge terms arise from the parabolics, which are like the original group $\mathbf{SL}_n$ but of lower dimension or rank. Thus the fudge terms, not factors because we are carrying on in an additive setting, correspond to zeta objects associated with lower steps in the ladder, in the present case the $\mathbf{SL}_n$ ladder.

We view spec($\mathbf{Z}[\mathbf{i}]$) as lying at the bottom of the geometric ladder, that is at the first level. The existence of such a ladder shifts the focus of attention from a single level to the way all levels affect each other, especially the way the bottom level interacts with all levels (providing a fudge factor), emphasizing the number-theoretic significance.

The G_n/K_n having negative curvature, it is fruitful to view spec($\mathbf{Z}[\mathbf{i}]$), or spec($\mathfrak{o}$) for the ring of integers of a number field, as having "negative curvature," whatever that means. This point of view led us to conjecture that every abelian subgroup of the Galois group of the algebraic closure of **Q** is topologically cyclic, in analogy with Preissmann's theorem in differential geometry. It then turned out that the above statement is a theorem proved by Geyer in the late 60s [Gey 69], thus giving a strong indication that the heuristic point of view may have a more substantial content.

In any case, we may summarize the above pattern with

$$G_n = \mathbf{SL}_n(\mathbf{C}), \quad \mathbf{X}_n = G_n/K_n, \quad \Gamma_n = \mathbf{SL}_n(\mathbf{Z}[\mathbf{i}])$$

assigning zeta object to $\Gamma_n\backslash\mathbf{X}_n = \Gamma_n\backslash G_n/K_n$ by the five steps procedure:

- Start with the heat kernel on $\mathbf{X}_n$.
- Periodize by Γ_n to $\Gamma_n\backslash\mathbf{X}_n$.
- Give explicitly the eigenfunction expansion of the heat kernel.
- Regularize the divergent terms to get a theta inversion relation, by integrating over $\Gamma_n\backslash G_n$.
- Apply the Gauss transform.

6 Further connections

Representation theory per se forms a motivating force for a whole establishment. We are motivated differently, namely by the development of zeta functions via theta inversion relations, and their analysis via regularized products, regularized series, and explicit formulas. It has been realized in different contexts (Selberg trace formula [Sel 56], Gangolli's construction of the heat kernel on general G/K's with co-compact discrete Γ) that theta inversion formulas can be viewed as part of a much larger context, stemming from the theory of semisimple Lie groups, symmetric spaces, and the heat kernel. Roughly speaking, the general setting for what we described above is that of such groups which have a structure like that of $\mathbf{SL}_n$ but more (or occasionally much more) complicated. If G is such a group, its associated symmetric space is G/K, where K is the (compact) subgroup in the Iwasawa decomposition $G = UAK$. One essential aspect of these spaces is that they are Cartan–Hadamard: complete, simply connected, with seminegative curvature in the language of differential geometry. The subgroup Γ is a discrete subgroup, which may or may not be such that $\Gamma\backslash G$ is compact, but always volume$(\Gamma\backslash G)$ is assumed finite. Of special interest are "“arithmetic" subgroups such as $\mathbf{SL}_n(\mathbf{Z})$ in $\mathbf{SL}_n(\mathbf{R})$ and $\mathbf{SL}_n(\mathbf{Z}[\mathbf{i}])$ in $\mathbf{SL}_n(\mathbf{C})$, which are not co-compact and introduce number theory in various ways, from the bottom up, as we saw in Section 5.

In the Iwasawa decomposition, U is unipotent (its elements are exponentials of nilpotent elements in the Lie algebra), and in particular solvable. The group A is isomorphic to a product of positive real multiplicative groups. The subgroup K is maximal compact, or better unitary for an appropriate scalar product just as on $\mathbf{SL}_n$. If $\mathfrak{n}, \mathfrak{a}, \mathfrak{k}$ are the respective Lie algebras, then $\mathfrak{n}$ is stable under the conjugation action by A, and thus U is normal in UA. The Lie algebra $\mathfrak{n}$ decomposes semismply into 1-dimensional subspaces, giving rise to eigencharacters α just as for $\mathbf{SL}_n$. A positive definite scalar product is given a priori on $\mathfrak{a}$, and there is a polar decomposition $G = KAK$, just as for $\mathbf{SL}_n$, so that for $b \in A$, one has $\log b \in \mathfrak{a}$ and the norm $|\log b|$ is defined.

Gaussians. One may thus define gaussians when G is a complex group just as we did in the case of $\mathbf{SL}_n(\mathbf{C})$. Gangolli [Gan 68] showed that the heat kernel in this complex case is a gaussian with the formula we gave in Section 4. Gaussians actually serve (at least) three purposes:

- One of them is to serve as test functions in a new general development of the theory of semisimple or reductive Lie groups. Indeed, we expect the space generated by the gaussians to be dense in anything one wants (for the case of $\mathbf{SL}_2$, see [JoL 03b]).
- Another is to provide the basis for explicit formulas in this theory.
- The third is to lead immediately into the heat kernel.

The controlling effect of the complex case. Gelfand–Naimark first treated the representation theory in the complex case of the classical groups [GeN 50/57], and Harish-Chandra completed this for all complex groups, and then all real groups [Har 54], [Har 58a] and [Har 58b], by means of the Harish-Chandra series, taking his motivation from linear differential equations. This method was followed in the standard references [Hel 84], [GaV 88], and also in [JoL 01a] for $\mathbf{SL}_n$. However, in Chapter XII of [JoL 01a] for spherical inversion, we suggested the possibility of an entirely different approach to the general case, having its origins in the Flensted-Jensen method of reduction to the complex case [FlJ 78], [FlJ 86]. This program is in the process of being carried out, using the normal transform and its relation to spherical inversion and the heat gaussian, as investigated in collaboration with A. Sinton. As we developed, the Flensted-Jensen method can be placed in a much more general context of a totally geodesic embedding or an Iwasawa embedding (compatible with the Iwasawa decomposition up to conjugacy) of one space G_1/K_1 into another G/K. Then the analysis on G_1/K_1 is related to the analysis on G/K by a commutative diagram, via a projection operator from the larger space to the smaller, namely integrating over the normal directions. For instance, the heat kernel of G_1/K_1 is so obtained form the heat kernel on G/K, and so is the theory of spherical inversion. This also gives a rapid insight into the real case as a "homomorphic image" of the complex case.

Since any semisimple Lie group can be naturally embedded into some $\mathbf{SL}_n(\mathbf{C})$ via its Killing representation (conjugation representation on its Lie algebra), it thus appears that $\mathbf{SL}_n(\mathbf{C})$ is not only an example but a controlling object for all semisimple Lie groups. Thus the pedagogically effective approach of working out the concrete case of $\mathbf{SL}_n(\mathbf{C})$ first is also mathematically effective to get at the general case.

Ladders. The natural way groups $\mathbf{SL}_n$ are embedded in each other is part of a more general system giving rise to other kinds of what we call ladders. Each symmetric space G_n/K_n is embedded in a G_{n+1}/K_{n+1}, in a totally geodesic way, preserving the Iwasawa decomposition (after possibly a conjugation). Thus we may have a sequence of geometric objects which can be displayed vertically as a ladder:

$$\begin{array}{c} | \\ G_{n+1}/K_{n+1} \\ | \\ G_n/K_n \\ | \\ G_{n-1}/K_{n-1} \\ | \end{array}$$

On the other hand, we have the ladder of associated zeta functions $L_n(s)$, according to the general five steps summarized at the end of Section 5. The L-notation suggests logarithmic derivatives as well as classical L-functions. The fudge terms, written additively in our set up so we do not say fudge factors, in the additive functional equation of our zeta objects (logarithmic derivatives), will conjecturally come mainly from the zeta functions of lower level in the ladder. Other fudge factors will include higher-dimensional versions of gamma functions. Thus we have a zeta ladder in parallel to the ladder of spaces. The spaces G_n/K_n are not compact, but can be compactified by the spaces G_m/K_m with $m < n$. A special case arises when these come from parabolic subgroups. Thus the occurrence of functions $L_m(s)$ as fudge terms for $L_n(s)$ reflects the geometric construction of compactification.

Furthermore, a new connection arises between classical number-theoretic objects and objects coming from geometry and analysis, because classical Dedekind zeta functions will occur as fudge factors of geometric zetas. In particular, to so-called "trivial" zeros or poles at a given level, i.e. those belonging to the fudge factors, are the main zeros or poles of lower levels. Thus the zeros of the ordinary Riemann zeta are "trivial" zeros for higher level zetas. More appropriately, they might be called **fudge zeros** for higher levels.

Connections with geometry. Similar ladders should occur with certain types of spaces arising from algebraic geometry and differential geometry, in various ways.

(a) First, moduli spaces in algebraic geometry have a tendency to be of type $\Gamma\backslash G/K$ or to be naturally embeddable to some $\Gamma\backslash G/K$ (G reductive or semisimple, and K the unitary subgroup). For instance, the moduli space of curves of genus $\geqq 2$ is embedded in the moduli space of its associated jacobian because of Torelli's theorem (two curves are isomorphic if and only if their jacobians are isomorphic). As for abelian varieties, their moduli spaces are obtained as quotients of G/K where G is a symplectic group Sp_{2g} (with g equal to the dimension).

Another possible ladder over the moduli space of curves of genus 1 starts with the moduli space of $K3$-surfaces, which is a $\Gamma\backslash G/K$ corresponding to the group $G = \mathrm{SO}_0(2,19)$, followed by Calabi–Yau manifolds moduli spaces, with their more complicated moduli structure embeddable in $\Gamma\backslash G/K$'s, or also possibly involving (b) below. Finally, we mention the moduli ladder of forms of higher degree as in a paper of Jordan [Jor 1880]. Thus analytic properties of Mellin(θ) (essentially a spectral zeta) or of Gauss(θ) (a broader context for Selberg's zeta), associated to such spaces will conjecturally get related to the algebraic geometry and differential geometry of such spaces in ways in which both complement those in the past, and new ways in the future.

In any case, the geometric ladder and the ladder of zeta functions reflect each other, thereby interlocking the theory of spaces coming from algebraic and differential geometry, with analysis and a framework whose origins to a large extent stem from analytic number theory. On the other hand, for some purposes, and in any case as a necessary preliminary for everything else, the purely analytic aspects have to be systematically available.

(b) There are other manifestations of ladder-like stratifications. A theorem of Griffiths [Gri 71] states that given a projective variety V over $\mathbf{C}$, there is a Zariski open subset which is a quotient of a bounded domain (of holomorphy), and this bounded domain is C^∞ isomorphic to a cell (euclidean space). We propose to go further, namely that Zariski open subset can be chosen so that its universal covering space is real analytically a G/K with G semisimple or reductive, so the Zariski open subset is a $\Gamma\backslash G/K$ with discrete Γ. Thus any projective variety could be stratified by a $\Gamma\backslash G/K$. This gives rise to the possibility of considering the classification of varieties or manifolds via stratification structures by such $\Gamma\backslash G/K$. In particular, what is the minimal Zariski closed subset of the moduli space of Calabi–Yau manifolds which one has to delete to get the complement expressible as a $\Gamma\backslash G/K$? How does this symmetric space relate to the symmetric space in which the moduli space i naturally embedded? Same question for the moduli space of curves of genus ≥ 2.

Topologists have concentrated on the classification problem via connected sums, but we find indications that the stratification structure and its connection with eigenexpansion analysis deserves greater attention. Thurston's conjecture for 3-manifolds fits into this ladder scheme.

Having a stratification as suggested above, one may then define an associated zeta function following the five steps listed previously, and related the analytic properties of these zeta functions with the algebraic-differential geometry of the variety.

(c) The trace formula and the spectral zeta function in connection with index theorems, have some history dating back to the seventies and eighties. We mention only a few papers: Atiyah–Bott–Patodi [AtBP 73], Atiyah–Donnelly–Singer [AtDS 83], Barbasch–Moscovici [BaM 83], Müller [Mul 83], [Mul 84], [Mul 87]. For a more complete bibliography, cf. Müller's Springer Lecture Notes [Mul 87]. These papers are partly directed toward index theorems and the connection with number-theoretic invariants, as in the proof of a conjecture of Hirzebruch in [AtDS 83] and [Mul 87]. In retrospect, we would interpret the Atiyah–Donnelly–Singer paper as working on two steps of a ladder, with the compactification of one space by another, and the index theorem being applied on this compact manifold. A reconsideration of the above mentioned papers in light of the present perspective is now in order.

Readers can compare Müller's formula for the heat kernel [Mul 84] Theorem 4.8 and [Mul 87], (9.5), obtained in the context of functional analysis, with our explicit formula. Working as we do in the complex case, where it is possible to use an explicit gaussian representation for the heat kernel, and using the Gauss transform rather than the Mellin transform, puts a very different slant on the whole subject, and allows us to go in a very different direction, starting with the explicit theta inversion relation and its Gauss transform.

Towers of ladders. The structure goes still further. To concentrate on certain aspects of analysis in the simplest case of G/K, we already picked the number field $\mathbf{Q}(\mathbf{i})$ instead of $\mathbf{Q}$ itself, so we used $\mathbf{Z}[\mathbf{i}]$ instead of $\mathbf{Z}$. However, one may consider an arbitrary number field, and the Hilbert–Asai symmetric space associated with it [Asa 70], [JoL 99]. Thus we may go up a tower of number fields $\{F_m\}$ (finite extensions), and then ladders over these $\{G_{m,n}/K_{m,n}\}$, giving rise to a tower of ladders, so a quarter lattice combining even more extensively geometric structures with classical number theoretic ones.

Going up a tower ipso facto introduces questions of number theory. Already for quadratic fields, the Eisenstein series gives rise to a whole direction as in [EGM 85], [EGM 87], [EGM 98], with the theory of special values.

However, we place emphasis on the extent to which the geometric structure is affecting the number theory by having number-theoretic objects at the bottom of geometric ladders. The main new question, as we see it, is how does the existence of a rigid sequence of zeta functions up the ladder force certain regularities of behavior on the arithmetic bottom?

References

[Art 74] J. Arthur, The Selberg trace formula for groups of Γ-rank one, *Ann. of Math.* **100** No. 2 (1974), 326–385.

[Art 78] J. Arthur, A Trace Formula for Reductive Groups I: Terms Associated to Classes in $G(\mathbf{Q})$, *Duke Math. J.* **45** No. 4 (1978), 911–952.

[Art 80] J. Arthur, A Trace Formula for Reductive Groups II: Applications to a Truncation Operator, *Compositio Mathematica* **40** No. 1 (1980), 87–121.

[Art 89] J. Arthur, *The Trace Formula and Hecke Operators, Number Theory, Trace Formulas, and Discrete Groups*, Academic Press (1989), 11–27.

[Asa 70] T. Asai, On a certain function analogous to $\log \eta(z)$, *Nagoya Math. J.* **40** (1970), 193–211.

[AtBP 73] M. Atiyah, R. Bott, V. Patodi, On the heat equation and the index theorem, *Inventiones Math.* **19** (1973), 279–330.

[AtDS 83] M. Atiyah, H. Donnelly, I. Singer, Eta invariants, signature defects of cusps, and values of L-functions, *Ann. of Math.* **118** (1983), 131–177.

[BaM 83] D. Barbasch, H. Moscovici, L^2 index and the Selberg trace formula, *J. Functional Analysis* **53** (1983), 151–201.

[Bea 83] A. F. Beardon, *The Geometry of Discrete Groups*, Springer Verlag, GTM, 1983.

[BeG 04] J. Bernstein and S. Gelbart, eds, *Introduction to the Langlands Program*, Birkhäuser Boston, 2004.

[Bor 62] A. Borel, Arithmetic properties of linear algebraic groups, *Proceedings International Congress of Mathematicians*, Stockholm (1962), 10–22.

[Bor 69] A. Borel, *Introduction aux Groupes Arithmétiques*, Hermann, Paris, 1969.

[BoG 83] A. Borel and H. Garland, Laplacian and the Discrete Spectrum of an Arithmetic Group, *Am. J. Math.* **105** No. 2 (1983), 309–335.

[BuO 94] U. Bunke and M. Olbrich, The wave kernel for the Laplacian on the classical locally symmetric space of rank one, theta functions, trace formulas, and the Selberg zeta function, with an appendix by Andreas Juhl. *Ann. Global. Analysis Geom.* **12** No. 4 (1994), 357–401; Appendix: 402–403 (1994).

[Cr 19] H. Cramér, Studien über die Nullstellen der Riemannschen Zetafunktion. *Math. Zeit.* **4**, (1919), 104–130.

[DeI 82] J.-M. Deshouillers and H. Iwaniec, Kloosterman sums and Fourier coefficients of cusp forms, *Invent. Math.* **70** (1982), 219–288.

[Dod 83] J. Dodziuk, Maximum principle for parabolic inequalities and the heat flow on open manifolds, *Indiana U. Math. J.* **32** (1983), 703–716.

[DoJ 98] J. Dodziuk and J. Jorgenson, *Spectral Asymptotics on Degenerating Hyperbolic 3-Manifolds*, Memoirs of the AMS No. 643, Vol. 135, 1998.

[Efr 87] I. Efrat, *The Selberg trace formula for* $\mathrm{PSL}_2(\mathbf{R})$, Mem. AMS 359 (1987).

[EfS 85] I. Efrat and P. Sarnak, The determinant of the Eisenstein matrix and Hilbert class fields, *Trans. AMS* **290** (1985), 815–824.

[EGM 85] J. Elstrodt, E. Grunewald, J. Mennicke, Eisenstein series on three dimensional hyperbolic spaces and imaginary quadratic fields, *J. reine angew. Math.* **360** (1985), 160–213.

[EGM 87] J. Elstrodt, E. Grunewald, J. Mennicke, Zeta Functions of binary hermitian forms and special values of Eisenstein series on three-dimensional hyperbolic space, *Math. Ann.* **277** (1987), 655–708.

[EGM 98] J. Elstrodt, E. Grunewald, J. Mennicke, *Groups acting on hyperbolic space*, Monograph in Mathematics, Springer Verlag, 1998.

[FlJ 78] M. Flensted-Jensen, Spherical functions on semisimple Lie groups: A method of reduction to the complex case, *J. Funct. Anal.* **30** (1978), 106–146.

[FlJ 86] M. Flensted-Jensen, *Analysis on Non-Riemannian Symmetric Spaces*, CBMS 61, 1986.

[Fre 04] E. Frenkel, Recent advances in the Langlands program, *Bull. AMS* **41** No. 2 (2004), p 151–184.

[Gaf 59] M. Gaffney, The conservation property of the heat equation on Riemannian manifolds, *Comm. Pure and Applied Math.* **12** (1959), 1–11.

[Gan 68] R. Gangolli, Asymptotic Behaviour of Spectra of Compact Quotients of Certain Symmetric Spaces, *Acta Math.* **121** (1968), 151–192.

[Gan 77] R. Gangolli, Zeta functions of Selberg's type for compact space forms of symmetric spaces of rank 1, *Illinois J. Math.* **21** (1977), 1–42.

[GaV 88] R. Gangolli and V.S. Varadarajan, *Harmonic Analysis of Spherical Functions on Real Reductive Groups*, Ergebnisse Math. **101**, Springer Verlag, 1988.

[GaW 80] R. Gangolli and G. Warner, Zeta functions of Selberg's type for some noncompact quotients of symmetric spaces of rank one, *Nagoya Math J.* **78** (1980), 1–44.

[Gar 60] L. Garding, Vecteurs analytiques dans les représentations des groupes de Lie, *Bull. Soc. Math. France* **88** (1960), 73–93.

[GeM 03] S. Gelbart and S. Miller, Riemann's zeta function and beyond, *Bulletin AMS* **41** No. 1 (2003), 50–112.

[GGP 66] I. Gelfand, M. Graev, I. Piatetski-Shapiro, *Representation theory and automorphic functions* (Generalized Functions Vol. 6), Moscow 1966, Translation, Saunders, 1969.

[GeN 50/57] I. M. Gelfand and M. A. Naimark, *Unitäre Darstellungen der klassischen Gruppen*, Akademie Verlag, Berlin, 1957; German translation of Unitary representations of the classical groups, (in Russian), Trudy Mat. Inst. Steklova 36 (1950), 1–288.

[GePS 63] I. Gelfand, I. Piatetski-Shapiro, *Representation theory and theory of automorphic functions,* Am. Math. Soc. Transl. ser 2 26 (1963), 173–200.

[Gey 69] W-D. Geyer, Unendliche algebraische Zahlkrper, bei denen jede Gleichung auflösbar von beschränkter Stufe ist, *Journal of Number Theory* **1** (1969), 346–374.

[God 66] R. Godement, The spectral decomposition of cusp forms, *Proc. Symp. Pure Math. AMS* **9** (1966), 225–234.

[Gri 71] P. Griffiths, Complex analytic properties of certain Zariski open sets on algebraic varieties, *Ann. of Math.* **94** (1971), 21–51.

[Har 54] Harish-Chandra, Representations of semisimple Lie groups III, *Trans. Am. Math. Soc.* **76** (1954), 234–253.

[Har 58a] Harish-Chandra, Spherical functions on a semisimple Lie group I, *Amer. J. Math.* **79** (1958), 241–310.

[Har 58b] Harish-Chandra, Spherical functions on a semisimple Lie group II, *Amer. J. Math.* **80** (1958), 533–613.

[Har 65] Harish-Chandra, Invariant distributions on semisimple Lie groups, *Pub. IHES* **27** (1965), 5–54.

[Har 68] Harish-Chandra, *Automorphic Forms on Semisimple Lie Groups*, Springer Lecture Notes 62 (1968); Notes by J.G.M. Mars.

[Hel 59] S. Helgason, Differential operators on homogeneous spaces, *Acta Math.* **102** (1959), 239–299.

[Hel 84] S. Helgason, *Groups and Geometric Analysis*, Academic Press, 1984.

[Hum 1884] G. Humbert, Sur la mesure de classes d'Hermite de discriminant donné dans un corps quadratique imaginaire, *C.R. Acad. Sci. Paris* Vol. **169** (1919), 448–454.

[Iwa 95] H. Iwaniec, *Introduction to the Spectral Theory of Automorphic Forms*, Biblioteca de la Revista Matematica Iberoamericana, Madrid, 1995.

[Jor 1880] C. Jordan, Mémoire sur l'équivalence des formes. *J. École Polytechnique* **XLVIII** (1880), 112-150.

[JoL 93] J. Jorgenson and S. Lang, *Basic analysis of regularized series and products*, Springer Lecture Notes **1564**, 1993.

[JoL 93b] J. Jorgenson and S. Lang, On Cramér's theorem for general Euler products with functional equation. *Math. Ann.* **297** (1993), 383–416.

[JoL 94] J. Jorgenson and S. Lang, *Explicit Formulas for regularized products and series*, Springer Lecture Notes **1593**, 1994.

[JoL 96] J. Jorgenson and S. Lang, Extension of analytic number theory and the theory of regularized harmonic series from Dirichlet series to Bessel series, *Math. Ann.* **306** (1996), 75–124.

[JoL 99] J. Jorgenson and S. Lang, Hilbert-Asai Eisenstein series, regularized products, and heat kernels, *Nagoya Math. J.* Vol. **153** (1999), 155–188.

[JoL 01a] J. Jorgenson and S. Lang, *Spherical Inversion on* $\mathbf{SL}_n(\mathbf{R})$, Springer Verlag MIM, 2001.

[JoL 01b] J. Jorgenson and S. Lang, The Ubiquitous Heat Kernel, *Mathematics Unlimited: 2001 and Beyond*, Vol I, Engquist and Schmid eds, Springer Verlag 2001, 665–683.

[JoL 03] J. Jorgenson and S. Lang, Heat Eisenstein series on $\mathbf{SL}_n(\mathbf{C})$, to appear. Note in proof: This article as appeared as Memoirs of the AMS No. 946, Vol. 201, 2009.

[JoL 03a] J. Jorgenson and S. Lang, Spherical inversion on $\mathbf{SL}_2(\mathbf{C})$, in *Heat Kernels and Analysis on manifolds, Graphs, and Metric Spaces, Contemporary Mathematics* **338**,

AMS (2003), pp. 241–270 edited by P. Auscher, T. Coulhon, A. Grigoryan

[JoL 03b] J. Jorgenson and S. Lang, Gaussian spaces of test functions, to appear. Note in proof: This article as appeared as *Math. Nachr.* **278** (2005), 824–832.

[JoL 04] J. Jorgenson and S. Lang, Heat Kernel and Theta Inversion on $\mathbf{SL}_2(\mathbf{C})$, to appear. Note in proof: This article as appeared as Springer Verlag MIM, 2008.

[JoLS 03] J. Jorgenson, S. Lang and A. Sinton, Spherical inversion and totally geodesic embeddings of non-compact G/K's, in preparation.

[JoLu 95] J. Jorgenson and R. Lundelius, Convergence of the heat kernel and the resolvant kernel on degenerating hyperbolic Riemann surfaces of finite volume, *Quaestiones Mathematicae* **18** (1995), 345–363.

[JoLu 97a] J. Jorgenson and R. Lundelius, Convergence of the normalized spectral function on degenerating hyperbolic Riemann surfaces of finite volume, *J. Func. Analysis* **149** (1997), 25–57.

[JoLu 97b] J. Jorgenson and R. Lundelius, A regularized heat trace for hyperbolic Riemann surfaces of finite volume, *Comment. Math. Helv.* **72** (1997), 636–659.

[Kat 92] S. Katok, *Fuchsian Groups*, University of Chicago press, 1992.

[Kub 68] T. Kubota, Über diskontinuierlicher Gruppen Picardschen Typus und zugehörige Eisensteinsche Reihen, *Nagoya Math. J.* **32** (1968), 259–271.

[Kub 73] T. Kutoba, *Elementary Theory of Eisenstein series*, Kodansha and John Wiley, Tokyo-New York, 1973,

[Lan 73/87] S. Lang, *Elliptic functions*, Addison Wesley, 1973; Second Edition, Springer Verlag, 1987.

[Lan 75/85] S. Lang, $\mathbf{SL}_2(\mathbf{R})$, Addison Wesley 1973, Springer Verlag 1985.

[Lan 93] S. Lang, *Real and Functional Analysis*, Springer Verlag, 1993.

[Lan 70/94] S. Lang, *Algebraic Number Theory*, Addison Wesley 1970; Second Edition, Springer Verlag, 1994.

[Lan 97] S. Lang, *Undergraduate Analysis*, Second Edition, Springer Verlag, 1997.

[Lan 99] S. Lang, *Math Talks for Undergraduates*, Springer Verlag, 1999.

[Lan 02] S. Lang, *Introduction to Differentiable Manifolds*, Second Edition, Springer Verlag 2002.

[LanJo 01] S. Lang, *Collected Papers, Volume V, with Jay Jorgenson, 1993–1999*, Springer Verlag, 2001.

[Lgld 66] R. P. Langlands, Eisenstein Series, *Proc. Symposium in Pure Mathematics, AMS, Boulder Colorado* 1966, *Algebraic Groups and Discontinuous Subgroups*, Borel and Mostow, editors, 235–252.

[Lgld 76] R. P. Langlands, *On the functional equations satisfied by Eisenstein series*, Springer Lecture Notes 544, 1976.

[Maa 49] H. Maass, Über eine neue Art von Nichtanalytischen automorphen Funktionen und die Bestimmung Dirichletscher Reihen durch Funtionalgleichungen, *Math. Ann.* **121** (1949), 141–183.

[McK 72] H.P. McKean, Selberg's Trace Formula as Applied to a Compact Riemann Surface, *Comm. Pure and Applied Math.* **XXV** (1972), 225–246.

[MoW 94] C. Moeglin and J.-L. Waldspurger, *Décomposition spectrale et séries d'Eisenstein*, Birkhäuser Progress in Mathematics **113** Boston, 1994.

[Mul 83] W. Müller, Spectral theory for Riemannian manifolds with cusps and a related trace formula, *Math. Nachrichten* **111** (1983), 197–288.

[Mul 84] W. Müller, Signature defects of cusps of Hilbert modular varieties and values of L-series at $s = 1$, *J. Diff. Geom.* **20** (1984), 55–119.

[Mul 87] W. Müller, *Manifolds with Cusps of Rank One*, Springer Lecture Notes **1244**, Springer Verlag 1987.

[Nel 59] E. Nelson, Analytic vectors, *Annals of Math.* **70** (1959), 572–615.

[Pic 1884] E. Picard, Sur un groupe de transformations des points de l'espace situés du même coté d'un plan, *Bull. Soc. Math. France* **12** (1884), 43–47.

[Roe 56] W. Roelcke, Uber die Wellengleichung bei Grenzkreisgruppen erster Art, *Sitz. Ber. Heidelberger Ak. der Wiss., Math. Nat. Kl.* 1956, **4** Abh.

[Roe 66] W. Roelcke, Das Eigenwertproblem der automorphen Formen in der hyperbolischer Ebene I, *Math. Ann.* **167** (1966), 292–337.

[Roe 67] W. Roelcke, Das Eigenwertproblem der automorphen Formen in der hyperbolisches Ebene II, *Math. Ann.* **168** (1967), 261–324.

[Sar 83] P. Sarnak, The arithmetic and geometry of some hyperbolic three manifolds, *Acta Math.* **151** (1983), 253–295.

[Sar 03] P. Sarnak, Spectra of Hyperbolic Surfaces, *Bull. Amer. Math. Soc.* **40** (2003), no. 4, 441478.

[Sel 56] A. Selberg, Harmonic Analysis and Discontinuous Groups in Weakly Symmetric Riemannian Spaces with Applications to Dirichlet Series, *International Colloquium on Zeta Functions, J. Indian Math. Soc.* (1956), 47–87.

[Sel 62] A. Selberg, Discontinuous Groups and Harmonic Analysis, *Proc. International Congress of Mathematicians*, Stockholm (1962), 177–189.

[Sel 89] A. Selberg, *Harmonic analysis. Introduction to the Göttingen lecture notes*, Collected Papers Vol. I, Springer, 1989.

[Szm 83] J. Szmidt, The Selberg trace formula for the Picard group $\mathbf{SL}_2(\mathbf{Z}[\mathbf{i}])$, *Acta Arith.* **42** (1983), 291–424.

[Szm 87] J. Szmidt, *The Selberg trace formula and imaginary quadratic fields*, Schriftenreihe des Sonderforschungsbereichs Geometrie und Analysis #52, Mathematics, University of Göttingen, 1987.

[Tam 60] T. Tamagawa, On Selberg's trace formula, *J. Faculty of Science*, University of Tokyo, Sec. I, VIII, Part 2, 363–386.

[Tit 51] E. C. Titchmarsh, *The Theory of the Riemann Zeta Function*, Oxford, 1951.

[Ven 73] A.B. Venkov, Expansion in automorphic eigenfunctions of the Laplace-Beltrami operator in classical symmetric spaces of rank one, and the Selberg trace formula. *Proc. Steklov Inst. Math.* **125** (1973), 1–48.

[War 79] G. Warner, Selberg's trace formula for non-uniform lattices: The **R**-rank one case, *Advance in Math. Studies* **6** (1979), 1–142.

[Wei 1885] K. Weierstrass, Über die analytische Darstellbarkeit sogenannter willkürlicher Funktionen einer reellen Veränderlichen, *Sitzungsbericht Königl. Akad. Wiss.*, 2 and 30 July 1885, 633–639 and 789–805.

[Yos 88] E. Yoshida, On an Application of Zagier's Method in the Theory of Selberg's Trace Formula, *Advanced Studies in Pure Mathematics* **13** (1988), Investigations in Number Theory, 193–214.

[Zag 79] D. Zagier, Eisenstein series and the Selberg trace formula, *in Automorphic Forms, representation theory and arithmetic, Tata Institute, Bombay* (1979), 303–355.

[Zag 82] D. Zagier, The Rankin-Selberg method for automorphic functions which are not of rapid decay, *J. Fac. Sci. Univ. Tokyo* I A **28** (1981), 415–437.

[Zog 82] P. Zograf, Selberg trace formula for the Hilbert modular group of a real quadratic number field, *J. Soviet Math.* **19** (1982), 1637–1652.

Applications of heat kernels on abelian groups: $\zeta(2n)$, quadratic reciprocity, Bessel integrals

Anders Karlsson

In memory of Serge Lang

Abstract The discussion centers around three applications of heat kernel considerations on $\mathbb{R}$, $\mathbb{Z}$ and their quotients. These are Euler's formula for $\zeta(2n)$, Gauss' quadratic reciprocity law, and the evaluation of certain integrals of Bessel functions. Some further applications are mentioned, including the functional equation of Riemann's ζ-function, the reflection formula for the Γ-function, and certain infinite sums of Bessel functions.

Key words heat kernels • Bessel functions • theta functions

Mathematics Subject Classification (2010): 11A, 33A, 33A40

1 Introduction

It was a well-known open problem at the beginning of the 18th century to determine the value of

$$\sum_{k=1}^{\infty} \frac{1}{k^2}.$$

In fact, Wallis and Leibniz failed in their attempts and the question was much discussed among the Bernoullis. It was therefore a sensation when the solution came

A. Karlsson (✉)
Section de Mathématiques, Université de Genéve, 2-4 rue de Lièvre, case postate 64, 1211 Genève, Switzerland
e-mail: anders.Karlsson@unige.ch

D. Goldfeld et al. (eds.), *Number Theory, Analysis and Geometry: In Memory of Serge Lang*, DOI 10.1007/978-1-4614-1260-1_14,

in 1734 from the young Euler, who later also found the general formula for $\zeta(2n)$, see Theorem 1 below.

Now consider instead the problem of solving quadratic equations mod p. A general quadratic equation reduces to studying

$$x^2 = q \bmod p$$

for any two distinct primes p and q. The main theorem for answering when this equation has a solution is the quadratic reciprocity law proved by Gauss in 1796; see Theorem 3 below.

It is a striking fact that both these two classic theorems of number theory, on the surface so different in character, can be deduced from one single analytical formula. We will see this in Sections 3 and 4. The analytical formula in question is the classical Poisson–Jacobi theta inversion identity which expresses the heat kernel on $\mathbb{R}/2\pi\mathbb{Z}$ in two ways. This proof of Gauss's theorem is known and can be found in e.g., [4], while the deduction of Euler's evaluation of $\zeta(2n)$ appears to be new (this proof is analogous to how Selberg's zeta function with functional equation is derived in [17]).

In Section 6 we will moreover see how to evaluate integrals of Bessel functions such as

$$\int_0^x J_n(t)dt \text{ or } \int_0^x J_n(t)J_m(x-t)dt$$

through a determination of the heat kernel on the space consisting of two(!) points.

Of course there are several other extraordinary applications of heat kernels and theta inversion, even on $\mathbb{R}$; see e.g., [15]. Here I selected the evaluation of $\zeta(2n)$ because the proof is both appealing and suggestive, and I chose to include the case of quadratic reciprocity because Lang liked it particularly much and he told me that one should try to do the same to every theta inversion in sight.

As will be clear, the approach is influenced by the ideas of Jorgenson and Lang. In Sections 2 and 5 I try to put the material in the framework of their program, where $\mathbb{R}$ and $\mathbb{Z}$ correspond to the lowest (or next to lowest) levels in the ladder structures. See [12] and [13] for more details on this.

Acknowledgements Support from the Swedish Research Council (VR) grant 2002-4771 and from the Göran Gustafsson Foundation is gratefully acknowledged.

2 Theta inversion on $\mathbb{R}$

The Poisson summation formula is usually stated in the following way:

$$\sum_{n=-\infty}^{\infty} f(n) = \sum_{n=-\infty}^{\infty} \widehat{f}(n), \tag{1}$$

where $\widehat{f}$ is the Fourier transform of f, an infinitely differentiable function such that f and all its derivatives decrease rapidly at infinity, which means that $\lim_{|x|\to\infty} |x|^m f^{(k)}(x) = 0$ for every $m,\ k \geq 0$. Although elegant as this formula no doubt is, it comes especially alive when one takes f to be the heat kernel on $\mathbb{R}$,

$$K^{\mathbb{R}}(t,x) := \frac{1}{\sqrt{4\pi t}} e^{-x^2/4t}.$$

Actually, as Lang pointed out to me, two important features are left out in the "roof formula" (1) as compared to e.g. (2) below: first, the spectral expansion on the quotient present in the proof is hidden, and second, the crucial t-variable structure is missing. In more detail, we start with $K^{\mathbb{R}}(t,x)$, which we periodize to make it 2π-periodic in x (see e.g., [15]). As such it has a Fourier series expansion, and one has after a computation of Fourier coefficients that

$$\frac{1}{\sqrt{4\pi t}} \sum_{n=-\infty}^{\infty} e^{-(x+2\pi n)^2/4t} = \frac{1}{2\pi} \sum_{n=-\infty}^{\infty} e^{-n^2 t} e^{inx}. \tag{2}$$

Now specializing by letting $x = 0$ one gets the *Poisson–Jacobi theta inversion formula*:

$$\frac{1}{\sqrt{4\pi t}} \sum_{n=-\infty}^{\infty} e^{-\pi^2 n^2/t} = \frac{1}{2\pi} \sum_{n=-\infty}^{\infty} e^{-n^2 t}, \tag{3}$$

proved for $t > 0$, but a posteriori valid for $\mathrm{Re}(t) > 0$ since both sides are analytic in that region. Riemann attributes this formula to Jacobi, who in turn attributes it to Poisson (see [7, p. 15]). Note that (3) is what we would get from (1) with $f(x) = K^{\mathbb{R}}(t, 2\pi x)$.

Define the theta function $\theta(t) = \sum_{k=-\infty}^{\infty} e^{-\pi k^2 t}$. Then the identity (3) becomes in a more compact form

$$\frac{1}{\sqrt{t}}\theta\left(\frac{1}{t}\right) = \theta(t), \tag{4}$$

which explains the name *theta inversion.*

In the many applications of these formulas t plays a crucial role. The theorems of Euler and Gauss are discussed below, and then there is also the original Riemann's meromorphic continuation and functional equation of his zeta function, which is recalled without the proof in Section 3.

Note that although these theorems of Euler, Gauss, and Riemann are discussed in most basic textbooks on number theory (e.g., [3], [7], [11], and [19]), it seems that nowhere is it pointed out that, remarkably, all three are consequences of the Poisson-Jacobi theta inversion formula. Considering this, we may be well-advised to study analogs of this formula more closely, which is what we do in Section 5 although only to a modest extent.

3 Special values of Riemann's zeta function

What follows is a proof of the following theorem:

Theorem 1 (Euler). *For any $k > 0$,*

$$\zeta(2k) := \sum_{n=1}^{\infty} \frac{1}{n^{2k}} = (-1)^{k-1} \frac{(2\pi)^{2k} B_{2k}}{2(2k)!},$$

where B_n denotes the Bernoulli numbers.

Recall that the Bernoulli numbers B_k are defined via

$$\frac{x}{e^x - 1} = \sum_{k=0}^{\infty} B_k \frac{x^k}{k!} = 1 - \frac{1}{2}x + \sum_{k=1}^{\infty} B_{2k} \frac{x^{2k}}{(2k)!}.$$

See [11, Ch. 15] for more information on Bernoulli numbers, and in this reference it is also remarked that the theorem above "constitutes one of [Euler's] most remarkable calculations," which in Euler's case does not mean little. Since $B_2 = 1/6$, $B_4 = -1/30$, and $B_6 = 1/42$, we get for example that $\zeta(2) = \pi^2/6$, $\zeta(4) = \pi^4/90$, and $\zeta(6) = \pi^6/945$.

Let f be a measurable function such that $|f(t)| = O(e^{bt})$ for some b as $t \to \infty$. The *Gauss transform* of $f(t)$ following Jorgenson–Lang, see e.g., [16, p. 301] or [13, p. 1], is

$$Gf(s) = 2s \int_0^{\infty} f(t) e^{-s^2 t} dt$$

and is an analytic function in s for $\mathrm{Re}(s^2) > b$. From [6, p. 25] one has that the Laplace transform of

$$\frac{1}{\sqrt{\pi t}} e^{-a^2/4t}, \text{ for } a \geq 0, \text{ is } \frac{1}{\sqrt{\sigma}} e^{-a\sqrt{\sigma}}.$$

Now if we take the Gauss transform of the left-hand side LHS of (3), we get for $s > 0$, by repeatedly interchanging the order of sums and integrals (justified by absolute and uniform convergence of the series and integrals in question), that

$$G(\mathrm{LHS})(s) = 2s \sum_{n=-\infty}^{\infty} \int_0^{\infty} \frac{1}{\sqrt{4\pi t}} e^{-\pi^2 n^2/t} e^{-s^2 t} dt = \frac{2s}{2} \sum_{n=-\infty}^{\infty} \frac{1}{s} e^{-2\pi|n|s}$$

$$= \sum_{n=-\infty}^{\infty} e^{-2\pi|n|s} = 1 + 2\frac{e^{-2\pi s}}{1 - e^{-2\pi s}} = \frac{1 + e^{-2\pi s}}{1 - e^{-2\pi s}}.$$

The right-hand side RHS becomes

$$G(\text{RHS})(s) = \frac{2s}{2\pi}\sum_{n=-\infty}^{\infty}\int_0^{\infty} e^{-tn^2}e^{-s^2t}dt = \frac{2s}{2\pi}\sum_{n=-\infty}^{\infty}\left[\frac{e^{-(n^2+s^2)t}}{-(n^2+s^2)}\right]_0^{\infty}$$
$$= \frac{1}{2\pi}\sum_{n=-\infty}^{\infty}\frac{2s}{s^2+n^2}.$$

Therefore we have that the Gauss transform of the theta identity on $\mathbb{R}$ gives:

Proposition 1. *For real $s \neq 0$,*

$$\frac{1+e^{-2\pi s}}{1-e^{-2\pi s}} = \frac{1}{2\pi}\sum_{n=-\infty}^{\infty}\frac{2s}{s^2+n^2}.$$

We now expand both sides in series expansions in s, for small $s > 0$:

$$\frac{1}{2\pi}\sum_{n=-\infty}^{\infty}\frac{2s}{s^2+n^2} = \frac{1}{\pi s}+\frac{2}{\pi s}\sum_{n=1}^{\infty}\frac{s^2}{n^2+s^2} = \frac{1}{\pi s}+\frac{2}{\pi s}\sum_{n=1}^{\infty}\frac{(s/n)^2}{1+(s/n)^2}$$
$$= \frac{1}{\pi s}+\frac{2}{\pi s}\sum_{n=1}^{\infty}\sum_{k=1}^{\infty}(-1)^{k-1}\left(\frac{s}{n}\right)^{2k}$$
$$= \frac{1}{\pi s}+\frac{2}{\pi s}\sum_{k=1}^{\infty}\left(\sum_{n=1}^{\infty}\frac{1}{n^{2k}}\right)(-1)^{k-1}s^{2k}.$$

On the other hand, in view of the definition of B_n, the left-hand side becomes

$$\frac{1+e^{-2\pi s}}{1-e^{-2\pi s}} = -1-\frac{2}{e^{-2\pi s}-1} = -1+\frac{1}{\pi s}\frac{-2\pi s}{e^{-2\pi s}-1}$$
$$= -1+\frac{1}{\pi s}+1+\frac{1}{\pi s}\sum_{k=1}^{\infty}B_{2k}\frac{(-2\pi s)^{2k}}{(2k)!}$$
$$= \frac{1}{\pi s}+\frac{1}{\pi s}\sum_{k=1}^{\infty}B_{2k}\frac{(2\pi)^{2k}s^{2k}}{(2k)!}.$$

Hence for integers $k > 0$

$$\sum_{n=1}^{\infty}\frac{1}{n^{2k}} = \frac{(-1)^{k-1}(2\pi)^{2k}}{2(2k)!}B_{2k},$$

which proves the theorem.

Let $\xi(s) = \pi^{-s/2}\Gamma(s/2)\zeta(s)$. It is worth recalling here that Riemann applied the Mellin transform to the theta inversion on $\mathbb{R}$, see [7, pp. 15–16] or [15], proving:

Theorem 2 (Riemann). *The function $\xi(s)$ admits an analytic continuation for all $s \neq 0, 1$ and*

$$\xi(s) = \xi(1-s).$$

From this and Theorem 1, it follows that

$$\zeta(1-2n) = -\frac{B_{2n}}{2n},$$

and, because of the poles of Γ, that at the negative even integers $\zeta(-2n) = 0$. These special values were found by Euler in 1749.

4 Quadratic reciprocity

We consider the following equation:

$$x^2 = q \bmod p$$

for any two distinct primes p and q. The Legendre symbol

$$\left(\frac{q}{p}\right)$$

is defined to be 1 if the above equation has a solution for some integer x and -1 otherwise unless $q = 0 \bmod p$ in which case the symbol is 0. The quadratic reciprocity law is:

Theorem 3 (Gauss). *For any two distinct odd primes p and q,*

$$\left(\frac{p}{q}\right)\left(\frac{q}{p}\right) = (-1)^{(p-1)(q-1)/4}.$$

Euler stated the theorem in 1783 but without proof. Legendre wrote only a partial proof, and the first correct proof was published by Gauss in 1796. This theorem was perhaps Gauss's favorite in number theory, which is also indicated by the name he attached to it: theorema aureum — the golden theorem.

The proof we present is based on a beautiful formula, due to Schaar from 1848, and which is of independent interest. It will here arise as the asymptotics expansion in the theta inversion formula for $t = \varepsilon + ip/q$, $\varepsilon \to 0$. We follow Bellman [4], who attributes this proof to Landsberg. A similar method of proof was employed by Hecke [10] to establish quadratic reciprocity for an arbitrary number field; see also [8], [3], and [18]. This might indicate that it is one of the better proofs out of the hundred or so published proofs of Gauss's theorem.

Let

$$S(p,q) := \sum_{r=0}^{q-1} e^{-i\pi r^2 p/q}.$$

Proposition 2. *Let p and q be two relatively prime integers. Then*

$$\frac{1}{\sqrt{q}} S(p,q) = \frac{e^{-i\pi/4}}{\sqrt{p}} \overline{S(q,p)},$$

or written out in full,

$$\frac{1}{\sqrt{q}} \sum_{k=0}^{q-1} e^{-i\pi k^2 p/q} = \frac{e^{-i\pi/4}}{\sqrt{p}} \sum_{l=0}^{p-1} e^{i\pi l^2 q/p}.$$

Proof. Let

$$\theta(t) = \sum_{k=-\infty}^{\infty} e^{-\pi k^2 t} = 1 + 2\sum_{k=1}^{\infty} e^{-\pi k^2 t}.$$

For $\varepsilon > 0$ we have

$$\theta(\varepsilon + ip/q) = 1 + 2\sum_{k=1}^{\infty} e^{-\pi k^2 \varepsilon} e^{-i\pi k^2 p/q} = 1 + 2\sum_{k=0}^{q-1} \left(e^{-i\pi k^2 p/q} \sum_{l=0}^{\infty} e^{-\pi (k+lq)^2 \varepsilon} \right).$$

The inner sum can be interpreted as a Riemann sum as $\varepsilon \to 0$ so that

$$\sum_{l=0}^{\infty} e^{-\pi(k+lq)^2\varepsilon} = \int_0^{\infty} e^{-\pi(k+xq)^2\varepsilon} dx + o(1) = \frac{1}{\pi q \sqrt{\varepsilon}} \int_{\pi k \sqrt{\varepsilon}}^{\infty} e^{-w^2} dw + o(1)$$

$$= \frac{1}{\pi q \sqrt{\varepsilon}} \left(\frac{\sqrt{\pi}}{2} + o(1) \right).$$

Hence

$$\theta(\varepsilon + ip/q) = 1 + 2\frac{1}{\pi q \sqrt{\varepsilon}} \left(\frac{\sqrt{\pi}}{2} + o(1) \right) S(p,q) = \frac{1}{q\sqrt{\pi\varepsilon}} (S(p,q) + o(1))$$

as $\varepsilon \to 0$.

On the other hand, start by noting that

$$\frac{1}{t} = \frac{1}{\varepsilon + ip/q} = \frac{\varepsilon}{\varepsilon^2 + p^2/q^2} - i\frac{p/q}{\varepsilon^2 + p^2/q^2} = \varepsilon\frac{q^2}{p^2} - i\frac{q}{p} + O(\varepsilon^2).$$

Therefore by the same argument, although with some extra care due to the presence of $O(\varepsilon^2)$ above, we get the asymptotics as $\varepsilon \to 0$ for

$$\theta\left(\frac{1}{\varepsilon + ip/q}\right) = \frac{1}{p\sqrt{\pi\varepsilon q^2/p^2}}(S(-q,p) + o(1)) = \frac{1}{q\sqrt{\pi\varepsilon}}(\overline{S(q,p)} + o(1)).$$

Finally, in view of that

$$\frac{1}{\sqrt{t}} = \frac{1}{\sqrt{\varepsilon + ip/q}} = e^{-i\pi/4}\sqrt{\frac{q}{p}} + o(1)$$

and comparing the two asymptotics in the theta inversion formula (4), the proposition is proved. □

Let the quadratic Gauss sum be

$$G(n,m) = \overline{S(2n,m)} = \sum_{r=0}^{m-1} e^{i2\pi r^2 n/m}.$$

We have:

Lemma 1. *Let p and q be two distinct primes. Then*

$$G(1,pq) = G(p,q)G(q,p).$$

Proof. Note that $k^2p^2 + l^2q^2$ equals $(kp + lq)^2$ mod pq, so we see that

$$G(p,q)G(q,p) = \sum_{k=0}^{q-1} e^{i2\pi k^2 p/q} \sum_{l=0}^{p-1} e^{i2\pi r^2 q/p} = \sum_{l=0}^{p-1}\left(\sum_{k=0}^{q-1} e^{i2\pi k^2 p/q}\right) e^{i2\pi l^2 q/p}$$
$$= \sum_{l=0}^{p-1}\sum_{k=0}^{q-1} e^{i2\pi(k^2p^2 + l^2q^2)/pq} = G(1,pq),$$

since $kp + lq$ runs through all the values 0 to $pq - 1$ mod pq exactly once. □

The connection to the Legendre symbol comes next:

Lemma 2. *Let p be an odd prime and assume that p does not divide n. Then*

$$G(n,p) = \left(\frac{n}{p}\right) G(1,p).$$

Proof. This is a simple calculation keeping in mind that as r runs from 1 to $p-1$, r^2 mod p goes through all the quadratic residues Q, exactly twice because $(p-r)^2 = r^2$ mod p:

$$G(n,p) = 1 + 2\sum_{k\in Q} e^{i2\pi kn/p}.$$

Now if n is a quadratic residue then clearly kn is a quadratic residue, and so

$$G(n,p) = 1 + 2\sum_{m\in Q} e^{i2\pi m/p} = G(1,p) = \left(\frac{n}{p}\right) G(1,p).$$

On the other hand, if n is a quadratic nonresidue then kn runs through the quadratic nonresidues Q', and we get

$$G(n,p) = 1 + 2\sum_{m\in Q'} e^{i2\pi m/p} = -1 - 2\sum_{l\in Q} e^{i2\pi l/p} = \left(\frac{n}{p}\right) G(1,p),$$

where the second equality comes from the evaluation of a geometric series:

$$1 + \sum_{m\in Q} e^{i2\pi m/p} + \sum_{m\in Q'} e^{i2\pi m/p} = \sum_{m=0}^{p-1} e^{i2\pi m/p} = 0. \qquad \square$$

We now prove Gauss's theorem. First we have using Proposition 2 for an odd number m that

$$G(1,m) = \overline{S(2,m)} = \frac{\sqrt{m}}{\sqrt{2}} e^{i\pi/4}(1 + e^{-i\pi m/2}) = i^{(m-1)^2/4}\sqrt{m}.$$

In view of the two lemmas we finally get

$$\left(\frac{p}{q}\right)\left(\frac{q}{p}\right) = \frac{G(p,q)\,G(q,p)}{G(1,q)\,G(1,p)} = \frac{G(1,pq)}{G(1,q)G(1,p)} = (-1)^{(p-1)(q-1)/4}$$

as required.

Finally, note that also the two so-called supplements come out directly from Proposition 2:

$$\left(\frac{-1}{p}\right) = \frac{\overline{G(1,p)}}{G(1,p)} = \frac{(-i)^{(p-1)^2/4}\sqrt{p}}{i^{(p-1)^2/4}\sqrt{p}} = (-1)^{(p-1)^2/4} = (-1)^{(p-1)/2}$$

and for an odd prime p,

$$\left(\frac{2}{p}\right) = \frac{G(2,p)}{G(1,p)} = \frac{\frac{\sqrt{p}}{2} e^{i\pi/4}\left(1 + e^{i\pi p/4} + e^{i\pi p} + e^{i\pi 9p/4}\right)}{i^{(p-1)^2/4}\sqrt{p}}$$

$$= \frac{e^{i\pi(p+1)/4} + e^{i\pi(9p+1)/4}}{2i^{(p-1)^2/4}} = (-1)^{(p^2-1)/8}.$$

5 Theta inversion on $\mathbb{Z}$

The heat kernel $K^{\mathbb{Z}}(t,x)$ on $\mathbb{Z}$ is the fundamental solution of

$$\left(\Delta + \frac{\partial}{\partial t}\right) f(t,x) = 0,$$

where

$$\Delta g(x) = g(x) - \frac{1}{2}(g(x-1) + g(x+1)).$$

It is easily verified that

$$K^{\mathbb{Z}}(t,x) = e^{-t} I_x(t),$$

where $x \in \mathbb{Z}$, $t \geq 0$, and I is the Bessel function

$$I_\nu(z) = \sum_{k=0}^{\infty} \frac{z^{\nu+2k}}{2^{\nu+2k} k! \Gamma(\nu+k+1)}.$$

(The relation to the more standard J-Bessel funtion is $I_n(z) = (-i)^n J_n(iz)$.) This can basically be found in Feller [9, pp. 58–60], see also my paper with Neuhauser [14] for a discussion. When passing to a quotient $\mathbb{Z}/m\mathbb{Z}$, we obtain the analogy of (2), except for a cancellation of the factor e^{-z},

$$\sum_{k=-\infty}^{\infty} I_{km+x}(z) = \frac{1}{m} \sum_{j=0}^{m-1} e^{\cos(2\pi j/m)z + 2\pi i j x/m}, \tag{5}$$

for any $z \in \mathbb{C}$ and integers x and $m > 0$, as was proved in [14]. Specializing to $x = 0$, we have the *theta inversion formula on* $\mathbb{Z}$,

$$\sum_{k=-\infty}^{\infty} I_{km}(z) = \frac{1}{m} \sum_{j=0}^{m-1} e^{\cos(2\pi j/m)z}, \tag{6}$$

or in more perfect analogy with (3),

$$e^{-t} \sum_{k=-\infty}^{\infty} I_{km}(t) = \frac{1}{m} \sum_{j=0}^{m-1} e^{-2\sin^2(\pi j/m)t}.$$

The beautiful formula (6) was in fact established earlier by Al-Jarrah, Dempsey, and Glasser [2] (compare also with Theorem 9 in [5]) by a very different method. Note however that these formulas do not seem to have been noticed previously in the vast classical literature on Bessel functions.

If we take the Gauss transform on this identity (now again multiplied by e^{-t}) it is possible, see [14], to get an explicit formula, which is thus the analog of Proposition 1:

Proposition 3. *For real $s \neq 0$, and $m > 0$ an integer,*

$$\frac{2s}{\sqrt{s^4+2s^2}} \frac{1+\left(s^2+1-\sqrt{s^4+2s^2}\right)^m}{1-\left(s^2+1-\sqrt{s^4+2s^2}\right)^m} = \frac{1}{m}\sum_{j=0}^{m-1} \frac{2s}{s^2+2\sin^2(\pi j/m)}.$$

This is the logarithmic derivative (up to the factor m) of a Selberg-type zeta function $Z^{\mathbb{Z}/m\mathbb{Z}}$ (the analogy coming from [17]). In this way we obtain the following [14]:

$$2^{2-m}\sinh^2\left(\frac{m}{2}\operatorname{arccosh}(s^2+1)\right) = ms^2\prod_{n=1}^{m-1}\left(1+\frac{s^2}{2\sin^2(\pi n/m)}\right) =: Z^{\mathbb{Z}/m\mathbb{Z}}(s)$$

which holds for any $s \in \mathbb{C}$.

In the case of $\mathbb{R}$, one gets (cf. the remarks on p. 5 in [13]) in an analogous fashion

$$2\sinh \pi s = 2\pi s \prod_{n=1}^{\infty}\left(1+\frac{s^2}{n^2}\right) =: Z^{\mathbb{R}/2\pi\mathbb{Z}}(s). \tag{7}$$

This in turn can be recast into the well-known reflection formula due to Euler:

Proposition 4. *For $z \in \mathbb{C}\setminus\mathbb{Z}$,*

$$\frac{\pi}{\sin \pi z} = \Gamma(z)\Gamma(1-z).$$

Proof. Recall that the gamma function can be defined through a Weierstrass product (where γ is Euler's constant):

$$\frac{1}{\Gamma(z)} = ze^{\gamma z}\prod_{n=1}^{\infty}\left(1+\frac{z}{n}\right)e^{-z/n}.$$

In view of this and using $\Gamma(w+1) = w\Gamma(w)$, the formula (7) with $s = iz$ becomes the desired identity. □

I hope this brief discussion further illustrates the wealth of identities which come out of formulas like (3) or (6).

6 Integrals of Bessel functions

Already the fact that $e^{-t}I_x(t)$ is the heat kernel on $\mathbb{Z}$ gives an alternative way of looking at Bessel functions; for example the addition theorem

$$J_n(t+s) = \sum_{k=-\infty}^{\infty} J_{n-k}(t)J_k(s)$$

becomes obvious if one thinks probabilistically.

From one of the basic recurrence formulas for J,

$$J_{\nu-1}(z) - J_{\nu+1}(z) = 2J_\nu'(z),$$

one can deduce that [1, 11.1.2] for $\nu > -1$,

$$\int_0^x J_\nu(t)dt = 2\sum_{k=0}^{\infty} J_{2k+\nu+1}(x).$$

In the cases where $\nu = l$ an integer, this latter sum can be simplified to a finite sum from the theta inversion formula for $\mathbb{Z}$ with $m = 2$, recalling that $I_{-n} = I_n$. One gets that

$$\int_0^x J_{2l}(t)dt = \int_0^x J_0(t)dt - 2\sum_{k=0}^{l-1} J_{2k+1}(x),$$

$$\int_0^x J_{2l+1}(t)dt = 1 - J_0(x) - 2\sum_{k=1}^{l} J_{2k}(x),$$

which are [1, 11.1.3] and [1, 11.1.4] respectively.

Other examples, even more adapted to our formula, are the convolution-type integrals

$$\int_0^x J_l(t)J_n(x-t)dt.$$

Here the following formula holds [1, 11.3.37]:

$$\int_0^x J_l(t)J_n(x-t)dt = 2\sum_{k=0}^{\infty}(-1)^k J_{2k+l+n+1}(x),$$

for integers $l, n \geq 0$. We now carry out an example of how to compute this in detail. First we rewrite the sum in terms of I-Bessel functions:

$$\begin{aligned}\int_0^x J_l(t)J_n(x-t)dt &= 2\sum_{k=0}^{\infty}(-1)^k i^{2k+l+n+1} I_{2k+l+n+1}(-ix)\\ &= 2i^{l+n+1}\sum_{k=0}^{\infty} I_{2k+l+n+1}(-ix).\end{aligned}$$

We continue, but now assuming that $l + n$ is even, and then using (5) with $m = 2$,

$$\begin{aligned}
\int_0^x J_l(t)J_n(x-t)dt &= (-1)^{\frac{l+n}{2}}\left(i\sum_{k=-\infty}^{\infty} I_{2k+1}(-ix) - 2i\sum_{k=0}^{(l+n)/2-1} I_{2k+1}(-ix)\right)\\
&= (-1)^{\frac{l+n}{2}}\left(\frac{i}{2}\left(e^{-ix}+e^{ix+i\pi}\right) - 2i\sum_{k=0}^{(l+n)/2-1} I_{2k+1}(-ix)\right)\\
&= (-1)^{\frac{l+n}{2}}\left(\sin x - 2\sum_{k=0}^{(l+n)/2-1}(-1)^k J_{2k+1}(x)\right).
\end{aligned}$$

Similarily, if $l+n$ is odd, one gets

$$\int_0^x J_l(t)J_n(x-t)dt = (-1)^{\frac{l+n+1}{2}}\left(\cos x + J_0(x) - 2\sum_{k=0}^{(l+n-1)/2}(-1)^k J_{2k}(x)\right).$$

In the special cases $n=-l$, or $n=1-l$ with $l=0$ these formulas become (compare with [1, 11.3.38] and [1, 11.3.39] which hold also for nonintegers $-1<l<1$)

$$\begin{aligned}
\int_0^x J_0(t)J_0(x-t)dt &= \sin x,\\
\int_0^x J_0(t)J_1(x-t)dt &= J_0(x) - \cos x.
\end{aligned}$$

7 Personal remarks

I had the great privilege to attend in total around ten semesters of mathematics courses taught by Serge Lang. Like many others I am grateful to him for his teaching, generosity, and constant encouragement. During the spring semester of 2005, when I was his office neighbor at Yale — a very special and interesting experience in itself — we often had conversations on topics related to the present paper. I was struck by the sad news of his death on September 12, 2005. I miss Serge, in particular his great sense of humor, and I feel fortunate for having known him.

References

1. *Handbook of Mathematical Functions with Formulas, Graphs, and Mathematical Tables*, eds: M. Abramowitz, I.A. Stegun, U.S. Department of Commerce, 1972.
2. A. Al-Jarrah, K. M. Dempsey, and M. L. Glasser, *Generalized series of Bessel functions*, J. Comput. Appl. Math. 143 (2002) 1–8.

3. T. Apostol, *Introduction to analytic number theory,* Undergraduate Texts in Mathematics, Springer-Verlag, New York – Heidelberg, 1976.
4. R. Bellman, *A brief introduction to theta functions*, Athena Series: Selected Topics in Mathematics, Holt, Rinehart and Winston, New York 1961.
5. F.R.K. Chung and S.-T. Yau, A combinatorial trace formula, In: *Tsing Hua lectures on geometry and analysis* (S.-T. Yau, ed.) International Press, Cambridge, MA, 1997, pp. 107–116.
6. G. Doetsch, *Theorie und Anwendung der Laplace-Transformation*, Dover Publications, New York, 1943.
7. H.M. Edwards, *Riemann's Zeta Function*, Academic Press, 1974.
8. M. Eichler, *Einführung in die Theorie der algebraischen Zahlen und Funktionen*, Birkhäuser Verlag, Basel–Stuttgart, 1963.
9. Feller, W., *An introduction to probability theory*, Vol. 2, 2nd ed., Wiley, 1971.
10. Hecke, E., *Lectures on the Theory of Algebraic Numbers*, Graduate Texts in Mathematics 77, Springer-Verlag, New York, 1981.
11. K. Ireland and M. Rosen, *A classical introduction to modern number theory*, Graduate Texts in Mathematics 84, Springer-Verlag, New York, 1990.
12. J. Jorgenson and S. Lang, The ubiquitous heat kernel, In: *Mathematics Unlimited — 2001 and beyond,* Springer-Verlag, Berlin, 2001, pp. 655–683.
13. J. Jorgenson and S. Lang, The heat kernel, theta inversion and zetas on $\Gamma\backslash G/K$, Preprint.
14. A. Karlsson and M. Neuhauser, Heat kernels, theta identities, and zeta functions on cyclic groups, In: *Topological and Asymptotic Aspects of Group Theory*, (Grigorchuk et al., eds.), *Contemp. Math.* 394 (2006) pp. 177–189.
15. S. Lang, Die Wärmeleitung auf dem Kreis und Thetafunktionen, *Elem. Math.* 51 (1996) 17–27 (Transl. in *Math Talks for Undergraduates,* Springer-Verlag, 1999).
16. Lang, S., *Collected papers. Vol. V. 1993–1999. With Jay Jorgenson.* Springer-Verlag, 2001.
17. H.P. McKean, Selberg's trace formula as applied to a compact Riemann surface, *Comm. Pure Appl. Math.* 25 (1972) 225–246.
18. M. R. Murty and A. Pacelli, Quadratic reciprocity via theta functions. In: *Number theory*, Mysore , 2005, 107–116, Ramanujan Math. Soc. Lect. Notes Ser., 1, 107–116.
19. I. Niven, H.S. Zuckerman, H.L. Montgomery, *An Introduction to the Theory of Numbers*, 5th edition, Wiley, New York, 1991.

Report on the irreducibility of L-functions

Nicholas M. Katz

Dedicated to the memory of Serge Lang

Abstract In this paper, in honor of the memory of Serge Lang, we apply ideas of Chavdarov and work of Larsen to study the $\mathbb{Q}$-irreducibility, or lack thereof, of various orthogonal L-functions, especially L-functions of elliptic curves over function fields in one variable over finite fields. We also discuss two other approaches to these questions, based on work of Matthews, Vaserstein, and Weisfeller, and on work of Zalesskii-Serezkin.

Key words Q-irreducibility • L-function • elliptic curves

Mathematics Subject Classification (2010): 11M38, 11M50, 14D10, 14D05

1 Introduction

By the pioneering work of Dwork [Dw-Rat] and Grothendieck [Gr-Rat], we know that zeta functions of varieties over finite fields, as well as L-functions attached to quite general algebro-geometric situations over finite fields, are rational functions. In many cases, either this function or its "interesting part" is a polynomial with $\mathbb{Q}$-coefficients. In such cases, it is natural to wonder about the factorization properties of this $\mathbb{Q}$-polynomial. This question was first investigated by Chavdarov [Chav, Theorems 2.1, 2.3, 2.5], who used monodromy techniques to show that for

N.M. Katz (✉)
Princeton University, Department of Mathematics, Fine Hall, Princeton, NJ 08544-1000, USA
e-mail: nmk@math.princeton.edu

D. Goldfeld et al. (eds.), *Number Theory, Analysis and Geometry: In Memory of Serge Lang*, DOI 10.1007/978-1-4614-1260-1_15,

a fixed genus $g \geq 1$, most genus g curves over a large finite field $\mathbb{F}_q$ have the numerator of their zeta function $\mathbb{Q}$-irreducible, i.e., the fraction of the genus g curves over $\mathbb{F}_q$ with this irreducibility property tends to 1 as q grows. [Strictly speaking, Chavdarov's literal result requires q to be a power of a fixed prime p.] Recently Kowalski [Kow-LSM] combined Chavdarov's monodromy methods with large sieve techniques to give quantitative refinements of Chavdarov's results.

It occurred to the author in the fall of 2001 that one might apply Chavdarov's ideas to study the irreducibility properties of L-functions of elliptic curves E over one-variable function fields K over finite fields $\mathbb{F}_q$. Here one knows that, as long as the j-invariant is non-constant, the L-function is a polynomial with $\mathbb{Z}$-coefficients, of known degree d, of the form

$$L(T) = \det(1 - qTA)$$

for a (necessarily unique up to conjugacy) element A in the compact real orthogonal group $O(d, \mathbb{R})$. The unitarized L-function,

$$L_u(T) := L(T/q) = \det(1 - TA)$$

thus has coefficients in $\mathbb{Z}[1/q]$. Being the reversed characteristic polynomial of an element A in $O(d, \mathbb{R})$, it satisfies the functional equation

$$T^d L_u(1/T) = \det(-A) L_u(T).$$

Here $\det(-A) = \pm 1$ is the "sign in the functional equation". With this normalization, the point $T = 1$ is the Birch and Swinnerton-Dyer point. The Birch and Swinnerton-Dyer conjecture states that the Mordell–Weil rank of E/K, $MWrk.(E/k)$, is equal to the multiplicity of $T = 1$ as a zero of $L_u(T)$ (which has come to be called the "analytic rank" of E/K, $an.rk.(E/k)$. One has (in the function field case) the a priori inequality

$$MWrk.(E/k) \leq an.rk.(E/k).$$

The analytic rank is odd if and only if the sign in the functional equation is -1, in which case the analytic rank, being odd, is at least 1. On the other hand, if the sign in the functional equation is $+1$, then the analytic rank, being even, has "no reason" to be nonzero. There is a general expectation that, in any reasonable enumeration sense, "most" elliptic curves will have the lowest possible analytic rank, i.e., 0 or 1, that is compatible with the sign in their functional equations. We refer the reader to [deJ-Ka, 9.7], [Ka-TLFM, 8.3, 9.11, 10.3] and [Ka-MMP, 13.1.7] for one approach to this sort of question.

One knows that, depending on the parity of d and on the sign in the functional equation, either 1 or -1 or both or neither necessarily occur as "imposed" eigenvalues of an element A in $O(d, \mathbb{R})$. More precisely, for d odd, $-\det(-A)$ is always an eigenvalue of A. For d even and $\det(-A) = -1$, both ± 1 are always eigenvalues of A. So it is natural to introduce the "reduced" polynomial

$$Rdet(1 - TA) := \det(1 - TA)/(1 - T); \;\; d \text{ odd, sign} - 1,$$

$$Rdet(1 - TA) := \det(1 - TA)/(1 + T); \;\; d \text{ odd, sign} + 1,$$

$$Rdet(1 - TA) := \det(1 - TA)/(1 - T^2); \;\; d \text{ even, sign} - 1,$$

$$Rdet(1 - TA) := \det(1 - TA); \;\; d \text{ even, sign} + 1,$$

and the reduced (unitarized) L-function

$$L_{u,red}(T) := \text{Rdet}(1 - TA)$$

We propose to show that in various settings, "most" elliptic curves have their reduced L-functions $\mathbb{Q}$-irreducible. The relevance to the Birch and Swinnerton-Dyer Conjecture is simply this: so long as the reduced L-function has degree ≥ 2, if it is $\mathbb{Q}$-irreducible, then it cannot have $T = 1$ as a root, and hence its analytic rank is as low as possible. This consequence for analytic rank gives nothing better than the already cited results [deJ-Ka, 9.7], [Ka-TLFM, 8.3, 9.11, 10.3] and [Ka-MMP, 13.1.7], the only interest is in the methods. [Work of Emmanuel Kowalski [Kow-RQT], Chris Hall [Ha], and Florent Jouve [Jo], using related ideas together with large sieve technology, allows one to do better.] It would be interesting to understand what is the analogue, if any, in the number field case, of the irreducibility of the reduced L-function.

To end this introduction, let us mention briefly a natural question that we do not discuss at all; given that "most" elliptic curves have their reduced L-functions $\mathbb{Q}$-irreducible, what are the galois groups (of the splitting fields, over $\mathbb{Q}$, of) the $\mathbb{Q}$-irreducible polynomials which arise? A natural guess is that for d odd, say $d = 2n + 1$, we should "usually" get the Weyl group of the root system B_n, independent of the sign in the functional equation. For d even and sign $+1$, say $d = 2n$, we should "usually" get the Weyl group of the root system D_n. But for d even and sign -1, say $d = 2n + 2$, we should "usually" get[1] the Weyl group of the root system C_n. The analogous question for families of curves of genus g, where we have symplectic monodromy, was posed and answered by Chavdarov [Chav] and made more quantitative by Kowalski [Kow-LSM]; here the galois group is "usually" the Weyl group of the root system C_g.

[1]The reason we expect this Weyl group is the fact [Weyl, (9.15) on p. 226] that in the compact orthogonal group $O(2n + 2, \mathbb{R})$, the space of conjugacy classes of sign (here sign = determinant) -1 is, with its "Hermann Weyl measure" of total mass one, isomorphic to the space of conjugacy classes in the compact symplectic group $USp(2n)$, with its "Hermann Weyl measure" of total mass one.

These results were worked out in the author's Princeton graduate courses of Fall, 2001 and of 2004–2005, and were presented in lectures at the University of Minnesota (2001), NYU (2001), the Newton Institute (2004), the University of Tokyo (2004), and Brown University (2005). It is a pleasure to thank the listeners for their stimulating questions.

2 The general setup, and the basic examples

We work over an integral domain R which is normal, finitely generated as a $\mathbb{Z}$-algebra, and whose fraction field has characteristic zero. Typically, R will simply be $\mathbb{Z}[1/N]$ for some integer $N \geq 1$. Over R, we are given a smooth R-scheme M/R of relative dimension $\nu \geq 1$ with geometrically connected fibres. Over M, we are given a proper smooth curve C/M and a closed subscheme $D \subset C$ which is finite etale over M. We denote by U/M the open curve

$$U := C - D.$$

Finally, over U we are given a relative elliptic curve E/U.

Before going further, let us give the two basic examples we have in mind.

The first example is the universal family of good degree d polynomial twists of the Legendre curve. Here R is $\mathbb{Z}[1/2]$. We fix an integer $d \geq 3$, and take for M the open set $Twist_d$ in the affine space $\mathbb{A}^d_R$ of all monic, degree d polynomials in one variable λ consisting of those polynomials $f(\lambda)$ for which the product $f(0)f(1)\mathrm{Discrim}(f)$ is invertible. Over this Twist_d we have the universal such polynomial, f_{univ}, and we have the constant curve $\mathbb{P}^1/\mathrm{Twist}_d$, with coordinate λ, in which we take for D the disjoint union of the sections $\infty, 0, 1$ and the zero locus of f_{univ}. So D is finite etale over Twist_d of degree $d+3$. Here we have

$$U = \mathbb{A}^1_{\mathrm{Twist}_d}[1/\lambda(\lambda-1)f_{\mathrm{univ}}(\lambda)].$$

Over this U, we take for E/U the twisted Legendre curve in $\mathbb{P}^2_U$ whose affine equation is

$$y^2 = f_{\mathrm{univ}}(\lambda)x(x-1)(x-\lambda).$$

For each finite field k of odd characteristic, and for each k-valued point f in $\mathrm{Twist}_d(k)$, we obtain a relative elliptic curve $E_{k,f}$ over the punctured λ-line $\mathbb{A}^1_k[1/\lambda(\lambda-1)f(\lambda)]$, namely the twisted Legendre curve $y^2 = f(\lambda)x(x-1)(x-\lambda)$. Its L-function is a polynomial of degree $2d$ if d is even, and of degree $2d-1$ if d is odd. We will show that as $\#k$ grows, the fraction of twisting polynomials f in $\mathrm{Twist}_d(k)$ for which the reduced L-function of the twisted Legendre curve is $\mathbb{Q}$-irreducible tends to 1. On the other hand, we have at present no means of addressing the following extremely natural question. Fix a finite field k of odd characteristic, and consider, as the integer d grows, the fraction of twisting polynomials f in

$\mathrm{Twist}_d(k)$ for which the reduced L-function of the twisted Legendre curve is $\mathbb{Q}$-irreducible. Does this fraction tend to 1 as d grows but k stays fixed? To some other nonzero limit (cf. [Poonen] for an analogous situation)? To any limit?

The second example is the universal family of good Weierstrass curves with g_2 and g_3 of at most specified degrees d_2 and d_3 respectively. Here R is $\mathbb{Z}[1/6]$. We fix integers $d_2 \geq 3$ and $d_3 \geq 3$, and we suppose that either $d_2 \geq 5$ or that $d_3 \geq 7$. We take for M the open set $W(d_2, d_3)$ in the affine space $\mathbb{A}_R^{1+d_2} \times \mathbb{A}_R^{1+d_3}$ consisting of those pairs of polynomials $(g_2(t), g_3(t))$ of degrees at most (d_2, d_3), for which the auxiliary polynomial $\Delta(g_2, g_3) := g_2(t)^3 - 27g_3(t)^2$ has degree exactly $Max(3d_2, 2d_3)$ and has *its* discriminant invertible. Over $W(d_2, d_3)$ we have the universal pair $(g_{2,univ}(t), g_{3,univ}(t))$, the constant curve $\mathbb{P}^1/W(d_2, d_3)$ with coordinate t, and the divisor D which is the disjoint union of the section ∞ and the zero locus of $\Delta(g_{2,univ}(t), g_{3,univ}(t))$. So D is finite étale over $W(d_2, d_3)$ of degree $1 + Max(3d_2, 2d_3)$. Here we have

$$U = \mathbb{A}^1_{W(d_2,d_3)}[1/\Delta(g_{2,\mathrm{univ}}(t), g_{3,univ}(t))].$$

Over this U, we take for E/U the relative elliptic curve given in $\mathbb{P}^2_U$ whose affine equation is the universal Weierstrass equation

$$y^2 = 4x^3 - g_{2,univ}(t)x - g_{3,univ}(t).$$

For each finite field k in which 6 is invertible, and for each k-valued point $(g_2(t), g_3(t))$ in $W(d_2, d_3)(k)$, we obtain the relative elliptic curve E_{k,g_2,g_3} over the punctured t-line $\mathbb{A}^1_k[1/\Delta(g_2, g_3)]$, namely the Weierstrass curve $y^2 = 4x^3 - g_2(t)x - g_3(t)$. Its L-function is a polynomial of degree $Max(3d_2, 2d_3) - 2$ if 12 divides $\mathrm{Max}(3d_2, 2d_3)$, otherwise of degree $Max(3d_2, 2d_3) - 4$. We will show that as $\#k$ grows, the fraction of points $(g_2(t), g_3(t))$ in $W(d_2, d_3)(k)$ for which the reduced L-function of the corresponding Weierstrass curve is $\mathbb{Q}$-irreducible tends to 1. Just as in the first example, if we fix a finite field k in which 6 is invertible, and vary the integers (d_2, d_3) in such a way that, say, $\mathrm{Min}(d_2, d_3)$ grows, we have no understanding of the limiting behavior, if any, of the fraction of points in $W(d_2, d_3)(k)$ whose reduced L-function is $\mathbb{Q}$-irreducible.

3 Back to the general setup; axiomatics

We return to the general setup. Thus R is an integral domain which is normal, finitely generated as a $\mathbb{Z}$-algebra, and whose fraction field has characteristic zero, and M/R is smooth of relative dimension $\nu \geq 1$ with geometrically connected fibres. Over M, we are given a proper smooth curve C/M and a closed subscheme $D \subset C$ which is finite étale over M. U/M is the open curve

$$U := C - D,$$

and over U we are given a relative elliptic curve E/U. So our picture is

$$E \to U \subset C \to M \to \mathrm{Spec}(R).$$

Let us name these morphisms, say

$$f : E \to U,$$
$$j : U \subset C,$$
$$\pi : C \to M.$$

If k is a finite field and $m \in M(k)$ is a k-valued point of M, then by base change we obtain from $E/U/M$ an open curve $U_{k,m}/k$ and a relative elliptic curve $E_{k,m}/U_{k,m}/k$. Let us recall the cohomological genesis of its unitarized L-function.

For a prime number ℓ, and A any of the rings $\mathbb{F}_\ell$, $\mathbb{Z}_\ell$, $\mathbb{Q}_\ell$ or $\overline{\mathbb{Q}_\ell}$, consider the lisse sheaf on $U[1/\ell]$ given by

$$\mathcal{F}_A := R^1 f_\star A.$$

It is a sheaf of free A-modules of rank 2, whose determinant is canonically the Tate-twisted constant sheaf $A(-1)$. So we have a canonical symplectic autoduality paring

$$\mathcal{F}_A \times \mathcal{F}_A \to A(-1).$$

Because R and hence M are normal and connected of generic characteristic zero, any lisse A-sheaf on $U[1/\ell]$ (here $\mathcal{F}_A$) is tamely ramified along the finite étale divisor $D[1/\ell]$. We next consider its extension by direct image,

$$\mathcal{G}_A := j_\star \mathcal{F}_A,$$

on $C[1/\ell]$. The autoduality pairing on $\mathcal{F}_A$ extends by direct image to a pairing

$$\mathcal{G}_A \times \mathcal{G}_A \to j_\star A(-1) \cong A(-1).$$

The formation of $\mathcal{G}_A$ on $C[1/\ell]$ commutes with arbitrary base change on $M[1/\ell]$, and its restriction to $D[1/\ell]$ is a lisse sheaf of free A-modules on $D[1/\ell]$. We then form the Tate-twisted higher direct image sheaf

$$\mathcal{H}_A := R^1 \pi_\star \mathcal{G}_A(1)$$

on $M[1/\ell]$. This is a lisse sheaf of (not necessarily free, when A is $\mathbb{Z}_\ell$) A-modules of finite type. Its formation commutes with arbitrary base change on $M[1/\ell]$. It is endowed with an A-linear cup product pairing

$$\mathcal{H}_A \times \mathcal{H}_A \to R^2 \pi_\star A(1) \cong A.$$

When A is a field, this pairing makes $\mathcal{H}_A$ orthogonally self-dual. When A is $\mathbb{Q}_\ell$ or $\overline{\mathbb{Q}_\ell}$, then $\mathcal{H}_A$ is, in addition, pure of weight zero. We view the lisse sheaf $\mathcal{H}_A$ as a representation of $\pi_1(M[1/\ell])$.

Theorem 3.1 *In the general setup $E/U/M/R$ as above, there exist integers $d \geq 0$ and $N \geq 1$ such that for ℓ not dividing N, $\mathcal{H}_{\mathbb{Z}_\ell}$ is a lisse sheaf of free $\mathbb{Z}_\ell$-modules of rank d on $M[1/\ell]$ which, by the cup product pairing*

$$\mathcal{H}_{\mathbb{Z}_\ell} \times \mathcal{H}_{\mathbb{Z}_\ell} \to \mathbb{Z}_\ell,$$

is orthogonally self-dual over $\mathbb{Z}_\ell$.

Proof. Pick an embedding of R into $\mathbb{C}$, and make the extension of scalars from R to $\mathbb{C}$. We denote by the superscript *an* the corresponding analytic objects. Thus we have the locally constant sheaf $\mathcal{H}_{\mathbb{Z}}^{\mathrm{an}}$ of finitely generated abelian groups on M^{an}, endowed with the cup product pairing to $\mathbb{Z}^{\mathrm{an}}$. If we tensor it with $\mathbb{Q}$, we obtain the locally constant sheaf $\mathcal{H}_{\mathbb{Q}}^{\mathrm{an}}$ on M^{an}, which by cup product is orthogonally self-dual. We take for d the rank of $\mathcal{H}_{\mathbb{Q}}^{\mathrm{an}}$. If we invert a suitable integer $N \geq 1$, and tensor $\mathcal{H}_{\mathbb{Z}}^{\mathrm{an}}$ with $\mathbb{Z}[1/N]$ to obtain (by the flatness of $\mathbb{Z}[1/N]$ over $\mathbb{Z}$) $\mathcal{H}_{\mathbb{Z}[1/N]}^{\mathrm{an}}$, we find that $\mathcal{H}_{\mathbb{Z}[1/N]}^{\mathrm{an}}$ is a locally constant sheaf of free $\mathbb{Z}[1/N]$-modules of rank d which under cup product is orthogonally self-dual over $\mathbb{Z}[1/N]$. We can take this N to be the N of the theorem. Indeed, for any ℓ not dividing N, we can make the flat extension of scalars from $\mathbb{Z}[1/N]$ to $\mathbb{Z}_\ell$ and infer that $\mathcal{H}_{\mathbb{Z}_\ell}^{\mathrm{an}}$ is a lisse sheaf of free $\mathbb{Z}_\ell$-modules of rank d on M^{an} which is orthogonally self-dual over $\mathbb{Z}_\ell$. By the comparison theorem, the restriction to $M_{\mathbb{C}}$ of $\mathcal{H}_{\mathbb{Z}_\ell}$ is therefore a lisse sheaf of free $\mathbb{Z}_\ell$-modules on $M_{\mathbb{C}}$ which is orthogonally self dual over $\mathbb{Z}_\ell$. It follows that the lisse sheaf $\mathcal{H}_{\mathbb{Z}_\ell}$ on $M[1/\ell]$ itself is torsion-free and $\mathbb{Z}_\ell$-autodual under the cup product pairing. Indeed, it suffices to check both the torsion-freeness of the lisse sheaf in question, namely $\mathcal{H}_{\mathbb{Z}_\ell}$, and the $\mathbb{Z}_\ell$-nondegeneracy of the pairing, at a single geometric point of $M[1/\ell]$.

We now consider two fibrewise conditions that may or may not hold in our general setup. Both of these conditions do hold in both of the examples given above (Legendre twists and Weierstrass families), cf. [Ka-MMP, 8.2.3 and 10.2.13] respectively for these two cases.

(1) For every finite field k, and for every k-valued point m in $M(k)$, the relative elliptic curve $E_{k,m}/U_{k,m}/k$ has non-constant j-invariant.

(2strong) For every finite field k and every ring homomorphism $\phi : R \to k$, denote by $M_{k,\phi}/k$ the fibre of M/R above (k, ϕ). For every ℓ invertible in k, consider the restriction to $M_{k,\phi}$ of the lisse sheaf $\mathcal{H}_{\mathbb{Q}_\ell}$ on $M[1/\ell]$. View this lisse sheaf as a representation $\rho_{k,\phi,\ell} : \pi_1(M_{k,\phi}) \to O(d, \mathbb{Q}_\ell)$. Under every such homomorphism $\rho_{k,\phi,\ell}$, the image in $O(d, \mathbb{Q}_\ell)$ of the geometric fundamental group

$$\pi_1^{\mathrm{geom}}(M_{k,\phi}) := \pi_1(M_{k,\phi} \otimes_k \overline{k}) \lhd \pi_1(M_{k,\phi})$$

is Zariski dense in $O(d, \overline{\mathbb{Q}_\ell})$.

In certain applications, cf. [Ka-MMP, 7.2.7, 8.2.5, 10.2.15] and [Ka-TLFM, 8.5.7, 8.6.7], one knows only that the following weaker version of the second condition holds.

(2weak) For every finite field k and every ring homomorphism $\phi : R \to k$, denote by $M_{k,\phi}/k$ the fibre of M/R above (k, ϕ). For every ℓ invertible in k, consider the restriction to $M_{k,\phi}$ of the lisse sheaf $\mathcal{H}_{\mathbb{Q}_\ell}$ on $M[1/\ell]$. View this lisse sheaf as a representation $\rho_{k,\phi,\ell} : \pi_1(M_{k,\phi}) \to O(d, \mathbb{Q}_\ell)$. Under each such homomorphisms $\rho_{k,\phi,\ell}$, the image in $O(d, \mathbb{Q}_\ell)$ of the geometric fundamental group

$$\pi_1^{\mathrm{geom}}(M_{k,\phi}) := \pi_1(M_{k,\phi} \otimes_k \overline{k}) \lhd \pi_1(M_{k,\phi})$$

is Zariski dense in *either* $SO(d, \overline{\mathbb{Q}}_\ell)$ *or in* $O(d, \overline{\mathbb{Q}}_\ell)$.

If the first condition holds, then for every ℓ invertible in k, the lisse sheaf $\mathcal{F}_{\mathbb{Q}_\ell}$ on $U_{k,m}/k$ is geometrically irreducible, and (hence) the unitarized L-function is given by the action of the Frobenius conjugacy class $Frob_{k,m}$ in $\pi_1(M[1/\ell])$ on the lisse sheaf $\mathcal{H}_{\mathbb{Q}_\ell}$:

$$L_u(E_{k,m}/U_{k,m}, T) = \det(1 - TFrob_{k,m}|\mathcal{H}_{\mathbb{Q}_\ell}).$$

[If we do not impose the first condition, the lisse sheaf $\mathcal{F}_{\mathbb{Q}_\ell}$ on $U_{k,m}/k$ could be geometrically constant (e.g., if E/U were a constant elliptic curve), in which case the unitarized L-function would not be a polynomial, but rather a rational function whose numerator is given by the right hand side.] Since these unitarized L-function have rational coefficients which "do not know about ℓ", we see that the sheaves $\mathcal{H}_{\mathbb{Q}_\ell}$ on $M[1/\ell]$ form, as ℓ varies, a "compatible system of orthogonal ℓ-adic representations" on M. Moreover, and this is the import of the previous theorem, there exists a single orthogonal group $O(d)/\mathbb{Z}[1/N]$, corresponding to a quadratic form over $\mathbb{Z}[1/N]$ in d variables whose discriminant is invertible in $\mathbb{Z}[1/N]$, such that for every ℓ we land in its $\mathbb{Q}_\ell$-points, and such that for ℓ prime to N, we land in its $\mathbb{Z}_\ell$ points. What is essential here is"only" the following (apparently weak) consequence of this last fact: for almost all ℓ (namely those ℓ prime to N), we are landing in the $\mathbb{Z}_\ell$ points of an orthogonal group over $\mathbb{Z}_\ell$ corresponding to a quadratic form over $\mathbb{Z}_\ell$ in d variables whose discriminant is invertible in $\mathbb{Z}_\ell$.

For each finite field k and each homomorphism $\phi : R \to k$, denote by $\mathrm{IrrFrac}(k, \phi) \in \mathbb{Q}$ the fraction of the k-valued points m in the fibre $M_{k,\phi}/k$ for which the reduced unitarized L-function $L_{u,\mathrm{red}}(E_{k,m}/U_{k,m}, T)$ is $\mathbb{Q}$-irreducible.

4 Statement of the main theorem

Theorem 4.1 *In the general setup $E/U/M/R$, suppose that the fibrewise conditions* (1) *and* (2weak) *of the previous section hold. Suppose also that d, the common degree of the L-functions, is ≥ 3. Given a real number $\epsilon > 0$, there exists a real* *constant* $X = X(\epsilon, E/U/M/R)$ *such that for any finite field* k *with* $\#k > X$, *and any homomorphism* $\phi : R \to k$, *we have*

$$\mathrm{IrrFrac}(k, \phi) \geq 1 - \epsilon.$$

5 Statement of an abstract version of the main theorem

Let us now consider an abstract version of our situation. We are given a finitely generated $\mathbb{Z}$-algebra R. Over R, we are given a smooth R-scheme M/R of relative dimension $\nu \geq 1$ with geometrically connected fibres. We are given an integer $d \geq 3$. For every prime ℓ such that $M[1/\ell]$ is nonempty, we are given a lisse $\mathbb{Q}_\ell$-sheaf $\mathcal{H}_{\mathbb{Q}_\ell}$ on $M[1/\ell]$ of rank d, together with a symmetric autoduality pairing

$$\mathcal{H}_{\mathbb{Q}_\ell} \times \mathcal{H}_{\mathbb{Q}_\ell} \to \mathbb{Q}_\ell.$$

These sheaves are assumed to form a compatible system of ℓ-adic representations on M (in the sense that each characteristic polynomial of Frobenius has rational coefficients which are independent of the auxiliary choice of allowed ℓ). Each sheaf $\mathcal{H}_{\mathbb{Q}_\ell}$ is assumed pure of weight zero. For all but finitely many ℓ, say for all ℓ outside a finite set S of primes, we are given a lisse $\mathbb{Z}_\ell$-sheaf $\mathcal{H}_{\mathbb{Z}_\ell}$ on $M[1/\ell]$ of free $\mathbb{Z}_\ell$ modules of rank d, together with a symmetric autoduality pairing over $\mathbb{Z}_\ell$,

$$\mathcal{H}_{\mathbb{Z}_\ell} \times \mathcal{H}_{\mathbb{Z}_\ell} \to \mathbb{Z}_\ell,$$

which is an integral form of $\mathcal{H}_{\mathbb{Q}_\ell}$ with its autoduality pairing.

For each finite field k and each homomorphism $\phi : R \to k$, denote by $\mathrm{IrrFrac}(k, \phi) \in \mathbb{Q}$ the fraction of the k-valued points m in the fibre $M_{k,\phi}/k$ for which the reduced characteristic polynomial $\mathrm{Rdet}(1 - T\mathrm{Frob}_{k,m}|\mathcal{H})$ is $\mathbb{Q}$-irreducible.

Theorem 5.1 *In the abstract version given above, with* $d \geq 3$, *suppose that the fibrewise condition* (2weak) *of the previous section holds. Given a real number* $\epsilon > 0$, *there exists a real constant* $X = X(\epsilon, R)$ *such that for any finite field* k *with* $\#k > X$, *and any homomorphism* $\phi : R \to k$, *we have*

$$\mathrm{IrrFrac}(k, \phi) \geq 1 - \epsilon.$$

We will fix $\epsilon > 0$, and prove the theorem for this value of ϵ. We reduce immediately to the case when R is reduced. If we have a finite decomposition of $\mathrm{Spec}(R)$ as the disjoint union of finitely many locally closed, reduced affine subschemes $\mathrm{Spec}(R_i)$, it suffices to prove the theorem (for our fixed $\epsilon > 0$), over each $\mathrm{Spec}(R_i)$ separately. Indeed, then we can take $X(\epsilon, R)$ to be $\mathrm{Max}_i(X(\epsilon, R_i))$. So by noetherian induction on $\mathrm{Spec}(R)$, it suffices to prove that the theorem holds, for our fixed $\epsilon > 0$, in some affine open neighborhood of some maximal point of

Spec(R). Any sufficiently small such open neighborhood is of the form Spec(R_1), with R_1 a normal integral domain which is a finitely generated $\mathbb{Z}$-algebra. Making the extension of scalars from R to R_1, we are reduced to proving the following "generic" version of the theorem.

Theorem 5.2 *In the abstract version given above, with $d \geq 3$, suppose that the fibrewise condition* (2weak) *of the previous section holds. Suppose in addition that R is a normal integral domain which is a finitely generated $\mathbb{Z}$-algebra. Given a real number $\epsilon > 0$, there exists a real constant $X = X(\epsilon, R)$ and a nonzero element $r = r(\epsilon) \in R$, such that for any finite field k with $\#k > X$, and any homomorphism $\phi : R \to k$ for which $\phi(r) \neq 0$, we have*

$$\mathrm{IrrFrac}(k, \phi) \geq 1 - \epsilon.$$

6 Interlude: Review of orthogonal groups over finite fields of odd characteristic

In this section, we fix an integer $d \geq 1$, a finite field $E = \mathbb{F}_q$ of odd characteristic, and a nondegenerate quadratic form in d variables over E, i.e., a d-dimensional E vector space V endowed with a symmetric E-bilinear form $\Psi : V \times V \to E$ which makes V autodual. We denote by $O(V, \Psi) := Aut_E(V, \Psi)$ the corresponding finite orthogonal group.

One knows that for fixed d and E, there are precisely two isomorphism classes of nondegenerate quadratic form, distinguished by whether or not the discriminant is a square in $E^\times$. When d is odd, the two isomorphism classes give rise to the same orthogonal group; indeed if (V, Ψ) represents one class, then for any nonsquare $\alpha \in E^\times$, $(V, \alpha\Psi)$ represents the other, while visibly their orthogonal groups coincide. So we may speak unambiguously of the group $O(d, E)$ when d is odd.

When $d = 2n$ is even, then the two cases are called the split case and the nonsplit case. The standard model for the split case is given by the quadratic form $\sum_{i=1}^{n} x_i x_{n+i}$ (so here $(-1)^n$Discriminant is a square), which we will denote $(split_{2n}, std)$. A convenient model for the nonsplit case is to take $V := \mathbb{F}_{q^{2n}}$ as our $\mathbb{F}_q$ vector space, endowed with the symmetric bilinear form

$$\Psi(x, y) := (1/2)\mathrm{Trace}_{\mathbb{F}_{q^{2n}}/\mathbb{F}_q}(xy^{q^n}),$$

and quadratic form

$$\Psi(x, x) = \mathrm{Trace}_{\mathbb{F}_{q^n}/\mathbb{F}_q}(\mathrm{Norm}_{\mathbb{F}_{q^{2n}}/\mathbb{F}_{q^n}}(x)).$$

For ease of later reference, we will refer to this model as the standard nonsplit model, and denote it $(\mathbb{F}_{q^{2n}}, std)$. The split and nonsplit orthogonal groups are *not* isomorphic; they even have different orders. We will denote them $O_{\text{spl}}(d, E)$ and $O_{\text{nonspl}}(d, E)$ respectively when we need to distinguish them.

On the Clifford algebra $Cl := Cl(V, \Psi)$ attached to (V, Ψ), we have the E-algebra involution I which is $v \mapsto -v$ on V, and the E-algebra anti-automorphism $x \mapsto t(x)$ which is the identity on V. We have its unit group $Cl^{\times}$. The unit group acts on the Clifford algebra by the sign-twisted conjugation action: $u \in Cl^{\times}$ acts as $x \mapsto I(u)xu^{-1}$. Inside $Cl^{\times}$ we have the (twisted) Clifford group, namely the subgroup $C^{\times}$ consisting of those elements which map V to itself. Every nonisotropic $v \in V$ lies in $C^{\times}$, for the map $x \mapsto I(v)xv^{-1}$ is then, for $x \in V$, reflection in v. Moreover, one knows that every element of $C^{\times}$ is a nonzero scalar times a (possibly empty) product of nonisotropic vectors $v \in V$; this corresponds to the fact that in the orthogonal group $O(V, \Psi)$, every element is a product of reflections in nonisotropic vectors. For $u \in C^{\times}$, its "norm" $N(u) := t(u)u$ lies in $E^{\times}$, and $x \mapsto N(x)$ is a group homomorphism. Its kernel is the group $\text{Pin}(V, \Psi)$:

$$\text{Pin}(V, \Psi) := \text{Ker}(N : C^{\times} \to E^{\times}).$$

The subgroup of $\text{Pin}(V, \Psi)$ consisting of the elements fixed by the involution I is the group $\text{Spin}(V, \Psi)$.

Remarks 6.1 The reader should be warned of a possible source of serious confusion. In the older literature, e.g., [Artin-GA], [Bour-AlgIX] and [Chev-Spin], the unit group is made to act on the Clifford algebra by the literal conjugation action: $u \in Cl^{\times}$ acts as $x \mapsto uxu^{-1}$, and one takes the (untwisted) Clifford group, denoted Γ in [Chev-Spin, 2.3], accordingly. This leads to unpleasant difficulties, centered on the fact that when V is odd-dimensional, there are nonscalar elements of Γ which act trivially on V, and the "norm" of an element of Γ need not be a scalar. Contorsions are adopted to get around these difficulties; one obtains the group $\text{Spin}(V, \Psi)$, but there is no $\text{Pin}(V, \Psi)$ in the older theory. The sign-twisted approach, and the group Pin, first appeared in [AtBS-Clif, 1.7, 3.1], cf. also [Kar-Clif, 1.1.4-8].

We have an exact sequence

$$\{1\} \to \pm 1 \to \text{Pin}(V, \Psi) \to O(V, \Psi) \to \pm 1,$$

in which the last map is the spinor norm, denoted

$$\text{sp} : O(V, \Psi) \to \pm 1.$$

The spinor norm is determined by its value on reflections Rfl_v in nonisotropic vectors $v \in V$ (since these elements generate $O(V, \Psi)$). For these, we have the explicit formula

$$sp(Rfl_v) = \text{the class of } \Psi(v, v) \text{ in } E^{\times}/(E^{\times})^2 \cong \pm 1.$$

If $d \geq 2$, the spinor norm is surjective, and we have a short exact sequence

$$\{1\} \to \pm 1 \to \mathrm{Pin}(V, \Psi) \to O(V, \Psi) \to \pm 1 \to \{1\},$$

under which the inverse image of $SO(V, \Psi)$ in $\mathrm{Pin}(V, \Psi)$ is $\mathrm{Spin}(V, \Psi)$. So we also have the more standard short exact sequence

$$\{1\} \to \pm 1 \to \mathrm{Spin}(V, \Psi) \to SO(V, \Psi) \to \pm 1 \to \{1\}.$$

We also have the determinant homomorphism

$$\det : O(V, \Psi) \to \pm 1.$$

The simultaneous kernel of these two homomorphisms, sp and det, is denoted $\Omega(V, \Psi)$.

When $d \geq 5$, or when $d = 4$ and we are in the nonsplit case, or when $d = 3$ and the characteristic is ≥ 5, the group $\Omega(V, \Psi)$ is, modulo its center, a nonabelian simple group, cf. [Artin-GA, Theorems 4.9, 5.20, 5.21, 5.27]. Moreover, in these cases, the only proper normal subgroups of $\Omega(V, \Psi)$ are subgroups of its center, and consequently $\Omega(V, \Psi)$ is its own commutator subgroup. The center of $\Omega(V, \Psi)$ is trivial if either d is odd or if the discriminant is a nonsquare, otherwise it is ± 1. When $d = 4$ and the characteristic is ≥ 5 and we are in the split case, then $\Omega(V, \Psi)/\pm 1$ is the product $PSL(2, E) \times PSL(2, E)$ of the simple group $PSL(2, E)$ with itself, cf. [Artin-GA, Theorem 5.22], and $\Omega(V, \Psi)$ is its own commutator subgroup [being a quotient of $Spin(V, \Psi) \cong SL(2, E) \times SL(2, E)$, which is its own commutator subgroup].

One knows [Artin-GA, Theorems 5.14, 5.17] that $\Omega(V, \Psi)$ is the commutator subgroup of $O(V, \Psi)$; indeed this was its *definition* before Chevalley introduced the use of Clifford algebras in these questions, cf. [Die-GC, Chpt. III, Section 12, p.23]. For $d \geq 2$, the quotient group $O(V, \Psi)/\Omega(V, \Psi)$ is, by the pair of maps (det, sp) the group $\{\pm 1\} \times \{\pm 1\}$. We will need to know, in each of the four cosets of $\Omega(V, \Psi)$ in $O(V, \Psi)$, lower bounds for the numbers of elements A whose reduced reversed characteristic polynomials $\mathrm{Rdet}(1 - TA)$ have, as E-polynomials, certain imposed factorization patterns. For each $(\alpha, \beta) \in \{\pm 1\} \times \{\pm 1\}$, we denote by

$$O(V, \Psi)(\det = \alpha, \mathrm{sp} = \beta)$$

the corresponding coset.

There is a further cautionary remark we need to make at this point. Suppose $d \geq 2$; we are given a subgroup H of $GL(V) := \mathrm{Aut}_E(V)$, and we are told that $H = O(V, \Psi)$ for some symmetric autoduality Ψ. Then the subgroup $\Omega(V, \Psi)$ is an intrinsic subgroup of H, namely its commutator subgroup. The det homomorphism

$$\det : H \to \pm 1$$

is intrinsic on H as a subgroup of $GL(V)$. However, the spinor norm homomorphism

$$\mathrm{sp} : H \to \pm 1$$

depends on the choice of Ψ. Indeed, if we replace Ψ by a nonzero scalar multiple $\alpha\Psi$ with α a nonsquare, the orthogonal group does not change, but the two spinor norms are related by

$$\mathrm{sp}_{(V,\alpha\Psi)}(h) = \det(h)\mathrm{sp}_{(V,\Psi)}(h).$$

On the other hand, since the quotient group $H/\Omega(V, \Psi)$ is of type $(2, 2)$, with det and $\mathrm{sp}_{(V,\Psi)}$ an $\mathbb{F}_2$-basis of its character group, we see that for any Ψ_1 on V with $O(V, \Psi) = O(V, \Psi_1)$, we have either $\mathrm{sp}_{(V,\Psi_1)}(h) = \mathrm{sp}_{(V,\Psi)}(h)$ for every $h \in H$, or we have $\mathrm{sp}_{(V,\Psi_1)}(h) = \det(h)\mathrm{sp}_{(V,\Psi)}(h)$ for every $h \in H$. Thus each of the two cosets of $\Omega(V, \Psi)$ in $H \cap SL(V) = SO(V, \Psi)$ is intrinsic, e.g., one is a subgroup and one isn't, but the two cosets of $\Omega(V, \Psi)$ in $H \setminus H \cap SL(V) = O(V, \Psi) \setminus SO(V, \Psi)$ may be interchanged by different choices of Ψ. In the discussion below, we work with particular models of our orthogonal groups, i.e., we make specific choices of Ψ. But we prove only statements which are invariant under replacing sp by det $\times$ sp.

Lemma 6.2 *Fix $d = 2n \geq 2$. Suppose $q := \#E \geq 7$. In each of the two cosets of $\Omega(d, E)$ in $SO_{\mathrm{nonspl}}(d, E)$, the fraction of elements A for which* $\mathrm{Rdet}(1 - TA)$ *is E-irreducible is at least $1/2n$.*

Proof. In the standard nonsplit model $(\mathbb{F}_{q^{2n}}, std)$, the group $\mu_{1+q^n} := \mu_{1+q^n}(\mathbb{F}_{q^{2n}})$, acting by homothety on $\mathbb{F}_{q^{2n}}$, lies in $SO_{\mathrm{nonspl}}(2n, E)$. Moreover, we know [Saito-sign, Lemma 1, parts 4 and 5] that the spinor norm, restricted to μ_{1+q^n}, is trivial precisely on the subgroup $\mu_{(1+q^n)/2}$ of squares. We also remark that every $\mathbb{F}_{q^{2n}}$-homothety which lies in the orthogonal group lies in μ_{1+q^n}. It follows that if $\zeta \in \mu_{1+q^n}$ is an element such that the field $\mathbb{F}_q(\zeta)$ is $\mathbb{F}_{q^{2n}}$, then its characteristic polynomial is an $\mathbb{F}_q$-irreducible palindromic polynomial, and its centralizer in $O_{\mathrm{nonspl}}(2n, E)$ is the subgroup μ_{1+q^n} [simply because any $\mathbb{F}_q$-linear endomorphism A of $\mathbb{F}_{q^{2n}}$ which commutes with ζ is $\mathbb{F}_{q^{2n}}$-linear, so an $\mathbb{F}_{q^{2n}}$-homothety]. So for any such ζ, its conjugacy class in $O_{\mathrm{nonspl}}(2n, E)$ contains $\#O_{\mathrm{nonspl}}(2n, E)/(1 + q^n)$ elements, all of which have the same $\mathbb{F}_q$-irreducible palindromic characteristic polynomial as ζ, as well as the same spinor norm and determinant as ζ. If we take a second such element ζ_1 which is not one of the $2n$ Galois conjugates of ζ, then its characteristic polynomial is a different $\mathbb{F}_q$-irreducible palindromic polynomial, so certainly its conjugacy class in $O_{\mathrm{nonspl}}(2n, E)$ is disjoint from that of ζ. [Conversely, Galois conjugate elements of μ_{1+q^n} are $O_{\mathrm{nonspl}}(2n, E)$-conjugate, since the Galois automorphisms of $\mathbb{F}_{q^{2n}}/\mathbb{F}_q$ lie in the orthogonal group, and their conjugation action on elements of μ_{1+q^n} is the same as their Galois action.] Denote temporarily by $N_\pm$ the number of elements $\zeta \in \mu_{1+q^n}$ of spinor norm ± 1 such that the field $\mathbb{F}_q(\zeta)$ is $\mathbb{F}_{q^{2n}}$. Taking the union of their conjugacy classes, we obtain $\#O_{\mathrm{nonspl}}(2n, E)N_\pm/2n(1 + q^n)$ elements in $O_{\mathrm{nonspl}}(2n, E)(\det = 1, \mathrm{sp} = \pm 1)$ with an E-irreducible palindromic characteristic polynomial.

One sees easily if $\zeta \in \mu_{1+q^n}$ is such that $\mathbb{F}_q(\zeta)$ is a proper subfield of of $\mathbb{F}_{q^{2n}}$, then either $\zeta = \pm 1$, or $\mathbb{F}_q(\zeta)$ is $\mathbb{F}_{q^{2a}}$ for some divisor $a < n$ of n, $\zeta \in \mu_{1+q^a}$ and n/a is odd. Thus, denoting $[x] :=$ Floor(x), at most $2 + [n/3]q^{[n/3]}$ of the elements in μ_{1+q^n} fail to generate $\mathbb{F}_{q^{2n}}$ over $\mathbb{F}_q$. So we have the estimates

$$N_{\pm} \geq (1+q^n)/2 - 2 - [n/3]q^{[n/3]}.$$

Treating separately the cases $[n/3] = 0$ and $[n/3] \geq 1$, we see that so long as $q \geq 7$, we have

$$N_{\pm} \geq (1+q^n)/4.$$

Thus we obtain at least

$$\#O_{\text{nonspl}}(2n, E)/8n = \#O_{\text{nonspl}}(2n, E)(\det = 1, \text{sp} = \pm 1)/2n$$

elements in $O_{\text{nonspl}}(2n, E)(\det = 1, \text{sp} = \pm 1)$ with an E-irreducible palindromic characteristic polynomial.

Lemma 6.3 *Fix $d = 2n \geq 4$. Suppose $q := \#E \geq 7$. In each of the two cosets $O(d, E)(\det = -1, sp = \pm 1)$, the fraction of elements A for which* Rdet$(1 - TA)$ *is E-irreducible is at least $1/4(2n-2)$.*

Proof. We take as model of our quadratic space

$$(\mathbb{F}_{q^{2n-2}}, std) \oplus (\mathbb{F}_{q^2}, \text{std})$$

in the *split* case, and

$$(\mathbb{F}_{q^{2n-2}}, std) \oplus (\text{split}_2, \text{std})$$

in the *nonsplit* case. Corresponding to these direct sum decompositions, we have inclusions of the corresponding orthogonal groups

$$O(2n-2, E) \times O(2, E) \subset O(2n, E).$$

In the orthogonal group of the first factor, take an element $\zeta \in \mu_{1+q^{n-1}}$ which generates $\mathbb{F}_{q^{2n-2}}$ over $\mathbb{F}_q$. In the orthogonal group of the second factor, take a reflection R of spinor norm one, e.g., take the reflection in a vector of square length one. The centralizer in $O(2n, E)$ of the element (ζ, R) is the product group $\mu_{1+q^{n-1}} \times \{\pm 1, \pm R\}$. [Indeed, if an element A in an orthogonal group over a field of characteristic not 2 has a (reversed or not, the two agree up to sign) characteristic polynomial which is a product $\prod_i f_i(T)$ of pairwise prime polynomials, each of which has its roots stable by $x \mapsto 1/x$, then the decomposition of the ambient vector space V as the direct sum of the spaces $V_i := \text{Ker}(f_i(A))$ is an orthogonal decomposition. Any endomorphism B of V which commutes with A preserves this decomposition, say $B = \oplus_i(B_i \ on \ V_i)$, and on each V_i, B_i commutes with $A|V_i$.

Moreover, if B is orthogonal, then so is each B_i.] The counting argument used to prove the lemma above then gives the asserted result. □

Lemma 6.4 *Fix $d = 2n + 1 \geq 3$. Suppose $q := \#E \geq 7$. In each of the four cosets $O(d, E)(\det = \pm 1, \mathrm{sp} = \pm 1)$, the fraction of elements A for which* $\mathrm{Rdet}(1 - TA)$ *is E-irreducible is at least $1/4n$.*

Proof. We take as model of our quadratic space

$$(\mathbb{F}_{q^{2n}}, \mathrm{std}) \oplus (\mathbb{F}_q, x^2),$$

and repeat the previous argument, now using elements of the form $(\zeta, \pm 1)$. □

Lemma 6.5 *Fix $d = 2n \geq 6$. Suppose $q := \#E \geq 7$. Fix a partition of n as $n = a + b$ with $1 \leq a < b$. In each of the two cosets of $\Omega(d, E)$ in $SO_{\mathrm{spl}}(d, E)$, the fraction of elements A for which* $\mathrm{Rdet}(1 - TA)$ *is of the form*

$$(\mathrm{E - irreducible\ of\ degree\ 2a})(\mathrm{E - irreducible\ of\ degree\ 2b})$$

is at least $1/32ab$.

Proof. We take as model of our quadratic space

$$(\mathbb{F}_{q^{2a}}, \mathrm{std}) \oplus (\mathbb{F}_{q^{2b}}, std),$$

and repeat the previous argument, now using elements of the form $(\zeta_a \in \mu_{1+q^a}, \zeta_b \in \mu_{1+q^b})$. If ζ_a (respectively ζ_b) has full degree $2a$ (respectively $2b$) over $\mathbb{F}_q$, the centralizer of this element in $O(2n, E)$ is the product group $\mu_{1+q^a} \times \mu_{1+q^b}$, and the argument concludes as before. □

Lemma 6.6 *Fix $d = 2n \geq 4$. Suppose $q := \#E \geq 7$. In each of the two cosets of $\Omega(d, E)$ in $SO_{\mathrm{spl}}(d, E)$, the fraction of elements A for which* $\mathrm{Rdet}(1 - TA)$ *is of the form*

$$(\mathrm{E - irreducible\ } P(T) \mathrm{\ of\ degree\ } n)(\mathrm{E - irreducible\ } Q(T) \mathrm{\ of\ degree\ } n),$$

with P and Q relatively prime, and with

$$Q(T) = (\mathrm{some\ constant\ in\ E}^{\times})T^n P(1/T),$$

is at least $1/2n$.

Proof. We take as model

$$V := \mathbb{F}_{q^n} \oplus \mathbb{F}_{q^n},$$

with the split quadratic form

$$\Psi(x \oplus y, x \oplus y) := \mathrm{Trace}_{\mathbb{F}_{q^n}/\mathbb{F}_q}(xy).$$

The group $\mathbb{F}_{q^n}^{\times}$ is then a subgroup of $SO(V, \Psi)$, with $\zeta \in \mathbb{F}_{q^n}^{\times}$ acting as

$$(\zeta, \zeta^{-1}) : x \oplus y \mapsto \zeta x \oplus \zeta^{-1} y.$$

By [Saito-sign, Lemma 1.1, part 2], the spinor norm, restricted to this $\mathbb{F}_{q^n}^{\times}$ subgroup, is trivial precisely on the squares. Take an element (ζ, ζ^{-1}) such that ζ has full degree n over $\mathbb{F}_q$, and such that ζ and ζ^{-1} have different irreducible polynomials over $\mathbb{F}_q$ (i.e., such that ζ and ζ^{-1} are not Galois conjugate). Then the centralizer of (ζ, ζ^{-1}) in $O(V, \Psi)$ is precisely the subgroup $\mathbb{F}_{q^n}^{\times}$. Moreover, knowing the characteristic polynomial of (ζ, ζ^{-1}) determines ζ up to replacing it by either one of its n conjugates or by one of the n conjugates of ζ^{-1}.

The ζ's which fail the first condition are those which lie in a proper subfield $\mathbb{F}_{q^a}$ for some divisor $a < n$ of n. Those which fail the second condition are those which lie in some subgroup μ_{1+q^a}, with $2a|n$. For $q \geq 7$, a routine counting shows that the number of ζ's which fail one or both of the two conditions is at most $(q^n - 1)/4$. The argument now concludes as before.

With these preliminary lemmas established, we get the following product theorems.

Theorem 6.7 *Fix an odd integer $d = 2n + 1 \geq 3$, an integer $r \geq 1$, and a list of r primes*

$$7 \leq \ell_1 < \ell_2 < \cdots < \ell_r.$$

Denote by G the subgroup of the product group $\prod_i O(d, \mathbb{F}_{\ell_i})$ consisting of those elements $(A_1, \ldots, A_r)$ all of whose determinants, viewed in ± 1, coincide. In any coset of $\prod_i \Omega(d, \mathbb{F}_{\ell_i})$ in G, the fraction of elements $(A_1, \ldots, A_r)$ such that $Rdet(1 - TA_i)$ is $\mathbb{F}_{\ell_i}$-irreducible for some i is at least

$$1 - (1 - 1/4n)^r.$$

Proof. The point is that the quotient $G/\prod_i \Omega(d, \mathbb{F}_{\ell_i})$ is naturally the product of $r + 1$ copies of ± 1, by means of the common value of the determinant and the spinor norms of the factors. So any coset is a product, either of cosets $O(d, \mathbb{F}_{\ell_i})(\det = 1, \mathrm{sp} = \alpha_i)$, or of cosets $O(d, \mathbb{F}_{\ell_i})(\det = -1, \mathrm{sp} = \alpha_i)$. The assertion is now immediate from Lemma 6.4.

Theorem 6.8 *Fix an even integer $d = 2n \geq 4$, an integer $r \geq 1$, and a list of r primes*

$$7 \leq \ell_1 < \ell_2 < \cdots < \ell_r.$$

Denote by G the subgroup of the product group $\prod_i O_{\mathrm{nonspl}}(d, \mathbb{F}_{\ell_i})$ consisting of those elements $(A_1, \ldots, A_r)$ all of whose determinants, viewed in ± 1, coincide. In any coset of $\prod_i \Omega(d, \mathbb{F}_{\ell_i})$ in G, the fraction of elements $(A_1, \ldots, A_r)$ such that $\mathrm{Rdet}(1 - TA_i)$ *is $\mathbb{F}_{\ell_i}$-irreducible for some i is at least*

$$1 - (1 - 1/8n)^r.$$

Proof. By the product structure of the coset, the assertion is immediate from Lemmas 6.2 and 6.3.

Theorem 6.9 *Fix an even integer $d = 2n \geq 4$, an integer $r \geq 1$, and a list of r primes*

$$7 \leq \ell_1 < \ell_2 < \cdots < \ell_r.$$

Denote by G the subgroup of the product group $\prod_i O(d, \mathbb{F}_{\ell_i})$ consisting of those elements $(A_1, \ldots, A_r)$ all of whose determinants, viewed in ± 1, coincide; the factor groups may be separately split or nonsplit at will. In any coset of $\prod_i \Omega(d, \mathbb{F}_{\ell_i})$ in G for which the common value of the determinant is -1, the fraction of elements $(A_1, \ldots, A_r)$ such that $\mathrm{Rdet}(1 - TA_i)$ *is $\mathbb{F}_{\ell_i}$-irrreducible for some i is at least*

$$1 - (1 - 1/8n)^r.$$

Proof. By the product structure of the coset, the assertion is immediate from Lemma 6.3.

Theorem 6.10 *Fix an even integer $d = 2n \geq 6$, an integer $r \geq 1$, and a list of r primes*

$$7 \leq \ell_1 < \ell_2 < \cdots < \ell_r.$$

Choose a partition of n, say $n = a + b$ with $1 \leq a < b$. Denote by G the subgroup of the product group $\prod_i O_{\mathrm{spl}}(d, \mathbb{F}_{\ell_i})$ consisting of those elements $(A_1, \ldots, A_r)$ all of whose determinants, viewed in ± 1, coincide. In any coset of $\prod_i \Omega(d, \mathbb{F}_{\ell_i})$ in G for which the common value of the determinant is $+1$, the fraction of elements $(A_1, \ldots, A_r)$ such that $Rdet(1 - TA_i)$ is of the form

$$(\mathrm{E} - \text{irreducible of degree 2a})(\mathrm{E} - \text{irreducible of degree 2b})$$

for some i, AND such that $\mathrm{Rdet}(1 - TA_j)$ *is of the form*

$$(\mathrm{E} - \text{irreducible of degree } n)(\text{a different E} - \text{irreducible of degree } n)$$

for some j, is at least

$$1 - (1 - 1/32ab)^r - (1 - 1/2n)^r.$$

Proof. By the product structure of the coset, the assertion is immediate from Lemmas 6.5 and 6.6. □

Theorem 6.11 *Fix $d = 4$, an integer $r \geq 1$, and a list of r primes*

$$7 \leq \ell_1 < \ell_2 < \cdots < \ell_r.$$

Denote by G the subgroup of the product group $\prod_i O_{\mathrm{spl}}(4, \mathbb{F}_{\ell_i})$ *consisting of those elements* $(A_1, \ldots, A_r)$ *all of whose determinants, viewed in* ± 1, *coincide. In any coset of* $\prod_i \Omega(4, \mathbb{F}_{\ell_i})$ *in G for which the common value of the determinant is* $+1$, *the fraction of elements* $(A_1, \ldots, A_r)$ *such that* $\mathrm{Rdet}(1 - TA_i)$ *is, for some i, of the form*

$$P(T)Q(T)$$

with $P(T)$ *and* $Q(T)$ *relatively prime* $\mathbb{F}_{\ell_i}$*-irreducibles of degree 2, neither of which is palindromic, and such that*

$$Q(T) = (\text{some constant in } \mathrm{E}^\times)T^2 P(1/T),$$

is at least

$$1 - (1 - 1/4)^r.$$

Proof. If we omitted the requirement that neither $P(T)$ nor $Q(T)$ be palindromic, the assertion would be immediate from the product structure of the coset, and Lemma 6.6. But the nonpalindromicity is automatic. Indeed, the fact that

$$Q(T) = (\text{some constant in } \mathrm{E}^\times)T^2 P(1/T),$$

tells us that if $\zeta \in \mathbb{F}_{\ell_i^2}^\times$ is a root of $P(T)$, then $1/\zeta$ is a root of $Q(T)$, hence cannot be a root of $P(T)$, since $P(T)$ and $Q(T)$ are relatively prime. But the two roots of a palindromic polynomial of degree two are reciprocals. Thus $P(T)$ is not palindromic, and similarly for $Q(T)$. □

7 Proof of Theorem 5.2, via a theorem of Larsen

Let us put ourselves in the situation which Theorem 5.2 purports to treat. Choose a finite field k and a ring homomorphism $\phi : R \to k$ (for instance, take a maximal ideal $\mathcal{I}$ of R, take k to be $R/\mathcal{I}$, and take ϕ to be canonical map of R onto $R/\mathcal{I}$). Making the extension of scalars $\phi : R \to k$, we get the space $M_{k,\phi}$. On $M_{k,\phi}$, we have the restrictions of the sheaves $\mathcal{H}_{\mathbb{Q}_\ell}$, for all ℓ invertible in k, as well as of the restrictions of the sheaves $\mathcal{H}_{\mathbb{Z}_\ell}$, for all such ℓ not in the finite set S. For each ℓ invertible in k, let us denote by Γ_ℓ the image in $O(d, \mathbb{Q}_\ell)$ of the arithmetic fundamental group $\pi_1(M_{k,\phi})$ under the homomorphism which "is" $\mathcal{H}_{\mathbb{Q}_\ell} | M_{k,\phi}$. Meanwhile, consider the composite map

$$\mathrm{Spin}(d, \mathbb{Q}_\ell) \to SO(d, \mathbb{Q}_\ell) \subset O(d, \mathbb{Q}_\ell).$$

According to a striking theorem of Larsen [Lar-Max, 3.17], the inverse image of Γ_ℓ in $\mathrm{Spin}(d, \mathbb{Q}_\ell)$ is, for a set of primes ℓ of Dirichlet density one, a "hyperspecial" maximal compact subgroup of $\mathrm{Spin}(d, \mathbb{Q}_\ell)$. Now for all ℓ invertible in k and not in S, Γ_ℓ lies in $O(d, \mathbb{Z}_\ell)$, and so its inverse image lies in $\mathrm{Spin}(d, \mathbb{Z}_\ell)$. Whenever this inverse image is a maximal compact subgroup of $\mathrm{Spin}(d, \mathbb{Q}_\ell)$, it must, by its maximality, be equal to the possibly larger compact subgroup $\mathrm{Spin}(d, \mathbb{Z}_\ell)$. Thus we infer that among the primes ℓ invertible in k and not in S, there is a set of Dirichlet density one, the "good primes" over (k, ϕ), for which the inverse image of Γ_ℓ in $\mathrm{Spin}(d, \mathbb{Z}_\ell)$ is the entire group $\mathrm{Spin}(d, \mathbb{Z}_\ell)$.

For each of these good ℓ, which we may take to all be ≥ 5, let us denote by $\Gamma_{\text{mod } \ell}$ the image in $O(d, \mathbb{F}_\ell)$ of the arithmetic fundamental group $\pi_1(M_{k,\phi})$ under the homomorphism which "is" $\mathcal{H}_{\mathbb{Z}_\ell} \otimes_{\mathbb{Z}_\ell} \mathbb{F}_\ell | M_{k,\phi}$. Then by Larsen's theorem, $\Gamma_{\text{mod } \ell}$ contains $\Omega(d, \mathbb{F}_\ell)$ for these good ℓ. Thus we have

$$\Omega(d, \mathbb{F}_\ell) \subset \Gamma_{\text{mod } \ell} \subset O(d, \mathbb{F}_\ell).$$

Let us denote by $\Gamma_{\text{geom, mod } \ell}$ the image in $O(d, \mathbb{F}_\ell)$ of the geometric fundamental group $\pi_1^{\text{geom}}(M_{k,\phi})$. Then

$$\Gamma_{\text{geom, mod } \ell} \lhd \Gamma_{\text{mod } \ell},$$

and the quotient is cyclic, being a quotient of $Gal(\overline{k}/k)$. We claim that for each good ℓ, we have

$$\Omega(d, \mathbb{F}_\ell) \subset \Gamma_{\text{geom, mod } \ell}.$$

Indeed, the intersection $\Omega(d, \mathbb{F}_\ell) \cap \Gamma_{\text{geom, mod } \ell}$ inside $\Gamma_{\text{mod } \ell}$ is a normal subgroup of $\Omega(d, \mathbb{F}_\ell)$ with cyclic quotient. As $d \geq 3$, $\Omega(d, \mathbb{F}_\ell)$ is its own commutator subgroup, so it has no proper normal subgroup which gives a cyclic quotient. Thus for each good ℓ we have

$$\Omega(d, \mathbb{F}_\ell) \subset \Gamma_{\text{geom, mod } \ell} \subset \Gamma_{\text{mod } \ell} \subset O(d, \mathbb{F}_\ell).$$

Suppose we are given an integer $r \geq 1$, and a list of r good primes

$$7 \leq \ell_1 < \ell_2 < \cdots < \ell_r.$$

Denote by G the subgroup of the product group $\prod_i O(d, \mathbb{F}_{\ell_i})$ consisting of those elements $(A_1, \ldots, A_r)$ all of whose determinants, viewed in ± 1, coincide. Denote by

$$\Gamma_{\text{geom, mod } \ell_1, \ell_2, \ldots \ell_r} \subset G \subset \prod_i O(d, \mathbb{F}_{\ell_i})$$

the image of the geometric fundamental group $\pi_1^{\text{geom}}(M_{k,\phi})$ under the direct sum of the various mod ℓ_i representations.

A key point is the following result of Goursat–Ribet type, cf. [Ribet-Gal, 5.2.2].

Lemma 7.1 *The group* $\Gamma_{\text{geom,mod } \ell_1, \ell_2, \ldots, \ell_r}$ *contains* $\prod_i \Omega(d, \mathbb{F}_{\ell_i})$*:*

$$\prod_i \Omega(d, \mathbb{F}_{\ell_i}) \subset \Gamma_{\text{geom, mod } \ell_1, \ell_2, \ldots, \ell_r} \subset G \subset \prod_i O(d, \mathbb{F}_{\ell_i}).$$

Proof. The projection of $\Gamma_{\text{geom,mod } \ell_1,\ell_2,\ldots,\ell_r} \subset \prod_i O(d,\mathbb{F}_{\ell_i})$ to each $O(d,\mathbb{F}_{\ell_i})$ factor contains $\Omega(d,\mathbb{F}_{\ell_i})$, as we have noted above. Now consider the commutator subgroup $D := D\Gamma_{\text{geom,mod } \ell_1,\ell_2,\ldots\ell_r}$ of $\Gamma_{\text{geom,mod } \ell_1,\ell_2,\ldots,\ell_r}$. As each group $\Omega(d,\mathbb{F}_{\ell_i})$ is its own commutator subgroup, D is a subgroup of $\prod_i \Omega(d,\mathbb{F}_{\ell_i})$ which maps onto each factor.

Suppose first that either $d \neq 4$, or that $d = 4$ and all our groups are nonsplit. Then the individual groups $\Omega(d,\mathbb{F}_{\ell_i})$ are simple modulo their centers, and the corresponding simple groups are pairwise non-isomorphic. So by Goursat's lemma [Ribet-Gal, 5.2.1], D maps onto each pair of factors $\Omega(d,\mathbb{F}_{\ell_i}) \times \Omega(d,\mathbb{F}_{\ell_j}), i < j$. Since each $\Omega(d,\mathbb{F}_{\ell_i})$ has no nontrivial abelian quotients, Ribet's lemma [Ribet-Gal, 5.2.2] shows that D is the full product $\prod_i \Omega(d,\mathbb{F}_{\ell_i})$.

In the remaining case, when $d = 4$ and all the groups are split, each $\Omega(4,\mathbb{F}_{\ell_i})/\pm 1$ is $PSL(2,\mathbb{F}_{\ell_i}) \times PSL(2,\mathbb{F}_{\ell_i})$. We first note that as $PSL(2,\mathbb{F}_{\ell_i})$ is simple and nonabelian, the only quotient groups of $\Omega(4,\mathbb{F}_{\ell_i})/\pm \cong PSL(2,\mathbb{F}_{\ell_i}) \times PSL(2,\mathbb{F}_{\ell_i})$ are the four obvious ones $(\{1\}\times\{1\}, \{1\}\times PSL(2,\mathbb{F}_{\ell_i}), PSL(2,\mathbb{F}_{\ell_i})\times\{1\}, PSL(2,\mathbb{F}_{\ell_i})\times PSL(2,\mathbb{F}_{\ell_i}))$. So the only quotient groups of $\Omega(4,\mathbb{F}_{\ell_i})$ are either these groups or, possibly, double covers of them. There is no quotient of order 2, since $\Omega(4,\mathbb{F}_{\ell_i})$ is its own commutator subgroup. Thus the only quotient groups $H_i \neq \{1\}$ of $\Omega(4,\mathbb{F}_{\ell_i})$ have the property that ℓ_i is the largest prime dividing the order of H_i (since ℓ_i is the largest prime dividing the order of $PSL(2,\mathbb{F}_{\ell_i})$). Therefore if $\ell_i \neq \ell_j$, then no quotient $H_i \neq \{1\}$ of $\Omega(4,\mathbb{F}_{\ell_i})$ is isomorphic to any quotient $H_j \neq \{1\}$ of $\Omega(4,\mathbb{F}_{\ell_j})$, simply because these quotients have different orders. So by Goursat's lemma [Ribet-Gal, 5.2.1], D maps onto each pair of factors $\Omega(4,\mathbb{F}_{\ell_i}) \times \Omega(4,\mathbb{F}_{\ell_j}), i < j$, and the proof then concludes as before, by invoking Ribet's lemma [Ribet-Gal, 5.2.2]. □

We now make a choice of the integer $r \geq 1$, and of the list of r good primes

$$7 \leq \ell_1 < \ell_2 < \ldots < \ell_r.$$

Recall the real $\epsilon > 0$ in the statement of the theorem we are to prove.
There are three separate cases to consider.

If $d = 2n + 1$ is odd, we choose r large enough that

$$(1 - 1/4n)^r < \epsilon/2,$$

and we take any list of r good primes

$$7 \leq \ell_1 < \ell_2 < \cdots < \ell_r.$$

If $d = 2n$ is even, we first look to see whether or not there are infinitely many good primes ℓ where our orthogonal group $O(d,\mathbb{F}_\ell)$ is nonsplit. If there are, we choose r large enough that

$$(1 - 1/8n)^r < \epsilon/2,$$

and we take any list of r good primes

$$7 \leq \ell_1 < \ell_2 < \cdots < \ell_r$$

at which the corresponding orthogonal group is nonsplit.

If $d = 2n$ is even, and there are at most finitely many good primes ℓ where our orthogonal group $O(d, \mathbb{F}_\ell)$ is nonsplit, then we choose r large enough that

$$(1 - 1/32n)^r < \epsilon/4,$$

and we take any list of r good primes

$$7 \le \ell_1 < \ell_2 < \cdots < \ell_r$$

at which the corresponding orthogonal group is split.

We now study what happens in the geometric generic fibre of M/R. Denote by K the fraction field of R, by $\overline{K}$ an algebraic closure of K, and by $M_{\overline{\eta}}$ the $\overline{K}$-scheme obtained by the extension of scalars $R \subset \overline{K}$.

Denote by

$$\Gamma_{\overline{\eta},\text{geom,mod } \ell_1,\ell_2,\ldots,\ell_r} \subset G \subset \prod_i O(d, \mathbb{F}_{\ell_i})$$

the image of the geometric fundamental group $\pi_1^{\text{geom}}(M_{\overline{\eta}})$ under the direct sum of the various mod ℓ_i representations. By a fundamental specialization result of Pink [Ka-ESDE, 8.18.2, (1)], this group *contains* (an $\prod_i O(d, \mathbb{F}_{\ell_i})$-conjugate of) the group $\Gamma_{\text{geom,mod } \ell_1,\ell_2,\ldots,\ell_r}$ we obtained by looking at the image of $\pi_1^{\text{geom}}(M_{k,\phi})$. As $\prod_i \Omega(d, \mathbb{F}_{\ell_i})$ is a normal subgroup of $\prod_i O(d, \mathbb{F}_{\ell_i})$, being its commutator subgroup, we therefore have

$$\prod_i \Omega(d, \mathbb{F}_{\ell_i}) \subset \Gamma_{\overline{\eta},\text{geom,mod } \ell_1,\ell_2,\ldots,\ell_r} \subset G \subset \prod_i O(d, \mathbb{F}_{\ell_i}).$$

By this same result of Pink [Ka-ESDE, 8.18.2, (2)], there is a dense open set U in $\mathrm{Spec}(R)$ such that for any geometric point $\overline{s}$ in U, the group

$$\Gamma_{\overline{s},\text{geom,mod } \ell_1,\ell_2,\ldots,\ell_r} \subset G \subset \prod_i O(d, \mathbb{F}_{\ell_i}),$$

obtained by looking at the image of $\pi_1^{\text{geom}}(M_{\overline{s}})$, is equal to (an $\prod_i O(d, \mathbb{F}_{\ell_i})$-conjugate of) $\Gamma_{\overline{\eta},\text{geom,mod } \ell_1,\ell_2,\ldots,\ell_r}$. As every subgroup of $\prod_i O(d, \mathbb{F}_{\ell_i})$ containing $\prod_i \Omega(d, \mathbb{F}_{\ell_i})$ is normal, we therefore have equality:

$$\Gamma_{\overline{s},geom,mod\ \ell_1,\ell_2,\ldots,\ell_r} = \Gamma_{\overline{\eta},\text{geom,mod } \ell_1,\ell_2,\ldots,\ell_r}$$

for every geometric point $\overline{s}$ in U. Each of the primes $\ell_1, \ldots, \ell_r$ is nonzero in R (because each is invertible in k under ϕ), so by shrinking U we may further assume that each of them is invertible on U.

We will show that the theorem holds, for the fixed $\epsilon > 0$, if we take for $r \in R$ any nonzero element such that $\mathrm{Spec}(R[1/r]) \subset U$. Denote by

$$\Gamma_{\text{arith,mod } \ell_1,\ell_2,\ldots,\ell_r} \subset G \subset \prod_i O(d, \mathbb{F}_{\ell_i})$$

the image of the arithmetic fundamental group $\pi_1(M[1/r])$. We now apply the Chebotarev density theorem in the uniform version given in [Ka-Sar, 9.7.13] to this situation, our $M[1/r]/R[1/r]$ taken as the X/S there, and with our groups

$$\Gamma_{\overline{\eta},\text{geom,mod } \ell_1,\ell_2,\ldots,\ell_r} \subset \Gamma_{\text{arith,mod } \ell_1,\ell_2,\ldots,\ell_r}$$

taken as the groups $K \subset K_{\text{arith}}$ there. In our situation, the quotient K_{arith}/K is abelian, so the sets $K_{\text{arith},\gamma}$ there are just the cosets of K in K_{arith}. For a given pair (k, ϕ) consisting of a finite field k and a ring homomorphism $\phi : R[1/r] \to k$, all the Frobenius conjugacy classes attached to the k-points of $M_{k,\phi}$ lie in a single coset of K in K_{arith}, say $K_{\text{arith},\gamma}$. Inside this coset $K_{\text{arith},\gamma}$, take the subset W which is defined as follows.

If $d = 2n + 1$ is odd, or if $d = 2n$ and all the r orthogonal groups $O(d, \mathbb{F}_{\ell_i})$ are nonsplit, W consists of all elements $(A_1, \ldots, A_r)$ in the coset $K_{\text{arith},\gamma}$ such that for some i, $\text{Rdet}(1 - TA_i)$ is $\mathbb{F}_{\ell_i}$-irreducible.

If $d = 2n \geq 6$ and all the r orthogonal groups $O(d, \mathbb{F}_{\ell_i})$ are split, then W is the disjoint union of two sets, W_- and W_+, defined as follows. The set W_- consists of those elements $(A_1, \ldots, A_r)$ in the coset $K_{\text{arith},\gamma}$ such that the common value of their determinant is -1 and such that for some i, $\text{Rdet}(1 - TA_i)$ is $\mathbb{F}_{\ell_i}$-irreducible. The set W_+ consists of those elements $(A_1, \ldots, A_r)$ in the coset $K_{\text{arith},\gamma}$ such that the common value of their determinant is $+1$ and such that for some i, $\text{Rdet}(1 - TA_i)$ is of the form

$$(\mathbb{F}_{\ell_i} - \text{irreducible of degree 2})(\mathbb{F}_{\ell_i} - \text{irreducible of degree } 2n - 2),$$

AND such that for some j, $\text{Rdet}(1 - TA_j)$ is of the form

$$(\mathbb{F}_{\ell_j} - \text{irreducible of degree } n)(\text{a different } \mathbb{F}_{\ell_j} - \text{irreducible of degree } n).$$

If $d = 4$ and all the r orthogonal groups $O(4, \mathbb{F}_{\ell_i})$ are split, then W is again the disjoint union of two sets, W_- and W_+. The set W_- is defined exactly as in the paragraph above. The set W_+ consists of those elements $(A_1, \ldots, A_r)$ in the coset $K_{arith,\gamma}$ such that the common value of their determinant is $+1$ and such that for some i, $\text{Rdet}(1 - TA_i)$ is of the form

$$P(T)Q(T)$$

with $P(T)$ and $Q(T)$ relatively prime $\mathbb{F}_{\ell_i}$-irreducibles of degree 2, neither of which is palindromic, and such that

$$Q(T) = (\text{some constant in } \mathbb{F}_{\ell_i}^{\times})T^2 P(1/T).$$

Decompose the K-coset $K_{\text{arith},\gamma}$ into cosets under the smaller group $\prod_i \Omega(d, \mathbb{F}_{\ell_i})$, say

$$K_{\text{arith},\gamma} = \coprod_a \text{Coset}_a.$$

In view of Theorems 6.7 through 6.10, we see that in each such coset, we have

$$\#(W \cap \text{Coset}_a)/\#\text{Coset}_a \geq 1 - \epsilon/2.$$

Summing over the cosets, we find that

$$\#W/\#K_{\text{arith},\gamma} \geq 1 - \epsilon/2.$$

By the Chebotarev density theorem in the uniform version given in [Ka-Sar, 9.7.13], there exist constants C and A such that if $\#k \geq 4A^2$, then

$$|\#W/\#K_{\text{arith},\gamma} - \#\{m \in M_{k,\phi}(k)|\text{Frob}_{k,m} \in W\}/\#M_{k,\phi}(k)|$$
$$\leq 2C\#K_{\text{arith}}/\text{Sqrt}(\#k).$$

For $\#k$ sufficiently large, we obviously have

$$2C\#K_{\text{arith}}/\text{Sqrt}(\#k) \leq \epsilon/2,$$

and hence for $\#k$ sufficiently large we have

$$\#\{m \in M_{k,\phi}(k)|\text{Frob}_{k,m} \in W\}/\#M_{k,\phi}(k) \geq 1 - \epsilon.$$

It remains only to show that whenever $\text{Frob}_{k,m}$ lies in W, then $\text{Rdet}(1-\text{TFrob}_{k,m})$ is $\mathbb{Q}$-irreducible. To see this, we argue as follows. This polynomial has coefficients in $\mathbb{Z}[1/\#k]$. If either d is odd, or d is even and each $O(d, \mathbb{F}_{\ell_i})$ is nonsplit, or d is even and the sign in the functional equation is -1, then for some i the reduction mod ℓ_i of this polynomial $\text{Rdet}(1 - \text{TFrob}_{k,m})$ is $\mathbb{F}_{\ell_i}$-irreducible, this being the defining property of W, and hence $\text{Rdet}(1 - \text{TFrob}_{k,m})$ is $\mathbb{Q}$-irreducible.

It remains to treat the case in which d is even, each $O(d, \mathbb{F}_{\ell_i})$ is split and the sign in the functional equation is $+1$. Suppose first that $d = 2n \geq 6$. Then for some i the reduction mod ℓ_i of $\text{Rdet}(1 - T\text{Frob}_{k,m})$ is the product of two $\mathbb{F}_{\ell_i}$-irreducibles, of degrees 2 and $d - 2$, while for some j the reduction mod ℓ_j of $\text{Rdet}(1 - T\text{Frob}_{k,m})$ is the product of two $\mathbb{F}_{\ell_j}$-irreducibles, both of degree n, this being the defining property of W in this case. So once again $\text{Rdet}(1 - \text{TFrob}_{k,m})$ must be $\mathbb{Q}$-irreducible. [For if it were $\mathbb{Q}$-reducible, its $\mathbb{Q}$-factorization would simultaneously be of the form (degree 2 irred.)(degree $d - 2$ irred.) and of the form (degree n irred.)(degree n irred.).] Suppose now that $d = 4$. Then for some i, $\text{Rdet}(1 - T\text{Frob}_{k,m})$ is the product

$$P(T)Q(T)$$

with $P(T)$ and $Q(T)$ relatively prime $\mathbb{F}_{\ell_i}$-irreducibles of degree 2, neither of which is palindromic, and such that

$$Q(T) = (\text{some constant in } \mathbb{F}_{\ell_i}^{\times})T^2 P(1/T).$$

This implies that $\text{Rdet}(1-T\text{Frob}_{k,m})$ is $\mathbb{Q}$-irreducible. For if it were $\mathbb{Q}$-reducible, its $\mathbb{Q}$-factorization would be as the product of two relatively prime $\mathbb{Q}$-irreducibles of degree 2, neither of which is palindromic. But $\text{Rdet}(1 - T\text{Frob}_{k,m})$ has all its eigenvalues on the unit circle (because pure of weight zero), hence both its $\mathbb{Q}$-irreducible quadratic factors have roots stable by inversion $\zeta \mapsto 1/\zeta$. Since these $\mathbb{Q}$-irreducible factors have degree 2, none of their roots is fixed by inversion (i.e., no root is ± 1), and hence each $\mathbb{Q}$-irreducible factor has roots of the form $(\zeta, 1/\zeta)$, hence is palindromic.

8 Another application of Theorem 5.1: Universal families of hypersurface sections

Let R be a finitely generated $\mathbb{Z}$-algebra, $\mathbb{P} = \mathbb{P}^N/R$ the projective space of some dimension N, and $X \subset \mathbb{P}$ a closed subscheme which is smooth over R with geometrically connected fibres, all of some common odd dimension $\nu = 2n+1 \geq 3$. Fix an integer $d \geq 1$. Denote by M/R the parameter space for smooth, degree d hypersurfaces in the ambient $\mathbb{P}$ which are transversal to X, by $\mathcal{H}_d/M \subset \mathbb{P}/M$ the universal family of these hypersurfaces, and by $\pi : X \cap \mathcal{H}_d \to M$ the corresponding universal family of smooth, degree d hypersurface sections of X. Concretely, if k is a field and $\phi : R \to k$ is a ring homomorphism, then $M_{k,\phi}$ is the parameter space for smooth, degree d hypersurfaces which are transversal to $X_{k,\phi}$. For each prime ℓ, we have the lisse (but not necessarily torsion-free) $\mathbb{Z}_\ell$-sheaf $R^{2n}\pi_\star\mathbb{Z}_\ell(n)$ on $M[1/\ell]$, together with its cup product pairing

$$R^{2n}\pi_\star\mathbb{Z}_\ell(n) \times R^{2n}\pi_\star\mathbb{Z}_\ell(n) \to R^{4n}\pi_\star\mathbb{Z}_\ell(2n) \cong \mathbb{Z}_\ell,$$

which is an orthogonal autoduality modulo torsion. Let us denote by $\rho : X \to Spec(R)$ the structural morphism of X/R. On $Spec(R[1/\ell])$, we have the lisse $\mathbb{Z}_\ell$-sheaf $R^{2n}\rho_\star\mathbb{Z}_\ell(n)$, and we denote by $R^{2n}\rho_\star\mathbb{Z}_\ell(n)_{M[1/\ell]}$ its pullback to $M[1/\ell]$. The canonical restriction map on cohomology gives an inclusion

$$R^{2n}\rho_\star\mathbb{Z}_\ell(n)_{M[1/\ell]} \subset R^{2n}\pi_\star\mathbb{Z}_\ell(n).$$

We denote by

$$Ev_{\mathbb{Z}_\ell} \subset R^{2n}\pi_\star\mathbb{Z}_\ell(n)$$

the orthogonal to $R^{2n}\rho_\star\mathbb{Z}_\ell(n)_{M[1/\ell]}$ under the cup product pairing. The lisse sheaves $Ev_{\mathbb{Z}_\ell}$ on $M[1/\ell]$,carry the induced cup product pairing

$$Ev_{\mathbb{Z}_\ell} \times Ev_{\mathbb{Z}_\ell} \to \mathbb{Z}_\ell.$$

If we tensor this situation with $\mathbb{Q}_\ell$, the Hard Lefschetz Theorem [De-Weil II, 4.1.2] tells us that this pairing on $Ev_{\mathbb{Q}_\ell} := Ev_{\mathbb{Z}_\ell} \otimes_{\mathbb{Z}_\ell} \mathbb{Q}_\ell$ is an orthogonal autoduality. By the Riemann Hypothesis for projective smooth varieties over finite fields [De-Weil I, 1.6], we know that the sheaves $Ev_{\mathbb{Q}_\ell}$ are pure of weight zero, and form a compatible system. By Gabber's theorem [Gab-Tors], for all but finitely many primes ℓ, the sheaves $R^{2n}\rho_\star\mathbb{Z}_\ell(n)$, $R^{2n}\rho_\star\mathbb{Z}_\ell(n)_{M[1/\ell]}$, and $Ev_{\mathbb{Z}_\ell}$ are all torsion free, and the cup product pairing makes $Ev_{\mathbb{Z}_\ell}$ orthogonally self dual over $\mathbb{Z}_\ell$. A fundamental result of Deligne [De-Weil II, 4.4.1,4.4.2s,and 4.4.9], amplified by [Ka-LAMM, 2.2.4] and [Ka-Pan, Corollaries 2 and 3], tells us that the condition (2 strong) holds for the compatible system given by the $Ev_{\mathbb{Q}_\ell}$, provided that $d \geq 3$ if $2n \geq 4$, or that $d \geq 4$ if $2n = 2$, and that, under these conditions, the common rank of the sheaves $Ev_{\mathbb{Q}_\ell}$ is ≥ 9. So Theorem 5.1 applies to this situation.

Let us spell out the simplest case of this situation. We take R to be $\mathbb{Z}$, and we take $X = \mathbb{P} = \mathbb{P}^{2n+1}/\mathbb{Z}$, $2n \geq 2$, with the identical embedding of $\mathbb{P}$ into itself. We fix an integer d with $d \geq 3$ if $2n \geq 4$, or $d \geq 4$ if $2n = 2$. For a finite field $k = \mathbb{F}_q$, and a smooth hypersurface H_d/k of degree d and dimension $2n$ over k, we know that its Zeta function is of the form

$$Zeta(H_d/k, T) = 1/P(H_d/k, T)\prod_{i=0}^{2n}(1 - q^i T).$$

Here $P(H_d/k, T) \in \mathbb{Z}[T]$ is the polynomial whose unitarization

$$P_u(H_d/k, T) := P(H_d/k, T/q^n)$$

is given by

$$P_u(H_d/k, T) = \det((1 - T\mathrm{Frob}_{k,H_d}|Ev_{\mathbb{Q}_\ell}),$$

for any prime ℓ invertible in k. The reduced unitarization $P_{u,\mathrm{red}}$ is defined by

$$P_{u,\mathrm{red}}(H_d/k, T) := \mathrm{Rdet}((1 - T\mathrm{Frob}_{k,H_d}|Ev_{\mathbb{Q}_\ell}).$$

For each finite field k, we denote by $\mathrm{IrrFrac}(k, d, 2n)$ the fraction of the smooth hypersurfaces H_d/k of degree d and dimension $2n$ over k for which the polynomial $P_{u,\mathrm{red}}(H_d/k, T)$ is $\mathbb{Q}$-irreducible. Then Theorem 5.1 gives us the following result.

Theorem 8.1 *In any sequence $i \mapsto k_i$ of finite fields whose cardinalities are strictly increasing, the sequence of fractions $i \mapsto \mathrm{IrrFrac}(k_i, d, 2n)$ tend to* 1.

9 Alternative approaches

The Chavdarov approach to studying irreducibility requires knowledge of mod ℓ monodromy for infinitely many primes ℓ. We have used Larsen's theorem [Lar-Max, 3.17] to infer, from information about the ℓ-adic monodromy for all (invertible on the base) ℓ, information about mod ℓ monodromy for a set of primes ℓ of Dirichlet density one. There are two other approaches, which, when they apply, give information about mod ℓ monodromy for all but finitely many primes ℓ.

The first is based on the theorem of Mathews, Vaserstein, and Weisfeller [MVW][2], which concerns a smooth groupscheme $G/\mathbb{Z}[1/N]$ whose complex fibre $G_{\mathbb{C}}$ is a connected, semisimple, simply connected group, and a finitely generated subgroup $\Gamma \subset G(\mathbb{Z}[1/N])$ which is Zariski dense in $G_{\mathbb{C}}$. For any ℓ which is prime to N, we have a "reduction mod ℓ" homomorphism

$$\Gamma \subset G(\mathbb{Z}[1/N]) \to G(\mathbb{F}_\ell).$$

The theorem asserts that Γ maps onto $G(\mathbb{F}_\ell)$ for all sufficiently large ℓ.

Let us explain an instance of when we can apply this method, and what kind of result it gives. Let us put ourselves in the general setup $E/U/M/R$ of Section 3, and assume that conditions (1) and (2weak) of that section hold, and that $d \geq 3$. Pick an embedding of R into $\mathbb{C}$, and make the corresponding extension of scalars. As explained in the proof of Theorem 3.1, we have, for some integer $N \geq 1$, an orthogonally self-dual lisse sheaf $\mathcal{H}^{\mathrm{an}}_{\mathbb{Z}[1/N]}$ on M^{an}. Enlarging N, we will further suppose that N is even. Let us denote by

$$\rho^{\mathrm{an}}_{\mathbb{C}} : \pi_1(M^{\mathrm{an}}) \to O(d, \mathbb{Z}[1/N])$$

the corresponding "transcendental" monodromy representation attached to $\mathcal{H}^{\mathrm{an}}_{\mathbb{Z}[1/N]}$. For every prime ℓ not dividing N, we also have the algebro-geometric ℓ-adic monodromy of $\mathcal{H}_{\mathbb{Z}_\ell}|M_{\mathbb{C}}$,

$$\rho_{\mathbb{C},\ell} : \pi_1(M_{\mathbb{C}}) \to O(d, \mathbb{Z}_\ell).$$

By the comparison theorem, the algebro-geometric fundamental group $\pi_1(M_{\mathbb{C}})$ is the profinite completion of $\pi_1(M^{\mathrm{an}})$. For every ℓ not dividing N, the ℓ-adic image $\rho_{\mathbb{C},\ell}(\pi_1(M_{\mathbb{C}})) \subset O(d, \mathbb{Z}_\ell)$ is the closure (in $O(d, \mathbb{Z}_\ell)$ with its profinite topology) of the topological image $\rho^{\mathrm{an}}_{\mathbb{C}}(\pi_1(M^{\mathrm{an}})) \subset O(d, \mathbb{Z}[1/N])$. By Pink's specialization theorem [Ka-ESDE, 8.18.2, (2)] applied to $\mathcal{H}_{\mathbb{Z}_\ell}$ on $M[1/\ell]$, we may infer from condition (2weak) that the ℓ-adic image $\rho_{\mathbb{C},\ell}(\pi_1(M_{\mathbb{C}}))$ is Zariski dense in either $O(d, \overline{\mathbb{Q}}_\ell)$ or in $SO(d, \overline{\mathbb{Q}}_\ell)$. By the ℓ-adic continuity of polynomial functions, it then

[2] Unlike the Larsen result or the Zalesskii–Serezkin result to be discussed below, this result [MVW] depends upon the classification of finite simple groups.

follows that the topological image $\rho^{\mathrm{an}}_{\mathbb{C}}(\pi_1(M^{\mathrm{an}})) \subset O(d, \mathbb{Z}[1/N])$ is Zariski dense in the same group, either $O(d, \overline{\mathbb{Q}}_\ell)$ or $SO(d, \overline{\mathbb{Q}}_\ell)$. Picking an embedding of fields $\overline{\mathbb{Q}}_\ell \subset \mathbb{C}$, we see that the topological image is Zariski dense in either $O(d, \mathbb{C})$ or in $SO(d, \mathbb{C})$. Since the topological fundamental group $\pi_1(M^{\mathrm{an}})$ is finitely generated, its image

$$\Gamma_1 := \rho^{\mathrm{an}}_{\mathbb{C}}(\pi_1(M^{\mathrm{an}})) \subset O(d, \mathbb{Z}[1/N])$$

is a finitely generated subgroup of $O(d, \mathbb{Z}[1/N])$ which is Zariski dense in either $O(d)$ or $SO(d)$. We cannot yet apply [MVW], because the orthogonal group $O(d)$ is not connected and its identity component $SO(d)$ is not simply connected. We get around this difficulty following an argument of Ron Livne. First, replace Γ_1 by the subgroup $\Gamma_2 \subset \Gamma_1$ of index 1 or 2 consisting of the elements of determinant $+1$. Then Γ_2 is a finitely generated, Zariski dense subgroup of $SO(d, \mathbb{Z}[1/N])$. Next consider the Spin group attached to our orthogonal group. The spinor norm gives an exact sequence

$$\{1\} \to \pm 1 \to \mathrm{Spin}(d, \mathbb{Z}[1/N]) \to SO(d, \mathbb{Z}[1/N]) \to \mathbb{Z}[1/N]^\times/(\mathbb{Z}[1/N]^\times)^2,$$

in which the last term, $\mathbb{Z}[1/N]^\times/(\mathbb{Z}[1/N]^\times)^2$, is finite, generated by -1 and by the primes dividing N. Now consider the composite homomorphism

$$\Gamma_2 \subset SO(d, \mathbb{Z}[1/N]) \to \mathbb{Z}[1/N]^\times/(\mathbb{Z}[1/N]^\times)^2.$$

Its image is finite. So the subgroup

$$\Gamma_3 := \mathrm{Ker}(\Gamma_2 \to \mathbb{Z}[1/N]^\times/(\mathbb{Z}[1/N]^\times)^2) \subset \Gamma_2$$

is a subgroup of finite index in Γ_2, so is still Zariski dense in SO, and still finitely generated. Every element of Γ_3 lifts, in two different ways, to $\mathrm{Spin}(d, \mathbb{Z}[1/N])$. Denote by

$$\Gamma \subset \mathrm{Spin}(d, \mathbb{Z}[1/N])$$

the complete inverse image of Γ_3. This group Γ is finitely generated (because Γ_3 is), and, as it maps onto Γ_3, it is Zariski dense in Spin. We may now apply the theorem of Mathews, Vaserstein, and Weisfeller [MVW], to $\Gamma \subset \mathrm{Spin}(d, \mathbb{Z}[1/N])$, to conclude that for all sufficiently large ℓ prime to N, say for all ℓ not in the finite set S, Γ maps onto $\mathrm{Spin}(d, \mathbb{F}_\ell)$. For any such ℓ, Γ_3 maps onto the image of $\mathrm{Spin}(d, \mathbb{F}_\ell)$ in $SO(d, \mathbb{F}_\ell)$, i.e., Γ_3 maps onto $\Omega(d, \mathbb{F}_\ell)$. So for any such ℓ, the image of Γ_1 in $O(d, \mathbb{F}_\ell)$ contains $\Omega(d, \mathbb{F}_\ell)$. Because the algebro-geometric fundamental group $\pi_1(M_{\mathbb{C}})$ is the profinite completion of $\pi_1(M^{\mathrm{an}})$, this last image is also the image of $\pi_1(M_{\mathbb{C}})$ in $O(d, \mathbb{F}_\ell)$. Thus we find that the image of $\pi_1(M_{\mathbb{C}})$ in $O(d, \mathbb{F}_\ell)$ contains $\Omega(d, \mathbb{F}_\ell)$ for every ℓ not in S.

So far, all of this is taking place on the complex fibre of M/R. Let us say that M/R is nicely compactifiable if there exists a proper smooth R-scheme $M^\wedge/R$ and a divisor $D \subset M^\wedge$ which has normal crossings relative to R, such that $M \cong$

$M^\wedge \setminus D$. By resolution over the characteristic zero fraction field of R, we know that there exists a nonzero $r \in R$ such that $M[1/r]/R[1/r]$ is nicely compactifiable. [This passage, from R to some $R[1/r]$, is not entirely harmless. For instance, in the second example, of Weierstrass families, where we start, for a given (d_2, d_3), with $R = \mathbb{Z}[1/6]$ and the corresponding $M = M_{d_2,d_3}/\mathbb{Z}[1/6]$, we do not know which, if any, other primes p we need to invert to get a nice compactification, nor do we know how this set of p depends on (d_2, d_3). In our 2004–2005 course, we followed the [MVW] method when M/R was nicely compactifiable, as explained in the next paragraph, but then invoked the Larsen method to handle separately each of the finitely many unknown bad p.]

When M/R is nicely compactifiable, with R a normal integral domain whose fraction field has characteristic zero, Abhyankar's lemma [SGA 1, XIII, 5.5] assures us that for any lisse sheaf on $M[1/\ell]$, and any geometric point s of $\mathrm{Spec}(R[1/\ell])$, its restriction to the geometric fibre M_s of $M[1/\ell]/R[1/\ell]$ is tamely ramified at each maximal point of D_s. We apply this to the lisse sheaf $\mathcal{H}_{\mathbb{Z}_\ell} \otimes \mathbb{F}_\ell$ on $M[1/\ell]$, for each ℓ not in S. The Tame Specialization Theorem [Ka-ESDE, 8.17.14] then tells us that for every ℓ not in S, and for every geometric point s of $\mathrm{Spec}(R[1/\ell])$, the image of $\pi_1(M_s)$ in $O(d, \mathbb{F}_\ell)$ contains $\Omega(d, \mathbb{F}_\ell)$.

We now turn to a second approach[3] to controlling the mod ℓ monodromy for all but finitely many ℓ. This approach is based on the Zalesskii-Serezkin classification [Zal-Ser, Theorem, page 478] of irreducible subgroups of $GL(n, \overline{\mathbb{F}_\ell})$, $\ell \geq 3, n \geq 3$, which are generated by reflections and which contain no transvections (:=unipotent pseudoreflections). We can apply this to describe all irreducible subgroups of orthogonal groups in odd characteristic generated by reflections because such orthogonal groups contain no transvections. Here is a baby version of their result in this case.

Theorem 9.1 (Zalesskii–Serezkin) *Given an integer $n \geq 3$, there exists a constant $C(n)$ with the following property. Let $\ell \geq 3$, and (V, Ψ) an n-dimensional $\mathbb{F}_\ell$-vector space with a symmetric autoduality Ψ. Let $G \subset O(V, \Psi)$ be an irreducible subgroup generated by reflections. Denote by $N_O(G)$ the normalizer of G in $O(V, \Psi)$. Then either $\Omega(V, \Psi) \subset G$, or we have the divisibility estimate $\#N_O(G)|C(n)$. Moreover, if $n \geq 9$, we can take $C(n) = 2^n(n+2)!$.*

Proof. We begin by recalling that if $G \subset O(V, \Psi)$ is an irreducible subgroup generated by reflections, then G is absolutely irreducible (i.e., G acts irreducibly after extending scalars from $\mathbb{F}_\ell$ to $\overline{\mathbb{F}_\ell}$). Now for any absolutely irreducible subgroup $G \subset O(V, \Psi)$, we have the divisibility estimate

$$\#N_O(G)|2\#\mathrm{Aut}(G),$$

simply because the kernel of the conjugation action homomorphism

---

[3] A third approach would be to appeal to the results of Hall [Ha].

$$N_O(G) \to \mathrm{Aut}(G)$$

lies in the subgroup of scalars in $O(V, \Psi)$, which is ± 1.

It is immediate from [Zal-Ser, Theorem, page 478] that for $n \geq 9$, there are at most two primitive such groups G which fail to contain $\Omega(V, \Psi)$, namely the symmetric group S_{n+1}, if ℓ is prime to $n + 1$, and the symmetric group S_{n+2}, if ℓ divides $n + 2$. For these G, every automorphism is inner. For $3 \leq n \leq 8$, there are finitely many more such primitive G, and these we handle by the $\#N_O(G)|2\#\mathrm{Aut}(G)$ divisibility.

We now consider the imprimitive such G. For $n \geq 5$, any imprimitive such group has a *unique* [Zal-Ser, 4.1] system of imprimitivity consisting of the lines L_i spanned by-linearly independent vectors e_i, and the induced homomorphism maps G onto the symmetric group S_n. For each of $n = 3$ and $n = 4$, there is at most one imprimitive G for which the system of imprimitivity is not unique [Zal-Ser, 4.1], and these cases are handled by the $\#N_O(G)|2\#\mathrm{Aut}(G)$ divisibility.

It remains to treat the case of an imprimitive such G which admits a unique system of imprimitivity. By uniqueness, the system of imprimitivity is respected by $N_O(G)$, so we have a homorphism of $N_O(G)$ onto S_n. It remains only to show the following claim: in the basis given by the vectors e_i, any element $g \in N_O(G)$ which lies in the kernel of this homomorphism, i.e., which is diagonal, has entries each ± 1. Indeed, we will show that any element $g \in O(V, \Psi)$ which is diagonal in this basis has entries ± 1. Let us denote by λ_i the diagonal entries of g.

From the fact that G induces every possible permutation of the lines L_i, we see that

(1) Either all square lengths $\Psi(e_i, e_i)$ are nonzero, or they are all zero.
(2) Either all cross terms $\Psi(e_i, e_j)$ are nonzero, for all $i \neq j$, or they are all zero.

If all $\Psi(e_i, e_i)$ are nonzero, our claim is obvious, since

$$\lambda_i^2 \Psi(e_i, e_i) = \Psi(g(e_i), g(e_i)) = \Psi(e_i, e_i).$$

If all $\Psi(e_i, e_i)$ vanish, then by nondegeneracy all $\Psi(e_i, e_j)$ are nonzero, for all $i \neq j$. From the identity

$$\lambda_i \lambda_j \Psi(e_i, e_j) = \Psi(g(e_i), g(e_j)) = \Psi(e_i, e_j),$$

we then infer that for every $i \neq j$, we have $\lambda_i \lambda_j = 1$, which in turn forces all λ_i to be equal to each other, with common value ± 1. □

Armed with this result, we can prove an "almost all ℓ" result about the mod ℓ monodromy of Lefschetz pencil of even fibre dimension $2n \geq 2$. Let us put ourselves in the situation of Section 8, but taking now the base ring R to be a finite field k. We take the degree d of the hypersurface sections large enough that the common rank N of the sheaves $Ev_{\mathbb{Q}_\ell}$, for every ℓ invertible in k, is ≥ 3. We suppose that the condition (2 strong) holds, and that there exist, over $\overline{k}$, Lefschetz pencils on

X of hypersurface sections of degree d for which (2 strong) holds as well. [As noted above, the first condition is automatic if $d \geq 3$ and $d + 2n \geq 6$, in which case we have $N \geq 9$. Moreover, in this case Lefschetz pencils exist, and sufficiently general ones will satisify (2 strong).]

Theorem 9.2 *For all sufficiently large primes ℓ, the image of the geometric fundamental group $\pi_1(M_{\overline{k}})$ in $O(N, \mathbb{F}_\ell)$ under the monodromy representation of $Ev_{\mathbb{Z}_\ell} \otimes \mathbb{F}_\ell$ contains $\Omega(N, \mathbb{F}_\ell)$. More precisely, it is the following subgroup of $O(N, \mathbb{F}_\ell)$: if $(-1)^n 2$ is a square in $\mathbb{F}_\ell$, it is the subgroup of elements of spinor norm one. If not, it is the subgroup of elements having $sp = \det$. Moreover, for any (sufficiently general, if char$(k) = 2$) Lefschetz pencil satisfying (2 strong), we have the same results for the image of its geometric monodromy, with a possibly larger set of "bad" ℓ.*

Proof. By Gabber's theorem [Gab-Tors], applied both to X and to any single smooth hypersurface section $X \cap H_d$ of degree d, we know that for all but finitely many ℓ, both spaces have their $\mathbb{Z}_\ell$-cohomology torsion-free, and the hard Lefschetz theorem holds mod ℓ on both. These are the "good ℓ" for the theorem. Because the fibre dimension $2n$ is even, we know, by [SGA 7 II, XV 3.4, XVIII 6.2 and 6.3], that "condition A" of [SGA 7 II, XVIII 5.3.5] holds for any Lefschetz pencil on X. An attentive reading of the entire exposé [SGA 7 II, XVIII] then shows that for all these good ℓ, the mod ℓ geometric monodromy of any (sufficiently general, if char$(k) = 2$) Lefschetz pencil is an irreducible subgroup of $O(N, \mathbb{F}_\ell)$ (this uses the conjugacy of the vanishing cycles [De-Weil II, 4.2.7]) which is generated by reflections in various vectors δ_i with square length $\delta_i \cdot \delta_i = (-1)^n 2$ (this is the Picard-Lefschetz formula [SGA 7 II, XV 3.4]).

Let us begin with a Lefschetz pencil, defined over $\overline{k}$ and hence over some finite extension E/k, for which (2 strong) holds. Since the statements to be proven are geometric, we may extend scalars, and reduce to the case when our Lefschetz pencil satisfying (2 strong) is defined over k. From the theorem of Zalesskii–Serezkin above, we see for a given good ℓ, there are only two possibilities: either the image $\Gamma_{\text{geom, mod } \ell}$ of the geometric monodromy group of our Lefschetz pencil is the asserted group, or its normalizer $N_O(\Gamma_{\text{geom, mod } \ell})$, which *contains* the mod ℓ image $\Gamma_{\text{arith, mod } \ell}$ of the arithmetic monodromy group, is a group whose order divides $C(N)$. We will show that this can happen for only finitely many good ℓ. Indeed, we will show that the inequality $\#\Gamma_{\text{arith, mod } \ell} \leq C(N)$ can hold for only finitely many good ℓ. For this, we argue as follows.

Because our pencil satisfies (2 strong), we know by Deligne's equidistribution theorem, cf. [Ka-GKM, 3.6], that as we run over larger and larger finite extensions E/k, and consider all the smooth, degree d hypersurface sections $X \cap H_d$ defined over E, the (unique conjugacy classes having) reversed characteristic polynomials

$$\det(1 - T\text{Frob}_{k, X \cap H_d} | Ev^n) \in \mathbb{Z}[1/\#k][T]$$

become equidistributed, for (the direct image of) Haar measure, in the space $O(N, \mathbb{R})^\#$ of conjugacy classes in the compact orthogonal group $O(N, \mathbb{R})$. The space $O(N,\mathbb{R})^{\#}$ is a compact metric space [namely, the set of all degree N polynomials in $1+T\mathbb{R}[T]$ all of whose roots lie on the unit circle], every nonempty open set has strictly positive measure, and it is infinite. So if we take $1+C(N)$ distinct points A_i in $O(N,\mathbb{R})^{\#}$, and tiny open balls B_i around A_i which are pairwise disjoint, then for E sufficiently large, we can find $1+C(N)$ different smooth, degree d hypersurface sections $X\cap H_{d,i}$ defined over E such that the reversed characteristic polynomial of $\mathrm{Frob}_{k,X\cap H_{d,i}}$ lands in B_i. So these reversed characteristic polynomials are pairwise distinct. Let us enumerate these polynomials, say $P_0(T), P_1(T), \ldots, P_{C(N)}(T)$. Now consider the product polynomial

$$R(T) := \prod_{0\le i<j\le C(N)} (P_i(T)-P_j(T)).$$

This is a nonzero polynomial in $\mathbb{Z}[1/\#k][T]$, hence it is nonzero mod all sufficiently large primes ℓ. For any good prime ℓ mod which it is nonzero, the $1+C(N)$ Frobenius conjugacy classes $\mathrm{Frob}_{k,X\cap H_{d,i}}$ must have distinct images in $\Gamma_{\mathrm{arith,\,mod}\ell}$, since they have distinct mod ℓ characteristic polynomials. So certainly we have $\#\Gamma_{\mathrm{arith,\,mod}\ell}\ge 1+C(N)$ for these good ℓ.

To treat the situation over M itself, we note that our single Lefschetz pencil above shows us for all sufficiently large good primes ℓ, the image of the geometric fundamental group $\pi_1(M_{\overline{k}})$ in $O(N,\mathbb{F}_\ell)$ under the monodromy representation of $Ev_{\mathbb{Z}_\ell}\otimes\mathbb{F}_\ell$ contains the asserted subgroup of $O(N,\mathbb{F}_\ell)$. To see that this image can be no bigger, use the fact that for any given good prime ℓ, Bertini's theorem says that already a sufficiently general Lefschetz pencil will have the same mod ℓ geometric monodromy as does M itself. Since other choices of Lefschetz pencils may require omitting fewer good ℓ than did our initial choice, the result over M may have fewer bad ℓ that the result for some particular choice of Lefschetz pencil. □

References

[Artin-GA] Artin, E., Geometric Algebra, Interscience Publishers, 1957. Reprinted in Wiley Classics Library, John Wiley, 1988.

[AtBS-Clif] Atiyah, M. F.; Bott, R.; Shapiro, A., Clifford modules. Topology 3 1964 suppl. 1, 3–38.

[Bour-AlgIX] Bourbaki, N. Eléments de mathématique. Première partie: Les structures fondamentales de l'analyse. Livre II: Algèbre. Chapitre 9: Formes sesquilinéaires et formes quadratiques. Actualités Sci. Ind. no. 1272 Hermann, 1959.

[Chav] Chavdarov, N., The generic irreducibility of the numerator of the zeta function in a family of curves with large monodromy. Duke Math. J. 87 (1997), no. 1, 151–180.

[Chev-Spin] Chevalley, C., The algebraic theory of spinors. Columbia University Press, New York, 1954. Reprinted in The Algebraic Theory of Spinors and Clifford Algebras. Collected Works. Vol. 2. Edited and with a Foreword by Pierre Cartier and Catherine Chevalley. With a postface by J.-P. Bourguignon. Springer-Verlag, Berlin, 1997.

[deJ-Ka] de Jong, A. Johan, Katz, Nicholas M., Monodromy and the Tate conjecture: Picard numbers and Mordell-Weil ranks in families. Israel J. Math. 120 (2000), part A, 47–79.

[De-Weil I] Deligne, P., La conjecture de Weil. Publ. Math. IHES 43 (1974), 273–307.

[De-Weil II] Deligne, P., La conjecture de Weil II. Publ. Math. IHES 52 (1981), 313–428.

[Die-GC] Dieudonné, J. Sur les groupes classiques. Troisième édition revue et corrigée. Publications de l'Institut de Mathématique de l'Université de Strasbourg, VI. Actualités Scientifiques et Industrielles, No. 1040. Hermann, Paris, 1959.

[Dw-Rat] Dwork, B., On the rationality of the zeta function of an algebraic variety. Amer. J. Math. 82 (1960), 631-648.

[Gab-Tors] Gabber, O., Sur la torsion dans la cohomologie l-adique d'une variété. C. R. Acad. Sci. Paris Sér. I Math. 297 (1983), no. 3, 179–182.

[Gr-Rat] Grothendieck, A., Formule de Lefschetz et rationalité des fonctions L. Séminaire Bourbaki, Vol. 9, Exp. No. 279, 41-55, Soc. Math. France, 1995.

[Ha] Hall, C., Big symplectic or orthogonal monodromy modulo ℓ, Duke Math. J. 141 (2008), no. 1, 179–203.

[Jo] Jouve, F., Méthodes de crible et sommes d'exponentielles, thesis, Univ. Bordeaux I, Déc., 2008.

[Kar-Clif] Karoubi, M., Algèbres de Clifford et K-théorie. Ann. Sci. Ecole Norm. Sup. (4) 1 1968 161–270.

[Ka-ESDE] Katz, N., Exponential sums and differential equations, Annals of Math. Study 124, Princeton Univ. Press, 1990.

[Ka-GKM] Katz, N., Gauss sums, Kloosterman sums, and monodromy groups, Annals of Math. Study 116, Princeton Univ. Press, 1988.

[Ka-LAMM] Katz, N., Larsen's alternative, moments, and the monodromy of Lefschetz pencils. Contributions to automorphic forms, geometry, and number theory, 521–560, Johns Hopkins Univ. Press, Baltimore, MD, 2004.

[Ka-MMP] Katz, N., Moments, monodromy, and perversity: a Diophantine perspective. Annals of Math. Study 159. Princeton University Press, 2005.

[Ka-Pan] Katz, N.; Pandharipande, R., Inequalities related to Lefschetz pencils and integrals of Chern classes. Geometric aspects of Dwork theory. Vol. I, II, 805–818, Walter de Gruyter GmbH & Co. KG, Berlin, 2004.

[Ka-Sar] Katz, N.; Sarnak, P., Random matrices, Frobenius eigenvalues, and monodromy. American Mathematical Society Colloquium Publications, 45. American Mathematical Society, Providence, RI, 1999.

[Ka-TLFM] Katz, N., Twisted L-functions and monodromy. Annals of Math. Study 150. Princeton University Press, 2002.

[Kow-LSM] Kowalski, E., The large sieve, monodromy and zeta functions of curves, Crelle 601 (2006), 29–69.

[Kow-RQT] Kowalski, E., On the rank of quadratic twists of elliptic curves over function fields, Int'l. J. of Number Theory 2 (2006), 267–288.

[Lar-Max] Larsen, M. Maximality of Galois actions for compatible systems. Duke Math. J. 80 (1995), no. 3, 601–630.

[MVW] Matthews, C. R.; Vaserstein, L. N.; Weisfeiler, B. Congruence properties of Zariski-dense subgroups. I. Proc. London Math. Soc. (3) 48 (1984), no. 3, 514–532.

[Poonen] Poonen, B., Bertini theorems over finite fields. Ann. of Math. (2) 160 (2004), no. 3, 1099-1127.

[Ribet-Gal] Ribet, K., Galois action on division points of Abelian varieties with real multiplications. Amer. J. Math. 98 (1976), no. 3, 751–804.

[Saito-sign] Saito, T., The sign of the functional equation of the L-function of an orthogonal motive. Invent. Math. 120 (1995), no. 1, 119–142.

[SGA 1] Revêtements étales et groupe fondamental. Séminaire de Géométrie Algébrique du Bois Marie 1960-1961 (SGA 1). Dirigé par Alexandre Grothendieck. Augmenté de deux exposés de M. Raynaud. Lecture Notes in Mathematics, Vol. 224, Springer-Verlag, 1971.

[SGA 7 II] Groupes de monodromie en géométrie algébrique. II. Séminaire de Géométrie Algébrique du Bois-Marie 1967-1969 (SGA 7 II). Dirigé par P. Deligne et N. Katz. Lecture Notes in Mathematics, Vol. 340. Springer-Verlag, 1973.

[Weyl] Weyl, H., Classical Groups, Princeton University Press, 1946.

[Zal-Ser] Zalesskiĭ, A. E.; Serežkin, V. N. Finite linear groups generated by reflections. (Russian) Izv. Akad. Nauk SSSR Ser. Mat. 44 (1980), no. 6, 1279-1307, 38, translated in Math. USSR. Izvestijia 17 (1981), No. 3, 477–503.

Remark on fundamental groups and effective Diophantine methods for hyperbolic curves

Minhyong Kim

Dedicated to the memory of Serge Lang

Abstract In a letter from Grothendieck to Faltings, it was suggested that a positive answer to the *section conjecture* should imply finiteness of points on hyperbolic curves over number fields. In this paper, we point out instead the analogy between the section conjecture and the finiteness conjecture for the Tate-Shafarevich group of elliptic curves. That is, the section conjecture should provide a terminating algorithm for finding all rational points on a hyperbolic curve equipped with a rational point.

Key words Diophantine geometry • fundamental group • Diophantine decidability

Mathematics Subject Classification (2010): 11G30; 14H25

In earlier articles [8–10] attention was drawn to the parallel between the ideas surrounding the well-known conjecture of Birch and Swinnerton-Dyer (BSD) for elliptic curves, and the mysterious *section conjecture* of Grothendieck [6] that concerns hyperbolic curves. We wish to explain here some preliminary ideas for "effective non-abelian descent" on hyperbolic curves equipped with at least one rational point. We again follow in an obvious manner the method of descent on elliptic curves and therefore rely on conjectures. In fact, the main point is to

M. Kim (✉)
Mathematical Instititute, University of Oxford, Oxford, OX1 3LB, United Kingdom

Department of Mathematics, Pohang University of Science and Technology, San 31 Hyoja-dong, Nam-gu Pohang, Gyungbuk 790-784, Korea
e-mail: minhyong.kim@ucl.ac.uk

D. Goldfeld et al. (eds.), *Number Theory, Analysis and Geometry: In Memory of Serge Lang*, DOI 10.1007/978-1-4614-1260-1_16,

substitute the section conjecture for the finiteness of the Tate–Shafarevich group. That is to say, the input of the section conjecture is of the form

section conjecture $\Rightarrow$ termination of descent.

At a number of different lectures delivered by the author on the topic of fundamental groups and Diophantine geometry, the question was raised about the role of *surjectivity* in the section conjecture as far as Diophantine applications are concerned. The demonstration of this implication is intended as something of a reply.

To *start* the descent, on the other hand, requires the use of p-adic Hodge theory and the unipotent Albanese map. In this process, in general, another conjecture is unfortunately needed. It could be, for example, the Bloch–Kato conjecture on surjectivity of the p-adic Chern class map that has been referred to in [9]. In other words, via the construction of Selmer varieties and Albanese maps, one deduces an implication

Bloch–Kato conjecture $\Rightarrow$ beginning of descent.

The main caveat here arises from the lack of actual knowledge of computational issues on the part of the author. To avoid misleading anyone about what is being achieved here, we have in the following section separated out the questionable portions as hypotheses [H] and [H′]. That is to say, the objects that mediate this process, namely Galois cohomology groups/varieties and maps between them, seem in principle to be computable. But even to the algorithmically illiterate perspective, it is obvious that actual computation would be daunting to the point of impossibility given the technology of the present day. Nevertheless, it is perhaps not entirely devoid of value to point out one direction of investigation in effective methods, in the hope that even incompetent strategies may eventually be improved through the focusing of sharper skills obviously available in the community. Hence the present paper.

One point of some theoretical interest concerns the comparison with “effective Mordell conjectures” in the usual sense where upper bounds for heights are proposed. If we fix a point b on the curve and measure heights with respect to the corresponding divisor, the height of another point measures the inverse distance from b at all places. So an upper bound for the height corresponds to a lower bound for the distance from b at all places. On the other hand, what the p-adic Hodge theory provides (in principle) is a lower bound for the p-adic distance between all pairs of points at one place. This lower bound is exactly what is required to start the descent.

Finally, we make the obvious point that the use of conjectures is probably not a serious obstacle from the computational perspective (that is, in comparison to the problem of feasibility). This is in the same spirit as the standard algorithms for computing Mordell–Weil groups of elliptic curves where the BSD conjecture is employed with just a few misgivings [3].

Acknowledgements The author was supported in part by a grant from the National Science Foundation and a visiting professorship at RIMS. He is grateful to Kazuya Kato, Shinichi Mochizuki, and Akio Tamagawa for a continuing stream of discussions on topics related to this paper, and for their generous hospitality during the Fall of 2006.

1 Brief review

Here we will be intentionally brief, referring the reader to [4] and [9] for a more thorough discussion.

Let $X/\mathbb{Q}$ be a proper smooth hyperbolic curve of genus g with a point $b \in X(\mathbb{Q})$ and let S be the set of primes of bad reduction for X. In the following, we shall be a bit sloppy and mostly omit separate notation for an integral model of X. Choose a prime $p \notin S$ and let $U^{\text{ét}} = \pi_1^{\text{ét},\mathbb{Q}_p}(X, b)$ be the $\mathbb{Q}_p$-unipotent étale fundamental group of $\bar{X} := X \times_{\text{Spec}(\mathbb{Q})} \text{Spec}(\bar{\mathbb{Q}})$ and $U_n^{\text{ét}} = (U^{\text{ét}})^{n+1} \backslash U^{\text{ét}}$ its quotient by the $(n+1)$th level of the descending central series normalized so that $(U^{\text{ét}})^1 = U^{\text{ét}}$. Let Γ be the Galois group of $\bar{\mathbb{Q}}$ over $\mathbb{Q}$. We have defined the Selmer varieties

$$H^1_f(\Gamma, U_n^{\text{ét}})$$

[9, 10] classifying Γ-equivariant torsors for $U_n^{\text{ét}}$ that are unramified at all places not in $\{p\} \cup S$ and crystalline at p. (This is $H^1_f(\Gamma_T, U_{n-1}^{\text{ét}})$ in the notation of [8] and [9], where $T = S \cup \{p\}$.) These are affine algebraic varieties over $\mathbb{Q}_p$ whose algebraic structures are defined inductively starting from

$$H^1_f(\Gamma, U_1^{\text{ét}}) \simeq H^1_f(\Gamma, H_1^{\text{ét}}(\bar{X}, \mathbb{Q}_p)),$$

which is a $\mathbb{Q}_p$ vector space. Recall the fundamental diagram ([9], end of section 2)

$$\begin{array}{ccccc}
X(\mathbb{Q}) & \hookrightarrow & X(\mathbb{Q}_p) & & \\
\downarrow{\scriptstyle \kappa_n^{\text{ét,glob}}} & & \downarrow{\scriptstyle \kappa_n^{\text{ét,loc}}} & \searrow{\scriptstyle \kappa_n^{\text{dr/cr}}} & \\
H^1_f(\Gamma, U_n^{\text{ét}}) & \xrightarrow{\text{loc}_p} & H^1_f(G_p, U_n^{\text{ét}}) & \xrightarrow{D} & U_n^{\text{dr}}/F^0
\end{array}$$

Here, $H^1_f(G_p, U_n^{\text{ét}})$ classifies $G_p := \text{Gal}(\bar{\mathbb{Q}}_p/\mathbb{Q}_p)$-equivariant torsors for $U_n^{\text{ét}}$ that are crystalline, while U_n^{dr}/F^0 classifies compatible pairs $T_n^{\text{dr}} \simeq T_n^{\text{cr}}$ of torsors for the de Rham and crystalline fundamental groups U_n^{dr} and U_n^{cr} equipped with Hodge filtrations and Frobenius endomorphisms compatible with the torsor structures. The maps associate to each point $x \in X(\mathbb{Q})$ the class of the torsor of paths from b to x in the appropriate category. So

$$\kappa_n^{\text{ét,glob}}(x) = [\pi_1^{\text{ét},\mathbb{Q}_p}(\bar{X}; b, x)_n]$$

with Γ-action,

$$\kappa_n^{\text{ét,loc}}(x) = [\pi_1^{\text{ét},\mathbb{Q}_p}(\bar{X}; b, x)_n]$$

with G_p-action, and

$$\kappa_n^{\mathrm{dr/cr}}(x) = [\pi_1^{\mathrm{dr}}(X \otimes \mathbb{Q}_p; b, x) \simeq \pi_1^{\mathrm{cr}}(Y; \bar{b}, \bar{x})],$$

where Y is the reduction mod p of a smooth $\mathbb{Z}[1/S]$ model for X.

In contrast to this mass of notation, the section conjecture considers just one map

$$\hat{\kappa} : X(\mathbb{Q}) \rightarrow H^1(\Gamma, \hat{\pi}_1(\bar{X}, b))$$

that sends a point $x \in X(\mathbb{Q})$ to the class of the profinite torsor of paths

$$\hat{\pi}_1(\bar{X}; b, x)$$

with Γ-action. It proposes that this map should be a bijection. The injectivity is already known as a consequence of the Mordell–Weil theorem for the Jacobian J of X, while the surjectivity seems to be a very deep problem. The question mentioned in the introduction arises exactly because the injectivity appears, at first glance, to be more relevant for finiteness than the surjectivity. The idea for using the *bijectivity* seems to have been to create a tension between the compact profinite topology of $H^1(\Gamma, \hat{\pi}_1(\bar{X}, b))$ and the "discrete nature" of $X(\mathbb{Q})$. At present it is unclear how this intuition is to be realized. But as mentioned, when the finiteness is obtained through a different approach, we wish to explain the use of the surjectivity for *finding* the full set of points.

Using the exact sequence

$$0 \rightarrow U^{n+1} \backslash U^n \rightarrow U_{n+1} \rightarrow U_n \rightarrow 0$$

for each of the fundamental groups, the global Selmer variety is fibered according to the sequence

$$0 \rightarrow H^1_f(\Gamma, (U^{\text{ét}})^{n+1} \backslash (U^{\text{ét}})^n) \rightarrow H^1_f(\Gamma, U^{\text{ét}}_{n+1}) \rightarrow H^1_f(\Gamma, U^{\text{ét}}_n),$$

which means that the kernel acts on the variety in the middle with orbit space a subset of the third object. If we denote by r_n the dimension of U_n, there is a recursive formula [11]

$$\Sigma_{i|n} i r_i = (g + \sqrt{g^2 - 1})^n + (g - \sqrt{g^2 - 1})^n$$

which implies in particular that

$$r_n = (g + \sqrt{g^2 - 1})^n / n + O(g^{n/2})$$

for some $O(g^{n/2})$ that can be explicitly computed.

The global Selmer variety has its dimension controlled by the Euler characteristic formula for the cohomology of the group $\Gamma_T = \mathrm{Gal}(\mathbb{Q}_T/\mathbb{Q})$, where $T = S \cup \{p\}$ and $\mathbb{Q}_T$ is the maximal extension of $\mathbb{Q}$ unramified outside T ([9], Section 3). It reads

$$\begin{aligned}&\dim H^1(\Gamma_T, (U^{\text{ét}})^{n+1}\backslash(U^{\text{ét}})^n) - \dim H^2(\Gamma_T, (U^{\text{ét}})^{n+1}\backslash(U^{\text{ét}})^n)\\ &\quad = \dim[(U^{\text{ét}})^{n+1}\backslash(U^{\text{ét}})^n]^-,\end{aligned}$$

where the minus in the superscript refers to the sign for the action of complex conjugation. The dimension of this minus part can be estimated as follows. The action of complex conjugation on the étale fundamental group $U^{\text{ét}}$ is compatible with its action on the Betti realization U^B of the motivic fundamental group [4], according to which

$$(U^{\mathrm{B}})^{n+1}\backslash(U^{\mathrm{B}})^n$$

has a pure Hodge structure of weight n. So when n is odd, we get

$$\dim[(U^{\text{ét}})^{n+1}\backslash(U^{\text{ét}})^n]^- = r_n/2.$$

But when $n = 2m$ is even, there is the contribution from the (m,m) component to consider, which can be complicated. This (m,m) component is a quotient of the (m,m)-part of

$$H_1(X(\mathbb{C}), \mathbb{C})^{\otimes 2m},$$

which has dimension $\binom{2m}{m}g^{2m}$. So for simplicity, we will just use the tautological estimate

$$\dim[(U^{\text{ét}})^{n+1}\backslash(U^{\text{ét}})^n]^- \leq r_n$$

for n even.

In [9], Section 3, we analyzed the use of the corresponding Tate–Shafarevich groups

$$\begin{aligned}Ш^2((U^{\text{ét}})^{n+1}\backslash(U^{\text{ét}})^n)) &:= \mathrm{Ker}[H^2(\Gamma_T, (U^{\text{ét}})^{n+1}\backslash(U^{\text{ét}})^n)\\ &\qquad \to \oplus_{v\in T} H^2(G_v, (U^{\text{ét}})^{n+1}\backslash(U^{\text{ét}})^n)],\end{aligned}$$

which is dual to

$$\begin{aligned}Ш^1(((U^{\text{ét}})^{n+1}\backslash(U^{\text{ét}})^n)^*(1)) &:= \mathrm{Ker}[H^1(\Gamma_T, ((U^{\text{ét}})^{n+1}\backslash(U^{\text{ét}})^n)^*(1))\\ &\qquad \to \oplus_{v\in T} H^1(G_v, ((U^{\text{ét}})^{n+1}\backslash(U^{\text{ét}})^n)^*(1))].\end{aligned}$$

There is a Chern class map [1]

$$\mathrm{ch}_{n,1} : K^{(1)}_{2-n-1}(X^n) \otimes \mathbb{Q}_p \to H^1(\Gamma, H^n(\bar{X}^n, \mathbb{Q}_p(1)))$$

for $n \neq 1$ whose image lies in a "geometric" subspace

$$H^1_g(\mathrm{Gal}, H^n(\bar{X}^n, \mathbb{Q}_p(1)))$$

that contains

$$Ш^1(H^n(\bar{X}^n, \mathbb{Q}_p(1))).$$

In fact,

$$Ш^1([(U^{\text{ét}})^{n+1} \backslash (U^{\text{ét}})^n)]^*(1))$$

is a subspace of $Ш^1(H^n(\bar{X}^n, \mathbb{Q}_p(1)))$ because the representation $(U^{\text{ét}})^{n+1} \backslash (U^{\text{ét}})^n$ is a direct summand of $H_1^{et}(\bar{X}, \mathbb{Q}_p)^{\otimes n}$, which, in turn, is a direct summand of $(H^n(\bar{X}^n, \mathbb{Q}_p))^*$. But Bloch and Kato conjecture that

$$\mathrm{ch}_{n,1} : K^{(1)}_{2-n-1}(X^n) \otimes \mathbb{Q}_p \to H^1_g(\mathrm{Gal}(\bar{\mathbb{Q}}/\mathbb{Q}), H^n(\bar{X}^n, \mathbb{Q}_p(1)))$$

is an isomorphism. Thus, when $n \geq 2$, we get

$$Ш^1([(U^{\text{ét}})^{n+1} \backslash (U^{\text{ét}})^n)]^*(1)) = 0.$$

We recall the explicit bound for the local H^2 ([9], Section 3). For $v \neq p$, we have

$$\dim H^2(G_v, (U^{\text{ét}})^{n+1} \backslash (U^{\text{ét}})^n) \leq ng^n + \frac{n(n-1)}{2}(2g)^2 g^{n-2},$$

while

$$\dim H^2(G_p, (U^{\text{ét}})^{n+1} \backslash (U^{\text{ét}})^n) \leq ng^n.$$

Finally, as regards the contribution of the Hodge filtration, we saw in [9] that

$$F^0((U^{\text{dr}})^{n+1} \backslash (U^{\text{dr}})^{n+1}) \leq g^n,$$

so that

$$\dim (U^{\text{dr}})^{n+1} \backslash (U^{\text{dr}})^{n+1} / F^0 \geq r_n - g^n = (g + \sqrt{g^2 - 1})^n / n - g^n + O(g^{n/2}).$$

2 Beginning the descent

Since it costs very little extra work to define, we will in fact consider the refined Selmer variety

$$H^1_{f,0}(\Gamma, U_n^{\text{ét}}) \subset H^1_f(\Gamma, U_n^{\text{ét}})$$

consisting of classes whose images in

$$H^1_f(\Gamma, U_1^{\text{ét}})$$

go to zero under all localization maps

$$H^1_f(\Gamma, U_1^{\text{ét}}) \overset{\text{loc}_v}{\to} H^1(G_v, U_1^{\text{ét}})$$

for $v \neq p$. As explained in [10], the image of $X(\mathbb{Q})$ under $\kappa_n^{\text{ét,glob}}$ lies in $H^1_{f,0}(\Gamma, U_n^{\text{ét}})$. From the estimates of the previous section, it is obvious that

assuming the Bloch–Kato conjecture, there is an effectively computable t such that

$$\dim H^1_{f,0}(\Gamma, U_n^{ét}) < \dim U_n^{\text{dr}}/F^0$$

for $n \geq t$.

Of course the computation starts out with an estimate for $\dim H^1_{f,0}(\Gamma, U_1^{\text{ét}})$ which according to the usual BSD is the same as the Mordell–Weil rank of J. After that, the dimension of $\dim H^1_{f,0}(\Gamma, U_n^{\text{ét}})$ grows as a function of n with an explicit upper bound, while the dimension of U_n^{dr}/F^0 grows with an explicit (and eventually bigger) lower bound. Written out, the estimate for growth looks like

$$\begin{aligned}\dim H^1_{f,0}(\Gamma, U_{2n+1}^{\text{ét}}) &\leq \dim H^1_{f,0}(\Gamma, U_{2n}^{\text{ét}}) \\ &+ r_{2n}/2 + |S|[(2n)g^{2n} + \frac{(2n)(2n-1)}{2}(2g)^2 g^{2n-2}] + (2n)g^{2n}\end{aligned}$$

and

$$\begin{aligned}\dim H^1_{f,0}(\Gamma, U_{2n+2}^{\text{ét}}) &\leq \dim H^1_{f,0}(\Gamma, U_{2n+1}^{\text{ét}}) + r_{2n+1} \\ &+ |S|[(2n+1)g^{2n+1} + \frac{(2n+1)(2n)}{2}(2g)^2 g^{2n-1}] + (2n+1)g^{2n+1},\end{aligned}$$

while

$$\dim U_{n+1} \geq \dim U_n + r_n - g^n.$$

We eventually get an inequality in the right direction because of the asymptotic behavior of r_n. In this regard, note that $g+\sqrt{g^2-1}>g$ for $g\geq 2$.

As a consequence of the discrepancy in dimension, the image of

$$D\circ \mathrm{loc}_p : H^1_{f,0}(\Gamma, U_t^{\text{ét}})\to U_t^{\mathrm{dr}}/F^0$$

is *not* Zariski dense. In contrast to difficult sets like $X(\mathbb{Q})$, the classifying spaces for torsors and the maps between them are algebro-geometric objects which can be computed in principle. This should work in the manner of computations with the usual method of Chabauty as appears, for example, in [7] (cf. the discussion of θ in the introduction). In case this is not convincing, we will adopt it as an additional hypothesis:

$[H]$: The map

$$D\circ \mathrm{loc}_p : H^1_{f,0}(\Gamma, U_t^{\text{ét}})\to U_t^{\mathrm{dr}}/F^0$$

can be computed.

The end result of this is that assuming B-K and [H], we can find an algebraic function α on U_t^{dr}/F^0 that vanishes on the image of $H^1_{f,0}(\Gamma, U_t^{\text{ét}})$. Now, when we restrict α to $X(\mathbb{Q}_p)$ it becomes a linear combination of p-adic iterated integrals. To elaborate on this point a little more, recall ([9], Section 1) the description of the coordinate ring of the de Rham fundamental group $U^{\mathrm{dr},0}$ for an affine curve X^0 obtained by deleting some rational divisor from X. In this case, when we choose a collection $a_1, a_2, \ldots, a_k$ of algebraic differential forms on X^0 inducing a basis of $H^1_{\mathrm{dr}}(X^0)$, the coordinate ring of $U^{\mathrm{dr},0}$ has the form

$$\mathbb{Q}_p\langle a_w\rangle,$$

the $\mathbb{Q}_p$ vector space generated by symbols a_w, one for each finite sequence w of numbers from $\{1,2,\ldots,k\}$. Furthermore, on $X^0(\mathbb{Z}_p)$, there is a lifting (depending on the previous choice of basis)

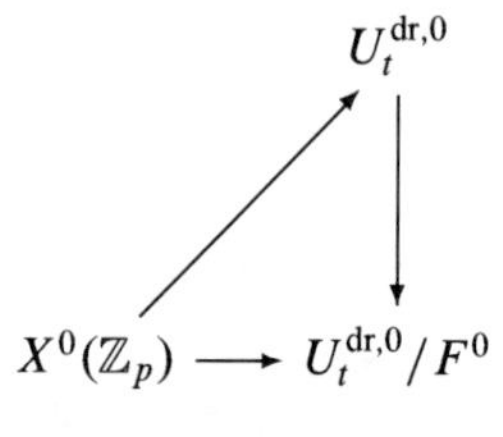

such that the restriction of a_w for $w=(i_1,i_2,\ldots,i_l)$ to $X^0(\mathbb{Z}_p)$ has the form

$$a_w(z)=\int_b^z a_{i_1}a_{i_2}\cdots a_{i_l}.$$

Also, there is a functorial map

$$U_t^{\mathrm{dr},0} \to U_t^{\mathrm{dr}}$$

compatible with the Hodge filtration so that the function α on U_t^{dr}/F^0 can be lifted to $U_t^{\mathrm{dr},0}$. That is to say, one can construct a diagram

$$\begin{array}{ccc} & & U_t^{\mathrm{dr},0} \\ & \nearrow & \downarrow \\ X^0(\mathbb{Z}_p) & \longrightarrow & U_t^{\mathrm{dr}}/F^0 \end{array}$$

enabling us to compute the restriction of α to $X^0(\mathbb{Z}_p)$ in terms of the a_w. The idea would be to carry this process out for two separate affine X^0 so as to cover $X(\mathbb{Z}_p)$ and then to express α in terms of iterated integrals on each affine open set. Of course, the problem of explicitly computing the local liftings is also a daunting task, although possible in theory. The author makes no pretense of knowing, as yet, how to reduce this to a tractable process. Perhaps it is safer to state it also explicitly as a hypothesis:

[H′]: The map

$$U_t^{\mathrm{dr},0} \longrightarrow U_t^{\mathrm{dr}}/F^0$$

can be computed.

Choose a representative $y \in X(\mathbb{Q}_p)$ for each point in $Y(\mathbb{F}_p)$ $(= X \bmod p)$ and a coordinate z_y centered at y. We must then approximate the zeros of α on $X(\mathbb{Q}_p)$ by expressing it as a power series in the z_y. This needs to be carried out to a sufficiently high degree of accuracy so that we can find an M and a finite collection $y_i \in X(\mathbb{Q}_p)$ for which

$$]y_i[_M := \{x \in X(\mathbb{Q}_p) | z_{y_i}(x) \leq p^{-M}\}$$

contains at most one zero of α. That is to say, we need to separate the zeros of α modulo p^M. Note that even at this point, since all expressions will be approximate, there would be no way to determine which of the y_i relate to actual points of $X(\mathbb{Q})$, even though an upper bound for the *number* of points may be available, as was emphasized by Coleman [2]. In fact, the process of separating the points using small disks already seems to occur, at least implicitly, in the method of Coleman–Chabauty. In the next section we will see how to combine that separation with the section conjecture.

We summarize the preceding passages as follows:

Observation 1 *Assuming the Bloch–Kato conjecture and the hypotheses [H] and [H′], there is an effectively computable M such that the map*

$$X(\mathbb{Q}) \hookrightarrow X(\mathbb{Q}_p) \to X(\mathbb{Z}/p^M)$$

is injective.

In our view, this statement is one rather essential justification for studying the Selmer varieties and unipotent Albanese maps. That is, Faltings' theorem as it stands does not seem to give, even in principle, a way of getting at this sort of effectivity. To belabor the obvious, the point is that the map

$$X(\mathbb{Q}) \to X(\mathbb{Q}_p)$$

is not a priori (i.e., before finding $X(\mathbb{Q})$) computable, even in principle, in contrast to the algebraic map

$$H^1_{f,0}(\Gamma, U_t^{\text{ét}}) \to U_t^{\text{dr}}/F^0.$$

When we embed $X(\mathbb{Q})$ inside $J(\mathbb{Q})$ using the base point b, we see then that we have an injection

$$X(\mathbb{Q}) \hookrightarrow J(\mathbb{Z}/p^M).$$

But the kernel of the reduction map

$$J(\mathbb{Q}) \to J(\mathbb{Z}/p^M)$$

is of finite index, and hence contains $NJ(\mathbb{Q})$ for some N. For example, one could take $N = |J(\mathbb{Z}/p^M)|$, which, in turn, can be computed from the formula

$$|J(\mathbb{Z}/p^M)| = p^{2g(M-1)}|J(\mathbb{F}_p)|,$$

since p is a prime of good reduction. So finally, we arrive at an effectively computable N such that

$$X(\mathbb{Q}) \to J(\mathbb{Q}) \to J(\mathbb{Q})/NJ(\mathbb{Q})$$

is injective. Let T_0 be S together with the set of primes dividing N, and Γ_{T_0} the fundamental group of $\mathrm{Spec}(\mathbb{Z}[1/T_0])$ with base point given by $\mathbb{Z}[1/T_0] \hookrightarrow \mathbb{Q} \hookrightarrow \bar{\mathbb{Q}}$. Thus we get an injection

$$X(\mathbb{Q}) \hookrightarrow H^1(\Gamma_{T_0}, J[N])$$

that allows us to begin descent.

3 Non-abelian descent and its termination

Once we have the final conclusion of the previous section, we can dispense entirely with the unipotent machinery and start to deal with the profinite formalism. There are many ways to construct a cofinal system for

$$\Delta := \hat{\pi}_1(\bar{X}, b),$$

of which we will use one described in a letter from Deligne to Thakur [5]. Let $K_n \subset \Delta$ be the intersection of all open subgroups of index $\leq n$. It is a characteristic open subgroup, and hence we can form the finite quotient $\Delta(n) := \Delta/K_n$. The order of this quotient has all prime divisors $\leq n$. Let Γ_n denote the fundamental group of $\mathrm{Spec}(\mathbb{Z}[1/n!])$. We also denote by $\pi(n)$ the quotient of $\hat{\pi}_1(X, b)$ by K_n, a group that fits into the exact sequence

$$0 \to \Delta(n) \to \pi(n) \to \Gamma \to 0.$$

For n larger than any prime in S, there is a pull-back diagram ([12], proof of Theorem 2.8)

$$\begin{array}{ccccccccc} 0 & \longrightarrow & \Delta(n) & \longrightarrow & \pi(n) & \longrightarrow & \Gamma & \longrightarrow & 0 \\ & & \Big\downarrow = & & \Big\downarrow & & \Big\downarrow & & \\ 0 & \longrightarrow & \Delta(n) & \longrightarrow & \hat{\pi}_1(\mathcal{X}_n)/K_n & \longrightarrow & \Gamma_n & \longrightarrow & 0 \end{array}$$

where $\mathcal{X}_n$ is a proper smooth model for X over $\mathrm{Spec}(\mathbb{Z}[1/n!])$. Therefore, we see that any point $x \in X(\mathbb{Q})$ defines a class in

$$H^1(\Gamma_n, \Delta(n))$$

and that we have a commutative diagram

$$\begin{array}{ccc} X(\mathbb{Q}) & \xhookrightarrow{\hat{\kappa}} & H^1(\Gamma, \Delta) \\ \Big\downarrow & & \Big\downarrow \\ H^1(\Gamma_n, \Delta(n)) & \hookrightarrow & H^1(\Gamma, \Delta(n)) \end{array} \tag{1}$$

There is a sequence of subsets containing $X(\mathbb{Q})$,

$$H^1(\Gamma, \Delta)_i \subset H^1(\Gamma, \Delta),$$

consisting of those classes whose projection to $H^1(\Gamma, \Delta(i))$ lies in the image of

$$H^1(\Gamma_i, \Delta(i)) \hookrightarrow H^1(\Gamma, \Delta(i)).$$

Let n_0 be larger than the primes in T_0. Then we have diagrams

$$\begin{array}{ccccc} & & H^1(\Gamma, \Delta)_i & \hookrightarrow & H^1(\Gamma, \Delta) \\ & & \downarrow & & \downarrow \\ & & H^1(\Gamma_i, \Delta(i)) & \hookrightarrow & H^1(\Gamma, \Delta(i)) \\ & & \downarrow & & \\ H^1(\Gamma_{T_0}, J[N]) & \hookrightarrow & H^1(\Gamma_i, J[N]) & & \end{array} \tag{2}$$

for $i \geq n_0$. Using this, we can define a decreasing sequence of subsets

$$H^1(\Gamma_{T_0}, J[N])_n \subset H^1(\Gamma_{T_0}, J[N])$$

for $n \geq n_0$ consisting of those classes whose images in $H^1(\Gamma_i, J[N])$ lift to $H^1(\Gamma_i, \Delta(i))$ for all $n_0 \leq i \leq n$. For $n \geq n_0$, we also have a commutative diagram

$$\begin{array}{ccc} X(\mathbb{Q}) & \hookrightarrow & H^1(\Gamma_n, \Delta(n)) \\ \cap\downarrow & & \downarrow \\ H^1(\Gamma_{T_0}, J[N]) & \hookrightarrow & H^1(\Gamma_n, J[N]) \end{array} \tag{3}$$

Meanwhile, there is an increasing sequence

$$X(\mathbb{Q})_n \subset X(\mathbb{Q}) \subset H^1(\Gamma_{T_0}, J[N])$$

consisting of the points with height (in some projective embedding) $\leq n$. We visualize the situation using the sort of filtration familiar from the arithmetic theory of elliptic curves:

$$\begin{aligned} \cdots X(\mathbb{Q})_n \subset X(\mathbb{Q})_{n+1} \subset \cdots \subset H^1(\Gamma_{T_0}, J[N])_{m+1} \\ \subset H^1(\Gamma_{T_0}, J[N])_m \subset \cdots \subset H^1(\Gamma_{T_0}, J[N]). \end{aligned} \tag{4}$$

Observation 2 *The section conjecture implies that*

$$X(\mathbb{Q})_n = H^1(\Gamma_{T_0}, J[N])_m$$

for n, m *sufficiently large. At this point,* $X(\mathbb{Q}) = X(\mathbb{Q})_n$.

That is to say, we know when to stop searching. The simple proof is written out just to make sure the author is not confused.

Proof. Assume the section conjecture. Then by diagrams (1) and (2), we have

$$H^1(\Gamma, \Delta)_i = H^1(\Gamma, \Delta)$$

for all i and we actually have maps

$$H^1(\Gamma, \Delta) \to H^1(\Gamma_i, \Delta(i))$$

for each i. Therefore,

$$H^1(\Gamma, \Delta) = \varprojlim H^1(\Gamma, \Delta(i)) = \varprojlim H^1(\Gamma_i, \Delta(i)).$$

Claim: Suppose $c \in H^1(\Gamma_{T_0}, J[N])$ is not in $X(\mathbb{Q})$. Then $c \notin H^1(\Gamma_{T_0}, J[N])_m$ for some m.

Proof of claim. If $c \in H^1(\Gamma_{T_0}, J[N])_m$ for each m, then $H^1(\Gamma, \Delta(m))_c \supset H^1(\Gamma_m, \Delta(m))_c$, the classes in $H^1(\Gamma, \Delta(m))$ that lift $c \in H^1(\Gamma_{T_0}, J[N]) \subset H^1(\Gamma, J[N])$ are non-empty for each m. Thus, the inverse limit $\varprojlim_m H^1(\Gamma, \Delta(m))_c \supset \varprojlim_m H^1(\Gamma_m, \Delta(m))_c$, containing an inverse limit of non-empty finite sets, is itself non-empty. Therefore, c would be in the image of $H^1(\Gamma, \Delta)$, and hence, in the image of $X(\mathbb{Q})$. □

Thus, eventually, $X(\mathbb{Q}) = H^1(\Gamma_{T_0}, J[N])_m$. Of course eventually $X(\mathbb{Q})_n = X(\mathbb{Q})$. Now suppose

$$X(\mathbb{Q})_n = H^1(\Gamma_{T_0}, J[N])_m$$

at any point. Then classes not in $H^1(\Gamma_{T_0}, J[N])_m$ cannot lift to $H^1(\Gamma, \Delta)_m$. And hence, they are not in $X(\mathbb{Q})$. That is to say, $X(\mathbb{Q})_n = X(\mathbb{Q})$. □

All the cohomology sets occurring in diagram (3) are finite and thereby have the nature of being computable through explicit Galois theory. Thus, the filtration (4) can be computed in principle. As mentioned in the introduction, the actual implementation of such an algorithm is obviously an entirely different matter.

References

1. Bloch, Spencer; Kato, Kazuya. L-functions and Tamagawa numbers of motives. *The Grothendieck Festschrift, Vol. 1*, 333–400, Prog. Math. 86, Birkhäuser, Boston, MA 1990.
2. Coleman, Robert F. Effective Chabauty. Duke Math. J. 52 (1985), no. 3, 765–770.
3. Cremona, John. *Algortihms for elliptic curves*. Online edition available at http://www.maths.nott.ac.uk/personal/jec/book/fulltext/index.html
4. Deligne, Pierre. Le groupe fondamental de la droite projective moins trois points. Galois groups over $\mathbb{Q}$ (Berkeley, CA, 1987), 79–297, Math. Sci. Res. Inst. Publ., 16, Springer, New York, 1989.
5. Deligne, Pierre. Letter to Dinesh Thakur. March 7, 2005.
6. Grothendieck, Alexandre. Brief an G. Faltings, *Geometric Galois Actions, 1*, 49–58, London Math. Soc. Lecture Note Ser., 242, Cambridge University Press, Cambridge, 1997.
7. Flynn, Victor A. Flexible Method for Applying Chabauty's Theorem. Compositio Math. 105 (1997), 79–94.
8. Kim, Minhyong. The motivic fundamental group of $\mathbf{P}^1 \setminus \{0, 1, \infty\}$ and the theorem of Siegel. Invent. Math. 161 (2005), no. 3, 629–656.
9. Kim, Minhyong. The unipotent Albanese map and Selmer varieties for curves, Publ. Res. Inst. Math. Sci. Volume 45, Number 1 (2009), 89-133.
10. Kim, Minhyong; Tamagawa, Akio: The l-component of the unipotent Albanese map. Math. Ann. 340 (2008), no. 1, 223–235.
11. Labute, John P. On the descending central series of groups with a single defining relation. J. Algebra 14 1970 16–23.
12. Wildeshaus, Jörg. Realizations of polylogarithms. Lecture Notes in Mathematics, 1650. Springer-Verlag, Berlin, 1997.

Ranks of elliptic curves in cubic extensions

Hershy Kisilevsky

Dedicated to the memory of S. Lang

Abstract Let $E/\mathbb{Q}$ be an elliptic curve defined over the rational field $\mathbb{Q}$. We examine the rank of the Mordell–Weil group $E(K)$ as K ranges over cubic extensions of $\mathbb{Q}$.

Key words Elliptic curves • cubic fields • Mordell–Weil groups

Mathematics Subject Classification (2010): 11G05, 11R16, 14H52

1 Introduction

Let $E/\mathbb{Q}$ be an elliptic curve defined over the rational field $\mathbb{Q}$. For any finite extension $K/\mathbb{Q}$, the Mordell–Weil group $E(K)$ of K-rational points of E is a finitely generated abelian group and we denote by $r_E(K)$ the $\mathbb{Z}$-rank of $E(K)$, i.e., the number of copies of $\mathbb{Z}$ in its standard decomposition as a direct sum of cyclic groups. In [F-K-K] we studied the behaviour of $r_E(K)$ as K ranged over *cyclic* cubic extensions of $\mathbb{Q}$ and showed that the condition $r_E(K) > r_E(\mathbb{Q})$ is controlled by the rational points on a certain K3-surface defined over $\mathbb{Q}$. For many curves E, we showed that $r_E(K) > r_E(\mathbb{Q})$ for an infinite number of cyclic cubic extensions $K/\mathbb{Q}$, and asked whether this was the case for every elliptic curve E. Assuming the Birch & Swinnerton-Dyer conjecture, this question translates to the following:

H. Kisilevsky (✉)
Department of Mathematics and Statistics and CICMA, Concordia University,
1455 de Maisonneuve Blvd. West, Montréal, Quebec, H3G 1M8, Canada
e-mail: kisilev@mathstat.concordia.ca

D. Goldfeld et al. (eds.), *Number Theory, Analysis and Geometry: In Memory of Serge Lang*, DOI 10.1007/978-1-4614-1260-1_17,

If $L(E,s)$ is the L-function of $E/\mathbb{Q}$, then does the twisted L-series $L(E,\chi,s)$ vanish at $s=1$ for an infinity of Dirichlet characters χ of order three?

In the present paper we consider this question as $K/\mathbb{Q}$ ranges over all cubic extensions.

We define the families $\mathscr{F}$ of fields we will consider. Let k be a global field and let $\overline{k}$ be an algebraic closure of k. Suppose that F/k is a separable extension of degree 1 or 2.

Definition 1.1. Define $\mathscr{F}(F)$ to be the collection of all separable cubic extensions K/k (in $\overline{k}$) whose Galois closures are $M = K \cdot F$.

We will consider all number fields to be subfields of the complex numbers $\mathbb{C}$, so that we consider $\overline{\mathbb{Q}} \subset \mathbb{C}$.

In this paper, given a cubic extension K/k, we will always write M for the Galois closure of K/k, and F for the quadratic resolvent field (so $F = k$ is permitted). In the case that $k = \mathbb{Q}$, then either $F = \mathbb{Q}$ or $F = \mathbb{Q}(\sqrt{D})$, where $D \neq 1$ is a fundamental discriminant. We write $\mathscr{F}(F) = \mathscr{F}(1)$ if $F = \mathbb{Q}$, and $\mathscr{F}(F) = \mathscr{F}(D)$ if $F = \mathbb{Q}(\sqrt{D})$, with $F/\mathbb{Q}$ a quadratic extension.

The family $\mathscr{F}(1)$ of *cyclic* cubic extensions $K/\mathbb{Q}$ was studied in [F-K-K] and the family $\mathscr{F}(-3)$ has been considered by T. Dokchitser [DoT], where

$$\mathscr{F}(-3) = \{K = \mathbb{Q}(\omega^i m^{\frac{1}{3}}) \mid m \in \mathbb{Z}, m \text{ not a cube}, i = 0, \pm 1\}$$

and where ω is a primitive cube root of unity.

The results of this paper are contained in the following theorem.

Theorem 1.1. *Let E be an elliptic curve defined over $\mathbb{Q}$, and let $D = 1$ or let $D \neq 1$ be a fundamental discriminant. Let E^D denote the (quadratic) twist of E by D. Then $r_E(K) > r_E(\mathbb{Q})$ for an infinite number of $K \in \mathscr{F}(D)$ if any of the following hold:*

(a) *Both* $r_E(\mathbb{Q}) \geq 1$ *and* $r_{E^D}(\mathbb{Q}) \geq 1$.
(b) $D > 0$ *and* $\operatorname{rank}_{\mathbb{Z}} E_{\mathrm{Tr}}(K_0) \geq 1$ *for some* $K_0 \in \mathscr{F}(D)$.
(c) $D < 0$ *and* $\operatorname{rank}_{\mathbb{Z}} E_{\mathrm{Tr}}(K_0) \geq 1$ *for some* $K_0 \in \mathscr{F}(D)$. *We must also assume that the density hypothesis holds (and in the case that E has CM, we assume also that $\mathbb{Q}(\sqrt{D})$ is distinct from $F' = \mathrm{End}(E) \otimes_{\mathbb{Z}} \mathbb{Q}$).*
(d) *(T. Dokchitser)* $D = -3$.
(e) $D < 0$, *and E is a semistable elliptic curve of conductor N_E, prime to D, such that* $\operatorname{sign}(E, \mathbb{Q}(\sqrt{D})) = -1$ *and* $L'(E/\mathbb{Q}(\sqrt{D}), 1) \neq 0$.

Here $E_{\mathrm{Tr}}(K_0)$ is the group of points of E defined over K_0 and with trace zero to $E(\mathbb{Q})$, $\operatorname{sign}(E, F)$ is the sign of the functional equation of the L-function of E viewed as an elliptic curve over the field F, and the density hypothesis is defined in Section 3.

Conjecture 1. For every elliptic curve $E/\mathbb{Q}$ and for every family $\mathscr{F}(D)$, there is an infinite number of $K \in \mathscr{F}(D)$ for which $r_E(K) > r_E(\mathbb{Q})$.

Question 1. Let k be a global field. Then for every elliptic curve E/k and for every family $\mathscr{F}(F)$ as above, is there an infinite number of $K \in \mathscr{F}(F)$ for which $r_E(K) > r_E(k)$?

When $[F : \mathbb{Q}] = 2$, the corresponding analytic question concerns the vanishing at $s = 1$ of the twist $L(E, \rho, s)$ of $L(E, s)$ by the character of the (unique) irreducible two-dimensional Artin representation associated with M. The situation for these L-functions differs from the cyclic case treated in [F-K-K] because the signs of the functional equations of the $L(E, \rho, s)$ are (to a large degree) independent of ρ in the families $\mathscr{F}(D)$, $D \neq 1$.

Acknowledgments This work was supported in part by a grant from NSERC.

2 Cubic families

Fix a fundamental discriminant D, i.e., either $D = 1$, or D is an integer which is either square-free and $\equiv 1 \pmod 4$ or 4 times a square-free integer $\equiv 2, 3 \pmod 4$, and let $F = \mathbb{Q}(\sqrt{D})$. We consider the family $\mathscr{F}(D)$ of all cubic extension fields $K/\mathbb{Q}$ whose Galois closure M contains F, i.e., $M = K \cdot F = K(\sqrt{D})$.

Unless otherwise stated, we assume, for the remainder of the paper, that D is a fundamental discriminant and that $D \neq 1$, i.e., we assume that $F = \mathbb{Q}(\sqrt{D})$ is a quadratic extension of the rational field $\mathbb{Q}$.

Let $\mathcal{M}_3(D)$ be the composite of all cyclic cubic extensions of F. Then $\mathcal{M}_3(D)/\mathbb{Q}$ is a Galois extension. Let $\mathcal{G}_3(D)$ denote the Galois group $\mathrm{Gal}(\mathcal{M}_3(D)/F)$ and let τ be an automorphism of $\mathcal{M}_3(D)$ of order 2 which lifts the non-trivial element of $\mathrm{Gal}(F/\mathbb{Q})$. Then $\mathcal{G}_3(D)$ splits under the natural action of τ as a direct sum

$$\mathcal{G}_3(D) = \mathcal{G}_3(D)^+ \oplus \mathcal{G}_3(D)^-.$$

Here $\mathcal{G}_3(D)^\pm = \{\sigma \in \mathcal{G}_3(D) \mid \tau(\sigma) = \tau \cdot \sigma \cdot \tau^{-1} = \sigma^\pm\}$. If $\mathcal{M}_3(D)^\pm$ is the subfield of $\mathcal{M}_3(D)$ fixed by $\mathcal{G}_3(D)^\mp$, then $\mathcal{M}_3(D)^+$ is the composite with F of all cyclic cubic extensions of $\mathbb{Q}$ and is the maximal abelian (over $\mathbb{Q}$) subfield of $\mathcal{M}_3(D)$, and $\mathcal{M}_3(D)^-$ is the composite of all S_3-extensions of $\mathbb{Q}$ having F as quadratic subfield. It then follows that $K \in \mathscr{F}(D)$ if and only if its Galois closure M is contained in $\mathcal{M}_3(D)^-$.

Suppose that $K/\mathbb{Q}$ is a finite extension and let $\mathrm{Tr} : E(K) \longrightarrow E(\mathbb{Q})$ denote the trace map. We denote by $E_{\mathrm{Tr}}(K)$ the kernel of the trace map.

We record the following lemmas.

Lemma 2.1. *let E be an elliptic curve defined over $\mathbb{Q}$. The following are equivalent:*

(i) $r_E(K) > r_E(\mathbb{Q})$;
(ii) $E(K)$ *contains a trace zero point of infinite order;*
(iii) $\#(E_{\mathrm{Tr}}(K)) = \infty$.

Proof. Since $E(K)$ is a finitely generated abelian group, it follows from the exact sequence

$$0 \longrightarrow E_{\mathrm{Tr}}(K) \longrightarrow E(K) \longrightarrow \mathrm{Tr}(E(K)) \longrightarrow 0$$

that $r_E(K) = \mathrm{rank}_{\mathbb{Z}}\, \mathrm{Tr}(E(K)) + \mathrm{rank}_{\mathbb{Z}}\, E_{\mathrm{Tr}}(K)$. But since $[E(\mathbb{Q}) : \mathrm{Tr}(E(K))] < \infty$, it follows that $r_E(\mathbb{Q}) = \mathrm{rank}_{\mathbb{Z}}\, \mathrm{Tr}(E(K))$, and hence that

$$r_E(K) = r_E(\mathbb{Q}) + \mathrm{rank}_{\mathbb{Z}}\, E_{\mathrm{Tr}}(K).$$

The statement of the lemma follows. □

In light of Lemma 2.1, the conjecture is equivalent to the statement that for every elliptic curve $E/\mathbb{Q}$, and for every fundamental discriminant D, $\#(E_{\mathrm{Tr}}(K)) = \infty$ for infinitely many $K \in \mathscr{F}(D)$.

Suppose now that E is an elliptic curve defined over $\mathbb{Q}$, and that $K/\mathbb{Q}$ is a non-Galois cubic extension whose Galois closure is M. Let F be the subfield of M of degree 2 such that $M = K \cdot F$.

Lemma 2.2. *The following are equivalent:*

(i) $r_E(K) > r_E(\mathbb{Q})$;
(ii) $r_E(M) > r_E(F)$.

Proof. Let $G = \mathrm{Gal}(M/\mathbb{Q}) = \langle \tau, \sigma \rangle \simeq S_3$, where $\tau^2 = \sigma^3 = 1$ and $\tau\sigma = \sigma^{-1}\tau$. Let V be a $\mathbb{Q}$-vector space which has an action by $G \simeq S_3$. Then there is a decomposition

$$V = V^0 \oplus V^\chi \oplus V^\rho,$$

where V^0, V^χ, and V^ρ are the subspaces of V corresponding to the trivial representation χ_0 of G, the non-trivial abelian representation χ of G, and the irreducible two-dimensional representation ρ of G, respectively. The subspaces which are pointwise fixed by the groups G, $\langle \sigma \rangle$ and $\langle \tau \rangle$ are V^0, $V^0 \oplus V^\chi$, and $V^0 \oplus V^\rho$ respectively. Applying this decomposition to $V = E(M) \otimes_{\mathbb{Z}} \mathbb{Q}$, we see that

$$\begin{aligned} r_E(K) > r_E(\mathbb{Q}) &\Leftrightarrow \dim_{\mathbb{Q}}(V^0 \oplus V^\rho) > \dim_{\mathbb{Q}}(V^0) \\ &\Leftrightarrow \dim_{\mathbb{Q}}(V^0 \oplus V^\chi \oplus V^\rho) > \dim_{\mathbb{Q}}(V^0 \oplus V^\chi) \\ &\Leftrightarrow r_E(M) > r_E(F). \end{aligned}$$

□

3 Associated surfaces

Let a Weierstrass model for $E/\mathbb{Q}$ be given by

$$y^2 = x^3 + Ax + B$$

with $A, B \in \mathbb{Z}$.

Now suppose that K is an extension of $\mathbb{Q}$ of degree 3, and that $P = (\alpha, \beta)$ is a point in $E(K)$. Since $[K : \mathbb{Q}]$ is odd, we see that $\beta \in \mathbb{Q}(\alpha)$ and therefore $\mathbb{Q}(\alpha) = \mathbb{Q}(P) = \mathbb{Q}$ or K. In either case, there exist $a, b, c \in \mathbb{Q}$ with $\beta = a + b\alpha + c\alpha^2$. We note that the condition that $c = 0$ is equivalent to the statement that P lies on the intersection of E with the line $L_{a,b}$ with equation $y = a + bx$. This is equivalent to the fact that α is a root of the cubic polynomial

$$f_{a,b}(x) = x^3 + Ax + B - (a + bx)^2.$$

The discriminant of $f_{a,b}$ is a *square* in $\mathbb{Q}$ if and only if either a root of $f_{a,b}(x)$ generates a *cyclic* cubic extension of $K/\mathbb{Q}$, or $f_{a,b}(x)$ has three rational roots. This is the case $D = 1$, i.e., $K \in \mathscr{F}(1)$, and this was treated in [F-K-K].

Let $S_E^D = S_E^D(d, a, b)$ be the affine surface defined by the equation

$$\begin{aligned} D \cdot d^2 &= \text{discriminant}(x^3 + Ax + B - (a + bx)^2) \\ &= -27a^4 - 4b^3a^3 + (54B - 30Ab^2)a^2 + (36Bb^3 + 24A^2b - 4Ab^5)a \\ &\quad + A^2b^4 - 4A^3 - 27B^2 + 4Bb^6 - 18ABb^2. \end{aligned}$$

Recall that for $K \in \mathscr{F}(D)$, $M = K(\sqrt{D})$ denotes the Galois closure of K and $\tau \in \mathrm{Gal}(M/\mathbb{Q})$ the non-trivial automorphism of order 2 of M fixing K. Also $F = \mathbb{Q}(\sqrt{D}) \subset M$ is the quadratic subfield of M and let σ be an automorphism of order 3 which generates $\mathrm{Gal}(M/F)$, so $\mathrm{Gal}(M/\mathbb{Q}) = \langle \sigma, \tau \mid \tau^2 = \sigma^3 = 1 \rangle \simeq S_3$, and $\tau\sigma = \sigma^{-1}\tau$.

Suppose that (d, a, b) is a $\mathbb{Q}$-rational point on S_E^D. If α is a root of $f_{a,b}(x)$, and $\beta = a + b\alpha$, then $P = (\alpha, \beta) \in E(\mathbb{Q}(\alpha))$. Since the discriminant of $f_{a,b}(x)$ is Dd^2, it follows that either $\mathbb{Q}(\alpha) = \mathbb{Q}$ or F and $f_{a,b}(x)$ has a rational root, and two conjugate roots in F, or $\mathbb{Q}(\alpha) = K$ is an extension of $\mathbb{Q}$ of degree 3, with $K \in \mathscr{F}(D)$ and $P \in E_{\mathrm{Tr}}(K)$, a point of trace zero. Thus $(d, a, b) \in S_E^D(\mathbb{Q})$ determines either a triple of points $\{P, Q, Q^\tau\}$ with Q, Q^τ conjugate points in $E(F)$ and $P = -(Q + Q^\tau) \in E(\mathbb{Q})$, or a point $P \in E_{\mathrm{Tr}}(K)$ of trace zero.

Conversely, suppose that $K/\mathbb{Q}$ is a cubic extension with $K \in \mathscr{F}(D)$ and that $P \in E_{\mathrm{Tr}}(K)$ is a point of trace zero. Then the points P, P^σ, and P^{σ^2} lie on a line L. Since the automorphism τ fixes P and interchanges P^σ and P^{σ^2}, the set $\{P, P^\sigma, P^{\sigma^2}\}$ is invariant under the action of $\mathrm{Gal}(M/\mathbb{Q})$, and therefore the line $L = L_{a,b}$ is defined over $\mathbb{Q}$. It follows that $P = (\alpha, \beta)$ with $\beta = a + b\alpha$ and with $a, b \in \mathbb{Q}$. Similarly, if we have a triple of points $\{P, Q, Q^\tau\}$ with Q, Q^τ conjugate points in $E(F)$ and $P = -(Q + Q^\tau) \in E(\mathbb{Q})$, then this determines a rational line $L_{a,b}$ intersecting E in these three points.

Therefore we have the following result:

Proposition 3.1. *The $\mathbb{Q}$-rational points of S_E^D are in one-to-one correspondence with rational lines intersecting E either in a triple of points $\{P, Q, Q^\tau\}$ with Q, Q^τ conjugate points in $E(F)$ and $P = -(Q + Q^\tau) \in E(\mathbb{Q})$, or in a point $P \in E_{\mathrm{Tr}}(K)$ of trace zero on a cubic field $K \in \mathscr{F}(D)$.*

Let E^D denote the quadratic twist of E by D. If E^D is given by the model $Dy^2 = x^3 + Ax + B$, then the map $\phi : E^D(\mathbb{Q}) \longrightarrow E(F)$ which takes a point $Q = (x, y)$ to $\phi(Q) = (x, \sqrt{D}y)$ is an injective homomorphism of groups such that τ acts as -1 on the image $\phi(E^D(\mathbb{Q})) \subseteq E(F)$. In fact $\phi(E^D(\mathbb{Q})) = \{P \in E(F) | P^\tau = -P\}$.

Let $\pi : S_E^D \longrightarrow \mathbb{A}^2$ be the projection onto affine 2-space given by $\pi((d, a, b)) = (a, b) \in \mathbb{A}^2$. We give conditions under which we can prove that $\pi(S_E^D(\mathbb{Q}))$ is Zariski dense in $\mathbb{A}^2$.

Suppose first that $D > 0$.

Proposition 3.2. *Suppose that $D > 0$. Assume either that* rank$_{\mathbb{Z}}$ $E_{\mathrm{Tr}}(K_0) \geq 1$ *for some $K_0 \in \mathcal{F}(D)$ or that both $r_E(\mathbb{Q}) \geq 1$ and $r_{E^D}(\mathbb{Q}) \geq 1$. Then the image of $\mathbb{Q}$–rational points $S_E^D(\mathbb{Q})$ under π is Zariski dense in $\mathbb{A}^2$. (In fact, $\pi(S_E^D(\mathbb{Q}))$ is dense in the usual topology in an open subset of $\mathbb{A}^2(\mathbb{R})$.)*

Proof. Suppose first that $K_0 \in \mathcal{F}(D)$ is a cubic field and that $P_0 \in E_{\mathrm{Tr}}(K_0)$ is a point of trace zero and of infinite order. Let M_0 be the Galois closure of K_0. Since $\mathrm{Tr}(P_0) = 0$, it follows that $P = P_0$ and $P' = P_0^\sigma$ generate a subgroup of $E(M_0)$ of rank 2 over $\mathbb{Z}$ which is stable under $\mathrm{Gal}(M_0/\mathbb{Q})$. Then for any $n \in \mathbb{Z}$, the line joining nP, nP' also contains nP^{σ^2} and so is defined over the rational field. In the case that both $r_E(\mathbb{Q}) \geq 1$ and $r_{E^D}(\mathbb{Q}) \geq 1$, let $M_0 = F = \mathbb{Q}(\sqrt{D})$. If P_0 and P_1 are points of infinite order on $E(\mathbb{Q})$ and $E^D(\mathbb{Q})$ respectively, let $P = P_0 + \phi(P_1)$ and $P' = P_0 - \phi(P_1)$ in $E(M_0)$. Then $P' = P^\tau$, and P and P' again generate a rank-two subgroup of $E(M_0)$ which is stable under $\mathrm{Gal}(M_0/\mathbb{Q})$. In this case for any $n \in \mathbb{Z}$, the line joining nP, nP' contains $-2nP_0 \in E(\mathbb{Q})$. Therefore it intersects E in two conjugate points of $E(F)$ and a point in $E(\mathbb{Q})$ and so is also defined over the rational field.

The assumption that $D > 0$ ensures that M_0 is a totally real field.

Let $\Lambda_E \subset \mathbb{C}$ be the period lattice associated to E and let

$$\mathbb{C}/\Lambda_E \longrightarrow E(\mathbb{C})$$
$$z(\mathrm{mod}\, \Lambda_E) \longmapsto P = (\wp(z), \wp'(z))$$

be the analytic parametrization of E.

Let $\mathcal{O}(D)$ denote the interior of the set $\pi(S_E^D(\mathbb{R}))$,

$$\mathcal{O}(D) = \{(a, b) \in \mathbb{A}^2(\mathbb{R}) \mid \mathrm{discriminant}(f_{a,b}(x)) > 0\}.$$

Then $\mathcal{O}(D)$ consists of those $(a, b) \in \mathbb{A}^2(\mathbb{R})$ such that the line $L_{a,b}$ intersects E in three distinct real points, and is a non-empty open subset of $\mathbb{R}^2$. Let $\mathcal{O}(D)^+ \subseteq \mathcal{O}(D)$ be the set of points corresponding to lines $L_{a,b}$ intersecting $E(\mathbb{R})$ at three (distinct) real points P_1, P_2, and P_3, on the connected component of the identity. Then $\mathcal{O}(D)^+$ is open in $\mathcal{O}(D)$.

Let z and $z' \in \mathbb{R}$ be real lifts of P and P' with respect to the analytic parametrization, and let ω_1 be a (non-zero) real period of E. Since P and P' are independent over $\mathbb{Z}$, it follows that z, z', and ω_1 are $\mathbb{Q}$-linearly independent in $\mathbb{R}$. Kronecker's theorem then implies that given any $x, y \in \mathbb{R}$ and $\epsilon > 0$, there are integers $m, n_1, n_2 \in \mathbb{Z}$ such that $|x - mz - n_1\omega_1| < \epsilon$ and $|y - mz' - n_2\omega_1| < \epsilon$ simultaneously (see Hardy & Wright Theorem 442, [H-W]). It follows that the multiples of (z, z') (mod Λ_E^2) are dense in $\mathbb{R}^2/(\mathbb{Z}\omega_1)^2$. Therefore given any two points Q_1, Q_2 in the connected component of the identity of $E(\mathbb{R})$ there is an integer $m \in \mathbb{Z}$ such that mP and mP' are arbitrarily close to Q_1 and Q_2, and so the line joining Q_1 and Q_2 can be approximated arbitrarily closely by the line joining mP and mP' which is defined over $\mathbb{Q}$ by the above. It now follows that $\pi(S_E^D(\mathbb{Q}))$ is dense (in the Euclidean topology) in $\mathcal{O}(D)^+$and is therefore Zariski dense in $\mathbb{A}^2$. □

Suppose now that $D < 0$. In this case the above proof fails because we cannot apply Kronecker's theorem. We appeal to the following:

Definition 3.1. (Density hypothesis.) Let $E/\mathbb{Q}$ be an elliptic curve defined over the rational field $\mathbb{Q}$. Suppose that P is an *algebraic* point on $E(\mathbb{C})$ such that no integer multiple of mP is real ($mP = \overline{mP}$) or purely imaginary (i.e., $mP = -\overline{mP}$), and if E has CM, then P is not the division of a real or a purely imaginary algebraic point by any (complex) endomorphism. Then the cyclic subgroup of $E(\mathbb{C})$ generated by P is dense (in the usual topology) in the complex points $E(\mathbb{C})$ of E. The density hypothesis can be restated more compactly as follows: Suppose $P \in E(\overline{\mathbb{Q}})$ is such that $\lambda(P) \notin E^{\pm}(\mathbb{C})$ for any $0 \neq \lambda \in \mathrm{End}(E)$. Then the cyclic subgroup generated by P is dense in $E(\mathbb{C})$. Here $E^+(\mathbb{C})$ and $E^-(\mathbb{C})$ are the subgroups of $E(\mathbb{C})$ on which complex conjugation acts by $+1$ and -1 respectively.

In this regard, Waldschmidt had already introduced the "density property" for commutative algbraic groups in [Wa1] and [Wa2]. He has shown for elliptic curves E defined over $\overline{\mathbb{Q}}$, which are not isogenous to their complex conjugates, that a subgroup of rank ≥ 3 of $E(\overline{\mathbb{Q}})$ is dense in $E(\mathbb{C})$. He has also pointed out that the density hypothesis above follows from the "elliptico-toric" conjecture of Cristiana Bertolin [Be].

Proposition 3.3. *Suppose that $D < 0$. Assume either that* $\mathrm{rank}_{\mathbb{Z}}\, E_{\mathrm{Tr}}(K_0) \geq 1$ *for some $K_0 \in \mathcal{F}(D)$ and that the density hypothesis holds (and in the case that E has CM, we assume also that $\mathbb{Q}(\sqrt{D})$ is distinct from $F' = \mathrm{End}(E) \otimes_{\mathbb{Z}} \mathbb{Q}$), or that both $r_E(\mathbb{Q}) \geq 1$ and $r_{E^D}(\mathbb{Q}) \geq 1$. Then the image of $\mathbb{Q}$-rational points $S_E^D(\mathbb{Q})$ under π is Zariski dense in $\mathbb{A}^2$. (In fact, $\pi(S_E^D(\mathbb{Q}))$ is dense in the usual topology in an open subset of $\mathbb{A}^2(\mathbb{R})$.)*

Proof. Suppose that $P_0 \in E_{\mathrm{Tr}}(K_0)$ is a point of infinite order for some cubic field $K_0 \in \mathcal{F}(D)$. We may suppose that P_0 is a real point and that its conjugates P_0^{σ} and $P_0^{\sigma^2}$ are complex conjugate points. Then we show that $P = P_0^{\sigma}$ satisfies the conditions of the density hypothesis. We note that the subgroup of $E(M_0)$ generated by $\{P_0, P_0^{\sigma}, P_0^{\sigma^2}\}$ is a group of $\mathbb{Z}$-rank 2 with the relation $P_0 + P_0^{\sigma} + P_0^{\sigma^2} = 0$. If a multiple of P_0^{σ} were real or purely imaginary, there would be a second independent relation of the form $mP_0^{\sigma} = \pm mP_0^{\sigma^2}$. Since this would imply that the group had

$\mathbb{Z}$-rank at most 1, it follows that $mP_0 \notin E^{\pm}(M)$ for any $0 \neq m \in \mathbb{Z}$. Furthermore, if E has CM by the quadratic field $F' \neq F$, let $M' = M_0 \cdot F'$, so that

$$\mathrm{Gal}(M'/\mathbb{Q}) \simeq \mathrm{Gal}(F'/\mathbb{Q}) \times \mathrm{Gal}(M_0/\mathbb{Q}) = \langle \tau' \rangle \times \langle \tau, \sigma \rangle.$$

Then τ fixes K_0F' and complex conjugation on M' is given by $\tau\tau'$. Now $P_0 \in E(K_0)$ is a real point and its conjugates P_0^{σ} and $P_0^{\sigma^2}$ are complex conjugate points in $E(M')$. Suppose that $\lambda(P_0) \in E^{\pm}(M')$ for some endomorphism $\lambda \in \mathrm{End}(E) \simeq \mathcal{O}_{F'}$. Then $\pm\lambda(P_0^{\sigma}) = \overline{\lambda(P_0^{\sigma})} = \overline{\lambda}(P_0^{\sigma^2})$, so that applying σ^2, we see that $\lambda(P_0) = \pm\overline{\lambda}(P_0^{\sigma})$. Since $\lambda(P_0) + \lambda(P_0^{\sigma}) + \lambda(P_0^{\sigma^2}) = 0$ we find that

$$(\lambda \pm \overline{\lambda})(P_0^{\sigma}) + \lambda(P_0^{\sigma^2}) = 0.$$

Applying τ and adding, we see that

$$(2\lambda \pm \overline{\lambda})(P_0^{\sigma} + P_0^{\sigma^2}) = 0.$$

But since $P_0^{\sigma} + P_0^{\sigma^2} = -P_0$ has infinite order, it follows that $2\lambda \pm \overline{\lambda} = 0$, and therefore $\lambda = 0$. Hence the density hypothesis implies that the multiples of P_0^{σ} are dense in $E(\mathbb{C})$.

In the case that both $r_E(\mathbb{Q}) \geq 1$ and $r_{E^D}(\mathbb{Q}) \geq 1$, then $E(\mathbb{Q})$ is dense in the real points E and $\phi(E^D(\mathbb{Q}))$ is dense in the purely imaginary points of E and so their sum $E(F) = E(\mathbb{Q}) + \phi(E^D(\mathbb{Q}))$ is dense in $E(\mathbb{C})$.

As before, let $(a,b) = \pi((d,a,b)) \in \mathbb{A}^2(\mathbb{R})$ for some $(d,a,b) \in \pi(S_E^D(\mathbb{Q}))$. If $L_{a,b}$ is the corresponding line, then $L_{a,b}$ intersects E in one real point (P_1, say) and a pair of conjugate complex points (Q_1 and $\overline{Q}_1$). Then the density hypothesis assumed above implies that there is a point Q either in $E(M_0)$ or in $E(F)$ arbitrarily close to Q_1, and so Q^{τ} is close to $\overline{Q}_1$ and hence the line L_0 joining Q, Q^{τ} and $-(Q + Q^{\tau})$ is close to $L_{a,b}$. But since the set $\{Q, Q^{\tau}, -(Q + Q^{\tau})\}$ is stable under the action of $\mathrm{Gal}(M_0/\mathbb{Q})$ or $\mathrm{Gal}(F/\mathbb{Q})$, the line L_0 is defined over $\mathbb{Q}$ and so $\pi(S_E^D(\mathbb{Q}))$ is dense (in the Euclidean topology) in $\mathcal{O}(D)$ and is therefore Zariski dense in $\mathbb{A}^2$. □

We can now prove Conjecture 1 under the assumptions of Propositions 3.2 and 3.3.

For fixed $b \in \mathbb{Q}$, the fibre over b, $X_{E,b} = X_{E,b}(d,a)$ is a curve of genus at most 1. Our strategy is to try to find $b \in \mathbb{Q}$ such that $X_{E,b}$ is either an elliptic curve of positive rank or a rational curve with infinitely many $\mathbb{Q}$-rational points. We then show that for almost all such b, only a finite number of rational points on any $X_{E,b}$ can correspond to a given cubic extension $K \in \mathscr{F}(D)$, so in the event that $X_{E,b}(\mathbb{Q})$ is infinite, we obtain the conclusion of Conjecture 1 that there is an infinite number of fields $K \in \mathcal{F}(D)$ for which $r_E(K) > r_E(\mathbb{Q})$.

Recall that a $\mathbb{Q}$-rational point $Q = (d,a,b) \in S_E^D(\mathbb{Q})$ corresponds to the three roots $\alpha, \alpha_1, \alpha_2$ (distinct if $d \neq 0$) of the polynomial $f_{a,b}(x) = x^3 + Ax + B - (a + bx)^2$. In addition there are two possibilities: either one of the roots is rational

and the other pair are conjugate elements in $\mathbb{Q}(\sqrt{D})$, or $\mathbb{Q}(\alpha), \mathbb{Q}(\alpha_1)$, and $\mathbb{Q}(\alpha_2)$ are conjugate cubic fields in $\mathscr{F}(D)$.
The following results appear in Section 3 of [F-K-K].

Proposition 3.4. *Fix a field K for which Faltings' theorem holds. Then for any fixed $b \neq 0$, there are only finitely many $a \in \mathbb{Q}$ such that the polynomial $f_{a,b}(x) = x^3 + Ax + B - (a + bx)^2$ has three distinct roots in K.*

Proof. Fix b. Suppose that

$$f_{a,b}(x) = x^3 + Ax + B - (a + bx)^2 = (x - \alpha_1)(x - \alpha_2)(x - \alpha_3)$$

with α_1, α_2, and $\alpha_3 \in K$. Hence α_1, α_2, and α_3 satisfy the following equations:

$$x_1 + x_2 + x_3 = b^2;$$
$$x_1x_2 + x_2x_3 + x_3x_1 = A - 2ab;$$
$$x_1x_2x_3 = a^2 - B.$$

This is a system of three equations in the four variables x_1, x_2, x_3, and a which defines a curve in (say) x_1 and x_2. A (Maple) calculation shows this to be a curve $C_b{}' = C_b{}'(x_1, x_2)$ of degree 4 in both x_1 and x_2 for which an affine equation is given by

$$\begin{aligned} C_b{}' : 0 = &-x_1^4 + (-2x_2 + 2b^2)x_1^3 + (-3x_2^2 - b^4 - 2A)x_1^2 \\ &+ (-2x_2A + 2x_2b^4 + 2b^2A - 2x_2^3)x_1 \\ &- A^2 - 2x_2^2A - x_2^4 + 2x_2b^2A - x_2^2b^4 + 2x_2^3b^2 + 4Bb^2. \end{aligned}$$

For generic values of b, the curve $C_b{}'$ has genus 3 . The set of b for which the genus can be less than three is finite and is a subset of the set of roots of the polynomial (in b)

$$-b^4(27B^2 + 4A^3)(-27A^2 + 18b^4A + 108Bb^2 + b^8)$$

(the ramification locus of $C_b{}'$). The root $b = 0$ gives rise to $C_0{}' : 0 = -(x_1^2 + x_1x_2 + x_2^2 + A)^2$. The roots of $-27A^2 + 18b^4A + 108Bb^2 + b^8$ are distinct for all A and B with $27B^2 + 4A^3 \neq 0$, and correspond to the slopes of the tangents to E at the non-trivial 3-torsion points of E and result in curves $C_b{}'$ of genus 2. It follows that for all $b \neq 0$, $C_b{}'$ has genus at least 2, and so by Faltings' theorem there is only a finite number of K-rational points $(x_1, x_2) \in C_b{}'(K)$. Therefore, for $b \neq 0$, the number of a such that $f_{a,b}(x)$ has three (distinct if $d \neq 0$) roots in K is finite. □

Proposition 3.5. *Fix $b_0 \in \mathbb{Q}$, $b_0 \neq 0$, and let $K/\mathbb{Q}$ be a fixed finite extension. Then there is only a finite number of points $Q = (d, a, b_0) \in S_E^D(\mathbb{Q})$ for which the points in $E(\overline{\mathbb{Q}})$ corresponding to Q (by Proposition 3.1) are defined over K.*

Proof. This follows immediately from Proposition 3.4.

We now prove the following:

Theorem 3.6. *Assume either that* $\mathrm{rank}_{\mathbb{Z}}\, E_{\mathrm{Tr}}(K_0) \geq 1$ *for some* $K_0 \in \mathscr{F}(D)$ *or that both* $r_E(\mathbb{Q}) \geq 1$ *and* $r_{E^D}(\mathbb{Q}) \geq 1$*. If* $D < 0$*, assume also the density hypothesis in the first case (and in the case that* E *has CM, we assume also that* F *is distinct from* $F' = \mathrm{End}(E) \otimes_{\mathbb{Z}} \mathbb{Q}$*). Then* $\mathrm{rank}_{\mathbb{Z}}\, E(K) > \mathrm{rank}_{\mathbb{Z}}\, E(\mathbb{Q})$ *for infinitely many cubic fields* $K \in \mathscr{F}(D)$.

Proof. Let $\psi : S_E^D \longrightarrow \mathbb{A}^1$ denote the map taking $Q = (d, a, b) \in S_E^D$ to $\psi(Q) = b$ with fibres, the curves $X_{E,b}$, generically of genus one. The desingularization $\widehat{S}_E^D$ of S_E^D is obtained by blowing up (twice) the nine rational double points of S_E^D, corresponding to nine 3-torsion points of E. Let J be the associated Jacobian fibration with corresponding map $\psi' : J \longrightarrow \mathbb{A}^1$. Then for generic $b \in \mathbb{Q}$, $\psi'^{-1}(b)$ is an elliptic curve defined over $\mathbb{Q}$ which is isomorphic over $\mathbb{Q}$ to $X_{E,b}$ if $X_{E,b}$ has a $\mathbb{Q}$-rational point. There is a degree-four map defined over $\mathbb{Q}$ (see [F-K-K], Lemma 5.5) from $f : \widehat{S}_E^D \longrightarrow J$ which is a dominant map. By Propositions 3.2 and 3.3, $S_E^D(\mathbb{Q})$ is Zariski dense in S_E^D, so $f(\widehat{S}_E^D(\mathbb{Q}))$ is Zariski dense in J since f is dominant. By Mazur's theorem [Ma] on the boundedness of the torsion over $\mathbb{Q}$, the set of $\mathbb{Q}$-rational points of J which are torsion points on the fibres is a Zariski closed subset of J. Since $f(\widehat{S}_E^D(\mathbb{Q}))$ is Zariski dense in J, there are infinitely many $Q \in \widehat{S}_E^D(\mathbb{Q})$ such that $f(Q)$ is a non-torsion point in infinitely many fibres in J. For such a point Q, $\psi^{-1}(Q)$ is isomorphic over $\mathbb{Q}$ to the elliptic curve $\psi'^{-1}(\mathbb{Q})$, which contains a $\mathbb{Q}$-rational point of infinite order. Thus, $\psi^{-1}(Q)$ also contains infinitely many $\mathbb{Q}$-rational points. It follows that for infinitely many $b \in \mathbb{Q}$, the curves $X_{E,b}$, have an infinite number of rational points. Proposition 3.4 then implies that there is an infinite number of fields $K \in \mathcal{F}(D)$ for which $r_E(K) > r_E(\mathbb{Q})$. □

Remark 3.1. In [F-K-K] we use an alternative argument proving the Zariski density of Propositions 3.2 and 3.3 and we get the same result without appealing to the density hypothesis.

4 Rational three torsion

If E has a 3-torsion point rational over $\mathbb{Q}$, then E has a Weierstrass equation of the form $E : y^2 + 3uxy + ty = x^3$ with $u, v \in \mathbb{Q}$ (see Knapp, p. 146, [Kn]). In this case, the surface S_E^D can be described by the affine equation

$$Dd^2 = \mathrm{discriminant}(x^3 - (a + bx)^2 - 3u(a + bx)x - t(a + bx)).$$

Then $S_E^D = S_E^D(d, a, b)$ has a degenerate fiber over $b = -3u$ which is the rational conic $X_{E,-3u}(d, a)$,

$$Dd^2 = -27(a + t)^2(a^2 + 4u^3 a + 4tu^3),$$

and which becomes

$$Dz^2 + 3w^2 = 12u^6 - 12tu^3$$

in the variables $z = d/3(a+t)$ and $w = a + 2u^3$.

This conic has a rational point if and only if $3(u^4 - ut)$ is a norm from $\mathbb{Q}(\sqrt{-3D})$. If (w_0, z_0) is a solution to this norm equation, we obtain a parametrization given by

$$a = a(r) = w - 2u^3 = (Dr^2 w_0 - 2Drz_0 - 3w_0 - 2u^3 Dr^2 - 6u^3)/(Dr^2 + 3)$$

and

$$d = -3(Dr^2 z_0 + 6w_0 r - 3z_0)(a(r) + t)/(3 + Dr^2).$$

Substituting this value $a = a(r)$ into the original elliptic curve gives the family of cubic equations

$$(a + bx)^2 + 3ux(a + bx) + t(a + bx) = x^3,$$

or

$$x^3 + 3u(t + a)x - a(a + t) = 0,$$

whose root $\alpha = \alpha(r)$ is the x-coordinate of the point $P = (\alpha, a + b\alpha) \in E(\mathbb{Q}(\alpha(r)))$. These expressions provide points of infinite order on the parametrized family of cubic fields $K_r = \mathbb{Q}(\alpha(r)) \in \mathcal{F}(D)$ for rational values of r.

In the case $D = -3$ the norm condition is trivially satisfied and we obtain points of infinite order on a parametrized family of fields K_r of the form $K_r = \mathbb{Q}(m^{\frac{1}{3}})$, where

$$m = \frac{2(r+1)(r-1)^2}{tr - t + 2u^3}$$

and where

$$x = \frac{-2(r - 1 - um^{\frac{1}{3}})}{m^{\frac{2}{3}}}$$

and

$$y = \frac{4u^3 - t(r-1)^2}{r^2 - 1} - 3ux.$$

Having such parametrized families allows us to give lower bounds for the number of $K \in \mathcal{F}(D)$ (counted by increasing discriminant) over which E acquires points. See [F-K-K] §7 for such results for certain elliptic curves $E/\mathbb{Q}$ and for the family $\mathcal{F}(1)$.

5 Theorems of T. Dokchitser and V. Dokchitser

In this section we take note of some work of T. Dokchitser [DoT] and V. Dokchitser [DoV] which tends to support Conjecture 1. In fact, the result of T. Dokchitser provides a proof of Conjecture 1 for $D = -3$.

Let E be an elliptic curve defined over $\mathbb{Q}$ and suppose that $L/\mathbb{Q}$ is a Galois extension. Let ρ be an Artin representation of $\mathrm{Gal}(L/K) = G$, and let ρ^* denote the contragredient representation. V. Dokchitser [DoV] has computed the global root number of $L(E, s, \rho)$ (the twist of the L-function $L(E/\mathbb{Q}, s)$ by ρ) when $\rho \simeq \rho^*$ and no prime of additive reduction for E is bad for ρ.

He shows (Corollary 2, [DoV]) that if ρ is the irreducible 2-dimensional Artin representation associated to $K \in \mathcal{F}(D)$, $D \neq 1$, and if E is an elliptic curve defined over $\mathbb{Q}$ whose conductor N_E is coprime to D, then

$$w_{\rho,E} = \mathrm{sign}(D)\left(\frac{D}{N_E}\right),$$

where $w_{\rho,E}$ is the sign of the functional equation for $L(E/\mathbb{Q}, s, \rho)$ and where $\left(\frac{\cdot}{\cdot}\right)$ is the Jacobi symbol.

This implies that for an elliptic curve $E/\mathbb{Q}$ of conductor N_E, if $D \neq 1$ is coprime to N_E, there are infinitely many $K \in \mathcal{F}(D)$ for which $w_{\rho,E} = -1$, and so the Birch & Swinnerton-Dyer conjecture would then predict that for such fields K we have $r_E(K) > r_E(\mathbb{Q})$.
In particular, he shows that for $E = E_{19A}$ (the elliptic curve over $\mathbb{Q}$ of conductor 19 labelled 19A in Cremona [Cr]), $L(E/\mathbb{Q}, s, \rho_m) = 0$ for *every* m prime to 19, where ρ_m is the irreducible 2-dimensional Artin representation associated to the field $\mathbb{Q}(m^{\frac{1}{3}})$. Since $r_{E_{19}}(\mathbb{Q}) = 0$, the Birch & Swinnerton-Dyer conjecture predicts that $r_{E_{19A}}(\mathbb{Q}(m^{\frac{1}{3}})) \geq 1$ for all integers m.

T. Dokchitser [DoT] proves the following theorem:

Theorem 5.1. *(Dokchitser, T.) For any number field k and elliptic curve E defined over k there is an infinite number of cubic extensions $K = k(\alpha^{\frac{1}{3}})/k$, $\alpha \in k$, such that $r_E(K) > r_E(k)$.*

This provides a proof of Conjecture 1 for $D = -3$.

6 Heegner points

In this section we follow a suggestion of Darmon to use Heegner points to verify Conjecture 1 in certain cases. Fix an elliptic curve $E/\mathbb{Q}$ of conductor N_E and fix a negative fundamental discriminant $D < 0$ with D and N_E relatively prime. Then $F = \mathbb{Q}(\sqrt{D})$ is an imaginary quadratic field, and we denote by $\mathcal{O}_F$ and $\mathcal{O}_c$ the maximal order of F and its suborder of conductor $c, c \in \mathbb{Z}$ respectively. We use the notation of Darmon [Da], and let $\mathrm{sign}(E, F)$ be the sign of the functional equation satisfied by $L(E/F, s)$. Then we use the following ([Da] Theorem 4.18):

Theorem 6.1. *Let E be a semistable elliptic curve of conductor N_E, and F an imaginary quadratic field of discriminant D prime to N_E. If* $\operatorname{sign}(E, F) = -1$, *then there is a non-trivial Heegner system $\{P_n\}$ attached to (E, F).*

We will need the following lemma, which is proved in [K-S], Lemma 2.1.

Let L be a number field, $\mathfrak{p}$ a finite prime of L, $I_{\mathfrak{p}}$ the group of fractional ideals prime to $\mathfrak{p}$, $P_{\mathfrak{p}}$ the group of principal fractional ideals in $I_{\mathfrak{p}}$, $P_{\mathfrak{p},1}$ the group of principal fractional ideals in $P_{\mathfrak{p}}$ generated by elements congruent to 1 mod $\mathfrak{p}$. Then $\mathcal{C}_L = I_{\mathfrak{p}}/P_{\mathfrak{p}}$ is the class group of L, $\mathcal{C}_{L,\mathfrak{p}} = I_{\mathfrak{p}}/P_{\mathfrak{p},1}$ is the ray class group with conductor $\mathfrak{p}$, and $\overline{P}_{\mathfrak{p}} = P_{\mathfrak{p}}/P_{\mathfrak{p},1}$ is the principal ray with conductor $\mathfrak{p}$. We have a short exact sequence

$$1 \longrightarrow \overline{P}_{\mathfrak{p}} \longrightarrow \mathcal{C}_{L,\mathfrak{p}} \longrightarrow \mathcal{C}_L \longrightarrow 1. \tag{1}$$

Let q be a prime, and consider the exact sequence of q-primary components

$$1 \longrightarrow \overline{P}_{\mathfrak{p}}^{(q)} \longrightarrow \mathcal{C}_{L,\mathfrak{p}}^{(q)} \longrightarrow \mathcal{C}_L^{(q)} \longrightarrow 1. \tag{2}$$

We are interested in primes $\mathfrak{p}$ for which the sequence (2) splits. Let $\mathfrak{a}_1, \ldots, \mathfrak{a}_s \in I_{\mathfrak{p}}$ be such that their images $\overline{\mathfrak{a}}_i$ in $\mathcal{C}_L^{(q)}$ form a basis of the finite abelian q-group $\mathcal{C}_L^{(q)}$. Let q^{m_i} be the order of $\overline{\mathfrak{a}}_i$, $i = 1, \ldots, s$. Then $\mathfrak{a}_i^{q^{m_i}} = (a_i) \in P_{\mathfrak{p}}$, $i = 1, \ldots, s$. Let $L_1 = L(\zeta_{q^m}, \sqrt[q^{m_i}]{a_i} | 1 \leq i \leq s)$ with ζ_{q^m} a primitive q^mth root of unity, and $m = \max\{1, m_1, \ldots, m_s\}$.

Lemma 6.2. *(Splitting lemma) In order that the sequence* (2) *split, it is sufficient that $\mathfrak{p}$ split completely in L_1.*

Let E be a curve satisfying the hypotheses of Theorem 6.1, and let $\{P_n\}$ be the (non-trivial) Heegner system attached to (E, F).

Theorem 6.3. *Let E be an elliptic curve of conductor N_E, and F an imaginary quadratic field of discriminant D prime to N_E. If there is a non-trivial Heegner system $\{P_n\}$ attached to (E, F), and if $L'(E/F, 1) \neq 0$, then there is an infinite number of cubic fields $K \in \mathcal{F}(D)$ for which $r_E(K) > r_E(\mathbb{Q})$.*

Proof. Let $\{P_n\}$ be a non-trivial Heegner system attached to (E, F), so $P_n \in E(H_n)$ where H_n is the ring class field over F of conductor n (n prime to N_E.) The Galois group $G_n = \operatorname{Gal}(H_n/F)$ is described by the exact sequence

$$0 \longrightarrow \frac{(\mathcal{O}_F/n\mathcal{O}_F)^*}{(\mathbb{Z}/n\mathbb{Z})^*(\mathcal{O}_F)^*} \longrightarrow G_n \longrightarrow \mathcal{C}_F \longrightarrow 0, \tag{3}$$

where $\mathcal{C}_F$ is the ideal class group of the ring of integers $\mathcal{O}_F$ of F. Taking $q = 3$ and $L = F$, the splitting lemma shows that there is set S of positive density of primes $\ell \in \mathbb{Z}$ satisfying the following properties:

$\ell \equiv 1 \pmod 3$;

the 3-primary part of the exact sequence (3) splits for $n = \ell$;

$\ell \cdot \mathcal{O}_F = \mathfrak{l} \cdot \bar{\mathfrak{l}}$ splits into *principal* ideals in $\mathcal{O}_F$.

In addition, since the set of primes $p \in \mathbb{Z}$ for which $a_p(E) = 2$ has density 0 (Serre [Se]), we may also suppose that the primes $\ell \in S$ have $a_\ell \neq 2$. For such primes ℓ, H_ℓ contains a subfield M_ℓ which is a cyclic cubic extension of F, disjoint from the Hilbert class field H_1, and Galois over $\mathbb{Q}$ with $\mathrm{Gal}(M_\ell/\mathbb{Q}) \simeq S_3$. Then from the properties of Heegner systems (cf. Darmon, Proposition 3.10, [Da]) we have

$$\mathrm{Trace}_{H_\ell/H_1}(P_\ell) = (a_\ell - \sigma_{\mathfrak{l}} - \sigma_{\mathfrak{l}}^{-1})P_1 = (a_\ell - 2)P_1,$$

since $\mathfrak{l}$ is a principal ideal in $\mathcal{O}_F$. Therefore

$$\mathrm{Trace}_{H_\ell/F}(P_\ell) = (a_\ell - 2)\mathrm{Trace}_{H_1/F}(P_1).$$

If we set $Q_\ell = \mathrm{Trace}_{H_\ell/M_\ell}(P_\ell)$, then $\mathrm{Trace}_{M_\ell/F}(Q_\ell) = \mathrm{Trace}_{H_1/F}(P_1)$. Since we assumed that $L'(E/F,1) \neq 0$, the Gross–Zagier theorem implies that $\mathrm{Trace}_{H_1/F}(P_1)$ has infinite order. Hence $Q_\ell \in E(M_\ell)$ is also a point of infinite order. Finally, if we had $r_E(M_\ell) = r_E(F)$, then $\mathrm{Gal}(M_\ell/F)$ would act trivially on $E(M_\ell)$ modulo torsion, and we would have that Q_ℓ is the 3-division of a point of $E(F)$ modulo torsion. Since the torsion in a cubic extension of F is bounded, this would account for only a finite number of fields. But the M_ℓ are distinct for different primes ℓ, so we see that there must be infinitely many fields M_ℓ/F for which $r_E(M_\ell) > r_E(F)$. By Lemma 2.1, we conclude that there is an infinite number of cubic fields $K_\ell \in \mathcal{F}(D)$ for which $r_E(K_\ell) > r_E(\mathbb{Q})$. □

Corollary 6.4. *Let E be a semistable elliptic curve of conductor N_E, and F an imaginary quadratic field of discriminant D prime to N_E. If* $\mathrm{sign}(E,F) = -1$, *and $L'(E/F,1) \neq 0$, then there is an infinite number of cubic fields $K \in \mathcal{F}(D)$ for which $r_E(K) > r_E(\mathbb{Q})$.*

References

[Be] C. Bertolin, *Périods de* 1*-motifs et transcendance*, J. Number Theory no. 2, **97** (2002), 204–221.

[Cr] J. Cremona, *Algorithms for modular elliptic curves*, Cambridge University Press, Cambridge, England, 1992.

[Da] H. Darmon, *Rational Points on Modular Elliptic Curves*, AMS (CBMS Regional Conference Series in Mathematics, Number 101), Providence R.I, 2004.

[DoT] T. Dokchitser, *Ranks of elliptic curves in cubic extensions*, Acta Arithmetica (2007), 357–360.

[Do T2] T. Dokchitser, *Computing special values of motivic L-functions*, Exper. Math, **13** (2004), 137–150.

[DoV] V. Dokchitser, *Root numbers of non-abelian twists of elliptic curves*, Proc. London Math. Soc. (3), **91** (2005), 300–324.

[H-W] G. H. Hardy and E. M. Wright, *An Introduction to the Theory of Numbers*, Clarenden Press, Oxford, 3rd ed., 1954.

[F-K-K] J. Fearnley, H. Kisilevsky, and M. Kuwata, *Vanishing and non-vanishing Dirichlet twists of L-functions of elliptic curves*, submitted, http://arxiv.org/abs/0711.1771.

[K-S] H. Kisilevsky and J. Sonn, *Abelian extensions of global fields with constant local degrees*, Math. Research Letters (4) **13** (2006), 599–605.

[Kn] A. Knapp, *Elliptic Curves*, Mathematical Notes **40**, Princeton University Press, 1992.

[Ma] B. Mazur, *Modular Curves and the Eisenstein Ideal*, Publ. math de l'IHES, **47**(1976), 33–186.

[Se] J.-P. Serre, *Quelques applications du théorème de densité de Chebotarev*, Publ. math. de l'IHES **54** (1981), 323–401.

[Wa1] M. Waldschmidt, *Densité de points rationnels sur un groupe algébrique*, Experiment. Math., **3** no. 4 (1994), 329–352, Errata. ibid., **4** (1995), no. 3, 255.

[Wa2] M. Waldschmidt, *Topologie des points rationnels*, Cours de Troisime Cycle 1994/95, Preprint, Univ. P. et M. Curie, 1994/95.

On effective equidistribution of expanding translates of certain orbits in the space of lattices

D. Y. Kleinbock and G. A. Margulis

In memory of Serge Lang

Abstract We prove an effective version of a result obtained earlier by Kleinbock and Weiss [KW] on equidistribution of expanding translates of orbits of horospherical subgroups in the space of lattices.

Key words homogeneous flows • exponential mixing • equidistribution

Mathematics Subject classification (2010): 37A17; 37A25

1 Introduction

The motivation for this work is a result obtained recently in [KW]. Fix $m, n \in \mathbb{N}$, set $k = m + n$ and let

$$G = \mathrm{SL}_k(\mathbb{R}),\ \Gamma = \mathrm{SL}_k(\mathbb{Z}),\ u_Y = \begin{pmatrix} I_m & Y \\ 0 & I_n \end{pmatrix},\ H = \{u_Y \,|\, Y \in M_{m,n}\}, \quad (1.1)$$

where $M_{m,n}$ stands for the space of $m \times n$ matrices with real entries. Then H is a unipotent abelian subgroup of G which is *expanding horospherical* with respect to

$$g_t = \mathrm{diag}(e^{t/m}, \ldots, e^{t/m}, e^{-t/n}, \ldots, e^{t/n}), \quad t > 0. \quad (1.2)$$

D.Y. Kleinbock (✉)
Department of Mathematics, Brandeis University, Waltham, MA 02454
e-mail: kleinboc@brandeis.edu

G.A. Margulis
Department of Mathematics, Yale University, New Haven, CT 06520
e-mail: margulis@math.yale.edu

D. Goldfeld et al. (eds.), *Number Theory, Analysis and Geometry: In Memory of Serge Lang*, DOI 10.1007/978-1-4614-1260-1_18,

The latter, by definition, means that the Lie algebra of H is the span of eigenspaces of $\mathrm{Ad}(g_t)$, $t > 0$, with eigenvalues bigger than 1 in absolute value.

The space $X \stackrel{\mathrm{def}}{=} G/\Gamma$ can be identified with the space of unimodular lattices in $\mathbb{R}^k$, on which G acts by left translations. Denote by π the natural projection $G \to X$, $g \mapsto g\Gamma$, and for any $z \in X$ let $\pi_z : G \to X$ be defined by $\pi_z(g) = gz$. Also denote by $\bar{\mu}$ the G-invariant probability measure on X and by μ the Haar measure on G such that $\pi_*\mu = \bar{\mu}$. Fix a Haar measure ν on H. Note that the H-orbit foliation is unstable with respect to the action of g_t, $t > 0$. It is well known that for any Borel probability measure ν' on H absolutely continuous with respect to ν and for any $z \in X$, g_t-translates of $(\pi_z)_*\nu'$ become equidistributed, that is, weak-$*$ converge to $\bar{\mu}$ as $t \to \infty$. An effective version of this statement was obtained in [KM1, Proposition 2.4.8]. In order to state that result, it will be convenient to introduce the following notation: for $f \in L^1(H,\nu)$, a bounded continuous function ψ on X, $z \in X$ and $g \in G$ define

$$I_{f,\psi}(g,z) \stackrel{\mathrm{def}}{=} \int_H f(h)\psi(g_t hz)\,d\nu(h).$$

In other words, $I_{f,\psi}(g,z)$ is the result of evaluation of the g-translate of $(\pi_z)_*\nu'$ at ψ, where $d\nu' = f\,d\nu$. Then equidistribution of g_t-translates of $(\pi_z)_*\nu'$ amounts to the convergence of $I_{f,\psi}(g_t,z)$ to $\int_H f \cdot \int_X \psi$ as $t \to \infty$ (unless it causes confusion, we will omit measures in the integration notation for the sake of brevity).

The following is a slightly simplified form of [KM1, Proposition 2.4.8]:

Theorem 1.1. *There exists $\gamma > 0$ such that for any $f \in C^\infty_{\mathrm{comp}}(H)$, $\psi \in C^\infty_{\mathrm{comp}}(X)$ and for any compact subset L of X there exists a constant $C = C(f,\psi,L)$ such that for all $z \in L$ and any $t \geq 0$*

$$\left| I_{f,\psi}(g_t,z) - \int_H f \int_X \psi \right| \leq Ce^{-\gamma t}. \tag{1.3}$$

The proof used the exponential decay of correlations of the G-action on X (called "condition (EM)" in [KM1]). See Section 2 for more detail.

Motivated by some questions in simultaneous Diophantine approximation, the first named author and Barak Weiss considered translates of H-orbits on X by diagonal elements of G other than g_t. Specifically, following [KW], let us denote by $\mathfrak{a}^+$ the set of k-tuples $\mathbf{t} = (t_1,\dots,t_k) \in \mathbb{R}^k$ such that

$$t_1,\dots,t_k > 0 \quad \text{and} \quad \sum_{i=1}^m t_i = \sum_{j=1}^n t_{m+j},$$

and for $\mathbf{t} \in \mathfrak{a}_+$ define

$$g_{\mathbf{t}} \stackrel{\mathrm{def}}{=} \mathrm{diag}\left(e^{t_1},\dots,e^{t_m},e^{-t_{m+1}},\dots,e^{-t_k}\right) \in G \tag{1.4}$$

and

$$\lfloor \mathbf{t} \rfloor \stackrel{\text{def}}{=} \min_{i=1,\dots,k} t_i$$

(the latter, roughly speaking, measures the distance between $\mathbf{t}$ and the walls of the cone $\mathfrak{a}^+ \subset \mathbb{R}^k$).

The theorem below is a reformulation of [KW, Theorem 2.2].

Theorem 1.2. *For any $f \in L^1(H, \nu)$, any continuous compactly supported $\psi : X \to \mathbb{R}$, any compact subset L of X and any $\varepsilon > 0$ there exists $T > 0$ such that*

$$\left| I_{f,\psi}(g_{\mathbf{t}}, z) - \int_H f \int_X \psi \right| < \varepsilon$$

for all $z \in L$ and $\mathbf{t} \in \mathfrak{a}^+$, $\lfloor \mathbf{t} \rfloor \geq T$.

That is, $g_{\mathbf{t}}$-translates of H-orbits become equidistributed as $\lfloor \mathbf{t} \rfloor \to \infty$ uniformly in z when the latter is restricted to compact subsets of X. The proof relies on S. G. Dani's classification of measures invariant under horospherical subgroups and the so-called *linearization method.* The purpose of the present paper is to prove an effective version of the above theorem:

Theorem 1.3. *There exists $\tilde{\gamma} > 0$ such that for any $f \in C^\infty_{\text{comp}}(H)$, $\psi \in C^\infty_{\text{comp}}(X)$ and for any compact $L \subset X$ there exists $\tilde{C} = \tilde{C}(f, \psi, L)$ such that for all $z \in L$ and all $\mathbf{t} \in \mathfrak{a}^+$*

$$\left| I_{f,\psi}(g_{\mathbf{t}}, z) - \int_H f \int_X \psi \right| \leq \tilde{C} e^{-\tilde{\gamma} \lfloor \mathbf{t} \rfloor}.$$

Note that the above statement follows from Theorem 1.1 when $k = 2$, that is, $G = \mathrm{SL}_2(\mathbb{R})$, but is new for $k > 2$. The proof uses the "exponential mixing" approach of [KM1, KM3] together with effective nondivergence estimates obtained in [KM2]. We will describe these two parts Sections 2 and 3 respectively, and then proceed with the proof of Theorem 1.3 in Section 4. We remark that the method of proof readily extends to the set-up more general than (1.1). Note also that, as observed by N. Shah in [S, Remark 1.0.2], Theorem 1.3 can be used to strengthen one of the main results of [KW], that is, [KW, Theorem 1.4], which constitutes a diophantine application of Theorem 1.2.

Acknowledgements The authors are grateful to the Fields Institute for Research in Mathematical Sciences (Toronto, Canada), where this project has commenced, and to the referee for useful remarks. The work of the first named author was supported in part by NSF Grants DMS-0239463 and DMS-0801064, and that of the second author by NSF Grants DMS-0244406 and DMS-0801195.

2 Exponential mixing and g_t-translates

Notation: We will fix a right-invariant metric *dist* on G, giving rise to the corresponding metric on X. $B(x, r)$ will stand for an open ball of radius r centered at x. If a metric space is G or its subgroups, we will abbreviate $B(e, r)$ to $B(r)$. When necessary, we will use subscripts denoting the ambient metric spaces. $\|\cdot\|_\ell$ and $\|\cdot\|_{C^\ell}$ will stand for the $(2,\ell)$-Sobolev and C^l norms respectively. We define

$$W^{2,\infty}(X) = \{\psi \in C^\infty(X) : \|\psi\|_\ell < \infty \ \forall\, \ell \in \mathbb{N}\};$$

clearly $C^\infty_{\text{comp}}(X) \subset W^{2,\infty}(X)$. In fact, $W^{2,\infty}(X)$ coincides with the set of functions $\psi \in C^\infty(X)$ such that $D\psi \in L^2(X)$ for any D from the universal enveloping algebra of Lie(G). We let $\langle\cdot,\cdot\rangle$ stand for the inner product in $L^2(X)$. We also denote by $\|\psi\|_{\text{Lip}}$ the Lipschitz constant of a function ψ on X,

$$\|\psi\|_{\text{Lip}} \stackrel{\text{def}}{=} \sup_{x,y\in X,\ x\neq y} \frac{|\psi(x) - \psi(y)|}{\text{dist}(x, y)},$$

and let $\text{Lip}(X) \stackrel{\text{def}}{=} \{\psi : \|\psi\|_{\text{Lip}} < \infty\}$.

The following property of the G-action on X is deduced in [KM3] from the spectral gap on $L^2(X)$:

Theorem 2.1 (KM3, Corollary 3.5). *There exist $\gamma > 0$ and $\ell \in \mathbb{N}$ such that for any two functions $\varphi, \psi \in W^{2,\infty}(X)$ and for any $t \geq 0$ one has*

$$\left|\langle g\varphi, \psi\rangle - \int_X \varphi \int_X \psi\right| \ll \|\varphi\|_\ell \|\psi\|_\ell \cdot e^{-\gamma\, \text{dist}(g,e)}.$$

Here and hereafter the implicit constants in $\ll$ depend only on the dimensions of the corresponding spaces and the choices of the metric. Taking $g = g_t$ as in (1.2), it follows that

$$\left|\langle g_t\varphi, \psi\rangle - \int_X \varphi \int_X \psi\right| \ll \|\varphi\|_\ell \|\psi\|_\ell \cdot e^{-\gamma t}. \tag{2.1}$$

An estimate analogous to (2.1) was used in [KM1] to derive Theorem 1.1. In this section we apply Theorem 2.1 to prove a statement similar to Theorem 1.1, providing some information as to how C in (1.3) depends on f and L. The argument follows the lines of the proof in [KM1]; in fact, the statement below is basically an intermediate step in the proof of [KM1, Proposition 2.4.8]. However we have decided to include details for the sake of making this paper self-contained.

To pass from $\langle g_t\varphi, \psi\rangle$ to $I_{f,\psi}(g_t, z)$, we need to thicken f into a suitable function φ on X. To explain this process, we need to introduce some more notation. Let

$$H^- = \left\{ \begin{pmatrix} I_m & 0 \\ Y & I_n \end{pmatrix} \middle| \, Y \in M_{n,m} \right\}$$

and

$$H^0 = \left\{ \begin{pmatrix} A & 0 \\ 0 & B \end{pmatrix} \middle| \, A \in \mathrm{GL}_m(\mathbb{R}),\ B \in \mathrm{GL}_n(\mathbb{R}),\ \det(A)\det(B) = 1 \right\}.$$

The product map $H^- \times H^0 \times H \to G$ is a local diffeomorphism; we will choose r_0 so that the inverse of this map is well defined on $B_G(r_0)$. Note that H^- is expanding horospherical with respect to g_{-t}, $t > 0$, while H^0 is centralized by $\{g_t\}$. Thus, the inner automorphism Φ_t of G given by $\Phi_t(g) \stackrel{\mathrm{def}}{=} g_t h(g_t)^{-1}$ is non-expanding on the group

$$\tilde{H} \stackrel{\mathrm{def}}{=} H^- H^0 = \left\{ \begin{pmatrix} A & 0 \\ Y & B \end{pmatrix} \right\};$$

in fact, one has

$$\forall\, r > 0\ \forall\, t > 0 \quad \Phi_t\big(B_{\tilde{H}}(r)\big) \subset B_{\tilde{H}}(r). \tag{2.2}$$

Let us choose Haar measures ν^-, ν^0 on H^-, H^0 respectively, normalized so that μ is locally almost the product of ν^-, ν^0 and ν. By the latter, in view of [B, Ch. VII, §9, Proposition 13], we mean that μ can be expressed via ν^-, ν^0 and ν in the following way: for any $\varphi \in L^1(G)$

$$\int_{H^- H^0 H} \varphi(g)\, d\mu(g) = \int_{H^- \times H^0 \times H} \varphi(h^- h^0 h)\Delta(h^0)\, d\nu^-(h^-)\, d\nu^0(h^0)\, d\nu(h), \tag{2.3}$$

where Δ is the modular function of (the non-unimodular group) $\tilde{H}$.

The "thickening" will be based on the following properties of the Sobolev norm, cf. [KM1, Lemma 2.4.7]:

Lemma 2.2. *(a) For any $r > 0$, there exists a nonnegative function $\theta \in C^\infty_{\mathrm{comp}}(\mathbb{R}^N)$ such that* $\mathrm{supp}(\theta)$ *is inside $B(r)$, $\int_{\mathbb{R}^N} \theta = 1$, and $\|\theta\|_\ell \ll r^{-(\ell+N/2)}$.*

(b) Given $\theta_1 \in C^\infty_{\mathrm{comp}}(\mathbb{R}^N)$, $\theta_2 \in C^\infty_{\mathrm{comp}}(\mathbb{R}^N)$, define $\theta \in C^\infty_{\mathrm{comp}}(\mathbb{R}^N)$ by $\theta(x) = \theta_1(x)\theta_2(x)$. Then $\|\theta\|_\ell \ll \|\theta_1\|_\ell \|\theta_2\|_{C^\ell}$.

(c) Given $\theta_1 \in C^\infty_{\mathrm{comp}}(\mathbb{R}^{N_1})$, $\theta_2 \in C^\infty_{\mathrm{comp}}(\mathbb{R}^{N_2})$, define $\theta \in C^\infty_{\mathrm{comp}}(\mathbb{R}^{N_1+N_2})$ by $\theta(x_1, x_2) = \theta_1(x_1)\theta_2(x_2)$. Then $\|\theta\|_\ell \ll \|\theta_1\|_\ell \|\theta_2\|_\ell$.

We will apply the above lemma to functions supported on small enough balls centered at identity elements in G, H, H^0, H^-.

Theorem 2.3. *Let $f \in C^\infty_{\text{comp}}(H)$, $0 < r < r_0/2$ and $z \in X$ be such that*

(i) $\operatorname{supp} f \subset B_H(r)$, *and*
(ii) π_z *is injective on* $B_G(2r)$.

Then for any $\psi \in W^{2,\infty}(X) \cap \operatorname{Lip}(X)$ with $\int_X \psi = 0$ there exists $E = E(\psi)$ such that for any $t \geq 0$ one has

$$\left|I_{f,\psi}(g_t, z)\right| \leq E\left(r\int_H |f| + r^{-(2\ell+N/2)}\|f\|_\ell e^{-\gamma t}\right), \tag{2.4}$$

where γ and ℓ are as in Theorem 2.1 and $N = m^2 + mn + n^2 - 1 = \dim \tilde{H}$.

Proof. Using Lemma 2.2, one can choose nonnegative functions $\theta^- \in C^\infty_{\text{comp}}(H^-)$, $\theta^0 \in C^\infty_{\text{comp}}(H^0)$ with

$$\int_{H^-} \theta^- = \int_{H^0} \theta^0 = 1 \tag{2.5}$$

such that

$$\operatorname{supp}(\theta^-) \cdot \operatorname{supp}(\theta^0) \subset B_{\tilde{H}}(r), \tag{2.6}$$

and at the same time

$$\|\tilde{\theta}\|_\ell \ll r^{(2\ell+N/2)}, \tag{2.7}$$

where $\tilde{\theta} \in C^\infty_{\text{comp}}(\tilde{H})$ is defined by

$$\tilde{\theta}(h^- h^0) \stackrel{\text{def}}{=} \theta^-(h^-)\theta^0(h^0)\Delta(h^0)^{-1}.$$

Also define $\varphi \in C^\infty_{\text{comp}}(X)$ by $\varphi(h^- h^0 hz) = \tilde{\theta}(h^- h^0) f(h)$; the definition makes sense because of (2.6) and assumptions (i), (ii) of the theorem. Then $I_{f,\psi}(g_t, z)$ can be reasonably well approximated by $\langle g_t\varphi, \psi\rangle = \langle \varphi, g_{-t}\psi\rangle$:

$$\begin{aligned}
&\left|I_{f,\psi}(g_t, z) - \langle \varphi, g_{-t}\psi\rangle\right| \\
&\underset{(2.3)}{=} \left|\int_H f(h)\psi(g_t hx)\, d\nu(h) - \int_G \tilde{\theta}(h^- h^0) f(h)\psi(g_t h^- h^0 hx)\, d\mu(h^- h^0 h)\right| \\
&\underset{(2.5)}{=} \left|\int_G \theta^-(h^-)\theta^0(h^0) f(h)\Big(\psi(g_t hx) - \psi\big(\Phi_t(h^- h^0) g_t hx\big)\Big)\Delta(h^0)^{-1}\, d\mu(h^- h^0 h)\right| \\
&\underset{(2.2),(2.6)}{\leq} \sup_{g \in B_{\tilde{H}}(r),\, y \in X} |\psi(gy) - \psi(y)| \int_G \left|\theta^-(h^-)\theta^0(h^0) f(h)\Delta(h^0)^{-1}\right| d\mu(h^- h^0 h) \\
&\underset{(2.3)}{\leq} \|\psi\|_{\text{Lip}} \cdot r \cdot \int_H |f|.
\end{aligned}$$

On the other hand, in view of Lemma 2.2 and π_z being a local isometry,

$$\|\varphi\|_\ell = \|\tilde{\theta} \cdot f\|_\ell \ll \|\tilde{\theta}\|_\ell \|f\|_\ell \underset{(2.7)}{\ll} r^{-(2\ell+N/2)} \|f\|_\ell,$$

hence by (2.1)

$$|\langle g_t\varphi, \psi\rangle| \ll r^{(2\ell+N/2)} \|f\|_\ell \|\psi\|_\ell e^{-\gamma t},$$

finishing the proof. □

Remark 2.4. In order to derive Theorem 1.1 from Theorem 2.3 it suffices to choose $r = e^{-\beta t}$ for some suitable β. The same trick will help us in the proof of Theorem 1.3. Note that t needs to be taken large enough so that condition (ii) of Theorem 2.3 is satisfied for all $z \in L$. The latter is possible because, in view of the compactness of L and discreteness of Γ in G, the value

$$r(L) \stackrel{\text{def}}{=} \inf_{z\in L} \sup\{r > 0 \mid \pi_z : G \to X \text{ is injective on } B(r)\}$$

is positive; we will call it the *injectivity radius of L*.

Remark 2.5. It is worthwhile to point out that H being the expanding horospherical subgroup relative to g_t, $t > 0$, was crucially important for the proof. When g_t is replaced with $g_{\mathbf{t}}$ where $\mathbf{t}$ is an arbitrary element of $\mathfrak{a}^+$, one can still talk about $\Phi_{\mathbf{t}}$, the inner automorphism of H given by

$$\Phi_{\mathbf{t}}(h) \stackrel{\text{def}}{=} g_{\mathbf{t}}h(g_{\mathbf{t}})^{-1}. \tag{2.8}$$

It is expanding on H, since the latter is contained in the expanding horospherical subgroup relative to $g_{\mathbf{t}}$; however it is not non-expanding on $\tilde{H}$ in the sense of (2.2); thus there is no guarantee that $I_{f,\psi}(g_t, z)$ is close to $\langle \varphi, g_{-t}\psi\rangle$ for φ constructed as in the above proof. We bypass this difficulty by means of an additional step, based on the nondivergence phenomenon, to be described in the next section.

3 Quantitative nondivergence

For any $\varepsilon > 0$ consider

$$K_\varepsilon \stackrel{\text{def}}{=} \pi\left(\left\{g \in G \;\middle|\; \|g\mathbf{v}\| \geq \varepsilon \quad \forall\, \mathbf{v} \in \mathbb{Z}^k \setminus \{0\}\right\}\right).$$

In other words, K_ε consists of lattices in $\mathbb{R}^k$ with no nonzero vector of length less than ε. These sets are compact by virtue of Mahler's Compactness Criterion (see [R, Corollary 10.9] or [BM]). Here $\|\cdot\|$ can be any norm on $\mathbb{R}^k$ which we will from now on take to be the standard Euclidean norm.

It was proved in [KM2], refining previous work on nondivergence of unipotent flows [M, D], that certain polynomial maps from balls in Euclidean spaces to X cannot take values outside of K_ε on a set of big measure. Namely, the following is a special case of [BKM, Theorem 6.2] (see also [KLW, KT, K] for further generalizations):

Theorem 3.1. *For $d \in \mathbb{N}$, let φ be a map $\mathbb{R}^d \to \mathrm{GL}_k(\mathbb{R})$ such that*

(i) all coordinates (matrix elements) of $\varphi(\cdot)$ are affine (degree 1 polynomials), and let a ball $B \subset \mathbb{R}^d$ and $0 < \rho \le 1$ be such that
(ii) for any $j = 1, \dots, k-1$ and any $\mathbf{w} \in \bigwedge^j(\mathbb{Z}^k) \smallsetminus \{0\}$ one has

$$\|\varphi(x)\mathbf{w}\| \ge \rho \quad \text{for some } x \in B.$$

Then for any positive $\varepsilon \le \rho$ one has

$$\lambda\left(\left\{x \in B \mid \pi\big(\varphi(x)\big) \notin K_\varepsilon\right\}\right) \ll \left(\frac{\varepsilon}{\rho}\right)^{1/d(k-1)} \lambda(B). \tag{3.1}$$

Here λ is Lebesgue measure on $\mathbb{R}^d$, and the Euclidean[1] norm $\|\cdot\|$ is naturally extended from $\mathbb{R}^k$ to its exterior powers. We remark that the way assumption (i) is used in the proof is by verifying that all the functions $x \mapsto \|\varphi(x)\mathbf{w}\|$, where $\mathbf{w} \in \bigwedge^j(\mathbb{Z}^k)$, are (C,α)-good on $\mathbb{R}^d$, with some fixed $C = C(d,k) > 0$ and $\alpha = 1/d(k-1)$, the exponent appearing in (3.1). See [KM2] for more detail.

Our plan is to apply Theorem 3.1 with $\varphi : M_{m,n} \to G$ given by

$$\varphi(Y) = g_{\mathbf{t}} u_Y g \tag{3.2}$$

for some $g \in G$ and $\mathbf{t} \in \mathfrak{a}^+$. It is clear that assumption (i) holds. As for (ii), we will need to have uniformity in $\mathbf{t} \in \mathfrak{a}^+$ and in g such that $\pi(g)$ belongs to a compact subset of X. This can be extracted from the next lemma, which is immediate from [KW, Proposition 2.4] applied to the representations of G on $\bigwedge^j(\mathbb{R}^k)$, $j = 1, \dots, k-1$:

Lemma 3.2. *There exists $\alpha > 0$ with the following property. Let B be a ball centered at 0 in $M_{m,n}$. Then one can find $b > 0$ such that for any $j = 1, \dots, k-1$, any $\mathbf{w} \in \bigwedge^j(\mathbb{R}^k)$ and any $\mathbf{t} \in \mathfrak{a}^+$ one has*

$$\sup_{Y \in B} \left\|g_{\mathbf{t}} u_Y \mathbf{w}\right\| \ge b e^{\alpha \lfloor \mathbf{t} \rfloor} \|\mathbf{w}\|.$$

[1] In [KM2] the statement of Theorem 5.2 involved the sup norm instead of the Euclidean one, which resulted in a restriction for ρ to be not greater than $1/k$; thus we chose to refer to [BKM] for the Euclidean norm version.

Corollary 3.3. *Let B be a neighborhood of 0 in $M_{m,n}$ and let $L \subset X$ be compact. Then there exists $b > 0$ such that for any $j = 1, \ldots, k-1$, any $\mathbf{w} \in \bigwedge^j(\mathbb{Z}^k) \smallsetminus \{0\}$, any $g \in \pi^{-1}(L)$ and any $\mathbf{t} \in \mathfrak{a}^+$ one has*

$$\sup_{Y \in B} \left\| g_{\mathbf{t}} u_Y g \mathbf{w} \right\| \geq b e^{\alpha \lfloor \mathbf{t} \rfloor}.$$

Proof. Apply the above lemma with $\mathbf{w}$ replaced by $g\mathbf{w}$; it follows from the compactness of L and discreteness of $\bigwedge^j(\mathbb{Z}^k)$ in $\bigwedge^j(\mathbb{R}^k)$ that

$$\inf\left\{\|g\mathbf{w}\| \mid \pi(g) \in L,\ \mathbf{w} \in \bigwedge^j(\mathbb{Z}^k) \smallsetminus \{0\}, j = 1, \ldots, k-1\right\}$$

is positive. □

Corollary 3.4. *Let $L \subset X$ be compact and let $B \subset H$ be a ball centered at $e \subset H$. Then there exists $T = T(B, L)$ such that for every $0 < \varepsilon < 1$, any $z \in L$ and any $\mathbf{t} \in \mathfrak{a}^+$ with $\lfloor \mathbf{t} \rfloor \geq T$ one has*

$$\nu\left(\{h \in B \mid g_{\mathbf{t}} h z \notin K_\varepsilon\}\right) \ll \varepsilon^{\frac{1}{mn(k-1)}} \nu(B).$$

Proof. Define T by $be^{\alpha T} = 1$, where α is given by Lemma 3.2 and b by Corollary 3.3 applied to $\log(B) \subset M_{m,n}$ and L. (Note that the exponential map from $M_{m,n}$ to H is an isometry.) Take φ as in (3.2) with $g \in \pi^{-1}(L)$. Then, in view of Corollary 3.3, assumption (ii) of Theorem 3.1, with $d = mn$, will be satisfied with $\rho = 1$ as long as $\lfloor \mathbf{t} \rfloor \geq T$. □

We conclude this section by an estimate of the injectivity radius of K_ε, to make it possible to combine the above corollary with Theorem 2.3. Observe that any lattice $\Lambda \in K_\varepsilon$ can be generated by vectors of norm $\ll 1/\varepsilon^{k-1}$; if $g\Lambda = \Lambda$ and $g \neq e$, then for one of those vectors $\mathbf{v}$ one has $\|g\mathbf{v} - \mathbf{v}\| \geq \varepsilon$. This implies that $\mathrm{dist}(e, g) \gg \varepsilon^k$. We arrive at

Proposition 3.5. *There exists positive $c = c(k)$ such that $r(K_\varepsilon) \geq c \cdot \varepsilon^k\ \forall\, \varepsilon > 0$.*

4 Proof of Theorem 1.3

Our goal in this section will be to find $\tilde{\gamma} > 0$ such that for any $f \in C^\infty_{\mathrm{comp}}(H)$, $\psi \in W^{2,\infty}(X) \cap \mathrm{Lip}(X)$ with $\int_X \psi = 0$ and compact $L \subset X$ there exists $\tilde{C} > 0$ such that for all $z \in L$ and all $\mathbf{t} \in \mathfrak{a}^+$ one has

$$\left| I_{f,\psi}(g_{\mathbf{t}}, z) \right| \leq \tilde{C} e^{-\tilde{\gamma} \lfloor \mathbf{t} \rfloor}. \tag{4.1}$$

Then Theorem 1.3 will follow by applying (4.1) with ψ replaced by $\psi - \int_X \psi$. Note also that, by increasing $\tilde{C}$ if needed, it is enough to prove (4.1) for $\mathbf{t}$ with large enough $\lfloor \mathbf{t} \rfloor$.

Given $\mathbf{t} \in \mathfrak{a}^+$, define $t \stackrel{\text{def}}{=} \lfloor \mathbf{t} \rfloor / 2$, and let

$$\mathbf{u} = \mathbf{u}(\mathbf{t}) \stackrel{\text{def}}{=} \mathbf{t} - \left(\frac{t}{m}, \dots, \frac{t}{m}, \frac{t}{n}, \dots, \frac{t}{n} \right). \tag{4.2}$$

Note that $\mathbf{u} \in \mathfrak{a}^+$, $\lfloor \mathbf{u} \rfloor \geq \lfloor \mathbf{t} \rfloor / 2 = t$, and $g_{\mathbf{t}} = g_t g_{\mathbf{u}}$ (here $g_{\mathbf{t}}$ and $g_{\mathbf{u}}$ are defined via (1.4), and g_t is as in (1.2)).

Take a function θ supported on $B_H(r)$ as in Lemma 2.2(a), with $r = e^{-\beta t}$ where β is to be specified later; since $\int_H \theta = 1$ and ν is translation-invariant, one can write

$$\begin{aligned} I_{f,\psi}(g_{\mathbf{t}}, z) &= \int_H f(h)\psi(g_{\mathbf{t}} hz)\, d\nu(h) \int_H \theta(y)\, d\nu(y) \\ &= \int_H \int_H f\big(\Phi_{\mathbf{u}}^{-1}(y)h\big)\theta(y)\psi\big(g_t g_{\mathbf{u}} \Phi_{\mathbf{u}}^{-1}(y)hz\big)\, d\nu(y)\, d\nu(h) \\ &= \int_H \int_H f\big(\Phi_{\mathbf{u}}^{-1}(y)h\big)\theta(y)\psi\big(g_t y g_{\mathbf{u}} hz\big)\, d\nu(y)\, d\nu(h). \end{aligned}$$

Note that $\Phi_{\mathbf{u}}^{-1}$ is a contracting automorphism of H, in fact, one has

$$\operatorname{dist}\big(e, \Phi_{\mathbf{u}}^{-1}(h)\big) \leq e^{-2\lfloor \mathbf{u} \rfloor} \operatorname{dist}(e, h) \leq e^{-2t} \operatorname{dist}(e, h)$$

for any $h \in H$. Choose $B = B(r)$ containing $\operatorname{supp} f$. Then the supports of all functions of the form $h \mapsto f\big(\Phi_{\mathbf{u}}^{-1}(y)h\big)$ are contained in

$$\tilde{B} \stackrel{\text{def}}{=} B\big(r + e^{-(2+\beta)t}\big).$$

By taking t large enough it is safe to assume that

$$e^{-\beta t} < r_0/2, \tag{4.3}$$

$\nu(\tilde{B}) \leq 2\nu(B)$, and also that $t > T \stackrel{\text{def}}{=} T(\tilde{B}, L)$ as in Corollary 3.4. Now define ε by

$$\varepsilon = \left(\frac{2}{c} e^{-\beta t} \right)^{1/k}, \tag{4.4}$$

where c is from Proposition 3.5, and denote

$$A \stackrel{\text{def}}{=} \big\{ h \in \tilde{B} \mid g_{\mathbf{u}} hz \notin K_\varepsilon \big\}.$$

Then for any $\mathbf{u} \in \mathfrak{a}^+$ with $\lfloor \mathbf{u} \rfloor \geq T$ and any $z \in L$ one knows, in view of Corollary 3.4, that

$$\nu(A) \ll \varepsilon^{\frac{1}{mn(k-1)}} \nu(\tilde{B}).$$

Hence the absolute value of

$$\int_A \int_H f\big(\Phi_{\mathbf{u}}^{-1}(y)h\big)\theta(y)\psi\big(g_t y g_{\mathbf{u}} hz\big)\,d\nu(y)\,d\nu(h)$$

is

$$\ll \varepsilon^{\frac{1}{mn(k-1)}}\nu(\tilde{B})\sup|f|\sup|\psi|\int_H \theta \ll \sup|f|\sup|\psi|\nu(B)\cdot e^{-\frac{\beta}{mnk(k-1)}t}.$$

Now let us assume that $h \in \tilde{B} \smallsetminus A$, and apply Theorem 2.3 with $r = e^{-\beta t}$, $g_{\mathbf{u}}hz$ in place of z and

$$f_h(y) \stackrel{\mathrm{def}}{=} f\big(\Phi_{\mathbf{u}}^{-1}(y)h\big)\theta(y)$$

in place of f. Clearly condition (i) follows from (4.3), and, since $g_{\mathbf{u}}hz \in K_\varepsilon$ whenever $h \notin A$, condition (ii) is satisfied in view of Proposition 3.5 and (4.4). Also, because $\Phi_{\mathbf{u}}^{-1}|_H$ is contracting, partial derivatives of f will not increase after precomposition with $\Phi_{\mathbf{u}}^{-1}$, and thus

$$\|f_h\|_\ell \underset{\text{Lemma 2.2(b)}}{\ll} \|f\|_{C^\ell}\|\theta\|_\ell \underset{\text{Lemma 2.2(a)}}{\ll} r^{-(\ell+mn/2)}\|f\|_{C^\ell}. \tag{4.5}$$

This way one gets

$$\begin{aligned}
&\left|\int_{\tilde{B}\smallsetminus A}\int_H f\big(\Phi_{\mathbf{u}}^{-1}(y)h\big)\theta(y)\psi\big(g_t y g_{\mathbf{u}} hz\big)\,d\nu(y)\,d\nu(h)\right| \\
&\le \int_{\tilde{B}\smallsetminus A}\left|I_{f_h,\psi}(g_t, g_{\mathbf{u}}hz)\right|\,d\nu(h) \\
&\underset{(2.4)}{\le} E(\psi)\left(r\int_H |f_h| + r^{-(2\ell+N/2)}\|f_h\|_\ell e^{-\gamma t}\right)\nu(\tilde{B}) \\
&\underset{(4.5)}{\ll} E(\psi)\left(\sup|f|\cdot e^{-\beta t} + \|f\|_{C^\ell}\cdot e^{-(\gamma-(2\ell+N/2)\beta)t}\right)\nu(B).
\end{aligned}$$

Combining the two estimates above, one can conclude that

$$\begin{aligned}
\left|I_{f,\psi}(g_{\mathbf{t}}, z)\right| &\ll C_1 e^{-\frac{\beta}{mnk(k-1)}t} + C_2 e^{-\beta t} + C_3 e^{-(\gamma-(2\ell+N/2)\beta)t} \\
&\le \max(C_1, C_2)e^{-\frac{\beta}{mnk(k-1)}t} + C_3 e^{-(\gamma-(2\ell+N/2)\beta)t},
\end{aligned}$$

where C_i, $i = 1, 2, 3$, depend on f, ψ and L. An elementary computation shows that choosing β equalizing the two exponents above will produce

$$\tilde{\gamma} = \frac{\gamma}{1 + mnk(k-1)(2\ell + N/2)}$$

such that (4.1) will hold with $\tilde{C} \ll \max(C_1, C_2, C_3)$. □

References

[B] N. Bourbaki *Éléments de mathématique*, Livre VI: Intégration, Chapitre 7: Mesure de Haar, Chapitre 8: Convolution et représentations, Hermann, Paris, 1963.

[BKM] V. Bernik, D. Kleinbock, and G. A. Margulis, *Khintchine-type theorems on manifolds: the convergence case for standard and multiplicative versions*, Internat. Math. Res. Notices, (2001), no. 9, 453–486.

[BM] B. Bekka and M. Mayer, *Ergodic theory and topological dynamics of group actions on homogeneous spaces*, Cambridge University Press, Cambridge, 2000.

[D] S. G. Dani, *On orbits of unipotent flows on homogeneous spaces*, II, Ergodic Theory Dynamical Systems **6** (1986), 167–182.

[K] D. Kleinbock, *An extension of quantitative nondivergence and applications to Diophantine exponents*, Trans. Amer. Math. Soc. **360** (2008), 6497–6523.

[KLW] D. Kleinbock, E. Lindenstrauss and B. Weiss, *On fractal measures and Diophantine approximation*, Selecta Math. **10** (2004), no. 4, 479–523.

[KM1] D. Kleinbock and G. A. Margulis, *Bounded orbits of nonquasiunipotent flows on homogeneous spaces*, Amer. Math. Soc. Transl. **171** (1996), 141–172.

[KM2] ———, *Flows on homogeneous spaces and Diophantine approximation on manifolds*, Ann. Math. **148** (1998), 339–360.

[KM3] ———, *Logarithm laws for flows on homogeneous spaces*, Inv. Math. **138** (1999), 451–494.

[KT] D. Kleinbock and G. Tomanov, *Flows on S-arithmetic homogeneous spaces and applications to metric Diophantine approximation*, Comm. Math. Helv. **82** (2007), 519–558.

[KW] D. Kleinbock and B. Weiss, *Dirichlet's theorem on diophantine approximation and homogeneous flows*, J. Mod. Dyn. **2** (2008), 43–62.

[M] G. A. Margulis, *On the action of unipotent group in the space of lattices*, In: Proceedings of the Summer School on group representations, (Budapest 1971), Académiai Kiado, Budapest, 1975, pp. 365–370.

[R] M. S. Raghunathan, *Discrete subgroups of Lie groups*, Springer-Verlag, Berlin and New York, 1972.

[S] N. A. Shah, *Equidistribution of expanding translates of curves and Dirichlet's theorem on diophantine approximation*, Invent. Math. **177**, no. 3, 509–532.

Elliptic Eisenstein series for $\mathrm{PSL}_2(\mathbb{Z})$

Jürg Kramer and Anna-Maria von Pippich

To the memory of Serge Lang

Abstract Let $\Gamma \subset \mathrm{PSL}_2(\mathbb{R})$ be a Fuchsian subgroup of the first kind acting by fractional linear transformations on the upper half-plane $\mathbb{H}$, and let $\Gamma\backslash\mathbb{H}$ be the associated finite volume hyperbolic Riemann surface. Associated to any cusp of $\Gamma\backslash\mathbb{H}$, there is the classically studied non-holomorphic (parabolic) Eisenstein series. In [11], Kudla and Millson studied non-holomorphic (hyperbolic) Eisenstein series associated to any closed geodesic on $\Gamma\backslash\mathbb{H}$. Finally, in [9], Jorgenson and the first named author introduced so-called elliptic Eisenstein series associated to any elliptic fixed point of $\Gamma\backslash\mathbb{H}$. In this article, we study elliptic Eisenstein series for the full modular group $\mathrm{PSL}_2(\mathbb{Z})$. We explicitly compute the Fourier expansion of the elliptic Eisenstein series and derive from this its meromorphic continuation.

Key words Eisenstein series • automorphic functions • Fourier coefficients • meromorphic continuation

Mathematics Subject Classification (2010): 11F03, 11F30, 11M36, 30F35

J. Kramer (✉)
Institut für Mathematik, Humboldt-Universität zu Berlin, D-10099 Berlin, Germany
e-mail: kramer@math.hu-berlin.de

A.-M. von Pippich
Institut für Mathematik, Humboldt-Universität zu Berlin D-10099 Berlin, Germany
e-mail: apippich@math.hu-berlin.de

D. Goldfeld et al. (eds.), *Number Theory, Analysis and Geometry: In Memory of Serge Lang*, DOI 10.1007/978-1-4614-1260-1_19,

1 Introduction

1.1 Background and main results

The theory of Eisenstein series plays a prominent role in the theory of automorphic functions and automorphic forms. Classically, in the theory of holomorphic modular forms, the Eisenstein series of weight $2k$ ($k \in \mathbb{N}, k \geq 2$) for the full modular group $\mathrm{PSL}_2(\mathbb{Z})$ are defined by

$$\mathcal{E}_{2k}(z) := \frac{1}{2} \sum_{\substack{(c,d) \in \mathbb{Z}^2 \\ (c,d)=1}} \frac{1}{(cz+d)^{2k}} \qquad (z = x + iy \in \mathbb{C},\ y > 0).$$

The arithmetic significance of these series is reflected by the fact that their Fourier coefficients are given by certain divisor sums.

More generally, in the theory of automorphic functions for Fuchsian subgroups Γ of the first kind of $\mathrm{PSL}_2(\mathbb{R})$, Eisenstein series are defined by

$$\mathcal{E}(z,s) := \sum_{\gamma \in \Gamma_\infty \backslash \Gamma} \mathrm{Im}(\gamma z)^s \qquad (s \in \mathbb{C},\ \mathrm{Re}(s) > 1);$$

here Γ_∞ denotes the stabilizer of the cusp $i\infty$ in the group Γ. For $s \in \mathbb{C}$ with $\mathrm{Re}(s) > 1$, the Eisenstein series $\mathcal{E}(z,s)$ are C^∞-functions in x, y. For $z \in \mathbb{C}$ with $\mathrm{Im}(z) > 0$, the series $\mathcal{E}(z,s)$ are holomorphic functions in s as long as $\mathrm{Re}(s) > 1$. It can be shown that the Eisenstein series $\mathcal{E}(z,s)$ admit a meromorphic continuation to the whole s-plane. The significance of $\mathcal{E}(z,s)$ relies on the fact that these series are eigenfunctions of the hyperbolic Laplacian Δ_{hyp} for the continuous spectrum. The classical approach to establishing the meromorphic continuation is based on the explicit knowledge of the Fourier expansion of $\mathcal{E}(z,s)$. Other approaches rely on the meromorphic continuation of the resolvent kernel of Δ_{hyp} or Colin de Verdière's method given in [3].

Observing that the series $\mathcal{E}(z,s)$ are associated to the cusp $i\infty$, S. Kudla and J. Millson introduced in [11] so-called hyperbolic Eisenstein series $\mathcal{E}_{\mathrm{hyp}}(z,s)$ associated to geodesics in the upper half-plane $\mathbb{H}$, and proved a partial meromorphic continuation and a Kronecker limit-type formula for these series. Following this point of view, J. Jorgenson and the first author were led to consider so-called elliptic Eisenstein series $\mathcal{E}_{\mathrm{ell}}(z,s)$ associated to elliptic fixed points $z_0 \in \mathbb{H}$ for Γ. In fact, these series were introduced in [9] (see also the unpublished paper [8]) in order to derive optimal sup-norm bounds for cusp forms of weight 2 for the subgroup Γ. An alternative, more elementary proof for these sup-norm bounds avoiding elliptic Eisenstein series is given in [7].

The elliptic Eisenstein series associated to an elliptic fixed point $z_0 \in \mathbb{H}$ for the subgroup Γ is defined by

$$\mathcal{E}_{\mathrm{ell}}(z,s) = \sum_{\gamma \in \Gamma_{z_0} \backslash \Gamma} \sinh\big(\varrho(\sigma_{z_0}^{-1}\gamma z)\big)^{-s} \qquad (z \neq z_0),$$

where Γ_{z_0} denotes the stabilizer of z_0 in Γ, $\sigma_{z_0} \in \mathrm{PSL}_2(\mathbb{R})$ is a scaling matrix for z_0, i.e., $\sigma_{z_0}(i) = z_0$, and $\varrho(z)$ denotes the hyperbolic distance from z to i. In the Ph.D. thesis [13] by the second named author, the meromorphic continuation of the elliptic Eisenstein series $\mathcal{E}_{\mathrm{ell}}(z,s)$ for any Fuchsian subgroup Γ of the first kind to the whole s-plane is proven using a variation of Colin de Verdière's method mentioned above. Moreover, various expansions of the series $\mathcal{E}_{\mathrm{ell}}(z,s)$ are computed and a Kronecker limit type formula is established there.

In this paper we study elliptic Eisenstein series in the special case $\Gamma = \mathrm{PSL}_2(\mathbb{Z})$ and $z_0 = i$. Following the classical approach, the main goal of this paper is to establish the meromorphic continuation of the series $\mathcal{E}_{\mathrm{ell}}(z,s)$ by means of its Fourier expansion thereby complementing work carried out in [13] in the special case $\Gamma = \mathrm{PSL}_2(\mathbb{Z})$. In order to achieve our goal, the Fourier expansion of $\mathcal{E}_{\mathrm{ell}}(z,s)$ has to be explicitly computed and the growth of the Fourier coefficients has to be controlled.

1.2 *Outline of the paper*

The paper is organized as follows. In Section 2, we recall and summarize basic notation and definitions used in this article.

In Section 3, we recall the classical Poincaré series $P_m(z,s)$ and relate them to the more recent Poincaré-type series $V_m(z,s)$ studied in [14]. We review how the meromorphic continuation of $P_m(z,s)$ can be obtained via its spectral expansion. Via the aforementioned relation, we obtain the meromorphic continuation of $V_m(z,s)$ to the whole s-plane.

In Section 4, we define the elliptic Eisenstein series $\mathcal{E}_{\mathrm{ell}}(z,s)$ associated to the elliptic fixed point i of $\mathrm{PSL}_2(\mathbb{Z})$. We show that it is holomorphic for $\mathrm{Re}(s) > 1$ and an automorphic function for $\mathrm{PSL}_2(\mathbb{Z})$. In contrast to the parabolic situation, the elliptic Eisenstein series fails to be an eigenfunction of Δ_{hyp}; instead it satisfies the differential equation

$$\big(\Delta_{\mathrm{hyp}} - s(1-s)\big)\mathcal{E}_{\mathrm{ell}}(z,s) = -s^2\mathcal{E}_{\mathrm{ell}}(z,s+2).$$

In Section 5, we calculate the Fourier coefficients of $\mathcal{E}_{\mathrm{ell}}(z,s)$. In order to simplify the exposition, we restrict our study to the case $z \in \mathbb{H}$ with $\mathrm{Im}(z) > 1$.

In Section 6, we obtain the meromorphic continuation of $\mathcal{E}_{\mathrm{ell}}(z,s)$ via its Fourier expansion. The main task here is to first meromorphically continue the mth Fourier coefficients $a_m(y,s)$ of $\mathcal{E}_{\mathrm{ell}}(z,s)$ and then to achieve suitable bounds for $a_m(y,s)$ with respect to m. The main result is stated in Theorem 6.10.

Acknowledgements We would like to express our thanks to J. Jorgenson for his valuable advice in the course of the write-up of this article. Furthermore, we would like to thank J. Funke, O. Imamoglu, and U. Kühn for helpful discussions.

Both authors acknowledge support from the DFG Graduate School *Berlin Mathematical School* and the DFG Research Training Group *Arithmetic and Geometry.*

2 Preliminaries

2.1 Basic notation

Let $\Gamma := \mathrm{PSL}_2(\mathbb{Z})$ be the modular group acting by fractional linear transformations on the upper half-plane $\mathbb{H} := \{z = x + iy \in \mathbb{C} \,|\, y > 0\}$, i.e., for $\gamma = \left(\begin{smallmatrix} a & b \\ c & d \end{smallmatrix}\right) \in \Gamma$ and $z \in \mathbb{H}$, we have

$$\gamma z := \frac{az + b}{cz + d}.$$

We denote by $\mathcal{F}_\Gamma$ a fundamental domain of Γ in $\mathbb{H}$. By $\Gamma_z := \mathrm{Stab}_\Gamma(z)$ we denote the stabilizer of $z \in \mathbb{H}$ in Γ, and we set

$$\Gamma_\infty := \left\{ \begin{pmatrix} 1 & n \\ 0 & 1 \end{pmatrix} \,\middle|\, n \in \mathbb{Z} \right\}.$$

As usual, we put $e(z) := \exp(2\pi i z)$ and denote by $\zeta(s)$ the Riemann zeta function.

In the rectangular coordinates x, y, the hyperbolic line element ds^2_{hyp}, the hyperbolic volume element μ_{hyp}, and the hyperbolic Laplacian Δ_{hyp} on $\mathbb{H}$ are given by

$$ds^2_{\mathrm{hyp}} = \frac{dx^2 + dy^2}{y^2}, \quad \mu_{\mathrm{hyp}} = \frac{dx\,dy}{y^2}, \quad \Delta_{\mathrm{hyp}} = -y^2\left(\frac{\partial^2}{\partial x^2} + \frac{\partial^2}{\partial y^2}\right).$$

We recall that the hyperbolic volume $\mathrm{vol}_{\mathrm{hyp}}(\mathcal{F}_\Gamma)$ of $\mathcal{F}_\Gamma$ is given by

$$\mathrm{vol}_{\mathrm{hyp}}(\mathcal{F}_\Gamma) = \int_{\mathcal{F}_\Gamma} \mu_{\mathrm{hyp}}(z) = \frac{\pi}{3}.$$

By $d_{\mathrm{hyp}}(z, w)$ we denote the hyperbolic distance from $z \in \mathbb{H}$ to $w \in \mathbb{H}$.

2.2 *Hyperbolic polar coordinates*

For $z = x+iy \in \mathbb{H}$, we define the hyperbolic polar coordinates $\varrho = \varrho(z), \vartheta = \vartheta(z)$ centered at $i \in \mathbb{H}$ by

$$\varrho(z) := d_{\mathrm{hyp}}(i, z), \quad \vartheta(z) := \measuredangle(L, T_z),$$

where L denotes the positive y-axis and T_z is the tangent at the unique geodesic passing through i and z at the point i.

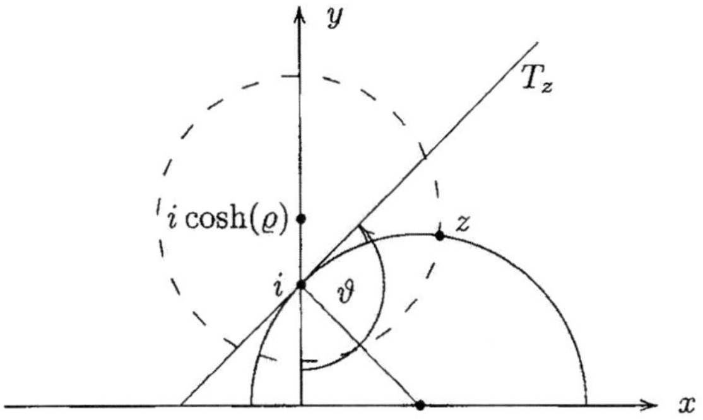

The relation between the x, y-coordinates and the ϱ, ϑ-coordinates is expressed through the formulas

$$x = \frac{\sinh(\varrho)\sin(\vartheta)}{\cosh(\varrho) + \sinh(\varrho)\cos(\vartheta)}, \quad y = \frac{1}{\cosh(\varrho) + \sinh(\varrho)\cos(\vartheta)}. \tag{1}$$

Using the above formulas, the hyperbolic line element and the hyperbolic Laplacian in terms of the hyperbolic polar coordinates take the form

$$ds^2_{\mathrm{hyp}} = \sinh^2(\varrho)d\vartheta^2 + d\varrho^2, \quad \Delta_{\mathrm{hyp}} = -\frac{\partial^2}{\partial\varrho^2} - \frac{1}{\tanh(\varrho)}\frac{\partial}{\partial\varrho} - \frac{1}{\sinh^2(\varrho)}\frac{\partial^2}{\partial\vartheta^2}.$$

From the well-known formula for the hyperbolic distance (see [Bea91], p. 131)

$$\cosh\big(d_{\mathrm{hyp}}(z, w)\big) = 1 + \frac{|z-w|^2}{2\mathrm{Im}(z)\mathrm{Im}(w)},$$

we obtain for $z = x+iy \in \mathbb{H}$ and $\gamma = \left(\begin{smallmatrix} a & b \\ c & d \end{smallmatrix}\right) \in \Gamma$,

$$\cosh\big(\varrho(\gamma z)\big) = \cosh\big(d_{\mathrm{hyp}}(z, \gamma^{-1}i)\big) = \frac{1}{2y}\Big(2y + (a^2+c^2)|z-\gamma^{-1}i|^2\Big).$$

A straightforward computation yields

$$\cosh\big(\varrho(\gamma z)\big) = \frac{1}{2y}\big((a^2+c^2)(x^2+y^2)+2(ab+cd)x+(b^2+d^2)\big). \quad (2)$$

2.3 Hypergeometric functions

For $a, b, c \in \mathbb{C}$, $c \neq -n$ ($n \in \mathbb{N}$), and $w \in \mathbb{C}$, we denote Gauss's hypergeometric function by $F(a,b;c;w)$. For $w \in \mathbb{C}$ with $|w| < 1$ it is defined by the series

$$F(a,b;c;w) := \sum_{k=0}^{\infty} \frac{(a)_k \cdot (b)_k}{(c)_k \cdot k!} \cdot w^k,$$

where $(\lambda)_k := \Gamma(\lambda+k)/\Gamma(\lambda)$ ($\lambda \in \mathbb{C}$, $k \in \mathbb{N}$) is the Pochhammer symbol; for $k \in \mathbb{N}$ with $k > 0$, we note the alternative formula $(\lambda)_k = \prod_{j=0}^{k-1}(\lambda+j)$. For $\mathrm{Re}(c) > \mathrm{Re}(b) > 0$, the hypergeometric function $F(a,b;c;w)$ has the integral representation (see [1], formula 15.3.1)

$$F(a,b;c;w) = \frac{\Gamma(c)}{\Gamma(b)\Gamma(c-b)} \int_0^1 t^{b-1}(1-t)^{c-b-1}(1-tw)^{-a}\, dt. \quad (3)$$

2.4 Parabolic Eisenstein series

For $z \in \mathbb{H}$ and $s \in \mathbb{C}$, the parabolic Eisenstein series $\mathcal{E}_{\mathrm{par}}$ is given by

$$\mathcal{E}_{\mathrm{par}}(z,s) := \sum_{\gamma \in \Gamma_\infty \backslash \Gamma} \mathrm{Im}(\gamma z)^s.$$

The parabolic Eisenstein series is known to be holomorphic for $s \in \mathbb{C}$ with $\mathrm{Re}(s) > 1$ with Fourier expansion given by

$$\mathcal{E}_{\mathrm{par}}(z,s) = y^s + \varphi(s)y^{1-s} + \sum_{n\neq 0} \varphi(n,s) y^{1/2} K_{s-1/2}(2\pi|n|y)e(nx), \quad (4)$$

where $K_{s-1/2}(\cdot)$ is the modified Bessel function of the second kind,

$$\varphi(s) = \frac{\sqrt{\pi}\,\Gamma(s-1/2)}{\Gamma(s)} \cdot \frac{\zeta(2s-1)}{\zeta(2s)} = \frac{\Lambda(2s-1)}{\Lambda(2s)},$$

and

$$\varphi(n,s) = \frac{2\pi^s |n|^{s-1/2}}{\Gamma(s)\zeta(2s)} \sum_{d|n} d^{-2s+1} = \frac{2}{\Lambda(2s)} \sum_{ab=|n|} \left(\frac{a}{b}\right)^{s-1/2};$$

here we set $\Lambda(s) := \pi^{-s/2}\Gamma(s/2)\zeta(s)$. The Fourier expansion (4) provides the meromorphic continuation of $\mathcal{E}_{\mathrm{par}}(z,s)$ to the whole s-plane with a simple pole at $s = 1$ with residue $\mathrm{res}_{s=1}\mathcal{E}_{\mathrm{par}}(z,s) = 1/\mathrm{vol}_{\mathrm{hyp}}(\mathcal{F}_\Gamma) = 3/\pi$, and other poles contributed by the non-trivial zeros of $\zeta(2s)$ in the strip $0 < \mathrm{Re}(s) < 1/2$. From the functional equation $\Lambda(s) = \Lambda(1-s)$, we get $\varphi(s)\varphi(1-s) = 1$, and hence the relation

$$\begin{aligned}\varphi(s)\varphi(n,1-s) &= \frac{2\,\Lambda(2s-1)}{\Lambda(2s)\Lambda(-2s+2)} \sum_{ab=|n|} \left(\frac{b}{a}\right)^{s-1/2} \\ &= \frac{2}{\Lambda(2s)} \sum_{ab=|n|} \left(\frac{a}{b}\right)^{s-1/2} = \varphi(n,s),\end{aligned} \tag{5}$$

which, using (4), proves the functional equation

$$\mathcal{E}_{\mathrm{par}}(z,s) = \varphi(s)\,\mathcal{E}_{\mathrm{par}}(z,1-s). \tag{6}$$

3 Poincaré series

In this section we recall results for two types of Poincaré series that are mostly known to the experts. However, for lack of complete reference, some proofs have to be elaborated.

Definition 3.1. For $z \in \mathbb{H}$, $s \in \mathbb{C}$, and $m \in \mathbb{Z}$, the *Poincaré series* P_m is defined by

$$P_m(z,s) := \sum_{\gamma\in\Gamma_\infty\backslash\Gamma} \mathrm{Im}(\gamma z)^s \exp\bigl(-2\pi|m|\mathrm{Im}(\gamma z)\bigr)e\bigl(m\mathrm{Re}(\gamma z)\bigr).$$

The Poincaré series is known to be holomorphic for $s \in \mathbb{C}$ with $\mathrm{Re}(s) > 1$, since it can be majorized by $P_0(z,\mathrm{Re}(s)) = \mathcal{E}_{\mathrm{par}}(z,\mathrm{Re}(s))$.

Remark 3.2. For $m \neq 0$, the Poincaré series $P_m(z,s)$ is bounded on $\mathbb{H}$ (see [10], p. 83) and hence admits a spectral expansion in terms of the eigenfunctions ψ_j associated to the discrete eigenvalues λ_j of Δ_{hyp} and the parabolic Eisenstein series $\mathcal{E}_{\mathrm{par}}$, namely

$$P_m(z,s) = \sum_{j=0}^{\infty} a_{j,m}(s)\,\psi_j(z) + \frac{1}{4\pi}\int_{-\infty}^{\infty} a_{1/2+ir,m}(s)\,\mathcal{E}_{\mathrm{par}}(z,1/2+ir)\,dr, \tag{7}$$

where the coefficients $a_{j,m}(s)$, resp. $a_{1/2+ir,m}(s)$, are given by

$$a_{j,m}(s) = \int_{\mathcal{F}_\Gamma} P_m(z,s)\overline{\psi}_j(z)\mu_{\mathrm{hyp}}(z), \text{ resp.}$$

$$a_{1/2+ir,m}(s) = \int_{\mathcal{F}_\Gamma} P_m(z,s)\overline{\mathcal{E}}_{\mathrm{par}}(z,1/2+ir)\mu_{\mathrm{hyp}}(z).$$

The expansion (7) is absolutely and locally uniformly convergent for $s \in \mathbb{C}$ with $\mathrm{Re}(s) > 1$.

As ususal, we enumerate the eigenvalues of the discrete spectrum by $0 = \lambda_0 < \lambda_1 \leq \lambda_2 \leq \ldots$; since $\Gamma = \mathrm{PSL}_2(\mathbb{Z})$, we have $\lambda_j = 1/4 + t_j^2 = s_j(1-s_j)$, i.e., $s_j = 1/2 + it_j$ with $t_j > 0$, as long as $j > 0$. For $j = 0$, the eigenfunction is given by $\psi_0(z) = \sqrt{3/\pi}$. For $j > 0$, the eigenfunction ψ_j is a cusp form and admits a Fourier expansion of the form

$$\psi_j(z) = \sum_{n\neq 0} \rho_j(n) y^{1/2} K_{s_j-1/2}(2\pi|n|y)e(nx). \tag{8}$$

The eigenvalues of the continuous spectrum are of the form $\lambda = 1/4 + r^2 = s(1-s)$, i.e., $s = 1/2 + ir$ with $r \in \mathbb{R}$. The corresponding eigenfunctions are given by the parabolic Eisenstein series $\mathcal{E}_{\mathrm{par}}(z, 1/2+ir)$.

Proposition 3.3. *For $z \in \mathbb{H}$, $s \in \mathbb{C}$ with $\mathrm{Re}(s) > 1$, and $m \neq 0$, the Poincaré series $P_m(z,s)$ has the following explicit spectral expansion:*

$$P_m(z,s)\, 2^{2s-1}\pi^{s-1}\Gamma(s)|m|^{s-1/2} = \sum_{j=1}^{\infty} \Gamma(s-s_j)\Gamma(s+s_j-1)\overline{\rho}_j(m)\,\psi_j(z)$$
$$+ \frac{1}{4\pi}\int_{-\infty}^{\infty} \Gamma(s-1/2-ir)\Gamma(s-1/2+ir)\overline{\varphi}(m, 1/2+ir)\, \mathcal{E}_{\mathrm{par}}(z,1/2+ir)\, dr. \tag{9}$$

Proof. The proof can easily be deduced from the spectral expansion given for the function $\widetilde{P}_m(z,s) = \pi^{s-1/2}\Gamma(s+1/2)^{-1}|m|^{s-1/2}P_m(z,s)$ in [12], p. 58. □

Proposition 3.4. *For $z \in \mathbb{H}$ and $m \neq 0$, the Poincaré series $P_m(z,s)$ admits a meromorphic continuation to the whole s-plane with simple poles at $s = s_j - N$ and $s = -s_j - N + 1$ ($N \in \mathbb{N}$) with residues*

$$\mathrm{res}_{s=s_j-N} P_m(z,s)$$
$$= \frac{(-1)^N 2^{-2s_j+2N+1}\pi^{-s_j+N+1}\Gamma(2s_j-N-1)}{N!\,\Gamma(s_j-N)|m|^{s_j-N-1/2}} \sum_{s_\ell = s_j} \overline{\rho}_\ell(m)\,\psi_\ell(z)$$

and

$$\begin{aligned}&\operatorname{res}_{s=-s_j-N+1} P_m(z,s)\\&\quad=\frac{(-1)^N 2^{2s_j+2N-1}\pi^{s_j+N}\Gamma(-2s_j-N+1)}{N!\,\Gamma(-s_j-N+1)|m|^{-s_j-N+1/2}}\sum_{s_\ell=s_j}\overline{\rho}_\ell(m)\,\psi_\ell(z),\end{aligned}$$

respectively.

Proof. Due to the lack of reference for the claimed residues, we have to discuss the proof briefly. In order to obtain the desired meromorphic continuation we will follow closely [12] and [10], and base our argument on the spectral expansion (9).

We start by discussing the meromorphic continuation of the discrete part

$$D(s) := \sum_{j=1}^{\infty}\Gamma(s-s_j)\Gamma(s+s_j-1)\overline{\rho}_j(m)\,\psi_j(z)$$

of the spectral expansion (9). The argument given in [10], p. 87, shows that $D(s)$ has a meromorphic continuation to the whole s-plane with simple poles at $s=s_j-N$ and $s=-s_j-N+1$ ($N\in\mathbb{N}$) arising from the Γ-factors. For later purposes, we note the bound (see [10], p. 87, adapted to the present situation)

$$\left|D(s)\right| \ll y^{-3/2}, \tag{10}$$

where the implied constant depends only on s (not a pole), but is independent of z and m. The dependence of the implied constant on s is uniform as long as s is contained in a compact set not containing s_j-N or $-s_j-N+1$ for some $N\in\mathbb{N}$. For the residues we compute

$$\operatorname{res}_{s=s_j-N} D(s) = \frac{(-1)^N}{N!}\Gamma(2s_j-N-1)\sum_{s_\ell=s_j}\overline{\rho}_\ell(m)\,\psi_\ell(z)$$

and

$$\operatorname{res}_{s=-s_j-N+1} D(s) = \frac{(-1)^N}{N!}\Gamma(-2s_j-N+1)\sum_{s_\ell=s_j}\overline{\rho}_\ell(m)\,\psi_\ell(z),$$

respectively.

We now turn to the meromorphic continuation of the continuous part

$$Q(s) := \frac{1}{4\pi}\int_{-\infty}^{\infty}\Gamma(s-1/2-ir)\Gamma(s-1/2+ir)\overline{\varphi}(m,1/2+ir)\,\mathcal{E}_{\mathrm{par}}(z,1/2+ir)\,dr \tag{11}$$

of the spectral expansion (9). By substituting $t := 1/2 + ir$, the integral (11) can be rewritten as

$$Q(s) = \frac{1}{4\pi i} \int_{1/2-i\infty}^{1/2+i\infty} \Gamma(s-t)\Gamma(s-1+t)\varphi(m,1-t)\, \mathcal{E}_{\mathrm{par}}(z,t)\, dt. \tag{12}$$

By construction, the integral (12) exists for $s \in \mathbb{C}$ with $\mathrm{Re}(s) > 1$ and represents a holomorphic function in this range. The argument given in [10], p. 89, shows that $Q(s)$ extends to a holomorphic function for $s \in \mathbb{C}$ with $\mathrm{Re}(s) \neq -N + 1/2$, and for $s = -N + 1/2$, where $N \in \mathbb{N}$. In order to extend $Q(s)$ to the whole s-plane, we rewrite the integral (12) by means of a different path of integration (see [12], p. 51) using the residue theorem as follows. Let $s_0 \in \mathbb{C}$ with $\mathrm{Re}(s_0) = -N + 1/2$ for some $N \in \mathbb{N}$ and $\mathrm{Im}(s_0) > 0$, and let $C(s_0)$ denote the integration path, which runs on the vertical line with $\mathrm{Re}(t) = 1/2$ from $-\infty$ to ∞ as before, but passes on the left-hand side around $-s_0 - N + 1$ and on the right-hand side around $s_0 + N$ in such a way that the only poles of the integrand being encircled by this new integration path are located at $t = -s_0 - N + 1$ and $t = s_0 + N$. For s with $\mathrm{Re}(s) > -N + 1/2$ being sufficiently close to s_0 such that $-s - N + 1$ and $s + N$ are still encircled by the path $C(s_0)$, we set

$$\widetilde{Q}(s) = \frac{1}{4\pi i} \int_{C(s_0)} \Gamma(s-t)\Gamma(s-1+t)\varphi(m,1-t)\, \mathcal{E}_{\mathrm{par}}(z,t)\, dt,$$

which is well defined by construction. Using the residue theorem and recalling (6) and (5), we then compute

$$\begin{aligned} Q(s) =& \frac{1}{4\pi i} \int_{C(s_0)} \Gamma(s-t)\Gamma(s-1+t)\varphi(m,1-t)\, \mathcal{E}_{\mathrm{par}}(z,t)\, dt \\ &- \frac{(-1)^N}{2\,N!} \Gamma(2s+N-1)\varphi(m,-s-N+1)\, \mathcal{E}_{\mathrm{par}}(z,s+N) \\ &+ \frac{(-1)^N}{2\,N!} \Gamma(2s+N-1)\varphi(m,s+N)\, \mathcal{E}_{\mathrm{par}}(z,-s-N+1) = \widetilde{Q}(s). \end{aligned}$$

For s with $\mathrm{Re}(s) < -N + 1/2$ being sufficiently close to s_0, we define $\widetilde{Q}(s)$ as above and verify again $Q(s) = \widetilde{Q}(s)$, now using Cauchy's theorem. By the choice of the integration path $C(s_0)$ it turns out that the integral

$$\widetilde{Q}(s_0) = \frac{1}{4\pi i} \int_{C(s_0)} \Gamma(s_0-t)\Gamma(s_0-1+t)\varphi(m,1-t)\, \mathcal{E}_{\mathrm{par}}(z,t)\, dt$$

is also well defined. Proceeding in an analogous way for $s_0 \in \mathbb{C}$ with $\mathrm{Re}(s_0) = -N + 1/2$ for some $N \in \mathbb{N}$, but $\mathrm{Im}(s_0) < 0$, we obtain the analytic continuation of $Q(s)$ to the whole s-plane.

All in all, these considerations show that $P_m(z,s)$ admits a meromorphic continuation to the whole s-plane with simple poles at $s = s_j - N$ and $s = -s_j - N + 1$ ($N \in \mathbb{N}$). The stated formulas for the residues are easily obtained from the residue computations for $D(s)$, taking into account that the factor $2^{-2s+1}\pi^{-s+1}\Gamma(s)^{-1}|m|^{-s+1/2}$ does not contribute further poles.

Before finishing the proof, we recall for later purposes that for $s \in \mathbb{C}$, we have the bound (see [10], p. 90)

$$\big|Q(s)\big| \ll y^{1/2}, \tag{13}$$

where the implied constant depends only on s (not a pole), but is independent of z and m. □

Definition 3.5. For $z \in \mathbb{H}$, $s \in \mathbb{C}$, and $m \in \mathbb{Z}$, the *Poincaré series* V_m is defined by

$$V_m(z,s) := \sum_{\gamma \in \Gamma_\infty \backslash \Gamma} \mathrm{Im}(\gamma z)^s e\big(m\mathrm{Re}(\gamma z)\big). \tag{14}$$

The Poincaré series is known to be holomorphic for $s \in \mathbb{C}$ with $\mathrm{Re}(s) > 1$, since it can be majorized by $V_0(z, \mathrm{Re}(s)) = \mathcal{E}_{\mathrm{par}}(z, \mathrm{Re}(s))$.

Lemma 3.6. *For $z \in \mathbb{H}$, $s \in \mathbb{C}$ with $\mathrm{Re}(s) > 1$, and $m \neq 0$, we have the relation*

$$V_m(z,s) = \sum_{k=0}^{\infty} \frac{(2\pi|m|)^k}{k!} P_m(z, s+k).$$

Proof. We first check the absolute and local uniform convergence of the series in the claimed relation for fixed $z \in \mathbb{H}$ and $s \in \mathbb{C}$ with $\mathrm{Re}(s) > 1$. Using the estimate

$$|cz + d| \geq C|ci + d|,$$

where $C = C(z)$ is a positive constant depending on z but which is independent of $(c,d) \in \mathbb{R}^2$, we obtain the bound

$$\sum_{k=0}^{\infty} \left| \frac{(2\pi|m|)^k}{k!} P_m(z, s+k) \right| \leq \sum_{k=0}^{\infty} \frac{(2\pi|m|)^k}{k!} \sum_{\gamma \in \Gamma_\infty \backslash \Gamma} \mathrm{Im}(\gamma z)^{\mathrm{Re}(s)+k}$$
$$= \sum_{k=0}^{\infty} \frac{(2\pi|m|)^k}{k!} \sum_{\gamma \in \Gamma_\infty \backslash \Gamma} \frac{y^{\mathrm{Re}(s)}}{|cz+d|^{2\mathrm{Re}(s)}} \cdot \frac{y^k}{|cz+d|^{2k}}$$
$$\leq \sum_{k=0}^{\infty} \frac{(2\pi|m|yC^{-2})^k}{k!} \sum_{\gamma \in \Gamma_\infty \backslash \Gamma} \frac{y^{\mathrm{Re}(s)}}{|cz+d|^{2\mathrm{Re}(s)}} \cdot \frac{1}{(c^2+d^2)^k}$$
$$\leq \exp(2\pi|m|yC^{-2}) \cdot \mathcal{E}_{\mathrm{par}}\big(z, \mathrm{Re}(s)\big).$$

This proves that the series in question converges absolutely and locally uniformly for $s \in \mathbb{C}$ with $\mathrm{Re}(s) > 1$.

Now the claimed relation can easily be derived by changing the order of summation; namely, we compute

$$\sum_{k=0}^{\infty} \frac{(2\pi|m|)^k}{k!} P_m(z, s+k)$$

$$= \sum_{\gamma \in \Gamma_\infty \backslash \Gamma} \mathrm{Im}(\gamma z)^s \exp\big(-2\pi|m|\mathrm{Im}(\gamma z)\big) e\big(m\mathrm{Re}(\gamma z)\big) \sum_{k=0}^{\infty} \frac{(2\pi|m|)^k}{k!} \mathrm{Im}(\gamma z)^k$$

$$= \sum_{\gamma \in \Gamma_\infty \backslash \Gamma} \mathrm{Im}(\gamma z)^s \exp\big(-2\pi|m|\mathrm{Im}(\gamma z)\big) e\big(m\mathrm{Re}(\gamma z)\big) \exp\big(2\pi|m|\mathrm{Im}(\gamma z)\big)$$

$$= V_m(z, s).$$

This completes the proof of the lemma. □

Proposition 3.7. *For $z \in \mathbb{H}$ and $m \neq 0$, the Poincaré series $V_m(z, s)$ admits a meromorphic continuation to the whole s-plane with simple poles at $s = s_j - 2N$ and $s = -s_j - 2N + 1$ ($N \in \mathbb{N}$) with residues*

$$\mathrm{res}_{s=s_j-2N} V_m(z, s) = \frac{2^{2N-1} \pi^{-s_j+2N+1} \Gamma(s_j - N - 1/2)}{(2N)!\, \Gamma(-N + 1/2) |m|^{s_j - 2N - 1/2}} \sum_{s_\ell = s_j} \overline{\rho}_\ell(m)\, \psi_\ell(z) \quad (15)$$

and

$$\mathrm{res}_{s=-s_j-2N+1} V_m(z, s) = \frac{2^{2N-1} \pi^{s_j+2N} \Gamma(-s_j - N + 1/2)}{(2N)!\, \Gamma(-N + 1/2) |m|^{-s_j - 2N + 1/2}} \sum_{s_\ell = s_j} \overline{\rho}_\ell(m)\, \psi_\ell(z), \quad (16)$$

respectively.

Proof. We start by proving that the Poincaré series $V_m(z, s)$ has a meromorphic continuation to the half-plane

$$\mathcal{H}'_N := \{ s \in \mathbb{C} \,|\, \mathrm{Re}(s) > -N \}$$

for any $N \in \mathbb{N}$. By Lemma 3.6, we can write

$$V_m(z, s) = \sum_{k=0}^{N} \frac{(2\pi|m|)^k}{k!} P_m(z, s+k) + \sum_{k=N+1}^{\infty} \frac{(2\pi|m|)^k}{k!} P_m(z, s+k). \quad (17)$$

We show that the series

$$\sum_{k=N+1}^{\infty} \frac{(2\pi|m|)^k}{k!} P_m(z, s+k)$$

is a holomorphic function on the half-plane $\mathcal{H}'_N$. For this we estimate as in the proof of Lemma 3.6, assuming $s \in \mathbb{C}$ with $\mathrm{Re}(s) > -N$,

$$\begin{aligned}
&\sum_{k=N+1}^{\infty} \left| \frac{(2\pi|m|)^k}{k!} P_m(z, s+k) \right| \\
&= \sum_{k=N+1}^{\infty} \left| \frac{(2\pi|m|)^{N+1}}{k!/(k-N-1)!} \cdot \frac{(2\pi|m|)^{k-N-1}}{(k-N-1)!} P_m\big(z, (s+N+1) + (k-N-1)\big) \right| \\
&\leq (2\pi|m|)^{N+1} \sum_{k=0}^{\infty} \left| \frac{(2\pi|m|)^k}{k!} P_m(z, s+N+1+k) \right| \\
&\leq (2\pi|m|)^{N+1} \cdot \exp(2\pi|m|yC^{-2}) \cdot \mathcal{E}_{\mathrm{par}}\big(z, \mathrm{Re}(s)+N+1\big).
\end{aligned}$$

This proves that the series in question converges absolutely and locally uniformly for $s \in \mathbb{C}$ with $\mathrm{Re}(s) > -N$, and hence the holomorphicity statement.

Since the finite sum $\sum_{k=0}^{N} (2\pi|m|)^k/k!\, P_m(z, s+k)$ is a meromorphic function on the whole s-plane by Proposition 3.4, we conclude that $V_m(z, s)$ has a meromorphic continuation to the half-plane $\mathcal{H}'_N$. Since N was chosen arbitrarily, this proves the meromorphic continuation of $V_m(z, s)$ to the whole s-plane.

In order to determine the poles of $V_m(z, s)$, we calculate its poles in the strip

$$\mathcal{S}'_N := \{s \in \mathbb{C} \mid -N < \mathrm{Re}(s) \leq -N+1\}$$

for any $N \in \mathbb{N}$. By considering $V_m(z, s)$ with its decomposition (17) in the strip $\mathcal{S}'_N$, we see that the poles come from the finite sum $F_N(z, s) := \sum_{k=0}^{N} (2\pi|m|)^k/k!\, P_m(z, s+k)$. By Proposition 3.4, $F_N(z, s)$ has poles in the strip $\mathcal{S}'_N$ at $s = s_j - N$ and $s = -s_j - N + 1$. The explicit formula for the residues of $P_m(z, s)$ given in Proposition 3.4 now leads to the following residue of $F_N(z, s)$ at $s = s_j - N$:

$$\begin{aligned}
\mathrm{res}_{s=s_j-N} F_N(z, s) &= \sum_{k=0}^{N} \frac{(2\pi|m|)^k}{k!} \mathrm{res}_{s=s_j-(N-k)} P_m(z, s) \\
&= \sum_{k=0}^{N} \frac{(2\pi|m|)^k}{k!} \frac{(-1)^{N-k} 2^{-2s_j+2(N-k)+1} \pi^{-s_j+(N-k)+1} \Gamma(2s_j-(N-k)-1)}{(N-k)!\,\Gamma(s_j-(N-k))|m|^{s_j-(N-k)-1/2}} \\
&\quad \times \sum_{s_\ell = s_j} \overline{\rho}_\ell(m)\, \psi_\ell(z)
\end{aligned}$$

$$= \sum_{k=0}^{N} \frac{(-1)^{N-k} 2^{-2s_j+2N-k+1} \pi^{-s_j+N+1} \Gamma(2s_j - N + k - 1)}{k!\,(N-k)!\,\Gamma(s_j - N + k) |m|^{s_j - N - 1/2}} \sum_{s_\ell = s_j} \overline{\rho}_\ell(m)\, \psi_\ell(z)$$

$$= \frac{(-1)^N\, 2^{N-1} \pi^{-s_j+N+1} \Gamma(s_j - N/2 - 1/2)}{N!\,\Gamma(-N/2 + 1/2) |m|^{s_j - N - 1/2}} \sum_{s_\ell = s_j} \overline{\rho}_\ell(m) \psi_\ell(z).$$

This shows that the residue in question vanishes if N is odd, and that the residue of $V_m(z,s)$ at $s = s_j - 2N$ is given by (15). Analogously, it is shown that the residue of $F_N(z,s)$ at $s = -s_j - N + 1$ is zero if N is odd, and that the residue of $V_m(z,s)$ at $s = -s_j - 2N + 1$ is given by (16). □

4 Elliptic Eisenstein series

Definition 4.1. For $z \in \mathbb{H}$ with $z \neq \gamma i$ for any $\gamma \in \Gamma$, and $s \in \mathbb{C}$, the *elliptic Eisenstein series* $\mathcal{E}_{\mathrm{ell}}$ is defined by

$$\mathcal{E}_{\mathrm{ell}}(z,s) = \sum_{\gamma \in \Gamma_i \backslash \Gamma} \sinh\big(\varrho(\gamma z)\big)^{-s}.$$

Lemma 4.2. *(i) For $z \in \mathbb{H}$ with $z \neq \gamma i$ for any $\gamma \in \Gamma$, the elliptic Eisenstein series $\mathcal{E}_{\mathrm{ell}}(z,s)$ converges absolutely and locally uniformly for $s \in \mathbb{C}$ with* $\mathrm{Re}(s) > 1$, *and hence defines a holomorphic function.*

(ii) The elliptic Eisenstein series $\mathcal{E}_{\mathrm{ell}}(z,s)$ is invariant under the action of Γ, i.e., we have $\mathcal{E}_{\mathrm{ell}}(\gamma z,s) = \mathcal{E}_{\mathrm{ell}}(z,s)$ for any $\gamma \in \Gamma$.

(iii) For fixed $s \in \mathbb{C}$ with $\mathrm{Re}(s) > 1$, *the elliptic Eisenstein series $\mathcal{E}_{\mathrm{ell}}(z,s)$ converges absolutely and uniformly for z in compacta $K \subseteq \mathbb{H}$ not containing any translate γi of i by $\gamma \in \Gamma$.*

Proof. (i) To ease notation, we write $s = \sigma + it \in \mathbb{C}$; we assume that $\sigma = \mathrm{Re}(s) > 1$. We fix $z \in \mathbb{H}$ such that $z \neq \gamma i$ for any $\gamma \in \Gamma$. Since Γ acts properly discontinuously on $\mathbb{H}$ and $z \neq \gamma i$ for any $\gamma \in \Gamma$, the minimum

$$R_1(z) := \min_{\gamma \in \Gamma} d_{\mathrm{hyp}}(i, \gamma z)$$

exists and is strictly positive. Introducing the quantity

$$C_1(z) := \frac{1 - \exp\big(-2R_1(z)\big)}{2} > 0,$$

we derive the inequality

$$\frac{1 - \exp\big(-2\varrho(\gamma z)\big)}{2} \geq C_1(z)$$

for all $\gamma \in \Gamma$. From this we obtain the estimate

$$\sinh\bigl(\varrho(\gamma z)\bigr) = \exp\bigl(\varrho(\gamma z)\bigr) \cdot \frac{1 - \exp\bigl(-2\varrho(\gamma z)\bigr)}{2} \geq C_1(z) \cdot \exp\bigl(\varrho(\gamma z)\bigr),$$

again for all $\gamma \in \Gamma$. From this we derive the estimate

$$\sum_{\gamma \in \Gamma_i \backslash \Gamma} \left| \sinh\bigl(\varrho(\gamma z)\bigr)^{-s} \right| = \sum_{\gamma \in \Gamma_i \backslash \Gamma} \sinh\bigl(\varrho(\gamma z)\bigr)^{-\sigma}$$
$$\leq C_1(z)^{-\sigma} \cdot \sum_{\gamma \in \Gamma_i \backslash \Gamma} \exp\bigl(-\sigma \varrho(\gamma z)\bigr).$$

In order to complete the proof of (i), we are left to show the local uniform convergence of the series

$$\sum_{\gamma \in \Gamma_i \backslash \Gamma} \exp\bigl(-\sigma \varrho(\gamma z)\bigr)$$

for $\sigma > 1$. To do this, we introduce for $r \in \mathbb{R}_{\geq 0}$ the quantities

$$G(r) := \bigl\{\gamma \in \Gamma_i \backslash \Gamma \bigm| \varrho(\gamma z) < r\bigr\}, \quad N(r) := \sharp G(r).$$

We note that the number $N(r)$ is finite, since Γ acts properly discontinously on $\mathbb{H}$ and $z \neq \gamma i$ for any $\gamma \in \Gamma$; in particular, we have $N(r) = 0$ for $0 \leq r \leq R_1(z)$.

For fixed $r \in \mathbb{R}_{>0}$, we are next going to estimate the number $N(r)$. Let $\mathcal{B}_r(i)$ denote the open hyperbolic disk of radius r centered at i containing the finitely many translates γz of z for $\gamma \in G(r)$. Then there exists a constant $\varepsilon(z) > 0$, depending on z, such that the open hyperbolic disks $\mathcal{B}_{\varepsilon(z)}(\gamma z)$ of radius $\varepsilon(z)$ centered at γz do not intersect for all $\gamma \in G(r)$ and are contained in $\mathcal{B}_r(i)$. Consequently, we obtain

$$N(r) \cdot \mathrm{vol}_{\mathrm{hyp}}\bigl(\mathcal{B}_{\varepsilon(z)}(\gamma z)\bigr) \leq \mathrm{vol}_{\mathrm{hyp}}\bigl(\mathcal{B}_r(i)\bigr) \quad \bigl(\gamma \in G(r)\bigr).$$

This yields the estimate

$$N(r) \leq \frac{4\pi \sinh^2\bigl(r/2\bigr)}{4\pi \sinh^2\bigl(\varepsilon(z)/2\bigr)} = \frac{\cosh(r) - 1}{2\sinh^2\bigl(\varepsilon(z)/2\bigr)} = \frac{\exp(r) + \exp(-r) - 2}{4\sinh^2\bigl(\varepsilon(z)/2\bigr)}$$
$$< \exp(r) \cdot \frac{1 + \exp(-2r)}{4\sinh^2\bigl(\varepsilon(z)/2\bigr)} < C_2(z) \cdot \exp(r) \tag{18}$$

with a suitable constant $C_2(z) > 0$ depending on z.

For fixed $R \in \mathbb{R}_{>0}$, the monotone increasing step function $N : [0, R] \longrightarrow \mathbb{N}$ induces a Stieltjes measure $dN(r)$ on the interval $[0, R]$. Since the function $\exp(-\sigma r) : [0, R] \longrightarrow \mathbb{R}_{>0}$ is continuous and the function $N(r)$ is of bounded variation, the function $\exp(-\sigma r)$ is Riemann–Stieltjes integrable with respect to $N(r)$ on the interval $[0, R]$. Furthermore, since $N(r)$ and $\exp(-\sigma r)$ are bounded on $[0, R]$, the theorem of partial integration can be applied to give

$$\begin{aligned}\sum_{\substack{\gamma\in\Gamma_i\backslash\Gamma\\ \gamma\in G(R)}} \exp\Big(-\sigma\varrho(\gamma z)\Big) &= \int_0^R \exp(-\sigma r)dN(r)\\ &= \Big[N(r)\exp(-\sigma r)\Big]_0^R - \int_0^R N(r)d\big(\exp(-\sigma r)\big)\\ &= \Big[N(r)\exp(-\sigma r)\Big]_0^R + \int_0^R \sigma N(r)\exp(-\sigma r)dr. \qquad (19)\end{aligned}$$

Using (18), the first summand of (19) can be bounded as

$$\Big[N(r)\exp(-\sigma r)\Big]_0^R = N(R)\exp(-\sigma R) < C_2(z)\exp\big((1-\sigma)R\big).$$

On the other hand, again using (18), the integral in (19) can be bounded as

$$\begin{aligned}\int_0^R \sigma N(r)\exp(-\sigma r)dr &< \sigma C_2(z)\int_0^R \exp\big((1-\sigma)r\big)dr\\ &= \frac{\sigma C_2(z)}{1-\sigma}\Big(\exp\big((1-\sigma)R\big)-1\Big).\end{aligned}$$

Summing up, we arrive at

$$\begin{aligned}\sum_{\gamma\in\Gamma_i\backslash\Gamma} \exp\big(-\sigma\varrho(\gamma z)\big) &= \lim_{R\to\infty} \sum_{\substack{\gamma\in\Gamma_i\backslash\Gamma\\ \gamma\in G(R)}} \exp\big(-\sigma\varrho(\gamma z)\big)\\ &\leq \lim_{R\to\infty}\Bigg[C_2(z)\exp\big((1-\sigma)R\big)\\ &\qquad + \frac{\sigma C_2(z)}{1-\sigma}\Big(\exp\big((1-\sigma)R\big)-1\Big)\Bigg]\\ &= \frac{\sigma C_2(z)}{\sigma-1},\end{aligned}$$

keeping in mind that $\sigma > 1$. The absolute and local uniform convergence of the elliptic Eisenstein series $\mathcal{E}_{\mathrm{ell}}(z, s)$ now follows for $s \in \mathbb{C}$ with $\mathrm{Re}(s) > 1$.

(ii) From definition 4.1 we immediately deduce for $s \in \mathbb{C}$ with $\mathrm{Re}(s) > 1$,

$$\mathcal{E}_{\mathrm{ell}}(\gamma z, s) = \mathcal{E}_{\mathrm{ell}}(z, s)$$

for all $\gamma \in \Gamma$, provided that $z \neq \gamma i$ for any $\gamma \in \Gamma$.

(iii) Finally, let $K \subseteq \mathbb{H}$ be a compact subset not containing any translate γi of i by $\gamma \in \Gamma$. Then the constants $C_1(z)$ and $C_2(z)$ constructed in the first part of the proof can be chosen uniformly for all $z \in K$. For fixed $s \in \mathbb{C}$ with $\mathrm{Re}(s) > 1$, the series $\mathcal{E}_{\mathrm{ell}}(z, s)$ therefore converges absolutely and uniformly on $K \subseteq \mathbb{H}$. □

Lemma 4.3. *For $z = x + iy \in \mathbb{H}$ with $z \neq \gamma i$ for any $\gamma \in \Gamma$, and $s \in \mathbb{C}$ with $\mathrm{Re}(s) > 1$, the elliptic Eisenstein series $\mathcal{E}_{\mathrm{ell}}(z, s)$ is twice continuously differentiable with respect to x, y.*

Proof. In order to prove the claim, we have to show in a first step that the series of partial derivatives

$$\sum_{\gamma \in \Gamma_i \backslash \Gamma} \frac{\partial}{\partial x} \sinh\big(\varrho(\gamma z)\big)^{-s}, \qquad \sum_{\gamma \in \Gamma_i \backslash \Gamma} \frac{\partial}{\partial y} \sinh\big(\varrho(\gamma z)\big)^{-s} \tag{20}$$

converge absolutely and uniformly on compacta $K \subseteq \mathbb{H}$ not containing any translate γi of i by $\gamma \in \Gamma$ provided that $\sigma = \mathrm{Re}(s) > 1$. To do this, we introduce for functions $f \in C^1(\mathbb{H})$ the notation

$$\nabla_{\mathrm{hyp}} f(z) := y^2 \Big(\Big(\frac{\partial f(z)}{\partial x} \Big)^2 + \Big(\frac{\partial f(z)}{\partial y} \Big)^2 \Big).$$

Letting $\varphi(x, y) := (a^2 + c^2)(x^2 + y^2) + 2(ab + cd)x + (b^2 + d^2)$, we have by (2),

$$\sinh\big(\varrho(\gamma z)\big) = \sqrt{\cosh^2\big(\varrho(\gamma z)\big) - 1} = \sqrt{\Big(\frac{\varphi(x, y)}{2y} \Big)^2 - 1},$$

from which we derive

$$\begin{aligned} \frac{\partial}{\partial x} \sinh\big(\varrho(\gamma z)\big) &= \frac{\cosh\big(\varrho(\gamma z)\big)}{\sinh\big(\varrho(\gamma z)\big)} \cdot \frac{\partial}{\partial x} \frac{\varphi(x, y)}{2y} \\ &= \coth\big(\varrho(\gamma z)\big) \cdot \frac{(a^2 + c^2)x + (ab + cd)}{y}, \end{aligned}$$

and

$$\frac{\partial}{\partial y} \sinh\big(\varrho(\gamma z)\big) = \coth\big(\varrho(\gamma z)\big) \cdot \Big((a^2 + c^2) - \frac{\varphi(x, y)}{2y^2} \Big).$$

A straightforward computation yields

$$\nabla_{\mathrm{hyp}} \sinh\big(\varrho(\gamma z)\big) = \cosh^2\big(\varrho(\gamma z)\big), \tag{21}$$

from which we deduce

$$\left|\frac{\partial}{\partial x} \sinh\big(\varrho(\gamma z)\big)\right| \leq y^{-1}\sqrt{\nabla_{\mathrm{hyp}} \sinh\big(\varrho(\gamma z)\big)} = y^{-1}\cosh\big(\varrho(\gamma z)\big),$$
$$\left|\frac{\partial}{\partial y} \sinh\big(\varrho(\gamma z)\big)\right| \leq y^{-1}\sqrt{\nabla_{\mathrm{hyp}} \sinh\big(\varrho(\gamma z)\big)} = y^{-1}\cosh\big(\varrho(\gamma z)\big).$$

By the choice of the compact set K, there is a positive constant C_K such that the inequality $\cosh\big(\varrho(\gamma z)\big) \leq C_K \cdot \sinh\big(\varrho(\gamma z)\big)$ holds for all $z \in K$. Therefore, we obtain for $z \in K$,

$$\left|\frac{\partial}{\partial x} \sinh\big(\varrho(\gamma z)\big)^{-s}\right| \leq C_K \cdot |s| \cdot y^{-1} \cdot \sinh\big(\varrho(\gamma z)\big)^{-\sigma},$$
$$\left|\frac{\partial}{\partial y} \sinh\big(\varrho(\gamma z)\big)^{-s}\right| \leq C_K \cdot |s| \cdot y^{-1} \cdot \sinh\big(\varrho(\gamma z)\big)^{-\sigma}.$$

The absolute and locally uniform convergence for the series (20) now follows from Lemma 4.2 provided that $\sigma > 1$.

To ease notation, we put for the second step $x_1 := x$ and $x_2 := y$. We will then show that for $j, k = 1, 2$, the series

$$\sum_{\gamma \in \Gamma_i \backslash \Gamma} \frac{\partial^2}{\partial x_j \partial x_k} \sinh\big(\varrho(\gamma z)\big)^{-s} \tag{22}$$

converges absolutely and uniformly on compacta $K \subseteq \mathbb{H}$ not containing any translate γi of i by $\gamma \in \Gamma$ provided that $\sigma = \mathrm{Re}(s) > 1$. Setting $f(z) := \sinh\big(\varrho(\gamma z)\big)$, we estimate for $z \in K$,

$$\begin{aligned}
&\left|\frac{\partial^2}{\partial x_j \partial x_k} \sinh\big(\varrho(\gamma z)\big)^{-s}\right| \\
&= \left|(-s)(-s-1) \cdot f(z)^{-(s+2)} \cdot \frac{\partial f(z)}{\partial x_j} \cdot \frac{\partial f(z)}{\partial x_k} + (-s)\, f(z)^{-(s+1)} \cdot \frac{\partial^2 f(z)}{\partial x_j \partial x_k}\right| \\
&\leq |s^2 + s| \cdot f(z)^{-(\sigma+2)} \cdot \left|\frac{\partial f(z)}{\partial x_j}\right| \cdot \left|\frac{\partial f(z)}{\partial x_k}\right| + |s| \cdot f(z)^{-(\sigma+1)} \cdot \left|\frac{\partial^2 f(z)}{\partial x_j \partial x_k}\right| \\
&\leq C_K^2 \cdot |s^2 + s| \cdot x_2^{-2} \cdot f(z)^{-\sigma} + |s| \cdot f(z)^{-(\sigma+1)} \cdot \left|\frac{\partial^2 f(z)}{\partial x_j \partial x_k}\right|.
\end{aligned}$$

We are left to estimate the term $|\partial^2 f(z)/\partial x_j\,\partial x_k|$. For this, we use the fact that for real functions $g(z) = g(x_1, x_2)$ defined on $\mathbb{H}$ with continuous first- and second-order partial derivatives, the inequality

$$\left|\frac{\partial^2 g(z)}{\partial x_j\,\partial x_k}\right| \leq x_2^{-2}\left(\sqrt{\nabla_{\mathrm{hyp}} g(z)} + \frac{\sqrt{\nabla^2_{\mathrm{hyp}} g(z)}}{\sqrt{\nabla_{\mathrm{hyp}} g(z)}} + |\Delta_{\mathrm{hyp}} g(z)|\right)$$

holds for all $z = x_1 + ix_2 \in \mathbb{H}$ provided that $\nabla_{\mathrm{hyp}} g(z) \neq 0$ (see [5]). Using (21), we obtain

$$\nabla^2_{\mathrm{hyp}} \sinh\big(\varrho(\gamma z)\big) = \nabla_{\mathrm{hyp}} \cosh^2\big(\varrho(\gamma z)\big) = 4\cosh^2\big(\varrho(\gamma z)\big)\sinh^2\big(\varrho(\gamma z)\big),$$

which yields

$$\frac{\nabla^2_{\mathrm{hyp}} \sinh\big(\varrho(\gamma z)\big)}{\nabla_{\mathrm{hyp}} \sinh\big(\varrho(\gamma z)\big)} = 4\sinh^2\big(\varrho(\gamma z)\big).$$

This, together with the relation $|\Delta_{\mathrm{hyp}} \sinh\big(\varrho(\gamma z)\big)| = 2\sinh\big(\varrho(\gamma z)\big) + \sinh\big(\varrho(\gamma z)\big)^{-1}$, leads to

$$\sqrt{\nabla_{\mathrm{hyp}} f(z)} + \frac{\sqrt{\nabla^2_{\mathrm{hyp}} f(z)}}{\sqrt{\nabla_{\mathrm{hyp}} f(z)}} + |\Delta_{\mathrm{hyp}} f(z)|$$
$$= \cosh\big(\varrho(\gamma z)\big) + 4\sinh\big(\varrho(\gamma z)\big) + \sinh\big(\varrho(\gamma z)\big)^{-1}.$$

Therefore, by the choice of the compact set K, there is a positive constant C'_K such that the inequality

$$\left|\frac{\partial^2 f(z)}{\partial x_j\,\partial x_k}\right| \leq C'_K \cdot x_2^{-2} \cdot \sinh\big(\varrho(\gamma z)\big)$$

holds for $z \in K$. Again, the absolute and locally uniform convergence for the series (22) now follows from Lemma 4.2 provided that $\sigma > 1$. This concludes the proof of the lemma. □

Lemma 4.4. *For $z \in \mathbb{H}$ with $z \neq \gamma i$ for any $\gamma \in \Gamma$, and $s \in \mathbb{C}$ with* $\mathrm{Re}(s) > 1$, *the elliptic Eisenstein series $\mathcal{E}_{\mathrm{ell}}(z, s)$ satisfies the differential equation*

$$\big(\Delta_{\mathrm{hyp}} - s(1-s)\big)\mathcal{E}_{\mathrm{ell}}(z, s) = -s^2\mathcal{E}_{\mathrm{ell}}(z, s+2).$$

Proof. Since the differential operator

$$\Delta_{\text{hyp}} = -\frac{\partial^2}{\partial \varrho^2} - \frac{1}{\tanh(\varrho)} \frac{\partial}{\partial \varrho} - \frac{1}{\sinh^2(\varrho)} \frac{\partial^2}{\partial \vartheta^2}$$

is invariant under the action of Γ, it suffices by Lemma 4.3 to prove the equality

$$\big(\Delta_{\text{hyp}} - s(1-s)\big) \sinh(\varrho)^{-s} = -s^2 \sinh(\varrho)^{-(s+2)}.$$

This follows immediately from the subsequent calculation

$$\begin{aligned}
\Delta_{\text{hyp}} \sinh(\varrho)^{-s} &= s(-s-1) \sinh(\varrho)^{-(s+2)} \cosh^2(\varrho) + s \sinh(\varrho)^{-s} \\
&\quad + s \sinh(\varrho)^{-(s+2)} \cosh^2(\varrho) \\
&= (-s^2 - s + s) \sinh(\varrho)^{-(s+2)} \big(1 + \sinh^2(\varrho)\big) + s \sinh(\varrho)^{-s} \\
&= -s^2 \sinh(\varrho)^{-(s+2)} + s(1-s) \sinh(\varrho)^{-s}.
\end{aligned}$$

□

5 Fourier expansion of the elliptic Eisenstein series

Lemma 5.1. *For* $z \in \mathbb{H}$ *with* $\text{Im}(z) \neq \text{Im}(\gamma^{-1} i)$ *for any* $\gamma \in \Gamma$, *and* $s \in \mathbb{C}$ *with* $\text{Re}(s) > 1$, *the elliptic Eisenstein series* $\mathcal{E}_{\text{ell}}(z,s)$ *admits the Fourier expansion*

$$\mathcal{E}_{\text{ell}}(z,s) = \sum_{m \in \mathbb{Z}} a_m(y,s) e(mx),$$

where

$$\begin{aligned}
a_m(y,s) = &\sum_{\gamma = \left(\begin{smallmatrix} a & b \\ c & d \end{smallmatrix}\right) \in \Gamma_i \backslash \Gamma / \Gamma_\infty} e\left(m \frac{ab+cd}{a^2+c^2}\right) \\
&\times \int_{-\infty}^{\infty} \left(-1 + \left(\frac{a^2+c^2}{2y} t^2 + \frac{(a^2+c^2)^2 y^2 + 1}{2(a^2+c^2)y}\right)^2\right)^{-s/2} e(-mt) dt.
\end{aligned}$$

Proof. Since $\mathcal{E}_{\text{ell}}(z+1,s) = \mathcal{E}_{\text{ell}}(z,s)$, the series $\mathcal{E}_{\text{ell}}(z,s)$ admits the Fourier expansion

$$\mathcal{E}_{\text{ell}}(z,s) = \sum_{m \in \mathbb{Z}} a_m(y,s) e(mx),$$

where

$$a_m(y,s) = \int_0^1 \mathcal{E}_{\mathrm{ell}}(z,s)e(-mx)dx = \sum_{\gamma=\left(\begin{smallmatrix} a & b \\ c & d \end{smallmatrix}\right)\in\Gamma_i\backslash\Gamma} \int_0^1 \sinh\bigl(\varrho(\gamma z)\bigr)^{-s} e(-mx)dx$$

$$= \sum_{\gamma=\left(\begin{smallmatrix} a & b \\ c & d \end{smallmatrix}\right)\in\Gamma_i\backslash\Gamma/\Gamma_\infty} \sum_{n\in\mathbb{Z}} \int_0^1 \sinh\bigl(\varrho(\gamma(z+n))\bigr)^{-s} e(-mx)dx$$

$$= \sum_{\gamma=\left(\begin{smallmatrix} a & b \\ c & d \end{smallmatrix}\right)\in\Gamma_i\backslash\Gamma/\Gamma_\infty} \int_{-\infty}^{\infty} \sinh\bigl(\varrho(\gamma z)\bigr)^{-s} e(-mx)dx.$$

Now, writing $\sinh^2\bigl(\varrho(\gamma z)\bigr) = -1 + \cosh^2\bigl(\varrho(\gamma z)\bigr)$, using (2), and substituting $t := x + \frac{ab+cd}{a^2+c^2}$, we obtain

$$\cosh\bigl(\varrho(\gamma z)\bigr) = \frac{1}{2y}\Bigl((a^2+c^2)t^2 + (a^2+c^2)y^2 + \frac{1}{a^2+c^2}\Bigr)$$

$$= \frac{a^2+c^2}{2y}t^2 + \frac{(a^2+c^2)^2y^2+1}{2(a^2+c^2)y}.$$

From this the claimed formula for $a_m(y,s)$ follows immediately. □

Proposition 5.2. *For $z \in \mathbb{H}$ with* $\mathrm{Im}(z) > 1$, *and $s \in \mathbb{C}$ with* $\mathrm{Re}(s) > 1$, *we have*

$$a_0(y,s) = \frac{2^s\sqrt{\pi}\,\Gamma(s-1/2)}{\Gamma(s)} \cdot y^{1-s} \sum_{k=0}^{\infty} \frac{(s-\frac{1}{2})_k \cdot (\frac{s}{2})_k}{(\frac{s}{2}+\frac{1}{2})_k \cdot k!} \cdot y^{-2k} \cdot V_0(s+2k),$$

where

$$V_0(s) := \sum_{\gamma=\left(\begin{smallmatrix} a & b \\ c & d \end{smallmatrix}\right)\in\Gamma_i\backslash\Gamma/\Gamma_\infty} \frac{1}{(a^2+c^2)^s}.$$

Proof. Letting $m = 0$, we derive from Lemma 5.1

$$a_0(y,s) = \sum_{\gamma=\left(\begin{smallmatrix} a & b \\ c & d \end{smallmatrix}\right)\in\Gamma_i\backslash\Gamma/\Gamma_\infty} b_{0,\gamma}(y,s),$$

where

$$b_{0,\gamma}(y,s) := 2\int_0^{\infty} \Bigl(-1 + \Bigl(\frac{a^2+c^2}{2y}t^2 + \frac{(a^2+c^2)^2y^2+1}{2(a^2+c^2)y}\Bigr)^2\Bigr)^{-s/2} dt$$

$$= \int_0^{\infty} \Bigl(-1 + \Bigl(\frac{a^2+c^2}{2y}t + \frac{(a^2+c^2)^2y^2+1}{2(a^2+c^2)y}\Bigr)^2\Bigr)^{-s/2} \frac{dt}{\sqrt{t}}.$$

Substituting

$$r := \frac{((a^2+c^2)y+1)^2}{(a^2+c^2)^2}\left(t + \frac{((a^2+c^2)y+1)^2}{(a^2+c^2)^2}\right)^{-1},$$

we obtain

$$b_{0,\gamma}(y,s) = \frac{2^s y^s (a^2+c^2)^{s-1}}{((a^2+c^2)y+1)^{2s-1}} \int_0^1 r^{s-3/2}(1-r)^{-1/2} \times \left(1 - \frac{4(a^2+c^2)y}{((a^2+c^2)y+1)^2}\cdot r\right)^{-s/2} dr.$$

Now using the integral representation (3) of Gauss's hypergeometric function $F(a', b'; c'; w)$ with

$$a' := \frac{s}{2}, \quad b' := s - \frac{1}{2}, \quad c' := s, \quad \text{and} \quad w := \frac{4(a^2+c^2)y}{((a^2+c^2)y+1)^2},$$

which is justified since $\mathrm{Re}(c') > \mathrm{Re}(b') = \mathrm{Re}(s) - 1/2 > 0$, we obtain

$$b_{0,\gamma}(y,s) = \frac{2^s y^s (a^2+c^2)^{s-1}}{((a^2+c^2)y+1)^{2s-1}} \cdot \frac{\sqrt{\pi}\,\Gamma(s-1/2)}{\Gamma(s)} \times F\left(\frac{s}{2}, s - \frac{1}{2}; s; \frac{4(a^2+c^2)y}{((a^2+c^2)y+1)^2}\right).$$

Since $\Gamma = \mathrm{PSL}_2(\mathbb{Z})$ and $y > 1$, we have

$$\frac{4(a^2+c^2)y}{((a^2+c^2)y+1)^2} < 1,$$

and so the hypergeometric function in question can be represented as a series, which shows that

$$F\left(\frac{s}{2}, s - \frac{1}{2}; s; \frac{4(a^2+c^2)y}{((a^2+c^2)y+1)^2}\right) = F\left(s - \frac{1}{2}, \frac{s}{2}; s; \frac{4(a^2+c^2)y}{((a^2+c^2)y+1)^2}\right).$$

Now the hypergeometric function under consideration is of the form $F(b', a'; 2a'; w)$, which allows us to apply the following formula (see [1], formula 15.3.17):

$$F(b',a';2a';w)$$
$$= 2^{2b'}(1+\sqrt{1-w})^{-2b'} F\left(b', b'-a'+\frac{1}{2}; a'+\frac{1}{2}; \left(\frac{1-\sqrt{1-w}}{1+\sqrt{1-w}}\right)^2\right). \tag{23}$$

Again, since $y > 1$, we have

$$\sqrt{1-\frac{4(a^2+c^2)y}{((a^2+c^2)y+1)^2}} = \frac{(a^2+c^2)y-1}{(a^2+c^2)y+1},$$

which leads to

$$F\left(s-\frac{1}{2},\frac{s}{2};s;\frac{4(a^2+c^2)y}{((a^2+c^2)y+1)^2}\right)$$
$$= 2^{2s-1}\left(\frac{2(a^2+c^2)y}{(a^2+c^2)y+1}\right)^{-2s+1} F\left(s-\frac{1}{2},\frac{s}{2};\frac{s}{2}+\frac{1}{2};\frac{1}{(a^2+c^2)^2y^2}\right)$$
$$= ((a^2+c^2)y)^{-2s+1}((a^2+c^2)y+1)^{2s-1} F\left(s-\frac{1}{2},\frac{s}{2};\frac{s}{2}+\frac{1}{2};\frac{1}{(a^2+c^2)^2y^2}\right). \tag{24}$$

Adding up, we obtain

$$b_{0,\gamma}(y,s) = \frac{2^s\sqrt{\pi}\,\Gamma(s-1/2)}{\Gamma(s)}\cdot\frac{y^{1-s}}{(a^2+c^2)^s}\cdot F\left(s-\frac{1}{2},\frac{s}{2};\frac{s}{2}+\frac{1}{2};\frac{1}{(a^2+c^2)^2y^2}\right).$$

Introducing the notation

$$g(s) := \frac{2^s\sqrt{\pi}\,\Gamma(s-1/2)}{\Gamma(s)},$$

we arrive at

$$a_0(y,s)$$
$$= g(s)\cdot y^{1-s} \sum_{\gamma=\left(\begin{smallmatrix} a & b \\ c & d\end{smallmatrix}\right)\in\Gamma_i\backslash\Gamma/\Gamma_\infty} \frac{1}{(a^2+c^2)^s} F\left(s-\frac{1}{2},\frac{s}{2};\frac{s}{2}+\frac{1}{2};\frac{1}{(a^2+c^2)^2y^2}\right)$$

$$= g(s) \cdot y^{1-s} \sum_{\gamma = \left(\begin{smallmatrix} a & b \\ c & d \end{smallmatrix}\right) \in \Gamma_i \backslash \Gamma / \Gamma_\infty} \frac{1}{(a^2+c^2)^s} \sum_{k=0}^{\infty} \frac{(s-\frac{1}{2})_k \cdot (\frac{s}{2})_k}{(\frac{s}{2}+\frac{1}{2})_k \cdot k!} \left(\frac{1}{(a^2+c^2)^2 y^2} \right)^k$$

$$= g(s) \cdot y^{1-s} \sum_{k=0}^{\infty} \frac{(s-\frac{1}{2})_k \cdot (\frac{s}{2})_k}{(\frac{s}{2}+\frac{1}{2})_k \cdot k!} \cdot y^{-2k} \cdot V_0(s+2k).$$

This completes the proof of the proposition. □

Remark 5.3. The statement of Proposition 5.2 can easily be generalized to the case $z \in \mathbb{H}$ with $\mathrm{Im}(z) \neq \mathrm{Im}(\gamma^{-1} i)$ for any $\gamma \in \Gamma$ as follows: When applying formula (23) in the case $y < (a^2+c^2)^{-1}$, formula (24) becomes

$$F\left(s - \frac{1}{2}, \frac{s}{2}; s; \frac{4(a^2+c^2)y}{((a^2+c^2)y+1)^2}\right)$$
$$= ((a^2+c^2)y+1)^{2s-1} F\left(s-\frac{1}{2}, \frac{s}{2}; s+\frac{1}{2}; (a^2+c^2)^2 y^2\right).$$

Therefore, we arrive at

$$a_0(y,s)$$
$$= g(s) \cdot y^{1-s} \sum_{\substack{\gamma = \left(\begin{smallmatrix} a & b \\ c & d \end{smallmatrix}\right) \in \Gamma_i \backslash \Gamma / \Gamma_\infty \\ y > (a^2+c^2)^{-1}}} (a^2+c^2)^{-s} \sum_{k=0}^{\infty} \frac{(s-\frac{1}{2})_k \cdot (\frac{s}{2})_k}{(\frac{s}{2}+\frac{1}{2})_k \cdot k!} \cdot \left((a^2+c^2)y\right)^{-2k}$$

$$+ g(s) \cdot y^{s} \sum_{\substack{\gamma = \left(\begin{smallmatrix} a & b \\ c & d \end{smallmatrix}\right) \in \Gamma_i \backslash \Gamma / \Gamma_\infty \\ y < (a^2+c^2)^{-1}}} (a^2+c^2)^{s-1} \sum_{k=0}^{\infty} \frac{(s-\frac{1}{2})_k \cdot (\frac{s}{2})_k}{(s+\frac{1}{2})_k \cdot k!} \cdot \left((a^2+c^2)y\right)^{2k}.$$

Proposition 5.4. *For $z \in \mathbb{H}$ with $\mathrm{Im}(z) > 1$, $s \in \mathbb{C}$ with $\mathrm{Re}(s) > 1$, and $m \neq 0$, we have*

$$a_m(y,s) = 2^s y^s \sum_{k_1=0}^{\infty} \sum_{k_2=0}^{\infty} \frac{(\frac{s}{2})_{k_1} \cdot (\frac{s}{2})_{k_2}}{k_1! \cdot k_2!} \cdot I_{-m}(y,s;k_1,k_2) \cdot V_{-m}(s+2k_1+2k_2),$$

where

$$I_m(y,s;k_1,k_2) := \int_{-\infty}^{\infty} (y+it)^{-s-2k_1} (y-it)^{-s-2k_2} e(mt)\, dt$$

and

$$V_m(s) = \sum_{\gamma \in \Gamma_i \backslash \Gamma / \Gamma_\infty} \mathrm{Im}(\gamma^{-1} i)^s e\big(m\mathrm{Re}(\gamma^{-1} i)\big).$$

Proof. For $m \neq 0$, we derive from Lemma 5.1

$$a_m(y,s) = \sum_{\gamma = \left(\begin{smallmatrix} a & b \\ c & d \end{smallmatrix}\right) \in \Gamma_i \backslash \Gamma / \Gamma_\infty} e\Big(m\frac{ab+cd}{a^2+c^2}\Big) b_{m,\gamma}(y,s),$$

where

$$b_{m,\gamma}(y,s) := \int_{-\infty}^{\infty} \bigg(-1 + \Big(\frac{a^2+c^2}{2y}t^2 + \frac{(a^2+c^2)^2y^2+1}{2(a^2+c^2)y}\Big)^2\bigg)^{-s/2} e(-mt)\, dt. \tag{25}$$

We write

$$\begin{aligned}
&-1 + \Big(\frac{a^2+c^2}{2y}t^2 + \frac{(a^2+c^2)^2y^2+1}{2(a^2+c^2)y}\Big)^2 \\
&= \frac{(a^2+c^2)^2}{(2y)^2}\bigg(t^2 + \Big(y + \frac{1}{a^2+c^2}\Big)^2\bigg)\bigg(t^2 + \Big(y - \frac{1}{a^2+c^2}\Big)^2\bigg) \\
&= \frac{(a^2+c^2)^2}{(2y)^2}\Big(it + y + \frac{1}{a^2+c^2}\Big)\Big(-it + y + \frac{1}{a^2+c^2}\Big)\Big(it + y - \frac{1}{a^2+c^2}\Big) \\
&\quad \times \Big(-it + y - \frac{1}{a^2+c^2}\Big) \\
&= \frac{(a^2+c^2)^2}{(2y)^2}(y+it)^2(y-it)^2\Big(1 - \frac{1}{(a^2+c^2)^2(y+it)^2}\Big) \\
&\quad \times \Big(1 - \frac{1}{(a^2+c^2)^2(y-it)^2}\Big).
\end{aligned} \tag{26}$$

Since $y > 1$, we have the estimate

$$\max_{-\infty < t < \infty}\Big(\frac{1}{|(a^2+c^2)^2(y \pm it)^2|}\Big) = \frac{1}{(a^2+c^2)^2y^2} < 1,$$

and hence we can write

$$\Big(1 - \frac{1}{(a^2+c^2)^2(y \pm it)^2}\Big)^{-s/2} = \sum_{k=0}^{\infty} \frac{(\frac{s}{2})_k}{k!(a^2+c^2)^{2k}} \cdot (y \pm it)^{-2k}.$$

Therefore, we obtain

$$b_{m,\gamma}(y,s) = \frac{(2y)^s}{(a^2+c^2)^s} \sum_{k_1=0}^{\infty} \sum_{k_2=0}^{\infty} \frac{(\frac{s}{2})_{k_1} \cdot (\frac{s}{2})_{k_2}}{k_1! \cdot k_2! \cdot (a^2+c^2)^{2(k_1+k_2)}}$$
$$\times \int_{-\infty}^{\infty} (y+it)^{-s-2k_1} (y-it)^{-s-2k_2} e(-mt)\, dt, \qquad (27)$$

from which the statement follows. □

Remark 5.5. The statement of Proposition 5.4 can be generalized to the case $z \in \mathbb{H}$ with $\operatorname{Im}(z) \neq \operatorname{Im}(\gamma^{-1}i)$ for any $\gamma \in \Gamma$. In this case the Fourier coefficient in question becomes

$$a_m(y,s) = \sum_{\substack{\gamma = \left(\begin{smallmatrix} a & b \\ c & d \end{smallmatrix}\right) \in \Gamma_i \backslash \Gamma / \Gamma_\infty \\ y > (a^2+c^2)^{-1}}} e\left(m \frac{ab+cd}{a^2+c^2}\right) b_{m,\gamma}^{(>)}(y,s)$$
$$+ \sum_{\substack{\gamma = \left(\begin{smallmatrix} a & b \\ c & d \end{smallmatrix}\right) \in \Gamma_i \backslash \Gamma / \Gamma_\infty \\ y < (a^2+c^2)^{-1}}} e\left(m \frac{ab+cd}{a^2+c^2}\right) b_{m,\gamma}^{(<)}(y,s),$$

where

$$b_{m,\gamma}^{(>)}(y,s) = \frac{2^s y^s}{(a^2+c^2)^{1-s}} \sum_{k_1=0}^{\infty} \sum_{k_2=0}^{\infty} \frac{(\frac{s}{2})_{k_1} \cdot (\frac{s}{2})_{k_2}}{k_1! \cdot k_2!} \cdot I_{-m/(a^2+c^2)}\Big((a^2+c^2)y, s; k_1, k_2\Big), \qquad (28)$$

$$b_{m,\gamma}^{(<)}(y,s) = \frac{2^s y^{1-s}}{(a^2+c^2)^s} \sum_{k_1=0}^{\infty} \sum_{k_2=0}^{\infty} \frac{(\frac{s}{2})_{k_1} \cdot (\frac{s}{2})_{k_2}}{k_1! \cdot k_2!} \cdot I_{-my}\Big(\frac{1}{(a^2+c^2)y}, s; k_1, k_2\Big). \qquad (29)$$

Here $b_{m,\gamma}^{(>)}(y,s)$ is obtained as in the proof of Proposition 5.4 (after a suitable substitution in (27)), whereas $b_{m,\gamma}^{(<)}(y,s)$ is obtained by rewriting (26) as

$$-1 + \left(\frac{a^2+c^2}{2y} t^2 + \frac{(a^2+c^2)^2 y^2 + 1}{2(a^2+c^2)y}\right)^2$$
$$= \frac{(a^2+c^2)^2 y^2}{4} \left(\frac{1}{(a^2+c^2)y} + \frac{it}{y}\right)^2 \left(\frac{1}{(a^2+c^2)y} - \frac{it}{y}\right)^2$$
$$\times \left(1 - \left(\frac{1}{(a^2+c^2)y} + \frac{it}{y}\right)^{-2}\right) \left(1 - \left(\frac{1}{(a^2+c^2)y} - \frac{it}{y}\right)^{-2}\right),$$

which, after using the expansion

$$\left(1-\left(\frac{1}{(a^2+c^2)y}\pm\frac{it}{y}\right)^{-2}\right)^{-s/2}=\sum_{k=0}^{\infty}\frac{(\frac{s}{2})_k}{k!}\cdot\left(\frac{1}{(a^2+c^2)y}\pm\frac{it}{y}\right)^{-2k},$$

yields the claimed formula (again after a suitable substitution in the corresponding integral).

Remark 5.6. The series $V_m(s)$ ($m \in \mathbb{Z}$) of Propositions 5.2 and 5.4 can be rewritten as follows: Consider the anti-isomorphism $\phi : \Gamma \to \Gamma$ given by $\gamma \mapsto \gamma^{-1}$. Since $\phi(\Gamma_\infty) = \Gamma_\infty$ and $\phi(\Gamma_i) = \Gamma_i$, we have $\phi(\Gamma_\infty\backslash\Gamma/\Gamma_i) = \Gamma_i\backslash\Gamma/\Gamma_\infty$. Therefore, we obtain

$$\begin{aligned}V_m(s) &= \sum_{\gamma\in\Gamma_i\backslash\Gamma/\Gamma_\infty}\mathrm{Im}(\gamma^{-1}i)^s e\big(m\mathrm{Re}(\gamma^{-1}i)\big)=\sum_{\gamma\in\Gamma_\infty\backslash\Gamma/\Gamma_i}\mathrm{Im}(\gamma i)^s e\big(m\mathrm{Re}(\gamma i)\big)\\ &=\frac{1}{2}V_m(i,s)\end{aligned}$$

with the Poincaré series (14) evaluated at $z = i$. Note that the series $V_0(s)$ multiplied by $\zeta(2s)$ equals the Dedekind zeta function associated to the field of $\mathbb{Q}(i)$.

6 Meromorphic continuation of the elliptic Eisenstein series

Lemma 6.1. *The series*

$$V_0(s)=\sum_{\gamma=\left(\begin{smallmatrix}a&b\\c&d\end{smallmatrix}\right)\in\Gamma_i\backslash\Gamma/\Gamma_\infty}\frac{1}{(a^2+c^2)^s}$$

converges absolutely and locally uniformly for $s \in \mathbb{C}$ with $\mathrm{Re}(s) > 1$, and hence defines a holomorphic function. It has a meromorphic continuation to the whole s-plane with a simple pole at $s = 1$ and poles at $s = \rho/2$, where ρ is a non-trivial zero of $\zeta(s)$. Furthermore, we have $V_0(1/2) = 0$.

Proof. Since $V_0(s) = \mathcal{E}_{\mathrm{par}}(i,s)/2$, the claimed assertions immediately follow from the known properties of the parabolic Eisenstein series $\mathcal{E}_{\mathrm{par}}(z,s)$ recalled in Section 2.4. In particular, the vanishing of $V_0(s)$ at $s = 1/2$ follows from the functional equation (6) by observing that $\varphi(1/2) = -1$. □

Lemma 6.2. *For $z \in \mathbb{H}$ with $\mathrm{Im}(z) > 1$, and $N \in \mathbb{N}$, the series*

$$\sum_{k=N+1}^{\infty}\frac{(s-\frac{1}{2})_k\cdot(\frac{s}{2})_k}{\Gamma(\frac{s}{2}+\frac{1}{2}+k)\cdot k!}\cdot y^{-2k}\cdot V_0(s+2k)$$

converges absolutely and locally uniformly for $s \in \mathbb{C}$ *with* $\mathrm{Re}(s) > -2N - 1$*, and hence defines a holomorphic function.*

Proof. Fix $N \in \mathbb{N}$, and let $s \in \mathbb{C}$ with $\mathrm{Re}(s) > -2N - 1$. Then, for $k \in \mathbb{N}$, we define the functions

$$f_k(y,s) := g_k(y,s) \cdot V_0(s+2k), \text{ where } g_k(y,s) := \frac{(s-\frac{1}{2})_k \cdot (\frac{s}{2})_k}{\Gamma(\frac{s}{2}+\frac{1}{2}+k)\cdot k!} \cdot y^{-2k}.$$

If $k \geq N+1$, we have $\mathrm{Re}(s+2k) \geq \mathrm{Re}(s) + 2N + 2 > 1$, whence the functions $V_0(s+2k)$ are holomorphic. Since the functions $g_k(y,s)$ are also holomorphic in the range under consideration, the functions $f_k(y,s)$ are holomorphic for $s \in \mathbb{C}$ with $\mathrm{Re}(s) > -2N - 1$ as long as $k \geq N+1$. We now estimate

$$\sum_{k=N+1}^{\infty} \left|f_k(y,s)\right| \leq V_0\big(\mathrm{Re}(s)+2N\big) \sum_{k=N+1}^{\infty} \left|g_k(y,s)\right|.$$

Since the ratio of successive terms in the latter series has limit

$$\lim_{k\to\infty} \frac{|g_{k+1}(y,s)|}{|g_k(y,s)|} = \lim_{k\to\infty} \left| \frac{\big(s-\frac{1}{2}+k\big)\big(\frac{s}{2}+k\big)}{\big(\frac{s}{2}+\frac{1}{2}+k\big)(1+k)} \cdot \frac{1}{y^2} \right| = \frac{1}{y^2} < 1,$$

we derive from d'Alembert's criterion that the series $\sum_{k=N+1}^{\infty} f_k(y,s)$ converges absolutely and locally uniformly for $s \in \mathbb{C}$ with $\mathrm{Re}(s) > -2N - 1$, which proves the claim. □

Proposition 6.3. *For* $z \in \mathbb{H}$ *with* $\mathrm{Im}(z) > 1$*, the function* $a_0(y,s)$ *has a meromorphic continuation to the whole* s*-plane with possible poles at* $s = 1 - 2N$*,* $s = \rho/2 - 2N$*,* $s = 1/2 - 2N$*, and* $s = -1/2 - 2N$ *(*$N \in \mathbb{N}$*), where* ρ *is a non-trivial zero of* $\zeta(s)$*.*

Proof. We start by proving that the function $a_0(y,s)$ has a meromorphic continuation to the half-plane

$$\mathcal{H}_N := \{s \in \mathbb{C} \,|\, \mathrm{Re}(s) > -2N - 1\}$$

for any $N \in \mathbb{N}$. By Proposition 5.2 and the duplication formula for the Γ-function, we can write, using the notation from the proof of Lemma 6.2,

$$\begin{aligned} a_0(y,s) &= \frac{2^s\sqrt{\pi}\,\Gamma(s-\frac{1}{2})\Gamma(\frac{s}{2}+\frac{1}{2})}{\Gamma(s)} \cdot y^{1-s} \left(\sum_{k=0}^{N} f_k(y,s) + \sum_{k=N+1}^{\infty} f_k(y,s) \right) \\ &= \frac{2\pi\,\Gamma(s-\frac{1}{2})}{\Gamma(\frac{s}{2})} \cdot y^{1-s} \left(\sum_{k=0}^{N} f_k(y,s) + \sum_{k=N+1}^{\infty} f_k(y,s) \right). \end{aligned} \tag{30}$$

Since $\mathrm{Re}(s) > -2N-1$ by assumption, Lemma 6.2 proves that the series $\sum_{k=N+1}^{\infty} f_k(y,s)$ is a holomorphic function on the half-plane $\mathcal{H}_N$. Since the finite sum $\sum_{k=0}^{N} f_k(y,s)$ is a meromorphic function on the whole s-plane by Lemma 6.1, we conclude that $a_0(y,s)$ has a meromorphic continuation to the half-plane $\mathcal{H}_N$. Since N was chosen arbitrarily, this proves the meromorphic continuation of $a_0(y,s)$ to the whole s-plane.

In order to determine the poles of $a_0(y,s)$, we calculate its poles in the strip

$$\mathcal{S}_N := \{s \in \mathbb{C} \mid -2N-1 < \mathrm{Re}(s) \leq -2N+1\}$$

for any $N \in \mathbb{N}$. By considering $a_0(y,s)$ with its decomposition (30) in the strip $\mathcal{S}_N$, we see that the poles come from the finite sum $\sum_{k=0}^{N} f_k(y,s)$, which has poles in the strip $\mathcal{S}_N$ arising from the function $f_N(y,s)$, more precisely from the factor $V_0(s+2N)$ at $s=1-2N$ and $s=\rho/2-2N$, where ρ is a non-trivial zero of $\zeta(s)$, and from the Γ-factor $\Gamma(s-1/2)$ at $s=1/2-2N$ and $s=-1/2-2N$. Therefore, the possible poles of $a_0(y,s)$ in the strip $\mathcal{S}_N$ are located at $s=1-2N$, $s=\rho/2-2N$, $s=1/2-2N$, and $s=-1/2-2N$, as claimed. □

Remark 6.4. Using Remark 5.3, one can establish the meromorphic continuation of $a_0(y,s)$ to the whole s-plane in the more general case $\mathrm{Im}(z) \neq \mathrm{Im}(\gamma^{-1}i)$ for any $\gamma \in \Gamma$, using the same techniques as in Lemma 6.2 and Proposition 6.3 applied accordingly to the modified situation. The poles of $a_0(y,s)$ turn out to be same as in the case $\mathrm{Im}(z) > 1$.

Lemma 6.5. *For $y > 1$, $s \in \mathbb{C}$, $m \neq 0$, and $k_1, k_2 \in \mathbb{N}$, let $I_m(y,s;k_1,k_2)$ denote the integral*

$$I_m(y,s;k_1,k_2) := \int_{-\infty}^{\infty} (y+it)^{-s-2k_1}(y-it)^{-s-2k_2} e(mt)\,dt.$$

Then the following assertions hold:

(i) The integral $I_m(y,s;k_1,k_2)$ converges absolutely and locally uniformly for $s \in \mathbb{C}$ with $\mathrm{Re}(s) > 1/2-k_1-k_2$, and hence defines a holomorphic function.
(ii) The integral $I_m(y,s;k_1,k_2)$ admits a holomorphic continuation to the whole s-plane.
(iii) Let $\Omega \subseteq \mathbb{C}$ be a compact subset and let $d \in \mathbb{N}$ be such that $\Omega \subseteq \{s \in \mathbb{C} \mid \mathrm{Re}(s) > 1/2-k_1-k_2-d/2\}$. Then, we have for all $s \in \Omega$ the bound

$$\left|I_m(y,s;k_1,k_2)\right| \ll \frac{(k_1+k_2)^d}{|m|^d} \cdot y^{-2(\mathrm{Re}(s)+k_1+k_2+d/2)+1},$$

where the implied constant depends on Ω and d, but is independent of m and k_1, k_2.

Proof. (i) For $s \in \mathbb{C}$ with $\mathrm{Re}(s) > 1/2 - k_1 - k_2$, we have the estimate

$$\begin{aligned}
\left|I_m(y,s;k_1,k_2)\right| &\leq \int_{-\infty}^{\infty} \left|(y+it)^{-s-2k_1}(y-it)^{-s-2k_2}\, e(mt)\right| dt \\
&= \int_{-\infty}^{\infty} (y^2+t^2)^{-\mathrm{Re}(s)-k_1-k_2}\, dt \\
&= y^{-2(\mathrm{Re}(s)+k_1+k_2)+1} \int_{-\infty}^{\infty} \left(1+\frac{t^2}{y^2}\right)^{-\mathrm{Re}(s)-k_1-k_2} \frac{dt}{y} \\
&= y^{-2(\mathrm{Re}(s)+k_1+k_2)+1}\, \frac{\pi\, \Gamma\big(\mathrm{Re}(s)-1/2+k_1+k_2\big)}{\Gamma\big(\mathrm{Re}(s)+k_1+k_2\big)} \\
&\leq \pi \cdot y^{-2(\mathrm{Re}(s)+k_1+k_2)+1}. \qquad (31)
\end{aligned}$$

For all $s \in \Omega$, where $\Omega \subseteq \{s \in \mathbb{C} \,|\, \mathrm{Re}(s) > 1/2 - k_1 - k_2\}$ is a compact subset, we therefore obtain the bound

$$\left|I_m(y,s;k_1,k_2)\right| \leq \pi,$$

which shows that the integral $I_m(y,s;k_1,k_2)$ converges absolutely and locally uniformly for $s \in \mathbb{C}$ with $\mathrm{Re}(s) > 1/2 - k_1 - k_2$.

(ii) Let $\mathrm{Re}(s) > 1/2 - k_1 - k_2$. Integration by parts yields

$$\begin{aligned}
I_m(y,s;k_1,k_2) &= \left[(y+it)^{-s-2k_1}(y-it)^{-s-2k_2}\frac{e(mt)}{2\pi i m}\right]_{t=-\infty}^{\infty} \\
&+ \frac{1}{2\pi m}\int_{-\infty}^{\infty} (s+2k_1)(y+it)^{-s-2k_1-1}(y-it)^{-s-2k_2}\, e(mt)\, dt \\
&- \frac{1}{2\pi m}\int_{-\infty}^{\infty} (s+2k_2)(y+it)^{-s-2k_1}(y-it)^{-s-2k_2-1}\, e(mt)\, dt.
\end{aligned}$$

In absolute values, the boundary term equals

$$\left|(y+it)^{-s-2k_1}(y-it)^{-s-2k_2}\frac{e(mt)}{2\pi i m}\right| = \frac{(y^2+t^2)^{-\mathrm{Re}(s)-k_1-k_2}}{2\pi|m|}.$$

Since $\mathrm{Re}(s) > 1/2 - k_1 - k_2$ and $y^2 + t^2 > 1$ for $t \in (-\infty, \infty)$, we have

$$(y^2+t^2)^{-\mathrm{Re}(s)-k_1-k_2} < (y^2+t^2)^{-1/2},$$

from which we conclude that the boundary term vanishes. Therefore, we obtain the recurrence formula

$$\begin{aligned} I_m(y,s;k_1,k_2) = &\frac{(s+2k_1)}{2\pi m} I_m\Big(y,s;k_1+\frac{1}{2},k_2\Big) \\ &- \frac{(s+2k_2)}{2\pi m} I_m\Big(y,s;k_1,k_2+\frac{1}{2}\Big). \end{aligned} \tag{32}$$

By part (i), both terms on the right-hand side are holomorphic for $s \in \mathbb{C}$ with $\mathrm{Re}(s) > -k_1 - k_2$. In this way we obtain the holomorphic continuation of $I_m(y,s;k_1,k_2)$ to the half-plane $\{s \in \mathbb{C} \mid \mathrm{Re}(s) > -k_1 - k_2\}$.

Let $d \in \mathbb{N}$. Applying relation (32) d times, we arrive at a formula of the type

$$\begin{aligned} I_m(y,s;k_1,k_2) = &\frac{1}{(2\pi m)^d} \sum_{j=0}^{d} P_{d,j}(s;k_1,k_2) \\ &\times I_m\Big(y,s;k_1+\frac{j}{2},k_2+\frac{d-j}{2}\Big), \end{aligned} \tag{33}$$

where $P_{d,j}(s;k_1,k_2)$ is a polynomial in s and k_1,k_2 of degree d. In fact, one can prove by induction on d that

$$P_{d,j}(s;k_1,k_2) = (-1)^{d-j} \cdot \binom{d}{d-j} \cdot (s+2k_1)_j \cdot (s+2k_2)_{d-j} \qquad (0 \leq j \leq d).$$

Now all the terms in (33) are holomorphic for $s \in \mathbb{C}$ with

$$\mathrm{Re}(s) > 1/2 - (k_1 + j/2) - (k_2 + (d-j)/2) = 1/2 - k_1 - k_2 - d/2.$$

Therefore formula (33) yields the holomorphic continuation of $I_m(y,s;k_1,k_2)$ to the half-plane $\{s \in \mathbb{C} \mid \mathrm{Re}(s) > 1/2 - k_1 - k_2 - d/2\}$. Since $d \in \mathbb{N}$ was chosen arbitrarily, this proves the holomorphic continuation of $I_m(y,s;k_1,k_2)$ to the whole s-plane.

(iii) Let $\Omega \subseteq \mathbb{C}$ be a compact subset and let $d \in \mathbb{N}$ be such that $\Omega \subseteq \{s \in \mathbb{C} \mid \mathrm{Re}(s) > 1/2 - k_1 - k_2 - d/2\}$. For $s \in \Omega$, the function $I_m(y,s;k_1,k_2)$ is given by formula (33). Since $\mathrm{Re}(s) > 1/2 - k_1 - k_2 - d/2 = 1/2 - (k_1 + j/2) - (k_2 + (d-j)/2)$, the bound (31) provides the estimate

$$\left| I_m\Big(y,s;k_1+\frac{j}{2},k_2+\frac{d-j}{2}\Big) \right| \ll y^{-2(\mathrm{Re}(s)+k_1+k_2+d/2)+1},$$

where the implied constant is universal. Furthermore, letting $s \in \Omega$, we have the bound

$$\left| P_{d,j}(s;k_1,k_2) \right| \ll k_1^j \cdot k_2^{d-j} \ll (k_1+k_2)^d \qquad (0 \leq j \leq d),$$

where the implied constant depends on Ω and d, but is independent of m and k_1, k_2. Altogether, as long as $s \in \Omega$, we have the bound

$$\begin{aligned}\left|I_m(y,s;k_1,k_2)\right| &\ll (d+1)\cdot(k_1+k_2)^d \cdot \frac{y^{-2(\mathrm{Re}(s)+k_1+k_2+d/2)+1}}{(2\pi|m|)^d} \\ &\ll \frac{(k_1+k_2)^d}{|m|^d}\cdot y^{-2(\mathrm{Re}(s)+k_1+k_2+d/2)+1},\end{aligned}$$

where the implied constant depends on Ω and d, but is independent of m and k_1, k_2. □

Lemma 6.6. *For $m \neq 0$, the following assertions hold:*

(i) The function $V_m(s)$ admits a meromorphic continuation to the whole s-plane with possible simple poles at $s = s_j - 2N$ and $s = -s_j - 2N + 1$ ($N \in \mathbb{N}$).

(ii) Let $N \in \mathbb{N}$ and $\Omega \subseteq \{s \in \mathbb{C} \mid \mathrm{Re}(s) > -2N - 1\}$ a compact subset not containing any pole of $V_m(s)$. Then, for all $s \in \Omega$, we have the bound

$$\left|V_m(s)\right| \ll |m|^{2N+2},$$

where the implied constant depends on Ω and N, but is independent of m.

(iii) Let $N \in \mathbb{N}$ and $\tilde{s}$ a pole of $V_m(s)$ with $\mathrm{Re}(\tilde{s}) = -2N + 1/2$. Then, the residue of $V_m(s)$ at $\tilde{s}$ is bounded by

$$\left|\mathrm{res}_{s=\tilde{s}} V_m(s)\right| \ll |m|^{2N},$$

where the implied constant depends on $\tilde{s}$ and N, but is independent of m.

Proof. (i) Since we have $V_m(s) = V_m(i,s)/2$ by Remark 5.5, the claim follows immediately from Proposition 3.7.

(ii) We will prove the claim more generally for the Poincaré series $V_m(z,s)$ for any $z \in \mathbb{H}$. For $s \in \Omega$, we then consider the decomposition

$$V_m(z,s) = \sum_{k=0}^{2N+1} \frac{(2\pi|m|)^k}{k!} P_m(z,s+k) + \sum_{k=2N+2}^{\infty} \frac{(2\pi|m|)^k}{k!} P_m(z,s+k). \quad (34)$$

From the proof of Proposition 3.7 we recall that the series on the right-hand side converges absolutely for $s \in \Omega$. Hence, we can rearrange the summation and find for $s \in \Omega$,

$$\left|\sum_{k=2N+2}^{\infty}\frac{(2\pi|m|)^k}{k!}P_m(z,s+k)\right|$$

$$=\left|\sum_{k=0}^{\infty}\frac{(2\pi|m|)^{2N+k+2}}{(2N+k+2)!}P_m(z,s+2N+k+2)\right|$$

$$=(2\pi|m|)^{2N+2}\left|\sum_{k=0}^{\infty}\frac{(2\pi|m|)^k}{(2N+k+2)!}\sum_{\gamma\in\Gamma_\infty\backslash\Gamma}\mathrm{Im}(\gamma z)^{s+2N+k+2}\right.$$

$$\left.\times\exp\bigl(-2\pi|m|\mathrm{Im}(\gamma z)\bigr)e\bigl(m\mathrm{Re}(\gamma z)\bigr)\right|$$

$$=(2\pi|m|)^{2N+2}\left|\sum_{\gamma\in\Gamma_\infty\backslash\Gamma}\mathrm{Im}(\gamma z)^{s+2N+2}\right.$$

$$\left.\times\exp\bigl(-2\pi|m|\mathrm{Im}(\gamma z)\bigr)e\bigl(m\mathrm{Re}(\gamma z)\bigr)\sum_{k=0}^{\infty}\frac{(2\pi|m|\mathrm{Im}(\gamma z))^k}{(2N+k+2)!}\right|$$

$$\leq(2\pi|m|)^{2N+2}\sum_{\gamma\in\Gamma_\infty\backslash\Gamma}\mathrm{Im}(\gamma z)^{\mathrm{Re}(s)+2N+2}\exp\bigl(-2\pi|m|\mathrm{Im}(\gamma z)\bigr)$$

$$\times\sum_{k=0}^{\infty}\frac{(2\pi|m|\mathrm{Im}(\gamma z))^k}{k!}$$

$$=(2\pi|m|)^{2N+2}\cdot\mathcal{E}_{\mathrm{par}}(z,\mathrm{Re}(s)+2N+2)\ll|m|^{2N+2},$$

where the implied constant depends on z, Ω, and N, but is independent of m.

In order to estimate the finite sum in the decomposition (34), we multiply the bounds (10) and (13) by the factor $2^{-2s+1}\pi^{-s+1}\Gamma(s)^{-1}|m|^{-s+1/2}$, and derive from the spectral expansion (9) of $P_m(z,s)$ for all $s\in\Omega$ the bound

$$\bigl|P_m(z,s+k)\bigr|\ll|m|^{-\mathrm{Re}(s)-k+1/2}\ll|m|^{2N-k+3/2}\qquad(k=0,\ldots,2N+1),$$

where the implied constant depends on z and Ω, but is independent of m. Hence, for all $s\in\Omega$, we obtain

$$\bigl|V_m(z,s)\bigr|\ll\sum_{k=0}^{2N+1}|m|^k\cdot|m|^{2N-k+3/2}+|m|^{2N+2}\ll|m|^{2N+2},$$

where the implied constant depends on z, Ω, and N, but is independent of m. This proves the second claim.

(iii) In order to prove the third claim, we recall formulas (15), resp. (16), together with the bound (see [10], p. 86, adapted to our situation)

$$\left|\rho_\ell(m)\right|^2 \ll |t_j|\exp(\pi|t_j|) \qquad (\ell \in \mathbb{N}:\, s_\ell = s_j = 1/2 + it_j),$$

where the implied constant is universal. Then we obtain

$$\left|\mathrm{res}_{s=s_j-2N} V_m(z,s)\right| \ll |m|^{-\mathrm{Re}(s_j)+2N+1/2} \ll |m|^{2N},$$

resp.

$$\left|\mathrm{res}_{s=-s_j-2N+1} V_m(z,s)\right| \ll |m|^{\mathrm{Re}(s_j)+2N-1/2} \ll |m|^{2N},$$

where the implied constants depend on z, s_j, and N, but are independent of m. □

Lemma 6.7. *For $z \in \mathbb{H}$ with $\mathrm{Im}(z) > 1$, $m \neq 0$, and $N \in \mathbb{N}$, the series*

$$2^s y^s \sum_{n=N+1}^{\infty} \sum_{k_1+k_2=n} \frac{(\frac{s}{2})_{k_1} \cdot (\frac{s}{2})_{k_2}}{k_1! \cdot k_2!} \cdot I_{-m}(y,s;k_1,k_2) \cdot V_{-m}(s+2k_1+2k_2)$$

converges absolutely and locally uniformly for $s \in \mathbb{C}$ with $\mathrm{Re}(s) > -2N-1$, and hence defines a holomorphic function.

Proof. Let $\Omega \subseteq \{s \in \mathbb{C} \mid \mathrm{Re}(s) > -2N-1\}$ be a compact subset. For $s \in \Omega$ and $k_1, k_2 \in \mathbb{N}$, we define the functions

$$f_{m;k_1,k_2}(y,s) := 2^s y^s \frac{(\frac{s}{2})_{k_1} \cdot (\frac{s}{2})_{k_2}}{k_1! \cdot k_2!} \cdot I_{-m}(y,s;k_1,k_2) \cdot V_{-m}(s+2k_1+2k_2).$$

If $k_1 + k_2 \geq N+1$, we have $\mathrm{Re}(s+2k_1+2k_2) \geq \mathrm{Re}(s) + 2N + 2 > 1$, whence the functions $V_{-m}(s+2k_1+2k_2)$ are holomorphic for $s \in \Omega$. By Lemma 6.5 (ii), the functions $I_{-m}(y,s;k_1,k_2)$ are holomorphic for $s \in \mathbb{C}$. Therefore, the functions $f_{m;k_1,k_2}(y,s)$ are holomorphic for $s \in \Omega$, as long as $k_1+k_2 \geq N+1$. Now choose $d \in \mathbb{N}$ with $d > 2N+1$; then we have $\Omega \subseteq \{s \in \mathbb{C} \mid \mathrm{Re}(s) > 1/2 - k_1 - k_2 - d/2\}$, as long as $k_1 + k_2 \geq N+1$. Using Lemma 6.5 (iii), we estimate for $s \in \Omega$,

$$\sum_{n=N+1}^{\infty} \sum_{k_1+k_2=n} \left|f_{m;k_1,k_2}(y,s)\right|$$

$$\ll V_0(\mathrm{Re}(s)+2N+2) \frac{2^{\mathrm{Re}(s)} y^{-\mathrm{Re}(s)-d+1}}{|m|^d} \sum_{n=N+1}^{\infty} \sum_{k_1+k_2=n} \frac{|(\frac{s}{2})_{k_1}| \cdot |(\frac{s}{2})_{k_2}|}{k_1! \cdot k_2!} \cdot \frac{(k_1+k_2)^d}{y^{2(k_1+k_2)}}$$

$$\ll \frac{y^{-\mathrm{Re}(s)-d+1}}{|m|^d} \sum_{n=N+1}^{\infty} \frac{n^d}{y^{2n}} \sum_{k_1=0}^{n} \frac{(|\frac{s}{2}|)_{k_1} \cdot (|\frac{s}{2}|)_{n-k_1}}{k_1! \cdot (n-k_1)!} = \frac{y^{-\mathrm{Re}(s)-d+1}}{|m|^d} \sum_{n=N+1}^{\infty} \frac{n^d}{y^{2n}} \frac{(|s|)_n}{n!},$$

where the implied constants depend on Ω, d, and N, but are independent of m and k_1, k_2. Since the ratio of successive terms in the latter series has limit

$$\lim_{n\to\infty}\left|\frac{(n+1)^d\cdot(n+|s|)}{n^d\cdot(n+1)}\cdot\frac{1}{y^2}\right|=\frac{1}{y^2}<1,$$

we derive from d'Alembert's criterion that the series in question converges absolutely and locally uniformly for $s\in\mathbb{C}$ with $\mathrm{Re}(s)>-2N-1$, which proves the claim.

For later purposes, we note for $s\in\Omega$ the bound

$$\left|\sum_{n=N+1}^{\infty}\sum_{k_1+k_2=n} f_{m;k_1,k_2}(y,s)\right|\ll|m|^{-d}, \tag{35}$$

where $d\in\mathbb{N}$ with $d>2N+1$, and where the implied constant depends on z, Ω, d, and N, but is independent of m. □

Proposition 6.8. *For $z\in\mathbb{H}$ with $\mathrm{Im}(z)>1$, and $m\neq 0$, the following assertions hold:*

(i) The function $a_m(y,s)$ admits a meromorphic continuation to the whole s-plane with possible simple poles at $s=s_j-2N$ and $s=-s_j-2N+1$ $(N\in\mathbb{N})$.

(ii) Let $N\in\mathbb{N}$ and $\Omega\subseteq\{s\in\mathbb{C}\,|\,\mathrm{Re}(s)>-2N-1\}$ a compact subset not containing any pole of $a_m(y,s)$. Then, for all $s\in\Omega$, we have the bound

$$\left|a_m(y,s)\right|\ll|m|^{-d},$$

where $d\in\mathbb{N}$ with $d>2N+1$, and where the implied constant depends on z, Ω, d, and N, but is independent of m.

(iii) Let $N\in\mathbb{N}$ and $\tilde{s}$ a pole of $a_m(y,s)$ with $\mathrm{Re}(\tilde{s})=-2N+1/2$. Then, the residue of $a_m(y,s)$ at $\tilde{s}$ is bounded by

$$\left|\mathrm{res}_{s=\tilde{s}}a_m(y,s)\right|\ll|m|^{-d},$$

where $d\in\mathbb{N}$ with $d>2N+3$, and where the implied constant depends on z, $\tilde{s}$, d, and N, but is independent of m.

Proof. (i) As before, we obtain the meromorphic continuation of $a_m(y,s)$ to the whole s-plane by constructing its meromorphic continuations to the half-planes

$$\mathcal{H}_N:=\{s\in\mathbb{C}\,|\,\mathrm{Re}(s)>-2N-1\}$$

for any $N \in \mathbb{N}$. Applying Proposition 5.4 and using the notation from the proof of Lemma 6.7, we can write

$$a_m(y,s) = \sum_{n=0}^{N} \sum_{k_1+k_2=n} f_{m;k_1,k_2}(y,s) + \sum_{n=N+1}^{\infty} \sum_{k_1+k_2=n} f_{m;k_1,k_2}(y,s). \tag{36}$$

Since $\mathrm{Re}(s) > -2N-1$ by assumption, Lemma 6.7 proves that the series

$$\sum_{n=N+1}^{\infty} \sum_{k_1+k_2=n} f_{m;k_1,k_2}(y,s)$$

is a holomorphic function on the half-plane $\mathcal{H}_N$. Since the first double sum in (36) is a meromorphic function on the whole s-plane, we conclude that $a_m(y,s)$ has a meromorphic continuation to the half-plane $\mathcal{H}_N$.

In order to determine the poles of $a_m(y,s)$, we calculate its poles in the strip

$$\mathcal{S}_N := \{s \in \mathbb{C} \mid -2N-1 < \mathrm{Re}(s) \leq -2N+1\}$$

for any $N \in \mathbb{N}$. By considering $a_m(y,s)$ with its decomposition (36) in the strip $\mathcal{S}_N$, we see that the poles come from the finite sum

$$\sum_{n=0}^{N} \sum_{k_1+k_2=n} f_{m;k_1,k_2}(y,s) = 2^s y^s \sum_{n=0}^{N} V_{-m}(s+2n) \sum_{k_1=0}^{n} \frac{(\frac{s}{2})_{k_1} \cdot (\frac{s}{2})_{n-k_1}}{k_1! \cdot (n-k_1)!} \times I_{-m}(y,s;k_1,n-k_1),$$

which has possible simple poles at $s = s_j - 2N$ and $s = -s_j - 2N + 1$ in the strip $\mathcal{S}_N$ arising from the factors $V_{-m}(s+2n)$ ($n = 0, \ldots, N$). Therefore, the possible poles of $a_m(y,s)$ in the strip $\mathcal{S}_N$ are located at $s = s_j - 2N$ and $s = -s_j - 2N + 1$ ($N \in \mathbb{N}$).

(ii) In order to prove the second claim, we let $s \in \Omega$, where $\Omega \subseteq \mathcal{H}_N$ is a compact subset not containing any pole of $a_m(y,s)$, and we decompose $a_m(y,s)$ as in (36). Choosing now $d' \in \mathbb{N}$ with $d' > 4N+3$ and applying the bounds obtained in Lemma 6.5 (iii) (note that $\mathrm{Re}(s) > 1/2 - n - d'/2$ for $n = 0, \ldots, N$) and Lemma 6.6 (ii) (note that $\mathrm{Re}(s) + 2n > -2(N-n) - 1$ for $n = 0, \ldots, N$) to the finite double sum in (36) and the bound (35) to the remaining series in (36), we obtain the estimate

$$\begin{aligned} \left|a_m(y,s)\right| &\ll \sum_{n=0}^{N} \left|V_{-m}(s+2n)\right| \sum_{k_1=0}^{n} \left|I_{-m}(y,s;k_1,n-k_1)\right| + |m|^{-d'} \\ &\ll \sum_{n=0}^{N} |m|^{2(N-n)+2} \cdot |m|^{-d'} + |m|^{-d'} \ll |m|^{-(d'-2N-2)}, \end{aligned}$$

where the implied constants depend on z, Ω, d', and N, but are independent of m. Setting $d := d' - 2N - 2$ and observing that $d > 2N + 1$, the proof of part (ii) is complete.

(iii) As in the proof of (ii), we work from the decomposition (36). We let $\tilde{s}$ be a pole of $a_m(y,s)$ with $\tilde{s} = -2N + 1/2$, i.e., $\tilde{s} \in \mathcal{S}_N$. As before, choosing $d' \in \mathbb{N}$ with $d' > 4N + 3$, the bounds obtained in Lemmas 6.5 (iii) and 6.6 (iii) give the estimate

$$\left|\mathrm{res}_{s=\tilde{s}} a_m(y,s)\right| \ll \sum_{n=0}^{N} \left|\mathrm{res}_{s=\tilde{s}} V_{-m}(s+2n)\right| \sum_{k_1=0}^{n} \left|I_{-m}(y,\tilde{s};k_1,n-k_1)\right|$$

$$\ll \sum_{n=0}^{N} |m|^{2(N-n)} \cdot |m|^{-d'} \ll |m|^{-(d'-2N)},$$

where the implied constants depend on z, $\tilde{s}$, d, and N, but are independent of m. Setting $d := d' - 2N$ and observing that $d > 2N + 3$, the proof of part (iii) is also complete. □

Remark 6.9. By means of Remark 5.5, one can establish the meromorphic continuation of $a_m(y,s)$ $(m \neq 0)$ to the whole s-plane in the more general case $\mathrm{Im}(z) \neq \mathrm{Im}(\gamma^{-1}i)$ for any $\gamma \in \Gamma$ by applying Lemma 6.5 (noting that this lemma also holds for $y > 0$ and $m \in \mathbb{R}$, $m \neq 0$) as well as by using the same techniques as in Lemma 6.7 and Proposition 6.8 applied according to the modified situation. The poles of $a_m(y,s)$ and their residues turn out to be same as in the case $\mathrm{Im}(z) > 1$. Moreover, also the statements (ii) and (iii) of Proposition 6.8 generalize to the case $\mathrm{Im}(z) \neq \mathrm{Im}(\gamma^{-1}i)$ for any $\gamma \in \Gamma$.

Theorem 6.10. *For $z \in \mathbb{H}$ with $\mathrm{Im}(z) > 1$, the elliptic Eisenstein series $\mathcal{E}_{\mathrm{ell}}(z,s)$ has a meromorphic continuation to the whole s-plane with possible poles at $s = s_\Gamma - 2N$, $s = s_j - 2N$, and $s = -s_j - 2N + 1$ $(N \in \mathbb{N})$, where s_Γ is a pole of $\Gamma(s-1/2)\mathcal{E}_{\mathrm{par}}(i,s)$, and $s_j = 1/2 + it_j$ with $t_j > 0$ and $s_j(1-s_j) = \lambda_j$ a discrete eigenvalue of Δ_{hyp}.*

Proof. Let $z \in \mathbb{H}$ with $\mathrm{Im}(z) > 1$. For $s \in \mathbb{C}$ with $\mathrm{Re}(s) > 1$, we represent the elliptic Eisenstein series $\mathcal{E}_{\mathrm{ell}}(z,s)$ by its Fourier expansion

$$\mathcal{E}_{\mathrm{ell}}(z,s) = \sum_{m \in \mathbb{Z}} a_m(y,s) e(mx), \tag{37}$$

where the coefficients $a_m(y,s)$ are explicitly given by Propositions 5.2 and 5.4 for $m = 0$ and $m \neq 0$, respectively. By Propositions 6.3 and 6.8, the functions $a_m(y,s)$ admit a meromorphic continuation to the whole s-plane.

In order to prove the meromorphic continuation of $\mathcal{E}_{\mathrm{ell}}(z,s)$ to the whole s-plane, let $s \in \Omega$, where $\Omega \subseteq \{s \in \mathbb{C} \,|\, \mathrm{Re}(s) > -2N - 1\}$ for some $N \in \mathbb{N}$ is acompact subset not containing any pole of $a_m(y,s)$ for all $m \in \mathbb{Z}$. Choosing $d \in \mathbb{N}$, $d > 2N+2$, we have by Proposition 6.8 (ii) the bound

$$\sum_{\substack{m\in\mathbb{Z}\\ m\neq 0}} \left|a_m(y,s)e(mx)\right| \ll \sum_{\substack{m\in\mathbb{Z}\\ m\neq 0}} |m|^{-2},$$

where the implied constant depends on z, Ω, d, and N, but is independent of m. Therefore, the Fourier expansion (37) converges absolutely and uniformly in Ω. This proves that $\mathcal{E}_{\mathrm{ell}}(z,s)$ is holomorphic in $s \in \mathbb{C}$ away form the poles of $a_m(y,s)$ for $m \in \mathbb{Z}$.

Now let $\tilde{s} \in \mathbb{C}$ be a pole of $a_m(y,s)$ for $m \neq 0$ as in Proposition 6.8 (i); then, $\mathrm{Re}(\tilde{s}) = -2N + 1/2$ for some $N \in \mathbb{N}$. Choosing $d \in \mathbb{N}$, $d > 2N+3$, we estimate using Proposition 6.8 (iii),

$$\lim_{s\to\tilde{s}}(s-\tilde{s}) \sum_{\substack{m\in\mathbb{Z}\\ m\neq 0}} a_m(y,s)e(mx) \ll \sum_{\substack{m\in\mathbb{Z}\\ m\neq 0}} \left|\mathrm{res}_{s=\tilde{s}} a_m(y,s)\right| \ll \sum_{\substack{m\in\mathbb{Z}\\ m\neq 0}} |m|^{-2},$$

where the implied constant depends on z, $\tilde{s}$, d, and N, but is independent of m.

In this way we obtain the meromorphic continuation of $\mathcal{E}_{\mathrm{ell}}(z,s)$ to the whole s-plane with possible poles at $s = s_\Gamma - 2N$, $s = s_j - 2N$, and $s = -s_j - 2N + 1$ ($N \in \mathbb{N}$). The poles at $s = s_\Gamma - 2N$ are contributed by $a_0(y,s)$; here s_Γ denotes a pole of $\Gamma(s-1/2)\mathcal{E}_{\mathrm{par}}(i,s)$. □

Remark 6.11. Using Remark 6.9, one can establish the meromorphic continuation of $\mathcal{E}_{\mathrm{ell}}(z,s)$ to the whole s-plane in the more general case $\mathrm{Im}(z) \neq \mathrm{Im}(\gamma^{-1}i)$ for any $\gamma \in \Gamma$. The poles of $\mathcal{E}_{\mathrm{ell}}(z,s)$ and their residues turn out the be same as in the case $\mathrm{Im}(z) > 1$.

Remark 6.12. The elliptic Eisenstein series $\mathcal{E}_{\mathrm{ell}}(z,s)$ has a simple pole at $s = 1$ with residue

$$\begin{aligned}\mathrm{res}_{s=1}\mathcal{E}_{\mathrm{ell}}(z,s) &= \mathrm{res}_{s=1}a_0(y,s) = 2\pi\,\mathrm{res}_{s=1}\left(\frac{V_0(s)}{\Gamma(s/2+1/2)}\right)\\ &= \pi\,\mathrm{res}_{s=1}\mathcal{E}_{\mathrm{par}}(i,s) = \frac{2\pi}{\sharp\Gamma_i}\cdot\frac{1}{\mathrm{vol}_{\mathrm{hyp}}(\mathcal{F}_\Gamma)} = 3;\end{aligned}$$

here we used the decomposition (30) for $a_0(y,s)$ with $N = 0$.

References

[1] M. Abramowitz, I.A. Stegun, *Handbook of Mathematical Functions*, Volume I, McGraw-Hill, 1965.

[2] A.F. Beardon, *The Geometry of Discrete Groups*, Graduate Texts in Mathematics, Springer-Verlag, 1995.

[3] Y. Colin de Verdière, *Une nouvelle démonstration du prolongement méromorphe des séries d'Eisenstein*, C. R. Acad. Sci. Paris Sér. I Math. 293 (1981), 361–363.
[4] I.S. Gradshteyn, I.M. Ryzhik, *Table of Integrals, Series, and Products*, Academic Press, 1980.
[5] H. Huber, *Über eine neue Klasse automorpher Funktionen und ein Gitterpunktproblem in der hyperbolischen Ebene. I*, Comment. Math. Helv. 30 (1956), 20–62.
[6] H. Iwaniec, *Spectral Methods of Automorphic Forms*, Graduate Studies in Mathematics, Vol. 53, Amer. Math. Soc., 2002.
[7] J. Jorgenson, J. Kramer, *Bounding the sup-norm for automorphic forms*, Geom. Funct. Anal. 14 (2004), 1267–1277.
[8] J. Jorgenson, J. Kramer, *Canonical metrics, hyperbolic metrics and Eisenstein series for* $PSL_2(\mathbb{R})$, unpublished preprint.
[9] J. Jorgenson, J. Kramer, *Sup-norm bounds for automorphic forms and Eisenstein series*, in *Arithmetic Geometry and Automorphic Forms*, J. Cogdell et al. (eds.), ALM 19, 407–444, Higher Education Press and International Press, Beijing-Boston, 2011.
[10] J. Jorgenson, C. O'Sullivan, *Convolution Dirichlet series and a Kronecker limit formula for second-order Eisenstein series*, Nagoya Math. J. 179 (2005), 47–102.
[11] S.S. Kudla, J.J. Millson, *Harmonic Differentials and Closed Geodesics on a Riemann Surface*, Invent. Math. 54 (1979), 193–211.
[12] H. Neunhöffer, *Über die analytische Fortsetzung von Poincaréreihen*, S.-B. Heidelberger Akad. Wiss. Math.-Natur. Kl. (1973), 33–90.
[13] A.-M. v. Pippich, *The arithmetic of elliptic Eisenstein series*, Ph.D. thesis, Humboldt-Universität zu Berlin (2010).
[14] P. Sarnak, *Estimates for Rankin–Selberg L-functions and quantum unique ergodicity*, J. Funct. Anal. 184 (2001), 419–453.

Consequences of the Gross–Zagier formulae: Stability of average L-values, subconvexity, and non-vanishing mod p

Philippe Michel and Dinakar Ramakrishnan

In memory of Serge Lang

Abstract Applying the celebrated results of Gross and Zagier for central values of L-series of holomorphic forms of prime level, we deduce an exact average formula for suitable twists of such L-values, with a relation to the class number of associated imaginary quadratic fieds, thereby strengthening a result of Duke. We also obtain a stability result, as well as subconvexity (in this setting), and certain non-vanishing assertions.

Key words Average L-values • Gross–Zagier formulae • Non-vanishing • Subconvexity • Stability

Mathematics Subject Classification (2010): 11F11; 11F12; 11F67

1 Introduction

In this paper we investigate some consequences of the Gross–Zagier type formulae which were introduced by Gross and Zagier and then generalized in various directions by Hatcher, Zhang, Kudla, and others [1, 11, 12, 14, 40]. Let us now recall these formulae in the classical context. Denote by K an imaginary quadratic field of

D. Ramakrishnan (✉)
Department of Mathematics, California Institute of Technology, Pasadena, CA 91125, USA
e-mail: dinakar@caltech.edu

P. Michel
EPFL, 1015 Lausanne, Switzerland
e-mail: philippe.michel@epfl.ch

D. Goldfeld et al. (eds.), *Number Theory, Analysis and Geometry: In Memory of Serge Lang*, DOI 10.1007/978-1-4614-1260-1_20,

discriminant $-D$, associated quadratic character $\chi_{-D} = (\frac{-D}{\cdot})$, and ring of integers $\mathcal{O}_K$. For any character Ψ of the ideal class group $\mathrm{Pic}(\mathcal{O}_K)$ of K, let g_Ψ be the associated weight one theta series attached to Ψ on the upper half plane $\mathcal{H}$ given by

$$g_\Psi(z) = \sum_{m\geq 0} r_\Psi(m) q^m, \quad q = \exp(2\pi\iota z), z \in \mathcal{H},$$

where, for $m \geq 1$,

$$r_\Psi(m) = \sum_{N(\mathfrak{a})=m} \Psi(\mathfrak{a})$$

and $\mathfrak{a} \subset \mathcal{O}_K$ ranges over the $\mathcal{O}_K$-ideals of norm m. We will denote the trivial character of $\mathrm{Pic}(\mathcal{O}_K)$ by 1_K.

Now let f be a holomorphic new cusp form of level N coprime with D, trivial nebentypus and weight $2k$:

$$f(z) = \sum_{m\geq 1} a_m(f) q^m.$$

Depending on how the primes dividing N split in K, the Gross–Zagier formula expresses the central value at $s = k$ (or the derivative there) of the Rankin–Selberg L-function

$$L(s, f, \Psi) := L(2s - 2k + 1, \chi_{-D}) \sum_{m\geq 1} a_m(f) r_\Psi(m) m^{-s}$$

in terms of an intersection/height pairing of the f-isotypic component $e_{\Psi,f}$ of a cycle e_Ψ living in some Hecke module $M = M_{k,N}$: Denoting this pairing by $\langle\cdot,\cdot\rangle_M$ and the Petersson inner product on $S_{2k}(N)$ by

$$\langle f, g\rangle = \int_{Y_0(N)} f(z)\overline{g(z)} y^{2k} \frac{dxdy}{y^2},$$

where $Y_0(N)$ denotes the open modular curve $\Gamma_0(N)\backslash\mathcal{H}$, one has

$$c_{k,K} \frac{L^{(i)}(k, f, \Psi)}{\langle f, f\rangle} = \langle e_{\Psi,f}, e_{\Psi,f}\rangle_M \tag{1}$$

for some constant $c_{k,K} > 0$ and the order of derivative $i = i_{K,N}$ is 0 or 1 (depending on the sign of the functional equation). Originally the formula was proven as follows (for $i = 0$): let $M_{2k}(N)$ (resp. $S_{2k}(N)$) denote the space of holomorphic forms (resp. cusp forms) of weight $2k$ level N and trivial nebentypus. The map

$$f \mapsto L(k, f, \Psi),$$

being linear on $S_{2k}(N)$, can be represented by a kernel $f \mapsto \langle f, G_\Psi \rangle$ for some $G_\Psi \in M_{2k}(N)$ (same for the first derivative). By the Rankin–Selberg theory

$$L(k, f, \Psi) = \int_{Y_0(N)} f(z)\overline{g_\Psi(z)}\,\overline{E_{2k-1}(z)}\,y^{(2k+1)/2}\frac{dxdy}{y^2}$$

for a suitable holomorphic Eisenstein series E_{2k-1} of weight $2k-1$. The determination of G_Ψ amounts to first taking the trace from level $N' = lcm(4, N)$ to N, and then computing the projection of $g_\Psi(z)E_{2k-1}(z)$ on $M_{2k}(N)$. This can be done, and one infers from the computation of the Fourier expansion of $g_\Psi(z)E_{2k-1}(z)$ that the Fourier coefficients $a_m(G_\Psi)$ of G_Ψ are relatively elementary expressions involving the arithmetical functions r_Ψ and variants thereof: see below for an example. One the other hand, using the theory of complex multiplication, Gross and Zagier, and subsequently other people, showed by an auxiliary computation that

$$G_\Psi(z) = a_0(G_\Psi) + \sum_{m\geq 1}\langle T_m e_\Psi, e_\Psi\rangle_M\, q^m,$$

where T_m denotes the m-th Hecke operator acting on the module M. The final result then follows from a formal argument involving the multiplicity one theorem. The main observation underlying this paper is that the above computation provides formally an expression for the *average* of the central values $L(k, f, \Psi)$. Namely, if $\mathcal{F}_{2k}(N)$ denotes the set of arithmetically normalized new forms, then $\{f/\langle f, f\rangle^{1/2}\}_{f\in\mathcal{F}_{2k}(N)}$ may be completed to an orthonormal basis of $S_{2k}(N)$. Then decomposing G_Ψ along such an orthonormal basis, and taking the m-th Fourier coefficient in the above decomposition, one deduces, for any $m \geq 1$,

$$\sum_{f\in\mathcal{F}_{2k}(N)} \frac{L(k, f, \Psi)}{\langle f, f\rangle} a_m(f) \;=\; a_m(G_\Psi) + \mathcal{A}_{\rm old}(m) + \mathcal{A}_{\rm Eis}(m),$$

where $\mathcal{A}_{\rm old}(m)$, resp. $\mathcal{A}_{\rm Eis}(m)$, is the contribution from the old forms, resp. the Eisenstein series, of weight $2k$ and level N. In principle, the *Eisenstein series contribution* could be evaluated explicitly, while the *old forms contribution* could be computed by induction on N by following the same scheme, although there is an added complication of finding a suitable orthonormal basis. We shall consider here the nicest possible situation in which these additional contributions have a particularly simple expression, in fact where the old part vanishes! Therefore we obtain, by the first step of the proof of the Gross–Zagier type formulae, a simple expression for the first moment

$$\sum_{f\in\mathcal{F}_{2k}(N)} \frac{L(k, f, \Psi)}{\langle f, f\rangle} a_m(f).$$

Let us now turn to a more specific statement.

For $K = \mathbb{Q}(\sqrt{-D})$, denote by $\mathcal{O}_K$ its ring of integers, by $\mathrm{Pic}(\mathcal{O}_K)$ its ideal class group,

$$h = h_K = |\mathrm{Pic}(\mathcal{O}_K)|,$$

its class number, and

$$u = |\mathcal{O}_K^{\times}/\{\pm 1\}|.$$

Given any ideal class group character Ψ, recall that we have set for $m \geq 1$

$$r_\Psi(m) = \sum_{N(\mathfrak{a})=m} \Psi(\mathfrak{a});$$

for $\Psi = 1_K$ the trivial character, we shall also denote $r_{1_K}(m)$ by

$$R(m) = r_{1_K}(m) = \sum_{\substack{\mathfrak{a}\subset\mathcal{O}_K \\ N(\mathfrak{a})=m}} 1.$$

We extend the definition of $r_\Psi(m)$ to $m = 0$ by setting

$$r_\Psi(0) = \begin{cases} 0, & \text{if } \Psi \neq 1_K \\ h/2u, & \text{if } \Psi = 1_K. \end{cases}$$

We also set

$$\sigma_N(m) = \sum_{\substack{d|m \\ (d,N)=1}} d.$$

Specializing the formulas of Gross–Zagier and Gross [11, 12], and averaging over newforms f, we obtain

Theorem 1. *Let $-D < 0$ be an odd fundamental discriminant; let N be a prime which is inert in $K = \mathbb{Q}(\sqrt{-D})$ and let $k \geq 1$ be an integer. For Ψ a character of* $\mathrm{Pic}(\mathcal{O}_K)$*, and for any positive integer m, we have the following exact identity:*

$$\frac{2(2k-2)!D^{1/2}u^2}{(4\pi)^{2k}} \sum_{f\in\mathcal{F}_{2k}(N)} \frac{L(k,f,\Psi)}{\langle f,f\rangle} a_m(f)$$

$$= -\delta\frac{12h^2}{N-1}\sigma_N(m) + um^{k-1}r_\Psi(mD)h + u^2m^{k-1}\sum_{n=1}^{\frac{mD}{N}} \Phi_k(n,\Psi,N). \quad (2)$$

Here

$$\Phi_k(n,\Psi,N) = d((n,D))\delta_1(\Psi)R(n)r_\Psi(mD-nN)P_{k-1}(1-\frac{2nN}{mD}),$$

with P_{k-1} denoting the $(k-1)$-th Legendre polynomial; $d(n)$ is the number of divisors of n; $\delta \in \{0,1\}$ is 1 if and only if $(k,\Psi)=(1,1_K)$; $\delta_1(\Psi)\in\{0,1\}$ is 1 if D is prime, and when D is composite, it is 1 if and only if $\Psi^2=1_K$ and there exist ideals $\mathfrak{a},\mathfrak{b}$, of respective norms $mD-nN$ and n, such that, for a prime ideal $\mathfrak{q}$ of norm q congruent to $-N$ mod D, the class of $\mathfrak{abq}$ is a square in $\mathrm{Pic}(\mathcal{O}_K)$.

An asymptotic formula involving the average on the left was first established for $k=1,\Psi=1_K$ by W. Duke ([6]), which spurred a lot of other work, including that of Iwaniec and Sarnak ([19]) relating it to the problem of Siegel zeros for $L(s,\chi_{-D})$. In the work of the second named author with J. Rogawski ([32]), a different proof of Duke's result was given (for all weights), using Jacquet's relative trace formula involving the integration of the kernel over the square of the split torus, and in addition, the intervening measure was identified.

It is important to note that one obtains a *stability theorem* when N is sufficiently large compared with D and m, and this could perhaps be considered the most unexpected consequence of our approach. Indeed, when $N>mD$, the sum on the far right of the identity furnished by Theorem 1 becomes zero, and our exact average simplifies as follows:

Corollary 1 (Stability). *With the above notations and assumptions, suppose, moreover, that $N>mD$. Then one has*

$$\frac{2(2k-2)!D^{1/2}u^2}{(4\pi)^{2k}}\sum_{f\in\mathcal{F}_{2k}(N)}\frac{L(k,f,\Psi)}{\langle f,f\rangle}a_m(f)$$
$$=-\delta\frac{12h^2}{N-1}\sigma_N(m)+um^{k-1}r_\Psi(mD)h.$$

We call the range $N>mD$ the *stable range*. As one can check with other instances of the Gross–Zagier formulas, such as for the derivative in the case of odd order of vanishing, this phenomenon appears to be quite general. It has been recently generalized to Hilbert modular forms of square-free level by B. Feigon and D. Whitehouse ([9]), using the relative trace formula, now by integrating the kernel over a non-split torus; they are also able to treat more general characters Ψ.

When $\Psi=1_K$, we have the factorization

$$L(s,f,1_K)=L(s,f_K)=L(s,f)L(s,f\otimes\chi_{-D}),$$

where f_K denotes the base change of f to K, $L(s,f)$ the Hecke L-function of f, and $f\otimes\chi_{-D}$ the twist of f by χ_{-D}. Thus for $m=1$ and $N>D$, we get the following explicit identity involving the class number of K:

$$\frac{2(2k-2)!D^{1/2}u}{(4\pi)^{2k}}\sum_{f\in\mathcal{F}_{2k}(N)}\frac{L(k,f)L(k,f\otimes\chi_{-D})}{\langle f,f\rangle}=h\left(1-\delta\frac{12h}{u(N-1)}\right).$$

In the weight 2 case, as N is taken to be a prime here, the cardinality of $\mathcal{F}_2(N)$ is just the genus $g_0(N)$ of the compactification $X_0(N)$ of $Y_0(N)$. It is amusing to note that when $g_0(N)$ is zero one finds that

$$h = \frac{(N-1)u}{12},$$

implying that $h = 1$ when $(-D, N)$ is $(-3, 5)$, $(-7, 13)$, $(-8, 13)$ or $(-11, 13)$, agreeing with known data. Similarly, $X_0(11)$ is an elliptic curve $E/\mathbb{Q}$, and if we denote by E_{-D} the $-D$-twist of E, we see, for $D = 3$, that the algebraic special value $A(1, E)A(1, E_{-3})$ is just $1/5$. (We recall that the algebraic special value $A(1, E)$ of an elliptic curve E over $\mathbb{Q}$ is the ratio of $L(1, E)$ by the real fundamental period of E.) In general one gets more complicated identities, involving average central values, which are all compatible with the Birch and Swinnerton-Dyer conjecture for E, E_{-D}, and the Shafarevich–Tate groups $\mathrm{Sh}(E)$, $\mathrm{Sh}(E_{-D})$.

1.1 Application to the subconvexity problem

We will now discuss some simple applications of the above exact average formula, the first one being a subconvex estimate for the central values $L(k, f, \Psi)$. We refer to [20] for a general discussion on the subconvexity problem. By the work of Waldspurger, the central value $L(k, f, \Psi)$ is non-negative and the convexity bound is given by

$$0 \leq L(k, f, \Psi) \ll_{\varepsilon} (kND)^{\varepsilon} kN^{1/2} D^{1/2},$$

for any $\varepsilon > 0$. We prove here.

Corollary 2 (Subconvexity). *Preserve the notations of Theorem 1. Then for any* $\varepsilon > 0$, *we have*

$$L(k, f, \Psi) \ll_{\varepsilon} (kDN)^{\varepsilon} kN^{1/2} D^{1/2} \Big(\frac{1}{N^{1/2}} + \frac{N^{1/2}}{D^{1/2}}\Big).$$

In particular this improves on convexity as long as

$$(kD)^{\delta} \leq N \leq D(kD)^{-\delta}$$

for some fixed $\delta > 0$.

Note that this breaks convexity when N is in a suitable range depending on k, D in which case the bound is subconvex in all parameters: such bounds are called *hybrid*. We refer to [29] of other hybrid subconvex bounds valid in quite general circumstances. The present bound however does not seem to be implied by [29].

At this point we do not know of any application of these subconvex estimates, but we are intrigued by them because they come for free and seem to be hard to prove with the current methods of analytic number theory (e.g., see [5, 24]). Note also that such bounds are fundamentally limited to the critical center $s = k$. For a generalization to the Hilbert modular case, where Ψ is allowed to be any ray class character, see [9].

1.2 Application to non-vanishing problems

Another line of application addresses the existence of f for which $L(k, f, \Psi)$ does not vanish. Indeed several variants of such problems have been considered in the past by various methods [6,19,23,30,37]. Here we obtain non-vanishing results that are valid with a fairly large uniformity in the parameters, and again such uniformity seems hard to achieve by purely analytic methods.

Theorem 2. *Preserve the hypotheses of Theorem 1. Suppose further that*

$$N \gg_\delta D^{1/2+\delta}$$

for some $\delta > 0$. Then there exists $f \in \mathcal{F}_{2k}(N)$ such that

$$L(k, f, \Psi) \neq 0.$$

The same conclusion holds as long as $N > D$ and either $k \neq 1$ or $\Psi \neq 1_K$.

When $\Psi = 1_K$, we also obtain a non-vanishing result in a somewhat greater range:

Theorem 3. *Suppose $\Psi = 1_K$, $k = 1$ and*

$$h < \frac{N-1}{12}.$$

Then there exist f such that

$$L(k, f)L(k, f \otimes \chi_{-D}) \neq 0.$$

Non-vanishing theorems of this kind, with an *explicit* dependence between N and D (like $N > D$ or $N - 1 > 12h$), are of some interest. For instance, in the paper [27], Merel needs to consider the following problem: Given a prime p and a character χ of conductor p which is not even and quadratic, does there exist an $f \in \mathcal{F}_2(p)$ such that $L(1, f \otimes \chi) \neq 0$? In the appendix of that paper, the first named author and E. Kowalski prove that this is the case when p is greater than an explicit but very large number. In particular, it has so far not been possible to answer the problem numerically in the finitely many remaining cases; this has been answered, however, for $p < 1000$ [26].

Closer to the main concern of the present paper, Ellenberg [7, 8] uses analytic methods to prove the non-vanishing of the twisted L-function $L(1, f \otimes \chi_{-4})$ for some f in $\mathcal{F}_2(N)$ for N of the form p^2 or $2p^2$ (p an odd prime) and with prescribed eigenvalues at the Atkin–Lehner operators w_2, w_p, subject to an *explicit* lower bound on p. Ellenberg concludes from this the non-existence of primitive integral solutions to the generalized Fermat equation $A^4 + B^2 = C^p$ as long as $p > 211$; that this equation has only a finite number of primitive solutions is a theorem of Darmon and Granville. (Since this article was written, there has been further progress in "The Diophantine equation $A^4 + 2^d B^2 = C^n$" by M.A. Bennett, J. Ellenberg, and N.C. Ng, to appear in Int. J. Number Theory.) Another related set of examples is in the work of Dieulefait and Urroz ([4]). In a sequel to this paper under preparation ([28]), we will develop a suitable generalization of the exact average formula to a class of composite levels N, and investigate similar questions by modifying the method. This extension is subtle for three reasons: N is not square-free, D is not odd, and N, D are not relatively prime.

1.3 Nonvanishing modulo p

The exactness of the Gross–Zagier formulae even enable us to obtain *average non-vanishing results* for the *algebraic part* of the $L(k, f, \Psi)$ modulo suitable primes p. Again, such a question has been considered in the past, see for example [2, 37]. However, these earlier works addressed the question of the existence of the non-vanishing of $L(k, f, \Psi)$ mod p when the form f is *fixed* and when the character Ψ varies. Here our results go in the other direction as we fix p and let N and f vary. Given $f \in \mathcal{F}_{2k}(N)$ and g_Ψ as above, we denote by $L^{\mathrm{alg}}(k, f, \Psi)$ the algebraic part of $L(k, f, \Psi)$ (see Section 6, (11), for a precise definition). It follows from the work of Shimura that $L^{\mathrm{alg}}(k, f, \Psi)$ is an algebraic number satisfying the reciprocity law

$$L^{\mathrm{alg}}(k, f, \Psi)^\sigma = L^{\mathrm{alg}}(k, f^\sigma, \Psi^\sigma)$$

for any σ automorphism of $\mathbb{C}$ [33].

Theorem 4. *Let $p > 2k + 1$ be a prime, let $\mathcal{P}$ be a chosen place in $\overline{\mathbb{Q}}$ above p, and let N, D be as in Theorem 1. Suppose, moreover, that p does not divide $h = h_{-D}$, that $N > D$, and that N is greater than some absolute constant. Then there exists $f \in \mathcal{F}_{2k}(N)$ such that*

$$L^{\mathrm{alg}}(k, f, \Psi) \not\equiv 0 \pmod{\mathcal{P}}.$$

The question of the integrality of $L^{\mathrm{alg}}(k, f, \Psi)$ is quite subtle, and our result only concerns the numerator of the L-value. When $\Psi = 1_K$, we also prove the following variant:

Theorem 5. *Notations and assumptions as in Theorem 4. Suppose moreover that* $\Psi = 1$ *and* $N > pD$. *Then there exists* $f \in \mathcal{F}_{2k}(N)$ *such that*

$$\sqrt{D}(2\pi)^{-2k} \frac{L(k, f)L(k, f \otimes \chi_{-D})}{\langle f, f \rangle} a_p(f) \not\equiv 0 \pmod{\mathcal{P}^{2k-1}}.$$

The assertion makes sense because the left-hand side is (see Section 6.1) a p-unit times $a_p(f)$ times $L^{\text{alg}}(k, f, 1_K)$.

There are two fundamental periods $c^+(f)$ and $c^-(f)$ associated to f such that for any Dirichlet character ν, the special value $L^{\text{alg}}(k, f \otimes \nu)$, defined as $L(k, f \otimes \nu)/c^{\text{sgn}(\nu(-1))}(f)$ times a simple factor (see Section 6, (12)) is an algebraic number. One gets the near-factorization

$$\eta_f L^{\text{alg}}(k, f, 1_K) = L^{\text{alg}}(k, f) L^{\text{alg}}(k, f \otimes \chi_{-D}),$$

where η_f is essentially the order of the congruence module considered by Hida, Wiles, Taylor, Flach, Diamond, and others, which measures the congruences f has with other modular forms modulo p. The needed non-divisibility properties of η_f (for suitable p) are understood (at least) if f is ordinary or $k = 1$. Now finally, let us suppose we are in the classical weight 2 situation, i.e., with $\Psi = 1_K$ and $k = 1$.

Theorem 6. *Let* p *an odd prime not dividing* Dh_{-D}, *with* D *odd. Then there exist infinitely many newforms of* f *of prime level* N *and weight* 2 *such that*

$$\text{num}\left(\frac{L^{\text{alg}}(1, f \otimes \chi_{-D})}{\eta_f}\right) \not\equiv 0 \pmod{p},$$

where η_f *is the order of the congruence module of* f.

See Section 6 for a discussion of η_f, which measures the congruences that f may have with other modular forms of the same weight and level. An analogue of Theorem 6 should also hold in a suitable range of p, for forms of higher weight, and this question will be taken up elsewhere.

1.4 Acknowledgements

Serge Lang always conveyed infectious excitement about mathematics to anyone he came into contact with, and he will be missed. He was quite interested in the values of L-functions and in the *divisibility properties* of arithmetic invariants, and it is a pleasure to dedicate this article to him. We would like to thank B. Gross for discussions concerning the occurrence of a factorial in Theorems 5.5, 5.6 of [11], and also M. Flach, H. Hida, K. Prasanna, V. Vatsal, and D. Whitehouse for helpful conversations concerning parts of the paper. Thanks are also due to Paul Nelson and the anonymous referee for reading the paper carefully and finding a number of typos and small inaccuracies.

The first author would like to thank Caltech for its hospitality during the preparation of this work, and acknowledge the partial support he received from the ERC advanced research grant no. 228304. The second author was partially supported by the National Science Foundation through the grants DMS-0402044 and DMS-0701089.

2 The weight 2 case

It may be instructive to explain why the exact average formula holds in the weight 2 case when $\Psi = 1$. Let B be a quaternion division algebra over $\mathbb{Q}$, ramified only at (a prime) N and ∞, with maximal order $\mathcal{O}$. Let Y be the associated rational curve such that $\mathrm{Aut}(Y) = \mathrm{B}^\times/\mathbb{Q}^\times$. Put

$$X = \mathrm{B}^\times \backslash Y \times \hat{\mathrm{B}}^\times / \hat{\mathcal{O}}^\times = \cup_{j=1}^n \Gamma_j \backslash Y,$$

where $\hat{\mathrm{B}}^\times = \prod_p{}' \mathrm{B}_p^\times$ and $\hat{\mathcal{O}}^\times = \prod_p \mathcal{O}_p^\times$, with each Γ_j being a finite group. Then $\mathrm{Pic}(X)$ is isomorphic to $\mathbb{Z}^n$ with natural basis $\{e_1, e_2, \ldots, e_n\}$, where each e_j is the class of $\Gamma_j \backslash Y$. Since N is inert in $K = \mathbb{Q}(\sqrt{-D})$, there is an embedding $f \in \mathrm{Hom}(K, \mathrm{B}) = Y(K)$. It results in certain *Heegner points*,

$$x = \mathrm{B}^\times \backslash \mathrm{B}^\times .(f, b) \in X$$

of discriminant $-D$, with $b \in \hat{\mathrm{B}}^\times / \hat{\mathcal{O}}^\times$. The set of these Heegner points is acted on transitively by $\mathrm{Pic}(\mathcal{O}_K)$, and for $A \in \mathrm{Pic}(\mathcal{O}_K)$ and x a Heegner point, we denote by x_A the corresponding translate; let

$$c = \sum_A x_A,$$

for A running over ideal classes of K. For any weight 2 eigenform f, let c_f denote the f-component c. Then by a beautiful theorem of B. Gross ([12, Prop.11.2]), providing an analogue for the L-value of the Gross-Zagier theorem for the first derivative, one has

$$\langle c_f, c_f \rangle_X = \frac{u^2\sqrt{D}}{8\pi^2} \frac{L(1, f)L(1, f \otimes \chi_{-D})}{\langle f, f \rangle},$$

where $\langle \cdot, \cdot \rangle_X$ is a natural *height pairing* on $\mathrm{Pic}(X)$. We have by orthogonality,

$$\langle c, T_m c \rangle_X = \langle c_E, T_m c_E \rangle_X + \sum_f \langle c_f, T_m c_f \rangle_X,$$

where T_m is the operator corresponding to the m-th Hecke operator on $M_2(N)$, f runs over newforms in $M_2(N)$, and E denotes the unique (holomorphic) Eisenstein series (of weight 2 and level N). Using the fact that f and E are Hecke eigenforms, and that $\langle c_E, c_E\rangle_X = \frac{12h^2}{N-1}$ (cf. [12, (11.7)]), we get, by averaging Gross's formula,

$$\frac{u^2\sqrt{D}}{8\pi^2}\sum_f \frac{L(1,f)L(1,f\otimes\chi_{-D})}{\langle f,f\rangle} = -\sigma_N(m)\frac{12h^2}{N-1} + \langle c, T_m c\rangle_X.$$

One has

$$\langle c, T_m c\rangle_X = \sum_A\sum_B \langle x_B, T_m x_{AB}\rangle_X,$$

and ([12, Prop. 10.8])

$$\sum_B \langle x_B, T_m x_{AB}\rangle_X = uhr_A(m) + \sum_{n=1}^{mD/N} r_A(mD-nN)d((n,D))R_{\{-NA\}}(n);$$

here $r_A(n)$ is the number of $\mathcal{O}_K$-ideals of norm n in the class A and $R_{\{-NA\}}(n)$ is a variant of $r_A(n)$ whose definition is recalled in (3) below; when D is prime, $R_{\{-NA\}}(n)$ is just $r_A(n)$.

The assertion of Theorem 1 now follows by summing over $A \in \mathrm{Pic}(\mathcal{O}_K)$. Moreover, when mD is less than N, $\sum_B \langle x_B, T_m x_{AB}\rangle_X$ simply equals $uhR_A(m)$, and this furnishes Corollary 1 (stability) in the weight 2 case.

3 Proof of the main identity for all $2k \geq 2$

3.1 Preliminaries

For $N \geq 1$, let $M_{2k}(N)$ (resp $S_{2k}(N)$) denote, as usual, the space of holomorphic modular forms (resp. cuspforms) of weight $2k$, level N and trivial character. For $f \in M_{2k}(N)$, we write the Fourier expansion at the infinite cusp as

$$f(z) = \sum_{m\geq 0} a_m(f)q^m, q = \exp(2\pi\imath z).$$

We denote by $\mathcal{F}_{2k}(N)$, the set of cuspidal newforms f (normalized in the usual way, so that the first Fourier coefficient $a_1(f)$ is 1.) Whenever it converges, we denote the Petersson inner product on $M_{2k}(N)$ by

$$\langle f, g\rangle = \int_{Y_0(N)} f(z)\overline{g(z)}y^{2k}\frac{dxdy}{y^2}.$$

With notations as in the introduction, define, for any ideal class $A \in \mathrm{Pic}(\mathcal{O}_K)$,

$$r_A(m) = \begin{cases} |\{\mathfrak{a} \subset \mathcal{O}_K, N(\mathfrak{a}) = m, \mathfrak{a} \in A\}| & \text{if } m \geq 1 \\ \frac{1}{2u} & \text{if } m = 0. \end{cases}$$

We also need a slight variant $R_{\{-NA\}}(m)$ defined as follows: given q a prime such that $q \equiv -N(\mathrm{mod}\ D)$, when q splits and writing $q\mathcal{O}_K = \mathfrak{q}\bar{\mathfrak{q}}$, we set

$$R_{\{-NA\}}(n) = |\{I : N(I) = n, \mathfrak{q}AI \in \mathrm{Pic}(\mathcal{O}_K)^2\}|. \tag{3}$$

Observe that this definition is independent of the choice of q and that when D is prime, $R_{\{-NA\}}(n)$ is just $R_A(n)$.

The theta series

$$\theta_A(z) = \sum_{m \geq 0} r_A(m) q^m, q = \exp(2\pi \imath z)$$

is a modular form of weight 1, level D and central character χ_{-D}. Moreover, for any $\Psi \in \widehat{\mathrm{Pic}(\mathcal{O}_K)}$, put

$$\theta_\Psi(z) = \sum_A \overline{\Psi}(A)\theta_A(z),$$

whose Fourier coefficients are then given by

$$a_m(\theta_\Psi) = \sum_A \overline{\Psi}(A) r_A(m) = r_{\overline{\Psi}}(m) = r_\Psi(m).$$

In particular, the constant term $a_0(\theta_\Psi)$ equals $\frac{1}{2u}\sum_A \overline{\Psi}(A)$, which is, by orthogonality, zero if and only if $\Psi \neq 1_K$, when θ_Ψ is a cusp form. Setting

$$L(s, f, A) := L^{(N)}(1 + 2(s-k), \chi_{-D}) \sum_{m \geq 1} \frac{a_m(f) r_A(m)}{m^s},$$

with

$$L^{(N)}(s, \chi_{-D}) \sum_{\substack{n>1 \\ (n,N)=1}} \frac{\chi_{-D}(n)}{n^s},$$

one has

$$L(s, f, \Psi) = \sum_{A \in \mathrm{Pic}(\mathcal{O}_K)} \Psi(A) L(s, f, A).$$

Define a holomorphic function G_A on the upper half plane $\mathcal{H}$, invariant under $z \to z+1$, by means of its Fourier expansion at infinity:

$$G_A(z) := \sum_{m=0}^{\infty} b_{m,A} q^m, \tag{4}$$

where

$$b_{m,A} = m^{k-1} \frac{h}{u} r_A(mD) \tag{5}$$
$$+ m^{k-1} \sum_{n=1}^{mD/N} \delta(n) r_A(mD - nN) R_{\{-NA\}}(n) P_{k-1}\left(1 - \frac{2nN}{mD}\right).$$

In this definition, u and $R(n) = \sum_A r_A(n)$ are as in the introduction, $R_{\{-NA\}}$ as in (3), $\delta(n)$ is 1 (resp. 2) if $(m, D) = 1$ (resp. $\neq 1$), and for $r \geq 0$, P_r is the r-th Legendre polynomial defined by

$$P_r(x) := \frac{1}{2^r} \sum_{m=1}^{[r/2]} (-1)^m \binom{r}{m} \binom{2r-2m}{r} x^{r-2m}.$$

The following result, due to Gross and Zagier, is crucial here:

Theorem 7. *G_A is a modular form of weight $2k$, level N, and trivial character; it is cuspidal if $k > 1$, and for every newform f of weight $2k$ and level N, we have*

$$L(k, f, A) = \frac{(4\pi)^{2k}}{2(2k-2)!D^{1/2}} \langle f, G_A \rangle.$$

For $k = 1$, see [12, p.291, Prop. 9.1], and for general k, this is in [11, Thm (5.6), p. 291].

3.2 Correction of a small error in [11], Theorem 5.6

We take this opportunity to correct a small error in [11]: The factor $(k-1)!$ should not be present in the numerator of the right-hand side of the second formula of Theorem (5.6) on p. 291 of [11]; it is "canceled" by the $(k-1)!$ term in the denominator on the right-hand side of the formula in Prop. (4.4) on p. 283 of *loc. cit.* This does not affect anything concerning the weight 2 case (when $k = 1$), which was the main thrust of [11]. Even for general k, it is of little consequence when k is fixed. It is however crucial for us, especially since we would like to study the behavior on average when k becomes large, and so we give a detailed discussion here.

To begin, in the statement of Theorem 5.5 (on page 290 of [11]), the formula should read

$$L_{\mathcal{A}}(f, 2k-1-r) = \frac{(-1)^{k-r}(2\pi)^{2(2k-1-r)}}{(2k-2-r)!} \frac{2^{2k-1}}{(2k-2)!} \frac{\varepsilon(N)r!}{|D|^{2k-r-1/2}} \times \langle f, \sum b_{m,r} q^m \rangle, \qquad (*)$$

with

$$b_{m,r} = \sum_{n=0}^{m|D|/N} r_{\mathcal{A}}(m|D| - nN) P_{k,r}(Nn, m|D|) \sigma_{2k-2r-2,\mathcal{A}}(n).$$

The only difference from the formula of Gross–Zagier is the appearance of $(2k-2-r)!$ in the denominator, instead of what they have in [11], namely $(2k-2-2r)!$.

To see that $(*)$ is the right formula, note that by Proposition 1.2 of [11], which is used in the proof (see p. 272),

$$(4\pi)^{-s-2k+1} N^s \Gamma(s+2k-1) L_{\mathcal{A}}(f, 2k-1-r) = \langle f, \tilde{\Phi}_{\bar{s}} \rangle.$$

To get Theorem 5.5 (of [11]), Gross and Zagier apply their Proposition 1.2 with $s = -r$, and so, using $\Gamma(2k-1-r) = (2k-2-r)!$, one obtains

$$L_{\mathcal{A}}(f, 2k-1-r) = \frac{(4\pi)^{2k-1-r} N^r}{(2k-2-r)!} \langle f, \tilde{\Phi}_{-r} \rangle. \qquad (**)$$

By Corollary 3.4 on page 281 (of *loc. cit.*),

$$\tilde{\Phi}_{-r}(z) = \sum_{m=0}^{\infty} a_{m,r}(y) e^{2\pi i m z},$$

where

$$a_{m,r}(y) = \sum_{n=0}^{m\delta/N} e_{n,r}(y) r_{\mathcal{A}}(m\delta - nN),$$

with $e_{n,r}(y)$ being given by the formulae on the top of page 282.

When $r = 0$, the form $\tilde{\Phi}_{-r}(z)$ is holomorphic. For $k > 1$ and $r > 0$, we may apply Proposition 5.1 on page 288 (of [11]) to conclude that

$$\langle f, \tilde{\Phi}_{-r} \rangle = \langle f, \Phi_{-r} \rangle,$$

where Φ_{-r} is the holomorphic projection given by

$$\Phi_{-r}(z) = \sum_{m=1}^{\infty} a_{m,r} e^{2\pi i m z}$$

with

$$a_{m,r} = \frac{(4\pi m)^{2k-1}}{(2k-2)!} \int_0^\infty a_{m,r}(y) e^{-4\pi m y} y^{2k-2} dy.$$

Working through the calculation of Gross and Zagier on the bottom of page 289 and the top of page 290, making use of Prop. 5.1, Cor. 3.4 and Prop. 4.4, we get exactly the same expression for the coefficients of Φ_{-r} as Gross and Zagier do on page 290, namely,

$$a_{m,r} = \frac{(-1)^{k-r} 2^{2k-1} \varepsilon(N) r! \pi^{2k-1-r}}{(2k-2)! N^r |D|^{2k-r-1/2}} b_{m,r},$$

with

$$b_{m,r} = \sum_{n=0}^{m|D|/N} r_{\mathcal{A}}(m|D| - nN) P_{k,r}(Nn, m|D|) \sigma_{2k-2r-2,\mathcal{A}}(n),$$

where $P_{k,r}(x, y)$, resp. $\sigma_{2\ell,\mathcal{A}}(n)$, is given by (5.3), resp. (5.4), on page 290.

It is important to pause and note that the expression for $b_{m,r}$ correctly involves $\sigma_{2k-2r-2,\mathcal{A}}(n)$ (and not $\sigma_{2k-r-2,\mathcal{A}}(n)$). The source of this is that $e_{n,r}(y)$ (occurring in the definition of $a_{m,r}(y)$) is $e_{-r}(n, \frac{Ny}{\delta}) e^{2\pi N n y/\delta}$ (see the top of page 282 of [11]), and for $n \neq 0$, we have from page 279,

$$e_s(n, y) = i \delta^{-s-2k+1/2} L_n(s) y^{-s-2k+2} V_s(ny),$$

where

$$L_n(s) = \sum_{D=D_1 D_2, D_2 | n} \varepsilon_{D_1}(-N) \chi_{D_1 D_2}(\mathcal{A}) \sum_{m | n/\delta_2, m>0} \frac{\varepsilon_{D_1}(m\delta_2) \varepsilon_{D_2}(n/m\delta_2)}{(m\delta_2)^{2s+2k-2}}.$$

Thus $L_n(s)$ is a sum of Dirichlet series at the argument $2s + 2k - 2$, and we are letting $s = -r$.

In any case, from the expression above for $a_{m,r}$, we get

$$\langle f, \tilde{\Phi}_{-r} \rangle = \frac{(-1)^{k-r} 2^{2k-1} \varepsilon(N) r! \pi^{2k-1-r}}{(2k-2)! N^r |D|^{2k-r-1/2}} \langle f, \sum b_{m,r} q^m \rangle.$$

Plugging this in (∗∗), canceling N^r, and consolidating the powers of 2π, we get (∗) as asserted.

These calculations extend to the case $r = k - 1$, when we get a fortuitous *cancelation* as the $r!$ in the numerator and the $(2k - 2 - r)!$ in the denominator are the same in (∗). If one had $(2k - 2 - 2r)!$ in the denominator, it would become 1 when $r = k - 1$, and this is why the formula in Theorem 5.6 (on page 291) has a spurious $(k - 1)!$ in the numerator. Again, this was not of major concern to Gross and Zagier as they were mainly considering $k = 1$ with a dramatic application to elliptic curves. It is amusing that it also doesn't matter for the $k = 2$ example Gross and Zagier work out at the bottom of page 291. We care only because the appearance of the unwanted $(k-1)!$ in the numerator would contradict the general RH.

It follows, after the simplifications Gross and Zagier indicate in [11], that Theorem 5.6 holds when the expression for $L_{\mathcal{A}}(f,k)$ is corrected to be the following:

$$L_{\mathcal{A}}(f,k) = \frac{(2\pi)^{2k}2^{2k-1}}{(2k-2)!|D|^{k-1/2}}\langle f, \sum b_{m,\mathcal{A}}q^m\rangle,$$

where $b_{m,\mathcal{A}}$ is as in the beginning of Theorem 5.6 of *loc. cit.*

3.3 The exact average formula

Let

$$E = E_{2,N} = \sum_{n=0}^{\infty} a_n(E)q^n$$

denote a holomorphic Eisenstein series for $\Gamma_0(N)$ of weight 2. Since N is prime, the modular curve $Y_0(N)$ has only two cusps, namely ∞ and 0. It then follows that E is unique up to scalar multiple, and so $E(z)/a_0(E)$ is well defined with constant term 1 at ∞. To be specific, we will take

$$E(z) = \frac{N-1}{12} + \sum_{m=1}^{\infty} \sigma_N(m)q^m,$$

where $\sigma_N(m) = \sum_{d|m,(d,N)=1} d$.

For $A \in \mathrm{Pic}(\mathcal{O}_K)$, with G_A being as in the previous section, put

$$G_A^{\mathrm{cusp}}(z) := G_A(z) - \delta_{k=1}\frac{b_{0,A}}{a_0(E)}E(z). \tag{6}$$

Then G_A^{cusp} is a holomorphic cuspform of level N, weight $2k$, and trivial character, with coefficients $a_m(G_A^{\mathrm{cusp}})$.

Lemma 3.1. *For $-D$ an odd fundamental discriminant and N a prime inert in K, we have, for any $m \geq 1$,*

$$\frac{2(2k-2)!D^{1/2}}{(4\pi)^{2k}} \sum_{f\in\mathcal{F}_{2k}(N)} \frac{L(k,f,A)}{\langle f,f\rangle}a_m(f)$$
$$= a_m(G_A^{\mathrm{cusp}}) = b_{m,A} - \delta_{k=1}\frac{b_{0,A}}{a_0(E)}a_m(E).$$

In order to prove this, we first need the following

Lemma 3.2. *Assume that N is a prime that is inert in $K = \mathbb{Q}(\sqrt{-D})$. Let f be any old form in $S_{2k}(N)$. Then we have, for every $A \in \mathrm{Pic}(\mathcal{O}_K)$,*

$$\langle f, G_A^{\mathrm{cusp}}\rangle = 0.$$

Such a lemma will not in general hold for composite N.

3.4 Proof of Lemma 3.2

There is nothing to prove when $k < 6$, since $S_{2k}(1)$ is zero in that case. Suppose that $k \geq 6$; since f is cuspidal, it suffices to prove that $\langle f, G_A\rangle = 0$. Put

$$G_\Psi := \sum_{A \in \mathrm{Pic}(\mathcal{O}_K)} \Psi(A) G_A.$$

It is sufficient to show that $\langle f, G_\Psi\rangle = 0$ for all ideal class characters Ψ of K. If $f = \sum_{n=1}^{\infty} a_n(f) q^n$, put

$$D(s, f \times \theta_\Psi) = L^{(N)}(1 + 2(s-k), \chi_{-D}) \sum_{n=1}^{\infty} \frac{a_n(f)\overline{a}_n(\theta_\Psi)}{n^s}. \tag{7}$$

Then the Rankin–Selberg method ([11, IV, §1]) gives the identity

$$(4\pi)^{-k}\Gamma(k) D(k, f \times \theta_\Psi) = \langle f, \mathrm{Tr}_{ND/N}(\theta_\Psi \mathcal{E}_{2k-1,N})\rangle \tag{8}$$

where $\mathcal{E}_{2k-1,N}$ is the result of slashing a holomorphic Eisenstein series of weight $2k-1$ (and character χ_{-D}) with the Atkin involution w_N, and $\mathrm{Tr}_{ND/D}$ denotes the trace from $\tilde{M}_{2k}(ND)$ to $\tilde{M}_{2k}(N)$, where $\tilde{M}_r(M)$ denotes the space of non-holomorphic modular forms of level M and weight r. Moreover, the calculations of Gross and Zagier ([11, IV, §5]) show that G_Ψ is the holomorphic projection of $\mathrm{Tr}_{ND/N}(\theta_\Psi \mathcal{E}_{2k-1,N})$, so that

$$\langle f, G_\Psi\rangle = \langle f, \mathrm{Tr}_{ND/N}(\theta_\Psi \mathcal{E}_{2k-1,N})\rangle.$$

Let f be a newform of level 1 (and weight $2k$). Then since N is prime, it defines two old forms of level N, namely $f_1(z) = f(z)$ and $f_N(z) = f(Nz)$, so that $a_m(f_N)$ is zero unless $N|m$, and $a_{mN}(f_N) = a_m(f)$. Since the new and old forms are orthogonal to each other under $\langle\cdot,\cdot\rangle$, and since the space of old forms of level N are spanned by $\{f_d, d = 1, N\}$ with f running over all the cuspforms of level 1, it suffices to prove that each $D(k, f_d \times \theta_\Psi) = 0$. For $d = 1$, one has

$$D(s, f_1 \times \theta_\Psi) = \left(1 - \frac{\chi_{-D}(N)}{N^s}\right) L(s, f \times \theta_\Psi). \tag{9}$$

This reduces to checking the vanishing of the right-hand side. Since f has level 1, the root number of $L(s, f \times \theta_\Psi)$ is -1, yielding the requisite vanishing. When $d = N$, $D(k, f_2 \times \theta_\Psi)$ is still a multiple of $L(k, f \times \theta_\Psi)$ and is therefore zero. □

3.5 *Proof of Lemma 3.1*

We may choose an orthogonal basis $\mathcal{B}$ of $S_{2k}(N)$ to be of the form $\mathcal{F}_{2k}(N) \cup \mathcal{B}'$, where $\mathcal{B}'$ consists of old forms. Clearly we have

$$\sum_{f \in \mathcal{B}} \frac{\langle f, G_A^{\mathrm{cusp}} \rangle}{\langle f, f \rangle} f = G_A^{\mathrm{cusp}}. \tag{10}$$

In view of Lemma 3.2, the sum on the left-hand side needs to run only over newforms f. Applying Theorem 7, and using (10), we obtain

$$\frac{2(2k-2)!D^{1/2}}{(4\pi)^{2k}} \sum_{f \in \mathcal{F}_{2k}(N)} \frac{L(k, f, A)}{\langle f, f \rangle} f = G_A^{\mathrm{cusp}}.$$

The lemma now follows by taking the m-th coefficient of the above identity. □

3.6 *Proof of Theorem 1*

The exact average formula follows by performing the averaging process $\sum_{A \in \mathrm{Pic}(\mathcal{O}_K)} \Psi(A) \ldots$ on both sides of the formula in Lemma 3.1, using the formula (5) for the coefficients $b_{m,A}$, and by noting that

$$\frac{a_m(E)}{a_0(E)} = \frac{12}{N-1} \sigma_N(m)$$

and that, when $k = 1$, $b_{0,A} = \frac{h}{2u^2}$. □

4 Subconvex Bounds

In this section, we prove Corollary 2. By the work of Waldspurger, Guo and Jacquet [13, 38] (see also [22] for $\Psi = 1_K$),

$$L(k, f, \Psi) \geq 0.$$

Thus from formula (2) for $m = 1$, we have

$$\frac{2(2k-2)!D^{1/2}}{(4\pi)^{2k}}\frac{L(k,f,\Psi)}{\langle f,f\rangle} \le \frac{h}{u} + \sum_{n=1}^{\frac{D}{N}} |\Phi_k(n,\Psi,N)|.$$

Since $|P_{k-1}(x)| \le 1$ for $|x| \le 1$ and $R(n), |r_\Psi(n)| \le d(n)$, so that

$$R(n)|r_\Psi(D-nN)| \le d(n)^2 + d(D-nN)^2,$$

we see that the n-sum on the right side is bounded by $\frac{D}{N}(\log D)^3$. From the class number formula, we have

$$h \ll D^{1/2}\log D$$

and

$$\langle f,f\rangle \ll (4\pi)^{-2k}(2k-1)!N(\log N)^3$$

as follows from [21, (2.31)]. Unlike the corresponding bound for Maass forms ([18]), this upper bound is elementary since f is holomorphic and by Deligne its Fourier coefficients satisfy the Ramanujan–Petersson bound. Thus we see that

$$L(k,f,\Psi) \ll (\log kN \log D)^3 k(N + D^{1/2}).$$

□

5 Application to non-vanishing

We prove here Theorem 2. Arguing exactly as above we have

$$\frac{2(2k-2)!D^{1/2}}{(4\pi)^{2k}} \sum_{f\in\mathcal{F}_{2k}(N)} \frac{L(k,f,\Psi)}{\langle f,f\rangle} = \frac{h}{u} - \delta\frac{6(h/u)^2}{N-1} + O\left(\frac{D}{N}(\log D)^3\right)$$

$$= \frac{h}{u} + O\left(\frac{D}{N}(\log D)^3\right).$$

By Siegel's theorem, which gives $h = D^{1/2+o(1)}$, we see that the right side is positive as soon as $N > D^{1/2+\delta}$ for some $\delta > 0$. If $N > D$, then we are in the stable range and we have

$$\frac{2(2k-2)!D^{1/2}}{(4\pi)^{2k}} \sum_{f\in\mathcal{F}_{2k}(N)} \frac{L(k,f,\Psi)}{\langle f,f\rangle} = \frac{h}{u}\left(1 - \delta\frac{6(h/u)}{N-1}\right). \tag{11}$$

When $\delta = 0$, this concludes the proof of Theorem 2 since $h \ge 1$. □

Suppose now that $\delta = 1$ (i.e., $k = 1, \Psi = 1_K$). Then we remark that

$$\sum_{n=1}^{\frac{D}{N}} \Phi_1(n, 1, N) \geq 0,$$

so that

$$\frac{D^{1/2}}{8\pi^2} \sum_{f \in \mathcal{F}_{2k}(N)} \frac{L(k, f, \Psi)}{\langle f, f \rangle} \geq \frac{h}{u}\left(1 - \frac{6(h/u)}{N-1}\right)$$

completing the proof of Theorem 3. □

6 Non-vanishing mod p

6.1 Algebraic Parts of L-values

Let us put

$$L^{\mathrm{alg}}(k, f, \Psi) = (-1)^k (2\pi)^{-2k} (k-1)!^2 g(\chi_{-D}) \frac{L(k, f, \Psi)}{\langle f, f \rangle}, \tag{12}$$

where $g(\chi_{-D}) = \sum_{x \pmod D} \chi_{-D}(x) \exp(2\pi i \frac{x}{D})$ is the Gauss sum. Then it is known by Shimura ([33], see also [16]), that $L^{\mathrm{alg}}(k, f, \psi)$ is an algebraic number obeying the reciprocity law:

$$L^{\mathrm{alg}}(k, f^\sigma, \Psi^\sigma) = L^{\mathrm{alg}}(k, f, \Psi)^\sigma,$$

for every automorphism σ of $\mathbb{C}$.

Next recall that for $\Psi = 1_K$, we have the factorization $L(k, f, \Psi) = L(k, f)$ $L(k, f \otimes \chi_{-D})$. For any Dirichlet character ν, the algebraic part of $L(k, f \otimes \nu)$ is given by

$$L^{\mathrm{alg}}(k, f \otimes \nu) = g(\overline{\nu})(k-1)! \frac{L(k, f \otimes \nu)}{(-2\pi i)^k c_\pm(f)}, \tag{13}$$

where $c_\pm(f)$ is a fundamental period of f, with $\pm = \nu(-1)$. Again, one has for any automorphism σ of $\mathbb{C}$, $L^{\mathrm{alg}}(k, f^\sigma \otimes \nu^\sigma)$ is $L^{\mathrm{alg}}(k, f \otimes \nu)^\sigma$.

This leads to the near-factorization

$$\eta_f L^{\mathrm{alg}}(k, f, 1_K) = L^{\mathrm{alg}}(k, f) L^{\mathrm{alg}}(k, f \otimes \chi_{-D}), \tag{14}$$

where η_f equals – thanks to a series of papers of Hida (cf. [16], [17]), Wiles ([39]), Taylor–Wiles ([35]), and Diamond–Flach–Guo ([3]) – the order of the congruence module of f, i.e., the number which counts the congruences of f with other modular forms of the same weight and level.

6.2 Proof of Theorems 4 and 5

From the definition of the algebraic part, the hypothesis of Theorem 4 and the formula (11), used in conjunction with $\delta = 0$, we have (up to multiplication by a p-unit)

$$\sum_{f \in \mathcal{F}_{2k}(N)} L^{\mathrm{alg}}(k, f, \Psi) = \frac{h}{u}.$$

The conclusion of Theorem 4 is immediate.

For the proof of Theorem 5, we have, assuming that $N > pD$,

$$\sum_{f \in \mathcal{F}_{2k}(N)} L^{\mathrm{alg}}(k, f, 1_K) = \frac{h}{u}\left(1 - \delta_{k=1}\frac{12(h/u)}{N-1}\right).$$

Therefore the conclusion holds except possibly if $p|(1 - \frac{12(h/u)}{N-1})$. Suppose we are in that latter case. Then we apply the exact formula of Corollary 1 with $m = p$ and get

$$\sum_{f \in \mathcal{F}_{2k}(N)} L^{\mathrm{alg}}(k, f, 1_K)a_p(f) = \frac{h}{u}\left(R(p) - \frac{12(h/u)}{N-1}(p+1)\right).$$

$R(p)$ is either 0 or 2, if it is zero, then the left-hand side of the previous formula is not divisible by p. If $R(p) = 2$, then $2 - \frac{12(h/u)}{N-1}$ is not divisible by p since by assumption $p|(1 - \frac{12(h/u)}{N-1})$. So we are done in all cases. □

6.3 *Proof of Theorem 6*

Here we are restricting to the weight 2 case, and by the theory of modular symbols, cf. Stevens [34] and Vatsal [36] (see also Prasanna [31]), we know that for any Dirichlet character ν, the special value $L^{\mathrm{alg}}(1, f \otimes \nu)$ is integral except possibly at the Eisenstein primes; these are the primes dividing

$$\tilde{N} := \prod_{q|N} q(q^2 - 1),$$

which is related to the order of the cuspidal divisor class group, studied for modular curves, among others, by Kubert and Lang.

We may, and we will, choose N to lie in the infinite family of primes that are inert in K and are such that $p \nmid \tilde{N}$.

Now Theorem 6 follows by the near-factorization (14) of $L^{\mathrm{alg}}(1, f, 1_K)$. It may be useful to note that when f has $\mathbb{Q}$-coefficients, with associated elliptic curve E over $\mathbb{Q}$, one knows (cf. Flach [10]) that any prime dividing η_F also divides the degree of the modular parametrization $X_0(N) \to E$.

References

1. H. Darmon and S.-W. Zhang (eds.), *Heegner points and Rankin L-series*, Mathematical Sciences Research Institute Publications, Vol. 49, Cambridge University Press, 2004. Papers from the Workshop on Special Values of Rankin L-Series held in Berkeley, CA, December 2001.
2. J. H. Bruinier, K. James, W. Kohnen, K. Ono, C. Skinner, and V. Vatsal, *Congruence properties of values of L-functions and applications*, Topics in Number Theory (University Park, PA), Math. Appl., Vol. 467, Kluwer Acad. Publ., Dordrecht, 1999, pp. 115–125.
3. F. Diamond, M. Flach, and L. Guo, *The Bloch-Kato conjecture for adjoint motives of modular forms*, Math. Res. Lett. **8** (2001), no. 4, 437–442.
4. L. Dieulefait and J.J. Urroz, *Solving Fermat-type equations via modular curves over polyquadratic fields*, Journal für die Reine und Angewandte Mathematik (Crelles Journal) (2009), no. 633, 183–195.
5. W. Duke, J. B. Friedlander, and H. Iwaniec, *Bounds for automorphic L-functions. II*, Invent. Math. **115** (1994), no. 2, 219–239.
6. W. Duke, *The critical order of vanishing of automorphic L-functions with large level*, Invent. Math. **119** (1995), no. 1, 165–174.
7. J. S. Ellenberg, *Galois representations attached to $\mathbb{Q}$-curves and the generalized Fermat equation $A^4 + B^2 = C^p$*, Amer. J. Math. **126** (2004), no. 4, 763–787.
8. J. S. Ellenberg, *On the error term in Duke's estimate for the average special value of L-functions*, Canad. Math. Bull. **48** (2005), no. 4, 535–546.
9. B. Feigon and D. Whitehouse, *Averages of central L-values of Hilbert modular forms with an application to subconvexity*, Duke Math. J. **149** (2009), no. 2, 347–410.
10. M. Flach, *On the degree of modular parametrizations*, Séminaire de Théorie des Nombres, Paris, 1991–92, Progr. Math., Vol. 116, Birkhäuser Boston, Cambridge, MA, 1993, pp. 23–36.
11. B.H. Gross, and D. B. Zagier, *Heegner points and derivatives of L-series*, Invent. Math. **84** (1986), no. 2, 225–320.
12. B. H. Gross, *Heights and the special values of L-series*, Number theory (Montreal, Que., 1985), CMS Conf. Proc., vol. 7, Amer. Math. Soc. Providence, RI, 1987, pp. 115–187.
13. J. Guo, *On the positivity of the central critical values of automorphic L-functions for* GL(2), Duke Math. J. **83** (1996), no.1, 157–190.
14. R. L. Hatcher, *Heights and L-series*, Canad. J. Math. **42** (1990), no.3, 533–560.
15. R. L. Hatcher, *Special values of L-series*, Proc. Amer. Math. Soc. **114** (1992), no.2, 337–343.
16. H. Hida, *A p-adic measure attached to the zeta functions associated with two elliptic modular forms. I*, Invent. Math. **79** (1985), no.1, 159–195.
17. H. Hida, *On the search of genuine p-adic modular L-functions for* GL(n), Mém. Soc. Math. Fr. (N.S.) (1996), no. 67, vi+110 (English, with English and French summaries). With a correction to: "On p-adic L-functions of GL(2) × GL(2) over totally real fields" [Ann. Inst. Fourier (Grenoble) **41** (1991), no. 2, 311–391; MR1137290 (93b:11052)].
18. J. Hoffstein and P. Lockhart, *Coefficients of Maass forms and the Siegel zero*, Ann. of Math. (2) **140** (1994), no.1, 161–181. With an appendix by Dorian Goldfeld, Jeffrey Hoffstein and Daniel Lieman.
19. H. Iwaniec and P. Sarnak, *The non-vanishing of central values of automorphic L-functions and Landau-Siegel zeros*, Israel J. Math. **120** (2000), 155–177.
20. H. Iwaniec and P. Sarnak, *Perspectives on the analytic theory of L-functions*, Geom. Funct. Anal. (2000), no. Special Volume, 705–741. GAFA 2000 (Tel Aviv, 1999).
21. H. Iwaniec, W. Luo, and P. Sarnak, *Low lying zeros of families of L-functions*, Inst. Hautes Études Sci. Publ. Math. (2000), no. 91, 55–131 (2001).
22. W. Kohnen and D. Zagier, *Values of L-series of modular forms at the center of the critical strip*, Invent. Math. **64** (1981), no. 2, 175–198.
23. E. Kowalski and P. Michel, *The analytic rank of $J_0(q)$ and zeros of automorphic L-functions*, Duke Math. J. **100** (1999), no.3, 503–542.

24. E. Kowalski P. Michel, and J. VanderKam, *Rankin–Selberg L-functions in the level aspect*, Duke Math. J. **114** (2002), no.1, 123–191.
25. S. Lang, *Introduction ot modular forms*, Grundlehren der Mathematischen Wissenschaften [Fundamental Principles of Mathematical Sciences], Vol. 222, Springer-Verlag, Berlin, 1995. With appendixes by D. Zagier and Walter Feit; Corrected reprint of the 1976 original.
26. L. Merel and W. A. Stein, *The field generated by the points of small prime order on an elliptic curve*, Internat. Math. Res. Notices (2001), no.20, 1075–1082.
27. L. Merel, *Sur la nature non-cyclotomique des points d'ordre dini des courbes elliptiques*, Duke Math. J. **110** (2001), no.1, 81–119 (French, with French summary); with and appendix by E. Kowalski and Ph. Michel.
28. Ph. Michel and D. Ramakrishnan, *Exact average of centeral L-values, class numbers, and a diophantine application*, (in preparation).
29. P. Michel and A. Venkatesh, *The subconvexity problem for* GL_2, Publ. Math. Inst. Hautes Études Sci., posted on 2010, no. 111, 171–271, DOI 10.1007/s10240-010-0025-8, (to appear in print). MR 2653249.
30. K. Ono and C. Skinner, *Fourier coefficients of half-integral weight modular forms modulo l*, Ann. of Math. (2) 147 (1998), no.2, 453–470.
31. K. Prasanna, *Arithmetic aspects of theta correspondence and periods of modular forms*, Eisenstein series and applications, Progr. Math., Vol. 258, Birkhäuser Boston, Cambridge, MA, 2008, pp. 251–269.
32. D. Ramakrishnan and J. Rogawski, *Average values of modular L-series via the relative trace formula*, Pure Appl. Math. Q. **1** (2005), no.4, 701–735.
33. G. Shimura, *The special values of the zeta functions associated with cusp forms*, Comm. Pure Appl. Math. **29** (1976), no.6, 783–804.
34. G. Stevens, *The cuspidal group and special values of L-functions*, Trans. Amer. Math. Soc. **291** (1985), no.2, 519–550.
35. R. Taylor and A. Wiles, *Ring-theoretic properties of certain Hecke algebras*, Ann. of Math. (2) **141** (1995), no.3, 553–572.
36. V. Vatsal, *Canonical periods and congruence formulae*, Duke Math. J. **98** (1999), no.2, 397–419.
37. V. Vatsal, *Special values of anticyclotomic L-functions* Duke Math. J. **116** (2003), no.2, 219-261.
38. J.-L. Waldspurger, *Sur les valeurs de certaines fonctions L automorphes en leur centre de symétrie*, Compositio Math. **54** (1985), no.2, 173–242 (French).
39. A. Wiles, *Modular elliptic curves and Fermat's last theorem*, Ann. of Math. (2) **141** (1995), no.3, 443–551.
40. S. Zhang, *Heights of Heegner cycles and derivatives of L-series*, Invent. Math. **130** (1997), no.1, 99–152.

A variant of the Lang–Trotter conjecture

M. Ram Murty and V. Kumar Murty

in memory of Serge Lang

Abstract In 1976, Serge Lang and Hale Trotter formulated general conjectures about the value distribution of traces of Frobenius automorphisms acting on an elliptic curve. In this paper, we study a modular analog. More precisely, we consider the distribution of values of Fourier coefficients of Hecke eigenforms of weight $k \geq 4$.

Key words Lang-Trotter conjecture • *abc* conjecture • Ramanujan τ-function • Atkin-Serre conjecture

Mathematics Subject Classification (2010): 11F03, 11F30

1 Introduction

Let E be an elliptic curve over a number field K. If $\mathfrak{p}$ is a prime of $\mathcal{O}_K$ and E has good reduction at $\mathfrak{p}$, denote by $a_{\mathfrak{p}}(E)$ the integer

$$N\mathfrak{p} + 1 - |E(\mathbb{F}_{\mathfrak{p}})|.$$

---

*Research of both authors partially supported by NSERC grants.

M.R. Murty (✉)
Department of Mathematics Queen's University, Kingston, Ontario, K7L 3N6, Canada
e-mail: murty@mast.queensu.ca

V.K. Murty
Department of Mathematics, University of Toronto, Toronto, Ontario, M5S 2E4, Canada
e-mail: murty@math.toronto.edu

D. Goldfeld et al. (eds.), *Number Theory, Analysis and Geometry: In Memory of Serge Lang*, DOI 10.1007/978-1-4614-1260-1_21,

In 1976, Lang and Trotter [4] formulated some conjectures about how often $a_{\mathfrak{p}}(E)$ takes a fixed value. More precisely, they conjectured that there is a constant $c_{E,a}$ (possibly zero) such that for $x \to \infty$,

$$\pi_{E,a}(x) := \#\{\mathfrak{p} : N\mathfrak{p} \leq x \text{ and } a_{\mathfrak{p}}(E) = a\} \sim c_{E,a} \frac{\sqrt{x}}{\log x},$$

provided we are in the generic case, that is, $a \neq 0$ or E does not have complex multiplication. The constant $c_{E,a}$ depends on the Galois representation attached to E. In 1981, Serre [13] proved that for any $\epsilon > 0$,

$$\pi_{E,a}(x) \ll_\epsilon x/(\log x)^{5/4-\epsilon},$$

in the generic case. The exponent $5/4$ was improved to 2 by Daqing Wan [17]. A further refinement was obtained by the second author in [5] where it is shown that

$$\pi_{E,a}(x) \ll \frac{x(\log \log x)^2}{(\log x)^2}.$$

The case $a_{\mathfrak{p}}(E) = 0$ corresponds to E having supersingular reduction at $\mathfrak{p}$. A classical result of Deuring shows that if E has complex multiplication by an order in an imaginary quadratic field F, the set of supersingular primes of K has density $1/2$ if F is not contained in K and zero if $F \subseteq K$. If E does not have complex multiplication, then Elkies, Kaneko, and R. Murty (see [1]) showed that

$$\pi_{E,0}(x) \ll x^{3/4}.$$

Recently, R. Taylor has announced the meromorphic continuation of symmetric power L-series attached to E (in the case that K is totally real and E has multiplicative reduction at some prime $\mathfrak{p}$). It is conjectured that these symmetric power L-functions extend to entire functions. If we assume this, together with an analogue of the Riemann hypothesis for them, K. Murty [6] has shown that

$$\pi_{E,a}(x) \ll x^{3/4}$$

if $a \neq 0$ or E does not have CM. A substantial generalization and reinterpretation of the Lang–Trotter conjecture can be found in [7], where a more general formulation in terms of Galois representations is made.

In this paper, we consider a normalized Hecke eigenform of weight $k \geq 4$ for the full modular group. We write

$$f(z) = \sum_{n=1}^{\infty} \lambda_f(n) e^{2\pi i n z}$$

for its Fourier expansion at $i\infty$. The field K_f generated by the values $\lambda_f(n)$ as n ranges over all positive integers is of finite degree over $\mathbb{Q}$. We write $\mathcal{O}_f$ for the ring of integers of K_f. In an earlier paper [9], we showed that if $\alpha \in \mathcal{O}_f$ is coprime to 2, then the number of solutions of the equation

$$\lambda_f(n) = \alpha \tag{1}$$

is bounded. Moreover, there is an effectively computable constant $c = c(\alpha) > 0$ such that all solutions n of the equation satisfy

$$n \leq \exp(|N(\alpha)|^c),$$

where $N(\alpha)$ is the norm of α from K_f to $\mathbb{Q}$. This means that for any given α, all the solutions of (1) can be effectively determined. If, in addition, we assume the *abc* conjecture for the number field K_f, then it was shown that the exponential bound can be improved to a polynomial bound of the form $c_1|N(\alpha)|^c$, for some constant $c_1 > 0$ and the same c as before. In the special case of the Ramanujan τ-function, we deduced that the number of solutions of the equation $\tau(n) = a$ with a odd is finite, a result obtained earlier in our joint work with Shorey [11]. Our methods are sufficiently versatile to be applied to related problems. For example, in [10], we study the greatest prime ideal factor of the ideal generated by $\lambda_f(p^n)$ for fixed p and varying n using similar techniques.

In this paper, we want to study the number $\nu_f(a)$ of solutions of the equation

$$|N(\lambda_f(n))| = a$$

for a given natural number a. We prove the following Theorem:

Theorem 1 *Let f be a normalized Hecke eigenform of weight $k \geq 4$ for the full modular group. Assume the abc conjecture for K_f. Let $d = [K_f : \mathbb{Q}]$. Then, for any $\epsilon > 0$,*

$$\sum_{a \leq x}{}' \nu_f(a) \ll x^{2/d(k-3)+\epsilon},$$

where the dash on the summation indicates that we sum over odd, positive a.

We immediately deduce the following corollary:

Corollary 2 *For any normalized Hecke eigenform f of weight $k \geq 4$ for the full modular group,*

$$\nu_f(a) \ll a^{2/d(k-3)+\epsilon},$$

provided a is odd and the abc conjecture holds for K_f.

What is interesting about this corollary is that it is consistent with the Atkin–Serre conjecture (see p.244 of [14]). This conjecture predicts that if f is of weight $k \geq 4$ and is not of CM type, then for sufficiently large primes p,

$$|\lambda_f(p)| \gg p^{(k-3)/2-\epsilon}. \tag{2}$$

As (2) is conjectured to hold for all conjugates f^σ of f, it implies that

$$|N(\lambda_f(p))| \gg p^{\frac{d(k-3)}{2}-\epsilon}$$

and so

$$\nu_f(a) \ll |a|^{\frac{2}{d(k-3)}+\epsilon}.$$

As was shown in [9], $\lambda_f(p)$ is divisible by 2 for all odd primes p in the level-one case. This is a key fact, since it implies that for α coprime to 2, the equation $\lambda_f(n) = \alpha$ forces n to be a perfect square (see [9]). Thus, Theorem 1 can be extended to higher levels, provided this property holds for all sufficiently large primes. Indeed, Ono and Taguchi [12] have shown that this is the case for all forms of level $2^a N_0$ with a arbitrary and $N_0 = 1, 3, 5, 15$, or 17. We record this observation in the following.

Theorem 3 *Let f be a normalized Hecke eigenform of weight $k \geq 4$ and level N. Suppose that for all primes sufficiently large, $\lambda_f(p)$ is divisible by 2. Assuming the abc conjecture for K_f, we have for any $\epsilon > 0$,*

$$\sum_{a \leq x}{}' \nu_f(a) \ll x^{2/d(k-3)+\epsilon},$$

where the dash on the summation indicates that we sum over a coprime to 2 and $d = [K_f : \mathbb{Q}]$.

Acknowledgements We would like to thank the referee for useful comments on an earlier version of this paper.

2 Preliminaries

We begin by reviewing results proved in an earlier paper [9].

Proposition 4 *Let f be a normalized cuspidal eigenform of weight $k \geq 4$ and level N. There is an effectively computable constant $c_1 > 0$ such that for $m \geq 2$ and every prime p, we have*

$$|\lambda_f(p^m)| \geq |\gamma_f(p,m)| p^{\frac{k-1}{2}(m - c_1 \log m)},$$

where $\gamma_f(p,m) = 1$ if m is even and $\lambda_f(p)$ if m is odd.

Proof. This is Proposition 2.2 of [9]. □

In particular, we see from this proposition that $\lambda_f(p^m) \neq 0$ when m is even and sufficiently large.

Proposition 5 *Let f be a Hecke eigenform of weight k and level N. Then, for all p sufficiently large, either $\lambda_f(p) = 0$ or $\lambda_f(p^a) \neq 0$ for all $a \geq 1$. Moreover,* *for each m, there is a binary form f_m of degree $[m/2]$, with integeral coefficients such that*

$$\lambda_f(p^m) = \gamma_f(p,m) f_m(\lambda_f(p)^2, p^{k-1}).$$

Proof. The first part of the assertion follows from the previous proposition or from Lemma 2.3 of [9]. The second part follows from the proof of the same lemma. The binary form $f_m(x, y)$ is

$$\prod_{r=1}^{[m/2]} (x - 4y\cos^2(\pi r/(m+1))),$$

which is easily seen to have integer coefficients by simple field-theoretic considerations. □

We will also have need of a version of Roth's theorem, which we record in the following lemma.

Lemma 6 (Roth's theorem) *Let f be a binary form with integer coefficients and degree $d \geq 3$. If f has distinct irrational roots, then,*

$$|f(x, y)| \gg \max(|x|, |y|)^{d-2-\epsilon},$$

where the implied constant depends only on the coefficients of f.

Proof. This essentially follows from Roth's theorem. See also [8]. □

A number-field version of this lemma will also be needed in the later sections, and this will be recalled in Section 4.

Our line of argument has its origins in [9] and [11]. In [11], it was observed that the Ramanujan τ-function has the fortuitous property that $\tau(p)$ is even for every prime p. By an analogue of Proposition 5 for the τ-function, we see that $\tau(p^m)$ is even for every odd m. Hence, if we are interested in the equation

$$\tau(n) = a$$

for a odd, it follows that n must be a perfect square, by virtue of the multiplicativity of τ. This was the key fact that enabled the application of results from Baker's theory to establish that the number of solutions to the equation $\tau(n) = a$, with a odd, is finite. This argument was extended to any normalized eigenform for the full modular group in [9]. As indicated in [9], results of Tate [15] imply that $\lambda_f(p)$ is divisible by 2 for every prime p. This enabled us to extend the results of [11] to the full modular case. As indicated in [9], the method can be generalized to arbitrary level provided that $\lambda_f(p)$ is divisible by 2 for all primes p sufficiently large. With this background information in place, we now outline our basic strategy.

We fix a positive integer a coprime to 2 and study the equation

$$|N(\lambda_f(n))| = a.$$

As $\lambda_f(n)$ is multiplicative, we see that $\lambda_f(p^m)$ is coprime to 2 for $p^m||n$. Now suppose that $\lambda_f(p)$ is divisible by 2 for all primes $p \geq c_0$. Then by Proposition 5, we see that $\lambda_f(p^m)$ is divisible by 2 for all *odd* m and $p \geq c_0$. Thus, if we write $n = n_0 n_1 n_2$, where the prime factors of n_1 are $< c_0$ satisfying $\lambda_f(p) \neq 0$, the prime factors p of n_0 are $< c_0$ with $\lambda_f(p) = 0$, and the prime factors of n_2 are $\geq c_0$, then we see that n_2 is a perfect square. For primes $p|n_1$, we have $p < c_0$ and $\lambda_f(p) \neq 0$, so that Proposition 4 shows that

$$|\lambda_f(p^m)| \geq |\gamma_f(p,m)| p^{\frac{k-1}{2}(m - c_1 \log m)}.$$

This means that n_1 is bounded, since the primes and prime powers that divide it are bounded. If we look at n_0, then $\lambda_f(p) = 0$ for each $p|n_0$. Since $p^m||n$, m must be even, for otherwise $\lambda_f(n) = 0$. Thus, n_0 is a perfect square. In any case, n has the form ab^2 with a, b coprime and a bounded and $\lambda_f(b^2) \neq 0$. Thus, we are motivated to study the Dirichlet series

$$D_f(s) = \sum_{n=1}^{\infty}{}' |N(\lambda_f(n^2))|^{-s},$$

where the dash in the summation means we go over those n such that $\lambda_f(n^2) \neq 0$. Since $\lambda_f(n^2)$ is multiplicative, we may write this as an Euler product:

$$D_f(s) = \prod_p{}' \left(\sum_{m=0}^{\infty} \frac{1}{|N(\lambda_f(p^{2m}))|^s} \right),$$

where the dash on the product indicates we go over primes p such that $\lambda_f(p^{2m}) \neq 0$ for any $m \geq 0$. Our objective is to determine a half-plane in which this series converges absolutely.

We remark that if the series

$$\sum_{a=1}^{\infty} \frac{\nu_f(a)}{a^s}$$

converges absolutely for $\Re(s) > c$, then

$$\sum_{n \leq x} \nu_f(a) \ll \sum_{n \leq x} \nu_f(a)(x/n)^{c+\epsilon} \ll x^{c+\epsilon},$$

for any $\epsilon > 0$. We will use this remark in our discussion below.

Let us note also that as

$$|N(\lambda_f(n^2))| \leq n^{(k-1)d} d(n^2),$$

where $d(n)$ denotes the number of divisors of n, the series does not converge for

$$\Re(s) \leq \frac{1}{d(k-1)}.$$

Moreover, as $D_f(s)$ is a Dirichlet series with non-negative coefficients, it must have a singularity at its abscissa of convergence, by a celebrated theorem of Landau. In particular, we have

$$\sum_{a \leq x} \nu_f(a) = \Omega(x^{1/d(k-1)}).$$

3 The special case of Ramanujan's τ-function

For the sake of clarity, we will first consider a special case, namely, the study of the Dirichlet series

$$D_\Delta(s) = \sum_{n=1}^{\infty}{}' \frac{1}{|\tau(n^2)|^s}.$$

Since $\tau(n^2)$ is a multiplicative function, we can expand the series as an infinite product over the primes:

$$D_\Delta(s) = \prod_p{}' \left(\sum_{m=0}^{\infty} |\tau(p^{2m})|^{-s} \right).$$

Our goal is to determine a region of convergence for this series. By Proposition 4, we see that

$$|\tau(p^{2m})| \geq p^{11m(1-\epsilon)}$$

for $m \geq m_0$ (say). This means that the series

$$\sum_{m \geq m_0} |\tau(p^{2m})|^{-\Re(s)} \ll \sum_{m \geq m_0} p^{-11m(1-\epsilon)\Re(s)}$$

converges for $\Re(s) > 0$. To deal with the other part of the series, we need to estimate $\tau(p^{2m})$ for $2 \leq m \leq m_0$. We can use Propostion 5 combined with Roth's theorem to derive a lower bound for $|\tau(p^{2m})|$ for $6 \leq m \leq m_0$. Indeed, Roth's theorem allows us to deduce that

$$|f_m(\tau(p)^2, p^{11})| \gg p^{11(m/2-2-\epsilon)}.$$

We need to discuss lower bounds for $\tau(p^2)$ and $\tau(p^4)$. For this, we need to invoke the *abc* conjecture. To this end, let us define the *radical* of a natural number n, denoted by $\text{rad}(n)$, to be the product of the distinct primes dividing n. The *abc* conjecture predicts that for any two coprime integers a, b,

$$\text{rad}(ab(a+b)) \gg \max(|a|, |b|)^{1-\epsilon},$$

for any $\epsilon > 0$. The implied constant will depend on ϵ but not on a, b.

Lemma 7 *Suppose that $\tau(p) \neq 0$. The abc conjecture implies that for any $\epsilon > 0$,*

$$|\tau(p^2)| \gg p^{9/2-\epsilon}$$

and

$$|\tau(p^4)| \gg p^{10-\epsilon}.$$

Proof. We first apply the *abc* conjecture to the equation

$$\tau(p^2) = \tau(p)^2 - p^{11}.$$

Suppose first that p is coprime to $\tau(p)$. From the *abc* conjecture, we deduce that

$$\text{rad}(\tau(p)^2 \tau(p^2) p^{11}) \gg p^{11(1-\epsilon)}.$$

Using $|\tau(p)| \leq 2p^{11/2}$, we obtain

$$|\tau(p^2)| \geq \text{rad}(|\tau(p^2)|) \gg p^{9/2(1-\epsilon)},$$

as desired. If $p|\tau(p)$, write $\tau(p) = p^a v_p$ with v_p coprime to p. As $\tau(p^2) \neq 0$, we deduce that

$$\text{rad}(v_p^2 p^{11-2a}(v_p^2 - p^{11-2a})) \gg p^{11-2a-\epsilon},$$

so that

$$\tau(p^2) = p^{2a}(v_p^2 - p^{11-2a}) \gg p^{9/2+a-\epsilon}.$$

This completes the proof of the first part. For the second part, consider

$$(2\tau(p)^2 - 3p^{11})^2 = 4\tau(p^4) - 5p^{22}.$$

Assuming first that p is coprime to $\tau(p)$, we can apply the *abc* conjecture to this equation to deduce

$$|\tau(p^4)| \gg p^{10(1-\epsilon)}.$$

If $p|\tau(p)$, then we write, as before, $\tau(p) = p^a v_p$ with v_p coprime to p. Then, we have

$$4\tau(p^4) = p^{4a}[(2v_p^2 - 3p^{11-2a})^2 + 5p^{22-4a}].$$

Applying the abc conjecture to the term in the square brackets, we obtain

$$|\tau(p^4)| \gg p^{10+2a-\epsilon},$$

so that the result is proved in this case also. □

We are now in a position to study the convergence of

$$\sum_{m \le m_0} |\tau(p^{2m})|^{-s}.$$

We break the sum into three parts:

$$|\tau(p^2)|^{-s} + |\tau(p^4)|^{-s} + \sum_{3 \le m \le m_0} |\tau(p^{2m})|^{-s}.$$

By our earlier discussion, the last sum is bounded by $p^{-33\Re(s)}$. By the previous lemma, the first two terms are

$$\ll p^{-\frac{9}{2}(1-\epsilon)\Re(s)}.$$

This result immediately implies that $D_\Delta(s)$ converges for $\Re(s) > 2/9$. Thus,

$$\sum_{a \le x}{}' v_\Delta(a) \ll x^{2/9+\epsilon}.$$

We record the following corollary for its own intrinsic interest.

Corollary 8 *If a is an odd number, the number of solutions of $\tau(n) = a$ is bounded by $O(|a|^{2/9+\epsilon})$, assuming the abc conjecture.*

4 The abc conjecture for number fields

Let K be an algebraic number field. Suppose $a, b, c \in K^*$ such that $a + b + c = 0$ Define

$$\text{rad}_K(a, b, c) = \prod_{\mathfrak{p}} N_{K/\mathbb{Q}}(\mathfrak{p}),$$

where the product is over those prime ideals for which the numbers

$$||a||_{\mathfrak{p}},\ ||b||_{\mathfrak{p}},\ ||c||_{\mathfrak{p}}$$

are unequal. We will also write rad (a) to be the product of norms of the distinct prime ideal divisors of (a). We define

$$H_K(a,b,c) = \prod_v \max(||a||_v, ||b||_v, ||c||_v),$$

where the product is over all valuations of K (both finite and infinite and we normalize the archimedean valuations by $||x||_v = |x|_v^{d_v}$ with $d_v = 1$ or 2 according as v is real or complex, and the nonarchimedean valuations by $||x||_v = N_{K/\mathbb{Q}}(\mathfrak{p})^{-v(x)}$). The abc conjecture for K is the following assertion. For any $\epsilon > 0$, there is a constant $C_{K,\epsilon}$ such that

$$H_K(a,b,c) \leq C_{K,\epsilon}(\text{rad}_K(a,b,c))^{1+\epsilon}.$$

A stronger version predicts that one may replace $C_{K,\epsilon}$ by

$$C_\epsilon^{[K:\mathbb{Q}]} D_K^{1+\epsilon},$$

where D_K is the absolute value of the discriminant of K. We will not be using this stronger version of the abc conjecture in our discussion below. We refer the reader to Vojta [16] for further details.

We first derive a consequence of the abc conjecture for number fields that will be applied in the subsequent discussion.

Lemma 9 *Let K be an algebraic number field and suppose that $\mathfrak{d} =$ gcd $((a),(b))$. Suppose for all finite primes $\mathfrak{p}$, $||a||_{\mathfrak{p}} \neq ||b||_{\mathfrak{p}}$ Assuming the abc conjecture for K, we have*

$$\text{rad}(a)\text{rad}(b)\text{rad}(a+b)/(\text{rad}(\mathfrak{d}))^2 \gg \left(\max(|N(a)|, |N(b)|, |N(a+b)|)/N(\mathfrak{d})^2\right)^{1-\epsilon},$$

where N stands for $N_{K/\mathbb{Q}}$ and the implied constant depends on K and ϵ.

Proof. Suppose first that $\mathfrak{d} = 1$. From the definition, we have

$$\text{rad}_K(a,b,a+b) = \prod_{\mathfrak{p}|ab(a+b)} N(\mathfrak{p}),$$

since $a, b, (a+b)$ are mutually coprime. Let us note that for every finite v, we also have that one of

$$||a||_v,\ ||b||_v,\ ||a+b||_v,$$

is 1, so that

$$H_K(a, b, a + b) \geq \max(|N(a)|, |N(b)|, |N(a + b)|).$$

The *abc* conjecture now implies the result in this case. If $\mathfrak{d} \neq 1$, let $\mathfrak{p}$ be a prime ideal dividing $\mathfrak{d}$. By our assumption, $\mathfrak{p}$ enters into the radical. $N(\mathfrak{p})$ enters three times into the product $\mathrm{rad}(a)\mathrm{rad}(b)\mathrm{rad}(a + b)$, and to remove two of the occurences, we can divide by $N(\mathfrak{p})^2$. This completes the proof. □

In our estimations below, we will need a number field version of Lemma 6, and this we record here.

Lemma 10 *Let K be an algebraic number field and f a binary form in $\mathcal{O}_K[x, y]$ with no repeated factors. Then, assuming the abc conjecture for K, we have*

$$\mathrm{rad}_K(f(u, v)) \gg H_K(u, v)^{d-2-\epsilon},$$

where d is the degree of f and $u, v \in K^*$.

Proof. This is proved on page 105 of [2]. □

We remark that if we replace $\mathrm{rad}_K(f(u, v))$ by $|f(u, v)|$, this is essentially Roth's theorem for number fields. Thus, the *abc* conjecture is making a stronger assertion than that implied by Roth's theorem. Indeed, since $|N(f(u, v))| \geq \mathrm{rad}_K(f(u, v))$, we deduce the following:

Corollary 11 *Let K be an algebraic number field and f a binary form in $\mathcal{O}_K[x, y]$. Then,*

$$|N(f(u, v))| \gg H_K(u, v)^{d-2-\epsilon},$$

where d is the degree of f and $u, v \in K^*$, assuming the abc conjecture for K.

Lemma 12 *Suppose that $\lambda_f(p) \neq 0$. Assume the abc conjecture for K_f. Then,*

$$|N(\lambda_f(p^2))| \gg p^{d(k-3)/2-\epsilon}$$

and

$$|N(\lambda_f(p^4))| \gg p^{d(k-2)-\epsilon},$$

where $d = [K_f : \mathbb{Q}]$ and p is unramified in K_f.

Proof. As before, we apply the *abc* conjecture to the equation

$$\lambda_f(p^2) = \lambda_f(p)^2 - p^{k-1}.$$

First suppose that $\lambda_f(p)$ and p are coprime. By Lemma 9 applied to the field K_f, we obtain

$$\mathrm{rad}_{K_f}(\lambda_f(p)^2, p^{k-1}, \lambda_f(p^2)) \gg p^{d(k-1)-\epsilon},$$

where $d = [K_f : \mathbb{Q}]$. We obtain

$$p^d |N(\lambda_f(p))N(\lambda_f(p^2))| \gg p^{d(k-1)-\epsilon},$$

from which we deduce, using the Ramanujan bound $|N(\lambda_f(p))| \leq 2^d p^{d(k-1)/2}$, that

$$|N(\lambda_f(p^2))| \gg p^{d(k-3)/2-\epsilon}.$$

Now suppose that $\mathfrak{p}^a || (\lambda_f(p))$, with $a \geq 1$. Then by taking norms, we obtain the inequality

$$p^{da} \leq p^{d(k-1)/2},$$

implying $a \leq (k-1)/2$. Since k is even, this is a strict inequality. Thus, $a < (k-1)/2$. Since p is unramified,

$$||p^{k-1}||_{\mathfrak{p}} = N(\mathfrak{p})^{-(k-1)} \neq ||\lambda_f(p)^2||_{\mathfrak{p}} = N(\mathfrak{p})^{-2a}.$$

By Lemma 9, we obtain as before,

$$|N(\lambda_f(p^2))| \gg p^{d(k-3)/2-\epsilon}.$$

The lower bound for $|N(\lambda_f(p^4))|$ is derived similarly. We apply the *abc* conjecture to the equation

$$(2\lambda_f(p)^2 - 3p^{k-1})^2 = 4\lambda_f(p^4) - 5p^{2k-2}.$$ □

5 The Dirichlet series $D_f(s)$

We will now study the series $D_f(s)$ and determine where it converges. Since $N(\lambda_f(n^2))$ is multiplicative, we have the Euler product

$$D_f(s) = \prod_p \left(\sum_{m=0}^{\infty} \frac{1}{|N(\lambda_f(p^{2m}))|^s} \right).$$

Our goal is to determine the region where the Euler product converges absolutely. We split the product into two parts: $p \leq c_0$ and $p > c_0$, for which we have that $\lambda_f(p)$ is divisible by 2. The first product is finite and is over those $\mathfrak{p}$ for which the $\lambda_f(p^m)$ are all coprime to 2. This product converges for $\Re(s) > 0$. Let us now consider the other product. We proceed as in the case of the τ-function. By Proposition 4, we see that for $m \geq m_0$ (say),

$$|\lambda_f(p^{2m})| \gg p^{m(k-1)(1-\epsilon)}.$$

A similar estimate holds with f replaced by any conjugate form f^σ. Thus the series in the Euler product converges for $\Re(s) > 0$ if we restrict $m \geq m_0$. By Corollary 11, we have

$$|f_m(\lambda_f(p)^2, p^{k-1})| \gg p^{(k-1)(m/2-2-\epsilon)}$$

for $6 \leq m \leq 2m_0$. Thus,

$$|\lambda_f(p^{2m})| \gg p^{(k-1)(m-2-\epsilon)}$$

for $3 \leq m \leq m_0$. We deduce that

$$|N(\lambda_f(p^{2m}))| \gg p^{(k-1)d(m-2-\epsilon)},$$

for $3 \leq m \leq m_0$. To complete our estimates, we need lower bounds for $|\lambda_f(p^2)|$ and $|\lambda_f(p^4)|$, which are provided by Lemma 12. From that lemma, we get that

$$|N(\lambda_f(p^2))| \gg p^{d(k-3)/2-\epsilon}, \quad |N(\lambda_f(p^4))| \gg p^{d(k-2)-\epsilon}.$$

Putting all this together shows the following:

Theorem 13 *Assume the abc conjecture for K_f. Let $d = [K_f : \mathbb{Q}]$. Then, the Dirichlet series $D_f(s)$ converges absolutely for $\Re(s) > 2/d(k-3)$. In particular,*

$$\sum_{a \leq x}{}' \nu_f(a) \ll x^{2/d(k-3)+\epsilon},$$

for any $\epsilon > 0$, where the summation is over odd, positive a.

References

1. N. Elkies, Distribution of supersingular primes, Journées Arithmétiques (Luminy, 1989), *Astérisque,* **198–200** (1991), 127–132.
2. N. Elkies, ABC implies Mordell, *International Math. Research Notices,* **1991** (1991), No. 7, 99–109.
3. M. Hindry and J. Silverman, Diophantine Geometry, an Introduction, *Graduate Texts in Mathematics,* **201**, Springer-Verlag, 2000.
4. S. Lang and H. Trotter, Frobenius Distributions in GL_2-extensions, *Lecture Notes in Mathematics,* **504** (1976), Springer.
5. V. Kumar Murty, Modular forms and the Chebotarev density theorem, II, in Analytic Number Theory, edited by Y. Motohashi, *London Math. Society Lecture Notes,* **247** (1997), 287–308, Cambridge University Press.
6. V. Kumar Murty, Explicit formulae and the Lang–Trotter conjecture, *Rocky Mountain Journal,* **15** (1985), 535–551.
7. V. Kumar Murty, Frobenius distributions and Galois representations, *Proc. Symp. Pure Math.,* **66.1** (1999), 193–211.

8. D.J. Lewis and K. Mahler, On the representation of integers by binary forms, *Acta Arith.*, **6** (1960/61), 333–363.
9. M. Ram Murty and V. Kumar Murty, Odd values of Fourier coefficients of certain modular forms, *International Journal of Number Theory,* **3** (2007), no. 3, 455–470.
10. M. Ram Murty and V. Kumar Murty, On a conjecture of Shorey, in Diophantine Equations, edited by N. Saradha, pp. 167–176, Narosa, 2008.
11. M. Ram Murty, V. Kumar Murty, and T.N. Shorey, Odd values of the Ramanujan τ-function, *Bulletin Soc. Math. France,* **115** (1987), no. 3, 391–395.
12. K. Ono and T. Taguchi, 2-adic properties of certain modular forms and their applications to arithmetic functions, *International Journal of Number Theory,* **1** (2005), no. 1, 75–101.
13. J.-P. Serre, Quelques applications du théorème de densité de Chebotarev, *Publ. Math. IHES,* **54** (1981), 123–201.
14. J.-P. Serre, Divisibilité de certaines fonctions arithmétiques, in *Séminaire Delange-Pisot-Poitou, 16e année* (1974/75), Théorie des nombres, Fasc. 1, Exp. No. 20, 28p., Secrétariat Mathématique, Paris, 1975.
15. J. Tate, The non-existence of certain Galois extensions of $\mathbb{Q}$ unramified outside 2, *Contemporary Mathematics,* **174** (1994), 153–156, American Math. Society, Providence, Rhode Island.
16. P. Vojta, Diophantine approximations and value distribution theory, *Lecture notes in mathematics,* **1239**, Springer-Verlag, Berlin, 1987.
17. D. Wan, On the Lang–Trotter conjecture, *Journal of Number Theory,* **35**(1990), 247–268.

Multiplicity estimates, interpolation, and transcendence theory

Michael Nakamaye

In loving memory of my mentor and friend Serge Lang

Abstract We discuss the problems of interpolation and multiplicity estimates on compactifications of commutative algebraic groups. We consider two extremal cases: one where multiplicity is imposed at a single point and the other where the conditions are imposed on an asymptotically growing set of points. Some conjectures and new results are given in both cases.

Key words multiplicity estimates • Seshadri constant • algebraic group

Mathematics Subject Classification (2010): 11J81, 14L40, 14C20

1 Introduction

One of Serge Lang's most memorable qualities was his incredible vision of mathematical structures. Whether studying diophantine approximation, rational points on algebraic varieties, or the far-reaching consequences of the heat kernel in number theory, Serge always had a simple, elegant vision of how, and often why, the mathematical universe behaves, what objects one might hope to find, and how to go about finding them. In this article we would like, to the best of our ability, to adopt a similarly broad point of view and study the global geometric properties of algebraic varieties which govern the behavior of multiplicity estimates and interpolation estimates on them.

Suppose that G is a commutative algebraic group of dimension d defined over the complex numbers. Let X be an equivariant compactification of G and A an

M. Nakamaye (✉)
Department of Mathematics, University of New Mexico, Albuquerque, NM 87131
e-mail: nakamaye@math.unm.edu

D. Goldfeld et al. (eds.), *Number Theory, Analysis and Geometry: In Memory of Serge Lang*, DOI 10.1007/978-1-4614-1260-1_22,

ample line bundle on X. If $\Gamma \subset G$ is a finite set, then we let

$$\Gamma(S) = \{x_1 + \ldots + x_S : x_i \in \Gamma \text{ for all } i\}.$$

For simplicity, we will assume that Γ contains the identity e_G of G. Interpolation asks to find the maximal order jets which a multiple of A can simultaneously generate at the points of $\Gamma(S)$. Multiplicity estimates, on the other hand, seek the maximal order to which a non-zero section of a multiple of A can vanish at the set of points $\Gamma(S)$. We will consider only the two extreme cases, where $\Gamma = e_G$ and where Γ contains non-torsion elements of G and S is large.

We consider first the case where Γ is reduced to a point. In this case, the fact that X is an equivariant compactification of a group variety is irrelevant, except insofar as the open subset $G \subset X$ is a homogeneous space for the action of G, and consequently $e_G \in G \subset X$ may be viewed as a very general point. Suppose then that X is a smooth projective variety and $\eta \in X$ is a general point. We now introduce formally the two numerical invariants of the pair (X, A) which will be studied.

Definition 1.1.1. Suppose X is a smooth projective variety, $x \in X$, and A an ample line bundle on X. Then

$$\epsilon(x, A) = \inf_{C \ni x} \frac{\deg_A(C)}{\mathrm{mult}_x(C)};$$

here the infimum runs over all integral curves $C \subset X$ passing through x.

The invariant $\epsilon(x, A)$ is called the *Seshadri constant* of A at x. The reader is encouraged to consult [L], Volume I, Chapter 5 for the basic properties of Seshadri constants, the geometric information they encode, as well as some very interesting problems, both solved and unsolved, regarding Seshadri constants. This numerical definition is equivalent (see [L] Theorem 5.1.17 for details) to a more intuitive geometric definition. In particular, $\epsilon(x, A)$ is the supremum of all non-negative rational numbers α such that the linear series $|nA|$ separates $n\alpha$-jets at x for n sufficiently large and divisible. Any irreducible subvariety $V \subset X$, of positive dimension and maximal with respect to the relation of inclusion, satisfying

$$\left(\frac{\deg_A(V)}{\mathrm{mult}_x(V)}\right)^{\frac{1}{\dim(V)}} = \epsilon(x, A)$$

is called *Seshadri exceptional* at x relative to A. An important result of Campana and Peternell [CP] asserts that a Seshadri exceptional subvariety always exists. Note finally that solving the interpolation problem for X and A at a point η is equivalent to bounding the Seshadri constant $\epsilon(\eta, A)$ from below.

Next, for an ample divisor A on X we let

$$m(A) = \sup_{D \equiv A} \left\{\mathrm{mult}_\eta(D) \mid D \in \mathrm{Div}(X) \otimes \mathbf{Q} \text{ effective}\right\} :$$

here $\equiv$ denotes numerical equivalence and η is a very general point of X. To see that the invariant $m(A)$ is well-defined, choose a very ample divisor B. For any $\mathbf{Q}$-divisor $D \equiv A$, we can choose general divisors $B_1, \ldots, B_{d-1}$ from the linear system $|B|$, containing η, so that the intersection

$$\text{support}(D) \cap B_1 \cap \cdots \cap B_{d-1}$$

is proper. It follows that

$$\text{mult}_\eta(D) \leq c_1(A) \cap c_1(B)^{d-1},$$

establishing that the supremum exists in the definition of $m(A)$.

The two invariants $m(A)$ and $\epsilon(\eta, A)$ are closely related. Indeed, as we will establish at the end of §1.2

$$\epsilon(\eta, A) \leq \sqrt[d]{\deg_A(X)} \leq m(A) \tag{1}$$

with equality holding on the one side if and only if it holds on the other. More generally, the smaller $\epsilon(\eta, A)$ is the more sparse will be the jets generated by $|kA|$ at η, and consequently the larger $m(A)$ will be. We can quantify this relationship precisely on a surface.

Theorem 1.1.2. *Suppose X is a smooth surface and $\eta \in X$ a very general point. Then for any ample line bundle A on X we have*

$$\frac{\deg_A(X)}{2} \leq m(A)\epsilon(\eta, A) \leq \deg_A(X).$$

For higher dimensional varieties, the analogue of Theorem 1.1.2 requires the introduction of new invariants.

Definition 1.1.3. For $1 \leq i \leq d$ let

$$\alpha_i(A) = \sup\left\{\alpha \in \mathbf{Q} : \dim\left(\text{BS}\left|kA \otimes m_\eta^{k\alpha}\right|\right) < i \ \text{ for } k \gg 0\right\};$$

here BS denotes the base locus and $m_\eta \subset \mathcal{O}_X$ the maximal ideal associated to the point η.

Note that $\alpha_1(A) = \epsilon(\eta, A)$ and $\alpha_d(A) = m(A)$. As in Theorem 1.1.2 one can establish the upper bound

$$\prod_{i=1}^{d} \alpha_i(A) \leq \deg_A(X).$$

A lower bound is harder to come by, however, since the proof of Theorem 1.1.2 uses in a fundamental way in order to establish (4) below the fact that the Seshadri exceptional curve C_η is a divisor. Along the lines of Theorem 1.1.2, we make the following conjecture:

Conjecture 1.1.4. *Suppose A is an ample line bundle on a smooth projective variety X of dimension d. Then*

$$\prod_{i=1}^{d} \alpha_i(A) \geq \frac{\deg_A(X)}{d!}.$$

As in the case of surfaces, Conjecture 1.1.4 is the best possible result. We will see that Conjecture 1.1.4 is true, as in the surface case, if there is a divisorial Seshadri exceptional subvariety of A at η. The interest of Conjecture 1.1.4 is that it gives a connection between $m(A)$ and $\epsilon(\eta, A)$. In particular, Conjecture 1.1.4 says that if $m(A)$ is small, then $\epsilon(\eta, A)$ cannot be too small. Thus any upper bound on $m(A)$ entails a lower bound on $\epsilon(\eta, A)$ and, conversely, if $\epsilon(\eta, A)$ is small, then $m(A)$ cannot be too small.

We will now consider non-zero values of the parameter S. Here the most refined results are due to Philippon [P1,P2]. From the geometric viewpoint, the one manner in which Philippon's zero estimates might be strengthened is to distinguish the cases where S is small from those where S is large. Indeed, these two cases are qualitatively very different. When $S = 1$, since no hypothesis is made concerning the finite set $\Gamma \subset G$, no meaningful conclusion can be drawn without imposing vanishing along $\Gamma(r)$ for $r > 1$. For example, if $G = \mathbf{A}^2$ and Γ consists of 1,000,000 points on a smooth conic $C \subset G$, then of course there is a regular function of degree two, namely the function defining the conic C, which vanishes to order one along all 1,000,000 points. Meanwhile, if the 1,000,000 points were general points of G, there would be no polynomial of degree smaller than roughly $\sqrt{2{,}000{,}000}$ vanishing at each of the points.

Thus the main goal, in this case, of Philippon's construction is to replace Γ with $\Gamma(2)$, thus allowing the group law on $\mathbf{A}^2$ to spread the points on C around. If, however, as may sometimes be assumed in applications, the set along which one imposes vanishing is already of the form $\Gamma(S)$ for $S > 1$, then a much stronger result can be expected to hold since regardless of the position of the points of Γ, they have already been distributed by the group law on G.

In order to evaluate how well the points of $\Gamma(S)$ will be distributed inside G, we will use a line of reasoning which grows out of the work of Faltings and Wüstholz [FW] on the Schmidt subspace theorem.

Definition 1.1.5. Suppose $H \subset G$ is a non-trivial subgroup. Let $\Gamma(\infty)$ be the subgroup of G generated by Γ and set

$$\rho(H) = \text{rank}(\Gamma(\infty) \cap H).$$

Setting $d(H) = \dim(H)$ we also define

$$\nu(\Gamma, H) = \frac{\rho(H)}{d(H)}.$$

A subgroup H is called the Γ-exceptional subgroup if $\nu(\Gamma, H) \geq \nu(\Gamma, H')$ for all non-trivial subgroups $H' \subset G$ and if H is maximal among such subgroups.

The same notion is defined, with the notation $\mu^*(\Gamma, H)$, and used in a similar fashion by Masser [M] and, not suprisingly given the relationship with [FW], it is closely related to a second invariant used by Masser and Wüstholz [MW] and Waldschmidt [Wal] to study related diophantine problems. In §2 we will show that a Γ-exceptional subgroup H always exists and, moreover, that there is an increasing sequence of subgroups

$$H_1 \subset \cdots \subset H_r$$

with H_1 the Γ-exceptional subgroup, $H_r = G$, and $\nu(\Gamma, H_1) > \ldots > \nu(\Gamma, H_r)$. Moreover if $H \subset G$ is a subgroup with $\nu(H) = \nu(H_i)$ then, $H \subset H_i$.

In order to study the generation of jets along $\Gamma(S)$ when S is large, we need a notion of Seshadri exceptional subvariety for the set of points $\Gamma(S)$. First, the Seshadri constant of A along a finite subset $Z \subset X$ is defined by

$$\epsilon(Z, A) = \inf_C \left\{ \frac{\deg_A(C)}{\sum_{x \in Z} \text{mult}_x(C)} \right\}$$

where the infimum is taken over all irreducible curves $C \subset X$ containing at least one point of Z. An irreducible subvariety $V \subset X$ of positive dimension is called Seshadri exceptional for A relative to Z if

$$\left(\frac{\deg_A(V)}{\sum_{x \in Z} \text{mult}_x(V)} \right)^{\frac{1}{\dim(V)}} = \epsilon(Z, A)$$

and if V is maximal with respect to the relation of inclusion.

Theorem 1.1.6. *Suppose the Γ-exceptional subgroup $H \subset G$ is proper. Then for all S sufficiently large, a translate of each Seshadri exceptional subvariety of A relative to $\Gamma(S)$ is contained in H.*

Theorem 1.1.6 is nearly optimal but once S is sufficiently large one might ask whether or not there are criteria for when the subgroup H *is* the Seshadri exceptional subvariety:

Question 1.1.7. *With hypotheses as in Theorem 1.1.6, is the subgroup H the Seshadri exceptional subvariety for A relative to $\Gamma(S)$ for all S sufficiently large?*

Quantitatively, an affirmative answer to Question 1.1.7 leads to an improvement of interpolation estimates involving factors of dim(G) and the rank of Γ, as we will see in Theorem 1.5.1 below. Qualitatively, however, an affirmative answer to Question 1.1.7 is much stronger than Theorem 1.5.1 as it identifies the obstruction to the interpolation problem in terms of the subgroups of G.

We are left to address the interpolation question in the case where

$$\nu(\Gamma, G) \geq \nu(\Gamma, H)$$

for all subgroups $H \subset G$. In this case, there are no obstruction subgroups and thus the points of $\Gamma(S)$ are well distributed with respect to subgroups of G.

Conjecture 1.1.8. *Suppose X is the compactification of a commutative algebraic group G and $\Gamma \subset G$ is not contained in G_{tors}. Suppose G is Γ-exceptional. Then for all S sufficiently large*

$$\epsilon(\Gamma(S), A) \geq \frac{\sqrt[d]{\frac{\deg_A(X)}{|\Gamma(S)|}}}{d}.$$

More generally, it is reasonable to look for the best possible lower bound for

$$\liminf_{S \to \infty} \frac{\epsilon(\Gamma(S), A)\sqrt[d]{|\Gamma(S)|}}{\sqrt[d]{\deg_A(X)}},$$

and attempt to characterize when the limit is close to 1. If the limit were always 1, then Question 1.1.7 would have an affirmative answer, in asymptotic form at least, without the hypothesis that the Γ-exceptional subgroup H is a proper subgroup of G: in other words, as S becomes larger and larger the Seshadri constant for A relative to $\Gamma(S)$ is closer and closer to maximal and thus X itself is becoming closer and closer to being Seshadri exceptional.

Finally we would like to address zero estimates when the parameter S is large. Let $h_i = \dim(H_i)$ for $1 \leq i \leq r$. As in Definition 1.1.3 we also introduce a constant to measure the dimension of the base locus of $H^0\left(kA \otimes \mathcal{I}_{\Gamma(S)}^{k\alpha}\right)$ for different values of α: for $1 \leq i \leq r$

$$\alpha_i(S, A) = \sup\left\{\alpha > 0 : \dim\left(\mathrm{BS}\left(A^{\otimes k} \otimes \mathcal{I}_{\Gamma(S)}^{k\alpha}\right)\right) < h_i \text{ for some } k > 0\right\}.$$

Only the dimensions of the subgroups H_i are considered in the definition of $\alpha_i(S, A)$ since, conjecturally, when $S \gg 0$ only translates of these subgroups will appear as obstructions:

Conjecture 1.1.9. *Suppose X is the compactification of a commutative group variety and Γ is not contained in G_{tors}. Suppose G is not Γ-exceptional so that $r \geq 2$. Then for $1 \leq j \leq r-1$ and $1 \gg \epsilon > 0$ there exists S sufficiently large so that*

$$\mathrm{BS}\left(kA \otimes \mathcal{I}_{\Gamma(S)}^{k(\alpha_{j+1}(S,A)-\epsilon)}\right)$$

consists of translates of H_k *where* $0 \leq k \leq j$. *Moreover, given* $0 < \beta < 1$, *for each* $1 \leq j \leq r-1$ *and for all* S *sufficiently large,*

$$(\alpha_{j+1}(S, A) - \alpha_j(S, A))^{c_j} \operatorname{card}(\Gamma(\lfloor S^\beta \rfloor) + H_j/H_j) \deg_A(\overline{H}_j) \leq \deg_A(X).$$

where $c_j = \operatorname{codim}(H_j, G)$ *and* $\lfloor S^\beta \rfloor$ *is the greatest integer less than or equal to* S^β.

Both parts of Conjecture 1.1.9 can be cleaned up and strengthened by enlarging the set $\Gamma(S)$ and this will be examined in §2. As stated, the degree inequality of Conjecture 1.1.9 would be an improvement of joint work with Ratazzi [NR]. The arguments of the article [NR] prove the best known results in the direction of Conjecture 1.1.9.

The organization of the paper is as follows. In §1 we prove Theorem 1.1.2 and its analogue in the higher dimensional case when there is a divisorial Seshadri exceptional subvariety for A at η. In §2 we discuss the basic properties of the invariant $\nu(\Gamma, H)$ and the filtration which it induces on G. We then study one example of this filtration carefully and relate this to Conjecture 1.1.9. In §3 we prove Theorem 1.1.6 while §4 is devoted to a discussion of Conjecture 1.1.8 and Conjecture 1.1.9. In particular, Conjecture 1.1.8 can be established up to explicit factors involving $\dim(G)$ and $\rho(G)$ and similarly for Conjecture 1.1.9, except that we cannot necessarily isolate each subgroup H_j and thus the inequality obtained involves an unknown subgroup containing H_j. Due to space constraints, we can only sketch the main ideas of the arguments in some cases but we have tried to include all important ideas. One aspect of this story which we do not have time to address here but which, particularly for applications in transcendence theory, merits serious attention is the issue of giving effective bounds, *as a function of* S, for zero estimates and interpolation estimates.

Acknowledgements As is clear from the testimony in §5, my greatest mathematical debt is to Serge Lang and it is with great affection that I dedicate this work to his memory. Others to whom I am endebted for many discussions involving the material of this paper are D. Bertrand, S. Fischler, P. Philippon, and N. Ratazzi.

1.2 Surfaces

Proof of Theorem 1.1.2.

One of the inequalities of Theorem 1.1.2 is trivial, namely

$$m(A)\epsilon(\eta, A) \leq \deg_A(X).$$

Indeed, choose an effective **Q**-divisor $D \equiv A$ with multiplicity at least $m(A) - \delta$ at η. Choose a general effective **Q**-divisor $E \equiv A$ with multiplicity at least

$\epsilon(\eta, A)-\delta$ at η. By definition of $\epsilon(\eta, A)$ we may assume that D and E meet properly at η and, by Bézout's theorem,

$$(\epsilon(\eta, A) - \delta)(m(A) - \delta) \leq \deg_A(X).$$

Allowing δ to approach zero establishes the upper bound on $m(A)\epsilon(\eta, A)$.

For the lower bound we use the counting methods of [N2] and a key observation of [EKL] also used in [N2]. Suppose that C_η is a Seshadri exceptional curve for A at η: if the Seshadri exceptional subvariety were X itself, then we would be in the situation where $\epsilon(\eta, A) = m(A) = \sqrt{\deg_A(X)}$ and there would be nothing to prove.

Suppose $D \in \left|kA \otimes m_\eta^{k\alpha}\right|$. Then D vanishes along C_η if $\alpha > \epsilon(\eta, A)$. In fact, using [N2] Lemma 1.3, which is in turn a variant of [EKL] Proposition 2.3, we have

$$\text{mult}_{C_\eta}(D) \geq k\alpha - k\epsilon(\eta, A) - 1. \tag{2}$$

We now define a sequence of vector spaces as follows. For each integer $a \geq 0$ let

$$V_a = H^0\left(X, kA \otimes m_\eta^a/m_\eta^{a+1}\right).$$

We have the trivial upper bound

$$\dim(V_a) \leq a + 1, \quad 1 \leq a \leq \lceil k\epsilon(\eta, A) + 1\rceil. \tag{3}$$

On the other hand, for $a > \lceil k\epsilon(\eta, A) + 1\rceil$ we can apply (2) to conclude that

$$\text{mult}_{C_\eta}(D) \geq a - \lceil k\epsilon(\eta, A) + 1\rceil.$$

In a worst case scenario, from the point of view of the counting, C_η is smooth at η. In this case, writing

$$D = (a - \lceil k\epsilon(\eta, A) + 1\rceil)C + D',$$

and then removing the multiple of C gives an inclusion

$$V_a \subset H^0\left(X, kA - (a - \lceil k\epsilon(\eta, A) + 1\rceil)\, C \otimes m_\eta^{\lceil k\epsilon(\eta,A)+1\rceil}/m_\eta^{\lceil k\epsilon(\eta,A)+2\rceil}\right).$$

In particular, for all $a > \lceil k\epsilon(\eta, A) + 1\rceil$,

$$\dim(V_a) \leq k\epsilon(\eta, A) + 2. \tag{4}$$

We have

$$h^0\left(X, kA \otimes m_\eta^{k\alpha}\right) = h^0(X, kA) - \sum_{i=0}^{k\alpha} h^0\left(X, kA \otimes m_\eta^i/m_\eta^{i+1}\right).$$

Applying (3) for each $1 \leq i \leq \lceil k\epsilon(\eta, A) + 1\rceil$, and (4) for each $i \geq \lceil k\epsilon(\eta, A) + 2\rceil$ we find, up to factors of $O(k)$,

$$\begin{aligned} h^0\left(X, kA \otimes m_\eta^{k\alpha}\right) &= h^0(X, kA) - \frac{k^2\epsilon(\eta, A)^2}{2} - (k\alpha - k\epsilon(\eta, A))k\epsilon(\eta, A) \\ &= h^0(X, kA) - k^2\alpha\epsilon(\eta, A) + \frac{k^2\epsilon(\eta, A)^2}{2}. \end{aligned}$$

By Riemann-Roch $h^0(X, kA) = \frac{k^2 \deg_A(X)}{2} + O(k)$. In particular, for $k \gg 0$ we see that $h^0\left(X, kA \otimes m_\eta^{k\alpha}\right) > 0$, and consequently $m(A) > \alpha$, as long as

$$\alpha\epsilon(\eta, A) - \frac{\epsilon(\eta, A)^2}{2} < \frac{\deg_A(X)}{2}.$$

Thus we conclude that

$$m(A) \geq \frac{\deg_A(X)}{2\epsilon(\eta, A)} + \frac{\epsilon(\eta, A)}{2}.$$

This concludes the proof of Theorem 1.1.2. Note that when $\epsilon(\eta, A)$ is small the lower bound becomes sharper.

Note that Theorem 1.1.2 is the best possible result of this type. Indeed, the right-hand side can be an equality as we see by taking $X = \mathbf{P}^2$ and $A = \mathcal{O}_{\mathbf{P}^2}(1)$. On the other hand the left-hand side becomes closer and closer to being an equality with $X = \mathbf{P}^1 \times \mathbf{P}^1$ and $A = \mathcal{O}(a, b)$ with $a \ll b$. Indeed, in this case $m(A) = a + b$ and $\epsilon(\eta, A) = a$ with Seshadri exceptional subvariety being $\mathbf{P}^1 \times \pi_2(\eta)$. We have $\deg_A(X) = 2ab$ and thus $\frac{\deg_A(X)}{2} = ab$ is only slightly less than $m(A)\epsilon(\eta, A) = ab + a$. According to [N1] this scenario only arises when the Seshadri exceptional curve is a fibre of a surjective map $\pi : X \to Y$ where Y is a curve.

We now discuss the analogue of Theorem 1.1.2 for the case where $\dim(X) > 2$ and where there is a Seshadri exceptional divisor D_η of A relative to $\Gamma(S)$ at η. In this case we have direct analogues of the three key estimates (2), (3), and (4). Suppose $D \in H^0\left(X, A \otimes m_\eta^{k\alpha}\right)$. Then we have as before

$$\text{mult}_{D_\eta}(D) \geq k\alpha - k\epsilon(\eta, A) - 1. \tag{5}$$

With V_a defined precisely as in the surface case we have

$$\dim(V_a) \leq \binom{a + \dim(X) - 1}{\dim(X) - 1}, \quad 1 \leq a \leq \lceil k\epsilon(\eta, A) + 1\rceil. \tag{6}$$

Lastly, for $a > \lceil k\epsilon(\eta, A) + 1\rceil$ we have

$$\dim(V_a) \leq \binom{\lceil k\epsilon(\eta, A) + 1\rceil + \dim(X) - 1}{\dim(X) - 1}. \tag{7}$$

Combining (6) and (7) we find, up to $O(k^{d-1})$,

$$h^0\left(X, ka \otimes m_\eta^{k\alpha}\right) \geq \frac{k^d \deg_A(X)}{d!} - \frac{k^d \epsilon(\eta, A)^d}{d!} - (k\alpha - k\epsilon(\eta, A))\frac{k^{d-1}\epsilon(\eta, A)^{d-1}}{(d-1)!}.$$

Arguing, as in the surface case, we deduce that

$$m(A) \geq \frac{\deg_A(X)}{d\epsilon(\eta, A)^{d-1}} + \epsilon(\eta, A)\frac{d-1}{d}. \tag{8}$$

Since the Seshadri exceptional subvariety D_η is a divisor, we have $\alpha_i(A) = \epsilon(\eta, A)$ for $1 \leq i \leq d-1$ and thus (8) implies that

$$\prod_{i=1}^{d} \alpha_i(d) = \epsilon(\eta, A)^{d-1} m(A) \geq \frac{\deg_A(X)}{d} + \epsilon(\eta, A)^d \left(\frac{d-1}{d}\right).$$

To conclude this section, we prove (1) from the introduction, generalizing the argument given on surfaces in the first paragraph of this section. The fact that $m(A) \geq \epsilon(\eta, A)$ is clear as $m(A)$ measures the maximum asymptotic multiplicity of a section of kA at η, while $\epsilon(\eta, A)$ measures the maximum asymptotic separation of jets of kA at η. Moreover, [L] Volume I, Proposition 5.1.9 asserts that

$$\epsilon(\eta, A) \leq \sqrt[d]{\deg_A(X)}. \tag{9}$$

The fact that $m(A) \geq \sqrt[d]{\deg_A(X)}$ is a simple counting argument using the Riemann-Roch theorem on X. We will show that

$$\epsilon(\eta, A)^{d-1} \cdot m(A) \leq \deg_A(X). \tag{10}$$

Combining (9) and (10) shows that if $\epsilon(\eta, A) = \sqrt[d]{\deg_A(X)}$, then $m(A) = \sqrt[d]{\deg_A(X)}$. The opposite implication, that if $m(A) = \sqrt[d]{\deg_A(X)}$, then $\epsilon(\eta, A) = \sqrt[d]{\deg_A(X)}$ is more subtle and is the content of the last section of [N2]. As for (10), given $\delta > 0$, choose $k > 0$ and sections $s_1, \ldots, s_d \in H^0(X, kA)$ with the following properties:

(a.) $\text{mult}_\eta(s_i) \geq (\epsilon(\eta, A) - \delta)k, \ 1 \leq i \leq d-1,$
(b.) $\text{mult}_\eta(s_d) \geq (m(A) - \delta)k,$
(c.) the tangent cones of the sections s_i at η meet properly.

Letting D_i be the zeroes of s_i, by (c.) the intersection $D_1 \cap \cdots D_d$ is proper and by (a) and (b) the intersection multiplicity at η will be at least

$$(\epsilon(\eta, A) - \delta)^{d-1}(m(A) - \delta)k^d.$$

Since this intersection multiplicity is at most the total degree of the intersection, $k^d \deg_A(X)$, (10) follows taking the limit as δ approaches zero.

1.3 Filtrations

In order to justify why, in the language of Definition 1.1.5, there is a Γ-exceptional subgroup $H \subset G$, suppose $H_1, H_2 \subset G$ are two distinct subgroups with $\nu(\Gamma, H_1) = \nu(\Gamma, H_2)$ both maximal. We may assume that neither subgroup contains the other. It is sufficient then to show that

$$\nu(\Gamma, H_1 + H_2) = \nu(\Gamma, H_1) = \nu(\Gamma, H_2). \tag{11}$$

If $H_1 \cap H_2 = \{0\}$, then $d(H_1 + H_2) = d(H_1) + d(H_2)$ and $\rho(H_1 + H_2) = \rho(H_1) + \rho(H_2)$ and thus (11) is true. Thus we may assume that $H_1 \cap H_2$ is non-trivial.

Since $\nu(\Gamma, H_1)$ and $\nu(\Gamma, H_2)$ are maximal we have

$$\nu(\Gamma, H_1 + H_2) \leq \nu(\Gamma, H_1).$$

If $\nu(\Gamma, H_1 + H_2) < \nu(\Gamma, H_1)$, then we claim that

$$\frac{\rho(H_1 \cap H_2)}{d(H_1 \cap H_2)} > \frac{\rho(H_1)}{d(H_1)}, \tag{12}$$

contradicting the maximality assumption on H_1. Note that

$$\nu(\Gamma, H_1 + H_2) = \frac{\rho(H_1 + H_2)}{d(H_1 + H_2)} = \frac{\rho(H_1) + \rho(H_2) - \rho(H_1 \cap H_2)}{d(H_1) + d(H_2) - d(H_1 \cap H_2)}. \tag{13}$$

If a, b, c, d are positive numbers with $a > c$ and $b > d$, then

$$\frac{a-c}{b-d} < \frac{a}{b} \text{ if and only if } \frac{a}{b} < \frac{c}{d}. \tag{14}$$

Combining (14) with (13) establishes (12) and concludes the proof that there is a unique maximal subgroup $H \subset G$ with $\nu(H)$ maximal.

Let H_1 be the Γ-exceptional subgroup of G. If $H_1 = G$, then we are done. If not, let ν_2 be the largest value, strictly smaller than $\nu(\Gamma, H_1)$, such that $\nu(\Gamma, H) = \nu_2$ for some subgroup $H \subset G$. Among all subgroups, choose H_2 maximal so that $\nu(\Gamma, H_2) = \nu_2$. As above we consider $H_1 + H_2$. If $H_1 \cap H_2 = \{0\}$; then $\nu(H_1) > \nu(H_1 + H_2) > \nu(H_2)$, contradicting the choice of H_2. Thus $H_1 \cap H_2$ is non-trivial. By choice of H_1 we have $\nu(H_1 \cap H_2) \leq \nu(H_1)$. If $\nu(H_1 \cap H_2) = \nu(H_1)$, then (13) gives the contradiction

$$\nu(H_1) > \nu(H_1 \cap H_2) > \nu(H_1 + H_2)$$

unless $H_1 \cap H_2 = H_1$. But if $H_1 \cap H_2 = H_1$, then we are done since this implies that $H_1 \subset H_2$. Thus we may assume that $\nu(H_1 \cap H_2) < \nu(H_1)$ and so, by the choice of H_2, $\nu(H_1 \cap H_2) \leq \nu(H_2)$. Using (13) again will show that

$$\nu(H_1 + H_2) > \nu(H_2), \tag{15}$$

unless $H_1 \cap H_2 = H_1$. In either case we are finished since (15) contradicts the choice of H_2.

If $H_2 = G$ we are done. If not, choose the largest value $\nu_3 < \nu_2$ so that there is a subgroup $H_3 \subset G$ with $\nu(H_3) = \nu_3$. We choose H_3 maximal and arguing precisely as in the previous paragraph shows that $H_2 \subset H_3$. The chain terminates when we find $H_r = G$.

In the rest of this section, we would like, in order to motivate Conjecture 1.1.9, to consider a special case of G and Γ where we can compute base loci explicitly. Let $G = \mathbf{A}^3$ and $X = \mathbf{P}^3$. Let $v_1 = (1,0,0)$, $v_2 = (0,1,0)$, and $v_3 = (0,0,1)$. Choose

$$\Gamma = \{0, a_{11}v_1, \ldots, a_{1r_1}v_1, a_{21}v_2, \ldots, a_{2r_2}v_2, a_{31}v_3, \ldots, a_{3r_3}v_3\},$$

where the a_{ij} are algebraically independent over $\mathbf{Q}$ and $r_1 \gg r_2 \gg r_3$. In this scenario we have

$$\begin{aligned} H_1 &= \mathbf{A}^1 \times \{0\} \times \{0\}, \\ H_2 &= \mathbf{A}^1 \times \mathbf{A}^1 \times \{0\}, \\ H_3 &= \mathbf{A}^1 \times \mathbf{A}^1 \times \mathbf{A}^1. \end{aligned}$$

We also have

$$\begin{aligned} \nu(\Gamma, H_1) &= r_1, \\ \nu(\Gamma, H_2) &= \frac{r_1 + r_2}{2}, \\ \nu(\Gamma, H_3) &= \frac{r_1 + r_2 + r_3}{3}. \end{aligned}$$

Finally, let

$$\begin{aligned} \Gamma_1 &= \{0, a_{11}v_1, \ldots, a_{1r_1}v_1\}, \\ \Gamma_2 &= \{0, a_{21}v_2, \ldots, a_{2r_2}v_2\}, \\ \Gamma_3 &= \{0, a_{31}v_3, \ldots, a_{3r_3}v_3\}. \end{aligned}$$

For this example it is useful to introduce a slightly larger set than $\Gamma(S)$, namely

$$\overline{\Gamma}(S) = \{a_1x_1 + \ldots + a_lx_l : 0 \leq a_i \leq S \text{ for all } i,\ x_i \in \Gamma\}.$$

Conjectures 1.1.8 and 1.1.9 should still hold with $\overline{\Gamma}(S)$ in place of $\Gamma(S)$. Consider, for $S > 0$ fixed and k large, the linear series

$$\left|\mathcal{O}_X(k) \otimes \mathcal{I}^{k\alpha}_{\overline{\Gamma}(S)}\right|. \tag{16}$$

Suppose $\alpha < (S+1)^{-r_1}$. We claim that the base locus of (16) is $\overline{\Gamma}(S)$. Choose coordinates x_1, x_2, x_3 on $\mathbf{A}^3$ where x_i is the coordinate from the ith factor $\mathbf{A}^1$. Let $\pi_i : \mathbf{A}^3 \to \mathbf{A}$ be the projection to the ith factor. Note that $\pi_i(\overline{\Gamma}(S)) = \overline{\Gamma}_i(S)$. Let $f_i(x_i)$ be a polynomial of degree k which vanishes along $\overline{\Gamma}_i(S)$ to order $k\alpha$. In order to verify that such a non-zero polynomial $f_i(x_i)$ exists, note that $|\overline{\Gamma}_i(S)| = (S+1)^{r_i}$. The cost of vanishing to order at least $k\alpha$ at a point is $\lceil k\alpha \rceil$. On the other hand, polynomials of degree at most k in x_i form a vector space of dimension $k+1$. Thus as long as

$$\lceil k\alpha \rceil (S+1)^{-r_i} \leq k+1$$

a non-zero polynomial $f_i(x_i)$ exists satisfying the required vanishing conditions. Using the hypothesis that $r_1 \gg r_2 \gg r_3$ and looking at the common zeroes of the polynomials $f_1(x_1)$, $f_2(x_2)$, and $f_3(x_3)$ shows that the base locus of (16) is exactly $\overline{\Gamma}(S)$ when $\alpha < (S+1)^{-r_1}$.

When $\alpha > (S+1)^{-r_1}$ we claim that $\mathbf{A}^1 \times \{P_2\} \times \{P_3\}$ is in the base locus of (16) for all $P_2 \in \overline{\Gamma}_2(S)$ and $P_3 \in \overline{\Gamma}_3(S)$. Indeed, if $\sigma \in H^0(X, \mathcal{O}_X(k))$ vanishes to order at least $k\alpha$ along $\overline{\Gamma}(S)$, then $\sigma|\mathbf{A}^1 \times \{P_2\} \times \{P_3\}$ is a polynomial of degree at most k with order of vanishing larger than $k(S+1)^{-r_1}$ at $(S+1)^{r_1}$ points, and this is impossible unless $\sigma|\mathbf{A}^1 \times \{P_2\} \times \{P_3\} = 0$. Using the functions $f_2(x_2)$, $f_3(x_3)$ defined above, we see that when $(S+1)^{-r_2} > \alpha > (S+1)^{-r_1}$ the base locus of (16) is precisely

$$\bigcup_{P_2 \in \overline{\Gamma}_2(S), P_3 \in \overline{\Gamma}_3(S)} \overline{\mathbf{A}^1 \times \{P_2\} \times \{P_3\}},$$

where $\overline{\mathbf{A}^1 \times \{P_2\} \times \{P_3\}}$ denotes the Zariski closure of $\mathbf{A}^1 \times \{P_2\} \times \{P_3\}$ in $\mathbf{P}^3$.

Suppose now that $\alpha > (S+1)^{-r_2}$. We claim that for $P_3 \in \overline{\Gamma}_3(S)$ the subvariety $\mathbf{A}^1 \times \mathbf{A}^1 \times \{P_3\}$ is in the base locus of (16) for each $P_3 \in \overline{\Gamma}_3(S)$. Suppose f is a polynomial of degree k vanishing to order at least $\lceil k\alpha \rceil$ at $\overline{\Gamma}(S)$. Suppose that f does not vanish identically along $\mathbf{A}^1 \times \mathbf{A}^1 \times \{P_3\}$ and let

$$g = f|\mathbf{A}^1 \times \mathbf{A}^1 \times \{P_3\}.$$

We claim that for each $P_1 \in \overline{\Gamma}_1(S)$ g vanishes at $\{P_1\} \times \mathbf{A}^1 \times \{P_3\}$. Indeed, if not, then $g|\{P_1\} \times \mathbf{A}^1 \times \{P_3\}$ has degree k and order of vanishing at least α at $(S+1)^{r_2}$ points, and this is impossible. Thus g is divisible by

$$h(x_1) = \prod_{P_1 \in \overline{\Gamma}_1(S)} (x_1 - P_1).$$

Consider the quotient $g(x_1, x_2)/h(x_1)$. This polynomial has degree $k - |\overline{\Gamma}_1(S)|$ and order of vanishing at least $\lceil k\alpha \rceil - 1$ at (x, y) for $x \in \overline{\Gamma}_1(S)$ and $y \in \overline{\Gamma}_2(S)$. Note that the argument applied to establish that g is divisible by h applies more generally. Indeed, if $r(x_1, x_2)$ is any polynomial with order of vanishing at least t along $\overline{\Gamma}_1(S) \times \overline{\Gamma}_2(S)$, with $t|\overline{\Gamma}_2(S)| > \deg(r)$, then r vanishes at $\{P_1\} \times \mathbf{A}^1$ for each $P_1 \in \overline{\Gamma}_1(S)$. Dividing $g(x_1, x_2)$ by $h(x_1)$ drops the degree of g by $|\Gamma_1(S)|$, while the order of vanishing along $\overline{\Gamma}_1(S) \times \overline{\Gamma}_2(S)$ has gone down by 1. Thus we conclude, setting $r(x_1, x_2) = g(x_1, x_2)/h(x_1)$, that $r(x_1, x_2)$ is also divisible by $h(x_1)$ and the argument can then be repeated until we reach a contradiction. We conclude that for $\alpha > (S+1)^{-r_2}$ the base locus of (16) contains

$$\bigcup_{P_3 \in \overline{\Gamma}_3(S)} \mathbf{A}^1 \times \mathbf{A}^1 \times \{P_3\}.$$

Using the function $f_3(x_3)$ defined above, we see that the base locus of (16) for $(S+1)^{-r_2} < \alpha < (S+1)^{-r_3}$ is precisely

$$\bigcup_{P_3 \in \overline{\Gamma}_3(S)} \overline{\mathbf{A}^1 \times \mathbf{A}^1 \times \{P_3\}}.$$

Suppose now that $\alpha > (S+1)^{-r_3}$ and that f is a polynomial of degree k vanishing at $\overline{\Gamma}(S)$ to order at least $k\alpha$. Using the argument of the previous case we will establish that f is divisible by the polynomial

$$h(x_1) = \prod_{P_1 \in \overline{\Gamma}_1(S)} (x_1 - P_1).$$

For $P_1 \in \overline{\Gamma}_1(S)$ consider $f|\{P_1\} \times \mathbf{A}^1 \times \mathbf{A}^1$. This has order of vanishing at least α at $\overline{\Gamma}_2(S) \times \overline{\Gamma}_3(S)$ and so, repeating the argument of the previous paragraph, f vanishes identically on $\{P_1\} \times \mathbf{A}^1 \times \mathbf{A}^1$. We then repeat the same argument for $f(x)/h(x_1)$ which has degree $k - |\overline{\Gamma}_1(S)|$ and order of vanishing at least $\lceil k\alpha \rceil - 1$ along $\overline{\Gamma}(S)$. We eventually reach a contradiction, concluding that f does not exist.

In this particular case it is easy to verify that Conjecture 1.1.9 holds with $\overline{\Gamma}(S)$ in place of $\Gamma(S)$. Indeed, we have provided explicit polynomials, for any $\epsilon > 0$, with the specified base locus. As indicated after the statement of Conjecture 1.1.9 the base locus of the linear series (16) is exactly translates of H_j, where j will depend on α.

What happens in this example for Conjecture 1.1.9 when $\Gamma(S)$ is used instead of $\overline{\Gamma}(S)$? Qualitatively we will find the same result since we have the inclusions

$$\Gamma(S) \subset \overline{\Gamma}(S) \subset \Gamma(dS),$$

but the quantitative sharpness of the estimate will be lost. To better understand what goes wrong, we will consider the same example but with $\Gamma(S)$ in place of the enlarged $\overline{\Gamma}(S)$.

When $\alpha > (S+1)^{-r_1}$ the subvariety $\mathbf{A}^1 \times \{0\} \times \{0\}$ will enter the base locus precisely as before. For other values of $P_2, P_3 \in \Gamma(S)$, however, the curve $\mathbf{A}^1 \times \{P_2\} \times \{P_3\}$ enters the base locus at a later point. For example, if $P_2 = 0$ and $P_3 = x$ for some $x \in \Gamma$ the argument above will only establish that $\mathbf{A}^1 \times \{P_2\} \times \{P_3\}$ enters the base locus once $\alpha > S^{-r_3}$. Thus as α grows, more and more curves of the form $\mathbf{A}^1 \times \{P_2\} \times \{P_3\}$ will enter the base locus. Eventually, the surface $\mathbf{A}^1 \times \mathbf{A}^1 \times \{0\}$ enters the base locus quickly to be followed by fibres over other points $P_3 \in \overline{\Gamma}_3(S)$. The values where the base locus jumps in dimension will be comparable asymptotically in S, but the structure is much more involved and does not lend itself to a simple statement as in Conjecture 1.1.9.

1.4 The Γ-exceptional subgroup

In this section we will prove Theorem 1.1.6. Note first that for S large, the existence of a proper Γ-exceptional subgroup H forces the Seshadri constant $\epsilon(\Gamma(S), A)$ to be highly submaximal. Indeed, for S large, using the subgroup H to estimate the value of $\epsilon(\Gamma(S), A)$, we have by [L] Volume I, Proposition 5.1.9

$$\epsilon(\Gamma(S), A) \leq \left(\frac{\deg(H)}{|\Gamma(S) \cap H|}\right)^{\frac{1}{\dim(H)}}.$$

Recalling that $\rho_H = \operatorname{rank}(\Gamma(\infty) \cap H)$, the right-hand side grows like $O\left(S^{-\frac{\rho_H}{\dim(H)}}\right)$, while the maximal value possible for $\epsilon(\Gamma(S), A)$ is

$$\left(\frac{\deg(G)}{|\Gamma(S)|}\right)^{\frac{1}{\dim(G)}} = O\left(S^{-\frac{\rho_G}{\dim(G)}}\right).$$

Since H is Γ-exceptional and is a proper subgroup of G we have $\frac{\rho_H}{\dim(H)} > \frac{\rho_G}{\dim(G)}$ and consequently for S large

$$\epsilon(\Gamma(S), A) \ll \left(\frac{\deg(G)}{|\Gamma(S)|}\right)^{\frac{1}{\dim(G)}}. \tag{17}$$

We claim that for k sufficiently large the linear series

$$\left| kA \otimes \mathcal{I}_{\Gamma(dS)-\Gamma(dS)}^{dk\epsilon(\Gamma(S),A)+d} \right| \tag{18}$$

is non-empty. Note that $\Gamma - \Gamma$ generates the same subgroup of G as Γ and in particular $|\Gamma(S) - \Gamma(S)|$ and $|\Gamma(S)|$ both grow like polynomials in S of degree ρ_G. Thus, according to (17), for $k, S \gg 0$ the linear series (18) is non-empty.

We will also need to study the growth of $|\Gamma(S) - \Gamma(S)|$ on subgroups. Suppose $H \subset G$ is an irreducible non-trivial subgroup. Choose a basis $x_1, \ldots, x_{\rho_G}$ for the free part of $\Gamma(\infty)$. Choose $r > 0$ so that each non-torsion element $x \in \Gamma$ can be expressed as

$$x = t_x + \sum_{i=1}^{\rho_G} a_i x_i,$$

where $t_x \in \Gamma(\infty)_{\text{tors}}$ and $|a_i| \leq r$. Let

$$A(S) = \left\{ \sum_{i=1}^{\rho_G} a_i x_i \, : \, |a_i| \leq Sr \right\}.$$

For all $S > 0$ if $x \in \Gamma(S)$, then there is an element $y \in \Gamma(\infty)_{\text{tors}}$ and $z \in A(S)$ so that

$$x = y + z.$$

Thus we have

$$\begin{aligned} |H \cap \Gamma(S)| &\leq |\Gamma(\infty)_{\text{tors}}||H \cap \Gamma(S) \cap \Gamma(\infty)_{\text{free}}| \\ &\leq |\Gamma(\infty)_{\text{tors}}||H \cap A(rS)| \end{aligned}$$

Let ρ_H be the rank of H. Then $|H \cap A(rS)| \leq (2rS + 1)^{\rho_H}$. In particular it is possible to choose S sufficiently large so that the linear series (18) is non-empty restricted to any subgroup H which satisfies $\nu(H) < \nu(G)$.

Proof of Theorem 1.1.6. Suppose that V is Seshadri exceptional for A relative to $\Gamma(S)$ for some $S \gg 0$. Thus by definition

$$\epsilon(\Gamma(S), A) = \left(\frac{\deg_A(V)}{\sum_{y \in \Gamma(S)} \text{mult}_y(V)} \right)^{\frac{1}{\dim(V)}}. \tag{19}$$

Let σ be a non-zero section of the linear series (18). We will outline a proof that

$$\sigma \text{ vanishes on } t_{-x}(V) + \ldots + t_{-x}(V) \subset Z(\sigma) \text{ with } d \text{ summands} \tag{20}$$

for each $x \in \Gamma(S)$. We will review here the argument to establish (20) for one and two summands and indicate how it extends to the general case.

We first claim that for $g \in G$ if V is Seshadri exceptional for A relative to $\Gamma(S)$ then $V + g$ is Seshadri exceptional for A relative to $\Gamma(S) + g$. Note that for all positive dimensional $W \subset X$

$$\frac{\deg_A(g + W)}{\sum_{y \in \Gamma(S)} \text{mult}_{g+y}(g + W)} = \frac{\deg_A(W)}{\sum_{y \in \Gamma(S)} \text{mult}_y(W)} \tag{21}$$

because the degree is translation invariant. As W varies over all curves, the infimum of the right-hand side is $\epsilon(\Gamma(S), A)$ by definition, while the infimum of the left hand side is $\epsilon(g + \Gamma(S), A)$ and thus

$$\epsilon(\Gamma(S), A) = \epsilon(g + \Gamma(S), A).$$

Using (19) and (21) we conclude that $V + g$ is Seshadri exceptional for A relative to $\Gamma(S) + g$ as claimed.

Suppose now that $x \in \Gamma(S)$. Since $V - x$ is Seshadri exceptional for A relative to $\Gamma(S) - x$, and since σ vanishes to order $> k\epsilon(\Gamma(S), A)$ at $\Gamma(S) - \Gamma(S)$ it follows from [NR] Proposition 3.2 that σ vanishes on $V - x$ for each $x \in \Gamma(S)$, establishing (20) for one summand. To establish (20) for two summands, note that for all $x, y \in \Gamma(S)$, $y + V - 2x$ is Seshadri exceptional for A relative to $y + \Gamma(S) - 2x$. Since σ vanishes to order $> 2k\epsilon(\Gamma(S), A) + 2$ along $y + \Gamma(S) - 2x$ we deduce that σ must vanish along $y + V - 2x$. Using the differentiation argument of [NR] Proposition 3.2 shows that in fact σ has order of vanishing at least $k\epsilon(\Gamma(S), A) + 1$ along $\Gamma(S) + V - 2x$. In particular, σ vanishes along $V + V - 2x$ as desired. This process can be iterated, using the fact that σ vanishes to order at least $kd\epsilon(\Gamma(S), A) + d$ along $\Gamma(dS) - \Gamma(dS)$, to establish (20).

Choose $x \in \Gamma(S) \cap V$: we know such an x exists since V is Seshadri exceptional for A relative to $\Gamma(S)$. Let $W = t_{-x}(V)$ so that W contains e_G and let $W + \ldots + W$, with d summands, be the Zariski closure in X of $(W \cap G) + \ldots + (W \cap G)$. Then from the previous paragraph we conclude that σ vanishes on $W + \ldots + W$. Since $\sigma \neq 0$ it follows that $W \cap G$ is degenerate. Let $H_W \subset G$ be the smallest subgroup containing W. Then $V \subset x + W \subset x + \overline{H}_W$ is still Seshadri exceptional for $A|(x + \overline{H}_W)$ relative to $\Gamma(S) \cap x + \overline{H}_W$. Thus $V - x$ is Seshadri exceptional relative to $\Gamma(S) - x$ for $A|\overline{H}_W$. If $\nu(H_W) < \nu(H)$ then, according to (18), for S sufficiently large, but not depending on W, the linear series

$$\left| kA|\overline{H}_W \otimes \mathcal{I}_{\Gamma(dS)-\Gamma(dS)}^{dk\epsilon(\Gamma(S),A)+d} \right|$$

is non-empty. Repeating the above argument will show that $V - x$ is degenerate inside $\overline{H}_W$ which is impossible by definition of H_W. Thus $\nu(H_W) = \nu(H)$, and so we have $H_W \subset H$. This concludes the proof of Theorem 1.1.6 since we have shown that $V - x \subset \overline{H}_W$.

Note that if $\dim(G) = 2$ then Theorem 1.1.6 implies that for all S sufficiently large a translate of the Γ-exceptional subgroup is Seshadri exceptional. This generalizes [NR] Théorème 1.5.

1.5 What can be shown

The obstruction to the interpolation problem is the subgroup H where $\nu(\Gamma, H)$ is maximal or in other words where $\Gamma(\infty) \cap H$ is most densely distributed. For multiplicity estimates, on the other hand, there are in general several possible choices for an obstruction subgroup, namely $H_i, 1 \leq i \leq r-1$. The best choice among the possible obstruction subgroups, from the point of view of the inequality in Conjecture 1.1.9, is the subgroup H_j for which $\alpha_{j+1} - \alpha_j$ is largest. Stated qualitatively, the best choice for the obstruction to multiplicity estimates is the subgroup K in which $\Gamma(\infty) \cap K$ is most densely distributed by comparison to $\Gamma(\infty) \cap K'$ for all subgroups K' properly containing K.

Here we would like to discuss what can be shown in the direction of Question 1.1.7 and Conjectures 1.1.8 and 1.1.9. First, regarding Question 1.1.7, Theorem 1.1.6 is, at least qualitatively, already a large step in the right direction, for even if the Γ-exceptional subgroup H is not itself Seshadri exceptional relative to $\Gamma(S)$ for large S, it remains the source of the interpolation problem and provides a good lower bound for the interpolation exponent. Concerning Conjecture 1.1.8 we have

Theorem 1.5.1. *Suppose G is a commutative algebraic group with compactification X and A an ample line bundle on X. Let $\Gamma \subset G$ be a finite set and let H be the Γ-exceptional subgroup and d_H its dimension. For all S sufficiently large,*

$$\epsilon(\Gamma(S), A) \geq \frac{1}{d_H}\left(\frac{\deg_A(H)}{|\Gamma(d_H S) \cap H|}\right)^{\frac{1}{d_H}}.$$

Similarly we have

$$\epsilon(\overline{\Gamma}(S), A) \geq \frac{1}{d_H}\left(\frac{\deg_A(H)}{|\overline{\Gamma}(d_H S) \cap H|}\right)^{\frac{1}{d_H}}.$$

The penalty to be paid for not knowing that H is Seshadri exceptional is that $|\Gamma(S)|$ must be replaced with $|\Gamma(d_H S)|$. Otherwise this would be Conjecture 1.1.8.

Proof of Theorem 1.5.1. Let V be the Seshadri exceptional subvariety for A relative to $\Gamma(S)$ for some $S \gg 0$. The proof of Theorem 1.1.6 shows that for some $x \in \Gamma(S)$ we have $V \subset x + H'$ where $H' \subset H$ and $\nu(H', \Gamma) = \nu(H, \Gamma)$. We assume that H' is the smallest subgroup containing a translate of V. We know that V is Seshadri exceptional for A relative to $\Gamma(S)$ and in particular V is still Seshadri exceptional for $A|x + H'$ relative to $\Gamma(S) \cap H'$. If $\epsilon(\Gamma(S), A) < \frac{1}{d_H}\left(\frac{\deg_A(H)}{|\Gamma(d_H S) \cap H|}\right)^{\frac{1}{d_H}}$, then the same inequality holds with $d_{H'}$ in place of d_H:

$$\epsilon(\Gamma(S), A) < \frac{1}{d_{H'}}\left(\frac{\deg_A(H)}{|\Gamma(d_{H'} S) \cap H|}\right)^{\frac{1}{d_{H'}}}. \tag{22}$$

The argument from the proof of Theorem 1.1.6, which we will repeat here without translations by a multiple of $-x$, will now provide a contradiction, giving a non-zero section $\sigma \in H^0(d_{H'}x + H', kA)$ which vanishes on $d_{H'}x + H' = (x + V) + \ldots (x + V)$ with $d_{H'}$ summands. Note that if $d_{H'} = 1$, then $V = x + H'$ with

$x \in \Gamma(S)$. Since V is Seshadri exceptional for A relative to $\Gamma(S)$ we have

$$\epsilon(\Gamma(S), A) = \frac{\deg_A(x + H')}{|\Gamma(S) \cap (x + H')|}$$

and this violates (22) for $S \gg 0$. Thus we may assume that $d_{H'} \geq 2$.

Suppose $0 \neq s \in H^0\left(2x + H', kA \otimes \mathcal{I}^{2k\epsilon(\Gamma(S),A)+2}_{2x+(\Gamma(2S)\cap H')}\right)$: such a section exists, for $k \gg 0$, by (22). We claim that for each $y \in \Gamma(S) \cap H'$ the section s vanishes on $2x + y + V$. If not, then, since $2x + y + (\Gamma(S) \cap H') \subset 2x + (\Gamma(2S) \cap H')$, we have

$$s \in H^0\left(2x + y + H', kA \otimes \mathcal{I}^{2k\epsilon(\Gamma(S),A)+2}_{2x+y+(\Gamma(S)\cap H')}\right).$$

By hypothesis V is Seshadri exceptional for A relative to $\Gamma(S)$, and so $2x + y + V$ is Seshadri exceptional for $A|2x + y + H'$ relative to $2x + y + (\Gamma(S) \cap H')$. In particular, $s|2x + y + V = 0$. The same argument will show that all derivatives of s up to order $k\epsilon(\Gamma(S), A) + 1$ vanish along $2x + \Gamma(S) + V$. Since the order of vanishing of s along $2x + \Gamma(S) + V$ is at least $k\epsilon(\Gamma(S), A) + 1$ we conclude, using again the fact that V is Seshadri exceptional for A relative to $\Gamma(S)$, that s vanishes along $2x + V + V$. This argument extends to show that if $0 \neq \sigma \in H^0\left(d_{H'}x + H', kA \otimes \mathcal{I}^{d_{H'}\epsilon(\Gamma(d_{H'}S),A)+d_{H'}}_{d_{H'}x+(\Gamma(d_{H'}S)\cap H')}\right)$, then σ vanishes along $V + \ldots + V$ with $d_{H'}$ summands of V. By hypothesis, however, $V + \ldots + V = d_{H'}x + H'$ and thus σ vanishes identically on $d_{H'}x + H'$. This is a contradiction, violating (22), and this establishes Theorem 1.5.1. The same proof can be copied line for line with $\overline{\Gamma}(S)$ in place of $\Gamma(S)$.

We now consider multiplicity estimates. There are two cases to consider, according to whether or not the Γ-exceptional subgroup H is a proper subgroup of G. When $H = G$ this means that the points in $\Gamma(S)$ for S large are well-distributed, and consequently one expects a very strong upper bound on the maximum multiplicity a section $\sigma \in H^0(X, kA)$ can have along $\Gamma(S)$.

Theorem 1.5.2. *Suppose X is an equivariant compactification of a commutative group variety G. Let $\Gamma \subset G$ be a finite set and assume that G is the Γ-exceptional subgroup. Then for $S \gg 0$*

$$m(\Gamma(S), A) \leq \left(\frac{\deg_A(X)}{|\Gamma(S)|}\right)^{\frac{1}{d}} \left(\frac{|\Gamma(dS)|^{\frac{d-1}{d}}}{|\Gamma(S)|}\right) d^{d-1}.$$

Proof of Theorem 1.5.2. This is a straightforward consequence of Theorem 1.5.1. Indeed, Theorem 1.5.1 implies that

$$\epsilon(\Gamma(S), A) \geq \frac{1}{d}\left(\frac{\deg_A(X)}{|\Gamma(dS)|}\right)^{\frac{1}{d}}.$$

On the other hand we always have, by the multipoint analogue of (10),

$$m(\Gamma(S), A)\epsilon(\Gamma(S), A)^{d-1}|\Gamma(S)| \leq \deg_A(X).$$

Thus

$$m(\Gamma(S), A) \leq \frac{\deg_A(X)}{|\Gamma(S)|\epsilon(\Gamma(S), A)^{d-1}}.$$

Plugging in the lower bound for $\epsilon(\Gamma(S), A)$ from Theorem 1.5.1 gives Theorem 1.5.2.

Note that the first part of the right-hand side of Theorem 1.5.2

$$\left(\frac{\deg_A(X)}{|\Gamma(S)|}\right)^{\frac{1}{d}}$$

represents the minimal value possible for $m(\Gamma(S), A)$. The second term is presumably unnecessary, while the third term, though certainly not optimal, is likely necessary in some form.

We will next study the case where the Γ-exceptional subgroup H is a proper subgroup of G so that the ultimate goal is Conjecture 1.1.9. As is normally the case with multiplicity estimates, one falls rather short of Conjecture 1.1.9 as, although it is easy to establish, using Theorem 1.5.2 for example, that there are many translates of subgroups in a given base locus, it is very difficult to show that these are not properly contained in some other base components.

We will outline, following [NR], what can be done toward improving the classical multiplicity esimates when we assume that S is large and that Γ is not contained inside the torsion subgroup of G. We begin with the key tool that will be used to enforce vanishing along subgroups and their translates.

Lemma 1.5.3. *Suppose $H \subset G$ is the Γ-exceptional subgroup. Suppose $1 \gg \epsilon > 0$ is fixed and let*

$$\sigma \in H^0\left(X, kA \otimes \mathcal{I}_{\Gamma(S)}^{kS^{-\nu(\Gamma,H)+\epsilon}}\right).$$

Then $\sigma|H = 0$ for $S \gg 0$.

Proof of Lemma 1.5.3. We must show that, given $\epsilon > 0$, for S sufficiently large

$$m(\Gamma(S), A|H) \leq S^{-\nu(\Gamma,H)+\epsilon}.$$

By Theorem 1.5.2 for all $S \gg 0$,

$$\begin{aligned} m(\Gamma(S), A|H) &\leq \left(\frac{\deg_A(X)}{|\Gamma(S) \cap H|}\right)^{\frac{1}{d_H}} \left(\frac{|\Gamma(dS) \cap H|^{\frac{d_H-1}{d_H}}}{|\Gamma(S) \cap H|}\right) d_H^{d_H-1} \\ &\leq \frac{1}{S^{\nu(\Gamma,H)}} c(G, \Gamma), \end{aligned} \tag{23}$$

where the constant $c(G, \Gamma)$ does not depend on S. If

$$\sigma \in H^0\left(X, kA \otimes \mathcal{I}_{\Gamma(S)}^{kS^{-\nu(\Gamma,H)+\epsilon}}\right)$$

as in the statement of Lemma 1.5.3 and $\sigma|H \neq 0$, then $m(\Gamma(S), A|H) \geq S^{-\nu(\Gamma,H)+\epsilon}$ which is impossible for $S \gg 0$ according to (23).

Lemma 1.5.3 also holds with $t_x^*(A)$ in place of A since the proof depends only on the numerical equivalence class of A. Suppose

$$0 \neq \sigma \in H^0\left(X, kA \otimes \mathcal{I}_{\Gamma(S)}^{kS^{-\nu(\Gamma,H_2)+\epsilon}}\right)$$

with $0 < \epsilon \ll 1$ as in Lemma 1.5.3. According to Lemma 1.5.3 and the differentiation process of [NR], Proposition 3.2, σ vanishes to order at least $k(S^{-\nu(\Gamma,H_2)-\epsilon} - S^{-\nu(\Gamma,H_1)+\epsilon})$ along H where $H = H_1$. Choose $0 < \beta < 1$ and suppose $x \in \Gamma\left(cS^\beta\right)$ for $c = \text{codim}(H, G)$. Note that for each $y \in \Gamma(\lfloor S - cS^\beta \rfloor)$ we have $x + y \in \Gamma(S)$. Using the fact that $m(\Gamma(S), A)$ does not depend on the numerical equivalence class of A, we have for $x \in \Gamma\left(cS^\beta\right)$

$$\begin{aligned}
m(\Gamma(S), A|x + H) &= m(\Gamma(S), t_{-x}^*(A)|x + H) \\
&\leq m\left(\Gamma\left(x + \lfloor S - cS^\beta \rfloor\right), t_{-x}^*(A)|x + H\right) \\
&= m\left(\Gamma\left(\lfloor S - cS^\beta \rfloor\right), A|H\right).
\end{aligned}$$

Thus

$$\sigma|x + H = 0 \text{ if } m\left(\Gamma\left(\lfloor S - cS^\beta \rfloor\right), A|H\right) \leq S^{-\nu(\Gamma,H)+\epsilon}.$$

This is deduced from Theorem 1.5.2 just as Lemma 1.5.3. Provided $S \gg 0$ then, using [NR] Proposition 3.2, σ vanishes on $x + H$ for all $x \in \Gamma(c\lfloor S^\beta \rfloor)$ to order at least

$$k\left(S^{-\nu(\Gamma,H_2)-\epsilon} - S^{-\nu(\Gamma,H_1)+\epsilon}\right).$$

Running the normal argument for zero estimates, as in [NR] §4, shows that there exists a subvariety $V \subset X$ so that $\Gamma\left(\lfloor S^\beta \rfloor\right) + V$ can be cut out by derivatives and translates of σ, each having order of vanishing at least

$$\frac{k\left(S^{-\nu(\Gamma,H_2)-\epsilon} - S^{-\nu(\Gamma,H_1)+\epsilon}\right)}{c}$$

along $\Gamma\left(\lfloor S^\beta \rfloor\right) + V$. The same is true for $\Gamma\left(\lfloor S^\beta \rfloor\right) + H_V$ where H_V is the connected component of the stabilizer of V containing the origin. Thus we find an inequality of the form

$$\alpha^c \text{card}\left(\Gamma\left(\lfloor S^\beta \rfloor\right) + H_V/H_V\right) \deg_A(\overline{H}_V) \leq \deg_A(X), \tag{24}$$

with $\alpha = \frac{k(S^{-\nu(\Gamma,H_2)-\epsilon} - S^{-\nu(\Gamma,H_1)+\epsilon})}{c}$. Moreover, we claim that $H_V \supset H$. Since any section of kA vanishing to order at least

$$\frac{k\left(S^{-\nu(\Gamma,H_2)-\epsilon} - S^{-\nu(\Gamma,H_1)+\epsilon}\right)}{c}$$

along $\Gamma\left(\lfloor S^\beta \rfloor\right) + y$ vanishes along $H + y$, it is sufficient to apply Lemma 1.5.3 to $\Gamma(\lfloor S^\beta \rfloor) + x$ for each $x \in V$. Thus $H \subset H_V$, as desired.

In order to apply this argument to force vanishing along H_i and its translates for $i \geq 2$ there is a problem as one does not necessarily control $m(\Gamma(S), A|H_i)$ in this case: indeed, for $i = 1$ we have Lemma 1.5.3 but this requires that H be the Γ-exceptional subgroup. On the other hand, one can certainly apply the same argument with $m(\Gamma(S), A) - \epsilon$ in place of $S^{-\nu(\Gamma,H_2)-\epsilon}$. The conclusion would then be that there is a subgroup H', containing H, so that $\Gamma\left(\lfloor S^\beta \rfloor\right) + H'$ can be cut out by translates and derivatives of σ and so

$$\beta^{\mathrm{codim}(H',G)} \mathrm{card}\left(\Gamma\left(\lfloor S^\beta \rfloor\right) + H'/H'\right) \deg_A(\overline{H}') \leq \deg_A(X), \tag{25}$$

with $\beta = \frac{m(\Gamma(S),A)-\epsilon}{d}$. Formulas (24) and (25) fall short of Conjecture 1.1.9 in two respects. First is the superfluous factor of involving the dimension of X on the left-hand side which is inherent to all of the arguments. Secondly, the subgroup in (25) can not necessarily be identified with one of the obstruction subgroups H_i.

1.6 In Memoriam

"Hi this is Serge Lang. So what are you doing?" These were the first words ever spoken to me by Serge, with a thick French accent, on the phone during the winter of 1990. I was applying to graduate schools and had written to Serge, whom I knew about through his many books, to inquire about the program at Yale. I never could have imagined the consequences of the small, innocent letter I sent to Serge. Indeed, as I would later learn, one of Serge's most remarkable qualities was that he did not recognize the legitimacy of any form of "innocent" interchange. Every conversation, every statement, engages us in a commitment to some world view, to some normative judgment. Serge demanded that we recognize this density of discourse and consequently take full responsibility for all implications and presuppositions implied in the statements and claims that we make.

While at other prestigious institutions I was readily dismissed (after all, how could an individual from a small liberal arts college with a degree in French brazenly pretend to do serious mathematics?), Serge made every effort humanly possible to stimulate and nourish my love for mathematics. He sent books and called several times each week to answer questions and help to guide my studies. Serge even lobbied on my behalf, to no avail, at the NSF as well as a few top graduate programs.

Once I decided to begin graduate school at Yale, Serge paid out of his own pocket to bring me down to Berkeley for the summer so that I could get started on my Ph.D. work.

Shortly after my arrival in Berkeley, I was given a preliminary version of Faltings' paper [F1] proving a generalization of the Mordell Conjecture. My initial project was to generalize this result to all subvarieties of abelian varieties, as was to be accomplished by Faltings [F2], and eventually to what Serge so fondly called pseudocanonical varieties. This did not seem like a very reasonable project at the time but, not knowing any better, I studied Faltings' paper [F1] closely and set to work trying to generalize it to the case of arbitrary subvarieties of abelian varieties. Being more of a geometer than a number theorist, I tried to solve the problem with more sophistocated methods from higher dimensional geometry, not realizing that the necessary tools did not exist (or required much greater ingenuity than I was capable of), and so began the road toward becoming an algebraic geometer.

Modern algebraic geometry was not Serge's area of expertise and, caring first and foremost about the development of his students, he sent me away to Berkeley for my third year of graduate school, in order to profit from a year long program in algebraic geometry at MSRI, and then to Harvard for my fourth and final year. Thus it is that I spent a grand total of a mere 13 months in New Haven and saw Serge only three times after graduating in 1994. Yet the impact of the few moments I spent with Serge has been immense.

For me, Serge was first and foremost the only person who ever truly believed in my mathematical abilities and consequently he always encouraged me to set lofty goals. Perhaps even more importantly in practice, he will always be a professional role model for me. Serge Lang was the embodiment of *integrity*, in academics, and more generally in life. He did not, like so many others in the same position, make judgments based on external trappings: Serge always looked beneath the surface and supported those whose talent he could sense, regardless of their particular circumstances.

It is my sincerest hope that, in memory of Serge and the ideals he stood for, those in positions of influence and authority will exercise their power honestly and without prejudice to help all those who desire to contribute to the growth of mathematics. When education and knowledge become the monopoly of an elite class, the entire world, academic and otherwise, suffers and it should be a moral imperative for every teacher to reach out to the larger community without discrimination.

Though in recent years I have largely withdrawn from the mathematical community, I have not abandoned my dream. One day in January 1997 I received a call from Serge. After his usual greeting, "Hi, it's me," I told him that I had just been through surgery on a catastrophic brain tumor. Serge's reply: "So can you still do math?" I did not have the heart to tell Serge the truth, namely that it did not matter and had not mattered since the first day I got up in front of a wonderful group of Math 122 students at Harvard. At that moment I had discovered my most fulfilling passion, namely teaching mathematics. I never told Serge, as I felt he would have been disappointed since he had such great hopes for my mathematical career and since he devoted so much time and energy to helping me.

Perhaps I was mistaken as Serge undoubtedly would have appreciated my desire to help others as opposed to pursuing greater success as a mathematician. On the other hand, Serge would certainly have chastised me for "chickening out" and with reason as I have assuredly not put my heart and soul into doing mathematics. In the end, there are simply too many other beautiful things in the world which I am not willing to sacrifice in order to prove better theorems. My admiration, however, for those who are willing to sacrifice is in no way lessened, particularly for someone like Serge who was always fully conscious of his intentions, desires, and needs, and was still willing to put everything aside in order to develop his mathematical vision. Having enjoyed a life of passionate pursuit of truth and equity, I hope that Serge rests now in peace, admiring the immense beauty of the mathematical world which he loved so much.

References

[CP] F. Campana and T. Peternell, *Algebraicity of the ample cone of projective varieties*, J. reine angew. Math., **404**, 1990, 160–166.

[EKL] L. Ein, O. Kuchle, R. Lazarsfeld, *Local positivity of ample line bundles*, Journal of Differential Geometry, **42**, 1995, 193–219.

[F1] G. Faltings, *Diophantine Approximation on Abelian Varieties*, Annals of Math., **133**, 1991, 549–576.

[F2] G. Faltings, *The general case of S. Lang's conjecture*, in: Christante and Messing (eds.), Barsotti symposium in algebraic geometry, Academic Press, 1994, 175–182.

[FW] G. Faltings and G. Wüstholz, *Diophantine approximations on projective spaces*, Inv. Math., **116**, 1994, 109–138.

[Fi] S. Fischler, *Interpolation on algebraic groups*, Compositio, **141**, 2005, 907–925.

[L] R. Lazarsfeld, *Positivity in Algebraic Geometry*, 2 Volumes, Springer, 2004.

[M] D. Masser, *Interpolation on group varieties*, in *Approximations diophantiennes et nombres transcendants*, Birkhaüser, 1983, 151–171.

[MW] D.W. Masser and G. Wüstholz, *Zero estimates on group varieties I*, Inv. Math., **64**, 1981, 489–516.

[N1] M. Nakamaye, *Seshadri constants and the geometry of surfaces*, Crelle, **564**, 2003, 205–214.

[N2] M. Nakamaye *Seshadri constants at very general points*, Transactions AMS, **357**, 3285–3297, 2004.

[N3] M. Nakamaye, *Multiplicity Estimates on Commutative Algebraic Groups*, Crelle, **607**, 2007, 217–235.

[NR] M. Nakamaye and N. Ratazzi, *Lemmes de multiplicités et constante de Seshadri*, Mathematische Zeitschrift, **259**, 2008, 915–933.

[P1] P. Philippon, *Lemmes de zéros dans les groupes algébriques commutatifs*, Bull. Soc. Math. France, **114**, 1986, 355–383.

[P2] P. Philippon, *Nouveaux lemmes de zéros dans les groupes algébriques commutatifs*, Rocky Mountain Journal of Math, **26 No. 3**, 1996, 1069–1088.

[Wal] M. Waldschmidt, *Nombres Transcendants et groupes algébriques*, Astérisque, **69-70**, 1979.

Sampling spaces and arithmetic dimension

Catherine O'Neil

To Serge. It was an honor to be your favorite sophomore.

Abstract This paper introduces the twin concepts of sampling spaces and arithmetic dimension, which together address the question of how to count the number, or measure the size of, families of objects over a number field or global field. It can be seen as an alternative to coarse moduli schemes, with more attention to the arithmetic properties of the ambient base field, and which leads to concrete algorithmic applications and natural height functions. It is compared to the definition of essential dimension.

Key words essential dimension • coarse moduli • elliptic curves

Mathematics Subject Classification (2010): 11G35

1 Introduction

This paper is the result of fomalizing a concept which was introduced to the author by her thesis advisor, Barry Mazur. Namely, it is the idea of a kind of parameter space which includes every example of a certain arithmetic type, and which can be studied in its own right. This leads directly to the definition of a "sampling space." Here are a few motivations for this definition. First, one wishes to formalize and standardize the many discussions surrounding the 'average behavior' of arithmetic objects. Second, one wishes to find comparison theorems of various

C. O'Neil (✉)
15 Claremont Avenue #91, New York, NY 10027
e-mail: cathy.oneil@gmail.com

D. Goldfeld et al. (eds.), *Number Theory, Analysis and Geometry: In Memory of Serge Lang*, DOI 10.1007/978-1-4614-1260-1_23,

different families of arithmetic objects. Third, one can ask for efficiency in these families; this question leads to the definition of *arithmetic dimension*.

When dealing with classes of objects such as Galois extensions, elliptic curves, class numbers, or some kind of cohomology classes over a number field, one often feels somewhat stuck between a geometric point of view and a discrete point of view, owing to the inherent arithmetic. One may want to ask for the growth of a function related to a kind of complexity, such as the rank of a Mordell–Weil group or the size of a class group. However, the answer may well depend on how you count. What is inherent in the discussion is the concept of listing all examples of a certain type, usually with some equivalence relation in effect, and taking averages. However, the way you list the examples matters. For example, in listing elliptic curves over $\mathbb{Q}$ (up to isomorphism) you may list all curves up to a given discriminant, or you may list all curves with coefficients smaller than a given size. Each of these is natural but the answer you get will typically depend on which one you chose.

Comparison theorems are intended to help us sort out the issue raised above of counting the same things in different ways. We shall see that many sampling spaces have natural height functions, as they are sometimes projective or quasi-projective schemes. Because of this, if we have two different sampling spaces for the same arithmetic objects, we can directly compare their heights. Since we are interested ultimately in listing all objects of a certain kind, then it obviously in our interest to make sure we have as little redundancy as possible. That is why we would like our sampling spaces to have minimal dimension. Indeed this can lead us to far more interesting and nuanced sampling spaces, as we will see in Section 5.

We are primarily interested in functors which are not representable. One standard way around this problem is to introduce stacks. Although many of the same issues will arise as in the theory of stacks (such as group actions), one can think of sampling spaces as being much more algorithmically motivated. Indeed, sampling spaces are intended to be concrete and, when possible, explicit.

One way to view a sampling space for a non-representable functor H is as an alternative to a coarse moduli scheme. In some sense, sampling spaces trade efficiency for completeness. A sampling space lives ***above*** its functor (i.e., maps to H) and loses no arithmetic information. By contrast, a coarse moduli scheme lives ***below*** its functor and does lose arithmetic information. The arithmetic dimension of a functor is the minimal dimension of a sampling space for that functor. The difference between the arithmetic dimension of H and the dimension of a coarse moduli scheme for H, if both exist, can be viewed as an arithmetic "bloating factor," i.e., the number of extra parameters that one needs to, say, program a computer to completely list all arithmetic objects of a given type over a number field versus over the complex numbers. A good example is given by the functor $\mathcal{E}ll$, which associates to a base scheme S in the category $\mathcal{C}_K$ of finite-type schemes over a field K, the elliptic curves over S up to isomorphism. The coarse moduli scheme for $\mathcal{E}ll$ is the j-line, a curve. However, to actually list all elliptic curves (up to isomorphism) over $\mathbb{Q}$, for example, one needs both the j-invariant and a separate parameter to take into account all the quadratic twists of a fixed elliptic curve (see the example in the next section). It is not hard to see that this forces any sampling space for $\mathcal{E}ll$ with respect to the field $\mathbb{Q}$ to have dimension at least 2. Therefore there is one dimension of arithmetic bloat for elliptic curves over $\mathbb{Q}$.

This is, as described above, not a new idea. There have been many interesting definitions which try to measure the complexity of a functor based on the dimension of a parameter space for the objects involved. However, the definition we give below is new, in that it is more specific to the arithmetic of the base field. In particular, in Section 3.1 we compare our definition with that of the 'essential dimension' of a group, and we show that the two are measuring different phenomena; in particular, our definition is sensitive to the existence of rational points on quadratic hypersurfaces in large projective spaces over number fields.

Acknowledgements The author would like to acknowledge the many conversations with Mira Bernstein which led to the definitions contained in this paper and conversations with Barry Mazur which motivated the investigation of these ideas.

2 Definitions

Let K be a field. Let $\mathcal{C}_K$ be the category of schemes of finite-type over K.

Definition 2.1. *An* arithmetic object *is a contravariant functor $\mathcal{F}$ from $\mathcal{C}_K$ to the category of sets.*

Definition 2.2. *A* sampling space *for an arithmetic object $\mathcal{F}$ is the data of a scheme T in $\mathcal{C}_K$ with a natural transformation Φ from the functor $Hom_{\mathcal{C}_K}(\cdot, T)$ to $\mathcal{F}$ which is required to be surjective on L-points whenever L is a finite extension of K.*

Remark. We denote by $\Phi : T \twoheadrightarrow \mathcal{F}$ the fact that T is a sampling space for $\mathcal{F}$ via the map Φ. When the map Φ is obvious we simply write $T \twoheadrightarrow \mathcal{F}$.

Definition 2.3. *The* arithmetic dimension *of an arithmetic object is the minimum dimension of a sampling space for that object.*

Example. We will prove that the functor $\mathcal{E}ll$ over a number field has arithmetic dimension 2 and over a finite field has arithmetic dimension at most 2. When K is a field of characteristic away from 2 and 3, every elliptic curve over K can be modelled (non-uniquely) by a Weierstrass equation of the form $E : y^2 = x^3 + Ax + B$, where $A, B \in K$ and where we assume $\Delta = 27B^2 + 4A^3 \neq 0$. In the case that K has characteristic 3, when E is ordinary then one can always write $E : y^2 = x^3 + Ax^2 + B$; we can deal with the supersingular case separately with a finite union of points. Similarly, when K has characteristic 2 and when E is ordinary, one can write $E : y^2 + xy = x^3 + Ax^2 + B$.[1] Therefore $\mathrm{Spec}(K[A, B, 1/\Delta]) \twoheadrightarrow \mathcal{E}ll$, and so the arithmetic dimension of $\mathcal{E}ll$ is bounded above by 2. We will now show that there is no 1-dimensional sampling space for $\mathcal{E}ll$ over a number field. Suppose we have a

[1] Thanks to René Schoof for these models.

curve sampling space T; then T will map (by composition with the map $\mathcal{E}ll \to \mathbb{P}^1_j$) to the j-line: $T \twoheadrightarrow \mathcal{E}ll$. The size of a fiber of this map is bounded on the one hand by the degree d of this map of curves. On the other hand, above a given j-invariant away from $j = 0$ or 1728, the fiber above a K-rational j-value corresponds to the set of all quadratic twists of a single elliptic curve with that j-invariant, i.e., with the set $H^1(K, \{\pm 1\}) \cong K^*/K^{*2}$. By the functoriality condition this would imply that for all finite separable extensions L/K the set L^*/L^{*2} is bounded (by d). This is clearly untrue for number fields.

As stated in the introduction, the above definitions allow us to make precise a measurement of the "dimension of pure arithmetic" of the functor $\mathcal{E}ll$, namely the discrepancy between the arithmetic dimension of $\mathcal{E}ll$ and that of its coarse moduli scheme, 1. Also, the example of elliptic curves nicely illustrates some of the following more general observations:

Remarks. (1) When H is representable by a scheme X in $\mathcal{C}_K$, then X is a sampling space for H. Moreover, the arithmetic dimension of H is just the dimension of X: any sampling space Y of H comes with a morphism $\phi : Y \to X$ which is surjective on $\overline{K}$ points. Since the closure of the image of ϕ has dimension at most $\dim(Y)$, and since the image has dimension equal to that of the dimension of X we know $\dim(Y) \geq \dim(X)$.

(2) A sampling space T of $\mathcal{F}$ comes with a tautological object: namely, take the image of the identity map $\mathrm{Id}_T \in T(T)$ to the corresponding T-valued point of $\mathcal{F}$.

(3) If there is a scheme "below the functor", then the arithmetic dimension of the functor is bounded below by the dimension of that scheme. For example if there is a finite-type K-scheme M and a transformation of functors $\mathcal{F} \to \mathrm{Hom}_{\mathcal{C}_K}(\cdot, M)$ whose values on L-points $[L : K] < \infty$ form a Zariski dense collection of points in M, then the arithmetic dimension of $\mathcal{F}$ is at least the dimension of M. This will typically be the case if M is a coarse moduli scheme of $\mathcal{F}$.

(4) Say we have a "arithmetic surjection" from an object F to an object H; this just means that there is a transformation of functors which is surjective on L-points whenever L is a finite extension of K. Then the arithmetic dimension of H is bounded above by the arithmetic dimension of F.

(5) We do need to assume the surjectivity for *all* finite field extensions. For example, in [7], it is proven that for any finite field $\mathbb{F}$ there exists a curve over $\mathbb{F}$ which is "space-filling," i.e., which has as many $\mathbb{F}$-valued points as its ambient space $\mathbb{A}^n_{\mathbb{F}}$ for $n > 1$. However, it is impossible to produce a curve which is a sampling space for the functor associated to $\mathbb{A}^n_{\mathbb{F}}$, since the number of points in affine n-space over a finite field grows asymptotically much faster than the number of points of a curve.

(6) In particular, an infinite disjoint union of copies of $\mathrm{Spec}(K)$ has no arithmetic dimension.

3 Important example: cohomology groups

Let G be a group scheme over K. Let $H^1_K(G)$ denote the functor taking a scheme X of finite type over K to the pointed set (which under certain restriction on G will have a natural group structure) $H^1_{\text{ét}}(X, G)$. When X is connected and G is a finite étale group scheme we can identify elements of this set with G-torsors over X, i.e., Galois coverings $Y \to X$ with an isomorphism $G \cong \text{Aut}_X(Y)$. When $X = \text{Spec}(L)$ for L a separable extension of K, this is just Galois cohomology and is denoted by $H^1(L, G)$.

Lemma 3.1. *A sampling space T/K of $H^1_K(G)$ has a "tautological" G-torsor $Y \to T$. Moreover, whenever L/K is an algebraic extension, the surjective map $\Phi_L : T(L) \to H^1(L, G)$ is given by pullback as follows, for $p \in T(L)$ thought of as a morphism $p : \text{Spec}(L) \to T$ mapping to the G-torsor $\Phi_L(p) : Y_p \to \text{Spec}(L)$:*

$$\begin{array}{ccc} Y_p & \to Y \\ \Phi_L(p) \downarrow & \square \downarrow . \\ Spec(L) & \stackrel{p}{\to} T \end{array}$$

Proof. By Remark 2 above, we take the image under this map of the identity map on T to get the G-torsor $Y \to T$. Next, given $p : \text{Spec}(L) \to T$ we get a map $T(T) \to T(L)$ by composition: $\alpha : T \to T$ maps to $\alpha \circ p$ (note that id_T map to p itself). The above diagram then follows from functoriality and the fact that the map $H^1_{\text{ét}}(T, G) \to H^1(L, G)$ is defined by pullback. □

3.1 Comparison with essential dimension

We refer to the many papers of Zinovy

Reichstein available at http://www.math.ubc.ca/$\sim$reichst/pub.html. The essential dimension of a finite group G with respect to a field K is defined in [2] as the least dimension of a K-variety Y which is the target of a dominant G-equivariant rational map of a faithful linear representation of G. This definition is generalized in [13] to algebraic groups and in [1] is given a functorial perspective. Namely it can be shown using the results in that paper that the following definition is equivalent (at least when K is infinite).

Definition 3.2. *Let $X \to Y$ be a morphism in $\mathcal{C}_K$ such that X is a G-torsor over Y. This gives a transformation of functors* $\text{Hom}(\cdot, Y) \to H^1(\cdot, G)$. *The essential dimension of G over K is the least dimension of an Y as above such that the map* $\text{Hom}(\text{Spec}(K'), Y) \to H^1(K', G)$ *is surjective for all field extensions K'/K.*

The above immediately implies:

Theorem 3.3. *The essential dimension of G is an upper bound for the arithmetic dimension of $H^1_K(G)$.*

Remark. The essential dimension of G over K is sometimes unequal to the arithmetic dimension of $H^1_K(G)$.

Remark. The above theorem is obvious when $K = \overline{K}$, since the arithmetic dimension of $H^1_K(G)$ is zero if $K = \overline{K}$ and G is finite. We show that even over number fields these two concepts are essentially different. In fact we show the following:

Theorem 3.4. *Let K be a totally imaginary number field and let $n \geq 4$; the essential dimension of (μ_2^5, K) is 5 but the arithmetic dimension of $H^1_K(\mu_2^5)$ is at most 4.*

Sketch of proof. We want to show that there is a four-dimensional scheme which maps onto $(K^*/K^{*2})^5$. So take $(a_i)_{1\leq i\leq 5} \in (K^*/K^{*2})^5$. Form the equation $\sum_{i=1}^5 a_i x_i^2 = 0$. This always has a point over K since we assumed K to be a totally imaginary number field. But then we can basically take the 4-dimensional sampling space to be

$$Spec(K[X_i, X_i^{-1}]_{1\leq i\leq 5}/(\sum_{i=1}^{5} a_i x_i^2 = 0)).$$

To be complete, we would need to take into consideration the fact that the solution to $\sum_{i=1}^5 a_i x_i^2 = 0$ may have some $x_i = 0$. This will add a finite number of 4-dimensional components to our sampling space.

Question. What is the actual arithmetic dimension of $H^1_K(\mu_2^5)$ for a number field K? How does the arithmetic dimension of $H^1_K(\mu_2^n)$ grow as a function of n? Note that when the base field K has a trivial Brauer group the arithmetic dimension of $H^1(\mu_n^d)$ is probably much smaller than d; see the section below on Brauer groups for the relationship and the section below on finite fields for such an example.

3.2 *Brauer groups*

As a special case of the above, when K is a number field, we have an identification $H^1_K(\mathrm{PGL}_n) \cong \mathrm{Br}(K)[n]$, where $Br(K)[n]$ is the n-torsion of the Brauer group of the field K. Consider the functor on $\mathcal{C}_K$ given by $X \mapsto H^2_{\text{ét}}(X, \mathbf{G}_m)[n]$ (the n-torsion in the cohomological Brauer group of X). We will denote this functor by $\mathrm{Br}_K[n]$.

Theorem 3.5. *Let K be a number field which contains the nth roots of unity. Then the arithmetic dimension of* $\mathrm{Br}_K[n]$ *is 2.*

Proof. The arithmetic dimension of $\mathrm{Br}_K[n]$ is at most two, since every division algebra D of order n in the Brauer group of a number field containing a primitive nth root of 1 is cyclic. Namely, it can be written as $D \cong \langle x, y | x^n = a, y^n = b, xyx^{-1}y^{-1} = \zeta\rangle$. Therefore $\mathbb{G}_m \times \mathbb{G}_m$ is a sampling space for $\mathrm{Br}_K[n]$. Next, say there is a 1-dimensional finite type K-scheme C which is a sampling space for $\mathrm{Br}_K[n]$. Denote by ξ the corresponding tautological object in $\mathrm{Br}(C)[n]$. This Brauer group element will manifest every cyclic algebra over any finite extension L of K as some fiber, since it is the tautological object of a sampling space. However, Tsen's theorem implies (with a bit of work) that the Brauer group of the base change $C \times_{\mathrm{Spec}(K)} \mathrm{Spec}(\overline{K})$ is trivial. This means there is some finite extension L of K over which ξ becomes trivial. It is clear that there doesnot exist such an L which simultaneously trivializes all cyclic algebras. □

Corollary 3.6. *Let K be a number field containing the nth roots of unity. Then the arithmetic dimension of $H^1_K(\mu_n^2)$ is 2.*

Proof. First, $\mathbb{G}_m \times \mathbb{G}_m$ is clearly a sampling space for $H^1_K(\mu_n^2)$ over any field. Next, use remark 4 on page 502 above and the fact that the "norm symbol map" $H^1_K(\mu_n^2) \to H^2(K, \mu_n) = \mathrm{Br}(K)[n]$ is surjective. □

Notice that the above proof is specific to number fields- in general, an Azumaya algebra is only a finite product of cyclic algebras even if we assume roots of unity.

Question. What if I don't assume any roots of unity? By Observation (3) below, since adjoining a pth root of unity to a field induces a prime-to-p extension, there is an upper bound on the arithmetic dimension of the functor $\mathrm{Br}_{\mathbb{Q}}[p]$ of $2(p-1)$. Can this be improved?

Observations.

1. There is no sampling space for the entire Brauer group over a number field. Namely, a given element of the Brauer group (cohomological or not) of a finite-type scheme over a field has a given (finite) order. But elements of the Brauer groups of number fields have arbitrary order.
2. When G is a finite group scheme over K of degree n the arithmetic dimension of $H^1_K(G)$ is bounded above by n. Namely, the regular representation (which is the action of G on $\Gamma(G, \mathcal{O}_G)$) is faithful, and then the results of [2] imply the essential dimension of the group scheme G over K is bounded by its dimension, which is n. Hence also the arithmetic dimension is bounded by n.
3. Let G be a finite commutative group scheme of order n over K. Let K'/K be a finite separable extension of degree d prime to n. For any finite extension L/K we get a sequence $H^1(L,G) \overset{\mathrm{res}}{\to} H^1(L \otimes_K K', G) \overset{\mathrm{cor}}{\to} H^1(L,G)$ whose composition is multiplication by d. On the other hand, suppose that T' is a finite type K'-scheme with a transformation $\mathrm{Hom}_{\mathcal{C}_{K'}}(\cdot, T') \to H^1_{K'}(G_{K'})$ which is a sampling space over K'. Set $T = \mathrm{Res}_{K'/K}(T')$. For a finite field extension $K \subset L$ consider the composition

$$T(L) = T'(L \otimes_K K') \to H^1(L \otimes_K K', G) \overset{\mathrm{cor}}{\to} H^1(L,G).$$

By what was said above and since $(d,n) = 1$ this is surjective. Hence we see that T is a sampling space for $H^1_K(G)$, provided we can show these maps fit together and come from a transformation of functors $\mathrm{Hom}(\cdot, T) \to H^1_K(G)$. Granted that this is indeed the case we see that $dim(T) = d\,dim(T')$. In other words we see that the arithmetic dimension of $H^1_K(G)$ is at most d times the arithmetic dimension of $H^1_{K'}(G_{K'})$. To construct the transformation $\mathrm{Hom}(\cdot, T) \to H^1_K(G)$ we have to construct a G-torsor over T. Let $X' \to T'$ be the G-torsor corresponding to the transformation $\mathrm{Hom}(\cdot, T') \to H^1_{K'}(G_{K'})$. Consider the morphism $\mathrm{Res}_{K'/K}(X') \to T = \mathrm{Res}_{K'/K}(T')$. This is a $\mathrm{Res}_{K'/K}(G_{K'})$-torsor. Since G is abelian there is a "norm" map $\mathrm{Res}_{K'/K}(G_{K'}) \to G$ which is a homomorphism of group schemes over K. The push-out of $\mathrm{Res}_{K'/K}(X')$ by this norm map gives the desired torsor.

4. If G is a group scheme and if we can embed G into an "H^1-trivial group scheme" H (such as GL_n or SL_n), then we have $0 \to G \to H \to H/G \to 0$. There is no guarantee that the scheme H/G is a group, but it still makes sense to take the long exact sequence of cohomology to the extent that we get $0 \to G(K) \to H(K) \to (H/G)(K) \to H^1(K, G) \to H^1(K, H) = 0$. That is, we get the sampling space H/G. For example, if $G = \mu_n$, we can take $H = \mathbb{G}_m$ to recover sampling space $\mathbb{G}_m$ for $H^1_K(\mu_n)$.

4 Over a finite field

Let p be an odd prime, let $\mathbb{F}_p$ be the finite field with p elements, and let d be a positive integer. For i ranging between 0 and d, define the map $f_i : \mathbb{F}_p \backslash \{-i\} \to \mathbb{F}_p^*/\mathbb{F}_p^{*2}$, sending the element x to $[x+i]$. Katz proves (Corollary 1.4.2.2, page 25 of [8]) that as p goes to infinity, the maps f_i have "equidistributed" and independent values. In other words, the map $(f_0, f_1, \ldots, f_d) : \mathbb{F}_p \backslash \{0, -1, -2, \ldots, -d\} \to (\mathbb{F}_p^*/\mathbb{F}_p^{*2})^{d+1}$ will asymptotically send x to any given element of $\mathbb{Z}/2\mathbb{Z}^{d+1}$ with equal likelihood.

The following theorem says nothing about equidistribution, but it generalises Katz's result in that it works modulo nth powers, and also because the conditions on the functions are very weak.

Theorem 4.1. *Let $\mathbb{F}$ be a finite field of characteristic p. For an integer n prime to p, the arithmetic dimension of $H^1_{\mathbb{F}}(\mu_n^d)$ is 1. In other words, there is a curve sampling space defined over $\mathbb{F}$ which parameterizes d "independent" nonzero elements of $\mathbb{F}$ mod nth powers.*

Lemma 4.2. *Let $f_1, f_2, \ldots, f_d \in \mathbb{F}[t]$ be functions with the property that the ring*

$$R = \mathbb{F}(t)[Y_1, Y_2, \ldots, Y_d]/(Y_1^n = f_1, Y_2^n = f_2, \ldots, Y_d^n = f_d) \otimes_{\mathbb{F}} \overline{\mathbb{F}}$$

is actually a field. Define the set $S = \{\alpha \in \overline{\mathbb{F}} | f_i(\alpha) = 0 \text{ for some } 1 \le i \le d\}$. The curve $\mathbb{A}^1 \backslash S$ is a scheme defined over $\mathbb{F}$, and becomes a sampling space for $H^1_L(\mu_n^d)$ whenever L is a sufficiently large field extension of $\mathbb{F}$.

Proof. Fix f_i as above, and for any field extension $L/\mathbb{F}$, define the map

$$\mathbb{A}^1_L \backslash S \to (L^*/L^{*n})^d$$

sending α to the element $([f_1(\alpha)], \ldots, [f_d(\alpha)])$. This will be a sampling space if, for an arbitrary element $([a_1], [a_2], \ldots, [a_d]) \in (L^*/L^{*n})^d$, there is an α and there are y_i, $1 \leq i \leq d$, such that for all i, $a_i y_i^n = f_i(\alpha)$. These equations define the function field of a curve $C_{([a_1],[a_2],\ldots,[a_d])}$ which is a cover, say by the map π, of $\mathbb{P}^1$ given by sending the pair $(\alpha, (y_1, y_2, \ldots, y_d))$ to α. Moreover, since the ring R above is a field, this curve is irreducible (a geometric condition, and so we can assume that all of the a_i are perfect nth powers). In order to get a sampling space, we need to show that the curve $C_{([a_1],[a_2],\ldots,[a_d])}$ has an L-rational point outside the fibers above S whenever L is large enough. We will make use of the Riemann Hypothesis for curves over finite fields:

$$\#C(\mathbb{F}) = q + 1 - \sum \alpha_i,$$

where the α_i are the eigenvalues of the Frobenius operator. The absolute value of each α_i is bounded by $\sqrt{q}$, and the number of eigenvalues is $2g$, where g is the genus of C. Moreover, the genus of $C_{([a_1],[a_2],\ldots,[a_d])}$ depends only on the degrees of the f_i's and the degree, n^d, of the map π. Finally, the number of rational points "hiding" above points in S is bounded again by the degrees of the f_i and by n and d. Thus when the size of L is large enough, we are assured of a choice of α and y_i as above. □

We can view the above results as saying that over number fields, we have a stronger version of Kummer theory, and over finite fields we have a very strong version of Kummer theory.

5 Models of genus one curves

Let $n \geq 3$ be an odd integer, and let K be a field of characteristic prime to n. For an elliptic curve E with origin $\mathcal{O}_E$ defined over K, fix a "base diagram" $f_E : E \to \mathbb{P}^{n-1}$ given by the full linear series associated to the line bundle $n \cdot \mathcal{O}_E$. A choice of f_E induces, by passing to automorphism group schemes (see [11]), the following commutative diagram of group schemes, where each the row is exact and the upper row is a theta group as defined in [9], page 221:

$$\begin{array}{ccccccccc} 0 & \to & \mathbb{G}_m & \to & \mathcal{G}_n & \to & E[n] & \to & 0 \\ & & \downarrow & & \downarrow & & \downarrow & & \\ 0 & \to & \mathbb{G}_m & \to & \mathrm{GL}_n & \to & \mathrm{PGL}_n & \to & 0. \end{array} \tag{5.1}$$

Definition 5.1. Define the map $Ob : H^1_K(E[n]) \to H^1_K(\mathrm{PGL}_n) \to H^2_K(\mathbb{G}_m)$ by composing two induced maps on cohomology from diagram (5.1) above.

Definition 5.2. The functor H_{Ob} is the set-theoretic kernel of the map Ob.

Remarks. (1) An element of $H_{\mathrm{Ob}}(K)$ can be seen to correspond to a certain kind of twist of the base diagram f, namely a diagram "of index n," $C \to \mathbb{P}^{n-1}$ where C is a genus-one curve whose Jacobian is isomorphic to E. A general element of $H^1(K, E[n])$, not necessarily lying in $H_{\mathrm{Ob}}(K)$, would be a more general twist, namely a diagram "of period n," $C \to S$, where S is a Brauer–Severi variety of dimension $n-1$. Moreover, every index-n object will be represented in H_{Ob}.

(2) For example, when n is 3, the elements in H_{Ob} correspond to all cubic curves (up to PGL_3-action) whose Jacobian elliptic curve is isomorphic to E.

(3) The map Ob is quadratic, so $H_{\mathrm{Ob}}(X)$ is not generally a group.

(4) All of the elements of the n-Selmer group for E will appear in H_{Ob}. For this reason we are interested in finding small sampling spaces for H_{Ob}; it would be the first step towards finding a sampling space for the Selmer group of E.

(5) Assume that n is odd and the characteristic of K is prime to n. Assume that E is an elliptic curve over K with $E[n](\overline{K}) \subset E(K)$, i.e., as group schemes, $E[n] \cong \mathbb{Z}/n\mathbb{Z} \times \mathbb{Z}/n\mathbb{Z}$. Then $H_{\mathrm{Ob}} \subset H^1_K(E[n]) \subset H^1_K(\mathbb{Z}/n\mathbb{Z} \times \mathbb{Z}/n\mathbb{Z})$. In [12], it is proved that the arithmetic dimension of H_{Ob} is at most n, and in the cases of $n = 3$ and $n = 5$ an explicit sampling space of dimension n has been found for this situation (see also Corollary 5.12 below for an improvement when $n = 2$). The image of the map $H_{\mathrm{Ob}} \to H^1_K(\mathbb{Z}/n\mathbb{Z}) \times H^1_K(\mathbb{Z}/n\mathbb{Z})$ is the set of trivial norm symbols of level n.

5.1 The general case: lower bound

Theorem 5.3. *Let K be a number field, $n \geq 2$ an integer, and E an elliptic curve. The arithmetic dimension of $H_{Ob,n}$ is at least 2.*

Proof. First we reduce to the case that we have "rational n-torsion," i.e., that $E[n] \cong \mathbb{Z}/n\mathbb{Z} \times \mathbb{Z}/n\mathbb{Z}$: this is so after a finite extension, and any 1-dimensional sampling space for $H_{\mathrm{Ob},n}$ would base change to a sampling space for the rational n-torsion case, so it is enough to show that no such sampling space exists. We will make use of the fact that in the case of rational n-torsion, Ob is a norm symbol map.

Lemma 5.4. *Let K be a number field. Given a nontrivial element $a \in K^*/K^{*n}$, the set of $b \in K^*/K^{*n}$ such that the corresponding norm symbol (a, b) is trivial is the infinite set $\mathbb{N}_{K(\sqrt[n]{a})/K}(K(\sqrt[n]{a})^*)/K^{*n}$.*

Proof of lemma. That the set *described in the lemma is as claimed* is the definition of the norm symbol. That it is infinite follows from local considerations. □

Now assume that we have a one-dimensional sampling space T for $H_{\mathrm{Ob},n}$. It consists of a finite union of schemes of dimension 0 and 1. By the pigeonhole principle and the above lemma, there exists some irreducible component of T, a curve C, such that the rational points of C are responsible for many of the trivial norm symbols, i.e., that for infinitely many classes $a \in K^*/K^{*n}$ and infinitely many $b \in \mathbb{N}_{K(\sqrt[n]{a})/K}(K(\sqrt[n]{a})^*)/K^{*n}$ there exist rational points $x_{(a,b)} \in C(K)$ mapping to their respective points in $H_{\mathrm{Ob},n}$.

The tautological object of T is an element of $H^1_{\text{ét}}(T, \mu_n \times \mu_n)$, and as such gives rise to a $\mu_n \times \mu_n$ field extension $L(T)/K(T)$, where $K(T)$ is the function field of T. Similarly we get a field extension $L(C)/K(C)$. Since $\zeta_n \in K$, by Kummer theory there exist elements d_1 and $d_2 \in K(C)$ whose nth roots give the extension $L(C)$. Moreover, the functions d_i define maps f of C onto $\mathbb{G}_m$. We may now construct the map $C \to \mathbb{G}_m \times \mathbb{G}_m \stackrel{\pi_2}{\to} \mathbb{G}_m$ where the second part is the second projection map. But by construction, the composite map is both nontrivial and has infinite fibers, so C cannot be a curve of finite type, a contradiction. □

5.2 *The general case: upper bound*

As above, fix a "base diagram," an embedding $f : E \to \mathbb{P}^{n-1}$. Then the finite group scheme $E[n]$ acts on E and extends to a faithful action on $\mathbb{P}^{n-1}$; in other words, we get an injective map of group schemes $i : E[n] \to \mathrm{PGL}_n$, and we can define the scheme $\mathbb{P}^{n-1}/E[n]$. This will not be smooth, because although the $E[n]$ action is faithful, there are points with nontrivial stabilizer. Denote by $\mathbb{P}^{n-1}/E[n]^{\mathrm{sm}}$ the smooth (open) part of this quotient. A lift $\tilde{v} \in \mathbb{P}^{n-1}(\overline{K})$ of a point $v \in \mathbb{P}^{n-1}/E[n]^{\mathrm{sm}}(K)$ is "almost rational," meaning that for each σ, there exists $T(\sigma) \in E[n](\overline{K})$ such that $\tilde{v}^\sigma = i(T(\sigma))(\tilde{v})$. In this way we can form a cocycle ξ_v for any such v, and we can map ξ_v to $H^1(K, E[n])$. Moreover, a different choice of a lift of v will differ by a coboundary. Thus we have defined a map of functors $\xi : (\mathbb{P}^{n-1}/E[n])^{\mathrm{sm}} \to H_{\mathrm{Ob}}$.

Theorem 5.5. *The scheme* $(\mathbb{P}^{n-1}/E[n])^{\mathrm{sm}}$ *is a sampling space for* H_{Ob}*, via* ξ *:*

$$\xi : (\mathbb{P}^{n-1}/E[n])^{\mathrm{sm}} \twoheadrightarrow H_{\mathrm{Ob}}.$$

Moreover, the following diagram commutes:

$$\begin{array}{ccccccc} E & \to & E/E[n] & \cong & E & \twoheadrightarrow & E/nE \\ \downarrow & & \downarrow & & \downarrow & & \downarrow \\ \mathbb{P}^{n-1} & \to & \mathbb{P}^{n-1}/E[n] & \twoheadrightarrow & H_{\mathrm{Ob}} & \to & H^1_K(E[n]) \end{array}, \tag{5.2}$$

where the rightmost column comes from cohomology, sending $P \in E(K)/nE(K)$ *to the cocycle* $Q^\sigma - Q$ *if* $nQ = P$.

Proof. We know that $E[n] \subset \mathrm{PGL}_n$ (see diagram (5.1) above) and we can extend this to an exact sequence of schemes $E[n] \to \mathrm{PGL}_n \to \mathrm{PGL}_n/E[n]$, since the action is fixed-point free. Note that since $E[n]$ is not normal inside PGL_n, the quotient scheme is not a group scheme. However, we can still make sense of the first part of an exact sequence of cohomology:

$$\mathrm{PGL}_n(K) \to \mathrm{PGL}_n/E[n](K) \to H^1(K, E[n]) \stackrel{\mathrm{Ob}}{\to} H^1(K, \mathrm{PGL}_n).$$

We immediately see that the scheme $\mathrm{PGL}_n/E[n]$ is a sampling space for H_{Ob}. We can cut down the dimension considerably by noting that the group PGL_n acts faithfully on $\mathbb{P}^{n-1}$, so we can replace $\mathrm{PGL}_n/E[n]$ by its compression $\mathbb{P}^{n-1}/E[n]$.[2] Concretely, any lift of a rational point of the scheme $\mathrm{PGL}_n/E[n]$ to $\mathrm{PGL}_n(\overline{K})$ is a matrix M which is almost rational, just as above: for every element $\sigma \in G_K$, we have $M^\sigma = i(T(\sigma)) \cdot M$. The image of M in $\mathrm{PGL}_n/E[n](K)$ maps to the cocycle $\sigma \mapsto T_\sigma$. Moreover, such an M can be thought of as moving C to E in the sense that, starting with equations $F_i(x)$ for E, the equations for C will be $F_i(Mx)$. For more on this perspective see [10]. The above compression amounts to the fact that one column of such an M will suffice to represent the cocycle $\sigma \mapsto T_\sigma$. Or indeed, if v is any rational point in C-space, then $M \cdot v$, the image of v in E-space, will suffice. See the examples worked out below for an explicit representation of the matrix M starting only with one of its columns.

Now for the commutativity of the diagram. The composite of the first two maps on the first line is multiplication by n, a rational point of $E/E[n]$ is just the class of a point $Q \in E(\overline{K})$ such that $nQ = P$ is rational. The third map sends the class of Q to $nQ = P$, and the connecting map from cohomology on the right divides P by n again. On the other hand, the map $E/E[n] \to \mathbb{P}^{n-1}/E[n]$ is defined over K and sends the class of Q to some rational point of $\mathbb{P}^{n-1}/E[n]$ and then associates to that point a cocycle which is clearly equivalent. □

Corollary 5.6. *The arithmetic dimension of $H_{\mathrm{Ob},n}$ is at most $n-1$.*

Corollary 5.7. *Let K be a number field. The arithmetic dimension of $H_{\mathrm{Ob},3}$ is 2.*

5.3 Visualizing elements of H_{Ob}

We take a definition from Section 3 of [5]: Let $\xi \in H^1(G_K, E)$ for an elliptic curve over the number field K. Suppose we have an exact sequence of abelian varieties $0 \to E \to J \to B \to 0$. We define ξ to be *visible* when ξ is in the kernel of the induced map $H^1(G_K, E) \to H^1(G_K, J)$.

We will specialize the situation a bit. Since $H^1(G_K, E)$ is a torsion group, the element ξ has some order n; we will assume that n is odd and at least 3. Moreover, assume that we have found another elliptic curve E' such that $E[n] \cong E'[n]$. Then we can construct the following exact sequence: $0 \to E \to (E \times E')/E[n] \to E' \to 0$, where we have glued E and E' together along their common subgroup scheme $E[n]$ and we have identified E' with $E'/E'[n]$.

[2]This idea of compression in the case of a faithful representation comes from the papers of Reichstein.

Then ξ is visualised by the abelian variety $(E \times E')/E[n]$ exactly when ξ lifts to a point P of $E'(K) = H^0(G_K, E')$. This means that there is a point $Q \in E'(\overline{K})$ where $nQ = P$ and where $\xi(\sigma) = Q^\sigma - Q$. This in turn is exactly the condition that the following diagram holds:

$$\begin{array}{ccccccccc} & P & \mapsto & (\sigma \mapsto Q^\sigma - Q) & & & & \\ 0 \to & E'(K)/nE'(K) & \to & H^1(G_K, E'[n]) & \to & H^1(G_K, E')[n] & \to 0 \\ & & & \cong\downarrow & & & \\ 0 \to & E(K)/nE(K) & \to & H^1(G_K, E[n]) & \to & H^1(G_K, E)[n] & \to 0 \\ & & & (\sigma \mapsto Q^\sigma - Q) & \mapsto & \xi & \end{array}$$

We will abuse notation slightly by saying that in this case, ξ is *visualised by* E'.

Remark. The above approach to visualisation was explained to the author in a talk by Tom Fisher at the "Explicit Arithmetic Geometry" conference in Paris, December 2004.

Now let E be an elliptic curve defined over K, and let $X(n)^E$ denote the twisted modular curve parameterizing pairs $(E'/S, E[n]_S \cong E'[n])$, where E' is an elliptic curve over a base S and $E[n]_S \to E'[n]$ is an isomorphism of group schemes over S respecting the Weil pairings. Let $\mathcal{E}$ denote the total space of the universal elliptic curve over $X(n)^E$. Then $E[n]$ acts on $\mathcal{E}$ in a natural way.

Theorem 5.8. *The base diagram* $f : E \to \mathbb{P}^{n-1}$ *gives rise to a map* $\mathcal{E} \to \mathbb{P}^{n-1}$, *which for* $n = 3$ *is a birational map and which for larger odd* n *is an embedding of* $\mathcal{E}$. *Moreover, the following diagram (of functors) commutes:*

$$\begin{array}{ccccccc} \mathcal{E} & \to & \mathcal{E}/E[n] & \cong & \mathcal{E} & \twoheadrightarrow & \mathcal{E}/n\mathcal{E} \\ \downarrow & & \downarrow & & \downarrow & & \downarrow \\ \mathbb{P}^{n-1} & \to & \mathbb{P}^{n-1}/E[n] & \twoheadrightarrow & H_{Ob} & \to & H^1_K(E[n]). \end{array} \tag{5.3}$$

Proof. First assume that the map $\mathcal{E} \to \mathbb{P}^{n-1}$ exists and fits in the above diagram. Assume that there is a geometric point x in the intersection of two elliptic curves E' and E'' in the family of elliptic curves $\mathcal{E}$. The first map on the top row above sends x to the intersection of $E'/E[n]$ and $E''/E[n]$; since this is essentially the multiplication by n map, a degree n^2 map, we know that the n^2 preimages of the image of x all lie on the intersection. Now we will show that two curves intersecting in n^2 points will be forced to be the same curve when n is at least 4. If there is a point y lying on E' but not on E'', we can project away from that point to land in $\mathbb{P}^{n-2}$ space, and E' will now be a degree-$n-1$ curve and E'' will still be a degree n curve. Now we project away from points lying off of E' and E'' until we get down to the projective plane; then the two curves will intersect in $n \cdot (n-1)$ points by Bézout's theorem, a contradiction. Note that this argument does not work for $n = 3$, and indeed it is well known that in this case $\mathcal{E}$ is birational to $\mathbb{P}^2$; with our construction, the elliptic curves in $\mathcal{E}$ will all intersect in their 9 flex points.

Now we need to show such a map exists. We sketch how to produce the morphism $\mathcal{E} \to \mathbb{P}^{n-1}$. Namely, let E' be an elliptic curve over a field extension K'/K endowed with an isomorphism $\alpha : E[n]_{K'} \to E'[n]$ compatible with the Weil pairings. Recall that we have fixed the base diagram $f_E : E \to \mathbb{P}^{n-1}$ which gives rise to an action of $E[n]$ on $\mathbb{P}^{n-1}$ by the associated map $i : E[n] \to \mathrm{PGL}_n$ which reproduces the Weil pairing via (5.1). Choosing a basis of $\Gamma(E', \mathcal{O}_{E'}(n \cdot \mathcal{O}))$ gives rise to a similar diagram $f_{E'} : E' \to \mathbb{P}^{n-1}_{K'}$ and $i' : E'[n] \to \mathrm{PGL}_{n,K'}$ reproducing the Weil pairing on $E'[n]$. We claim that there is a unique automorphism $\beta : \mathbb{P}^{n-1}_{K'} \to \mathbb{P}^{n-1}_{K'}$ such that β intertwines the action of $E[n]_{K'}$ via i and base change with the action of $E[n]_{K'}$ via α and i'. Although we leave this to the reader, we point out that it is straightforward to prove this over the algebraic closure and then deduce the general case using that a solution β is unique since there are no automorphisms of $\mathbb{P}^{n-1}$ commuting with the action i. The upshot is that there is a unique choice of the embedding $f_{E'} : E' \to \mathbb{P}^{n-1}_{K'}$ so that the resulting action i' agrees with the action i via α. Doing this procedure over the base $X(n)^E$ gives a canonical morphism $\mathcal{E} \to \mathbb{P}^{n-1}$.

Commutativity of the diagram. By construction the morphism $\mathcal{E} \to \mathbb{P}^{n-1}$ is $E[n]$-equivariant, which gives us the first commutative square. The last commutative square is obvious. To see that the middle square commutes, consider a fibre E' of $\mathcal{E}$ over some K' point of $X(n)^E$. By (5.2) the diagram commutes with E replaced by E'. Hence it suffices to see that the obstruction map Ob only cares about the isomorphism class of the pair $(E[n], e)$ where e is the Weil pairing. This follows from the results in Section 4 of [11]. □

Definition 5.9. *Denote by $H^1_K(E[n])^{\mathcal{E}}$ the image of $\mathcal{E}$ in $H^1_K(E[n])$. This is the visualised part of $H^1_K(E[n])$.*

An open question is to what extent $H^1_K(E[n])^{\mathcal{E}}$ contains the n-Selmer group of E. To this end we can recover one result from [5]:

Corollary 5.10. *The 3-part of the Selmer group of an elliptic curve is always visualized.*

Proof. In fact the above proves that all of H_{Ob} is visualised, and we know that the 3-torsion in the Selmer group lies inside H_{Ob}. □

Corollary 5.11. *When $n = 3$, the universal elliptic curve $\mathcal{E}$ as defined above can be seen as a sampling space for H_{Ob}; it is a rational surface.*

Remark. We already knew that $\mathbb{P}^2/E[3]$ is a sampling space for H_{Ob} when $n = 3$. It is a unirational surface, and it is known to be rational over $\overline{K}$. The above corollary tells us that it is in fact rational over K, at least when the modular curve $X(n)^E$ has a rational point, as it does here by construction, since we constructed it so that the point corresponding to E itself is rational. See the next section for explicit equations for $n = 3$ in the case where $E[3] \cong \mathbb{Z}/3\mathbb{Z} \times \mu_3$.

5.3.1 Example

When $n = 3$ and $E[3](\overline{K}) = E[3](K)$, we will explicitly compute a rational parameterization of the sampling space $\mathbb{T}_3$ for H_{Ob} and the tautological curve above it. We will also compute the *generic* trivial level-3 norm symbol.

Let ζ denote a fixed primitive third root of unity. We may standardize the action of the 3-torsion as follows: take a basis $\langle S, T\rangle$ of $E[3]$ such that S sends the coordinate x to x, y to ζy, and z to $\zeta^2 z$, and such that T sends the coordinate x to y, y to z, and z to x. Define $s = xyz, t = x^3 + y^3 + z^3, u = x^3y^3 + y^3z^3 + z^3x^3, v = x^6y^3+y^6z^3+z^6x^3$, and $\overline{v} = x^3y^6+y^3z^6+z^3x^6$. These are clearly invariant under the above action, and they satisfy the relations $f_1 : v+\overline{v} = tu-3s^3$ and $f_2 : v\overline{v} = 9s^6 - 6tus^3 + u^3 + t^3s^3$. Our surface $\mathbb{T}_3$ is an open part of $Proj(K[s,t,u,v,\overline{v}]/(f_1, f_2))$, where s and t are of weight 1, u is of weight 2, and v and $\overline{v}$ are of weight 3.

Note the above action of $E[3]$ forces $E = E_\lambda$ to lie in the family

$$E_\lambda : X^3 + Y^3 + Z^3 - 3\lambda XYZ = 0$$

for some $\lambda \in K$. Fix $\mathcal{O} = (1;-1;0)$.

We will find a rational parameterization of S by identifying S birationally with $\mathcal{E}$ as in Section 5.3. To be exact, note that if we choose a rational point of $\mathbb{P}^2$, we can (almost always) uniquely solve for λ above: $\lambda = \frac{X^3+Y^3+Z^3}{3XYZ} = \frac{t}{3s}$.[3]

Note that $\mathcal{E}$ also has a natural elliptic curve family in $\mathbb{P}^2_K$ sitting over it: a rational point P of $\mathcal{E}$ lies on an elliptic curve E'; over such a point P the elliptic curve is simply E' embedded in $\mathbb{P}^2$ by the line bundle corresponding to $2 \cdot \mathcal{O} + P$. This is not the family we are looking for; rather, we are looking for the family of genus-one curves all of which have Jacobian isomorphic to the fixed E. The fact that *two* such curves lie above each point means that their corresponding cocycles are visualised (see below).

From Theorem 5.8 we see that if we identify $\mathcal{E}$ birationally with $\mathbb{P}^2$, we can express the multiplication by 3 map on $\mathbb{P}^2$, denoted by [3], as a function on $\mathbb{P}^2/E[3]$ and this will give us its birational model. On the level of points, we are taking a rational point P of $\mathbb{P}^2$ and finding the elliptic curve E_P it lies on as above; we then extract a Q on E_P such that $3 \cdot Q = P$, and form the cocycle $Q^\sigma - Q$.

Note that even with the general Weierstrass equations for elliptic curves, this makes sense, but would require knowing how to extract this cube root in general.

The function [3] vanishes on the hyperplane $Z = 0$ exactly at those points Q such that $3 \cdot Q = \mathcal{O}, S$, or $-S$. Similarly, [3] vanishes on the hyperplane $X = 0$

[3] What if we can't solve for λ? This happens when one and more of X, Y, and Z is zero. Two being zero would mean the point is rational, which gives us the trivial cocycle. Acting by T above if necessary (which only changes the cocycle mod coboundaries) we can assume our point looks like $(0;1;\alpha)$, and so the corresponding cocycle ξ sends σ to $(0;1;\sigma(\alpha))$, but on the other hand sends it to itself acted on by $\xi(\sigma) \in E[3]$. We see then that $\xi(\sigma) = i \cdot S$ for some i, so ξ pulls back to $H^1(G_K, \mathbb{Z}/3\mathbb{Z}) \cong K^*/K^{*3}$. However these cocycles are also covered by the points $(1;\alpha;\alpha^2)$, which have well-defined λ.

exactly at those points Q such that $3 \cdot Q = T, T + S$, or $T - S$, and [3] vanishes on the hyperplane $Y = 0$ exactly at those point Q such that $3 \cdot Q = -T, -T + S$, or $-T - S$.

For every $T \in E[3]$, there is a cubic form a_T which intersects E_λ in exactly those points Q such that $3 \cdot Q = T$. By work in [4] we can easily compute such a_T, well-defined up to (the same) scalar, when $T \neq \mathcal{O}$ as $T_v \cdot M_T \cdot v$, where v is the point $(x; y; z) \in \mathbb{P}^2$ represented by the column vector $(x \quad y \quad z)^\tau$, and T_v is the tangent plane to v represented by the row vector $(x^2 - \lambda yz \quad y^2 - \lambda xz \quad z^2 - \lambda xy)$. Moreover we can take $a_\mathcal{O}$ to be the Hessian cubic taking v to $-2\lambda^2(x^3 + y^3 + z^3) + (8 - 2\lambda^3)xyz$. Then the composite birational map $[3] : \mathbb{P}^2 \to \mathbb{P}^2/E[n] \cong \mathbb{P}^2$ is given by

$$[3] : x \mapsto ((a_T a_{T+S} a_{T-S})(x); (a_{-T} a_{-T+S} a_{-T-S})(x); (a_\mathcal{O} a_S a_{-S})(x)).$$

We know that these functions can be written as functions on S, i.e., as functions of s, t, u, v, and $\overline{v}$ as above, and indeed we have the following:

$$\begin{aligned} (a_T a_{T+S} a_{T-S})(x) &= (1 - \lambda^3)(\overline{v} - 3s^3) \\ (a_{-T} a_{-T+S} a_{-T-S})(x) &= (1 - \lambda^3)(v - 3s^3) \\ (a_\mathcal{O} a_S a_{-S})(x) &= (t^2 - 3u)(-2\lambda^2 t + (8 - 2\lambda^3)s). \end{aligned} \tag{5.4}$$

We can substitute $\lambda = \frac{t}{3s}$, and the map simplifies to

$$[3] : (x, y, z) \mapsto (\overline{v} - 3s^3; v - 3s^3; 8s(t^2 - 3u)).$$

This map factors through $\mathbb{T}_3 \subset \operatorname{Proj}(K[s, t, u, v, \overline{v}]/(f_1, f_2))$, namely by sending a point $(s; t; u; v; \overline{v})$ to $(\overline{v} - 3s^3; v - 3s^3; 8s(t^2 - 3u))$. Since all the scheme maps in Theorem 5.8 in the case $n = 3$ are birational, this gives us a birational model of our sampling space; indeed we will now compute the inverse map. Starting with $(a; b; c) \in \mathbb{P}^2$, we will compute its inverse image in $\mathbb{T}_3$. First, over an open part of our surface we may assume $s = 1$, in which case we can recover $v = b + 3$ and $\overline{v} = c + 3$. Moreover, since both $(a; b; c)$ and its preimage $(x; y; z)$ lie on the same curve E_λ, we know that $\lambda = t(a, b, c)/3s(a, b, c) = \frac{a^3+b^3+c^3}{3abc} = \frac{t}{3s} = t/3$. In other words, $t = \frac{a^3+b^3+c^3}{abc}$. Finally, we can solve for u to get $u = -a/24 + 1/3(\frac{a^3+b^3+c^3}{abc})^2$.

Next we will explicitly compute the tautological genus-one curve lying over $\mathbb{T}_3$. To do so, start with a rational point $P \in \mathbb{T}_3(K)$ and a lift $(x; y; z)$ in $\mathbb{P}^2(\overline{K})$. The matrix M_P which takes $f : E \to \mathbb{P}^2$ to the fiber above P, $f_P : C \to S$, satisfies the following property: for any $\sigma \in G_K$,

$$M_P^\sigma = M_{T(\sigma)} M_P, \iff \begin{pmatrix} x \\ y \\ z \end{pmatrix}^\sigma = M_{T(\sigma)} \begin{pmatrix} x \\ y \\ z \end{pmatrix},$$

and any such matrix $M \in \mathrm{PGL}_n$ which satisfies that property will bring $f : E \to \mathbb{P}^2$ to some diagram equivalent to $f_P : C \to S$. With that in mind, we will define

$$M_P = \begin{pmatrix} x & 0 & 0 \\ 0 & y & 0 \\ 0 & 0 & z \end{pmatrix} (x^3 I + y^3 M_T^{-1} + z^3 M_T).$$

A calculation shows that it satisfies the above property. To find the tautological curve over $\mathbb{T}_3$ it remains to act by M_P, that is, to expand $F_C(x) = F(M_P(x))$, where $F(X, Y, Z) = X^3 + Y^3 + Z^3 - 3\lambda XYZ$ is the cubic giving E. By construction the coefficients will be functions on $\mathbb{T}_3$. A calculation gives us:

$$\begin{aligned} F_C(X, Y, Z) = {} & (t^4 - 4ut^2 + 2u^2 + 3s^3t)X^3 + (vt - u^2)Y^3 + (\overline{v}t - u^2)Z^3 \\ & +(2\overline{v}t - 3u^2 + 3s^3t)X^2Y + (3s^3t - \overline{v}t)Y^2Z + (3u^2 - \overline{v}t - 6s^3t) \\ & \times Z^2X + (2vt - 3u^2 + 3s^3t)X^2Z + (3u^2 - vt - 6s^3t)Y^2X \\ & +(3s^3t - vt)Z^2Y + (3t^2u - t^4)XYZ. \end{aligned}$$

Corollary 5.12. *The generic level-three trivial norm symbol is given by a rational map $\mathbb{P}^2/E[3] \cong \mathbb{T}_3 \to \mathbb{G}_m^2$ which sends the image of $x \in \mathbb{P}^2$ to $(a_S(x)^3, a_T(x)^3)$, where S and T generate $E[n]$. Explicitly, it sends $(s; t; u; v; \overline{v})$ to*

$$(3\zeta^2 v + 3\zeta\overline{v} + t^3 - 2ut - v - \overline{v} + 6s^3, v + 3su + 3s^2t + 6s^3).$$

Proof. The function a_T can be computed as $a_T = T_v \cdot M_T \cdot v$. For any $\sigma \in G_K$, we then have $a_T^\sigma = (T_v)^\sigma \cdot M_T^\sigma \cdot v^\sigma = T_{v^\sigma} \cdot M_T \cdot v^\sigma$, since the G_K action on $E[n]$ and thus M_T is trivial, and since the entries of T_v are polynomials in the coordinates of v. By construction, $v^\sigma = M_{T(\sigma)}v$, and a calculation yields $T_{v^\sigma} = T_v M_{T(\sigma)}^{-1}$, so $a_T^\sigma = [M_T, M_{T(\sigma)}] \cdot T_v \cdot M_T \cdot v$. By work in [10], $[M_T, M_{T(\sigma)}] = e(T, T(\sigma))$. □

What if we assume a slightly less stringent condition, namely that $E[n] \cong \mathbb{Z}/n\mathbb{Z} \times \mu_n$? By previous observations, we can "base change" to $K(\zeta_n)$ and use the corestriction map to get a sampling space over K. However, the dimension gets multiplied by the degree of the field extension $K(\zeta_n)/K$, which divides $\phi(n)$.

In fact, when $E[n] \cong \mathbb{Z}/n\mathbb{Z} \times \mu_n$, then $H_K^1(E[n]) = H_K^1(\mathbb{Z}/n\mathbb{Z}) \times H_K^1(\mu_n)$; the arithmetic dimension of $H_K^1(\mu_n)$ is of course 1 when K is a number field (since its set of K points is infinite but a sampling space is $\mathbb{G}_m$). However, the arithmetic dimension of $H_{\mathbb{Q}}^1(\mathbb{Z}/n\mathbb{Z})$ is bounded by the essential dimension of $\mathbb{Z}/n\mathbb{Z}$, which is expected to be unbounded (and related to $\phi(n)$). Note that this is equivalent to asking for the dimension of a versal family of $\mathbb{Z}/n\mathbb{Z}$-extensions of $\mathbb{Q}$, a classical problem from inverse Galois theory. In other words, the functor H_{Ob}, which is carved out of the functor $H^1(E[n])$, seems to get more complicated as the Galois action on $E[n]$ becomes more complicated. The conjecture says that in spite of this, H_{Ob} becomes no more complicated than growing linearly with n.

6 Further remarks

- I would like to compare arithmetic dimensions over number fields with those over p-adic fields and finite fields. Since many schemes automatically have points over such fields, the arithmetic dimensions of functors such as $H^1_K(G)$ over such fields should generally speaking be smaller. Of course I would like to make this precise. Certainly such things as the (functor associated to the) Brauer group over such fields is understood to be smaller.
- A major motivation for the definitions in this paper is to explore the question of finding a sampling space for the Tate–Shavarevich group of a fixed elliptic curve E over a number field K (suitably extended to a functor on the category $\mathcal{C}_K$) or prove that no such sampling space exists. Note we have some freedom in extending the functor to the full category $\mathcal{C}_K$. However, the most obvious approach would have $X \in \mathcal{C}_K$ map to $Ker(H^1_{\text{ét}}(X, E) \to \prod_{v \in K} H^1_{ét}(X \otimes_K K_v, G)$. In this case, we would have a tautological genus-one curve lying over any sampling space. It is not hard to see that this would imply that there is a (uniform) bound for the order of elements of the Tate-Shafarevich group of E over all finite extensions of K, a very strong result. However, there may be more nuanced ways to extend the functor. What is probably an easier task is to find a series of sampling spaces for the p-power torsion of the Tate–Shavarevich group, for all primes p.

  In an analogy for Hilbert symbols, if the ground field K is large enough to admit rational p^n torsion for all n (in which case K is not a number field), then an element of $\text{Ш}(E)$ would correspond to a coherent sequence of trivial Hilbert symbols of level p^n for all n. The above assumption on the ground field is Iwasawa-like; if possible we will work without such an assumption, so as to stay over number fields.
- One major goal is to not only have a sampling space for a functor but to have a *nice* sampling space, namely one that is as close as possible to $\mathbb{A}^n$. As we have seen, this is often attainable. We want nice sampling spaces primarily because then the points are extremely accessible (i.e., so we really have an analogy with the concept of a "parameter space"), but also because we have a well-defined notion of height of a point. This could be useful in quantifying the arithmetic objects that we are parameterizing. Often we have a different notion of height on the objects already, for example by looking at the size of the coefficients of the objects. We would want to prove "comparison theorems" in order to justify the height measurements coming from our sampling space. An example of a comparison theorem would state that if the maximal height of a coefficient of a model of a cubic equation C is D, then there is a point of the height at most $f(D)$ on the sampling space of cubics which maps to C, where $f(D)$ is a simple function of D. With the aid of a comparison theorem we would be able to compute the asymptotic behavior of our arithmetic objects.

Another benefit of having a good sampling space for objects, especially with an accompanying explicit tautological model of the objects, is simply that we can test for, say, local points on the models.

- There are numerous simple questions one can ask about sampling spaces which give rise to interesting and basic questions about elliptic curves and the arithmetic of number fields. For example, we know that the arithmetic dimension of $H^1_K(\mu_n)$ is 1 when K is a number field, but is $\mathbb{G}_m$ (with the standard map) the *only* sampling space? Is there an elliptic curve, an open part of which acts (with an appropriate map) as a sampling space? First note that any sampling space would have to map (rationally) through $\mathbb{G}_m$ with the standard map after we adjoin the nth roots of unity, since Kummer theory will supply us with an element in the function field of the sampling space. Then (an open part of) an elliptic curve sampling space would map through $\mathbb{G}_m$; clearly it would not be possible for the map to be a *homomorphism* of the group schemes, since the Mordell–Weil group is finitely generated. However, it is possible it could be some other map, like sending a point $(x, y) \in E(K)$ to say $y \in \mathbb{G}_m(K)$. This raises the question of "growth" of rational points as K grows: the functor $H^1_K(\mu_n)(K) \cong K^*/K^{*n}$ enlarges with *any* non-trivial extension L/K, and so would any sampling space. Therefore we couldn't possibly have an elliptic curve sampling space unless we found an elliptic curve whose rank grew with every extension. This is very unlikely but I don't know a proof that no such elliptic curve exists except over $\mathbb{Q}$. Note finally that by Faltings' theorem no genus-2 (or higher) curve could be a sampling space for $H^1_K(\mu_n)$; another related question would be, is there another map (whose degree is larger than 1) making $\mathbb{G}_m$ a sampling space in a different way?

References

[1] Berhuy, G. and Favi, G., *Essential Dimension: A Functorial Point of View*, Doc. Math. 8 (2003), 279–330 (electronic).
[2] Buhler, J., and Reichstein, Z., *On the Essential Dimension of a Finite Group*, Compositio Math. 106 (1997), 159–179.
[3] Cassels, J. W. S., *Lectures on Elliptic Curves,* Cambridge University Press, Cambridge, 1991.
[4] Cremona, J. E.; Fisher, T. A.; O'Neil, C.; Simon, D.; Stoll, M., *Explicit n-descent on elliptic curves. I. Algebra,* J. Reine Angew. Math. 615 (2008), 121–155.
[5] Cremona, J., Mazur, B., *Visualizing elements in the Shafarevich-Tate group,* Experiment. Math. 9 (2000), no. 1, 13–28.
[6] Hartshorne, R., *Algebraic Geometry,* Springer-Verlag, New York, 1977.
[7] Katz, N. M., *Space filling curves over finite fields,* Math. Res. Lett. 6 (1999), no. 5–6, 613–624.
[8] Katz, N. M., *Sommes exponentielles,* Astérisque, 79 (1980), Société Mathématique de France, Paris.
[9] Mumford, D., *Abelian Varieties*, Oxford University Press, Oxford, 1985.
[10] O'Neil, C., *Jacobians of Genus One Curves,* Math. Res. Lett. 8 (2001), no. 1–2, pp. 125–140.
[11] O'Neil, C., *The Period-Index Obstruction for Elliptic Curves,* J. of Number Theory, 95 (2002), no. 2, pp. 329–339.

[12] O'Neil, C., *Models of Some Genus One Curves with Applications to Descent,* J. Number Theory 112 (2005), no. 2, 369–385.

[13] Reichstein, Z., *On the Notion of Essential Dimension for Algebraic Groups*, Transformation Groups, 5, no. 3 (2000), pp. 265–304.

[14] Schaefer, E. F.; Stoll, M., *How to do a p-descent on an elliptic curve,* Trans. Amer. Math. Soc. 356 (2004), no. 3, 1209–1231 (electronic).

Recovering function fields from their decomposition graphs

Florian Pop

In memory of Serge Lang

Abstract We develop the *global theory* of a strategy to tackle a program initiated by Bogomolov in 1990. That program aims at giving a group-theoretical recipe by which one can reconstruct function fields $K|k$ with $\mathrm{td}(K|k) > 1$ and k algebraically closed from the maximal pro-ℓ abelian-by-central Galois group Π_K^c of K, where ℓ is any prime number $\neq \mathrm{char}(k)$.

Key words anabelian geometry • pro-ℓ groups • Galois theory • function fields • valuations theory • (Riemann) space of prime divisors • Hilbert decomposition theory • Parshin chains • decomposition graphs

Mathematics Subject Classification (2010): Primary 12E, 14E, 14H, 14J Secondary 12E30, 14E99, 14H30, 14J99

1 Introduction

Recall that the birational anabelian conjecture originating in ideas presented in Grothendieck's *Esquisse d'un Programme* [11] and *Letter to Faltings* [12] asserts roughly the following: First, there should exist a group-theoretical recipe by which one can recognize the absolute Galois groups G_K of finitely generated infinite fields K among all profinite groups. Second, if $G = G_K$ is such an absolute Galois group, then the group-theoretical recipe should recover the field K from G_K in a functorial way. Third, the recipe should be invariant under open homomorphisms of absolute

F. Pop (✉)
Dept of Mathematics, University of Pennsylvania, 209 S 33rd St, Philadelphia, PA 19104, USA
e-mail: pop@math.upenn.edu

D. Goldfeld et al. (eds.), *Number Theory, Analysis and Geometry: In Memory of Serge Lang*, DOI 10.1007/978-1-4614-1260-1_24,

Galois groups. In particular, the category of finitely generated infinite fields (up to Frobenius twist) should be equivalent to the category of their absolute Galois groups and open outer homomorphisms between these groups. A first instance of this situation is the celebrated Neukirch–Uchida theorem, which says that global fields are characterized by their absolute Galois groups. I will not go into further detail about the results concerning Grothendieck's (birational) anabelian geometry, but the interested reader can find more about this in Szamuely's Bourbaki Séminaire talk [34], Faltings' Séminaire Bourbaki talk [10], Stix [35], and newer results by Mochizuki [19], Saidi–Tamagawa [33], Minhyong Kim [13], and Koenigsmann [15] concerning the (birational) section conjecture.

The idea behind Grothendieck's anabelian geometry is that the *arithmetical Galois action* on rich geometric fundamental groups (such as the geometric absolute Galois group) makes objects very rigid, so that there is no room left for non-geometric open morphisms between such rich fundamental groups endowed with arithmetical Galois action.

On the other hand, Bogomolov [2] advanced at the beginning of the 1990s the idea that one should have anabelian-type results in total absence of an arithmetical action as follows: Let ℓ be a fixed rational prime number. Consider function fields $K|k$ over algebraically closed fields k of characteristic $\neq \ell$. For each such function field $K|k$, let $\Pi_K^{c} := \mathrm{Gal}(K''|K)$ be the Galois group of a maximal pro-ℓ abelian-by-central Galois extension $K''|K$. Note that if $G^{(1)} = G_K$ and $G^{(i+1)} := [G^{(i)}, G_K](G^{(i)})^{\ell^\infty}$ for $i \geq 1$ are the central ℓ^∞ terms of the absolute Galois group G_K of K, then we have that $\Pi_K = G^{(1)}/G^{(2)}$ is the Galois group of the maximal pro-ℓ abelian subextension $K'|K$ of $K''|K$, and $\Pi_K^{c} = G^{(1)}/G^{(3)}$; and denoting by $G^{(\infty)}$ the intersection of all the $G^{(i)}$, it follows that $G_K(\ell) := G_K/G^{(\infty)}$ is the maximal pro-ℓ quotient of G_K. Now the program initiated by Bogomolov [2] has as ultimate goal to recover function fields $K|k$ with $\mathrm{td}(K|k) > 1$ as above from Π_K^{c} in a functorial way. (Note that Bogomolov denotes Π_K^{c} by PGal_K^{c}.) If successful, this program would go far beyond Grothendieck's birational anabelian conjectures, as k being algebraically closed implies that there is no arithmetical action in the game. The program initiated by Bogomolov is not completed yet, and this paper is a contribution towards trying to settle that program; see the historical note below for more about this.

Since this paper is quite abstract, let me announce the following "concrete" result, whose proof relies in an essential way on the Main Theorem of this paper (see Pop [30] for a complete proof):

Theorem I *Let $K|k$ be a function field with* $\mathrm{td}(K|k) > 1$ *and k an algebraic closure of a finite field. Then the following hold:*

(1) There exists a group-theoretical recipe which recovers $K|k$ from Π_K^{c}.
(2) The above group-theoretical recipe is functorial in the following sense: Let $L|l$ be a function field with l an algebraically closed field, and let $\Phi : \Pi_K \to \Pi_L$ be the abelianization of some isomorphism $\Phi^{c} : \Pi_K^{c} \to \Pi_L^{c}$. Then denoting by L^{i} and K^{i} the perfect closures, there exist an isomorphism of field extensions $\imath : L^{\mathrm{i}}|l \to K^{\mathrm{i}}|k$ and an ℓ-adic unit $\varepsilon \in \mathbb{Z}_\ell^\times$ such that $\varepsilon \cdot \Phi$ is induced by $\imath$.

Moreover, the isomorphism $\imath$ is unique up to Frobenius twists, and the ℓ-adic unit ε is unique up to multiplication by p-powers, where $p = \mathrm{char}(k)$.

(3) *For a function field $L|l$ as above, let $\mathrm{Isom}^{\mathrm{F}}(L, K)$ be the set of isomorphisms of field extensions $\imath : L^{\mathrm{i}}|l \to K^{\mathrm{i}}|k$ up to Frobenius twists, and let $\mathrm{Isom}^{\mathrm{c}}(\Pi_K, \Pi_L)$ be the set of abelianizations of continuous group isomorphisms $\Pi^{\mathrm{c}}_K \to \Pi^{\mathrm{c}}_L$ up to multiplication by ℓ-adic units $\varepsilon \in \mathbb{Z}_\ell^\times$. Then there is a canonical bijection*

$$\mathrm{Isom}^{\mathrm{F}}(L|l, K|k) \to \mathrm{Isom}^{\mathrm{c}}(\Pi_K, \Pi_L).$$

A sketch of a strategy to functorially recover $K|k$ from pro-ℓ Galois information, in particular to prove the above Theorem I, can be found essentially already in (the notes of) Pop [25], and has as starting point the following simple idea: Let $\widehat{K}$ be the ℓ-adic completion of the multiplicative group $K^\times$ of $K|k$.[1] Since the cyclotomic character of K is trivial, one can identify the ℓ-adic Tate module $\mathbb{T}_{K,\ell}$ of K with $\mathbb{Z}_\ell$ (non-canonically), and let $\imath_K : \mathbb{T}_{K,\ell} \to \mathbb{Z}_\ell$ be a fixed identification. Via Kummer theory, one has isomorphisms of ℓ-adically complete groups:

$$\widehat{K} = \mathrm{Hom}_{\mathrm{cont}}(\Pi_K, \mathbb{T}_{K,\ell}) \xrightarrow{\imath_K} \mathrm{Hom}_{\mathrm{cont}}(\Pi_K, \mathbb{Z}_\ell),$$

i.e., $\widehat{K}$ can be recovered from Π_K, hence from Π^{c}_K via the projection $\Pi^{\mathrm{c}}_K \to \Pi_K$. On the other hand, since $k^\times$ is divisible, $\widehat{K}$ equals the ℓ-adic completion of the free abelian group $K^\times/k^\times$. Now the idea of recovering $K|k$ is as follows:

(a) First, give a recipe to recover the image $\jmath_K(K^\times) = K^\times/k^\times$ of the ℓ-adic completion functor $\jmath_K : K^\times \to K^\times/k^\times \subset \widehat{K}$ inside the "known" $\widehat{K} = \mathrm{Hom}_{\mathrm{cont}}(\Pi_K, \mathbb{Z}_\ell)$.

(b) Second, interpreting $K^\times/k^\times =: \mathcal{P}(K)$ as the projectivization of the infinite-dimensional k-vector space $(K, +)$, give a recipe to recover the projective lines $\mathfrak{l}_{x,y} := (kx + ky)^\times/k^\times$ inside $\mathcal{P}(K)$, where $x, y \in K$ are k-linearly independent.

(c) Third, apply the *fundamental theorem of projective geometries* of Artin [1], and deduce that $K|k$ can be recovered from $\mathcal{P}(K)$ endowed with all the lines $\mathfrak{l}_{x,y}$.

(d) Finally, show that the recipes above are functorial, i.e., they are invariant under isomorphisms of profinite groups $\Pi_K \to \Pi_L$ which are abelianizations of isomorphisms $\Pi^{\mathrm{c}}_K \to \Pi^{\mathrm{c}}_L$. In particular, such isomorphisms $\Pi_K \to \Pi_L$ originate actually from geometry.

The strategy from Pop [25] to tackle the above problems (a), (b), (c), (d), above is in principle similar to the strategies (initiated by Neukirch and Uchida) for tackling Grothendieck's anabelian conjectures. It has two main parts, as follows, the terminology being as introduced later:

---

[1]Recall that for an abelian group A, its ℓ-adic completion is by definition $\widehat{A} := \varprojlim_e A/\ell^e$.

Local theory: It has as input the Galois/group-theoretical information Π_K^c. It should be a recipe which in a first approximation recovers from Π_K^c the decomposition/inertia groups of prime divisors of $K|k$ in Π_K (N.B., not in Π_K^c). The final output of the local theory should be the *total decomposition graph* $\mathcal{G}_{\mathcal{D}_K^{\rm tot}}$ of $K|k$. This recipe should be invariant under isomorphisms $\Pi_K \to \Pi_L$ which are induced by some isomorphisms $\Pi_K^c \to \Pi_L^c$.

Global theory: Its input is the total decomposition graph $\mathcal{G}_{\mathcal{D}_K^{\rm tot}}$ of $K|k$. It should be a recipe which in a first approximation recovers the *geometric decomposition graphs* $\mathcal{G}_{\mathcal{D}_K}$ (together with some of their special properties) for $K|k$ from $\mathcal{G}_{\mathcal{D}_K^{\rm tot}}$ together with their sets of *rational quotients* $\mathfrak{A}_K = \{\Phi_{\kappa_x}\}_{\kappa_x}$. In a second approximation, this recipe should recover $\mathscr{P}(K)$ and its projective lines from the $\mathcal{G}_{\mathcal{D}_K}$ endowed with their rational quotients $\mathfrak{A}_K = \{\Phi_{\kappa_x}\}_{\kappa_x}$. It thus should finally recover the function field $K|k$. Moreover, this recipe should be functorial, i.e., invariant under isomorphisms of total decomposition graphs $\mathcal{G}_{\mathcal{D}_K^{\rm tot}} \to \mathcal{G}_{\mathcal{D}_L^{\rm tot}}$.

This paper deals mainly with questions of the above *global theory,* precisely, recovering the geometric decomposition graphs (together with some of their special properties) from the total decomposition graph, and finally proving the main result of the paper, which is to show that morphisms of (total) decomposition graphs that are compatible with rational projections originate in a precise way from geometry.

Before announcing the main result here, let us briefly introduce the main concepts and objects, which will be discussed/studied in detail later.

• **Prime divisor graphs** (see Section 3 for more details)

Recall that in the context above, a (Zariski) prime divisor of a function field $K|k$ is a discrete valuation v of K whose valuation ring is the local ring $\mathcal{O}_{X,x_1}$ of the generic point x_1 of some Weil prime divisor of some normal model $X \to k$ of $K|k$. If so, then the residue field Kv of v is the function field $Kv = \kappa(x_1)$, and therefore, $\mathrm{td}(Kv|k) = \mathrm{td}(K|k) - 1$. A set of Zariski prime divisors D of $K|k$ is called a *geometric set* if there exists a quasiprojective normal model $X \to k$ of $K|k$ such that $D = D_X$ is the set of valuations v_{x_1} defined by the generic points x_1 of all the Weil prime divisors of X. We next generalize the prime divisors of $K|k$ as follows: First, for a valuation $\mathfrak{v}$ of K the following are equivalent:

(i) $\mathfrak{v}$ is trivial on k, the residue field has $\mathrm{td}(K\mathfrak{v}|k) = \mathrm{td}(K|k) - r$, and there exists a chain of valuations $\mathfrak{v}_1 < \cdots < \mathfrak{v}_r := \mathfrak{v}$.

(ii) $\mathfrak{v}$ is the valuation-theoretical composition $\mathfrak{v} = v_r \circ \cdots \circ v_1$, where v_1 is a prime divisor of K, and inductively, v_{i+1} is a prime divisor of the residue function field $\kappa(v_i)\,|\,k$.

A valuation $\mathfrak{v}$ of K which satisfies the above equivalent conditions is called a *prime r-divisor* of $K|k$; and a sequence of prime divisors $(v_r, \ldots, v_1)$ as above will be called a *Parshin r-chain* of $K|k$. By definition, the trivial valuation will be considered a generalized prime divisor of rank zero, and the corresponding Parshin chain is the trivial Parshin chain. Finally, note that $r \le \mathrm{td}(K|k)$, and that in the above notation, one has $\mathfrak{v}_i = v_i \circ \cdots \circ v_1$ for all $i \ge 1$.

The *total prime divisor graph* $\mathscr{D}_K^{\text{tot}}$ of K is the following half-oriented graph:

(a) the vertices of $\mathscr{D}_K^{\text{tot}}$ are the residue fields $K\mathfrak{v}$ of all generalized prime divisors $\mathfrak{v}$ of $K|k$ viewed as distinct function fields.
(b) For given $\mathfrak{v} = v_r \circ \cdots \circ v_1$ and $\mathfrak{w} = w_s \circ \cdots \circ w_1$, the edges from $K\mathfrak{v}$ to $K\mathfrak{w}$ are as follows:

   (i) If $\mathfrak{v} = \mathfrak{w}$, i.e., $K\mathfrak{v} = K\mathfrak{w}$, then the trivial valuation $\mathfrak{v}/\mathfrak{w} = \mathfrak{w}/\mathfrak{v}$ is the only edge from $K\mathfrak{v} = K\mathfrak{w}$ to itself; and it is by definition a non-oriented edge.
   (ii) If $K\mathfrak{v} \neq K\mathfrak{w}$, then the set of edges from $K\mathfrak{v}$ to $K\mathfrak{w}$ is non-empty iff $s = r+1$ and $v_i = w_i$ for $1 \le i \le r$; and if so, then $w_s = \mathfrak{w}/\mathfrak{v}$ is the only edge from $K\mathfrak{v}$ to $K\mathfrak{w}$, and it is by definition an oriented edge.

A *geometric prime divisor graph* for $K|k$ is any connected subgraph $\mathscr{D}_K$ of $\mathscr{D}_K^{\text{tot}}$ which satisfies the following conditions: First, for each vertex $K\mathfrak{v}$ of $\mathscr{D}_K$, the set $D_\mathfrak{v}$ of all non-trivial edges of $\mathscr{D}_K$ originating from $K\mathfrak{v}$ is a geometric set of prime divisors of $K\mathfrak{v}\,|\,k$. Second, all maximal branches of non-trivial edges of $\mathscr{D}_K$ originate at K and have length equal to $\text{td}(K|k)$. Equivalently, $\mathscr{D}_K$ is a half-oriented connected graph having $K = K_0$ as origin and satisfying:

(a) the vertices of $\mathscr{D}_K$ are distinct function fields $K_i|k$ over k;
(b) for every vertex K_i, the trivial valuation of K_i is the only edge from K_i to itself. And the set of non-trivial edges v_i originating at K_{i^*} is a geometric set of prime divisors of $K_{i^*}|k$, and if v_i is a non-trivial edge from K_{i^*} to K_i, then $K_i = K_{i^*}v_i$;
(c) the only cycles of the graph are the non-oriented edges, and all the maximal branches consisting of oriented edges only have length equal to $\text{td}(K|k)$.

The functorial behavior of geometric prime divisor graphs is as follows:

(1) *Embeddings.* Let $L|l \hookrightarrow K|k$ be an embedding of function fields which maps l onto k. Then the canonical restriction map of valuations $\text{Val}_K \to \text{Val}_L$, $v \mapsto v|_L$, gives rise to a morphism of the total prime divisor graphs $\varphi_\imath : \mathscr{D}_K^{\text{tot}} \to \mathscr{D}_L^{\text{tot}}$, which moreover is surjective. The relation between *geometric prime divisor graphs* $\mathscr{D}_K$ and $\mathscr{D}_L$ is a little bit more subtle; see Proposition 37: Given geometric prime divisor graphs $\mathscr{D}_K$ and $\mathscr{D}_L$, there exist geometric prime divisor graphs $\mathscr{D}_K^0$ and $\mathscr{D}_L^0$ containing $\mathscr{D}_K$, respectively $\mathscr{D}_L$, such that $\varphi_\imath$ defines a surjective morphism of geometric prime divisor graphs:

$$\varphi_\imath : \mathscr{D}_K^0 \to \mathscr{D}_L^0.$$

(2) *Restrictions.* Given a generalized prime divisor $\mathfrak{v}$ of $K|k$, let $\mathscr{D}_\mathfrak{v}^{\text{tot}}$ be the set of all generalized prime divisors $\mathfrak{w}$ of $K|k$ with $\mathfrak{v} \le \mathfrak{w}$. Then the map

$$\mathscr{D}_\mathfrak{v}^{\text{tot}} \to \mathscr{D}_{K\mathfrak{v}}^{\text{tot}}, \quad \mathfrak{w} \mapsto \mathfrak{w}/\mathfrak{v},$$

is an isomorphism of $\mathscr{D}_{\mathfrak{v}}^{\rm tot}$ onto $\mathscr{D}_{K\mathfrak{v}}^{\rm tot}$. Moreover, if $K\mathfrak{v}$ is a vertex of some geometric prime divisor graph $\mathscr{D}_K$ for $K|k$, then one has that the maximal subgraph $\mathscr{D}_{K\mathfrak{v}}$ of $\mathscr{D}_K$ whose initial vertex is $K\mathfrak{v}$ is a geometric graph of prime divisors of $K\mathfrak{v}$.

• **Decomposition graphs** (see Section 3 for more details)

Let $K|k$ be as considered above. Then we have the following, see e.g., Pop [28], Introduction, for a discussion of these facts: For every prime divisor v of $K|k$ one has $T_v \cong \mathbb{T}_{\ell,K}$, and for every prime r-divisor $\mathfrak{v}$ one has $T_{\mathfrak{v}} \cong \mathbb{T}_{\ell,K}^r$. Further, for generalized prime divisors $\mathfrak{v}$ and $\mathfrak{w}$ one has $Z_{\mathfrak{v}} \cap Z_{\mathfrak{w}} \neq 1$ if and only if $\mathfrak{v}, \mathfrak{w}$ are not independent as valuations, i.e., $\mathscr{O} := \mathscr{O}_{\mathfrak{v}}\mathscr{O}_{\mathfrak{w}} \neq K$; and if so, then $\mathscr{O}$ is the valuation ring of a generalized prime divisor $\mathfrak{u}$ of $K|k$ which turns out to be the unique generalized prime divisor with $T_{\mathfrak{u}} = T_{\mathfrak{v}} \cap T_{\mathfrak{w}}$, and also the unique generalized prime divisor of $K|k$ maximal with the property $Z_{\mathfrak{v}}, Z_{\mathfrak{w}} \subseteq Z_{\mathfrak{u}}$.

In particular, $\mathfrak{v} = \mathfrak{w}$ iff $T_{\mathfrak{v}} = T_{\mathfrak{w}}$ iff $Z_{\mathfrak{v}} = Z_{\mathfrak{w}}$. Further, $\mathfrak{v} < \mathfrak{w}$ iff $T_{\mathfrak{v}} \subset T_{\mathfrak{w}}$ strictly iff $Z_{\mathfrak{v}} \supset Z_{\mathfrak{w}}$ strictly, and $T_{\mathfrak{w}}/T_{\mathfrak{v}} \cong \mathbb{Z}_\ell^{s-r}$ if $\mathfrak{v}$ is a prime r-divisor and $\mathfrak{w}$ is a prime s-divisor.

We conclude that the partial ordering of the set of all generalized prime divisors $\mathfrak{v}$ of $K|k$ *is encoded in the set* of their inertia/decomposition groups $T_{\mathfrak{v}} \subseteq D_{\mathfrak{v}}$. In particular, the existence of the trivial, respectively a non-trivial, edge from $K\mathfrak{v}$ to $K\mathfrak{w}$ in $\mathscr{D}_K^{\rm tot}$ is equivalent to $T_{\mathfrak{v}} = T_{\mathfrak{w}}$, respectively to $T_{\mathfrak{v}} \subset T_{\mathfrak{w}}$ and $T_{\mathfrak{w}}/T_{\mathfrak{v}} \cong \mathbb{Z}_\ell$.

Via the Galois correspondence and the functorial properties of the Hilbert decomposition theory for valuations, we attach to the total prime divisor graph $\mathscr{D}_K^{\rm tot}$ of $K|k$ a graph $\mathscr{G}_{\mathscr{D}_K^{\rm tot}}$ whose vertices and edges are in bijection with those of $\mathscr{D}_K^{\rm tot}$ as follows:

(a) The vertices of $\mathscr{G}_{\mathscr{D}_K^{\rm tot}}$ are $\Pi_{K\mathfrak{v}}$, viewed as distinct pro-ℓ groups (all $\mathfrak{v}$).
(b) If the edge from $K\mathfrak{v}$ to $K\mathfrak{w}$ exists, the corresponding edge from $\Pi_{K\mathfrak{v}}$ to $\Pi_{K\mathfrak{w}}$ is endowed with the pair of groups $T_{\mathfrak{w}/\mathfrak{v}} \subseteq Z_{\mathfrak{w}/\mathfrak{v}}$ viewed as subgroups of $\Pi_{K\mathfrak{v}}$; thus $\Pi_{K\mathfrak{w}} = Z_{\mathfrak{w}/\mathfrak{v}}/T_{\mathfrak{w}/\mathfrak{v}}$.

The graph $\mathscr{G}_{\mathscr{D}_K^{\rm tot}}$ will be called the *total decomposition graph* of $K|k$, or of Π_K. If $\mathscr{D}_K \subseteq \mathscr{D}_K^{\rm tot}$ is a geometric graph of prime divisors of $K|k$, the corresponding subgraph $\mathscr{G}_{\mathscr{D}_K} \subseteq \mathscr{G}_{\mathscr{D}_K^{\rm tot}}$ will be called a *geometric decomposition graph* for $K|k$, or for Π_K.

Next recall that the isomorphy type of (the maximal abelian pro-ℓ quotient of) the fundamental group $\Pi_1(X) := \pi_1^{{\rm ab},\ell}(X)$ of complete regular models $X \to k$, if such models exist, depends on $K|k$ only, and not on $X \to k$. Moreover, one can recover $\Pi_1(X)$ as being $\Pi_1(X) = \Pi_K/T_K$, where T_K is the subgroup of G_K generated by all the inertia groups T_v with v a prime divisor of $K|k$. This justifies calling the group $\Pi_{1,K} := \Pi_K/T_K$ the *birational fundamental group for* $K|k$. As discussed at Fact 57, there always exist quasiprojective normal models $X \to k$ for $K|k$ such that $T_K = T_{D_X}$, where T_{D_X} is the closed subgroup of Π_K generated by all T_v with $v \in D_X$. We will say that a model $X \to k$ of $K|k$ and/or that D_X is *complete regular-like* if $T_K = T_{D_X}$ and the rational rank $\mathrm{rr}(\mathfrak{Cl}(X))$ of the divisor class group $\mathfrak{Cl}(X)$ is positive, and for every normal quasiprojective model $\tilde{X}$ with $D_X \subseteq D_{\tilde{X}}$ one has that $\mathrm{rr}\big(\mathfrak{Cl}(\tilde{X})\big) = \mathrm{rr}(\mathfrak{Cl}(X)) + |D_{\tilde{X}} \backslash D_X|$. Note that a complete regular like curve is a complete normal curve and viceversa. We say that a geometric decomposition graph $\mathcal{G}_{\mathcal{D}_K}$ is *complete regular-like* if, for all vertices $\mathfrak{v}$ of $\mathcal{D}_K$ with $\mathrm{td}(K\mathfrak{v}|k) > 0$, one has that the set $D_{\mathfrak{v}}$ of 1-edges of $\mathcal{G}_{\mathcal{D}_{K\mathfrak{v}|k}}$ is complete regular-like.

As shown in Proposition 22, there exists a group-theoretical recipe by which one can recover the geometric decomposition graphs (and the property of being *complete regular-like*) from the total decomposition graph $\mathcal{G}_{\mathcal{D}_K^{\mathrm{tot}}}$. Further, by Proposition 39, that recipe is invariant under isomorphisms $\Phi : \mathcal{G}_{\mathcal{D}_K^{\mathrm{tot}}} \to \mathcal{G}_{\mathcal{D}_L^{\mathrm{tot}}}$, i.e., every such isomorphism gives rise by restriction to isomorphisms of the (complete regular-like) decomposition graphs for $K|k$ onto the (complete regular-like) ones for $L|l$.

The functorial properties of the graphs of prime divisors translate to the following functorial properties of the decomposition graphs:

(1) *Embeddings.* Let $\imath : L|l \hookrightarrow K|k$ be an embedding of function fields which maps l onto k. Then the canonical projection homomorphism $\Phi_\imath : \Pi_K \to \Pi_L$ is an open homomorphism, and for every generalized prime divisor $\mathfrak{v}$ of $K|k$ and its restriction $\mathfrak{v}_L$ to L, one has that $\Phi_\imath(Z_{\mathfrak{v}}) \subseteq Z_{\mathfrak{v}_L}$ is an open subgroup, and $\Phi_\imath(T_{\mathfrak{v}}) \subseteq T_{\mathfrak{v}_L}$ satisfies $\Phi_\imath(T_{\mathfrak{v}}) = 1$ iff $\mathfrak{v}_L$ is the trivial valuation. Therefore, $\Phi_\imath$ gives rise to a *morphism* of total decomposition graphs, which we denote by the same symbol

$$\Phi_\imath : \mathcal{G}_{\mathcal{D}_K^{\mathrm{tot}}} \to \mathcal{G}_{\mathcal{D}_L^{\mathrm{tot}}}.$$

In turn, for given geometric decomposition graphs $\mathcal{D}_K$ and $\mathcal{D}_L$, for which $\imath$ gives rise to a morphism of geometric decomposition graphs $\mathcal{D}_K \to \mathcal{D}_L$, the above $\Phi_\imath$ morphism of total decomposition graphs gives rise to a morphisms of geometric decomposition graphs $\Phi_\imath : \mathcal{G}_{\mathcal{D}_K} \to \mathcal{G}_{\mathcal{D}_L}$, as defined later in Sections 4 and 5.

(2) *Restrictions.* Given a generalized prime divisor $\mathfrak{v}$ of $K|k$, let $\mathrm{pr}_{\mathfrak{v}} : Z_{\mathfrak{v}} \to \Pi_{K\mathfrak{v}}$ be the canonical projection. Then for eveary $\mathfrak{w} \geq \mathfrak{v}$ we have that $T_{\mathfrak{w}} \subseteq Z_{\mathfrak{w}}$ are mapped onto $T_{\mathfrak{w}/\mathfrak{v}} \subseteq Z_{\mathfrak{w}/\mathfrak{v}}$. Therefore, the total decomposition graph of $K\mathfrak{v}|k$ can be recovered from that of $K|k$ in a canonical way via $\mathrm{pr}_{\mathfrak{v}} : Z_{\mathfrak{v}} \to \Pi_{K\mathfrak{v}}$.

• **Rational quotients** (see Section 5 for more details). Let $K|k$ be a function field as above satisfying $\mathrm{td}(K|k) > 1$. For every non-constant function $t \in K$, let κ_t be the relative algebraic closure of $k(t)$ in K. Since $\mathrm{td}(\kappa_t|k) = 1$, it follows that κ_t has a unique complete normal model $X_t \to k$, which is a projective smooth curve. Therefore, the set of prime divisors of $\kappa_t|k$ is actually in bijection with the (local rings at the) closed points of X_t, thus with the set of Weil prime divisors of X_t. Therefore, the total prime divisor graph $\mathcal{D}_{\kappa_t}^{\mathrm{tot}}$ for $\kappa_t|k$ is actually the *unique maximal* geometric prime divisor graph for $\kappa_t|k$. We denote $\mathcal{D}_{\kappa_t}^{\mathrm{tot}}$ simply by $\mathcal{D}_{\kappa_t}$.

Let $\imath_t : \kappa_t \to K$ be the canonical embedding, and $\Phi_{\kappa_t} : \Pi_K \to \Pi_{\kappa_t}$ the (surjective) canonical projection. Then by the functoriality of embeddings, Φ_{κ_t} gives rise canonically to a morphism $\Phi_{\kappa_t} : \mathcal{G}_{\mathcal{D}_K^{\mathrm{tot}}} \to \mathcal{G}_{\kappa_t}$. Moreover, if $\mathcal{G}_{\mathcal{D}_K}$ is a geometric decomposition graph for $K|k$, then Φ_{κ_t} restricts to a morphism of geometric decomposition graphs $\Phi_{\kappa_t} : \mathcal{G}_{\mathcal{D}_K} \to \mathcal{G}_{\kappa_t}$.

In the above context, if $\kappa_t = k(t)$, we say that Φ_{κ_t} is a *rational quotient* of $\mathcal{G}_{\mathcal{D}_K^{\rm tot}}$ as well as of every geometric decomposition graph $\mathcal{G}_{\mathcal{D}_K}$ for $K|k$. We call such $t \in K$ "general elements" of K, and usually denote general elements of K by x, in order to distinguish them from the "usual" non-constant $t \in K$. A "birational" Bertini-type argument shows that there are "many" general elements in K; see Lang [18], Ch. VIII, and/or Roquette [30], § 4, respectively Fact 43 in Section 5: For any given algebraically independent functions $t, t' \in K$, not both inseparable, $t_{a',a} := t/(a't' + a)$ is a general element of K for almost all $a', a \in k$. A set of general elements $\Sigma \subset K$ is a *Bertini set* if Σ contains almost all elements $t_{a',a}$ for all t, t' as above. We denote by $\mathfrak{A}_K = \{\Phi_{\kappa_x}\}_{\kappa_x}$ the set of all the rational quotients of $K|k$, and consider subsets $\mathfrak{A} \subset \mathfrak{A}_K$ containing all the $\Phi_{\kappa_x} \in \mathfrak{A}$, $x \in \Sigma$, with Σ some Bertini set of general elements, and call them, for short, *Bertini-type sets* of rational quotients.

The relation between rational projections and morphisms of geometric decomposition graphs is as follows: Let $\imath : L|l \hookrightarrow K|k$ be an embedding of function fields with $\imath(l) = k$, such that $K|\imath(L)$ a separable field extension, and $\mathrm{td}(L|l) > 1$. Then there exists a Bertini-type set $\mathfrak{B} = \{\Phi_{\kappa_y}\}_{\kappa_y}$ for $L|l$ such that $\kappa_x := \imath(\kappa_y)$ is relatively algebraically closed in K for all κ_y. Hence for all $\Phi_{\kappa_y} \in \mathfrak{B}$ and the corresponding $\Phi_{\kappa_x} \in \mathfrak{A}_K$, $\kappa_x := \imath(\kappa_y)$, we get that the isomorphism $\Phi_{\kappa_x\kappa_y} : \mathcal{G}_{\kappa_x} \to \mathcal{G}_{\kappa_y}$ defined by $\imath_{\kappa_x\kappa_y} := \imath|_{\kappa_y}$ satisfies the condition

$$\Phi_{\kappa_y} \circ \Phi_\imath = \Phi_{\kappa_x\kappa_y} \circ \Phi_{\kappa_x}.$$

Because of this property, we will say that $\Phi_\imath$ is *compatible with rational quotients*.

- **Abstract decomposition graphs**

It is one of our main tasks in the present manuscript to define and study *abstract decomposition graphs*, which resemble the geometric decomposition graphs $\mathcal{G}_{\mathcal{D}_K}$ (this will be done in Section 2) and to define *proper morphisms* of such abstract decomposition graphs, in particular their *rational quotients* (which will be done in Section 4). The abstract decomposition graphs, which endowed with families of rational quotients resemble the complete regular-like geometric decomposition graphs as introduced above, will be called *complete regular-like abstract decomposition graphs.*

The main result of this manuscript is the following; see Theorem 45 for a more general assertion, and Definition 21, Fact/Definition 43 (2), Definition 33 (and Definitions 12 and 9), and Definition/Remark 34 for the definitions of all the terms:

Main Theorem *Let $K|k$ and $L|l$ be function fields with* $\mathrm{td}(K|k) > 1$. *Let $\mathcal{G}_{\mathcal{D}_K^{\rm tot}}$ and $\mathcal{H}_{\mathcal{D}_L^{\rm tot}}$ be their total decomposition graphs, which we endow with Bertini-type sets of rational quotients $\mathfrak{A}$, respectively $\mathfrak{B}$. Then the following hold:*

(1) There exists a group-theoretical recipe which recovers $K|k$ from $\mathcal{G}_{\mathcal{D}_K^{\rm tot}}$ endowed with $\mathfrak{A}$. Moreover, this recipe is invariant under isomorphisms in the following

sense: Up to multiplication by ℓ-adic units and composition with automorphisms $\Phi_\imath$ of $\mathcal{G}_{\mathcal{D}_K^{\rm tot}}$ defined by automorphisms $\imath : K^{\rm i}|k \to K^{\rm i}|k$, there exists at most one isomorphism $\Phi : \mathcal{G}_{\mathcal{D}_K^{\rm tot}} \to \mathcal{H}_{\mathcal{D}_L^{\rm tot}}$ of abstract decomposition graphs *which is compatible with the sets of rational quotients $\mathfrak{A}$ and $\mathfrak{B}$.*

(2) *The following more precise assertion holds: Suppose that* $\mathrm{td}(L|l) > 1$. *Let $\mathcal{G}_{\mathcal{D}_K}$ and $\mathcal{H}_{\mathcal{D}_L}$ be geometric complete regular-like decomposition graphs for $K|k$, which endowed with $\mathfrak{A}$, respectively $\mathfrak{B}$, are viewed as* complete regular-like abstract decomposition graphs. *Then for every morphism*

$$\Phi : \mathcal{G}_{\mathcal{D}_K} \to \mathcal{H}_{\mathcal{D}_L}$$

which is compatible with the sets of rational quotients $\mathfrak{A}$ and $\mathfrak{B}$, there exist an ℓ-adic unit $\varepsilon \in \mathbb{Z}_\ell^\times$ and an embedding of field extensions

$$\imath : L^{\rm i}|l \to K^{\rm i}|k$$

such that $\Phi = \varepsilon \cdot \Phi_\imath$, where $\Phi_\imath : \mathcal{G}_{\mathcal{D}_K} \to \mathcal{H}_{\mathcal{D}_L}$ is the canonical morphism defined by $\imath$ as above.

Further, $\imath(l) = k$, and $\imath$ is unique up to Frobenius twists, and ε is unique up to multiplication by powers of p, where $p =$ $\mathrm{char}(k)$.

We notice that the Main Theorem above (together with Propositions 22 and 39) reduces the problem of functorially recovering $K|k$ from $\Pi_K^{\rm c}$, thus completing the proof of the above Theorem I, to recovering the *total decomposition graph* $\mathcal{G}_{\mathcal{D}_K^{\rm tot}}$ of $K|k$ and its *rational quotients.* In the case that k is an algebraic closure of a finite field, both these problems were solved in Pop [27], but working with the full pro-ℓ Galois group $G_K(\ell)$ instead of $\Pi_K^{\rm c}$. Nevertheless, the methods of Pop [27] to recover $\mathcal{G}_{\mathcal{D}_K^{\rm tot}}$ and its rational quotients used only the set of all divisorial groups $T_v \subset Z_v$ inside Π_K. Using the local theory developed in Pop [28] instead of the local theory of Pop [27], a complete proof of Theorem I above is given in Pop [30].

Historical note

The idea to recover $K|k$ from $\Pi_K^{\rm c}$ originates from Bogomolov [2], and a first attempt to do so can be found in his fundamental paper [2]. Although that paper is too sketchy to make clear what the author precisely proposes, a thorough inspection shows that it provides a fundamental tool for recovering inertia elements of valuations v of K (which nevertheless may be non-trivial on k). This is Bogomolov's theory of *commuting liftable pairs*; see Bogomolov–Tschinkel [3] for detailed proofs. On the other hand, it is not at all clear how and whether one could develop a "global theory" along the lines (vaguely) suggested in [2], and there was virtually no progress on the problem for about a decade.

A sketch of a viable global theory —at least in the case that *k is an algebraic closure of a finite field*— was proposed in the notes of my MSRI talk in the fall of

1999; see Pop [25]. In the second part of Pop [26], the technical details concerning the global theory hinted at in Pop [25] were worked out. Actually, this paper is an elaboration of parts of Pop [26], and the main theorem here, more precisely Theorem 45, is the *Hom-form* of the *Isom-form* of Theorem 5.11 of [26]. However, I should mention that in [26] the mixed "arithmetic + geometric situation" was considered as well as non-abelian Galois groups, which is of interest in the case that k is not algebraically closed.

In the case that k *is an algebraic closure of a finite field,* let me finally mention:

- In the paper Pop [27], a recipe to functorially recover $K|k$ from $G_K(\ell)$, in particular a proof of (a slightly stronger form of) the above target result was given. First, the assertion one proves using $G_K(\ell)$ instead of Π_K^c is stronger, namely, if $\Phi : \Pi_K \to \Pi_L$ is the abelianization of an isomorphism $\Phi(\ell) : G_K(\ell) \to G_L(\ell)$, then there exists an isomorphism $\imath : L^{\rm i}|l \to K^{\rm i}|k$ (unique up to Frobenius twists) which defines Φ; thus one does not need to "adjust" Φ by multiplying by an ℓ-adic unit $\varepsilon \in \mathbb{Z}_\ell^\times$. I should also observe that the full $G_K(\ell)$ was used in Pop [27] essentially only in order to recover the divisorial subgroups of Π_K via the canonical projection $G_K(\ell) \to \Pi_K$, whereas all the other steps of the local and global theory are virtually identical with the ones in the case of Π_K^c. (The recipe to recover the divisorial subgroups of Π_K via $\Pi_K^c \to \Pi_K$ is given in Pop [28] and uses Bogomolov's theory of commuting liftable pairs as a "black box." That recipe is used in Pop [30].)
- Bogomolov–Tschinkel [4], [5], consider the case $K = k(X)$, where X is a projective smooth surface over k. In the initial variant of their manuscript [4], they considered only the case that $\pi_1(X)$ is finite, and proved that if Π_K^c and Π_L^c are isomorphic, then $K|k$ and $L|l$ are isomorphic up to pure inseparable closures, provided k and l are algebraic closures of finite fields with chat $\neq 2$ (which is less precise than what the above target result gives in this case). Nevertheless, in the published version [5] of their earlier manuscript [4], they announce their main result for surfaces in a form almost identical with the target result above and use a strategy of proof which is in many ways very similar to that announced in Pop [25], and used in Pop [27].

Acknowledgements First I would like to thank P. Deligne for several useful discussions we had during my visits to IAS Princeton. Actually, the theory of abstract decomposition graphs presented here is inspired by a suggestion of his (letter from September 1995) for an axiomatic approach to the birational anabelian conjecture in the arithmetical case. I also want to thank J.-L. Colliot-Thélène for several discussions we had at MSRI in the fall of 1999, as well as M. Saidi, T. Szamuely, and J. Stix for "discussion sessions" at Bonn and Penn. Finally, I would like to thank a few others for their interest in this work, including D. Harbater, J. Koenigsmann, P. Lochak, J. Mínáč, H. Nakamura, and A. Tamagawa.

It is a great honor for me to to contribute to this volume in memory of Serge Lang. The present manuscript is an expanded version of a manuscript initially submitted at the beginning of 2007.

The author was supported by NSF grants DMS-0401056 and DMS-0801144.

2 Pro-ℓ abstract decomposition graphs

In this section we develop an abelian pro-ℓ prime divisor decomposition theory for "abstract function fields" which is similar in some sense to the abstract class field theory. Throughout ℓ is a fixed prime number, and $\delta \geq 0$ is a non-negative integer.

2.1 Axioms and definitions

Definition 1. A *level-δ* (pro-ℓ) *abstract decomposition graph* is a connected half-oriented graph $\mathcal{G}$ whose vertices are endowed with pro-ℓ abelian groups G_i and whose edges v_i are endowed with pairs of pro-ℓ abelian groups $T_{v_i} \subseteq Z_{v_i}$ satisfying the following:
Axiom (I): The vertices of $\mathcal{G}$ are pro-ℓ abelian free groups G_i, and $\mathcal{G}$ has an origin, which we denote by $G_0 = G$.
Axiom (II): The edges v_i and the corresponding $T_{v_i} \subseteq Z_{v_i}$ satisfy the following:

(i) For every vertex G_i there exists a unique *non-oriented edge* v_{i0} from G_i to itself, and the corresponding pair of pro-ℓ groups is $\{1\} =: T_{v_{i0}} \subseteq Z_{v_{i0}} := G_i$. For all other vertices $G_{i^*} \neq G_i$ there exists at most one edge v_i from G_{i^*} to G_i. If v_i exists, we say that v_i is the *oriented edge* from G_{i^*} to G_i, and v_i is endowed with a pair $T_{v_i} \subseteq Z_{v_i}$ of subgroups of G_{i^*} such that $T_{v_i} \cong \mathbb{Z}_\ell$ and $G_i = Z_{v_i}/T_{v_i}$.
The edges of $\mathcal{G}$ are also called *valuations* of $\mathcal{G}$; in particular, the edges originating from G_i are called *valuations* of G_i. The non-oriented edge v_{i0} from G_i to itself is called the *trivial valuation* of G_i, whereas the oriented edges v_i originating from G_{i^*} are called *non-trivial valuations* of G_{i^*}.
The groups $T_{v_i} \subseteq Z_{v_i}$ are called the *inertia*, respectively *decomposition*, groups of v_i; and $G_i := Z_{v_i}/T_{v_i}$ is called the *residue group* of v_i.

(ii) For distinct non-trivial edges $v_i \neq v_{i'}$ originating from G_{i^*}, one has $Z_{v_i} \cap Z_{v_{i'}} = 1$, hence $T_{v_i} \cap T_{v_{i'}} = 1$ holds as well.
For every cofinite subset $\mathfrak{U}_i$ of the set of non-trivial edges v_i originating from G_{i^*}, let $T_{\mathfrak{U}_i}$ be the closed subgroup of G_{i^*} generated by all the T_{v_i}, $v_i \in \mathfrak{U}_i$. A system $(\mathfrak{U}_{i,\alpha})_\alpha$ of such cofinite subsets is called *cofinal*, if every finite set of valuations v_i as above is contained in the complement of $\mathfrak{U}_{i,\alpha}$ for some α.

(iii) There exist cofinal systems $(\mathfrak{U}_{i,\alpha})_\alpha$ such that $T_{v_i} \cap T_{\mathfrak{U}_{i,\alpha}} = 1$ for all α and all $v_i \notin \mathfrak{U}_{i,\alpha}$.

Axiom (III): The non-oriented edges v_{i0} are the only cycles of the graph $\mathcal{G}$, and all maximal branches of non-trivial edges of $\mathcal{G}$ have length equal to δ.

Definition/Remark 2. Let $\mathcal{G}$ be an abstract decomposition graph of level-$\delta_{\mathcal{G}}$ on a pro-ℓ group $G = G_0$. We will say that $\mathcal{G}$ is a level-$\delta_{\mathcal{G}}$ *abstract decomposition* *graph* on G. A valuation of $G = G_0$ will be called a 1-*edge* of $\mathcal{G}$. If no confusion is possible, we will denote the 1-edges of $\mathcal{G}$ simply by v; thus the corresponding pro-ℓ groups involved are denoted by $T_v \subseteq Z_v$ and $G_v := Z_v/T_v$.

(1) Consider any δ such that $0 \leq \delta \leq \delta_{\mathcal{G}}$. By induction on δ it is easy to see that $\mathcal{G}$ has a unique maximal connected abstract decomposition subgraph containing the origin G of $\mathcal{G}$ and having all branches of oriented edges of length δ.

(2) Let $\mathfrak{v} = (v_r, \ldots, v_1)$ be a path of length $\delta_{\mathfrak{v}} := r$ of non-trivial valuations originating at $G = G_0$. This means by definition that v_1 is a non-trivial valuation of G_0, and if $r > 1$, then for all $i < r$ one has inductively that G_i is the residue group of v_i, and v_{i+1} is a non-trivial valuation of G_i. In particular, G_r is the residue group of v_r. Then there exists a unique maximal connected subgraph $\mathcal{G}_{\mathfrak{v}}$ of $\mathcal{G}$ having $G_{\mathfrak{v}} := G_r$ as origin. Clearly, $\mathcal{G}_{\mathfrak{v}}$ is in a natural way an abstract decomposition graph of level $\delta_{\mathcal{G}} - \delta_{\mathfrak{v}}$ on $G_{\mathfrak{v}}$.

We say that $\mathcal{G}_{\mathfrak{v}}$ is an r-*residual* abstract decomposition graph of $\mathcal{G}$. In particular, the unique 0-residual abstract decomposition graph of $\mathcal{G}$ is $\mathcal{G}$ itself.

(3) For every path $\mathfrak{v} = (v_r, \ldots, v_1)$ of length $\delta_{\mathfrak{v}} = r$ as above, we will say that $G_{\mathfrak{v}}$ is an r-*residual group* of $\mathcal{G}$, precisely that $G_{\mathfrak{v}}$ is the $\mathfrak{v}$-*residual group* of $\mathcal{G}$. One can further elaborate as follows: For $r > 1$ we set $\mathfrak{w} = (v_{r-1}, \ldots, v_1)$, and suppose that the *inertia/decomposition groups* $T_{\mathfrak{w}} \subseteq Z_{\mathfrak{w}} \subseteq G_0$ of $\mathfrak{w}$ have been defined inductively such that the *residue group* $G_{\mathfrak{w}} := Z_{\mathfrak{w}}/T_{\mathfrak{w}}$ of $\mathfrak{w}$ is $G_{\mathfrak{w}} = G_{v_{r-1}}$. We then define the inertia/decomposition groups $T_{\mathfrak{v}} \subseteq Z_{\mathfrak{v}}$ of $\mathfrak{v}$ in G_0 as being the preimages of $T_{v_r} \subseteq Z_{v_r} \subseteq G_{v_{r-1}}$ via $Z_{\mathfrak{w}} \to Z_{\mathfrak{w}}/T_{\mathfrak{w}} = G_{v_{r-1}}$. Note that by definition we have $Z_{\mathfrak{v}}/T_{\mathfrak{v}} =: G_{\mathfrak{v}}$ and $T_{\mathfrak{v}} \cong \mathbb{Z}_\ell^{\delta_{\mathfrak{v}}}$.

We call $\mathfrak{v} = (v_r, \ldots, v_1)$ a *generalized valuation* of $G = G_0$, or a *multi-index* of length $\delta_{\mathfrak{v}} := r$ of $\mathcal{G}$. And we will say that $\delta_{\mathfrak{v}}$ is the rank of $\mathfrak{v}$ or that $\mathfrak{v}$ is a generalized r-valuation if $r = \delta_{\mathfrak{v}}$.

Given generalized valuations $\mathfrak{v} = (v_r, \ldots, v_1)$, $\mathfrak{w} = (w_s, \ldots, w_1)$, we will say that $\mathfrak{w} \leq \mathfrak{v}$ if $s \leq r$, and $v_i = w_i$ for all $i \leq s$. From the definitions one gets that if $\mathfrak{w} \leq \mathfrak{v}$, then $Z_{\mathfrak{v}} \subseteq Z_{\mathfrak{w}}$ and $T_{\mathfrak{w}} \subseteq T_{\mathfrak{v}}$. On the other hand, by Axiom II (ii), it immediately follows that the converse of (any of) these assertions is also true. We will say that $\mathfrak{v}$ and $\mathfrak{w}$ are *dependent* if there exists some $q > 0$ such that $v_i = w_i$ for $i \leq q$. For dependent generalized valuations $\mathfrak{v}$ and $\mathfrak{w}$ as above, the following are equivalent:

(a) q is maximal such that $v_i = w_i$ for $i \leq q$.
(b) $T_{\mathfrak{v}} \cap T_{\mathfrak{w}} \cong \mathbb{Z}_\ell^q$.
(c) q is maximal such that $Z_{\mathfrak{v}}, Z_{\mathfrak{w}}$ are both contained in the decomposition group of some generalized q-valuation of $G = G_0$.

(4) In order to have a uniform notation, we take $\mathfrak{v} = v_0$ to be the *trivial multi-index*, or the *trivial path*, of $\mathcal{G}$ as the unique one having length equal to 0. We further set $Z_{v_0} := G_0$ and $T_{v_0} = \{1\}$. In particular, one has $G_{v_0} = Z_{v_0}/T_{v_0} = G_0$, which is compatible with the other notations/conventions. Further, $v_0 \leq \mathfrak{v}$ for all multi-indices $\mathfrak{v}$.

Definition/Remark 3. Let $\mathcal{G}$ be a level-$\delta_{\mathcal{G}}$ abstract decomposition graph on the abelian pro-ℓ group $G = G_0$. In notation as above, we consider the following:

(1) Define $\widehat{\Lambda}_{\mathcal{G}} := \operatorname{Hom}(G, \mathbb{Z}_\ell)$. Since G is a pro-ℓ free abelian group, $\widehat{\Lambda}_{\mathcal{G}}$ is a free ℓ-adically complete $\mathbb{Z}_\ell$-module (in ℓ-adic duality with G).

From now on suppose that $\delta_{\mathcal{G}} > 0$. Recall that $T_v \subset Z_v$ and $G_v = Z_v/T_v$ denote respectively the inertia, the decomposition, and the residue groups at the 1-edges v of $\mathcal{G}$, i.e., at the valuations v of G.

(2) Denote by $T \subseteq G$ the closed subgroup generated by all the inertia groups T_v (all v as above). We set $\Pi_{1,\mathcal{G}} := G/T$ and call it the abstract fundamental group of $\mathcal{G}$. One has a canonical exact sequence

$$1 \to T \to G \to \Pi_{1,\mathcal{G}} \to 1.$$

Taking continuous $\mathbb{Z}_\ell$-Homs, we get an exact sequence of the form

$$0 \to \widehat{U}_{\mathcal{G}} := \operatorname{Hom}(\Pi_{1,\mathcal{G}}, \mathbb{Z}_\ell) \xrightarrow{\text{can}} \widehat{\Lambda}_{\mathcal{G}} := \operatorname{Hom}(G, \mathbb{Z}_\ell) \xrightarrow{\jmath^{\mathcal{G}}} \widehat{\Lambda}_T := \operatorname{Hom}(T, \mathbb{Z}_\ell).$$

We will call $\widehat{U}_{\mathcal{G}} := \operatorname{Hom}(\Pi_{1,\mathcal{G}}, \mathbb{Z}_\ell)$ the *unramified part* of $\widehat{\Lambda}_{\mathcal{G}}$. And if no confusion is possible, we will identify $\widehat{U}_{\mathcal{G}}$ with its image in $\widehat{\Lambda}_{\mathcal{G}}$.

(3) Next we have a closer look at the structure of $\widehat{\Lambda}_{\mathcal{G}}$. For every 1-edge v as above, the inclusions $T_v \hookrightarrow Z_v \hookrightarrow G$ give rise to restriction homomorphisms as follows:

$$\jmath^v : \widehat{\Lambda}_{\mathcal{G}} \xrightarrow{\operatorname{res}_{Z_v}} \widehat{\Lambda}_{Z_v} := \operatorname{Hom}(Z_v, \mathbb{Z}_\ell) \xrightarrow{\operatorname{res}_v} \widehat{\Lambda}_{T_v} := \operatorname{Hom}(T_v, \mathbb{Z}_\ell).$$

(a) We set $\widehat{U}^1_v = \ker(\operatorname{res}_{Z_v})$ and $\widehat{U}_v = \ker(\jmath^v)$ and call them the *principal v-units*, respectively the *v-units*, in $\widehat{\Lambda}_{\mathcal{G}}$. And observe that the unramified part of $\widehat{\Lambda}_{\mathcal{G}}$ is exactly $\widehat{U}_{\mathcal{G}} = \cap_v \ker(\jmath^v)$.

(b) The family $(\jmath^v)_v$ gives rise canonically to a continuous homomorphism $\widehat{\oplus}_v\, \jmath^v$ of ℓ-adically complete $\mathbb{Z}_\ell$-modules

$$\widehat{\oplus}_v\, \jmath^v : \widehat{\Lambda}_{\mathcal{G}} \to \widehat{\Lambda}_T \hookrightarrow \widehat{\oplus}_v\, \widehat{\Lambda}_{T_v}$$

Thus identifying $\widehat{\Lambda}_T$ with its image inside $\widehat{\oplus}_v\, \widehat{\Lambda}_{T_v}$, one has $\jmath^{\mathcal{G}} = \widehat{\oplus}_v\, \jmath^v$ on $\widehat{\Lambda}_{\mathcal{G}}$. We define $\widehat{\operatorname{Div}}_{\mathcal{G}} := \widehat{\oplus}_v\, \widehat{\Lambda}_{T_v}$ and call it the *ℓ-adic abstract divisor group* of $\mathcal{G}$.

(c) Finally, we set $\widehat{\mathfrak{Cl}}_{\mathcal{G}} = \operatorname{coker}(\jmath^{\mathcal{G}})$ and call it the ℓ-adic abstract divisor class group of $\mathcal{G}$. And observe that we have a canonical exact sequence

$$0 \to \widehat{U}_{\mathcal{G}} \hookrightarrow \widehat{\Lambda}_{\mathcal{G}} \xrightarrow{\jmath^{\mathcal{G}}} \widehat{\operatorname{Div}}_{\mathcal{G}} \xrightarrow{\text{can}} \widehat{\mathfrak{Cl}}_{\mathcal{G}} \to 0\,.$$

(4) Let $\widehat{\Lambda}_{\mathscr{G},\mathrm{fin}} := \{x \in \widehat{\Lambda}_{\mathscr{G}} \mid \jmath^{v}(x) = 0 \text{ for almost all } v\}$. We notice that by Axiom II (iii), the $\mathbb{Z}_\ell$-module $\widehat{\Lambda}_{\mathscr{G},\mathrm{fin}}$ is dense in $\widehat{\Lambda}_{\mathscr{G}}$. Indeed, let $(\mathfrak{U}_\alpha)_\alpha$ be a cofinal system of 1-edges v. Then setting $G_\alpha = G/T_{\mathfrak{U}_\alpha}$ and $T_\alpha = T/T_{\mathfrak{U}_\alpha}$, we have a canonical exact sequence

$$1 \to T_\alpha \to G_\alpha \to \Pi_{1,\mathscr{G}} \to 1\,,$$

and T_α is generated by the images $T_{v,\alpha}$ of T_v (all $v \notin \mathfrak{U}_\alpha$) in G_α. Clearly, the image of the inflation homomorphism $\inf_\alpha : \mathrm{Hom}\big(G_\alpha, \mathbb{Z}_\ell\big) \to \mathrm{Hom}\big(G, \mathbb{Z}_\ell\big)$ is exactly

$$\Delta_\alpha := \{x \in \widehat{\Lambda}_{\mathscr{G}} \mid \jmath^{v}(x) = 0 \text{ for all } v \in \mathfrak{U}_\alpha\} = \cap_{v \in \mathfrak{U}_\alpha} \ker(\jmath^{v}).$$

Taking inductive limits over the cofinal system $(\mathfrak{U}_\alpha)_\alpha$, the density assertion follows.

We observe that $\jmath^{\mathscr{G}}(\widehat{\Lambda}_{\mathscr{G},\mathrm{fin}}) \cong \widehat{\Lambda}_{\mathscr{G}}/\widehat{U}_{\mathscr{G}}$ is a $\mathbb{Z}_\ell$-submodule of the $\mathbb{Z}_\ell$-free module $\oplus_v \widehat{\Lambda}_{T_v} \cong \oplus_v \mathbb{Z}_\ell v$; hence $\jmath^{\mathscr{G}}(\widehat{\Lambda}_{\mathscr{G},\mathrm{fin}})$ is a free $\mathbb{Z}_\ell$-module too. Therefore, for every $\mathbb{Z}_\ell$-submodule $\Delta \subseteq \widehat{\Lambda}_{\mathscr{G}}$, its image $\jmath^{\mathscr{G}}(\Delta)$ under $\jmath^{\mathscr{G}}$ is a free $\mathbb{Z}_\ell$-module. The rank of $\jmath^{\mathscr{G}}(\Delta)$ will be called the *corank* of Δ.

We notice that a $\mathbb{Z}_\ell$-submodule $\Delta \subset \widehat{\Lambda}_{\mathscr{G}}$ has *finite corank* iff Δ is contained in $\ker(\jmath^{v})$ for almost all v. Clearly, the sum of two finite corank submodules of $\widehat{\Lambda}_{\mathscr{G}}$ is again of finite corank. Thus the set of such submodules is inductive, and one has

$$\widehat{\Lambda}_{\mathscr{G},\mathrm{fin}} = \cup_\Delta \text{ (all finite corank } \Delta) = \cup_\alpha \Delta_\alpha.$$

(5) We say that $\mathscr{G}$ is *complete curve-like* if the following holds: There exist generators τ_v of T_v such that $\prod_v \tau_v = 1$, and this is the only pro-relation satisfied by the system of elements $\mathfrak{T} = (\tau_v)_v$. We call such a system $\mathfrak{T} = (\tau_v)_v$ a *distinguished system of inertia generators.*

We notice the following: Let $\mathscr{G}$ be complete curve-like, and let $\mathfrak{T} = (\tau_v)_v$ and $\mathfrak{T}' = (\tau'_v)_v$ be distinguished systems of inertia generators. Then $\tau'_v = \tau_v^{\varepsilon_v}$ for some ℓ-adic units $\varepsilon_v \in \mathbb{Z}_\ell$, because both τ_v and τ'_v are generators of T_v. Hence we have $1 = \prod_v \tau'_v = \prod_v \tau_v^{\varepsilon_v}$. By the uniqueness of the relation $\prod_v \tau_v = 1$, it follows that $\varepsilon_v = \varepsilon$ for some fixed ℓ-adic unit $\varepsilon \in \mathbb{Z}_\ell$.

Next consider some δ with $0 < \delta \le \delta_{\mathscr{G}}$. We say that $\mathscr{G}$ is *level-δ complete curve-like* if all the $(\delta - 1)$-residual abstract decomposition graphs $\mathscr{G}_{\mathfrak{v}}$ are residually complete curve-like. In particular, "level 1 complete curve-like" is the same as "complete curve-like."

(6) For every 1-vertex v consider the exact sequence $1 \to T_v \to Z_v \to G_v \to 1$ given by Axiom II (i). Let $\inf_v : \mathrm{Hom}\big(G_v, \mathbb{Z}_\ell\big) \to \mathrm{Hom}\big(Z_v, \mathbb{Z}_\ell\big)$ be the resulting inflation homomorphism. Since $T_v = \ker(Z_v \to G_v)$, it follows that $\mathrm{res}_{Z_v}(\widehat{U}_v)$ is the image of the inflation map infl_v. Therefore there exists a canonical exact sequence

$$0 \to \widehat{U}_v^{1} \longrightarrow \widehat{U}_v \xrightarrow{\jmath_v} \mathrm{Hom}\big(G_v, \mathbb{Z}_\ell\big) = \widehat{\Lambda}_{\mathscr{G}_v} \to 0,$$

and we call $\jmath_v$ the *v-reduction homomorphism.*

(7) In particular, if $\delta_{\mathcal{G}} > 1$, then $\delta_{\mathcal{G}_v} = \delta_{\mathcal{G}} - 1 > 0$ for every 1-vertex v, and we have the corresponding exact sequence for the residual abstract decomposition graph $\mathcal{G}_v$

$$0 \to \widehat{U}_{\mathcal{G}_v} \hookrightarrow \widehat{\Lambda}_{\mathcal{G}_v} \xrightarrow{\jmath^{\mathcal{G}_v}} \widehat{\mathrm{Div}}_{\mathcal{G}_v}.$$

We will say that $\mathcal{G}$ is *ample* if $\delta_{\mathcal{G}} > 0$ and the following conditions are satisfied:

(i) $\jmath^{\Sigma} : \widehat{\Lambda}_{\mathcal{G}} \longrightarrow \oplus_{v\in\Sigma} \Lambda_{T_v}$ is surjective for every finite set Σ, where $\jmath^{\Sigma} := \oplus_{v\in\Sigma}\jmath^{v}$.

(ii) If $\delta_{\mathcal{G}} > 1$, then the following hold:

(a) $\jmath_v(\widehat{U}_{\mathcal{G}}) \subseteq \widehat{U}_{\mathcal{G}_v}$ and $\widehat{U}_{\mathcal{G}_v} + \jmath_v(\widehat{\Lambda}_{\mathcal{G}\mathrm{,fin}} \cap \widehat{U}_v) = \widehat{\Lambda}_{\mathcal{G}_v\mathrm{,fin}}$ for every v.

(b) For every finite-corank submodule $\Delta \subseteq \widehat{\Lambda}_{\mathcal{G}}$, there exists v such that $\Delta \subseteq \widehat{U}_v$, and Δ and $\jmath_v(\Delta)$ have equal coranks.

Notice that the condition (ii) above is empty in the case $\delta_{\mathcal{G}} = 1$. Thus if $\delta_{\mathcal{G}} = 1$, then condition (i) is necessary and sufficient for $\mathcal{G}$ to be ample.

Next consider $0 < \delta \leq \delta_{\mathcal{G}}$. We say that $\mathcal{G}$ is *ample up to level* δ if all the residual abstract decomposition graphs $\mathcal{G}_{\mathfrak{v}}$ for $\mathfrak{v}$ such that $0 \leq \delta_{\mathfrak{v}} < \delta$ are ample. In particular, "ample up to level 1" is the same as "ample."

2.2 Abstract $\mathbb{Z}_{(\ell)}$ divisor groups

Definition 4. (1) Let M be the ℓ-adic completion of a free $\mathbb{Z}$-module. A $\mathbb{Z}_{(\ell)}$-submodule $\mathcal{M}_{(\ell)} \subseteq M$ of M is called a $\mathbb{Z}_{(\ell)}$*-lattice* in M (for short, a lattice) if $\mathcal{M}_{(\ell)}$ is a free $\mathbb{Z}_{(\ell)}$-module, it is ℓ-adically dense in M, and it satisfies the following equivalent conditions:

(a) $M/\ell = \mathcal{M}_{(\ell)}/\ell$.

(b) $\mathcal{M}_{(\ell)}$ has a $\mathbb{Z}_{(\ell)}$-basis $\mathfrak{B}$ which is ℓ-adically independent in M.

(c) Every $\mathbb{Z}_{(\ell)}$-basis of $\mathcal{M}_{(\ell)}$ is ℓ-adically independent in M.

(2) Let $N \subset \mathcal{M}_{(\ell)} \subseteq M$ be $\mathbb{Z}_{(\ell)}$-submodules of M such that N and M/N are ℓ-adically complete and torsion-free. We call $\mathcal{M}_{(\ell)}$ an N*-lattice* in M, if $\mathcal{M}_{(\ell)}/N$ is a lattice in M/N.

(3) In the context above, a *true lattice* in M is a free abelian subgroup $\mathcal{M}$ of M such that $\mathcal{M}_{(\ell)} := \mathcal{M} \otimes \mathbb{Z}_{(\ell)}$ is a lattice in M in the above sense. And we will say that a $\mathbb{Z}$-submodule $\mathcal{M} \subseteq M$ is a *true* N*-lattice* in M if $N \subset \mathcal{M}$ and $\mathcal{M}/N$ is a true lattice in M/N.

(4) Let M be an arbitrary $\mathbb{Z}_\ell$-module. We say that subsets M_1, M_2 of M are ℓ*-adically equivalent* if there exists an ℓ-adic unit $\varepsilon \in \mathbb{Z}_\ell$ such that $M_2 = \varepsilon \cdot M_1$ inside M. Further, given systems $S_1 = (x_i)_i$ and $S_2 = (y_i)_i$ of elements of M, we will say that S_1 and S_2 are ℓ*-adically equivalent* if there exists an ℓ-adic unit $\varepsilon \in \mathbb{Z}_\ell$ such that $x_i = \varepsilon\, y_i$ (all i).

(5) We define correspondingly the ℓ-adic N-equivalence of N-lattices, etc.

Construction 5. Let $\mathcal{G}$ be an abstract decomposition graph on G which is *level-δ complete curve-like* and *ample up to level δ* for some given $\delta > 0$. Recall the last exact sequence from point (4) from Definition/Remark 3:

$$0 \to \widehat{U}_{\mathcal{G}} \hookrightarrow \widehat{\Lambda}_{\mathcal{G}} \xrightarrow{\jmath^{\mathcal{G}}} \widehat{\mathrm{Div}}_{\mathcal{G}} \xrightarrow{\mathrm{can}} \widehat{\mathfrak{Cl}}_{\mathcal{G}} \to 0.$$

The aim of this subsection is to describe the ℓ-adic equivalence class of a lattice $\mathrm{Div}_{\mathcal{G}}$ in $\widehat{\mathrm{Div}}_{\mathcal{G}}$, in case it exists, which will be called an *abstract divisor group* of $\mathcal{G}$. In case the lattice $\mathrm{Div}_{\mathcal{G}} \subset \widehat{\mathrm{Div}}_{\mathcal{G}}$ exists, it satisfies

$$\mathrm{Div}_{\mathcal{G}} \otimes \mathbb{Z}_\ell = \oplus_v \Lambda_{T_v}.$$

Further, the existence (of the equivalence class) of the lattice $\mathrm{Div}_{\mathcal{G}}$ will turn out to be equivalent to the existence (of the equivalence class) of a $\widehat{U}_{\mathcal{G}}$-lattice $\Lambda_{\mathcal{G}}$ in $\widehat{\Lambda}_{\mathcal{G}}$, which will turn out to be the preimage of $\mathrm{Div}_{\mathcal{G}}$ in $\widehat{\Lambda}_{\mathcal{G}}$. In particular, if $\Lambda_{\mathcal{G}}$ exists, it satisfies

$$\Lambda_{\mathcal{G}} \otimes \mathbb{Z}_\ell = \widehat{\Lambda}_{\mathcal{G}\mathrm{,fin}}.$$

The case $\delta = 1$, i.e., $\mathcal{G}$ is complete curve-like and ample.
In the notation from Definition/Remark 3 (5) above, let $\mathfrak{T} = (\tau_v)_v$ be a distinguished system of inertia generators. Further, let $\mathcal{F}_{\mathfrak{T}}$ be the abelian pro-ℓ free group on the system $\mathfrak{T}$ (written multiplicatively). Then one has a canonical exact sequence of pro-ℓ groups

$$1 \to \tau^{\mathbb{Z}_\ell} \to \mathcal{F}_{\mathfrak{T}} \to T \to 1,$$

where $\tau = \prod_v \tau_v$ in $\mathcal{F}_{\mathfrak{T}}$ is the pro-ℓ product of the generators τ_v (all v). Observing that $\mathrm{Hom}\big(\mathcal{F}_{\mathfrak{T}}, \mathbb{Z}_\ell\big) \cong \widehat{\mathrm{Div}}_{\mathcal{G}}$ in a canonical way, and taking ℓ-adically continuous Homs, we get an exact sequence

$$0 \to \widehat{\Lambda}_T = \mathrm{Hom}\big(T, \mathbb{Z}_\ell\big) \to \widehat{\mathrm{Div}}_{\mathcal{G}} = \mathrm{Hom}\big(\mathcal{F}_{\mathfrak{T}}, \mathbb{Z}_\ell\big) \to \mathbb{Z}_\ell = \mathrm{Hom}\big(\tau^{\mathbb{Z}_\ell}, \mathbb{Z}_\ell\big) \to 0,$$

where the last homomorphism maps each φ to its "trace": $\varphi \mapsto \big(\tau \mapsto \sum_v \varphi(\tau_v)\big)$. Thus $\widehat{\Lambda}_T$ consists of all the homomorphisms $\varphi \in \mathrm{Hom}\big(\mathcal{F}_{\mathfrak{T}}, \mathbb{Z}_\ell\big)$ with trivial trace.

Consider the system $\mathfrak{B} = (\varphi_v)_v$ of all the functionals $\varphi_v \in \mathrm{Hom}\big(\mathcal{F}_{\mathfrak{T}}, \mathbb{Z}_\ell\big) = \widehat{\mathrm{Div}}_{\mathcal{G}}$ defined by $\varphi_v(\tau_w) = 1$ if $v = w$, and $\varphi_v(\tau_w) = 0$ for all $v \neq w$. We denote by

$$\mathrm{Div}_{\mathfrak{T}} = \langle \mathfrak{B} \rangle_{(\ell)} \subset \widehat{\mathrm{Div}}_{\mathcal{G}}$$

the $\mathbb{Z}_{(\ell)}$-submodule of $\mathrm{Hom}\big(\mathcal{F}_{\mathfrak{T}}, \mathbb{Z}_\ell\big) = \widehat{\mathrm{Div}}_{\mathcal{G}}$ generated by $\mathfrak{B}$. Then $\mathrm{Div}_{\mathfrak{T}}$ is a lattice in $\widehat{\mathrm{Div}}_{\mathcal{G}}$, and $\mathfrak{B}$ is an *ℓ-adic basis* of $\widehat{\mathrm{Div}}_{\mathcal{G}}$. We next set

$$\mathrm{Div}^0_{\mathfrak{T}} := \{\textstyle\sum_v a_v \varphi_v \in \mathrm{Div}_{\mathfrak{T}} \mid \sum_v a_v = 0\} = \mathrm{Div}_{\mathfrak{T}} \cap \widehat{\Lambda}_T.$$

Clearly, $\mathrm{Div}^0_{\mathfrak{T}}$ is a lattice in $\widehat{\Lambda}_T$. And moreover, the system $(e_w = \varphi_w - \varphi_v)_{w \neq v}$ is an ℓ-adic $\mathbb{Z}_{(\ell)}$-basis of $\mathrm{Div}^0_{\mathfrak{T}}$ for every fixed v.

The dependence of $\mathrm{Div}_{\mathfrak{T}}$ on $\mathfrak{T} = (\tau_v)_v$ is as follows. Let $\mathfrak{T}' = (\tau'_v)_v = \mathfrak{T}^\varepsilon$ with $\varepsilon \in \mathbb{Z}_\ell^\times$ be another distinguished system of inertia generators. If $\mathfrak{B}' = (\varphi'_v)_v$ is the dual basis to $\mathfrak{T}'$, then $\varepsilon \cdot \mathfrak{B}' = \mathfrak{B}$. Thus $\mathfrak{B}$ and $\mathfrak{B}'$ are ℓ-adically equivalent, and we have $\mathrm{Div}_{\mathfrak{T}} = \varepsilon \cdot \mathrm{Div}_{\mathfrak{T}'}$ and $\mathrm{Div}^0_{\mathfrak{T}} = \varepsilon \cdot \mathrm{Div}^0_{\mathfrak{T}'}$.

Therefore, all the subgroups of $\widehat{\mathrm{Div}}_{\mathcal{G}}$ of the form $\mathrm{Div}_{\mathfrak{T}}$, respectively $\mathrm{Div}^0_{\mathfrak{T}}$, are ℓ-adically equivalent (for all distinguished $\mathfrak{T}$). Hence the ℓ-adic equivalence classes of $\mathrm{Div}_{\mathfrak{T}}$ and $\mathrm{Div}^0_{\mathfrak{T}}$ do not depend on $\mathfrak{T}$, but only on $\mathcal{G}$.

Fact 6. *In the above context, denote by $\Lambda_{\mathfrak{T}}$ the preimage of $\mathrm{Div}^0_{\mathfrak{T}}$, thus of $\mathrm{Div}_{\mathfrak{T}}$, in $\widehat{\Lambda}_{\mathcal{G}}$. Consider all the finite-corank submodules $\Delta \subset \Lambda_{\mathcal{G},\mathrm{fin}}$ with $\widehat{U}_{\mathcal{G}} \subset \Delta$. Then the following hold:*

(i) $\Lambda_{\mathfrak{T}}$ is a $\widehat{U}_{\mathcal{G}}$-lattice in $\widehat{\Lambda}_{\mathcal{G}}$, and $\Lambda_{\mathfrak{T}} \subset \widehat{\Lambda}_{\mathcal{G},\mathrm{fin}}$.
(ii) $\Delta \cap \Lambda_{\mathfrak{T}}$ is a $\widehat{U}_{\mathcal{G}}$-lattice in Δ (all Δ as above).

Moreover, $\jmath^v(\Lambda_{\mathfrak{T}}) = \mathbb{Z}_{(\ell)}\varphi_v$ (all v).

Proof. Clear. □

Definition 7. In the context of Fact 6 above, we define objects as follows:

(1) A lattice of the form $\mathrm{Div}_{\mathfrak{T}} \subset \widehat{\mathrm{Div}}_{\mathcal{G}}$ will be called an *abstract divisor group* of $\mathcal{G}$. We will further say that $\mathrm{Div}^0_{\mathfrak{T}}$ is the *abstract divisor group of degree* 0 in $\mathrm{Div}_{\mathfrak{T}}$.
(2) The $\widehat{U}_{\mathcal{G}}$-lattice $\Lambda_{\mathfrak{T}}$ is called a *divisorial $\widehat{U}_{\mathcal{G}}$-lattice* for $\mathcal{G}$ in $\widehat{\Lambda}_{\mathcal{G}}$. And we will say that $\Lambda_{\mathfrak{T}}$ and $\mathrm{Div}_{\mathfrak{T}}$ *correspond* to each other, and that $\mathfrak{T}$ defines them.
- Note that $\Lambda_{\mathcal{G}} \subset \widehat{\Lambda}_{\mathcal{G},\mathrm{fin}}$ and $\Lambda_{\mathcal{G}} \otimes \mathbb{Z}_\ell = \widehat{\Lambda}_{\mathcal{G},\mathrm{fin}}$. Indeed, if $x \in \Lambda_{\mathcal{G}}$, then $\jmath^v(x) = 0$ for almost all v, etc.

The case $\delta > 1$.
We begin by mimicking the construction from the case $\delta = 1$, and then conclude the construction by induction on δ. Thus let $\mathfrak{T} = (\tau_v)_v$ be any system of generators for the inertia groups T_v (all 1-edges v). Further let $\mathcal{F}_{\mathfrak{T}}$ be the abelian pro-ℓ free group on the system $\mathfrak{T}$ (written multiplicatively). Then T is a quotient $\mathcal{F}_{\mathfrak{T}} \to T \to 1$ in a canonical way. Observing that $\mathrm{Hom}(\mathcal{F}_{\mathfrak{T}}, \mathbb{Z}_\ell) \cong \widehat{\mathrm{Div}}_{\mathcal{G}}$ in a canonical way, by taking ℓ-adic Homs we get an exact sequence

$$0 \to \mathrm{Hom}(T, \mathbb{Z}_\ell) \to \mathrm{Hom}(\mathcal{F}_{\mathfrak{T}}, \mathbb{Z}_\ell) = \widehat{\mathrm{Div}}_{\mathcal{G}}.$$

Next let $\mathfrak{B} = (\varphi_v)_v$ be the system of all the functionals $\varphi_v \in \mathrm{Hom}(\mathcal{F}_{\mathfrak{T}}, \mathbb{Z}_\ell)$ defined by $\varphi_v(\tau_w) = 1$ if $v = w$, and $\varphi_v(\tau_w) = 0$ for all $v \neq w$. We denote by

$$\mathrm{Div}_{\mathfrak{T}} = \langle \mathfrak{B} \rangle_{(\ell)} \subset \mathrm{Hom}(\mathcal{F}_{\mathfrak{T}}, \mathbb{Z}_\ell)$$

the $\mathbb{Z}_{(\ell)}$-submodule generated by $\mathfrak{B}$. Then $\mathfrak{B}$ *is an ℓ-adic basis* of $\mathrm{Hom}(\mathcal{F}_{\mathfrak{T}}, \mathbb{Z}_\ell)$, i.e., $\mathrm{Div}_{\mathfrak{T}}$ is ℓ-adically dense in $\widehat{\mathrm{Div}}_{\mathcal{G}} = \mathrm{Hom}(\mathcal{F}_{\mathfrak{T}}, \mathbb{Z}_\ell)$, and there are no non-trivial

ℓ-adic relations between the elements of $\mathfrak{B}$. We will call $\mathfrak{B} = (\varphi_v)_v$ the "dual basis" to $\mathfrak{T}$, and remark that $\mathrm{Div}_{\mathfrak{T}}$ is a lattice in $\mathrm{Hom}(T, \mathbb{Z}_\ell)$.
Finally, let $\mathfrak{T}' = (\tau'_v)_v$ be another system of inertia generators, and suppose that $\mathfrak{T}' = \mathfrak{T}^\varepsilon$ for some $\varepsilon \in \mathbb{Z}_\ell$. If $\mathfrak{B}' = (\varphi'_v)_v$ is the dual basis to $\mathfrak{T}'$, then $\varepsilon\,\varphi'_v = \varphi_v$ inside $\mathrm{Hom}(T, \mathbb{Z}_\ell)$. Thus $\varepsilon \cdot \mathfrak{B}' = \mathfrak{B}$. In other words, $\mathfrak{B}$ and $\mathfrak{B}'$ are ℓ-adically equivalent, and we have $\mathrm{Div}_{\mathfrak{T}} = \varepsilon \cdot \mathrm{Div}_{\mathfrak{T}'}$.

Fact 8. *In the notations from above let a $\widehat{U}_{\mathcal{G}_v}$-lattice $\Lambda_{\mathcal{G}_v} \subset \widehat{\Lambda}_{\mathcal{G}_v}$ with $\widehat{U}_{\mathcal{G}_v} \subset \Lambda_{\mathcal{G}_v}$ be given for every valuation v of $\mathcal{G}$. Then the following hold:*

(1) Up to ℓ-adic equivalence, there exists at most one $\widehat{U}_{\mathcal{G}}$-lattice $\Lambda_{\mathcal{G}}$ in $\widehat{\Lambda}_{\mathcal{G}}$ such that first, $\widehat{U}_{\mathcal{G}} \subset \Lambda_{\mathcal{G}} \subset \widehat{\Lambda}_{\mathcal{G},\mathrm{fin}}$, and second, for every finite-corank submodule Δ of $\widehat{\Lambda}_{\mathcal{G},\mathrm{fin}}$ with $\widehat{U}_{\mathcal{G}} \subset \Delta$ and the corresponding $\Delta_v := \jmath_v(\Delta \cap \widehat{U}_v) + \widehat{U}_{\mathcal{G}_v} \subset \widehat{\Lambda}_{\mathcal{G}_v,\mathrm{fin}}$ the following hold:

(i) $\Lambda_\Delta := \Delta \cap \Lambda_{\mathcal{G}}$ is a $\widehat{U}_{\mathcal{G}}$-lattice in Δ.
(ii) $\jmath_v(\Lambda_\Delta \cap \widehat{U}_v) + \widehat{U}_{\mathcal{G}_v}$ is a $\widehat{U}_{\mathcal{G}_v}$-lattice in Δ_v, which is ℓ-adically $\widehat{U}_{\mathcal{G}_v}$-equivalent to $\Lambda_{\mathcal{G}_v} \cap \Delta_v$.

Moreover, if the $\widehat{U}_{\mathcal{G}}$-lattice $\Lambda_{\mathcal{G}}$ exists, then its ℓ-adic equivalence class depends only on the ℓ-adic equivalence classes of the $\widehat{U}_{\mathcal{G}_v}$-lattices $\Lambda_{\mathcal{G}_v}$ (all v).

(2) In the above context, suppose that $\mathcal{G}$ is ample, and that the $\widehat{U}_{\mathcal{G}}$-lattice $\Lambda_{\mathcal{G}}$ satisfying the conditions (i), (ii), *exists. Then $\widehat{U}_{\mathcal{G}_v} + \jmath_v(\Lambda_{\mathcal{G}} \cap \widehat{U}_v)$ is a $\widehat{U}_{\mathcal{G}_v}$-lattice, which moreover is ℓ-adically $\widehat{U}_{\mathcal{G}_v}$-equivalent to $\Lambda_{\mathcal{G}_v}$ (all v).*

Proof. To (1): Let $\Lambda_{\mathcal{G}}, \Lambda'_{\mathcal{G}}$ be $\widehat{U}_{\mathcal{G}}$-lattices in $\widehat{\Lambda}_{\mathcal{G}}$ satisfying the conditions from (1) above. Let $\Delta \in \widehat{\Lambda}_{\mathcal{G},\mathrm{fin}}$ have finite non-zero corank, and satisfy $\widehat{U}_{\mathcal{G}} \subset \Delta$. By the ampleness of $\mathcal{G}$, it follows that there exists v such that, first, $\Delta \subseteq \widehat{U}_v$, and second, Δ and $\Delta_v := \jmath_v(\Delta) + \widehat{U}_{\mathcal{G}_v}$ have equal coranks. Therefore, $\jmath_v$ defines an isomorphism of $\Delta/\widehat{U}_{\mathcal{G}}$ onto $\Delta_v/\widehat{U}_{\mathcal{G}_v}$, and one has

$$(*) \qquad \ker(\jmath_v) \cap \Delta \subseteq \widehat{U}_{\mathcal{G}}, \quad \jmath_v(\Delta) \cap \widehat{U}_{\mathcal{G}_v} \subseteq \jmath_v(\widehat{U}_{\mathcal{G}}).$$

For Δ as above, set $\Lambda'_\Delta = \Delta \cap \Lambda'_{\mathcal{G}}$. Then by hypothesis (i), it follows that Λ_Δ and Λ'_Δ are both $\widehat{U}_{\mathcal{G}}$-lattices in Δ. Further, by hypothesis (ii), both $\Lambda_{\Delta_v} := \widehat{U}_{\mathcal{G}_v} + \jmath_v(\Lambda_\Delta)$ and $\Lambda'_{\Delta_v} := \widehat{U}_{\mathcal{G}_v} + \jmath_v(\Lambda'_\Delta)$ are $\widehat{U}_{\mathcal{G}_v}$ lattices in Δ_v, which are both equivalent to the $\widehat{U}_{\mathcal{G}_v}$-lattice $\Lambda_{\mathcal{G}_v} \cap \Delta_v$. Therefore, there exists $\varepsilon \in \mathbb{Z}_\ell^\times$ such that $\Lambda'_{\Delta_v} = \varepsilon \cdot \Lambda_{\Delta_v}$.

Claim. $\Lambda'_\Delta = \varepsilon \cdot \Lambda_\Delta$.

Indeed, $\Lambda'_{\Delta_v} = \varepsilon \cdot \Lambda_{\Delta_v}$ implies that $\jmath_v(\Lambda'_\Delta) \subseteq \varepsilon \cdot \jmath_v(\Lambda_\Delta) + \widehat{U}_{\mathcal{G}_v}$. Hence for every $e' \in \Lambda'_\Delta$ there exist $e \in \Lambda_\Delta$ and $u_v \in \widehat{U}_{\mathcal{G}_v}$ such that $\jmath_v(e') = \varepsilon\,\jmath_v(e) + u_v$. Therefore we have $u_v = \jmath_v(e' - \varepsilon\,e) \in \jmath(\Delta)$, and hence $u_v \in \jmath_v(\Delta) \cap \widehat{U}_{\mathcal{G}_v}$. Hence by assertion $(*)$ above, there exists $u \in \widehat{U}_{\mathcal{G}}$ such that $\jmath_v(u) = u_v$; thus $\jmath_v(u) = \jmath_v(e' - \varepsilon\,e)$. But then we have $e' - (\varepsilon\,e + u) \in \ker(\jmath_v) \cap \Delta$, thus $e' - (\varepsilon\,e + u) \in \widehat{U}_{\mathcal{G}}$ by assertion $(*)$. We conclude that $e' \in \varepsilon\,e + \widehat{U}_{\mathcal{G}}$. Since $e' \in \Lambda'_\Delta$

was arbitrary, we have $\Lambda'_\Delta \subseteq \varepsilon \cdot \Lambda_\Delta + \widehat{U}_{\mathcal{G}}$. On the other hand, by hypothesis we have $\widehat{U}_{\mathcal{G}} \subset \Lambda_\Delta$ and $\widehat{U}_{\mathcal{G}} \subset \Lambda'_\Delta$. Hence the above inclusion is actually equivalent to $\Lambda'_\Delta \subseteq \varepsilon \cdot \Lambda_\Delta$. By symmetry, the other inclusion also holds, and we finally get $\Lambda'_\Delta = \varepsilon \cdot \Lambda_\Delta$.

We also observe that ε is unique up to multiplication by rational ℓ-adic units, because $\Lambda'_\Delta/\widehat{U}_{\mathcal{G}} = \varepsilon \cdot \Lambda'_\Delta/\widehat{U}_{\mathcal{G}}$ are ℓ-adically equivalent lattices in the non-trivial $\mathbb{Z}_\ell$-module $\Delta/\widehat{U}_{\mathcal{G}}$. Hence recalling that $\Lambda_{\mathcal{G}} = \cup_\Delta \Lambda_\Delta$ and $\Lambda'_{\mathcal{G}} = \cup_\Delta \Lambda'_\Delta$, and taking into account the uniqueness of ε, one immediately gets that $\Lambda'_{\mathcal{G}} = \varepsilon \cdot \Lambda'_{\mathcal{G}}$, as claimed.

To (2): First, since $\Lambda_{\mathcal{G}} = \cup_\Delta \Lambda_\Delta$ as mentioned above, it follows from hypotheses (i), (ii), that $\widehat{U}_{\mathcal{G}_v} + \jmath_v(\Lambda_{\mathcal{G}} \cap \widehat{U}_v)$ is ℓ-adically equivalent to some $\widehat{U}_{\mathcal{G}_v}$-sublattice of $\Lambda_{\mathcal{G}_v}$, as this is the case for all the $\widehat{U}_{\mathcal{G}_v} + \jmath_v(\Lambda_\Delta \cap \widehat{U}_v)$. After replacing $\Lambda_{\mathcal{G}_v}$ by some properly chosen ℓ-adic multiple, say $\varepsilon \cdot \Lambda_{\mathcal{G}_v}$ with $\varepsilon \in \mathbb{Z}_\ell^\times$, without loss of generality, we can suppose that $\jmath_v(\Lambda_{\mathcal{G}} \cap \widehat{U}_v) \subseteq \Lambda_{\mathcal{G}_v}$, and thus $\widehat{U}_{\mathcal{G}_v} + \jmath_v(\Lambda_{\mathcal{G}} \cap \widehat{U}_v) \subseteq \Lambda_{\mathcal{G}_v}$. For the converse inclusion, let $\Gamma \subseteq \widehat{\Lambda}_{\mathcal{G}_v}$ be a finite-corank submodule. Then by the ampleness of $\mathcal{G}$, see Definition/Remark 3 (7) (ii), there exists a finite-corank submodule $\Delta \subseteq \widehat{\Lambda}_{\mathcal{G}}$ such that $\Gamma \subseteq \widehat{U}_{\mathcal{G}_v} + \jmath_v(\Delta \cap \widehat{U}_v)$. But then by properties (i), (ii), we get $\Gamma \cap \Lambda_{\mathcal{G}_v} \subseteq \widehat{U}_{\mathcal{G}_v} + \jmath_v(\Lambda_\Delta \cap \widehat{U}_v) \subseteq \widehat{U}_{\mathcal{G}_v} + \jmath_v(\Lambda_{\mathcal{G}} \cap \widehat{U}_v)$. Since Γ was arbitrary and $\Lambda_{\mathcal{G}_v} = \widehat{U}_{\mathcal{G}_v} + \cup_\Gamma(\Gamma \cap \Lambda_{\mathcal{G}_v})$, the converse inclusion follows. □

Let $\mathcal{G}$ be an abstract decomposition graph which is both level-δ complete curve-like and ample up to level δ for some $\delta > 1$. In particular, all residual abstract decomposition graphs $\mathcal{G}_{\mathfrak{v}}$ to non-trivial indices $\mathfrak{v}$ of length $\delta_{\mathfrak{v}} < \delta$ are both *level-$(\delta - \delta_{\mathfrak{v}})$ complete curve-like and ample up to level* $(\delta - \delta_{\mathfrak{v}})$; and if $\delta_{\mathfrak{v}} = \delta - 1$, then $\mathcal{G}_{\mathfrak{v}}$ is complete curve-like and ample. Hence if $\delta_{\mathfrak{v}} = \delta - 1$, then $\mathcal{G}_{\mathfrak{v}}$ has an abstract divisor group $\mathrm{Div}_{\mathcal{G}_{\mathfrak{v}}}$ as defined/introduced in Definition 7. In the above context, let us fix notation as follows:

Definition 9. In the above context, we define an *abstract divisor group* of $\mathcal{G}$ (if it exists) to be the lattice defined by any system $\mathfrak{T}$ of inertia generators as above,

$$\mathrm{Div}_{\mathcal{G}} := \mathrm{Div}_{\mathfrak{T}} \subset \widehat{\mathrm{Div}}_{\mathcal{G}},$$

which together with its preimage $\Lambda_{\mathcal{G}}$ in $\widehat{\Lambda}_{\mathcal{G}}$ satisfies inductively on δ the following:

(i) Abstract divisor groups $\mathrm{Div}_{\mathcal{G}_v}$ exist for all residual abstract decomposition graphs $\mathcal{G}_v$. Let $\Lambda_{\mathcal{G}_v}$ be the preimage of $\mathrm{Div}_{\mathcal{G}_v}$ in $\widehat{\Lambda}_{\mathcal{G}_v}$ (all v).

(ii) $\Lambda_{\mathcal{G}}$ satisfies conditions (i), (ii) from Fact 8 for all finite corank submodules $\Delta \subset \widehat{\Lambda}_{\mathcal{G}}$ with respect to the preimages $\Lambda_{\mathcal{G}_v}$ defined at (i) above.

- Note that if $\Lambda_{\mathcal{G}}$ exists, then $\Lambda_{\mathcal{G}} \subset \widehat{\Lambda}_{\mathcal{G},\mathrm{fin}}$ and $\Lambda_{\mathcal{G}} \otimes \mathbb{Z}_\ell = \widehat{\Lambda}_{\mathcal{G},\mathrm{fin}}$. Indeed, if $x \in \Lambda_{\mathcal{G}}$, then $\jmath^v(x) = 0$ for almost all v, etc.

Remarks 10. Let $\mathcal{G}$ be an abstract decomposition graph which is level-δ complete curve-like and ample up to level δ for some $\delta > 0$. Suppose that an abstract divisor group $\mathrm{Div}_{\mathcal{G}} := \mathrm{Div}_{\mathfrak{T}}$ for $\mathcal{G}$ exists, and let $\Lambda_{\mathcal{G}}$ be its preimage in $\widehat{\Lambda}_{\mathcal{G}}$. Then one has:

(1) The homomorphism $\jmath^v : \widehat{\Lambda}_{\mathcal{G}} = \mathrm{Hom}\big(G, \mathbb{Z}_\ell\big) \xrightarrow{\mathrm{res}_v} \mathrm{Hom}\big(T_v, \mathbb{Z}_\ell\big) = \mathbb{Z}_\ell \varphi_v$ gives rise by restriction to a surjective homomorphism

$$\jmath^v : \Lambda_{\mathcal{G}} \to \mathbb{Z}_{(\ell)}\, \varphi_v.$$

Indeed, by condition (i) of the ampleness, see Definition/Remark 3 (7), it follows that $\jmath^v(\widehat{\Lambda}_{\mathcal{G}}) = \mathbb{Z}_\ell\, \varphi_v$. Further, since $\Lambda_{\mathcal{G}}$ is ℓ-adically dense in $\widehat{\Lambda}_{\mathcal{G}}$, it follows that $\jmath^v(\Lambda_{\mathcal{G}})$ is dense in $\mathbb{Z}_\ell\, \varphi_v$. Thus the assertion.

(2) Moreover, the $\widehat{U}_{\mathcal{G}}$-lattice $\Lambda_{\mathcal{G}}$ endowed with all the homomorphisms $\jmath^v$ determines $\mathrm{Div}_{\mathcal{G}}$ as the additive subgroup

$$\mathrm{Div}_{\mathcal{G}} = \textstyle\sum_v \mathbb{Z}_{(\ell)}\varphi_v = \sum_v \jmath^v(\Lambda_{\mathcal{G}}) \subset \widehat{\mathrm{Div}}_{\mathcal{G}}$$

generated by the $\jmath^v(\Lambda_{\mathcal{G}})$ for all the v. Therefore, giving an abstract divisor group $\mathrm{Div}_{\mathcal{G}}$ is *equivalent* to giving a $\widehat{U}_{\mathcal{G}}$-lattice $\Lambda_{\mathcal{G}}$ in $\widehat{\Lambda}_{\mathcal{G}}$ such that inductively we have:

   (i) $\Lambda_{\mathcal{G}}$ satisfies conditions (i), (ii) from Fact 8 with respect to the preimages $\Lambda_{\mathcal{G}_v}$ of some abstract divisor groups $\mathrm{Div}_{\mathcal{G}_v}$ (all v).

   (ii) $\jmath^v(\Lambda_{\mathcal{G}}) \cong \mathbb{Z}_{(\ell)}$ (all v), and $\Lambda_{\mathcal{G}}$ is the preimage of $\oplus_v \jmath^v(\Lambda_{\mathcal{G}})$ via $\jmath^{\mathcal{G}}$.

(3) Finally, for an abstract divisor group $\mathrm{Div}_{\mathcal{G}}$ for $\mathcal{G}$ and its preimage $\Lambda_{\mathcal{G}}$ in $\widehat{\Lambda}_{\mathcal{G}}$, we set $\mathfrak{Cl}_{\Lambda_{\mathcal{G}}} = \mathrm{Div}_{\mathcal{G}}/\jmath^{\mathcal{G}}(\Lambda_{\mathcal{G}})$ and call it the abstract ideal class group of $\Lambda_{\mathcal{G}}$. Thus one has a commutative diagram of the form

$$(*)\qquad \begin{array}{ccccccccc} 0 \to & \widehat{U}_{\mathcal{G}} & \hookrightarrow & \Lambda_{\mathcal{G}} & \xrightarrow{\jmath^{\mathcal{G}}} & \mathrm{Div}_{\mathcal{G}} & \xrightarrow{\mathrm{can}} & \mathfrak{Cl}_{\mathcal{G}} & \to 0 \\ & \downarrow & & \downarrow & & \downarrow & & \downarrow & \\ 0 \to & \widehat{U}_{\mathcal{G}} & \hookrightarrow & \widehat{\Lambda}_{\mathcal{G}} & \xrightarrow{\jmath^{\mathcal{G}}} & \widehat{\mathrm{Div}}_{\mathcal{G}} & \xrightarrow{\mathrm{can}} & \widehat{\mathfrak{Cl}}_{\mathcal{G}} & \to 0 \end{array}$$

where the first three vertical morphisms are the canonical inclusions, and the last one is the ℓ-adic completion homomorphism.

Proposition 11. *Let $\mathcal{G}$ be an abstract decomposition graph which is level-δ complete curve-like and ample up to level $\delta > 0$. Then any two abstract divisor groups $\mathrm{Div}_{\mathcal{G}}$ and $\mathrm{Div}'_{\mathcal{G}}$ for $\mathcal{G}$ are ℓ-adically equivalent as lattices in $\widehat{\mathrm{Div}}_{\mathcal{G}}$. Equivalently, their preimages $\Lambda_{\mathcal{G}}$ and $\Lambda'_{\mathcal{G}}$ in $\widehat{\Lambda}_{\mathcal{G}}$ are ℓ-adically equivalent $\widehat{U}_{\mathcal{G}}$-lattices in $\widehat{\Lambda}_{\mathcal{G}}$. In particular, there exist distinguished systems of inertia generators $\mathfrak{T}$ and $\mathfrak{T}'$ defining $\mathrm{Div}_{\mathcal{G}}$, respectively $\mathrm{Div}'_{\mathcal{G}}$, which are ℓ-adically equivalent, i.e., $\mathfrak{T}' = \mathfrak{T}^\varepsilon$ for some ℓ-adic unit $\varepsilon \in \mathbb{Z}_\ell^\times$.*

Proof. We prove this assertion by induction on δ. For $\delta = 1$, the uniqueness is already shown, see Fact 6, and Definition 7 in case $\delta = 1$. Now suppose that $\delta > 1$. Let $\mathrm{Div}_{\mathcal{G}_v}$ and $\mathrm{Div}'_{\mathcal{G}_v}$ be abstract divisor groups for $\mathcal{G}$ used for the definition of $\mathrm{Div}_{\mathcal{G}}$, respectively $\mathrm{Div}'_{\mathcal{G}}$ (all v). By the induction hypothesis, $\mathrm{Div}_{\mathcal{G}_v}$ and $\mathrm{Div}'_{\mathcal{G}_v}$ are ℓ-adically equivalent. Thus their preimages Λ_v and Λ'_v in $\widehat{\Lambda}_{\mathcal{G},v}$ are ℓ-adically equivalent $\widehat{U}_{\mathcal{G},v}$-lattices. Therefore, by Fact 8, the lattices $\Lambda_{\mathcal{G}}$ and $\Lambda'_{\mathcal{G}}$ (which are the preimages of $\mathrm{Div}_{\mathcal{G}}$ respectively $\mathrm{Div}'_{\mathcal{G}}$ in $\widehat{\Lambda}_{\mathcal{G}}$) are ℓ-adically equivalent. Finally, use Remark 10 (2), above to conclude. □

Definition 12. Let $\mathcal{G}$ be an abstract decomposition graph which is level-δ complete curve-like and ample up to level δ. We will say that $\mathcal{G}$ is a *divisorial abstract decomposition graph* if it has abstract divisor groups $\mathrm{Div}_{\mathcal{G}} = \mathrm{Div}_{\mathfrak{T}}$ as introduced above. If this is the case, we will denote by $\Lambda_{\mathcal{G}}$ the preimage of $\mathrm{Div}_{\mathcal{G}}$ in $\widehat{\Lambda}_{\mathcal{G}}$, and call it a *divisorial $\widehat{U}_{\mathcal{G}}$-lattice* in $\widehat{\Lambda}_{\mathcal{G}}$.

3 Abstract decomposition graphs arising from algebraic geometry

3.1 Some general valuation-theoretical nonsense

Let K be an arbitrary field. The space of all equivalence classes of valuations Val_K of K is in a canonical way a partially ordered set by $v \le w$ iff $\mathcal{O}_w \subseteq \mathcal{O}_v$ iff $\mathfrak{m}_v \subseteq \mathfrak{m}_w$, and if so, then $\mathfrak{m}_v \subset \mathcal{O}_w$ is a prime ideal of $\mathcal{O}_w$, and $\mathcal{O}_v$ is the localization $\mathcal{O}_v = (\mathcal{O}_w)_{\mathfrak{m}_v}$. The unique minimal element of Val_K is the trivial valuation v_0 which has $\mathcal{O}_{v_0} = K$ as valuation ring. Further, the minimal *non-trivial* elements of Val_K are exactly the rank-one valuation rings of K (which then correspond to the equivalence classes of non-archimedean absolute values of K). Note that if $v \le w$, then $\mathcal{O}_w/\mathfrak{m}_v$ is a valuation ring in the residue field Kv of v. We denote the corresponding valuation of Kw by w/v, and call it the *quotient* of w by v. Conversely, given $v \in \mathrm{Val}_K$ and a valuation $\overline{w}$ of Kv, the preimage $\mathcal{O}$ of $\mathcal{O}_{\overline{w}}$ under $\mathcal{O}_v \to Kv$ is a valuation ring of a valuation $w \ge v$ such that $w/v = \overline{w}$. We define $\overline{w} \circ v := w$, and call it the composition of $\overline{w}$ and v. Val_K has in a canonical way the structure of a (half-oriented) graph with origin $K = Kv_0$ as follows:

(a) The vertices are the residue fields Kv indexed by $v \in \mathrm{Val}_K$.
(b) The set of edges from Kv to Kw is non-empty if and only if $v \le w$ and $\mathrm{rank}(w/v) \le 1$. If so, then w/v is the unique edge from Kv to Kw. We say that w/v is a non-trivial oriented edge if $\mathrm{rank}(w/v) = 1$, respectively we call w/v a trivial non-oriented edge if $v = w$, i.e., w/v is the trivial valuation of Kv.

We will call the graph defined above the *valuation graph* for K. There are two functorial constructions one should mention here:

(1) *Embeddings.* Let $\imath : L \hookrightarrow K$ be a field embedding and $\varphi_\imath : \mathrm{Val}_K \to \mathrm{Val}_L$, $v \mapsto v_L := v|_L$, the canonical restriction map. Then $\varphi_\imath$ is surjective and compatible with the ordering of valuations. And if $v \le w$ in Val_K, then $v_L \le w_L$ in Val_L, and $\mathrm{rank}(w_L/v_L) \le \mathrm{rank}(w/v)$. Hence if the edge w/v from Kv to Kw exists, then the edge w_L/v_L from Lv_L to Lw_L exists too. Therefore, $\varphi_\imath$ defines a canonical projection from the valuation graph of K onto the valuation graph of L, under which Kv is mapped to Lv_L, and the edge w/v from Kv to Kw (if it exists) is mapped to the edge w_L/v_L from Lv_L to Lw_L. Note that if w/v is a non-trivial oriented edge such that $w_L = v_L$, then w/v is mapped to the trivial non-oriented edge of $Lw_L = Lv_L$.
(2) *Restrictions.* Let Kv be the residue field of v, and let $\mathrm{Val}_v = \{w \in \mathrm{Val}_K \mid w \ge v\}$ be the set of all refinements of v. Then $\mathrm{Val}_v \to \mathrm{Val}_{Kv}$, $w \mapsto w/v$, is a canonical bijection which respects the ordering, thus defines an isomorphism of the subgraph Val_v of the valuation graph for K onto the valuation graph Val_{Kv} for Kv.

• The Galois decomposition theoretical side

Let ℓ be a fixed prime number as above. For every field K which contains the ℓ^∞ roots of unity, let $K'|K$ be a maximal pro-ℓ abelian extension, and we denote by $\Pi_K = \mathrm{Gal}(K'|K)$ its Galois group. For $v \in \mathrm{Val}_K$ and prolongations v' of v to K', we have that the inertia/decomposition groups $T_{v'} \subseteq Z_{v'}$ of the several prolongations $v'|v$ are conjugated under Π_K; hence these groups are equal, as Π_K is commutative. We will denote them by $T_v \subseteq Z_v$, and call them the *inertia/decomposition groups* at v. Recall that $\Pi_{Kv} = Z_v/T_v$ canonically.

Via the Galois correspondence and using the functorial properties of Hilbert decomposition theory, we attach to Val_K a graph $\mathcal{G}_{\mathrm{Val}_K}$ which is in bijection with Val_K and has vertices and edges as follows: The vertices of $\mathcal{G}_{\mathrm{Val}_K}$ are indexed by the (distinct) pro-ℓ abelian groups Π_{Kv}. Concerning edges, if v/w is the unique edge from some Kw to some Kv (hence, either $w = v$ and v/w is the trivial valuation on $Kv = Kw$, or $v < w$ and $\mathrm{rank}(w/v) = 1$ on Kv), then the unique edge from Π_{Kv} to Π_{Kw} is the pair of groups $T_{w/v} \subseteq Z_{w/v}$ viewed as subgroups of Π_{Kv}. Note that in case w/v is the trivial valuation, we have merely by definition that $T_{w/v} = 1$ and $Z_{w/v} = \Pi_{Kw}$.

We will call $\mathcal{G}_{\mathrm{Val}_K}$ the *valuation decomposition graph* of K, or of Π_K.

Note that the above functorial constructions concerning embeddings and restrictions give rise functorially to corresponding functorial constructions on the Galois side as follows:

(1) *Embeddings.* Let $\imath : L \hookrightarrow K$ be an embedding of fields, and consider a prolongation $\imath' : L' \hookrightarrow K'$ of $\imath$. Then $\imath'$ gives rise to a projection $\Phi_\imath : \Pi_K \to \Pi_L$, which in turn gives rise canonically to a *morphism of valuation decomposition graphs*, which we denote by $\Phi_\imath$ again:

$$\Phi_\imath : \mathcal{G}_{\mathrm{Val}_K} \to \mathcal{G}_{\mathrm{Val}_L}.$$

Note that $\Phi_\imath$ maps the profinite group Π_{Kv} at the vertex Kv into the profinite group Π_{Lv_L} at the corresponding vertex Lv_L. And concerning edges, $\Phi_\imath$ maps $T_{w/v} \subseteq Z_{w/v}$ into the pair $T_{w_L/v_L} \subseteq Z_{w_L/v_L}$ of the corresponding inertia/decomposition subgroups of w_L/v_L in Π_{Lv_L}.

(2) *Restrictions.* For $w \in \mathrm{Val}_v$, one has $Z_w \subseteq Z_v$ and $T_v \subseteq T_w$. And under the canonical projection $Z_v \to \Pi_{Kv}$, every $T_w \subseteq Z_w$ is mapped onto $T_{w/v} \subseteq Z_{w/v}$ in Π_{Kv}, etc.

3.2 *Recovering the geometric decomposition graphs from the total decomposition graph*

Let $K|k$ be a function field as introduced in the introduction. We notice that the total graph of prime divisors $\mathscr{D}_K^{\mathrm{tot}}$ of $K|k$, as defined in the introduction, is the subgraph of Val_K whose vertices are the generalized prime divisors of $K|k$ and whose non-trivial edges are of the form $\mathfrak{w}/\mathfrak{v}$ with $\mathfrak{w} > \mathfrak{v}$ generalized prime divisors. (If so, then $\mathfrak{w}/\mathfrak{v}$ is a prime divisor of $K\mathfrak{v}\,|\,k$.) We also recall that a subgraph $\mathscr{D}_K$ of $\mathscr{D}_K^{\mathrm{tot}}$ was called a *geometric graph of prime divisors* for $K|k$ if for every vertex $\mathfrak{v}$ of $\mathscr{D}_K$, the following hold: First, the trivial edge from $K\mathfrak{v}$ to itself is an edge of $\mathscr{D}_K$, and second, the set of non-trivial edges $D_\mathfrak{v}$ originating from $K\mathfrak{v}$ form a geometric set of prime divisors of $K\mathfrak{v}\,|\,k$.

Concerning the Galois theoretical side, the *total decomposition graph* $\mathscr{G}_{\mathscr{D}_K^{\mathrm{tot}}}$ of $K|k$, or of Π_K, is the subgraph of the valuation decomposition graph $\mathscr{G}_{\mathrm{Val}_K}$ which is defined by the total prime divisors graph $\mathscr{D}_K^{\mathrm{tot}}$. And a *geometric decomposition graph* for $K|k$, or for Π_K, is any subgraph $\mathscr{G}_{\mathscr{D}_K}$ of $\mathscr{G}_{\mathscr{D}_K^{\mathrm{tot}}}$ which corresponds to a geometric graph $\mathscr{D}_K$ of prime divisors.

In this subsection we give a recipe to recover/describe the geometric decomposition graphs $\mathscr{G}_{\mathscr{D}_K}$ for $K|k$ inside the total decomposition graph $\mathscr{G}_{\mathscr{D}_K^{\mathrm{tot}}}$ of $K|k$ using only the Galois theoretical information encoded in $\mathscr{G}_{\mathscr{D}_K^{\mathrm{tot}}}$.

We begin by recalling a criterion for the description of the geometric sets of prime divisors of a function field $K|k$, as presented in Pop [24], Subsection 2 D), an idea which was used in essence already in Pop [27].

Let $K|k$ be a function field over an algebraically closed base field as usual. For every normal model $X \to k$ of $K|k$, we denote by D_X the set of all the prime divisors of $K|k$ defined by the Weil prime divisors of X.

Fact 13. *For a set D of prime divisors of $K|k$, the following conditions are equivalent:*

(i) *For all normal models $X \to k$ of $K|k$ one has that D and D_X are almost equal. (Recall that two sets are almost equal if their symmetric difference is finite.)*

(ii) *D is geometric, i.e., there exists a quasi-projective normal model $X \to k$ of $K|k$ such that $D = D_X$.*

Recall that a *line* on a k-variety X is an integral k-subvariety of X, which is a curve of geometric genus equal to 0. We denote by X^{line} the union of all the lines on X.

We will say that a variety $X \to k$ is *very unruly* if the set X^{line} is not dense in X. In particular, a curve X is very unruly iff its geometric genus g_X is positive. Further recall that being very unruly is a birational notion. In particular, it makes sense to say that a function field $K|k$ with $\text{td}(K|k) = d > 0$ is *very unruly* if $K|k$ has models $X \to k$ which are very unruly.

Suppose that $d > 1$. We call a prime divisor v of $K|k$ *very unruly* if $Kv\,|\,k$ is very unruly as a function field over k. A prime divisor v of $K|k$ is very unruly iff there exist a normal model $X \to k$ of $K|k$ and a very unruly prime Weil divisor X_1 of X such that $v = v_{X_1}$.

The following is a more precise form of Proposition 2.6 from Pop [27], but see rather Pop [24], Section 3, for details:

Proposition 14. *With the usual notation, the following hold:*

(1) A set D of prime divisors of $K|k$ is geometric iff there exists a finite ℓ-elementary subextension $K_0|K$ of $K'\,|\,K$ of degree ℓ^d such that for every ℓ-elementary subextension $K_1|K$ of $K'\,|\,K$ of degree ℓ^{2d} containing K_0, one has that D consists of almost all prime divisors v of $K|k$ whose prolongations $v_1|v$ to $K_1|L$ are very unruly prime divisors of K_1.

(2) Let $L|K$ be a finite subextension of $K'\,|\,K$. Then a set D_L of prime divisors of $L|k$ is geometric iff there exists a geometric set of prime divisors D of $K|k$ such that D_L is almost equal to the prolongation of D to L.

Proof. The proof is more or less identical with the one from Pop [24]; thus we refer the reader to that work for the details: Choose some transcendence basis $(t_1, \ldots, t_d)$ of $K|k$ and a "sufficiently general" separable polynomial $p(T) \in k[T]$ of degree ≥ 3. For $m = 1, \ldots, d$, consider $u_m \in K'$ with $u_m^\ell = p(t_m)$. Then $K_0 = K[u_1, \ldots, u_d]$ has degree ℓ^d over K, and it does the job; see [24] for details. □

Using the proposition above, one deduces the following inductive procedure on $d = \text{td}(K|k)$ for deciding whether a given set D of prime divisors of $K|k$ is geometric, respectively whether a finite subextension $L|K$ of $K'|K$ viewed as a function field $L|k$ is very unruly.

Criterion 15. By induction on d, we consider criteria $\mathscr{P}^{(d)}_{\text{geom}}(D)$ and $\mathscr{P}^{(d)}_{\text{v.u.}}(L|K)$ for sets of prime divisors D of $K|k$ to be geometric sets of prime divisors, respectively for finite subextensions $L|K$ of $K'\,|\,K$ to be very unruly, as follows:

(1) Case $d = 1$:

- $\mathscr{P}^{(1)}_{\text{geom}}(D)$: D is almost equal to the set of all prime divisors of $K|k$.
- $\mathscr{P}^{(1)}_{\text{v.u.}}(L|K)$: The genus of the complete normal model of $L|k$ satisfies $g_{L|k} > 0$.

(2) Case $d > 1$:

- $\mathscr{P}^{(d)}_{\text{geom}}(D)$: With $K_0|K$ and $K_1|K$ as in Proposition 14, the set D is almost equal to the set of all prime divisors v of $K|k$ whose prolongations $v_1|v$ to $K_1|K$ satisfy $\mathscr{P}^{(d-1)}_{\text{v.r}}(K_1v_1|Kv)$.
- $\mathscr{P}^{(d)}_{\text{v.u.}}(L|K)$: There exists a set D of prime divisors of $K|k$ such that $\mathscr{P}^{(d)}_{\text{geom}}(D)$ holds, and for almost all $v \in D$, the prolongations $w|v$ of v to $L|K$ satisfy $\mathscr{P}^{(d-1)}_{\text{v.u.}}(Lw\,|\,Kv)$.

Remarks 16. (1) As mentioned in the introduction, if $\mathfrak{v}$ is a generalized prime divisor of $K|k$, then via the canonical projection $pr_{\mathfrak{v}} : Z_{\mathfrak{v}} \to \Pi_{K\mathfrak{v}}$, one can recover the total decomposition graph of $K\mathfrak{v}\,|\,k$ as follows: The generalized prime divisors of $K\mathfrak{v}\,|\,k$ are precisely the valuations of the from $\mathfrak{w}/\mathfrak{v}$ with $\mathfrak{w}$ a generalized prime divisor satisfying $\mathfrak{v} \leq \mathfrak{w}$. In turn, these are exactly the generalized prime divisors $\mathfrak{w}$ such that $T_{\mathfrak{v}} \subseteq T_{\mathfrak{w}}$, or equivalently $Z_{\mathfrak{w}} \subseteq Z_{\mathfrak{v}}$. If so, then $T_{\mathfrak{w}/\mathfrak{v}} \subseteq Z_{\mathfrak{w}/\mathfrak{v}}$ are the images of $T_{\mathfrak{w}} \subseteq Z_{\mathfrak{w}}$; thus the total decomposition graph of $K\mathfrak{v}\,|\,k$ can be recovered from the total decomposition graph of $K|k$ via $pr_{\mathfrak{v}}$.

(2) The finite subextensions $L|K$ of $K'|K$ are in bijection with all the open subgroups $\Delta \subseteq \Pi_K$. And if $\mathfrak{v}$ is a generalized prime divisor of $K|k$, and $\mathfrak{w}$ is a prolongation of $\mathfrak{v}$ to L, then under the canonical projection $\mathrm{pr}_{\mathfrak{v}} : Z_{\mathfrak{v}} \to \Pi_{K\mathfrak{v}}$ we have that if $L|K$ corresponds to $\Delta \subseteq \Pi_K$, then the finite residual subextension $L\mathfrak{w}\,|\,K\mathfrak{v}$ of $K\mathfrak{v}'|K\mathfrak{v}$ corresponds to the open subgroup $\Delta_v := \mathrm{pr}_{\mathfrak{v}}(Z_{\mathfrak{v}} \cap \Delta)$ of $\Pi_{K\mathfrak{v}}$.

(3) Let $\mathscr{G} \subset \mathscr{G}_{\mathscr{D}_K^{\text{tot}}}$ be a connected full subgraph containing the origin Π_K of $\mathscr{G}_{\mathscr{D}_K^{\text{tot}}}$ and having all maximal oriented branches of length $d = \mathrm{td}(K|k)$. (Here "full" means that for all vertices $\Pi_{K\mathfrak{v}}$ and $\Pi_{K\mathfrak{w}}$ of $\mathscr{G}$ one has that if the edge $\mathfrak{w}/\mathfrak{v}$ from $K\mathfrak{v}$ to $K\mathfrak{w}$ exists, then this edge endowed with $T_{\mathfrak{w}/\mathfrak{v}} \subset Z_{\mathfrak{w}/\mathfrak{v}}$ is contained in $\mathscr{G}$.) In particular, the following hold:

   (a) For for every vertex $\Pi_{K\mathfrak{v}}$ of $\mathscr{G}$, the trivial edge from $\Pi_{K\mathfrak{v}}$ to itself endowed with the inertia/decomposition group of the trivial valuation $\{1\} \subset \Pi_{K\mathfrak{v}}$ belongs to $\mathscr{G}$.
   (b) If $\Pi_{K\mathfrak{v}}$ and $\Pi_{K\mathfrak{w}}$ belong to $\mathscr{G}$, and $\mathfrak{w}/\mathfrak{v}$ is a prime divisor of $K\mathfrak{v}$, then the edge $\mathfrak{w}/\mathfrak{v}$ endowed with $T_{\mathfrak{w}/\mathfrak{v}} \subseteq Z_{\mathfrak{w}/\mathfrak{v}}$ belongs to $\mathscr{G}$.
   (c) All maximal branches of non-trivial edges have length $d := \mathrm{td}(K|k)$.

(4) Let $\mathscr{D} \subset \mathscr{D}_K^{\text{tot}}$ be the (connected full) subgraph defined by $\mathscr{G}$. For every vertex $K\mathfrak{v}$ of $\mathscr{D}$, or equivalently a vertex $\Pi_{K\mathfrak{v}}$ of $\mathscr{G}$, let $D_{\mathfrak{v}}$ be the set of prime divisors v of $K\mathfrak{v}\,|\,k$ which are the non-trivial edges of $\mathscr{D}$ originating from $K\mathfrak{v}$. Then by the definitions one has the following:
$\mathscr{G}$ is a geometric decomposition graph iff $D_{\mathfrak{v}}$ is a geometric set of prime divisors of $K\mathfrak{v}\,|\,k$ for every vertex $K\mathfrak{v}$ of $\mathscr{D}$, and all maximal oriented branches of $\mathscr{G}$ have length $\mathrm{td}(K|k)$.

(5) We conclude that recovering/describing the geometric decomposition graphs inside $\mathcal{G}_{\mathcal{D}_K^{\mathrm{tot}}}$ is equivalent to recovering/describing the geometric sets of prime divisors of the function fields $K\mathfrak{v}\,|\,k$ for all generalized prime divisors $\mathfrak{v}$.

We do this by showing that the geometric Criterion 15 can be recovered from, respectively interpreted in, the group-theoretical information encoded in $\mathcal{G}_{\mathcal{D}_K^{\mathrm{tot}}}$.

Gal-Criterion 17.
By induction on $d_{\mathfrak{v}} := \mathrm{td}(K\mathfrak{v}\,|\,k)$, we give criteria $\mathrm{Gal}\mathscr{P}^{(d)}_{\mathrm{geom}}(D)$ and $\mathrm{Gal}\mathscr{P}^{(d)}_{\mathrm{v.u.}}(L|K\mathfrak{v})$ for sets of prime divisors D of $K\mathfrak{v}|k$ to be geometric sets of prime divisors, respectively for finite subextensions $L|K\mathfrak{v}$ of $(K\mathfrak{v})'\,|\,K\mathfrak{v}$ to be very unruly, as follows:
Case $d_{\mathfrak{v}} = 1$:
Then $K\mathfrak{v}\,|\,k$ is the function field of a complete smooth curve $X_{\mathfrak{v}} \to k$ with function field $\kappa(X_{\mathfrak{v}}) = K\mathfrak{v}$. And the set of all non-trivial generalized prime divisors equals the set of prime divisors of $K\mathfrak{v}\,|\,k$, which is $D_{X_{\mathfrak{v}}}$. Let $(T_v)_v$ be the system of all divisorial inertia groups in $\Pi_{K\mathfrak{v}}$ (which is part of the hypothesis, as $\Pi_{K\mathfrak{v}}$ comes endowed with the total decomposition graph of $K\mathfrak{v}\,|\,k$, hence encodes the set of all the T_v, $v \in D_{X_{\mathfrak{v}}}$), and let $T_{K\mathfrak{v}}$ be the closed subgroup of $\Pi_{K\mathfrak{v}}$ generated by all T_v. Then $\Pi_{K\mathfrak{v}}/T_{K\mathfrak{v}} = \pi_1^{\ell,\mathrm{ab}}(X_{\mathfrak{v}})$ is the pro-ℓ abelian fundamental group of $X_{\mathfrak{v}}$. Since $\mathrm{char}(k) \neq \ell$, it follows that $\pi_1^{\ell,\mathrm{ab}}(X_{\mathfrak{v}}) \cong \mathbb{Z}_\ell^{2g_{\mathfrak{v}}}$, where $g_{\mathfrak{v}}$ is the genus of $X_{\mathfrak{v}}$. For every non-empty set $D \subset D_{X_{\mathfrak{v}}}$ of prime divisors of $K\mathfrak{v}\,|\,k$, let T_D be the closed subgroup of $\Pi_{K\mathfrak{v}}$ generated by T_v, $v \in D$. Then $\Pi_{K\mathfrak{v}}/T_D$ is a pro-ℓ abelian free group on $2g_{\mathfrak{v}} + r - 1$ generators, where $r = |D_{X_{\mathfrak{v}}} \backslash D|$. Since D is geometric iff r is finite, we get that D is geometric iff $\Pi_{K\mathfrak{v}}/T_D$ is topologically finitely generated. Hence $\mathscr{P}^{(1)}_{\mathrm{geom}}(D)$ is equivalent to:

- $\mathrm{Gal}\mathscr{P}^{(1)}_{\mathrm{geom}}(D)$: *$\Pi_{K\mathfrak{v}}/T_D$ is finitely generated.*

Let $L|K\mathfrak{v}$ be a finite subextension of $K\mathfrak{v}'\,|\,K\mathfrak{v}$ corresponding to $\Delta \subseteq \Pi_{K\mathfrak{v}}$ as above, and let $\pi_L : \Pi_{K\mathfrak{v}} \to \Pi_{K\mathfrak{v}}/\Delta = \mathrm{Gal}(L|K\mathfrak{v})$ be the corresponding finite quotient. For every prime divisor v of $K\mathfrak{v}\,|\,k$ and a prolongation $w|v$ of v to L, we have that the inertia group of $w|v$ in $G = \mathrm{Gal}(L|K\mathfrak{v})$ is precisely $T_w := \pi_L(T_v)$; hence the ramification index of $w|v$ is $e_{L,v} := |\pi_L(T_v)|$. Further, by the fundamental equality, the number $n_{L,v}$ of prolongations of v to L can be computed as $[L : K\mathfrak{v}] = n_{L,v} \cdot e_{L,v}$. Thus applying the Hurwitz genus formula, one has $2g_L - 2 = [L : K\mathfrak{v}](2g_{\mathfrak{v}} - 2) + \sum_v n_{L,v}\big(e_{L,v} - 1\big)$. And since $n_{L,v} = [L : K\mathfrak{v}]/e_{L,v}$, we see that $g_L > 0$ iff either $g_{\mathfrak{v}} > 0$ or $\sum_v (1 - 1/e_{L,v}) \geq 2$. Therefore, taking into account that $e_{L,v} = |\pi_L(T_v)|$, and that $g_{L|k} = 0$ iff $\Pi_{K\mathfrak{v}} = T_{K\mathfrak{v}}$, we get that the geometric criterion $\mathscr{P}^{(1)}_{\mathrm{v.r}}(D)$ is equivalent to

- $\mathrm{Gal}\mathscr{P}^{(1)}_{\mathrm{v.r}}(L|K\mathfrak{v})$: *Either $\Pi_{K\mathfrak{v}} \neq T_{K\mathfrak{v}}$ or otherwise $\sum_v \big(1 - 1/|\pi_L(T_v)|\big) \geq 2$.*

Case $d_{\mathfrak{v}} > 1$:
Recall that the total decomposition graph of $K\mathfrak{v}\,|\,k$ can be recovered from $\mathcal{G}_{\mathcal{D}_K^{\mathrm{tot}}}$ as mentioned above. Hence without loss of generality (and in order to simplify notions), we can suppose that $K\mathfrak{v} = K$. In particular, we will denote by v the

prime divisors of $K|k$, and for finite subextensions $L|K$ of $K'|K$, we denote by $w|v$ the prolongations of v to L. Note that since $L|K$ is abelian, thus Galois, all the prolongations $w|v$ are conjugated under $\mathrm{Gal}(L|K)$, and thus have the same ramification indices and residue function fields $Lw|k$, which are isomorphic over $Kv|k$.

Now let D be a set of prime divisors v of $K|k$. Then $\mathscr{P}^{(d)}_{\mathrm{geom}}(D)$ is equivalent to

- $\mathrm{Gal}\mathscr{P}^{(d)}_{\mathrm{geom}}(D)$: *With $K_0|K$ and $K_1|K$ as in Proposition 14, the set D is almost equal to the set of all prime divisors v of $K|k$ whose prolongations $v_1|v$ to $K_1|K$ satisfy* $\mathrm{Gal}\mathscr{P}^{(d-1)}_{\mathrm{v.r}}(K_1v_1|Kv)$.

Similarly, let $L|K$ be a finite subextension of $K'|K$. Then the geometric criterion $\mathscr{P}^{(d)}_{\mathrm{v.r}}(L|K)$ is equivalent to the following:

- $\mathrm{Gal}\mathscr{P}^{(d)}_{\mathrm{v.r}}(L|K)$: *There is a set D of prime divisors of $K|k$ satisfying* $\mathrm{Gal}\mathscr{P}^{(d)}_{\mathrm{geom}}(D)$ *such that for almost all prime divisors $v \in D$, all the prolongations $w|v$ of v to $L|K$ satisfy* $\mathrm{Gal}\mathscr{P}^{(d-1)}_{\mathrm{v.r}}(Lw|Kv)$.

This concludes the proof of the claim that the geometric sets of prime divisors of each $K\mathfrak{v}|k$ can be recovered from the total decomposition graph of $K\mathfrak{v}|k$, thus from that of $K|k$.

3.3 Geometric decomposition graphs as abstract decomposition graphs

Let $K|k$ be a function field over an algebraically closed field k with $\mathrm{char}(k) \neq \ell$. Generalizing the divisor graphs of prime divisors from the introduction, we define a *level-δ geometric prime divisor graph* for $K|k$ as being a (half) oriented graph $\mathscr{D}_K$ defined as follows:

(I) The vertices of $\mathscr{D}_K$ are distinct function fields $K_i|k$ over k. And $\mathscr{D}_K$ has an origin which is $K_0 := K$.

(II) For every vertex K_i, the trivial valuation v_{i0} of K_i is the only edge from K_i to itself, and we view this edge as a non-oriented one, or a trivial edge. Further, the set of all the oriented edges starting at K_{i^*} is a geometric set D_{i^*} of prime divisors v_i of K_{i^*}. We call these edges non-trivial, and if $v_i \in D_{i^*}$ is such a non-trivial edge from K_{i^*} to K_i, then $K_i = K_{i^*}v_i$. In particular, $\mathrm{td}(K_i|k) = \mathrm{td}(K_{i^*}|k) - 1$.

(III) The trivial valuations are the only cycles of $\mathscr{D}_K$, and all the maximal branches of non-trivial edges of $\mathscr{D}_K$ have length equal to δ, hence $\delta \leq \mathrm{td}(K|k)$.

As indicated above, we attach to $\mathscr{D}_K$ the corresponding subgraph $\mathscr{G}_{\mathscr{D}_K} \subset \mathscr{G}_{\mathrm{Val}_K}$. Hence by definition one has:

(I) The vertices of $\mathcal{G}_{\mathcal{D}_K}$ are in bijection with the vertices of $\mathcal{D}_K$, via the Galois correspondence, i.e., the vertices of $\mathcal{G}_{\mathcal{D}_K}$ are the pro-ℓ groups Π_{K_i} with K_i vertex of $\mathcal{D}_K$. In particular, $\Pi_{K_0} := \Pi_K$ is the origin of $\mathcal{G}_{\mathcal{D}_K}$.

(II) The edges of $\mathcal{G}_{\mathcal{D}_K}$ are in bijection with the edges of $\mathcal{D}_K$. The trivial edge v_{i0} from K_i to itself is endowed with $\{1\} =: T_{v_{i0}} \subset Z_{v_{i0}} := \Pi_{K_i}$, i.e., with $\{1\} \subset \Pi_{K_i}$. Every non-trivial edge v_i is endowed with the inertia/decomposition groups $T_{v_i} \subseteq Z_{v_i}$. In particular, if v_i is an edge from $K_{i'}$ to $K_i = K_{i'}v_i$, then $\Pi_{K_i} = Z_{v_i}/T_{v_i}$.

(III) The trivial valuations are the only cycles of $\mathcal{G}_{\mathcal{D}_K}$, and all the maximal branches originating from Π_{K_0} and consisting of non-trivial edges of $\mathcal{G}_{\mathcal{D}_K}$ have length δ.

Proposition 18. *With the above notation, $\mathcal{G}_{\mathcal{D}_K}$ is a level-δ abstract decomposition graph.*

Proof. Indeed, all the axioms of an abstract decomposition graph are more or less well-known facts concerning Hilbert decomposition theory for valuations. For instance, if v_i is a prime divisor of $K_{i*}|k$, then all the prolongations v_i' of v_i to K_{i*}' are conjugated under $\Pi_{K_{i*}}$; therefore, their inertia, respectively decomposition, groups are equal, say equal to $T_{v_i} \subseteq Z_{v_i}$. Further, $T_{v_i} \cong \mathbb{Z}_\ell$, and the residue field $K_{i*}'v_i'$ equals $(K_{i*}v_i)'$, thus $(K_{i*}v_i)' = Z_{v_i}/T_{v_i}$, etc. Moreover, for prime divisors $v_i \neq w_i$ of $K_{i*}|k$ one has the following; see e.g., Pop [28], Introduction, and especially Proposition 2.5 (2): The decomposition groups Z_{v_i} and Z_{w_i} have trivial intersection. And finally, if $X_{i*} \to k$ is any quasiprojective normal variety, and $D_{X_{i*}}$ is the set of Weil prime divisors of X_{i*}, then every open subgroup of $\Pi_{K_{i*}}$ contains almost all inertia groups T_{v_i}. Indeed, in every finite separable extension of K_{i*} only finitely many Weil prime divisors of X_{i*} are ramified, etc. □

Remarks 19. Let $\mathcal{G}_{\mathcal{D}_K}$ be a level-δ abstract decomposition graph as above, and to simplify notation a little bit, set $\widehat{\Lambda}_{\mathcal{D}_K} := \widehat{\Lambda}_{\mathcal{G}_{\mathcal{D}_K}}$, $\widehat{U}_{\mathcal{D}_K} := \widehat{U}_{\mathcal{G}_{\mathcal{D}_K}}$, $\widehat{\mathrm{Div}}_{\mathcal{D}_K} := \widehat{\mathrm{Div}}_{\mathcal{G}_{\mathcal{D}_K}}$, and $\widehat{\mathfrak{Cl}}_{\mathcal{D}_K} := \widehat{\mathfrak{Cl}}_{\mathcal{G}_{\mathcal{D}_K}}$. We next analyze/describe the abstract objects $\widehat{U}_{\mathcal{D}_K}$, $\widehat{\Lambda}_{\mathcal{D}_K}$, $\widehat{\mathrm{Div}}_{\mathcal{D}_K}$ and $\widehat{\mathfrak{Cl}}_{\mathcal{D}_K}$ and relate the abstract exact sequence

$$1 \to \widehat{U}_{\mathcal{D}_K} \to \widehat{\Lambda}_{\mathcal{D}_K} \xrightarrow{\jmath^{\mathcal{D}_K}} \widehat{\mathrm{Div}}_{\mathcal{D}_K} \to \widehat{\mathfrak{Cl}}_{\mathcal{D}_K} \to 0$$

to the geometry of $K|k$ as reflected in the geometric information encoded in the prime divisor graph $\mathcal{D}_K$. In order to do so, let us consider some normal model $X \to k$ of $K|k$ such that $D := D_X$ is the set of 1-edges of the given $\mathcal{D}_K$. Without loss of generality, we can and will suppose that X is quasi-projective. Then by Krull's Hauptidealsatz, $\mathcal{U}_D := \Gamma(X, \mathcal{O}_X)^\times$ depends on D only, and not on X; $\mathcal{H}_D(K) := K^\times/\mathcal{U}_D$ is isomorphic to the group of principal divisors on X. Hence since $\mathrm{Div}(D) := \mathrm{Div}(X)$ depends on D only, and not on X, it follows that $\mathfrak{Cl}(D) := \mathfrak{Cl}(X)$ depends on D only, and not on X; one has a canonical exact sequence

$$0 \to \mathcal{H}_D(K) \xrightarrow{\mathrm{div}_D} \mathrm{Div}(D) \xrightarrow{\mathrm{pr}} \mathfrak{Cl}(D) \to 0,$$

and the resulting exact sequence of ℓ-adically complete groups:

$$0 \to \mathbb{T}_{\ell,\mathfrak{Cl}(D)} \longrightarrow \widehat{\mathcal{H}_D}(K) \xrightarrow{\mathrm{div}_D} \widehat{\mathrm{Div}}(D) \longrightarrow \widehat{\mathfrak{Cl}}(D) \to 0,$$

where $\mathbb{T}_{\ell,\mathfrak{Cl}(D)} = \varprojlim_n {}_n\mathfrak{Cl}(D)$, with $n = \ell^e$ and $e \geq 0$, is the ℓ-adic Tate module of $\mathfrak{Cl}(D)$. Since $1 \to \mathcal{U}_D/k^\times \to K^\times/k^\times \to \mathcal{H}_D(K) \to 1$ is an exact sequence of free abelian groups, so is $1 \to \widehat{\mathcal{U}}_D \to \widehat{K} \to \widehat{\mathcal{H}_D}(K) \to 1$. Hence if $\widehat{U}_D \subset \widehat{K}$ is the preimage of $\mathbb{T}_{\ell,\mathfrak{Cl}(D)} \hookrightarrow \widehat{\mathcal{H}_D}(K)$ under $\widehat{K} \to \widehat{\mathcal{H}_D}(K)$, we finally get an exact sequence of the form $0 \to \widehat{U}_D \to \widehat{K} \to \widehat{\mathrm{Div}}(D) \to \widehat{\mathfrak{Cl}}(D) \to 0$, and therefore, $\widehat{U}_D$ fits canonically into an exact sequence $1 \to \widehat{\mathcal{U}}_D \to \widehat{U}_D \to \mathbb{T}_{\ell,\mathfrak{Cl}(D)} \to 0$.

(1) By Kummer theory we have an identification: $\widehat{\Lambda}_{\mathcal{D}_K} := \mathrm{Hom}(\Pi_K, \mathbb{Z}_\ell) = \widehat{K}$.

(2) Concerning/describing $\widehat{U}_{\mathcal{D}_K}$: Recall that we defined $\widehat{U}_{\mathcal{D}_K} := \mathrm{Hom}(\Pi_{1,\mathcal{D}_K}, \mathbb{Z}_\ell)$, where $\Pi_{1,\mathcal{D}_K} := \Pi_K/T_{\mathcal{D}_K}$ and $T_{\mathcal{D}_K}$ is the group generated by all the inertia groups T_v with v all the 1-edges of $\mathcal{D}_K$. By the definitions, we have $T_{\mathcal{D}_K} = T_D$ and $\Pi_{1,\mathcal{D}_K} = \Pi_{1,D}$. Further, in the notations from Fact 55, it follows that $\Pi_{1,D}$ is the Pontryagin dual of Δ_∞, which fits canonically in the exact sequence

$$0 \to \mathcal{U}_D \otimes \mathbb{Q}_\ell/\mathbb{Z}_\ell \to \Delta_\infty \to {}_{\ell^\infty}\mathfrak{Cl}(D) \to 0.$$

Let $\Delta_0 \subset \Delta_\infty$ be the maximal divisible subgroup. Since $\mathcal{U}_D \otimes \mathbb{Q}_\ell/\mathbb{Z}_\ell$ is divisible, it follows by Fact 54 that Δ_0 fits into an exact sequence of the form

$$0 \to \mathcal{U}_D \otimes \mathbb{Q}_\ell/\mathbb{Z}_\ell \to \Delta_0 \to {}_{\ell^\infty}A_0(X) \to 0,$$

and Δ_∞/Δ_0 finite. Hence $\Pi_{1,D}$ has finite torsion, $\widehat{U}_{\mathcal{D}_K} := \mathrm{Hom}(\Pi_{1,\mathcal{D}_K}, \mathbb{Z}_\ell)$ is the Pontryagin dual of Δ_0, and we get an exact sequence $1 \to \widehat{\mathcal{U}}_D \to \widehat{U}_{\mathcal{D}_K} \to \mathbb{T}_{\ell,\mathfrak{Cl}(D)} \to 0$. Finally $\widehat{U}_{\mathcal{D}_K} = \widehat{U}_D$, and this gives the precise description of $\widehat{U}_{\mathcal{D}_K}$ in geometric terms.

(3) Concerning $\widehat{\mathrm{Div}}_{\mathcal{D}_K}$: For every prime divisor v one has a commutative diagram

$$(*)_v \qquad \begin{array}{ccc} \widehat{K} & \xrightarrow{v} & \widehat{vK} \\ \downarrow & & \downarrow{\scriptstyle \theta^v} \\ \mathrm{Hom}(\Pi_K, \mathbb{Z}_\ell) & \xrightarrow{\jmath^v} & \mathrm{Hom}(T_v, \mathbb{Z}_\ell) \end{array}$$

The diagrams $(*)_v$ with $v \in D$ give rise canonically to a commutative diagram

$$\begin{array}{ccccc} \widehat{K} & \to \widehat{\mathrm{Div}}(D) = \widehat{\oplus}_v vK & \to & \widehat{\mathfrak{Cl}}(D) & \to 0 \\ \downarrow & \downarrow{\scriptstyle \widehat{\oplus}\theta^v} & & \downarrow & \\ \mathrm{Hom}(\Pi_K, \mathbb{Z}_\ell) & \xrightarrow{\jmath^{\mathcal{G}}} \widehat{\oplus}_v \mathrm{Hom}(T_v, \mathbb{Z}_\ell) & \xrightarrow{\mathrm{can}} & \widehat{\mathcal{P}}_{\mathcal{D}_K} & \to 0 \end{array}$$

where the vertical maps are isomorphisms, and $\widehat{\mathscr{P}}_{\mathscr{D}_K}$ is simply the quotient of the middle group by the first one. Therefore, the resulting canonical identification $\widehat{K} \to \mathrm{Hom}(\Pi_K, \mathbb{Z}_\ell) =: \widehat{\Lambda}_{\mathscr{D}_K}$ gives rise a canonical isomorphism $\widehat{\mathrm{Div}}(D) \to \widehat{\oplus}_v \mathrm{Hom}(T_v, \mathbb{Z}_\ell) =: \widehat{\mathrm{Div}}_{\mathscr{D}_K}$.

(4) Finally, the above identifications $\widehat{K} \to \mathrm{Hom}(\Pi_K, \mathbb{Z}_\ell)$ and $\widehat{\mathrm{Div}}(D) \to \widehat{\mathrm{Div}}_{\mathscr{D}_K}$ give rise to an identification $\widehat{\mathfrak{Cl}}(D) \to \widehat{\mathscr{P}}_{\mathscr{D}_K} =: \widehat{\mathfrak{Cl}}_{\mathscr{D}_K}$. Hence by the structure of $\mathfrak{Cl}(D) := \mathfrak{Cl}(X)$ given in Fact 54, it follows that $\widehat{\mathfrak{Cl}}_{\mathscr{D}_K} \cong \widehat{\mathfrak{Cl}}(D) = \widehat{A}_1(X)$, and is thus a finite $\mathbb{Z}_\ell$-module.

Fact 20. *With the above notation, let* $\mathrm{Div}'(D)$ *be the preimage in* $\mathrm{Div}(D) := \mathrm{Div}(X)$ *of the maximal ℓ-divisible subgroup* $\mathfrak{Cl}'(D)$ *of* $\mathfrak{Cl}(D) := \mathfrak{Cl}(X)$. *Then one has:*

(1) $\mathrm{Div}'(D) \hookrightarrow \mathrm{Div}(D)$ *gives rise to an embedding* $\widehat{\mathrm{Div}'}(D) \hookrightarrow \widehat{\mathrm{Div}}(D)$.

(2) $\widehat{\mathrm{Div}'}(D) = \ker\big(\widehat{\mathrm{Div}}(D) \to \widehat{\mathfrak{Cl}}(D)\big) = \mathrm{div}_D(\widehat{K})$, *and* $\mathrm{Div}'(D) = \widehat{\mathrm{Div}'}(D) \cap \mathrm{Div}(D)$.

(3) *Let $\tilde{D} \supseteq D$ be geometric sets with $\Pi_{1,D} = \Pi_{1,\tilde{D}}$. Then* $\mathrm{Div}'(\tilde{D}) \subseteq \mathrm{Div}'(D)$ *has finite bounded index. Finally, for every D large enough,* $\mathrm{Div}'(D) \subset \mathrm{Div}(X)$ *depends on $K|k$ only, and* $\widehat{\mathfrak{Cl}}(\tilde{D}) \cong \widehat{\mathfrak{Cl}}(D) \oplus \mathbb{Z}_\ell^{|\tilde{D}\backslash D|}$.

Proof. To (1) and (2): We get a commutative diagram of the form

$$\begin{array}{ccccccccc}
0 \to & \mathscr{H}_D(K) & \to & \mathrm{Div}'(D) & \to & \mathfrak{Cl}'(D) & \to 0 \\
 & \| & & \downarrow & & \downarrow & \\
0 \to & \mathscr{H}_D(K) & \to & \mathrm{Div}(D) & \to & \mathfrak{Cl}(D) & \to 0 \\
 & & & \downarrow & & \downarrow & \\
 & & & \mathscr{D} & \to & \mathscr{C} &
\end{array}$$

where $\mathscr{D} = \mathrm{Div}(D)/\mathrm{Div}'(D)$, $\mathscr{C} = \mathfrak{Cl}(D)/\mathfrak{Cl}'(D)$, and $\mathscr{D} \to \mathscr{C}$ is an isomorphism. By the structure of $\mathfrak{Cl}(D) = \mathfrak{Cl}(X)$ in Fact 54, $\mathscr{C} := \mathfrak{Cl}(X)/\mathfrak{Cl}'(X)$ is of the form $\mathscr{C} = C_1 + C_2$ with C_1 a finite abelian ℓ-group, and C_2 a finitely generated free abelian group. Hence we get embeddings of the ℓ-adic completions $\mathscr{C} \hookrightarrow \widehat{\mathscr{C}}$ and $\mathscr{D} \hookrightarrow \widehat{\mathscr{D}}$. Further, since $\mathfrak{Cl}(D)'$ is ℓ-divisible, its ℓ-adic completion is trivial, hence $\widehat{\mathfrak{Cl}}(D) \to \widehat{\mathscr{C}}$ is an isomorphism. And finally, the middle (exact) column defines an exact sequence $\widehat{\mathrm{Div}'}(X) \hookrightarrow \widehat{\mathrm{Div}}(X) \to \widehat{\mathscr{D}} \to 0$. The remaining assertions follow easily in the same way, by chasing in the commutative diagram above.

To (3): Let $D = D_X$ and $\tilde{D} = D_Y$ for some normal models $X \to k$ and $Y \to k$ of $K|k$. By Fact 55, especially assertion (4), it follows that $\mathscr{U}_X = \mathscr{U}_Y$, and with $Y \to k$ as in Fact 55 we have that $A_{\mathrm{tors}}(Y) \subseteq A_0(Y) \subseteq A_\tau(Y)$ are birational invariants of $K|k$, etc. □

Definition/Remark 21. Consider notation as above.

(1) A geometric set $D := D_X$ of prime divisors for $K|k$ is called *complete regular-like*, if $\Pi_{1,D} = \Pi_{1,K}$ and $\widehat{\mathfrak{Cl}}(D)$ has positive rational rank and for every geometric set of prime divisors $\tilde{D} \supseteq D$ one has $\widehat{\mathfrak{Cl}}(\tilde{D}) \cong \widehat{\mathfrak{Cl}}(D) \oplus \mathbb{Z}_\ell^{|\tilde{D}\setminus D|}$.

- Note that if $X \to k$ is a complete regular variety, then D_X is complete regular-like, but the converse is not true in general. Nevertheless, if X is a curve, then D_X is complete regular-like iff X is a complete regular curve.

(2) Let $\mathscr{D}_K$ be a level-δ geometric graph of prime divisors for $K|k$. For each vertex $\mathfrak{v}$ of $\mathscr{D}_K$, let $D_\mathfrak{v}$ is the set of non-trivial 1-edges of $\mathscr{D}$ with origin $K\mathfrak{v}$. We say that $\mathscr{D}_K$ is *complete regular-like*, if $D_\mathfrak{v}$ is complete regular-like for all $\mathfrak{v}$ with $\mathrm{td}(K\mathfrak{v}|k) > 0$.

(3) The complete regular-like prime divisor graphs for $K|k$ are abundant. Moreover, for every geometric prime divisor graph $\mathscr{D}'_K \subset \mathscr{D}_K^{\mathrm{tot}}$ there exist complete regular-like decomposition graphs $\mathscr{D}_K$ with $\mathscr{D}'_K \subseteq \mathscr{D}_K$. Indeed, we proceed by induction on the transcendence degree as follows: Let D' be the set of 1-indices of $\mathscr{D}'_K$. Let $D \supseteq D'$ be any complete regular-like set of prime divisors of K. Then $\mathrm{td}(Kv|k) < \mathrm{td}(K|k)$ for all $v \in D$, hence for every $v \in D$ by induction one has the following: There exists complete regular-like prime divisor graphs $\mathscr{D}_{Kv}$ for $Kv|k$. Moreover, if $v \in D'$, then there exists a complete regular-like prime divisor graph $\mathscr{D}_{Kv}$ which contains the residual prime divisor graph $\mathscr{D}'_{Kv}$ of $\mathscr{D}'_K$. Then the resulting prime divisor graph $\mathscr{D}_K$ having D as set of 1-indices and $\mathscr{D}_{Kv}$ as residual prime divisor graph at each $v \in D$ is by definition complete regular-like.

Combining the above discussion with the one in the previous subsection, we get:

Proposition 22. *In the above notations and context, the following hold:*

(1) *The geometric decomposition graphs $\mathscr{G}_{\mathscr{D}_K}$ for $K|k$ can be recovered by a group-theoretical recipe from the group-theoretical information encoded in $\mathscr{G}_{\mathscr{D}_K^{\mathrm{tot}}}$.*
(2) *Moreover, given a geometric decomposition graph $\mathscr{G}_{\mathscr{D}_K}$, the fact that $\mathscr{G}_{\mathscr{D}_K}$ is complete regular-like can be recovered from the total decomposition graph $\mathscr{G}_{\mathscr{D}_K^{\mathrm{tot}}}$ endowed with $\mathscr{G}_{\mathscr{D}_K}$.*
(3) *In particular, the complete regular-like decomposition graphs $\mathscr{G}_{\mathscr{D}_K}$ for $K|k$ can be recovered from the group-theoretical information encoded in $\mathscr{G}_{\mathscr{D}_K^{\mathrm{tot}}}$.*

Proof. To (1): This is more or less the Gal-Criterion 17 combined with Remarks 16.

To (2): Let $\mathscr{D}_K$ be a geometric graph of prime divisors for $K|k$, and $\mathscr{G}_{\mathscr{D}_K}$ the corresponding geometric decomposition graph for $K|k$. For every vertex $K\mathfrak{v}$ of $\mathscr{D}_K$, let $X_\mathfrak{v} \to k$ be a normal model of $K\mathfrak{v}|k$ such that $D_\mathfrak{v} = D_{X_\mathfrak{v}}$. Let $T_{D_\mathfrak{v}} \subseteq \Pi_{K\mathfrak{v}}$ be the closed subgroup generated by all the inertia groups T_v, $v \in D_\mathfrak{v}$. Further, by Remark 16 (1), the total decomposition graph of $K\mathfrak{v}|k$ can be recovered from $\mathscr{G}_{\mathscr{D}_K^{\mathrm{tot}}}$. In particular, the set of the inertia groups T_w of all the prime divisors w of $K\mathfrak{v}|k$, hence the closed subgroup $T_{K\mathfrak{v}}$ generated by all these inertia groups, can be recovered from $\mathscr{G}_{\mathscr{D}_K^{\mathrm{tot}}}$ endowed with $\mathscr{G}_{\mathscr{D}_K}$. Therefore, from $\mathscr{G}_{\mathscr{D}_K^{\mathrm{tot}}}$ endowed with $\mathscr{G}_{\mathscr{D}_K}$ one can check whether the following hold:

(a) $\{T_v \mid v \in D_{\mathfrak{v}}\}$ equals the set $\{T_w \mid w$ all the prime divisors of $K\mathfrak{v}\,|\,k\}$.
(b) $T_{K\mathfrak{v}} = T_{D_{\mathfrak{v}}}$.
(c) $\widehat{\mathfrak{Cl}}(D_{\mathfrak{v}})$ is not finite.
(d) $\widehat{\mathfrak{Cl}}(\tilde{D}_{\mathfrak{v}}) \cong \widehat{\mathfrak{Cl}}(D_{\mathfrak{v}}) \oplus \mathbb{Z}_\ell^{|\tilde{D}_{\mathfrak{v}}\setminus D_{\mathfrak{v}}|}$ for every geometric set $\tilde{D}_{\mathfrak{v}} \supseteq D_{\mathfrak{v}}$ of prime divisors of $K\mathfrak{v}|k$.

Note that (a) holds iff $X_{\mathfrak{v}}$ is a complete normal curve. Indeed, if $\dim(X_{\mathfrak{v}}) > 1$, then for every given normal model $Y_{\mathfrak{v}} \to k$ of $K\mathfrak{v}\,|\,k$ there exist infinitely many prime divisors of $K\mathfrak{v}\,|\,k$ which are not Weil prime divisors of $Y_{\mathfrak{v}}$. And if $X_{\mathfrak{v}}$ is a normal curve over k, then the set of prime divisors of $K\mathfrak{v}\,|\,k$ is in bijection with the set of all the points in the normal completion of $X_{\mathfrak{v}}$, etc. Further, the conditions (b), (c), (d), are by definition the necessary conditions for $D_{\mathfrak{v}}$ to be complete regular-like. This concludes the proof of assertion (2).
To (3): This follows immediately by combining assertions (1) and (2) above. □

Proposition 23. *With the above notation, let $\mathcal{G}_{\mathcal{D}_K}$ be a complete regular-like decomposition graph for $K|k$. Then the following hold:*

(1) *$\mathcal{G}_{\mathcal{D}_K}$ viewed as an abstract decomposition graph of level $\delta := \mathrm{td}(K|k)$ is δ complete curve like and ample up to level δ. Thus $\mathcal{G}_{\mathcal{D}_K}$ is divisorial.*
(2) *Let D be the set of 1-edges of $\mathcal{G}_{\mathcal{D}_K}$. Then the following hold:*

 (a) *$\widehat{U}_{\mathcal{D}_K}$ is the ℓ-adic dual of $\Pi_{1,K} = \Pi_{1,D}$, and thus it depends on $K|k$ only, and not on $\mathcal{G}_{\mathcal{D}_K}$. Therefore we denote $\widehat{U}_K := \widehat{U}_{\mathcal{D}_K}$.*
 (b) *$\mathrm{Div}(D)_{(\ell)} := \mathrm{Div}(D) \otimes \mathbb{Z}_{(\ell)}$ is an abstract divisor group of $\mathcal{G}_{\mathcal{D}_K}$, which we call the canonical abstract divisor group of $\mathcal{G}_{\mathcal{D}_K}$.*
 (c) *The preimage $\Lambda_{\mathcal{D}_K}$ of $\mathrm{Div}(D)_{(\ell)}$ in $\widehat{K}$ will be called the canonical divisorial $\widehat{U}_{\mathcal{D}_K}$-lattice of $\mathcal{G}_{\mathcal{D}_K}$, and it has the following description:*

- *Let $\mathrm{Div}'(D) \subseteq \mathrm{Div}(D)$ be the preimage of the ℓ-divisible subgroup $\mathfrak{Cl}'(D)$ of $\mathfrak{Cl}(D)$. Then $\mathrm{Div}'(D)_{(\ell)} = \mathrm{div}_D(\Lambda_{\mathcal{D}_K})$ and $\mathrm{div}_D(\Lambda_{\mathcal{D}_K})/\mathcal{H}_D(K)_{(\ell)} = \mathfrak{Cl}'(D)_{(\ell)}$.*

(3) *Up to multiplication by ℓ-adic units, the canonical $\widehat{U}_K$-lattice $\Lambda_K := \Lambda_{\mathcal{D}_K}$ depends only on $K|k$, and not on $\mathcal{D}_K$.*

Proof. To (1): First, the fact that $\mathcal{G}_{\mathcal{D}_K}$ is complete curve-like follows from the fact that all the $(\delta - 1)$ residual function fields $K\mathfrak{v}|k$ have $\mathrm{td}(K\mathfrak{v}|k) = 1$, and the facts that $\mathcal{G}_{\mathcal{D}_K}$ is complete regular-like. In order to show that $\mathcal{G}_{\mathcal{D}_K}$ is ample up to level $\delta = \mathrm{td}(K|k)$, we have to show that conditions (i), (ii), from Definition/Remark 3 (7) are satisfied. First, condition (i) follows immediately from the weak Approximation Lemma. To check condition (ii) is a little bit more technical though. Since $\mathcal{G}_{\mathcal{D}_K}$ is complete regular-like, for every multi-index $\mathfrak{v}$ of $\mathcal{G}_{\mathcal{D}_K}$, there exists a complete regular-like model $X_{\mathfrak{v}} \to k$ of $K\mathfrak{v}|k$ such that $D_{X_{\mathfrak{v}}}$ is the set of 1-vertices of $\mathcal{G}_{\mathcal{D}_{K\mathfrak{v}}}$; hence in particular, one has $\Pi_{1,K\mathfrak{v}} = \Pi_{1,D_{X_{\mathfrak{v}}}}$. And we will denote by $X \to k$ the corresponding model of $K|k$; thus $\Pi_{1,K} = \Pi_{1,D_X}$.

We carry out induction on $d = \mathrm{td}(K|k) > 1$ as follows.

Let $\Delta \subseteq \widehat{K}$ be an ℓ-adically closed submodule. Then Δ gives rise functorially to a subextension $K_\Delta|K$ of $K'|K$ by setting $K_\Delta := \cup_n K_n$, where $K_n := K[\sqrt[n]{\Delta_n}\,]$, and $\Delta_n \subseteq K^\times/n$ is the image of Δ in $K^\times/n = \widehat{K}/n$ for all $n = \ell^e$. (Note that since $\Delta \subseteq \widehat{K}$ is closed, Δ is the projective limit of the Δ_n's inside $\widehat{K}$.) We notice that $K_n|K$ is $\mathbb{Z}/n$ elementary abelian with Galois group equal to $\mathrm{Hom}(\Delta_n, \mu_n)$; thus $K_\Delta|K$ has Galois group $\mathrm{Hom}_{\mathrm{cont}}(\Delta, \mathbb{T}_\ell)$, where $\mathbb{T}_\ell$ is the Tate module of the μ_{ℓ^∞} roots of unity, which we have identified with $\mathbb{Z}_\ell$.

Further, let v be an *arbitrary* valuation of K, and v_Δ a prolongation to K_Δ. Then under the above correspondence one has the following: The decomposition field of $v_\Delta|v$ is K_{Δ_0}, where $\Delta_0 := \Delta \cap \ker(\jmath_v)$, and the inertia field of $v_\Delta|v$ is K_{Δ_1}, where $\Delta_1 := \Delta \cap \widehat{U}_v$. In particular, v is unramified in $K_\Delta|K$ iff $\Delta \subset \widehat{U}_v$.

Checking condition (ii) (a) from Definition/Remark 3 (7): The main technical tool for the proof is Theorem B from Pop [29], which implies the following: Since $\Delta := \widehat{U}_{\mathcal{D}_K}$ is the ℓ-adic dual of $\Pi_{1,K}$, it follows by the definition of $\Pi_{1,K}$ that the corresponding subextension $K_\Delta|K$ is the maximal subextension of $K'|K$ in which all prime divisors v of $K|k$ are unramified. But then by [29] it follows that all the k-valuations of $K|k$ are unramified in $K_\Delta|K$. And correspondingly, the same is true for all the residue function fields $K\mathfrak{v}|k$ of $\mathcal{D}_K$. Now for v a fixed prime divisor of $K|k$, let $\mathcal{V}$ be the set of the k-valuations $\mathfrak{v} = w \circ v$ of $K|k$, with $w \in D_{X_v}$ the set of prime divisors defined by the complete regular-like model $X_v \to k$ mentioned at the beginning of the proof. Since $\widehat{U}_\mathfrak{v} \subseteq \widehat{U}_v$ and $\jmath_v(\widehat{U}_\mathfrak{v}) = \widehat{U}_w$ for $\mathfrak{v} = w \circ v$, it follows that setting $\Delta_v := \cap_{\mathfrak{v}\in\mathcal{V}} \widehat{U}_\mathfrak{v}$, we have

$$\Delta_v \subseteq \widehat{U}_v \text{ and } \jmath_v(\Delta_v) = \jmath_v\big(\cap_\mathfrak{v} \widehat{U}_\mathfrak{v}\big) \subseteq \cap_\mathfrak{v} \jmath_v(\widehat{U}_\mathfrak{v}) = \cap_w \widehat{U}_w = \widehat{U}_{\mathcal{D}_{Kv}}.$$

On the other hand, by the discussion above, all the k-valuations of $K|k$ are unramified in $K_\Delta|K$. Hence in particular so are all the $\mathfrak{v} \in \mathcal{V}$; thus $\Delta \subseteq \widehat{U}_\mathfrak{v}$ for all $\mathfrak{v} \in \mathcal{V}$. We conclude that $\Delta \subseteq \Delta_v$. Therefore, $\jmath_v$ maps $\widehat{U}_{\mathcal{D}_K} =: \Delta$ into $\jmath_v(\Delta_v) \subseteq \widehat{U}_{\mathcal{D}_{Kv}}$, as claimed.

For the second assertion of condition (ii) (a) from Definition/Remark 3 (7), let $\widehat{K}_{\mathrm{fin}} \subset \widehat{K}$ be the union of all the finite-corank submodules of $\widehat{K}$, and define $\widehat{Kv}_{\mathrm{fin}} \subset \widehat{Kv}$ correspondingly. We then have to show that $\widehat{U}_{\mathcal{D}_{Kv}} \cdot \jmath_v(\widehat{K}_{\mathrm{fin}} \cap \widehat{U}_v) = \widehat{Kv}_{\mathrm{fin}}$. Let $\Delta \subset \widehat{K}_{\mathrm{fin}}$ be a finite-corank $\mathbb{Z}_\ell$-submodule. Since Δ has finite corank, there exits a cofinite subset $D' \subset D_X$ such that $v'(\Delta) = 0$ for every $v' \in D$. Therefore, if x_v is the center of v on X, for every sufficiently small open neighborhood $X' \subset X$, we have $v'(\Delta) = 0$ for all $v \in D_{X'}$, $v' \neq v$. In particular, since X is normal, thus smooth at x_v, we can choose $X' \subset X$ to be smooth such that $x_v \in X'$ and $w(\Delta) = 0$ for all $w \in D_{X'}$, $w \neq v$. Since $\Delta \cap \widehat{U}_v$ is contained in $\widehat{U}_{D_{X'}}$, it is sufficient to show that $\jmath_v(\widehat{U}_{D_{X'}})$ is contained in $\widehat{Kv}_{\mathrm{fin}}$, hence mutatis mutandis, we can suppose that $\Delta := \widehat{U}_{D_{X'}}$. If so, $K_\Delta|K$ is the maximal subextension of $K'|K$ in which all $v' \in D_{X'}$ are unramified. Let $\tilde{X} \to k$ be a projective normal completion of X (note that $\tilde{X} \to k$ exists, because $X' \subset X$, and X is normal quasi-projective), and set $\tilde{S} := \tilde{X}\backslash X'$; hence S is a proper closed subset of X which does not contain x_v. Let further $X_{x_v} \subset \tilde{X}$ be the closure of x_v in $\tilde{X}$, and set $S := \tilde{S} \cap X_{x_v}$, thus $S \subset X_{x_v}$ is a proper closed subset. Further, we view $X_{x_v} \to k$ as a projective, thus proper (not necessarily normal) model of $Kv|k$. For every k-valuation w of $Kv|k$ we claim the following:

Claim. Suppose that $w(\jmath_v(\Delta)) \neq 0$. Then the center x_w of w on X_{x_v} lies in S.

Indeed, since $w(\jmath_v(\Delta)) \neq 0$, it follows that setting $\mathfrak{v} = w \circ v$ as a valuation of $K|k$, we have $\mathfrak{v}(\Delta) \neq 0$. Therefore, by the introductory discussion above, it follows that $\mathfrak{v}$ is ramified in $K_\Delta|K$. We claim that the center $x_\mathfrak{v}$ of $\mathfrak{v}$ on $\tilde{X}$ lies in $\tilde{S}$. By contradiction, let $x_\mathfrak{v} \in X'$. Since X' is smooth, hence regular, by the purity of the branch locus, one has $\Pi_{1,D_{X'}} = \Pi_1(X)$. Hence every finite cover $Y' \to X'$ defined by some open subgroup of $\Pi_{1,X'}$ is étale. Hence the cover $Y' \times_X \operatorname{Spec} \mathcal{O}_\mathfrak{v} \to \operatorname{Spec} \mathcal{O}_\mathfrak{v}$ is étale, thus unramified. Therefore, $\mathfrak{v}$ is unramified in $K_\Delta|K$, contradiction! Since $\mathcal{O}_\mathfrak{v} \subset \mathcal{O}_v$, by the valuative criterion for properness, we have $\mathcal{O}_{\tilde{X},x_\mathfrak{v}} \subset \mathcal{O}_{\tilde{X},x_v}$, and $x_\mathfrak{v}$ lies in the closure of $\{x_v\}$ in $\tilde{X}$, hence in X_{x_v}. In particular, since $\mathfrak{v}$ has no center on X', it follows that $x_\mathfrak{v} \in S = X_{x_v} \cap \tilde{S}$. Finally, using the valuative criterion for properness again, it follows that viewing X_{x_v} as a projective (not necessarily normal) model of $Kv|k$, the center of w on X_{x_v} is precisely $x_\mathfrak{v}$. This concludes the proof of the Claim.

Using the claim above, we finish the proof of (ii) (a) from Definition/Remark 3 (7) as follows: For every geometric set of prime divisors D_v of $Kv|k$, only finitely many $w \in D_v$ have center in S. Hence by the claim, only finitely many $w \in D_v$ satisfy $w(\jmath_v(\Delta)) \neq 0$. From this we conclude that $\jmath_v(\Delta)$ has finite corank.

Checking (ii) (b) from Definition/Remark 3 (7): Let $\Delta \subset \widehat{K}_{\text{fin}}$ be a finite corank $\mathbb{Z}_\ell$-module. Further let $X' \subset X$ be a smooth open subvariety such that $\Delta \subset \widehat{U}_{D_{X'}}$. As in the proof of (ii) (a), mutatis mutandis, it sufficient to check (ii) (b) for the $\mathbb{Z}_\ell$-submodule of finite-corank $\Delta := \cap_{v \in D_{X'}} \widehat{U}_v$.

In order to do so, let $X' \subset \tilde{X}$ be a normal projective completion of X', and $\tilde{X} \hookrightarrow \mathbb{P}^N_k$ a projective embedding. If H is a general hyperplane, and $Z := \tilde{X} \cap H$ and $Z' := X' \cap H$ are the corresponding hyperplane sections, it follows that $Z' \hookrightarrow X'$ is a prime Weil divisor such that $Z' \to k$ is smooth, because $X' \to k$ was so. Further, $Z' \hookrightarrow X'$ gives rise to a surjective group homomorphism $\Pi_1(Z') \to \Pi_1(X')$, which is an isomorphism if $\dim(X') > 2$. Hence since X' and Z' are smooth, thus regular, by the purity of the branch locus we have $\Pi_{1,D_{X'}} = \Pi_1(X')$ and $\Pi_{1,D_{Z'}} = \Pi_1(Z')$; thus we get a surjective projection $\Pi_{1,D_{Z'}} \to \Pi_{1,D_{X'}}$. Let $v := v_{Z'}$ be the prime divisor of $K|k$ defined by the Weil prime divisor Z' of X'. Then taking ℓ-adic duals, it follows as in the proof of (ii) (a) above that the surjectivity of the projection $\Pi_{1,D_{Z'}} \to \Pi_{1,D_{X'}}$ implies that $\jmath_v : \widehat{U}_{X'} \to \widehat{U}_{Z'}$ is injective, as claimed.

To (3): It follows immediately from (the proof of) assertion (1) above, together with Remarks 19 (2), (3), and (4) and Fact 20 and Fact 55 (4). □

4 Morphisms and rational quotients of abstract decomposition graphs

4.1 Morphisms

Let $\mathcal{G}$ and $\mathcal{H}$ be given abstract decomposition graphs of levels $\delta_{\mathcal{G}}$ and $\delta_{\mathcal{H}}$, based on $G = G_0$, respectively $H = H_0$. We denote as usual by $T_v \subset Z_v$ and $G_v = Z_v/T_v$ the 1-edges, respectively the 1-vertices, of $\mathcal{G}$, and correspondingly by $T_w \subset Z_w$ and $G_w = Z_w/T_w$ those for $\mathcal{H}$. Further, $\mathcal{G}_v$ and $\mathcal{H}_w$ are the corresponding 1-residual abstract decomposition graphs, which have then level $\delta_{\mathcal{G}} - 1$, respectively $\delta_{\mathcal{H}} - 1$. We also recall that v_0 and w_0 are the trivial valuations of G, respectively H, and that their inertia groups are trivial by definition.

Definition/Remark 24. In the above context we define:

(1) Let $\Phi : G_0 \to H_0$ be a (continuous) group homomorphism. Let $\mathfrak{v}$ and $\mathfrak{w}$ be multi-indices for $\mathcal{G}$ and $\mathcal{H}$. We define inductively on the length of $\mathfrak{v}$ the fact that $\mathfrak{w}$ *corresponds to* $\mathfrak{v}$ *via* Φ as follows; see Definition/Remark 2, especially points (3) and (4), to recall notation:

  (i) The trivial multi-index $\mathfrak{w} = w_0$ corresponds to $\mathfrak{v}$ if and only if $\Phi(T_{\mathfrak{v}}) = 1$ and $\Phi(Z_{\mathfrak{v}})$ is open in H_0. And the only $\mathfrak{w}$ which corresponds to the trivial multi-index $\mathfrak{v} = v_0$ is the trivial multi-index $\mathfrak{w} = w_0$.

  (ii) Suppose that $\mathfrak{w} = (w_s, \ldots, w_1)$ and $\mathfrak{v} = (v_r, \ldots, v_1)$ are both non-trivial, and let us set $\mathfrak{v} = (\mathfrak{v}_1, v_1)$ and $\mathfrak{w} = (\mathfrak{w}_1, w_1)$ with $\mathfrak{v}_1$ and $\mathfrak{w}_1$ the corresponding multi-indices for the residual abstract decomposition graphs $\mathcal{G}_{v_1}$, respectively $\mathcal{H}_{w_1}$. (Note that $\mathfrak{v}_1$ and/or $\mathfrak{w}_1$ might be trivial.) Then we say that $\mathfrak{w}$ corresponds to $\mathfrak{v}$ if and only if one of the following hold:

    (a) If $\Phi(T_{v_1}) = 1$, then under $\Phi_{v_1} : G_{v_1} = Z_{v_1}/T_{v_1} \to H_0$, inductively one has that $\mathfrak{w}$ corresponds to $\mathfrak{v}_1$.

    (b) If $\Phi(T_{v_1}) \neq 1$, then $\Phi(T_{v_1}) \subseteq T_{w_1}$ and $\Phi(Z_{v_1}) \subseteq T_{w_1}$ are open subgroups, and under $\Phi_{v_1} : G_{v_1} = Z_{v_1}/T_{v_1} \to Z_{w_1}/T_{w_1} = H_{w_1}$, inductively one has that $\mathfrak{w}_1$ corresponds to $\mathfrak{v}_1$.

(2) Let $\mathfrak{w}$ correspond to some $\mathfrak{v}$. Then for every $\mathfrak{w}' \leq \mathfrak{w}$, there exists $\mathfrak{v}' \leq \mathfrak{v}$ such that $\mathfrak{w}'$ corresponds to $\mathfrak{v}'$. The proof of this assertion follows easily by induction on the length of $\mathfrak{v}$, and we will omit it.

(3) Finally, let $\mathcal{V}_{\mathcal{G}}$ and $\mathcal{V}_{\mathcal{H}}$ be the sets of the multi-indices $\mathfrak{v}$ of $\mathcal{G}$, respectively $\mathfrak{w}$ of $\mathcal{H}$, and let $\mathcal{V}_{\mathcal{G},\Phi} \subseteq \mathcal{V}_{\mathcal{G}}$ be the set of all $\mathfrak{v} \in \mathcal{V}_{\mathcal{G}}$ such that there exists some $\mathfrak{w}_{\mathfrak{v}} \in \mathcal{V}_{\mathcal{H}}$ which corresponds to $\mathfrak{v}$. Then the correspondence defined at (1) above gives rise to a map $\varphi_\Phi : \mathcal{V}_{\mathcal{G},\Phi} \to \mathcal{V}_{\mathcal{H}}$, $\mathfrak{v} \mapsto \varphi_\Phi(\mathfrak{v}) = \mathfrak{w} := \mathfrak{w}_{\mathfrak{v}}$.

(4) If $\varphi_\Phi(\mathfrak{v}) = \mathfrak{w}$, we say that Φ *maps* $\mathfrak{v}$ *to* $\mathfrak{w}$, or that $\mathfrak{w}$ *is the image of* $\mathfrak{v}$ under Φ.

Definition 25. With the above notation, let $\delta \leq \delta_{\mathcal{G}}, \delta_{\mathcal{H}}$ be a non-negative integer. We define a *level-δ morphism* $\Phi : \mathcal{G} \to \mathcal{H}$ inductively on δ and $\delta_{\mathcal{G}}$ as follows:

(1) A *level-zero morphism* $\Phi : \mathcal{G} \to \mathcal{H}$ is any group homomorphism $\Phi : G \to H$ under which w_0 corresponds to v_0. Equivalently, Φ is open.
(2) A *level-δ morphism* $\Phi : \mathcal{G} \to \mathcal{H}$ is any level-zero morphism $\Phi : G \to H$ which inductively on $\delta_{\mathcal{G}}$ and on $\delta > 0$ satisfies the following:

(i) Almost all 1-vertices of $\mathcal{H}$ correspond to some 1-vertices of $\mathcal{G}$, and every 1-vertex of $\mathcal{H}$ corresponds to only finitely many (maybe to none) of the 1-vertices of $\mathcal{G}$.
(ii) If the trivial valuation w_0 corresponds to a 1-edge v, then $\delta_{\mathcal{G}_v} = \delta_{\mathcal{G}} - 1 \geq \delta$, and the canonical group homomorphism $\Phi_v : G_v = Z_v/T_v \to H_0$ defines a level-δ morphism of the corresponding residual abstract decomposition graphs $\mathcal{G}_v$ and $\mathcal{H}$.
(iii) If w is a 1-edge corresponding to the 1-edge v, then the group homomorphism $\Phi_v : G_v = Z_v/T_v \to Z_w/T_w = H_w$ defines a level-$(\delta-1)$ morphism of the corresponding residual abstract decomposition graphs $\mathcal{G}_v$ and $\mathcal{H}_w$.

Remarks 26. In the above context, let $\Phi : \mathcal{G} \to \mathcal{H}$ be a level-δ morphism of abstract decomposition graphs.

(1) The morphism Φ gives rise to a *Kummer homomorphism*

$$\widehat{\Lambda}_{\mathcal{H}} := \mathrm{Hom}\big(H, \mathbb{Z}_\ell\big) \xrightarrow{\hat{\phi}} \mathrm{Hom}\big(G, \mathbb{Z}_\ell\big) =: \widehat{\Lambda}_{\mathcal{G}}, \quad \varphi \mapsto \varphi \circ \Phi.$$

Since Φ has an open image, and $\widehat{\Lambda}_{\mathcal{G}}$ and $\widehat{\Lambda}_{\mathcal{H}}$ are torsion-free, $\hat{\phi}$ is *injective.*

From now on *suppose that* $\delta > 0$, and that $\mathfrak{v}$ and $\mathfrak{w}$ are the multi-indices of $\mathcal{G}$, respectively $\mathcal{H}$, which correspond to each other. Let $\delta_{\mathfrak{v}}$ and $\delta_{\mathfrak{w}}$ be their lengths, and suppose that $\delta_{\mathfrak{w}} < \delta$.

(2) $\Phi_{\mathfrak{v}} : \mathcal{G}_{\mathfrak{v}} \to \mathcal{H}_{\mathfrak{w}}$ has level $(\delta - \delta_w)$ and the resulting *residual Kummer homomorphism* $\hat{\phi}_{\mathfrak{v}} : \widehat{\Lambda}_{\mathcal{H}_{\mathfrak{w}}} \to \widehat{\Lambda}_{\mathcal{G}_{\mathfrak{v}}}$ is injective by the remark above applied to $\Phi_{\mathfrak{v}}$.
(3) To simplify notation, let us set $\widehat{\Lambda}_{Z_{\mathfrak{v}}} = \mathrm{Hom}\big(Z_{\mathfrak{v}}, \mathbb{Z}_\ell\big)$ and $\widehat{\Lambda}_{T_{\mathfrak{v}}} = \mathrm{Hom}\big(T_{\mathfrak{v}}, \mathbb{Z}_\ell\big)$, thus in particular, $\widehat{\Lambda}_{T_{\mathfrak{v}}} \cong \mathbb{Z}_\ell^{\delta_{\mathfrak{v}}}$. The inclusions $T_{\mathfrak{v}} \hookrightarrow Z_{\mathfrak{v}} \hookrightarrow G$ and the canonical exact sequence $1 \to T_{\mathfrak{v}} \to Z_{\mathfrak{v}} \to G_{\mathfrak{v}} \to 1$ give rise in the same way as at Definition/Remark 3, points (3) and (6), to morphisms of ℓ-adically complete $\mathbb{Z}_\ell$-modules as follows:

$$\jmath^{\mathfrak{v}} : \widehat{\Lambda}_{\mathcal{G}} \xrightarrow{\mathrm{res}_Z} \widehat{\Lambda}_{Z_{\mathfrak{v}}} \xrightarrow{\mathrm{res}_T} \widehat{\Lambda}_{T_{\mathfrak{v}}} \quad \text{and} \quad 0 \to \widehat{\Lambda}_{\mathcal{G}_{\mathfrak{v}}} \xrightarrow{\inf} \widehat{\Lambda}_{Z_{\mathfrak{v}}} \xrightarrow{\mathrm{res}_T} \widehat{\Lambda}_{T_{\mathfrak{v}}} \to 0.$$

In particular, setting $\widehat{U}^1_{\mathfrak{v}} := \ker(\mathrm{res}_Z)$ and $\widehat{U}_{\mathfrak{v}} = \ker(\jmath^{\mathfrak{v}})$, we get exact sequences

$$0 \to \widehat{U}_{\mathfrak{v}} \longrightarrow \widehat{\Lambda}_{\mathcal{G}} \xrightarrow{\jmath^{\mathfrak{v}}} \widehat{\Lambda}_{T_{\mathfrak{v}}} \to 0 \quad \text{and} \quad 0 \to \widehat{U}^1_{\mathfrak{v}} \hookrightarrow \widehat{U}_{\mathfrak{v}} \xrightarrow{\jmath_{\mathfrak{v}}} \widehat{\Lambda}_{\mathcal{G}_{\mathfrak{v}}} \to 0.$$

The surjective morphism $\jmath_{\mathfrak{v}} : \widehat{U}_{\mathfrak{v}} \to \widehat{\Lambda}_{\mathcal{G}_{\mathfrak{v}}}$ is called the canonical $\mathfrak{v}$-*reduction homomorphism*.

(4) By induction on $\delta_{\mathfrak{v}}$ and $\delta_{\mathfrak{w}}$, one gets the following: $\Phi(Z_{\mathfrak{v}}) \subseteq Z_{\mathfrak{w}}$ and $\Phi(Z_{\mathfrak{v}})$ is open in $Z_{\mathfrak{w}}$, and $\Phi(T_{\mathfrak{v}}) \subseteq T_{\mathfrak{w}}$ and $\Phi(T_{\mathfrak{v}})$ is open in $T_{\mathfrak{w}}$. Hence since Φ is open and restricts to an open homomorphism $Z_{\mathfrak{v}} \to Z_{\mathfrak{w}}$ and $T_{\mathfrak{v}} \to T_{\mathfrak{w}}$, by taking ℓ-adic duals we get commutative diagrams with *injective columns and exact rows* as follows:

$$\begin{array}{ccccc} \widehat{U}_{\mathfrak{w}} & \longrightarrow & \widehat{\Lambda}_{Z_{\mathfrak{w}}} & \longrightarrow & \widehat{\Lambda}_{T_{\mathfrak{w}}} \\ \downarrow\hat{\phi} & & \downarrow\hat{\phi} & & \downarrow\hat{\phi}^{\mathfrak{v}} \\ \widehat{U}_{\mathfrak{v}} & \longrightarrow & \widehat{\Lambda}_{Z_{\mathfrak{v}}} & \longrightarrow & \widehat{\Lambda}_{T_{\mathfrak{v}}} \end{array} \quad \text{and} \quad \begin{array}{ccccc} \widehat{U}^{1}_{\mathfrak{w}} & \hookrightarrow & \widehat{U}_{\mathfrak{w}} & \xrightarrow{\jmath_{\mathfrak{w}}} & \widehat{\Lambda}_{\mathcal{H}_{\mathfrak{w}}} \\ \downarrow\hat{\phi} & & \downarrow\hat{\phi} & & \downarrow\hat{\phi}_{\mathfrak{v}} \\ \widehat{U}^{1}_{\mathfrak{v}} & \hookrightarrow & \widehat{U}_{\mathfrak{v}} & \xrightarrow{\jmath_{\mathfrak{v}}} & \widehat{\Lambda}_{\mathcal{H}_{\mathfrak{v}}} \end{array}$$

(5) A special case of the above discussion is that $\mathfrak{v} = v$ and $\mathfrak{w} = w$ are 1-vertices. If τ_v and τ_w are inertia generators at v, respectively w, there exists a unique $a_{vw} \in \mathbb{Z}_\ell$ such that $\Phi(\tau_v) = \tau_w^{a_{vw}}$. And we have commutative diagrams dual to each other:

$$\begin{array}{ccc} T_v & \longrightarrow & G \\ \downarrow\Phi & & \downarrow\Phi \\ T_w & \longrightarrow & H \end{array} \quad \text{and} \quad \begin{array}{ccc} \widehat{\Lambda}_{\mathcal{H}} & \xrightarrow{\jmath^w} & \mathbb{Z}_\ell\varphi_w \\ \downarrow\hat{\phi} & & \downarrow a_{vw} \\ \widehat{\Lambda}_{\mathcal{G}} & \xrightarrow{\jmath^v} & \mathbb{Z}_\ell\varphi_v \end{array}$$

where φ_w and φ_v are as in Construction 5. Further, the horizontal maps in the first diagram are the inclusions, and the last vertical map in the second diagram denotes the $\mathbb{Z}_\ell$-morphism defined by $\varphi_w \mapsto a_{vw}\varphi_v$.

Definition/Remark 27. Let $\Phi : \mathcal{G} \to \mathcal{H}$ be a level-δ morphism of abstract decomposition graphs. We will say that:

(1) Φ is *proper* if first each $\mathfrak{w}$ corresponds to some $\mathfrak{v}$ and every $\mathfrak{v}$ has an image $\mathfrak{w}$, and second, inductively on $\delta_{\mathcal{G}}$, for every 1-edge v of $\mathcal{G}$ and the corresponding edge w of $\mathcal{H}$ (which could be the trivial edge), the residual morphism $\Phi_v : \mathcal{G}_v \to \mathcal{G}_w$ is a proper one.
(2) Φ defines $\mathcal{H}$ as a *level-δ quotient* of $\mathcal{G}$, or that $\mathcal{H}$ is a *level-δ quotient* of $\mathcal{G}$ via Φ, if Φ is proper, and we have $\Phi(G) = H$.
(3) We notice that a level δ morphism $\Phi : \mathcal{G} \to \mathcal{H}$ is a proper morphism iff in the notations from Definition/Remark 24 (3), for every $\mathfrak{w}$ and $\mathfrak{v}$ which correspond to each other one has $\mathcal{V}_{\mathcal{G}_{\mathfrak{v}},\Phi_{\mathfrak{v}}} = \mathcal{V}_{\mathcal{G}_{\mathfrak{v}}}$ and the residual map $\varphi_{\Phi_{\mathfrak{v}}} : \mathcal{V}_{\mathcal{G}_{\mathfrak{v}}} \to \mathcal{V}_{\mathcal{H}_{\mathfrak{w}}}$ is onto.
(4) If $\Phi : \mathcal{G} \to \mathcal{H}$ is a proper morphism, and $\mathfrak{w}$ and $\mathfrak{v}$ correspond to each other, then the corresponding residual morphism $\Phi_{\mathfrak{v}} : \mathcal{G}_{\mathfrak{v}} \to \mathcal{G}_{\mathfrak{w}}$ is proper.

Remarks 28. Let $\Phi : \mathcal{G} \to \mathcal{H}$ be a level $\delta > 0$ proper morphism.

(1) Let $\mathfrak{v}$ be a multi-index of $\mathcal{G}$, and let $T_{\mathfrak{v}} \cong \mathbb{Z}_\ell^{\delta_{\mathfrak{v}}}$ be the inertia group at $\mathfrak{v}$, as defined in Remark/Definition 2 (3). Then $\Phi(T_{\mathfrak{v}})$ is a free $\mathbb{Z}_\ell$-module of rank $\delta' \leq \delta_{\mathfrak{v}}$. Suppose that $\delta' \leq \delta$. Then using the *properness* of Φ, one checks by

induction on $\delta_{\mathfrak{v}}$ that there exists a unique multi-index $\mathfrak{w}$ such that the following hold: $\Phi(Z_{\mathfrak{v}}) \subseteq Z_{\mathfrak{w}}$ and $\Phi(T_{\mathfrak{v}}) \subseteq T_{\mathfrak{w}}$ are open subgroups. Thus in particular, $\mathfrak{w}$ corresponds to $\mathfrak{v}$.

(2) Denote by $T_{\mathcal{G}}$ the subgroup of G generated by all the inertia elements of G, and define $T_{\mathcal{H}} \subseteq H$ correspondingly. Then in the notation from Definition/Remark 3 (2), Φ gives rise to a commutative diagram as follows:

$$\begin{array}{ccccccc} 1 \to & T_{\mathcal{G}} & \longrightarrow & G & \longrightarrow & \Pi_{1,\mathcal{G}} & \to 1 \\ & \downarrow \Phi & & \downarrow \Phi & & \downarrow & \\ 1 \to & T_{\mathcal{H}} & \longrightarrow & H & \longrightarrow & \Pi_{1,\mathcal{H}} & \to 1. \end{array}$$

Next suppose that $\mathcal{G}$ and $\mathcal{H}$ are divisorial, and let $\mathfrak{T}_{\mathcal{G}} = (\tau_v)_v$ and $\mathfrak{T}_{\mathcal{H}} = (\tau_w)_w$ be distinguished systems of generators for $\mathcal{G}$, respectively $\mathcal{H}$, which give rise to abstract divisor groups $\mathrm{Div}_{\mathfrak{T}_{\mathcal{G}}}$ and $\mathrm{Div}_{\mathfrak{T}_{\mathcal{H}}}$ for $\mathcal{G}$, respectively $\mathcal{H}$, and abstract divisorial lattices $\Lambda_{\mathcal{G}}$ and $\Lambda_{\mathcal{H}}$.

(3) For every w, denote by X_w the set of all the v to which w corresponds. Then X_w is finite non-empty (by the fact that Φ is proper). For every w and $v \in X_w$, there exists a unique $a_{vw} \in \mathbb{Z}_\ell$ such that $\Phi(\tau_v) = \tau_w^{a_{vw}}$. Equivalently, denoting $\mathfrak{B}_{\mathcal{G}} = (\varphi_v)_v$ and $\mathfrak{B}_{\mathcal{H}} = (\varphi_w)_w$ the dual bases to $\mathfrak{T}_{\mathcal{G}} = (\tau_v)_v$ and $\mathfrak{T}_{\mathcal{H}} = (\tau_w)_w$ as defined/introduced at Construction 5, by Remark 26 (4) above via $\hat{\phi}$ we have

$$\varphi_w \mapsto \textstyle\sum_{v \in X_w} a_{vw} \varphi_v,$$

and therefore $\hat{\phi}$ gives rise to a morphism

$$\mathrm{div}_\Phi : \widehat{\mathrm{Div}}_{\mathcal{H}} \to \widehat{\mathrm{Div}}_{\mathcal{G}}$$

which maps $\mathrm{Div}_{\mathfrak{T}_{\mathcal{H}}} \otimes \mathbb{Z}_\ell$ into $\mathrm{Div}_{\mathfrak{T}_{\mathcal{G}}} \otimes \mathbb{Z}_\ell$ and fits into the following commutative diagram:

$$(*) \qquad \begin{array}{ccccccccc} 0 \to & \widehat{U}_{\mathcal{H}} & \longrightarrow & \widehat{\Lambda}_{\mathcal{H}} & \xrightarrow{\jmath^{\mathcal{H}}} & \widehat{\mathrm{Div}}_{\mathcal{H}} & \longrightarrow & \widehat{\mathfrak{Cl}}_{\mathcal{H}} & \to 0 \\ & \downarrow \hat{\phi} & & \downarrow \hat{\phi} & & \downarrow \mathrm{div}_\Phi & & \downarrow \mathrm{can} & \\ 0 \to & \widehat{U}_{\mathcal{G}} & \longrightarrow & \widehat{\Lambda}_{\mathcal{G}} & \xrightarrow{\jmath^{\mathcal{G}}} & \widehat{\mathrm{Div}}_{\mathcal{G}} & \longrightarrow & \widehat{\mathfrak{Cl}}_{\mathcal{G}} & \to 0 \end{array}$$

(4) Recall that for every divisorial $\widehat{U}_{\mathcal{G}}$-lattice $\Lambda_{\mathfrak{T}_{\mathcal{G}}}$ in $\widehat{\Lambda}_{\mathcal{G}}$ one has the following: $\widehat{\Lambda}_{\mathcal{G},\mathrm{fin}} = \Lambda_{\mathfrak{T}_{\mathcal{G}}} \otimes \mathbb{Z}_\ell$, and therefore $\widehat{\Lambda}_{\mathcal{G},\mathrm{fin}}$ is exactly the preimage of $\mathrm{Div}_{\mathfrak{T}_{\mathcal{G}}} \otimes \mathbb{Z}_\ell$ under $\jmath^{\mathcal{G}}$. Hence from the commutative diagram $(*)$ above it follows that

$$(**)\ \widehat{\Lambda}_{\mathcal{H},\mathrm{fin}} = \hat{\phi}^{-1}(\widehat{\Lambda}_{\mathcal{G},\mathrm{fin}}),\ \hat{\phi}(\widehat{\Lambda}_{\mathcal{H},\mathrm{fin}}) \cap \widehat{U}_{\mathcal{G}} = \hat{\phi}(\widehat{U}_{\mathcal{H}}),\ \hat{\phi}(\widehat{\Lambda}_{\mathcal{H},\mathrm{fin}}) = \hat{\phi}(\widehat{\Lambda}_{\mathcal{H}}) \cap \widehat{\Lambda}_{\mathcal{G},\mathrm{fin}}.$$

In particular, $\hat{\phi}$ maps finite-corank submodules into such, and preimages of finite-corank submodules under $\hat{\phi}$ are again such.

(5) With the above notation, the following are equivalent:

(a) There exist $\mathfrak{T}_{\mathcal{G}} = (\tau_v)_v$, $\mathfrak{T}_{\mathcal{H}} = (\tau_w)_w$ such that $a_{vw} \in \mathbb{Z}_{(\ell)}$ for all $w, v \in X_w$.
(b) $\mathrm{div}_\Phi(\mathrm{Div}_{\mathfrak{T}_{\mathcal{H}}}) \subseteq \mathrm{Div}_{\mathfrak{T}_{\mathcal{G}}}$.
(c) $\hat{\phi}(\Lambda_{\mathcal{H}}) \subseteq \Lambda_{\mathcal{G}}$.

And if the above equivalent conditions are satisfied, one has equalities as follows:

$$(***)\ \mathrm{Div}_{\mathfrak{T}_{\mathcal{H}}} = \mathrm{div}_\Phi^{-1}(\mathrm{Div}_{\mathfrak{T}_{\mathcal{G}}}),\ \Lambda_{\mathfrak{T}_{\mathcal{H}}} = \hat{\phi}^{-1}(\Lambda_{\mathfrak{T}_{\mathcal{G}}}),\ \hat{\phi}(\Lambda_{\mathfrak{T}_{\mathcal{H}}}) = \hat{\phi}(\widehat{\Lambda}_{\mathcal{H}}) \cap \Lambda_{\mathfrak{T}_{\mathcal{G}}}.$$

Thus in particular, $\Lambda_{\mathfrak{T}_{\mathcal{H}}}$ can be recovered from $\Lambda_{\mathfrak{T}_{\mathcal{G}}}$ via $\hat{\phi}$.

Proof of (5): The implication (a) $\Rightarrow$ (b) follows immediately from the definition of div_Φ, and the implication (b) $\Rightarrow$ (c) follows from the definition of $\Lambda_{\mathcal{H}}$ and $\Lambda_{\mathcal{G}}$. In order to prove (c) $\Rightarrow$ (a), let $v \in X_w$ be given. Then combining Remark 10 (1), with the second diagram from Remark 26 (4), we get a commutative diagram of the form

$$\begin{array}{ccc} \Lambda_{\mathfrak{T}_{\mathcal{H}}} & \xrightarrow{\jmath^w} & \mathbb{Z}_{(\ell)}\varphi_w \\ \downarrow \hat{\phi} & & \downarrow a_{vw} \\ \Lambda_{\mathfrak{T}_{\mathcal{G}}} & \xrightarrow{\jmath^v} & \mathbb{Z}_{(\ell)}\varphi_v; \end{array}$$

hence it follows that $a_{vw} \in \mathbb{Z}_{(\ell)}$, as claimed. Finally, let us show that in case the equivalent conditions (a), (b), (c), are satisfied, the equalities $(**)$ hold. First observe that since $\hat{\phi}$ and div_Φ are injective, all the above equalities are equivalent. Thus it is enough to prove one of them, say the first one: Recall that by point (3) above, $\mathrm{Div}_{\mathfrak{T}_{\mathcal{H}}} \otimes \mathbb{Z}_\ell$ and $\mathrm{Div}_{\mathfrak{T}_{\mathcal{G}}} \otimes \mathbb{Z}_\ell$ are free $\mathbb{Z}_\ell$-modules on the bases $\mathfrak{B}_{\mathcal{H}} = (\varphi_w)_w$, respectively $\mathfrak{B}_{\mathcal{G}} = (\varphi_v)_v$, and that div_Φ maps the former $\mathbb{Z}_\ell$-module into the latter one. Hence $\mathrm{div}_\Phi^{-1}(\mathrm{Div}_{\mathfrak{T}_{\mathcal{G}}}) \subseteq \mathrm{Div}_{\mathfrak{T}_{\mathcal{H}}} \otimes \mathbb{Z}_\ell$. Now let $x = \sum_w b_w\varphi_w$ with $b_w \in \mathbb{Z}_\ell$ be an element of $\mathrm{div}_\Phi^{-1}(\mathrm{Div}_{\mathfrak{T}_{\mathcal{G}}})$. Then $\mathrm{div}_\Phi(x) = \sum_w \sum_{v \in X_w} b_w a_{vw}\varphi_v$ lies in $\mathrm{Div}_{\mathfrak{T}_{\mathcal{G}}}$; hence $b_w a_{vw} \in \mathbb{Z}_{(\ell)}$ for all w and $v \in X_w$. On the other hand, since $a_{vw} \in \mathbb{Z}_{(\ell)}$, it follows that b_w are rational numbers. Since they lie in $\mathbb{Z}_\ell$ as well, it follows that $b_w \in \mathbb{Z}_{(\ell)}$. But then we finally get that $x = \sum_w b_w\varphi_w$ lies in $\Lambda_{\mathfrak{T}_{\mathcal{H}}}$ as claimed.

Definition/Remark 29. Let $\Phi : \mathcal{G} \to \mathcal{H}$ be a level-δ proper morphism of divisorial abstract decomposition graphs with $\delta > 0$.

(1) We say that Φ is *divisorial*, if all residual morphisms $\Phi_{\mathfrak{v}} : \mathcal{G}_{\mathfrak{v}} \to \mathcal{H}_{\mathfrak{w}}$ with $\mathfrak{w}$ of length $< \delta$ satisfy the equivalent conditions (a), (b), (c) from (5) above.
(2) If $\Phi : \mathcal{G} \to \mathcal{H}$ is divisorial, the residual Kummer morphism $\hat{\phi}_{\mathfrak{v}}$ maps divisorial lattices for $\mathcal{H}_{\mathfrak{w}}$ into divisorial lattices for $\mathcal{G}_{\mathfrak{v}}$. And the commutative diagram $(*)$ from Remarks 28 above gives rise to a commutative sub-diagram:

$$(**) \qquad \begin{array}{ccccccccc} 0 \to & \widehat{U}_{\mathcal{H}} & \to & \Lambda_{\mathcal{H}} & \xrightarrow{\jmath^{\mathcal{H}}} & \mathrm{Div}_{\mathcal{H}} & \to & \mathfrak{Cl}_{\mathcal{H}} & \to 0 \\ & \downarrow \hat{\phi} & & \downarrow \hat{\phi} & & \downarrow \mathrm{div}_\Phi & & \downarrow \mathrm{can} & \\ 0 \to & \widehat{U}_{\mathcal{G}} & \to & \Lambda_{\mathcal{G}} & \xrightarrow{\jmath^{\mathcal{G}}} & \mathrm{Div}_{\mathcal{G}} & \to & \mathfrak{Cl}_{\mathcal{G}} & \to 0 \end{array}$$

(3) It is not too difficult to give examples of proper morphisms Φ of divisorial abstract decomposition graphs such that Φ are not divisorial. Indeed, one can give such examples even in the case that both $\mathcal{G}$ and $\mathcal{H}$ are complete curve like, and Φ is a proper morphism of level $\delta = 1$. The next proposition shows that actually the case $\delta = 1$ is the "generic" source for proper non-divisorial morphisms.

Proposition 30. *Let $\Phi : \mathcal{G} \to \mathcal{H}$ be a level-δ proper morphism of divisorial abstract decomposition graphs, where $\delta = \delta_{\mathcal{H}} > 0$. Then the following hold:*

(1) *Let $\mathfrak{v}$, $\mathfrak{w}$ be all pairs of multi-indices of $\mathcal{G}$, respectively of $\mathcal{H}$, such that $\mathfrak{w}$ has length $\delta_{\mathcal{H}} - 1$ and corresponds to $\mathfrak{v}$. Suppose that* for all such pairs $\mathfrak{v}, \mathfrak{w}$ *the residual morphism $\Phi_{\mathfrak{v}} : \mathcal{G}_{\mathfrak{v}} \to \mathcal{H}_{\mathfrak{w}}$ is divisorial. Then $\Phi : \mathcal{G} \to \mathcal{H}$ is divisorial.*

(2) *If Φ is an isomorphism, then $\hat{\phi}$ is an isomorphism too, and Φ is a divisorial morphism of abstract decomposition graphs. Hence for any abstract divisor groups* $\mathrm{Div}_{\mathcal{G}}$ *and* $\mathrm{Div}_{\mathcal{H}}$ *of $\mathcal{G}$, respectively $\mathcal{H}$, there exists $\varepsilon \in \mathbb{Z}_\ell^\times$ such that the diagram below is commutative:*

$$(***)\qquad \begin{array}{ccccccccc} 0 \to & \widehat{U}_{\mathcal{H}} & \longrightarrow & \Lambda_{\mathcal{H}} & \xrightarrow{\jmath^{\mathcal{H}}} & \mathrm{Div}_{\mathcal{H}} & \longrightarrow & \mathfrak{Cl}_{\mathcal{H}} & \to 0 \\ & \downarrow{\scriptstyle \varepsilon\cdot\hat{\phi}} & & \downarrow{\scriptstyle \varepsilon\cdot\hat{\phi}} & & \downarrow{\scriptstyle \varepsilon\cdot\mathrm{div}_{\Phi}} & & \downarrow{\scriptstyle \varepsilon\cdot\mathrm{can}} & \\ 0 \to & \widehat{U}_{\mathcal{G}} & \longrightarrow & \Lambda_{\mathcal{G}} & \xrightarrow{\jmath^{\mathcal{G}}} & \mathrm{Div}_{\mathcal{G}} & \longrightarrow & \mathfrak{Cl}_{\mathcal{G}} & \to 0 \end{array}$$

Proof. To (1): One carries out induction on $\delta_{\mathcal{G}}$.

Case (1) $\delta_{\mathcal{G}} = 1$. Then $1 = \delta_{\mathcal{G}} \geq \delta = \delta_{\mathcal{H}} > 0$; hence all these numbers equal 1, and the assertion follows from/by the definitions and the hypothesis of the proposition.

Case (2) $\delta > 1$ arbitrary. Let v be some 1-index of $\mathcal{G}$, and w the image of v under Φ. Note that w is either the trivial valuation w_0, or otherwise w is a 1-index of $\mathcal{H}$. We show that the resulting residual morphism $\Phi_v : \mathcal{G}_v \to \mathcal{H}_w$ satisfies the hypothesis of the proposition: First $\Phi_v : \mathcal{G}_v \to \mathcal{H}_w$ is a proper morphism, as $\Phi : \mathcal{G} \to \mathcal{H}$ was so by hypothesis. Second, let $\mathfrak{w}_w$ be a multi-index of $\mathcal{H}_w$ of length $\delta_{\mathcal{H}_w} - 1$, and $\mathfrak{v}_w$ a multi-index of $\mathcal{G}_v$ such that $\mathfrak{w}_w$ corresponds to $\mathfrak{v}_w$ under Φ_v.

Claim 1. $\Phi_{\mathfrak{v}_w} : \mathcal{G}_{\mathfrak{v}_w} \to \mathcal{H}_{\mathfrak{w}_w}$ is divisorial.

Indeed, let us first suppose that $w = w_0$ is the trivial valuation. Then $\mathcal{H}_w = \mathcal{H}$, and $\mathfrak{w} := \mathfrak{w}_w$ is a multi-index of $\mathcal{H}$ of length $\delta - 1$. Further, $(\mathfrak{v}_w, v)$ is a multi-index of $\mathcal{G}$ which corresponds to $\mathfrak{w}$. And we have that $\mathcal{G}_{\mathfrak{v}_w} = \mathcal{G}_{(\mathfrak{v}_w,v)}$, $\mathcal{H}_{\mathfrak{w}_w} = \mathcal{H}_{\mathfrak{w}}$, and $\Phi_{\mathfrak{v}_w} : \mathcal{G}_{\mathfrak{v}_w} \to \mathcal{H}_{\mathfrak{w}_w}$ is actually the same as $\Phi_{(\mathfrak{v}_w,v)} : \mathcal{G}_{(\mathfrak{v}_w,v)} \to \mathcal{H}_{\mathfrak{w}}$. But then the claim follows from the hypothesis of the Proposition. Next suppose that $w \neq w_0$ is a 1-index of $\mathcal{H}$. Then $\Phi_{\mathfrak{v}}$ has level $\delta - 1 = \delta_{\mathcal{H}} - 1 = \delta_{\mathcal{H}_w}$. Moreover, if $\mathfrak{w}_w$ is a multi-index of $\mathcal{H}_w$ of length $\delta_{\mathcal{H}_w} - 1$, then $(\mathfrak{w}_w, w)$ is a multi-index of $\mathcal{H}$ of length

$$(\delta_{\mathcal{H}_w} - 1) + 1 = \delta_{\mathcal{H}_w} = \delta_{\mathcal{H}} - 1.$$

And since $\mathfrak{w}_w$ corresponds to $\mathfrak{v}_w$, and w to v, it follows that $(\mathfrak{w}_w, w)$ corresponds to $(\mathfrak{v}_w, v)$. But then by the hypothesis of the proposition, the residual morphism

$$\Phi_{(\mathfrak{v}_w,v)} : \mathcal{G}_{(\mathfrak{v}_w,v)} \to \mathcal{H}_{(\mathfrak{w}_w,w)}$$

is divisorial. On the other hand, by definitions we have identifications $\mathcal{G}_{(\mathfrak{v}_w,v)} = \mathcal{G}_{\mathfrak{v}_w}$ and $\mathcal{H}_{(\mathfrak{w}_w,w)} = \mathcal{H}_{\mathfrak{w}_w}$, and $\Phi_{(\mathfrak{v}_w,v)} = \Phi_{\mathfrak{v}_w}$. This completes the proof of the Claim 1.

Coming back to the proof of assertion 1 of the proposition, let $\Lambda_{\mathcal{H}}$ and $\Lambda_{\mathcal{G}}$ be divisorial lattices in $\widehat{\Lambda}_{\mathcal{H}}$, respectively $\widehat{\Lambda}_{\mathcal{G}}$. For $\Gamma \subset \widehat{\Lambda}_{\mathcal{H}}$ of finite corank and satisfying $\Gamma \cap \widehat{U}_{\mathcal{H}} = (0)$, set $\Delta := \hat{\phi}(\Gamma)$.

Claim 2. $\hat{\phi} : \widehat{\Lambda}_{\mathcal{H}} \to \widehat{\Lambda}_{\mathcal{G}}$ maps Γ isomorphically onto its image Δ, and further one has that $\Delta \cap \widehat{U}_{\mathcal{G}} = (0)$, and Δ is a finite-corank $\mathbb{Z}_\ell$-submodule of $\widehat{\Lambda}_{\mathcal{G}}$.

Indeed, by the diagram $(*)$ from Remark 28 (3), and in the notation from there, we have that $\jmath^{\mathcal{H}}$ is injective on Γ, since $\Gamma \cap \widehat{U}_{\mathcal{H}} = (0)$. Since $\widehat{\mathrm{div}_\Phi}$ is injective, it finally follows that $\jmath^{\mathcal{H}}(\Gamma)$ is mapped injectively into $\widehat{\mathrm{Div}}_{\mathcal{G}}$. Therefore, $\hat{\phi}$ maps Γ injectively into $\widehat{\Lambda}_{\mathcal{G}}$, and $\Delta := \hat{\phi}(\Gamma)$ has trivial intersection with $\widehat{U}_{\mathcal{G}}$. Now let us check that Δ has finite corank in $\widehat{\Lambda}_{\mathcal{G}}$: First let v be a 1-edge of $\mathcal{G}$ such that $\jmath^v(\Gamma) \neq (0)$. Equivalently, $\Delta = \hat{\phi}(\Gamma)$ has a non-trivial image under

$$\jmath^v \circ \hat{\phi} : \widehat{\Lambda}_{\mathcal{H}} \xrightarrow{\hat{\phi}} \widehat{\Lambda}_{\mathcal{G}} \xrightarrow{\jmath^v} \mathbb{Z}_\ell \varphi_v.$$

Hence $\jmath^v \circ \hat{\phi}(\Delta)$ is non-trivial. Since the above sequence is ℓ-adically dual to $T_v \hookrightarrow G \xrightarrow{\Phi} H$, it follows that $\Phi(T_v) \neq 1$ in H. Since Φ is proper by hypothesis, it follows that there exists w such that $\Phi(T_v) \subseteq T_w$, and $\Phi(T_v)$ is open in T_w. Hence finally w corresponds to v. Therefore, if $\jmath^v(\Delta) \neq (0)$, then there exists some $w \neq w_0$ corresponding to v.

Next, by the commutativity of the second diagram in Remark 26 (4), it follows that $\jmath^v(\Delta) \neq (0)$ if and only if $\jmath^w(\Gamma) \neq (0)$. Now since Γ has finite corank, there exist only finitely many valuations w of H such that $\jmath^w(\Gamma) \neq (0)$. Finally, for each such w there exist only finitely many v's such that w corresponds to one of the v's. Thus finally there are only finitely many valuations v of G such that $\jmath^v(\Delta) \neq (0)$. This completes the proof of Claim 2.

Now suppose that Γ is non-trivial. Then we have the following situation: Γ and its isomorphic image Δ are non-trivial finite-corank submodules of $\widehat{\Lambda}_{\mathcal{H}}$, respectively $\widehat{\Lambda}_{\mathcal{G}}$. Since $\Delta \cap \widehat{U}_{\mathcal{G}} = (0)$, it follows that $\Delta \cap \Lambda_{\mathcal{G}}$ is a lattice in Δ which completely determines the divisorial lattice $\Lambda_{\mathcal{G}}$ in the ℓ-adic equivalence class of all the divisorial lattices of $\mathcal{G}$. Correspondingly, the same is true for $\Gamma \cap \widehat{U}_{\mathcal{H}}$ and $\Lambda_{\mathcal{H}}$, etc. On the other hand, since Δ has finite corank, by the ampleness of $\mathcal{G}$, there exist valuations v of G such that the following are satisfied:

(j) $\Delta \subset \widehat{U}_v$ and $\jmath_v$ maps Δ injectively into $\widehat{\Lambda}_{\mathcal{G}_v}$, and for such v set $\Delta_v := \jmath_v(\Delta)$.

(jj) $\Delta_v \cap \widehat{U}_{\mathcal{G}_v} = (0)$, because $\Delta \cap \widehat{U}_{\mathcal{G}} = (0)$ by the discussion above.

For such a valuation v, the lattice $\Delta \cap \Lambda_{\mathcal{G}}$ is mapped by $\jmath_v$ isomorphically onto a lattice in Δ_v. Hence by the properties (i), (ii), from Fact 8, we get that there exists a *unique* divisorial $\widehat{U}_{\mathcal{G}_v}$-lattice $\Lambda_{\mathcal{G}_v}$ of $\mathcal{G}_v$ such that $\jmath_v(\Delta \cap \Lambda_{\mathcal{G}}) = \Delta_v \cap \Lambda_{\mathcal{G}_v}$. For v as above we analyze the following cases:

Case (a): $\Phi(T_v) = 1$. Then the trivial valuation w_0 corresponds to v, and for the residual morphism $\Phi_v : \mathcal{G}_v \to \mathcal{H}$ we have that Φ_v is divisorial by Claim 1. Hence by Remark 28 (5), there exists a unique divisorial $\widehat{U}_{\mathcal{H}}$-lattice $\Lambda_{\mathcal{H}}$ of $\mathcal{H}$ such that the Kummer homomorphism $\hat{\phi}_v : \widehat{\Lambda}_{\mathcal{H}} \to \widehat{\Lambda}_{\mathcal{G}_v}$ maps $\Lambda_{\mathcal{H}}$ into $\Lambda_{\mathcal{G}_v}$.

Case (b): $\Phi(T_v) \neq 1$. Then there is a non-trivial valuation w corresponding to v, and for the corresponding residual morphism $\Phi_v : \mathcal{G}_v \to \mathcal{H}_w$ we have that Φ_v is divisorial by Claim 1. Hence by Remark 28 (5), there exists a unique divisorial $\widehat{U}_{\mathcal{H}_w}$-lattice $\Lambda_{\mathcal{H}_w}$ of $\mathcal{H}_w$ such that the Kummer homomorphism $\hat{\phi}_v : \widehat{\Lambda}_{\mathcal{H}_w} \to \widehat{\Lambda}_{\mathcal{G}_v}$ maps $\Lambda_{\mathcal{H}_w}$ into $\Lambda_{\mathcal{G}_v}$. Moreover, since $\hat{\phi}$ maps Γ isomorphically onto its image Δ, and $\jmath_v$ maps Δ isomorphically onto its image Δ_v, we get that since $\jmath_v \circ \hat{\phi}$ and $\jmath_w \circ \hat{\phi}_v$ coincide on Γ, it follows that $\jmath_w$ maps Γ isomorphically onto its image Γ_w, and that $\hat{\phi}_v$ maps Γ_w isomorphically onto Δ_v. Therefore, w satisfies mutatis mutandis the conditions (j), (jj), above with respect to Γ. Hence $\Gamma_w \cap \Lambda_{\mathcal{H}_w}$ is a lattice in Γ_w. Hence there exists a unique divisorial $\widehat{U}_{\mathcal{H}}$-lattice $\Lambda_{\mathcal{H}}$ of $\mathcal{H}$ such that $\Gamma \cap \Lambda_{\mathcal{H}}$ is mapped isomorphically onto $\Gamma_w \cap \Lambda_{\mathcal{H}_w}$.

Claim 3. In both cases above, $\hat{\phi}$ maps $\Lambda_{\mathcal{H}}$ into $\Lambda_{\mathcal{G}}$.

First, with the notation from above, it is clear by the discussion above that $\hat{\phi}$ maps $\Gamma \cap \Lambda_{\mathcal{H}}$ isomorphically onto $\Delta \cap \Lambda_{\mathcal{G}}$. Now let Γ' be a finite-corank $\mathbb{Z}_\ell$-module such that $\Gamma' \cap \widehat{U}_{\mathcal{H}} = 1$ and $\Gamma \subseteq \Gamma'$. Let $\Lambda'_{\mathcal{H}}$ be the divisorial $\widehat{U}_{\mathcal{H}}$-lattice given by the construction above when starting with Γ' instead of Γ. Then we have that $\Gamma \cap \Lambda'_{\mathcal{H}}$ is a lattice in Γ, which is ℓ-adically equivalent to $\Gamma \cap \Lambda_{\mathcal{H}}$. Hence $\hat{\phi}(\Gamma \cap \Lambda_{\mathcal{H}})$ and $\hat{\phi}(\Gamma \cap \Lambda'_{\mathcal{H}})$ are ℓ-adically equivalent lattices in $\Delta = \hat{\phi}(\Gamma)$, and both of them are contained in $\Lambda_{\mathcal{G}}$. Hence $\hat{\phi}(\Gamma \cap \Lambda_{\mathcal{H}}) = \hat{\phi}(\Gamma \cap \Lambda'_{\mathcal{H}})$, thus $\Gamma \cap \Lambda_{\mathcal{H}} = \Gamma \cap \Lambda'_{\mathcal{H}}$. Hence finally $\Lambda_{\mathcal{H}}$ and $\Lambda'_{\mathcal{H}}$ are equal. In other words, for every finite-corank $\mathbb{Z}_\ell$-module Γ' of $\mathcal{H}$ as above we have that if $\Gamma \subseteq \Gamma'$, then $\Gamma' \cap \Lambda_{\mathcal{H}}$ is mapped into $\Lambda_{\mathcal{G}}$. But then $\Lambda_{\mathcal{H}} = \widehat{U}_{\mathcal{H}} + \cup_{\Gamma'}(\Gamma' \cap \Lambda_{\mathcal{H}})$ is mapped into $\Lambda_{\mathcal{G}}$, as claimed.

To (2): Since Φ is an isomorphism, it follows that $\delta_{\mathcal{G}} = \delta = \delta_{\mathcal{H}}$, hence Φ gives rise to a bijection of the multi-indices $\mathfrak{v}$ and $\mathfrak{w}$ of length $\delta - 1$ of $\mathcal{G}$, respectively of $\mathcal{H}$; and if $\mathfrak{v}$ and $\mathfrak{w}$ are such indices, then the residual morphism $\Phi_{\mathfrak{v}} : \mathcal{G}_{\mathfrak{v}} \to \mathcal{H}_{\mathfrak{w}}$ is by definition an isomorphism of compete curve-like abstract decomposition graphs. Thus by assertion 1, it is sufficient to prove that all $\Phi_{\mathfrak{v}} : \mathcal{G}_{\mathfrak{v}} \to \mathcal{H}_{\mathfrak{w}}$ as above are divisorial. Let $(\sigma_v)_v$ be a distinguished system of inertia generators for $G_{\mathfrak{v}}$, where the v are the 1-edges of $\mathcal{G}_{\mathfrak{v}}$. If w is the 1-edge of $\mathcal{G}_{\mathfrak{w}}$ corresponding to v, then setting $\tau_w := \Phi_{\mathfrak{v}}(\sigma_v)$, the system $(\tau_w)_w$ is a distinguished system of inertia generators of $\mathcal{H}_{\mathfrak{w}}$. In particular, condition (a) from Remark/Definition 28 (5), is satisfied. Thus $\Phi_{\mathfrak{v}}$ is divisorial by definition, etc. □

4.2 *Rational quotients and geometric like abstract decomposition graphs*

We begin by first defining rational quotients of divisorial abstract decomposition graphs. The point is that (divisorial) abstract decomposition graphs that arise from geometry have "sufficiently many" rational quotients; and morphisms of (divisorial) abstract decomposition graphs arising from geometry are compatible with the rational quotients. This suggests that for applications, one should consider/study divisorial abstract decomposition graphs endowed with "sufficiently many" rational quotients, and morphisms of such enriched structures.

To begin with, let $\mathcal{G}_\alpha$ be a level-one complete curve-like abstract decomposition graph. Recall the notation from Construction 5, Case $\delta = 1$: For every distinguished system of generators $\mathfrak{T}_\alpha = (\tau_v)_v$ of $\mathcal{G}_\alpha$, we have an exact sequence

$$0 \to \widehat{U}_{\mathcal{G}_\alpha} \hookrightarrow \Lambda_{\mathfrak{T}_\alpha} \xrightarrow{\jmath^{\mathcal{G}_\alpha}} \mathrm{Div}_{\mathfrak{T}_\alpha} \xrightarrow{\text{can}} \mathfrak{Cl}_{\mathfrak{T}_\alpha} \cong \mathbb{Z}_{(\ell)} \to 0\,.$$

Definition/Remark 31. With the notation from above we define:

(1) A level-one divisorial abstract decomposition graph $\mathcal{G}_\alpha$ is called *rational* if $\widehat{U}_{\mathcal{G}_\alpha} = (0)$ for some (thus every) distinguished system of inertia generators $\mathfrak{T}_\alpha$ of $\mathcal{G}_\alpha$, as introduced in Construction 5, Case $\delta = 1$.

We notice the following: Since $\widehat{U}_{\mathcal{G}_\alpha} = (0)$, every $\widehat{U}_{\mathcal{G}_\alpha}$-lattice in $\mathcal{G}_\alpha$ is actually a lattice in $\widehat{\Lambda}_{\mathcal{G}_\alpha}$. Let $\mathfrak{T}_\alpha = (\tau_v)_v$ be a distinguished system of inertia generators, and $\mathfrak{B}_\alpha = (\varphi_v)_v$ the corresponding $\mathbb{Z}_{(\ell)}$-basis of $\mathrm{Div}_{\mathfrak{T}_\alpha}$. An element of the form

$$\boldsymbol{x} = \varphi_{v'} - \varphi_v$$

is called a *generating element* of $\Lambda_{\mathcal{G}_\alpha}$. We set $(\boldsymbol{x})_0 := v'$ and $(\boldsymbol{x})_\infty := v$, and call these the *zero*, respectively the *pole*, of $\boldsymbol{x}$. Further, we define

$$\mathcal{P}_v = \{\boldsymbol{x} \in \Lambda_{\mathfrak{T}_\alpha} \mid \boldsymbol{x} \text{ generating, and } (\boldsymbol{x})_\infty = v\} = \{\,\varphi_{v'} - \varphi_v \mid \text{ all } v' \neq v\},$$

and call it a *generating set* at v for $\Lambda_{\mathfrak{T}_\alpha}$. Clearly, $\mathcal{P}_v$ defines a $\mathbb{Z}_{(\ell)}$-basis of $\Lambda_{\mathfrak{T}_\alpha}$ for every v. And if $\mathfrak{T}'_\alpha = \mathfrak{T}^\varepsilon_\alpha$ is another distinguished system of inertia generators, and $\mathcal{P}'_v$ is correspondingly defined, then $\varepsilon \in \mathbb{Z}^\times_\ell$ is the unique ℓ-adic unit such that $\varepsilon \cdot \mathcal{P}'_v = \mathcal{P}_v$.

(2) Let $\mathcal{G}$ be a level-δ divisorial abstract decomposition graph, where $\delta > 0$. We will consider quotients $\Phi_\alpha : \mathcal{G} \to \mathcal{G}_\alpha$ of $\mathcal{G}$ together with their Kummer homomorphisms $\hat{\phi}_\alpha : \widehat{\Lambda}_{\mathcal{G}_\alpha} \to \widehat{\Lambda}_{\mathcal{G}}$, and in order to simplify notation we set

$$\widehat{\Lambda}_\alpha := \hat{\phi}_\alpha(\widehat{\Lambda}_{\mathcal{G}_\alpha}) \subseteq \widehat{\Lambda}_{\mathcal{G}}.$$

Further, for every multi-index $\mathfrak{v}$ of $\mathcal{G}$, let $\jmath_\mathfrak{v} : \widehat{U}_\mathfrak{v} \to \widehat{\Lambda}_\mathfrak{v}$ be the canonical reduction homomorphism; see Remark 26 for definitions.

With the above notation, we say that $\Phi_\alpha : \mathcal{G} \to \mathcal{G}_\alpha$ is a *rational quotient* of $\mathcal{G}$, if $\mathcal{G}_\alpha$ is rational, and Φ_α is divisorial and satisfies the following:

(i) For all multi-indices $\mathfrak{v}$ the following hold: If $\jmath_{\mathfrak{v}}$ is non-trivial on $\widehat{\Lambda}_\alpha \cap \widehat{U}_{\mathfrak{v}}$, then $\widehat{\Lambda}_\alpha \subset \widehat{U}_{\mathfrak{v}}$, and $\jmath_{\mathfrak{v}}$ is injective on $\widehat{\Lambda}_\alpha$, or equivalently, $\jmath_{\mathfrak{v}} \circ \hat{\phi}_\alpha$ is injective on $\widehat{\Lambda}_{\mathcal{G}_\alpha}$.

(ii) For every finite $\mathbb{Z}_\ell$-module $\Delta \subset \widehat{\Lambda}_{\mathcal{G},\mathrm{fin}}$ with $\widehat{U}_{\mathcal{G}} \subseteq \Delta$, there exist 1-edges v such that $\jmath^v(\widehat{\Lambda}_\alpha) \neq 0$ and $\ker(\Delta \xrightarrow{\jmath_v} \widehat{\Lambda}_{\mathcal{G}_v}) = \Delta \cap \widehat{\Lambda}_\alpha$.

Fact 32. *Let $\Phi_\alpha : \mathcal{G} \to \mathcal{G}_\alpha$ be a rational quotient. Then $\widehat{U}_{\mathcal{G}} \cap \widehat{\Lambda}_\alpha = 0$, and one has:*

(1) *Let $\Lambda_{\mathcal{G}}$ be a divisorial $\widehat{U}_{\mathcal{G}}$-lattice in $\widehat{\Lambda}$. Then there exists a unique divisorial lattice $\Lambda_{\mathcal{G}_\alpha}$ in $\widehat{\Lambda}_{\mathcal{G}_\alpha}$ such that $\hat{\phi}_\alpha(\Lambda_{\mathcal{G}_\alpha})$ is contained in $\Lambda_{\mathcal{G}}$. Moreover, the images $\Lambda_\alpha := \hat{\phi}_\alpha(\Lambda_{\mathcal{G}_\alpha})$ can be recovered from $\widehat{\Lambda}_\alpha = \hat{\phi}_\alpha(\widehat{\Lambda}_{\mathcal{G}_\alpha})$ and $\Lambda_{\mathcal{G}}$ as follows:*

$$(*) \qquad \Lambda_\alpha := \hat{\phi}_\alpha(\Lambda_{\mathcal{G}_\alpha}) = \widehat{\Lambda}_\alpha \cap \Lambda_{\mathcal{G}}.$$

(2) *One can recover Λ_α from $\Lambda_{\mathcal{G}}$ using the maps $\jmath^v$ and $\jmath_v$ as follows:*

$(**)\, \Lambda_\alpha = \{x \in \Lambda_{\mathcal{G}} \mid$ *For all v with $\jmath^v(\widehat{\Lambda}_\alpha) \neq 0$ and $\jmath^v(x) = 0$, one has $\jmath_v(x) = 0$*$\}$.

Proof. First, since $\mathcal{G}_\alpha$ is rational, by definition we have $\widehat{U}_{\mathcal{G}_\alpha} = 0$. But then by Remark 28 (3), one has $0 = \hat{\phi}(\widehat{U}_{\mathcal{G}_\alpha}) = \widehat{U}_{\mathcal{G}} \cap \widehat{\Lambda}_\alpha$, as claimed.
To (1): Since Φ_α defines $\mathcal{G}_\alpha$ as a rational quotient of $\mathcal{G}$, it is divisorial (by definition), and $\widehat{U}_{\mathcal{G}_\alpha} = 0$. Hence we can conclude by applying Remark 28 (5).
To (2): Clearly, if $x \in \Lambda_\alpha$, then it satisfies the hypothesis from $(**)$, i.e., for all v with $\jmath^v(\widehat{\Lambda}_\alpha) \neq 0$ and $\jmath^v(x) = 0$ one has $\jmath_v(x) = 0$. For the converse, let $x \in \Lambda_{\mathcal{G}}$ satisfy hypothesis $(**)$, i.e., be such that for all v with $\jmath^v(\widehat{\Lambda}_\alpha) \neq 0$ and $\jmath^v(x) = 0$ one has $\jmath_v(x) = 0$. Since $\widehat{U}_{\mathcal{G}} \cap \widehat{\Lambda}_\alpha = 0$, by condition (ii) in the definition of Φ_α, it follows that there exist v such that $\jmath^v(\widehat{\Lambda}_\alpha) \neq 0$ and $\jmath_v$ is injective on $\widehat{U}_{\mathcal{G}}$. Therefore, by the hypothesis $(**)$, it follows that $x \notin \widehat{U}_{\mathcal{G}}$. By contradiction, suppose that $x \notin \Lambda_\alpha$. Let $\Delta = \widehat{U}_{\mathcal{G}} + \mathbb{Z}_\ell x$. Since $x \in \widehat{\Lambda}_{\mathcal{G},\mathrm{fin}}$, we have $\Delta \subset \widehat{\Lambda}_{\mathcal{G},\mathrm{fin}}$, and since $x \notin \widehat{U}_{\mathcal{G}}$, the inclusion $\widehat{U}_{\mathcal{G}} \subset \Delta$ is strict.

Case (a). $\Delta \cap \widehat{\Lambda}_\alpha = (0)$.

Then by property (ii) of Φ_α it follows that there exists v such that $\jmath^v(\widehat{\Lambda}_\alpha) \neq 0$, and $\Delta \subseteq \widehat{U}_v$ and $\jmath_v$ is injective on Δ. In particular, $x \in \widehat{U}_v$ and $\jmath_v(x)$ is non-trivial. Contradiction!

Case (b). $\Delta \cap \widehat{\Lambda}_\alpha \neq (0)$.

Then there exist $u \in \widehat{U}_{\mathcal{G}}$, $b \in \mathbb{Z}_\ell$ and $z \in \widehat{\Lambda}_\alpha$, $z \neq 0$, such that $u + bx = z$. In particular, since $z \in \Delta \subset \Lambda_{\mathcal{G}} \subset \widehat{\Lambda}_{\mathcal{G},\mathrm{fin}}$, we have $z \in \widehat{\Lambda}_{\mathcal{G},\mathrm{fin}}$. Further, $b \neq 0$, because $\widehat{U}_{\mathcal{G}} \cap \widehat{\Lambda}_\alpha = 0$. Setting $\Delta_0 := \widehat{U}_{\mathcal{G}}$, we have $\Delta_0 \subset \Delta$, and $\Delta_0 \cap \widehat{\Lambda}_\alpha = 0$. Hence there exists v such that $\jmath^v(\widehat{\Lambda}_\alpha) \neq 0$, and $\Delta \subseteq \widehat{U}_v$, and $\jmath_v$ is injective on $\Delta_0 = \widehat{U}_{\mathcal{G}}$. Thus we have

$$\jmath_v(-u) = \jmath_v(z - u) = b\,\jmath_v(x),$$

hence $\jmath_v(u) \neq 0$ iff $\jmath_v(x) \neq 0$, because $b \neq 0$. First, if $u \neq 0$, then $\jmath_v(u) \neq 0$, hence $\jmath_v(x) \neq 0$. Since $\jmath^v(\widehat{\Lambda}_\alpha) \neq 0$, this contradicts the hypothesis $(**)$. Second, if $u = 0$, then $bx = z \in \widehat{\Lambda}_\alpha$. Since $\widehat{\Lambda}_{\mathcal{G}}/\widehat{\Lambda}_\alpha$ is torsion-free, it follows that $x \in \widehat{\Lambda}_\alpha$, as claimed. □

Definition 33. Let $\mathcal{G}$ be a divisorial abstract decomposition graph, and let $\Lambda_{\mathcal{G}} \subset \widehat{\Lambda}_{\mathcal{G}}$ be a fixed divisorial $\widehat{U}_{\mathcal{G}}$-lattice. Let $\mathfrak{A}_0 = \{\Phi_\alpha\}_\alpha$ be the set of rational quotients of $\mathcal{G}$. For every subset $\mathfrak{A} \subseteq \mathfrak{A}_0$, we define

$$\Lambda_{\mathfrak{A}} = \sum\nolimits_{\Phi_\alpha \in \mathfrak{A}} \Lambda_\alpha$$

as the $\mathbb{Z}_{(\ell)}$-submodule of $\Lambda_{\mathcal{G}} \subset \widehat{\Lambda}_{\mathcal{G}}$ generated by all the Λ_α with $\Phi_\alpha \in \mathfrak{A}$.

(1) We say that $\mathfrak{A}$ is an *ample set of rational quotients* of $\mathcal{G}$, if the following hold:

  (i) For all α, α' one has that if $\Phi_\alpha \neq \Phi_{\alpha'}$, then $\widehat{\Lambda}_\alpha \cap \widehat{\Lambda}_{\alpha'} = (0)$.
  (ii) $\Lambda_{\mathfrak{A}} \cap \widehat{U}_{\mathcal{G}} = (0)$ and $\Lambda_{\mathfrak{A}}$ is ℓ-adically dense in $\widehat{\Lambda}_{\mathcal{G}}$.

(2) Suppose that $\mathfrak{A}$ is an ample set of rational quotients of $\mathcal{G}$. We will say that $\mathcal{G}$ is *geometric like with respect to* $\mathfrak{A}$ if for every α, α' there exists a multi-index $\mathfrak{v}$ of $\mathcal{G}$ such that:

  (j) $\widehat{\Lambda}_\alpha$ and $\widehat{\Lambda}_{\alpha'}$ are contained in $\widehat{U}_{\mathfrak{v}}$.
  (jj) $\jmath_{\mathfrak{v}}$ maps $\widehat{\Lambda}_\alpha$ and $\widehat{\Lambda}_{\alpha'}$ injectively into $\Lambda_{\mathcal{G}_{\mathfrak{v}}}$, and $\jmath_{\mathfrak{v}}(\widehat{\Lambda}_\alpha) = \jmath_{\mathfrak{v}}(\widehat{\Lambda}_{\alpha'})$.

(3) In the above context, we will call $\Lambda_{\mathfrak{A}}$ an $\mathfrak{A}$*-arithmetical lattice*. Its ℓ-adic equivalence class depends in general on $\mathfrak{A}$, and not only on equivalence class of $\Lambda_{\mathcal{G}}$. Further,

$$\widehat{U}_{\mathcal{G}} + \Lambda_{\mathfrak{A}} \subseteq \Lambda_{\mathcal{G}}$$

is a $\widehat{U}_{\mathcal{G}}$-lattice in $\widehat{\Lambda}_{\mathcal{G}}$, and therefore $\Lambda_{\mathcal{G}} \,/\, (\widehat{U}_{\mathcal{G}} + \Lambda_{\mathfrak{A}})$ is a torsion free divisible group, hence a $\mathbb{Q}$-vector space. But in general, $\widehat{U}_{\mathcal{G}} + \Lambda_{\mathfrak{A}}$ is not necessarily a divisorial $\widehat{U}_{\mathcal{G}}$-lattice.

Definition/Remark 34. Let $\mathcal{G}$ and $\mathcal{H}$ be geometric-like abstract decomposition graphs with respect to some sets of rational quotients $\mathfrak{A}_0 = \{\Phi_\alpha\}_\alpha$, respectively $\mathfrak{B}_0 = \{\Phi_\beta\}_\beta$, and let a proper morphism $\Phi : \mathcal{G} \to \mathcal{H}$ of level $\delta := \delta_H$ be given.

(1) We say that Φ is *compatible with rational quotients* if there exist ample subsets $\mathfrak{A} \subseteq \mathfrak{A}_0$ and $\mathfrak{B} \subseteq \mathfrak{B}_0$ satisfying the following: First, $\mathcal{G}$ and $\mathcal{H}$ are geometric-like with respect to $\mathfrak{A}$, respectively $\mathfrak{B}$. Second, for each $\Psi_\beta \in \mathfrak{B}$ there exist $\Phi_\alpha \in \mathfrak{A}$ and an isomorphism $\Phi_{\alpha\beta} : \mathcal{G}_\alpha \to \mathcal{H}_\beta$ such that the following diagram is commutative:

$$(*) \qquad \begin{array}{ccc} \mathcal{G} & \xrightarrow{\Phi} & \mathcal{H} \\ \downarrow{\scriptstyle \Phi_\alpha} & & \downarrow{\scriptstyle \Psi_\beta} \\ \mathcal{G}_\alpha & \xrightarrow{\Phi_{\alpha\beta}} & \mathcal{H}_\beta \end{array}$$

(2) We observe that in the above context, for every $\Psi_\beta \in \mathfrak{B}$ there exists a unique Φ_α satisfying hypothesis (∗). Indeed, if $\Phi_{\alpha'}$ together with $\Phi_{\alpha'\beta}$ also satisfy hypothesis (∗), then $\widehat{\Lambda}_{\mathcal{G}_\alpha} = \hat{\phi}_{\alpha\beta}(\widehat{\Lambda}_{\mathcal{H}_\beta})$, and therefore we get

$$\widehat{\Lambda}_\alpha := \hat{\phi}_\alpha(\widehat{\Lambda}_{\mathcal{G}_\alpha}) = \hat{\phi}_\alpha(\hat{\phi}_{\alpha\beta}(\widehat{\Lambda}_{\mathcal{H}_\beta})) = \hat{\phi}(\hat{\psi}_\beta(\widehat{\Lambda}_{\mathcal{H}_\beta})) = \hat{\phi}(\widehat{\Lambda}_\beta).$$

Since the same is true correspondingly for α', we finally get $\widehat{\Lambda}_\alpha = \hat{\phi}(\widehat{\Lambda}_\beta) = \widehat{\Lambda}_{\alpha'}$. But then by Definition 33 (1) (i) it follows that $\Phi_\alpha = \Phi_{\alpha'}$, as claimed.

In the above context, we say that α *corresponds to* β if the hypothesis (∗) is satisfied for Ψ_β and Φ_α. Thus α corresponds to β if and only if $\hat{\phi}(\widehat{\Lambda}_\beta) = \widehat{\Lambda}_\alpha$.

Proposition 35. *In the above context, let $\Phi : \mathcal{G} \to \mathcal{H}$ be a level-δ proper morphism of geometric-like abstract decomposition graphs which is compatible with the rational quotients $\mathfrak{A}$ and $\mathfrak{B}$, where $\delta := \delta_H$. Then Φ is divisorial.*

(1) *More precisely, let $\hat{\phi} : \widehat{\Lambda}_{\mathcal{H}} \to \widehat{\Lambda}_{\mathcal{H}}$ be the Kummer homomorphism of Φ. Let $\Lambda_{\mathfrak{B}}$ be an arithmetical lattice for $\mathcal{H}$ defined by $\mathfrak{B}$. Then there exists a unique arithmetical lattice $\Lambda_{\mathfrak{A}}$ for $\mathcal{G}$ defined by $\mathfrak{A}$ such that $\hat{\phi}(\Lambda_{\mathfrak{B}}) \subseteq \Lambda_{\mathfrak{A}}$, and one has*

$$\hat{\phi}(\Lambda_{\mathfrak{B}}) = \hat{\phi}(\widehat{\Lambda}_{\mathcal{H}}) \cap \Lambda_{\mathfrak{A}}.$$

(2) *Suppose that $\hat{\phi}(\Lambda_{\mathfrak{B}}) \subseteq \Lambda_{\mathfrak{A}}$; and for each α and β consider the unique divisorial lattices $\Lambda_{\mathcal{H}_\beta} \subset \widehat{\Lambda}_{\mathcal{H}_\beta}$ and $\Lambda_{\mathcal{G}_\alpha} \subset \widehat{\Lambda}_{\mathcal{G}_\alpha}$ such that $\Lambda_\beta := \hat{\phi}_\beta(\Lambda_{\mathcal{H}_\beta}) = \hat{\phi}_\beta(\widehat{\Lambda}_{\mathcal{H}_\beta}) \cap \Lambda_{\mathfrak{B}}$ and $\Lambda_\alpha := \hat{\phi}_\alpha(\Lambda_{\mathcal{G}_\alpha}) = \hat{\phi}_\alpha(\widehat{\Lambda}_{\mathcal{G}_\alpha}) \cap \Lambda_{\mathfrak{A}}$. Then for all α, β it follows that $\Phi_\alpha \in \mathfrak{A}$ corresponds to $\Psi_\beta \in \mathfrak{B}$ if and only if $\hat{\phi}_{\alpha\beta}(\Lambda_{\mathcal{H}_\beta}) = \Lambda_{\mathcal{G}_\alpha}$ and $\hat{\phi}(\Lambda_\beta) = \Lambda_\alpha$.*

Proof. It is clear that assertion (2) follows from assertion (1) and previous discussion. Therefore we will concentrate on the proof of assertion (1).
First recall that by Definition/Remark 34 (2), we have that α corresponds to β if and only if $\hat{\phi}(\widehat{\Lambda}_\beta) = \widehat{\Lambda}_\alpha$. Using this we deduce the following:

- $\hat{\phi}$ maps $\widehat{\Lambda}_{\mathfrak{B}} := \sum_{\Psi_\beta \in \mathfrak{B}} \widehat{\Lambda}_\beta$ into $\widehat{\Lambda}_{\mathfrak{A}} := \sum_{\Phi_\alpha \in \mathfrak{A}} \widehat{\Lambda}_\alpha$.
- Let $\Lambda_{\mathfrak{B}}$ and $\Lambda_{\mathfrak{A}}$ be fixed arithmetical lattices of $\mathcal{H}$, respectively $\mathcal{G}$. For a given β, choose α corresponding to it. By Definition/Remark 31 (3) above, and with the notation from there we have that there exists a unique divisorial lattice $\Lambda_{\mathcal{G}_\alpha}$ in $\widehat{\Lambda}_{\mathcal{G}_\alpha}$ such that $\hat{\phi}_\alpha$ maps $\Lambda_{\mathcal{G}_\alpha}$ into $\Lambda_{\mathfrak{A}}$, and actually $\hat{\phi}_\alpha(\Lambda_{\mathcal{G}_\alpha}) = \widehat{\Lambda}_\alpha \cap \Lambda_{\mathfrak{A}}$. And correspondingly, the same is true for β, i.e., there exists a unique $\Lambda_{\mathcal{H}_\beta}$ in $\widehat{\Lambda}_{\mathcal{H}_\beta}$ such that $\hat{\psi}_\beta(\Lambda_{\mathcal{H}_\beta}) = \widehat{\Lambda}_\beta \cap \Lambda_{\mathfrak{B}}$.

Since α corresponds to β, with the notation from Definition 34, let $\hat{\phi}_{\alpha\beta}$ be the Kummer isomorphism defined by $\Phi_{\alpha\beta}$. Then $\hat{\phi}_{\alpha\beta}(\Lambda_{\mathcal{H}_\beta})$ is a divisorial lattice in $\widehat{\Lambda}_{\mathcal{G}_\alpha}$. Thus there exists an ℓ-adic unit $\varepsilon_{\alpha\beta}$ such that

$$\hat{\phi}_{\alpha\beta}(\Lambda_{\mathcal{H}_\beta}) = \varepsilon_{\alpha\beta} \cdot \Lambda_{\mathcal{G}_\alpha}.$$

On the other hand, the commutativity of the diagram (∗) from Definition/Remark 34 translated in terms of Kummer homomorphisms means that the above equality is equivalent to the following: For all β and its corresponding α one has

$$(\alpha\beta) \qquad \hat{\phi}(\Lambda_\beta) = \varepsilon_{\alpha\beta} \cdot \Lambda_\alpha \,.$$

Let β and β', and the corresponding α and α' be given. Hence $\hat{\phi}$ maps $\widehat{\Lambda}_\beta$ and $\widehat{\Lambda}_{\beta'}$ isomorphically onto $\widehat{\Lambda}_\alpha$, respectively $\widehat{\Lambda}_{\beta'}$. Since $\mathcal{G}$ is geometric-like with respect to the family of rational projections $\mathfrak{A}$, it follows that there exists some multi-index $\mathfrak{v}$ of $\mathcal{G}$ which has the properties (j), (jj), of Definition 33 (2).

Before moving on, we recall that by Fact 8 (2), the fixed divisorial $\widehat{U}_{\mathcal{G}}$-lattice $\Lambda_{\mathcal{G}}$ of $\mathcal{G}$ defines uniquely a $\mathfrak{v}$-residual $\widehat{U}_{\mathcal{G}_\mathfrak{v}}$-lattice $\Lambda_{\mathcal{G}_\mathfrak{v}}$ by setting

$$\Lambda_{\mathcal{G}_\mathfrak{v}} := \widehat{U}_{\mathcal{G}_\mathfrak{v}} + \jmath_\mathfrak{v}(\Lambda_{\mathcal{G}} \cap \widehat{U}_\mathfrak{v}).$$

We further remark that condition (j) from Definition 33 (2) implies that $\Phi_\alpha(T_\mathfrak{v}) = 1$. Hence Φ_α gives rise to a residual morphism $\Phi_{\mathfrak{v}\alpha} : \mathcal{G}_\mathfrak{v} \to \mathcal{G}_\alpha$. And if $\hat{\phi}_{\mathfrak{v}\alpha} : \widehat{\Lambda}_{\mathcal{G}_\alpha} \to \Lambda_{\mathcal{G}_\mathfrak{v}}$ is the Kummer homomorphism of $\Phi_{\mathfrak{v}\alpha}$, then $\jmath_\mathfrak{v} \circ \hat{\phi}_\alpha = \hat{\phi}_{\mathfrak{v}\alpha}$. Therefore we have

$$\jmath_\mathfrak{v}(\widehat{\Lambda}_\alpha) = \hat{\phi}_{\mathfrak{v}\alpha}(\Lambda_{\mathcal{G}_\alpha}), \quad \jmath_\mathfrak{v}(\Lambda_\alpha) = \hat{\phi}_{\mathfrak{v}\alpha}(\Lambda_{\mathcal{G}_\alpha}).$$

Now since Φ_α is divisorial, $\Phi_{\mathfrak{v}\alpha}$ is so by definition. Hence by Remark 28, 5), we have:

$$\hat{\phi}_{\mathfrak{v}\alpha}(\Lambda_{\mathcal{G}_\alpha}) = \hat{\phi}_{\mathfrak{v}\alpha}(\widehat{\Lambda}_{\mathcal{G}_\alpha}) \cap \Lambda_{\mathcal{G}_\mathfrak{v}} \,.$$

Thus combining the assertions above, we finally get

$$\jmath_\mathfrak{v}(\Lambda_\alpha) = \jmath_\mathfrak{v}(\widehat{\Lambda}_\alpha) \cap \Lambda_{\mathcal{G}_\mathfrak{v}} \,.$$

On the other hand, both α and α' satisfy condition j) from Definition 33 (2). Hence by symmetry, the equalities above hold correspondingly for α' too. And since by condition jj) of Definition 33 (2), one has $\hat{\phi}_{\mathfrak{v}\alpha}(\widehat{\Lambda}_{\mathcal{G}_\alpha}) =: \widehat{\Lambda}_{\mathfrak{v},\alpha\alpha'} := \hat{\phi}_{\mathfrak{v}\alpha'}(\widehat{\Lambda}_{\alpha'})$, we get

$$(\alpha) \qquad \jmath_\mathfrak{v}(\Lambda_\alpha) = \widehat{\Lambda}_{\mathfrak{v},\alpha\alpha'} \cap \Lambda_{\mathcal{G}_\mathfrak{v}} = \jmath_\mathfrak{v}(\Lambda_{\alpha'}) \,.$$

On the other hand, since Φ is proper, there exists some $\mathfrak{w}$ corresponding to $\mathfrak{v}$. Recall the second diagram in Remark 26 (3), from which we bring forward

$$\begin{array}{ccc} \widehat{U}_\mathfrak{w} & \xrightarrow{\jmath_\mathfrak{w}} & \widehat{\Lambda}_{\mathcal{H}_\mathfrak{w}} \\ \downarrow\hat{\phi} & & \downarrow\hat{\phi}_\mathfrak{v} \\ \widehat{U}_\mathfrak{v} & \xrightarrow{\jmath_\mathfrak{v}} & \widehat{\Lambda}_{\mathcal{G}_\mathfrak{v}} \end{array}$$

and recall that $\hat{\phi}$, $\hat{\phi}_{\mathfrak{v}}$ are injective. Since $\widehat{\Lambda}_\beta = \hat{\phi}(\widehat{\Lambda}_{\mathcal{G}_\alpha})$, $\widehat{\Lambda}_{\beta'} = \hat{\phi}(\widehat{\Lambda}_{\mathcal{G}_{\alpha'}})$, we get:

(c) $\widehat{\Lambda}_\beta, \widehat{\Lambda}_{\beta'} \subset \widehat{U}_{\mathfrak{w}}$.
(d) $\widehat{\Lambda}_\beta$ and $\widehat{\Lambda}_{\beta'}$ are mapped by $\jmath_{\mathfrak{w}} : \widehat{U}_{\mathfrak{w}} \to \widehat{\Lambda}_{\mathcal{H}_{\mathfrak{w}}}$ injectively into $\widehat{\Lambda}_{\mathcal{H}_{\mathfrak{w}}}$, and have equal images $\jmath_{\mathfrak{w}}(\widehat{\Lambda}_\beta) =: \widehat{\Lambda}_{\mathfrak{w},\beta\beta'} := \jmath_{\mathfrak{w}}(\widehat{\Lambda}_{\beta'})$.

And note that $\hat{\phi}_{\mathfrak{v}}$ maps $\Lambda_{\mathfrak{w},\beta\beta'}$ isomorphically onto $\Lambda_{\mathfrak{v},\alpha\alpha'}$. Then going through the same steps as above and using notation correspondingly, we get as above

$$(\beta) \qquad \jmath_{\mathfrak{w}}(\Lambda_\beta) = \jmath_{\mathfrak{w}}(\Lambda_{\beta'}) .$$

We conclude the proof of the proposition as follows: For β, β' and α, α' corresponding to them, with the notation from above, we have by relation (α) above,

$$\varepsilon_{\alpha\beta} \cdot \Lambda_\alpha = \hat{\phi}(\Lambda_\beta) \quad \text{and} \quad \varepsilon_{\alpha'\beta'} \cdot \Lambda_{\alpha'} = \hat{\phi}(\Lambda_{\beta'})$$

for some ℓ-adic units $\varepsilon_{\alpha\beta}$ and $\varepsilon_{\alpha'\beta'}$. Applying $\jmath_{\mathfrak{v}}$ to the above equalities, and taking into account that by the commutativity of the diagram above one has $\jmath_{\mathfrak{v}} \circ \hat{\phi} = \hat{\phi}_{\mathfrak{v}} \circ \jmath_{\mathfrak{w}}$ on $\widehat{U}_{\mathfrak{w}}$, thus on Λ_β, $\Lambda_{\beta'} \subset \widehat{U}_{\mathfrak{w}}$, we finally get

$$\jmath_{\mathfrak{v}}(\varepsilon_{\alpha\beta} \cdot \Lambda_\alpha) = \jmath_{\mathfrak{v}}\big(\hat{\phi}(\Lambda_\beta)\big) = (\jmath_{\mathfrak{v}} \circ \hat{\phi})(\Lambda_\beta)\big) = (\hat{\phi}_{\mathfrak{v}} \circ \jmath_{\mathfrak{w}})(\Lambda_\beta)\big) = \hat{\phi}_{\mathfrak{v}}\big(\jmath_{\mathfrak{w}}(\Lambda_\beta)\big)$$

and correspondingly

$$\jmath_{\mathfrak{v}}(\varepsilon_{\alpha'\beta'} \cdot \Lambda_{\alpha'}) = \jmath_{\mathfrak{v}}\big(\hat{\phi}(\Lambda_{\beta'})\big) = (\jmath_{\mathfrak{v}} \circ \hat{\phi})(\Lambda_{\beta'})\big) = (\hat{\phi}_{\mathfrak{v}} \circ \jmath_{\mathfrak{w}})(\Lambda_{\beta'})\big) = \hat{\phi}_{\mathfrak{v}}\big(\jmath_{\mathfrak{w}}(\Lambda_{\beta'})\big).$$

On the other hand, $\jmath_{\mathfrak{w}}(\Lambda_\beta) = \jmath_{\mathfrak{w}}(\Lambda_{\beta'})$ by remark (β) above; hence the last two terms of the equalities above are equal. Thus we get

$$\jmath_{\mathfrak{v}}(\varepsilon_{\alpha\beta} \cdot \Lambda_\alpha) = \jmath_{\mathfrak{v}}(\varepsilon_{\alpha'\beta'} \cdot \Lambda_{\alpha'}), \quad \text{hence} \quad \varepsilon_{\alpha\beta} \cdot \jmath_{\mathfrak{v}}(\Lambda_\alpha) = \varepsilon_{\alpha'\beta'} \cdot \jmath_{\mathfrak{v}}(\Lambda_{\alpha'}).$$

On the other hand, $\jmath_{\mathfrak{v}}(\Lambda_\alpha) = \jmath_{\mathfrak{v}}(\Lambda_{\alpha'})$, by equalities ($\alpha$) above. Thus finally

$$\varepsilon_{\alpha\beta} \cdot \jmath_{\mathfrak{v}}(\Lambda_\alpha) = \varepsilon_{\alpha'\beta'} \cdot \jmath_{\mathfrak{v}}(\Lambda_\alpha) .$$

Next recall that if $\hat{\phi}_{\mathfrak{v}\alpha} : \widehat{\Lambda}_{\mathcal{G}_\alpha} \to \widehat{\Lambda}_{\mathcal{G}_{\mathfrak{v}}}$ is the Kummer homomorphism of the residual morphism $\Phi_{\mathfrak{v}\alpha} : \mathcal{G}_{\mathfrak{v}} \to \mathcal{G}_\alpha$, then we have $\jmath_{\mathfrak{v}}(\Lambda_\alpha) = \hat{\phi}_{\mathfrak{v}\alpha}(\Lambda_{\mathcal{G}_\alpha})$, and the latter is a $\widehat{U}_{\mathcal{G}_{\mathfrak{v}}}$-sublattice of $\Lambda_{\mathcal{G}_{\mathfrak{v}}}$. Hence finally $\varepsilon_{\alpha\beta}/\varepsilon_{\alpha'\beta'}$ must be a rational ℓ-adic unit. Since β, β' were arbitrary, we conclude that for every fixed β_0 and the corresponding α_0, after setting $\varepsilon := \varepsilon_{\alpha_0\beta_0}$, one has $\hat{\phi}(\Lambda_\beta) = \varepsilon \cdot \Lambda_\alpha$. Equivalently, $\hat{\phi}$ maps $\Lambda_{\mathfrak{B}} = \sum_\beta \Lambda_\beta$ into $\varepsilon \cdot \Lambda_{\mathfrak{A}} = \varepsilon \cdot \sum_\beta \Lambda_\alpha$. □

5 Morphisms arising from algebraic geometry

5.1 Morphisms

Let k and l be algebraically closed fields of characteristic $\neq \ell$. Let $K|k$ and $L|l$ be function fields, and let

$$\imath : L|l \hookrightarrow K|k$$

be an embedding of function fields such that l is mapped isomorphically onto k, and $K|\imath(L)$ is a separable field extension; see e.g., Lang [18] for a thorough discussion of this situation.

As defined in the introduction, let $\mathcal{D}_K^{\mathrm{tot}}$ and $\mathcal{D}_L^{\mathrm{tot}}$ be the total graphs of prime divisors on K, respectively on L. Then $\imath$ gives rise in a canonical way to a morphism of the total prime divisor graphs

$$\varphi_\imath : \mathcal{D}_K^{\mathrm{tot}} \to \mathcal{D}_L^{\mathrm{tot}}.$$

The precise definition of $\varphi_\imath$ is as follows: First let v be a prime divisor of $K|k$. Then either the restriction $v_L := v|_L$ of v to $L|l$ is the trivial valuation w_0 of $L|l$, or v_L is a prime divisor of $L|l$ otherwise. In both cases, $\imath$ gives rise to an embedding of the residue function fields

$$\imath_v : Lv_L|l \hookrightarrow Kv|k\,.$$

Inductively, we deduce from this that if $\mathfrak{v} = v_r \circ \cdots \circ v_1$ is a prime r-divisor of $K|k$ as defined in the Introduction, then $\mathfrak{w} := \mathfrak{v}|_L$ is a prime s-divisor of $L|l$ for some non-negative integer $s \leq r$. Moreover, by general valuation theory, it follows that every generalized prime divisor of $L|l$ is the restriction of some generalized prime divisor of $K|k$; hence $\varphi_\imath$ is surjective, etc.

The situation will become clearer after we analyze in more detail how *geometric prime divisor graphs* $\mathcal{D}_K$ of $K|k$ behave under $\varphi_\imath$.

First, observe that if $K|\imath(L)$ is finite, then for every generalized prime divisor $\mathfrak{w}$ of $L|l$, its fiber is finite of cardinality bounded by $[K : \imath(L)]$. From this one immediately deduces that the image of every geometric decomposition graph for $K|k$ under $\varphi_\imath$ is a geometric decomposition graph for $L|l$, etc.

Therefore, let us assume from now on that $K|\imath(L)$ is not algebraic. Then denoting by $K_1|k$ the relative algebraic closure of $\imath(L)$ in K, we have that $K_1|\imath(L)$ is finite separable, and $K|K_1$ is a regular function field extension. The situation of $L|l \hookrightarrow K_1|k$ was explained above. Thus mutatis mutandis, let $K|\imath(L)$ be a *regular field extension.*

Lemma 36. *Let X be a projective normal model for $K|k$, and $D \supseteq D_X$ a set of prime divisors with $D \backslash D_X$ finite. Then there exist a projective normal model $\tilde{X}$ for $K|k$ and a dominant k morphism $\phi : \tilde{X} \to X$ such that $D \subset D_{\tilde{X}}$; hence D is geometric.*

Proof. Clear. □

Using the lemma above, we have that there exist projective normal models $X \to k$ for $K|k$ such that D_X contains the 1-edges of $\mathscr{D}_K$ and X is complete regular-like. And correspondingly, the same holds for $L|l$ and $\mathscr{D}_L$. On the other hand, the regular embedding of function fields $\imath : L|l \to K|k$ is the generic fiber of a dominant rational map $f : X \dashrightarrow Y$ which factors through $\imath : l \to k$. And note that since X and Y are normal, f is defined at all points x_1 of codimension one of X. Moreover, replacing $X \to k$ by a properly chosen blowup, and normalizing the resulting k-variety, we can suppose that $f : X \to Y$ is a k-morphism of projective normal varieties. And since $K|\imath(L)$ is a regular field extension, it follows that $f : X \to Y$ has geometric generic integral fibers. Hence by the characterization of (the dimension of) the fibers the following hold:

- At almost all points x_1 of codimension one in X, $f(x_1)$ is either the generic point of Y, or $y_1 = f(x_1)$ is a point of codimension one of Y otherwise.
- On a Zariski open subset $V \subset Y$, the fiber X_y at $y \in V$ is irreducible, and if $\overline{X}_y \subseteq X$ is the Zariski closure, one has the following:

$$\operatorname{codim}(y) + \dim(\overline{X}_y) = \dim(X)\,.$$

Hence for almost all points y_1 of codimension one in Y, the closure of the fiber $\overline{X}_{y_1}$ is irreducible and has $\dim(\overline{X}_{y_1}) = \dim(X) - 1$. Equivalently, $\overline{X}_{y_1}$ is a Weil prime divisor of X, and its generic point x_1 has codimension one in X and is mapped to y_1.

In birational terms this means the following: For every prime divisor $v = v_{x_1} \in D_X$ let $w := v|_L = \varphi_\imath(v)$ be its restriction to L. Then the center of w on Y is $y_1 = f(x_1)$, and one of the following holds:

(a) w is the trivial valuation of $L|l$. This is so iff y_1 is the generic point of Y.
(b) w is a prime divisor of $L|l$. Then either y_1 has codimension one in Y, and if so, then w is the Weil prime divisor defined by y_1, or y_1 has codimension > 1.

In particular, we see that the following hold: First, all $w \in D_Y$ have preimages v in D_X, and for almost all w the preimage v is unique. Second, there are at most finitely many "exceptional" $v \in D_X$ for which $\varphi_\imath(v)$ does lie in D_Y. Let Σ_f be that set.

We now claim that for the given projective models $X \to k$ and $Y \to l$ as above, there exist quasi-projective normal models $\tilde{X} \to k$ and $\tilde{Y} \to l$ dominating $X \to k$ and $Y \to l$, and a morphism $\tilde{f} : \tilde{X} \to \tilde{Y}$ above $f : X \to Y$, and having the following property:

$$(*) \qquad \varphi_\imath(D_{\tilde{X}} \cup \{v_0\}) = D_{\tilde{Y}} \cup \{w_0\},$$

where v_0 and w_0 are the trivial valuations. In particular, $D_X \subseteq D_{\tilde{X}}$ and $D_Y \subseteq D_{\tilde{Y}}$. Indeed, if $\varphi_\imath(D_X \cup \{v_0\}) = D_Y \cup \{w_0\}$, i.e., if the exceptional set Σ_f is empty, then there is nothing to prove. Hence consider some $v := v_{x_1} \in \Sigma_f$ such that the center y_v of $w = \varphi_\imath(v)$ has codimension > 1. Let $Y_v \subset Y$ be the closure of y_v in Y. Setting $Y_1 := Y$, and $Z_1 := Y_v$, we consider a sequence of blowups

$\dots \to Y_{n+1} \to Y_n \to \dots$ as follows: $Z_n \subset Y_n$ is the closure of the center of w on Y_n. We stop if Z_n has codimension 1, and blow up Z_n otherwise. Then the above sequence is finite. Moreover, if $\mathrm{codim}(Z_n) > 1$, then $Y_{n+1} \to Y_n$ is an isomorphism outside Z_n. But then if the process above stops say at Y_n, it follows that Z_n is the center of w on Y_n, and $\mathrm{codim}(Z_n) = 1$. An easy Noether induction shows that one gets models $\tilde{Y}'$ dominating Y such that $\varphi_\imath(D_X) \subseteq D_{\tilde{Y}'} \cup \{w_0\}$. On the other hand, $f : X \to Y$ can be interpreted as a dominant rational map $\tilde{f} : X \dashrightarrow \tilde{Y}'$. Since X is normal, and $\tilde{Y}'$ is complete, $\tilde{f}$ is defined at all points $v \in D_X$ and maps these points into $D_{\tilde{Y}}$ by the discussion above. To conclude, let $S_Y \subset \tilde{Y}'$ be the Zariski closure of the (finite) complement of $D_{\tilde{Y}'} \backslash \varphi_\imath(D_X)$, and S_X the preimage of S_Y under f. Finally, set $\tilde{Y} := \tilde{Y}' \backslash S_Y$, and $\tilde{X} = U(f) \backslash S_X$, where $U(f)$ is the the domain of f. Then by the choices made, it follows that f defines a dominant morphism $\tilde{f} : \tilde{X} \to \tilde{Y}$ which has the required property $(*)$.

Now using the fact $(*)$ above and proceeding by induction on the transcendence degree of the residual function fields $L\mathfrak{w}|l \hookrightarrow K\mathfrak{v}|k$, a straightforward Noether induction argument shows finally the following.

Proposition 37. *In the above context, let $\mathcal{D}_K \subset \mathcal{D}_K^{\mathrm{tot}}$ and $\mathcal{D}_L \subset \mathcal{D}_L^{\mathrm{tot}}$ be geometric graphs of prime divisors for $K|k$, respectively $L|l$. Then there exists a unique maximal geometric subgraph $\mathcal{D}'_K \subset \mathcal{D}_K$ such that $\varphi_\imath$ defines by restriction a morphism of graphs of prime divisors*

$$\varphi_\imath : \mathcal{D}'_K \to \mathcal{D}_L.$$

Moreover, for given geometric graphs $\mathcal{D}_K \subset \mathcal{D}_K^{\mathrm{tot}}$ and $\mathcal{D}_L \subset \mathcal{D}_L^{\mathrm{tot}}$ as above, there exist geometric graphs of prime divisors $\mathcal{D}_K^0 \supseteq \mathcal{D}_K$ and $\mathcal{D}_L^0 \supseteq \mathcal{D}_L$ for $K|k$, respectively $L|l$, such that $\varphi_\imath$ defines by restriction a surjective morphism of graphs of prime divisors

$$\varphi_\imath : \mathcal{D}_K^0 \to \mathcal{D}_L^0.$$

Using Galois theory and decomposition theory of valuations, the above facts have the following translation in terms of abstract decomposition graphs: Let $\imath' : L' \to K'$ be a prolongation of $\imath : L|l \to K|k$ to L', and let

$$\Phi_\imath : \Pi_K \to \Pi_L$$

be the corresponding canonical projection of Galois groups. Then since $\imath : L|l \to K|k$ is a morphism of function fields, it follows that the relative algebraic closure L_1 of $L|l$ in $K|k$ is a finite extension of L, thus a function field over l. But then it follows that $\Phi_\imath$ is an open homomorphism.

Moreover, if $\varphi_\imath(v) = w$, and v' is a prolongation of v to K', then the restriction w' of v' to L' satisfies, first, that w' is a prolongation of w to L'. Second, let $T_v \subset Z_v$ and $T_w \subset Z_w$ be the corresponding decomposition groups. Then $\Phi_\imath(Z_v) \subset Z_w$ and

$\Phi_\imath(T_v) \subset T_w$ are open subgroups. (This discussion includes the case that w is the trivial valuation of L.) Moreover, if w is non-trivial, then $wL \subset vK$ has finite index $e(v|w)$. Hence we have commutative diagrams of the form

$$\begin{array}{ccc} L & \xrightarrow{w} & wL \subset \widehat{wL} = \mathrm{Hom}\big(T_w, \mathbb{Z}_\ell\big) \\ \downarrow{\scriptstyle\imath} & & \downarrow{\scriptstyle e(v|w)} \\ K & \xrightarrow{v} & vK \subset \widehat{vK} = \mathrm{Hom}\big(T_v, \mathbb{Z}_\ell\big) \end{array}$$

Therefore, if γ_w and γ_v are the unique positive generators of vK, respectively wL, then γ_w is mapped to $e(v|w) \cdot \gamma_v$. Thus if $\tau_v \in T_v$ and $\tau_w \in T_w$ are the arithmetical inertia generators as defined/introduced at Remark 19 (2), then from the commutativity of the above diagrams and definitions it follows that $\Phi_\imath(\tau_v) = \tau_w^{e(v|w)}$.

Now combining these observations with Proposition 37 above and Remark 28, especially (5), we obtain the following by merely applying the definitions:

Proposition 38. *In the notation from Proposition 37, the embedding of function fields $\imath : L|l \hookrightarrow K|k$ and the resulting canonical homomorphism $\Phi_\imath : \Pi_K \to \Pi_L$ give rise in a natural way to a level-*td$(L|l)$ *morphism $\Phi_\imath : \mathcal{G}_{\mathcal{D}_K} \to \mathcal{G}_{\mathcal{D}_L}$ of the corresponding abstract decomposition graphs.*

(1) *Moreover, if $\varphi_\imath : \mathcal{D}_K \to \mathcal{D}_L$ is a proper morphism of graphs of prime divisors, then the corresponding $\Phi_\imath : \mathcal{G}_{\mathcal{D}_K} \to \mathcal{G}_{\mathcal{D}_L}$ is a proper morphism of abstract decomposition graphs.*
(2) *Further, if $\Phi_\imath : \mathcal{G}_{\mathcal{D}_K} \to \mathcal{G}_{\mathcal{D}_L}$ is proper, and both $\mathcal{G}_{\mathcal{D}_K}$ and $\mathcal{G}_{\mathcal{D}_L}$ are complete regular-like, hence divisorial by Proposition 23, then $\Phi_\imath$ is divisorial.*
(3) *The Kummer homomorphism $\hat{\phi} : \widehat{L} \to \widehat{K}$ of Φ is actually the ℓ-adic completion of the embedding of function fields $\imath : L|l \hookrightarrow K|k$. In particular, $\imath$ defines $\Phi_\imath$ uniquely.*
(4) *Moreover, $\Phi_\imath$ defines $\imath$ uniquely up to Frobenius twists.*

Proof. Assertions (1), (2), and (3) follow from the discussion above.

To (4): Recall that in the Introduction we considered an identification $\imath_K : \mathbb{T}_{\ell,K} \to \mathbb{Z}_\ell$ of the ℓ-adic Tate module of K with $\mathbb{Z}_\ell$, and via that identification one gets the identification $\widehat{K} = \mathrm{Hom}_{\mathrm{cont}}(\Pi_K, \mathbb{Z}_\ell)$. Explicitly, this identification works as follows: For each $x \in K^\times$, let $\delta(x) : \Pi_K \to \mathbb{T}_{\ell,K}$ be the corresponding character defined in Kummer theory. Then $\delta_x := \imath_K \circ \delta(x)$ is the homomorphism $\delta_x : \Pi_K \to \mathbb{Z}_\ell$ defined by x. Given the embedding $\imath : L|l \hookrightarrow K|k$, by the functoriality of Kummer theory one has $\delta\big(\imath(y)\big) = \imath \circ \delta(x) \circ \Phi$. Therefore, if we choose the identifications $\imath_K : \mathbb{T}_{\ell,K} \to \mathbb{Z}_\ell$, $\imath_L : \mathbb{T}_{\ell,L} \to \mathbb{Z}_\ell$ compatible with $\imath$, i.e., such that $\imath_L = \imath_K \circ \imath$, it follows that one has $\delta_{\imath(y)} = \delta_u \circ \Phi$; hence the Kummer homomorphism defined by $\Phi_\imath$ is

$$\hat{\phi} : \widehat{L} = \mathrm{Hom}(\Pi_L, \mathbb{Z}_\ell) \to \mathrm{Hom}(\Pi_K, \mathbb{Z}_\ell) = \widehat{K}, \quad \delta_y \mapsto \delta_{\imath(y)}$$

and therefore, $\hat{\phi}$ is exactly the ℓ-adic completion of the embedding $\imath : L^\times \to K^\times$.

Now let $\imath' : L|l \hookrightarrow K|k$ be a further embedding of function fields such that $\Phi_{\imath'} = \Phi_\imath$. Then choosing $\imath'_K : \mathbb{T}_{\ell,L} \to \mathbb{Z}_\ell$ such that $\imath'_L = \imath_K \circ \imath'$, it follows that the Kummer homomorphism $\hat{\phi}'$ of $\Phi_{\imath'} = \Phi_\imath$ in this new setting is the ℓ-adic completion of $\imath'$. On the other hand, there exists an ℓ-adic unit $\varepsilon \in \mathbb{Z}_\ell^\times$ such that $\imath'_L = \varepsilon \cdot \imath_L$. If so, then we have $\hat{\phi}' = \varepsilon \cdot \hat{\phi}$ on $\widehat{L}$. Since $\hat{\phi}$ is the ℓ-adic completion of $\imath$, and $\hat{\phi}'$ is the ℓ-adic completion of $\imath'$, if we denote by $\jmath_K : K^\times \to \widehat{K}$ the ℓ-adic completion homomorphisms, we have

$$\jmath_K\big(\imath'(y)\big) = \varepsilon \cdot \jmath_K\big(\imath(y)\big), \quad y \in L^\times .$$

Therefore, ε must be a rational ℓ-adic unit, say $\varepsilon = m/n$ with n, m natural numbers relatively prime to ℓ. Equivalently, there exists $a_y \in k$ such that $\imath'(y) = a_y \imath(y)^{m/n}$ in K, hence $\imath'(y)$ is of the form $\imath'(y) = u^{m/n}$ in K, as k is algebraically closed. But then $\imath'(y)$ is an nth power in K. Since this is the case for all $\imath'(y) \in \imath'(L)$, it finally follows that $n = p^k$ is a power of the characteristic exponent p of k and l. By symmetry, the same is true for m. Hence finally ε is a power of the characteristic exponent of k and l. Equivalently, $\imath'$ is a Frobenius twist of $\imath$. □

Before studying the rational quotients in more detail in the next subsection, we mention the following weak version of the main result mentioned in the introduction. By Proposition 23, if $\mathcal{G}_{\mathcal{D}_K}$ is a complete regular-like decomposition graph for $K|k$, the canonical sequence $1 \to \widehat{U}_{\mathcal{D}_K} \to \Lambda_{\mathcal{D}_K} \to \mathrm{Div}(D_K)_{(\ell)} \to \mathfrak{Cl}_{\mathcal{D}_K} \to 0$ can be recovered from $\mathcal{G}_{\mathcal{D}_K}$ up to multiplication by ℓ-adic units $\varepsilon \in \mathbb{Z}_\ell^\times$.

Proposition 39. *Let $K|k$ and $L|l$ be function fields over algebraically closed fields of characteristic $\neq \ell$, and $\Phi : \mathcal{G}_{\mathcal{D}_K^{\mathrm{tot}}} \to \mathcal{G}_{\mathcal{D}_L^{\mathrm{tot}}}$ be an isomorphism. The following hold:*

(1) *For every geometric decomposition graph $\mathcal{G}_{\mathcal{D}_K}$ for $K|k$ there exists a geometric decomposition graph $\mathcal{G}_{\mathcal{D}_L}$ for $L|l$ such that Φ defines an isomorphism $\mathcal{G}_{\mathcal{D}_K} \to \mathcal{G}_{\mathcal{D}_L}$, and $\mathcal{G}_{\mathcal{D}_K}$ is complete regular-like (hence abstract divisorial) iff $\mathcal{G}_{\mathcal{D}_L}$ is so.*

(2) *Let $\Phi : \mathcal{G}_{\mathcal{D}_K} \to \mathcal{G}_{\mathcal{D}_L}$ be an isomorphism as above, $\mathcal{G}_{\mathcal{D}_K}$ and $\mathcal{G}_{\mathcal{D}_L}$ be complete regular-like decomposition graphs with sets of 1-edges D_K and D_L. Then Φ is divisorial, and there exists $\varepsilon \in \mathbb{Z}_\ell^\times$ such that the Kummer isomorphism $\hat{\phi}$ of Φ makes the diagram below commutative*

$$\begin{array}{ccccccccc}
0 \to & \widehat{U}_{\mathcal{D}_L} & \longrightarrow & \Lambda_{\mathcal{D}_L} & \xrightarrow{\mathrm{div}_{D_L}} & \mathrm{Div}(D_L)_{(\ell)} & \longrightarrow & \mathfrak{Cl}_{\mathcal{D}_L} & \to 0 \\
& \downarrow{\scriptstyle \varepsilon\cdot\hat{\phi}} & & \downarrow{\scriptstyle \varepsilon\cdot\hat{\phi}} & & \downarrow{\scriptstyle \varepsilon\cdot\mathrm{div}_\Phi} & & \downarrow{\scriptstyle \varepsilon\cdot\mathrm{can}} & \\
0 \to & \widehat{U}_{\mathcal{D}_K} & \longrightarrow & \Lambda_{\mathcal{D}_K} & \xrightarrow{\mathrm{div}_{D_K}} & \mathrm{Div}(D_K)_{(\ell)} & \longrightarrow & \mathfrak{Cl}_{\mathcal{D}_K} & \to 0.
\end{array}$$

Proof. Assertion (1) follows immediately by sorting through the proof of Propositions 22, as the group-theoretical recipe given there is invariant under group isomorphisms. Assertion (2) follows immediately from assertion (1) above, combined with Propositions 23 and 30. □

5.2 Rational quotients

Next we turn our attention to rational projections of abstract decomposition graphs $\mathcal{G}_{\mathcal{D}_K}$ as above. Let $t \in K$ be an arbitrary non-constant function, and let κ_t be the relative algebraic closure of $k(t)$ in K. Then $\kappa_t|k$ is a function field in one variable. We endow $\kappa_t|k$ with its unique complete normal model $X_t \to k$, which is also projective, and consider the corresponding graph of prime divisors $\mathcal{D}_{\kappa_t}$ for κ_t and the resulting complete regular-like decomposition graph $\mathcal{G}_{\kappa_t}$ for Π_{κ_t}. Then $\mathcal{G}_{\kappa_t}$ has level $\delta = 1$ and is divisorial, by Proposition 23. Moreover, if g_t is the geometric genus of X_t, then we have:

(a) $\widehat{\mathfrak{Cl}}_{\mathcal{G}_{\kappa_t}} \cong \mathbb{Z}_\ell$.
(b) $\widehat{U}_{\mathcal{G}_{\kappa_t}} \cong \mathbb{Z}_\ell^{2g_t}$ as the ℓ-adic dual of $\Pi_1(X_t) \cong \mathbb{Z}_\ell^{2g}$, thus g_t is encoded in $\mathcal{G}_{\kappa_t}$.
(c) The canonical (surjective) projection $\Phi_{\kappa_t} : \Pi_K \to \Pi_{\kappa_t}$ defines a level-one morphism of abstract decomposition graphs $\Phi_{\kappa_t} : \mathcal{G}_{\mathcal{D}_K} \to \mathcal{G}_{\kappa_t}$, by Proposition 38.

Before going into the details of characterizing the rational projections, let us mention the following fact for later use:

Proposition 40. *With the above notation, suppose that $\kappa_t \hookrightarrow K$ is a regular field extension, i.e., K is separably generated over κ_t, and $K \cap \overline{\kappa_t} = \kappa_t$. For normal models X and X_t of $K|k$, respectively of $\kappa_t|k$, let $f : X \dashrightarrow X_t$ be a rational map defining $\kappa_t \hookrightarrow K$. Then:*

(1) *There exists on open subset $U \subset X_t$ such that the fibers $f_x : X_x \dashrightarrow \kappa(s) = k$ of f at $s \in U(k)$ are integral, and the generic point x_s of X_s is the center of the unique prime divisor $v_s \in D_X$ which restricts to $s \in X_t$. In particular, $v_s(K) = v_s(\kappa_t)$.*
(2) *Let $L|K$ be a finite separable extension of K which is linearly disjoint from $\overline{\kappa_t}$ over K, i.e., $L \cap \overline{\kappa_t} = \kappa_t$. Then for almost all $s \in U(k)$ the prime divisor v_s has a unique prolongation w_s to L, and moreover, $w_s|v_s$ is totally inert, i.e., $Lw_s \,|\, Kv_s$ is separable and $[Lw_s : Kv_s] = [L : K]$.*
(3) *Let $\Delta \subset \widehat{K}$ be a $\mathbb{Z}_\ell$-submodule of finite corank such that $\Delta \cap \widehat{\kappa_t} = 1$. Then for almost all $s \in U(k)$ one has that $\Delta \subset \widehat{U}_{v_s}$ and $\jmath_{v_s}$ maps Δ injectively into $\widehat{Kv}$.*

Proof. To (1): Since $K|\kappa_t$ is a regular, it follows that the generic fiber $f_{\kappa_t} : X_{\kappa_t} \dashrightarrow \kappa_t$ of f is geometrically integral. Hence the fiber $f_s : X_s \to \kappa(s)$ of f is geometrically integral for s in a Zariski open subset $s \in U \subset X_t$. The remaining facts are just the (valuation-theoretical) birational interpretation of this fact.

To (2): This is just a souped-up version of assertion (1), using the fact that since $L|K$ is separable, the fundamental equality $[L : K] = \sum_i e(w_i|v) f(w_i|v)$ is satisfied for every discrete valuation v of K and the set of its prolongations $w_i|v$ to L.

To (3): Choose some projective normal model $X \to k$ of $K|k$ such that the rational map $X \dashrightarrow X_t$ is defined on the whole X, and $\Pi_{1,K} = \Pi_{1,D_X}$. Then $K^\times/\kappa_t^\times$ embeds in the divisor group $\mathrm{Div}(X_{\kappa_t})$ of the generic fiber $X_{\kappa_t} \to \kappa_t$ of $X \to X_t$. Hence $K^\times/\kappa_t^\times$ is a free abelian group, and we have an exact sequence of free abelian groups

$$1 \to \kappa_t^\times/k^\times \to K^\times/k^\times \to K^\times/\kappa_t^\times \to 1,$$

and its ℓ-adic completion $1 \to \widehat{\kappa_t} \to \widehat{K} \to \widehat{\mathscr{K}} \to 1$, where $\mathscr{K} := K^\times/\kappa_t^\times$. Note that since $\Delta \cap \widehat{\kappa_t} = 1$ by hypothesis, the map $\widehat{K} \to \widehat{\mathscr{K}}$ is injective on Δ. For $n = \ell^e$, consider the exact sequence $1 \to \kappa_t^\times/n \to K^\times/n \to \mathscr{K}/n \to 1$, and let $\Delta_n \subset K^\times/n$ be the image of Δ in $K^\times/n$. Then Δ is the projective limit of $(\Delta_n)_n$. Further, setting $\Xi_n := \Delta_n \cap (\kappa_t^\times/n)$, the projective limit of $(\Xi_n)_n$ equals $\Delta \cap \widehat{\kappa_t} = 1$. Hence for every n_0 there exists $n > n_0$ such that the image of $\Xi_n \to \Xi_{n_0}$ is trivial.

The Kummer theory interpretation of the facts above is: Let $K_n := K[\sqrt[n]{\Delta_n}\,]$ be the corresponding $\mathbb{Z}/n$ elementary abelian extension of K. Then $\mathrm{Gal}(K_n|K)$ is isomorphic to $\mathrm{Hom}(\Delta_n, \mu_n)$, and setting $\kappa_n := K_n \cap \overline{\kappa_t}$ one actually has $\kappa_n = \kappa_t[\sqrt[n]{\Xi_n}\,]$. And further, $\mathrm{Gal}(\kappa_n|\kappa_t)$ is canonically isomorphic to $\mathrm{Hom}(\Xi_n, \mu_n)$, and the canonical projection $\mathrm{Gal}(K_n|K) \to \mathrm{Gal}(\kappa_n|\kappa_t)$ is given by $\mathrm{Hom}(\Delta_n, \mu_n) \to \mathrm{Hom}(\Xi_n, \mu_n)$, which is defined by the inclusion $\Xi_n \hookrightarrow \Delta_n$. In particular, setting $M_n := K\kappa_n$, it follows that $K_n|M_n$ is a $\mathbb{Z}/n$ elementary abelian extension with $\mathrm{Gal}(K_n|M_n)$ canonically isomorphic to $\mathrm{Hom}(\Delta_n/\Xi_n, \mu_n)$.

Now recall that Δ is the projective limit of $(\Delta_n)_n$; hence for n_0 sufficiently large, the map $\Delta \to \Delta/\ell$ factors through $\Delta \to \Delta_{n_0}$. Second, for any fixed n_0, if $n > n_0$ is sufficiently large, the image of $\Xi_n \to \Xi_{n_0}$ is trivial. Therefore, the canonical map $\Delta_n \to \Delta_{n_0}$ factors through Δ_n/Ξ_n; and therefore, the canonical map $\Delta \to \Delta/\ell$ factors through Δ_n/Ξ_n. We conclude that if $\delta > 0$ is the rank of the finite free $\mathbb{Z}_\ell$-module Δ, i.e., $\Delta \cong \mathbb{Z}_\ell^\delta$, then $\mathrm{Gal}(K_n|M_n)$ has $(\mathbb{Z}/\ell)^\delta$ as a quotient.

In order to simplify and fix notation, for $n > n_0$ as above, set $M := M_n$, $L := K_n$, and $\kappa := \kappa_n$; hence $M = K\kappa$ and $L|\kappa$ is a regular field extension, because $\kappa_n = K_n \cap \overline{\kappa_t}$. And denoting by $\Delta_M \subset M^\times/n$ the image of Δ in $M^\times/n$, one has $L = M[\sqrt[n]{\Delta_M}\,]$, and in particular $\Delta_M \cong \Delta_n/\Xi_n$ by the fact that $\mathrm{Gal}(L|M)$ is canonically isomorphic to both $\mathrm{Hom}(\Delta_n/\Xi_n, \mu_n)$ and $\mathrm{Hom}(\Delta_M, \mu_n)$. In particular, Δ_M has $(\mathbb{Z}/\ell)^\delta$ as a quotient.

Changing gears, let $Z_t \to X_t$ be the normalization of X_t in the function field extension $\kappa_t \hookrightarrow \kappa$, and $Z \to X$ the normalization of X in the field extension $K \hookrightarrow M$. Then the morphism $X \to X_t$ is dominated by $Z \to Z_t$, and the following holds: Since $M|\kappa$ is a regular field extension, the generic fiber of $Z \to Z_t$ is geometrically integral. Therefore, there exists an open subvariety $V \subset Z_t$ such that for all $s \in V(k)$, the fiber $Z_s \to \kappa(s) = k$ is integral. Consequently, the prime divisor v_s of M defined by the Weil prime divisor $Z_s \subset Z$ is the unique prime divisor in D_Z which restricts to the point $s \in V(k)$.

Further, let $Y \to Z$ be the normalization of Z in $M \hookrightarrow L$. Then arguing as above, it follows that for almost all $s \in V(k)$, the fiber $Y_s \to \kappa(s) = k$ of $Y \to Z_t$ at $s \in V(k)$ is integral, and the prime divisor w_s of $L|k$ defined by the Weil prime divisor $Y_s \subset Y$ is the only prime divisor in D_Y which restricts to the point $s \in V(k)$.

But then w_s must restrict to v_s too, and moreover, w_s is the only prolongation of v_s from M to L and $w_s|v_s$ is inert. By the fundamental equality we conclude that

$$\mathrm{Gal}(L|M) = Z_{w_s|v_s} \to \mathrm{Gal}\big(Lw_s|Mv_s\big)$$

is an isomorphism. By general valuation theory one has $Lw_s = Mv_s[\sqrt[n]{\Delta_M v_s}]$, where $\Delta_M v_s$ is the image of $\Delta_M \subset M^\times/n$ under the residue map $U_{v_s}/n \to Mv_s/n$ induced by $\jmath_{v_s} : U_{v_s} \to Mv_s$. By Kummer theory applied to both $L|M$ and $Lw_s\,|\,Mv_s$, we conclude that $\mathrm{Gal}(L|M) \to \mathrm{Gal}(Lw_s|Mv_s)$ is an isomorphism iff $\Delta_M \to \Delta_M v_s$ is an isomorphism. From this we finally conclude that $(\mathbb{Z}/\ell)^\delta$ is a quotient of $\Delta_M v_s$. Therefore from the commutativity of the diagram of surjective morphisms

$$\begin{array}{ccccc} \Delta & \to & \Delta_n & \to & \Delta_M \\ \downarrow{\scriptstyle \jmath_v} & & \downarrow{\scriptstyle \jmath_v} & & \downarrow{\scriptstyle \jmath_{v_s}} \\ \jmath_v(\Delta) & \to & \Delta_n v & \to & \Delta_M v_s \end{array}$$

it follows that $\jmath_v(\Delta)$ has $(\mathbb{Z}/\ell)^\delta$ as a quotient, because $\Delta_M v_s$ does so. But then since $\Delta \cong \mathbb{Z}_\ell^\delta$, and $\jmath_v(\Delta)$ has no torsion, being a submodule of the torsion free $\mathbb{Z}_\ell$-module $\widehat{Kv}$, it follows that $\jmath_v$ maps Δ isomorphically onto $\jmath_v(\Delta)$. □

Proposition 41. *Let $\mathcal{G}_{\mathcal{D}_K}$ be a complete regular-like decomposition graph, which we view as a divisorial abstract decomposition graph, as indicated in Proposition 23. Then with the above notation, and that of Definition/Remark 31, for $t \in K$ the following are equivalent:*

(i) *$\Phi_{\kappa_t} : \mathcal{G}_{\mathcal{D}_K} \to \mathcal{G}_{\kappa_t}$ is a rational quotient of $\mathcal{G}_{\mathcal{D}_K}$.*
(ii) *κ_t is a rational function field.*

Proof. To (i) ⇒ (ii): First recall that $\widehat{U}_{\mathcal{G}_{\kappa_t}} \cong \mathbb{Z}_\ell^{2g_t}$, where g_t is the genus of X_t. Hence $\mathcal{G}_{\kappa_t}$ is rational if and only if $g = 0$, or equivalently, κ_t is a rational function field.

For (ii) ⇒ (i), we first claim that $\mathcal{G}_{\kappa_t}$ is rational. Indeed, by the discussion above, $\widehat{U}_{\mathcal{G}_{\kappa_t}} = 0$ and $\widehat{\mathfrak{Cl}}_{\mathcal{G}_{\kappa_t}} \cong \mathbb{Z}_\ell$, thus the claim. Next we claim that $\Phi_{\kappa_t} : \mathcal{G}_{\mathcal{D}_K} \to \mathcal{G}_{\kappa_t}$ defines $\mathcal{G}_{\kappa_t}$ as a quotient of $\mathcal{G}_{\mathcal{K}}$. Indeed, first $\Phi_{\kappa_t} : \Pi_K \to \Pi_{\kappa_t}$ is surjective by the definition of κ_t. Thus it is left to show that Φ_{κ_t} is proper, i.e., to show that if $\Phi_{\mathfrak{v}} : \mathcal{G}_{\mathfrak{v}} \to \mathcal{G}_{\kappa_t}$ is a level-one residual morphism for Φ_{κ_t}, then the following hold:

(a) Each 1-index of $\mathcal{G}_{\mathfrak{v}}$ is mapped under $\Phi_{\mathfrak{v}}$ to some multi-index of $\mathcal{G}_{\kappa_t}$.
(b) Each multi-index of $\mathcal{G}_{\kappa_t}$ corresponds to some multi-index of $\mathcal{G}_{\mathfrak{v}}$.

To prove (a), let $X_{\mathfrak{v}} \to k$ be a normal model of $K_{\mathfrak{v}}|k$ such that $D_{X_{\mathfrak{v}}}$ is the set of all the 1-vertices of $\mathcal{G}_{\mathfrak{v}}$. Then the embedding of k function fields $\imath : \kappa_t \hookrightarrow K_{\mathfrak{v}}$ is defined by some dominant rational map $f : X_{\mathfrak{v}} \dashrightarrow X_t$. Since $X_{\mathfrak{v}}$ is normal and X_t is complete, it follows that f is defined at all points of codimension 1. This means that for every $v \in D_{X_{\mathfrak{v}}}$, if v is trivial on κ_t, then $f(v)$ is the generic point of X_t, and hence v is mapped to the trivial valuation of $\mathcal{G}_{\kappa_t}$; and if v is non-trivial on κ_t, then $f(v)$ is a closed point in X_t.

To prove (b), we proceed by induction on $d_{\mathfrak{v}} = \mathrm{td}(K_{\mathfrak{v}}|k)$. If $d_{\mathfrak{v}} = 1$, then $X_{\mathfrak{v}}$ is a normal curve. Since $\mathcal{G}_{\mathcal{D}_K}$ was assumed to be divisorial, $\mathcal{G}_{\mathfrak{v}}$ is divisorial too by definition. Hence by Proposition 23 (1), $X_{\mathfrak{v}} \to k$ is a complete normal curve. But then the dominant rational map $f : X_{\mathfrak{v}} \dashrightarrow X_t$ is a surjective morphism. Finally, if $d_{\mathfrak{v}} > 1$, then there exist "many" $v \in D_{X_{\mathfrak{v}}}$ which are trivial on κ_t. But then $\Phi_{\mathfrak{v}}$ gives rise to a level-one residual morphism $\Phi_v : \mathcal{G}_v \to \mathcal{G}_{\kappa_t}$ of $\mathcal{G}_{\mathfrak{v}}$. Since $\mathrm{td}(Kv|k) < d_{\mathfrak{v}}$, by induction Φ_v is proper. On the other hand, the set of multi-indices $\mathcal{V}_{\mathcal{G}_v}$ of $\mathcal{G}_v$ is contained in the set of multi-indices $\mathcal{V}_{\mathcal{G}_{\mathfrak{v}}}$. Hence finally, every vertex of $\mathcal{G}_{\kappa_t}$ corresponds to some vertex of $\mathcal{G}_{\mathfrak{v}}$.

Finally, it is left to check properties (i), (ii), Remark/Definition 31 (2).

Checking property (i) form Remark/Definition 31 (2): If $\jmath_{\mathfrak{v}}(\widehat{U}_v \cap \widehat{\kappa_t})$ is non-trivial, then $\jmath_{\mathfrak{v}}$ maps $\widehat{\kappa_t}$ injectively into $\widehat{K_{\mathfrak{v}}}$. Indeed, let $\mathfrak{v} = v_r \circ \cdots \circ v_1$ with v_i prime divisors. Then if $\mathfrak{v}$ is not trivial on $\kappa_t^\times$, then $\kappa_t\mathfrak{v} = k$, and hence $\jmath_{\mathfrak{v}}$ is trivial on $\widehat{\kappa_t} \cap \widehat{U}_{\mathfrak{v}}$. Second, if $\mathfrak{v}$ is trivial on $\kappa_t^\times$ then $\kappa_t\mathfrak{v} = \kappa_t$; hence $\widehat{\kappa_t} \subseteq \widehat{U}_{\mathfrak{v}}$, and $\jmath_{\mathfrak{v}}$ is injective on $\widehat{\kappa_t}$.

Checking property (ii) form Remark/Definition 31 (2): Let us view K as a function field over the rational function field κ_t; hence $\mathrm{td}(K|\kappa_t) = \mathrm{td}(K|k) - 1 > 0$, and κ_t is relatively algebraically closed in K. Moreover, after replacing κ_t by a finite purely inseparable extension (which is of the form κ_y with $y^{p^e} = t$ for some power p^e of $p = \mathrm{char}(k)$, which does not change the Galois theory of the situation), we can suppose that $K|\kappa_t$ is actually a regular field extension. Let $X \to k$ be any normal model of $K|k$, and let $\mathbb{P}^1_k$ be the projective t-line over k. Since $K|\kappa_t$ is regular, by Fact 40 there exists a cofinite subset $S \subset k$ such that the prime divisor v_a of $K|k$ defined by the fiber $X_s \subset X$ at the point $s \in \mathbb{P}^1_k$ defined by $a \in S$ restricts to the $(t-a)$-adic valuation of κ_t, and $t-a$ is a uniformizing parameter of v_a. Now recall that $K|\kappa_t$ is a (regular) function field over κ_t with $\mathrm{td}(K|\kappa_t) = \mathrm{td}(K|k) - 1$ positive, and one has an exact sequence of the form

$$1 \to \kappa_t^\times \to K^\times \to K^\times/\kappa_t^\times \to 1.$$

And note that the last group is a free abelian group, since it is contained in the group of Weil prime divisors of any projective normal model $Y \to \kappa_t$ of $K|\kappa_t$. Therefore, $\kappa_t^\times$ has complements, say $\mathcal{K} \subset K^\times$ in $K^\times$, and hence we have $K^\times = \kappa_t^\times \cdot \mathcal{K}$ with $\mathcal{K}$ a free abelian subgroup with $\mathcal{K} \cap \kappa_t^\times = \{1\}$. Moreover, we can "adjust" $\mathcal{K}$ in such a way as to have $v_a(\mathcal{K}) = 0$ for all $a \in S$ as above. Indeed, if $(\alpha_i)_i$ is a $\mathbb{Z}$-basis of $\mathcal{K}$, then replacing each α_i by $\beta_i := \alpha_i \prod_{a \in S}(t-a)^{-v_a(\alpha_i)}$, the resulting system $(\beta_i)_i$ generates freely a $\mathbb{Z}$-submodule $\mathcal{K}_1$ of $\mathcal{K}$ such that $v_a(\mathcal{K}_1) = 0$ for all $a \in S$, and $\mathcal{K}_1$ is a complement of $\kappa_t^\times$ in $K^\times$.

Therefore, we may and will suppose that $\mathscr{K} \subset K^\times$ is a complement of $\kappa_t^\times$ in $K^\times$ such that $v_a(\mathscr{K}) = 0$, or equivalently $\mathscr{K} \subset U_{v_a}$, for all $a \in S$; and taking ℓ-adic completions, $\widehat{\mathscr{K}}$ is a complement of $\widehat{\kappa_t}$ in $\widehat{K}$ with the same property, i.e., $\widehat{\mathscr{K}} \subset \widehat{U}_{v_a}$ for all $a \in S$.

Now let $\Delta \subset \widehat{K}_{\text{fin}}$ be a $\mathbb{Z}_\ell$-submodule of finite corank, which means that $v(\Delta) = 0$ for almost all $v \in D_X$. In particular, this implies that $\Delta \subset \widehat{U}_{v_a}$ for almost all $a \in S$. Let $\Delta_1 \subset \widehat{\mathscr{K}}$ and $\Delta_2 \subset \widehat{\kappa_t}$ be the projections of Δ on $\widehat{\mathscr{K}}$, respectively $\widehat{\kappa_t}$. We claim that both Δ_1 and Δ_2 have finite corank, i.e., $v(\Delta_1) = 0$ and $v(\Delta_2) = 0$ for almost all v. Indeed, let $\Sigma \subset D_X$ be the finitely many $v \in D_X$ such that $v|_{\kappa_t}$ is non-trivial, and $v \neq v_a$ for all $a \in S$. Every $x \in \Delta$ has a unique presentation of the form $x = x_1 x_2$ with $x_i \in \Delta_i$, and clearly, $v(x) = v(x_1) + v(x_2)$ for all v. Then for $v \in D_X \backslash \Sigma$ satisfying $v(\Delta) = 0$, since $v(x_1) + v(x_2) = v(x) = 0$, we must have $v(x_1) \neq 0$ iff $v(x_2) \neq 0$ for $v \in D_X$. On the other hand, if $v(x_2) \neq 0$, then $v|_{\kappa_t}$ is non-trivial; hence $v = v_a$ for some $a \in S$, by the fact that $v \in D_X \backslash \Sigma$. And if $v = v_a$ for some $a \in S$, then $v(\mathscr{K}) = 0$; hence $v(x_1) = 0$. We conclude that for $v \in D_X \backslash \Sigma$ we have $v(\Delta) = 0$ iff $v(\Delta_i) = 0$ for $i = 1, 2$. Thus both Δ_1 and Δ_2 have finite corank.

Now since $\Delta_1 \cap \widehat{\kappa_t} = 1$, it follows by Proposition 40 (3) that $\jmath_{v_a}$ maps Δ_1 injectively into $\widehat{Kv_a}$ for almost all $a \in S$. Further, we notice that $\jmath_{v_a}$ is trivial on Δ_2 for almost all $a \in S$. (Indeed, $f \in \kappa_t$ is a v_a-unit iff a is neither a zero nor a pole of f; and if so, then $\jmath_{v_a}(f) = \jmath_{v_a}\big(f(a)\big) = 1$, because $\jmath_{v_a}\big(f(a)\big) \in k^\times$, and $\jmath_{v_a}$ is trivial on $k^\times$.) Therefore, for $x = x_1 x_2 \in \Delta$ with $x_i \in \Delta_i$ as above, one has $\jmath_{v_a}(x) = \jmath_{v_a}(x_1)$. Hence for $x \in \Delta$ as above we have $\jmath_{v_a}(x) = 1$ iff $\jmath_{v_a}(x_1) = 1$ iff $x \in \Delta_2$, as claimed. □

Notations 42. We introduce notation as follows:

(1) Let $\{\kappa_x = k(x) \mid x \text{ general element of } K\}$ be the set of all the subfields of K generated by general elements $x \in K$. Note that $\kappa_x = \kappa_{x'}$ if and only if x' is a linear transformation $x' = (ax + b)/(cx + d)$ of x.
Further, let $\mathfrak{A}_K = \{\Phi_{\kappa_x}\}_{\kappa_x}$ be the set of the corresponding rational quotients of $\mathscr{G}_{\mathscr{D}_K}$, and note that $\Phi_{\kappa_x} = \Phi_{\kappa_{x'}}$ if and only if $x' = (ax+b)/(cx+d)$ is a linear transformation of x.

(2) In order to simplify notation, we identify κ_x with the corresponding subfield of K. This identification defines a canonical embedding $\widehat{\kappa}_x \hookrightarrow \widehat{K}$ which turns out to be the inflation map defined by the canonical projection $\Phi_{\kappa_x} : \Pi_K \to \Pi_{\kappa_x}$. Therefore, the ℓ-adic completion homomorphism $\jmath_K : K^\times \to \widehat{K}$ then identifies $\jmath_{\kappa_x}(\kappa_x^\times)$ with $\jmath_K(\kappa_x^\times)$ inside $\widehat{K}$.

(3) Let $\imath : L|l \to K|k$ be an embedding of function fields such that $\imath(l) = k$ and $K|\imath(L)$ a separable field extension. Let $\mathfrak{A}_K = \{\Phi_{\kappa_x}\}_{\kappa_x}$ and $\mathfrak{B}_L = \{\Psi_{\kappa_y}\}_{\kappa_y}$ the sets of all rational quotients of $K|k$, respectively $L|l$. Finally, let $\mathfrak{B}_\imath \subseteq \mathfrak{B}_L$ be the set of all Ψ_{κ_y} such that $\imath(\kappa_y)$ is relatively algebraically closed in K. Thus in the context of Proposition 38, by taking into account Fact 41, for $\Psi_{\kappa_y} \in \mathfrak{B}_\imath$ and the corresponding $\Phi_{\kappa_x} \in \mathfrak{A}_K$, one has commutative diagrams in which $\Phi_{\kappa_x \kappa_y}$ is an isomorphism:

$$\begin{array}{ccc} \Pi_K & \xrightarrow{\Phi_\imath} & \Pi_L \\ \downarrow{\scriptstyle \Phi_{\kappa_x}} & & \downarrow{\scriptstyle \Psi_{\kappa_y}} \\ \Pi_{\kappa_x} & \xrightarrow{\Phi_{\kappa_x\kappa_y}} & \Pi_{\kappa_y} \end{array} \quad \text{and} \quad \begin{array}{ccc} \mathcal{G}_{\mathcal{D}_K} & \xrightarrow{\Phi_\imath} & \mathcal{G}_{\mathcal{D}_L} \\ \downarrow{\scriptstyle \Phi_{\kappa_x}} & & \downarrow{\scriptstyle \Psi_{\kappa_y}} \\ \mathcal{G}_{\kappa_x} & \xrightarrow{\Phi_{\kappa_x\kappa_y}} & \mathcal{G}_{\kappa_y} \end{array}$$

Fact/Definition 43. In the context from Notation 42 above, the following hold: *Birational Bertini*: Let $x, t \in K$ be algebraically independent over k, and let x be separable in K, i.e., x is not a p-power in K, where $p = \text{char}(k)$. Then for all but finitely many $a \in k$, the element $ax + t$ is a general element of K, i.e., $k(ax + t)$ is relatively algebraically closed in K; see e.g. Lang [18], Ch. VIII, Lemma in the proof of Theorem 7, or Roquette [32], §4.

(1) We will use the above "birational Bertini" repeatedly in the following form: Let $x, t \in K$ be fixed algebraically independent functions over k, with x separable, e.g., general. Then the following hold:

  (a) $t_a := ax + t$ is a general element of K for almost all $a \in k$.
  (b) $t_{a',a} := t/(a'x + a)$ is a general element of K for all $a' \in k^\times$ and almost all $a \in k$.
  (c) $t_{a'',a',a} := (a''t + a'x + a + 1)/(t + a'x + a)$ is a general element of K for all $a'' \in k$ and almost all $a', a \in k$.

(2) For $x, t \in K$ as above, the general elements of the form t_a, $t_{a',a}$, $t_{a'',a',a}$, will be called *general elements of Bertini type* defined by x, t. Further, a set $\Sigma \subset K^\times$ will be called a *Bertini set*, if for all $x, t \in K$ which are algebraically independent over k, and x is separable, one has $t_a, t_{a',a}, t_{a'',a',a} \in \Sigma$ for all $a'' \in k$, and almost all $a', a \in k$. Clearly, Σ generates the multiplicative group $K^\times$ by assertion (1) (b) above.

  We say that a set of rational quotients $\mathfrak{A} \subseteq \mathfrak{A}_K$ is of *Bertini type*, if $\mathfrak{A}$ has a subset of the form $\mathfrak{A}_\Sigma := \{\Phi_{\kappa_x} \mid x \in \Sigma\}$ for some Bertini set $\Sigma \subset K^\times$.

(3) Next let $\imath : L|l \to K|k$ be an embedding of function fields such that $\imath(l) = k$ and $K|\imath(L)$ separable. Then for every separable element $y \in L$, one has that $x := \imath(y)$ is a separable element of $K|k$. Further, directly from the definition of a general element of Bertini-type one gets the following: Let $u_b, u_{b',b}, u_{b'',b',b} \in L$ be general elements of Bertini type defined by some $y, u \in L$. Then for all $b'' \in l$ and almost all $b', b \in l$, the images $t_b := \imath(y_b)$, $t_{b',b} := \imath(u_{b',b})$, $t_{b'',b',b} = \imath(u_{b'',b',b})$ are general elements of Bertini type in $K|k$ defined by $x := \imath(y)$, $t := \imath(u)$.

(4) From this we deduce that there exist Bertini sets $\Delta \subset L^\times$ and $\Sigma \subset K^\times$ such that $\imath(\Delta) \subseteq \Sigma$. Therefore, for the corresponding Bertini-type sets of rational quotients $\mathfrak{B}_\Delta$ and $\mathfrak{A}_\Sigma$, we have that if $\kappa_y \in \mathfrak{B}_\Delta$, then $\kappa_x := \imath(\kappa_y)$ lies in $\mathfrak{A}_\Sigma$, etc.

Proof. The only assertions which are perhaps not obvious are (1) (b) and c).

To (1) (b): $t_{a',a}$ is general if and only if $1/t_{a',a} = a'(x/t) + a(1/t)$ is general. Now note that if $x/t, 1/t \in K$ are algebraically independent over k, and because

x is separable, it follows that at least one of the two elements is separable. Finally apply the "birational Bertini".

To (1) (c): Let $\alpha := 1 - a''$. Then $t_{a'',a',a} = a'' + (\alpha a'x + \alpha a + 1]/(t + a'x + a)$ is a general element if and and only if $t' := (\alpha a'x + \alpha a + 1)/(t + a'x + a)$ is so. Note that $t + a'x + a$ is a general element for all $a \in k^\times$ and almost all $a' \in k$ by the "birational Bertini." Hence if $\alpha = 0$, then $t' := 1/(t + a'x + a)$ is a general element. Finally, if $\alpha \neq 0$, then x' is a general element if and only if $1/t' = (t - \frac{1}{\alpha})/(\alpha a'x + \alpha a + 1) + \frac{1}{\alpha}$ is a general element, thus if and only if $(t - \frac{1}{\alpha})/(\alpha a'x + \alpha a + 1)$ is a general element. And the latter is a general element for all $\alpha a + 1 \in k^\times$ and almost all $a' \in k$, by Case (1) (b). □

Proposition 44. *With the above notation, the following hold:*

(1) *Suppose that* $\mathrm{td}(K|k) > 1$, *and let* $\mathcal{G}_{\mathcal{D}_K}$ *be a complete regular-like geometric decomposition graph, which we view as a divisorial abstract decomposition graph. Then endowing* $\mathcal{G}_{\mathcal{D}_K}$ *with a Bertini-type set* $\mathfrak{A} \subseteq \mathfrak{A}_K$ *of rational projections,* $\mathcal{G}_{\mathcal{D}_K}$ *becomes a geometric like abstract decomposition graph satisfying the following:* $K^\times_{(\ell)} := \jmath_K(K^\times)\otimes\mathbb{Z}_{(\ell)}$ *is an* arithmetical lattice *defined by* $\mathfrak{A}$ *inside* $\widehat{K}$, *which we call the canonical arithmetical lattice.*

(2) *Let* $\imath : L|l \to K|l$ *be an embedding of function fields such that* $\imath(l) = k$, *and* $K|\imath(L)$ *is separable. Let* $\mathcal{H}_{\mathcal{D}_L}$ *be a complete regular-like abstract decomposition graph for* $L|l$ *such that*

$$\Phi_\imath : \mathcal{G}_{\mathcal{D}_K} \to \mathcal{H}_{\mathcal{D}_L}$$

gives rise to a proper morphism of abstract decomposition graphs. Then there exist Bertini-type sets $\mathfrak{B}$ *of rational quotients for* $\mathcal{H}_{\mathcal{D}_L}$ *such that* $\Phi_\imath$ *is compatible with the rational projections* $\mathfrak{B}$ *and* $\mathfrak{A}$.

Proof. To (1): Let us check that $\mathfrak{A}$ satisfies the conditions from Definition 33. Let $X \to k$ be a quasiprojective normal model of $K|k$ such that D_X is the set of all 1-vertices of $\mathcal{G}_{\mathcal{X}}$.

Step 1. $\mathfrak{A}$ is an ample family of rational quotients for $\mathcal{G}_{\mathcal{D}_K}$. Indeed, first recall that by Fact 43 (2), the set $\Sigma_{\mathfrak{A}}$ generates $K^\times$. Therefore, with the notation from Definition 33 we have $\Lambda_{\mathfrak{A}_\Sigma} = K^\times_{(\ell)}$, hence $\Lambda_{\mathfrak{A}} = K^\times_{(\ell)}$ too. From this we deduce, first, that $\Lambda_{\mathfrak{A}}$ is ℓ-adically dense in $\widehat{K}$, as $\jmath_K(K^\times)$ itself is so. Second, since $\mathcal{D}_K$ was supposed to be complete regular-like, for every non-constant $x \in K$ there exists $v \in D_X$ such that $v(x) \neq 0$. Equivalently, for every non-trivial $x \in K^\times_{(\ell)}$, there exists $v \in D^1_{\mathcal{D}_K}$ such that $v(x) \neq 0$. But this means exactly that $K^\times_{(\ell)} \cap \widehat{U}_{\mathcal{G}_{\mathcal{X}}}$ is trivial. From this discussion, condition (ii) of Definition 33 follows. For condition (i), observe that $\Phi_{\kappa_x} \neq \Phi_{\kappa_{x'}}$ implies that $\kappa_x \neq \kappa_{x'}$. But then $\kappa_x \cap \kappa_{x'} = k$; hence $\widehat{\kappa}_x$ and $\widehat{\kappa}_{x'}$ have trivial intersection inside $\widehat{K}$.

Step 2. $\mathcal{G}_{\mathcal{D}_K}$ endowed with $\mathfrak{A}$ is geometric like.

Indeed, let κ_x and $\kappa_{x'}$ be given. If $\kappa_x = \kappa_{x'}$, then there is nothing to prove. Hence let $\kappa_x \neq \kappa_{x'}$. Since κ_x and $\kappa_{x'}$ are relatively algebraically closed in K, it follows that

x, x' are actually algebraically independent over k. Therefore, by the "birational Bertini," it follows that for almost all $a, a' \in k$ we have that $t := ax - a'x'$ gives rise to a dominant rational map $f : X \dashrightarrow \mathbb{P}^1_t$ such that for general points $t = b$, the fiber X_b is a Weil prime divisor of X, and x, x' are non-constant on X_b. The birational translation of this is the following: If $v := v_{X_b} \in D_X$ is the corresponding prime divisor of K, then x, x' are v-units such that $\jmath_v(x)$, $\jmath_v(x')$ are not constant in the residue field Kv of v. But then $\kappa_x^\times$ and $\kappa_{x'}^\times$ consist of non-principal v-units, and are mapped isomorphically into the residue field Kv. Moreover, since $t = ax - a'x'$ has $v(t) > 0$, it follows that $ax \equiv a'x' \,(\mathrm{mod}\, \mathfrak{m}_v)$, hence $\jmath_v(\kappa_x) = \jmath_v(\kappa_{x'})$. Taking ℓ-adic completions, we deduce from this that conditions j), jj) of Definition 33 are satisfied at v.

To (2): Apply Fact 43 (3), (4), and the commutative diagrams from Notations 42 (3). □

6 Proof of Main Theorem

In this section we will give a proof of the Main Theorem from the introduction. We will actually prove a slightly more general result than the Main Theorem announced in the introduction, in the sense that the first part of the theorem proved below compares complete regular-like geometric decomposition graphs with geometric-like abstract decomposition graphs.

Theorem 45. *Let $K|k$ be a function field with* $\mathrm{td}(K|k) > 1$*, and let $\mathcal{G}_{\mathcal{D}_K}$ be a complete regular-like geometric decomposition graph for $K|k$. We endow $\mathcal{G}_{\mathcal{D}_K}$ with a Bertini-type set $\mathfrak{A}$ of rational quotients, and view it as a* geometric like abstract decomposition graph.

(1) *Let $\mathcal{H}$ endowed with a family of rational quotients $\mathfrak{B}$ be a geometric like abstract decomposition graph. Then up to multiplication by ℓ-adic units and composition with automorphisms $\Phi_\imath : \mathcal{G}_{\mathcal{D}_K} \to \mathcal{G}_{\mathcal{D}_K}$ defined by embedding of function fields $\imath : K|k \to K|k$ such that $K|\imath(K)$ is purely inseparable, there exists at most one isomorphism $\Phi : \mathcal{G}_{\mathcal{D}_K} \to \mathcal{H}$ of abstract decomposition graphs which is compatible with the rational quotients $\mathfrak{A}$ and $\mathfrak{B}$.*

(2) *Let $L|l$ be a further function field with* $\mathrm{td}(L|l) > 1$*, and let $\mathcal{H}_{\mathcal{D}_L}$ be a complete regular-like abstract decomposition graph for $L|l$. We endow $\mathcal{H}_{\mathcal{D}_L}$ with a Bertini-type set $\mathfrak{B}$ of rational quotients, and view it as a* geometric like abstract decomposition graph. *Let*

$$\Phi : \Pi_K \to \Pi_L$$

be an open group homomorphism which defines a proper morphism $\Phi : \mathcal{G}_{\mathcal{D}_K} \to \mathcal{H}_{\mathcal{D}_L}$ of abstract decomposition graphs compatible with the rational quotients $\mathfrak{B}$ and $\mathfrak{A}$. Then there exist an ℓ-adic unit ε and an embedding of function fields

$$\imath : L|l \to K|k$$

such that $\Phi = \varepsilon \cdot \Phi_\imath$, where $\Phi_\imath : \mathcal{G}_{\mathcal{D}_K} \to \mathcal{H}_{\mathcal{D}_L}$ is the functorial morphism of decomposition graphs defined by $\imath$ as indicated above. Further, $\imath(l) = k$, $\imath$ is unique up to Frobenius twists, and ε is unique up to multiplication by p^n-powers, where $p = \mathrm{char}(k)$ and $n \in \mathbb{Z}$.

Proof. Since (1) follows from (2), it suffices to prove assertion (2).
Recall that by Proposition 44, $K^\times_{(\ell)} := \jmath_K(K^\times) \otimes \mathbb{Z}_{(\ell)}$ and $L^\times_{(\ell)} := \jmath_L(L^\times) \otimes \mathbb{Z}_{(\ell)}$ are arithmetical lattices for $\mathcal{G}_{\mathcal{D}_K}$ endowed with $\mathfrak{A}$, respectively for $\mathcal{H}_{\mathcal{D}_L}$ endowed with $\mathfrak{B}$. Now by Proposition 35, it follows that $\hat{\phi}(L^\times_{(\ell)})$ is contained in a unique arithmetical lattice of $\mathcal{G}_{\mathcal{D}_K}$. Since the arithmetical lattices of $\mathcal{G}_{\mathcal{D}_K}$ are ℓ-adically equivalent to $K^\times_{(\ell)}$, there exists an ℓ-adic unit ε such that

$$\hat{\phi}(L^\times_{(\ell)}) \subseteq \varepsilon \cdot K^\times_{(\ell)} .$$

Therefore, after replacing Φ by $\varepsilon \cdot \Phi$, without loss of generality we can make the following hypothesis:

Hypothesis I. *$\hat{\phi}$ maps $L^\times_{(\ell)}$ isomorphically into $K^\times_{(\ell)}$.*

We further recall that $\jmath_K(K^\times) = K^\times/k^\times$ and $\jmath_L(L^\times) = L^\times/l^\times$ are true lattices in $K^\times_{(\ell)}$, respectively $L^\times_{(\ell)}$. In order to simplify notation, we denote by

$$\mathsf{x} = \jmath_K(x) = k^\times x, \quad \mathsf{y} = \jmath_L(y) = l^\times y$$

the image of $x \in K^\times$ under $\jmath_K$, respectively that of $y \in L^\times$ under $\jmath_L$. Further, we will always denote elements of $K^\times_{(\ell)}$, respectively of $L^\times_{(\ell)}$, in boldface:

$$\boldsymbol{x} \in K^\times_{(\ell)}, \quad \boldsymbol{y} \in L^\times_{(\ell)} .$$

Next we want to remark the following: Let $\Phi_{\kappa_x} \in \mathfrak{A}$ correspond to some $\Phi_{\kappa_y} \in \mathfrak{B}$, and let $\kappa_{x,(\ell)} \subset \widehat{\kappa}_x$ and $\kappa_{y,(\ell)} \subset \widehat{\kappa}_y$ be the unique divisorial lattices such that $\hat{\phi}_{\kappa_x}(\kappa_{x,(\ell)}) \subset K^\times_{(\ell)}$, respectively $\hat{\phi}_{\kappa_y}(\kappa_{y,(\ell)}) \subset L^\times_{(\ell)}$. Then by Proposition 35 (2), we get $\hat{\phi}_{\kappa_x\kappa_y}(\kappa_{y,(\ell)}) = \kappa_{x,(\ell)}$, $\hat{\phi} \circ \hat{\phi}_{\kappa_y}(\kappa_{y,(\ell)}) = \hat{\phi}_{\kappa_x}(\kappa_{x,(\ell)})$. Hence taking into account the identifications from Notations 42, i.e., $\kappa_{x,(\ell)} = K^\times_{(\ell)} \cap \hat{\phi}_{\kappa_x}(\widehat{\kappa}_x)$ inside $\widehat{K} = \widehat{\Lambda}_{\mathcal{G}_{\mathcal{D}_K}}$, and $\kappa_{y,(\ell)} = L^\times_{(\ell)} \cap \hat{\phi}_{\kappa_y}(\widehat{\kappa}_y)$ inside $\widehat{L} = \widehat{\Lambda}_{\mathcal{H}_{\mathcal{D}_L}}$, the above assertion is equivalent to the fact that if $\Phi_{\kappa_x} \in \mathfrak{A}$ corresponds to $\Phi_{\kappa_y} \in \mathfrak{B}$, then

$$\hat{\phi}(\kappa_{y,(\ell)}) = \kappa_{x,(\ell)} .$$

Hence we have $\jmath_L(\kappa_y^\times) \subset \kappa_{y,(\ell)}$ and $\jmath_K(\kappa_x^\times) \subset \kappa_{x,(\ell)}$, and the task now is to understand the precise relation between $\hat{\phi} \circ \jmath_L(\kappa_y{}^\times)$ and $\jmath_K(\kappa_x^\times)$ inside $\kappa_{x,(\ell)}$.

Lemma 46. *Let $\Phi_{\kappa_x} \in \mathfrak{A}$ correspond to some $\Phi_{\kappa_y} \in \mathfrak{B}$. Then there exist unique relatively prime and prime to ℓ integers $m, n > 0$ such that the following hold:*

(1) $\hat{\phi}\big(n\cdot \jmath_L(ly+l)^\times\big) = m\cdot \jmath_K(kx+k)^\times$ *and* $\hat{\phi}\big(n\cdot \jmath_L(\kappa_y^\times)\big) = m\cdot \jmath_K(\kappa_x^\times)$ *in* $\kappa_{x,(\ell)}$.

(2) $\hat{\phi}\big(n\cdot \jmath_L(y)\big) = m\cdot \jmath_K(x)$*, provided* $\mathbb{Z}_{(\ell)}\cdot \jmath_L(y)$ *is mapped by* $\hat{\phi}$ *into* $\mathbb{Z}_{(\ell)}\cdot \jmath_K(x)$*, and such a choice of a generator* x *for* κ_x *is always possible.*

Proof. We begin by considering the systems of *arithmetical inertia generators* $\mathfrak{T}_{\kappa_y}$ of $\mathcal{G}_{\kappa_y}$, respectively $\mathfrak{T}_{\kappa_x}$ of $\mathcal{G}_{\kappa_x}$, as introduced at Definition/Remark 19 (2). We let $(y) = w' - w$ be the divisor of $y \in \kappa_y = l(y)$, and $(x) = v' - v$ be the divisor of $x \in \kappa_x = k(x)$. Then $\kappa_x = k(x)$, and in the notations from Definition/Remark 31 (1), we have:

(a) $\mathsf{y} := \jmath_L(y) = \varphi_{w'} - \varphi_w$, and $\mathscr{P}_{\mathsf{y}} := \mathscr{P}_w = \jmath_L(ly+l)^\times \subset \kappa_{y,(\ell)}$ is the generating set at w with respect to $\mathfrak{T}_{\kappa_y}$.

(b) $\mathsf{x} := \jmath_K(x) = \varphi_{v'} - \varphi_v$, and $\mathscr{P}_{\mathsf{x}} := \mathscr{P}_v = \jmath_K(kx+k)^\times \subset \kappa_{x,(\ell)}$ is the generating set at v with respect to $\mathfrak{T}_{\kappa_y}$.

Moreover, we can choose x from the beginning in such a way that v', v are the preimages of w', w under $\Phi_{\kappa_x\kappa_y}$. Equivalently, we have $\hat{\phi}(\mathbb{Z}_{(\ell)}{\cdot}\mathsf{y}) = \mathbb{Z}_{(\ell)}{\cdot}\mathsf{x}$. Therefore there exist unique relatively prime integers $m, n > 0$ such that

$$(*) \qquad \hat{\phi}(n\cdot \mathsf{y}) = m\cdot \mathsf{x}\,.$$

On the other hand, the image $\Phi_{\kappa_x\kappa_y}(\mathfrak{T}_{\kappa_y})$ of $\mathfrak{T}_{\kappa_y}$ under the isomorphism $\Phi_{\kappa_x\kappa_y}$ is a distinguished system of inertia generators for κ_x such that $\hat{\phi}(\mathscr{P}_w)$ is the generating set at v with respect to $\Phi_{\kappa_x\kappa_y}(\mathfrak{T}_{\kappa_y})$. By the uniqueness up to ℓ-adic equivalence of the distinguished systems of inertia generators we have $\Phi_{\kappa_x\kappa_y}(\mathfrak{T}_{\kappa_y}) = \mathfrak{T}^{\varepsilon}_{\kappa_x}$ for a unique ℓ-adic unit $\varepsilon \in \mathbb{Z}_\ell^\times$. Hence $\hat{\phi}(\mathscr{P}_w) = \varepsilon^{-1}\cdot \mathscr{P}_v$, and in particular, $\hat{\phi}(\mathsf{y}) = \varepsilon^{-1}\cdot \mathsf{x}$ inside $\kappa_{x,(\ell)}$. Then by the fact $(*)$ above, it follows that $\varepsilon = n/m$; hence both m, n are relatively prime to ℓ. Finally, we get

$$(*)' \qquad \hat{\phi}(n\cdot \mathscr{P}_{\mathsf{y}}) = m\cdot \mathscr{P}_{\mathsf{x}}\,.$$

Clearly, if m', n' are relatively prime integers such that $\hat{\phi}(n'\cdot \mathscr{P}_{\mathsf{y}}) = m'\cdot \mathscr{P}_{\mathsf{x}}$, then we must have $\hat{\phi}(n'\cdot \mathsf{y}) = m'\cdot \mathsf{x}$. Therefore, $(m,n) = (m',n')$ by the uniqueness of m, n.

Finally, since $\mathscr{P}_{\mathsf{y}}$ and $\mathscr{P}_{\mathsf{x}}$ generate $\kappa_y^\times/l^\times$ inside $\kappa_{y,(\ell)}$, respectively $\kappa_x^\times/k^\times$ inside $\kappa_{x,(\ell)}$, we deduce that for the unique m, n above, one has

$$(*)'' \qquad \hat{\phi}\big(n\cdot \jmath_L(\kappa_y^\times)\big) = m\cdot \jmath_K(\kappa_x^\times)\,.$$

This completes the proof of the lemma. □

Norming 47. In the context of Lemma 46 above, suppose that $\mathbb{Z}_{(\ell)}{\cdot}\jmath_L(y)$ is mapped by $\hat{\phi}$ into $\mathbb{Z}_{(\ell)}\cdot \jmath_K(x)$. Then we will say that $\hat{\phi}$ is y-*normed* if $\hat{\phi}\circ \jmath_L(y) = \jmath_K(x)$. Clearly, a priori, $\hat{\phi}$ might not be normed with respect to any $\Phi_{\kappa_y} \in \mathfrak{B}$ and the corresponding $\Phi_{\kappa_x} \in \mathfrak{A}$. Nevertheless, we can "artificially" remedy this as follows:

With the notation from Lemma 46 above, suppose that we have chosen the generator x such that $\mathbb{Z}_{(\ell)} \cdot \jmath_L(y)$ is mapped by $\hat{\phi}$ into $\mathbb{Z}_{(\ell)} \cdot \jmath_K(x)$. Hence we have $\hat{\phi} \circ \jmath_L(y) = (m/n) \cdot \jmath_K(x)$. Further, notice that $\eta_{\hat{\phi}} := m/n$ is an ℓ-adic unit. And replacing the morphism $\Phi : \Pi_K \to \Pi_L$ by its $\eta_{\hat{\phi}}$-multiple $\Phi' := \eta_{\hat{\phi}} \cdot \Phi$ amounts to replacing $\hat{\phi}$ by its $(1/\eta_{\hat{\phi}})$-multiple $\hat{\phi}' := (1/\eta_{\hat{\phi}}) \cdot \hat{\phi}$. In particular, we have $\eta_{\hat{\phi}'} = 1$; hence $\hat{\phi}'$ is y-normed.

Hence we have the following: Let $\kappa_y \in \mathfrak{B}$ and its corresponding $\kappa_x \in \mathfrak{A}$ be given such that $\mathbb{Z}_{(\ell)} \cdot \jmath_L(y)$ is mapped by $\hat{\phi}$ into $\mathbb{Z}_{(\ell)} \cdot \jmath_K(x)$. Then after replacing Φ by a properly chosen multiple $\eta \cdot \Phi$ with $\eta \in \mathbb{Z}_{(\ell)}$, the resulting Kummer homomorphism $(1/\eta) \cdot \hat{\phi}$ is y-normed. Hence mutatis mutandis, we can suppose that $\hat{\phi}$ satisfies the following *norming hypothesis*:

Hypothesis II. *$\kappa_x \in \mathfrak{A}$ corresponds to $\kappa_y \in \mathfrak{B}$, and $\hat{\phi} \circ \jmath_L(y) = \jmath_K(x)$, hence $\hat{\phi}$ is y-normed.*

Remark/Notation 48. If $\hat{\phi}$ is y-normed, then by Lemma 46, $\hat{\phi}$ defines bijections

$$(\dagger) \qquad \hat{\phi} : \jmath_L(ly + l)^\times \to \jmath_L(ky + k)^\times, \quad \hat{\phi} : \jmath_L(\kappa_y^\times) \to \jmath_L(\kappa_x^\times).$$

We set $M_K := \hat{\phi}\big(\jmath_L(L^\times)\big) \cap \jmath_K(K^\times)$, and let $M_L \subseteq \jmath_L(L^\times)$ be the preimage of M_K under $\hat{\phi}$. Then $\jmath_L(\kappa_y^\times) \subset M_L$ and $\jmath_K(\kappa_x^\times) \subset M_K$ by the fact (†) above, and notice that

$$\hat{\phi} : M_L \to M_K$$

is an isomorphism which maps $\jmath_L(\kappa_y^\times)$ isomorphically onto $\jmath_K(\kappa_x^\times)$. We will say that $u \in L^\times$ and $t \in K^\times$ *correspond to each other* if the following hold:

$$\jmath_L(u) \in M_L, \quad \jmath_K(t) \in M_K, \quad \text{and} \quad \hat{\phi} \circ \jmath_L(u) = \jmath_K(t)\,.$$

Finally we notice that $M_L \otimes \mathbb{Z}_{(\ell)} = L^\times_{(\ell)}$ inside $\widehat{L}$.

Lemma 49. *Suppose that $t \in K$ and $u \in L$ correspond to each other via $\hat{\phi}$. Then $\mathscr{P}_t := (kt{+}k)^\times\!/k^\times {=} \jmath_K(kt{+}k)^\times \subset M_K$, $\mathscr{P}_u := (l\,u{+}l)^\times\!/l^\times {=} \jmath_L(l\,u{+}l)^\times {\subset} M_L$.*

Proof. Case 1: $u \in \kappa_y$. Then $t \in \kappa_x$, and we are in the situation of Lemma 46 above with $m = n = 1$, from which the assertion follows.

Case 2: $u \notin \kappa_y$. Since $\kappa_y = l(y)$ is relatively algebraically closed in L, it follows that u, y are algebraically independent over l. Correspondingly, the same is true for t, x, i.e., t, x are algebraically independent over k. Then by the Fact 43 (1), we have:

(i) $t_{a',a} := t/(a'x + a)$ is a general element of K for almost all $a', a \in k$.
(ii) $u_{b',b} := u/(b'x + b)$ is a general element of L for almost all $b', b \in l$.

Hence by condition (†) of Remark/Notation 48, we conclude the following: For a', a as at (i), let $y_{a',a} \in (ly + l)^\times$ be such that $\hat{\phi} \circ \jmath_L(y_{a',a}) = \jmath_K(a'x + a)$. Then by (ii), $u_{a',a} := u/y_{a',a}$ is a general element of L for almost all $a', a \in k$. And note that

$$\hat{\phi} \circ \jmath_L(u_{a',a}) = \hat{\phi} \circ \jmath_L(u/y_{a',a}) = \jmath_K\big(t/(a'x + a)\big) = \jmath_K(t_{a',a}).$$

In particular, since $\mathfrak{A}$ and $\mathfrak{B}$ contain some Bertini-type subsets, we can suppose that $\kappa_{t_{a',a}} \in \mathfrak{A}$ and $\kappa_{u_{a',a}} \in \mathfrak{B}$, and $\kappa_{t_{a',a}}$ corresponds to $\kappa_{u_{a',a}}$ under Φ. On the other hand, since by hypothesis we have $\jmath_K(t) \in M_K$ and $\jmath_L(u) \in M_L$, and by Remarks/Notation 48 above, $\jmath_K(kx + k) \subset M_K$ and $\jmath_L(ly + l) \subset M_L$, it follows that for almost $a, a' \in k$, the following hold:

(a) $t_{a',a} \in K$ and $u_{a',a} \in L$, respectively $t_{a',a+1} \in K$ and $u_{a',a+1} \in L$, are general elements which correspond to each other under $\hat{\phi}$.
(a)′ Hence $\hat{\phi}$ is normed with respect to both $u_{a',a}$ and $u_{a',a+1}$.

For b, b' as at (ii), let $x_{b',b} \in kx + k$ be such that $\hat{\phi}\big(\jmath_L(b'y + b)\big) = \jmath_K(x_{b',b})$. Then by (i), for all b and almost all b', the element $t_{b',b} := t/x_{b',b}$ is general, and note that

$$\jmath_K(t_{b',b}) = \jmath_K(t/x_{b',b}) = \hat{\phi} \circ \jmath_L\big(u/(b'y + b)\big) = \hat{\phi} \circ \jmath_L(u_{b',b}).$$

In particular, $\kappa_{t_{b',b}} \in \mathfrak{A}$ and $\kappa_{u_{b',b}} \in \mathfrak{B}$, and $\kappa_{t_{b',b}}$ corresponds to $\kappa_{u_{b',b}}$ under Φ, and $\hat{\phi}$ is normed with respect $u_{b',b}$. Reasoning as above, for almost $b', b \in l$ one has:

(b) $t_{b',b} \in K^\times$ and $u_{b',b} \in L$, respectively $t_{b',b+1} \in K$ and $u_{b',b+1} \in L$, are general elements which correspond to each other under $\hat{\phi}$.
(b)′ Hence $\hat{\phi}$ is normed with respect to both $u_{b',b}$ and $u_{b',b+1}$.

But then by the fact (†) from Remark/Notation 48 applied to the functions $t_{a',a} \in K^\times$ and $u_{a',a} \in L$, it follows that $\jmath_K(\kappa^\times_{t_{a',a}}) = \hat{\phi} \circ \jmath_L(\kappa^\times_{u_{a',a}}) \subset \hat{\phi} \circ \jmath_L(L^\times)$, and therefore we also have $\jmath_K(\kappa^\times_{t_{a',a}}) \subset \hat{\phi} \circ \jmath_L(L^\times) \cap \jmath_K(K^\times) = M_K$; and the same holds correspondingly for the other three pairs of functions which correspond to each other under $\hat{\phi}$. Thus finally we get

$$\jmath_K(\kappa^\times_{t_{a',a}}) \subset M_K, \quad \jmath_K(\kappa^\times_{t_{a',a+1}}) \subset M_K, \quad \jmath_L(\kappa^\times_{u_{b',b}}) \subset M_L, \quad \jmath_K(\kappa^\times_{u_{b',b+1}}) \subset M_L.$$

Finally, for $a, a', a'' \in k$, consider the functions

$$t_{a'',a',a} = (a''t + a'x + a + 1)/(t + a'x + a).$$

Then by Fact 43 (1), it follows that for all a'', and almost all a', a, the function $t_{a'',a',a}$ is a general element of K too. On the other hand, a direct computation shows that

$$t_{a'',a',a} = \frac{a'x + a + 1}{a'x + a} \cdot \frac{a''t_{a',a+1} + 1}{t_{a',a} + 1}.$$

Since the images via $\jmath_K$ of both the denominators and the numerators of the fractions above lie in M_K, we get $\jmath_K(t_{a'',a',a}) \in M_K$. Reasoning as previously in the case of $t_{a',a}$, we find general elements $u_{a'',a',a} \in L$ such that $\kappa_{t_{a'',a',a}}$ corresponds to $\kappa_{u_{a'',a',a}}$, etc. And we further define correspondingly functions

$$u_{b'',b',b} = \frac{b'x+b+1}{b'x+b} \cdot \frac{b''u_{b',b+1}+1}{u_{b',b}+1},$$

and find functions $t_{b'',b',b} \in K$, etc. Finally one gets $\jmath_K(\kappa_{t_{a'',a',a}}) \subset M_K$ for all $a'' \in k$, and almost all $a', a \in k$. And correspondingly $\jmath_L(\kappa_{u_{b'',b',b}}) \subset M_L$ for all $b'' \in l$, and almost all $b', b \in l$.

Now we conclude the proof of the fact that $\jmath_K(kt+k)^\times \subseteq M_K$ as follows: First, since $\jmath_K(\kappa^\times_{t_{a'',a',a}}) \subset M_K$, we have $\jmath_K(t_{a'',a',a}-1) \in M_K$. On the other hand,

$$t_{a'',a',a} - 1 = [(a''-1)t+1]/(t+a'x+a)\,.$$

Now observe that $t + a'x + a = (a'x+a+1)/t_{0,a',a}$. Hence $\jmath_K(t+a'x+a) \in M_K$, as $\jmath_K(t_{0,a',a}), \jmath_K(a'x+a+1) \in M_K$. Thus we finally deduce that $\jmath_K\big((a''-1)t+1\big) \in M_K$ for all $a'' \in k$. Hence $\mathscr{P}_t = \jmath_K(kt+k)^\times \subset M_K$, as $\jmath_K(t) \in M_K$ by hypothesis. In a completely similar way, one concludes that $\mathscr{P}_u = \jmath_L(lu+l)^\times \subset M_L$. □

Lemma 50. *Let $K_0 = \jmath_K^{-1}(M_K) \cup \{0\} \subseteq K$ and $L_0 = \jmath_L^{-1}(M_L) \cup \{0\} \subseteq L$ be the preimages of M_K, respectively M_L, in K, respectively L, together with 0 added. Then $K_0 \subseteq K$ and $L_0 \subseteq L$ are function subfields.*

Proof. Indeed, since M_K is a subgroup of $\widehat{K}$, its preimage $\jmath_K^{-1}(M_K)$ in $K^\times$ is a subgroup too. We check that K_0 is closed with respect to addition: For $t, t' \in K_0$ non-zero, $t'' = t'/t \in K_0$, and $t+t' = t(t''+1)$. On the other hand, by Lemma 49 we have $t''+1 \in K_0$. Hence finally we get $t+t' = t(t''+1) \in K_0$. The proof of the assertion concerning L_0 is similar, and we omit it. □

Next we observe that $M_K = \jmath_K(K_0^\times) = K_0^\times/k^\times$ can be viewed in a canonical way as the projectivization $\mathscr{P}(K_0) := K_0^\times/k^\times$ of the infinite-dimensional k-vector space $(K_0,+)$. And correspondingly, $M_L = \jmath_L(L_0^\times) = L_0^\times/l^\times =: \mathscr{P}(L_0)$ is the projectivization of the infinite-dimensional l-vector space $(L_0,+)$. And since the Kummer homomorphism $\hat{\phi} : \widehat{L} \to \widehat{K}$ maps M_L bijectively onto M_K, the restriction of $\hat{\phi}$ defines a bijection:

$$\phi := \hat{\phi}|_{\mathscr{P}(L_0)} : \mathscr{P}(L_0) = M_L \to M_K = \mathscr{P}(K_0).$$

Notice that the lines in $\mathscr{P}(K_0)$ are subsets of the form $\mathfrak{l}_{t_0,t_1} := (kt_0+kt_1)^\times/k^\times$ with t_0, t_1 k-linearly independent functions in K_0. In particular, setting $t := t_1/t_0$, we see that $\mathfrak{l}_{t_0,t_1} = t_0 \cdot \mathscr{P}_t$, where $\mathscr{P}_t := (kt+k)^\times/k^\times = \jmath_K(kt+k)^\times$. Further note that $\mathfrak{l}_{t_0,t_1}$ depends only on $\mathfrak{t}_0 = \jmath_K(t_0)$ and $\mathfrak{t}_1 := \jmath_K(t_1)$, and not on the functions t_0, t_1 themselves. We will therefore also write $\mathfrak{l}_{\mathfrak{t}_0,\mathfrak{t}_1}$ for the line $\mathfrak{l}_{t_0,t_1}$, and $\mathscr{P}_\mathfrak{t}$ for $\mathscr{P}_t$.

Correspondingly, the same holds for lines in $\mathscr{P}(L_0)$.

Lemma 51. *The morphism $\phi : \mathscr{P}(L_0) \to \mathscr{P}(K_0)$ respects colineations; more precisely, ϕ maps each line $\mathfrak{l}_{\mathsf{u}_0,\mathsf{u}_1} \subset \mathscr{P}(L_0)$ bijectively onto $\mathfrak{l}_{\mathsf{t}_0,\mathsf{t}_1} \subset \mathscr{P}(K_0)$, where $\mathsf{t}_0 = \phi(\mathsf{u}_0)$, $\mathsf{t}_1 = \phi(\mathsf{u}_1)$.*

Proof. Setting $\mathsf{t} = \phi(\mathsf{u})$, we get $\mathfrak{l}_{\mathsf{t}_0,\mathsf{t}_1} = \mathsf{t}_0 \cdot \mathfrak{l}_{\mathsf{t}}$ and $\mathfrak{l}_{\mathsf{u}_0,\mathsf{u}_1} = \mathsf{u}_0 \cdot \mathfrak{l}_{\mathsf{u}}$. Hence taking into account that ϕ respects the multiplication, it follows that it is sufficient to show that ϕ maps $\mathscr{P}_{\mathsf{u}}$ bijectively onto $\mathscr{P}_{\mathsf{t}}$, provided $\mathsf{t} := \phi(\mathsf{u})$.

Recall that $\hat{\phi}$ is y-normed, and $\hat{\phi}(\mathsf{y}) = \mathsf{x}$, where $\mathsf{y} = \jmath_L(y)$, $\mathsf{x} = \jmath_K(x)$, for x and y corresponding to each other under $\hat{\phi}$. Moreover, by fact (†) from Remark/Notations 48, $\hat{\phi}$ maps $\mathscr{P}_{\mathsf{y}} = \mathscr{P}_y$ bijectively onto $\mathscr{P}_{\mathsf{x}} = \mathscr{P}_x$. Recall that for every 1-index v of $\mathscr{G}_{\mathscr{D}_K}$, and the corresponding 1-index w of $\mathscr{H}_{\mathscr{D}_L}$, one has commutative diagrams of the form, see Remark 26 (3), and (4)

$$\begin{array}{ccc} \widehat{U}_w & \xrightarrow{\jmath_w} & \widehat{Lw} \\ \downarrow\hat{\phi} & & \downarrow\hat{\phi}_v \\ \widehat{U}_v & \xrightarrow{\jmath_v} & \widehat{Kv} \end{array} \qquad \text{and} \qquad \begin{array}{ccc} \widehat{L} & \xrightarrow{\jmath^w} & \mathbb{Z}_\ell\varphi_w \\ \downarrow\hat{\phi} & & \downarrow a_{vw} \\ \widehat{K} & \xrightarrow{\jmath^v} & \mathbb{Z}_\ell\varphi_v. \end{array}$$

Let κ_t be the relative algebraic closure of $k(t)$ in K_0. We claim that $\phi(\mathscr{P}_u) \subset \jmath_K(\kappa_t)$. Indeed, let v be such that $v(\mathsf{t}') \neq 0$ for some $\mathsf{t}' = \phi(\mathsf{u}')$ with $\mathsf{u}' \in \mathscr{P}_{\mathsf{u}}$. Then by the commutativity of the second diagram above we get $w(\mathsf{u}') \neq 0$. But then it follows that $\jmath_w$ is trivial on $l(u)^\times \cap U_w$. Hence by the commutativity of the first diagram above, it follows that $\jmath_v$ is trivial on $\phi \circ \jmath_L\big(l(u)^\times\big)$, in particular on $\phi(\mathscr{P}_{\mathsf{u}})$. By contradiction, suppose that $\phi(\mathscr{P}_u) \not\subset \jmath_K(\kappa_t)$. Then $\exists$ $\mathsf{u}_1 \in \mathscr{P}_u$ and $t_1 \in K_0$ such that t and t_1 are algebraically independent over k, and $\jmath_K(t_1) =: \mathsf{t}_1 = \phi(\mathsf{u}_1)$. On the other hand, since t, t_1 are algebraically independent over k, there exist "many" v satisfying the following: v is not trivial on $k(t)$ and t is a v-unit, and v is trivial on $k(t_1)$. Note that v being non-trivial on $k(t)$ and t being a v-unit implies that the residue of t at v lies in k; hence $\jmath_v(\mathsf{t}) = 0$. Now let w correspond to v under Φ. Then by the commutativity of the above diagrams, $\mathsf{u} = \jmath_L(u)$ is a w-unit, and $\jmath_w(\mathsf{u}) = 0$. Therefore, w is non-trivial on $l(u)$. Further, $\mathsf{u}_1 = \jmath_L(u_1)$ is a w-unit, and $\jmath_w(\mathsf{u}_1) \neq 0$. Hence w satisfies both that w is non-trivial on $l(u)$ and that $\jmath_w$ is non-trivial on $U_w \cap l(u)^\times$. Contradiction!

Now choose a prime divisor v of $K|k$ such that the following are satisfied:

(i) v is trivial on κ_x, and t is a v-unit.
(ii) x and t have equal residues in Kv; hence $\jmath_v(x) = \jmath_v(t)$.

Note that (ii) implies that v is trivial on κ_t too, hence $\jmath_v$ maps both $\kappa_x^\times$ and $\kappa_t^\times$ injectively into the residue field Kv.
Now let w correspond to v under the proper morphism $\Phi : \mathscr{G}_{\mathscr{D}_K} \to \mathscr{G}_{\mathscr{D}_L}$. Then reasoning as in the proof of Proposition 35, it follows that the following hold:

(j) w is trivial on κ_y.
(jj) y and u have equal residues in Lw; hence $\jmath_w(u) = \jmath_w(y)$.

Further, since $\jmath_w$ and $\jmath_v$ respect addition and multiplication, the following hold:

$$\jmath_v(\mathscr{P}_t) = \mathscr{P}_{\jmath_v(t)}, \ \jmath_v(\mathscr{P}_x) = \mathscr{P}_{\jmath_v(x)}, \ \text{and} \ \jmath_w(\mathscr{P}_u) = \mathscr{P}_{\jmath_w(u)}, \ \jmath_w(\mathscr{P}_y) = \mathscr{P}_{\jmath_w(y)}.$$

Hence by (i), (ii), respectively (j), (jj), we get $\mathscr{P}_{\jmath_v(t)} = \mathscr{P}_{\jmath_v(x)}$ and $\mathscr{P}_{\jmath_w(u)} = \mathscr{P}_{\jmath_w(y)}$. Since $\hat{\phi}(\mathscr{P}_y) = \mathscr{P}_x$ by the choice of x and y, it follows from $\hat{\phi}_v \circ \jmath_w = \jmath_v \circ \hat{\phi}$ that

$$\hat{\phi}_v(\mathscr{P}_{\jmath_w(y)}) = \hat{\phi}_v\big(\jmath_w(\mathscr{P}_y)\big) = \jmath_v\big(\hat{\phi}(\mathscr{P}_y)\big) = \jmath_v(\mathscr{P}_x) = \mathscr{P}_{\jmath_v(x)}.$$

Since $\jmath_v \circ \hat{\phi} = \hat{\phi}_v \circ \jmath_w$, from the equalities above we finally get

$$\jmath_v\big(\hat{\phi}(\mathscr{P}_u)\big) = \hat{\phi}_v\big(\jmath_w(\mathscr{P}_u)\big) = \hat{\phi}_v\big(\mathscr{P}_{\jmath_w(u)}\big) = \hat{\phi}_v(\mathscr{P}_{\jmath_w(y)}) = \mathscr{P}_{\jmath_v(x)} = \mathscr{P}_{\jmath_v(t)} = \jmath_v(\mathscr{P}_t).$$

Hence $\jmath_v\big(\hat{\phi}(\mathscr{P}_u)\big) = \jmath_v(\mathscr{P}_t)$. Since both $\hat{\phi}(\mathscr{P}_u)$ and $\mathscr{P}_t$ are subsets of $\jmath_K(\kappa_t^\times)$, and $\jmath_v$ is injective on $\jmath_K(\kappa_t^\times)$, we get $\hat{\phi}(\mathscr{P}_u) = \mathscr{P}_t$, as claimed. □

In order to conclude the first part of the proof of Theorem 45 we proceed as follows: By Lemma 50 above, ϕ respects colineations. Therefore, by the *fundamental theorem of projective geometries,* see e.g, Artin [1], ϕ is the projectivization $\phi = \mathscr{P}(\phi')$ of some linear $\imath_0$-isomorphism $\phi' : (L_0, +) \to (K_0, +)$, i.e., there exists a field isomorphism $\imath_0 : l \to k$, and ϕ' is an isomorphism of abelian groups, such that $\phi'(au) = \imath_0(a)\phi'(u)$ for all $a \in l$ and $u \in L_0$. Moreover, ϕ' is unique up to composition by homotheties of the form $l_a \circ \phi' \circ l_b$ (all $a \in k$, $b \in l$). Further, since $k^\times = \ker(\jmath_K)$ and $l^\times = \ker(\jmath_L)$, it follows that $\phi'(l) = k$. We set

$$\phi_0 := \big(1/\phi'(1)\big)\,\phi',$$

and claim that ϕ_0 is a field isomorphism which maps l isomorphically onto k. Indeed, for a fixed $y \in L_0$, consider $\phi_y : L_0 \to K_0$ defined by $\phi_y(u) := \phi_0(yu)$. Then ϕ_y is a linear $\imath_0$-isomorphism. Set $x = \phi_0(y)$. Then considering projectivizations, and using the fact that $\phi = \mathscr{P}(\phi_0)$ is multiplicative, it follows that for all $u \in L_0$ we have

$$\mathscr{P}(\phi_y)(u) = \mathscr{P}(\phi_0)(yu) = \mathscr{P}(l_x) \circ \mathscr{P}(\phi_0)(u),$$

where l_x is the multiplication by x on K_0. Therefore, there exist $a \in k^\times$ and $b \in l^\times$ such that $l_b \circ \phi_y \circ l_b = l_x \circ \phi_0$. In other words, $a\,\phi_0(y\,b\,u) = x\,\phi_0(u)$ for all $u \in L_0$. Setting $u = 1$, and taking into account that $\phi_0(1) = 1$, we have $a\,\imath_0(b)\,x = x$. Thus $a\,\imath_0(b) = 1$, and hence the effects of l_a and l_b cancel each other. Hence we have

$$\phi_0(y\,u) = x\,\phi_0(u) = \phi_0(y)\,\phi_0(u), \quad (\text{all } u, y \in L_0),$$

hence ϕ_0 is a field morphism. And since $\phi_0(L_0) = K_0$ and $\phi_0(l) = k$, it follows that $\phi_0 : L_0|l \to K_0|k$ is an isomorphism of field extension, as claimed.

Finally, in order to conclude the proof of Theorem 45 we prove the following:

Lemma 52. *$L|L_0$ is a purely inseparable field extension.*

Proof. First, recall that by the last assertion mentioned at Remark/Notations 48, we have $M_L \otimes \mathbb{Z}_{(\ell)} = L^{\times}_{(\ell)}$ inside $\widehat{L}$. Equivalently, for every $u \in L^{\times}$ there exists a prime to ℓ integer $n_u > 0$ such that $u^{n_u} \in L_0$. This means in particular that $L|L_0$ is an algebraic extension. Since $L|l$ is a function field over the (algebraically closed) field l, it follows that $L_0|l$ is so, and $L|L_0$ is actually a finite field extension, of degree $[L : L_0] = n > 0$. From this we deduce that if n_u is minimal such that $u^{n_u} \in L$, then $n_u|n$. In particular, all the nth powers u^n, $u \in L$, are contained in L_0. But then $n = [L : L_0]$ must be a power of the characteristic, as claimed. □

For the uniqueness of $\imath$ up to Frobenius twists, one uses the last lemma above, and applies Proposition 38. This completes the proof of Theorem 45. □

7 Appendix

Here we recall a few facts concerning the pro-ℓ abelian quotient of the fundamental group, and facts concerning the structure of the divisor class group as an abstract group. All these seem to be folklore and might be well be known to the experts, but I cannot give a quick reference. Throughout this section, $K|k$ is a function field over an algebraically closed field k of characteristic $p \geq 0$.

7.1 The Kummer interpretation of ${}_n\mathfrak{Cl}(X)$

Let $K|k$ be a function field, and $X \to k$ a normal model of $K|k$. Recall that D_X is the set of prime divisors of $K|k$ which are defined by the Weil prime divisors of X, and $\text{div}_X : K^{\times} \to \text{Div}(X)$ the corresponding divisor map. Then denoting by $\mathcal{U}_X := \Gamma(X, \mathcal{O}_X)^{\times}$ the invertible global functions on X, one has $\ker(\text{div}_X) = \mathcal{U}_X$, and $\mathcal{U}_X/k^{\times}$ is a finitely generated free abelian group. Finally, one has an exact sequence of the form

$$1 \to \mathcal{U}_X \to K^{\times}/k^{\times} \to \mathcal{H}_X(K) \to 0,$$

where $\mathcal{H}_X(K) := \text{div}_X(K)$ is the group of principal (Weil) divisors of X, and the exact sequence defining the divisor class group is

$$1 \to \mathcal{H}_X(K) \to \text{Div}(X) \to \mathfrak{Cl}(X) \to 0.$$

We now want to recall the Kummer interpretation of the prime to the characteristic torsion of $\mathfrak{Cl}(X)$, which is as follows: Let $\text{char}(k) = p \geq 0$ be the characteristic of k, and n a positive integer not divisible by p. Then tensoring the last exact sequence with $\mathbb{Z}/n$, we get the exact sequence

$$0 \to {}_n\mathfrak{Cl}(X) \hookrightarrow \mathscr{H}_X(K)/n \to \mathrm{Div}(X)/n \to \mathfrak{Cl}(X)/n \to 0,$$

where ${}_n\mathfrak{Cl}(X)$ is the n-torsion of $\mathfrak{Cl}(X)$. In particular, if we denote by Δ_n the preimage of ${}_n\mathfrak{Cl}(X) \subset \mathscr{H}_X(F)/n$ in $\mathscr{H}_X(K)$, it follows that we have a canonical exact sequence

$$0 \to \Delta_n \hookrightarrow K^\times/n \to \mathrm{Div}(X)/n \to \mathfrak{Cl}(X)/n \to 0,$$

because $K^\times/n \cong (K^\times/k^\times)/n$ by the fact that k being algebraically closed. Notice also that Δ_n fits into an exact sequence of the form $1 \to \mathscr{U}_X/n \to \Delta_n \to {}_n\mathfrak{Cl}(X) \to 0$; thus it is a quotient of $(\mathbb{Z}/n)^I$ for some index set I.

Fact 53. *In the above context, set $K_n := K[\sqrt[n]{\Delta_n}]$. Then $K_n|K$ is an n-elementary abelian extension of K satisfying:*

(1) $\mathrm{Gal}(K_n|K) \cong \mathrm{Hom}(\Delta_n, \mu_n)$ *canonically.*
(2) *$K_n|K$ is the maximal n-elementary abelian extension of K in which all $v \in D_X$ are unramified.*
(3) *In particular, for $\ell \neq \mathrm{char}(k)$, let $K_{\ell^\infty}|K$ be the maximal pro-ℓ abelian extension in which all $v \in D_X$ are unramified. Then*

$$\mathrm{Gal}(K_{\ell^\infty}|K) = \Pi_{1,D_X} = \mathrm{Hom}(\Delta_\infty, \mu_{\ell^\infty}),$$

where $\Delta_\infty = \varinjlim_n \Delta_n$ fits canonically in $0 \to \mathscr{U}_X \otimes \mathbb{Q}_\ell/\mathbb{Z}_\ell \to \Delta_\infty \to {}_{\ell^\infty}\mathfrak{Cl}(X) \to 0$.
(4) *Finally, ${}_n\mathfrak{Cl}(X)$ and $K_n|K$ are finite by Fact 54 below.*

Proof. The first assertion is clear by Kummer theory. For the second assertion, consider the following commutative diagram:

$$\begin{array}{ccccccccc} 1 \to & \Delta_n & \to & K^\times/n & \to & \mathrm{Div}(X)/n & \to & \mathfrak{Cl}(X)/n & \to 0 \\ & \downarrow & & \downarrow & & \| & & \| & \\ 0 \to & {}_n\mathfrak{Cl}(X) & \to & \mathscr{H}_X(K) & \to & \mathrm{Div}(X)/n & \to & \mathfrak{Cl}(X)/n & \to 0 \end{array}$$

We first show that all $v \in D_X$ are unramified in $K_n|K$. Equivalently, we have to prove that for all $f \in K^\times$ whose images lie in Δ_n, all $v \in D_X$ are unramified in $K(\sqrt[n]{f}\,)|K$. Now if the image of $f \in K^\times$ lies in Δ_n, then the image of f in $\mathscr{H}_X(K)$ lies actually in ${}_n\mathfrak{Cl}(X)$. Therefore, the divisor (f) of f has trivial image in $\mathrm{Div}(X)/n$; in other words, there exists some divisor $P \in \mathrm{Div}(X)$ such that $nP = (f)$, i.e., $v(f) \in n \cdot vK$ for all $v \in D_X$. But then by Hilbert decomposition theory, it follows that every $v \in D_X$ is unramified in $K(\sqrt[n]{f}\,)|K$. Since f was arbitrary, it follows that all $v \in D_X$ are unramified in $K_n|K$.

Conversely, for $f \in K^\times$, suppose that all $v \in D_X$ are unramified in $K(\sqrt[n]{f}\,)|K$. Equivalently, $v(f) \in n \cdot vK$ for all $v \in D_X$; hence $\mathrm{div}_X(f) \in n \cdot \mathrm{Div}(X)$. Thus there exists some divisor $P \in \mathrm{Div}(X)$ such that $\mathrm{div}_X(f) = nP$ in $\mathrm{Div}(X)$. But then $\mathrm{div}_X(f) \in \mathscr{H}_X(K)$ has a trivial image in $\mathrm{Div}(X)/n$, and therefore, the image of f in $K^\times/n$ lies actually in Δ_n, as claimed.

Finally, the proof of assertion (3) follows from assertion (2) by "taking limits." □

7.2 On the Weil divisor class group $\mathfrak{Cl}(X)$

We want to say a few words about the divisor class group $\mathfrak{Cl}(X) := \mathrm{Div}(X)/\mathscr{H}_X(K)$ (as an abstract group) of a normal model $X \to k$ of $K|k$ in more detail.

Fact 54. *Let k be an algebraically closed field, and $X \to k$ be an integral normal variety. Then the divisor class group $\mathfrak{Cl}(X)$ can be written as a (direct) sum:*

$$(*) \qquad \mathfrak{Cl}(X) \cong A_0(X) + A_1(X),$$

where $A_0(X)$ is the maximal divisible subgroup of $\mathfrak{Cl}(X)$, and $A_1(X)$ is a finitely generated abelian group. Further:

(1) *The maximal torsion divisible subgroup $A_t(X) \subseteq A_0(X)$ of $\mathfrak{Cl}(X)$ is a quotient of $(\mathbb{Q}/\mathbb{Z})^r$ for some $r \geq 0$.*
(2) *If k is an algebraic closure of a finite field, then $A_0(X) = A_t(X)$.*

Proof. First, since $A_0(X) \subseteq \mathfrak{Cl}(X)$ is the maximal divisible group, it follows that it has complements $A(X)$; hence $\mathfrak{Cl}(X) = A_0(X) + A(X)$ as a direct sum. Let K be the function field of X. Let $Y \to k$ and $X \to k$ be normal models of $K|k$ such that $D_X \subseteq D_Y$. We first claim the following:

Claim 1. $\mathfrak{Cl}(X)$ satisfies $(*)$ iff $\mathfrak{Cl}(Y)$ satisfies $(*)$.

Indeed, let $A_0(X) \subset \mathfrak{Cl}(X)$ and $A_0(Y) \subset \mathfrak{Cl}(Y)$ be the (unique) maximal divisible subgroups. Setting $S := D_Y \backslash D_X$, we have that S is finite, $D_Y = D_X \cup S$, and the canonical projection map $\mathrm{pr} : \mathrm{Div}(Y) \to \mathrm{Div}(X)$ has as kernel $\Delta_S := \sum_{v \in S} \mathbb{Z}v$. Thus the canonical projection $\mathrm{pr} : \mathrm{div}_Y(K^\times) \to \mathrm{div}_X(K^\times)$ has kernel $\mathscr{U}_S := \mathrm{div}_Y(K^\times) \cap \Delta_S$, which a finitely generated group, and the canonical map $\overline{\mathrm{pr}} : \mathfrak{Cl}(Y) \to \mathfrak{Cl}(X)$ has kernel $\Delta_S/\mathscr{U}_S$. Further note that $\overline{\mathrm{pr}}\big(A_0(Y)\big) \subset A_0(X)$, because the former group is divisible, thus contained in the unique maximal divisible subgroup $A_0(X)$ of $\mathfrak{Cl}(X)$, and $\overline{\mathrm{pr}} : A_0(Y) \to A_0(X)$ has kernel $\Delta_{S,0} := (\Delta_S/\mathscr{U}_S) \cap A_0(Y)$, which is a finitely generated group, as $\Delta_S/\mathscr{U}_S$ is so. Hence setting $A_S := (\Delta_S/\mathscr{U}_S)/\Delta_{S,0}$, we finally get an exact sequence of the form:

$$(\dagger) \qquad 0 \to A_S \to A(Y) \to A(X) \to 0.$$

Since A_S is a finitely generated group, it follows that $A(Y)$ if finitely generated iff $A(X)$ is so. This concludes the proof of Claim 1.

We next notice that the assertion of the above Claim 1 holds in the same form for the Cartier divisor class group $\mathfrak{CaCl}(X)$, and the proof of this fact is word-by-word the same as in the case of $\mathfrak{Cl}(X)$. Further, if X is smooth (enough factorial), then $\mathfrak{CaCl}(X) = \mathfrak{Cl}(X)$.

We conclude the proof of Fact 54 as follows: Let $X \to k$ be an arbitrary normal integral variety. Further, let $\tilde{X} \to k$ be a projective normal model of the function field $K := \kappa(X)$ of $X \to k$. Then $\tilde{X}$ and X are birational, and hence they have isomorphic open subsets $U \subset X$ and $\tilde{U} \subset \tilde{X}$. Moreover, we can suppose that $U \cong \tilde{U}$ are actually smooth over k. By Claim 1 and its Cartier form, it follows that, first, $\mathfrak{Cl}(X)$ has the structure $(*)$ iff $\mathfrak{Cl}(U)$ does, and second, $\mathfrak{CaCl}(\tilde{X})$ has the structure $(*)$ iff $\mathfrak{CaCl}(\tilde{U})$ does. On the other hand, since $U \cong \tilde{U}$ is smooth over k, hence factorial, we have $\mathfrak{Cl}(U) \cong \mathfrak{CaCl}(\tilde{U})$. Thus finally, it is sufficient to show that $\mathfrak{CaCl}(\tilde{X})$ has the structure $(*)$. In order to conclude, recall that for $\tilde{X} \to k$ projective integral normal, the structure of $\mathfrak{CaCl}(\tilde{X})$ is known: Indeed, by Kleiman [14], Theorem 4.8, Theorem 5.4, Corollary 6.17, Remark 6.19, it follows that $\mathfrak{CaCl}(\tilde{X}) = \mathrm{Pic}(\tilde{X})(k)$ has the structure $(*)$. This concludes the proof of Fact 54. □

Fact 55. *In the above context and notation, let $X \to k$ and $Y \to k$ be normal models of $K|k$ such that $D_X \subseteq D_Y$. In the notations from the proof of Claim 1, let $A_{\mathrm{tor}}(X) \subseteq A_0(X)$ be the torsion subgroup, whence $A_{\mathrm{tor}}(X)$ is divisible too; and let $A_\tau(X) \subseteq \mathfrak{Cl}(X)$ be the preimage of the torsion group of $\mathfrak{Cl}(X)/A_0(X)$, hence $A_\tau/A_0(X)$ is finite, and $\mathfrak{Cl}(X)/A_\tau(X)$ is finitely generated free abelian. The projection $\overline{\mathrm{pr}} : \mathfrak{Cl}(Y) \to \mathfrak{Cl}(X)$ gives rise to homomorphisms $\mathrm{pr}_\tau : A_\tau(Y) \to A_\tau(X)$, $\mathrm{pr}_0 : A_0(Y) \to A_0(X)$, $\mathrm{pr}_{\mathrm{tor}} : A_{\mathrm{tor}}(Y) \to A_{\mathrm{tor}}(X)$, and the following hold:*

(1) *$\mathrm{pr}_{\mathrm{tor}}$ has finite kernel and cokernel isomorphic to $(\mathbb{Q}/\mathbb{Z})^r$, where r the rational rank of $(\Delta_S/\mathscr{U}_S) \cap A_0(Y)$. Therefore, $A_{\mathrm{tor}}(X) \cong A_{\mathrm{tor}}(Y) \oplus (\mathbb{Q}/\mathbb{Z})^r$ as abstract groups.*
(2) *The rational ranks of $A_1(Y)$, $A_1(X)$ satisfy $\mathrm{rr}(A_1(Y)) = \mathrm{rr}(A_1(X)) + |S| - r$. Hence one has $\mathrm{rr}(A_1(Y)) = \mathrm{rr}(A_1(X)) + |S|$ iff $\mathscr{U}_S$ is trivial iff $\mathscr{U}_Y = \mathscr{U}_X$.*
(3) *If the equivalent conditions from (2) above are satisfied, then $\mathrm{pr}_{\mathrm{tor}}$ and pr_0 are isomorphisms, and pr_τ maps $A_\tau(Y)/A_0(Y)$ injectively into $A_\tau(X)/A_0(X)$.*
(4) *Suppose that X has/satisfies the following equivalent properties:*
    (i) *$A_\tau(X)/A_0(X)$ and $A_{\mathrm{tor}}(X)$ are minimal.*
    (ii) *$A_1(Y) \cong A_1(X) \oplus \mathbb{Z}^{|D_Y \setminus D_X|}$ for all normal models $Y \to k$ of $K|k$ with $D_X \subset Y$.*
    (iii) *$|A_\tau(X)/A_0(X)|$ is minimal and $\mathrm{rr}(A_1(Y)) = \mathrm{rr}(A_1(X)) + |D_Y \setminus D_X|$ for all Y as above.*

Then $A_{\mathrm{tor}} \subseteq A_0(X) \subseteq A_\tau(X)$ are birational invariants of the function field $k(X)$.

Proof. Everything follows immediately from the proof of Claim 1. For assertions (2), (3) and (4), use the exact sequence (†). In particular, note that if the equivalent conditions from assertions (2) are satisfied, then $\mathscr{U}_S$ trivial, and therefore $\Delta_{S,0} = \Delta \cap A_0(X)$ is a torsion subgroup of the free abelian group Δ_S. Hence $\Delta_{S,0}$ is trivial, and $A_S = \Delta_S$ is a finite free $\mathbb{Z}$-module. Thus the exact sequence (†) becomes $0 \to \Delta_S \to A(Y) \to A(X) \to 0$, etc. □

7.3 On the (pro-ℓ abelian) "birational" fundamental group $\Pi_{1,K}$

Let $K|k$ be a function field. It is well known that if $K|k$ has regular complete models $X \to k$, then the "usual" fundamental group $\pi_1(X)$ whose open subgroups parametrize all the étale connected covers of X is a birational invariant of $K|k$, in the sense that $\pi_1(X)$ does not depend on the particular regular complete model $X \to k$. In general, one can consider the following replacement for the fundamental group of complete regular models: Let $D_{K|k}$ be the set of prime divisors of $K|k$. For every $v \in D_{K|k}$, let $\overline{v}$ be prolongations of v to an algebraic closure $\overline{K}$ of K, and $T_{\overline{v}} \subset G_K$ the inertia group of $\overline{v}$ in the separable closure $K^{\mathrm{s}}|K$ of K in $\overline{K}$. Then the prolongations $\overline{v}$ are conjugated under G_K; thus the set of all the inertia groups $T_{\overline{v}}$ is closed under G_K-conjugation. Therefore, the closed subgroup $T_K \subset G_K$ generated by all the $T_{\overline{v}}$ for all $v \in D_{K|k}$ and all their prolongations $\overline{v}$ to $\overline{K}$ is a normal subgroup of G_K. Moreover, setting

$$\pi_{1,K} := G_K/T_K,$$

we see by the functoriality of Hilbert decomposition theory that the fixed field $\tilde{K}$ of T_K in K^{s} is the maximal field extension of K in which all the prime divisors v of $K|k$ are not ramified. We will call $\pi_{1,K}$ the *birational fundamental group* for $K|k$.

More generally, let $D \subset D_{K|k}$ be any set of prime divisors of K, e.g., $D = D_X$ is the set of prime divisors defined by the Weil prime divisors of a normal model $X \to k$ of $K|k$. Then we denote by $T_D \subset G_K$ the subgroup generated by all the $T_{\overline{v}}$ with $v \in D$ and $\overline{v}$ all the prolongations of v to $\overline{K}$. As above, T_D is normal in G_K, and the quotient

$$\pi_{1,D} := G_K/T_D$$

will be called the *fundamental group* for D. We notice that open subgroup of $\pi_{1,D}$ parametrize all the finite extensions $L|K$ of K in which all $v \in D$ are not ramified.

In the same way, we introduce/define the pro-ℓ abelianizations of the fundamental groups introduced above,

$$\Pi_{1,K} := \Pi_K/T_{D_{K|k}} \quad \text{and} \quad \Pi_{1,D} := \Pi_K/T_D,$$

and call them the pro-ℓ abelian *birational fundamental group* for $K|k$, respectively for D.

Remarks 56. Let $X \to k$ be a normal model of $K|k$, and $\pi_1(X) \to \Pi_1(X) := \pi_1^{\ell,\mathrm{ab}}(X)$ the canonical projection. In the above context and notation, the following hold:

(a) For every $D \subseteq D_K$ there are canonical surjective projections $\pi_{1,D} \to \pi_{1,K}$, and $\pi_{1,D_X} \to \pi_1(X)$, which by the functoriality of Hilbert decomposition theory give rise to surjective projections $\Pi_{1,D} \to \Pi_{1,K}$ and $\Pi_{1,D_X} \to \Pi_1(X)$.

(b) Nevertheless, if X is regular, then $\pi_{1,D_X} = \pi_1(X)$, thus also $\Pi_{1,D_X} = \Pi_1(X)$, by the *purity of the branch locus.*
(c) And if X is complete and regular, then $\pi_{1,K} = \pi_1(X)$, thus also $\Pi_{1,K} = \Pi_1(X)$, by the *purity of the branch locus.*

Fact 57. *In the above context, the following hold:*

(1) *Let $X \to k$ be a complete normal model of $K|k$, and $U \subseteq X$ is a regular open subvariety. Then there are canonical surjective projections $\pi_1(U) \to \pi_{1,K} \to \pi_1(X)$.*
(2) *There exists a geometric set D_X such that $\Pi_{1,D_X} \to \Pi_{1,K}$ is an isomorphism. Hence $\Pi_{1,D_{X'}} \to \Pi_{1,K}$ is an isomorphism, provided $D_X \subseteq D_{X'}$.*
(3) *Let X be an affine normal curve with $\Pi_{1,D_X} = \Pi_{1,K}$. Then either $X \cong \mathbb{A}^1_k$, or X is isomorphic to $E \backslash \{\mathrm{pt}\}$ with E a complete curve of genus one.*
(4) *If X is a normal model of $K|k$ such that $\Pi_{1,D_X} = \Pi_{1,K}$, then the group of global invertible sections on X is $\mathcal{U}_X := k^\times$.*

Proof. To (1): The existence and surjectivity of $\pi_1(U) \to \pi_{1,K}$ follow from (a) and (b) above. For the existence and surjectivity of $\pi_{1,K} \to \pi_1(X)$, we notice first that since $X \to k$ is a complete variety, for every $v \in D^1_K$, there exists a dominant canonical k-morphism $\operatorname{Spec} \mathcal{O}_v \to X$. On the other hand, since a base change of an étale cover is étale, we have that if $Y \to X$ is some finite connected étale cover defined by some finite quotient of $\pi_1(X)$, then Y is integral, and $Y \times_X \operatorname{Spec} \mathcal{O}_v$ is étale over $\mathcal{O}_v$. Equivalently, v is unramified in the field extension $k(X) \hookrightarrow k(Y)$; hence the image of the inertia group T_v in $\pi_1(X)$ is trivial, etc.

For assertion (2), first consider a small enough affine open subset $X_0 \subset X'$ such that X_0 is regular. Then $X_0 \to k$ is a quasi-projective regular model for $K|k$; hence $\pi_{1,D_{X_0}} = \pi_1(X_0)$, by the purity of the branch locus. Therefore, $\pi_{1,D_{X_0}}$ is finitely generated, hence a finite module as $\mathbb{Z}_\ell$, as it is an abelian pro-ℓ group. Since $\Pi_{1,D_{X_0}} \to \Pi_{1,K}$ is surjective, $\Pi_{1,K}$ is a finite $\mathbb{Z}_\ell$-module too. But then

$$\Delta = \ker(\Pi_{1,D_{X_0}} \to \Pi_{1,K})$$

is also a finite $\mathbb{Z}_\ell$-module. Finally, by the definition of $\Pi_{1,K}$, it follows that for every $g \in \Delta$ there exists some $v \in D^1_K$ such that $g \in T_v$. Since Δ is finitely generated, there exists a finite set $\Sigma \subset D^1_K$ such that the images of T_v (all $v \in \Sigma$) in $\Pi_{1,D_{X_0}}$ generate Δ. In order to conclude, consider any quasi-projective normal model $X \to k$ such that $D_{X'}, \Sigma \subseteq D_X$ (hence in particular, $D_{X_0} \subseteq D_X$ too).

Assertion (3) follows immediately from the structure theorem for (the abelian pro-ℓ quotient of the) fundamental groups of a normal curve.

Finally, assertion (4) follows from the following: By contradiction, let $f \in U_X$ be a non-constant global invertible section. For every $n = \ell^e$ with $e \geq 0$, consider the normalization $X_n \to X$ of X in the finite subextension $K_n := K[\sqrt[n]{f}\,]$ of $K \hookrightarrow K'$. Then since f is a v-unit for all $v \in D_X$, it follows that v is unramified in $X_n \to X$, and therefore, $\mathrm{Gal}(K_n|K)$ is a quotient of Π_{1,D_X}. On the other hand, if w is a prime divisor of $K|k$ with $w(f) > 0$, then w is ramified in $K_n|K$ for $n \gg 0$; hence $\mathrm{Gal}(K_n|K)$ is not a quotient of $\Pi_{1,K}$. Contradiction! □

References

[1] Artin, E., *Geometric Algebra*, Interscience Publishers Inc., New York, 1957
[2] Bogomolov, F. A., *On two conjectures in birational algebraic geometry,* in: Algebraic Geometry and Analytic Geometry, ICM-90 Satellite Conference Proceedings, ed. A. Fujiki et all, Springer Verlag Tokyo 1991.
[3] Bogomolov, F. A. and Tschinkel, Y., *Commuting elements in Galois groups of function fields,* in: Motives, Polylogarithms and Hodge theory, eds F.A. Bogomolov, L. Katzarkov, International Press, 2002.
[4] Bogomolov, F. A. and Tschinkel, Y., *Reconstruction of function fields,* See arXiv:math/ 0303075v2[math.AG]16Oct2003,
[5] Bogomolov, F. A. and Tschinkel, Y., *Reconstruction of function fields,* Geometric And Functional Analysis, Vol **18** (2008), 400–462.
[6] Bourbaki, N., *Algèbre commutative*, Hermann Paris, 1964.
[7] Deligne, P., *Le groupe fondamental de la droite projective moins trois points,* in: Galois groups over **Q**, Math. Sci. Res. Inst. Publ. **16**, 79–297, Springer 1989.
[8] Deligne, P., Letter to Pop, September 1995.
[9] Faltings, G., *Curves and their fundamental groups (following Grothendieck, Tamagawa and Mochizuki),* Astérisque **252** (1998), Exposé 840.
[10] Geometric Galois Actions I, LMS LNS Vol **242**, eds. L. Schneps – P. Lochak, Cambridge Univ. Press 1998.
[11] Grothendieck, A., Letter to Faltings, June 1983. See [GGA].
[12] Grothendieck, A., Esquisse d'un programme, 1984. See [GGA].
[13] Kim, M., *The motivic fundamental group of* $\mathbb{P}^1\backslash\{0,1,\infty\}$ *and the theorem of Siegel,* Invent. Mathematicae **161** (2005), 629–656.
[14] Kleiman, S. L., *The Picard Scheme. Fundamental algebraic geometry,* Math. Surveys Monogr. **123**, 235–321.
[15] Koenigsmann, J., *On the "Section Conjecture" in anabelian geometry,* J. reine angew. Math. **588** (2005), 221–235.
[16] Kuhlmann, F.-V., Book on Valuation Theory.See: http://math.usask.ca/~fvk/Fvkbook.htm.
[17] Lang, S., *Algebra,* Revised third edition. Graduate Texts in Mathematics 211. Springer-Verlag, New York, 2002.
[18] Lang, S., Introduction to Algebraic geometry, Third printing, with corrections. Addison-Wesley Publishing Co., Inc., Reading, Mass., 1972.
[19] Mochizuki, Sh., *Absolute Anabelian Cuspidalizations of Proper Hyperbolic Curves,* See http://www.kurims.kyoto-u.ac.jp/~motizuki/papers-english.html.
[20] Mumford, D., *The Red Book of Varieties and Scheme*, LNM 1358, 2nd expanded edition, Springer-Verlag 1999.
[21] Neukirch, J., *Über eine algebraische Kennzeichnung der Henselkörper,* J. reine angew. Math. **231** (1968), 75–81.
[22] Neukirch, J., *Kennzeichnung der p-adischen und endlichen algebraischen Zahlkörper,* Invent. math. **6** (1969), 269–314.
[23] Parshin, A. N., *Finiteness Theorems and Hyperbolic Manifolds,* in: The Grothendieck Festschrift III, eds. P. Cartier et all, PM Series Vol 88, Birkhäuser Boston, 1990.
[24] Pop, F., *On Alterations and birational anabelian geometry,* in: Resolution of Singularities, Birkhäuser PM Series, Vol. **181**, p. 519–532; eds. Herwig Hauser et al., Birkhäuser Verlag, Basel 2000.
[25] Pop, F., *Recovering* $K|k$ *from* $G_K(\ell)$, MSRI Talk notes, Fall 1999. See http://www.msri.org/ publications/ln/msri/1999/gactions/pop/1/index.html
[26] Pop, F., *The birational anabelian conjecture —revisited—*, Manuscript, Princeton/Bonn 2002. See http://www.math.leidenuniv.nl/gtem/view.php
[27] Pop, F., *Pro-*ℓ *birational anabelian geometry over alg. closed fields I,* See arXiv:math/ 0307076v1[math.AG]5July2003.

[28] Pop, F., *Pro-ℓ abelian-by-central Galois theory of Zariski prime divisors,* Israel J. Math. **180** (2010), 43–68.
[29] Pop, F., *Inertia elements versus Frobenius elements,* Math. Annalen **438** (2010), 1005-1017.
[30] Pop, F., *On the birational anabelian program initiated by Bogomolov* I, Manuscript 2009 (submitted).
[31] Roquette, P., *Zur Theorie der Konstantenreduktion algebraischer Mannigfaltigkeiten,* J. reine angew. Math. **200** (1958), 1–44.
[32] Roquette, P., *Nonstandard aspects of Hilbert's irreducibility theorem,* in: Model theory and algebra (A memorial tribute to Abraham Robinson), LNM Vol. **498**, Springer, Berlin, 1975. 231–275.
[33] Saidi, M. and Tamagawa, T., *Prime to p version of Grothendieck's anabelian conjecture for Hyperbolic Curves over Finite Fields of Characteristic* $p > 0$, Publ. RIMS Kyoto University **45**, no. 1 (2009), 135–186.
[34] Szamuely, T., *Groupes de Galois de corps de type fini (d'après Pop),* Astérisque **294** (2004), 403–431.
[35] Stix, J., *Projective anabelian curves in positive characteristic and descent theory for log-étale covers,* thesis, Univ. of Bonn, 2002.
[36] Uchida, K., *Isomorphisms of Galois groups of solvably closed Galois extensions,* Tôhoku Math. J. **31** (1979), 359–362.
[37] Uchida, K., *Homomorphisms of Galois groups of solvably closed Galois extensions,* J. Math. Soc. Japan **33** (1981).
[38] Zariski, O. and Samuel, P., *Commutative Algebra*, Vol. II, Springer-Verlag, New York, 1975.

Irreducible spaces of modular units

David E. Rohrlich

In memory of Serge Lang

Abstract We give a representation-theoretic decomposition of the group of modular units of prime level. Apart from the formulation, the results obtained are contained in those of Gross [3].

Key words modular units • Siegel units • Kubert–Lang parametrization • Manin–Drinfeld theorem

Mathematics Subject Classification (2010): 11F03, 11F42, 14G35

Introduction

The group $\mathrm{SL}(2,\mathbb{Z})$ of 2×2 matrices with integer coefficients and determinant 1 acts on the complex upper half-plane H by fractional linear transformations, and if $\Gamma \subset \mathrm{SL}(2,\mathbb{Z})$ is a subgroup of finite index then the orbit space of H under Γ will be denoted $Y(\Gamma)$ and endowed with the unique complex structure for which the quotient map $\pi : H \to Y(\Gamma)$ is holomorphic. As a Riemann surface, $Y(\Gamma)$ is noncompact but of finite type: $Y(\Gamma) = X(\Gamma) \smallsetminus C$, where $X(\Gamma)$ is a compact Riemann surface and $C \subset X(\Gamma)$ the finite nonempty subset of *cusps* of Γ. A *modular function* for Γ is the pullback under π of the restriction to $Y(\Gamma)$ of a meromorphic function on $X(\Gamma)$. We denote the field of modular functions for Γ by $\mathfrak{M}^{\Gamma}$.

D.E. Rohrlich (✉)
Department of Mathematics and Statistics, Boston University, Boston, MA 02215
e-mail: rohrlich@math.bu.edu

D. Goldfeld et al. (eds.), *Number Theory, Analysis and Geometry: In Memory of Serge Lang*, DOI 10.1007/978-1-4614-1260-1_25,

Now fix Γ and let $\mathcal{O} \subset \mathfrak{M}^\Gamma$ be the subring of holomorphic functions. Its unit group $\mathcal{O}^\times$ will be denoted U, and an element of U will be called a *modular unit*. Thus a modular unit is a modular function which is holomorphic and nowhere vanishing on H. If we identify $\mathfrak{M}^\Gamma$ with the function field of $X(\Gamma)$ then U corresponds to the multiplicative group of meromorphic functions on $X(\Gamma)$ with divisorial support in C. Writing $\mathbb{C}^\times$ for the subgroup of constant functions in U, we deduce that the free abelian group $U/\mathbb{C}^\times$ has rank at most $|C|-1$.

A theorem of Manin [8] and Drinfeld [2] asserts that the rank is exactly $|C|-1$ if Γ is a *congruence subgroup* of $\mathrm{SL}(2,\mathbb{Z})$, in other words a subgroup containing the kernel of the reduction-mod-N map $\mathrm{SL}(2,\mathbb{Z}) \to \mathrm{SL}(2,\mathbb{Z}/N\mathbb{Z})$ for some integer $N \geqslant 1$. The kernel itself is called the *principal congruence subgroup of level N* and denoted $\Gamma(N)$, and the proof of the Manin–Drinfeld theorem immediately reduces to the case $\Gamma = \Gamma(N)$. It is at this juncture that the work of Kubert and Lang [6] enters the picture, providing a refinement of the Manin–Drinfeld theorem which is crucial if one is concerned, as we shall be here, with the action of $\mathrm{SL}(2,\mathbb{Z}/N\mathbb{Z})$ on U arising from the identification $\mathrm{SL}(2,\mathbb{Z}/N\mathbb{Z}) \cong \mathrm{SL}(2,\mathbb{Z})/\Gamma(N)$.

The refinement introduced by Kubert and Lang is a parametrization of the group $U/\mathbb{C}^\times$, or at least of some large subgroup of it, using the "Siegel functions"

$$g_a(z) = -q_\tau^b e^{\pi i a_2(a_1-1)}(1-q_z)\prod_{n\geqslant 1}(1-q_\tau^n q_z)(1-q_\tau^n/q_z) \tag{1}$$

([6], p. 29, formula K4), where the parameter $a = (a_1, a_2)$ is an ordered pair of rational numbers a_1 and a_2, not both integers, while $q_z = e^{2\pi i z}$ with $z \in H$ and $q_\tau = e^{2\pi i\tau}$ with $\tau = a_1 z + a_2$. Also $b = (a_1^2 - a_1 + 1/6)/2$ and $q_\tau^b = e^{2\pi i b\tau}$. If $a_1, a_2 \in N^{-1}\mathbb{Z}$ then g_a is in general a modular function only for $\Gamma(12N^2)$, but Kubert and Lang give a criterion for a product of such Siegel functions to be a modular function for $\Gamma(N)$, hence by virtue of the product expansion (1) a modular unit for $\Gamma(N)$. We state the criterion only when N is relatively prime to 6; for the general case see [6], pp. 76–78. Suppose that

$$g = \prod_a g_a^{m(a)}, \tag{2}$$

where a runs over $N^{-1}\mathbb{Z}^2 \smallsetminus \mathbb{Z}^2$ and $m(a) \in \mathbb{Z}$ with $m(a) = 0$ for all but finitely many a. (If one prefers, the product can be taken to run over a set of coset representatives for $\mathbb{Z}^2$ in $N^{-1}\mathbb{Z}^2 \smallsetminus \mathbb{Z}^2$: up to a constant factor, g_a depends only on $a + \mathbb{Z}^2$.) The Kubert–Lang criterion states that g is modular of level N if and only if

$$\sum_a m(a) \equiv 0 \pmod{12} \tag{3}$$

and

$$\sum_a m(a)(Na_1)^2 \equiv \sum_a m(a)(Na_2)^2 \equiv \sum_a m(a)(Na_1)(Na_2) \equiv 0 \pmod{N}. \tag{4}$$

Kubert and Lang refer to (4) as the "quadratic relations," but they are actually linear relations in the exponents $m(a)$. It is only the coefficients of these linear relations that are quadratic in the parameter. Be that as it may, let us use the term *Siegel group* for the subgroup of $U/\mathbb{C}^\times$ consisting of the cosets of all products (2) satisfying (3) and (4). Kubert and Lang prove that the Siegel group has the maximal rank $|C| - 1$, thereby recovering the Manin–Drinfeld theorem, and in the optimal case where $N = p^n$ with a prime $p \geqslant 5$ and an integer $n \geqslant 1$ they show that the Siegel group is the full group of modular units modulo constants, so that (2), (3), and (4) give a complete description of $U/\mathbb{C}^\times$ ([6], p. 83, Theorem 1.3).

The present note gives an application of their result. Henceforth p is a fixed prime $\geqslant 7$ and U is the group of modular units for $\Gamma(p)$. Also $G = \mathrm{PSL}(2, \mathbb{F}_p)$. We shall determine the structure of the natural representation of G on the vector space U/U^p over $\mathbb{F}_p$. Most of what we prove is already contained in the results of Gross [3], but this may not be obvious, because our point of view is quite different. In [3] the goal is to understand the Galois module structure of the subgroup of the Jacobian of $X(p)$ generated by the cusps. Furthermore, the Galois group at issue in [3] is the full arithmetic Galois group $\mathrm{GL}(2, \mathbb{F}_p)/\{\pm 1\}$ of the cover $X(p) \to \mathbf{P}^1$, where $X(p)$ is viewed as a curve over $\mathbb{Q}(e^{2\pi i/p})$ and $\mathbf{P}^1$ as the j-line over $\mathbb{Q}$. In the present note, by contrast, we consider only the geometric subgroup $G = \mathrm{PSL}(2, \mathbb{F}_p)$ of this Galois group, because the applications we have in mind involve regular Galois extensions, as in [7] and [9]. This point of view dictates the formulation of our final result, which can be summarized as follows: Every irreducible nontrivial representation of G over $\mathbb{F}_p$ occurs with multiplicity one in the maximal semisimple subspace of the "noncongruence part" of U/U^p (to be defined).

The proof of this assertion includes a purely representation-theoretic component, which is divided into three steps—Proposition 2, Propositions 3 and 4, and Proposition 5—and may appear inefficient, particularly in comparison to the proof of the corresponding facts for $\mathrm{GL}(2, \mathbb{F}_p)/\{\pm 1\}$ in [3] (p. 73, Proposition 5.1). On the other hand, the three steps do illustrate three different tools—Brauer–Nesbitt theory, Frobenius reciprocity, and orthogonal idempotents respectively—while Gross's argument rests on yet a fourth tool, namely highest weight vectors. The more important point, however, is that here as in [3], the really crucial input is not abstract representation theory but rather the explicit parametrization of modular units provided by Kubert and Lang. To appreciate this point it is helpful to recall that the natural representation of G on the space of modular forms of weight 2 and level p was decomposed into irreducibles in two papers of Hecke [4], [5]. As one would expect, most of the work in these papers goes into decomposing the space of cusp forms, but it is actually the space of Eisenstein series—dealt with by Hecke in a few lines—which has some bearing on the present note. The reason is simple: if $f \in U$ then $(d \log f)/dz$ is an Eisenstein series of weight 2 and level p. In fact the space of all such Eisenstein series is simply $\mathbb{C} \otimes_{\mathbb{Z}} (d \log U)/dz$. Furthermore, since the kernel of $f \mapsto (d \log f)/dz$ is the subgroup of constant functions $\mathbb{C}^\times \subset U^p$, we see that U/U^p is isomorphic as an $\mathbb{F}_p[G]$-module to $\mathbb{F}_p \otimes_{\mathbb{Z}} (d \log U)/dz$. Thus the representation of G on U/U^p arises via tensor product with $\mathbb{F}_p$ from a G-stable $\mathbb{Z}$-form of the space of Eisenstein series. It follows that the semisimplification of U/U^p can be computed directly from Hecke's decomposition of the space of Eisenstein series into irreducibles. But computing the structure of U/U^p itself is another matter entirely, and in attempting to do so we will find that the Kubert–Lang parametrization of modular units is an indispensable tool.

Whatever else it may accomplish, the real value of this note to its author is the opportunity it provides to acknowledge an enormous personal debt to Serge Lang, to whom I owe my career in mathematics. I also take this opportunity to thank the referee of [9], whose suggestion for simplifying the proof of Proposition 7 of [9] turned out to be an essential ingredient of the present work.

1 The module of parameters

The $\mathbb{Z}[G]$-module M introduced below is a first approximation to the domain of the Kubert–Lang map parametrizing U. Our goal is to decompose the associated representation of G on the vector space $V = M/pM$ over $\mathbb{F}_p$.

1.1 Preliminaries

The irreducible representations of G in characteristic p can be classified using a single invariant: their dimension. Indeed for each integer k satisfying $0 \leqslant k \leqslant (p-1)/2$ there is an absolutely irreducible representation σ_k of G over $\mathbb{F}_p$ of dimension $2k+1$, and σ_k is unique up to isomorphism. Furthermore, every irreducible representations of G in characteristic p is isomorphic to some σ_k. In order to work with an explicit model we shall take σ_k to be the $(2k)$th symmetric power of the tautological two-dimensional projective representation of G. Then the space of σ_k consists of binary homogeneous polynomials $f(x, y)$ of degree $2k$ over $\mathbb{F}_p$, and the action of G is given by the formula

$$(\sigma_k(g)f)(x, y) = f(ax + cy, bx + dy), \tag{5}$$

where g is the image in G of the element

$$\tilde{g} = \begin{pmatrix} a & b \\ c & d \end{pmatrix} \tag{6}$$

of $\mathrm{SL}(2, \mathbb{F}_p)$.

Put $R = \mathbb{F}_p^2 \setminus \{(0,0)\}$. We define M to be the free $\mathbb{Z}$-module of rank $(p^2-1)/2$ consisting of functions $m : R \to \mathbb{Z}$ such that $m(-r) = m(r)$ for $r \in R$. An action of G on M is given by the formula

$$(g \cdot m)(r) = m(r\tilde{g}), \tag{7}$$

where $\tilde{g}$ is either of the two lifts of g to $SL(2, \mathbb{F}_p)$ and $r\tilde{g}$ is the product of the 1×2 row vector r and the matrix $\tilde{g}$. Of course this action is formally the same as (5), except that m is now an even function $R \to \mathbb{Z}$ rather than a homogeneous polynomial over $\mathbb{F}_p$.

Given a field F, put $V_F = F \otimes_{\mathbb{Z}} M$ and extend the action (7) by linearity to a representation τ_F of G on V_F. We can identify V_F with the vector space of dimension $(p^2-1)/2$ over F consisting of even functions $m : R \to F$, and then the action of G is again formally the same as in (5) and (7). We are primarily interested in the case $F = \mathbb{F}_p$, and in this case we write V_F and τ_F simply as V and τ.

1.2 Irreducible constituents

Write B for the image in G of the upper triangular subgroup of $SL(2, \mathbb{F}_p)$ and $N \subset B$ for the image of the strictly upper triangular subgroup (i.e., the subgroup defined by the conditions $c = 0$, $a = d = 1$ in (6)). We denote the trivial one-dimensional character of any group by 1, leaving both the group and the implicit field of scalars to be inferred from the context. In the following proposition, for example, 1 is the trivial one-dimensional character of N with values in F, and $\mathrm{ind}_N^G 1$ is the representation of G over F which it induces.

Proposition 1. $\tau_F \cong \mathrm{ind}_N^G 1$.

Proof. Take the space of $\mathrm{ind}_N^G 1$ to consist of functions $f : G \to F$ satisfying $f(ng) = f(g)$ for $n \in N$ and $g \in G$, with G acting by right translation. As we have already noted, V_F is also a space of functions, namely the space of even functions $m : R \to F$. Furthermore, given f in the space of $\mathrm{ind}_N^G 1$, we obtain an element $m_f \in V_F$ by setting $m_f(r) = f(g)$ if $e\tilde{g} = \pm r$, where e is the row vector $(0, 1) \in R$. The map $f \mapsto m_f$ is readily verified to be G-equivariant and injective, and its domain and range both have dimension $(p^2 - 1)/2$. □

We now take $F = \mathbb{F}_p$ and compute the semisimplification of τ:

Proposition 2. *The multiplicity of σ_k as a constituent of τ is 1 if $k = 0$ or $k = (p-1)/2$ and 2 if $1 \leqslant k \leqslant (p-3)/2$.*

Proof. Given $t \in \mathbb{F}_p^\times$, let $a(t)$ denote the image in B of the diagonal matrix with diagonal entries t, t^{-1}. The map $t \mapsto a(t)$ induces an isomorphism of quotient groups $\mathbb{F}_p^\times/\{\pm 1\} \cong B/N$, and we can compose the inverse of this isomorphism with even powers of the Teichmüller character $\omega : \mathbb{F}_p^\times \to \mathbb{Z}_p^\times$ to obtain characters of B. More precisely, we define $\xi_k : B \to \mathbb{Q}_p^\times$ $(0 \leqslant k \leqslant (p-3)/2)$ by setting

$$\xi_k(a(t)n) = \omega(t)^{2k} \qquad (t \in \mathbb{F}_p^\times,\ n \in N).$$

Then $\mathrm{ind}_N^B 1 \cong \oplus_{k=0}^{(p-3)/2} \xi_k$, whence Proposition 1 and the identification $\mathrm{ind}_N^G 1 = \mathrm{ind}_B^G(\mathrm{ind}_N^B 1)$ give

$$\tau_{\mathbb{Q}_p} \cong \oplus_{k=0}^{(p-3)/2} \pi_k \tag{8}$$

with $\pi_k = \mathrm{ind}_B^G \xi_k$ (cf. formula (22) of [4]). We remark that $\pi_0 \cong 1 \oplus \eta$ with an absolutely irreducible representation η of dimension p over $\mathbb{Q}_p$, while if $p \equiv 1 \bmod 4$ then $\pi_{(p-1)/4}$ decomposes over $\overline{\mathbb{Q}}_p$ as the direct sum of two inequivalent irreducible representations ζ and ζ' of dimension $(p+1)/2$. Apart from these exceptions, the direct summands in (8) are absolutely irreducible (although not distinct, as $\pi_k \cong \pi_{(p-1-2k)/2}$ for $1 \leqslant k \leqslant (p-3)/2$).

Put $\mathcal{M} = \mathbb{Z}_p \otimes_{\mathbb{Z}} M$. Then $\mathcal{M}$ is a G-stable $\mathbb{Z}_p$-lattice in $V_{\mathbb{Q}_p}$ and $V = \mathbb{F}_p \otimes \mathcal{M}$. Hence the semisimplification of V can be read from (8) and the mod-p decomposition numbers of G. These decomposition numbers are implicit in Brauer–Nesbitt [1] (p. 590) and explicitly computed by Srinivasan [10] (pp. 107–108). In applying [10], note that for $n = 1$ her $\Phi(r_0)$ and $\varphi(r_0)$ coincide. Hence taking $r_0 = 2k$ in formula (3.5) of [10], we find that the character of our π_k coincides on p-regular conjugacy classes with the sum of the Brauer characters of our σ_k and $\sigma_{(p-1-2k)/2}$. In the first instance this conclusion holds only when $1 \leqslant k \leqslant (p-3)/2$ and $k \neq (p-1)/4$, but in fact it holds also when $k = 0$ (by the first three lines on p. 108 of [10]) and when $k = (p-1)/4$ (by formula (3.7) of [10]). The upshot is that in all cases, the semisimplification of the reduction modulo p of π_k coincides with $\sigma_k \oplus \sigma_{(p-1-2k)/2}$. Hence the proposition follows from (8). □

1.3 Irreducible subspaces and quotient spaces

Next we determine the multiplicity of σ_k as a quotient representation of τ. Given representations α and β of a group J on vector spaces W_α and W_β over a field F, write $\mathrm{Hom}_{F[J]}(\alpha, \beta)$ for $\mathrm{Hom}_{F[J]}(W_\alpha, W_\beta)$.

Proposition 3. *For* $0 \leqslant k \leqslant (p-1)/2$,

$$\dim_{\mathbb{F}_p} \mathrm{Hom}_{\mathbb{F}_p[G]}(\tau, \sigma_k) = 1.$$

Proof. Proposition 1 and Frobenius reciprocity give

$$\mathrm{Hom}_{\mathbb{F}_p[G]}(\tau, \sigma_k) \cong \mathrm{Hom}_{\mathbb{F}_p[N]}(1, \mathrm{res}_N^G \sigma_k).$$

Now N is generated by the element u corresponding to the choices $a = b = d = 1$ and $c = 0$ in (6), so it suffices to see that the subspace of vectors fixed by $\sigma_k(u)$ is one-dimensional. Let A be the matrix of $\sigma_k(u)$ relative to the ordered basis $x^{2k}, x^{2k-1}y, \dots, y^{2k}$, and let a_{ij} be the (i,j)-entry of A for $1 \leqslant i, j \leqslant 2k+1$. Using (5) to write $(\sigma_k(u)f)(x,y) = f(x, x+y)$, one readily verifies that A is upper triangular, that $a_{ii} = 1$ for all i, and that $a_{i,i+1} \neq 0$ for $1 \leqslant i \leqslant 2k$. It follows that the Jordan normal form of A consists of a single Jordan block, whence x^{2k} is the unique eigenvector of $\sigma_k(u)$ up to scalar multiples. □

A similar statement holds for subrepresentations:

Proposition 4. *For* $0 \leqslant k \leqslant (p-1)/2$,

$$\dim_{\mathbb{F}_p} \mathrm{Hom}_{\mathbb{F}_p[G]}(\sigma_k, \tau) = 1.$$

Proof. In view of Proposition 3 it suffices to see that both σ_k and τ are self-dual. The self-duality of σ_k follows from the fact that irreducible representations of G over $\mathbb{F}_p$ are determined up to isomorphism by their dimension. The self-duality of τ follows from the fact that the symmetric bilinear form

$$\langle m, m' \rangle = \sum_{r \in R} m(r) m'(r) \qquad (m, m' \in V) \tag{9}$$

is nondegenerate and G-invariant. □

1.4 Homogeneous components

Recall that $\mathcal{M} = \mathbb{Z}_p \otimes_{\mathbb{Z}} M$ and that $\omega : \mathbb{F}_p^\times \to \mathbb{Z}_p^\times$ is the Teichmüller character. We shall view the elements of $\mathcal{M}$ as even functions $m : R \to \mathbb{Z}_p$. We define $\mathcal{M}^{(k)} \subset \mathcal{M}$ to be the $\mathbb{Z}_p$-submodule consisting of all $m \in \mathcal{M}$ such that

$$m(tr) = \omega(t)^{2k} m(r)$$

for $t \in \mathbb{F}_p^\times$ and $r = (r_1, r_2) \in R$, where $tr = (tr_1, tr_2)$. The linear endomorphisms $e^{(k)}$ of $\mathcal{M}$ given by

$$(e^{(k)} m)(r) = \frac{1}{p-1} \sum_{t \in \mathbb{F}_p^\times} \omega^{-2k}(t) m(tr) \tag{10}$$

$(0 \leqslant k \leqslant (p-3)/2)$ form a family of orthogonal idempotents projecting $\mathcal{M}$ onto the respective submodules $\mathcal{M}^{(k)}$ and summing to the identity, so we have

$$\mathcal{M} = \oplus_{k=0}^{(p-3)/2} \mathcal{M}^{(k)}. \tag{11}$$

In fact (11) is a decomposition into $\mathbb{Z}_p[G]$-submodules, because the idempotents $e^{(k)}$ commute with the action of G. Hence the space of τ likewise decomposes into G-stable subspaces:

$$V = \oplus_{k=0}^{(p-3)/2} V^{(k)} \tag{12}$$

with $V^{(k)} = \mathbb{F}_p \otimes_{\mathbb{Z}_p} \mathcal{M}^{(k)}$. Let $\tau^{(k)}$ denote the representation of G on $V^{(k)}$. We have assumed that $p \geqslant 7$ rather than merely $p \geqslant 5$ to ensure the validity of the following statement (and several others hereafter):

Proposition 5. *If $1 \leqslant k \leqslant (p-3)/2$ then $\tau^{(k)}$ has a unique irreducible subrepresentation and a unique irreducible quotient representation, and they are equivalent to σ_k and $\sigma_{(p-1-2k)/2}$ respectively. On the other hand, $\tau^{(0)} \cong \sigma_0 \oplus \sigma_{(p-1)/2}$.*

Proof. The first point is that the free $\mathbb{Z}_p$-module $\mathcal{M}^{(k)}$ has rank $p+1$. Indeed for each of the $p+1$ lines ℓ through the origin in $\mathbb{F}_p^2$, fix an element $r_\ell \in R$ which spans ℓ, and define a function $f_{\ell,k} \in \mathcal{M}^{(k)}$ by

$$f_{\ell,k}(r) = \begin{cases} \omega(t)^{2k} & \text{if } r = tr_\ell \text{ with } t \in \mathbb{F}_p^\times \\ 0 & \text{if } r \notin \ell. \end{cases}$$

For fixed k the $p+1$ functions $f_{\ell,k}$ have pairwise disjoint supports and are therefore linearly independent over $\mathbb{Z}_p$. Hence $\mathcal{M}^{(k)}$ has rank at least $p+1$. But $\mathcal{M}$ has rank $(p+1)(p-1)/2$, so we deduce from (11) that $\mathcal{M}^{(k)}$ has rank exactly $p+1$, as claimed.

It follows that $V^{(k)}$ has dimension $p+1$ over $\mathbb{F}_p$. But an irreducible representation of G over $\mathbb{F}_p$ has dimension $\leqslant p$, so $V^{(k)}$ has a *proper* irreducible subspace and hence at least two irreducible constituents. On the other hand, V has exactly $p-1$ irreducible constituents (Proposition 2), so we deduce from (12) that $V^{(k)}$ has exactly two constituents.

To identify these constituents up to isomorphism, we introduce a $\mathbb{Z}[G]$-submodule $\mathcal{N}_k$ of $\mathcal{M}$ for $0 \leqslant k \leqslant (p-3)/2$. Given $m \in \mathcal{M}$, let $\overline{m} : R \to \mathbb{F}_p$ denote the reduction of m modulo p. We define $\mathcal{N}_k \subset \mathcal{M}$ to be the submodule consisting of all m such that $\overline{m}$ coincides with a binary homogeneous polynomial of degree $2k$ over $\mathbb{F}_p$. Strictly speaking, we should say "coincides with the function $R \to \mathbb{F}_p$ defined by" such a polynomial, but the distinction is moot: a homogeneous polynomial of degree $< p$ which vanishes on R is zero. Thus the map $m \mapsto \overline{m}$ determines an embedding of $\mathcal{N}_k/(\mathcal{N}_k \cap p\mathcal{M})$ into the space of σ_k. In fact this embedding is surjective and hence a G-isomorphism, because any even function $R \to \mathbb{F}_p$ can be lifted to an even function $R \to \mathbb{Z}_p$.

Now put $\mathcal{N}_k^{(l)} = e^{(l)}\mathcal{N}_k$ $(0 \leqslant l \leqslant (p-3)/2)$. It is readily verified that if $l \neq k$ then the image of $\mathcal{N}_k^{(l)}$ under $m \mapsto \overline{m}$ is $\{0\}$. On the other hand, we have just seen that the map $m \mapsto \overline{m}$ gives a G-isomorphism of $\mathcal{N}_k/(\mathcal{N}_k \cap p\mathcal{M})$ onto the space of σ_k. It follows that the domain of this G-isomorphism can be replaced by $\mathcal{N}_k^{(k)}/(\mathcal{N}_k^{(k)} \cap p\mathcal{M}^{(k)})$. But the latter can be viewed as a G-stable subspace $W^{(k)}$ of $V^{(k)}$, and the representation of G on $W^{(k)}$ is therefore equivalent to σ_k. Furthermore, we have seen that $V^{(k)}$ has exactly two irreducible constituents, so the quotient $V^{(k)}/W^{(k)}$ is also irreducible. Since its dimension is $(p+1)-(2k+1) = p-2k$, we deduce that the quotient representation is equivalent to $\sigma_{(p-1-2k)/2}$. In summary, the representation of G on $W^{(k)}$ and on $V^{(k)}/W^{(k)}$ is equivalent to σ_k and to $\sigma_{(p-1-2k)/2}$ respectively.

To see that $\tau^{(0)} \cong \sigma_0 \oplus \sigma_{(p-1)/2}$, we observe that the set of indices k satisfying $1 \leqslant k \leqslant (p-3)/2$ is stable under $k \mapsto (p-1-2k)/2$. It follows that σ_0 and $\sigma_{(p-1)/2}$

occur as constituents of $V^{(k)}$ if and only if $k = 0$. On the other hand, σ_0 and $\sigma_{(p-1)/2}$ occur not merely as constituents but as subrepresentations of τ (Proposition 4). It follows that they occur as subrepresentations of $\tau^{(0)}$, whence $\tau^{(0)} \cong \sigma_0 \oplus \sigma_{(p-1)/2}$.

Finally, suppose that $1 \leqslant k \leqslant (p-3)/2$. If W is an irreducible subspace of $V^{(k)}$ then the representation of G on W is equivalent to an irreducible constituent of $\tau^{(k)}$, hence either to σ_k or to $\sigma_{(p-1-2k)/2}$. But if $W \neq W^{(k)}$ then the first possibility is excluded, because σ_k occurs as a subrepresentation of τ with multiplicity one (Proposition 4). As for the second possibility, it coincides with the first (and is therefore excluded when $W \neq W^{(k)}$) if $k = (p-1)/4$. Otherwise it is excluded by Proposition 4 again, because $\sigma_{(p-1-2k)/2}$ already occurs as a subrepresentation of $\tau^{((p-1-2k)/2)}$, and the spaces $V^{((p-1-2k)/2)}$ and $V^{(k)}$ are linearly independent. We conclude that $W^{(k)}$ is the unique irreducible subspace of $V^{(k)}$, and since $V^{(k)}$ has just two irreducible constituents it follows that $V^{(k)}/W^{(k)}$ is the unique irreducible quotient. □

2 The quadratic relations

To move a step closer to U we turn from M to the $\mathbb{Z}[G]$-submodule Q of M defined by the quadratic relations of Kubert and Lang. As before, our primary concern is the representation of G on the associated vector space over $\mathbb{F}_p$, which is now the space $V' = Q/pQ$.

2.1 Preliminaries

To define Q, recall that given $m \in M$ we write $\overline{m} : R \to \mathbb{F}_p$ for the reduction of m modulo p. We will also let N denote the $\mathbb{Z}[G]$-submodule of M consisting of all n for which $\overline{n}$ has the form

$$\overline{n}(r) = ar_1^2 + br_1r_2 + cr_2^2 \tag{13}$$

with $a, b, c \in \mathbb{F}_p$, where $r = (r_1, r_2)$. Since N is a $\mathbb{Z}$-form of the $\mathbb{Z}_p[G]$-module previously denoted $\mathcal{N}_1$, it might be more logical to denote it N_1, but for simplicity we omit the subscript (and thereby void our previous use of N as a positive integer or as the subgroup of G corresponding to strictly upper triangular matrices). We define Q to consist of those $m \in M$ such that

$$\sum_{r \in R} \overline{m}(r)\overline{n}(r) = 0 \tag{14}$$

for all $n \in N$.

It is immediate from this description that Q contains pM. Thus M/Q is a quotient of the finite-dimensional vector space $V = M/pM$ over $\mathbb{F}_p$. In fact since Q is defined by the vanishing of three linearly independent linear forms on M/pM (namely those corresponding to the choices $(a,b,c) = (1,0,0)$, $(0,1,0)$, and $(0,0,1)$ in (13) and (14)) we see that M/Q has dimension three over $\mathbb{F}_p$. In particular Q has finite index in M, so by the Brauer–Nesbitt theorem, the representation τ' of G on the space $V' = Q/pQ$ has the same semisimplification as τ. Thus Proposition 2 holds with τ replaced by τ'. This is not quite true of Proposition 5, however, and our next proposition gives an analogue of Proposition 5 valid for τ'.

2.2 Homogeneous components

Put $\mathcal{Q} = \mathbb{Z}_p \otimes_{\mathbb{Z}} Q$. Then $\mathcal{Q}$ is stable under $e^{(k)}$ (cf. (10), (13), and (14)). Hence

$$\mathcal{Q} = \oplus_{k=0}^{(p-3)/2} \mathcal{Q}^{(k)}$$

with $\mathcal{Q}^{(k)} = e^{(k)}\mathcal{Q}$. Thus putting $V'^{(k)} = \mathcal{Q}^{(k)}/p\mathcal{Q}^{(k)}$ we have

$$V' = \oplus_{k=0}^{(p-3)/2} V'^{(k)}, \tag{15}$$

a decomposition of V' into G-stable subspaces. Let $\tau'^{(k)}$ denote the representation of G on $V'^{(k)}$.

Proposition 6. *If $1 \leqslant k \leqslant (p-5)/2$ then $\tau'^{(k)}$ has a unique irreducible subrepresentation and a unique irreducible quotient representation, and they are equivalent to σ_k and $\sigma_{(p-1-2k)/2}$ respectively. On the other hand, $\tau'^{(0)} \cong \sigma_0 \oplus \sigma_{(p-1)/2}$ and $\tau'^{((p-3)/2)} \cong \sigma_1 \oplus \sigma_{(p-3)/2}$.*

Proof. Suppose first that $k \neq (p-3)/2$. We claim that $\mathcal{M}^{(k)} \subset \mathcal{Q}$, whence $\mathcal{M}^{(k)} = \mathcal{Q}^{(k)}$. To see this, take $m \in \mathcal{M}^{(k)}$ and $n \in N$, and write

$$\sum_{r\in R} \overline{m}(r)\overline{n}(r) = \sum_{\ell\in\Lambda}\sum_{r\in R\cap\ell} \overline{m}(r)\overline{n}(r),$$

where Λ is the set of lines through the origin in $\mathbb{F}_p^2$. For each $\ell \in \Lambda$ choose a vector $r_\ell \in R$ spanning ℓ. Then the inner sum on the right-hand side can be written as a sum over $t \in \mathbb{F}_p^\times$, with $r = tr_\ell$. The homogeneity of $\overline{m}$ and $\overline{n}$ then gives

$$\sum_{r\in R} \overline{m}(r)\overline{n}(r) = \sum_{\ell\in\Lambda} \overline{m}(r_\ell)\overline{n}(r_\ell) \sum_{t\in\mathbb{F}_p^\times} t^{2k+2}.$$

Since $k \neq (p-3)/2$ the exponent of t on the right-hand side is $< p-1$ and consequently the inner sum is 0. Thus $\mathcal{M}^{(k)} \subset \mathcal{Q}$ and $\mathcal{M}^{(k)} = \mathcal{Q}^{(k)}$, as claimed.

It follows that if $k \neq (p-3)/2$ then $\tau'^{(k)} \cong \tau^{(k)}$, whence the assertions at hand reduce to those of Proposition 5. To handle the remaining case $k = (p-3)/2$, we recall that τ and τ' have isomorphic semisimplifications and are direct sums of their respective homogeneous components $\tau^{(k)}$ and $\tau'^{(k)}$. Since $\tau'^{(k)} \cong \tau^{(k)}$ for $k \neq (p-3)/2$, we deduce that the semisimplifications of $\tau'^{((p-3)/2)}$ and $\tau^{((p-3)/2)}$ are likewise isomorphic. Thus by Proposition 5, $\tau'^{((p-3)/2)}$ has exactly two irreducible constituents, namely $\sigma_{(p-3)/2}$ and σ_1.

Now $\mathcal{M}$ and $\mathcal{Q}$ are also the direct sums of their homogeneous components $\mathcal{M}^{(k)}$ and $\mathcal{Q}^{(k)}$, and we have seen that the vector space $\mathcal{M}/\mathcal{Q} = M/Q$ has dimension three over $\mathbb{F}_p$ (cf. (13) and (14)) while $\mathcal{M}^{(k)} = \mathcal{Q}^{(k)}$ for $k \neq (p-3)/2$. Consequently $\mathcal{M}^{((p-3)/2)}/\mathcal{Q}^{((p-3)/2)}$ is also three-dimensional over $\mathbb{F}_p$, as is therefore the subspace $Y = p\mathcal{M}^{((p-3)/2)}/p\mathcal{Q}^{((p-3)/2)}$ of $V'^{((p-3)/2)}$. Since $\tau'^{((p-3)/2)}$ has just the two irreducible constituents σ_1 and $\sigma_{(p-3)/2}$ of dimensions 3 and $p-2$ respectively, we deduce that the representation of G on Y is σ_1. Thus σ_1 is a subrepresentation of $\tau'^{((p-3)/2)}$ and $\sigma_{(p-3)/2}$ is the corresponding quotient representation.

It remains to see that σ_1 is also a quotient representation of $\tau'^{((p-3)/2)}$, whence $\sigma_{(p-3)/2}$ is a subrepresentation and $\tau'^{((p-3)/2)} \cong \sigma_1 \oplus \sigma_{(p-3)/2}$. To this end, consider the bilinear pairing $\prec *, * \succ: Q \times N \to \mathbb{Z}$ given by

$$\prec m, n \succ = \frac{1}{p} \sum_{r \in R} m(r) n(r) \qquad (m \in Q, \ n \in N).$$

Write L for the $\mathbb{Z}[G]$-submodule of Q consisting of those m such that

$$\prec m, n \succ \equiv 0 \pmod{p}$$

for all $n \in N$. Put $\mathcal{L} = \mathbb{Z}_p \otimes_{\mathbb{Z}} L$. Then $\mathcal{L}$ is stable under $e^{(k)}$, so putting $\mathcal{L}^{(k)} = e^{(k)}\mathcal{L}$ we have

$$\mathcal{L} = \oplus_{k=0}^{(p-3)/2} \mathcal{L}^{(k)}.$$

We claim that $\mathcal{L}^{((p-3)/2)}$ contains $p\mathcal{Q}^{((p-3)/2)}$ and that the quotient space $Z = \mathcal{Q}^{((p-3)/2)}/\mathcal{L}^{((p-3)/2)}$ of $V'^{(p-3)/2}$ is of positive dimension $\leqslant 3$. An immediate consequence of the claim is that the representation of G on Z is equivalent to σ_1, so verifying the claim will complete the proof.

It is immediate from the definitions that L contains pQ and hence that $\mathcal{L}$ contains $p\mathcal{Q}$. On the other hand, $\mathcal{L}$ does not contain $p\mathcal{M}$: for if $m \in M$ is the function taking the value 1 on $(\pm 1, 0)$ and 0 elsewhere then $\prec pm, n \succ \not\equiv 0 \bmod p$ for any $n \in N$ satisfying (13) with $a \neq 0$. It follows that for some k with $0 \leqslant k \leqslant (p-3)/2$ we have $p\mathcal{M}^{(k)} \not\subset \mathcal{L}^{(k)}$. But we have seen that $p\mathcal{Q} \subset \mathcal{L}$ and that $p\mathcal{Q}^{(k)} = p\mathcal{M}^{(k)}$ for $k \neq (p-3)/2$. Hence $\mathcal{L}^{((p-3)/2)}$ does not contain $p\mathcal{M}^{((p-3)/2)}$, and we deduce that $\mathcal{L}^{((p-3)/2)}/p\mathcal{Q}^{((p-3)/2)}$ is a subspace of $V'^{((p-3)/2)}$ of positive codimension. On the other hand, the codimension is $\leqslant 3$, because the subspace is defined by the vanishing of three linear forms on $V'^{((p-3)/2)}$, namely the forms sending $m + pQ^{((p-3)/2}$ to $\prec m, n \succ$ modulo p with n as in (13) and $(a, b, c) = (1, 0, 0)$, $(0, 1, 0)$, and $(0, 0, 1)$. Our claim follows. □

3 The Kubert–Lang map

Given a matrix $\tilde{\gamma} \in \mathrm{SL}(2, \mathbb{Z})$, we identify its image $\gamma \in \mathrm{PSL}(2, \mathbb{Z})$ with the fractional linear transformation of H defined by γ. Thus if f is a function on H and $\tilde{\gamma}$ is the right-hand side of (6) then $f \circ \gamma$ is the function $z \mapsto f((az+b)/(cz+d))$. Since the image of $\Gamma(p)$ in $\mathrm{PSL}(2, \mathbb{Z})$ has quotient G and fixes the elements of U we can make U into a $\mathbb{Z}[G]$-module by setting

$$g \cdot f = f \circ \gamma^{-1}$$

for $g \in G$ and $f \in U$, where $\gamma \in \mathrm{PSL}(2, \mathbb{Z})$ is any lift of g. The resulting representation of G on the vector space $V'' = U/U^p$ over $\mathbb{F}_p$ will be denoted τ''.

Given $a \in p^{-1}\mathbb{Z}^2 \setminus \mathbb{Z}^2$, define the Siegel function g_a as in (1). For $r \in R$ we put $f_r = g_a^{12}$, where $a \in p^{-1}\mathbb{Z}^2$ is chosen so that r coincides with the residue class of pa modulo $p\mathbb{Z}^2$. Since a can be replaced by any element of the coset $a + \mathbb{Z}^2$, the function g_a^{12} is determined only up to multiplication by a pth root of unity ([6], p. 28, Formula K2), but the coset $f_r U^p$ is uniquely determined by r because U^p contains $\mathbb{C}^\times$. Furthermore, if $m \in Q$ then the function

$$f^m := \prod_{r \in R} f_r^{m(r)}$$

belongs to U (cf. (3) and (4), or in other words [6], p. 76, Theorem 5.2). Hence the assignment $m + pQ \mapsto f^m U^p$ defines an $\mathbb{F}_p$-linear map $\Phi : V' \to V''$.

Proposition 7. *The map Φ is surjective with one-dimensional kernel, and it intertwines τ' with τ''.*

Proof. The argument echos the proof of Proposition 0 of [9], which in turn merely assembles a number of results from [6]. Let us at least recall the relevant citations: The surjectivity of Φ follows from [6], p. 83, Theorem 1.3, because p is prime to 12 and thus the map $fU^p \mapsto f^{12}U^p$ is an automorphism of U/U^p. That the kernel of Φ is one-dimensional follows from the surjectivity, because V' has dimension $(p^2 - 1)/2$ over $\mathbb{F}_p$ while V'' has dimension $(p^2 - 3)/2$ ([6], p. 42, Theorem 3.2). Finally, the G-equivariance of Φ follows from [6], p. 27, Formula K1. □

Put $V''^{(k)} = \Phi(V'^{(k)})$, so that

$$V'' = \oplus_{k=0}^{(p-3)/2} V''^{(k)}.$$

We write $\tau''^{(k)}$ for the representation of G on $V''^{(k)}$.

Proposition 8. *If* $1 \leqslant k \leqslant (p-5)/2$ *then* $\tau''^{(k)}$ *has a unique irreducible subrepresentation and a unique irreducible quotient representation, and they are equivalent to* σ_k *and* $\sigma_{(p-1-2k)/2}$ *respectively. On the other hand,* $\tau''^{(0)} \cong \sigma_{(p-1)/2}$ *and* $\tau''^{((p-3)/2)} \cong \sigma_1 \oplus \sigma_{(p-3)/2}$.

Proof. Combine Propositions 6 and 7 and observe that V' has exactly one G-stable subspace of dimension one. □

We conclude with some remarks which will lead to a slight reformulation of Proposition 8. Since $p \geqslant 7$, the two direct summands of $\tau''^{((p-3)/2)}$ are inequivalent, so there is a unique subspace $W''^{((p-3)/2)}$ of $V''^{((p-3)/2)}$ on which the representation of G is equivalent to $\sigma_{(p-3)/2}$. We shall refer to the subspace

$$V''_{\text{non}} = \left(\oplus_{k=0}^{(p-5)/2} V''^{(k)}\right) \oplus W''^{((p-3)/2)}$$

of V'' as the *noncongruence part* of V''. The *congruence part* of V'' is the unique subspace V''_{cong} of $V''^{((p-3)/2)}$ on which the representation of G is equivalent to σ_1. Thus

$$V'' = V''_{\text{non}} \oplus V''_{\text{cong}}. \tag{16}$$

To explain the terminology, write $\mathfrak{K}$ for the field $\mathfrak{M}^{\Gamma(p)}$ of modular functions for $\Gamma(p)$, and given a subspace W of V'' write $\mathfrak{K}_W$ for the Kummer extension of $\mathfrak{K}$ obtained by adjoining the pth roots of all $f \in U$ such that $fU^p \in W$. (Note that $\mathfrak{K}^{\times p} \cap U = U^p$, whence $\mathrm{Gal}(\mathfrak{K}_W/\mathfrak{K})$ is dual to W under the Kummer pairing. In particular, $[\mathfrak{K}_W : \mathfrak{K}] = |W|$.) Also put

$$\mathfrak{M}_{\text{cong}} = \bigcup_{N \geqslant 1} \mathfrak{M}^{\Gamma(N)},$$

so that $\mathfrak{M}_{\text{cong}}$ is the compositum of the modular function fields for all congruence subgroups of $\mathrm{SL}(2,\mathbb{Z})$. We claim that

$$\mathfrak{K}_{V''} \cap \mathfrak{M}_{\text{cong}} = \mathfrak{K}_{V''_{\text{cong}}}. \tag{17}$$

Together, (16) and (17) justify the designation "noncongruence part" for V''_{non}.

To prove (17), we recall from the proof of Proposition 6 that the subspace of $V'^{((p-3)/2)}$ on which G acts via σ_1 is pM/pQ (strictly speaking we should identify this subspace as $p\mathcal{M}^{((p-3)/2)}/p\mathcal{Q}^{((p-3)/2)}$, not pM/pQ, but $\mathcal{M}^{(k)} = \mathcal{Q}^{(k)}$ for $k \neq (p-3)/2$). Thus $\Phi(pM/pQ) = V''_{\text{cong}}$. It follows (see [9], Proposition 2, p. 12) that $\mathfrak{K}_{V''_{\text{cong}}}$ is the field of modular functions for $\Gamma(p^2)$, whence the right-hand side of (17) is contained in the left-hand side. For the reverse inclusion, put

$$\Gamma = \{\gamma \in \mathrm{SL}(2,\mathbb{Z}) : f \circ \gamma = f \text{ for all } f \in \mathfrak{K}_{V''} \cap \mathfrak{M}_{\text{cong}}\}.$$

Then the field of modular functions for Γ is the left-hand side of (17). In particular, since the left-hand side of (17) is a subfield of $\mathfrak{M}_{\text{cong}}$ it follows that Γ is a congruence subgroup. But the least common multiple of the cusp amplitudes of Γ divides p^2, because the field $\mathfrak{K}_{V''}$ is generated over $\mathfrak{K}$ by pth roots of elements of $\mathfrak{K}$. Thus the Wohlfahrt level of Γ divides p^2, and since Γ is a congruence subgroup its Wohlfahrt level equals its congruence level by the Fricke–Wohlfahrt theorem [11]: $\Gamma(p^2) \subset \Gamma$. Taking modular function fields of the two sides reverses the inclusion and thus gives the inclusion of the left-hand side of (17) in the right-hand side.

Now put $W''^{(0)} = V''^{(0)}$, and for $1 \leqslant k \leqslant (p-5)/2$ let $W''^{(k)}$ be the unique irreducible subspace of $V''^{(k)}$. Then the maximal semisimple subspace or socle of V''_{non} is $\oplus_{k=0}^{(p-3)/2} W''^{(k)}$, and we may describe its structure as follows:

Proposition 9. *The representation of G on the maximal semisimple subspace of V''_{non} is equivalent to $\oplus_{k=1}^{(p-1)/2} \sigma_k$.*

References

[1] R. Brauer and C. Nesbitt, *On the modular characters of groups,* Ann. of Math. **42** (1941), 556–590.

[2] V. G. Drinfeld, *Two theorems on modular curves,* Functional Analyis and its Applications **7** (1973), 155–156.

[3] B. H. Gross, *Representation theory and the cuspidal group of $X(p)$,* Duke Math. J. **54** (1987), 67–75.

[4] E. Hecke, *Über ein Fundamentalproblem aus der Theorie der elliptischen Modulfunktionen,* Abh. Math. Sem. Hamb. **6** (1928), 235–257 (= *Math. Werke* # 28, 525–547).

[5] E. Hecke, *Über das Verhalten der Integrale 1. Gattung bei Abbildungen, insbesondere in der Theorie der elliptischen Modulfunktionen,* Abh. Math. Sem. Hamb. **8** (1930), 271–281 (= *Math. Werke* # 29, 548–558).

[6] D. S. Kubert and S. Lang, *Modular Units*, Springer-Verlag, Grundlehren Math. Wissen. Vol. 244, 1981.

[7] Á. Lozano-Robledo, *On the surjectivity of Galois representations attached to elliptic curves over number fields*, Acta Arith. **117** (2005), 283–291.

[8] Yu. Manin, *Parabolic points and zeta functions of modular curves*, Izv. Akad. Nauk SSSR Ser. Math. **36** (1972), 19–66.

[9] D. E. Rohrlich, *Modular units and the surjectivity of a Galois representation,* J. of Number Theor. **107** (2004), 8–24.

[10] B. Srinivasan, *On the modular characters of the special linear group $SL(2, p^n)$*, Proc. London Math. Soc. **14** (1964), 101–114.

[11] K. Wohlfahrt, *An extension of F. Klein's level concept,* Ill. J. Math. **8** (1964), 529–535.

Equidistribution and generalized Mahler measures

L. Szpiro and T. J. Tucker

Dedicated to the memory of Serge Lang, who taught the world number theory for more than fifty years, through his research, lectures, and books

Abstract If K is a number field and $\varphi : \mathbb{P}^1_K \longrightarrow \mathbb{P}^1_K$ is a rational map of degree $d > 1$, then at each place v of K, one can associate to φ a generalized Mahler measure for polynomials $F \in K[t]$. These Mahler measures give rise to a formula for the canonical height $h_\varphi(\beta)$ of an element $\beta \in \overline{K}$; this formula generalizes Mahler's formula for the usual Weil height $h(\beta)$. In this paper, we use Diophantine approximation to show that the generalized Mahler measure of a polynomial F at a place v can be computed by averaging $\log |F|_v$ over the periodic points of φ.

Key words Height functions • Mahler measure • dynamical systems • periodic points • equidistribution • Diophantine approximation

Mathematics Subject Classification (2010): Primary 37P30; Secondary 11J68

L. Szpiro (✉)
Program in Mathematics, Graduate Center of CUNY, 365 Fifth Avenue, New York, NY 10016-4309
e-mail: lszpiro@gc.cuny.edu

T.J. Tucker
Department of Mathematics, Hylan Building, University of Rochester Rochester, NY 14627
e-mail: ttucker@math.rochester.edu

D. Goldfeld et al. (eds.), *Number Theory, Analysis and Geometry: In Memory of Serge Lang*, DOI 10.1007/978-1-4614-1260-1_26,

1 Introduction

The usual Weil height of a rational number x/y, where x and y are integers without a common prime factor, is defined as

$$h(x/y) = \log \max(|x|, |y|).$$

More generally, one can define the usual Weil height $h(\beta)$ of an algebraic number β in a number field K by summing $\log \max(|\beta|_v, |1|)$ over all of the absolute values v of K. Mahler [Mah60] has proved that if F is a nonzero irreducible polynomial in $\mathbb{Z}[t]$ with coprime coefficients such that $F(\beta) = 0$, then

$$\deg(F)h(\beta) = \int_0^1 \log|F(e^{2\pi i\theta})|d\theta. \tag{1.1}$$

The quantity $\int_0^1 \log|F(e^{2\pi i\theta})|d\theta$ is often referred to as the *Mahler measure* of F.

It is easy to see that $h(\beta^2) = 2h(\beta)$ for any algebraic number β. Similarly, it is easy to check that for any continuous function g on the unit circle, we have

$$\int_0^1 g((e^{2\pi i\theta})^2)d\theta = \int_0^1 g(e^{2\pi i\theta})d\theta.$$

Furthermore, the unit circle is the Julia set of $\varphi : x \to x^2$. Thus, Mahler's formula says that one obtains the height of an algebraic number by integrating its minimal polynomial against the unique measure μ such that $\varphi_*\mu = \mu$ and μ is supported on the Julia set of φ.

Now let $\varphi : \mathbb{P}^1_{\mathbb{C}} \longrightarrow \mathbb{P}^1_{\mathbb{C}}$ be any nonconstant rational map. Brolin [Bro65] and Lyubich [Lyu83] have constructed a totally φ-invariant probability measure μ_φ (that is, we have $\varphi_*\mu = \mu$) with support on the Julia set of φ; Freire, Lopes, and Mañe [FLM83] have demonstrated that this measure is the *unique* totally φ-invariant probability measure with support on the Julia set of φ. When φ is defined over a number field K, Call and Silverman [CS93] have constructed a height function h_φ with the properties that (1) $h_\varphi(\varphi(x)) = (\deg \varphi)h_\varphi(x)$ and (2) there is a constant C_φ such that $|h(x) - h_\varphi(x)| < C_\varphi$ for all $x \in \mathbb{P}^1(\overline{K})$. In [PST04], it is shown that Mahler's formula (1.1) generalizes to the adelic formula

$$(\deg F)h_\varphi(x) = \sum_{\text{places } v \text{ of } K} \int_{\mathbb{P}^1(\mathbb{C}_v)} \log|F|_v \, d\mu_{\varphi,v}, \tag{1.2}$$

where β is an algebraic point, F is a nonzero irreducible polynomial in $\mathbb{Q}[t]$ such that $F(\beta) = 0$, the measure $\mu_{\varphi,v}$ at an archimedean place is the totally φ-invariant probability measure constructed by Brolin and Lyubich, and the integral $\int_{\mathbb{P}^1(\mathbb{C}_v)} \log|F|_v \, d\mu_{\varphi,v}$ at a finite place v is defined so that its value is the v-adic analog of the value at an archimedean place (note that as defined in [PST04], these are not integrals per se). Favre and Rivera-Letelier have also given a proof of 1.2, using actual integrals on Berkovich spaces; Piñeiro [Piñ05] and Chambert-Loir and Thuillier [CLT04, Thu06] have recently proved higher-dimensional generalizations of 1.2.

Lyubich [Lyu83] has also proved that for any continuous function g and any archimedean place v, the integrals $\int_{\mathbb{P}^1(\mathbb{C}_v)} g \, d\mu_{\varphi,v}$ can be computed by averaging g on the periodic points of φ; that is,

$$\lim_{k\to\infty} \frac{1}{(\deg \varphi)^k} \sum_{\varphi^k(w)=w} g(w) = \int_{\mathbb{P}^1(\mathbb{C}_v)} g \, d\mu_{\varphi,v}. \tag{1.3}$$

Autissier [Aut01], Bilu [Bil97], Szpiro, Ullmo, and Zhang [SUZ97], and others have obtained generalizations and variations of this result. The most recent generalization, proved independently by Baker and Rumely [BR06], Chambert-Loir [CL06], and Favre and Rivera-Letelier [FRL04] and [FRL06] states that (1.3) continues to hold when the periodic points w such that $\varphi^k(w) = w$ are replaced by the conjugates of any infinite nonrepeating sequence of algebraic points with height tending to 0 and when the measure $\mu_{\varphi,v}$ is the totally φ-invariant measure without point masses at classical points on the v-adic Berkovich space (see [Ber90]) for a finite place v.

The function $\log|F|$, for F a nonconstant polynomial, is not continuous in general, of course. Thus, the equidistribution results cited above do not allow us to compute Mahler measures by averaging $\log|F|_v$ over points of small height. One can, however, show that for any $\beta \in \bar{\mathbb{Q}}$, we have

$$[\mathbb{Q}(\beta) : \mathbb{Q}]h(\beta) = \lim_{k\to\infty} \frac{1}{d^k} \sum_{\xi^n=1} \log|F(\xi)| = \int_0^1 \log|F(e^{2\pi i\theta})| d\theta, \tag{1.4}$$

where F is a nonzero irreducible polynomial in $\mathbb{Z}[t]$ with coprime coefficients such that $F(\beta) = 0$ (see [EW99, Chapter 1], [Sch74]). Everest, Ward, and Ní Fhlathúin have proved similar results for maps that come from multiplication on an elliptic curve [EW99, Chapter 6], [EF96]. The proofs of these results make use of the theory of linear forms in logarithms [Bak75], [Dav95], which is used to show that the periodic points of the maps in question have strong Diophantine properties. It is not clear how to apply the theory of linear forms in logarithms in the case of more general rational maps. In this paper, we use Roth's theorem [Rot55] from Diophantine approximation in place of the theory of linear forms in logarithms. This allows us to work in greater generality.

1.1 Statements of the main theorems

The main results of this paper extend (1.4) to a formula that holds for all rational maps. Let K be a number field or a function field of characteristic zero, let v be a place of K, and let $\varphi : \mathbb{P}^1_K \longrightarrow \mathbb{P}^1_K$ be a nonconstant rational map of degree $d > 1$. We prove the following equidistribution result for the periodic points of φ.

Theorem 5.7. *For any nonzero polynomial F with coefficients in $\overline{K}$, we have*

$$\int_{\mathbb{P}^1(\mathbb{C}_v)} \log|F|_v \, d\mu_{\varphi,v} = \lim_{k\to\infty} \frac{1}{d^k} \sum_{\substack{\varphi^k([w:1])=[w:1] \\ F(w)\neq 0}} \log|F(w)|_v.$$

This allows us to show that for any point $\beta \in \overline{K}$, the canonical height $h_\varphi(\beta)$ can be computed by taking the average of the log of the absolute value of a minimal polynomial for β over the periodic points of φ.

Theorem 5.10. *For any $\beta \in \overline{K}$ and any nonzero irreducible $F \in K[t]$ such that $F(\beta) = 0$, we have*

$$(\deg K)(\deg F)(h_\varphi(\beta) - h_\varphi(\infty)) = \sum_{\text{places } v \text{ of } K} \lim_{k\to\infty} \frac{1}{d^k} \sum_{\substack{\varphi^k([w:1])=[w:1] \\ F(w)\neq 0}} \log|F(w)|_v.$$

In both the theorems, the w are counted with multiplicity. We explain what multiplicity means in this context in Section 1.

We are also able to prove that $\int_{\mathbb{P}^1(\mathbb{C}_v)} \log|F|_v \, d\mu_{\varphi,v}$ is the limit as n goes to infinity of the average of $\log|F|_v$ on the points w for which $\varphi^n(w) = \alpha$, where α is an algebraic point that is not an exceptional point for φ. We state this in Theorem 5.6. This enables us to prove Theorem 5.9, which is the analog of Theorem 5.10 for the points w such that $\varphi^n(w) = \alpha$.

The strategy of the proof of the main theorems is fairly simple. By additivity, it suffices to prove our results for polynomials of the form $F(t) = t - \beta$ for $\beta \in \overline{K}$. After Section 3, we are reduced to showing that

$$\lim_{k\to\infty} \frac{1}{d^k} \sum_{\substack{\varphi^k([w:1])=[w:1] \\ w\neq\beta}} \log|w-\beta|_v = \lim_{k\to\infty} \frac{\log\max(|P_k(\beta,1)|_v, |Q_k(\beta,1)|_v)}{d^k} - \lim_{k\to\infty} \frac{\log\max(|P_k(1,0)|_v, |Q_k(1,0)|_v)}{d^k}, \tag{1.5}$$

where φ^k is written as

$$\varphi^k([T_0 : T_1]) = [P_k(T_0, T_1) : Q_k(T_0, T_1)]$$

for coprime homogeneous polynomials P_k and Q_k in $K[T_0, T_1]$. The points w for which $\varphi^k(w) = w$ are just the solutions to the equation $P_k(w, 1) - wQ_k(w, 1) = 0$.

Thus, we get the left-hand side of (1.5) by taking the limit of $\log|P_k(\beta,1) - \beta Q_k(\beta,1)|_v/d^k$ as k goes to ∞. For each k, we rewrite this as

$$\frac{\log|Q_k(\beta,1)|_v}{d^k} + \frac{\log\left|\frac{P_k(\beta,1)}{Q_k(\beta,1)} - \beta\right|_v}{d^k}$$

and use Diophantine approximation to show that the second term in the equation above usually goes to 0 as $k \to \infty$; our theorems then follow after a bit of calculation. The Diophantine approximation result we use is Roth's theorem, which we state in Section 4.3 as theorem 4.1. We use Roth's theorem to derive Lemma 5.2, which is the key lemma in our proofs of the main theorems. The idea for the proof of Lemma 5.2 comes from Siegel's famous paper [Sie29]. We should note that after writing this paper we discovered that Silverman [Sil93] has used methods very similar to those found here at the beginning of Section 5; we require a slight modification of his results along these lines, however, so we present the necessary argument here in full.

Propositions 5.4 and 5.5 deal with the additional complications that may arise when the β in (1.5) is preperiodic. These complications are overcome with somewhat lengthy – but essentially basic – calculations that are very similar to some of the computations carried out by Morton and Silverman in [MS95].

In Section 6, we construct a simple counterexample that shows that Theorem 5.7 will not hold in general when the polynomial F does not have algebraic coefficients (it is likely that the theorem will also fail if the point α is not algebraic). We construct a transcendental number β such that the limit $\lim_{k\to\infty} \frac{1}{2^k} \sum_{\xi^{2^k}=1} \log|\xi - \beta|$ does not exist. This means that there is no way to prove the main results of this paper without using some special properties of algebraic numbers.

Acknowledgements We would like to thank M. Baker, A. Chambert-Loir, L. DeMarco, C. Petsche, R. Rumely, and S. Zhang for many helpful conversations. In particular, we thank M. Baker, L. DeMarco, and R. Rumely for suggesting some of the applications mentioned in Section 7. The first author was partially supported by NSF Grant 0071921. The second author was partially supported by NSF Grant 0101636.

2 Notation and terminology

We fix the following notation:

- K is a number field or a function field of characteristic 0 (by function field we mean a finite algebraic extension of a field of the form $K_{\text{cons}}(T)$ where K_{cons} is algebraically closed in K);
- v is a place of K;
- K_v is the completion of K at v;
- $\mathbb{C}_v$ is the completion of an algebraic closure of K_v at v;

- $\overline{K}$ is the algebraic closure of K in $\mathbb{C}_v$ (note that this means that v extends to all of $\overline{K}$);
- $n_v = [K_v : \mathbb{Q}_v]$ if K is a number field;
- $n_v = 1$ if K is a function field;
- $\deg K = [K : \mathbb{Q}]$ if K is a number field;
- $\deg K = 1$ if K is a function field.

We let $|\cdot|_v$ be an absolute value on $\mathbb{C}_v$ corresponding to v. When K is a function field and π_v generates the maximal prime $\mathcal{M}_v$ in the local ring $\mathfrak{o}_v$ corresponding to v, we specify that

$$|\pi_v|_v = e^{-[(\mathfrak{o}_v/\mathcal{M}_v):K_{\text{cons}}]},$$

where K_{cons} is the field of constants in K. When K is a number field and v is nonarchimedean, we normalize $|\cdot|_v$ so that

$$|p|_v = p^{-n_v}$$

when v lies over p. When K is a number field and v is archimedean we normalize so that $|\cdot|_v = |\cdot|^{n_v}$ on $\mathbb{Q}$, where $|\cdot|$ is the usual archimedean absolute value on $\mathbb{Q}$.

Throughout this paper, we will work with a nonconstant morphism $\varphi : \mathbb{P}^1_K \longrightarrow \mathbb{P}^1_K$ of degree $d > 1$. We choose homogeneous polynomials $P, Q \in K[T_0, T_1]$ of degree d without a common factor along with a coordinate system $[s : t]$ for $\mathbb{P}^1_{\overline{K}}$ such that

$$\varphi([T_0 : T_1]) = [P(T_0, T_1) : Q(T_0, T_1)],$$

where P and Q have no common zero in $\mathbb{P}^1(\overline{K})$. We let $P_1 = P$ and $Q_1 = Q$, and for $k \geq 2$ we define P_k and Q_k recursively by

$$P_k(T_0, T_1) = P_{k-1}(P(T_0, T_1), Q(T_0, T_1))$$

and

$$Q_k(T_0, T_1) = Q_{k-1}(P(T_0, T_1), Q(T_0, T_1)).$$

Having chosen coordinates, we can define the usual Weil height as

$$h([a : b]) = \frac{1}{\deg K} \sum_{\text{places } v \text{ of } K} \log \max(|a|_v, |b|_v)$$

when $a, b \in K$. When a and b lie in an extension L of K, this definition extends to

$$h([a : b]) = \frac{1}{[L : K](\deg K)} \sum_{\text{places } w \text{ of } L} [L_w : K_v] \log \max(|a|_w, |b|_w), \tag{2.1}$$

where L_w is the completion of L at w and the absolute value $|\cdot|_w$ restricts to some $|\cdot|_v$ on K.

As in [CS93], we define the canonical height h_φ as

$$h_\varphi([a:b]) = \lim_{k\to\infty} \frac{h(\varphi^k([a:b]))}{d^k}. \tag{2.2}$$

We say that $\alpha \in \mathbb{P}^1(\overline{K})$ is a ***periodic*** point for φ if there exists a positive integer n such that $\varphi^n(\alpha) = \alpha$. If α is periodic, we define the ***period*** of α to be the smallest positive integer ℓ such that $\varphi^\ell(\alpha) = \alpha$. We say that α is ***perperiodic*** if there exists a positive integer n such that $\varphi^n(\alpha)$ is periodic.

We will use a small amount of the theory of dynamics on the projective plane; for a more thorough account of the subject, we refer the reader to Milnor's [Mil99] and Beardon's [Bea91] books on the subject. We say that $\alpha \in \mathbb{P}^1(\overline{K})$ is an ***exceptional*** point for φ if $\varphi^2(\alpha) = \alpha$ and φ^2 is totally ramified at α. This is equivalent to saying that the set $\bigcup_{k=1}^\infty (\varphi^k)^{-1}(\alpha)$ is finite (see [Bea91, Chapter 4.1]). If α is exceptional, then at each place v, there is a maximal v-adically open set $\mathcal{U}$ containing α such that the sequence $(\varphi^{\ell k}(\beta))_k$ converges to α for each $\beta \in \mathcal{U}$, where ℓ is the period of α (which is either 1 or 2). We call $\mathcal{U}$ the ***attracting basin*** of α (see [Bea91, Chapter 6.3], which uses the terminology "local basin").

We always count points with multiplicities in this paper. The multiplicity of a point $[z:1]$ in the multiset $\{w \mid \varphi^k(w) = w\}$ is the highest power of $t - z$ that divides the polynomial $P_k(t,1) - tQ_k(t,1)$. The multiplicity of a point $[z:1]$ in the multiset $\{w \mid \varphi^k(w) = [s:u]\}$ is the highest power of $t - z$ that divides the polynomial $uP_k(t,1) - sQ_k(t,1)$ (here s, u, and z are taken to be elements of $\overline{K}$, while t is taken to be a variable).

We note that everything done in this paper depends upon our choice of coordinates. In particular, our integrals are closely related to the canonical local heights (see [CG97]) for the point $[1:0]$ at infinitely, so our choice of the point at infinity affects all of our integrals. To emphasize the fact that we treat $[1:0]$ as the point at infinity, we denote it as ∞ where appropriate.

3 Brolin–Lyubich integrals and local heights

We will work with the limits

$$\lim_{k\to\infty} \frac{\log\max(|P_k(a,b)|_v, |Q_k(a,b)|_v)}{d^k} \tag{3.1}$$

for $(a,b) \in \mathbb{C}_v \setminus \{(0,0)\}$. For a proof that these limits exist, see [PST04], [BR06], or [CG97] (the proof is essentially an exercise in using telescoping sums and geometric series). Note that Call and Goldstine [CG97, Theorem 3.1] have shown that

$$\hat{h}_{\varphi,v}([\beta:1]) = \lim_{k\to\infty} \frac{\log\max(|P_k(\beta,1)|_v, |Q_k(\beta,1)|_v)}{d^k}$$

is the unique Weil function for [1 : 0] at v (see [Lan83, Chapter 10] for a definition of Weil functions) that satisfies

$$\hat{h}_{\varphi,v}(\varphi([a:b])) = d\hat{h}_{\varphi,v}([a:b]) + \log\left|Q\left(\frac{a}{b},1\right)\right|_v,$$

for any $[a:b] \neq [1:0]$ (see [CG97, Theorem 2.1]). The function $\hat{h}_{\varphi,v}(\cdot)$ is called a canonical local height for φ.

As noted in the introduction, Brolin [Bro65] and Lyubich [Lyu83] have constructed a totally φ-invariant measure $\mu_{\varphi,v}$ with support on the Julia set of φ, when v is an infinite place (see also [FLM83]). More recently, Baker and Rumely [BR06], Chambert-Loir [CL06], and Favre and Rivera-Letelier [FRL04] and [FRL06] have constructed a φ-invariant measure $\mu_{\varphi,v}$ on the Berkovich space associated to $\mathbb{P}^1(\mathbb{C}_v)$; this is the unique φ-invariant measure without point masses at classical points on the v-adic Berkovich space (see [Ber90]) for a finite place v.

Proposition 3.1. *Let v be a place of an algebraically closed field $\mathbb{C}_v$ that is complete with respect to v, and let $F(t) = t - \beta$ for $\beta \in \mathbb{C}_v$. Then*

$$\int_{\mathbb{P}^1(\mathbb{C}_v)} \log|F|_v d\mu_{\varphi,v} = \lim_{k\to\infty} \frac{\log\max(|P_k(\beta,1)|_v, |Q_k(\beta,1)|_v)}{d^k} - \lim_{k\to\infty} \frac{\log\max(|P_k(1,0)|_v, |Q_k(1,0)|_v)}{d^k}. \tag{3.2}$$

Proof. We will prove this following the methods of Baker and Rumely [BR06, BR10]. The proposition could also be proved using the work of Favre and Rivera-Letelier [FRL06] or Chambert-Loir and Thuillier [CLT04,Thu06], who proved more general Mahler formulas (but do not formulate them in terms of limits such as (3.1)). In [BR06], Baker and Rumely show that for $w \in \mathbb{C}_v$, the function H_w defined by

$$H_w([a:b]) = -\log|wb-a|_v + \lim_{k\to\infty} \frac{\log\max(|P_k(1,0)|_v, |Q_k(1,0)|_v)}{d^k}$$

is subharmonic on $\mathbb{P}^1(\mathbb{C}_v) \setminus \{[w:1]\}$. Furthermore, if Δ is a suitably normalized distributional Laplacian (i.e., a suitable multiple of $-dd^c$ considered in the distributional sense, which can be extended to the setting of Berkovich spaces as described in [BR10]), then

$$\Delta H_w = -\mu_{\varphi,v} + \delta_w \tag{3.3}$$

where δ_w is the usual Dirac point mass at w. Similarly, we have

$$\Delta \log|t-\beta|_v = \delta_{[1:0]} - \delta_\beta$$

(see [FRL06, Section 5.1] or the same reasoning that gives (3.3)). Now, since $\log|t-\beta|_v$ and H_w are both subharmonic on $\mathbb{P}^1(\mathbb{C}_v) \setminus \{[1:0],[w:1],[\beta:1]\}$, we have

$$\begin{aligned}\int_{\mathbb{P}^1(\mathbb{C}_v)} \log|t-\beta|_v d\mu_{\varphi,v} &= \left(\int_{\mathbb{P}^1(\mathbb{C}_v)} \log|t-\beta|_v (-\Delta H_w)\right) + \log|w-\beta|_v \\ &= \left(\int_{\mathbb{P}^1(\mathbb{C}_v)} H_w (-\Delta \log|t-\beta|_v)\right) + \log|w-\beta|_v \\ &= H_w([\beta:1]) - H_w([1:0]) + \log|w-\beta|_v, \qquad (3.4)\end{aligned}$$

by the self-adjoint property of Δ (which follows from [BR10, Proposition 5.28] when v is nonarchimedean and is simply integration by parts when v is archimedean). Thus, we have

$$\begin{aligned}&-\log|w-\beta|_v + \lim_{k\to\infty} \frac{\log\max(|P_k(\beta,1)|_v, |Q_k(\beta,1)|_v)}{d^k} + \log|1| \\ &\quad - \lim_{k\to\infty} \frac{\log\max(|P_k(1,0)|_v, |Q_k(1,0)|_v)}{d^k} + \log|w-\beta|_v \\ &= \lim_{k\to\infty} \frac{\log\max(|P_k(\beta,1)|_v, |Q_k(\beta,1)|_v)}{d^k} \\ &\quad - \lim_{k\to\infty} \frac{\log\max(|P_k(1,0)|_v, |Q_k(1,0)|_v)}{d^k},\end{aligned}$$

as desired. □

Note that although our integrals are defined for points in $\mathbb{C}_v$, the results we prove in Section 5 only apply to points in $\overline{K}$, where K is a number field or function field of characteristic 0.

4 Preliminaries from Diophantine approximation

The following well-known theorem of Roth [Rot55] is the principal tool from Diophantine approximation that is used in this paper.

Theorem 4.1. *(Roth). If $\alpha \in \mathbb{C}$ is algebraic over $\mathbb{Q}$, then for any $\epsilon > 0$, there is a constant C such that*

$$\left|\alpha - \frac{a}{b}\right| > \frac{C}{|b|^{2+\epsilon}},$$

for all $a/b \in \mathbb{Q}$ such that $a/b \neq \alpha$.

We will need to work in slightly greater generality. In the terminology of the previous section, Roth's theorem admits the following generalization (see [Lan83, Theorem 7.1.1]), which holds when K is number field or a function field of characteristic 0.

Theorem 4.2. *Let $\alpha_1, \ldots, \alpha_n$ be elements of $\overline{K}$ and let $L \subset \overline{K}$ be a finite extension of K. Then, for any $\epsilon > 0$ and any places v of K and w of L such that $w|v$, we have*

$$\frac{1}{[L:K](\deg K)} \sum_{i=1}^{n} \max(0, -\log |\alpha_i - \beta|_v^{[L_w:K_v]n_v}) \leq (2+\epsilon)h(\beta) + O(1),$$

for all $\beta \in L$ not in the set $\{\alpha_1, \ldots, \alpha_n\}$.

Let $[a:1]$ be a point in $\mathbb{P}^1(\overline{K})$. Then for any $[b:1] \neq [a:1]$ in $\mathbb{P}^1(\mathbb{C}_v)$, we let

$$\lambda_{[a:1],v}([b:1]) = \max(-\log |b-a|_v, 0).$$

We extend this definition to the point at $[1:0]$ by letting

$$\lambda_{[a:1],v}([1:0]) = 0$$

and

$$\lambda_{[1:0],v}([b:1]) = \max(0, \log |b|_v). \tag{4.1}$$

We will work with divisors on $\mathbb{P}^1_{\overline{K}}$ rather than elements of $\overline{K}$. Let $D = \sum_{i=1}^{n} m_i \alpha_i$, where $\alpha_i \in \mathbb{P}^1(\overline{K})$ and $m_i \in \mathbb{Z}$. We let

$$\lambda_{D,v}(\beta) = \sum m_i \lambda_{\alpha_i,v}(\beta)$$

for points $\beta \in \mathbb{P}^1(\mathbb{C}_v)$ that are not in Supp D. Then $\lambda_{D,v}$ is a **Weil function** for D at v as defined in [Lan83, Chapter 10]. It is easy to check that for any divisor D and any rational map φ on $\mathbb{P}^1$, we have

$$\lambda_{D,v}(\varphi(\beta)) = \lambda_{\varphi^* D,v}(\beta) + O(1), \tag{4.2}$$

for all $\beta \in \mathbb{P}^1(\overline{K})$ away from the support of D and $\varphi^* D$. This is a general functorial property of Weil functions, as explained in [Lan83, Chapter 10].

For a divisor $D = \sum_{i=1}^{n} m_i \alpha_i$, where $\alpha_i \in \mathbb{P}^1(\overline{K})$, we define

$$r(D) = \max_i (m_i).$$

With this terminology, it follows from Theorem 4.2 that for any $\epsilon > 0$, any finite extension L of K, and any positive divisor D on $\mathbb{P}^1(\overline{K})$ with $r(D) = 1$, we have

$$\frac{1}{[L:K](\deg K)} \lambda_{D,v}(\beta) \leq (2+\epsilon)h(\beta) + O(1)$$

for all $\beta \in \mathbb{P}^1(L)$ away from the support of D. Hence, for any positive divisor D we have

$$\frac{1}{[L:K](\deg K)}\lambda_{D,v}(\beta) \le r(D)(2+\epsilon)h(\beta) + O(1). \tag{4.3}$$

5 Main results

We begin with a simple lemma on how $r((\varphi^n)^*(D)$ behaves as $n \to \infty$ when D is a divisor that does not contain an exceptional point of φ. We recall that in general if $D = \sum_{i=1}^n m_i\alpha_i$ is a divisor on $\mathbb{P}^1$ and $\psi : \mathbb{P}^1 \longrightarrow \mathbb{P}^1$ is a nonconstant rational map, then

$$\psi^* D = \sum_{i=1}^{n} \sum_{\psi(\beta_i)=\alpha_i} m_i e(\beta_i/\alpha_i)\beta_i \tag{5.1}$$

where $e(\beta_i/\alpha_i)$ is the ramification index of ψ at β_i.

Lemma 5.1. *Let D be a divisor such that* Supp D *does not contain any exceptional points of* φ. *Then* $\lim_{k\to\infty} \frac{r((\varphi^k)^*D)}{d^k} = 0$.

Proof. Recall that α is an exceptional point if and only if $\varphi^2(\alpha) = \alpha$ and φ is totally ramified at both α and $\varphi(\alpha)$. Since φ has at most two totally ramified points, it follows that if α is not exceptional, then one of α, $\varphi(\alpha)$, and $\varphi^2(\alpha)$ is not a totally ramified point of φ. Since the degree of φ^3 is d^3, this means that for any divisor E such that Supp E does not contain an exceptional point, we have $r((\varphi^3)^*E) < d^3 r(E)$ (by (5.1)), so $r((\varphi^3)^*E) \le d^2(d-1)r(E)$ Now, since Supp D does not contain an exceptional point, $\mathrm{Supp}(\varphi^k)^*D$ does not contain an exceptional point for any k. Thus, for any $k \ge 3$, we see that $\frac{r((\varphi^k)^*D)}{d^k}$ is less than or equal to $((d-1)/d)^{(k-2)/3}r(D)$, which goes to zero as k goes to infinity. □

5.1 *Using Roth's theorem*

Roth's Theorem allows us to prove the following lemma. The idea of the proof is that if $\varphi^{k+\ell}(\beta)$ approximates D very closely, then $\varphi^k(\beta)$ approximates $(\varphi^\ell)^*D$ very closely. Since $\varphi^k(\beta)$ has height approximately equal to $1/d^\ell$ times the height of $\varphi^{k+\ell}(\beta)$, this makes $h(\varphi^k(\beta))$ small relative to $\lambda_{(\varphi^\ell)^*D}(\beta)$. Repeating this for infinitely many $\varphi^k(\beta)$ gives a contradiction to Roth's theorem. This idea is due to Siegel [Sie29]; similar arguments can be found in [Sil93].

Lemma 5.2. *Let D be a positive divisor on $\mathbb{P}^1$ such that* Supp D *does not contain any of the exceptional points of φ. Let β be a point in $\mathbb{P}^1(\overline{K})$ for which there is a strictly increasing sequence of integers $(e_i)_{i=1}^{\infty}$ such that $\varphi^{e_i}(\beta) \notin$* Supp D*. Then*

$$\lim_{i\to\infty} \frac{\lambda_{D,v}(\varphi^{e_i}(\beta))}{d^{e_i}} = 0. \tag{5.2}$$

Proof. Let L be a finite extension of K for which $\beta \in \mathbb{P}^1(L)$. Choose $\delta > 0$. By Lemma 5.1, we may pick an integer ℓ such that $\frac{r((\varphi^\ell)^* D)}{d^\ell} < \delta/2$. We may then write $\frac{r((\varphi^\ell)^* D)(2+\epsilon)}{d^\ell} = \delta$ for some $\epsilon > 0$. For any e_i, we have $\varphi^{e_i-\ell}(\beta) \notin \operatorname{Supp}(\varphi^\ell)^* D$ since $\varphi^{e_i}(\beta) \notin \operatorname{Supp} D$. Thus, applying Roth's Theorem (as expressed in (4.3)), we find that for all e_i we have

$$\frac{1}{[L:K](\deg K)} \lambda_{(\varphi^\ell)^* D,v}(\varphi^{e_i-\ell}(\beta)) \le r((\varphi^\ell)^* D)(2+\epsilon) h(\varphi^{e_i-\ell}(\beta)) + O(1).$$

Using (4.2) and the fact that $h(\varphi^{e_i}(\beta)) \le d^\ell h_\varphi(\varphi^{e_i-\ell}(\beta)) + O(1)$, we then obtain

$$\begin{aligned}
\frac{1}{[L:K](\deg K)} \lambda_{D,v}(\varphi^{e_i}(\beta)) &\le \frac{1}{[L:K](\deg K)} \lambda_{(\varphi^\ell)^* D,v}(\varphi^{e_i-\ell}(\beta)) + O(1) \\
&\le r((\varphi^\ell)^* D)(2+\epsilon) h(\varphi^{e_i-\ell}(\beta)) + O(1) \\
&\le \frac{r((\varphi^\ell)^* D)(2+\epsilon)}{d^\ell} h(\varphi^{e_i}(\beta)) + O(1) \\
&\le \delta h(\varphi^{e_i}(\beta)) + O(1) \\
&\le \delta d^{e_i} h(\beta) + O(1).
\end{aligned}$$

Dividing through by d^{e_i} gives

$$\limsup_{i\to\infty} \frac{\lambda_{D,v}(\varphi^{e_i}(\beta))}{d^{e_i}} \le [L:K](\deg K)\delta h(\beta).$$

Since $\lambda_{D,v}(\varphi^{e_i}(\beta)) \ge 0$, letting δ go to zero gives (5.2), as desired. □

This allows us to prove the following proposition, which will be used to prove Theorems 5.6 and 5.7.

Proposition 5.3. *Let $\alpha = [s:u]$ be a nonexceptional point in $\mathbb{P}^1(\overline{K})$. Then for any point $\beta = [a:b]$ in $\mathbb{P}^1(\overline{K})$ and any strictly increasing sequence of integers $(e_i)_{i=1}^{\infty}$ such that $\varphi^{e_i}(\beta) \ne \alpha$, we have*

$$\lim_{i\to\infty} \frac{\log |u P_{e_i}(a,b) - s Q_{e_i}(a,b)|_v}{d^{e_i}} = \lim_{i\to\infty} \frac{\log \max(|P_{e_i}(a,b)|_v, |Q_{e_i}(a,b)|_v)}{d^{e_i}}.$$

Proof. Note that we know that the limit on the right-hand side of the equation above exists by the discussion at the beginning of Section 3.

If $[1:0]$ is an exceptional point of φ, let $\mathcal{U}$ be its attracting basin; if $[1:0]$ is not exceptional, let $\mathcal{U}$ simply equal $\{[1:0]\}$. We will divide $(e_i)_{i=1}^{\infty}$ into two subsequences: one consisting of the e_i for which $\varphi^{e_i}(\beta) \notin \mathcal{U}$ and one consisting of the remaining integers in the sequence $(e_i)_{i=1}^{\infty}$. Let $(\ell_j)_{j=1}^{\infty}$ be the subsequence consisting of all integers ℓ_j in $(e_i)_{i=1}^{\infty}$ such that $\varphi^{\ell_j}(\beta) \notin \mathcal{U}$ (this subsequence may be empty). We have

$$\lim_{j\to\infty} \frac{\max(\log|P_{\ell_j}(a,b)/Q_{\ell_j}(a,b)|_v, 0)}{d^{\ell_j}} = 0. \tag{5.3}$$

If $[1:0]$ is not exceptional, this follows from Lemma 5.2 applied to $D = [1:0]$, along with (4.1). If $[1:0]$ is exceptional, the fact that $\varphi^{\ell_j}(\beta) \notin \mathcal{U}$ for all j implies that $|P_{\ell_j}(a,b)/Q_{\ell_j}(a,b)|_v$ is bounded for all j, so (5.3) clearly holds. It follows immediately from (5.3) that

$$\lim_{j\to\infty} \frac{\log\max(|P_{\ell_j}(a,b)|_v, |Q_{\ell_j}(a,b)|_v)}{d^{\ell_j}} = \lim_{j\to\infty} \frac{\log|Q_{\ell_j}(a,b)|_v}{d^{\ell_j}}. \tag{5.4}$$

Note that if $u = 0$, then

$$uP_{\ell_j}(a,b) - sQ_{\ell_j}(a,b) = sQ_{\ell_j}(a,b),$$

so we are done. Otherwise, by Lemma 4.2, we have

$$\lim_{j\to\infty} \frac{\max\left(0, -\log\left|\frac{P_{\ell_j}(a,b)}{Q_{\ell_j}(a,b)} - \frac{s}{u}\right|_v\right)}{d^{\ell_j}} = 0.$$

Combining this with (5.3), we see that

$$\lim_{j\to\infty} \frac{\log\left|\frac{P_{\ell_j}(a,b)}{Q_{\ell_j}(a,b)} - \frac{s}{u}\right|_v}{d^{\ell_j}} = 0.$$

Thus, using (5.4), we obtain

$$\begin{aligned}
&\lim_{j\to\infty} \frac{\log|uP_{\ell_j}(a,b) - sQ_{\ell_j}(a,b)|_v}{d^{\ell_j}} \\
&= \lim_{j\to\infty} \frac{\log\left(|Q_{\ell_j}(a,b)|_v |u|_v \left|\frac{P_{\ell_j}(a,b)}{Q_{\ell_j}(a,b)} - \frac{s}{u}\right|_v\right)}{d^{\ell_j}} \\
&= \lim_{j\to\infty} \frac{\log|Q_{\ell_j}(a,b)|_v}{d^{\ell_j}} + \lim_{j\to\infty} \frac{\log\left|\frac{P_{\ell_j}(a,b)}{Q_{\ell_j}(a,b)} - \frac{s}{u}\right|_v}{d^{\ell_j}} \\
&= \lim_{j\to\infty} \frac{\log\max(|P_{\ell_j}(a,b)|_v, |Q_{\ell_j}(a,b)|_v)}{d^{\ell_j}},
\end{aligned}$$

as desired.

Now let $(m_j)_{j=1}^{\infty}$ be the subsequence of $(e_i)_{i=1}^{\infty}$ consisting of all integers m_j in $(e_i)_{i=1}^{\infty}$ such that $\varphi^{m_j}(\beta) \in \mathcal{U}$ (this subsequence may also be empty). If $\alpha = [1:0]$, then $[1:0]$ is not exceptional by assumption, so there are no m_j and we are done. Otherwise, we have

$$\lim_{j\to\infty} \frac{|sQ_{m_j}(a,b)|_v}{|uP_{m_j}(a,b)|_v} = 0,$$

since $\frac{P_{m_j}(a,b)}{Q_{m_j}(a,b)}$ goes to infinity and $u \neq 0$. This implies that

$$\begin{aligned}\lim_{j\to\infty} &\frac{\log|uP_{m_j}(a,b) - sQ_{m_j}(a,b)|_v}{d^{m_j}} \\ &= \lim_{j\to\infty} \frac{\log|uP_{m_j}(a,b)|_v}{d^{m_j}} \\ &= \lim_{j\to\infty} \frac{\log\max(|P_{m_j}(a,b)|_v, |Q_{m_j}(a,b)|_v))}{d^{m_j}}.\end{aligned}$$

Since every element of the sequence $(e_i)_{i=1}^{\infty}$ is in $(\ell_j)_{j=1}^{\infty}$ or $(m_j)_{j=1}^{\infty}$, this completes our proof. □

5.2 *Preperiodic points*

Proposition 5.3 provides all the information we need when $\varphi^k([a:b]) = [s:u]$ for at most finitely many k; this will always be the case when $[s:u]$ is not periodic. When $[s:u]$ is periodic, however, there may be infinitely many k such that $\varphi^k([a:b]) = [s:u]$. New complications arise when this is the case; we treat these complications in Propositions 5.4 and 5.5.

Suppose that $(bT_0 - aT_1)^{w_k}$ is the highest power of $(bT_0 - aT_1)$ that divides $uP_k(T_0,T_1) - sQ_k(T_0,T_1)$ in $\overline{K}[T_0,T_1]$. We write

$$uP_k(T_0,T_1) - sQ_k(T_0,T_1) = (bT_0 - aT_1)^{w_k} G_k(T_0,T_1),$$

where G_k is a polynomial in $\overline{K}[T_0,T_1]$ such that $G_k(a,b) \neq 0$.

Proposition 5.4. *Let $[s:u]$ be a nonexceptional point of φ. Then, with notation as above, we have*

$$\lim_{k\to\infty} \frac{\log|G_k(a,b)|_v}{d^k} = \lim_{k\to\infty} \frac{\log\max(|P_k(a,b)|_v, |Q_k(a,b)|_v)}{d^k}. \tag{5.5}$$

Proof. By Proposition 5.3, equation (5.5) holds if we restrict to the k for which $\varphi^k([a:b]) \neq \alpha$. If there are only finitely many k such that $\varphi^k([a:b]) = \alpha$, we are therefore done. Otherwise, let j be the smallest positive integer such that $\varphi^j([\beta : 1]) = \alpha$ and let ℓ be the period of α. Then $\varphi^k([\beta : 1]) = \alpha$ precisely when k is of the form $j + m\ell$ for some integer $m \geq 0$. If $\varphi^\ell([s : u]) = [s : u]$, then $uT_0 - sT_1$ divides $uP_\ell(T_0, T_1) - sQ_\ell(T_0, T_1)$.

Suppose that $u \neq 0$. Then, expanding Q_ℓ in the variables $uT_0 - sT_1$ and T_1, we see that since $uT_0 - sT_1$ cannot divide $Q_\ell(T_0, T_1)$ (because if it did, then it would also divide $P_\ell(T_0, T_1)$ and we know that Q_ℓ and P_ℓ have no factors), we have

$$Q_\ell(T_0, T_1) = g_0 T_1^{d^\ell} + (uT_0 - sT_1) W(T_0, T_1)$$

for some nonzero $g_0 \in \overline{K}$ and some $W(T_0, T_1) \in \overline{K}[T_0, T_1]$. For any $m \geq 1$ we thus have

$$Q_{m\ell} = g_0 (Q_{(m-1)\ell})^{d^\ell} + (uP_{(m-1)\ell} - sQ_{(m-1)\ell}) W(P_{(m-1)\ell}, Q_{(m-1)\ell}).$$

Using induction, we see then that

$$Q_{m\ell}(T_0, T_1) = g_0^{\sum_{i=0}^{m-1} d^{i\ell}} T_1^{d^{m\ell}} + (uT_0 - sT_1) W_m(T_0, T_1), \tag{5.6}$$

for some polynomial $W_m(T_0, T_1) \in \overline{K}[T_0, T_1]$. Similarly, we may write

$$\begin{aligned} &uP_\ell(T_0, T_1) - sQ_\ell(T_0, T_1) \\ &\quad = (uT_0 - sT_1)^r f_r T_1^{d-r} + (uT_0 - sT_1)^{r+1} V(T_0, T_1), \end{aligned} \tag{5.7}$$

for some nonzero $f_r \in \overline{K}$, some integer $r > 0$, and some $V(T_0, T_1)$ in $\overline{K}[T_0, T_1]$. Since $[s : u]$ is not an exceptional point of φ, we have $r < d^\ell$ (note that if r were equal to d^ℓ, then φ would have to ramify totally at $\varphi([s : u]), \ldots, \varphi^\ell[s : u]$, which would imply that $\ell = 2$ and that $[s : u]$ is therefore an exceptional point, as explained in Section 2). Then for any m, we have

$$\begin{aligned} uP_{m\ell} - sQ_{m\ell} &= (uP_{(m-1)\ell} - sQ_{(m-1)\ell})^r f_r Q_{(m-1)\ell}^{d-r} \\ &\quad + (P_{(m-1)\ell} - sQ_{(m-1)\ell})^{r+1} V(P_{(m-1)\ell}, Q_{(m-1)\ell}), \end{aligned}$$

so, using (5.6), (5.7), and induction, we obtain

$$\begin{aligned} &uP_{m\ell}(T_0, T_1) - sQ_{m\ell}(T_0, T_1) \\ &\quad = (uT_0 - sT_1)^{r^m} f_r^{\sum_{i=0}^{m-1} r^i} T_1^{d^{m\ell} - r^m} g_0^{\sum_{i=0}^{m-1} (d^{i\ell} - r^i)} \\ &\quad + (uT_0 - sT_1)^{r^m + 1} Z_m(T_0, T_1), \end{aligned} \tag{5.8}$$

for Z_m a polynomial in $\overline{K}[T_0, T_1]$. Since $r < d^\ell$, we have

$$\lim_{m \to \infty} \frac{\log |f_r^{\sum_{i=0}^{m-1} r^i} g_0^{\sum_{i=0}^{m-1} (d^{i\ell} - r^i)}|_v}{d^{m\ell}} = \lim_{m \to \infty} \frac{\log |g_0^{\sum_{i=0}^{m-1} d^{i\ell}}|_v}{d^{m\ell}} = \frac{\log |g_0|_v}{d^\ell - 1}.$$

Now, let ϵ be the highest power of $aT_0 - bT_1$ that divides $uP_j - sQ_j$. Using (5.8), we see that we have

$$uP_{j+m\ell}(T_0, T_1) - sQ_{j+m\ell}(T_0, T_1) = (bT_0 - aT_1)^{\epsilon r^m} G_{j+m\ell}(T_0, T_1)$$

for a polynomial $G_{j+m\ell} \in \overline{K}[T_0, T_1]$. Letting m go to infinity, we see from (5.8) that

$$\lim_{m\to\infty} \frac{\log |G_{j+m\ell}(a,b)|_v}{d^{j+m\ell}} = \frac{\log |g_0|_v}{d^j(d^\ell - 1)} + \frac{\log |Q_j(a,b)|_v}{d^j}.$$

Similarly, (5.6) yields

$$\lim_{m\to\infty} \frac{\log |Q_{j+m\ell}(a,b)|_v}{d^{j+m\ell}} = \frac{\log |g_0|_v}{d^j(d^\ell - 1)} + \frac{\log |Q_j(a,b)|_v}{d^j}.$$

Moreover, since $uP_{j+m\ell}(a,b) = sQ_{j+m\ell}(a,b)$ for every m, we have

$$\lim_{m\to\infty} \frac{\log |P_{j+m\ell}(a,b)|_v}{d^{j+m\ell}} = \lim_{m\to\infty} \frac{\log |Q_{j+m\ell}(a,b)|_v}{d^{j+m\ell}}.$$

Hence

$$\lim_{m\to\infty} \frac{\log |G_{j+m\ell}(a,b)|_v}{d^{j+m\ell}} = \lim_{m\to\infty} \frac{\log \max(|P_{j+m\ell}(a,b)|_v, |Q_{j+m\ell}(a,b)|_v)}{d^{j+m\ell}},$$

which completes our proof in the case $u \neq 0$. The proof in the case $u = 0$ proceeds in exactly the same way, using T_0 in place of T_1. □

We have a similar result for the polynomials $T_0 P_k - T_1 Q_k$. We write

$$T_0 P_k(T_0, T_1) - T_1 Q_k(T_0, T_1) = (bT_0 - aT_1)^{n_k} H_k(T_0, T_1)$$

where H_k is a polynomial in $\overline{K}[T_0, T_1]$ such that $H_k(a,b) \neq 0$. The proof of the following proposition is similar to Morton's and Silverman's proof of [MS95, Lemma 3.4], but it requires a bit more detail since it yields information about $H_k(a,b)$ as well as n_k.

Proposition 5.5. *With notation as above, we have*

$$\lim_{k\to\infty} \frac{\log |H_k(a,b)|_v}{d^k} = \lim_{k\to\infty} \frac{\log \max(|P_k(a,b)|_v, |Q_k(a,b)|_v)}{d^k}. \tag{5.9}$$

Furthermore, n_k remains bounded as k goes to infinity.

Proof. If $(e_i)_{i=1}^{\infty}$ is a strictly increasing sequence of integers such that $\varphi^{e_i}([a:b]) \neq [a:b]$ for each e_i, then

$$H_{e_i}(T_0, T_1) = T_0 P_{e_i}(T_0, T_1) - T_1 Q_{e_i}(T_0, T_1)$$

for all e_i. Hence, by Proposition 5.3, we have

$$\lim_{i\to\infty} \frac{\log|H_{e_i}(a,b)|_v}{d^{e_i}} = \lim_{i\to\infty} \frac{\log\max(|P_{e_i}(a,b)|_v, |Q_{e_i}(a,b)|_v)}{d^{e_i}}.$$

If $[a:b]$ is not periodic, this finishes the proof. Thus, we may assume that $[a:b]$ is periodic. The rest of the proof is a computation. We divide it into three steps.

Step I. We begin by changing variables so that $[a:b]$ becomes $[0:1]$. If $b=0$, we write $U_0 = T_1/a$ and $U_1 = -T_0$. We then let

$$R(U_0,U_1) = \frac{1}{a}Q(T_0,T_1)$$

and

$$S(U_0,U_1) = -P(T_0,T_1)$$

(this is simply the inverse of the transformation we defined on T_0 and T_1; our change of variables is obtained by conjugation by a change-of-basis matrix). If $b \neq 0$, we write $U_1 = \frac{1}{b}T_1$ and

$$U_0 = bT_0 - aT_1.$$

We then let $S(U_0,U_1) = Q(T_0,T_1)/b$ and

$$R(U_0,U_1) = bP(T_0,T_1) - aQ(T_0,T_1).$$

We define R_k and S_k recursively by letting $R_1 = R$, $S_1 = S$, and setting

$$R_{k+1}(U_0,U_1) = R_k(R(U_0,U_1), S(U_0,U_1))$$

and

$$S_{k+1}(U_0,U_1) = S_k(R(U_0,U_1), S(U_0,U_1)).$$

By the construction of our change of variables, we have

$$U_1R_k(U_0,U_1) - U_0S_k(U_0,U_1) = T_0P_k(T_0,T_1) - T_1Q_k(T_0,T_1) \tag{5.10}$$

as polynomials in T_0 and T_1. Hence, if $U_0^{n_k}$ is the highest power of U_0 that divides $U_1R_k(U_0,U_1) - U_0S_k(U_0,U_1)$ and τ_k is the coefficient of the $U_0^{n_k}U_1^{d^k-n_k}$ term in $U_1R_k(U_0,U_1) - U_0S_k(U_0,U_1)$, then

$$\tau_k = H_k(a,b).$$

Now let ℓ be the smallest positive integer for which $\varphi^\ell([a:b]) = [a:b]$. Note that $|S_{m\ell}(1,0)|_v = \frac{|Q_{m\ell}(a,b)|_v}{|b|_v}$ if $b \neq 0$ and

$$|S_{m\ell}(1,0)|_v = |P_{m\ell}(a,b)|_v/|a|_v$$

otherwise. Since

$$[P_{m\ell}(a,b) : Q_{m\ell}(a,b)] = [a : b]$$

for every m, it follows that

$$\lim_{m\to\infty} \frac{\log |S_{m\ell}(0,1)|_v}{d^{m\ell}} = \lim_{m\to\infty} \frac{\log \max(|P_{m\ell}(a,b)|_v, |Q_{m\ell}(a,b)|_v)}{d^{m\ell}}.$$

Thus, it will suffice to show that

$$\lim_{m\to\infty} \frac{\log |\tau_{m\ell}|_v}{d^{m\ell}} = \lim_{m\to\infty} \frac{\log |S_{m\ell}(0,1)|_v}{d^{m\ell}}. \tag{5.11}$$

We write

$$R_\ell(U_0, U_1) = \sum_{i=1}^{d^\ell} f_i U_0^i U_1^{d^\ell - i}$$

(note that U_0 divides R_ℓ by our change of variables) and

$$S_\ell(U_0, U_1) = \sum_{i=0}^{d^\ell} g_i U_0^i U_1^{d^\ell - i}.$$

Using induction, we see that

$$R_{m\ell}(U_0, U_1) \equiv f_1^m g_0^{(\sum_{j=0}^{m-1} d^{j\ell})-m} U_0 U_1^{d^{m\ell}-1} \pmod{U_0^2}$$

and

$$S_{m\ell}(U_0, U_1) \equiv g_0^{\sum_{j=0}^{m-1} d^{j\ell}} U_1^{d^{m\ell}} \pmod{U_0^2}.$$

Thus, we have

$$\begin{aligned} &U_1 R_{m\ell}(U_0, U_1) - U_0 S_{m\ell}(U_0, U_1) \\ &\equiv g_0^{\sum_{j=0}^{m-1} d^{j\ell}} ((f_1/g_0)^m - 1) U_0 U_1^{d^{m\ell}} \pmod{U_0^2}. \end{aligned} \tag{5.12}$$

Step II. We will now treat the m for which $(f_1/g_0)^m \neq 1$. We have

$$|\log |(f_1/g_0)^m - 1|_v \leq h((f_1/g_0)^m - 1) \leq 2m[K(f_1/g_0) : K]h(f_1/g_0)$$

for all m such that $(f_1/g_0)^m \neq 1$ (this is a simple version of Liouville's theorem), so

$$\lim_{\substack{m\to\infty \\ (f_1/g_0)^m \neq 1}} \frac{\log |(f_1/g_0)^m - 1|_v}{d^{m\ell}} = 0.$$

Thus, dividing (5.12) through by U_0, we obtain

$$\lim_{\substack{m\to\infty\\(f_1/g_0)^m\neq 1}} \frac{\log|\tau_{m\ell}|_v}{d^{m\ell}} = \lim_{m\to\infty} \frac{\log|g_0^{\sum_{j=0}^{m-1} d^{j\ell}}|_v}{d^{m\ell}} = \lim_{m\to\infty} \frac{\log|S_{m\ell}(0,1)|_v}{d^{m\ell}},$$

as desired.

Step III. We are left with treating the m for which $(f_1/g_0)^m = 1$. Let ρ be the smallest positive integer m such that $(f_1/g_0)^m = 1$ and write $\omega = \rho\ell$. For $q \geq 1$ we write

$$R_{q\omega}(U_0, U_1) = \sum_{i=1}^{d^{q\omega}} x_i^{[q]} U_0^i U_1^{d^{q\omega}-i}$$

(the summation starts at 1 since U_0 divides $R_{q\omega}$) and

$$S_{q\omega}(U_0, U_1) = \sum_{i=0}^{d^{q\omega}} y_i^{[q]} U_0^i U_1^{d^{q\omega}-i}.$$

Since $f_1^\rho = g_0^\rho$ by assumption, we have $y_0^{[1]} = x_1^{[1]}$ by (5.12). Multiplying R_ω and S_ω through by a constant will change all of the limits we are calculating by the same fixed amount, so we may assume that $y_0^{[1]} = x_1^{[1]} = 1$. Let r be the smallest integer greater than 0 such that $x_r^{[1]} \neq y_{r-1}^{[1]}$ (we have $r \geq 2$ since $(f_1/g_0)^m = 1$). Then U_0^r divides $U_1 R_\omega - U_0 S_\omega$, which in turn divides $U_1 R_{q\omega} - U_0 S_{q\omega}$ for any q; hence U_0^r divides $U_1 R_{q\omega} - U_0 S_{q\omega}$ for every q, so $x_j^{[q]} = y_{j-1}^{[q]}$ for $j < r$. To calculate $x_r^{[q]} - y_{r-1}^{[q]}$, we introduce some notation: we let

$$\left(\sum_{i=0}^{M} t_i U_0^i U_1^{M-i}\right)_j = t_j$$

for any polynomial $\sum_{i=0}^{M} t_i U_0^i U_1^{M-i}$. We have

$$\begin{aligned} x_r^{[q]} - y_{r-1}^{[q]} &= \sum_{i=1}^{r} x_i^{[q-1]} \left((R_\omega)^i (S_\omega)^{d^{(q-1)\omega}-i}\right)_r \\ &\quad - \sum_{j=0}^{r-1} y_j^{[q-1]} \left((R_\omega)^j (S_\omega)^{d^{(q-1)\omega}-j}\right)_{r-1}. \end{aligned} \tag{5.13}$$

For any $i < r$, we have $x_i^{[1]} = y_{i-1}^{[1]}$, so $(U_0 R_\omega)_i = (U_1 S_\omega)_i$. Hence, we have

$$\left((R_\omega)^j (S_\omega)^{d^{(q-1)\omega}-j}\right)_{r-1} = \left((R_\omega)^{j+1} (S_\omega)^{d^{(q-1)\omega}-j-1}\right)_r$$

for $j > 0$. For $j = 0$, we have

$$\left(S_\omega^{d^{(q-1)\omega}}\right)_{r-1} = \left((R_\omega + (x_r^{[1]} - y_{r-1}^{[1]})U_0^r U_1^{d^\omega - r})S_\omega^{d^{(q-1)\omega}-1}\right)_r$$
$$= \left(R_\omega S_\omega^{d^{(q-1)\omega}-1}\right)_r + (x_r^{[1]} - y_{r-1}^{[1]}),$$

since $y_0^{[1]} = x_1^{[1]} = 1$.

Using equation (5.13), we see that

$$x_r^{[q]} - y_{r-1}^{[q]} = \sum_{i=1}^{r} x_i^{[q-1]} \left((R_\omega)^i (S_\omega)^{d^{q\omega}-i}\right)_r$$
$$- \sum_{j=0}^{r-1} x_{j+1}^{[q-1]} \left((R_\omega)^{j+1} (S_\omega)^{d^{(q-1)\omega}-j-1}\right)_r + (x_r^{[1]} - y_{r-1}^{[1]})(x_1^{[q-1]})$$
$$+ (x_r^{[q-1]} - y_{r-1}^{[q-1]}) \left((R_\omega)^r (S_\omega)^{d^{(q-1)\omega}-r}\right)_r$$
$$= (x_r^{[1]} - y_{r-1}^{[1]})(x_1^{[q-1]}) + (x_r^{[q-1]} - y_{r-1}^{[q-1]}),$$

We have $y_0^{[q-1]} = x_1^{[q-1]} = 1$, since $y_0^{[1]} = x_1^{[1]} = 1$. Thus, assuming inductively that

$$x_r^{[q-1]} - y_{r-1}^{[q-1]} = (q-1)(x_r^{[1]} - y_{r-1}^{[1]}),$$

we have

$$x_r^{[q]} - y_{r-1}^{[q]} = q(x_r^{[1]} - y_{r-1}^{[1]}). \tag{5.14}$$

Note in particular that $n_{q\omega} = r$ for all q, so n_k is bounded for all k, as desired.

Now

$$\lim_{q\to\infty} \frac{\log |q(x_r^{[1]} - y_{r-1}^{[1]})|_v}{d^{q\omega}} = 0$$

and $\tau_{q\omega} = x_r^{[q]} - y_{r-1}^{[q]}$. Since $S_{q\omega}(1,0)$ is simply $y_0^{[q-1]} = 1$, we have

$$\lim_{q\to\infty} \frac{\log |\tau_{q\omega}|_v}{d^{q\omega}} = 0 = \lim_{q\to\infty} \frac{\log |S_{q\omega}(0,1)|_v}{d^{q\omega}},$$

which gives us (5.11) and thus completes our proof. □

5.3 Proofs of the main theorems

Now we can show that the integral $\int_{\mathbb{P}^1(\mathbb{C}_v)} \log|t-\beta|_v d\mu_{\varphi,v}$ can be computed by taking the limit of the average of $\log|\beta - w|_v$ on the points in $\varphi^{-k}(\alpha)$, as $k \to \infty$, for any nonexceptional point α.

Theorem 5.6. *Let $\alpha = [s : u]$ be a nonexceptional point in $\mathbb{P}^1(\overline{K})$. Then for any nonzero polynomial $F(t) \in \overline{K}[t]$ we have*

$$\int_{\mathbb{P}^1(\mathbb{C}_v)} \log|F|_v \, d\mu_{\varphi,v} = \lim_{k\to\infty} \frac{1}{d^k} \sum_{\substack{\varphi^k([w:1])=\alpha \\ F(w)\neq 0}} \log|F(w)|_v,$$

where the $[w : 1]$ for which $\varphi^k([w : 1]) = \alpha$ are counted with multiplicity.

Proof. The polynomial F factors as $F(t) = \gamma\prod_{i=1}^n (t-\beta_i)$, where γ and $\beta_1, \ldots, \beta_n$ are elements of $\overline{K}$. For each β_i, the multiplicity of β_i in $(\varphi^k)^*\alpha$ is at most $r((\varphi^k)^*\alpha)$ (where $r((\varphi^k)^*\alpha)$ is defined as in Section 4). Since α is not exceptional, we have $\lim_{k\to\infty} \frac{r((\varphi^k)^*\alpha)}{d^k} = 0$, by Lemma 5.1. Thus,

$$\lim_{k\to\infty} \frac{1}{d^k} \sum_{\substack{\varphi^k([w:1])=\alpha \\ w\neq\beta_j}} \log|w-\beta_j|_v = \lim_{k\to\infty} \frac{1}{d^k} \sum_{\substack{\varphi^k([w:1])=\alpha \\ F(w)\neq 0}} \log|w-\beta_j|_v$$

for each β_j. Hence, it suffices to show that

$$\int_{\mathbb{P}^1(\mathbb{C}_v)} \log|t-\beta|_v \, d\mu_{\varphi,v} = \lim_{k\to\infty} \frac{1}{d^k} \sum_{\substack{\varphi^k([w:1])=\alpha \\ w\neq\beta}} \log|w-\beta|_v \tag{5.15}$$

for any $\beta \in \overline{K}$.

Note that $\varphi^k([w : 1]) = [s : u]$ if and only if $uP_k(w, 1) - sQ_k(w, 1) = 0$. Thus, as polynomials in t, we have

$$uP_k(t,1) - sQ_k(t,1) = \eta_k \prod_{\varphi^k([w:1])=[s:u]} (t-w),$$

where $\eta_k \in \overline{K}$. We write

$$uP_k(t,1) - sQ_k(t,1) = (t-\beta)^{w_k} G_k(t,1)$$

for a polynomial G_k such that $G_k(\beta, 1) \neq 0$, as in Proposition 5.4. Note that

$$G_k(t,1) = \eta_k \prod_{\substack{\varphi^k([w:1])=\alpha \\ w\neq\beta}} (t-w).$$

Plugging β in for t and taking logs of absolute values gives

$$\log|G_k(\beta,1)|_v = \log|\eta_k|_v + \sum_{\substack{\varphi^k([w:1])=[s:u] \\ w\neq\beta}} \log|w-\beta|_v. \tag{5.16}$$

Applying Proposition 4.4 therefore yields

$$\lim_{k\to\infty} \frac{1}{d^k} \sum_{\substack{\varphi^k([w:1])=\alpha \\ w\neq\beta}} \log|w-\beta|_v + \frac{\log|\eta_k|_v}{d^k}$$
$$= \lim_{k\to\infty} \frac{\log\max(|P_k(\beta,1)|_v, |Q_k(\beta,1)|_v)}{d^k}. \tag{5.17}$$

Now, writing

$$uP_k(T_0,T_1) - sQ_k(T_0,T_1) = T_1^{w_k} V_k(T_0,T_1)$$

for some polynomial V_k such that $V_k(1,0) \neq 0$, we see that $\eta_k = V_k(1,0)$. Applying Proposition 4.4, we obtain

$$\lim_{k\to\infty} \frac{\log|\eta_k|_v}{d^k} = \lim_{k\to\infty} \frac{\log\max(|P_k(1,0)|_v, |Q_k(1,0)|_v)}{d^k}.$$

Substituting this equality into (5.17) gives

$$\lim_{k\to\infty} \frac{1}{d^k} \sum_{\substack{\varphi^k([w:1])=\alpha \\ [w:1]\neq\beta}} \log|w-\beta|_v = \lim_{k\to\infty} \frac{\log\max(|P_k(\beta,1)|_v, |Q_k(\beta,1)|_v)}{d^k}$$
$$- \lim_{k\to\infty} \frac{\log\max(|P_k(1,0)|_v, |Q_k(1,0)|_v)}{d^k}. \tag{5.18}$$

Using Proposition 3.1, we obtain (5.15). □

Now we show that the same result holds when we average $\log|\beta - w|_v$ over periodic points rather than inverse images of a point.

Theorem 5.7. *For any any polynomial $F \in \overline{K}[t]$ we have*

$$\int_{\mathbb{P}^1(\mathbb{C}_v)} \log|F|_v \, d\mu_{\varphi,v} = \lim_{k\to\infty} \frac{1}{d^k} \sum_{\substack{\varphi^k([w:1])=[w:1] \\ F(w)\neq 0}} \log|F(w)|_v,$$

where the $[w:1]$ for which $\varphi^k([w:1]) = w$ are counted with multiplicity.

Proof. As in the proof of Theorem 5.6, it will suffice to show that

$$\int_{\mathbb{P}^1(\mathbb{C}_v)} \log|t-\beta|_v \, d\mu_{\varphi,v} = \lim_{k\to\infty} \frac{1}{d^k} \sum_{\substack{\varphi^k([w:1])=[w:1] \\ w\neq\beta}} \log|w-\beta|_v \tag{5.19}$$

for any $\beta \in \overline{K}$ (this follows from the fact that the multiplicity of each β_i as a k-periodic point is bounded for all k by Proposition 5.5).

We have $\varphi^k([w:1]) = [w:1]$ if and only if $P_k(w,1) - wQ_k(w,1) = 0$. Thus,

$$P_k(t,1) - tQ_k(t,1) = \gamma_k \prod_{\varphi^k([w:1])=[w:1]} (t-w),$$

for some $\gamma_k \in \overline{K}$. We write

$$P_k(t,1) - tQ_k(t,1) = (t-\beta)^{n_k} H_k(t,1)$$

for a polynomial H_k such that $H_k(\beta,1) \neq 0$. We have

$$H_k(t,1) = \gamma_k \prod_{\substack{\varphi^k([w:1])=[w:1] \\ w\neq\beta}} (t-w).$$

Then, plugging β in for t, taking logs of absolute values, and applying Proposition 4.5 gives

$$\lim_{k\to\infty} \frac{1}{d^k} \sum_{\substack{\varphi^k([w:1])=[w:1] \\ w\neq\beta}} \log|\beta-w|_v + \frac{\log|\gamma_k|_v}{d^k}$$
$$= \lim_{k\to\infty} \frac{\log\max(|P_k(\beta,1)|_v, |Q_k(\beta,1)|_v)}{d^k}. \tag{5.20}$$

Writing

$$T_1 P_k(T_0,T_1) - T_0 Q_k(T_0,T_1) = T_1^{n_k} W_k(T_0,T_1)$$

for a polynomial W_k such that $W_k(1,0) \neq 0$, we see that $\gamma_k = W_k(1,0)$. By Proposition 4.5, we have

$$\lim_{k\to\infty} \frac{\log|\gamma_k|_v}{d^k} = \lim_{k\to\infty} \frac{\log\max(|P_k(1,0)|_v, |Q_k(1,0)|_v)}{d^k}.$$

Combining this equality with (5.20) and Proposition 3 yields (5.19). □

We are now ready to prove the results regarding the computation of the canonical height $h_\varphi(\beta)$. First, we'll need a lemma. Note that the lemma does not follow directly from the work of Call and Goldstine [CG97], since they only prove that in a fixed number field, the local canonical heights sum to the global canonical height. What is required here is slightly different.

Lemma 5.8. *Let* $\beta = [a : b]$ *in* $\mathbb{P}^1(\overline{K})$. *Let* $[a_1 : b_1], \ldots, [a_n : b_n]$ *be the conjugates of* $[a : b]$ *under the action of* $Gal(\overline{K}/K)$. *Then*

$$[K(\beta) : K](\deg K) h_\varphi([a : b]) = \sum_{\text{places } v \text{ of } K} \lim_{k\to\infty} \sum_{i=1}^{n} \frac{\log \max(|P_k(a_i, b_i)|_v, |Q_k(a_i, b_i)|_v)}{d^k}. \tag{5.21}$$

Proof. For all but finitely many v, we have $|a_i|_v = |b_i|_v = 1$. Furthermore, for all but finitely many v, we have

$$\log \max(|P_k(s,t)|_v, |Q_k(s,t)|_v) = 0 \tag{5.22}$$

for all k whenever $|s|_v = |t|_v = 1$. This is true, for example, at all nonarchimedean v of good reduction for φ in the sense of [PST04]. Indeed, when v is a finite place, (5.22) will hold for all $|s|_v = |t|_v = 1$ unless either $|\mathrm{Res}(P(T_0, 1), Q(T_0, 1))|_v$ or $|\mathrm{Res}(P(1, T_1), Q(1, T_1))|_v$ is less than 1, where Res is the usual resultant of two polynomials (see [BK86, p. 279, Proposition 4]). Thus, we can interchange the limit and the sum on the right-hand side of (5.21) so that

$$\lim_{k\to\infty} \sum_{\text{places } v \text{ of } K} \sum_{i=1}^{n} \frac{\log \max(|P_k(a_i, b_i)|_v, |Q_k(a_i, b_i)|_v)}{d^k} = \sum_{\text{places } v \text{ of } K} \lim_{k\to\infty} \sum_{i=1}^{n} \frac{\log \max(|P_k(a_i, b_i)|_v, |Q_k(a_i, b_i)|_v)}{d^k}. \tag{5.23}$$

Now let L be the field $K(\beta)$ and let w be a place of L that extends the place v of K; we write $w \mid v$. The field L has n embeddings $i : L \hookrightarrow \mathbb{C}_v$; for exactly $[L_w : K_v]$ of these embeddings, we have $|i(x)|_v = |x|_w$ for all $x \in L$. This yields $[L_w : K_v]$ conjugates $[a' : b']$ of $[a : b]$ such that $|a|_w = |a'|_v$ and $|b|_w = |b'|_w$. Hence, we see that

$$\sum_{i=1}^{n} \log \max(|P_k(a_i, b_i)|_v, |Q_k(a_i, b_i)|_v) = \sum_{w|v} [L_w : K_v] \log \max(|P_k(a, b)|_w, |Q_k(a, b)|_v).$$

Thus, we have

$$\sum_{\text{places } v \text{ of } K} \sum_{i=1}^{n} \log \max(|P_k(a_i, b_i)|_v, |Q_k(a_i, b_i)|_v)$$
$$= [K(\beta) : K](\deg K) h(\varphi^k([a : b])),$$

by (1.0.6). It follows from (1.0.7) and (4.8.3) that we therefore have

$$\begin{aligned} &\sum_{\text{places } v \text{ of } K} \lim_{k\to\infty} \sum_{i=1}^{n} \frac{\log \max(|P_k(a_i, b_i)|_v, |Q_k(a_i, b_i)|_v)}{d^k} \\ &= [K(\beta) : K](\deg K) \lim_{k\to\infty} \frac{h(\varphi^k([a : b]))}{d^k} \\ &= [K(\beta) : K](\deg K) h_\varphi([a : b]). \end{aligned}$$

□

Theorem 5.9. *Let α be any point in $\mathbb{P}^1(\overline{K})$ that is not an exceptional point of φ. Then, for any $\beta \in \overline{K}$ and any nonzero irreducible $F \in K[t]$ such that $F(\beta) = 0$, we have*

$$\begin{aligned} &(\deg K)(\deg F)(h_\varphi(\beta) - h_\varphi(\infty)) \\ &= \sum_{\text{places } v \text{ of } K} \lim_{k\to\infty} \frac{1}{d^k} \sum_{\substack{\varphi^k([w:1])=\alpha \\ F(w)\neq 0}} \log |F(w)|_v, \end{aligned}$$

where the $[w : 1]$ for which $\varphi^k([w : 1]) = \alpha$ are counted with multiplicity.

Proof. Write $F(t) = \gamma \prod_{i=1}^{n}(t - \beta_i)$ where $\gamma \in K$ and the β_i are the conjugates of β under the action of $\mathrm{Gal}(\overline{K}/K)$. By the product formula, we have $\sum_{\text{places } v \text{ of } K} \log |\gamma|_v = 0$. Thus, using Theorem 4.6 and Proposition 2.1, we see that

$$\begin{aligned} &\sum_{\text{places } v \text{ of } K} \lim_{k\to\infty} \frac{1}{d^k} \sum_{\substack{\varphi^k([w:1])=\alpha \\ F(w)\neq 0}} \log |F(w)|_v \\ &= \sum_{\text{places } v \text{ of } K} \lim_{k\to\infty} \frac{1}{d^k} \sum_{\substack{\varphi^k([w:1])=\alpha \\ F(w)\neq 0}} \log \left| \prod_{i=1}^{n} (w - \beta_i) \right|_v \\ &= \sum_{i=1}^{n} \lim_{k\to\infty} \frac{\log \max(|P_k(\beta_i, 1)|_v, |Q_k(\beta_i, 1)|_v)}{d^k} \\ &\quad -(\deg F) \lim_{k\to\infty} \frac{\log \max(|P_k(1, 0)|_v, |Q_k(1, 0)|_v)}{d^k}. \end{aligned} \tag{5.24}$$

By Lemma 4.8, the quantity on the last two lines is equal to

$$(\deg F)(\deg K)(h_\varphi(\beta) - h_\varphi(\infty)),$$

as desired. □

Theorem 5.10. *For any $\beta \in \overline{K}$ and any nonzero irreducible $F \in K[t]$ such that $F(\beta) = 0$, we have*

$$(\deg K)(\deg F)(h_\varphi(\beta) - h_\varphi(\infty))$$
$$= \sum_{\text{places } v \text{ of } K} \lim_{k\to\infty} \frac{1}{d^k} \sum_{\substack{\varphi^k([w:1])=[w:1] \\ F(w)\neq 0}} \log|F(w)|_v,$$

where the $[w:1]$ for which $\varphi^k([w:1]) = w$ are counted with multiplicity.

Proof. The proof is the same as the proof of Theorem 4.9, using Theorem 5.7 in place of Theorem 5.6. □

6 A counterexample

The main theorems of this paper are *not* true when we work over the complex numbers $\mathbb{C}$ rather than $\overline{K}$. Let $K = \mathbb{Q}$ and let $\varphi([x : y]) = [x^2 : y^2]$ be the usual squaring map. Let v be the archimedean place of $\mathbb{Q}$, so that $\mathbb{C}_v$ is just the usual complex numbers $\mathbb{C}$. We define the function ψ on the positive integers recursively by $\psi(1) = 2$ and $\psi(n) = 2^{(n\psi(n-1))}$. Let $\alpha = \sum_{n=1}^{\infty} 1/\psi(n)$ and let $\beta = e^{2\pi i\alpha}$. Note that for any t, we have $|e^{2\pi i t} - 1| \le \pi(t - [t])$ (where $[t]$ is the greatest integer less than or equal to t). Letting $\ell_n = \log_2 \psi(n)$, we then have

$$\frac{1}{2^{\ell_n}} \sum_{w^{2^{\ell_n}}=1} \log|w-\beta|_v = \frac{\log|\beta^{\psi(n)} - 1|}{\psi(n)}$$

$$\le \frac{1}{\psi(n)} \log(\pi\,(\psi(n)\alpha - [\psi(n)\alpha]))$$

$$\le \frac{1}{\psi(n)} \log\left(\pi \frac{\psi(n)}{\psi(n+1)} \sum_{j=0}^{\infty} \frac{1}{2^{j\psi(n+1)}}\right)$$

$$\le \log\pi + 1 - n\log 2 + \log 2.$$

Thus, $\frac{1}{2^{\ell_n}} \sum_{w^{2^{\ell_n}}=1} \log|\beta - w|_v$ goes to $-\infty$ as $n \to \infty$, so

$$\lim_{k\to\infty} \frac{1}{2^k} \sum_{w^{2^k}=1} \log|w-\beta|_v$$

does not exist.

7 Applications and further questions

7.1 Lyapunov exponents

The Lyapunov exponent $L(\varphi)$ of a rational map $\varphi : \mathbb{P}^1_{\mathbb{C}} \longrightarrow \mathbb{P}^1_{\mathbb{C}}$ (see [Mañ88]) can be defined as follows. Choosing coordinates $[T_0 : T_1]$ for $\mathbb{P}^1_{\mathbb{C}}$, letting $t = T_0/T$, and writing $\varphi(t) = P(t)/Q(t)$ for polynomials P and Q, we define

$$L(\varphi) = \int_{\mathbb{P}^1(\mathbb{C})} \log|\varphi'(t)| d\mu_\varphi,$$

where μ_φ is the unique measure of maximal entropy measure for φ on $\mathbb{P}^1$; this measure of maximal entropy is the same as the Brolin–Lyubich measure discussed in Section 3 (see [Mañ83]).

The Lyapunov exponent can be computed via equidistribution on certain subsequences of inverse images of nonexceptional points in $\mathbb{P}^1(\mathbb{C})$ (see [DeM03], [Mañ88]). That is, given a nonexceptional point α in $\mathbb{P}^1(\mathbb{C})$, there is an infinite strictly increasing sequence of integers $(m_i)_{i=1}^{\infty}$ such that

$$L(\varphi) = \lim_{i\to\infty} \frac{1}{(\deg\varphi)^{m_i}} \sum_{\substack{\varphi^{m_i}(\beta)=\alpha \\ \varphi'(\beta)\neq 0 \\ \beta\neq\infty}} \log|\varphi'(\beta)|.$$

It is not known, however, whether $L(\varphi)$ can be computed by taking the limit of the average φ' on the periodic points of φ.

When φ is defined over a number field K, however, we obtain the following result as a corollary of 5.17 Theorem.

Corollary 7.1. *Let K be a number field and let $\varphi : \mathbb{P}^1_{\mathbb{C}} \longrightarrow \mathbb{P}^1_{\mathbb{C}}$ be a nonconstant rational map that is defined via base extension from a map $\varphi : \mathbb{P}^1_K \longrightarrow \mathbb{P}^1_K$. Let φ' be defined as above. Then*

$$L(\varphi) = \lim_{k\to\infty} \frac{1}{(\deg\varphi)^k} \sum_{\substack{\varphi^k(\xi)=\xi \\ \varphi'(\xi)\neq 0 \\ \xi\neq\infty}} \log|\varphi'(\xi)|.$$

Proof. We may write φ' as a quotient of polynomials $A(t)/B(t)$ with coefficients in K. This yields $\log|\varphi'(t)| = \log|A(t)| - \log|B(t)|$. The corollary then follows immediately from Theorem 5.7. □

This corollary says that if φ is a rational function defined over a number field, then the Lyapunov exponent of φ is completely determined by the derivative of φ at the periodic points of φ. This means that the derivative of φ at the periodic points of φ also determines the Hausdorff dimension of the Julia set (see [FLM83]).

7.2 Computing with points of small height

The results in [Bil97], [Aut01], [BR06], [FRL04], [FRL06], and [CL06] all apply not only to the periodic points and backwards iterates of a point that we treat in this paper but to all points of small height in the algebraic closure of a number field K. For example, one of the main theorems in [BR06], [FRL04], [FRL06], and [CL06] states that for any continuous function g on $\mathbb{P}^1(\mathbb{C}_v)$ and any infinite nonrepeating sequence of points (α_n) in $\mathbb{P}^1(\overline{K})$ such that $\lim_{n\to\infty} h_\varphi(\alpha_n) = 0$, one has

$$\lim_{n\to\infty} \frac{1}{|\mathrm{Gal}(\alpha_n)|} \sum_{\sigma\in\mathrm{Gal}(\alpha_n)} g(\alpha_n^\sigma) = \int_{\mathbb{P}^1(\mathbb{C}_v)} g\, d\mu_{v,\varphi}, \tag{7.1}$$

where $\mathrm{Gal}(\alpha_n)$ is the Galois group of the Galois closure of $K(\alpha_n)$ over K.

Baker, Ih, and Rumely [BIR08] and Autissier ([Aut06]) have produced counterexamples that show that (6.1.1) does not always hold when the function g is replaced with $\log|F|_v$ for F a polynomial. All of these examples involve infinite nonrepeating sequences of points $(\alpha_n) \in \bar{\mathbb{Q}}$ such that $\lim_{n\to\infty} h(\alpha_n) = 0$ and

$$\lim_{n\to\infty} \frac{1}{|\mathrm{Gal}(\alpha_n)|} \sum_{\sigma\in\mathrm{Gal}(\alpha_n)} \log|\alpha_n^\sigma - 2| \neq \int_0^1 \log|e^{2\pi i\theta} - 2|\, d\theta.$$

The points (α_n) are not preperiodic in any of these examples. Thus, it may be possible to prove that the main results of this paper continue to hold when we work with any nonrepeating sequence of Galois orbits of preperiodic points. This would imply the following conjectured generalization of Siegel's theorem for integral points.

Conjection 7.2 (Ih). *For any nonpreperiodic point $\beta \in \mathbb{P}^1_{\mathfrak{o}_K}(\overline{K})$, there are at most finitely many preperiodic points of φ in $\mathbb{P}^1_{\mathfrak{o}_K}(\overline{K})$ that are integral relative to β. (Here, $\mathfrak{o}_K$ is the ring of integers of K and α is said to be integral relative to β if the Zariski closure of α does not meet the Zariski closure of β in $\mathbb{P}^1_{\mathfrak{o}_K}$.)*

Baker, Ih, and Rumely have proven that this is true when φ is a Lattès map or the usual squaring map $x \mapsto x^2$. Using Theorem 5.10 and arguing as in [BIR08] (or as in [Sil93], which presents a related result), it is possible to derive the following weak version of Ih's conjecture in general.

Proposition 7.3. *For any nonpreperiodic point $\beta \in \mathbb{P}^1(\overline{K})$, there are at most finitely many n such that all $\alpha \in \mathbb{P}^1(\overline{K})$ of period n are β-integral.*

References

[Aut01] P. Autissier, *Points entiers sur les surfaces arithmétiques*, J. Reine. Angew. Math **531** (2001), 201–235.

[Aut06] P. Autissier, *Sur une question d'équirépartition de nombres algébriques*, C. R. Math. Acad. Sci. Paris **342** (2006), no. 9, 639–641.

[Bak75] A. Baker, *Transcendental Number Theory*, Cambridge University Press, Cambridge, 1975.

[BIR08] M. Baker, S.-I. Ih, and R. Rumely, *A finiteness property of torsion points*, Algebra Number Theory **2** (2008), no. 2, 217–248.

[BR06] M. Baker and R. Rumely, *Equidistribution of small points, rational dynamics, and potential theory*, Ann. Inst. Fourier (Grenoble) **56** (2006), 625–688.

[BR10] ———, *Potential theory and dynamics on the Berkovich projective line*, Mathemetical Surveys and Monographs, 159, AMS, 2010.

[Bea91] A. F. Beardon, *Iteration of rational functions*, Springer-Verlag, New York, 1991.

[Ber90] V. G. Berkovich, *Spectral theory and analytic geometry over nonarchimedean fields*, AMS Mathematical Surveys and Monographs, American Mathematical Society, Providence, 1990.

[Bil97] Y. Bilu, *Limit distribution of small points on algebraic tori*, Duke Math. J. **89** (1997), 465–476.

[BK86] E. Brieskorn and H. Knörrer, *Plane algebraic curves*, Birkhäuser, Basel, 1986, Translated by J. Stillwell.

[Bro65] H. Brolin, *Invariant sets under iteration of rational functions*, Ark. Mat. **6** (1965), 103–144.

[CG97] G. S. Call and S. Goldstine, *Canonical heights on projective space*, J. Number Theory **63** (1997), 211–243.

[CL06] A. Chambert-Loir, *Mesures et équidistribution sur les espaces de Berkovich*, J. Reine Angew. Math. **595**, (2006), 215–235.

[CLT04] A. Chambert-Loir and A. Thuillier, *Mesures de Mahler et équidistribution logarithmique*, Ann. Inst. Fourier (Grenoble) **59** (2009), no. 3, 977–1014.

[CS93] G. S. Call and J. Silverman, *Canonical heights on varieties with morphisms*, Compositio Math. **89** (1993), 163–205.

[Dav95] S. David, *Minorations de formes linéaires de logarithmes elliptiques*, Mém. Soc. Math. France **62** (1995), 143 pp.

[DeM03] L. DeMarco, *Dynamics of rational maps: Lyapunov exponents, bifurcations, and capacity*, Math. Ann. **203** (2003), 43–73.

[EF96] G. Everest and Bríd Ní Fhlathúin, *The elliptic Mahler measure*, Math. Proc. Cambridge Philos. Soc. **120** (1996), 13–25.

[EW99] G. Everest and T. Ward, *Heights of Polynomials and Entropy in Algebraic Dynamics*, Springer-Verlag, New York, 1999.

[FLM83] A. Freire, A. Lopes, and R. Mañé, *An invariant measure for rational maps*, Bol. Soc. Brasil. Mat. **14** (1983), no. 1, 45–62.

[FRL04] C. Favre and J. Rivera-Letelier, *Théorème d'équidistribution de Brolin en dynamique p-adique*, C. R. Math. Acad. Sci. Paris **339** (2004), no. 4, 271–276.

[FRL06] ———, *Équidistribution quantitative des points de petite hauteur sur la droite projective*, Math. Ann. **335** (2006), no. 2, 311–361.

[Lan83] S. Lang, *Fundamentals of Diophantine geometry*, Springer-Verlag, New York, 1983.

[Lyu83] M. Lyubich, *Entropy properties of rational endomorphisms of the Riemann sphere*, Ergodic Theory Dynam. Systems **3** (1983), 351–385.

[Mah60] K. Mahler, *An application of Jensen's formula to polynomials*, Mathematika **7** (1960), 98–100.

[Mañ83] R. Mañé, *On the uniqueness of the maximizing measure for rational maps*, Bol. Soc. Brasil. Mat. **14** (1983), no. 1, 27–43.

[Mañ88] R. Mañé, *The Hausdorff dimension of invariant probabilities of rational maps*, Dynamical Systems, Valparaiso 1986 (R. Bamon, R. Labarca, and J. Palis, eds.), Springer-Verlag, 1988, pp. 86–117.

[Mil99] J. Milnor, *Dynamics in one complex variable*, Vieweg, Braunschweig, 1999.

[MS95] P. Morton and J. H. Silverman, *Periodic points, multiplicities, and dynamical units*, J. Reine Angew. Math. **461** (1995), 81–122.

[Piñ05] J. Piñeiro, *Mahler formula for morphisms on* $\mathbb{P}^n$, Ph.D. thesis, City University of New York, 2005.

[PST04] J. Piñeiro, L. Szpiro, and T. Tucker, *Mahler measure for dynamical systems on* $\mathbb{P}^1$ *and intersection theory on a singular arithmetic surface*, Geometric methods in algebra and number theory (F. Bogomolov and Y. Tschinkel, eds.), Progress in Mathematics 235, Birkhäuser, 2004, pp. 219–250.

[Rot55] K. F. Roth, *Rational approximations to algebraic numbers*, Mathematika **2** (1955), 1–20, corrigendum, ibid. **2** (1955), 168.

[Sch74] A. Schinzel, *Primitive divisors of the expression* $a^n - b^n$ *in algebraic number fields*, J. Reine Angew. **268/269** (1974), 27–33, Collection of articles dedicated to Helmut Hasse on his seventy-fifth birthday, II.

[Sie29] C. L. Siegel, *Über einige Anwendungen diophantische Approximationen*, Abh. Preuss. Akad. Wiss. Phys. Math. Kl. (1929), 41–69.

[Sil93] J. H. Silverman, *Integer points, Diophantine approximation, and iteration of rational maps*, Duke Math. J. **71** (1993), no. 3, 793–829.

[SUZ97] L. Szpiro, E. Ullmo, and S. Zhang, *Equirépartition des petits points*, Invent. Math. **127** (1997), 337–347.

[Thu06] A. Thuillier, *Théorie de potential sur les courbes en géométrie non archimédienne, applications à la théorie d'arakelov*, Ph.D. thesis, Université de Rennes 1, 2006.

Représentations p-adiques de torsion admissibles

Marie-France Vignéras

A Serge Lang, un mathématicien et un homme de qualité, intègre et passionné.

Abstract We give some properties of admissible smooth representations of a reductive p-adic group G over a field of characteristic p, deduced from Pontryagin duality, Ollivier's equivalence of categories when $G = PGL(2, Q_p)$, and finiteness properties of the Hecke ring of a pro-p-Iwahori-Hecke algebra.

Key words Representations • p-adic reductive groups • pro-p-Iwahori-Hecke algebra • Pontryagin duality

Mathematics Subject Classification (2010): 20C08, 11F 33, 11E95, 33D80

1 Introduction

La théorie des représentations lisses des groupes réductifs p-adiques sur un corps de caractéristique p en est à ses débuts, et un expert des représentations sur un corps de caractéristique différente de p est désemparé sans mesure de Haar. Nous avons réuni ici quelques propriétés de l'admissibilité, qui sont des applications assez faciles de trois théories: la dualité de Pontryagin, bien connue des arithméticiens p-adiques, l'équivalence de catégories de Ollivier pour $GL(2, \mathbf{Q}_p)/p^{\mathbf{Z}}$, et les propriétés de finitude des $\mathbf{Z}$-algèbres de Hecke d'un pro-p-sous-groupe d'Iwahori.

M.-F. Vignéras (✉)
Institut de Mathématiques de Jussieu, Université de Paris 7-Denis Diderot

D. Goldfeld et al. (eds.), *Number Theory, Analysis and Geometry: In Memory of Serge Lang*, DOI 10.1007/978-1-4614-1260-1_27,

Soit L un corps commutatif localement compact de caractéristique 0, d'anneau des entiers O_L, d'uniformisante p_L, de corps résiduel k_L de caractéristique p. Posons $A := L, O_L$ ou k_L.

Soit **G** un groupe analytique p-adique, ou ce qui est équivalent: **G** contient un pro-p-sous-groupe ouvert isomorphe à un sous-groupe fermé de $GL_d(\mathbf{Z}_p)$ for some $d \geq 1$ [4] 8.33 page 201 et Interlude A (o),(n) page 97.

Une A-représentation V de **G** est un A-module topologique sur lequel **G** opère continûment. Selon que $A = L, O_L, k_L$, on dit que V est p-adique, p-adique entière, modulo p.

Pour toute partie H de **G**, on note V^H le A-sous-module des $v \in V$ fixes par chaque élément de H. Lorsque $V = \cup_H V^H$ où H parcourt les sous-groupes ouverts compacts de **G**, on dit que V est lisse.

Une A-représentation lisse V telle que V^H est un A-module de type fini pour tout sous-groupe ouvert compact H de **G**, est appelée admissible.

Si V est un O_L-module discret de torsion, l'action continue de **G** sur V est lisse; on dit que V est une représentation p-adique de torsion de **G**.

Soit

$$0 \to V_1 \to V_2 \to V_3 \to 0$$

une suite exacte de A-représentations lisses de **G**. Si V_2 est admissible, alors V_1 est admissible, et si V_1 et V_3 sont admissibles, alors V_2 est admissible, car l'anneau A est noetherien.

Si $A = L$, l'existence d'une mesure de Haar sur **G** à valeurs dans L implique l'exactitude du foncteur $V \to V^H$, et la stabilité de l'admissibilité par quotient. Pour $A = O_L, k_L$, le foncteur $V \to V^H$ n'est pas exact à droite et l'on se demande si la stabilité par quotient de l'admissibilité reste vraie.

Nous allons montrer que la réponse est oui si V est une représentation p-adique entière annulée par une puissance de p, en particulier si V est une représentation modulo p, en utilisant la dualité de Pontryagin et le lemme de Nakayama topologique [1].

Soit U un pro-p-sous-groupe ouvert fixé de **G** et soit V une représentation p-adique entière de **G** telle que $p^k V = 0$ pour un entier $k \geq 1$. Le dual de Pontryagin $V^\vee$ est toujours annulé par p^k, et par restriction à U c'est un module profini sur la O_L-algèbre d'Iwasawa de U. La dualité de Pontryagin échange les U-invariants V^U et les U-coinvariants $(V^\vee)_U$ et respecte la propriété d'être fini. En appliquant le lemme de Nakayama topologique, on obtient le théorème 4 qui implique les deux théorèmes suivants.

Théorème 1. *V est admissible si et seulement si V^U est fini.*

On comparera avec le résultat de Paskunas [6]: toute représentation lisse V sur un corps k de caractérisque p d'un groupe topologique G contenant un pro-p-sous-groupe ouvert U telle que $\dim_k V^U < \infty$ est admissible. La démonstration basée sur les enveloppes injectives, n'utilise ni la dualité de Pontryagin ni le lemme de Nakayama.

Théorème 2. *Si V est admissible, tout sous-quotient de V est admissible.*

Un groupe réductif p-adique est le groupe des points rationnels d'un groupe réductif connexe sur une extension finie de $\mathbf{Q}_p$; c'est un groupe analytique p-adique. Le groupe des points rationnels d'un groupe réductif connexe sur corps local de caractéristique p n'est pas un groupe analytique p-adique.

Corollaire 1. *Si V est une représentation admissible d'un sous-groupe de Levi d'un sous-groupe parabolique $\mathbf{P}$ d'un groupe réductif p-adique $\mathbf{G}$, alors tout sous-quotient de l'induite parabolique $\mathrm{ind}_{\mathbf{P}}^{\mathbf{G}} V$ est admissible.*

En effet [9] I.V.6, l'induction parabolique respecte l'admissibilité puisque $\mathbf{G}/\mathbf{P}$ est compact par la décomposition d'Iwasawa, la propriété d'être entière et annulée par p^k.

La seule représentation irréductible de U sur k_L est isomorphe à la représentation triviale de dimension 1, et le foncteur d'induction compacte $\mathrm{ind}_U^{\mathbf{G}}$ est exact. Cette remarque et le théorème 2 impliquent

Corollaire 2. *Une représentation de type fini V de $\mathbf{G}$ sur k_L admet une filtration G-équivariante finie de quotients cycliques engendrés par un vecteur U-invariant. Si V est admissible, les quotients sont aussi admissibles.*

Dans le cas très particulier où $\mathbf{G} = GL(2, \mathbf{Q}_p)/p^{\mathbf{Z}}$, l'on sait beaucoup plus. La raison est la suivante:

Soit I_1 le sous-groupe des matrices de $GL(2, \mathbf{Z}_p)$ congrues au sous-groupe strictement triangulaire supérieur de $GL(2, \mathbf{F}_p)$ modulo p, appelé un p-sous-groupe d'Iwahori de $GL(2, \mathbf{Q}_p)$, et qui tient lieu de p-sous-groupe de Sylow. On l'identifie à son image (isomorphe) dans $\mathbf{G}$. Sa $\mathbf{Z}$-algèbre de Hecke $H_{\mathbf{Z}}(GL(2, \mathbf{Z}_p), I_1)$ est un module de type fini sur un anneau commutatif de type fini ([10] th.3), c'est donc une algèbre noetherienne. Soit $\overline{\mathbf{F}}_p$ une clôture algébrique de k_L. Par le théorème d'équivalence de Ollivier [7], la catégorie des modules à droite de $H_{\overline{\mathbf{F}}_p}(\mathbf{G}, I_1)$ est équivalente à celle des représentations lisses de $\mathbf{G}$ sur $\overline{\mathbf{F}}_p$ qui sont engendrées par leur vecteurs invariants par I_1 [7].

En utilisant la filtration (corollaire 2) nous obtenons le résultat (bien connu pour les représentations complexes ou modulo ℓ d'un groupe réductif p-adique):

Théorème 3. *La catégorie des $\overline{\mathbf{F}}_p$-représentations de $GL(2, \mathbf{Q}_p)/p^{\mathbf{Z}}$ est noetherienne, et une $\overline{\mathbf{F}}_p$-représentation de $GL(2, \mathbf{Q}_p)/p^{\mathbf{Z}}$ est admissible de type fini si et seulement si elle est de longueur finie.*

On peut remplacer les données $\overline{\mathbf{F}}_p$ ou $GL(2, \mathbf{Q}_p)/p^{\mathbf{Z}}$ du théorème par k_L ou $GL(2, \mathbf{Q}_p)$, si le théorème d'équivalence de Ollivier est valable avec ces nouvelles données.

On ne sait pas si une représentation irréductible V d'un groupe réductif p-adique $\mathbf{G}$ sur un corps algébriquement clos C de caractéristique p, est admissible ou a un caractère central (c'est vrai si la caractéristique de C est différente de p). On sait que V irréductible a un caractère central si C est non dénombrable (preuve classique), ou si V est admissible (V^{I_1} de dimension finie contient une droite stable par le centre), ou si $\mathbf{G}$ est déployé et le $H_{\overline{\mathbf{F}}_p}(\mathbf{G}, I_1)$-module V^{I_1} contient un module simple M (car M a un caractère central ([11] 5.3 plus [10])).

2 Dualité de Pontryagin

On note $\mathrm{Mod}_{\mathrm{tor}}$ la catégorie abélienne des O_L-modules discrets de torsion et des applications O_L-linéaires, et $\mathrm{Mod}_{\mathrm{prof}}$ la catégorie abélienne des O_L-modules profinis et des applications O_L-linéaires continues.

Tout $V \in \mathrm{Mod}_{\mathrm{tor}}$ est la limite inductive de ses O_L-sous-modules finis W, muni de la topologie de la limite inductive [2] AII.93; les applications de transition étant les inclusions. Tout $M \in \mathrm{Mod}_{\mathrm{prof}}$ est la limite projective de ses quotients M/N par les O_L-sous-modules ouverts N, muni de la topologie de la limite projective; les applications de transition sont les surjections.

Une application linéaire continue dans $\mathrm{Mod}_{\mathrm{prof}}$ est fermée car un O_L-module profini est compact et séparé [3] I.63.

La dualité de Pontryagin

$$E^\vee := \mathrm{Hom}_{O_L}(E, L/O_L)$$

pour $E = V$ ou M, satisfait $(E^\vee)^\vee = E$, elle échange limites inductives et projectives (topologies comprises), L/O_L et O_L, et induit une équivalence contravariante entre $\mathrm{Mod}_{\mathrm{tor}}$ et $\mathrm{Mod}_{\mathrm{prof}}$.

On considère les sous-catégories abéliennes $\mathrm{Mod}_{\mathrm{tor}}(\mathbf{G}) \subset \mathrm{Mod}_{\mathrm{tor}}$ et $\mathrm{Mod}_{\mathrm{prof}}(\mathbf{G}) \subset \mathrm{Mod}_{\mathrm{prof}}$ des O_L-modules discrets de torsion ou profinis munis d'une application continue de $\mathbf{G}$. Le groupe $\mathbf{G}$ agit continuement sur $E = V$ ou M si l'application

$$(g, x) \to gx : \mathbf{G} \times E \to E$$

est continue comme fonction des deux variables. Alors $V \in \mathrm{Mod}_{\mathrm{tor}}(\mathbf{G})$ est la limite inductive de ses O_L-sous-modules finis W stables par $\mathbf{G}$, et $M \in \mathrm{Mod}_{\mathrm{prof}}(\mathbf{G})$ est la limite projective de ses quotients par les O_L-sous-modules ouverts N stables par $\mathbf{G}$. L'action contragrédiente du groupe $\mathbf{G}$ sur le dual de Pontryagin $E^\vee$ est définie par

$$(gx^*, gx) = (x^*, x) \quad (g \in \mathbf{G}, x \in E, x^* \in E^*).$$

La dualité de Pontryagin induit une équivalence contravariante entre $\mathrm{Mod}_{\mathrm{tor}}(\mathbf{G})$ et $\mathrm{Mod}_{\mathrm{prof}}(\mathbf{G})$.

Soit H un groupe profini topologiquement de type fini (par exemple un sous-groupe fermé de $\mathbf{G}$ [4] prop. 3.11, page 51). Soit $E = V \in \mathrm{Mod}_{\mathrm{tor}}(H)$ ou $E = M \in \mathrm{Mod}_{\mathrm{prof}}(H)$.

Le submodule $E^H \subset E$ des H-invariants de E est fermé car l'action de H est continue. Le module E_H des H-coinvariants de E est $E/E(H)$ où $E(H)$ est le O_L-sous-module de E engendré par

$$gx - x \quad (g \in H, x \in E).$$

Proposition 1. *(1) $E(H)$ est fermé dans E.*
(2) La dualité de Pontryagin échange invariants et coinvariants

$$(E^H)^\vee = (E^\vee)_H \,, \quad \text{et} \quad (E/E^H)^\vee = E^\vee(H)\,.$$

Preuve. (1) Tout sous-module de $E = V$ est fermé (la topologie est discrète) et le problème ne se pose que pour $E = M$. Soit X_H un ensemble fini engendrant un sous-groupe dense H' de H. Le O_L-sous-module M' de M engendré par $(h-1)M$ pour $h \in X_H$ est fermé car M est compact. On a $M(H') = M'$ car l'égalité

$$h_1 \dots h_r m - m = h_1 h_2 \dots h_r m - h_2 \dots h_r m + h_2 \dots h_r m - m$$

pour $h_1, \dots, h_r \in X_H$ et $m \in M$, montre par induction sur r que $h_1 \dots h_r m - m \in M'$. La continuité de l'action de H implique que $M(H') = M(H)$. Donc $M(H) = M'$ est fermé dans M.

(2) $f \in E^\vee$ s'annule sur $E(H)$ si et seulement si f est invariant par H,

$$(E^\vee)^H = (E_H)^\vee.$$

Remplaçons E par $E^\vee$, puis prenons le dual de Pontryagin et l'on obtient $(E^H)^\vee = (E^\vee)_H$. Ceci implique $(E/E^H)^\vee = E^\vee(H)$ car l'on a la suite exacte

$$1 \to (E/E^H)^\vee \to E^\vee \to (E^H)^\vee \to 0.$$

3 Le lemme de Nakayama

Module signifiera module à gauche. Si A est un anneau et N un A-module, on note $d_A(N) \in \mathbf{N} \cup \infty$ le nombre minimal de générateurs du A-module N.

Proposition 2. [1] §3 (Lemme de Nakayama)
Soient A un anneau topologique compact, N un A-module profini et I un idéal bilatère topologiquement nilpotent de A. Alors

$$d_A(N) = d_{A/I}(N/I).$$

Soit U un pro-p-sous-groupe de rang fini (par exemple un pro-p-sous-groupe fermé de $\mathbf{G}$). Un O_L-module profini muni d'une action continue de U s'identifie à un module profini sur la O_L-algèbre d'Iwasawa

$$\Lambda := O_L[[U]] = \varprojlim_{U'} O_L[U/U']$$

où U' parcourt les sous-groupes distingués d'indice fini de U. Comme U est de rang fini, un sous-groupe est ouvert si et seulement s'il est d'indice fini, l'anneau compact Λ est local, sa topologie est celle donnée par les puissances de son radical, et Λ est noetherien à gauche et à droite [8] §2.2.7. Tout idéal propre de Λ est topologiquement nilpotent. Le noyau I_U de l'homomorphisme d'augmentation $\Lambda \to O_L$ est l'idéal à gauche engendré par

$$u - 1 \quad (u \in X_U)$$

où X_U est un sous-ensemble fini de U engendrant topologiquement U. Le radical de Λ est l'idéal engendré par I_U et p_L [5] 2.1.8.

Proposition 3. *Soit* $N \in \mathrm{Mod}_{\mathrm{prof}}(U)$. *Pour tout sous-groupe distingué d'indice fini* $U' \subset U$, *on a*

$$d_\Lambda(N) = d_{O_L[U/U']}(N_{U'}).$$

En particulier $d_\Lambda(N) = d_{O_L}(N_U)$.

Preuve. Soit $I_{U/U'}$ le noyau du morphisme canonique surjectif $\Lambda \to O_L[U/U']$. Comme idéal à gauche, $I_{U/U'}$ est engendré par $u - 1$ pour u dans un sous-ensemble $X_{U'}$ fini de U' engendrant topologiquement U'. L'espace des U'-coinvariants de N est

$$N_{U'} = N/I_{U/U'}N.$$

On applique le lemme de Nakayama à $(\Lambda, N, I_{U/U'})$.

Corollaire 3. *Les propriétés suivantes sont équivalentes:*

i. N *est un* Λ*-module de type fini,*
ii. N_H *est un* O_L*-module de type fini pour un sous-groupe d'indice fini* H *de* U.
iii. N_H *est un* O_L*-module de type fini pour tout sous-groupe d'indice fini* H *de* U.

Preuve. Un sous-groupe d'indice fini H de U contient un sous-groupe distingué U' d'indice fini de U. La proposition 3 implique que N est un Λ-module de type fini si et seulement si $N_{U'}$ est un O_L-module de type fini. Il reste à montrer que N_H est un O_L-module de type fini si et seulement si $N_{U'}$ est un O_L-module de type fini. La O_L-algèbre d'Iwasawa de H est un module libre de type fini sur celle de U', engendrée par les images d'un système de représentants de H/U'. Donc

$$d_{O_L[[H]]}(N) \leq d_{O_L[[U']]}(N) \leq [H : U'] d_{O_L[[H]]}(N).$$

On applique le cas particulier de la proposition 3 à U' et à H et le corollaire est démontré.

Corollaire 4. *Soit* U' *un sous-groupe ouvert distingué de* U. *Alors*

$$d_{O_L}(N_U) \leq d_{O_L}(N_{U'}) \leq [U : U'] \, d_{O_L}(N_U).$$

4 Application à l'admissibilité

Soit V une O_L-représentation lisse de $\mathbf{G}$ annulée par p^k pour un entier $k \geq 1$. Soit U un pro-p-sous-groupe ouvert de $\mathbf{G}$.

Théorème 4. *(1) Les propriétés suivantes sont équivalentes:*

i. V *est admissible.*
ii. Le dual de Pontryagin $V^\vee$ *restreint à* H *est un* $O_L[[H]]$*-module de type fini, pour tout sous-groupe ouvert compact* H *de* $\mathbf{G}$.
iii. Le dual de Pontryagin $V^\vee$ *restreint à* U *est un* $O_L[[U]]$*-module de type fini.*
iv. V^U *est fini.*

(2) Si U' est un sous-groupe d'indice fini distingué dans U, alors

$$d_{O_L}(V^U) \leq d_{O_L}(V^{U'}) \leq [U : U']\, d_{O_L}(V^U).$$

Preuve. La dualité de Pontryagin respecte la propriété d'être fini (et non d'être un O_L-module de type fini), d'être annulé par p^k (et non d'être de torsion), et échange U-invariants et U-coinvariants (proposition 1). Un O_L-module de torsion est de type fini si et seulement s'il est fini. On en déduit que V est admissible si et seulement si $(V^\vee)_H$ est un O_L-module de type fini, pour tout sous-groupe ouvert compact H de $\mathbf{G}$. On applique alors les corollaires 3, 4.

Nous démontrons le théorème 2. Un quotient W de V est annulé par p^k. La dualité de Pontryagin est contravariante donc $W^\vee$ est un sous-module de $V^\vee$. L'anneau $O_L[[U]]$ est noetherien donc si $V^\vee$ est un $O_L[[U]]$-module de type fini, il en est de même du $O_L[[U]]$-sous-module $W^\vee$. Si V est admissible, W l'est aussi par le théorème 4.

5 Représentations de type fini

Soit U un pro-p-sous-groupe ouvert de $\mathbf{G}$ et soit V une représentation de type fini de $\mathbf{G}$ sur k_L. Nous allons montrer que V admet une filtration $\mathbf{G}$-équivariante finie de quotients de type fini engendrés par leur U-invariants.

Soient $v_1, \ldots, v_r$ un système fini de générateurs de V, et soit W la représentation de U engendrée par ces éléments dans V. Comme V est lisse, $\dim_k W$ est fini. La représentation V est quotient de la représentation induite compacte $\mathrm{ind}_U^{\mathbf{G}}(W)$. La seule représentation irréductible de U sur k_L est isomorphe à la représentation triviale de dimension 1, et le foncteur d'induction compacte $\mathrm{ind}_U^{\mathbf{G}}$ est exact. La représentation finie W de U sur k_L a une filtration U-equivariante de longueur $r = \dim_k W$,

$$0 \subset W_1 \subset \cdots \subset W_r = W \quad \text{avec} \quad \dim_k W_i/W_{i-1} = 1\,,$$

induisant une filtration $\mathbf{G}$-equivariante,

$$0 \subset \mathrm{ind}_H^{\mathbf{G}}(W_1) \subset \cdots \subset \mathrm{ind}_U^{\mathbf{G}}(W_r) = \mathrm{ind}_U^{\mathbf{G}}(W)$$

de quotients isomorphes

$$\mathrm{ind}_U^{\mathbf{G}}(W_i)/\,\mathrm{ind}_U^{\mathbf{G}}(W_{i-1}) \simeq \mathrm{ind}_U^{\mathbf{G}}(1_{k_L})$$

pour tout $1 \leq i \leq r$. L'image de cette filtration par l'application surjective canonique $\mathrm{ind}_U^{\mathbf{G}}(W) \to V$ est une filtration $\mathbf{G}$-equivariante de V,

$$0 \subset V_1 \subset \cdots \subset V_r = V$$

de quotients isomorphes à des quotients (éventuellement nuls) du "module universel" $\mathrm{ind}_U^{\mathbf{G}}(1_{k_L})$. Si de plus V est admissible, les quotients sont admissibles (théorème 2). Le corollaire 2 est démontré.

Nous remarquons que si V' est une sous-représentation de $\mathbf{G}$, la filtration induite par celle de V sur V',

$$0 \subset V'_1 \subset \cdots \subset V'_r = V' , \quad V'_i := V' \cap V_i ,$$

a ses quotients isomorphes à des sous-quotients du module universel $\mathrm{ind}_U^{\mathbf{G}}(1_{k_L})$. On déduit de la démonstration du corollaire 2 et de cette remarque:

Proposition 4. *Si tout sous-quotient de* $\mathrm{ind}_U^{\mathbf{G}}(1_{k_L})$ *est de type fini, alors la catégorie des représentations de* $\mathbf{G}$ *sur* k_L *est noetherienne.*

Si tout quotient admissible de $\mathrm{ind}_U^{\mathbf{G}}(1_{k_L})$ *est de longeur finie, alors toute représentation admissible de type fini de* $\mathbf{G}$ *sur* k_L *est de longeur finie.*

Ces propriétés sont vérifiées pour $\mathbf{G} = GL(2, \mathbf{Q}_p)/p^{\mathbf{Z}}$, par l'équi-valence de catégories [7] et la noetheriannité de $H_{k_L}(\mathbf{G}, I_1)$ [10]. Le théorème 3 est démontré.

References

1. Balister P.N. and Howson S. Note on Nakayama's lemma for compact Λ-modules. *Asian Journal of Mathematics* 1, (1997) 224–229.
2. Bourbaki Nicolas. *Algèbre* Ch. 1 à 3. Hermann (1970).
3. Bourbaki Nicolas. *Topologie générale* Ch. 1 à 4. Hermann (1971).
4. Dixon J.D. Du Sautoy M.P.F. Mann A. and Segal D. *Analytic pro-p-groups*. Cambridge studies in advanced mathematics 61. Second edition 2003.
5. Lazard Michel. *Groupes anlytiques p-adiques*. IHES (26) (1965) 5–219.
   Grundlheren der mathematischen Wissenschaften 323. Springer, 2000.
6. Paskunas Vytautas. Coefficient systems and supersingular representations of $GL_2(F)$. *Mémoires de la S.M.F.* 99 (2004).
7. Ollivier Rachel. Le foncteur des invariants sous l'action du pro-p-Iwahori de $GL_2(F)$. Jour. für die Reine and angewandt'e mathematik 635(2009), 149–185.
8. Venjakob Otmar. Characteristic Elements in Noncommutative Iwasawa Theory. *J. reine angew. Math.* 583 (2005).
9. Vignéras Marie-France. *Représentations ℓ-modulaires d'un groupe réductif p-adique avec $\ell \neq p$*. Progress in Math. 137. Birkhäuser, 1996.
10. Vignéras Marie-France. On a numerical Langlands correspondence modulo p with the pro-p-Iwahori Hecke ring. *Mathematische Annalen* 2004. Erratum volume 333, no. 3, du 28 octobre (2005) 699–701.
11. Vignéras Marie-France. Représentations irréductibles de $GL(2, F)$ modulo p. In *L-functions and Galois representations*, ed. Burns, Buzzard, Nekovar, LMS Lecture Notes 320 (2007).

Multiplier ideal sheaves, Nevanlinna theory, and Diophantine approximation

Paul Vojta

Dedicated to the memory of Serge Lang. No one could hope for a better mentor.

Abstract This paper states a conjecture for Nevanlinna theory or diophantine approximation, with a sheaf of ideals in place of the normal crossings divisor. This is done by using a correction term involving a multiplier ideal sheaf. This new conjecture trivially implies earlier conjectures in Nevanlinna theory or diophantine approximation, and in fact is equivalent to these conjectures. Although it does not provide anything new, it may be a more convenient formulation for some applications.

Key words Nevanlinna theory • multiplier ideal sheaf • Second Main Theorem

Mathematics Subject Classification (2010): 11J97 (primary); 14G25, 32H30, 14F18 (secondary)

Introduction

This paper discusses a formulation of the author's conjectures in diophantine approximation and Nevanlinna theory, in which the assumption that D be a normal crossings divisor is dropped, and the conjectured inequality is changed to

$$h_{K,k}(P) + m_S(\mathfrak{a}, P) - m_S(\mathscr{I}^-(\mathfrak{a}), P) \le \epsilon\, h_{A,k}(P) + d_k(P) + O(1)$$

P. Vojta (✉)
Department of Mathematics, University of California, 970 Evans Hall #3840, Berkeley, CA 94720-3840
e-mail: vojta@math.berkeley.edu

D. Goldfeld et al. (eds.), *Number Theory, Analysis and Geometry: In Memory of Serge Lang*, DOI 10.1007/978-1-4614-1260-1_28,

in the number field case, and a similar inequality is conjectured for Nevanlinna theory. Here $\mathfrak{a}$ is the ideal sheaf associated to D and $\mathscr{I}^{-}(\mathfrak{a})$ is a type of multiplier ideal sheaf. For other notations, see Sections 1 and 2 and Conjectures 4.1 and 4.2.

Dropping the condition on D is possible here because of the additional term $-m_S(\mathscr{I}^{-}(\mathfrak{a}), P)$ on the left-hand side of the inequality.

This may be a more convenient formulation for some applications. Also, this conjecture shows how multiplier ideal sheaves may have a role in Nevanlinna theory and diophantine approximation, and therefore may give more information on the structure of the situation.

Section 1 briefly describes multiplier ideal sheaves, and gives a variant definition specific to this situation. It also includes related definitions. Section 2 recalls the standard definitions in Nevanlinna theory, as well as their counterparts in number theory, that are needed in this paper. Section 3 describes proximity functions for sheaves of ideals, using work of Silverman and Yamanoi. Sections 4 and 5 form the heart of the paper, giving the conjectures and showing their equivalence to previous conjectures.

Throughout this paper, X is a smooth complete variety over $\mathbb{C}$ (in the case of Nevanlinna theory) or over a global field of characteristic zero (in the case of Diophantine approximation). For the purposes of this paper, a global field is a number field or a function field of dimension one.

Acknowledgements Supported by NSF grants DMS-0200892 and DMS-0500512.

1 Multiplier ideal sheaves

Definition 1.1. Let $f : X \to Y$ be a morphism of smooth varieties. Then the *relative canonical divisor* $K_{X/Y}$ of X over Y is defined to be the divisor class

$$K_{X/Y} = K_X - f^* K_Y,$$

where K_X and K_Y are the canonical divisor classes on X and Y, respectively. As Lazarsfeld notes [3, below (9.1)], if f is a proper birational morphism then $K_{X/Y}$ is represented by a unique effective divisor supported on the exceptional locus of f; a local equation is given by the determinant of the derivative, $\det(df)$. In other words, this is the ramification divisor.

Definition 1.2. A Weil divisor D on X is *reduced* if it is effective and all prime divisors occurring in it have multiplicity 1. If $D = \sum n_i D_i$ is a Weil divisor, written such that the D_i are distinct prime divisors and $n_i \neq 0$ for all i, then D_{red} denotes the divisor $\sum D_i$.

Definition 1.3. A subset of X has *normal crossings* if it is defined in a neighborhood of any point by an equation in local analytic coordinates of the form $z_1 \cdots z_r = 0$ (and therefore is of pure codimension 1). A divisor D on X is a *normal crossings divisor* if it is reduced and if its support has normal crossings.

Definition 1.4. Let $D = \sum n_i D_i$ be a Weil $\mathbb{Q}$-divisor, where $n_i \in \mathbb{Q}$ for all i and the D_i are distinct prime Weil divisors. Then the *floor* of D is the divisor

$$\lfloor D \rfloor = \sum \lfloor n_i \rfloor D_i \ .$$

Note that this definition does not respect linear or numerical equivalence.

Definition 1.5. Let $\mathfrak{a}$ be a nonzero sheaf of ideals on X, and let $c \in \mathbb{R}_{\geq 0}$. Let $\mu\colon X' \to X$ be a proper birational morphism such that X' is a smooth variety and

$$\mu^*(\mathfrak{a}) = \mathscr{O}_{X'}(-F)$$

for a divisor F on X' with normal crossings support. Then the *multiplier ideal sheaf* associated to $\mathfrak{a}$ and c is the ideal sheaf

$$\mathscr{I}(\mathfrak{a}^c) = \mu_* \mathscr{O}_{X'}\big(K_{X'/X} - \lfloor cF \rfloor\big).$$

By a theorem of Esnault and Viehweg [3, Thm. 9.2.18], this definition is independent of the choice of μ.

For our purposes we need a slightly different definition.

Definition 1.6. Let $\mathfrak{a}$ and c be as above. We then define

$$\mathscr{I}^-(\mathfrak{a}^c) = \lim_{\epsilon \to 0+} \mathscr{I}(\mathfrak{a}^{c-\epsilon}).$$

Here we use the discrete topology on the set of ideal sheaves on X, and note that the limit exists because there are only finitely many coefficients in $\lfloor (c-\epsilon)F \rfloor$.

We also write $\mathscr{I}(\mathfrak{a}) = \mathscr{I}(\mathfrak{a}^1)$ and $\mathscr{I}^-(\mathfrak{a}) = \mathscr{I}^-(\mathfrak{a}^1)$.

Example 1.7. Let D be a normal crossings divisor on X and let $\mathfrak{a} = \mathscr{O}(-D)$. Then we can take $X' = X$, in which case $F = D$ and

$$K_{X'/X} = \lfloor (1-\epsilon)F \rfloor = 0,$$

so $\mathscr{I}^-(\mathfrak{a}) = \mathscr{O}_X$ (the ideal sheaf corresponding to the empty closed subscheme). More generally, if D is effective and has normal crossings support but is not necessarily reduced, then $\lfloor (1-\epsilon)F \rfloor = D - D_{\text{red}}$, and therefore

$$\mathscr{I}^-(\mathscr{O}(-D)) = \mathscr{O}(-(D - D_{\text{red}})).$$

2 Definitions from Nevanlinna theory

This section gives, for the convenience of the reader, various standard definitions from Nevanlinna theory, as well as their counterparts in number theory.

We start with the standard definitions of Nevanlinna theory. First, a *Weil function* for a divisor D on X is a continuous function $\lambda_D \colon X \setminus \operatorname{Supp} D \to \mathbb{R}$ such that if D is locally represented by a principal divisor (f) on an open set U, then $\lambda_D + \log|f|$ extends to a continuous function on all of U.

For a holomorphic curve $f \colon \mathbb{C} \to X$ whose image does not lie in Supp D, the *proximity function* for f relative to D is then defined as

$$m_f(D,r) = \int_0^{2\pi} \lambda_D(f(re^{i\theta}))\frac{d\theta}{2\pi}$$

for all $r > 0$. This depends on the choice of Weil function λ_D for D, but this affects the proximity function only by a bounded amount (depending only on the two Weil functions).

To define the counting function of f relative to D, we first let $\operatorname{ord}_z \Delta$ denote the multiplicity of an analytic divisor Δ on $\mathbb{C}$ at a point z. The *counting function* for f with respect to D is then

$$N_f(D,r) = \sum_{0<|z|<r} (\operatorname{ord}_z f^*D)\log\frac{r}{|z|} + (\operatorname{ord}_0 f^*D)\log r, \tag{2.1}$$

and the *characteristic* (or *height*) function is defined to be

$$T_{D,f}(r) = m_f(D,r) + N_f(D,r).$$

It is well known that the latter depends up to $O(1)$ only on the linear equivalence class of D, so it is possible to define the characteristic function $T_{D,f}(r)$ for a divisor class D (or for a line sheaf).

Definition 2.2. A divisor D on X is *big* if there is a constant $c > 0$ such that $h^0(X, nD) \ge cn^{\dim X}$ for all sufficiently large and divisible integers n.

It is known, and easy to check, that the proximity, counting, and height functions are linear in D. Moreover, if D is a big divisor and f has Zariski-dense image, then the height is as big as possible, up to a multiplicative constant. If D is big and D' is another divisor, then $T_{D',f}(r) \le c\,T_{D,f}(r) + O(1)$ for some constant c. Thus, heights relative to big divisors occur frequently in error terms. For details on these assertions, see [7, § 11].

A slightly more general situation involves finite ramified coverings. These correspond to considering algebraic points instead of rational points in number theory. The general setup is that B is a connected Riemann surface, $p \colon B \to \mathbb{C}$ is a proper surjective holomorphic map, and $f \colon B \to X$ is holomorphic.

$$\begin{array}{ccc} B & \xrightarrow{\ f\ } & X \\ \big\downarrow{\scriptstyle p} & & \\ \mathbb{C} & & \end{array} \tag{2.3}$$

For real $r > 0$ let

$$B\langle r\rangle = \{b \in B : |p(b)| = r\}$$
$$\text{and} \qquad B(r) = \{b \in B : |p(b)| < r\},$$

and let σ be the measure

$$\sigma = \frac{1}{\deg p} p^*\left(\frac{d\theta}{2\pi}\right)$$

on $B\langle r\rangle$.

Then the proximity function can be defined in this context as

$$m_f(D,r) = \int_{B\langle r\rangle} \lambda_d \circ f \cdot \sigma,$$

and the counting function can be defined as

$$N_f(D,r) = N_{f^*D}(r),$$

where

$$N_\Delta(r) = \frac{1}{\deg p}\left(\sum_{b\in B(r)\setminus p^{-1}(0)} (\operatorname{ord}_b \Delta)\log\frac{r}{|p(b)|} + \sum_{b\in p^{-1}(0)} (\operatorname{ord}_b \Delta)\log r\right) \tag{2.4}$$

for an analytic divisor Δ on B. We also define the *ramification counting function* for p to be

$$N_{\mathrm{Ram}(p)}(r) = N_R(r)\ ,$$

where R is the ramification divisor of p.

For more details on these definitions, see [7, § 26].

The corresponding definitions in the number field (or global field) case are as follows.

Let k be a global field, and let M_k denote its set of places. If k is a number field, then M_k is the disjoint union of a set S_∞, which is in one-to-one correspondence with the set of embeddings $\sigma\colon k \hookrightarrow \mathbb{C}$ modulo complex conjugation, and a set which is in one-to-one correspondence with the set of nonzero prime ideals of the ring of integers $\mathscr{O}_k$ of k. Norms $\|\cdot\|_v$ associated to places $v \in M_k$ are normalized so that if $v \in S_\infty$, then $\|x\|_v = |\sigma(x)|^{[k_v:\mathbb{R}]}$, where v corresponds to $\sigma\colon k \hookrightarrow \mathbb{C}$ and k_v is the local field at v; and if $v \notin S_\infty$, then $\|\pi\|_v = (\mathscr{O}_k : \mathfrak{p})^{-1}$, where v corresponds to $\mathfrak{p} \subseteq \mathscr{O}_k$ and $\pi \in \mathfrak{p}\setminus\mathfrak{p}^2$. For function fields, the set of places and their associated norms is assumed to be given with the function field, and $S_\infty = \emptyset$. For either type of field, the *product formula*

$$\prod_{v\in M_k} \|x\|_v = 1$$

holds for all $x \in k^*$ (by assumption, in the case of function fields).

Let D be a divisor on X. A *Weil function* λ_D is a real-valued function on $\coprod_{v \in M_k} (X \setminus \operatorname{Supp} D)(\bar{k}_v)$; its restriction to $(X \setminus \operatorname{Supp} D)(\bar{k}_v)$ is denoted $\lambda_{D,v}$. Each $\lambda_{D,v}$ is similar to a Weil function in Nevanlinna theory, normalized like $-\log \| f \|_v$, but there are additional conditions as v varies, as well as extra considerations due to the fact that $\bar{k}_v$ is not locally compact. The details are too extensive to give here; see [2, Ch. 10]. Another (incomplete) reference is [7, § 8].

Now let $P \in X(k)$ be a point not lying on the support of D, and let $S \supseteq S_\infty$ be a finite set of places of k. Then the *proximity function* for P relative to D (and S) is defined as

$$m_S(D, P) = \sum_{v \in S} \lambda_{D,v}(P),$$

the *counting function* is defined as

$$N_S(D, P) = \sum_{v \notin S} \lambda_{D,v}(P),$$

and the *height* is defined as

$$h_{D,k}(P) = m_S(D, P) + N_S(D, P) = \sum_{v \in M_k} \lambda_{D,v}(P).$$

These are linear in D, and again the height depends up to $O(1)$ only on the linear equivalence class of D. These definitions depend on the choice of Weil function λ_D, but only up to a constant depending only on the Weil functions involved. The definitions also depend on k, but k may be omitted from the notation since it is usually fixed.

As in the case of Nevanlinna theory, if $\Sigma \subseteq X(k)$ is a set for which all infinite subsets are Zariski-dense, if D is a big divisor, and if D' is any other divisor, then $h_{D',k}(P) \le c\, h_{D,k}(P) + O(1)$ for some constant c. For more details, see [7, Sects. 9–10].

These definitions generalize to points $P \in X(\bar{k})$ as follows. Assume $P \notin \operatorname{Supp} D$, and let L be a finite extension of k containing $k(P)$. For places $w \in M_L$ we write $w \mid S$ if w lies over a place $v \in S$, and $w \nmid S$ otherwise. Noting that $\bar{L}_w = \bar{k}_v$ if $w \mid v$, we write $\lambda_{D,w} = [L_w : k_v]\lambda_{D,v}$. We then define the proximity, counting, and height functions as

$$m_S(D, P) = \frac{1}{[L : k]} \sum_{\substack{w \in M_L \\ w \mid S}} \lambda_{D,w}(P),$$

$$N_S(D, P) = \frac{1}{[L : k]} \sum_{\substack{w \in M_L \\ w \nmid S}} \lambda_{D,w}(P),$$

and

$$h_{D,k}(P) = m_S(D,P) + N_S(D,P) = \frac{1}{[L:k]} \sum_{w \in M_L} \lambda_{D,w}(P).$$

These expressions do not depend on the choice of L.

Corresponding to $N_{\mathrm{Ram}(p)}(r)$ in Nevanlinna theory, we consider the discriminant of $k(P)$ for an algebraic point:

$$d_k(P) = \frac{\log |D_{k(P)}|}{[k(P):k]} - \log |D_k|,$$

where D_k denotes the discriminant of the number field k. The corresponding definition in the function field case involves the ramification divisor of the morphism of curves corresponding to the field inclusion $k \hookrightarrow L$; it is left to the reader as an exercise. For more details on discriminants, see [7, § 23].

3 Proximity functions for ideal sheaves

Silverman [5, 2.2] introduced Weil functions associated to sheaves of ideals on X. By [5, Thm. 2.1], there is a unique way to associate to each ideal sheaf $\mathfrak{a} \neq (0)$ of X a Weil-like function $\lambda_{\mathfrak{a}}$ on $X \setminus Y$, where Y is the closed subscheme associated to $\mathfrak{a}$, such that $\lambda_{\mathfrak{a}} = \lambda_D$ is a Weil function in the usual sense if $\mathfrak{a} = \mathscr{O}(-D)$ for some effective Cartier divisor D, and $\lambda_{\mathfrak{a}+\mathfrak{b}} = \min\{\lambda_{\mathfrak{a}}, \lambda_{\mathfrak{b}}\}$ for all nonzero ideal sheaves $\mathfrak{a}$ and $\mathfrak{b}$ of X. Here uniqueness and equality are up to addition of functions bounded by M_k-constants. An M_k-constant is a function $v \mapsto c_v$ from M_k to $\mathbb{R}$ such that $c_v = 0$ for all but finitely many v.

Weil functions of ideal sheaves also satisfy the following conditions:

(3.1) They are functorial in the sense that if $f: X' \to X$ is a morphism of complete varieties with $f(X') \not\subseteq Y$, then $\lambda_{f^*\mathfrak{a}} = \lambda_{\mathfrak{a}} \circ f$.
(3.2) If $\mathfrak{a} \subseteq \mathfrak{b}$ are ideal sheaves on X, then $\lambda_{\mathfrak{a}} \geq \lambda_{\mathfrak{b}}$.

See also Noguchi [4] and Yamanoi [8, 2.2]. They used similar Weil functions to define proximity functions relative to ideal sheaves. These are defined as follows. Let $f: \mathbb{C} \to X$ be a holomorphic curve whose image is not entirely contained in Y. Then we define the *proximity function* $m_f(\mathfrak{a}, r)$ in the usual way:

$$m_f(\mathfrak{a}, r) = \int_0^{2\pi} \lambda_{\mathfrak{a}}(f(re^{i\theta})) \frac{d\theta}{2\pi},$$

with the obvious adaptations in the case of (2.3). Similarly, if X is a variety over a global field k, if $P \in X(L)$ for some finite extension L of k, and if $S \supseteq S_\infty$ is a finite set of places of k, then

$$m_S(\mathfrak{a}, P) = \frac{1}{[L:k]} \sum_{\substack{w \in M_L \\ w \mid S}} \lambda_{\mathfrak{a},w}(P).$$

This expression is independent of the choice of L. Again, these proximity functions agree (up to $O(1)$) with $m_f(D,r)$ and $m_S(D,r)$, respectively, when $\mathfrak{a} = \mathscr{O}(-D)$. They also satisfy (3.1) and (3.2) (again, up to $O(1)$).

4 Conjectures

In Nevanlinna theory, we make the following conjecture:

Conjecture 4.1. Let X be a nonsingular complete complex variety, let K be the canonical divisor class on X, let $\mathfrak{a} \neq (0)$ be an ideal sheaf on X, let A be a big divisor on X, and let $\epsilon > 0$. Then there is a proper Zariski-closed subset Z of X, depending only on X, $\mathfrak{a}$, A, and ϵ, such that if $p\colon B \to \mathbb{C}$ and $f\colon B \to X$ are as in (2.3) and the image of f is not contained in Z, then

$$T_{K,f}(r) + m_f(\mathfrak{a},r) - m_f(\mathscr{I}^-(\mathfrak{a}),r) \leq_{\text{exc}} \epsilon\, T_{A,f}(r) + N_{\text{Ram}(p)}(r) + O(1).$$

Here the subscript "exc" means that the inequality holds outside of a set of r of finite Lebesgue measure. The implicit constant in $O(1)$ is independent of r but depends on all other data.

The corresponding conjecture in number theory is:

Conjecture 4.2. Let k be a global field of characteristic zero, let $S \supseteq S_\infty$ be a finite set of places of k, let r be a positive integer, let X be a nonsingular complete variety over k, let K be the canonical divisor class of X, let $\mathfrak{a} \neq (0)$ be an ideal sheaf on X, let A be a big divisor on X, and let $\epsilon > 0$. Then there is a proper Zariski-closed subset Z of X, depending only on k, S, X, $\mathfrak{a}$, r, A, and ϵ, such that

$$h_{K,k}(P) + m_S(\mathfrak{a},P) - m_S(\mathscr{I}^-(\mathfrak{a}),P) \leq \epsilon\, h_{A,k}(P) + d_k(P) + O(1)$$

for all $P \in (X \setminus Z)(\bar{k})$ with $[k(P):k] \leq r$. Again, the implicit constant in $O(1)$ is independent of P but depends on all other data.

These conjectures obviously generalize earlier conjectures in each case. Indeed, let D be a normal crossings divisor and let $\mathfrak{a} = \mathscr{O}(-D)$. Then $m_f(\mathscr{I}^-(\mathfrak{a}),r) = O(1)$ and $m_f(\mathfrak{a},r) = m_f(D,r)$, and likewise in the diophantine case.

Proposition 4.3. *Conjectures 4.1 and 4.2 are equivalent to their respective special cases in which $\mathfrak{a} = \mathscr{O}(-D)$ with D as above.*

Proof. Let $\mu\colon X' \to X$, $K_{X'/X}$, and F be as in the definition of multiplier ideal sheaf, and choose $\eta > 0$ such that $\mathscr{I}^-(\mathfrak{a}) = \mathscr{I}(\mathfrak{a}^{1-\eta})$. In the Nevanlinna case, let $g\colon B \to X'$ be a lifting of f; then

$$
\begin{aligned}
T_{K_X,f}(r) &+ m_f(\mathfrak{a},r) - m_f(\mathscr{I}^-(\mathfrak{a}),r) \\
&\le T_{K_{X'},g}(r) - m_g(K_{X'/X},r) + m_g(F,r) - m_g(-K_{X'/X} + \lfloor (1-\eta)F \rfloor, r) \\
&\quad + O(1) \\
&= T_{K_{X'},g}(r) + m_g(F_{\text{red}},r) + O(1) \\
&\le_{\text{exc}} \epsilon\, T_{A,f}(r) + N_{\text{Ram}(p)}(r) + O(1).
\end{aligned}
$$

Here we use the fact that

$$\mu^*\mathscr{I}(\mathfrak{a}^{1-\eta}) = \mu^*\mu_*\mathscr{O}_{X'}(K_{X'/X} - \lfloor (1-\eta)F \rfloor) \subseteq \mathscr{O}_{X'}(K_{X'/X} - \lfloor (1-\eta)F \rfloor)$$

and therefore

$$
\begin{aligned}
m_f(\mathscr{I}(\mathfrak{a}^{1-\eta}),r) &\ge m_g(\mathscr{O}_{X'}(K_{X'/X} - \lfloor (1-\eta)F \rfloor),r) + O(1) \\
&= m_g(-K_{X'/X} + \lfloor (1-\eta)F \rfloor, r) + O(1)\,.
\end{aligned}
$$

The diophantine case is similar and is left to the reader. □

Remark 4.4. Although an arbitrary complete variety may not have a big line sheaf (or any nontrivial line sheaf) [1, pp. 25–26 and p. 72], a nonsingular complete variety always does. Indeed, let U be a nonempty open affine on a nonsingular complete variety X, pick generators $x_1,\dots,x_r$ for the affine ring $\mathscr{O}_X(U)$, and let D be a Weil divisor whose support contains the polar divisors of all x_i. Then D is big.

5 Truncated counting functions

Variations of the above conjectures using truncated counting functions can also be made. Truncated counting functions are defined in Nevanlinna theory as follows.

Definition 5.1. Let t be a positive integer, let D be an effective divisor on X, and let $f\colon \mathbb{C} \to X$ be a holomorphic curve whose image is not contained in the support of D. Then the *truncated counting function* for f relative to D is

$$N_f^{(t)}(D,r) = \sum_{0<|z|<r} \min\{\operatorname{ord}_z f^*D, t\} \log\frac{r}{|z|} + \min\{\operatorname{ord}_0 f^*D, t\}\log r$$

(cf. (2.1)). In the more general situation of (2.3), the truncated counting function is $N_f^{(t)}(D,r) = N_{f^*D}^{(t)}(r)$, where in place of (2.4) we have

$$
\begin{aligned}
N_\Delta^{(t)}(r) = \frac{1}{\deg p}\Biggl(&\sum_{b\in B\langle r\rangle \setminus p^{-1}(0)} \min\{\operatorname{ord}_b \Delta, t\} \log\frac{r}{|p(b)|} \\
&+ \sum_{b\in p^{-1}(0)} \min\{\operatorname{ord}_b \Delta, t\}\log r \Biggr).
\end{aligned}
$$

Still more generally, if $\mathfrak{a} \neq (0)$ is an ideal sheaf on X and if $f: B \to X$ is a holomorphic map whose image is not contained in the closed subscheme associated to $\mathfrak{a}$, then $f^*\mathfrak{a}$ is a nonzero analytic ideal sheaf on B. Since B is smooth and of dimension 1, any such ideal sheaf is of the form $\mathscr{O}(-\Delta)$ for an analytic divisor Δ on B, and we define

$$N_f(\mathfrak{a}, r) = N_\Delta(r) \qquad \text{and} \qquad N_f^{(t)}(\mathfrak{a}, r) = N_\Delta^{(t)}(r).$$

Finally, we also define a height function

$$T_{\mathfrak{a},f}(r) = m_f(\mathfrak{a}, r) + N_f(\mathfrak{a}, r).$$

The conjecture in Nevanlinna theory is then:

Conjecture 5.2. Let $p: B \to \mathbb{C}$ be as in (2.3), let X be a nonsingular complete complex variety, let K be the canonical divisor class on X, let $\mathfrak{a} \neq (0)$ be a sheaf of ideals on X, let A be a big divisor on X, and let $\epsilon > 0$. Then there is a proper Zariski-closed subset Z of X, depending only on $\deg p$, X, $\mathfrak{a}$, A, and ϵ, such that the inequality

$$N_f^{(1)}(\mathfrak{a}, r) + N_{\operatorname{Ram}(p)}(r) \geq_{\text{exc}} T_{K,f}(r) + T_{\mathfrak{a},f}(r) - T_{\mathscr{I}^-(\mathfrak{a}),f}(r) - \epsilon\, T_{A,f}(r) - O(1)$$

holds for all nonconstant holomorphic curves $f: B \to X$ whose images are not contained in Z. The implicit constant in $O(1)$ is as in Conjecture 4.1.

In the diophantine case, the corresponding definition and conjecture are as follows.

Definition 5.3. Let t be a positive integer, let $S \supseteq S_\infty$ be a finite set of places of k, let D be an effective divisor on X, let λ_D be a Weil function for D, let $P \in X(\bar{k}) \setminus \operatorname{Supp} D$, and let $L = k(P)$. Then the *truncated counting function* of P relative to D is

$$N_S^{(t)}(D, P) = \frac{1}{[L:k]} \sum_{\substack{w \in M_L \\ w \nmid S}} \min\{\lambda_{D,w}(P), -t \log \|\pi_w\|_w\},$$

where π_w denotes an element of $\mathscr{O}_L$ that lies in the prime ideal associated to w, but not in the square of that ideal. Note that, because ramification may affect the truncation, L is fixed in the above expression. More generally, if $\mathfrak{a}$ is a nonzero ideal sheaf on X and if $\lambda_{\mathfrak{a}}$ is a Weil function associated to it, then the truncated counting function relative to $\mathfrak{a}$ is defined as

$$N_S^{(t)}(\mathfrak{a}, P) = \frac{1}{[L:k]} \sum_{\substack{w \in M_L \\ w \nmid S}} \min\{\lambda_{\mathfrak{a},w}(P), -t \log \|\pi_w\|_w\},$$

and its non-truncated counterpart is defined analogously. Finally, we define a height relative to $\mathfrak{a}$ in the usual way as

$$h_{\mathfrak{a},k}(P) = m_S(\mathfrak{a}, P) + N_S(\mathfrak{a}, P) = \frac{1}{[L:k]} \sum_{w \in M_L} \lambda_{\mathfrak{a},w}(P).$$

Conjecture 5.4. Let k be a global field of characteristic zero, let $S \supseteq S_\infty$ be a finite set of places of k, let r be a positive integer, let X be a nonsingular complete variety over k, let K be the canonical divisor class of X, let $\mathfrak{a} \neq (0)$ be an ideal sheaf on X, let A be a big divisor on X, and let $\epsilon > 0$. Then there is a proper Zariski-closed subset Z of X, depending only on k, S, X, $\mathfrak{a}$, r, A, and ϵ, such that

$$N_S^{(1)}(\mathfrak{a}, P) + d_k(P) \geq h_{K,k}(P) + h_{\mathfrak{a},k}(P) - h_{\mathscr{I}^-(\mathfrak{a}),k}(P) - \epsilon\, h_{A,k}(P) - O(1)$$

for all $P \in (X \setminus Z)(\bar{k})$ with $[k(P) : k] \leq r$. The implicit constant is as in Conjecture 4.2.

In each case, if $\mathfrak{a} = \mathscr{O}(-D)$ with D a normal crossings divisor, then the above conjectures reduce to conjectures that have already been posed; see [6] for the diophantine case.

Again, we have a converse:

Proposition 5.5. *Conjectures 5.2 and 5.4 are equivalent to their respective special cases in which $\mathfrak{a} = \mathscr{O}(-D)$ with D as above.*

Proof. In the diophantine case this follows by the same argument as before. Indeed, let $\mu \colon X' \to X$, F, and η be as before; assuming that [6, Conj. 2.3] holds for F_{red} on X', we have

$$\begin{aligned}
N_S^{(1)}(\mathfrak{a}, P) + d_k(P) &= N_S^{(1)}(F_{\text{red}}, P') + d_k(P') \\
&\geq h_{K_{X'}+F_{\text{red}}}(P') - \epsilon\, h_{\mu^* A}(P') - O(1) \\
&= h_{K'_X}(P') - h_{K_{X'/X}}(P') + h_F(P') - h_{-K_{X'/X}+\lfloor(1-\eta)F\rfloor}(P') \\
&\quad - \epsilon h_A(P) - O(1) \\
&\geq h_{K_X}(P) + h_{\mathfrak{a}}(P) - h_{\mathscr{I}^-(\mathfrak{a})}(P) - \epsilon h_A(P) - O(1), \quad (5.5)
\end{aligned}$$

where $P' \in X'$ lies over $P \in X$. The proof in the Nevanlinna case is analogous. □

References

1. Fulton, W.: Introduction to toric varieties, *Annals of Mathematics Studies*, vol. 131. Princeton University Press, Princeton, NJ (1993). The William H. Roever Lectures in Geometry.
2. Lang, S.: Fundamentals of Diophantine Geometry. Springer-Verlag, New York, 1983.

3. Lazarsfeld, R.: Positivity in Algebraic Geometry. II, *Ergebnisse der Mathematik und ihrer Grenzgebiete. 3. Folge. A Series of Modern Surveys in Mathematics [Results in Mathematics and Related Areas. 3rd Series. A Series of Modern Surveys in Mathematics]*, Vol. 49. Springer-Verlag, Berlin, 2004. Positivity for vector bundles, and multiplier ideals
4. Noguchi, J.: Nevanlinna Theory in Several Variables and Diophantine Approximation [Japanese]. Kyoritsu Publ., Tokyo, 2003.
5. Silverman, J.H.: Arithmetic distance functions and height functions in Diophantine geometry. *Math. Ann.* **279**(2), 193–216 (1987). DOI 10.1007/BF01461718. URL http://dx.doi.org/10.1007/BF01461718.
6. Vojta, P.: A more general *abc* conjecture. Internat. Math. Res. Notices **1998**(21), 1103–1116 (1998)
7. Vojta, P.: Diophantine approximation and Nevanlinna theory. In: *Arithmetic Geometry* (Lectures given at the C.I.M.E. Summer School held in Cetraro, Italy, September 10–15, 2007), Lecture Notes in Math., vol. 2009, pp. 111–230. Springer, 2010.
8. Yamanoi, K.: Algebro-geometric version of Nevanlinna's lemma on logarithmic derivative and applications. *Nagoya Math.* J. **173**, 23–63, 2004.

Recent advances in Diophantine approximation

Michel Waldschmidt

Dedicated to the memory of Serge Lang

Abstract A basic question of Diophantine approximation, which is the first issue we discuss, is to investigate the rational approximations to a single real number. Next, we consider the algebraic or polynomial approximations to a single complex number, as well as the simultaneous approximation of powers of a real number by rational numbers with the same denominator. Finally we study generalisations of these questions to higher dimensions. Several recent advances have been made by B. Adamczewski, Y. Bugeaud, S. Fischler, M. Laurent, T. Rivoal, D. Roy, and W.M. Schmidt, among others. We review some of these works.

Key words Diophantine approximation • rational approximation • simultaneous approximation • approximation by algebraic numbers • approximation by linear forms • irrationality measures • transcendence criterion • criteria for algebraic independence • Dirichlet • Hurwitz • Thue-Siegel-Roth-Schmidt • Khintchine • Davenport • Sprindzuck • Laurent • Roy

Mathematics Subject Classification (2010): 11Jxx

M. Waldschmidt (✉)
Université Pierre et Marie Curie–Paris 6, UMR 7586 IMJ Institut de, Mathématiques de Jussieu, 4 Place Jussieu 75252 Paris Cedex 05 FRANCE
e-mail: miw@math.jussieu.fr
http://www.math.jussieu.fr/~miw/

D. Goldfeld et al. (eds.), *Number Theory, Analysis and Geometry: In Memory of Serge Lang*, DOI 10.1007/978-1-4614-1260-1_29,

Introduction

The history of Diophantine approximation is quite old: it includes, for instance, early estimates for π, computations related to astronomical studies, and the theory of continued fraction expansion.

There are positive results: *any irrational number has good rational approximations*. One of the simplest tools to see this is Dirichlet's box principle; other methods are continued fraction expansions, Farey series; and geometry of numbers (Minkowski's Theorem). There are negative results: *no number has too good (and at the same time too frequent) approximations*. Some results are valid for all (irrational) numbers, others only for restricted classes of numbers, like the class of algebraic numbers. There is a metric theory (§1.2) which deals with almost all numbers in the sense of the Lebesgue measure.

One main goal of the theory of Diophantine approximation is to compare, on the one hand, the distance between a given real number ξ and a rational number p/q, with, on the other hand, the denominator q of the approximant. An approximation is considered *sharp* if $|\xi - p/q|$ is *small* compared to q. This subject is a classical one; there are a number of surveys, including those by S. Lang [78, 80–82]. Further general references are [26, 36, 46, 59, 60, 68, 75, 124].

The works by J. Liouville, A. Thue, C.L. Siegel, F.J. Dyson, A.O. Gel'fond, Th. Schneider and K.F. Roth essentially solve the question for the case where ξ is algebraic. In a different direction, a lot of results are known which are valid for almost all numbers, after Khintchine and others.

Several questions arise in this context. One may consider either *asymptotic* or else *uniform* approximation. The former asks to only for infinitely many solutions to some inequality, while the latter requires that occurrences of such approximations be not too lacunary. As a consequence, one introduces in §1.1 two exponents for the rational approximation to a single real number ξ, namely $\omega(\xi)$ for the asymptotic approximation and $\widehat{\omega}(\xi)$ for the uniform approximation; a lower bound for such an exponent means that sharp rational approximations exist, an upper bound means that too sharp estimates do no exist. To indicate with a "hat" the exponents of *uniform* Diophantine approximation is a convention which originates in [41].

In this context a new exponent, $\nu(\xi)$, inspired by the pioneering work of R. Apéry in 1976 on $\zeta(3)$, has been introduced recently by T. Rivoal and S. Fischler (§1.3).

After rational approximation to a single real number, several other questions naturally arise. One may investigate, for instance, the *algebraic* approximation properties of real or complex numbers, replacing the set of rational numbers by the set of real or complex algebraic numbers. Again, in this context, there are two main points of view: either one considers the distance $|\xi - \alpha|$ between the given real or complex number ξ and algebraic numbers α, or else one investigates the smallness of $|P(\xi)|$ for P a non-zero polynomial with integer coefficients. In both cases there are two parameters, the degree and the height of the algebraic number or of the polynomial, in place of a single one in degree 1, namely q for $\xi - p/q$ or for $P(X) = qX - p$. Algebraic and polynomial approximations are related: on the one hand (Lemma 9), the irreducible polynomial of an algebraic number close to ξ takes a small value at ξ, while on the other hand (Lemma 10), a polynomial taking a small value at ξ is likely to have a root close to ξ. However, these connections are not completely understood yet: for instance, while it is easy (by means of Dirichlet's box principle – Lemma 15) to prove the existence of polynomials P having small values $|P(\xi)|$ at the point ξ, it is not so easy to show that sharp algebraic approximations exist (cf. Wirsing's conjecture 37).

The occurrence of two parameters raises more questions to investigate: often one starts by taking the degree fixed and looking at the behaviour of the approximations as the height tends to infinity; one might do the opposite, fix the height and let the degree tend to infinity: this is the starting point of a classification of complex numbers by V.G. Sprindžuk (see [36] Chap. 8 p. 166). Another option is to let the sum of the degree and the logarithm of the height tend to infinity: this is the choice of S. Lang who introduced the notion of *size* [79] Chap. V, in connection with questions of algebraic independence [78].

The approximation properties of a real or complex number ξ by polynomials of degree at most n (§2.3) will give rise to two exponents, $\omega_n(\xi)$ and $\widehat{\omega}_n(\xi)$, which coincide with $\omega(\xi)$ and $\widehat{\omega}(\xi)$ for $n = 1$. Gel'fond's transcendence criterion (§2.2) is related to an upper bound for the asymptotic exponent of polynomial approximation $\widehat{\omega}_n(\xi)$ valid for all transcendental numbers.

The approximation properties of a real or complex number ξ by algebraic numbers of degree at most n (§2.5) will give rise to two further exponents, an asymptotic $\omega_n^*(\xi)$ and a uniform $\widehat{\omega}_n^*(\xi)$, which also coincide with $\omega(\xi)$ and $\widehat{\omega}(\xi)$ for $n = 1$.

For a *real* number ξ, there is a third way of extending the investigation of rational approximation, which is the study of simultaneous approximation by rational numbers of the n-tuple $(\xi, \xi^2, \dots, \xi^n)$. Once more there are an asymptotic exponent $\omega_n'(\xi)$ and a uniform $\widehat{\omega}_n'(\xi)$; again they coincide with $\omega(\xi)$ and $\widehat{\omega}(\xi)$ for $n = 1$. These two new exponents suffice to describe the approximation properties of a real number by algebraic numbers of degree at most n (the star exponents), thanks to a transference result (Proposition 40) based on the theory of convex bodies of Mahler.

Several relations among these exponents are known, but a number of problems remain open: for instance, for fixed $n \geqslant 1$ the spectrum of the sextuple

$$(\omega_n(\xi),\ \widehat{\omega}_n(\xi),\ \omega_n'(\xi),\ \widehat{\omega}_n'(\xi),\ \omega_n^*(\xi),\ \widehat{\omega}_n^*(\xi)) \in \mathbf{R}^6$$

is far from being completely understood. As we shall see for almost all real numbers ξ and for all algebraic numbers of degree $> n$,

$$\omega_n(\xi) = \widehat{\omega}_n(\xi) = \omega_n^*(\xi) = \widehat{\omega}_n^*(\xi) = n \quad \text{and} \quad \omega_n'(\xi) = \widehat{\omega}_n'(\xi) = 1/n.$$

A review of the known properties of these six exponents is given in [41]. We shall repeat some of these facts here (beware that our notation for ω_n' and $\widehat{\omega}_n'$ are the λ_n and $\widehat{\lambda}_n$ from [41], which are the inverses of their w_n' and $\widehat{w}_n'$, also used by Y. Bugeaud in §3.6 of [36]; here we wish to be compatible with the notation of M. Laurent in [91] for the higher-dimensional case).

Among a number of new results in this direction, we shall describe those achieved by D. Roy and others. In particular, a number of results for the case $n = 2$ have been recently obtained.

Simultaneous approximations to a tuple of numbers is the next step. The subspace theorem of W.M. Schmidt (Theorem 1B in Chapter VI of [124]), which is a powerful generalisation of the Thue–Siegel–Roth Theorem, deals with the approximation of algebraic numbers. It says that tuples of algebraic numbers do behave like almost all tuples. It is a fundamental tool with a number of deep consequences [29]. Another point of view is the metrical one, dealing with almost all numbers. Further questions arise which should concern all tuples, and these considerations raise many open problems. We shall report on recent work by M. Laurent who introduces a collection of new exponents for describing the situation.

We discuss these questions mainly in the case of real numbers. Most results (so far as they are not related to the density of $\mathbf{Q}$ into $\mathbf{R}$) are valid also for complex numbers with some modifications (however see [40]), as well and for non-Archimedean valuations, especially p-adic numbers but also (to some extent) for function fields. We make no attempt to be exhaustive. There are a number of related issues which we do not study in detail here — sometimes we just give a selection of recent references. Among them are

- Questions of inhomogeneous approximation.
- Littlewood's Conjecture ([36], Chap. 10).
- Measures of irrationality, transcendence, linear independence, algebraic independence of specific numbers. Effective refinements of Liouville's Theorem are studied in [30] (see also Chap. 2 of [36]).
- Results related to the complexity of the development of irrational algebraic numbers, automata, normality of classical constants (including irrational algebraic numbers) — the Bourbaki lecture by Yu. Bilu [29] on Schmidt's subspace Theorem and its applications describes recent results on this topic and gives further references.
- Connection between Diophantine conditions and dynamical systems.
- Diophantine questions related to Diophantine geometry. Earlier surveys dealing extensively with this issue were written by S. Lang. A recent reference on this topic is [70].
- In a preliminary version of the present paper, the list of topics which were not covered included also refined results on Hausdorff dimension, Diophantine approximation of dependent quantities and approximation on manifolds, and hyperbolic manifolds, also the powerful approach initiated by Dani and Margulis, developed by many specialists. We quote here V.V. Beresnevich, V.I. Bernik, H. Dickinson, M.M. Dodson, D.Y. Kleinbock, É.I. Kovalevskaya, G.A. Margulis, F. Paulin, S.L. Velani. However, thanks to the contribution of Victor Beresnevich and Maurice Dodson who kindly agreed to write Sections 2.7 and 3.6 (and also to contribute by adding remarks, especially on §1.2), these topics are no longer excluded.

We discuss only briefly a few questions of algebraic independence; there is much more to say on this matter, especially in connection with Diophantine approximation. Although we quote some recent transcendence criteria as well as criteria for algebraic independence, we do not fully cover this topic either (and do not mention criteria for linear independence).

Acknowledgments Many thanks to Boris Adamczewski, Victor Beresnevich, Yann Bugeaud, Maurice Dodson, Michel Laurent, Claude Levesque, Damien Roy for their enlightening remarks and their comments on preliminary versions of this paper. Sections 2.7 and 3.6, as well as part of Section 1.2, have been written by Victor Beresnevich and Maurice Dodson. I wish also to thank Dinakar Ramakrishnan who completed the editorial work in a very efficient way.

1 Rational approximation to a real number

1.1 Asymptotic and uniform rational approximation: ω and $\widehat{\omega}$

Since $\mathbf{Q}$ is dense in $\mathbf{R}$, *for any $\xi \in \mathbf{R}$ and any $\epsilon > 0$ there exists $b/a \in \mathbf{Q}$ for which*

$$\left|\xi - \frac{b}{a}\right| < \epsilon.$$

Let us write the conclusion

$$|a\xi - b| < \epsilon a.$$

It is easy to improve this estimate: *Let $a \in \mathbf{Z}_{>0}$ and let b be the nearest integer to $a\xi$. Then*

$$|a\xi - b| \leqslant 1/2.$$

A much stronger estimate is due to Dirichlet (1842) and follows from the box or pigeonhole principle; see for instance [78], [60], [124] Chap. I, sTh. 1A:

Theorem 1 (Uniform Dirichlet's theorem). *For each real number $N > 1$, there exist q and p in $\mathbf{Z}$ with $1 \leqslant q < N$ such that*

$$|q\xi - p| < \frac{1}{N}\cdot$$

As an immediate consequence ([124] Chap. I Cor. 1B):

Corollary 2 (Asymptotic Dirichlet's theorem). *If $\xi \in \mathbf{R}$ is irrational, then there exist infinitely many $p/q \in \mathbf{Q}$ for which*

$$\left|\xi - \frac{p}{q}\right| < \frac{1}{q^2}\cdot$$

Our first concern is to investigate whether it is possible to improve the uniform estimate of Theorem 1 as well as the asymptotic estimate of Corollary 2.

We start with Corollary 2. Using either the theory of continued fractions or Farey series, one deduces a slightly stronger statement, proved by A. Hurwitz in 1891 (Theorem 2F in Chap. I of [124]):

Theorem 3 (Hurwitz). *For any real irrational number ξ, there exist infinitely many $p/q \in \mathbf{Q}$ such that*

$$\left|\xi - \frac{p}{q}\right| < \frac{1}{\sqrt{5}q^2}.$$

For the golden ratio $\gamma = (1+\sqrt{5})/2$ and for the numbers related to the golden ratio by a homographic transformation $(ax+b)/(cx+d)$ (where a, b, c, d are rational integers satisfying $ad - bc = \pm 1$), this asymptotic result is optimal. For all other irrational real numbers, Hurwitz proved that the constant $\sqrt{5}$ can be replaced by $\sqrt{8}$ ([124] Chap. I Cor. 6C). These are the first elements of the *Lagrange spectrum*: $\sqrt{5}, \sqrt{8}, \sqrt{221}/5, \sqrt{1517}/13, \ldots$, (references are given in Chap.I, §6 of [124]; the book [47] is devoted to the Lagrange and Markov spectra).

Lagrange noticed as early as 1767 (see [59] Chap. 1, Theorem 1.2) that for all irrational quadratic numbers, the exponent 2 in q^2 in the conclusion of Corollary 2 is optimal: more generally, Liouville's inequality (1844) produces, for each algebraic number ξ of degree $d \geqslant 2$, a constant $c(\xi)$ such that for all rational numbers p/q,

$$\left|\xi - \frac{p}{q}\right| > \frac{c(\xi)}{q^d}.$$

Admissible values for $c(\xi)$ are easy to specify (Th. 1 Chap. 1 §1 of [128], [60] p. 6, Th. 1E of [125], Th. 1.2 of [36]).

A *Liouville number* is a real number ξ for which the opposite estimate holds: *for any $\kappa > 0$, there exists a rational number p/q such that*

$$0 < \left|\xi - \frac{p}{q}\right| < \frac{1}{q^\kappa}. \tag{4}$$

A *very well approximable number* is a real number ξ for which there exists $\kappa > 2$ such that the inequality (4) has infinitely many solutions. A nice example of such a number is

$$\xi_\kappa := 2\sum_{n=1}^{\infty} 3^{-\lceil \kappa^n \rceil}$$

for κ a real number > 2. This number belongs to the middle-third Cantor set $\mathcal{K}$, which is the set of real numbers whose base-three expansions are free of the digit 1. In [95], J. Levesley, C. Salp, and S.L. Velani show that ξ_κ is an element of $\mathcal{K}$ with irrationality exponent $\mu(\xi_\kappa) = \kappa$ for $\kappa \geqslant (3+\sqrt{5})/2$ and $\geqslant \kappa$ for $2 < \kappa \leqslant (3+\sqrt{5})/2$. This example answers a question of K. Mahler on the existence of very well approximable numbers which are not Liouville numbers in $\mathcal{K}$. In [38], Y. Bugeaud shows that $\mu(\xi_\kappa) = \kappa$ for $\kappa \geqslant 2$, and more generally, that for $\kappa \geqslant 2$ and $\lambda > 0$, the number

$$\xi_{\lambda,\kappa} := 2\sum_{n=n_0}^{\infty} 3^{-\lceil \lambda\kappa^n \rceil}$$

has $\mu(\xi_{\lambda,\kappa}) = \kappa$.

Given $\kappa \geqslant 2$, denote by E_κ the set of real numbers ξ satisfying the following property: *the inequality (4) has infinitely many solutions in integers p and q, and for any $c < 1$ there exists q_0 such that, for $q \geqslant q_0$,*

$$\left|\xi - \frac{p}{q}\right| > \frac{c}{q^\kappa}\cdot \tag{5}$$

Then for any $\kappa \geqslant 2$ this set E_κ is not empty. Explicit examples were given by Jarník in 1931 (see [2] for a variant). In [20], V.V. Beresnevich, H. Dickinson and S.L. Velani raised the question of the Hausdorff dimension of the set E_κ. The answer is given by Y. Bugeaud in [35]: this dimension is $2/\kappa$.

We now consider the uniform estimate of Theorem 1. Let us show that for any irrational number ξ, Dirichlet's theorem is essentially optimal: one cannot replace $1/N$ by $1/(2N)$. This was already observed by Khintchine in 1926 [74]:

Lemma 6. *Let ξ be a real number. Assume that there exists a positive integer N_0 such that for each integer $N \geqslant N_0$, there exist $a \in \mathbf{Z}$ and $b \in \mathbf{Z}$ with $1 \leqslant a < N$ and*

$$|a\xi - b| < \frac{1}{2N}\cdot$$

Then ξ is rational and $a\xi = b$ for each $N \geqslant N_0$.

Proof. By assumption for each integer $N \geqslant N_0$ there exist $a_N \in \mathbf{Z}$ and $b_N \in \mathbf{Z}$ with $1 \leqslant a_N < N$ and

$$|a_N\xi - b_N| < \frac{1}{2N}\cdot$$

Our goal is to check $a_N\xi = b_N$ for each $N \geqslant N_0$.

Let $N \geqslant N_0$. Write (a,b) for (a_N, b_N) and (a', b') for (a_{N+1}, b_{N+1}):

$$|a\xi - b| < \frac{1}{2N}\ (1 \leqslant a \leqslant N-1), \quad |a'\xi - b'| < \frac{1}{2N+2}\ (1 \leqslant a' \leqslant N).$$

Eliminate ξ between $a\xi - b$ and $a'\xi - b'$: the rational integer

$$ab' - a'b = a(b' - a'\xi) + a'(a\xi - b)$$

satisfies $|ab' - a'b| < 1$; hence it vanishes and $ab' = a'b$.

Therefore the rational number $b_N/a_N = b_{N+1}/a_{N+1}$ does not depend on $N \geqslant N_0$. Since

$$\lim_{N\to\infty} b_N/a_N = \xi,$$

it follows that $\xi = b_N/a_N$ for all $N \geqslant N_0$. □

Remark. As pointed out to me by M. Laurent, an alternative argument is based on continued fraction expansions.

Coming back to Theorem 1 and Corollary 2, we associate to each real irrational number ξ two exponents ω and $\widehat{\omega}$ as follows.

Starting with Corollary 2, we introduce the *asymptotic irrationality exponent of a real number* ξ, which is denoted by $\omega(\xi)$:

$$\omega(\xi) = \sup\Big\{w;\ \text{there exist infinitely many } (p,q) \in \mathbf{Z}^2 \\ \text{with } q \geqslant 1 \text{ and } 0 < |q\xi - p| \leqslant q^{-w}\Big\}.$$

Some authors prefer to introduce the *irrationality exponent* $\mu(\xi) = \omega(\xi) + 1$ *of* ξ which is denoted by $\mu(\xi)$:

$$\mu(\xi) = \sup\Big\{\mu;\ \text{there exist infinitely many } (p,q) \in \mathbf{Z}^2 \\ \text{with } q \geqslant 1 \text{ and } 0 < \left|\xi - \frac{p}{q}\right| \leqslant q^{-\mu}\Big\}.$$

An upper bound for $\omega(\xi)$ or $\mu(\xi)$ is an *irrationality measure* for ξ, namely a lower bound for $|\xi - p/q|$ when $p/q \in \mathbf{Q}$.

Liouville numbers are the real numbers ξ with $\omega(\xi) = \mu(\xi) = \infty$.

Since no set E_κ (see property (5)) with $\kappa \geqslant 2$ is empty, the *spectrum* $\{\omega(\xi)\ ;\ \xi \in \mathbf{R} \setminus \mathbf{Q}\}$ of ω is the whole interval $[1, +\infty]$, while the spectrum $\{\mu(\xi)\ ;\ \xi \in \mathbf{R} \setminus \mathbf{Q}\}$ of μ is $[2, +\infty]$.

According to the theorem of Thue–Siegel–Roth [29], *for any real algebraic number* $\xi \in \mathbf{R} \setminus \mathbf{Q}$,

$$\omega(\xi) = 1.$$

We shall see (in §1.2) that the same holds for almost all real numbers.

The other exponent related to Dirichlet's Theorem 1 is the *uniform irrationality exponent of* ξ, denoted by $\widehat{\omega}(\xi)$:

$$\widehat{\omega}(\xi) = \sup\Big\{w;\ \text{for any } N \geqslant 1, \text{ there exists } (p,q) \in \mathbf{Z}^2 \\ \text{with } 1 \leqslant q \leqslant N \text{ and } 0 < |q\xi - p| \leqslant N^{-w}\Big\}.$$

In the singular case of a rational number ξ we set $\omega(\xi) = \widehat{\omega}(\xi) = 0$. It is plain from the definitions that for any $\xi \in \mathbf{R} \setminus \mathbf{Q}$,

$$\omega(\xi) \geqslant \widehat{\omega}(\xi) \geqslant 1.$$

In fact Lemma 6 implies that *for any* $\xi \in \mathbf{R} \setminus \mathbf{Q}$, $\widehat{\omega}(\xi) = 1$. Our motivation to introduce a notation for a number which is always equal to 1 is that it will become non–trivial in more general situations ($\widehat{\omega}_n$ in §2.3, $\widehat{\omega}'_n$ in §2.4, $\widehat{\omega}^*_n$ in §2.5).

1.2 Metric results

The metric theory of Diophantine approximation provides statements which are valid for almost all (real or complex) numbers, which means for all numbers outside a set of Lebesgue measure 0. Among many references on this topic, we quote [26, 36, 69, 129, 130]. See also §2.7 and §3.6 below.

One of the early results is due to Capelli: *for almost all* $\xi \in \mathbf{R}$,

$$\omega(\xi) = \widehat{\omega}(\xi) = 1 \quad \textit{and} \quad \mu(\xi) = 2.$$

This is one of many instances where irrational algebraic numbers behave like almost all numbers. However one cannot expect that *all* statements from Diophantine approximation which are satisfied by all numbers outside a set of measure 0 will be satisfied by all irrational algebraic numbers, just because such an intersection of sets of full measure is empty. As pointed out to me by B. Adamczewski, S. Schanuel (quoted by S. Lang in [78] p.184 and [83] Chap. II §2 Th. 6) gave a more precise formulation of such a remark as follows.

Denote by $\mathcal{K}$ (for Khintchine) the set of *non-increasing* functions Ψ from $\mathbf{R}_{\geqslant 1}$ to $\mathbf{R}_{>0}$. Set

$$\mathcal{K}_c = \left\{ \Psi \in \mathcal{K}\, ;\, \sum_{n \geqslant 1} \Psi(n) \text{ converges} \right\}, \quad \mathcal{K}_d = \left\{ \Psi \in \mathcal{K}\, ;\, \sum_{n \geqslant 1} \Psi(n) \text{ diverges} \right\}.$$

Hence $\mathcal{K} = \mathcal{K}_c \cup \mathcal{K}_d$.

A well-known theorem of A. Ya. Khintchine in 1924 (see [73], [75] and Th. 1.10 in [36]) has been refined as follows (an extra condition that the function $x \mapsto x^2\Psi(x)$ be decreasing has been dropped) – see Beresnevich, Dickinson, and Velani [21]:

Theorem 7 (Khintchine). *Let* $\Psi \in \mathcal{K}$. *Then for almost all real numbers* ξ, *the inequality*

$$|q\xi - p| < \Psi(q) \tag{8}$$

has

- *only finitely many solutions in integers* p *and* q *if* $\Psi \in \mathcal{K}_c$
- *infinitely many solutions in integers* p *and* q *if* $\Psi \in \mathcal{K}_d$.

S. Schanuel proved that the set of real numbers which behave like almost all numbers from the point of view of Khintchine's theorem in the convergent case has measure 0. More precisely, the set of real numbers ξ such that for any *smooth convex* function $\Psi \in \mathcal{K}_c$, the inequality (8) has only finitely many solutions is the set of real numbers with bounded partial quotients (*badly approximable numbers* – see [124] Chap. I §5; other characterisations of this set are given in [83] Chap. II §2 Th. 6).

Moreover B. Adamczewski and Y. Bugeaud noticed that given any irrational ξ, either there exists a $\Psi \in \mathcal{K}_d$ for which

$$|q\xi - p| < \Psi(q)$$

has no integer solutions or there exists a $\Psi \in \mathcal{K}_c$ for which

$$|q\xi - p| < \Psi(q)$$

has infinitely many integer solutions.

1.3 The exponent ν of S. Fischler and T. Rivoal

Let $\xi \in \mathbf{R} \setminus \mathbf{Q}$. In [63], S. Fischler and T. Rivoal introduce a new exponent $\nu(\xi)$ which they define as follows.

When $\underline{u} = (u_n)_{n\geqslant 1}$ is an increasing sequence of positive integers, define another sequence of integers $\underline{v} = (v_n)_{n\geqslant 1}$ by $|u_n\xi - v_n| < 1/2$ (i.e. v_n is the nearest integer to $u_n\xi$) and set

$$\alpha_\xi(\underline{u}) = \limsup_{n\to\infty} \frac{|u_{n+1}\xi - v_{n+1}|}{|u_n\xi - v_n|}, \quad \beta(\underline{u}) = \limsup_{n\to\infty} \frac{u_{n+1}}{u_n}.$$

Then

$$\nu(\xi) = \inf \log \sqrt{\alpha_\xi(\underline{u})\beta(\underline{u})},$$

where $\underline{u}$ ranges over the sequences which satisfy $\alpha_\xi(\underline{u}) < 1$ and $\beta(\underline{u}) < +\infty$. Here we agree that $\inf \emptyset = +\infty$.

They establish a connection with the irrationality exponent by proving

$$\mu(\xi) \leqslant 1 - \frac{\log \beta(\underline{u})}{\log \alpha_\xi(\underline{u})}.$$

As a consequence, *if* $\nu(\xi) < +\infty$, *then* $\mu(\xi) < +\infty$.

If ξ is quadratic, Fischler and Rivoal produce a sequence $\underline{u}$ with $\alpha_\xi(\underline{u})\beta(\underline{u}) = 1$, hence $\nu(\xi) = 0$.

This new exponent ν is motivated by Apéry-like proofs of irrationality and measures. Following the works of R. Apéry, A. Baker, F. Beukers, G. Rhin and C. Viola, and M. Hata among others, S. Fischler and T. Rivoal deduce

$$\nu(2^{1/3}) \leqslant (3/2)\log 2, \quad \nu(\zeta(3)) \leqslant 3, \quad \nu(\pi^2) \leqslant 2, \quad \nu(\log 2) \leqslant 1.$$

Also $\nu(\pi) \leqslant 21$.

The spectrum of $\nu(\xi)$ is not yet known. According to [63], *for any* $\xi \in \mathbf{R} \setminus \mathbf{Q}$, *the inequalities* $0 \leqslant \nu(\xi) \leqslant +\infty$ *hold.* Further, *for almost all* $\xi \in \mathbf{R}$, $\nu(\xi) = 0$. Furthermore, S. Fischler and T. Rivoal, completed by B. Adamczewski [1], proved that *any irrational algebraic real number* ξ *has* $\nu(\xi) < +\infty$.

There are examples of $\xi \in \mathbf{R} \setminus \mathbf{Q}$ for which $\nu(\xi) = +\infty$, but all known examples with $\nu(\xi) = +\infty$ so far have $\mu(\xi) = +\infty$.

Fischler and Rivoal ask whether *it is true that* $\nu(\xi) < +\infty$ *implies* $\mu(\xi) = 2$. Another related question they raise in [63] is whether *there are numbers* ξ *with* $0 < \nu(\xi) < +\infty$.

2 Polynomial, algebraic, and simultaneous approximation to a single number

We define the (usual) height $\mathrm{H}(P)$ of a polynomial

$$P(X) = a_0 + a_1X + \cdots + a_nX^n$$

with complex coefficients as the maximum modulus of its coefficients, while its length $\mathrm{L}(P)$ is the sum of the moduli of these coefficients:

$$\mathrm{H}(P) = \max_{0 \leqslant i \leqslant n} |a_i|, \quad \mathrm{L}(P) = \sum_{i=0}^{n} |a_i|.$$

The height $\mathrm{H}(\alpha)$ and length $\mathrm{L}(\alpha)$ of an algebraic number α are the height and length of its minimal polynomial over $\mathbf{Z}$.

2.1 Connections between polynomial approximation and approximation by algebraic numbers

Let ξ be a complex number. To produce a sharp *polynomial approximation* to ξ is to find a non-zero polynomial $P \in \mathbf{Z}[X]$ for which $|P(\xi)|$ is small. An *algebraic approximation* to ξ is an algebraic number α such that the distance $|\xi - \alpha|$ between

ξ and α is small. There are close connections between both questions. On the one hand, if $|P(\xi)|$ is small, then ξ is close to a root α of P. On the other hand, if $|\xi - \alpha|$ is small then the minimal polynomial of α assumes a small value at ξ. These connections explain that the classifications of transcendental numbers in S, T, and U classes by K. Mahler coincide with the classifications of transcendental numbers in S^*, T^*, and U^* classes by J.F. Koksma (see [128] Chap. III and [36] Chap. 3).

The easy part is the next statement (Lemma 15 Chap. III §3 of [128], §15.2.4 of [138], Prop. 3.2 §3.4 of [36]).

Lemma 9. *Let $f \in \mathbf{C}[X]$ be a non-zero polynomial of degree D and length L, let $\alpha \in \mathbf{C}$ be a root of f and let $\xi \in \mathbf{C}$ satisfy $|\xi - \alpha| \leqslant 1$. Then*

$$|f(\xi)| \leqslant |\xi - \alpha| L D (1 + |\xi|)^{D-1}.$$

The other direction requires more work (see Chap. III §3 of [128], §3.4 of [36]). The next result is due to G. Diaz and M. Mignotte [52] (cf. Lemma 15.13 of [138]).

Lemma 10. *Let $f \in \mathbf{Z}[X]$ be a non-zero polynomial of degree D. Let ξ be a complex number, α a root of f at minimal distance of ξ and k the multiplicity of α as a root of f. Then*

$$|\xi - \alpha|^k \leqslant D^{3D-2} \mathrm{H}(f)^{2D} |f(\xi)|.$$

Further similar estimates are due to M. Amou and Y. Bugeaud [7].

2.2 *Gel'fond's transcendence criterion*

The so-called *transcendence criterion,* proved by A.O. Gel'fond in 1949, is an auxiliary result in the method he introduced in [64–66] (see also [67] and [68]) for proving algebraic independence results. An example is the algebraic independence of the two numbers $2^{\sqrt[3]{2}}$ and $2^{\sqrt[3]{4}}$. More generally, he proved that *if α is a non–zero algebraic number, $\log \alpha$ a non-zero logarithm of α and β an algebraic number of degree $d \geqslant 3$, then at least 2 among the $d-1$ numbers*

$$\alpha^\beta,\ \alpha^{\beta^2},\ \ldots,\alpha^{\beta^{d-1}}$$

are algebraically independent. Here α^z stands for $\exp(z \log \alpha)$.

While the Gel'fond–Schneider transcendence method for solving Hilbert's seventh problem on the transcendence of α^β relies on a *Liouville type* estimate, namely a lower bound for a non-zero value $|P(\xi)|$ of a polynomial P at an algebraic point ξ, Gel'fond's method for algebraic independence requires a more sophisticated result, namely the fact that *there is no non-trivial uniform sequence of polynomials taking small values at a given transcendental number.*

Here is a version of this transcendence criterion [93, 131].

Theorem 11 (Gel'fond's transcendence criterion). *Let $\xi \in \mathbf{C}$. Assume there is a sequence $(P_N)_{N\geqslant N_0}$ of non–zero polynomials in $\mathbf{Z}[X]$, where P_N has degree $\leqslant N$ and height $\mathrm{H}(P_N) \leqslant e^N$, for which*

$$|P_N(\xi)| \leqslant e^{-6N^2}.$$

Then ξ is algebraic and $P_N(\xi) = 0$ for all $N \geqslant N_0$.

Proof (sketch of). The idea of the proof is basically the same as for Lemma 6 which dealt with degree-1 polynomials: one eliminates the variable using two consecutive elements of the sequence of polynomials. In degree 1 linear algebra was sufficient. For higher degrees the resultant of polynomials is a convenient substitute.

Fix $N \geqslant N_0$. Since $|P_N(\xi)|$ is small, ξ is close to a root α_N of P_N, hence P_N is divisible by a power Q_N of the irreducible polynomial of α_N and $|Q_N(\xi)|$ is small. The resultant of the two polynomials Q_N and Q_{N+1} has absolute value < 1; hence it vanishes and therefore α_N does not depend on N. □

In 1969, H. Davenport and W.M. Schmidt ([49] Theorem 2b) prove the next variant of Gel'fond's transcendence criterion, where now the degree is fixed.

Theorem 12 (Davenport and Schmidt). *Let ξ be a real number and $n \geqslant 2$ a positive integer. Assume that for each sufficiently large positive integer N there exists a non–zero polynomial $P_N \in \mathbf{Z}[X]$ of degree $\leqslant n$ and usual height $\leqslant N$ for which*

$$|P_N(\alpha)| \leqslant N^{-2n+1}.$$

Then ξ is algebraic of degree $\leqslant n$.

The next sharp version of Gel'fond's transcendence criterion 11, restricted to quadratic polynomials, is due to B. Arbour and D. Roy, 2004 [8].

Theorem 13 (Arbour and Roy). *Let ξ be a complex number. Assume that there exists $N_0 > 0$ such that for any $N \geqslant N_0$, there exists a polynomial $P_N \in \mathbf{Z}[X]$ of degree $\leqslant 2$ and height $\leqslant N$ satisfying*

$$|P_N(\xi)| \leqslant \frac{1}{4} N^{-\gamma-1}.$$

Then ξ is algebraic of degree $\leqslant 2$ and $P_N(\xi) = 0$ for all $N \geqslant N_0$.

Variants of the transcendence criterion have been considered by D. Roy in connection with his new approach towards Schanuel's conjecture [138] §15.5.3:

Conjecture 14 (Schanuel). *Let $x_1, \ldots, x_n$ be $\mathbf{Q}$-linearly independent complex numbers. Then n at least of the $2n$ numbers $x_1, \ldots, x_n, e^{x_1}, \ldots, e^{x_n}$ are algebraically independent.*

In [111, 112], D. Roy states a Diophantine approximation conjecture, which he shows to be equivalent to Schanuel's Conjecture 14.

Roy's conjecture involves polynomials for which bounds for the degree, height, and absolute values at given points are assumed. The first main difference with Gel'fond's situation is that the smallness of the values is not as strong as what is achieved by Dirichlet's box principle when a single polynomial is constructed. Hence elimination arguments cannot be used without further ideas. On the other hand, the assumptions involve not only one polynomial for each n as in Theorem 11, but a collection of polynomials, and they are strongly related. This new situation raises challenging questions on which some advances have already been achieved. In particular in [93] Laurent and Roy obtain variants of Gel'fond's criterion 11 involving multiplicities. Further progress has been subsequently made by D. Roy in [120, 121].

We shall consider extensions of the transcendence criterion to criteria for algebraic independence in §3.1.

2.3 Polynomial approximation to a single number

A simple application of Dirichlet's box principle (see the proof of Lemma 8.1 in [36]) yields the existence of polynomials with small values at a given real point:

Lemma 15. *Let ξ be a real number and n a positive integer. Set $c = (n+1)\max\{1, |\xi|\}^n$. Then, for each positive integer N, there exists a non-zero polynomial $P \in \mathbf{Z}[X]$, of degree $\leqslant n$ and usual height* $\mathrm{H}(P) \leqslant N$, *satisfying*

$$|P(\xi)| \leqslant cN^{-n}.$$

Variants of this lemma rely on the geometry of numbers: for instance from Th. B2 in [36] one deduces that in the case $0 < |\xi| < 1/2$, if $N \geqslant 2$, then the conclusion holds also with $c = 1$ (see the proof of Prop. 3.1 in [36]).

Theorem 1 in §1 yields a refined estimate for the special case $n = 1$. The statement is plain in the case where ξ is algebraic of degree $\leqslant n$ as soon as N exceeds the height of the irreducible polynomial of ξ; this is why when n is fixed we shall most often assume that ξ is either transcendental or else algebraic of degree $> n$.

The fact that the exponent n in Lemma 15 cannot be replaced by a larger number (even if we seek such a solution only for infinitely many N) was proved by Sprindžuk [129], who showed in 1965 that *for ξ outside a set of Lebesgue measure zero and for each $\epsilon > 0$, there are only finitely many non–zero integer polynomials of degree at most n with*

$$|P(\xi)| \leqslant \mathrm{H}(P)^{-n-\epsilon}.$$

We introduce, for each positive integer n and each real number ξ, two exponents $\omega_n(\xi)$ and $\widehat{\omega}_n(\xi)$ as follows.

The number $\omega_n(\xi)$ denotes the supremum of the real numbers w for which there exist infinitely many positive integers N for which the system of inequalities

$$0 < |x_0 + x_1\xi + \cdots + x_n\xi^n| \leqslant N^{-w}, \quad \max_{0 \leqslant i \leqslant n} |x_i| \leqslant N \tag{16}$$

has a solution in rational integers $x_0, x_1, \ldots, x_n$. The inequalities (16) can be written

$$0 < |P(\xi)| \leqslant \mathrm{H}(P)^{-w},$$

where P denotes a non-zero polynomial with integer coefficients and degree $\leqslant n$. A *transcendence measure* for ξ is a lower bound for $|P(\xi)|$ in terms of the height $\mathrm{H}(P)$ and the degree $\deg P$ of P. Hence one can view an upper bound for $\omega_n(\xi)$ as a transcendence measure for ξ.

These numbers arise in Mahler's classification of complex numbers ([128] Chap. III §1 and [36] §3.1).

A uniform version of this exponent is the supremum $\widehat{\omega}_n(\xi)$ of the real numbers w such that, for any sufficiently large integer N, the same system (16) has a solution. An upper bound for $\widehat{\omega}_n(\xi)$ is a *uniform transcendence measure* for ξ.

Clearly, from the definitions, we see that these exponents generalize those from §1.1: for $n = 1$, $\omega_1(\xi) = \omega(\xi)$ and $\widehat{\omega}_1(\xi) = \widehat{\omega}(\xi)$. From Lemma 15 one deduces, for any $n \geqslant 1$ and any $\xi \in \mathbf{R}$ which is not algebraic of degree $\leqslant n$,

$$n \leqslant \widehat{\omega}_n(\xi) \leqslant \omega_n(\xi). \tag{17}$$

Moreover, $\omega_n \leqslant \omega_{n+1}$ and $\widehat{\omega}_n \leqslant \widehat{\omega}_{n+1}$. As a consequence, Liouville numbers have $\omega_n(\xi) = +\infty$ for all $n \geqslant 1$.

The value of the exponents ω_n and $\widehat{\omega}_n$ for almost all real numbers and for all algebraic numbers of degree $> n$ is n. The following metric result is due to V.G. Sprindžuk [129]:

Theorem 18 (Sprindžuk). *For almost all numbers $\xi \in \mathbf{R}$,*

$$\omega_n(\xi) = \widehat{\omega}_n(\xi) = n \textit{ for all } n \geqslant 1.$$

As a consequence of W.M. Schmidt's subspace theorem one deduces (see [36] Th. 2.8 and 2.9) the value of $\omega_n(\xi)$ and $\widehat{\omega}_n(\xi)$ for ξ algebraic irrational:

Theorem 19 (Schmidt). *Let $n \geqslant 1$ be an integer and ξ an algebraic number of degree $d > n$. Then*

$$\omega_n(\xi) = \widehat{\omega}_n(\xi) = n.$$

The spectrum of the exponent ω_n is $[n, +\infty]$. For $\widehat{\omega}_n$ it is not completely known. From Theorem 12 one deduces:

Theorem 20 (Davenport and Schmidt). *For any real number ξ which is not algebraic of degree $\leqslant n$,*

$$\widehat{\omega}_n(\xi) \leqslant 2n - 1.$$

For the special case $n = 2$ a sharper estimate holds: from Theorem 13 of B. Arbour and D. Roy one deduces

$$\widehat{\omega}_2(\xi) \leqslant \gamma + 1 \tag{21}$$

(recall that γ denotes the golden ratio $(1 + \sqrt{5})/2$).

In [49], Davenport and Schmidt comment:

"We have no reason to think that the exponents in these theorems are best possible."

It was widely believed for a while that $\widehat{\omega}_2(\xi)$ would turn out to be equal to 2 for all $\xi \in \mathbf{R}$ which are not rational or quadratic irrationals. Indeed, otherwise, for κ in the interval $2 < \kappa < \widehat{\omega}_2(\xi)$, the inequalities

$$0 < |x_0 + x_1\xi + x_2\xi^2| \leqslant cN^{-\kappa}, \quad \max\{|x_0|, |x_1|, |x_2|\} \leqslant N, \tag{22}$$

would have, for a suitable constant $c > 0$ and for all sufficiently large N, a non-trivial solution in integers $(x_0, x_1, x_2) \in \mathbf{Z}^3$. However, these inequalities define a convex body whose volume tends to zero as N tends to infinity. In such circumstances one does not expect a non–trivial solution to exist.[1] In general $\widehat{\omega}_2(\alpha, \beta)$ may be infinite (Khintchine, 1926; see [46]). However, here we have the restriction $\beta = \alpha^2$.

Hence it came as a surprise in 2003 when D. Roy [113] showed that the estimate (21) is optimal, by constructing examples of real (transcendental) numbers ξ for which $\widehat{\omega}_2(\xi) = \gamma + 1 = 2.618\ldots$.

By means of a transference principle of Jarník (Th. 2 of [72]), Th. 1 of [115] can be reformulated as follows (see also Th. 1.5 of [116]).

Theorem 23 (Roy). *There exists a real number ξ which is neither rational nor a quadratic irrational and which has the property that for a suitable constant $c > 0$, for all sufficiently large integers N, the inequalities (22) have a solution $(x_0, x_1, x_2) \in \mathbf{Z}^2$ with $\kappa = \gamma + 1$. Any such number is transcendental over $\mathbf{Q}$ and the set of such real numbers is countable.*

In [118], answering a question of Y. Bugeaud and M. Laurent [41], D. Roy shows that the exponents $\widehat{\omega}_2(\xi)$, where ξ ranges through all real numbers which are not algebraic of degree $\leqslant 2$, form a dense subset of the interval $[2, 1 + \gamma]$.

D. Roy calls *extremal* a number which satisfies the conditions of Theorem 23; from the point of view of approximation by quadratic polynomials, these numbers present a closest behaviour to quadratic real numbers.

Here is the first example [113] of an extremal number ξ. Recall that *the Fibonacci word*

$$w = abaababaabaababaababaabaababaabaab\ldots$$

[1] Compare with the definition of *singular systems* in §7, Chap. V of [46].

is the fixed point of the morphism $a \mapsto ab, b \mapsto a$. It is the limit of the sequence of words starting with $f_1 = b$ and $f_2 = a$ and defined inductively by concatenation as $f_n = f_{n-1}f_{n-2}$. Now let A and B be two distinct positive integers and let $\xi \in (0,1)$ be the real number whose continued fraction expansion is obtained from the Fibonacci word w by replacing the letters a and b by A and B:

$$[0; A, B, A, A, B, A, B, A, A, B, A, A, B, A, B, A, A, \ldots].$$

Then ξ is extremal.

In [113, 115, 119], D. Roy investigates the approximation properties of extremal numbers by rational numbers, by quadratic numbers as well as by cubic integers.

Theorem 24 (Roy). *Let ξ be an extremal number. There exist positive constants $c_1, \ldots, c_5$ with the following properties:*

(1) For any rational number $\alpha \in \mathbf{Q}$ we have

$$|\xi - \alpha| \geqslant c_1 H^{-2}(\log H)^{-c_2}$$

with $H = \max\{2, \mathrm{H}(\alpha)\}$.

(2) For any algebraic number α of degree at most 2 *we have*

$$|\xi - \alpha| \geqslant c_3 \mathrm{H}(\alpha)^{-2\gamma-2}.$$

(3) There exist infinitely many quadratic real numbers α with

$$|\xi - \alpha| \leqslant c_4 \mathrm{H}(\alpha)^{-2\gamma-2}.$$

(4) For any algebraic integer α of degree at most 3 *we have*

$$|\xi - \alpha| \geqslant c_5 \mathrm{H}(\alpha)^{-\gamma-2}.$$

Moreover, in [114], he shows that for some extremal numbers ξ, property (4) holds with the exponent $-\gamma - 1$ in place of $-\gamma - 2$.

In [116] D. Roy describes the method of Davenport and Schmidt and he gives a sketch of proof of his construction of extremal numbers.

In [119] he gives a sufficient condition for an extremal number to have bounded quotients and constructs new examples of such numbers.

The values of the different exponents for the extremal numbers which are associated with Sturmian words (including the Fibonacci word) have been obtained by Y. Bugeaud and M. Laurent [41]. Furthermore, they show that the spectrum $\{\widehat{\omega}_2(\xi); \xi \in \mathbf{R}\setminus\overline{\mathbf{Q}}\}$ is not countable. See also their joint works [42,43]. Their method involves words with many palindromic prefixes. S. Fischler in [61, 62] defines new exponents of approximation which allow him to obtain a characterization of the values of $\widehat{\omega}_2(\xi)$ obtained by these authors.

In [3], B. Adamczewski and Y. Bugeaud prove that for any extremal number ξ, there exists a constant $c = c(\xi)$ such that for any integer $n \geqslant 1$,

$$\omega_n(\xi) \leqslant \exp\{c(\log(3n))^2(\log\log(3n))^2\}.$$

In particular, an extremal number is either a S-number or a T-number in Mahler's classification.

Recent results on simultaneous approximation to a number and its square, on approximation to real numbers by quadratic integers and on quadratic approximation to numbers associated with *Sturmian words* have been obtained by M. Laurent, Y. Bugeaud, S. Fischler, D. Roy, and other.

2.4 *Simultaneous rational approximation to powers of a real number*

Let ξ be a real number and n a positive integer.

We consider first the simultaneous rational approximation of successive powers of ξ. We denote by $\omega'_n(\xi)$ the supremum of the real numbers w for which there exist infinitely many positive integers N for which the system

$$0 < \max_{1\leqslant i\leqslant n} |x_i - x_0\xi^i| \leqslant N^{-w}, \quad \text{with} \quad \max_{0\leqslant i\leqslant n} |x_i| \leqslant N, \tag{25}$$

has a solution in rational integers $x_0, x_1, \ldots, x_n$.

An upper bound for $\omega'_n(\xi)$ yields a *simultaneous approximation measure* for $\xi, \xi^2, \ldots, \xi^n$.

Next the uniform simultaneous approximation measure is the supremum $\widehat{\omega}'_n(\xi)$ of the real numbers w such that for any sufficiently large integer N, the same system (25) has a solution in rational integers $x_0, x_1, \ldots, x_n$.

Notice that for $n = 1$, $\omega'_1(\xi) = \omega(\xi)$ and $\widehat{\omega}'_1(\xi) = \widehat{\omega}(\xi)$.

According to Dirichlet's box principle, *for all ξ and n,*

$$\frac{1}{n} \leqslant \widehat{\omega}'_n(\xi) \leqslant \omega'_n(\xi).$$

Khintchine's transference principle (see Th. B.5 in [36] and Theorem 61 below) yields relations between ω'_n and ω_n. As remarked in Theorem 2.2 of [41], the same proof yields similar relations between $\widehat{\omega}'_n$ and $\widehat{\omega}_n$.

Theorem 26. *Let n be a positive integer and ξ a real number which is not algebraic of degree $\leqslant n$. Then*

$$\frac{1}{n} \leqslant \frac{\omega_n(\xi)}{(n-1)\omega_n(\xi)+n} \leqslant \omega'_n(\xi) \leqslant \frac{\omega_n(\xi)-n+1}{n}$$

and

$$\frac{1}{n} \leqslant \frac{\widehat{\omega}_n(\xi)}{(n-1)\widehat{\omega}_n(\xi)+n} \leqslant \widehat{\omega}'_n(\xi) \leqslant \frac{\widehat{\omega}_n(\xi)-n+1}{n}.$$

The second set of inequalities follows from the inequalities (4) and (5) of V. Jarník in Th. 3 of [72], with conditional refinements given by the inequalities (6) and (7) of the same theorem.

In particular, $\omega_n(\xi) = n$ if and only if $\omega'_n(\xi) = 1/n$. Also, $\widehat{\omega}_n(\xi) = n$ if and only if $\widehat{\omega}'_n(\xi) = 1/n$.

The spectrum of $\omega'_n(\xi)$, where ξ ranges over the set of real numbers which are not algebraic of degree $\leqslant n$, is investigated by Y. Bugeaud and M. Laurent in [42]. Only the case $n = 2$ is completely solved.

It follows from Theorem 18 that *for almost all real numbers* ξ*,*

$$\omega'_n(\xi) = \widehat{\omega}'_n(\xi) = \frac{1}{n} \quad \text{for all } n \geqslant 1.$$

Moreover, a consequence of Schmidt's theorem 19 is that *for all* $n \geqslant 1$ *and for all algebraic real numbers* ξ *of degree* $d > n$*,*

$$\omega'_n(\xi) = \widehat{\omega}'_n(\xi) = \frac{1}{n} = \frac{1}{\omega_n(\xi)}.$$

Theorems 2a and 4a of the paper [49] by H. Davenport and W.M. Schmidt (1969) imply that upper bounds for $\widehat{\omega}'_n(\xi)$ are valid for all real numbers ξ which are not algebraic of degree $\leqslant n$. For instance,

$$\widehat{\omega}'_1(\xi) = 1, \quad \widehat{\omega}'_2(\xi) \leqslant 1/\gamma = 0.618\ldots, \quad \widehat{\omega}'_3(\xi) \leqslant 1/2.$$

A slight refinement was obtained by M. Laurent [88] in 2003 (for the odd values of $n \geqslant 5$).

Theorem 27 (Davenport and Schmidt, Laurent). *Let* $\xi \in \mathbf{R} \setminus \mathbf{Q}$ *and* $n \geqslant 2$*. Assume* ξ *is not algebraic of degree* $\leqslant \lceil n/2 \rceil$*. Then*

$$\widehat{\omega}'_n(\xi) \leqslant \lceil n/2 \rceil^{-1} = \begin{cases} 2/n & \text{if } n \text{ is even,} \\ 2/(n+1) & \text{if } n \text{ is odd.} \end{cases}$$

The definition of $\widehat{\omega}'_n$ with a supremum does not reflect the accuracy of the results in [49]; for instance, the upper bound $\widehat{\omega}'_2(\xi) \leqslant 1/\gamma$ is not as sharp as Theorem 1a of [49] which is the following:

Theorem 28 (Davenport and Schmidt). *Let* ξ *be a real number which is not rational or a quadratic irrational. There exists a constant* $c > 0$ *such that for arbitrarily large values of* N*, the inequalities*

$$\max\{|x_1 - x_0\xi|, |x_2 - x_0\xi^2|\} \leqslant cN^{-1/\gamma}, \quad |x_0| \leqslant N,$$

have no solution $(x_0, x_1, x_2) \in \mathbf{Z}^3$.

Before restricting ourselves to the small values of n, we emphasise that there is a huge lack in our knowledge of the spectrum of the set

$$\big(\omega_n(\xi),\ \widehat{\omega}_n(\xi),\ \omega'_n(\xi),\ \widehat{\omega}'_n(\xi)\big) \in \mathbf{R}^4,$$

where ξ ranges over the set of real numbers which are not algebraic of degree $\leqslant n$.

Consider the special case $n = 2$ and the question of quadratic approximation. As pointed out by Y. Bugeaud, a formula due to V. Jarník (1938) (Theorem 1 of [72]; see also Corollary A3 in [118] and [91]) relates $\widehat{\omega}_2$ and $\widehat{\omega}'_2$:

$$\widehat{\omega}'_2(\xi) = 1 - \frac{1}{\widehat{\omega}_2(\xi)}\cdot \tag{29}$$

Therefore the properties of $\widehat{\omega}_2$ which we considered in §2.3 can be translated into properties of $\widehat{\omega}'_2$. For instance, $\widehat{\omega}'_2(\xi) = 1/2$ if and only if $\widehat{\omega}_2(\xi) = 2$, and this holds for almost all $\xi \in \mathbf{R}$ (see Theorem 18) and for all algebraic real numbers ξ of degree $\geqslant 3$ (see Theorem 19). If $\xi \in \mathbf{R}$ is neither rational nor a quadratic irrational, Davenport and Schmidt have proved

$$\widehat{\omega}'_2(\xi) \leqslant 1/\gamma = 0.618\ldots. \tag{30}$$

The extremal numbers of D. Roy in Theorem 23 satisfy $\widehat{\omega}'_2(\xi) = 1/\gamma$. More precisely, they are exactly the numbers $\xi \in \mathbf{R}$ which are not rational or quadratic irrationals and satisfy the following property: *there exists a constant $c > 0$ such that for any sufficiently large number N, the inequalities*

$$\max\{|x_1 - x_0\xi|, |x_2 - x_0\xi^2|\} \leqslant cN^{-1/\gamma}, \quad 0 < \max\{|x_0|, |x_1|, |x_2|\} \leqslant N,$$

have a solution in rational integers x_0, x_1, x_2. (This was the original definition).

In [118], using Jarník's formula (29), D. Roy shows that the set of $(\widehat{\omega}'_2(\xi), \widehat{\omega}_2(\xi)) \in \mathbf{R}^2$, where ξ ranges over the set of real numbers which are not algebraic of degree $\leqslant 2$, is dense in the piece of curve

$$\{(1 - t^{-1}, t)\ ;\ 2 \leqslant t \leqslant \gamma + 1\}.$$

We conclude with the case $n = 3$ and the question of cubic approximation. When $\xi \in \mathbf{R}$ is not algebraic of degree $\leqslant 3$, the estimate for $\widehat{\omega}'_3(\xi)$ by Davenport and Schmidt [49] is

$$\frac{1}{3} \leqslant \widehat{\omega}'_3(\xi) \leqslant \frac{1}{2}\cdot$$

As we have seen, the lower bound is optimal (equality holds for almost all numbers and all algebraic numbers of degree $> n$). The upper bound has been improved by D. Roy in [119]

$$\widehat{\omega}'_3(\xi) \leqslant \frac{1}{2}(2\gamma + 1 - \sqrt{4\gamma^2 + 1}) = 0.4245\ldots.$$

2.5 Algebraic approximation to a single number

Let ξ be a real number and n a positive integer.

Denote by $\omega_n^*(\xi)$ the supremum of the real numbers w for which there exist infinitely many positive integers N with the following property: *there exists an algebraic number α of degree $\leqslant n$ and height $\leqslant N$ satisfying*[2]

$$0 < |\xi - \alpha| \leqslant N^{-w-1}. \tag{31}$$

An upper bound for $\omega_n^*(\xi)$ is a *measure of algebraic approximation* for ξ. These numbers arise in Koksma's classification of complex numbers (Chap. III §3 of [128] and §3.3 of [36]).

Next, denote by $\widehat{\omega}_n^*(\xi)$ the supremum of the real numbers w such that, *for any sufficiently large integer N, there exists an algebraic number α of degree $\leqslant n$ and height $\leqslant N$ satisfying*

$$|\xi - \alpha| \leqslant \mathrm{H}(\alpha)^{-1} N^{-w}.$$

An upper bound for $\widehat{\omega}_n^*(\xi)$ yields a *uniform measure of algebraic approximation* for ξ.

From Schmidt's subspace theorem one deduces, for a real algebraic number ξ of degree d and for $n \geqslant 1$,

$$\widehat{\omega}_n^*(\xi) = \omega_n^*(\xi) = \min\{n, d-1\}.$$

See [36] Th. 2.9 and 2.11.

That there are relations between ω_n and ω_n^* (and, for the same reason, between $\widehat{\omega}_n$ and $\widehat{\omega}_n^*$) can be expected from Lemmas 9 and 10. Indeed, a lot of information on these numbers has been devised in order to compare the classifications of Mahler and Koksma. The estimate

$$\omega_n(\xi) \geqslant \omega_n^*(\xi),$$

which follows from Lemma 9, was known by Koksma (see also Wirsing's paper [139]). In the reverse direction, the inequalities

$$\omega_n^*(\xi) \geqslant \omega_n(\xi) - n + 1, \quad \omega_n^*(\xi) \geqslant \frac{\omega_n(\xi) + 1}{2} \tag{32}$$

and

$$\omega_n^*(\xi) \geqslant \frac{\omega_n(\xi)}{\omega_n(\xi) - n + 1} \tag{33}$$

were obtained by E. Wirsing in 1960 [139] (see §3.4 of [36]).

---

[2] The occurrence of -1 in the exponent of the right-hand side of (31) is already plain for degree 1 polynomials, comparing $|\alpha - p/q|$ and $|q\alpha - p|$.

A consequence is that for a real number ξ which is not algebraic of degree $\leqslant n$, if $\omega_n(\xi) = n$ then $\omega_n^*(\xi) = n$.

The inequality (33) of Wirsing has been refined in Theorem 2.1 of [41] as follows.

Theorem 34 (Bugeaud and Laurent). *Let n be a positive integer and ξ a real number which is not algebraic of degree $\leqslant n$. Then*

$$\widehat{\omega}_n^*(\xi) \geqslant \frac{\omega_n(\xi)}{\omega_n(\xi) - n + 1} \quad \textit{and} \quad \omega_n^*(\xi) \geqslant \frac{\widehat{\omega}_n(\xi)}{\widehat{\omega}_n(\xi) - n + 1}.$$

A number of recent papers are devoted to this topic, including the survey given in the first part of [41] as well as Bugeaud's papers [5, 31, 34, 37, 39], where further references can be found.

We quote Proposition 2.1 of [41], which gives connections between the six exponents ω_n, $\widehat{\omega}_n$, ω'_n, $\widehat{\omega}'_n$, ω_n^*, $\widehat{\omega}_n^*$.

Proposition 35. *Let n be a positive integer and ξ a real number which is not algebraic of degree $\leqslant n$. Then*

$$\frac{1}{n} \leqslant \widehat{\omega}'_n(\xi) \leqslant \min\{1, \omega'_n(\xi)\}$$

and

$$1 \leqslant \widehat{\omega}_n^*(\xi) \leqslant \min\{\omega_n^*(\xi),\ \widehat{\omega}_n(\xi)\} \leqslant \max\{\omega_n^*(\xi),\ \widehat{\omega}_n(\xi)\} \leqslant \omega_n(\xi).$$

A further relation connecting ω_n^* and $\widehat{\omega}'_n$ has been discovered by H. Davenport and W.M. Schmidt in 1969 [49]. We discuss their contribution in §2.6. For our immediate concern here we only quote the following result:

Theorem 36. *Let n be a positive integer and ξ a real number which is not algebraic of degree $\leqslant n$, Then*

$$\omega_n^*(\xi)\widehat{\omega}'_n(\xi) \geqslant 1.$$

The spectral question for ω_n^* is one of the main challenges in this domain. Wirsing's conjecture states that for any integer $n \geqslant 1$ and any real number ξ which is not algebraic of degree $\leqslant n$, we have $\omega_n^*(\xi) \geqslant n$. In other terms:

Conjecture 37 (Wirsing). *For any $\epsilon > 0$ there is a constant $c(\xi, n, \epsilon) > 0$ for which there are infinitely many algebraic numbers α of degree $\leqslant n$ with*

$$|\xi - \alpha| \leqslant c(\xi, n, \epsilon)\mathrm{H}(\alpha)^{-n-1+\epsilon}.$$

In 1960, E. Wirsing [139] proved that for any real number which is not algebraic of degree $\leqslant n$, the lower bound $\omega_n^*(\xi) \geqslant (n+1)/2$ holds: it suffices to combine (32) with the lower bound $\omega_n(\xi) \geqslant n$ from (17) (see [124] Chap. VIII Th. 3B). More precisely, he proved that for such a $\xi \in \mathbf{R}$ *there is a constant* $c(\xi, n) > 0$ *for which there exist infinitely many algebraic numbers* α *of degree* $\leqslant n$ *with*

$$|\xi - \alpha| \leqslant c(\xi, n)\mathrm{H}(\alpha)^{-(n+3)/2}.$$

The special case $n = 2$ of this estimate was improved in 1967 when H. Davenport and W.M. Schmidt [48] replaced $(n+3)/2 = 5/2$ by 3. This is optimal for the approximation to a real number by quadratic algebraic numbers. This is the only case where Wirsing's conjecture is solved. More recent estimates are due to V.I. Bernik and K. Tishchenko [28,132–136]. This question is studied by Y. Bugeaud in his book [36] (§3.4) where he proposes the following *main problem*:

Conjecture 38 (Bugeaud). *Let* $(w_n)_{n\geqslant 1}$ *and* $(w_n^*)_{n\geqslant 1}$ *be two non–decreasing sequences in* $[1, +\infty]$ *for which*

$$n \leqslant w_n^* \leqslant w_n \leqslant w_n^* + n - 1 \quad \textit{for any } n \geqslant 1.$$

Then there exists a transcendental real number ξ *for which*

$$\omega_n(\xi) = w_n \quad \textit{and} \quad \omega_n^*(\xi) = w_n^* \quad \textit{for any } n \geqslant 1.$$

A summary of known results on this problem is given in §7.8 of [36].

The spectrum

$$\left\{\omega_n(\xi) - \omega_n^*(\xi)\ ;\ \xi \in \mathbf{R} \text{ not algebraic of degree} \leqslant n\right\} \subset [0, n-1]$$

of $\omega_n - \omega_n^*$ for $n \geqslant 2$ was studied by R.C. Baker in 1976 who showed that it contains $[0, 1-(1/n)]$. This has been improved by Y. Bugeaud in [34]: it contains the interval $[0, n/4]$.

Most results concerning $\omega_n^*(\xi)$ and $\widehat{\omega}_n^*(\xi)$ for $\xi \in \mathbf{R}$ have extensions to complex numbers, only the numerical estimates are slightly different. However, see [40].

2.6 *Approximation by algebraic integers*

An innovative and powerful approach was initiated in the seminal paper [49] by H. Davenport and W.M. Schmidt (1969). It rests on the transference principle arising from the geometry of numbers and Mahler's theory of *polar convex bodies* and allows one to deal with approximation by algebraic integers of bounded degree. The next statement includes a refinement by Y. Bugeaud and O. Teulié (2000) [45] who observed that one may treat approximations by algebraic integers of given degree; the sharpest results in this direction are due to M. Laurent [88].

From the estimate $\omega_n^*(\xi)\widehat{\omega}_n'(\xi) \geqslant 1$ in Theorem 36 one deduces the following statement. Let n be a positive integer and let ξ be a real number which is not algebraic of degree $\leqslant n$. Let λ satisfy $\widehat{\omega}'_n(\xi) < \lambda$. Then for $\kappa = (1/\lambda) + 1$, there is a constant $c(n, \xi, \kappa) > 0$ such that the equation

$$|\xi - \alpha| \leqslant c(n, \xi, \kappa)\mathrm{H}(\alpha)^{-\kappa} \tag{39}$$

has infinitely many solutions in algebraic numbers α of degree n. In this statement one may replace "algebraic numbers α of degree n" by "algebraic integers α of degree $n + 1$" and also by "algebraic units α of degree $n + 2$."

Proposition 40. *Let $\kappa > 1$ be a real number, n a positive integer and ξ be a real number which is not algebraic of degree $\leqslant n$. Assume $\widehat{\omega}'_n(\xi) < 1/(\kappa - 1)$. Then there exists a constant $c(n, \xi, \kappa) > 0$ such that there are infinitely many algebraic integers α of degree $n + 1$ satisfying (39) and there are infinitely many algebraic units α of degree $n + 2$ satisfying (39).*

Suitable values for κ are deduced from Theorem 27 and estimate (21). For instance, from Theorem 36 and the estimate $\widehat{\omega}'_2(\xi) \leqslant 1/\gamma$ of Davenport and Schmidt in (30) one deduces $\omega_2^*(\xi) \geqslant \gamma$. Hence for any $\kappa < 1 + \gamma$ the assumptions of Proposition 40 are satisfied. More precisely, the duality (or transference) arguments used by Davenport and Schmidt to prove Theorem 36 together with their Theorem 28 enabled them to deduce the next statement ([49], Th. 1).

Theorem 41 (Davenport and Schmidt). *Let $\xi \in \mathbf{R}$ be a real number which is neither rational nor a quadratic irrational. Then there is a constant $c > 0$ with the following property: there are infinitely many algebraic integers α of degree at most 3 which satisfy*

$$0 < |\xi - \alpha| \leqslant c\mathrm{H}(\alpha)^{-\gamma-1}. \tag{42}$$

Lemma 9 shows that under the same assumptions, for another constant $c > 0$ there are infinitely many monic polynomials $P \in \mathbf{Z}[X]$ of degree at most 3 satisfying

$$|P(\xi)| \leqslant c\mathrm{H}(P)^{-\gamma}. \tag{43}$$

Estimates (42) and (43) are optimal for certain classes of *extremal* numbers [114]. Approximation of extremal numbers by cubic integers is studied by D. Roy in [114, 115]. Further papers dealing with approximation by algebraic integers include [4, 45, 122, 135].

Another development of the general and powerful method of Davenport and Schmidt deals with the question of approximating simultaneously several numbers by conjugate algebraic numbers: this is done in [122] and refined in [117] by D. Roy. Also in [117] D. Roy gives variants of Gel'fond's transcendence criterion involving not only a single number ξ but sets $\{\gamma + \xi_1, \ldots, \gamma + \xi_m\}$ or $\{\gamma\xi_1, \ldots, \gamma\xi_m\}$. In two recent manuscripts [120, 121], D. Roy produces new criteria for the additive and for the multiplicative groups.

A different application of transference theorems is to link inhomogeneous Diophantine approximation problems with homogeneous ones [42].

2.7 Overview of metrical results for polynomials

Here we give a brief account of some significant results that have produced new ideas and generalisations, as well as some interesting problems and conjectures. We begin with the probabilistic theory (that is, Lebesgue measure statements) and continue with the more delicate Hausdorff measure/dimension results. Results for multivariable polynomials, in particular, the recent proof of a conjecture of Nesterenko on the measure of algebraic independence of almost all real m-tuples, will be sketched in § 3.6, as will metrical results on simultaneous approximation. Note that many of the results suggested here have been established in the far more general situation of Diophantine approximation on manifolds. However, for simplicity, we will only explain this Diophantine approximation for the case of integral polynomials.

Mahler's problem [96], which arose from his classification of real (and complex) numbers, remains a major influence over the metrical theory of Diophantine approximation. As mentioned in §2.3, the problem was settled by Sprindzuk in 1965. Answering a question posed by A. Baker in [9], Bernik [25] established a generalisation of Mahler's problem akin to Khintchine's one-dimensional convergence result in Theorem 7, involving the critical sum

$$\sum_{h=1}^{\infty} \Psi(h) \tag{44}$$

of values of the function $\Psi : \mathbb{N} \to \mathbb{R}^+$ that defines the error of approximation.

Theorem 45 (Bernik, 1989). *Given a monotonic Ψ such that the critical sum* (44) *converges, for almost all $\xi \in \mathbb{R}$ the inequality*

$$|P(\xi)| < H(P)^{-n+1}\Psi(H(P)) \tag{46}$$

has only finitely many solutions in $P \in \mathbb{Z}[x]$ with $\deg P \leqslant n$.

In the case $n = 1$, inequality (46) reduces to rational approximations of real numbers and is covered by Khintchine's Theorem 7 [73]. Khintchine's Theorem 7 also covers the solubility of (46) when $n = 1$ and (44) diverges. For arbitrary n the complementary divergence case of Theorem 45 has been established by Beresnevich, who has shown in [13] that if (44) diverges then for almost all real ξ inequality (46) has infinitely many solutions $P \in \mathbb{Z}[x]$ with $\deg P = n$. In fact the latter statement follows from the following analogue of Khintchine's Theorem 7 for approximation by algebraic numbers, also established in [13].

Theorem 47 (Beresnevich, 1999). *Let $n \in \mathbb{N}$, $\Psi : \mathbb{N} \to \mathbb{R}^+$ be a monotonic error function and $\mathcal{A}_n(\Psi)$ the set of real ξ such that*

$$|\xi - \alpha| < H(\alpha)^{-n}\Psi(H(\alpha)) \tag{48}$$

has infinitely many solutions in real algebraic numbers of degree $\deg \alpha = n$. *Then* $\mathcal{A}_n(\Psi)$ *has full Lebesgue measure if the sum (44) diverges and zero Lebesgue measure otherwise.*

Bugeaud [32] has proved an analogue of Theorem 47 for approximation by algebraic integers:

Theorem 49 (Bugeaud, 2002). *Let* $n \in \mathbb{N}$, $n \geqslant 2$, $\Psi : \mathbb{N} \to \mathbb{R}^+$ *be a monotonic error function and* $\mathcal{I}_n(\Psi)$ *be the set of real* ξ *such that*

$$|\xi - \alpha| < H(\alpha)^{-n+1}\Psi(H(\alpha)) \tag{50}$$

has infinitely many solutions in real algebraic integers of degree $\deg \alpha = n$. *Then* $\mathcal{I}_n(\Psi)$ *has full Lebesgue measure if the sum (44) diverges and zero Lebesgue measure otherwise.*

No analogue of Theorem 45 is known for the monic polynomial case; however, see the final section of [23].

Unlike Theorem 45, the convergence parts of Theorems 47 and 49 are rather trivial consequences of the Borel–Cantelli lemma, which also implies that the monotonicity condition is unnecessary in the case of convergence. The intriguing question now arises whether the monotonicity condition in Theorem 45 and in the divergence part of Theorems 47 and 49 can be dropped. Beresnevich [16] has recently shown that the monotonicity condition on Ψ can indeed be safely removed from Theorem 45. Regarding Theorems 47 and 49, removing the monotonicity condition is a fully open problem - [16]. In fact, in dimension $n = 1$, removing the monotonicity from Theorem 47 falls within the Duffin and Schaeffer problem [56]. Note, however, that the higher -dimensional Duffin–Schaeffer problem has been settled in the affirmative [108].

It is interesting to compare Theorems 47 and 49 with their global counterparts. In the case of approximation by algebraic numbers of degree $\leqslant n$, the appropriate statement is known as the Wirsing conjecture 37 (see § 2.5). The latter has been verified for $n = 2$ by Davenport and Schmidt but is open in higher dimensions. Theorem 47 implies that the statement of the Wirsing conjecture 37 holds for almost all real ξ – the actual conjecture states that it is true at least for all transcendental ξ. In the case of approximation by algebraic integers of degree $\leqslant n$, Roy has shown that the statement analogous to Wirsing's conjecture is false [114]. However, Theorem 49 implies that the statement holds for almost all real ξ, In line with the recent "metrical" progress on Littlewood's conjecture by Einsiedler, Katok, and Lindenstrauss [57], it would be interesting to find out whether the set of possible exceptions to the Wirsing–Schmidt conjecture is of Hausdorff dimension zero. A similar question can also be asked about approximation by algebraic integers; this would shed light on the size of the set of exceptions, shown to be non-empty by Roy.

A. Baker [10] suggested a strengthening of Mahler's problem in which the height $H(P) = \max\{|a_n|, \dots, |a_0|\}$ of the polynomial $P(x) = a_n x^n + \cdots + a_1 x + a_0$ is replaced by

$$H^{\times}(P) = \prod_{i=1}^{n} \max\{1, |a_i|\}^{1/n}.$$

The corresponding statement has been established by Kleinbock and Margulis [76] in a more general context of Diophantine approximation on manifolds. Specialising their result to polynomials gives the following.

Theorem 51 (Kleinbock & Margulis, 1998). *Let $\varepsilon > 0$. Then for almost all $\xi \in \mathbb{R}$ the inequality*

$$|P(\xi)| < H^{\times}(P)^{-n-\varepsilon} \tag{52}$$

has only finitely many solutions in $P \in \mathbb{Z}[x]$ with $\deg P \leqslant n$.

A multiplicative analogue of Theorem 45 with $H(P)$ replaced by $H^{\times}(P)$ has been obtained by Bernik, Kleinbock and Margulis [27] (also within the framework of manifolds). Note that in their theorem the convergence of (44) must be replaced by the stronger condition that $\sum_{h=1}^{\infty} \Psi(h)(\log h)^{n-1} < \infty$. This condition is believed to be optimal but it is not known if the multiplicative analogue of Theorem 47, when $H(P)$ is replaced by $H^{\times}(P)$, holds. In [23], Beresnevich and Velani have proved an inhomogeneous version of the theorem of Kleinbock and Margulis and, in particular, an inhomogeneous version of Theorem 51.

With [11], A. Baker and W.M. Schmidt pioneered the use of Hausdorff dimension in the context of approximation of real numbers by algebraic numbers with a natural generalisation of the Jarník–Besicovitch theorem:

Theorem 53 (Baker & Schmidt, 1970). *Let $w \geqslant n$. Then the set of $\xi \in \mathbb{R}$ for which*

$$|\xi - \alpha| < H(\alpha)^{-w} \tag{54}$$

holds for infinitely many algebraic numbers α with $\deg \alpha \leqslant n$ has Hausdorff dimension $(n+1)/(w+1)$.

In particular, Theorem 53 implies that the set

$$A(w) = \left\{\xi \in \mathbb{R} : |P(\xi)| < H(P)^{-w} \text{ for infinitely many } P \in \mathbb{Z}[x],\ \deg P \leqslant n\right\} \tag{55}$$

has Hausdorff dimension at least $(n+1)/(w+1)$. Baker and Schmidt conjectured that this lower bound is sharp, and this was established by Bernik in [24]:

Theorem 56 (Bernik, 1983). *Let $w \geqslant n$. Then* $\dim A(w) = \dfrac{n+1}{w+1}$.

This theorem has an important consequence for the spectrum of Diophantine exponents already discussed (see Chap. 5 of [36]). Bugeaud [32] has obtained an analogue of Theorem 53 in the case of algebraic integers. However, obtaining an analogue of Theorem 56 for the case of algebraic integers is as yet an open problem.

Recently, Beresnevich, Dickinson, and Velani have established a sharp Hausdorff measure version of Theorem 53, akin to a classical result of Jarník. In order to avoid introducing various related technicalities, we refer the reader to [21, §12.2]. Their result implies the corresponding divergent statement for the Hausdorff measure of the set of $\xi \in \mathbb{R}$ such that (46) holds infinitely often. Obtaining the corresponding convergent statement represents yet another open problem.

There are various generalisations of the above results to the case of complex and p-adic numbers and more generally to the case of S-arithmetic (for instance by D. Kleinbock and G. Tomanov in [77]).

3 Simultaneous Diophantine approximation in higher dimensions

In §2.3, we considered polynomial approximation to a complex number ξ, which is the study of $|P(\xi)|$ for $P \in \mathbf{Z}[X]$. As we have seen, negative results on the existence of polynomial approximations lead to *transcendence measures*. A more general situation is to fix several complex numbers $x_1, \dots, x_m$ and to study the smallness of polynomials in these number; negative results provide *measures of algebraic independence* to $x_1, \dots, x_m$.

This is again a special case, where $\xi_i = x_1^{a_1} \cdots x_m^{a_m}$, of the study of linear combinations in $\xi_1, \dots, \xi_n$, where $\xi_1, \dots, \xi_n$ are given complex numbers. Now negative results are *measures of linear independence* to $\xi_1, \dots, \xi_n$.

There are still more general situations which we are not going to consider thoroughly but which are worth mentioning, namely the study of *simultaneous approximation of dependent quantities* and *approximation on a manifold* (see for instance [26]).

We start with the question of algebraic independence (§3.1) in connection with extensions to higher dimensions of Gel'fond's Criterion 11. Next (§§3.2 and 3.3) we discuss a recent work by M. Laurent [91], who introduces further coefficients for the study of simultaneous approximation. The special case of two numbers (§3.4) is best understood so far.

There is a very recent common generalisation of the question of Diophantine approximation to a point in $\mathbf{R}^n$ which is considered in §3.3 on the one hand, and of the question of approximation to a real number by algebraic numbers of bounded degree considered in §2.5 on the other hand. It consists in the investigation of the approximation to a point in $\mathbf{R}^n$ by algebraic hypersurfaces, or more generally algebraic varieties defined over the rationals. This topic has been recently investigated by W. M. Schmidt in [127] and [126].

3.1 Criteria for algebraic independence

In §2.2 we quoted Gel'fond's algebraic independence results of two numbers of the form α^{β^i} $(1 \leqslant i \leqslant d-1)$. His method has been extended in the work of several mathematicians including A.O. Gel'fond, A.A. Smelev, W.D. Brownawell, G.V. Chudnovsky, P. Philippon, Yu.V. Nesterenko, G. Diaz (see [68], [79], [131], [101, 102], [104, 107], [59] Chap. 6 and [103]). So far the best known result, due to G. Diaz [50], proves "half" of what is expected.

Theorem 57. *Let β be an algebraic number of degree $d \geqslant 2$ and α a non–zero algebraic number. Moreover, let* $\log\alpha$ *be any non-zero logarithm of α. Write α^z in place of* $\exp(z\log\alpha)$. *Then among the numbers*

$$\alpha^{\beta},\ \alpha^{\beta^2},\ \ldots,\alpha^{\beta^{d-1}},$$

at least $\lceil (d+1)/2 \rceil$ are algebraically independent.

In order to prove such a result, as pointed out by S. Lang in [78], it would have been sufficient to replace the transcendence criterion theorem 11 by a criterion for algebraic independence. However, an example, going back to A.Ya. Khintchine in 1926 [74] and quoted in J.W.S. Cassels's book ([46] Chap. V, Th. 14; see also the appendix of [104] and Appendix A of [121]), shows that in higher dimensions, some extra hypothesis cannot be avoided (and this is a source of difficulty in the proof of Theorem 57). After the work of W.D. Brownawell and G.V. Chudnovsky, such criteria were proved by P. Philippon [104, 107], Yu.V. Nesterenko [101, 102], M. Ably, C. Jadot (further references are given in [59] and [138] §15.5). Reference [103] is an introduction to algebraic independence theory which includes a chapter on multihomogeneous elimination by G. Rémond [109] and a discussion of criteria for algebraic independence by P. Philippon [107].

Further progress was made by M. Laurent and D. Roy in 1999, who produced criteria with multiplicities [92, 93] (see also [94]) and considered questions of approximation by algebraic sets. Moreover, in [93] they investigate the approximation properties, by algebraic numbers of bounded degree and height, of a m-tuple which generates a field of transcendence degree 1. This means that the corresponding point in $\mathbf{C}^m$ belongs to an affine curve defined over $\mathbf{Q}$. For $m=1$ they proved in [92] the existence of approximation; this has been improved by G. Diaz in [51]. Further contributions are due to P. Philippon (see for instance [106]).

A very special case of the investigation of Laurent and Roy is a result related to Wirsing's lower bound for ω_n^* (see §2.5), with a weak numerical constant, but with a lower bound for the degree of the approximation. Their result (Corollary 1 of §2 of [93]) has been improved by Y. Bugeaud and O. Teulié (Corollary 5 of [45]) who prove that the approximations α can be required to be algebraic numbers of exact degree n or algebraic integers of exact degree $n+1$.

Theorem 58 (Bugeaud and Teulié). *Let $\epsilon > 0$ be a real number, $n \geqslant 2$ an integer and ξ a real number which is not algebraic of degree n. Then the inequality*

$$|\xi - \alpha| \leqslant \mathrm{H}(\alpha)^{-((n+3)/2)+\epsilon}$$

has infinitely many solutions in algebraic integers α of degree n.

The ϵ in the exponent can be removed by introducing a constant factor. Further, for almost all $\alpha \in \mathbf{R}$, the result holds with the exponent replaced by n, as shown by Y. Bugeaud in [33].

There are close connections between questions of algebraic independence and simultaneous approximation of numbers. We shall not discuss this subject thoroughly here; it would deserve another survey. We just quote a few recent papers.

Applications of Diophantine approximation questions to transcendental number theory are considered by P. Philippon in [105]. M. Laurent [87] gives heuristic motivations in any transcendence degree. Conjecture 15.31 of [138] on simultaneous approximation of complex numbers suggests a path towards results on large transcendence degree.

In [110] D. Roy shows some limitations of the conjectures of algebraic approximation by constructing points in $\mathbf{C}^m$ which do not have good algebraic approximations of bounded degree and height, when the bounds on the degree and height are taken from specific sequences. The coordinates of these points are Liouville numbers.

3.2 *Four exponents: asymptotic or uniform simultaneous approximation by linear forms or by rational numbers*

Let $\xi_1, \ldots, \xi_n$ be real numbers. Assume that the numbers $1, \xi_1, \ldots, \xi_n$ are linearly independent over $\mathbf{Q}$. There are (at least) two points of view for studying approximation to $\xi_1, \ldots, \xi_n$. On the one hand, one may consider linear forms (see for instance [78])

$$|x_0 + x_1\xi_1 + \cdots + x_n\xi_n|.$$

On the other hand, one may investigate the existence of simultaneous approximation by rational numbers

$$\max_{1 \leqslant i \leqslant n} \left| \xi_i - \frac{x_i}{x_0} \right|.$$

Each of these two points of view has two versions, an asymptotic one (with exponent denoted by ω) and a uniform one (with exponent denoted by $\widehat{\omega}$). This gives rise to four exponents introduced in [42] (see also [91]),

$$\omega(\theta), \quad \widehat{\omega}(\theta), \quad \omega({}^t\theta), \quad \widehat{\omega}({}^t\theta),$$

where

$$\theta = (\xi_1, \ldots, \xi_n) \quad \text{and} \quad {}^t\theta = \begin{pmatrix} \xi_1 \\ \vdots \\ \xi_n \end{pmatrix}.$$

We shall recover the situation of §§2.1 and 2.4 in the special case where $\xi_i = \xi^i$, $1 \leqslant i \leqslant n$: for $\theta = (\xi, \xi^2, \ldots, \xi^n)$,

$$\omega(\theta) = \omega_n(\xi), \quad \widehat{\omega}(\theta) = \widehat{\omega}_n(\xi), \quad \omega({}^t\theta) = \omega'_n(\xi), \quad \widehat{\omega}({}^t\theta) = \widehat{\omega}'_n(\xi).$$

Notice that the index n is implicit in the notation involving ω, since it is the number of components of θ.

We start with the question of *asymptotic approximation by linear forms*. We denote by $\omega(\theta)$ the supremum of the real numbers w for which there exist infinitely many positive integers N for which the system

$$|x_0 + x_1\xi_1 + \cdots + x_n\xi_n| \leqslant N^{-w}, \quad 0 < \max_{0 \leqslant i \leqslant n} |x_i| \leqslant N, \tag{59}$$

has a solution in rational integers $x_0, x_1, \ldots, x_n$. An upper bound for $\omega(\theta)$ is a *linear independence measure* for $1, \xi_1, \ldots, \xi_n$.

The hat version of $\omega(\theta)$ is, as expected, related to the study of *uniform approximation by linear forms*: we denote by $\widehat{\omega}(\theta)$ the supremum of the real numbers w such that for any sufficiently large integer N, the same system (59) has a solution.

Obviously $\widehat{\omega}(\theta) \leqslant \omega(\theta)$.

The second question is that of asymptotic simultaneous approximation by rational numbers. Following again [91], we denote by $\omega({}^t\theta)$ the supremum of the real numbers w for which there exist infinitely many positive integers N for which the system

$$\max_{1 \leqslant i \leqslant n} |x_i - x_0\xi_i| \leqslant N^{-w}, \quad \text{with} \quad 0 < \max_{0 \leqslant i \leqslant n} |x_i| \leqslant N \tag{60}$$

has a solution in rational integers $x_0, x_1, \ldots, x_n$. An upper bound for $\omega({}^t\theta)$ is a *simultaneous approximation measure* for $1, \xi_1, \ldots, \xi_n$.

The uniform simultaneous approximation by rational numbers is measured by the hat version of ω: we denote by $\widehat{\omega}({}^t\theta)$ the supremum of the real numbers w such that for any sufficiently large integer N, the same system (60) has a solution.

Again $\widehat{\omega}({}^t\theta) \leqslant \omega({}^t\theta)$.

Transference principles provide relations between $\omega(\theta)$ and $\omega({}^t\theta)$. The next result (Khintchine, 1929 [74]) shows that $\omega(\theta) = n$ if and only if $\omega({}^t\theta) = 1/n$.

Theorem 61 (Khintchine transference principle). *If we set $\omega = \omega(\theta)$ and ${}^t\omega = \omega({}^t\theta)$, then we have*

$$\omega \geqslant n\,{}^t\omega + n - 1 \quad \textit{and} \quad {}^t\omega \geqslant \frac{\omega}{(n-1)\omega + n}.$$

In order to study these numbers, M. Laurent introduces in [91] further exponents as follows.

3.3 Further exponents, following M. Laurent

For each d in the range $0 \leqslant d \leqslant n-1$, M. Laurent [91] introduces two exponents, one for asymptotic approximation $\omega_d(\theta)$ and one for uniform approximation $\widehat{\omega}_d(\theta)$, which measure the quality of simultaneous approximation to the given tuple $\theta = (\xi_1, \ldots, \xi_n)$ from various points of view. First embed $\mathbf{R}^n$ into $\mathbf{P}^n(\mathbf{R})$ by mapping $\theta = (\xi_1, \ldots, \xi_n)$ to $(\xi_1 : \cdots : \xi_n : 1)$.

Now for $0 \leqslant d \leqslant n-1$, define

$$\omega_d(\theta) = \sup\Bigl\{w;\ \text{there exist infinitely many vectors}$$

$$X = x_0 \wedge \cdots \wedge x_d \in \Lambda^{d+1}(\mathbf{Z}^{n+1}) \ \text{ for which } |X \wedge \theta| \leqslant |X|^{-w}\Bigr\}$$

and

$$\widehat{\omega}_d(\theta) = \sup\Bigl\{w;\ \text{for any sufficiently large } N, \text{ there exists}$$

$$X = x_0 \wedge \cdots \wedge x_d \in \Lambda^{d+1}(\mathbf{Z}^{n+1})$$

$$\text{such that } 0 < |X| \leqslant N \quad \text{and} \quad |X \wedge \theta| \leqslant N^{-w}\Bigr\}.$$

Hence $\omega_d(\theta) \geqslant \widehat{\omega}_d(\theta)$.

The multivector $X = x_0 \wedge \cdots \wedge x_d$ is a system of Plücker coordinates of the linear projective subvariety $L = \langle x_0, \ldots, x_d \rangle \subset \mathbf{P}^n(\mathbf{R})$. Then

$$\frac{|X \wedge \theta|}{|X||\theta|}$$

is essentially the distance $d(\theta, L) = \min_{x \in L} d(\theta, x)$ between the image of θ in $\mathbf{P}^n(\mathbf{R})$ to L. As a consequence, equivalent definitions are as follows, where $\mathrm{H}(L)$ denotes the Weil height of any system of Plücker coordinates of L.

$$\omega_d(\theta) = \sup\Bigl\{w;\ \text{there exist infinitely many } L, \text{ rational over } \mathbf{Q},$$

$$\dim L = d \ \text{ and } \ d(\theta, L) \leqslant \mathrm{H}(L)^{-w-1}\Bigr\}$$

and

$$\widehat{\omega}_d(\theta) = \sup\Big\{w;\ \text{for any sufficiently large } N, \text{ there exists } L, \text{ rational over } \mathbf{Q},$$
$$\dim L = d,\ \mathrm{H}(L) \leqslant N \text{ and } d(\theta, L) \leqslant \mathrm{H}(L)^{-1}N^{-w}\Big\}.$$

In the extremal cases $d = 0$ and $d = n-1$, one recovers the exponents of § 3.2:

$$\omega_0(\theta) = \omega({}^t\theta), \quad \widehat{\omega}_0(\theta) = \widehat{\omega}({}^t\theta), \quad \omega_{n-1}(\theta) = \omega(\theta), \quad \widehat{\omega}_{n-1}(\theta) = \widehat{\omega}(\theta).$$

The lower bound

$$\widehat{\omega}_d(\theta) \geqslant \frac{d+1}{n-d} \quad \text{for all} \quad d = 0, \ldots, n-1$$

valid for all θ (with $1, \xi_1, \ldots, \xi_n$ linearly independent over $\mathbf{Q}$) follows from the results of W.M. Schmidt in his foundational paper [123] (see [44]). In particular for $d = n-1$ and $d = 0$ respectively, this lower bound yields

$$\widehat{\omega}(\theta) \geqslant n \quad \text{and} \quad \widehat{\omega}({}^t\theta) \geqslant 1/n$$

and in the special case $\xi_i = \xi^i$ $(1 \leqslant i \leqslant n)$ one recovers the lower bounds

$$\widehat{\omega}_n(\xi) \geqslant n \quad \text{and} \quad \widehat{\omega}'_n(\xi) \geqslant 1/n,$$

which we deduced in § 2.3 and § 2.4 respectively from Dirichlet's box principle.

It was proved by Khintchine in 1926 [74] that $\omega(\theta) = n$ if and only if $\omega({}^t\theta) = 1/n$. In [91], M. Laurent slightly improves on earlier inequalities due to W.M. Schmidt [123], splitting the classical Khintchine's transference principle (Theorem 61) into intermediate steps.

Theorem 62 (Schmidt, Laurent). *Fix $n \geqslant 1$ and $\theta = (\xi_1, \ldots, \xi_n) \in \mathbf{R}^n$. Set $\omega_d = \omega_d(\theta)$, $0 \leqslant d \leqslant n-1$. The "going up transference principle" is*

$$\omega_{d+1} \geqslant \frac{(n-d)\omega_d + 1}{n-d-1}, \quad 0 \leqslant d \leqslant n-2,$$

while the "going down transference principle" is

$$\omega_{d-1} \geqslant \frac{d\omega_d}{\omega_d + d + 1}, \quad 1 \leqslant d \leqslant n-1.$$

Moreover, these estimates are optimal.

As a consequence of Theorem 62, one deduces that if $\omega_d = (d+1)/(n-d)$ for one value of d in the range $0 \leqslant d \leqslant n-1$, then the same equality holds for all $d = 0, 1, \ldots, n-1$. Hence, for almost all $\theta \in \mathbf{R}^n$,

$$\omega_d(\theta) = \widehat{\omega}_d(\theta) = \frac{d+1}{n-d} \quad \text{for} \quad 0 \leqslant d \leqslant n-1.$$

A complement to Theorem 62, involving the hat coefficients, is given in [91] Th. 3.

A problem raised in [91] is to *find the spectrum in* $(\mathbf{R} \cup \{+\infty\})^n$ *of the n-tuples*

$$\big(\omega_0(\theta), \ldots, \omega_{n-1}(\theta)\big),$$

where θ *ranges over the elements* $(\xi_1, \ldots, \xi_n)$ *in* $\mathbf{R}^n$ *with* $1, \xi_1, \ldots, \xi_n$ *linearly independent over* $\mathbf{Q}$. Partial results are given in [91].

In [42] Y. Bugeaud and M. Laurent define and study exponents of *inhomogeneous* Diophantine approximation. Further progress on this topic has been achieved by M. Laurent in [89].

3.4 Dimension 2

We consider the special case $n = 2$ of §3.3: we replace (ξ_1, ξ_2) by (ξ, η). So let ξ and η be two real numbers with $1, \xi, \eta$ linearly independent over $\mathbf{Q}$.

Khintchine's transference Theorem 61 reads in this special case

$$\frac{\omega(\xi,\eta)}{\omega(\xi,\eta)+2} \leqslant \omega\begin{pmatrix}\xi\\ \eta\end{pmatrix} \leqslant \frac{\omega(\xi,\eta)-1}{2}\cdot$$

V. Jarník studied these numbers in a series of papers from 1938 to 1959 (see [36,42, 90]). He proved that both sides are optimal. Also Jarník's formula (of which (29) is a special case) reads

$$\widehat{\omega}\begin{pmatrix}\xi\\ \eta\end{pmatrix} = 1 - \frac{1}{\widehat{\omega}(\xi,\eta)}\cdot \tag{63}$$

The spectrum of each of our four exponents is as follows:

$\omega(\xi,\eta)$ takes any value in the range $[2, +\infty]$,

$\omega\begin{pmatrix}\xi\\ \eta\end{pmatrix}$ takes any value in the range $[1/2, 1]$,

$\widehat{\omega}(\xi,\eta)$ takes any value in the range $[2, +\infty]$,

$\widehat{\omega}\begin{pmatrix}\xi\\ \eta\end{pmatrix}$ takes any value in the range $[1/2, 1]$.

Moreover, for almost all $(\xi,\eta)\in\mathbf{R}^2$,

$$\omega(\xi,\eta)=\widehat{\omega}(\xi,\eta)=2,\quad \omega\begin{pmatrix}\xi\\ \eta\end{pmatrix}=\widehat{\omega}\begin{pmatrix}\xi\\ \eta\end{pmatrix}=\frac{1}{2}\cdot$$

A more precise description of the spectrum of the quadruple is due to M. Laurent [90]:

Theorem 64 (Laurent). *Assume $1,\xi,\eta$ are linearly independent over $\mathbf{Q}$. The four exponents*

$$\omega=\omega(\xi,\eta),\quad \omega'=\omega\begin{pmatrix}\xi\\ \eta\end{pmatrix},\quad \widehat{\omega}=\widehat{\omega}(\xi,\eta),\quad \widehat{\omega}'=\widehat{\omega}\begin{pmatrix}\xi\\ \eta\end{pmatrix}$$

are related by

$$2\leqslant\widehat{\omega}\leqslant+\infty,\quad \widehat{\omega}'=\frac{\widehat{\omega}-1}{\widehat{\omega}},\quad \frac{\omega(\widehat{\omega}-1)}{\omega+\widehat{\omega}}\leqslant\omega'\leqslant\frac{\omega-\widehat{\omega}+1}{\widehat{\omega}}$$

with the obvious interpretation if $\omega=+\infty$. Conversely, for any $(\omega,\omega',\widehat{\omega},\widehat{\omega}')$ in $(\mathbf{R}_{>0}\cup\{+\infty\})^4$ satisfying the previous inequalities, there exists $(\xi,\eta)\in\mathbf{R}^2$, with $1,\xi,\eta$ linearly independent over $\mathbf{Q}$, for which

$$\omega=\omega(\xi,\eta),\quad \omega'=\omega\begin{pmatrix}\xi\\ \eta\end{pmatrix},\quad \widehat{\omega}=\widehat{\omega}(\xi,\eta),\quad \widehat{\omega}'=\widehat{\omega}\begin{pmatrix}\xi\\ \eta\end{pmatrix}.$$

As a consequence:

Corollary 65. *The exponents $\omega=\omega(\xi,\eta)$, $\widehat{\omega}=\widehat{\omega}(\xi,\eta)$ are related by*

$$\omega\geqslant\widehat{\omega}(\widehat{\omega}-1)\quad\text{and}\quad\widehat{\omega}\geqslant 2.$$

Conversely, for any $(\omega,\widehat{\omega})$ satisfying these conditions, there exists (ξ,η) for which

$$\omega(\xi,\eta)=\omega\quad\text{and}\quad\widehat{\omega}(\xi,\eta)=\widehat{\omega}.$$

Corollary 66. *The exponents $\omega'=\omega\begin{pmatrix}\xi\\ \eta\end{pmatrix}$, $\widehat{\omega}'=\widehat{\omega}\begin{pmatrix}\xi\\ \eta\end{pmatrix}$ are related by*

$$\omega'\geqslant\frac{\widehat{\omega}'^2}{1-\widehat{\omega}'}\quad\text{and}\quad\frac{1}{2}\leqslant\widehat{\omega}'\leqslant 1.$$

Conversely, for any $(\omega',\widehat{\omega}')$ satisfying these conditions, there exists (ξ,η) with

$$\omega\begin{pmatrix}\xi\\ \eta\end{pmatrix}=\omega'\quad\text{and}\quad\widehat{\omega}\begin{pmatrix}\xi\\ \eta\end{pmatrix}=\widehat{\omega}'.$$

The next open problem has been raised by M. Laurent:

Open Problem 67 (Laurent). *Is there an extension of Jarník's equality (63) in higher dimensions relating $\widehat{\omega}(\theta)$ and $\widehat{\omega}({}^t\theta)$ for $\theta \in \mathbf{R}^n$?*

3.5 Approximation by hypersurfaces

In dimension 1 an irreducible hypersurface is nothing else than a point. The exponents $\omega_n(\xi)$ and their hat companions in § 2.3 measure $|P(\xi)|$ for $P \in \mathbf{Z}[X]$, while $\omega_n^*(\xi)$ of §2.5 measure the distance between a point $\xi \in \mathbf{C}$ and algebraic numbers α.

A generalisation of these questions in higher dimensions, where $\xi \in \mathbf{C}^n$, is the study of $|P(\xi)|$ for $P \in \mathbf{Z}[X_1, \dots, X_n]$ and of $\min_\alpha |\xi - \alpha|$, where α runs over the set of zeros of such P. As already mentioned in the introduction of §3, a lower bound for $|P(\xi)|$ when the degree of P is fixed is nothing else than a linear independence measure for $(\xi, \dots, \xi^n)$. To consider such quantities also when the degree of P varies yields a generalisation of Mahler's classification to several variables, which has been considered by Yu Kunrui [140]. A generalisation to higher dimensions of both Mahler and Koksma classifications has been achieved by W.M. Schmidt in [126], who raises a number of open problems suggesting that the close connection between the two classifications in dimension 1 does not extend to the classification of tuples.

In [127] W.M. Schmidt deals with approximation to points ξ in $\mathbf{R}^n$ or in $\mathbf{C}^n$ by algebraic hypersurfaces, and more generally by algebraic varieties, defined over the rationals.

Let $\mathcal{M}$ be a nonempty finite set of monomials in $x_1, \dots, x_n$ with $|\mathcal{M}|$ elements. Denote by $\mathcal{P}(\mathcal{M})$ the set of polynomials in $\mathbf{Z}[x_1, \dots, x_n]$ which are linear combinations of monomials in $\mathcal{M}$. Using Dirichlet's box principle or Minkowski's theorem on linear forms, one shows the existence of nonzero elements in $\mathcal{P}(\mathcal{M})$ for which $|P(\xi)|$ is small. It is a much more difficult task to get the existence of nonzero elements in $\mathcal{P}(\mathcal{M})$ for which the distance $\delta\big(\xi, A(P)\big)$ between ξ and the hypersurface $A(P)$ defined by $P = 0$ is small.

W.M. Schmidt asks whether given ξ and $\mathcal{M}$, there exists $c = c(\xi, \mathcal{M}) > 0$ such that there are infinitely many $P \in \mathcal{P}(\mathcal{M})$ with $\delta\big(\xi, A(P)\big) \leqslant cH(P)^{-m}$, where $m = |\mathcal{M}|$ in the real case $\xi \in \mathbf{R}^n$ and $m = |\mathcal{M}|/2$ in the complex case $\xi \in \mathbf{C}^n$. He proves such an estimate when $|\mathcal{M}| = n + 1$, and also in the real case when $|\mathcal{M}| = n + 2$. In the case $|\mathcal{M}| = n + 1$ he proves a uniform result, in the sense of Y. Bugeaud and M. Laurent [41]: given $N \geqslant 1$, there is a $P \in \mathcal{P}(\mathcal{M})$ with height $H(P) \leqslant N$ for which $\delta\big(\xi, A(P)\big) \leqslant cN^{1-m}H(P)^{-1}$. A number of further results are proved in which the exponent is not the conjectured one. The author also investigates the approximation by algebraic hypersurfaces (another reference on this topic is [94]).

Special cases of the very general and deep results of this paper were due to F. Amoroso, W.D. Brownawell, M. Laurent and D. Roy, and P. Philippon. Further previous results related with Wirsing's conjecture were also achieved by V.I. Bernik and K.I. Tishchenko.

An upper bound for the distance $\delta(\xi, A)$ means that there is a point on the hypersurface A (or more generally the variety A) close to ξ. The author also investigates the "size" of the set of such elements. The auxiliary results proved in [127] on this question have independent interest.

3.6 Further metrical results

The answer to the question of Schmidt on approximation of points $\xi \in \mathbb{R}^m$ by algebraic hypersurfaces $A(P)$ is almost surely affirmative. This follows from a general theorem established by Beresnevich, Bernik, Kleinbock and Margulis. With reference to Section 3.5, let $\mathcal{M}$ be a set of monomials of cardinality $m = |\mathcal{M}|$ in variables $x_1, \dots, x_k$, where we naturally assume that $m \geqslant 2$. Further, let $P(\mathcal{M})$ be the set of polynomials in $\mathbb{Z}[x_1, \dots, x_k]$ which are linear combinations of monomials in $\mathcal{M}$. Given a function $\Psi : \mathbb{N} \to (0, +\infty)$, let

$$\mathcal{A}_k(\Psi, \mathcal{M}) = \Big\{(\xi_1, \dots, \xi_k) \in [0,1]^k : |P(\xi_1, \dots, \xi_k)| < H(P)^{-m+2}\Psi(H(P)) \\ \text{for infinitely many } P \in P(\mathcal{M})\Big\}.$$

We are interested in $|\mathcal{A}_k(\Psi, \mathcal{M})|$, the k-dimensional Lebesgue measure of $\mathcal{A}_k(\Psi, \mathcal{M})$.

Theorem 68 (Beresnevich, Bernik, Kleinbock and Margulis). *For any decreasing Ψ,*

$$|\mathcal{A}_k(\Psi, \mathcal{M})| = \begin{cases} 0 \ \textit{if} \sum_{h=1}^{\infty} \Psi(h) < \infty\,, \\ 1 \ \textit{if} \sum_{h=1}^{\infty} \Psi(h) = \infty\,. \end{cases}$$

The convergence case of this theorem has been independently established by Beresnevich [15] and Bernik, Kleinbock and Margulis [27] using different techniques. The multiplicative analogue of the convergence part of Theorem 68, where $H(P)$ is replaced with $H^{\times}(P)$, has also been obtained in [27]. In addition, Theorem 68 holds when $\mathcal{M}$ is a set of m analytic functions defined on $(0,1)^k$ and linearly independent over $\mathbb{R}$. The analyticity assumption can also be relaxed towards a non-degeneracy condition.

The divergence case is established in [19] in the following stronger form connected with §3.5, where the notation δ and $A(P)$ are explained.

Theorem 69 (Beresnevich, Bernik, Kleinbock and Margulis). *Let Ψ be decreasing and such that $\sum_{h=1}^{\infty}\Psi(h)$ diverges. Then for almost all $\xi = (\xi_1,\dots,\xi_k) \in [0,1]^k$,*

$$\delta(\xi, A(P)) < H(P)^{-m+1}\Psi(H(P))$$

has infinitely many solutions $P \in P(\mathcal{M})$.

Taking $\Psi(h) = h^{-1}\log^{-1} h$, we get the following corollary which answers Schmidt's question in §3.5 in the affirmative for almost all points:

Corollary 1. *For almost all $\xi = (\xi_1,\dots,\xi_k) \in \mathbb{R}^k$, the inequality*

$$\delta(\xi, A(P)) < H(P)^{-m}\log^{-1} H(P) \tag{70}$$

has infinitely many solutions $P \in P(\mathcal{M})$.

Another interesting corollary corresponds to the special case of $\mathcal{M}$ being the set of all monomials of degree at most d. In this case we simply have the case of approximation by multivariable polynomials of degree at most d, where now

$$m = \binom{k+d}{d}.$$

In the case of convergence in Theorem 68, a lower bound for the Hausdorff dimension of $\mathcal{A}_k(\Psi,\mathcal{M})$ is implied by a general theorem for manifolds of Dickinson and Dodson [53]. Obtaining the corresponding upper bound in general remains an open problem but see Theorem 56 and [12, 17, 54]. The Hausdorff measure version of Theorem 69 has been established in [21].

Yet another class of interesting problems concerns the measure of transcendence and algebraic independence of numbers. Recall that complex numbers $z_1,\dots,z_m$ are called algebraically independent if for any non-zero polynomial $P \in \mathbb{Z}[x_1,\dots,x_m]$, the value $P(z_1,\dots,z_m)$ is not 0. Actually, $P(z_1,\dots,z_m)$ can still get very small when $z_1,\dots,z_m$ are algebraically independent. Indeed, using Dirichlet's pigeonhole principle, one can readily show that there is a constant $c_1 > 0$ such that for any real numbers $x_1,\dots,x_m$, there are infinitely many polynomials $P \in \mathbb{Z}[x_1,\dots,x_m]$ such that

$$|P(x_1,\dots,x_m)| < e^{-c_1 t(P)^{m+1}}, \tag{71}$$

where $t(P) = \deg P + \log H(P)$ is called the type of P. A conjecture of Mahler [97] proved by Nesterenko [100] says that in the case $m = 1$ for almost all real numbers x_1 there is a constant $c_0 > 0$ such that $|P(x_1)| > e^{-c_0 t(P)^2}$ for all non-zero $P \in \mathbb{Z}[x]$. Nesterenko has also shown that for $\tau = m+2$, for almost all $(x_1,\dots,x_m) \in \mathbb{R}^m$ there is a constant $c_0 > 0$ such that

$$|P(x_1,\dots,x_m)| > e^{-c_0 t(P)^{\tau}} \qquad \text{for all non-zero } P \in \mathbb{Z}[x_1,\dots,x_m] \tag{72}$$

and conjectured that the latter is indeed true with the exponent $\tau = m + 1$. This has been verified by Amoroso [6] over $\mathbb{C}$, but the "real" conjecture has been recently established by Nesterenko's student Mikhailov [99]:

Theorem 73 (Mikhailov, 2007). *Let $\tau = m+1$. Then for almost all $(x_1, \ldots, x_m) \in \mathbb{R}^m$ there is a constant $c_0 > 0$ such that (72) holds.*

We conclude by discussing the interaction of metrical, analytic, and other techniques in the question of counting and distribution of rational points near a given smooth planar curve Γ. In what follows we will assume that the curvature of Γ is bounded between two positive constants. Let $N_\Gamma(Q, \delta)$ denote the number of rational points $(p_1/q, p_2/q)$, where $p_1, p_2, q \in \mathbb{Z}$ with $0 < q \leqslant Q$, within a distance at most δ from Γ.

Huxley [71] has proved that for any $\varepsilon > 0$, $N_\Gamma(Q, \delta) \ll Q^{3+\varepsilon}\delta + Q$. Until recently, this bound has remained the only non–trivial result. Furthermore, very little has been known about the existence of rational points near planar curves for $\delta < Q^{-3/2}$, that is whether $N_\Gamma(Q, \delta) > 0$ when $\delta < Q^{-3/2}$. An explicit question of this type motivated by Elkies [58] has been recently raised by Barry Mazur who asks: "*given a smooth curve in the plane, how near to it can a point with rational coordinates get and still miss?*" (Question (3) in [98, § 11]). When $\delta = o(Q^{-2})$, the rational points in question cannot miss Γ if Γ is a rational quadratic curve in the plane (see [26]). This leads to $N_\Gamma(Q, \delta)$ vanishing for some choices of Γ when $\delta = o(Q^{-2})$. For example, the curve Γ given by $x^2 + y^2 = 3$ has no rational points [22]. When $\delta \gg Q^{-2}$, a lower bound on $N_\Gamma(Q, \delta)$ can be obtained using Khintchine's transfer principle. However, such a bound would be far from being close to the heuristic count of $Q^3\delta$ (see [26]). The first sharp lower bound on $N_\Gamma(Q, \delta)$ has been given by Beresnevich [14], who has shown that for the parabola $\Gamma = (x, x^2)$, $N_\Gamma(Q, \delta) \gg Q^3\delta$ when $\delta \gg Q^{-2}$.

Recently, Beresnevich, Dickinson, and Velani [22] have shown that for an arbitrary smooth planar curve Γ with non-zero curvature $N_\Gamma(Q, \delta) \gg Q^3\delta$ when $\delta \gg Q^{-2}$. Moreover, they show that the rational points in question are uniformly distributed in the sense that they form a ubiquitous system (see [18] for a discussion on ubiquity and related notions). They further apply this to get various metric results about simultaneous approximation to points on Γ. These include a Khintchine-type theorem and its Hausdorff measure analogue. In particular, for any $w \in (1/2, 1)$ they explicitly obtain the Hausdorff dimension of the set of w-approximable points on Γ:

$$\dim\Big\{(x, y) \in \Gamma : \max\{\|qx\|, \|qy\|\} < q^{-w} \text{ for infinitely many } q \in \mathbb{N}\Big\} = \frac{2-w}{1+w}. \tag{74}$$

Here Γ is a smooth planar curve with non–vanishing curvature. Using analytic methods, Vaughan and Velani [137] have shown that $\varepsilon > 0$ can be removed from Huxley's estimate for $N_\Gamma(Q, \varepsilon)$. Combining the results of [22] and [137] gives the following natural generalisation of Khintchine's theorem.

Theorem 75 (Beresnevich, Dickinson, Vaughan, Velani). *Let $\psi : \mathbb{N} \to (0, +\infty)$ be monotonic. Let Γ be a $C^{(3)}$ planar curve of finite length ℓ with non–vanishing curvature and let*

$$\mathcal{A}_2(\psi, \Gamma) = \Big\{(x, y) \in \Gamma : \max\{\|qx\|, \|qy\|\} < \psi(q) \text{ holds for infinitely many } q \in \mathbb{N}\Big\}.$$

Then the arclength[3] $|\mathcal{A}_2(\psi, \Gamma)|$ of $\mathcal{A}_2(\psi, \Gamma)$ satisfies

$$|\mathcal{A}_2(\psi, \Gamma)| = \begin{cases} 0 \text{ if } \sum_{h=1}^{\infty} h\,\psi(h) < \infty, \\ \ell \text{ if } \sum_{h=1}^{\infty} h\,\psi(h) = \infty. \end{cases}$$

Furthermore, let $s \in (0, 1)$ and let $\mathcal{H}^s$ denote the s-dimensional Hausdorff measure. Then

$$\mathcal{H}^s(\mathcal{A}_2(\psi, \Gamma)) = \begin{cases} 0 & \text{if } \sum_{h=1}^{\infty} h^{2-s}\psi(h)^s < \infty, \\ +\infty & \text{if } \sum_{h=1}^{\infty} h^{2-s}\psi(h)^s = \infty. \end{cases} \tag{76}$$

Note that (74) is a consequence of (76).

In higher dimensions, Druţu [55] has studied the distribution of rational points on non-degenerate rational quadrics in $\mathbb{R}^n$ and obtained a result similar to (76) in the case $\psi(q) = o(q^{-2})$. However, simultaneous Diophantine approximation on manifolds as well as the distribution of rational points near manifolds (in particular algebraic varieties) is little understood.[4] In other words, the higher-dimensional version of the "near-misses" question of Mazur mentioned above has never been systematically considered.

References

1. B. Adamczewski – "Sur l'exposant de densité des nombres algébriques", Int. Math. Res. Not., article ID rnm024, 6 pages, 2007.
2. B. Adamczewski and Y. Bugeaud – "Palindromic continued fractions.", *Ann. Inst. Fourier* **57** (2007), no. 5, p. 1557–1574.
3. — , "Mesures de transcendance et aspects quantitatifs de la méthode de Thue–Siegel–Roth–Schmidt", Proc. London Math. Soc. 101 (2010), 1–31.
4. G. Alain – "Simultaneous approximation of a real number by all conjugates of an algebraic number.", *Acta Arith.* **127** (2007), no. 1, p. 63–70.

---

[3]One-dimensional Lebesgue measure on Γ.

[4]see Beresnevich: Rational points near manifolds and metric Diophantine approximation. Ann. of Math. (to appear). http://arxiv.org/abs/0904.0474.

5. K. ALNIAÇIK, Y. AVCI and Y. BUGEAUD – "On U_m-numbers with small transcendence measure", *Acta Math. Hungar.* **99** (2003), no. 4, p. 271–277.
6. F. AMOROSO – "Polynomials with high multiplicity", *Acta Arith.* **56** (1990), no. 4, p. 345–364.
7. M. AMOU and Y. BUGEAUD – "On integer polynomials with multiple roots", *Mathematika* **54** (2007), no. 1-2, p. 83–92.
8. B. ARBOUR and D. ROY – "A Gel′fond type criterion in degree two", *Acta Arith.* **111** (2004), no. 1, p. 97–103.
9. A. BAKER – "On a theorem of Sprindžuk", *Proc. Royal Soc. Series A* **292** (1966), p. 92–104.
10. — , *Transcendental number theory*, Cambridge University Press, London, 1975.
11. A. BAKER and W. M. SCHMIDT – "Diophantine approximation and Hausdorff dimension", *Proc. Lond. Math. Soc.* **21** (1970), p. 1–11.
12. R. C. BAKER – "Dirichlet's theorem on Diophantine approximation", *Math. Proc. Cam. Phil. Soc.* **83** (1978), p. 37–59.
13. V. BERESNEVICH – "On approximation of real numbers by real algebraic numbers", *Acta Arith.* **90** (1999), no. 2, p. 97–112.
14. — , "Distribution of rational points near a parabola", *Dokl. Nats. Akad. Nauk Belarusi* **45** (2001), no. 4, p. 21–23, 123.
15. — , "A Groshev type theorem for convergence on manifolds", *Acta Math. Hungar.* **94** (2002), no. 1-2, p. 99–130.
16. — , "On a theorem of V. Bernik in the metric theory of Diophantine approximation", *Acta Arith.* **117** (2005), no. 1, p. 71–80.
17. V. BERESNEVICH, V. I. BERNIK and M. M. DODSON – "On the Hausdorff dimension of sets of well-approximable points on nondegenerate curves", *Dokl. Nats. Akad. Nauk Belarusi* **46** (2002), no. 6, p. 18–20, 124.
18. — , "Regular systems, ubiquity and Diophantine approximation", A panorama of number theory or the view from Baker's garden (Zürich, 1999), Cambridge Univ. Press, Cambridge, 2002, p. 260–279.
19. V. BERESNEVICH, V. I. BERNIK, D. Y. KLEINBOCK and G. A. MARGULIS – "Metric Diophantine approximation: the Khintchine-Groshev theorem for nondegenerate manifolds", *Mosc. Math. J.* **2** (2002), no. 2, p. 203–225, Dedicated to Yuri I. Manin on the occasion of his 65th birthday.
20. V. BERESNEVICH, H. DICKINSON and S. VELANI – "Sets of exact 'logarithmic' order in the theory of Diophantine approximation", *Math. Ann.* **321** (2001), no. 2, p. 253–273.
21. — , "Measure theoretic laws for lim sup sets", *Mem. Amer. Math. Soc.* **179** (2006), no. 846, p. x+91.
22. — , "Diophantine approximation on planar curves and the distribution of rational points", *Ann. of Math. (2)* **166** (2007), no. 3, p. 367–426.
23. V. BERESNEVICH and S. VELANI – "An inhomogeneous transference principle and Diophantine approximation", Proc. Lond. Math. Soc. (3) 101 (2010), no. 3, 821–851.
24. V. I. BERNIK – "An application of Hausdorff dimension in the theory of Diophantine approximation", *Acta Arith.* **42** (1983), no. 3, p. 219–253, (In Russian). English transl. in *Amer. Math. Soc. Transl.* **140** (1988), 15–44.
25. — , "On the exact order of approximation of zero by values of integral polynomials", *Acta Arith.* **53** (1989), p. 17–28, (In Russian).
26. V. I. BERNIK and M. M. DODSON – *Metric Diophantine approximation on manifolds*, Cambridge Tracts in Mathematics, vol. 137, Cambridge University Press, Cambridge, 1999.
27. V. I. BERNIK, D. KLEINBOCK and G. A. MARGULIS – "Khintchine-type theorems on manifolds: the convergence case for standard and multiplicative versions", *Internat. Math. Res. Notices* (2001), no. 9, p. 453–486.
28. V. I. BERNIK and K. I. TISHCHENKO – "Integral polynomials with an overfall of the coefficient values and Wirsing's theorem", *Dokl. Akad. Nauk Belarusi* **37** (1993), no. 5, p. 9–11, 121 (1994).

29. Y. BILU – "The many faces of the subspace theorem [after Adamczewski, Bugeaud, Corvaja, Zannier...]", Séminaire Bourbaki, 59e année (2006–2007), No 967, 2006.
30. E. BOMBIERI and W. GUBLER – *Heights in Diophantine geometry*, New Mathematical Monographs 4. Cambridge: Cambridge University Press, 2006.
31. Y. BUGEAUD – "Approximation par des nombres algébriques", *J. Number Theory* **84** (2000), no. 1, p. 15–33.
32. —, "Approximation by algebraic integers and Hausdorff dimension", *J. Lond. Math. Soc.* **65** (2002), p. 547–559.
33. —, "Approximation by algebraic integers and Hausdorff dimension.", *J. Lond. Math. Soc., II. Ser.* **65** (2002), no. 3, p. 547–559.
34. —, "Mahler's classification of numbers compared with Koksma's", *Acta Arith.* **110** (2003), no. 1, p. 89–105.
35. —, "Sets of exact approximation order by rational numbers", *Math. Ann.* **327** (2003), no. 1, p. 171–190.
36. — , *Approximation by algebraic numbers*, Cambridge Tracts in Mathematics, vol. 160, Cambridge University Press, Cambridge, 2004.
37. —, "Mahler's classification of numbers compared with Koksma's. III", *Publ. Math. Debrecen* **65** (2004), no. 3-4, p. 305–316.
38. —, "Diophantine approximation and Cantor sets.", *Math. Ann.* **341** (2008), no. 3, p. 677–684.
39. — , "Mahler's classification of numbers compared with Koksma's. II", *Diophantine approximation. Festschrift for Wolfgang Schmidt*, vol. 16, 2008, Developments in Math, Eds: H. P. Schlickewei, K. Schmidt and R. Tichy, Springer-Verlag, p. 107–121.
40. Y. BUGEAUD and J.-H. EVERTSE – "Approximation of complex algebraic numbers by algebraic numbers of bounded degree", Ann. Scuola Normale Superiore di Pisa, **8** (2009), 333–368.
41. Y. BUGEAUD and M. LAURENT – "Exponents of Diophantine approximation and Sturmian continued fractions", *Ann. Inst. Fourier (Grenoble)* **55** (2005), no. 3, p. 773–804.
42. — , "On exponents of homogeneous and inhomogeneous Diophantine approximation", *Moscow Math. J.* **5** (2005), p. 747–766.
43. —, "Exponents of Diophantine approximation", *Diophantine Geometry proceedings, Scuola Normale Superiore Pisa, Ser. CRM* **4** (2007), p. 101–121.
44. — , "On transfer inequalities in Diophantine approximation. II", *Math. Zeitschrift*, **265** (2010), 249–262.
45. Y. BUGEAUD and O. TEULIÉ – "Approximation d'un nombre réel par des nombres algébriques de degré donné", *Acta Arith.* **93** (2000), no. 1, p. 77–86.
46. J. W. S. CASSELS – *An introduction to Diophantine approximation*, Cambridge Tracts in Mathematics and Mathematical Physics, No. 45, Cambridge University Press, New York, 1957.
47. T. W. CUSICK and M. E. FLAHIVE – *The Markoff and Lagrange spectra*, Mathematical Surveys and Monographs, vol. 30, American Mathematical Society, Providence, RI, 1989.
48. H. DAVENPORT and W. M. SCHMIDT – "Approximation to real numbers by quadratic irrationals", *Acta Arith.* **13** (1967), p. 169–176.
49. — , "Approximation to real numbers by algebraic integers", *Acta Arith.* **15** (1969), p. 393–416.
50. G. DIAZ – "Grands degrés de transcendance pour des familles d'exponentielles", *J. Number Theory* **31** (1989), no. 1, p. 1–23.
51. —, "Une nouvelle propriété d'approximation diophantienne", *C. R. Acad. Sci. Paris Sér. I Math.* **324** (1997), no. 9, p. 969–972.
52. G. DIAZ and M. MIGNOTTE – "Passage d'une mesure d'approximation à une mesure de transcendance", *C. R. Math. Rep. Acad. Sci. Canada* **13** (1991), no. 4, p. 131–134.
53. H. DICKINSON and M. M. DODSON – "Extremal manifolds and Hausdorff dimension", *Duke Math. J.* **101** (2000), no. 2, p. 271–281.
54. M. M. DODSON, B. P. RYNNE and J. A. G. VICKERS – "Metric Diophantine approximation and Hausdorff dimension on manifolds", *Math. Proc. Cam. Phil. Soc.* **105** (1989), p. 547–558.

55. C. DRUŢU – "Diophantine approximation on rational quadrics", *Math. Ann.* **333** (2005), no. 2, p. 405–469.
56. R. J. DUFFIN and A. C. SCHAEFFER – "Khintchine's problem in metric Diophantine approximation", *Duke Math. J.* **8** (1941), p. 243–255.
57. M. EINSIEDLER, A. KATOK and E. LINDENSTRAUSS – "Invariant measures and the set of exceptions to Littlewood's conjecture", *Ann. of Math. (2)* **164** (2006), no. 2, p. 513–560.
58. N. ELKIES – "Rational points near curves and small nonzero $|x^3 - y^2|$ via lattice reduction", *Algorithmic number theory (Leiden, 2000)*, Lecture Notes in Comput. Sci., vol. 1838, Springer, Berlin, 2000, http://arxiv.org/abs/math/0005139, p. 33–63.
59. N. I. FEL′DMAN and Y. V. NESTERENKO – "Transcendental numbers", in *Number Theory, IV*, Encyclopaedia Math. Sci., vol. **44**, Springer, Berlin, 1998, p. 1–345.
60. N. I. FEL′DMAN and A. B. ŠIDLOVSKIĬ – "The development and present state of the theory of transcendental numbers", *Uspehi Mat. Nauk* **22** (1967), no. 3 (135), p. 3–81.
61. S. FISCHLER – "Spectres pour l'approximation d'un nombre réel et de son carré", *C. R. Math. Acad. Sci. Paris* **339** (2004), no. 10, p. 679–682.
62. — , "Palindromic prefixes and Diophantine approximation", *Monatshefte Math.* **151** (2007), no. 1, p. 1–87.
63. S. FISCHLER and T. RIVOAL – "Un exposant de densité en approximation rationnelle", *International Mathematics Research Notices* (2006), no. 24, p. 48, Article ID 95418.
64. A. O. GEL′FOND – "On the algebraic independence of algebraic powers of algebraic numbers", *Doklady Akad. Nauk SSSR (N.S.)* **64** (1949), p. 277–280.
65. — , "On the algebraic independence of transcendental numbers of certain classes", *Doklady Akad. Nauk SSSR (N.S.)* **67** (1949), p. 13–14.
66. — , "On the algebraic independence of transcendental numbers of certain classes", *Uspehi Matem. Nauk (N.S.)* **4** (1949), no. 5(33), p. 14–48.
67. — , "The approximation of algebraic numbers by algebraic numbers and the theory of transcendental numbers", *Amer. Math. Soc. Translation* (1952), no. 65, p. 45.
68. — , *Transcendental and algebraic numbers*, Translated by Leo F. Boron from the first Russian edition "Transcendentnye i algebraičeskie čisla", Gosudarstv. Izdat. Tehn.-Teor. Lit., Moscow 1952, Dover Publications Inc., New York, 1960.
69. G. HARMAN – *Metric number theory.*, vol. 18, London Mathematical Society Monographs. New Series, Oxford: Clarendon Press, 1998.
70. P.-C. HU and C.-C. YANG – *Distribution theory of algebraic numbers*, de Gruyter Expositions in Mathematics, vol. 45, Walter de Gruyter GmbH & Co. KG, Berlin, 2008.
71. M. N. HUXLEY – "The rational points close to a curve", *Ann. Scuola Norm. Sup. Pisa Cl. Sci. (4)* **21** (1994), no. 3, p. 357–375.
72. V. JARNIK – "Zum Khintchineschen "Übertragungssatz", *Acad. Sci. URSS., Fil. Géorgienne, Trav. Inst. math., Tbilissi, 3* (1938), p. 193–216 (German).
73. A. Y. KHINTCHINE – "Einige Sätze über Kettenbrüche, mit Anwendungen auf die Theorie der Diophantischen Approximationen", *Math. Ann.* **92** (1924), p. 115–125.
74. — , "Über eine Klasse linearer Diophantischer Approximationen", *Rendiconti Palermo,* **50** (1926), p. 170–195.
75. — , *Continued fractions*, Dover Publications Inc., Mineola, NY, 1964.
76. D. Y. KLEINBOCK and G. A. MARGULIS – "Flows on homogeneous spaces and Diophantine approximation on manifolds", *Ann. of Math. (2)* **148** (1998), no. 1, p. 339–360.
77. D. KLEINBOCK and G. TOMANOV – "Flows on S-arithmetic homogeneous spaces and applications to metric Diophantine approximation.", *Comment. Math. Helv.* **82** (2007), no. 3, p. 519–581.
78. S. LANG – "Report on Diophantine approximations", *Bull. Soc. Math. France* **93** (1965), p. 177–192, = [84] p. 326–341.
79. — , *Introduction to transcendental numbers*, Addison-Wesley Publishing Co., Reading, Mass.-London-Don Mills, Ont., 1966, = [84] p. 396–506.
80. — , "Transcendental numbers and Diophantine approximations", *Bull. Amer. Math. Soc.* **77** (1971), p. 635–677, = [85] p. 1–43.

81. — , "Higher dimensional Diophantine problems", *Bull. Amer. Math. Soc.* **80** (1974), p. 779–787, = [85] p. 102–110.
82. — , "Old and new conjectured Diophantine inequalities", *Bull. Amer. Math. Soc. (N.S.)* **23** (1990), no. 1, p. 37–75, = [86] p. 355–393.
83. — , *Introduction to Diophantine approximations*, second éd., Springer-Verlag, New York, 1995.
84. — , *Collected papers. Vol. I*, Springer-Verlag, New York, 2000, 1952–1970.
85. — , *Collected papers. Vol. II*, Springer-Verlag, New York, 2000, 1971–1970.
86. — , *Collected papers. Vol. III*, Springer-Verlag, New York, 2000, 1978–1990.
87. M. LAURENT – "New methods in algebraic independence", *Győry, Kálmán (ed.) et al., Number theory. Diophantine, computational and algebraic aspects. Proceedings of the international conference, Eger, Hungary, July 29–August 2, 1996*, Berlin: de Gruyter. 311-330 , 1998.
88. — , "Simultaneous rational approximation to the successive powers of a real number", *Indag. Math. (N.S.)* **14** (2003), no. 1, p. 45–53.
89. — , "On inhomogeneous Diophantine approximation and Hausdorff dimension", preprint, Proceedings of a conference dedicated to Gelfond, Moscow, 9 pages, 2007.
90. — , "Exponents of Diophantine Approximation in dimension two", Michel Waldschmidt 6 octobre 2011 16:32.
91. — , "On transfer inequalities in Diophantine approximation", in "*Analytic Number Theory - Essays in Honour of Klaus Roth*", Cambridge University Press, 2009, p. 306–314.
92. M. LAURENT and D. ROY – "Criteria of algebraic independence with multiplicities and interpolation determinants", *Trans. Amer. Math. Soc.* **351** (1999), no. 5, p. 1845–1870.
93. — , "Sur l'approximation algébrique en degré de transcendance un", *Ann. Inst. Fourier (Grenoble)* **49** (1999), no. 1, p. 27–55.
94. — , "Criteria of algebraic independence with multiplicities and approximation by hypersurfaces", *J. reine angew. Math.* **536** (2001), p. 65–114.
95. J. LEVESLEY, C. SALP and S. VELANI – "On a problem of K. Mahler: Diophantine approximation and Cantor sets", *Math. Ann.* **338** (2007), no. 1, p. 97–118.
96. K. MAHLER – "Über das Maß der Menge aller S-Zahlen", *Math. Ann.* **106** (1932), p. 131–139.
97. — , "On the order function of a transcendental number", *Acta Arith.* **18** (1971), p. 63–76.
98. B. MAZUR – "Perturbations, deformations, and variations (and "near-misses") in geometry, physics, and number theory", *Bull. Amer. Math. Soc. (N.S.)* **41** (2004), no. 3, p. 307–336 (electronic).
99. S. V. MIKHAĬLOV – "A transcendence type for almost all points in an m-dimensional real space", *Mat. Sb.* **198** (2007), no. 10, p. 67–88.
100. Y. V. NESTERENKO – "An order function for almost all numbers", *Mat. Zametki* **15** (1974), p. 405–414.
101. — , "Algebraic independence of algebraic powers of algebraic numbers", *Mat. Sb. (N.S.)* **123(165)** (1984), no. 4, p. 435–459.
102. — , "The measure of algebraic independence of values of certain functions", *Mat. Sb. (N.S.)* **128(170)** (1985), no. 4, p. 545–568, 576.
103. Y. NESTERENKO and P. PHILIPPON (éds.) – *Introduction to algebraic independence theory*, Lecture Notes in Mathematics, vol. 1752, Springer-Verlag, Berlin, 2001.
104. P. PHILIPPON – "Critères pour l'indépendance algébrique", *Inst. Hautes Études Sci. Publ. Math.* **64** (1986), p. 5–52.
105. — , "Une approche méthodique pour la transcendance et l'indépendance algébrique de valeurs de fonctions analytiques", *J. Number Theory* **64** (1997), no. 2, p. 291–338.
106. — , "Approximations algébriques des points dans les espaces projectifs. I", *J. Number Theory* **81** (2000), no. 2, p. 234–253.
107. — , "Criteria for algebraic independence", Introduction to algebraic independence theory, Lecture Notes in Math., Vol. 1752, Springer, Berlin, 2001, Chap. 8 in [103], p. 133–141.

108. A. POLLINGTON AND R. VAUGHAN – "The k-dimensional Duffin and Schaeffer conjecture", *Mathematika* **37** (1990), p. 190–200.
109. G. RÉMOND – "Élimination multihomogène", in *Introduction to algebraic independence theory*, Lecture Notes in Math., Vol. 1752, Springer, Berlin, 2001, Chap. 5 in [103], p. 53–81.
110. D. ROY – "Approximation algébrique simultanée de nombres de Liouville", *Canad. Math. Bull.* **44** (2001), no. 1, p. 115–120.
111. —, "An arithmetic criterion for the values of the exponential function", *Acta Arith.* **97** (2001), no. 2, p. 183–194.
112. —, "Une formule d'interpolation en deux variables", *J. Théor. Nombres Bordeaux* **13** (2001), no. 1, p. 315–323, 21è Journées Arithmétiques (Rome, 2001).
113. —, "Approximation simultanée d'un nombre et de son carré", *C. R. Math. Acad. Sci. Paris* **336** (2003), no. 1, p. 1–6.
114. —, "Approximation to real numbers by cubic algebraic integers. II", *Ann. of Math. (2)* **158** (2003), no. 3, p. 1081–1087.
115. —, "Approximation to real numbers by cubic algebraic integers. I", *Proc. London Math. Soc. (3)* **88** (2004), no. 1, p. 42–62.
116. —, "Diophantine approximation in small degree", Number theory, CRM Proc. Lecture Notes, vol. 36, Amer. Math. Soc., Providence, RI, 2004, p. 269–285.
117. —, "Simultaneous approximation by conjugate algebraic numbers in fields of transcendence degree one", *Int. J. Number Theory* **1** (2005), no. 3, p. 357–382.
118. —, "On two exponents of approximation related to a real number and its square", *Canadian J. Math.* **59** (2007), no. 1, p. 211–224,
119. —, "On the continued fraction expansion of a class of numbers", *Diophantine approximation. Festschrift for Wolfgang Schmidt*, vol. 16, 2008, in: Developments in Math, Eds: H. P. Schlickewei, K. Schmidt and R. Tichy, Springer-Verlag, p. 347–361.
120. —, "Small value estimates for the mutiplicative group", *Acta Arith.* **135.4** (2008), p. 357–393.
121. —, "Small value estimates for the additive group", *Int. J. Number Theory*, **6** (2010), no. 4, 919–956.
122. D. ROY and M. WALDSCHMIDT – "Diophantine approximation by conjugate algebraic integers", *Compos. Math.* **140** (2004), no. 3, p. 593–612.
123. W. M. SCHMIDT – "On heights of algebraic subspaces and Diophantine approximations", *Ann. of Math. (2)* **85** (1967), p. 430–472.
124. —, *Diophantine approximation*, Vol. **785**, Lecture Notes in Mathematics. Berlin-Heidelberg-New York: Springer-Verlag, 1980.
125. —, *Diophantine approximations and Diophantine equations*, Lecture Notes in Mathematics, vol. **1467**, Springer-Verlag, Berlin, 1991.
126. —, "Mahler and Koksma classification of points in $\mathbf{R}^n$ and $\mathbf{C}^n$", *Functiones et Approximatio* **35** (2006), p. 307–319.
127. —, "Diophantine approximation by algebraic hypersurfaces and varieties", *Trans. Amer. Math. Soc.* **359** (2007), no. 5, p. 2221–2241.
128. T. SCHNEIDER – *Einführung in die transzendenten Zahlen*, Springer-Verlag, Berlin, 1957, French Transl., Introduction aux nombres transcendants, Gauthier-Villars, Paris (1959).
129. V. G. SPRINDŽUK – *Mahler's problem in metric number theory*, Translated from the Russian by B. Volkmann. Translations of Mathematical Monographs, Vol. 25, American Mathematical Society, Providence, R.I., 1969.
130. —, *Metric theory of Diophantine approximations*, V. H. Winston & Sons, Washington, D.C., 1979.
131. R. TIJDEMAN – "On the algebraic independence of certain numbers", *Nederl. Akad. Wetensch. Proc. Ser. A* **74**=*Indag. Math.* **33** (1971), p. 146–162.
132. K. I. TISHCHENKO – "On new methods in the problem of the approximation of real numbers by algebraic numbers of degree at most three", *Vests = ı Nats. Akad. Navuk Belarus = ı Ser. F=ız.-Mat. Navuk* **4** (2000), p. 26–31, 141.
133. —, "On some special cases of the Wirsing conjecture", *Vests i Nats. Akad. Navuk Belarus = ı Ser. F = ız.-Mat. Navuk* **3** (2000), p. 47–52, 140.

134. —, "On the application of linearly independent polynomials in the Wirsing problem", *Dokl. Nats. Akad. Nauk Belarusi* **44** (2000), no. 5, p. 34–36, 124.
135. —, "On the approximation of real numbers by algebraic integers of the third degree", *Dokl. Nats. Akad. Nauk Belarusi* **45** (2001), no. 1, p. 17–19, 136.
136. —, "On approximation of real numbers by algebraic numbers of bounded degree", *J. Number Theory* **123** (2007), no. 2, p. 290–314.
137. R. C. VAUGHAN and S. VELANI – "Diophantine approximation on planar curves: the convergence theory", *Invent. Math.* **166** (2006), no. 1, p. 103–124.
138. M. WALDSCHMIDT – *Diophantine approximation on linear algebraic groups Transcendence properties of the exponential function in several variables*, Grundlehren der Mathematischen Wissenschaften, vol. 326, Springer-Verlag, Berlin, 2000.
139. E. WIRSING – "Approximation mit algebraischen Zahlen beschränkten Grades", *J. reine angew. Math.* **206** (1960), p. 67–77.
140. K. R. YU – "A generalization of Mahler's classification to several variables", *J. Reine angew. Math.* **377** (1987), p. 113–126.